FOUNDATIONS of MATHEMATICAL PHYSICS

Sadri Hassani
Illinois State University

ALLYN and BACON
Boston London Toronto Sydney Tokyo Singapore

To Sarah, Dane, and Daisy

A Division of Simon & Schuster, Inc.
160 Gould Street
Needham Heights, MA 02194

Series Editor: Nancy Forsyth
Series Editorial Assistant: Christopher Rawlings
Production Administrator: Peter Petraitis
Production Coordinator: Jane Hoover/Lifland et al., Bookmakers
Editorial-Production Service: Lifland et al., Bookmakers
Cover Administrator: Linda Dickinson
Manufacturing Buyer: Megan Cochran

9 780205 123797 9000
ISBN 0-205-12379-1
H23799

Library of Congress Cataloging-in-Publication Data

Hassani, Sadri.
Foundations of mathematical physics / Sadri Hassani.
p. cm.
Includes bibliographical references and index.
ISBN 0-205-12379-1
1. Mathematical physics. I. Title.
QC20.H39 1990
530.1'5--dc20 90-42556
CIP

Printed in the United States of America

10 9 8 7 6 5 4 3 2 1 95 94 93 92 91 90

Contents

Preface

This is a book for physics students interested in the mathematics they use. It is also a book for mathematics students who wish to see some of the abstract ideas with which they are familiar come alive in an applied setting. The level of presentation is that of an advanced undergraduate or beginning graduate course (or sequence of courses) traditionally called "Mathematical Methods of Physics" or some variation on this theme. Since there are a variety of books available on the subject, a justification for yet another is probably in order.

The existing mathematical physics books fall into two categories. In the first, formalism is stressed, with little or no emphasis on its application in concrete examples. The second category, on the other hand, either completely lacks formalism or deemphasizes it to the point of extinction. The first category is inaccessible to the novice; the second appears as a collection of *ad hoc* facts and materials lacking coherence and underlying unity.

PHILOSOPHY

It is my belief that a mathematical physics book should be more than a lexicographic collection of facts about the diagonalization of matrices, tensor analysis, Legendre polynomials, contour integration, and solutions to differential equations. A mathematical physics book must also be more than a mere investigation of relations between abstract structures; that approach is suitable for a mathematics course. *An ideal mathematical physics book strikes a balance between formalism and application, between the abstract and the concrete.* This ideal may be hard to achieve, but, in my opinion, it is crucial, and has been the guiding criterion in writing this book.

An attempt has been made to include as much of the essential formalism as is necessary to render the book optimally coherent. This entails stating and proving a large number of theorems, propositions, lemmas, and corollaries. The benefit of such an approach is that the student will recognize clearly both the power and the limitation of a mathematical idea used in physics. There is a tendency on the part of the novice to universalize the mathematical methods and ideas encountered in physics courses because the limitations of these methods and ideas are not clearly pointed out.

FEATURES

To better understand theorems, propositions, and so forth, students need to see them in action. There are approximately *600 worked-out examples and exercises* in this book and the accompanying *Student Solutions Manual*, providing a vast arena in which students can watch the formalism unfold. The philosophy underlying this abundance can be summarized as "An example is worth a thousand words of explanation." Thus, whenever a statement is intrinsically vague or hard to grasp, worked-out examples and/ or exercises are provided to clarify it. The inclusion of such a large number of examples and exercises is the means by which the balance between formalism and application has been achieved. However, although examples and exercises are essential in understanding mathematical physics, they are only one side of the coin. The theorems, propositions, lemmas, and corollaries are equally important. In fact, I have to complement my earlier statement by saying "A thousand words of explanation are worth an example."

Another feature of the book, which is not conspicuously emphasized in other comparable books, is the attempt to exhibit—as much as is useful and applicable—*interrelationships between the various topics* discussed in the book. Thus, the underlying theme of a vector space (which, in my opinion, is the most primitive concept at this level of presentation) recurs throughout the book and alerts the reader to the connection between various seemingly unrelated topics.

To facilitate reference to them, all mathematical statements (definitions, theorems, propositions, lemmas, and corollaries) have been numbered consecutively within each section and are preceded by the section number. For example, **4.2.9 Definition** indicates the ninth mathematical statement (which happens to be a definition) in Section 4.2. The end of a proof is marked by a solid black box, ■, placed at the right margin. Examples are indented and numbered separately, but also consecutively, within each section. That is, **Example 3.6.12** refers to the twelfth example in Section 3.6. The end of an example is designated by a solid black bullet, •, again placed at the right margin. The equations in each example are numbered separately.

ORGANIZATION AND TOPICAL COVERAGE

The book is divided into five parts. Part I, the first five chapters, is devoted to a thorough study of vector spaces in their broadest possible sense. As the unifying theme of the book, vector spaces demand careful analysis, and Chapters 2, 3, and 5 provide this. Special mention of Chapter 4 is in order here. Recent years have seen a burgeoning interest in differential geometry because of its relevance in the study of the fundamental interactions. Through a multitude of worked-out examples and exercises, this otherwise difficult material has become fairly accessible to the beginner. Since no part of the book is dependent on this chapter, its inclusion or omission will not affect subsequent reading.

Part II is entirely devoted to complex variables. Chapter 6 deals with basic properties of complex numbers and functions, and Chapter 7 addresses the more advanced topics of complex analysis, such as complex series and their convergence, residues, dispersion relations, analytic continuation, and so forth.

Part III treats mainly ordinary differential equations. Chapter 8 shows how ordinary differential equations of second order arise in physical problems, and Chapter 9 discusses these differential equations thoroughly. Chapter 10 focuses on the Sturm-Liouville problem and contains a great number of examples solved by the method of expansion in terms of the eigenfunctions of Sturm-Liouville systems.

In Part IV, Chapters 11, 12, and 13 cover operators on Hilbert spaces, Green's functions, and integral equations. Special emphasis is placed on Green's functions in

Chapters 11 and 12. Chapter 12 may be of special interest because it discusses Green's functions in an m-dimensional Euclidean space, a treatment that is completely new and not found in any other mathematical physics book.

Part V, entitled "Special Topics," consists of Chapter 14, on gamma and beta functions, and Chapter 15, which treats numerical methods.

To avoid the uncontrollable expansion of the book—a tendency discovered during the writing of the second draft—I had to exclude some topics, such as group theory, calculus of variations, and nonlinear dynamics (chaos). Any of these topics, if treated in the spirit of the rest of the book, would have required a substantial thickening of an already voluminous book.

This book is suitable for either a one-semester course or a full-year sequence. Suggestions as to parts of the book suitable for various course structures are as follows:

Graduate, one-semester: Chapters 2, 3, 5, and 6, Sections 7.2 and 7.3, Chapter 8, Sections 9.2–9.5, Chapter 10, Sections 11.4 and 11.5, Sections 12.3, 12.4 and 12.5.1, Chapter 15

Graduate, two-semester: all *except* Chapter 1 and Sections 7.1, 11.2, 12.1, and 12.2

Undergraduate, one-semester: Chapter 1, Sections 2.1, 2.2, 2.3.1, 2.3.2, and 2.3.4, Sections 3.1–3.4 and 3.6.1–3.6.3, Sections 5.1, 5.2.1, 5.3, and 5.4, Chapter 6, Section 7.2, Sections 8.1–8.3 and 8.4.1, Sections 9.2.1, 9.2.2, 9.3, and 9.4, and Chapter 15

Undergraduate, two-semester: Chapters 1, 2, and 3, Sections 5.1, 5.2.1, 5.3, and 5.4, Chapter 6, Section 7.2, Chapter 8, Sections 9.2–9.4, Chapter 10, Section 11.4, Chapters 14 and 15

Of course, depending on the taste of the instructor and the composition of the class, the coverage can be altered. The book is designed to allow a great deal of flexibility. For instance, the instructor can discuss all the examples in a given section or choose among them, can go through the proofs in great detail or completely ignore them.

I should mention the acronyms used in this book. In the course of writing the later chapters, I found myself endlessly repeating long expressions such as "*i*nhomogeneous *s*econd-*o*rder *l*inear *d*ifferential *e*quations." In order to save space, I used acronyms such as ISOLDE.

This book is my attempt at bringing formalism and problem-solving skills together, since the two are interrelated, and one reinforces the other. It is my hope that this approach will be beneficial both to the students and to the instructors. *I will greatly appreciate any comments and suggestions for improvements.*

Although extreme care was taken to correct all the misprints, it is likely that I have missed some of them. I shall be most grateful to those readers kind enough to bring to my attention any remaining mistakes, typographical or otherwise, so that I can incorporate the corrections in future printings.

ACKNOWLEDGMENTS

Many excellent textbooks, too numerous to cite individually here, have influenced the writing of this book. The following, however, are noteworthy for both their excellence and the amount of their influence.

BIRKHOFF, G., and G.-C. ROTA, *Ordinary Differential Equations*, 3rd ed. New York: John Wiley, 1978.

BISHOP, R., and S. GOLDBERG, *Tensor Analysis on Manifolds*. New York: Dover, 1980.

DENERY, P., and A. KRZYWICK, *Mathematics for Physicists*. New York: Harper & Row, 1967.

HALMOS, P., *Finite-Dimensional Vector Spaces*, 2nd ed. Princeton, NJ: D. Van Nostrand, 1958.

It gives me great pleasure to thank all the people who contributed to the making of this book. Special thanks go to Deborah Shreffler for typing the manuscript, to Nancy Forsyth, Senior Editor at Allyn and Bacon, for her enthusiasm and support, and to Jane Hoover and Peter Petraitis for thoroughly and effectively coordinating the overall production of the book. I am also indebted to my wife Sarah for her assistance in preparing the index and for her endless support during the long and arduous writing process.

NOTE TO THE STUDENT

Mathematics and physics are like the game of chess (or, for that matter, like *any* game)—you will learn them only by "playing" them. No amount of reading about the game will make you a master. In this book you will find a large number of examples, exercises, and problems. Go through as many examples as possible, and try to reproduce them. At the ends of sections are exercises whose solutions can be found in the *Student Solutions Manual*, published separately. Trying to solve these exercises on your own is your chance to improve your problem-solving ability. Look at the solutions only as a check or the absolute last resort. The problems at the ends of chapters are also an integral part of the learning process. Try as many of them as you can. The exercises and problems often fill in missing steps, and in this respect they are essential for a thorough understanding of the book.

Sadri Hassani

Introduction: Mathematical Preliminaries

Modern mathematics starts with the basic (and undefinable) concept of *set*. We think of a set as a structureless family, or collection, of objects. We speak, for example, of the set of students in a college, of men in a city, of women working for a corporation, of vectors in space, of points in a plane, or of events in the continuum of space-time. Each member a of a set A is called an *element* of that set. This relation is denoted by $a \in \mathrm{A}$ (read "a is an element of A" or "a belongs to A"), and its negation by $a \notin \mathrm{A}$. Sometimes a is called a *point* of the set A to emphasize the geometric connotation.

0.1 OPERATIONS ON SETS

A set is usually designated by enumeration of its elements between braces. For example, $\{2, 4, 6, 8\}$ signifies the set consisting of the first four even natural numbers; $\{0, \pm 1, \pm 2, \pm 3, \ldots\}$ is the set of all integers; $\{1, x, x^2, x^3, \ldots\}$ is the set of all nonnegative powers of x; and $\{1, i, -1, -i\}$ is the set of the four fourth roots of unity. It is impossible to enumerate some sets, however. In such cases the set is described by either a verbal or a mathematical statement that holds for all of its elements. Such a statement is generally denoted by $\{x|P(x)\}$ and read "the set of all x's such that $P(x)$ is true." The foregoing examples of sets can be written alternatively as follows:

$$\{n|n \text{ is even and } 1 < n < 9\}$$

$$\{\pm n|n \text{ is a natural number}\}$$

$$\{y|y = x^n \text{ and } n \text{ is a natural number}\}$$

$$\{z|z^4 = 1 \text{ and } z \text{ is a complex number}\}$$

In a frequently used shorthand notation, the last two sets can be abbreviated as $\{x^n | n \geqslant 0$ and n is an integer$\}$ and $\{z | z^4 = 1\}$. Similarly, the unit circle can be denoted as $\{z \,|\, |z| = 1\}$, the closed interval $[a, b]$ as $\{x | a \leqslant x \leqslant b\}$, the open interval (a, b) as $\{x | a < x < b\}$, and the set of all powers of x as $\{x^n\}_{n=1}^{\infty}$. This last notation will be used frequently in this book.

If $a \in \mathrm{A}$ whenever $a \in \mathrm{B}$, we say that B is a *subset* of A and write $\mathrm{B} \subset \mathrm{A}$ or $\mathrm{A} \supset \mathrm{B}$. If $\mathrm{A} \subset \mathrm{B}$ and $\mathrm{B} \subset \mathrm{A}$, then $\mathrm{A} = \mathrm{B}$. The set defined by $\{a | a \neq a\}$ is called the *empty set* and denoted by $\varnothing$. Clearly $\varnothing$ contains no elements and is a subset of any arbitrary set. The collection of all subsets (including $\varnothing$) of a set A is denoted by 2^{A}. The reason for this notation is that the number of subsets of the set $\mathrm{F}_n = \{a_1, a_2, a_3, \ldots, a_n\}$ is 2^n. This can easily be seen by noting that there are n sets of the form $\{a_i\}$ (which are called *singletons*), $n(n-1)/2$ sets of the form $\{a_i, a_j\}$, where $i \neq j$, $[n(n-1)(n-2)]/3!$ sets of the form $\{a_i, a_j, a_k\}$, where $i \neq j \neq k$, and so forth. Including the empty set and the full set as subsets gives

$$1 + n + \frac{n(n-1)}{2} + \frac{n(n-1)(n-2)}{3!} + \cdots + n + 1 = (1+1)^n = 2^n$$

as the number of subsets of F_n.

If A and B are sets, their *union*, denoted by $\mathrm{A} \cup \mathrm{B}$, is the set containing all elements that belong to A or B or both. The *intersection* of the sets A and B, denoted by $\mathrm{A} \cap \mathrm{B}$, is the set containing all elements belonging to both A and B. The *complement* of a set A is denoted by $\sim\mathrm{A}$ and defined as

$$\sim\mathrm{A} \equiv \{a | a \notin \mathrm{A}\}$$

The complement of B in A (or their difference) is

$$\mathrm{A} \sim \mathrm{B} \equiv \{a | a \in \mathrm{A} \quad \text{and} \quad a \notin \mathrm{B}\}$$

In any application of set theory there is an underlying universal set whose subsets are the subject of study. This universal set is usually clear from the context. For example, in the study of the properties of integers, the set of integers, denoted by $\mathbb{Z}$, is the universal set. The set of complex numbers, $\mathbb{C}$, is the universal set in complex analysis, and the set of reals, $\mathbb{R}$, is the universal set in real analysis. With a universal set X in mind, one can write $\mathrm{X} \sim \mathrm{A}$ instead of $\sim\mathrm{A}$.

0.2 CARTESIAN PRODUCTS

From two given sets A and B, it is possible to form a new set by taking pairs of elements, one from A and one from B. The *Cartesian product* of A and B, denoted by $\mathrm{A} \times \mathrm{B}$, is the set of ordered pairs (a, b), where $a \in \mathrm{A}$ and $b \in \mathrm{B}$. This is expressed in set theoretic notation as follows:

$$\mathrm{A} \times \mathrm{B} = \{(a, b) | a \in \mathrm{A} \quad \text{and} \quad b \in \mathrm{B}\}$$

We can generalize this to an arbitrary number of sets. If $A_1, A_2, A_3, \ldots, A_n$ are sets, then the Cartesian product of these sets is

$$A_1 \times A_2 \times \cdots \times A_n = \{(a_1, a_2, \ldots, a_n) | a_1 \in A_1, a_2 \in A_2, \ldots, a_n \in A_n\}$$

which is the set of ordered n-tuples. If $A_1 = A_2 = \cdots = A_n = A$, then we write A^n instead of $A \times A \times \cdots \times A$, and

$$A^n = \{(a_1, a_2, \ldots, a_n) | a_i \in A_i, \quad i = 1, 2, \ldots, n\}$$

The most familiar example of a Cartesian product occurs when $A = \mathbb{R}$. Then $\mathbb{R}^2$ is the set of pairs (x_1, x_2) with $x_1 \in \mathbb{R}, x_2 \in \mathbb{R}$. This is simply the points in a plane. Similarly, $\mathbb{R}^3$ is the set of triplets (x_1, x_2, x_3), or the points in space, and $\mathbb{R}^n = \{(x_1, \ldots, x_n) | x_i \in \mathbb{R}\}$ is the set of n-tuples.

0.3 MAPPINGS

A very important concept is that of mapping, by which a connection is established between two sets. A *mapping*, f, from a set X to a set Y, denoted by $f: X \to Y$, is a correspondence between elements of X and those of Y in which all the elements of X participate and each element of X corresponds to only one element of Y. If $y \in Y$ is the element that corresponds to $x \in X$ via the mapping f, we write $y = f(x)$ and call y the *image* of x under f. Thus, by the definition of mapping, $x \in X$ cannot have two (or more) different images.

If A is a subset of X, we call $f(A) = \{f(x) | x \in A\}$ the *image* of A (see Fig. 0.1). Similarly, if $B \subset Y$, we call $f^{-1}(B) = \{x \in X | f(x) \in B\}$ the *inverse image* (or *preimage*) of B. In words, $f^{-1}(B)$ consists of all elements in X whose images are in $B \subset Y$. If B consists of a single element, b, then $f^{-1}(b) = \{x \in X | f(x) = b\}$ consists of all elements of X that are mapped to b. Note that it is possible for many points of X to have the same image in Y.

If $f(x_1) = f(x_2)$ implies that $x_1 = x_2$, we call f *injective* or *one-to-one* (denoted 1-1). For an injective map only one element of X corresponds to an element of Y, although there may exist elements of Y that do not participate in the process of mapping. If $f(X) = Y$, the mapping is said to be *surjective*, or *onto*. A map that is both injective and surjective is said to be *bijective*, or to be a *one-to-one correspondence*. Two sets that are in one-to-one correspondence with one another have the same

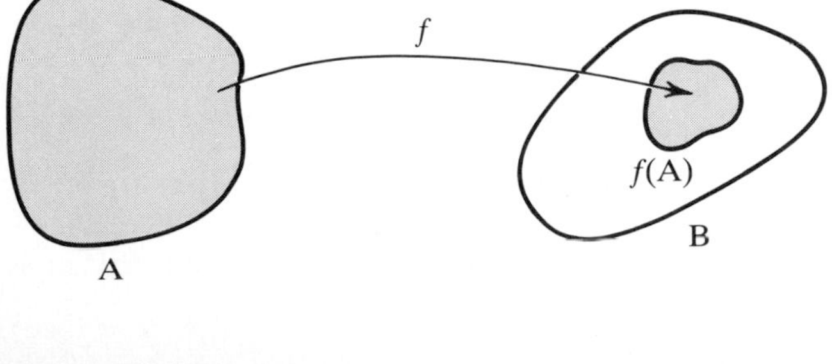

Figure 0.1 The function f maps the set A into the set B, and the range of f, $f(A)$, is shown as a subset of B.

number of elements. If $f: \mathrm{X} \to \mathrm{Y}$ is a bijection from X onto Y, then for each $y \in \mathrm{Y}$ there is one and only one element in X. Thus, there is a mapping $f^{-1}: \mathrm{Y} \to \mathrm{X}$ from Y to X given by $f^{-1}(y) = x$, where x is the unique element such that $f(x) = y$. This mapping is called the *inverse* of f.

In the definition of mapping the set X is called the *domain*, the set Y the *codomain*, and $f(\mathrm{X})$ the *range* of the mapping f. A mapping whose codomain is the set of real numbers, $\mathbb{R}$, or the set of complex numbers, $\mathbb{C}$, is commonly called a *function*.

For example, calculus involves the study of functions $f: \mathbb{R} \to \mathbb{R}$. The two functions $f: \mathbb{R} \to \mathbb{R}$ and $g: \mathbb{R} \to (-1, +1)$ given by $f(x) = x^3$ and $g(x) = \tanh x$ are bijective. The latter function, by the way, shows that there are as many points in the whole real line as there are in the interval $(-1, +1)$. If we denote the set of positive real numbers by $\mathbb{R}^+$, then the function $f: \mathbb{R} \to \mathbb{R}^+$ given by $f(x) = x^2$ is surjective but not injective [both x and $-x$ map to the same $f(x)$]. However, the function $g: \mathbb{R}^+ \to \mathbb{R}$ given by the same rule, $g(x) = x^2$, is injective but not surjective. On the other hand, $h: \mathbb{R}^+ \to \mathbb{R}^+$ given by $h(x) = x^2$ is bijective, but $u: \mathbb{R} \to \mathbb{R}$ given by $u(x) = x^2$ is neither injective nor surjective.

Let M_n denote the set of $n \times n$ real matrices. Define a function $f: \mathrm{M}_n \to \mathbb{R}$ by $f(\mathsf{A}) = \det \mathsf{A}$ for $\mathsf{A} \in \mathrm{M}_n$. This function is clearly surjective but not injective. Note that $f^{-1}(1)$ is the set of all matrices whose determinant is 1. Such matrices occur frequently in physical applications and will be encountered later in this book.

Another example of interest is $f: \mathbb{C} \to \mathbb{R}$ given by $f(z) = |z|$. This function is also neither injective nor surjective. Here $f^{-1}(1)$ is the unit circle, the circle of radius 1 in the complex plane.

There are many situations in which the domain of a mapping is a Cartesian product. Consider the mapping $f: \mathrm{X} \times \mathrm{X} \to \mathrm{Y}$ given by $f(x_1, x_2) = y$ in which X and Y are arbitrary sets. Two specific cases are worthy of mention. The first is when $\mathrm{Y} = \mathbb{R}$. An example is the dot product on vectors. Thus, if X is the set of vectors in the plane, we can define $f(\mathbf{x}_1, \mathbf{x}_2) = \mathbf{x}_1 \cdot \mathbf{x}_2$. The second important case is when $\mathrm{Y} = \mathrm{X}$. Then we speak of a *binary operation* on X by which an element in X is associated with two elements in X. For instance, let $\mathrm{X} = \mathbb{Z}$, the set of all integers; then the function $f: \mathbb{Z} \times \mathbb{Z} \to \mathbb{Z}$ defined by $f(m, n) = mn$ is the operation of multiplication of integers. Similarly, $g: \mathbb{R} \times \mathbb{R} \to \mathbb{R}$ given by $g(x, y) = x + y$ is the (binary) operation of addition of real numbers.

If $f: \mathrm{X} \to \mathrm{Y}$ and $g: \mathrm{Y} \to \mathrm{W}$, then the mapping $h: \mathrm{X} \to \mathrm{W}$ given by $h(x) = g(f(x))$ is called the *composition* of f and g, denoted as $h = g \circ f$ (see Fig. 0.2). If f and g are both bijections with inverses f^{-1} and g^{-1}, respectively, then h also has an inverse, and verifying that $h^{-1} = (g \circ f)^{-1} = f^{-1} \circ g^{-1}$ is straightforward.

0.4 METRIC SPACES

Although sets are at the roots of modern mathematics, by themselves they are of only formal and abstract interest. To make sets interesting and useful, it is necessary to introduce some structures on them. There are two general procedures for the imple-

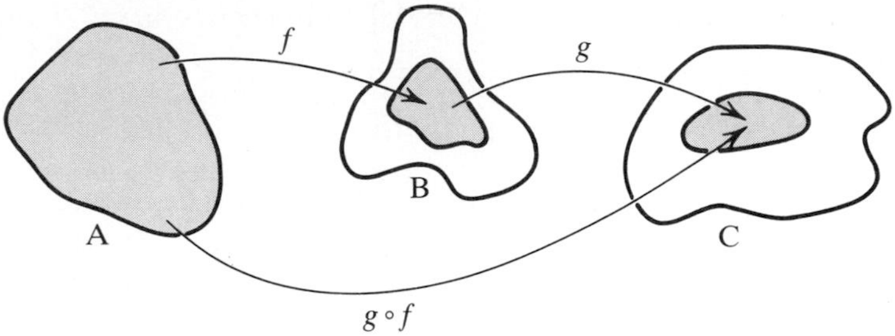

Figure 0.2 In the composition of two mappings, f and g, note the order in which f and g appear.

mentation of such structures. These are the abstractions of the two major branches of mathematics—algebra and analysis.

We can turn a set into an algebraic structure by introducing a binary operation on it. For example, a vector space, which is an algebraic structure that will be studied in detail in Chapters 2 and 3, consists, among other things, of the binary operation of vector addition. A group, which will not be studied in this book but which may be familiar to the reader, is a set together with the binary operation of multiplication. There are many other examples of algebraic systems, and they constitute the rich subject of abstract algebra.

When analysis, the other branch of mathematics, is abstracted using the concept of sets, it leads to topology, in which the concepts of limits and continuity play a central role. This is also a very rich subject with far-reaching consequences and applications.

Fascinating as the two above-mentioned areas of mathematics are, the real beauty and power of abstract mathematics are attained only when the two are combined. This opens up topics such as topological groups, Lie groups, algebraic topology, differentiable manifolds, differential geometry, functional analysis, and many others, each of which has revolutionized understanding of mathematical systems, which in turn has clarified many physical problems.

The aim of this book is not as grandiose as might be suggested by the foregoing. Although some algebraic systems will be discussed and the ideas of limit and continuity in the sequel will be used, this will be done in a very limited fashion, by introducing and employing the concepts when they are needed. On the other hand, some general concepts will be introduced when they require minimum prerequisites. One of these is the metric space, which, although very general in nature, is much less abstract than the topological space, which is the starting point of many branches of modern mathematics. Therefore, consider the following definition.

0.4.1 Definition A *metric space* is a set X together with a real-valued function $d: \mathrm{X} \times \mathrm{X} \to \mathbb{R}$, such that

(i) $d(x, y) \geqslant 0$, and $d(x, y) = 0$ if and only if $x = y$

(ii) $d(x, y) = d(y, x)$ (symmetry)

(iii) $d(x, y) \leqslant d(x, z) + d(z, y)$ (the triangle inequality)

It is important to note that X is a completely arbitrary set and needs no structure other than the metric structure defined above. In this respect Definition 0.4.1 is very general and encompasses many different situations, as the following examples will show.

Before examining the examples, note that the function d defined above is the abstraction of the notion of distance: (i) says that the distance between any two points is always positive and is zero only if the two points coincide; (ii) says that the distance between two points does not change if the two points are interchanged; (iii) states the known fact that the sum of the lengths of two sides of a triangle is always greater than or equal to the length of the third side.

Now consider these examples:

(1) Let $\mathrm{X} = \mathbb{Q}$, the set of rational numbers, and define $d(x, y) = |x - y|$.

(2) Let $\mathrm{X} = \mathbb{R}$, and define $d(x, y)$ as above.

(3) Let X consist of the points on the surface of a sphere. We can define two distance functions on X. Let $d_1(P, Q)$ be the straight line segment joining P and Q on the sphere. It is not hard to convince oneself that d satisfies all the properties of the metric function. We can also define $d_2(P, Q)$ as the length of the arc of the great circle passing through points P and Q on the surface of the sphere.

(4) Let $\mathrm{C}(a, b)$ denote the set of continuous real-valued functions on the interval $[a, b]$. We can define

$$d(f, g) = \int_a^b |f(x) - g(x)|\, dx$$

for $f, g \in \mathrm{C}(a, b)$.

(5) Let $\mathrm{C_B}(a, b)$ denote the set of bounded continuous real-valued functions on the interval $[a, b]$. We then define

$$d(f, g) = \max_{x \in [a, b]} \{|f(x) - g(x)|\}$$

for $f, g \in \mathrm{C_B}(a, b)$. Here we consider $|f(x) - g(x)|$ for all values of x in $[a, b]$, pick the maximum value, and christen it $d(f, g)$.

The metric function creates a natural setting in which to test the closeness of points in a metric space. One occasion on which the idea of closeness becomes essential is in the study of a sequence. A *sequence* is a mapping $s : \mathbb{N} \to \mathrm{X}$ from the set of natural numbers, $\mathbb{N}$, into the metric space X. Given a positive integer, n, such a mapping associates with it a point, $s(n)$, of the metric space X. It is customary to write s_n (or x_n to match the symbol X) instead of $s(n)$ and to enumerate the values of the function by writing $\{x_n\}_{n=1}^{\infty}$ instead of the mapping.

Knowledge of the behavior of a sequence for large values of n is of fundamental importance. In particular, it is important to know whether a sequence approaches a finite value as n increases. We say that a sequence $\{x_n\}_{n=1}^{\infty}$ *converges* to a value x if $\lim_{n\to\infty} d(x_n, x) = 0$. This convergence can also be denoted by $d(x_n, x) \to 0$, or simply $x_n \to x$. In other words, for any positive real number ε, there exists a natural number N such that $d(x_n, x) < \varepsilon$ whenever $n \geqslant N$.

With a given sequence, $\{x_n\}_{n=1}^{\infty}$, it may not be possible to test directly for convergence because this requires a knowledge of the limit point, x. However, it is possible to do the next best thing—to see whether the points of the sequence get closer and closer as n gets larger and larger. A *Cauchy sequence* is one for which $\lim_{\substack{m\to\infty \\ n\to\infty}} d(x_m, x_n) = 0$, or $d(x_m, x_n) \to 0$. We can test directly whether or not a sequence is Cauchy. However, the fact that a sequence is Cauchy does not guarantee that it converges. For example, let the metric space be the set of rational numbers, $\mathbb{Q}$, with the metric function $d(x, y) = |x - y|$. Consider the sequence $\{x_n\}_{n=1}^{\infty}$, where $x_n = \sum_{k=1}^{n} (1 + 1/k)^k$. It is clear that x_n is a rational number for any n. Also, to show that $|x_n - x_m| \to 0$ is an exercise in calculus. Thus, the sequence is Cauchy. However, it is well known that $\lim_{n\to\infty} x_n = e = 2.71828\ldots$, which is not a rational number.

A metric space in which every Cauchy sequence converges is called a *complete metric space*. Complete metric spaces play a crucial role in modern analysis. The preceding example shows that $\mathbb{Q}$ is not a complete metric space. However, if the limit points of all Cauchy sequences are added to $\mathbb{Q}$, the resulting space becomes complete. This complete space is, of course, the real number system, $\mathbb{R}$. A general property of all metric spaces is that any incomplete metric space can be "enlarged" to a complete metric space. The process of enlargement can be stated in a mathematically precise way, but that will not be done here.

0.5 CARDINALITY

Some counting terminology is currently used in mathematical physics and is therefore defined here.

The process of counting is a one-to-one comparison of one set with another. If two sets are in one-to-one correspondence, that is, if a bijection exists between them, they are said to have the same *cardinality*. Two sets with the same cardinality essentially have the same "number" of elements. The set $F_n = \{1, 2, \ldots, n\}$ is finite and has cardinality n. Any set from which there is a bijection onto F_n is said to be finite with n elements.

Now consider the set of natural numbers, $\mathbb{N} = \{1, 2, \ldots, n, \ldots\}$. If there exists a bijection between a set A and $\mathbb{N}$, then A is said to be infinite but *countable*. Some examples of infinitely countable sets are the set of all integers, the set of all rational numbers, the set of even natural numbers, the set of odd natural numbers, and the set of all prime numbers.

It may seem surprising that a subset (such as the set of all even numbers) can be put into one-to-one correspondence with its full set (the set of all natural numbers); however, this is a property shared by all infinite sets. In fact, sometimes they are defined as those sets that are in one-to-one correspondence with one of their proper subsets. It is also surprising to discover that there are as many rational numbers as there are natural numbers.

Sets that are neither finite nor countable are said to be *uncountable* or to form a *continuum* or to have the cardinality of the continuum. In some sense they are "more infinite" than any countable set. Examples of uncountable sets are points in the interval $[0, 1]$, the real number system, points in a plane, and points in space. It can be shown that all these sets have the same cardinality. Thus, there are as many points in three-dimensional space (the whole universe) as there are in the interval $[0, 1]$ or in any other finite interval.

Cardinality is a very intricate mathematical notion with many surprising results. One of these concerns the so-called *Cantor set*. Consider the interval $[0, 1]$. Remove the open interval $(\frac{1}{3}, \frac{2}{3})$ from its middle. This means that the points $\frac{1}{3}$ and $\frac{2}{3}$ will not be removed. From the remaining portion, $[0, \frac{1}{3}] \cup [\frac{2}{3}, 1]$, remove the two middle thirds; the remaining portion will then be $[0, \frac{1}{9}] \cup [\frac{2}{9}, \frac{1}{3}] \cup [\frac{2}{3}, \frac{7}{9}] \cup [\frac{8}{9}, 1]$. Do this indefinitely. What is the cardinality of the remaining set, which is called the Cantor set? Intuitively we expect hardly anything to be left. We might guess that the number of points remaining is at most infinite but countable. The surprising fact is that the cardinality is that of the continuum! Thus, after removal of infinitely many middle thirds, the set that remains has as many points as in the original set.

0.6 NOTATION

Certain standard notations in use in mathematics and mathematical physics save a lot of writing and will often be used in this book. These symbols and acronyms are listed in Table 0.1.

Negation is usually denoted by a slash (/) through a symbol. For example, $\not\Rightarrow$ means "does not imply," and $\nexists$ means "does not exist."

TABLE 0.1 SYMBOLS AND ACRONYMS USED IN THIS BOOK

$A \times B$	the Cartesian product of the sets A and B
(a, b)	the open interval with end points a and b, consisting of all real numbers x such that $a < x < b$
$[a, b]$	the closed interval with end points a and b, consisting of all real numbers x such that $a \leqslant x \leqslant b$
$A - \{a\}$	all points in the set A except the point a
$a < \infty$	a is finite
BC	boundary condition(s)
BVP	boundary value problem
$\mathbb{C}$	the set of complex numbers
$C(a, b)$	the set of real-valued continuous functions defined on the interval $[a, b]$
$\mathbb{C}(a, b)$	the set of complex-valued continuous functions defined on the interval $[a, b]$
$\mathbb{C}^n$	the space of complex n-tuples
$C^n(X)$	the collection of functions on the set X that have derivatives of all orders less than or equal to n
$C^\infty(X)$	the collection of functions on the set X having derivatives of all orders
$C_F^\infty(X)$	the subset of $C^\infty(X)$ whose members vanish outside some bounded (finite, compact) subset of X
CIF	Cauchy integral formula
DE	differential equation
DO	differential operator
$\hat{\mathbf{e}}$	the unit vector
$f: A \to B$	the mapping f from the set A into the set B
FODE	first-order differential equation
FOLDE	first-order linear differential equation
GF	Green's function
HDE	hypergeometric differential equation
HSOLDE	homogeneous second-order differential equation
IE	integral equation
iff	if and only if
IO	integral operator
ISOLDE	inhomogeneous second-order differential equation
IVP	initial value problem
LHS	left-hand side
$\mathbb{N}$	the set of natural numbers
ODE	ordinary differential equation
$\mathcal{P}_n^c$	the set of polynomials with degree n (or less) whose coefficients are complex
$\mathcal{P}_n^r$	the set of polynomials with degree n (or less) whose coefficients are real
PDE	partial differential equation
$\mathbb{Q}$	the set of rational numbers
$\mathbb{R}$	the set of real numbers
$\mathbb{R}^+$	the set of positive real numbers
$\mathbb{R}^n$	the n-dimensional Euclidean space; the space of real n-tuples
RHS	right-hand side
SODE	second-order differential equation
SODO	second-order differential operator
SOLDE	second-order linear differential equation
SOLDO	second-order linear differential operator
UHP	upper half-plane
$\mathcal{V}_N$	an N-dimensional vector space
$\mathcal{V} \otimes \mathcal{W}$	tensor product of the two vector spaces $\mathcal{V}$ and $\mathcal{W}$
wrt	with respect to
$\mathbb{1}$	the unit operator
1_N	the N-dimensional unit matrix
0	the generic symbol characterizing the real number zero, the complex zero, and the zero function
$\mathsf{0}$	the zero matrix
$\mathbb{O}$	the zero operator
$\mathbf{0}$ or $\|0\rangle$	the zero vector
$\approx$	approximately equal to
$\in$	belong(s) to; is an element of
$*$	complex conjugation; duality sign
$\bullet$	end of an example
$\blacksquare$	end of a proof
$\forall$	for all
$\dagger$	Hermitian conjugate (of an operator or a matrix)
$\Rightarrow$	implies
$\cap$	intersection
$\equiv$	is defined to be; is equal to by definition
$\Leftrightarrow$	is equivalent to; if and only if
$\\| \ \\|$	the norm function
$\leftrightarrow$	one-to-one correspondence
$\prod_{i=m}^{n} a_i$	the product $a_m \cdot a_{m+1} \cdot a_{m+2} \cdot \cdots \cdot a_{n-1} \cdot a_n$
$\propto$	is proportional to
$\{a_i\}_{i=1}^n$	the set containing $a_1, a_2, \ldots, a_n$
$\ni$	such that
$\sum_{i=m}^{n} a_i$	summation from m to n; $a_m + a_{m+1} + \cdots + a_n$
$\exists$	there exist(s)
$\sim$	transpose (of a matrix); Fourier transform (of a function)
$\cup$	union

Part I

SPACES OF VECTORS, TENSORS, AND FUNCTIONS

1

Vectors in the Plane and in Space

The basic theme on which most of mathematical physics is built is the notion of vectors. This concept will be the focus of this first part of this book. Chapters 2 and 3 will develop abstract vector spaces. However, to pave the way to abstraction, this chapter reviews some of the relevant properties of the more familiar planar and spatial vectors. The terminology normally reserved for abstract vector spaces is used here so that the novice will see the connection between the abstract structures and the more familiar objects.

1.1 BASIC PROPERTIES OF VECTORS IN THE PLANE

This chapter introduces vectors in ordinary space and discusses their properties with an eye to generalization to higher dimensions. We start with vectors in the plane and the most common definition of a vector as "something that points somewhere" (Fig. 1.1). Examples of vectors are displacement, **r**; velocity, **v**; momentum, **p**; electric field, **E**; and magnetic field, **B**.[1]

Vectors would be useless without some kind of operations on them. The most basic operation is changing the length of a vector. This is accomplished by multiplying the vector by a real positive number. For example, 3.2**r** is a vector in the same direction as **r** but 3.2 times as long. We can change the direction of a vector by 180° by multiplying it by -1. That is, $-1 \times \mathbf{r} = -\mathbf{r}$, a vector having the same length as **r** but pointing in the opposite direction. We can combine these two operations and

[1] Vectors will be denoted by letters printed in boldface type.

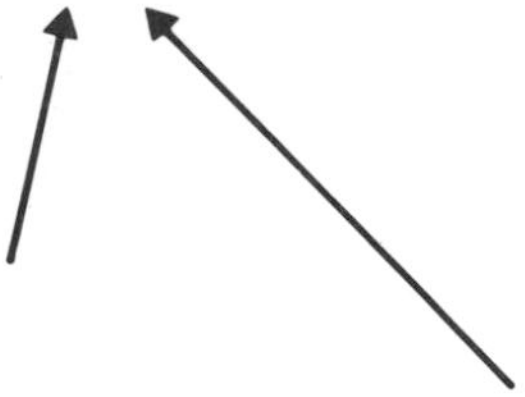

Figure 1.1 Two different vectors in the plane.

think of multiplying a vector by any real (positive or negative) number. Thus, $-0.732\mathbf{r}$ is a vector that is 0.732 times as long as $\mathbf{r}$ *and* points in the opposite direction. The combined operation is summarized as follows.

1.1.1 Operation A vector can be multiplied by any real number. The result is another vector.

Another operation is the addition of two vectors. This operation is inspired by the obvious addition law for displacements. In Fig. 1.2 two displacements, $\mathbf{r}_1$ and $\mathbf{r}_2$, are added to give the resultant, or the "sum," $\mathbf{R}$, which is also a vector:

$$\mathbf{r}_1 + \mathbf{r}_2 = \mathbf{R}$$

We can summarize this operation as follows.[2]

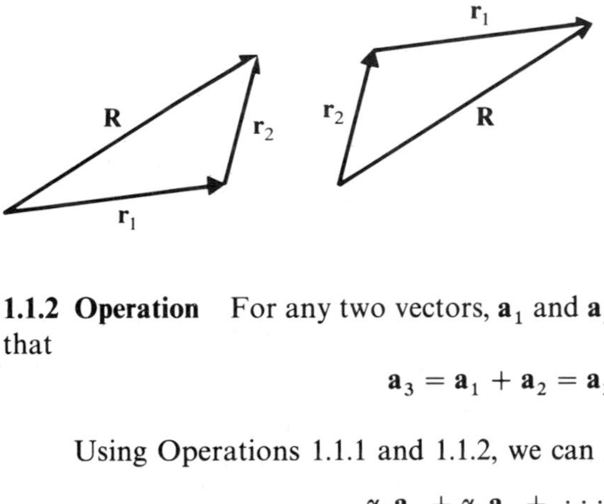

Figure 1.2 The vector $\mathbf{R}$ is the sum of two displacement vectors, $\mathbf{r}_1$ and $\mathbf{r}_2$. Note that $\mathbf{r}_1 + \mathbf{r}_2 = \mathbf{r}_2 + \mathbf{r}_1$.

1.1.2 Operation For any two vectors, $\mathbf{a}_1$ and $\mathbf{a}_2$, there exists a third vector, $\mathbf{a}_3$, such that

$$\mathbf{a}_3 = \mathbf{a}_1 + \mathbf{a}_2 = \mathbf{a}_2 + \mathbf{a}_1$$

Using Operations 1.1.1 and 1.1.2, we can form a sum,[3]

$$\alpha_1 \mathbf{a}_1 + \alpha_2 \mathbf{a}_2 + \cdots + \alpha_n \mathbf{a}_n \tag{1.1}$$

where $\alpha_1, \alpha_2, \ldots, \alpha_n$ are real numbers and $\mathbf{a}_1, \mathbf{a}_2, \ldots, \mathbf{a}_n$ are vectors. The sum in (1.1) is called a *linear combination* of the n vectors $\mathbf{a}_1, \mathbf{a}_2, \ldots, \mathbf{a}_n$. If we can find some set of

[2] The definition tacitly assumes that a vector can be moved parallel to itself without affecting it.

[3] To justify the expression in (1.1), one also has to postulate the associativity of addition, that is, that $\mathbf{a} + (\mathbf{b} + \mathbf{c}) = (\mathbf{a} + \mathbf{b}) + \mathbf{c}$.

real numbers, $\alpha_1, \alpha_2, \ldots, \alpha_n$ (not all of which are zero), such that the sum in (1.1) is zero, we say that the vectors $\mathbf{a}_1, \mathbf{a}_2, \ldots, \mathbf{a}_n$ are *linearly dependent*. If no such set of real numbers can be found, then the vectors $\mathbf{a}_1, \mathbf{a}_2, \ldots, \mathbf{a}_n$ are called *linearly independent*. In other words, if

$$\alpha_1 \mathbf{a}_1 + \alpha_2 \mathbf{a}_2 + \cdots + \alpha_n \mathbf{a}_n = 0 \tag{1.2}$$

implies that[4]

$$\alpha_1 = \alpha_2 = \cdots = \alpha_n = 0$$

then $\mathbf{a}_1, \mathbf{a}_2, \ldots, \mathbf{a}_n$ are linearly independent.

Example 1.1.1

(a) The two vectors $\hat{\mathbf{e}}_x$ and $\hat{\mathbf{e}}_y$ (sometimes denoted as $\hat{\mathbf{i}}$ and $\hat{\mathbf{j}}$) are linearly independent because[5]

$$\alpha \hat{\mathbf{e}}_x + \beta \hat{\mathbf{e}}_y = 0 \tag{1}$$

can be satisfied only if both α and β are zero. If one of them, say α, were different from zero, one could divide (1) by it and get one of the vectors as a multiple of the other,

$$\hat{\mathbf{e}}_x = -\frac{\beta}{\alpha} \hat{\mathbf{e}}_y$$

which is impossible.

(b) The two functions x^2 and x^3, where x *assumes arbitrary values*, are linearly independent. Because if

$$\alpha x^2 + \beta x^3 = 0 \tag{2}$$

for arbitrary values of x, then letting $x = 1$ and $x = -1$ gives, respectively,

$$\alpha + \beta = 0$$
$$\alpha - \beta = 0$$

which has the unique solution

$$\alpha = \beta = 0$$

Therefore, the only way (2) can hold is if both α and β are zero. This establishes the linear independence of x^2 and x^3. ●

It is easy to show that any three vectors in two dimensions are *linearly dependent*.[6] In Fig. 1.3 three arbitrary vectors are shown extended in both directions. From the tip of one of the vectors ($\mathbf{a}_3$ in the figure), a line is drawn parallel to one of the other two vectors such that it meets the third vector (or its extension) at point D.

[4] To be specific, we have to define a zero vector, $\mathbf{0}$, such that $\mathbf{a} + \mathbf{0} = \mathbf{a}$ for all $\mathbf{a}$. The zero in (1.2) is really this zero vector.

[5] The caret, ^, symbolizes a vector whose length is 1. The precise meaning of "length" will be given shortly. For now $\hat{\mathbf{e}}_x$ and $\hat{\mathbf{e}}_y$ are treated simply as vectors in the x and y directions, respectively.

[6] The precise definition of "dimension" will be given in Chapter 2. The intuitive meaning of the word is sufficient for the discussion here. A two-dimensional space means a plane.

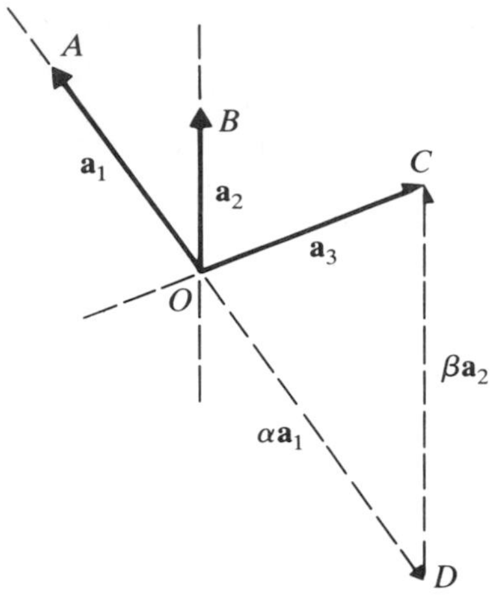

Figure 1.3 Any three vectors in a plane are linearly dependent.

The vectors $\overrightarrow{OD}$ and $\overrightarrow{DC}$ are proportional to $\mathbf{a}_1$ and $\mathbf{a}_2$, respectively, and their sum is equal to $\mathbf{a}_3$. So we can write

$$\mathbf{a}_3 = \overrightarrow{OD} + \overrightarrow{DC} = \alpha\mathbf{a}_1 + \beta\mathbf{a}_2$$

or

$$\alpha\mathbf{a}_1 + \beta\mathbf{a}_2 - \mathbf{a}_3 = 0$$

and $\mathbf{a}_1$, $\mathbf{a}_2$, and $\mathbf{a}_3$ are linearly dependent.

Clearly we cannot do the same with two arbitrary vectors. To see this, consider $\mathbf{a}_1$ and $\mathbf{a}_2$ and assume that there are nonzero real numbers α' and β' such that

$$\alpha'\mathbf{a}_1 + \beta'\mathbf{a}_2 = 0$$

Then

$$\mathbf{a}_1 = -\frac{\beta'}{\alpha'}\mathbf{a}_2$$

and $\mathbf{a}_1$ is proportional to $\mathbf{a}_2$, or $\mathbf{a}_1$ and $\mathbf{a}_2$ lie along the same line. This is contrary to the assumption that $\mathbf{a}_1$ and $\mathbf{a}_2$ are arbitrary. Thus, we have just proved the following theorem.

1.1.3 Theorem The maximum number of linearly independent vectors in a plane is two. Any three vectors in a plane are linearly dependent. ■

This brings us to the notion of a basis.

1.1.4 Definition A set of linearly independent vectors forms a *basis* if any vector can be written as a linear combination of the set.

The vectors $\mathbf{a}_1$ and $\mathbf{a}_2$ in Fig. 1.3 form a basis denoted as $\{\mathbf{a}_1, \mathbf{a}_2\}$. In fact, any two vectors that are not colinear (do not lie along the same line) form a basis in their plane.

With the notion of a basis comes the concept of components of a vector. Given a basis, there is a unique way (see Exercise 1.1.1) in which a particular vector can be written in terms of the vectors forming the basis. The unique coefficients of the basis vectors are called the *components* (or coordinates) of the particular vector in that basis. For instance, the components of $\mathbf{a}_3$ in the basis $\{\mathbf{a}_1, \mathbf{a}_2\}$ of Fig. 1.3 are (α, β).

Example 1.1.2

In Fig. 1.4 the vector $\mathbf{a}$ has components (3, 4) in the basis consisting of the linearly independent vectors $\hat{\mathbf{e}}_x$ and $\hat{\mathbf{e}}_y$. It is clear in the figure that

$$\mathbf{a} = 3\hat{\mathbf{e}}_x + 4\hat{\mathbf{e}}_y$$ •

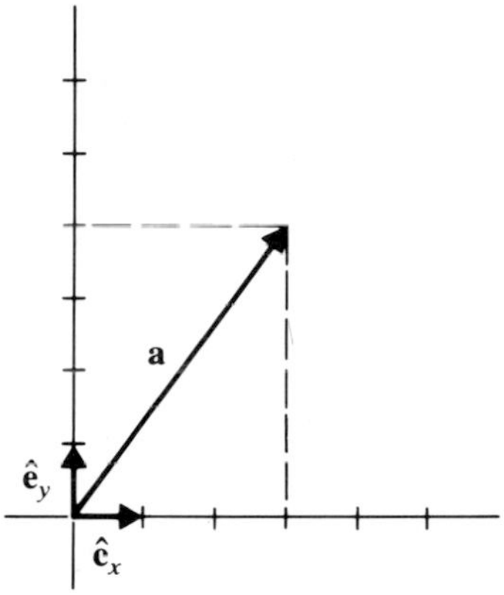

Figure 1.4 The components of the vector $\mathbf{a}$ in the basis $\{\mathbf{e}_x, \mathbf{e}_y\}$ are (3, 4).

Exercises

1.1.1 Show that the components of a vector in a given basis in a plane are unique.

1.1.2 What are the components of each vector in a basis $\{\mathbf{a}_1, \mathbf{a}_2\}$ with respect to that basis?

1.2 TRANSFORMATION OF COMPONENTS

There are infinitely many bases in a plane, because there are infinitely many pairs of vectors that can be chosen to form a basis. Therefore, there are infinitely many sets of components for any particular vector, and it is desirable to be able to find a relation between any two such sets. Such a relation employs the machinery of matrices, whose elementary properties should be familiar to the reader and on which Chapter 3 will elaborate (see, however, the exercises at the end of this section).

Consider a vector $\mathbf{c}$ with components (α_1, α_2) in the basis $\{\mathbf{a}_1, \mathbf{a}_2\}$ and components (α'_1, α'_2) in the basis $\{\mathbf{a}'_1, \mathbf{a}'_2\}$. We can write

$$\mathbf{c} = \alpha_1 \mathbf{a}_1 + \alpha_2 \mathbf{a}_2 \qquad \text{and} \qquad \mathbf{c} = \alpha'_1 \mathbf{a}'_1 + \alpha'_2 \mathbf{a}'_2$$

Since $\mathbf{a}'_1$ and $\mathbf{a}'_2$ form a basis, any vector, in particular, $\mathbf{a}_1$ or $\mathbf{a}_2$, can be written in terms of them:

$$\mathbf{a}_1 = \gamma_{11}\mathbf{a}'_1 + \gamma_{21}\mathbf{a}'_2$$

$$\mathbf{a}_2 = \gamma_{12}\mathbf{a}'_1 + \gamma_{22}\mathbf{a}'_2$$

where $(\gamma_{11}, \gamma_{21})$ and $(\gamma_{12}, \gamma_{22})$ are, respectively, components of $\mathbf{a}_1$ and $\mathbf{a}_2$ in the basis $\{\mathbf{a}'_1, \mathbf{a}'_2\}$. Combining the above equations, we obtain

$$\alpha_1(\gamma_{11}\mathbf{a}'_1 + \gamma_{21}\mathbf{a}'_2) + \alpha_2(\gamma_{12}\mathbf{a}'_1 + \gamma_{22}\mathbf{a}'_2) = \alpha'_1\mathbf{a}'_1 + \alpha'_2\mathbf{a}'_2$$

or

$$(\alpha'_1 - \gamma_{11}\alpha_1 - \gamma_{12}\alpha_2)\mathbf{a}'_1 + (\alpha'_2 - \gamma_{21}\alpha_1 - \gamma_{22}\alpha_2)\mathbf{a}'_2 = 0$$

The linear independence of $\mathbf{a}'_1$ and $\mathbf{a}'_2$ gives

$$\alpha'_1 = \gamma_{11}\alpha_1 + \gamma_{12}\alpha_2 \qquad (1.3a)$$

$$\alpha'_2 = \gamma_{21}\alpha_1 + \gamma_{22}\alpha_2 \qquad (1.3b)$$

These equations can be written concisely as

$$\begin{pmatrix} \alpha'_1 \\ \alpha'_2 \end{pmatrix} = \begin{pmatrix} \gamma_{11} & \gamma_{12} \\ \gamma_{21} & \gamma_{22} \end{pmatrix} \begin{pmatrix} \alpha_1 \\ \alpha_2 \end{pmatrix}$$

Alternatively, if we introduce the column vectors

$$\boldsymbol{a} \equiv \begin{pmatrix} \alpha_1 \\ \alpha_2 \end{pmatrix} \quad \text{and} \quad \boldsymbol{a}' \equiv \begin{pmatrix} \alpha'_1 \\ \alpha'_2 \end{pmatrix} \qquad (1.4a)$$

and the 2×2 matrix

$$\mathsf{\Gamma} \equiv \begin{pmatrix} \gamma_{11} & \gamma_{12} \\ \gamma_{21} & \gamma_{22} \end{pmatrix} \qquad (1.4b)$$

then Eqs. (1.3) take the form

$$\boldsymbol{a}' = \mathsf{\Gamma}\boldsymbol{a} \qquad (1.4c)$$

Let us now imagine that we have a third basis, $\{\mathbf{a}''_1, \mathbf{a}''_2\}$. Then we can write $\mathbf{a}'_1$ and $\mathbf{a}'_2$ as linear combinations,

$$\mathbf{a}'_1 = \gamma'_{11}\mathbf{a}''_1 + \gamma'_{21}\mathbf{a}''_2$$

and

$$\mathbf{a}'_2 = \gamma'_{12}\mathbf{a}''_1 + \gamma'_{22}\mathbf{a}''_2$$

where $(\gamma'_{11}, \gamma'_{21})$ and $(\gamma'_{12}, \gamma'_{22})$ are, respectively, the components of $\mathbf{a}'_1$ and $\mathbf{a}'_2$ in the new basis. These two equations lead to equations involving components,

$$\alpha''_1 = \gamma'_{11}\alpha'_1 + \gamma'_{12}\alpha'_2 \qquad (1.5a)$$

and

$$\alpha''_2 = \gamma'_{21}\alpha'_1 + \gamma'_{22}\alpha'_2 \qquad (1.5b)$$

where (α''_1, α''_2) are the components of $\mathbf{c}$ in the basis $\{\mathbf{a}''_1, \mathbf{a}''_2\}$. We can combine Eqs. (1.5) in matrix form, as before, to obtain

$$\boldsymbol{a}'' = \mathsf{\Gamma}'\boldsymbol{a}' \qquad (1.6a)$$

where $$\boldsymbol{\alpha}'' = \begin{pmatrix} \alpha_1'' \\ \alpha_2'' \end{pmatrix} \quad \text{and} \quad \Gamma' = \begin{pmatrix} \gamma_{11}' & \gamma_{12}' \\ \gamma_{21}' & \gamma_{22}' \end{pmatrix} \tag{1.6b}$$

and $\boldsymbol{\alpha}'$ is as defined before.

We can also discover how $\boldsymbol{\alpha}''$ and $\boldsymbol{\alpha}$ are related by substituting (1.3) in (1.5). This leads to the equations

$$\alpha_1'' = (\gamma_{11}'\gamma_{11} + \gamma_{12}'\gamma_{21})\alpha_1 + (\gamma_{11}'\gamma_{12} + \gamma_{12}'\gamma_{22})\alpha_2$$

and $$\alpha_2'' = (\gamma_{21}'\gamma_{11} + \gamma_{22}'\gamma_{21})\alpha_1 + (\gamma_{21}'\gamma_{12} + \gamma_{22}'\gamma_{22})\alpha_2$$

which, in matrix form, become

$$\boldsymbol{\alpha}'' = \Gamma''\boldsymbol{\alpha} \tag{1.7a}$$

where $$\Gamma'' \equiv \begin{pmatrix} \gamma_{11}'\gamma_{11} + \gamma_{12}'\gamma_{21} & \gamma_{11}'\gamma_{12} + \gamma_{12}'\gamma_{22} \\ \gamma_{21}'\gamma_{11} + \gamma_{22}'\gamma_{21} & \gamma_{21}'\gamma_{12} + \gamma_{22}'\gamma_{22} \end{pmatrix} \tag{1.7b}$$

On the other hand, the matrix equations (1.4c) and (1.6a) yield

$$\boldsymbol{\alpha}'' = \Gamma'(\Gamma\boldsymbol{\alpha})$$

This equation is consistent with Eq. (1.7a) only if matrix multiplication is defined in the usual manner. This consistency requirement is the motivation for the common definition of matrix multiplication. We thus have

$$\Gamma'' = \Gamma'\Gamma$$

Note the order of the product.

Equations (1.3), (1.5), and (1.7) relate the components of the *same* vector, **c**, in *different* bases. However, there is another, more physical, way of interpreting these equations. Consider (1.3) in matrix form. Here $\boldsymbol{\alpha}$ is a column vector representing the components of **c** in the basis $\{\mathbf{a}_1, \mathbf{a}_2\}$. Applying the matrix Γ to it yields a new column vector $\boldsymbol{\alpha}'$, which can be interpreted as the components of a *new* vector, **c**′, in the *same* basis. So, in essence we have changed the vector **c** into a new vector, **c**′, via the *transformation* Γ. The concept of transformation of a vector will be important later in this book. The first interpretation mentioned above is called a *passive transformation* (**c** is passively unchanged as basis vectors are altered); the second interpretation is called *active transformation* (**c** is actively changed into **c**′).

Exercises

1.2.1 What vector is obtained when the vector $\mathbf{a}_2$ in Exercise 1.1.2 is actively transformed with the matrix $\Gamma = \begin{pmatrix} 0 & 1 \\ 0 & 0 \end{pmatrix}$?

1.2.2 For the Pauli spin matrices

$$\sigma_1 = \begin{pmatrix} 0 & 1 \\ 1 & 0 \end{pmatrix}, \quad \sigma_2 = \begin{pmatrix} 0 & -i \\ i & 0 \end{pmatrix} \quad \sigma_3 = \begin{pmatrix} 1 & 0 \\ 0 & -1 \end{pmatrix}$$

show that

$$\sigma_1\sigma_2 - \sigma_2\sigma_1 = 2i\sigma_3$$
$$\sigma_2\sigma_3 - \sigma_3\sigma_2 = 2i\sigma_1$$
$$\sigma_3\sigma_1 - \sigma_1\sigma_3 = 2i\sigma_2$$

1.2.3 For the Dirac matrices

$$\alpha_0 = \begin{pmatrix} 1 & 0 & 0 & 0 \\ 0 & 1 & 0 & 0 \\ 0 & 0 & -1 & 0 \\ 0 & 0 & 0 & -1 \end{pmatrix} \qquad \alpha_1 = \begin{pmatrix} 0 & 0 & 0 & 1 \\ 0 & 0 & 1 & 0 \\ 0 & 1 & 0 & 0 \\ 1 & 0 & 0 & 0 \end{pmatrix}$$

$$\alpha_2 = \begin{pmatrix} 0 & 0 & 0 & -i \\ 0 & 0 & i & 0 \\ 0 & -i & 0 & 0 \\ i & 0 & 0 & 0 \end{pmatrix} \qquad \alpha_3 = \begin{pmatrix} 0 & 0 & 1 & 0 \\ 0 & 0 & 0 & -1 \\ 1 & 0 & 0 & 0 \\ 0 & -1 & 0 & 0 \end{pmatrix}$$

show that

$$(\alpha_i)^2 = 1_4 \qquad \text{for } i = 0, 1, 2, 3$$

and that

$$\alpha_i\alpha_j + \alpha_j\alpha_i = 0 \qquad \text{for } i \neq j$$

3 INNER PRODUCT

A vector space as defined and discussed so far is too limited for the exciting world of physics. Unless we add more substance to it, no physicist will be able to use it. One of the things physicists require is the concept of inner product (or scalar product, or dot product), which is closely related to the concept of metric.

1.3.1 Metric Function (Metric Tensor)

The metric function is defined as follows.[7]

1.3.1 Definition A *metric* is a bilinear mapping, g, from the set of vector pairs into the set of real numbers, $\mathbb{R}$, with the following properties:[8]

$$g(\mathbf{a}_1, \mathbf{a}_2) = g(\mathbf{a}_2, \mathbf{a}_1) \tag{1.8a}$$

$$g(\alpha_1\mathbf{a}_1 + \alpha_2\mathbf{a}_2, \mathbf{a}_3) = \alpha_1 g(\mathbf{a}_1, \mathbf{a}_3) + \alpha_2 g(\mathbf{a}_2, \mathbf{a}_3) \tag{1.8b}$$

$$g(\mathbf{a}_1, \alpha_2\mathbf{a}_2 + \alpha_3\mathbf{a}_3) = \alpha_2 g(\mathbf{a}_1, \mathbf{a}_2) + \alpha_3 g(\mathbf{a}_1, \mathbf{a}_3) \tag{1.8c}$$

for all real numbers $\alpha_1, \alpha_2, \alpha_3$ and all vectors $\mathbf{a}_1, \mathbf{a}_2, \mathbf{a}_3$.

[7] Do not confuse this metric function with the metric used in the Introduction. The two are intimately related, but the latter is defined on any set and the former is defined only on a set of vectors.

[8] For a partial motivation, see Exercise 1.3.1, in which (1.8c) is "proved" using plane geometry and trigonometry.

Note that $g(\mathbf{a}_1, \mathbf{a}_2)$ is a real number for all $\mathbf{a}_1$ and $\mathbf{a}_2$. Property (1.8a) says that the order of arguments is immaterial (symmetric metric); (1.8b) expresses the fact that g is a linear function in its first argument; and (1.8c) expresses the linearity of g in its second argument (hence the word "bilinear").

Definition 1.3.1 is the most general metric that can be constructed. The definition is made more specific by considering $g(\mathbf{a}, \mathbf{a})$ for an arbitrary vector $\mathbf{a}$. A metric is called *positive semidefinite* if

$$g(\mathbf{a}, \mathbf{a}) \geqslant 0 \qquad \forall\ \mathbf{a} \tag{1.8d}$$

If, in addition, $g(\mathbf{a}, \mathbf{a}) = 0$ implies that $\mathbf{a} = \mathbf{0}$, the metric is *positive definite*. Not all metrics are positive. For instance, it is possible for $g(\mathbf{a}, \mathbf{a})$ to be negative in relativity theory. This will be discussed in Chapter 2. For now, we consider only positive definite metrics and write

$$\|\mathbf{a}\| \equiv \sqrt{g(\mathbf{a}, \mathbf{a})} \geqslant 0 \tag{1.9}$$

where $\|\mathbf{a}\|$ is called the *norm* (or the *length*) of the vector $\mathbf{a}$.[9] Any vector $\mathbf{a}$ of nonzero length can be made a unit vector by dividing it by its norm; that is,

$$\hat{\mathbf{e}}_{\mathbf{a}} = \frac{\mathbf{a}}{\|\mathbf{a}\|}$$

1.3.2 Scalar Product (Inner Product)

All of the preceding considerations are abstract. To make contact with the real world, consider the vector

$$\mathbf{c} = \mathbf{a}_1 + \alpha \mathbf{a}_2$$

where α is any real number and $\mathbf{a}_1$ and $\mathbf{a}_2$ are two arbitrary vectors (in a plane). Using Definition 1.3.1 and Eq. (1.9), we have

$$0 \leqslant g(\mathbf{c},\mathbf{c}) = g(\mathbf{a}_1 + \alpha\mathbf{a}_2, \mathbf{a}_1 + \alpha\mathbf{a}_2) = \|\mathbf{a}_1\|^2 + \alpha^2\|\mathbf{a}_2\|^2 + 2\alpha g(\mathbf{a}_1, \mathbf{a}_2) \tag{1.10}$$

This inequality holds for all values of α. In particular, it holds for that value that makes the RHS minimum. Differentiating the RHS, setting the result equal to zero, and solving for the minimizing α, we get

$$\alpha = -\frac{g(\mathbf{a}_1, \mathbf{a}_2)}{\|\mathbf{a}_2\|^2}$$

Substituting this expression for α in Eq. (1.10), we obtain

$$\|\mathbf{a}_1\|^2 + \frac{[g(\mathbf{a}_1, \mathbf{a}_2)]^2}{\|\mathbf{a}_2\|^4}\|\mathbf{a}_2\|^2 - 2\frac{[g(\mathbf{a}_1, \mathbf{a}_2)]^2}{\|\mathbf{a}_2\|^2} \geqslant 0$$

[9] Actually the concept of length, or norm, is more primitive than the metric function, g. This is reflected in the way the scalar product is defined in introductory physics courses in terms of lengths and the cosine of the angle between two vectors. In fact, there are normed linear spaces that cannot be elevated to a space with a metric function (see Chapter 5).

or

$$\|\mathbf{a}_1\|^2 \|\mathbf{a}_2\|^2 \geqslant [g(\mathbf{a}_1, \mathbf{a}_2)]^2$$

Thus

$$\frac{|g(\mathbf{a}_1, \mathbf{a}_2)|}{\|\mathbf{a}_1\| \|\mathbf{a}_2\|} \leqslant 1 \tag{1.11}$$

if $\|\mathbf{a}_1\| \neq 0 \neq \|\mathbf{a}_2\|$.

The LHS of (1.11) is a positive expression that is *always* less than 1 (as long as $\mathbf{a}_1$ and $\mathbf{a}_2$ are of nonzero length). It is therefore natural to *define* an angle θ by

$$\cos\theta \equiv \frac{g(\mathbf{a}_1, \mathbf{a}_2)}{\|\mathbf{a}_1\| \|\mathbf{a}_2\|}$$

or

$$g(\mathbf{a}_1, \mathbf{a}_2) = \|\mathbf{a}_1\| \|\mathbf{a}_2\| \cos\theta$$

The RHS is recognizable as the elementary definition of the *scalar product* of two vectors. It is, therefore, customary to write[10]

$$g(\mathbf{a}_1, \mathbf{a}_2) \equiv \mathbf{a}_1 \cdot \mathbf{a}_2 = \|\mathbf{a}_1\| \|\mathbf{a}_2\| \cos\theta \tag{1.12}$$

Although Eq. (1.11), which motivated the concept of an angle, is true for all kinds of vectors (see Section 2.2), the concept of the angle as defined in (1.12) is restricted to vectors in a plane and in space, where θ is simply the angle between two vectors, as indicated in Fig. 1.5.

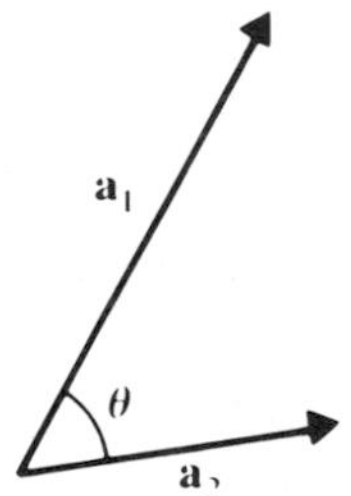

Figure 1.5 The vectors $\mathbf{a}_1$ and $\mathbf{a}_2$ participating in the inner product $\mathbf{a}_1 \cdot \mathbf{a}_2$.

Example 1.3.1

Let us investigate how the scalar product looks in terms of the components in some basis.

Consider the basis $\{\mathbf{a}_1, \mathbf{a}_2\}$ and two arbitrary vectors $\mathbf{b}$ and $\mathbf{c}$. We can write

$$\mathbf{b} = \beta_1 \mathbf{a}_1 + \beta_2 \mathbf{a}_2 \qquad \text{and} \qquad \mathbf{c} = \gamma_1 \mathbf{a}_1 + \gamma_2 \mathbf{a}_2 \tag{1}$$

where (β_1, β_2) are the components of $\mathbf{b}$ and (γ_1, γ_2) are those of $\mathbf{c}$ in the basis $\{\mathbf{a}_1, \mathbf{a}_2\}$. Now we compute the scalar product of $\mathbf{b}$ and $\mathbf{c}$:

$$\begin{aligned}\mathbf{b} \cdot \mathbf{c} \equiv g(\mathbf{b}, \mathbf{c}) &= g(\beta_1 \mathbf{a}_1 + \beta_2 \mathbf{a}_2, \gamma_1 \mathbf{a}_1 + \gamma_2 \mathbf{a}_2) \\ &= \beta_1\gamma_1 g(\mathbf{a}_1, \mathbf{a}_1) + \beta_1\gamma_2 g(\mathbf{a}_1, \mathbf{a}_2) + \beta_2\gamma_1 g(\mathbf{a}_2, \mathbf{a}_1) + \beta_2\gamma_2 g(\mathbf{a}_2, \mathbf{a}_2)\end{aligned} \tag{2}$$

[10] In elementary vector algebra (1.12) is the starting point of the discussion of the dot (scalar) product. However, the approach here, although more abstract, has the advantage of being more easily applied to more general cases.

We define a 2×2 matrix, G, with elements given as follows:

$$g_{ij} \equiv g(\mathbf{a}_i, \mathbf{a}_j) \equiv \mathbf{a}_i \cdot \mathbf{a}_j \qquad i, j = 1, 2 \tag{3}$$

That is,

$$\mathsf{G} = \begin{pmatrix} \mathbf{a}_1 \cdot \mathbf{a}_1 & \mathbf{a}_1 \cdot \mathbf{a}_2 \\ \mathbf{a}_2 \cdot \mathbf{a}_1 & \mathbf{a}_2 \cdot \mathbf{a}_2 \end{pmatrix} \equiv \begin{pmatrix} g_{11} & g_{12} \\ g_{21} & g_{22} \end{pmatrix} \tag{4}$$

Then (2) can be written as

$$\mathbf{b} \cdot \mathbf{c} = \beta_1 g_{11} \gamma_1 + \beta_1 g_{12} \gamma_2 + \beta_2 g_{21} \gamma_1 + \beta_2 g_{22} \gamma_2$$

or, in matrix form, as

$$\mathbf{b} \cdot \mathbf{c} = (\beta_1 \quad \beta_2) \begin{pmatrix} g_{11} & g_{12} \\ g_{21} & g_{22} \end{pmatrix} \begin{pmatrix} \gamma_1 \\ \gamma_2 \end{pmatrix} \equiv \tilde{\boldsymbol{\beta}} \mathsf{G} \boldsymbol{\gamma} \tag{5}$$

where $\boldsymbol{\beta}$ and γ are column vectors and $\tilde{\boldsymbol{\beta}}$ is the row vector associated with $\boldsymbol{\beta}$.

The matrix G is the matrix of the metric function g. In a general basis G may not take a simple form and may even be a function of position. However, in some special bases G may be considerably simplified. For instance, consider the familiar basis $\{\hat{\mathbf{e}}_x, \hat{\mathbf{e}}_y\}$ in the plane. These vectors give rise to a metric with elements

$$g_{11} = \mathbf{a}_1 \cdot \mathbf{a}_1 \equiv \hat{\mathbf{e}}_x \cdot \hat{\mathbf{e}}_x = 1 \qquad g_{12} = \mathbf{a}_1 \cdot \mathbf{a}_2 \equiv \hat{\mathbf{e}}_x \cdot \hat{\mathbf{e}}_y = 0$$

$$g_{21} = \mathbf{a}_2 \cdot \mathbf{a}_1 \equiv \hat{\mathbf{e}}_y \cdot \hat{\mathbf{e}}_x = 0 \qquad g_{22} = \mathbf{a}_2 \cdot \mathbf{a}_2 \equiv \hat{\mathbf{e}}_y \cdot \hat{\mathbf{e}}_y = 1$$

Thus, the matrix G will be

$$\mathsf{G} = \begin{pmatrix} 1 & 0 \\ 0 & 1 \end{pmatrix} = \mathsf{1}_2 \tag{6}$$

that is, the two-dimensional identity matrix. In the basis $\{\hat{\mathbf{e}}_x, \hat{\mathbf{e}}_y\}$ Eq. (5) takes the simple form

$$\mathbf{b} \cdot \mathbf{c} = (\beta_1 \quad \beta_2) \begin{pmatrix} 1 & 0 \\ 0 & 1 \end{pmatrix} \begin{pmatrix} \gamma_1 \\ \gamma_2 \end{pmatrix} = (\beta_1 \quad \beta_2) \begin{pmatrix} \gamma_1 \\ \gamma_2 \end{pmatrix} = \beta_1 \gamma_1 + \beta_2 \gamma_2 \tag{7}$$

which is the familiar form of the scalar product in terms of components. This simple form is one of the motivations for using orthonormal bases (described in the next subsection).

Note that for a vector $\mathbf{v}$ whose components in an orthonormal basis $\{\hat{\mathbf{e}}_x, \hat{\mathbf{e}}_y\}$ are (v_1, v_2), we can use (7) to write

$$\mathbf{v} \cdot \mathbf{v} = v_1^2 + v_2^2$$

and (1.9) gives

$$\|\mathbf{v}\| = \sqrt{\mathbf{v} \cdot \mathbf{v}} = \sqrt{v_1^2 + v_2^2} \tag{8}$$

As a concrete example, consider the vectors

$$\mathbf{a}_1 = \hat{\mathbf{e}}_x + \hat{\mathbf{e}}_y \qquad \text{and} \qquad \mathbf{a}_2 = 2\hat{\mathbf{e}}_x + \hat{\mathbf{e}}_y \tag{9}$$

The metric matrix elements in the basis $\{\mathbf{a}_1, \mathbf{a}_2\}$ are

$$g_{11} = \mathbf{a}_1 \cdot \mathbf{a}_1 = (\hat{\mathbf{e}}_x + \hat{\mathbf{e}}_y) \cdot (\hat{\mathbf{e}}_x + \hat{\mathbf{e}}_y) = 2 \qquad g_{12} = \mathbf{a}_1 \cdot \mathbf{a}_2 = (\hat{\mathbf{e}}_x + \hat{\mathbf{e}}_y) \cdot (2\hat{\mathbf{e}}_x + \hat{\mathbf{e}}_y) = 3$$

$$g_{21} = \mathbf{a}_2 \cdot \mathbf{a}_1 = \mathbf{a}_1 \cdot \mathbf{a}_2 = g_{12} = 3 \qquad g_{22} = (2\hat{\mathbf{e}}_x + \hat{\mathbf{e}}_y) \cdot (2\hat{\mathbf{e}}_x + \hat{\mathbf{e}}_y) = 5$$

or, in matrix form,

$$\mathsf{G} = \begin{pmatrix} 2 & 3 \\ 3 & 5 \end{pmatrix} \tag{10}$$

Now consider the vectors **b** and **c**, whose components in $\{\mathbf{a}_1, \mathbf{a}_2\}$ are, respectively, (1, 1) and (− 3, 2). We can compute the scalar product of **b** and **c** in terms of these components using (5) and (10):

$$\mathbf{b}\cdot\mathbf{c} = (1 \quad 1)\begin{pmatrix} 2 & 3 \\ 3 & 5 \end{pmatrix}\begin{pmatrix} -3 \\ 2 \end{pmatrix} = 1 \tag{11}$$

We can also write **b** and **c** in terms of $\hat{\mathbf{e}}_x$ and $\hat{\mathbf{e}}_y$ and use (7) to find $\mathbf{b}\cdot\mathbf{c}$. Since **b** has the components (1, 1) in $\{\mathbf{a}_1, \mathbf{a}_2\}$, it can be written as

$$\mathbf{b} = \mathbf{a}_1 + \mathbf{a}_2 = (\hat{\mathbf{e}}_x + \hat{\mathbf{e}}_y) + (2\hat{\mathbf{e}}_x + \hat{\mathbf{e}}_y) = 3\hat{\mathbf{e}}_x + 2\hat{\mathbf{e}}_y$$

Similarly,

$$\mathbf{c} = -3\mathbf{a}_1 + 2\mathbf{a}_2 = -3(\hat{\mathbf{e}}_x + \hat{\mathbf{e}}_y) + 2(2\hat{\mathbf{e}}_x + \hat{\mathbf{e}}_y) = \hat{\mathbf{e}}_x - \hat{\mathbf{e}}_y$$

Thus, in $\{\hat{\mathbf{e}}_x, \hat{\mathbf{e}}_y\}$ **b** has components (3,2), and **c** has components (1, − 1). Then (7) gives

$$\mathbf{b}\cdot\mathbf{c} = (3 \quad 2)\begin{pmatrix} 1 \\ -1 \end{pmatrix} = 3 - 2 = 1 \tag{12}$$

which agrees with (11). •

Example 1.3.2

Consider a metric represented by the matrix

$$\mathsf{G} = \begin{pmatrix} 1 & 0 \\ 0 & -1 \end{pmatrix} \tag{1}$$

For such a metric the results of Example 1.3.1 lead to a scalar product between two vectors $\mathbf{a} = (\alpha_1, \alpha_2)$ and $\mathbf{b} = (\beta_1, \beta_2)$; this product is given by

$$\mathbf{a}\cdot\mathbf{b} = \tilde{\boldsymbol{\alpha}}\mathsf{G}\boldsymbol{\beta} = (\alpha_1 \quad \alpha_2)\begin{pmatrix} 1 & 0 \\ 0 & -1 \end{pmatrix}\begin{pmatrix} \beta_1 \\ \beta_2 \end{pmatrix} = \alpha_1\beta_1 - \alpha_2\beta_2 \tag{2}$$

In particular,

$$\|\mathbf{a}\|^2 \equiv \mathbf{a}\cdot\mathbf{a} = \alpha_1^2 - \alpha_2^2 \tag{3}$$

It is clear that $\|\mathbf{a}\|^2$ could be positive, negative, or zero depending on whether $|\alpha_1| > |\alpha_2|$, $|\alpha_1| < |\alpha_2|$, or $|\alpha_1| = |\alpha_2|$. Such a metric is not positive definite. In fact, there can be a nonzero vector such as $\mathbf{a} = (\alpha, \alpha)$ that has a zero norm; such a vector is called a *null vector*.

The generalization of the metric in (1) to four dimensions (three for space and one for time) is at the root of the theory of relativity. In the special theory of relativity, the null vectors are the four-dimensional vectors corresponding to the space-time displacements of light signals. •

1.3.3 Orthogonality

Vectors that are mutually perpendicular are very convenient for describing physical situations. For example, choosing basis vectors that are perpendicular significantly simplifies the important operation of rotation.

1.3.2 Definition Two vectors are said to be *orthogonal* if their scalar product vanishes. A vector is *normal* if it is of unit norm (length). Two vectors are *orthonormal* if they are orthogonal and both are normal.

Consider the vector $\mathbf{a}$ in the plane and write it as a linear combination of two orthonormal basis vectors, $\hat{\mathbf{e}}_1$ and $\hat{\mathbf{e}}_2$,

$$\mathbf{a} = \alpha_1 \hat{\mathbf{e}}_1 + \alpha_2 \hat{\mathbf{e}}_2$$

as shown in Fig. 1.6. The orthogonality of $\hat{\mathbf{e}}_1$ and $\hat{\mathbf{e}}_2$ implies that

$$\hat{\mathbf{e}}_1 \cdot \hat{\mathbf{e}}_2 = 0 = \hat{\mathbf{e}}_2 \cdot \hat{\mathbf{e}}_1$$

This in turn can be used to obtain the components α_1 and α_2 very conveniently. We simply take the inner product of $\mathbf{a}$ with $\hat{\mathbf{e}}_1$ to obtain

$$\mathbf{a} \cdot \hat{\mathbf{e}}_1 = \alpha_1 \hat{\mathbf{e}}_1 \cdot \hat{\mathbf{e}}_1 + \alpha_2 \hat{\mathbf{e}}_2 \cdot \hat{\mathbf{e}}_1$$

But $\hat{\mathbf{e}}_1 \cdot \hat{\mathbf{e}}_1 = 1$, so

$$\mathbf{a} \cdot \hat{\mathbf{e}}_1 = \alpha_1$$

Similarly, the inner product of $\mathbf{a}$ with $\hat{\mathbf{e}}_2$ is

$$\mathbf{a} \cdot \hat{\mathbf{e}}_2 = \alpha_2$$

Note that we would not have these simple relations if $\hat{\mathbf{e}}_1$ and $\hat{\mathbf{e}}_2$ were not orthonormal.

Now suppose we rotate $\mathbf{a}$ (without changing its length) through an angle θ to a new position, as shown by $\mathbf{a}'$ in Fig. 1.6. What relation is there between the components (α'_1, α'_2) of $\mathbf{a}'$ and the components (α_1, α_2) of $\mathbf{a}$? The simplest way to answer this question is to note that

$$\alpha'_1 = \mathbf{a}' \cdot \hat{\mathbf{e}}_1 = \|\mathbf{a}'\| \cos(\theta + \varphi) = \|\mathbf{a}\| \cos(\theta + \varphi)$$

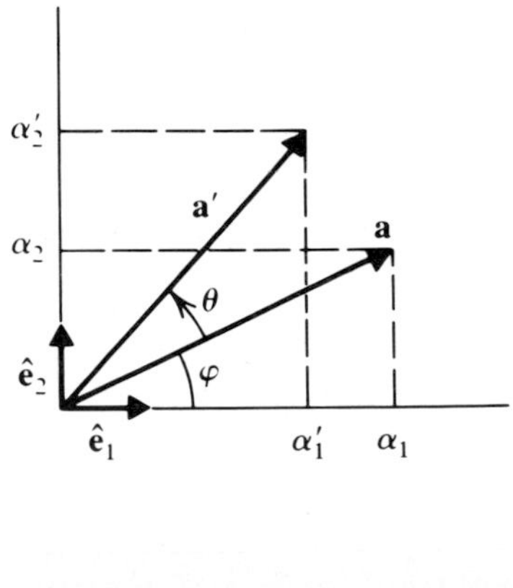

Figure 1.6 The vectors $\mathbf{a}$ and $\mathbf{a}'$ related by a rotation of the angle θ.

and $$\alpha_2' = \mathbf{a}' \cdot \hat{\mathbf{e}}_2 = \|\mathbf{a}'\| \sin(\theta + \varphi) = \|\mathbf{a}\| \sin(\theta + \varphi)$$

where we have used the fact that the lengths of **a** and **a**′ are the same. Now we expand the sine and cosine to get

$$\alpha_1' = \|\mathbf{a}\| \cos\theta \cos\varphi - \|\mathbf{a}\| \sin\theta \sin\varphi$$

and $$\alpha_2' = \|\mathbf{a}\| \sin\theta \cos\varphi + \|\mathbf{a}\| \cos\theta \sin\varphi$$

and use the fact that (see Fig. 1.6)

$$\|\mathbf{a}\| \cos\varphi = \alpha_1 \qquad \text{and} \qquad \|\mathbf{a}\| \sin\varphi = \alpha_2$$

to obtain

$$\alpha_1' = \alpha_1 \cos\theta - \alpha_2 \sin\theta$$

$$\alpha_2' = \alpha_1 \sin\theta + \alpha_2 \cos\theta$$

These equations can be written in matrix form as

$$\begin{pmatrix} \alpha_1' \\ \alpha_2' \end{pmatrix} = \begin{pmatrix} \cos\theta & -\sin\theta \\ \sin\theta & \cos\theta \end{pmatrix} \begin{pmatrix} \alpha_1 \\ \alpha_2 \end{pmatrix} \tag{1.13}$$

The matrix A that multiplies the column vector on the RHS of (1.13) has a special and important property. If we interchange its rows and columns so that the first row becomes the first column and the second row becomes the second column, we end up with the matrix

$$\tilde{\mathsf{A}} \equiv \begin{pmatrix} \cos\theta & \sin\theta \\ -\sin\theta & \cos\theta \end{pmatrix}$$

This operation is called *transposition*, and the matrix $\tilde{\mathsf{A}}$ is called the *transpose* of the matrix in (1.13). In general, the transpose of an $m \times n$ matrix whose element in the ith row and jth column is a_{ij} is an $n \times m$ matrix whose element in the ith row and jth column is a_{ji}.

For the particular matrix A in (1.13),

$$\mathsf{A}\tilde{\mathsf{A}} = \begin{pmatrix} \cos\theta & -\sin\theta \\ \sin\theta & \cos\theta \end{pmatrix} \begin{pmatrix} \cos\theta & \sin\theta \\ -\sin\theta & \cos\theta \end{pmatrix} = \begin{pmatrix} 1 & 0 \\ 0 & 1 \end{pmatrix} = \tilde{\mathsf{A}}\mathsf{A}$$

This shows that for this particular 2×2 matrix the transpose is the inverse, or

$$\tilde{\mathsf{A}} = \mathsf{A}^{-1} \tag{1.14}$$

Matrices for which the transpose is the inverse are called orthogonal for reasons that will become clear later. They, and their generalization to complex matrices, occupy a special place in quantum theory and other applications.

1.3.4 The Gram-Schmidt Process

Once the importance of orthonormal basis vectors has been demonstrated, the question that naturally arises is whether they exist at all. In other words, given a set of independent vectors that form a basis, can we find new orthonormal basis vectors from linear combinations of them? The answer is yes, and the systematic procedure for obtaining orthonormal basis vectors out of any set of linearly independent basis vectors is called the *Gram-Schmidt orthonormalization process.*

In the plane Gram-Schmidt orthonormalization is very simple. Suppose we have two basis vectors, $\mathbf{a}_1$ and $\mathbf{a}_2$ (see Fig. 1.7). To obtain two orthonormal vectors, we first make a unit vector, $\hat{\mathbf{e}}_1$, out of $\mathbf{a}_1$ by dividing it by its length, $\|\mathbf{a}_1\|$:

$$\hat{\mathbf{e}}_1 = \frac{\mathbf{a}_1}{\|\mathbf{a}_1\|}$$

It is clear that $\hat{\mathbf{e}}_1$ is a unit vector because

$$\hat{\mathbf{e}}_1 \cdot \hat{\mathbf{e}}_1 = \frac{\mathbf{a}_1}{\|\mathbf{a}_1\|} \cdot \frac{\mathbf{a}_1}{\|\mathbf{a}_1\|} = \frac{\mathbf{a}_1 \cdot \mathbf{a}_1}{\|\mathbf{a}_1\|^2} = 1$$

Now we take a linear combination of $\mathbf{a}_2$ and $\hat{\mathbf{e}}_1$ that is orthogonal to $\hat{\mathbf{e}}_1$. This can be done by writing

$$\mathbf{a}_2 = (\mathbf{a}_2)_{\|} + (\mathbf{a}_2)_{\perp}$$

where $(\mathbf{a}_2)_{\|}$ and $(\mathbf{a}_2)_{\perp}$ are, respectively, parallel and perpendicular to $\hat{\mathbf{e}}_1$. Since $(\mathbf{a}_2)_{\|}$ is in the direction of $\hat{\mathbf{e}}_1$, we can write

$$(\mathbf{a}_2)_{\|} = \|(\mathbf{a}_2)_{\|}\|\hat{\mathbf{e}}_1 = \|\mathbf{a}_2\| \cos\theta\, \hat{\mathbf{e}}_1 = (\mathbf{a}_2 \cdot \hat{\mathbf{e}}_1)\hat{\mathbf{e}}_1$$

From the two preceding equations, we obtain

$$(\mathbf{a}_2)_{\perp} = \mathbf{a}_2 - (\mathbf{a}_2 \cdot \hat{\mathbf{e}}_1)\hat{\mathbf{e}}_1$$

We see that $(\mathbf{a}_2)_{\perp}$ is a linear combination of $\mathbf{a}_2$ and $\hat{\mathbf{e}}_1$ ($\mathbf{a}_2 \cdot \hat{\mathbf{e}}_1$ is just a number) and is perpendicular to $\hat{\mathbf{e}}_1$ [it is easy to see that $(\mathbf{a}_2)_{\perp} \cdot \hat{\mathbf{e}}_1 = 0$]. If we divide $(\mathbf{a}_2)_{\perp}$ by its length, we get a unit vector that is perpendicular to $\hat{\mathbf{e}}_1$. We can call this unit vector $\hat{\mathbf{e}}_2$:

$$\hat{\mathbf{e}}_2 = \frac{(\mathbf{a}_2)_{\perp}}{\|(\mathbf{a}_2)_{\perp}\|} = \frac{\mathbf{a}_2 - (\mathbf{a}_2 \cdot \hat{\mathbf{e}}_1)\hat{\mathbf{e}}_1}{\|\mathbf{a}_2 - (\mathbf{a}_2 \cdot \hat{\mathbf{e}}_1)\hat{\mathbf{e}}_1\|}$$

We have obtained a pair of orthonormal basis vectors from two arbitrary linearly independent basis vectors. This procedure is applicable to more general cases, and Gram-Schmidt orthonormalization for three and more dimensions will be illustrated later.

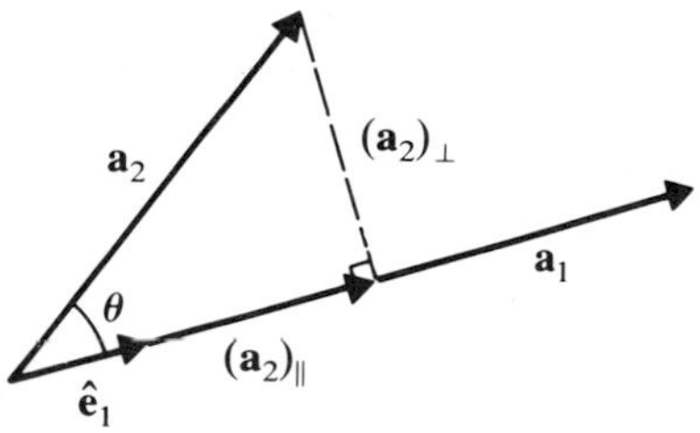

Figure 1.7 Gram-Schmidt orthonormalization in the plane.

As discussed in Example 1.3.1, in terms of the components of vectors in an orthonormal basis, the inner product becomes simple and symmetric. Using the linearity of the scalar product and the orthonormality of $\hat{\mathbf{e}}_1$ and $\hat{\mathbf{e}}_2$, we can write

$$\begin{aligned}\mathbf{a}\cdot\mathbf{b} &\equiv (\alpha_1\hat{\mathbf{e}}_1 + \alpha_2\hat{\mathbf{e}}_2)\cdot(\beta_1\hat{\mathbf{e}}_1 + \beta_2\hat{\mathbf{e}}_2) \\ &= \alpha_1\beta_1\hat{\mathbf{e}}_1\cdot\hat{\mathbf{e}}_1 + \alpha_1\beta_2\hat{\mathbf{e}}_1\cdot\hat{\mathbf{e}}_2 + \alpha_2\beta_1\hat{\mathbf{e}}_2\cdot\hat{\mathbf{e}}_1 + \alpha_2\beta_2\hat{\mathbf{e}}_2\cdot\hat{\mathbf{e}}_2 \qquad (1.15) \\ &= \alpha_1\beta_1 + \alpha_2\beta_2\end{aligned}$$

which is the expression in Example 1.3.1.

Exercises

1.3.1 Show that $\mathbf{a}\cdot(\mathbf{b}+\mathbf{c}) = \mathbf{a}\cdot\mathbf{b} + \mathbf{a}\cdot\mathbf{c}$ in two dimensions. (*Warning*! This exercise mostly involves geometry and trigonometry.)

1.3.2 Find the angle between two vectors **b** and **c**, whose components in an orthonormal basis are, respectively, (1,2) and (2, − 3).

1.3.3 Use the Gram-Schmidt process to find the orthonormal vectors obtained from the two vectors in Exercise 1.3.2.

1.3.4 Let A be a 2×2 antisymmetric matrix, that is, one with the property that

$$a_{ij} = -a_{ji} \qquad \forall\, i, j$$

In particular, the diagonal elements, a_{ii}, are zero. (Why?) Show that e^{A} is an orthogonal matrix, where, by definition,

$$e^{\mathsf{A}} = 1_2 + \mathsf{A} + \frac{1}{2!}\mathsf{A}^2 + \cdots = \sum_{k=0}^{\infty} \frac{1}{k!}\mathsf{A}^k$$

1.4 VECTORS IN SPACE

The ideas developed so far can be easily generalized to vectors in space. In particular, if $\mathbf{a}_1, \mathbf{a}_2, \ldots, \mathbf{a}_n$ are vectors in space and $\alpha_1, \alpha_2, \ldots, \alpha_n$ are real numbers, then the linear combination

$$\alpha_1\mathbf{a}_1 + \alpha_2\mathbf{a}_2 + \cdots + \alpha_n\mathbf{a}_n \equiv \sum_{i=1}^{n} \alpha_i\mathbf{a}_i$$

is also a vector in space.

It can be shown geometrically that any four vectors in space are linearly dependent. The proof involves the same ideas used for vectors in the plane (see Fig. 1.3), but their implementation is more complicated because it is harder to draw three-dimensional objects on paper. Although any four vectors are linearly dependent in three dimensions, any three noncoplanar (not lying in the same plane) vectors are linearly independent and form a basis in space.

We can also define a metric in space. It has all the properties expressed by (1.8a)–(1.8c). If we desire, we can make it positive definite by imposing (1.8d) and the fact that $g(\mathbf{a}, \mathbf{a}) = 0$ requires $\mathbf{a}$ to be zero. We can then define the angle θ between two vectors by (1.12). In other words, all the properties of vectors in the plane translate verbatim to vectors in space.

Example 1.4.1

The parametric equation of a line through two given points can be obtained in vector form by noting that any point in space defines a vector whose components are the coordinates of the given point. That is, if the components of the points P and Q in Fig. 1.8 are, respectively, (p_1, p_2, p_3) and (q_1, q_2, q_3), then we can define vectors $\mathbf{p}$ and $\mathbf{q}$ with those components. An arbitrary point X with components (x_1, x_2, x_3) [or (x, y, z)] will lie on the line PQ if and only if the vector $\mathbf{x} \equiv (x_1, x_2, x_3)$ has its tip on that line. This will happen if and only if the vector joining P and X, namely $\mathbf{x} - \mathbf{p}$, is proportional to the vector joining P and Q, namely $\mathbf{q} - \mathbf{p}$. Thus, for some real number t, we must have

$$\mathbf{x} - \mathbf{p} = t(\mathbf{q} - \mathbf{p}) \tag{1}$$

or

$$\mathbf{x} = \mathbf{p} + t(\mathbf{q} - \mathbf{p}) = (1 - t)\mathbf{p} + t\mathbf{q} \tag{2}$$

Equation (2) is the vector form of the equation of a line. We can write it in component form by noting that the equality of vectors implies the equality of corresponding components. Thus,

$$\begin{aligned} x_1 &= (1 - t)p_1 + tq_1 \\ x_2 &= (1 - t)p_2 + tq_2 \\ x_3 &= (1 - t)p_3 + tq_3 \end{aligned} \tag{3}$$

which is the usual parametric equation for a line. ●

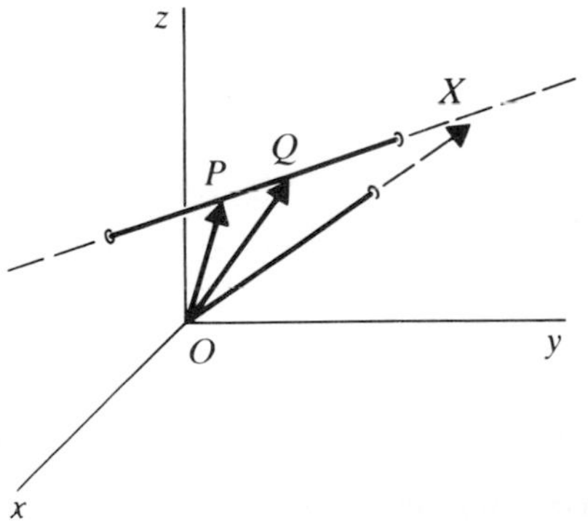

Figure 1.8 The line connecting P and Q.

1.4.1 Orthogonal Vectors in Space

It is convenient to construct orthonormal basis vectors in space. This can be done by the Gram-Schmidt orthonormalization process. Consider three noncoplanar vectors $\mathbf{a}_1$, $\mathbf{a}_2$, and $\mathbf{a}_3$, as shown in Fig. 1.9. In the plane of the two vectors $\mathbf{a}_1$ and $\mathbf{a}_2$, we can apply the Gram-Schmidt process and obtain two orthonormal vectors $\hat{\mathbf{e}}_1$ and $\hat{\mathbf{e}}_2$

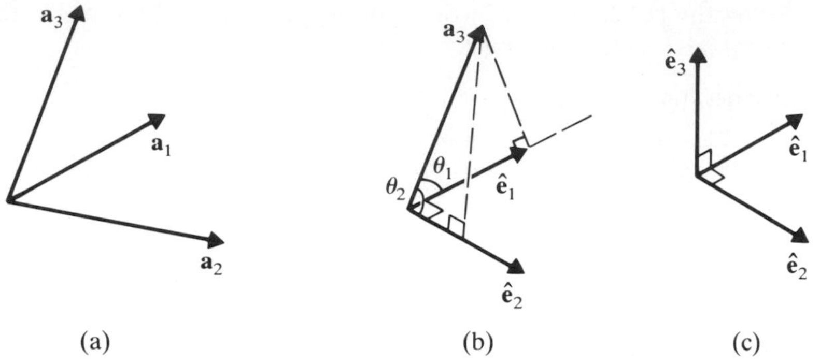

Figure 1.9 Gram-Schmidt orthonormalization in space: (a) three independent vectors, (b) $\hat{\mathbf{e}}_1$ and $\hat{\mathbf{e}}_2$ obtained by orthonormalization of $\mathbf{a}_1$ and $\mathbf{a}_2$ in the plane, and (c) the three orthonormal vectors.

[Fig. 1.9(b)]. To get the third vector in the set, we note that we can write

$$\mathbf{a}_3 = (\mathbf{a}_3)_\perp + (\mathbf{a}_3^{(1)})_{||} + (\mathbf{a}_3^{(2)})_{||}$$

where $(\mathbf{a}_3)_\perp$ is the component of $\mathbf{a}_3$ perpendicular to $\hat{\mathbf{e}}_1$ and $\hat{\mathbf{e}}_2$, $(\mathbf{a}_3^{(1)})_{||}$ is the component parallel to $\hat{\mathbf{e}}_1$, and $(\mathbf{a}_3^{(2)})_{||}$ is the component parallel to $\hat{\mathbf{e}}_2$. We can write $(\mathbf{a}_3^{(1)})_{||}$ and $(\mathbf{a}_3^{(2)})_{||}$ as follows:

$$(\mathbf{a}_3^{(1)})_{||} = \|\mathbf{a}_3\| \cos\theta_1\, \hat{\mathbf{e}}_1 = (\mathbf{a}_3 \cdot \hat{\mathbf{e}}_1)\hat{\mathbf{e}}_1$$

$$(\mathbf{a}_3^{(2)})_{||} = \|\mathbf{a}_3\| \cos\theta_2\, \hat{\mathbf{e}}_2 = (\mathbf{a}_3 \cdot \hat{\mathbf{e}}_2)\hat{\mathbf{e}}_2$$

Thus, we have

$$\mathbf{a}_3 = (\mathbf{a}_3)_\perp + (\mathbf{a}_3 \cdot \hat{\mathbf{e}}_1)\hat{\mathbf{e}}_1 + (\mathbf{a}_3 \cdot \hat{\mathbf{e}}_2)\hat{\mathbf{e}}_2$$

or

$$(\mathbf{a}_3)_\perp = \mathbf{a}_3 - (\mathbf{a}_3 \cdot \hat{\mathbf{e}}_1)\hat{\mathbf{e}}_1 - (\mathbf{a}_3 \cdot \hat{\mathbf{e}}_2)\hat{\mathbf{e}}_2$$

Now we define $\hat{\mathbf{e}}_3$:

$$\hat{\mathbf{e}}_3 = \frac{(\mathbf{a}_3)_\perp}{\|(\mathbf{a}_3)_\perp\|} = \frac{\mathbf{a}_3 - (\mathbf{a}_3 \cdot \hat{\mathbf{e}}_1)\hat{\mathbf{e}}_1 - (\mathbf{a}_3 \cdot \hat{\mathbf{e}}_2)\hat{\mathbf{e}}_2}{\|\mathbf{a}_3 - (\mathbf{a}_3 \cdot \hat{\mathbf{e}}_1)\hat{\mathbf{e}}_1 - (\mathbf{a}_3 \cdot \hat{\mathbf{e}}_2)\hat{\mathbf{e}}_2\|}$$

Once $\hat{\mathbf{e}}_1$, $\hat{\mathbf{e}}_2$, and $\hat{\mathbf{e}}_3$ are constructed, any vector can be written in terms of them. If **a** and **b** are vectors, we can write

$$\mathbf{a} = \alpha_1\hat{\mathbf{e}}_1 + \alpha_2\hat{\mathbf{e}}_2 + \alpha_3\hat{\mathbf{e}}_3$$

$$\mathbf{b} = \beta_1\hat{\mathbf{e}}_1 + \beta_2\hat{\mathbf{e}}_2 + \beta_3\hat{\mathbf{e}}_3$$

As with vectors in the plane, we can express the dot product in terms of the components of **a** and **b** (see, however, Exercise 1.4.1):

$$\mathbf{a} \cdot \mathbf{b} = \alpha_1\beta_1 + \alpha_2\beta_2 + \alpha_3\beta_3 \tag{1.16}$$

Equation (1.16) follows from the distributive property of the inner product [Eqs. (1.8b) and (1.8c)] and the orthonormality of $\hat{\mathbf{e}}_1$, $\hat{\mathbf{e}}_2$, and $\hat{\mathbf{e}}_3$.

1.4.2 Orthogonal 3 × 3 Matrices

If we represent **a** and **b** as the column vectors

$$\boldsymbol{\alpha} \equiv \begin{pmatrix} \alpha_1 \\ \alpha_2 \\ \alpha_3 \end{pmatrix} \quad \text{and} \quad \boldsymbol{\beta} \equiv \begin{pmatrix} \beta_1 \\ \beta_2 \\ \beta_3 \end{pmatrix}$$

then (1.16) becomes

$$\tilde{\boldsymbol{\alpha}}\boldsymbol{\beta} \equiv (\alpha_1 \quad \alpha_2 \quad \alpha_3)\begin{pmatrix} \beta_1 \\ \beta_2 \\ \beta_3 \end{pmatrix} = \alpha_1\beta_1 + \alpha_2\beta_2 + \alpha_3\beta_3 = \mathbf{a}\cdot\mathbf{b} \tag{1.17}$$

Now consider the same inner product in a different orthonormal basis. It is clear from earlier considerations in two dimensions that there must exist a 3 × 3 matrix $\mathsf{\Gamma}$ such that

$$\boldsymbol{\alpha}' = \begin{pmatrix} \alpha'_1 \\ \alpha'_2 \\ \alpha'_3 \end{pmatrix} = \mathsf{\Gamma}\boldsymbol{\alpha} \quad \text{and} \quad \boldsymbol{\beta}' = \begin{pmatrix} \beta'_1 \\ \beta'_2 \\ \beta'_3 \end{pmatrix} = \mathsf{\Gamma}\boldsymbol{\beta}$$

where $\alpha'_1, \alpha'_2, \alpha'_3$ and $\beta'_1, \beta'_2, \beta'_3$ are the components of **a** and **b**, respectively, in the new basis (passive transformation) *or* the components of the transformed vector in the original coordinate system (active transformation). We can now ask what the inner product would be in this new situation. In other words, if we apply the same transformation to both **a** and **b**, what happens to $\mathbf{a}\cdot\mathbf{b}$, or $\tilde{\boldsymbol{\alpha}}\boldsymbol{\beta}$?

Geometrically, since the lengths and angles do not change in going from one orthonormal basis to another, we conclude that

$$\mathbf{a}'\cdot\mathbf{b}' = \mathbf{a}\cdot\mathbf{b}$$

or

$$\tilde{\boldsymbol{\alpha}}'\boldsymbol{\beta}' = \tilde{\boldsymbol{\alpha}}\boldsymbol{\beta}$$

This means that in an orthonormal coordinate transformation (or rotation) the dot product does not change.[11] Substituting for $\boldsymbol{\alpha}'$ and $\boldsymbol{\beta}'$, we obtain

$$\widetilde{(\mathsf{\Gamma}\boldsymbol{\alpha})}(\mathsf{\Gamma}\boldsymbol{\beta}) = \tilde{\boldsymbol{\alpha}}\boldsymbol{\beta} \tag{1.18}$$

It can easily be shown (and will be in Chapter 3) that, generally, the transpose of the product of two matrices is equal to the product of the transposes *in reverse order*. That is, for any two matrices, A and B, we have

$$\widetilde{(\mathsf{AB})} = \tilde{\mathsf{B}}\tilde{\mathsf{A}}$$

[11] The term "coordinate transformation" is used rather loosely here. What is really meant is a *rigid rotation* that involves no change in distances or angles. In general, a coordinate transformation may involve a change in length or angles.

Therefore, Eq. (1.18) becomes

$$(\tilde{\alpha}\tilde{\Gamma})(\Gamma\beta) = \tilde{\alpha}\beta$$

or

$$\tilde{\alpha}(\tilde{\Gamma}\Gamma)\beta = \tilde{\alpha}\beta$$

Since this is true for all α and β, we must have

$$\tilde{\Gamma}\Gamma = 1_3 \tag{1.19}$$

which means that Γ must be an *orthogonal matrix*. This is a generalization to space of the result we obtained for the plane.

If we write Γ as

$$\Gamma = \begin{pmatrix} \gamma_{11} & \gamma_{12} & \gamma_{13} \\ \gamma_{21} & \gamma_{22} & \gamma_{23} \\ \gamma_{31} & \gamma_{32} & \gamma_{33} \end{pmatrix}$$

then Eq. (1.19) can be written as

$$\begin{pmatrix} \gamma_{11} & \gamma_{21} & \gamma_{31} \\ \gamma_{12} & \gamma_{22} & \gamma_{32} \\ \gamma_{13} & \gamma_{23} & \gamma_{33} \end{pmatrix}\begin{pmatrix} \gamma_{11} & \gamma_{12} & \gamma_{13} \\ \gamma_{21} & \gamma_{22} & \gamma_{23} \\ \gamma_{31} & \gamma_{32} & \gamma_{33} \end{pmatrix} = \begin{pmatrix} 1 & 0 & 0 \\ 0 & 1 & 0 \\ 0 & 0 & 1 \end{pmatrix}$$

The product on the LHS is a 3×3 matrix whose elements must equal the corresponding elements of the matrix on the RHS. That is, the one-one element of $\tilde{\Gamma}\Gamma$ must equal the one-one element of the RHS; this yields

$$\gamma_{11}^2 + \gamma_{21}^2 + \gamma_{31}^2 = 1$$

Similarly, the equality of the one-two elements on both sides gives

$$\gamma_{11}\gamma_{12} + \gamma_{21}\gamma_{22} + \gamma_{31}\gamma_{32} = 0$$

and so on. Thus we obtain nine equations. However, simple inspection of these equations reveals that only six of them are independent. Therefore, we can only solve for the nine unknowns in terms of three of them. These three parameters are completely arbitrary.

What this means physically is that three parameters are required to specify a rigid rotation of the axes. There are many ways to specify these three parameters. One of the most useful and convenient ways is by using *Euler angles* (see Fig. 1.10). In terms of these angles the matrix Γ can be shown to be (see Example 1.4.2)

$$\Gamma = \begin{pmatrix} \cos\psi\cos\varphi - \sin\psi\cos\theta\sin\varphi & -\cos\psi\sin\varphi - \sin\psi\cos\theta\cos\varphi & \sin\psi\sin\theta \\ \sin\psi\cos\varphi + \cos\psi\cos\theta\sin\varphi & -\sin\psi\sin\varphi + \cos\psi\cos\theta\cos\varphi & -\cos\psi\sin\theta \\ \sin\theta\sin\varphi & \sin\theta\cos\varphi. & \cos\theta \end{pmatrix}$$

It is straightforward to verify that

$$\tilde{\Gamma}\Gamma = \Gamma\tilde{\Gamma} = 1_3$$

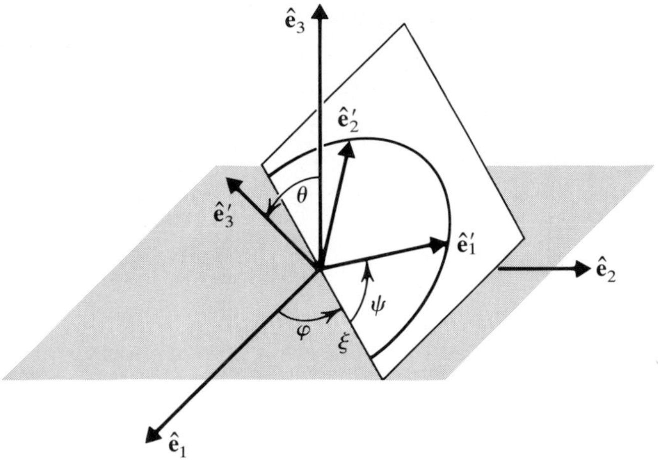

Figure 1.10 Euler angles and the relation between two coordinate systems, $\{\hat{\mathbf{e}}_1, \hat{\mathbf{e}}_2, \hat{\mathbf{e}}_3\}$ and $\{\hat{\mathbf{e}}'_1, \hat{\mathbf{e}}'_2, \hat{\mathbf{e}}'_3\}$.

This equation says that the rows (columns) of an orthogonal matrix form an orthonormal set of vectors (hence the name "orthogonal matrix"). The reader can easily verify that the inner product of any row (column) with any other row (column) is zero, and the norm of any row (column) is one. Euler angles are useful in describing the rotational motion of a rigid body in mechanics.

Example 1.4.2

From Fig. 1.10 it is clear that the system $\{\hat{\mathbf{e}}'_1, \hat{\mathbf{e}}'_2, \hat{\mathbf{e}}'_3\}$ is obtained from $\{\hat{\mathbf{e}}_1, \hat{\mathbf{e}}_2, \hat{\mathbf{e}}_3\}$ by the following three operations.

(a) Rotate the coordinate system about the $\hat{\mathbf{e}}_3$ axis through an angle φ. This corresponds to a rotation in the $\hat{\mathbf{e}}_1$-$\hat{\mathbf{e}}_2$ plane of angle φ, leaving the $\hat{\mathbf{e}}_3$ axis unchanged. The matrix corresponding to such a rotation is

$$\mathsf{\Gamma}_1 = \begin{pmatrix} \cos\varphi & -\sin\varphi & 0 \\ \sin\varphi & \cos\varphi & 0 \\ 0 & 0 & 1 \end{pmatrix} \tag{1}$$

It is clear that this matrix leaves the z coordinate of a column vector (x, y, z) unchanged while rotating the x and y coordinates by φ.

(b) Rotate the new coordinate system around the new $\hat{\mathbf{e}}_1$ axis (the ξ axis in the figure) through an angle θ. The corresponding matrix is

$$\mathsf{\Gamma}_2 = \begin{pmatrix} 1 & 0 & 0 \\ 0 & \cos\theta & -\sin\theta \\ 0 & \sin\theta & \cos\theta \end{pmatrix} \tag{2}$$

(c) Rotate the system about the new $\hat{\mathbf{e}}_3$ axis (the $\hat{\mathbf{e}}'_3$ axis in the figure) through an angle ψ. The corresponding matrix is

$$\Gamma_3 = \begin{pmatrix} \cos\psi & -\sin\psi & 0 \\ \sin\psi & \cos\psi & 0 \\ 0 & 0 & 1 \end{pmatrix} \tag{3}$$

It is easily verified that

$$\Gamma = \Gamma_3\Gamma_2\Gamma_1 \tag{4}$$

Note the order of multiplication. ●

1.4.3 Vector (Cross) Product

Another operation, which is defined only for vectors in space, is the *vector product.*[12] Given two space vectors, **a** and **b**, we can find a third space vector, **c**, such that

$$\mathbf{c} = \mathbf{a} \times \mathbf{b}$$

The magnitude of **c** is defined by

$$\|\mathbf{c}\| \equiv \|\mathbf{a}\|\,\|\mathbf{b}\| \sin\theta$$

where θ is the angle between **a** and **b**. The direction of **c** is given by the right-hand rule: if **a** is turned to **b** (note the order in which **a** and **b** appear here) through the smaller of the two angles between **a** and **b**, a right-handed screw that is perpendicular to **a** and **b** will advance in the direction of $\mathbf{a} \times \mathbf{b}$. This definition implies that

$$\mathbf{a} \times \mathbf{b} = -\mathbf{b} \times \mathbf{a} \tag{1.20}$$

and that

$$\mathbf{a} \cdot (\mathbf{a} \times \mathbf{b}) = \mathbf{b} \cdot (\mathbf{a} \times \mathbf{b}) = 0$$

That is, $\mathbf{a} \times \mathbf{b}$ is perpendicular to both **a** and **b**. This fact makes it clear why $\mathbf{a} \times \mathbf{b}$ is not defined in the plane. Although it is possible to define $\mathbf{a} \times \mathbf{b}$ for vectors **a** and **b** lying in a plane, $\mathbf{a} \times \mathbf{b}$ will not lie in that plane (it will be perpendicular to that plane). For the vector product, **a** and **b** must be considered space vectors.

The vector product has the following properties:

$$\mathbf{a} \times (\alpha\mathbf{b}) = \alpha\mathbf{a} \times \mathbf{b}$$

$$\mathbf{a} \times (\mathbf{b} + \mathbf{c}) = \mathbf{a} \times \mathbf{b} + \mathbf{a} \times \mathbf{c}$$

$$\mathbf{a} \times \mathbf{b} = -\mathbf{b} \times \mathbf{a}$$

$$\mathbf{a} \times \mathbf{a} = 0$$

Using these properties, we can write the vector product of two vectors in terms of their

[12] For a discussion of the connection between the cross product and antisymmetric tensors, see Example 4.1.18.

components in an orthonormal basis:

$$
\begin{aligned}
\mathbf{a} \times \mathbf{b} &= (\alpha_1 \hat{\mathbf{e}}_1 + \alpha_2 \hat{\mathbf{e}}_2 + \alpha_3 \hat{\mathbf{e}}_3) \times (\beta_1 \hat{\mathbf{e}}_1 + \beta_2 \hat{\mathbf{e}}_2 + \beta_3 \hat{\mathbf{e}}_3) \\
&= \alpha_1\beta_1 \hat{\mathbf{e}}_1 \times \hat{\mathbf{e}}_1 + \alpha_1\beta_2 \hat{\mathbf{e}}_1 \times \hat{\mathbf{e}}_2 + \alpha_1\beta_3 \hat{\mathbf{e}}_1 \times \hat{\mathbf{e}}_3 + \alpha_2\beta_1 \hat{\mathbf{e}}_2 \times \hat{\mathbf{e}}_1 \\
&\quad + \alpha_2\beta_2 \hat{\mathbf{e}}_2 \times \hat{\mathbf{e}}_2 + \alpha_2\beta_3 \hat{\mathbf{e}}_2 \times \hat{\mathbf{e}}_3 + \alpha_3\beta_1 \hat{\mathbf{e}}_3 \times \hat{\mathbf{e}}_1 + \alpha_3\beta_2 \hat{\mathbf{e}}_3 \times \hat{\mathbf{e}}_2 \\
&\quad + \alpha_3\beta_3 \hat{\mathbf{e}}_3 \times \hat{\mathbf{e}}_3
\end{aligned}
$$

But by the fourth property of the vector product, we have

$$\hat{\mathbf{e}}_1 \times \hat{\mathbf{e}}_1 = \hat{\mathbf{e}}_2 \times \hat{\mathbf{e}}_2 = \hat{\mathbf{e}}_3 \times \hat{\mathbf{e}}_3 = 0$$

Also, from the way $\hat{\mathbf{e}}_1$, $\hat{\mathbf{e}}_2$, and $\hat{\mathbf{e}}_3$ are oriented relative to one another in a so-called *right-handed coordinate system*, we get

$$
\begin{aligned}
\hat{\mathbf{e}}_1 \times \hat{\mathbf{e}}_2 &= -\, \hat{\mathbf{e}}_2 \times \hat{\mathbf{e}}_1 = \hat{\mathbf{e}}_3 \\
\hat{\mathbf{e}}_1 \times \hat{\mathbf{e}}_3 &= -\, \hat{\mathbf{e}}_3 \times \hat{\mathbf{e}}_1 = -\, \hat{\mathbf{e}}_2 \\
\hat{\mathbf{e}}_2 \times \hat{\mathbf{e}}_3 &= -\, \hat{\mathbf{e}}_3 \times \hat{\mathbf{e}}_2 = \hat{\mathbf{e}}_1
\end{aligned}
$$

Using these relations, we obtain

$$\mathbf{a} \times \mathbf{b} = (\alpha_2\beta_3 - \alpha_3\beta_2)\hat{\mathbf{e}}_1 + (\alpha_3\beta_1 - \alpha_1\beta_3)\hat{\mathbf{e}}_2 + (\alpha_1\beta_2 - \alpha_2\beta_1)\hat{\mathbf{e}}_3$$

This equation can be nicely written in a determinant form[13]

$$\mathbf{a} \times \mathbf{b} = \begin{vmatrix} \hat{\mathbf{e}}_1 & \hat{\mathbf{e}}_2 & \hat{\mathbf{e}}_3 \\ \alpha_1 & \alpha_2 & \alpha_3 \\ \beta_1 & \beta_2 & \beta_3 \end{vmatrix} \tag{1.21}$$

Strictly speaking, the elements of a determinant must be numbers, so the appearance of unit vectors in the first row is not completely "legal." However, we can consider the determinant form in (1.21) to be a mnemonic device useful for remembering the components of $\mathbf{a} \times \mathbf{b}$. Note the order in which the components of **a** and **b** appear in the determinant. A change in the ordering of **a** and **b** in the vector product corresponds to an interchange of two rows of the determinant, which results in a change in its sign.

There are other operations that can further enrich the set of vectors. For example, given a set of vectors, it may be possible to form other vectors through the binary operation of multiplication. If we define a product for vectors, the set of vectors turns into an algebra; this subject in its most general form is not dealt with in this book. However, a special type of product of vectors, namely tensors, will be discussed in Chapter 4.

[13]Here a basic familiarity with determinants is assumed. The theory of determinants and their relation to matrices will be developed in Chapters 3 and 4.

Example 1.4.3

From the definition of the vector product and Fig. 1.11, we note that

$$\|\mathbf{a} \times \mathbf{b}\| = \text{area of the parallelogram defined by } \mathbf{a} \text{ and } \mathbf{b}$$

So we can use (1.21) to find the area of a parallelogram defined by two vectors directly, in terms of their components.

For instance, the area defined by $\mathbf{a} = (1, 1, -2)$ and $\mathbf{b} = (2, 0, 3)$ can be found by calculating the vector product,

$$\mathbf{a} \times \mathbf{b} = \begin{vmatrix} \hat{\mathbf{e}}_1 & \hat{\mathbf{e}}_2 & \hat{\mathbf{e}}_3 \\ 1 & 1 & -2 \\ 2 & 0 & 3 \end{vmatrix} = 3\hat{\mathbf{e}}_1 - 7\hat{\mathbf{e}}_2 - 2\hat{\mathbf{e}}_3$$

and then computing its length,

$$\|\mathbf{a} \times \mathbf{b}\| = \sqrt{3^2 + (-7)^2 + (-2)^2} = \sqrt{62}$$

•

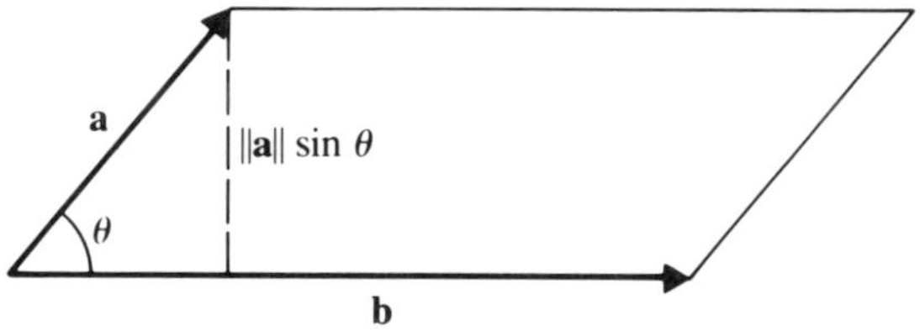

Figure 1.11 The area of a parallelogram is the product of the height and the base.

Example 1.4.4

The volume of a parallelopiped defined by three noncoplanar vectors, **a**, **b**, and **c**, is given by $|\mathbf{a} \cdot (\mathbf{b} \times \mathbf{c})|$. This can be seen from Fig. 1.12, where it is clear that the volume is

$$\text{volume} = (\text{area of base})(\text{altitude}) = \|\mathbf{b} \times \mathbf{c}\|(\|\mathbf{a}\| \cos\theta) = |(\mathbf{b} \times \mathbf{c}) \cdot \mathbf{a}|$$

The absolute value is taken to ensure the positivity of the area. In terms of components we have

$$\begin{aligned} \text{volume} &= (\mathbf{b} \times \mathbf{c})_1 \alpha_1 + (\mathbf{b} \times \mathbf{c})_2 \alpha_2 + (\mathbf{b} \times \mathbf{c})_3 \alpha_3 \\ &= (\beta_2\gamma_3 - \beta_3\gamma_2)\alpha_1 + (\beta_3\gamma_1 - \beta_1\gamma_3)\alpha_2 + (\beta_1\gamma_2 - \beta_2\gamma_1)\alpha_3 \end{aligned}$$

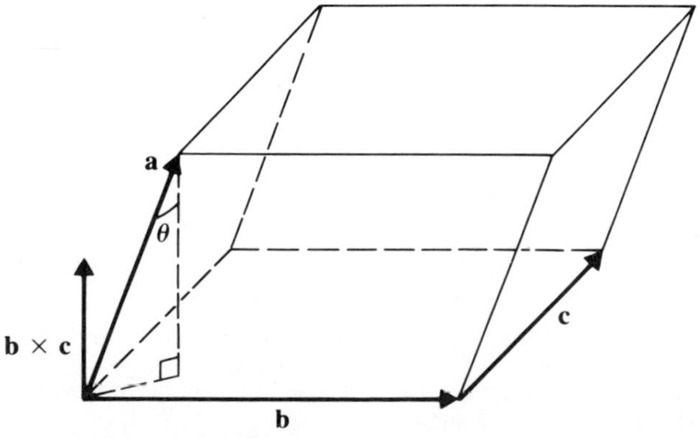

Figure 1.12 The volume of a parallelopiped is the area of the base times the height.

which can be written in determinant form as

$$\text{volume} = |\mathbf{a}\cdot(\mathbf{b}\times\mathbf{c})| = \begin{vmatrix} \alpha_1 & \alpha_2 & \alpha_3 \\ \beta_1 & \beta_2 & \beta_3 \\ \gamma_1 & \gamma_2 & \gamma_3 \end{vmatrix}$$

Note that the absolute value of the result must be taken. •

Exercises

1.4.1 Find the metric tensor associated with the basis vectors $\mathbf{a}_1 = \hat{\mathbf{e}}_x + \hat{\mathbf{e}}_y$, $\mathbf{a}_2 = \hat{\mathbf{e}}_x + \hat{\mathbf{e}}_z$, and $\mathbf{a}_3 = \hat{\mathbf{e}}_y + \hat{\mathbf{e}}_z$.

1.4.2 Calculate the inner product of the two vectors **b** and **c**, whose components in the basis $\{\mathbf{a}_1, \mathbf{a}_2, \mathbf{a}_3\}$ of Exercise 1.4.1 are, respectively, $(1, -1, 2)$ and $(0, 2, 3)$.

1.4.3 Use the Gram-Schmidt orthonormalization process to find three orthonormal vectors out of the basis $\{\mathbf{a}_1, \mathbf{a}_2, \mathbf{a}_3\}$ of Exercise 1.4.1.

1.4.4 Use Gram-Schmidt orthonormalization to show that the three vectors $(2, -1, 3)$, $(-1, 1, -2)$, and $(3, 1, 2)$ are linearly dependent.

1.4.5 Find the vector form of the equation of the plane defined by the three points P, Q, and R with coordinates (p_1, p_2, p_3), (q_1, q_2, q_3) and (r_1, r_2, r_3), respectively

1.4.6 Derive the law of sines for a triangle using vector methods.

1.4.7 Show that a necessary and sufficient condition for three vectors, $\mathbf{a}_1$, $\mathbf{a}_2$, and $\mathbf{a}_3$, to be in the same plane is

$$\mathbf{a}_1 \cdot (\mathbf{a}_2 \times \mathbf{a}_3) = 0 \tag{1}$$

1.5 VECTOR ANALYSIS IN CARTESIAN COORDINATES

Let us now leave the algebra of vectors and briefly discuss the subject of calculus of vectors, or *vector analysis*. The treatment of vector algebra in N dimensions will be continued in Chapter 2.

Basic to the study of vector analysis is the notion of *field*, which plays a key role in many areas of physics—in the motion of fluids, in conduction of heat, in electromagnetic theory, in gravitation, and so forth. All these situations involve a physical quantity that *varies from point to point* (also from time to time) and thus is a function of space coordinates (and time). This physical quantity can be either a scalar, in which case we speak of a *scalar field*, or a vector, in which case we speak of a *vector field*. For example, the temperature of the atmosphere is a scalar field because it is a function of space coordinates (equator versus the poles) and time (summer versus winter) and because temperature has no direction associated with it. On the other hand, wind velocity is a vector field because (1) it is a vector and (2) its magnitude *and* direction depend on space coordinates (and time). In general, when we talk of a vector field, we are dealing with *three* functions of space (along with time), corresponding to the three components of the vector.

As a specific example, consider a situation in which the wind is horizontal, and, depending on location, both the east-west components and the north-south components change. We can write (ignoring the time variable)

$$v_x = f_1(x, y, z)$$
$$v_y = f_2(x, y, z)$$
$$v_z = 0$$

where f_1 and f_2 are some functions of coordinates. Note that the x and y components are both functions of z (height above the ground in this example) although the velocity has no component up or down ($v_z = 0$). This simply means that as we go to higher altitudes, the two components of velocity will change (typically they increase). The point to emphasize here is that for a vector field we are dealing with three functions of coordinates (and time).[14]

Given scalar and vector fields, we can perform analytic operations (differentiation, integration, etc.) on them to obtain new scalar and vector fields. Let us consider differentiation first.

1.5.1 Time Differentiation

The preceding discussion of fields included the possibility of time-dependence. In general, fields are functions of time and the operation of differentiation with respect to time can be defined and used. For example, if $\mathbf{v}$ denotes the velocity of a particle, then the time derivative of $\mathbf{v}$ is acceleration.

Time differentiation for scalar fields is exactly the same as for ordinary functions. For vector fields we follow the same procedure as for ordinary functions to define the derivative:

$$\frac{d\mathbf{a}(t)}{dt} = \lim_{\Delta t \to 0} \frac{\mathbf{a}(t + \Delta t) - \mathbf{a}(t)}{\Delta t}$$

In terms of components, when $\hat{\mathbf{e}}_1, \hat{\mathbf{e}}_2$, and $\hat{\mathbf{e}}_3$ *do not change with time*, we have

$$\begin{aligned}
\frac{d\mathbf{a}(t)}{dt} &= \lim_{\Delta t \to 0} \frac{[\alpha_1(t + \Delta t) - \alpha_1(t)]\hat{\mathbf{e}}_1 + [\alpha_2(t + \Delta t) - \alpha_2(t)]\hat{\mathbf{e}}_2 + [\alpha_3(t + \Delta t) - \alpha_3(t)]\hat{\mathbf{e}}_3}{\Delta t} \\
&= \left(\lim_{\Delta t \to 0} \frac{\alpha_1(t + \Delta t) - \alpha_1(t)}{\Delta t} \right)\hat{\mathbf{e}}_1 + \left(\lim_{\Delta t \to 0} \frac{\alpha_2(t + \Delta t) - \alpha_2(t)}{\Delta t} \right)\hat{\mathbf{e}}_2 \\
&\quad + \left(\lim_{\Delta t \to 0} \frac{\alpha_3(t + \Delta t) - \alpha_3(t)}{\Delta t} \right)\hat{\mathbf{e}}_3 \\
&= \frac{d\alpha_1}{dt}\hat{\mathbf{e}}_1 + \frac{d\alpha_2}{dt}\hat{\mathbf{e}}_2 + \frac{d\alpha_3}{dt}\hat{\mathbf{e}}_3 \qquad (1.22)
\end{aligned}$$

[14] Also, it is assumed that the three functions representing the three components of a vector field have derivatives of all orders. This assumption will apply in all subsequent discussions.

which states that the ith component of the derivative of a vector is the derivative of the ith component of that vector. Note that this is not true when $\hat{\mathbf{e}}_1$, $\hat{\mathbf{e}}_2$, and $\hat{\mathbf{e}}_3$ are time-dependent, which typically happens when the motion of a particle is described in the so-called *curvilinear coordinates*, of which spherical and cylindrical coordinates are examples (see Example 1.5.1). In Cartesian coordinates $\hat{\mathbf{e}}_1$, $\hat{\mathbf{e}}_2$, and $\hat{\mathbf{e}}_3$ are fixed vectors, and, therefore, (1.22) is valid.

Example 1.5.1

Let us calculate the acceleration of a particle moving in a plane in an elementary fashion. (For a more elegant example, see Exercise 1.6.3.) We describe the motion in terms of the polar coordinates r and θ (Fig. 1.13). The position is described by the vector $\mathbf{r}$. We can write this vector as

$$\mathbf{r} = r\hat{\mathbf{e}}_r \tag{1}$$

where r is the length of $\mathbf{r}$ and $\hat{\mathbf{e}}_r$ is a unit vector in the direction of $\mathbf{r}$. Note that $\hat{\mathbf{e}}_r$ changes with time. Differentiating (1) gives the velocity:

$$\mathbf{v} = \frac{d\mathbf{r}}{dt} = \frac{dr}{dt}\hat{\mathbf{e}}_r + r\frac{d\hat{\mathbf{e}}_r}{dt} \tag{2}$$

Figure 1.13 shows that $\Delta\hat{\mathbf{e}}_r$ is in the direction of $\hat{\mathbf{e}}_\theta$, which is perpendicular to $\hat{\mathbf{e}}_r$ and in the direction in which θ is increasing. Thus we can write

$$\Delta\hat{\mathbf{e}}_r \approx \|\Delta\hat{\mathbf{e}}_r\|\,\hat{\mathbf{e}}_\theta \approx \|\hat{\mathbf{e}}_r\|\,\Delta\theta\,\hat{\mathbf{e}}_\theta$$

where $\|\Delta\hat{\mathbf{e}}_r\|$ is approximated by the arc of a circle of radius $\|\hat{\mathbf{e}}_r\|$ subtended by an angle $\Delta\theta$. However, since $\hat{\mathbf{e}}_r$ is of unit length, we have

$$\Delta\hat{\mathbf{e}}_r \approx \Delta\theta\,\hat{\mathbf{e}}_\theta$$

and

$$\frac{\Delta\hat{\mathbf{e}}_r}{\Delta t} \approx \frac{\Delta\theta}{\Delta t}\hat{\mathbf{e}}_\theta$$

In the limit as $\Delta t \to 0$,

$$\frac{d\hat{\mathbf{e}}_r}{dt} = \frac{d\theta}{dt}\hat{\mathbf{e}}_\theta \equiv \dot{\theta}\hat{\mathbf{e}}_\theta \tag{3}$$

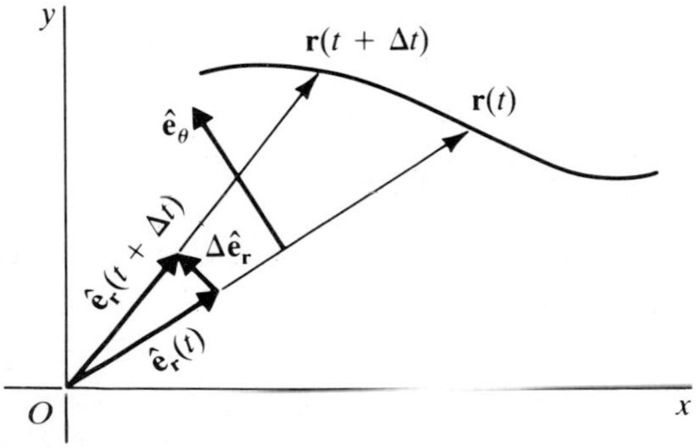

Figure 1.13 The unit vector $\hat{\mathbf{e}}_r$ changes as t varies.

where the dot signifies differentiation with respect to time. Thus, (2) becomes

$$\mathbf{v} = \dot{r}\hat{\mathbf{e}}_\mathbf{r} + r\dot{\theta}\hat{\mathbf{e}}_\theta \tag{4}$$

Since we want to find acceleration, $\mathbf{a}$, we need not only $d\hat{\mathbf{e}}_\mathbf{r}/dt$ but also $d\hat{\mathbf{e}}_\theta/dt$. To calculate the latter, note that $\hat{\mathbf{e}}_\mathbf{r}\cdot\hat{\mathbf{e}}_\theta = 0$. Thus

$$0 = \frac{d}{dt}(\hat{\mathbf{e}}_\mathbf{r}\cdot\hat{\mathbf{e}}_\theta) = \left(\frac{d\hat{\mathbf{e}}_\mathbf{r}}{dt}\right)\cdot\hat{\mathbf{e}}_\theta + \hat{\mathbf{e}}_\mathbf{r}\cdot\left(\frac{d\hat{\mathbf{e}}_\theta}{dt}\right)$$

and, using (3), we get

$$0 = \dot{\theta} + \hat{\mathbf{e}}_\mathbf{r}\cdot\left(\frac{d\hat{\mathbf{e}}_\theta}{dt}\right)$$

or

$$\hat{\mathbf{e}}_\mathbf{r}\cdot\left(\frac{d\hat{\mathbf{e}}_\theta}{dt}\right) = -\dot{\theta} \tag{5}$$

On the other hand, differentiating $\hat{\mathbf{e}}_\theta\cdot\hat{\mathbf{e}}_\theta = 1$ with respect to time gives

$$\left(\frac{d\hat{\mathbf{e}}_\theta}{dt}\right)\cdot\hat{\mathbf{e}}_\theta + \hat{\mathbf{e}}_\theta\cdot\left(\frac{d\hat{\mathbf{e}}_\theta}{dt}\right) = 0$$

or

$$\hat{\mathbf{e}}_\theta\cdot\left(\frac{d\hat{\mathbf{e}}_\theta}{dt}\right) = 0 \tag{6}$$

But $d\hat{\mathbf{e}}_\theta/dt$ is a vector in the plane and, therefore, can be written as

$$\frac{d\hat{\mathbf{e}}_\theta}{dt} = \alpha\hat{\mathbf{e}}_\mathbf{r} + \beta\hat{\mathbf{e}}_\theta \tag{7}$$

We take the inner product of (7) with $\hat{\mathbf{e}}_\mathbf{r}$ and $\hat{\mathbf{e}}_\theta$, use (5) and (6), and conclude that $\alpha = -\dot{\theta}$ and $\beta = 0$. Thus,

$$\frac{d\hat{\mathbf{e}}_\theta}{dt} = -\dot{\theta}\hat{\mathbf{e}}_\mathbf{r} \tag{8}$$

This result could also have been obtained geometrically by keeping track of the change in the direction of $\hat{\mathbf{e}}_\theta$ as $\hat{\mathbf{e}}_\mathbf{r}$ changes.

Having obtained $d\hat{\mathbf{e}}_\mathbf{r}/dt$ and $d\hat{\mathbf{e}}_\theta/dt$, we can now differentiate (4) to get the acceleration:

$$\begin{aligned}\mathbf{a} = \frac{d\mathbf{v}}{dt} &= \ddot{r}\hat{\mathbf{e}}_\mathbf{r} + \dot{r}\frac{d\hat{\mathbf{e}}_\mathbf{r}}{dt} + \dot{r}\dot{\theta}\hat{\mathbf{e}}_\theta + r\ddot{\theta}\hat{\mathbf{e}}_\theta + r\dot{\theta}\frac{d\hat{\mathbf{e}}_\theta}{dt} \\ &= \ddot{r}\hat{\mathbf{e}}_\mathbf{r} + \dot{r}\dot{\theta}\hat{\mathbf{e}}_\theta + \dot{r}\dot{\theta}\hat{\mathbf{e}}_\theta + r\ddot{\theta}\hat{\mathbf{e}}_\theta - r\dot{\theta}^2\hat{\mathbf{e}}_\mathbf{r} \\ &= (\ddot{r} - r\dot{\theta}^2)\hat{\mathbf{e}}_\mathbf{r} + (2\dot{r}\dot{\theta} + r\ddot{\theta})\hat{\mathbf{e}}_\theta\end{aligned} \tag{9}$$

This is the familiar form of the acceleration in polar coordinates, which is used in central force problems in mechanics. The use of polar coordinates is convenient in such situations, because if the force is central, then $\mathbf{a}$ will be in the direction of $\hat{\mathbf{e}}_\mathbf{r}$, and its $\hat{\mathbf{e}}_\theta$ component must therefore vanish. Thus, for a central force

$$2\dot{r}\dot{\theta} + r\ddot{\theta} = 0$$

Multiplying by r, we obtain

$$2r\dot{r}\dot{\theta} + r^2\ddot{\theta} = 0$$

or

$$\frac{d}{dt}(r^2\dot{\theta}) = 0 \tag{10}$$

which tells us that the quantity $r^2\dot{\theta}$ is a constant. But $mr^2\dot{\theta}$ is the angular momentum of the particle with respect to the origin. Thus, (10) is a statement of the conservation of angular momentum for situations involving a central force. ●

1.5.2 The Gradient

In many situations that arise in physics, rates of change of certain scalar functions with distance are of importance. For instance, the way potential energy changes with movement in space is directly related to the force producing the potential energy. Similarly, the rate of change (derivative) of electrostatic potential with respect to distance gives the electrostatic field. The concept of gradient makes precise the vague notion of a derivative with respect to distance.

Let us investigate the notion of differentiation with respect to distance, starting with one variable. In Fig. 1.14 a function, $f(x)$, has an increment, Δf, corresponding to a change in x, or Δx. If Δx is small enough, we can write

$$\Delta f \approx \left(\frac{df}{dx}\right)_{x=x_0} \Delta x$$

where $(df/dx)_{x=x_0}$ is a measure of how fast the function $f(x)$ is changing at the point x_0.

With one variable there is no ambiguity in defining the derivative, because there is only one way we can change x, the (only) coordinate.[15] With two or more variables the situation is completely different, as is illustrated in Fig. 1.15. A point $P_0 = (x_0, y_0)$

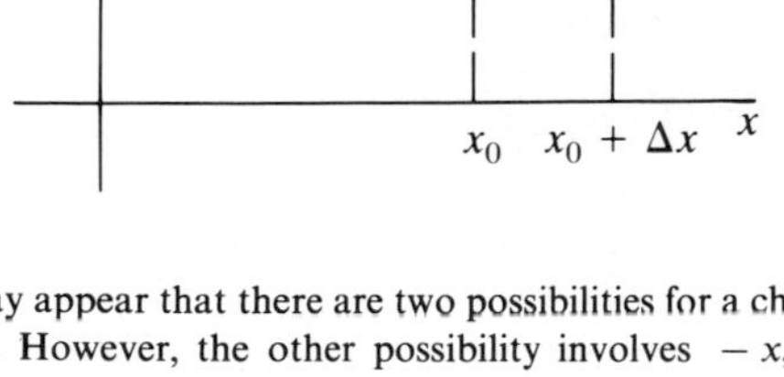

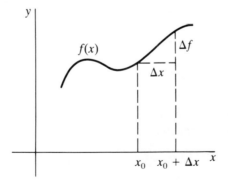

Figure 1.14 The derivative of a function $f(x)$ is a measure of how fast $f(x)$ changes in going from x_0 to a neighboring point, $x_0 + \Delta x$.

[15] It may appear that there are two possibilities for a change in x and, therefore, two possibilities for the derivative. However, the other possibility involves $-x$, the negative of the original variable. The derivative, of course, will remain the same regardless of the sign of x.

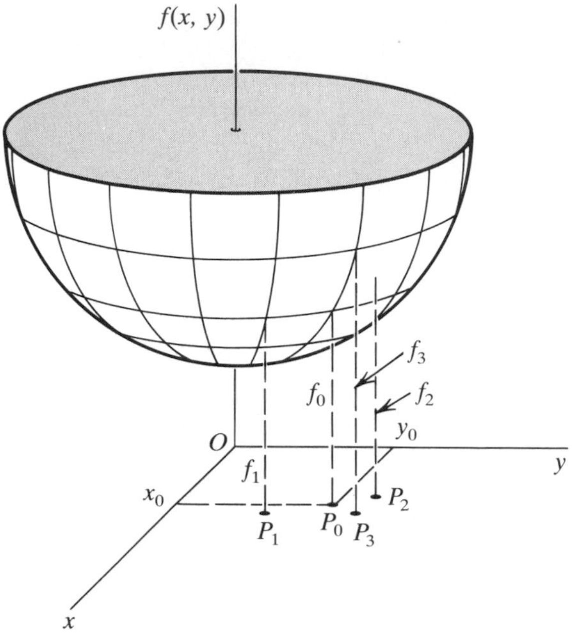

Figure 1.15 The change in $f(x, y)$ depends on the direction in which we move away from (x_0, y_0). The gradient is in the direction that maximizes this change.

in the xy-plane is shown with the corresponding value of the function $f(x_0, y_0) = f_0$. Out of the infinity of points that are close to P_0 and cause a change in the function, only three are shown. These indicate how the change in $f(x, y)$ depends on the direction in which the neighboring point is located with respect to P_0. For example, if we move in the direction P_0P_1, there is very little change in $f(x, y)$, but if we move in the direction P_0P_2, we notice more change in the function, and if we move in the direction P_0P_3, the change seems to be maximum. This maximum change, and the direction associated with it, is called the *gradient*.

Let us use $d\mathbf{r}_3$ to denote the (infinitesimal) displacement vector connecting P_0 to P_3 *in the xy-plane*. Then we conclude that out of infinitely many directions away from P_0, only $d\mathbf{r}_3$ has the property that the corresponding change in $f(x, y)$ is maximum. We also know from calculus that if $f(x, y)$ is differentiable, we can write very generally

$$df = \left(\frac{\partial f}{\partial x}\right)_{P_0} dx + \left(\frac{\partial f}{\partial y}\right)_{P_0} dy$$

where dx and dy are the components of the displacement from P_0 and df is the change in f corresponding to the increments dx and dy. We can rewrite this equation as

$$df = (\boldsymbol{\nabla} f)_{P_0} \cdot d\mathbf{r} = \| \boldsymbol{\nabla} f \| \, \| d\mathbf{r} \| \cos\theta$$

where, by definition,

$$\boldsymbol{\nabla} f \equiv \left(\frac{\partial f}{\partial x}, \frac{\partial f}{\partial y}\right)_{P_0} \tag{1.23}$$

is a vector in the xy-plane. It is clear that df will be maximum when $\cos\theta = 1$, that is, when $d\mathbf{r}$ is in the direction of ∇f. On the other hand, we saw above that $d\mathbf{r}_3$ maximizes df. We conclude, therefore, that ∇f must be along $d\mathbf{r}_3$. The vector in (1.23) is the gradient of f at P_0.

The preceding discussion makes it clear that the maximum change in a function at a point P_0 is obtained by moving in the direction of the gradient and is given by

$$df = \|\nabla f\| \, \|d\mathbf{r}\|$$

where

$$\|d\mathbf{r}\| = \sqrt{(dx)^2 + (dy)^2}$$

The notion of gradient can be generalized to three variables although it is harder to visualize than the two-variable case. In three dimensions we deal with a function $f(x, y, z)$ (which cannot be plotted as in Fig. 1.15!) and ask which direction, $d\mathbf{r} = (dx, dy, dz)$, maximizes the change in f. We have (assuming that f is differentiable, of course)

$$df = \frac{\partial f}{\partial x} dx + \frac{\partial f}{\partial y} dy + \frac{\partial f}{\partial z} dz$$

or

$$df = \nabla f \cdot d\mathbf{r} = \|\nabla f\| \, \|d\mathbf{r}\| \cos\theta$$

where

$$\nabla f = \left(\frac{\partial f}{\partial x}, \frac{\partial f}{\partial y}, \frac{\partial f}{\partial z}\right) \tag{1.24}$$

It is clear again that df is maximum when $\cos\theta = 1$, that is, when $d\mathbf{r}$ is in the direction of ∇f, the gradient of the function.

The symbol ∇ can be thought of as a vector operator (called "del") whose components are $\partial/\partial x$, $\partial/\partial y$, and $\partial/\partial z$. Thus, we can write

$$\nabla = \hat{\mathbf{e}}_x \frac{\partial}{\partial x} + \hat{\mathbf{e}}_y \frac{\partial}{\partial y} + \hat{\mathbf{e}}_z \frac{\partial}{\partial z} \tag{1.25}$$

In its most general form an operator is a mapping from one set into another. This vector operator maps differentiable functions into vector fields. That is, ∇ operates on differentiable functions and produces vector fields.

Example 1.5.2

As an example of the calculation of a gradient, let us consider the scalar field

$$V(x, y, z) = \frac{k}{\sqrt{x^2 + y^2 + z^2}}$$

where k is a constant, and find its gradient at the point (1, 1, 0).

In general, we have

$$\nabla V = \frac{\partial V}{\partial x} \hat{\mathbf{e}}_x + \frac{\partial V}{\partial y} \hat{\mathbf{e}}_y + \frac{\partial V}{\partial z} \hat{\mathbf{e}}_z \tag{1}$$

where $\partial V/\partial x$, $\partial V/\partial y$, and $\partial V/\partial z$ are given by

$$\frac{\partial V}{\partial x} = \frac{\partial}{\partial x}\left(\frac{k}{\sqrt{x^2 + y^2 + z^2}}\right) = -\frac{kx}{(x^2 + y^2 + z^2)^{3/2}}$$

$$\frac{\partial V}{\partial y} = -\frac{ky}{(x^2 + y^2 + z^2)^{3/2}}$$

$$\frac{\partial V}{\partial z} = -\frac{kz}{(x^2 + y^2 + z^2)^{3/2}}$$

Therefore, at the point (1, 1, 0) we get

$$\left(\frac{\partial V}{\partial x}\right)_{(1,1,0)} = -\frac{k}{2\sqrt{2}} \qquad \left(\frac{\partial V}{\partial y}\right)_{(1,1,0)} = -\frac{k}{2\sqrt{2}} \qquad \left(\frac{\partial V}{\partial z}\right)_{(1,1,0)} = 0$$

or

$$\nabla V = -\frac{k}{2\sqrt{2}}(1, 1, 0) \tag{2}$$

Note that ∇V is proportional to (1, 1, 0), the position vector of the point P. ●

1.5.3 Flux and Divergence

Another important differentiation associated with vector fields is divergence, which is closely related to the concept of flux. A concrete example of a vector field is the velocity field of a fluid. The situation is depicted in Fig. 1.16, where the fluid is flowing from left to right. An infinitesimal ring of area da is situated in the flow. How much fluid is passing through the ring per unit time? It is clear that the answer depends on the density of the fluid, its speed, the size of the area da, and also the relative orientation of the direction of the flow and the unit normal to the area, denoted by $\hat{\mathbf{e}}_n$. A little thought allows us to deduce that the amount of fluid of *constant unit density* passing through da is proportional to[16]

$$d\Phi = \mathbf{v}\cdot\hat{\mathbf{e}}_n\, da \equiv \mathbf{v}\cdot d\mathbf{a} \tag{1.26}$$

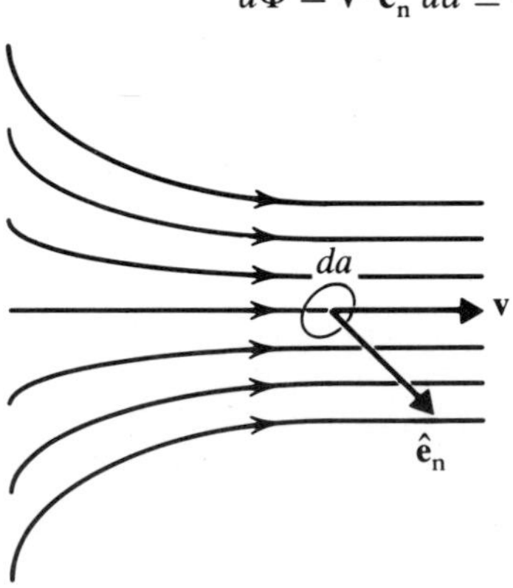

Figure 1.16 The flux of the velocity vector $\mathbf{v}$ through the area da is given as $d\Phi = \mathbf{v}\cdot\hat{\mathbf{e}}_n\, da$, where $\hat{\mathbf{e}}_n$ is the unit vector normal to da.

[16] The mass of fluid passing through da is given exactly by $\rho\mathbf{v}\cdot\hat{\mathbf{e}}_n\, da$, where ρ is the density of the fluid at da. Here, for simplicity, the density is taken as equal to 1.

where $d\Phi$ is called the *flux of* $\mathbf{v}$ *through da*. Sometimes it is useful to associate a vector, $d\mathbf{a} = \hat{\mathbf{e}}_n\, da$, with the element of area, as we have done in (1.26). If the ring is replaced by a large surface, S, then we have to integrate over all the area and write

$$\Phi = \iint_S \mathbf{v} \cdot \hat{\mathbf{e}}_n\, da = \iint_S \mathbf{v} \cdot d\mathbf{a} \tag{1.27}$$

where Φ is the total flux through S and $\iint_S$ denotes integration over the area S.

Example 1.5.3

Consider the flow of a river and assume that the velocity of the water is

$$\mathbf{v} = v_0\left(1 - \frac{x^2}{a^2}\right)\hat{\mathbf{e}}_z$$

where x is the distance from the midpoint of the river and a is half of the width of the river. Let us find the flux of the velocity, assuming that the cross section of the river is a rectangle with depth equal to h. The situation is depicted in Fig. 1.17.

The normal to the area da is perpendicular to the xy-plane and is in the same direction as the velocity. Thus, we have

$$d\Phi = \mathbf{v} \cdot d\mathbf{a} = v\, da = v\, dx\, dy$$

and

$$\Phi = \iint_S v\, dx\, dy$$

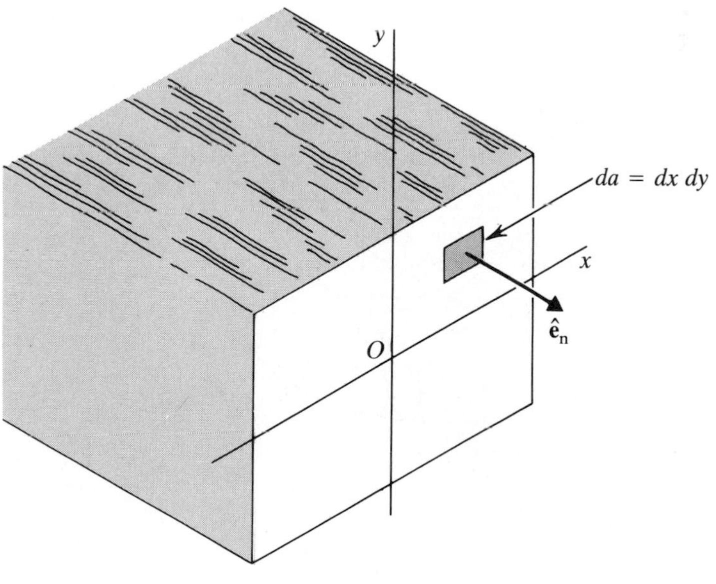

Figure 1.17 The river with its cross section.

where S is the cross-sectional area of the river, $S = 2ah$. For this particular situation the velocity is independent of y, and we can write

$$\Phi = \int_{-h/2}^{h/2} dy \int_{-a}^{a} v_0\left(1 - \frac{x^2}{a^2}\right)dx = hv_0 \int_{-a}^{a} \left(dx - \frac{x^2}{a^2}\,dx\right) = hv_0(2a - \tfrac{2}{3}a) = \tfrac{2}{3}Sv_0 \qquad \bullet$$

Example 1.5.4

As another example, let us consider the flux of the electric field of a point charge located at a distance d from the center of a circle of radius a (Fig. 1.18). The element of flux (in the cgs system) is given by

$$d\Phi = \mathbf{E}\cdot d\mathbf{a} = \|\mathbf{E}\| \cos\theta \, da = \|\mathbf{E}\| \cos\theta \, \rho \, d\rho \, d\varphi = \frac{q}{r^2}\frac{d}{r}\rho \, d\rho \, d\varphi = \frac{qd}{(d^2 + \rho^2)^{3/2}} \rho \, d\rho \, d\varphi$$

where the polar coordinates (ρ, φ) are used to specify a point in the plane of the circle at which the element of area is $\rho \, d\rho \, d\varphi$. To find the total flux, we integrate the last expression above:

$$\Phi = \iint_S \frac{qd}{(d^2 + \rho^2)^{3/2}} \rho \, d\rho \, d\varphi = qd \int_0^{2\pi} d\varphi \int_0^a \frac{\rho \, d\rho}{(d^2 + \rho^2)^{3/2}}$$

$$= 2\pi q d[-(d^2 + \rho^2)^{-1/2}|_0^a] = 2\pi q\left(1 - \frac{d}{\sqrt{d^2 + a^2}}\right) \qquad \bullet$$

It is often necessary to calculate the flux through a closed surface. For instance, the flux of the electric field due to a charge distribution inside a volume bounded by

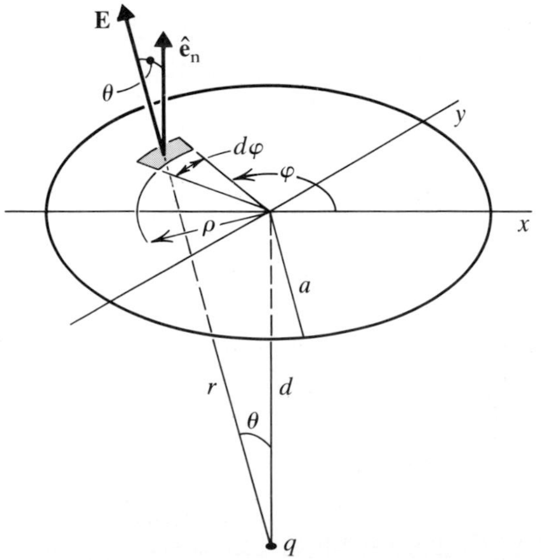

Figure 1.18 Electric flux through a circle.

a surface is of importance in electromagnetic theory. Let us calculate such a flux. To begin, consider a small volume, dV, bounded by a rectangular box of sides dx, dy, and dz (Fig. 1.19), and calculate the net *outward* flux of an arbitrary vector field, $\mathbf{A}(x, y, z)$. The six faces of the box are assumed to be so small that the angle between the normal to each face and the vector field $\mathbf{A}$ is constant over da. Since we are calculating the outward flux of $\mathbf{A}$, we must assume that $\hat{\mathbf{e}}_n$ is always *pointing out* of the volume.

To evaluate (1.26) for this particular case, we write

$$d\Phi = (d\Phi_1 + d\Phi_2) + (d\Phi_3 + d\Phi_4) + (d\Phi_5 + d\Phi_6)$$

where each pair of parentheses indicates one axis. For instance, $d\Phi_1$ is the flux through the face having a normal component along the positive x-axis, $d\Phi_2$ is the flux through the face having a normal component along the negative x-axis, and so on. Let us first look at $d\Phi_1$, which can be written as

$$d\Phi_1 = \mathbf{A}_1 \cdot \hat{\mathbf{e}}_{n_1}\, da_1$$

or, since $\hat{\mathbf{e}}_{n_1}$ is the same as $\hat{\mathbf{e}}_x$,

$$d\Phi_1 = \mathbf{A}_1 \cdot \hat{\mathbf{e}}_x\, da_1 = A_{1x}\, da_1$$

This requires some explanation. The subscript 1 in A_{1x} indicates the evaluation of the vector field at the midpoint (or any other point, since the area is infinitesimal) of the first face. The subscript x in A_{1x}, of course, means the x component. So, A_{1x} means the x component of $\mathbf{A}$ at the midpoint of face 1; da_1 is the area of face 1, which is merely $dy\, dz$ (Fig. 1.19). Since the midpoint of face 1 has coordinates $(x + dx/2, y, z)$ [note that the center of the box has coordinates (x, y, z)], we can write

$$d\Phi_1 = A_x\left(x + \frac{dx}{2}, y, z\right) dy\, dz$$

Since A_x is a (differentiable) function of x, y, and z, this equation implies an evaluation

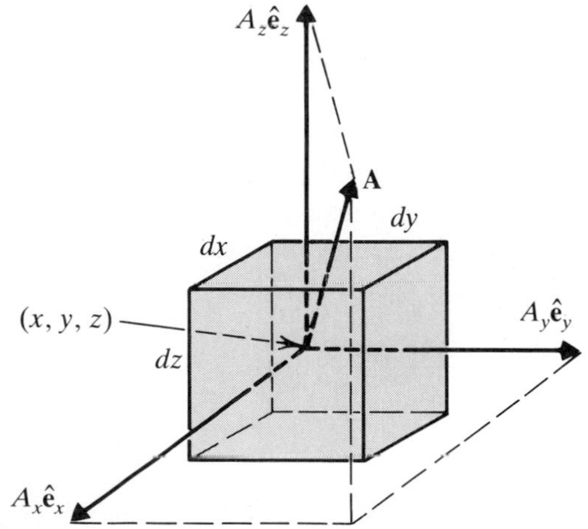

Figure 1.19 Diagram used in calculating the net outward flux of a vector field $\mathbf{A}(x, y, z)$ through a closed infinitesimal surface bounding the rectangular box of volume dV.

of the function A_x at the coordinates $x + dx/2$, y, and z. Using the Taylor-series expansion of a function and keeping terms only up to the first order, we can write

$$A_x\left(x + \frac{dx}{2}, y, z\right) = A(x, y, z) + \frac{dx}{2}\frac{\partial A_x}{\partial x} + \cdots$$

We thus have

$$d\Phi_1 = \left[A_x(x, y, z) + \frac{dx}{2}\frac{\partial A_x}{\partial x}\right] dy\, dz$$

Similarly, we write

$$d\Phi_2 = \mathbf{A}_2 \cdot \hat{\mathbf{e}}_{n_2}\, da_2 = \mathbf{A}_2 \cdot (-\hat{\mathbf{e}}_x)\, da_2 = -A_{2x}\, dy\, dz = -A_x\left(x - \frac{dx}{2}, y, z\right) dy\, dz$$

$$= -\left[A_x(x, y, z) - \frac{dx}{2}\frac{\partial A_x}{\partial x}\right] dy\, dz$$

Adding the expressions for $d\Phi_1$ and $d\Phi_2$, we obtain

$$d\Phi_1 + d\Phi_2 = dy\, dz\left[A_x(x, y, z) + \frac{dx}{2}\frac{\partial A_x}{\partial x} - A_x(x, y, z) + \frac{dx}{2}\frac{\partial A_x}{\partial x}\right]$$

$$= \frac{\partial A_x}{\partial x}\, dx\, dy\, dz = \frac{\partial A_x}{\partial x}\, dV$$

where dV is the volume of the little box. By the same reasoning, we obtain

$$d\Phi_3 + d\Phi_4 = \frac{\partial A_y}{\partial y}\, dV$$

and

$$d\Phi_5 + d\Phi_6 = \frac{\partial A_z}{\partial z}\, dV$$

The total flux is then

$$d\Phi = \left(\frac{\partial A_x}{\partial x} + \frac{\partial A_y}{\partial y} + \frac{\partial A_z}{\partial z}\right) dV$$

The expression in parentheses is called the *divergence* of the vector field **A**. The divergence can be written more compactly if we recall that the vector operator $\mathbf{\nabla}$ has components $(\partial/\partial x, \partial/\partial y, \partial/\partial z)$ and note that the expression in parentheses looks like a dot product of this operator with the vector **A**. Thus, we define

$$\mathbf{\nabla} \cdot \mathbf{A} \equiv \frac{\partial A_x}{\partial x} + \frac{\partial A_y}{\partial y} + \frac{\partial A_z}{\partial z} \tag{1.28}$$

and write

$$d\Phi = \mathbf{\nabla} \cdot \mathbf{A}\, dV \tag{1.29}$$

Equation (1.29) states that the outward flux of a vector **A** through an *infinitesimal* surface of a *closed* box is equal to the divergence of the vector evaluated at the center of the box (or any other point, since the cube is small) times the volume of the box. We thus have a definition of divergence.

1.5.1 Definition The outward flux per unit volume of a vector field through a small closed surface bounding a small volume located at a point P (in the limit that the volume goes to zero) is called the divergence of that vector field at the point P. That is,

$$\text{div}\,\mathbf{A} \equiv \nabla\cdot\mathbf{A} \equiv \lim_{\Delta V\to 0}\frac{\Delta\Phi}{\Delta V}$$

The obvious question is whether we can generalize (1.29) to larger and more arbitrarily shaped surfaces. Consider two boxes with one face in common (Fig. 1.20) and denote the volume on the left by a and the one on the right by b. The total flux is, of course, the sum of the fluxes through *all six faces of the composite box*:

$$d\Phi = (d\Phi_1 + d\Phi_2) + (d\Phi_3 + d\Phi_4) + (d\Phi_5 + d\Phi_6)$$

where as usual $d\Phi_1$ is the total flux through the face having a normal in the positive x direction, $d\Phi_2$ that through the face having a normal in the negative x direction, and so on.

Figure 1.20 shows all the faces involved. It is evident that

$$d\Phi_1 = d\Phi_{a1} + d\Phi_{b1}$$

where $d\Phi_{a1}$ is the flux through the positive x face of box a and $d\Phi_{b1}$ is the flux through the positive x face of box b. Using similar notation, we can write

$$d\Phi_2 = d\Phi_{a2} + d\Phi_{b2}$$

and

$$d\Phi_5 + d\Phi_6 = d\Phi_{a5} + d\Phi_{b5} + d\Phi_{a6} + d\Phi_{b6}$$

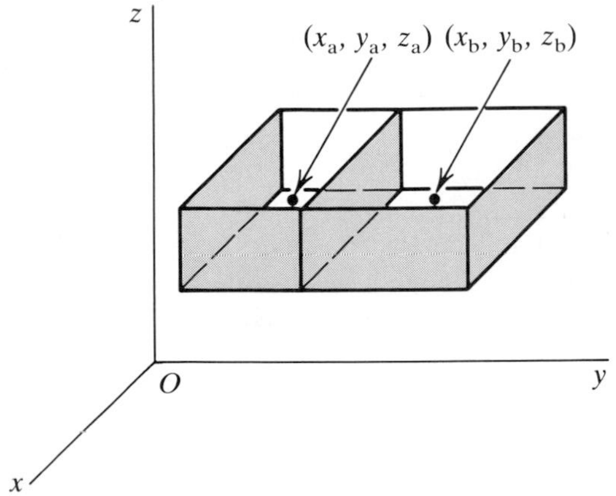

Figure 1.20 Illustration of the fact that the total flux through a surface composed of two boxes is equal to the sum of the divergence times the volume for each box.

However, for the y faces we have

$$d\Phi_3 = d\Phi_{b3}$$

$$d\Phi_4 = d\Phi_{a4}$$

because the face of the composite box in the positive y direction belongs to box b and that in the negative y direction to box a. Now, note that the outward flux through the left face of box b is the negative of the outward flux through the right face of box a; that is,

$$d\Phi_{b4} = -\, d\Phi_{a3}$$

or

$$d\Phi_{b4} + d\Phi_{a3} = 0$$

Thus, we obtain

$$d\Phi_3 + d\Phi_4 = d\Phi_{a3} + d\Phi_{a4} + d\Phi_{b3} + d\Phi_{b4}$$

Using all the above relations, we obtain

$$d\Phi = (d\Phi_{a1} + d\Phi_{a2}) + (d\Phi_{a3} + d\Phi_{a4}) + (d\Phi_{a5} + d\Phi_{a6}) + (d\Phi_{b1} + d\Phi_{b2}) + (d\Phi_{b3} + d\Phi_{b4}) + (d\Phi_{b5} + d\Phi_{b6})$$

or

$$d\Phi = d\Phi_a + d\Phi_b$$

where $d\Phi_a$ is the total flux through closed box a and $d\Phi_b$ is that through closed box b. Now we can use (1.29) for each of the closed boxes and write

$$d\Phi = (\nabla \cdot \mathbf{A})_a \, dV_a + (\nabla \cdot \mathbf{A})_b \, dV_b$$

where each subscript indicates evaluation at a point inside the appropriate volume.

It is now clear that we can generalize to any (small) number of boxes:

$$d\Phi = (\nabla \cdot \mathbf{A})_a \, dV_a + (\nabla \cdot \mathbf{A})_b dV_b + \cdots + (\nabla \cdot \mathbf{A})_z \, dV_z$$

It is worth emphasizing that the LHS of this equation is the outward flux through the *bounding surface only*. For example, $d\Phi$ could be the total flux through the six faces of a cubic box divided up into smaller infinitesimal cubic boxes.

For an arbitrary *closed* surface, we can divide up the volume into N (a large number) cubic boxes and write[17]

$$\Phi \approx \sum_{i=1}^{N} (\nabla \cdot \mathbf{A})_i \, dV_i \tag{1.30}$$

The use of the approximation sign reflects the facts that the shape of the closed surface is arbitrary (it is not a box), that N, although large, is not infinite, and that the sizes of boxes are not small enough. To attain equality we must make the sizes of the boxes smaller and smaller and their number larger and larger and write (1.30) in an exact

[17] We should really use ΔV_i here instead of dV_i, but it is customary in physics literature not to distinguish between the two.

integral form:

$$\Phi = \iiint\limits_V \nabla \cdot \mathbf{A}\, dV \tag{1.31}$$

Then, using (1.27), we can state an important theorem.

1.5.2 Theorem (The Divergence Theorem)

$$\iint\limits_S \mathbf{A} \cdot d\mathbf{a} = \iiint\limits_V \nabla \cdot \mathbf{A}\, dV$$ ■

The divergence theorem states that the surface integral (flux) of any vector field **A** through a *closed* surface S bounding the volume V is equal to the volume integral of the divergence of **A**.

To improve our physical intuition of divergence, let us consider the flow of a fluid of density $\rho(x, y, z, t)$ and velocity $\mathbf{v}(x, y, z, t)$. The flux of matter (the amount of matter crossing a surface area da per unit time) is obtained by multiplying (1.26) by ρ, giving

$$d\Phi_{\mathrm{m}} = (\rho\mathbf{v}) \cdot d\mathbf{a}$$

which, for a closed surface, leads to

$$\Phi_{\mathrm{m}} = \iint\limits_S (\rho\mathbf{v}) \cdot \hat{\mathbf{e}}_{\mathrm{n}}\, da = \iiint\limits_V \nabla \cdot (\rho\mathbf{v})\, dV$$

by the divergence theorem.

In particular, if the closed surface is infinitesimally small, we have

$$d\Phi_{\mathrm{m}} = \nabla \cdot (\rho\mathbf{v})\, dV$$

Conservation of matter, on the other hand, indicates that the rate of decrease in the amount of matter in dV must equal the flux, $d\Phi_{\mathrm{m}}$, or

$$d\Phi_{\mathrm{m}} = -\frac{\partial}{\partial t}(\rho\, dV) = -\frac{\partial \rho}{\partial t}\, dV$$

for fixed dV. Note the important minus sign. Combining the preceding two equations, we obtain

$$\frac{\partial \rho}{\partial t} + \nabla \cdot (\rho\mathbf{v}) = 0 \tag{1.32}$$

which is the differential form of the conservation of matter in any fluid flow. Equation (1.32) is sometimes written in an alternate form by first rewriting it as

$$\frac{\partial \rho}{\partial t} + (\nabla\rho) \cdot \mathbf{v} + \rho \nabla \cdot \mathbf{v} = 0$$

or

$$\frac{\partial \rho}{\partial t} + \left(\frac{\partial \rho}{\partial x}\frac{dx}{dt} + \frac{\partial \rho}{\partial y}\frac{dy}{dt} + \frac{\partial \rho}{\partial z}\frac{dz}{dt} \right) + \rho \nabla \cdot \mathbf{v} = 0 \tag{1.33}$$

By the definition of the total derivative, we have

$$\frac{d\rho\,(x, y, z, t)}{dt} = \frac{\partial \rho}{\partial t} + \frac{\partial \rho}{\partial x}\frac{dx}{dt} + \frac{\partial \rho}{\partial y}\frac{dy}{dt} + \frac{\partial \rho}{\partial z}\frac{dz}{dt}$$

Therefore, Eq. (1.33) becomes

$$\frac{d\rho}{dt} + \rho \nabla \cdot \mathbf{v} = 0 \tag{1.34}$$

Equation (1.32) or (1.34) is called the *continuity equation.* Note that the time derivative is a partial derivative in (1.32) and a total derivative in (1.34).

1.5.4 Line Integral and Curl

Section 1.5.3 introduced the concept of flux and the surface integral associated with it. This section discusses the notion of a line integral. The prime example of a line integral is the work done by a force. Consider the force field $\mathbf{F}(x, y, z, t)$ acting on an object and imagine the object being displaced by an infinitesimal distance, $d\mathbf{r}$. Then the work done by the force in effecting this displacement is defined as follows:

$$dW = \mathbf{F}(x, y, z, t) \cdot d\mathbf{r}$$

Note that $d\mathbf{r}$, being a displacement, has components (dx, dy, dz). Thus, we can also write

$$dW = F_x(x, y, z, t)dx + F_y(x, y, z, t)dy + F_z(x, y, z, t)dz$$

To calculate the work for a finite displacement, such as the one shown in Fig. 1.21, we break up the displacement into N small segments, calculate the work for each segment, and add all contributions to obtain

$$W \approx \mathbf{F}(x_1, y_1, z_1) \cdot d\mathbf{r}_1 + \mathbf{F}(x_2, y_2, z_2) \cdot d\mathbf{r}_2 + \cdots + \mathbf{F}(x_N, y_N, z_N) \cdot d\mathbf{r}_N$$

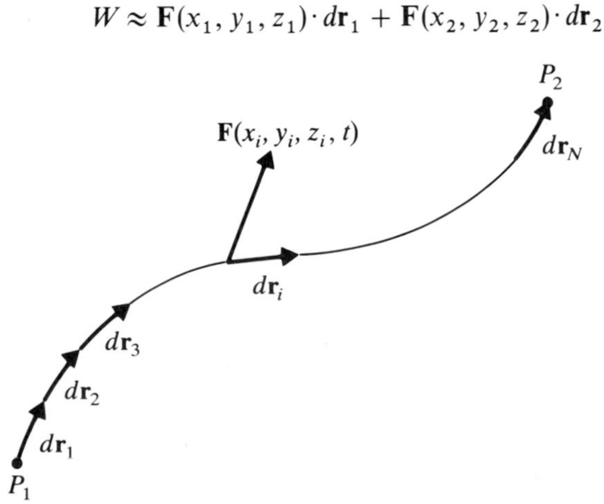

Figure 1.21 The line integral of a vector field $\mathbf{F}(x, y, z, t)$ from point P_1 to point P_2.

where we have ignored the dependence on time and denoted the midpoint (or any other point) of $d\mathbf{r}_i$ by (x_i, y_i, z_i). We can write the foregoing equation more compactly:[18]

$$W \approx \sum_{i=1}^{N} \mathbf{F}(x_i, y_i, z_i)\cdot d\mathbf{r}_i$$

The approximation sign can be removed by taking $d\mathbf{r}_i$ as small as possible and N as large as possible. Then we have

$$W = \int_{P_1}^{P_2} \mathbf{F}(x, y, z)\cdot d\mathbf{r} \equiv \int_{C} \mathbf{F}\cdot d\mathbf{r} \tag{1.35}$$

Equation (1.35) is by definition the *line integral* of the force field $\mathbf{F}$. In this particular case it is the work done by $\mathbf{F}$ in moving from P_1 to P_2. Of course, we can apply the line integral to any vector field, not just force. In electromagnetic theory, for example, the line integrals of the electric and magnetic fields play a central role.

Example 1.5.5

Let us evaluate the line integral of $\mathbf{F}_1 = \hat{\mathbf{e}}_x x^2 + \hat{\mathbf{e}}_y y^2$ along the semicircle shown in Fig. 1.22. Along this semicircle $x = a\cos\varphi$ and $y = a\sin\varphi$. Thus, *on the semicircle* $\mathbf{F}_1$ can be written as

$$\mathbf{F}_1 = \hat{\mathbf{e}}_x a^2\cos^2\varphi + \hat{\mathbf{e}}_y a^2\sin^2\varphi$$

Similarly, the element of displacement, $d\mathbf{r}$, has components

$$dx = d(a\cos\varphi) = -a\sin\varphi\, d\varphi$$

$$dy = d(a\sin\varphi) = a\cos\varphi\, d\varphi$$

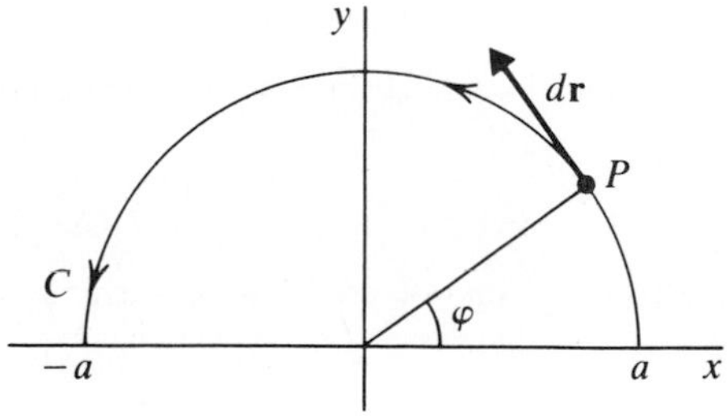

Figure 1.22 The semicircular path for calculating the line integral.

[18] Again we use $d\mathbf{r}_i$ instead of $\Delta\mathbf{r}_i$.

Thus,

$$\int_C \mathbf{F}_1 \cdot d\mathbf{r} = \int_0^{\pi} (\hat{\mathbf{e}}_x a^2 \cos^2 \varphi + \hat{\mathbf{e}}_y a^2 \sin^2 \varphi) \cdot (-\hat{\mathbf{e}}_x a \sin \varphi \, d\varphi + \hat{\mathbf{e}}_y a \cos \varphi \, d\varphi)$$

$$= a^3 \int_0^{\pi} (-\cos^2 \varphi \sin \varphi + \sin^2 \varphi \cos \varphi) d\varphi = a^3 (\tfrac{1}{3} \cos^3 \varphi |_0^{\pi} + \tfrac{1}{3} \sin^3 \varphi |_0^{\pi})$$

$$= -\tfrac{2}{3} a^3$$ ●

Example 1.5.6

We can also use Cartesian coordinates to solve the problem in Example 1.5.5. We note that *on the semicircle* $y = \sqrt{a^2 - x^2}$, so $dy = -x(a^2 - x^2)^{-1/2} \, dx$. Thus,

$$\mathbf{F}_1 \cdot d\mathbf{r} = [\hat{\mathbf{e}}_x x^2 + \hat{\mathbf{e}}_y (a^2 - x^2)] \cdot [\hat{\mathbf{e}}_x \, dx - \hat{\mathbf{e}}_y x (a^2 - x^2)^{-1/2} \, dx] = x^2 \, dx - x(a^2 - x^2)^{1/2} dx$$

and

$$\int_C \mathbf{F}_1 \cdot d\mathbf{r} = \int_a^{-a} [x^2 - x(a^2 - x^2)^{1/2}] dx = \tfrac{1}{3} x^3 |_a^{-a} - \tfrac{1}{3} (a^2 - x^2)^{3/2} |_a^{-a} = -\tfrac{2}{3} a^3$$

as expected. ●

Example 1.5.7

Let us consider a vector field that is a little different from that in the preceding examples. Let $\mathbf{F}_2 = \hat{\mathbf{e}}_x y^2 + \hat{\mathbf{e}}_y x^2$. Then

$$\begin{aligned}
\mathbf{F}_2 \cdot d\mathbf{r} &= (\hat{\mathbf{e}}_x a^2 \sin^2 \varphi + \hat{\mathbf{e}}_y a^2 \cos^2 \varphi) \cdot (-\hat{\mathbf{e}}_x a \sin \varphi \, d\varphi + \hat{\mathbf{e}}_y a \cos \varphi \, d\varphi) \\
&= a^3 (\cos^3 \varphi - \sin^3 \varphi) d\varphi \\
&= a^3 (\cos \varphi - \sin \varphi)(\cos^2 \varphi + \sin^2 \varphi + \sin \varphi \cos \varphi) d\varphi \\
&= a^3 (\cos \varphi - \sin \varphi)(1 + \sin \varphi \cos \varphi) d\varphi \\
&= a^3 (\cos \varphi - \sin \varphi + \cos^2 \varphi \sin \varphi - \sin^2 \varphi \cos \varphi) d\varphi
\end{aligned}$$

and the line integral becomes

$$\int_C \mathbf{F}_2 \cdot d\mathbf{r} = \int_0^{\pi} a^3 (\cos \varphi - \sin \varphi + \cos^2 \varphi \sin \varphi - \sin^2 \varphi \cos \varphi) d\varphi$$

$$= a^3 (\sin \varphi + \cos \varphi - \tfrac{1}{3} \cos^3 \varphi - \tfrac{1}{3} \sin^3 \varphi) |_0^{\pi} = -\tfrac{4}{3} a^3$$ ●

Curl of a vector field. Line integrals around a closed path are of special interest. For a vector field, **A**, and a closed path, *C*, we denote the line integral as

$\oint_C \mathbf{A} \cdot d\mathbf{r}$, where the circle on the integral sign indicates that the path is closed and C denotes the particular path taken.

Let us develop the analogue of the divergence theorem for closed line integrals. To begin, we consider a small closed rectangular path with a unit normal $\hat{\mathbf{e}}_n$, which is related to the direction of traversing the path by a right-hand rule.[19] Without loss of generality we assume that the rectangle is parallel to the xy-plane with sides parallel to the x-axis and the y-axis and that $\hat{\mathbf{e}}_n$ is parallel to the z-axis (see Fig. 1.23). The line integral can be written as

$$\oint_C \mathbf{A} \cdot d\mathbf{r} = \int_a^b \mathbf{A} \cdot d\mathbf{r} + \int_b^c \mathbf{A} \cdot d\mathbf{r} + \int_c^d \mathbf{A} \cdot d\mathbf{r} + \int_d^a \mathbf{A} \cdot d\mathbf{r}$$

Along ab the element of displacement, $d\mathbf{r}$, is always in the positive x direction and has magnitude dx, so it can be written as $d\mathbf{r} = \hat{\mathbf{e}}_x\, dx$. Thus, the first integral on the RHS above becomes

$$\int_a^b \mathbf{A} \cdot d\mathbf{r} = \int_a^b \mathbf{A}_1 \cdot d\mathbf{r}_1 = \int_a^b \mathbf{A}_1 \cdot (\hat{\mathbf{e}}_x\, dx) = \int_a^b A_{1x}\, dx$$

where, as before, the subscript 1 indicates that we have to evaluate $\mathbf{A}$ at the midpoint of ab and the subscript x denotes the x component. Now, since ab is small and *the cosine of the angle between* $\mathbf{A}$ *and* $d\mathbf{r}$ *does not change appreciably on ab,* we can omit the integral sign and write

$$\int_a^b \mathbf{A} \cdot d\mathbf{r} = A_{1x}\, dx = A_x\!\left(x, y - \frac{dy}{2}, z\right) dx$$

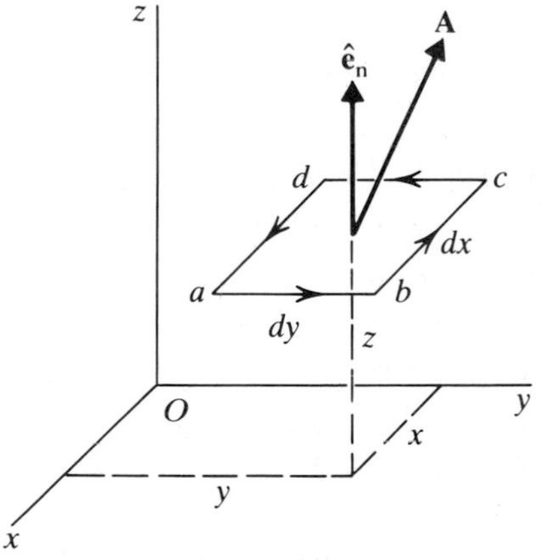

Figure 1.23 A closed rectangular path in a plane parallel to the xy-plane with center at (x, y, z).

[19]The following is the right-hand rule that, as explained in the text, relates the sense of integration to the direction of the unit normal to the surface: if your right-hand fingers are curled in the direction of integration along the curve, your thumb should point in the direction of $\hat{\mathbf{e}}_n$.

Using the Taylor expansion, as before, we obtain

$$\int_a^b \mathbf{A}\cdot d\mathbf{r} = \left[A_x(x, y, z) - \frac{dy}{2}\frac{\partial A_x}{\partial y}\right] dx$$

Similarly, we can write

$$\int_c^d \mathbf{A}\cdot d\mathbf{r} = \int_c^d \mathbf{A}_2\cdot d\mathbf{r}_2 = \int_c^d \mathbf{A}_2\cdot(-\hat{\mathbf{e}}_x\, dx)$$

$$= -\int_c^d A_{2x}\, dx = -A_{2x}\, dx = -A_x\left(x, y + \frac{dy}{2}, z\right) dx$$

$$= -\left[A_x(x, y, z) + \frac{dy}{2}\frac{\partial A_x}{\partial y}\right] dx$$

Adding the last two equations yields

$$\int_a^b \mathbf{A}\cdot d\mathbf{r} + \int_c^d \mathbf{A}\cdot d\mathbf{r} = -\frac{\partial A_x}{\partial y}\, dx\, dy$$

The contributions from the other two sides of the rectangle can also be calculated:

$$\int_b^c \mathbf{A}\cdot d\mathbf{r} + \int_d^a \mathbf{A}\cdot d\mathbf{r} = A_{3y}\, dy - A_{4y}\, dy$$

$$= A_y\left(x + \frac{dx}{2}, y, z\right) dy - A_y\left(x - \frac{dx}{2}, y, z\right) dy$$

$$= \left[A_y(x, y, z) + \frac{dx}{2}\frac{\partial A_y}{\partial x}\right] dy - \left[A_y(x, y, z) - \frac{dx}{2}\frac{\partial A_y}{\partial x}\right] dy$$

$$= \frac{\partial A_y}{\partial x}\, dx\, dy$$

The sum of these two equations gives the total contribution:

$$\oint_C \mathbf{A}\cdot d\mathbf{r} = \left(\frac{\partial A_y}{\partial x} - \frac{\partial A_x}{\partial y}\right) dx\, dy \tag{1.36}$$

Let us look at Eq. (1.36) more closely. The expression in parentheses can be interpreted as the z component of the cross product of the gradient operator, ∇, with $\mathbf{A}$. In fact, using the mnemonic determinant form of the vector product, we can write

$$\nabla \times \mathbf{A} = \begin{vmatrix} \hat{\mathbf{e}}_x & \hat{\mathbf{e}}_y & \hat{\mathbf{e}}_z \\ \dfrac{\partial}{\partial x} & \dfrac{\partial}{\partial y} & \dfrac{\partial}{\partial z} \\ A_x & A_y & A_z \end{vmatrix}$$

$$= \hat{\mathbf{e}}_x\left(\frac{\partial A_z}{\partial y} - \frac{\partial A_y}{\partial z}\right) + \hat{\mathbf{e}}_y\left(\frac{\partial A_x}{\partial z} - \frac{\partial A_z}{\partial x}\right) + \hat{\mathbf{e}}_z\left(\frac{\partial A_y}{\partial x} - \frac{\partial A_x}{\partial y}\right) \tag{1.37}$$

This cross product is called the *curl* of **A** and is an important quantity in vector analysis. We will look more closely at it later. At this point, however, we are interested only in its definition as applied in Eq. (1.36). The RHS of that equation can be written as

$$\left(\frac{\partial A_y}{\partial x} - \frac{\partial A_x}{\partial y}\right) dx\, dy = (\nabla \times \mathbf{A})_z\, dx\, dy = (\nabla \times \mathbf{A})\cdot \hat{\mathbf{e}}_z\, da$$

where da is the area of the rectangle, or $dx\, dy$. Noting that $\hat{\mathbf{e}}_z$ is in the direction normal to the area, we can replace $\hat{\mathbf{e}}_z$ with $\hat{\mathbf{e}}_{\mathrm{n}}$. Therefore, we can write (1.36) as

$$\oint_C \mathbf{A}\cdot d\mathbf{r} = (\nabla \times \mathbf{A})\cdot \hat{\mathbf{e}}_{\mathrm{n}}\, da = (\nabla \times \mathbf{A})\cdot d\mathbf{a} \tag{1.38}$$

Equation (1.38) states that for an infinitesimal rectangular path, C, the closed line integral is equal to the normal component of the curl of **A** evaluated at the center of the rectangle times the area of the rectangle. This result does not depend on the choice of coordinate system. In fact, any rectangle defines a plane and we are at liberty to designate that plane the xy-plane.

What happens with an arbitrary large closed path? Figure 1.24 shows an arbitrary closed path C with an arbitrary surface S, whose edge is the given curve. We divide S into small rectangular areas and assign a direction to their contours dictated by the direction of integration around C. If we sum all the contributions from the small rectangular paths, we will be left with the integration around C because the contributions from the common sides of adjacent rectangles cancel. This is because the sense of integration along their common side is opposite for two adjacent rectangles (see Fig. 1.24). Thus, the macroscopic version of (1.38) is

$$\oint_C \mathbf{A}\cdot d\mathbf{r} \approx \sum_{i=1}^{N} (\nabla \times \mathbf{A})_i \cdot \hat{\mathbf{e}}_{\mathrm{n}_i}\, da_i = \sum_{i=1}^{N} (\nabla \times \mathbf{A})_i \cdot d\mathbf{a}_i$$

where $(\nabla \times \mathbf{A})_i$ is the curl of **A** evaluated at the center of the ith rectangle, which has area da_i and normal $\hat{\mathbf{e}}_{\mathrm{n}_i}$, and N is the number of rectangles on the surface S. If the areas become smaller and smaller as N gets larger and larger, we can replace the summation by an integral and obtain the following theorem.

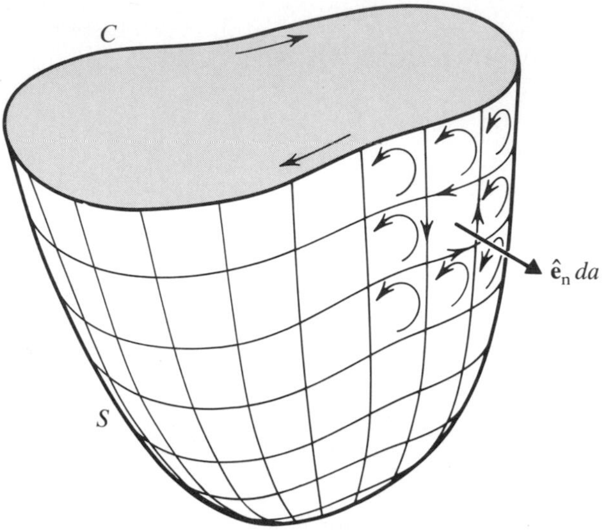

Figure 1.24 The curve C with an arbitrary surface whose edge is that curve. The sum of the line integrals around the rectangular paths shown is equal to the line integral around the curve C.

1.5.3 Theorem (Stokes' Theorem)

$$\oint_C \mathbf{A}\cdot d\mathbf{r} = \iint_S (\nabla \times \mathbf{A})\cdot d\mathbf{a} \qquad \blacksquare$$

In words, Stokes' theorem states that the line integral of a vector field **A** around a closed path C is equal to the surface integral of the curl of **A** on *any* surface whose only edge is path C. The direction of the normal to the infinitesimal area da of the surface S is related to the direction of integration around C by the right-hand rule.

Conservative vector fields. Of great importance are conservative vector fields, which are defined as follows.

1.5.4 Definition A *conservative vector field* is one that has a vanishing line integral around *every* closed path.

An immediate result of this definition is that the line integral of a conservative vector field **A** between two arbitrary points in space is independent of the path taken (see Problem 1.26). Thus, we can associate with **A** some function that depends only on the two points and not on how to go from one to the other. In particular, we can define a function $V(x, y, z)$ such that

$$V(x_1, y_1, z_1) - V(x_2, y_2, z_2) = \int_{P_1}^{P_2} \mathbf{A}\cdot d\mathbf{r}$$

where P_1 and P_2 have coordinates (x_1, y_1, z_1) and (x_2, y_2, z_2), respectively. The function V is well-defined because the integral is path-independent. For a general point P with coordinates (x, y, z), the definition of the function $V(x, y, z)$ gives

$$V(x, y, z) = V(x_0, y_0, z_0) - \int_{P_0}^{P} \mathbf{A} \cdot d\mathbf{r} \tag{1.39}$$

This $V(x, y, z)$ is called the *potential* associated with the vector field $\mathbf{A}$. Sometimes the potential at P_0 is taken to be zero, in which case we have

$$V(x, y, z) = -\int_{P_0}^{P} \mathbf{A} \cdot d\mathbf{r} \tag{1.40}$$

where P_0 is called the *potential reference point*.

When P_1 and P_2 are an infinitesimal distance from each other, we can write

$$dV = -\mathbf{A} \cdot d\mathbf{r}$$

where dV is the infinitesimal change in the potential of $\mathbf{A}$ due to the displacement $d\mathbf{r}$. On the other hand, V, being a scalar differentiable function of x, y, and z, has an infinitesimal increment dV,

$$dV = (\nabla V) \cdot d\mathbf{r}$$

so we have

$$(\nabla V) \cdot d\mathbf{r} = -\mathbf{A} \cdot d\mathbf{r}$$

But this is true for an arbitrary $d\mathbf{r}$. Taking $d\mathbf{r}$ to be $\hat{\mathbf{e}}_x\,dx$, $\hat{\mathbf{e}}_y\,dy$, and $\hat{\mathbf{e}}_z\,dz$ in turn, we obtain the equality of the three components of ∇V and $-\mathbf{A}$. Therefore, we have

$$\mathbf{A} = -\nabla V \tag{1.41}$$

which states that a conservative vector field can be written as the negative gradient of a potential.

Another property of a conservative vector field can be obtained by rewriting (1.38), which is true for an arbitrary infinitesimal closed path:

$$\oint_C \mathbf{A} \cdot d\mathbf{r} = (\nabla \times \mathbf{A}) \cdot \hat{\mathbf{e}}_n\, da$$

However, the LHS is zero because $\mathbf{A}$ is conservative. That is, we have

$$(\nabla \times \mathbf{A}) \cdot \hat{\mathbf{e}}_n\, da = 0$$

This is true for arbitrary da and $\hat{\mathbf{e}}_n$. Therefore, we have the important conclusion that

$$\nabla \times \mathbf{A} = 0$$

for a conservative vector field. It is important to note that although the integral $\oint_C \mathbf{A}\cdot d\mathbf{r}$ is zero and $d\mathbf{r}$ is small, we cannot deduce that $\mathbf{A}\cdot d\mathbf{r} = 0$ and, therefore, $\mathbf{A} = 0$. (Why?)

A conservative vector field demands the vanishing of the curl. But is $\nabla \times \mathbf{A} = 0$ *sufficient* for a vector field to be conservative? The answer, in general, is *no*! (Example 1.6.3 illustrates the point.) In general, if the vector field is well-defined and well-behaved (smoothly varying, differentiable, etc.) in a region of space, U, then $\nabla \times \mathbf{A} = 0$ in U implies that $\oint_C \mathbf{A}\cdot d\mathbf{r} = 0$ for all closed curves C lying entirely in U. In modern mathematical jargon such a region U is said to be *contractable to zero*, which means that any closed curve in U can be contracted to a point without encountering any singular point of the vector field (where it is not defined or well-behaved). We thus have the following theorem.

1.5.5 Theorem Let the region U in space be contractable to zero for the vector field $\mathbf{A}$. Then for any closed curve C in U the following two statements are equivalent:

$$\nabla \times \mathbf{A} = 0$$

and

$$\oint_C \mathbf{A}\cdot d\mathbf{r} = 0 \qquad \forall\ C \text{ in } V$$ ■

It should be clear that $\nabla \times \mathbf{A} \neq 0$ always implies that $\mathbf{A}$ is *not* conservative. However, $\nabla \times \mathbf{A} = 0$ implies that $\mathbf{A}$ is conservative *only* if the region in question is contractable to zero.

Example 1.5.8

Let us refer back to Examples 1.5.5 and 1.5.7 and consider a different path, the portion of the x-axis between points $(a, 0)$ and $(-a, 0)$, as shown in Fig. 1.25. Along this path $y = 0$ and $dy = 0$, so

$$\int_{C'} \mathbf{F}_1\cdot d\mathbf{r} = \int_{C'} [\hat{\mathbf{e}}_x x^2 + \hat{\mathbf{e}}_y(0)]\cdot[\hat{\mathbf{e}}_x\, dx + \hat{\mathbf{e}}_y(0)] = \int_a^{-a} x^2\, dx = -\tfrac{2}{3}a^3$$

which is the same as the result in Example 1.5.5. But

$$\int_{C'} \mathbf{F}_2\cdot d\mathbf{r} = \int_{C'} [\hat{\mathbf{e}}_x(0) + \hat{\mathbf{e}}_y x^2]\cdot[\hat{\mathbf{e}}_x\, dx + \hat{\mathbf{e}}_y(0)] = 0$$

is different from the result in Example 1.5.7.

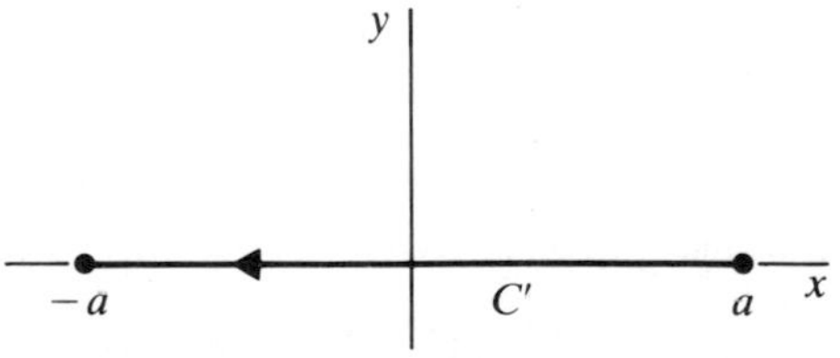

Figure 1.25 The straight path for evaluating the line integral.

We might suspect that $\mathbf{F}_1$ is a conservative field but $\mathbf{F}_2$ is definitely not. To see whether $\mathbf{F}_1$ is conservative, we take its curl:

$$\nabla \times \mathbf{F}_1 = \begin{vmatrix} \hat{\mathbf{e}}_x & \hat{\mathbf{e}}_y & \hat{\mathbf{e}}_z \\ \dfrac{\partial}{\partial x} & \dfrac{\partial}{\partial y} & \dfrac{\partial}{\partial z} \\ x^2 & y^2 & 0 \end{vmatrix} = 0$$

This result *and* the fact that $\mathbf{F}_1$ is well-behaved over all space lead, by Theorem 1.5.5, to the conclusion that $\mathbf{F}_1$ is indeed conservative. The potential associated with this vector field can be obtained by referring to Fig. 1.26. Since $\mathbf{F}_1$ is conservative, it does not matter which path is taken to reach $P(x_1, y_1)$, which is an arbitrary point in the xy-plane. Taking the path indicated in the figure makes the calculations simpler. Along (1) $y = 0$ and $dy = 0$, and along (2) $x = x_1$ and $dx = 0$. Therefore,

$$\int_0^P \mathbf{F}_1 \cdot d\mathbf{r} = \int_{(1)} \mathbf{F}_1 \cdot d\mathbf{r} + \int_{(2)} \mathbf{F}_1 \cdot d\mathbf{r} = \int_0^{x_1} x^2\, dx + \int_0^{y_1} y^2\, dy = \tfrac{1}{3} x_1^3 + \tfrac{1}{3} y_1^3$$

Taking the origin as the reference point for the potential means that $V(0, 0, 0) = 0$. Then

$$V(x_1, y_1) = -\int_0^P \mathbf{F}_1 \cdot d\mathbf{r} = -\tfrac{1}{3}(x_1^3 + y_1^3)$$

or, in general,

$$V(x, y) = -\tfrac{1}{3}(x^3 + y^3)$$

The gradient of this potential has components

$$\frac{\partial V}{\partial x} = -x^2 \quad \text{and} \quad \frac{\partial V}{\partial y} = -y^2$$

Thus,

$$\mathbf{F}_1 = -\nabla V$$

as expected of a conservative vector field.

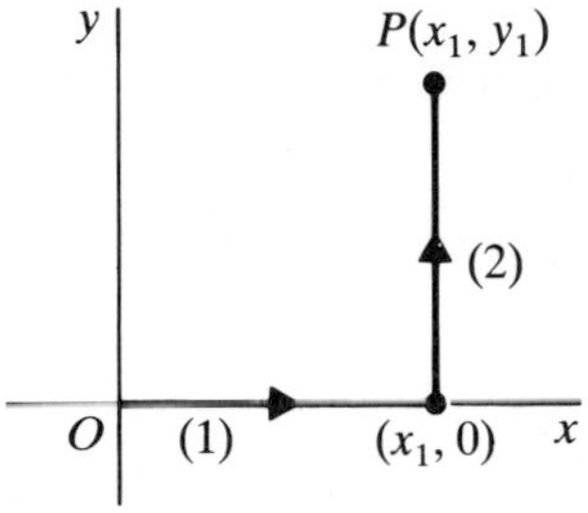

Figure 1.26 A broken path from O to P.

The curl of $\mathbf{F}_2$ is

$$\nabla \times \mathbf{F}_2 = \begin{vmatrix} \hat{\mathbf{e}}_x & \hat{\mathbf{e}}_y & \hat{\mathbf{e}}_z \\ \dfrac{\partial}{\partial x} & \dfrac{\partial}{\partial y} & \dfrac{\partial}{\partial z} \\ y^2 & x^2 & 0 \end{vmatrix} = \hat{\mathbf{z}}(2x - 2y) \neq 0$$

Thus, as expected, $\mathbf{F}_2$ is not a conservative field and has no potential function. •

1.5.5 Double Del Operations

This subsection briefly discusses different combinations of the vector operator ∇ with itself. By direct differentiation we can easily verify that

$$\nabla \times (\nabla f) \equiv 0$$

Similarly, we can show that

$$\nabla \cdot (\nabla \times \mathbf{A}) \equiv 0$$

Finally, the divergence of the gradient is an important and frequently occurring operator called the *Laplacian*:

$$\nabla \cdot (\nabla f) \equiv \nabla^2 f = \frac{\partial^2 f}{\partial x^2} + \frac{\partial^2 f}{\partial y^2} + \frac{\partial^2 f}{\partial z^2} \tag{1.42}$$

As will become evident, the Laplacian occurs throughout physics, in situations ranging from the waves on a drum to the diffusion of matter in space, the propagation of electromagnetic waves, and even the most basic behavior of matter on a subatomic scale, as governed by the Schrödinger equation of quantum mechanics.

Exercises

1.5.1 The total angular momentum, $\mathbf{L}$, of a system consisting of N point particles of masses $m_1, m_2, \ldots, m_N$ located instantaneously at $\mathbf{r}_1, \mathbf{r}_2, \ldots, \mathbf{r}_N$ relative to an arbitrary origin is defined as

$$\mathbf{L} = \sum_{i=1}^{N} \mathbf{r}_i \times \mathbf{p}_i \tag{1}$$

where $\mathbf{p}_i$ is the momentum of the ith particle. Show that

$$\frac{d\mathbf{L}}{dt} = \boldsymbol{\tau} \tag{2}$$

where

$$\boldsymbol{\tau} = \sum_{i=1}^{N} \mathbf{r}_1 \times (\mathbf{F}_i)_{\text{ext}} \tag{3}$$

and $(\mathbf{F}_i)_{\text{ext}}$ is the external force acting on the ith particle.

1.5.2 Using Cartesian coordinates, show that for a general function of the form $f(r)$ with $r = \sqrt{x^2 + y^2 + z^2}$, the gradient is

$$\nabla f = \frac{df}{dr}\hat{\mathbf{e}}_r$$

1.5.3 Consider the surface defined by $f(x, y, z) = C$, where C is a constant. Find a unit vector normal to this surface at the point (x_0, y_0, z_0).

1.5.4 Prove this vector identity:

$$\nabla\cdot(f\mathbf{A}) = (\nabla f)\cdot\mathbf{A} + f\nabla\cdot\mathbf{A}$$

1.5.5 Evaluate the line integral of

$$\mathbf{F}(x, y) = (x^2 + 3y)\hat{\mathbf{e}}_x + (y^2 + 2x)\hat{\mathbf{e}}_y$$

from the origin to the point (1, 2), first **(a)** along the line joining the two points and then **(b)** along the path going straight from the origin to the point (1, 0) and then straight from (1, 0) to (1, 2). (See Fig. 1.27.) **(c)** Is **F** conservative?

1.5.6 Repeat Exercise 1.5.5 for the vector field

$$\mathbf{F}(x, y) = (2x^2y + 4y)\hat{\mathbf{e}}_x + (3y^2x + 3x)\hat{\mathbf{e}}_y$$

1.5.7 Is the following vector field conservative?

$$\mathbf{A}(x, y, z) = (2xy + 3z^2)\hat{\mathbf{e}}_x + (x^2 + 4yz)\hat{\mathbf{e}}_y + (2y^2 + 6xz)\hat{\mathbf{e}}_z$$

If so, find the potential associated with it.

1.5.8 Show that

$$\iiint_V (\nabla\times\mathbf{A})\,dV = \iint_S \hat{\mathbf{e}}_n \times \mathbf{A}\,da$$

where **A** is an arbitrary vector field and S is the surface bounding the volume V.

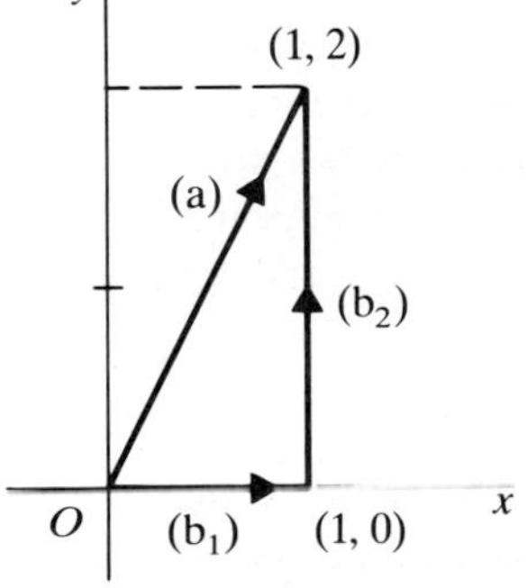

Figure 1.27 Broken and straight paths from the origin to (1, 2).

1.6 VECTOR ANALYSIS IN CURVILINEAR COORDINATES

In previous sections we have worked mostly with rectangular (Cartesian) coordinate systems in which the unit vectors $\hat{\mathbf{e}}_x$, $\hat{\mathbf{e}}_y$, and $\hat{\mathbf{e}}_z$ are fixed in direction. However, because of the symmetry of a physical situation, it is sometimes more convenient to use other coordinate systems. Two systems used very frequently in such cases are the spherical and cylindrical coordinate systems. It is necessary, therefore, to be familiar with expressions such as the gradient, divergence, curl, and Laplacian in these coordinate systems.

We will investigate these coordinate systems as special cases of more general systems generically called *curvilinear coordinates.* A point P_0 in space can be determined not only by the triple (x_0, y_0, z_0), but also by the values of any three independent functions of these coordinates. Let us denote these three functions as $f_1(x, y, z)$, $f_2(x, y, z)$ and $f_3(x, y, z)$ and assume that the values of these functions at P_0 are q_1, q_2, and q_3, respectively; that is,

$$f_1(x_0, y_0, z_0) = q_1 \tag{1.43a}$$

$$f_2(x_0, y_0, z_0) = q_2 \tag{1.43b}$$

$$f_3(x_0, y_0, z_0) = q_3 \tag{1.43c}$$

Then we say that the curvilinear coordinates of P_0 are (q_1, q_2, q_3). We can replace (x_0, y_0, z_0) with an arbitrary triple (x, y, z) and think of (1.43) as the equations of three surfaces that intersect at the point P_0. For example, the three equations

$$\sqrt{x^2 + y^2 + z^2} = r$$

$$\tan^{-1}\frac{\sqrt{x^2 + y^2}}{z} = \theta$$

and

$$\tan^{-1}\frac{y}{x} = \varphi$$

describe, respectively, a sphere of radius r, a cone of angle θ, and a plane passing through the z-axis and making an angle φ with the xz-plane. The intersection of all these surfaces is a point with curvilinear coordinates (r, θ, φ). We demand that the three surfaces of (1.43) be perpendicular to each other, and we take the unit vectors $\hat{\mathbf{e}}_1$, $\hat{\mathbf{e}}_2$, and $\hat{\mathbf{e}}_3$ to be perpendicular to the first, second, and third surface, respectively. We also assume that $\hat{\mathbf{e}}_1 \times \hat{\mathbf{e}}_2 = \hat{\mathbf{e}}_3$ and that these unit vectors are mutually perpendicular to one another. Obviously, these unit vectors do not have a fixed orientation in space; their orientation depends on the point being considered. Figure 1.28 shows the coordinates q_1, q_2, and q_3 of a general point, P, the unit vectors $\hat{\mathbf{e}}_1$, $\hat{\mathbf{e}}_2$, and $\hat{\mathbf{e}}_3$, and the surfaces perpendicular to them. For the special surfaces defined by

$$x = q_1 \qquad y = q_2 \qquad z = q_3$$

we obviously recover the rectangular coordinates.

Let us now consider the elements of length in the curvilinear coordinates, because all the important operations (gradient, divergence, and curl) involve changes

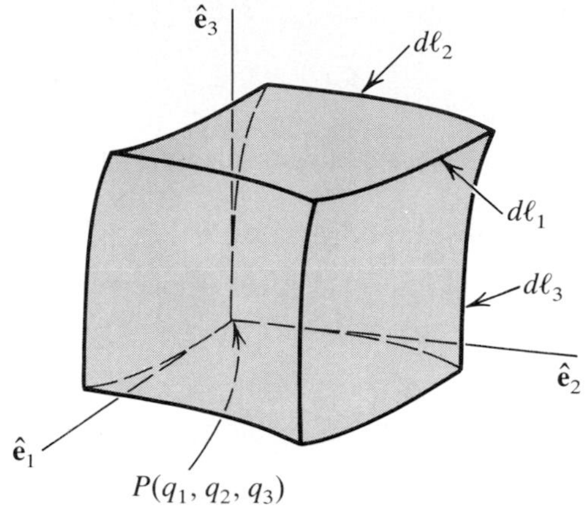

Figure 1.28 The curvilinear coordinates q_1, q_2, and q_3. The unit vectors $\hat{\mathbf{e}}_1, \hat{\mathbf{e}}_2$, and $\hat{\mathbf{e}}_3$ and the elements of length in the three directions are also shown. Note that the orientation of the unit vectors is such that $\hat{\mathbf{e}}_1 \times \hat{\mathbf{e}}_2 = \hat{\mathbf{e}}_3$.

of fields due to changes in distances. Let us call $d\ell_1$ an element of length perpendicular to the surface defined by (1.43a). This is the distance between the surface

$$f_1(x, y, z) = q_1$$

and the nearby parallel surfacc

$$f_1(x, y, z) = q_1 + dq_1$$

Since $d\ell_1$ is perpendicular to the surface of constant q_1, both q_2 and q_3 remain constant along $d\ell_1$ (see Fig. 1.28). Therefore, we can generally write this element of length as

$$d\ell_1 = h_1 dq_1$$

where h_1 is, in general, a function of the coordinates q_1, q_2, and q_3. Similarly,

$$d\ell_2 = h_2 dq_2 \qquad \text{and} \qquad d\ell_3 = h_3 dq_3$$

We can combine all the equations for the elements of length and write them as a single vector equation:

$$d\mathbf{r} = \hat{\mathbf{e}}_1 d\ell_1 + \hat{\mathbf{e}}_2 d\ell_2 + \hat{\mathbf{e}}_3 d\ell_3 = \sum_{i=1}^{3} \hat{\mathbf{e}}_i h_i dq_i \tag{1.44}$$

Equation (1.44) is useful; for example, we can obtain the unit vectors $\hat{\mathbf{e}}_i$ as follows. Dividing both sides of (1.44) by dq_1 and assuming that dq_2 and dq_3 are zero, we obtain

$$\frac{\partial \mathbf{r}}{\partial q_1} \equiv \left.\frac{d\mathbf{r}}{dq_1}\right|_{\substack{q_2 = \text{constant} \\ q_3 = \text{constant}}} = \hat{\mathbf{e}}_1 h_1$$

or

$$\hat{\mathbf{e}}_1 = \frac{1}{h_1} \frac{\partial \mathbf{r}}{\partial q_1}$$

More generally,

$$\hat{\mathbf{e}}_j = \frac{1}{h_j}\frac{\partial \mathbf{r}}{\partial q_j} \qquad \text{for } j = 1, 2, 3 \tag{1.45}$$

Equation (1.45) can be used to find $\hat{\mathbf{e}}_1, \hat{\mathbf{e}}_2$, and $\hat{\mathbf{e}}_3$ in terms of unit vectors of other coordinate systems.

The volume element in curvilinear coordinates is

$$dV = d\ell_1 d\ell_2 d\ell_3 = h_1 h_2 h_3 \, dq_1 dq_2 dq_3 \tag{1.46}$$

1.6.1 Cylindrical Coordinates

The cylindrical coordinate system is illustrated in Fig. 1.29. The position of any point P in space is specified by the three cylindrical coordinates ρ, φ, and z. The coordinate ρ is the perpendicular distance from the z-axis; φ is the azimuth angle of the plane containing P and the z-axis, measured from the xz-plane; and z is the distance from the xy-plane.

At the point P there are three mutually orthogonal directions specified by three unit vectors: $\hat{\mathbf{e}}_\rho$ is in the direction of the perpendicular from the z-axis extended through P; $\hat{\mathbf{e}}_\varphi$ is perpendicular to the plane containing the z-axis and P and pointing in the direction corresponding to increasing φ; and $\hat{\mathbf{e}}_z$ is in the positive z direction. These unit vectors do *not* maintain the same directions in space as the point P moves, but they always remain mutually orthogonal.

The vector $\mathbf{r}$ describing the position of P is

$$\mathbf{r} = \hat{\mathbf{e}}_\rho \rho + \hat{\mathbf{e}}_z z$$

This expression does not involve the angle φ and, therefore, does not determine a point uniquely. This is due to the fact that the unit vectors shown in Fig. 1.29 do not have a fixed orientation in space.

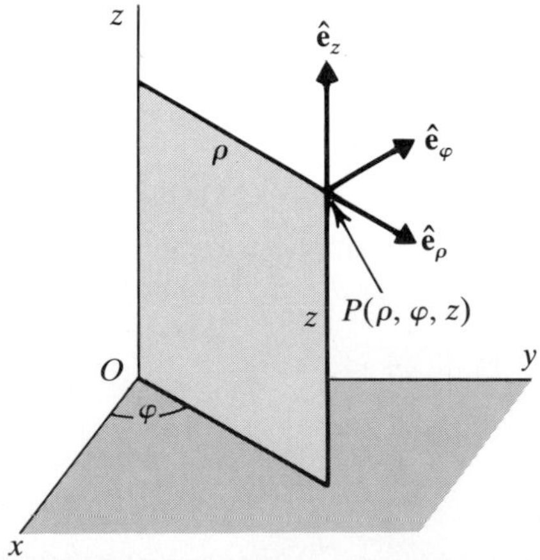

Figure 1.29 The cylindrical coordinate system.

Elements of length corresponding to infinitesimal changes in the coordinates of a point are important. If the coordinates φ and z of the point P are kept constant while ρ is allowed to increase by $d\rho$, then P is displaced by an amount $d\mathbf{r} = \hat{\mathbf{e}}_\rho \, d\rho$. On the other hand, if ρ and z are held fixed while φ is allowed to increase by $d\varphi$, then P is displaced by $d\mathbf{r} = \hat{\mathbf{e}}_\varphi \rho \, d\varphi$. Finally, if ρ and φ are held constant and z is allowed to increase by dz, then $d\mathbf{r} = \hat{\mathbf{e}}_z \, dz$. For arbitrary increments, $d\rho$, $d\varphi$, and dz, we have

$$d\mathbf{r} = \hat{\mathbf{e}}_\rho \, d\rho + \hat{\mathbf{e}}_\varphi \rho \, d\varphi + \hat{\mathbf{e}}_z \, dz \tag{1.47}$$

and the magnitude of $d\mathbf{r}$ is

$$\| d\mathbf{r} \| = \sqrt{(d\rho)^2 + (\rho \, d\varphi)^2 + (dz)^2}$$

Figure 1.30 shows these displacements as well as the volume element associated with them. This infinitesimal volume is

$$dV = \rho \, d\rho \, d\varphi \, dz$$

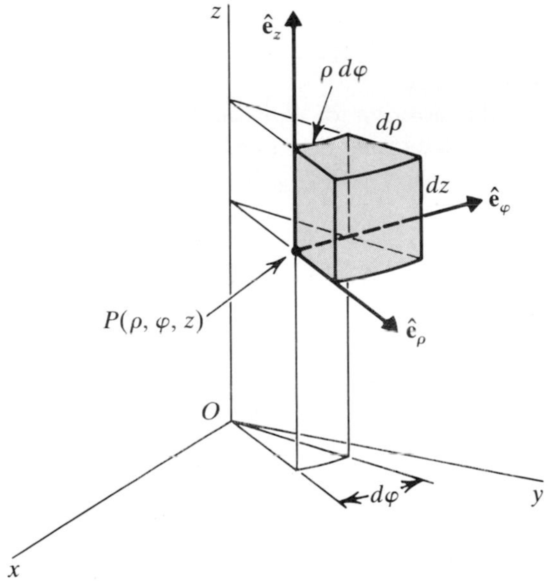

Figure 1.30 The distances involved in changing the coordinates ρ, φ, and z by infinitesimal amounts. The volume element corresponding to these distances is also shown.

1.6.2 Spherical Coordinates

The position of a point P is specified in spherical coordinates by r, θ, and φ, with r being the distance from the origin, θ the angle between the z-axis and the radius vector, and φ the azimuthal angle. At the point P the unit vectors are as shown in Fig. 1.31: $\hat{\mathbf{e}}_r$ is in the direction of the radius vector extended through P; $\hat{\mathbf{e}}_\theta$ is perpendicular to the radius vector in the plane containing the z-axis and the radius vector and pointing in the direction of increasing θ; $\hat{\mathbf{e}}_\varphi$ is perpendicular to both $\hat{\mathbf{e}}_r$ and

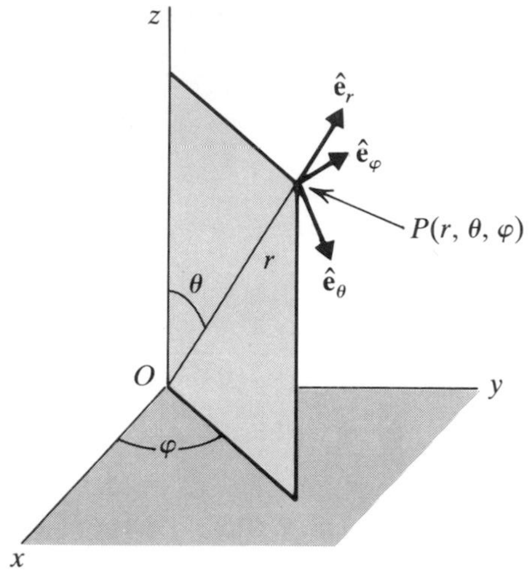

Figure 1.31 The spherical coordinate system.

$\hat{\mathbf{e}}_\theta$ and related to them by $\hat{\mathbf{e}}_\varphi = \hat{\mathbf{e}}_r \times \hat{\mathbf{e}}_\theta$. As before, these unit vectors do *not* maintain the same direction in space as P moves.

The vector $\mathbf{r}$ describing the position of P is simply $\mathbf{r} = \hat{\mathbf{e}}_r r$, and, as with cylindrical coordinates, it does *not* determine P uniquely.

The displacement element $d\mathbf{r}$ corresponding to arbitrary increments of the coordinates is

$$d\mathbf{r} = \hat{\mathbf{e}}_r\, dr + \hat{\mathbf{e}}_\theta r\, d\theta + \hat{\mathbf{e}}_\varphi r \sin\theta\, d\varphi \tag{1.48}$$

and its magnitude is

$$\| d\mathbf{r} \| = \sqrt{(dr)^2 + (r\, d\theta)^2 + (r \sin\theta\, d\varphi)^2}$$

Note that $\| d\mathbf{r} \| \neq dr$! The increments and the volume element associated with them are shown in Fig. 1.32. The volume element is

$$dV = r^2 \sin\theta\, dr\, d\theta\, d\varphi$$

The foregoing results are summarized in Table 1.1.

It is also useful to have a relation between the unit vectors in spherical or cylindrical coordinates and the Cartesian unit vectors $\hat{\mathbf{e}}_x$, $\hat{\mathbf{e}}_y$, and $\hat{\mathbf{e}}_z$. To obtain such relations we first note that

$$x = \rho \cos\varphi \qquad y = \rho \sin\varphi \qquad z = z$$

for cylindrical coordinates and

$$x = r \sin\theta \cos\varphi \qquad y = r \sin\theta \sin\varphi \qquad z = r\cos\theta$$

for spherical coordinates. Thus, we can write

$$\mathbf{r} = \hat{\mathbf{e}}_x x + \hat{\mathbf{e}}_y y + \hat{\mathbf{e}}_z z = \hat{\mathbf{e}}_x \rho \cos\varphi + \hat{\mathbf{e}}_y \rho \sin\varphi + \hat{\mathbf{e}}_z z$$

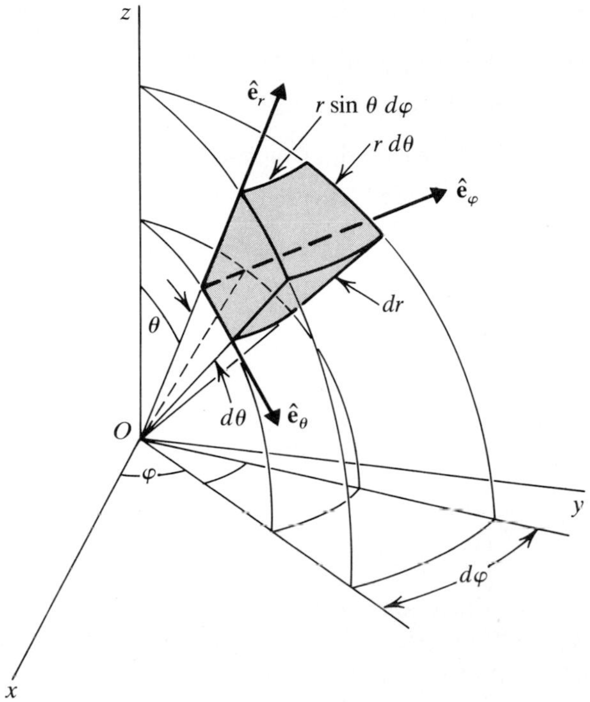

Figure 1.32 Displacements and volume element in the spherical coordinate system.

TABLE 1.1 THE MOST COMMONLY USED COORDINATE SYSTEMS

Curvilinear	Cartesian	Cylindrical	Spherical
q_1	x	ρ	r
q_2	y	φ	θ
q_3	z	z	φ
h_1	1	1	1
h_2	1	ρ	r
h_3	1	1	$r \sin\theta$

for cylindrical coordinates and

$$\mathbf{r} = \hat{\mathbf{e}}_x r \sin\theta \cos\varphi + \hat{\mathbf{e}}_y r \sin\theta \sin\varphi + \hat{\mathbf{e}}_z r \cos\theta$$

for spherical coordinates. We can now use Eq. (1.45) and Table 1.1 to obtain

$$\hat{\mathbf{e}}_\rho = \frac{1}{h_\rho}\frac{\partial \mathbf{r}}{\partial \rho} = \hat{\mathbf{e}}_x \cos\varphi + \hat{\mathbf{e}}_y \sin\varphi$$

$$\hat{\mathbf{e}}_\varphi = \frac{1}{h_\varphi}\frac{\partial \mathbf{r}}{\partial \varphi} = \frac{1}{\rho}[-\hat{\mathbf{e}}_x \rho \sin\varphi + \hat{\mathbf{e}}_y \rho \cos\varphi] = -\hat{\mathbf{e}}_x \sin\varphi + \hat{\mathbf{e}}_y \cos\varphi$$

$$\hat{\mathbf{e}}_z = \frac{1}{h_z}\frac{\partial \mathbf{r}}{\partial z} = \hat{\mathbf{e}}_z$$

Similarly,

$$\hat{\mathbf{e}}_r = \hat{\mathbf{e}}_x \sin\theta\cos\varphi + \hat{\mathbf{e}}_y \sin\theta\sin\varphi + \hat{\mathbf{e}}_z \cos\theta$$

$$\hat{\mathbf{e}}_\theta = \hat{\mathbf{e}}_x \cos\theta\cos\varphi + \hat{\mathbf{e}}_y \cos\theta\sin\varphi - \hat{\mathbf{e}}_z \sin\theta$$

$$\hat{\mathbf{e}}_\varphi = -\hat{\mathbf{e}}_x \sin\varphi + \hat{\mathbf{e}}_y \cos\varphi$$

It is easily verified that the LHS's of these equations represent orthonormal vectors.

We are now in a position to find the gradient, divergence, and curl operators in general curvilinear coordinates. Once these are found, finding the operators in cylindrical and spherical coordinates entails simply substituting the appropriate values of q_1, q_2, and q_3 and h_1, h_2, and h_3.

1.6.3 The Gradient Operator

Finding the form of the gradient operator requires the rate of change with distance of a scalar function f in each of the coordinate directions:

$$\nabla f = \hat{\mathbf{e}}_1 \frac{\partial f}{\partial \ell_1} + \hat{\mathbf{e}}_2 \frac{\partial f}{\partial \ell_2} + \hat{\mathbf{e}}_3 \frac{\partial f}{\partial \ell_3}$$

However,

$$\frac{\partial f}{\partial \ell_1} = \lim_{\Delta \ell_1 \to 0} \frac{\Delta f}{\Delta \ell_1} = \lim_{\Delta q_1 \to 0} \frac{\Delta f}{h_1 \Delta q_1} = \frac{1}{h_1}\frac{\partial f}{\partial q_1}$$

and similar relations for $\partial f/\partial \ell_2$ and $\partial f/\partial \ell_3$ lead to

$$\nabla f = \hat{\mathbf{e}}_1 \frac{1}{h_1}\frac{\partial f}{\partial q_1} + \hat{\mathbf{e}}_2 \frac{1}{h_2}\frac{\partial f}{\partial q_2} + \hat{\mathbf{e}}_3 \frac{1}{h_3}\frac{\partial f}{\partial q_3} \tag{1.49}$$

Using appropriate values for q_1, q_2, q_3, h_1, h_2, and h_3 from Table 1.1, we can immediately write the gradient in the cylindrical coordinate system:

$$\nabla f = \hat{\mathbf{e}}_\rho \frac{\partial f}{\partial \rho} + \hat{\mathbf{e}}_\varphi \frac{1}{\rho}\frac{\partial f}{\partial \varphi} + \hat{\mathbf{e}}_z \frac{\partial f}{\partial z}$$

and in the spherical coordinate system:

$$\nabla f = \hat{\mathbf{e}}_r \frac{\partial f}{\partial r} + \hat{\mathbf{e}}_\theta \frac{1}{r}\frac{\partial f}{\partial \theta} + \hat{\mathbf{e}}_\varphi \frac{1}{r\sin\theta}\frac{\partial f}{\partial \varphi}$$

1.6.4 The Divergence

To find the divergence of a vector $\mathbf{A}$, we consider the volume element of Fig. 1.33 and find the outward flux through the sides of the volume. For the front face we have

$$d\Phi_{\mathrm{F}} = \mathbf{A}_{\mathrm{F}} \cdot \hat{\mathbf{e}}_1 \, da_{\mathrm{F}}$$

where $\mathbf{A}_{\mathrm{F}}$ means the value of $\mathbf{A}$ at the center of the front face and da_{F} is the area of the front face. Following the arguments presented in Section 1.5.3 for the rectangular case, we write

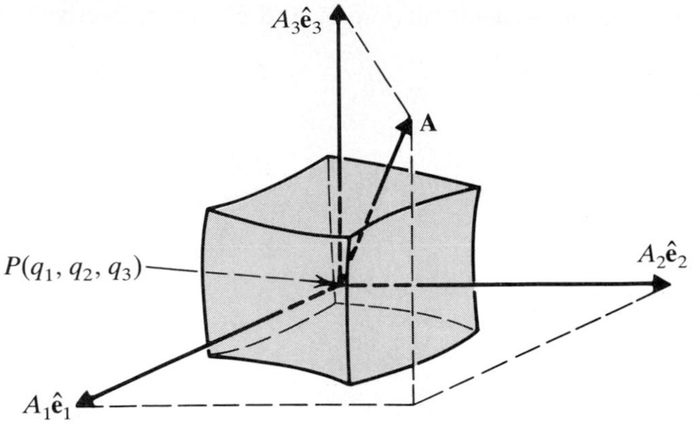

Figure 1.33 An element of volume in curvilinear coordinates, centered at $P(q_1, q_2, q_3)$, where the vector $\mathbf{A}$ has the value $\mathbf{A}(q_1, q_2, q_3)$.

$$\begin{aligned} d\Phi_F &= A_1\left(q_1 + \frac{dq_1}{2}, q_2, q_3\right)h_2\left(q_1 + \frac{dq_1}{2}, q_2, q_3\right)h_3\left(q_1 + \frac{dq_1}{2}, q_2, q_3\right)dq_2\,dq_3 \\ &= \left[A_1(q_1, q_2, q_3) + \frac{dq_1}{2}\frac{\partial A_1}{\partial q_1}\right]\left[h_2(q_1, q_2, q_3) + \frac{dq_1}{2}\frac{\partial h_2}{\partial q_1}\right] \\ &\quad \times \left[h_3(q_1, q_2, q_3) + \frac{dq_1}{2}\frac{\partial h_3}{\partial q_1}\right] \end{aligned}$$

Multiplying out and keeping terms up to the third order, we obtain

$$d\Phi_F = \left[A_1 h_2 h_3 + A_1 h_2 \frac{\partial h_3}{\partial q_1}\frac{dq_1}{2} + A_1 h_3 \frac{\partial h_2}{\partial q_1}\frac{dq_1}{2} + h_2 h_3 \frac{\partial A_1}{\partial q_1}\frac{dq_1}{2}\right]dq_2\,dq_3$$

Similarly, for the back face we have

$$\begin{aligned} d\Phi_B &= \mathbf{A}_B\cdot(-\hat{\mathbf{e}}_1)da_B \\ &= -\mathbf{A}_1\left(q_1 - \frac{dq_1}{2}, q_2, q_3\right)h_2\left(q_1 - \frac{dq_1}{2}, q_2, q_3\right)h_3\left(q_1 - \frac{dq_1}{2}, q_2, q_3\right)dq_2 dq_3 \\ &= -\left[A_1 - \frac{dq_1}{2}\frac{\partial A_1}{\partial q_1}\right]\left[h_2 - \frac{dq_1}{2}\frac{\partial h_2}{\partial q_1}\right]\left[h_3 - \frac{dq_1}{2}\frac{\partial h_3}{\partial q_1}\right]dq_2\,dq_3 \\ &= -\left[A_1 h_2 h_3 - A_1 h_2 \frac{\partial h_3}{\partial q_1}\frac{dq_1}{2} - A_1 h_3 \frac{\partial h_2}{\partial q_1}\frac{dq_1}{2} - h_2 h_3 \frac{\partial A_1}{\partial q_1}\frac{dq_1}{2}\right]dq_2\,dq_3 \end{aligned}$$

Adding the front and back contributions, we obtain

$$d\Phi_1 = d\Phi_F + d\Phi_B = \frac{\partial}{\partial q_1}(h_2 h_3 A_1)dq_1\,dq_2\,dq_3$$

Similarly, the fluxes through the faces perpendicular to $\hat{\mathbf{e}}_2$ and $\hat{\mathbf{e}}_3$ are, respectively,

$$d\Phi_2 = \frac{\partial}{\partial q_2}(h_1 h_3 A_2)dq_1\, dq_2\, dq_3$$

and

$$d\Phi_3 = \frac{\partial}{\partial q_3}(h_1 h_2 A_3)dq_1\, dq_2\, dq_3$$

The total flux through the infinitesimal volume can now be written as

$$d\Phi = \left[\frac{\partial}{\partial q_1}(h_2 h_3 A_1) + \frac{\partial}{\partial q_2}(h_1 h_3 A_2) + \frac{\partial}{\partial q_3}(h_1 h_2 A_3)\right] dq_1\, dq_2\, dq_3 \tag{1.50}$$

From Definition 1.5.1 we know that the divergence is flux per unit volume, that is,[20]

$$\text{div}\,\mathbf{A} \equiv \boldsymbol{\nabla}\cdot\mathbf{A} = \frac{d\Phi}{dV}$$

This definition of divergence is independent of the coordinate system used and is suitable for the present discussion. Using the equation for the element of volume [Eq. (1.46)] and Eq. (1.50), we obtain

$$\boldsymbol{\nabla}\cdot\mathbf{A} = \frac{1}{h_1 h_2 h_3}\left[\frac{\partial}{\partial q_1}(h_2 h_3 A_1) + \frac{\partial}{\partial q_2}(h_1 h_3 A_2) + \frac{\partial}{\partial q_3}(h_1 h_2 A_3)\right] \tag{1.51}$$

In particular, for cylindrical coordinates

$$\begin{aligned}\boldsymbol{\nabla}\cdot\mathbf{A} &= \frac{1}{\rho}\left[\frac{\partial}{\partial \rho}(\rho A_\rho) + \frac{\partial}{\partial \varphi}(A_\varphi) + \frac{\partial}{\partial z}(\rho A_z)\right] \\ &= \frac{1}{\rho}\frac{\partial}{\partial \rho}(\rho A_\rho) + \frac{1}{\rho}\frac{\partial A_\varphi}{\partial \varphi} + \frac{\partial A_z}{\partial z}\end{aligned}$$

and for spherical coordinates

$$\begin{aligned}\boldsymbol{\nabla}\cdot\mathbf{A} &= \frac{1}{r^2 \sin\theta}\left[\frac{\partial}{\partial r}(r^2 \sin\theta\, A_r) + \frac{\partial}{\partial \theta}(r \sin\theta\, A_\theta) + \frac{\partial}{\partial \varphi}(r A_\varphi)\right] \\ &= \frac{1}{r^2}\frac{\partial}{\partial r}(r^2 A_r) + \frac{1}{r \sin\theta}\left[\frac{\partial}{\partial \theta}(\sin\theta\, A_\theta) + \frac{\partial A_\varphi}{\partial \varphi}\right]\end{aligned}$$

Example 1.6.1

Consider the vector field defined by

$$\mathbf{A} = kr^2 \hat{\mathbf{e}}_r$$

[20] Strictly speaking, in the curvilinear coordinates the divergence (or curl) of a vector field **A** is not the dot (or cross) product of a simple vector operator with **A**. In this case $\boldsymbol{\nabla}\cdot\mathbf{A}$ (or $\boldsymbol{\nabla}\times\mathbf{A}$) should be viewed as a single entity whose definition is given by (1.51) [or (1.53)]. Some authors use div **A** (or curl **A**) instead of $\boldsymbol{\nabla}\cdot\mathbf{A}$ (or $\boldsymbol{\nabla}\times\mathbf{A}$). However, as long as $\boldsymbol{\nabla}\cdot\mathbf{A}$ (or $\boldsymbol{\nabla}\times\mathbf{A}$) is viewed as a single entity, no confusion will arise.

where k is a constant. Let us verify the divergence theorem for a spherical surface of radius R (see Fig. 1.34). The element of flux, $d\Phi$, is given by

$$d\Phi = \mathbf{A}\cdot\hat{\mathbf{e}}_n\,da = kR^2\hat{\mathbf{e}}_r\cdot\hat{\mathbf{e}}_n\,da = kR^2\,da$$
$$= kR^4\sin\theta\,d\theta\,d\varphi$$

where $\mathbf{A}$ is evaluated at $r = R$ (the surface of the sphere) and the element of area on the surface is

$$da = (R\,d\theta)(R\sin\theta\,d\varphi)$$

Integrating over the range of θ (0 to π) and φ (0 to 2π), we obtain

$$\Phi = \iint \mathbf{A}\cdot\hat{\mathbf{e}}_n\,da = kR^4\int_0^{\pi}\sin\theta\,d\theta\int_0^{2\pi}d\varphi = 4\pi kR^4$$

On the other hand, using the expression for divergence in the spherical coordinate system and noting that $A_\theta = 0 = A_\varphi$, we obtain

$$\nabla\cdot\mathbf{A} = \frac{1}{r^2}\frac{\partial}{\partial r}(r^2A_r) = \frac{1}{r^2}\frac{d}{dr}(kr^4) = 4kr$$

Therefore, $$\iiint_V \nabla\cdot\mathbf{A}\,dV = \iiint 4krr^2\sin\theta\,dr\,d\theta\,d\varphi$$
$$= 4k\int_0^{2\pi}d\varphi\int_0^{\pi}\sin\theta\,d\theta\int_0^{R}r^3\,dr$$
$$= 4k(2\pi)(2)\tfrac{1}{4}R^4 = 4\pi kR^4$$

This demonstrates the validity of the divergence theorem for this particular case. •

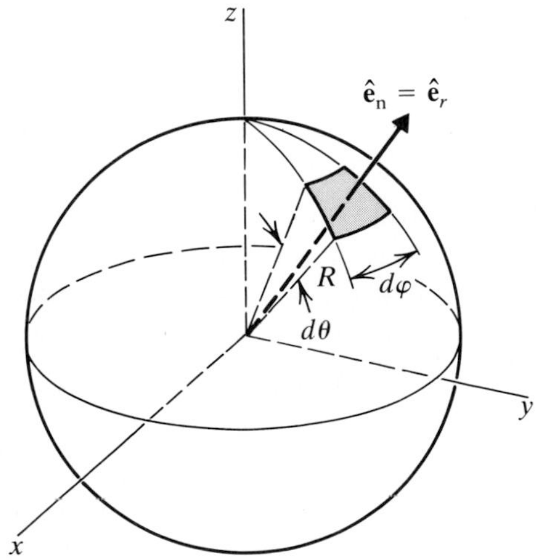

Figure 1.34 Flux through a sphere.

Example 1.6.2

Let us consider the vector field

$$\mathbf{A} = \frac{k}{r^2}\hat{\mathbf{e}}_r$$

and a spherical surface of radius R as in Example 1.6.1. First, we evaluate the flux through the surface:

$$\iint_S \mathbf{A}\cdot\hat{\mathbf{e}}_n\,da = \iint_S \frac{k}{R^2}\hat{\mathbf{e}}_r\cdot\hat{\mathbf{e}}_n R^2 \sin\theta\,d\theta\,d\varphi$$

$$= k\int_0^{2\pi} d\varphi \int_0^{\pi} \sin\theta\,d\theta = 4\pi k \tag{1}$$

On the other hand, $\nabla\cdot\mathbf{A} = 0$, as can easily be verified by noting that $A_r = k/r^2$ and $A_\theta = 0 = A_\varphi$ and using the expression for divergence in the spherical coordinate system. Thus, the RHS of the divergence theorem is zero, but the LHS is $4\pi k$. The problem has to do with the fact that at $r = 0$ neither the vector function nor its divergence is defined. When we integrate over the volume, we are including the origin, and since divergence blows up there, we cannot integrate properly. What is interesting is that $\nabla\cdot\mathbf{A} = 0$ for *all points* except at the origin. Thus, the contribution to the integral $\iiint_V \nabla\cdot\mathbf{A}\,dV$ must all come from the origin. This means that if we break up the volume V so that $V = V_\varepsilon + V_{\text{rest}}$, where V_ε is an infinitesimal volume around the origin and V_{rest} is the remainder of V, we get

$$\iiint_V \nabla\cdot\mathbf{A}\,dV = \iiint_{V_\varepsilon} \nabla\cdot\mathbf{A}\,dV + \iiint_{V_{\text{rest}}} \nabla\cdot\mathbf{A}\,dV = \iiint_{V_\varepsilon} \nabla\cdot\mathbf{A}\,dV$$

because $\nabla\cdot\mathbf{A}$ vanishes everywhere in V_{rest}. But the RHS must equal $4\pi k$, so

$$\iiint_{V_\varepsilon} \nabla\cdot\mathbf{A}\,dV \approx \nabla\cdot\mathbf{A}\,V_\varepsilon = 4\pi k$$

where the fact that V_ε is infinitesimal allows us to disregard the integral sign. This result indicates that $\nabla\cdot\mathbf{A}$ has the following properties:

(i) it is zero everywhere except at the origin,

(ii) it is infinite at the origin (so $\nabla\cdot\mathbf{A}\,V_\varepsilon$ is finite), and

(iii) its integral over all space (note that the sphere is arbitrary but the surface integral is constant independent of the sphere) is constant.

Such a function is called a *Dirac delta function* and will be considered further in Chapter 5. Here we merely denote it as $\delta(\mathbf{r})$ and note that

$$\iiint_V \delta(\mathbf{r})\,dV = 1 \tag{2}$$

The correct result, consistent with (1), is obtained if we let

$$\nabla \cdot \mathbf{A} = 4\pi k \delta(\mathbf{r})$$

or

$$\nabla \cdot \left(\frac{\hat{\mathbf{e}}_r}{r^2} \right) = 4\pi \delta(\mathbf{r}) \tag{3}$$

●

1.6.5 The Curl

First, we need a coordinate-independent definition of curl. Equation (1.38) suggests such a definition. We define the component of the curl in the $\hat{\mathbf{e}}_n$ direction by

$$(\nabla \times \mathbf{A}) \cdot \hat{\mathbf{e}}_n = \lim_{\Delta a \to 0} \frac{1}{\Delta a} \oint_C \mathbf{A} \cdot d\mathbf{r} \tag{1.52}$$

which states that the component of the curl of $\mathbf{A}$ in the direction of $\hat{\mathbf{e}}_n$ is equal to the line integral of $\mathbf{A}$ around an infinitesimal closed path, C, divided by the area of the closed path, Δa, whose normal is $\hat{\mathbf{e}}_n$. The direction of $\hat{\mathbf{e}}_n$ and the sense of integration are again related by the right-hand rule. This definition clearly makes no use of any coordinate system and can be used generally.

Next, we choose a closed path perpendicular to $\hat{\mathbf{e}}_1$ and calculate the line integral of $\mathbf{A}$ around it. The situation is depicted in Fig. 1.35. For the paths labeled (1), (2), (3), and (4), we have the following contributions to the line integral, in which terms higher

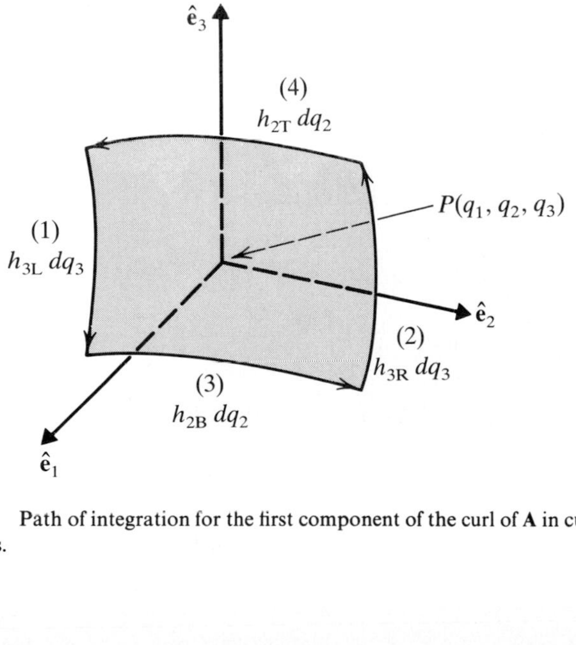

Figure 1.35 Path of integration for the first component of the curl of **A** in curvilinear coordinates.

than the second order have been omitted:

$$(1)\ \mathbf{A}_{\mathrm{L}}\cdot d\mathbf{r}_{\mathrm{L}} = \mathbf{A}_{\mathrm{L}}\cdot(-\hat{\mathbf{e}}_3\, d\ell_{\mathrm{L}}) = -A_{3\mathrm{L}}\, d\ell_{\mathrm{L}} = -A_{3\mathrm{L}}\, h_{3\mathrm{L}}\, dq_3$$

$$= -A_3\left(q_1, q_2 - \frac{dq_2}{2}, q_3\right) h_3\left(q_1, q_2 - \frac{dq_2}{2}, q_3\right) dq_3$$

$$= -\left[A_3 - \frac{dq_2}{2}\frac{\partial A_3}{\partial q_2}\right]\left[h_3 - \frac{dq_2}{2}\frac{\partial h_3}{\partial q_2}\right] dq_3$$

$$= -A_3 h_3 + \frac{1}{2}\frac{\partial}{\partial q_2}(h_3 A_3)\, dq_2\, dq_3$$

$$(2)\ \mathbf{A}_{\mathrm{R}}\cdot d\mathbf{r}_{\mathrm{R}} = \mathbf{A}_{\mathrm{R}}\cdot(\hat{\mathbf{e}}_3\, d\ell_{\mathrm{R}}) = A_{3\mathrm{R}}\, d\ell_{\mathrm{R}} = A_{3\mathrm{R}}\, h_{3\mathrm{R}}\, dq_3$$

$$= A_3\left(q_1, q_2 + \frac{dq_2}{2}, q_3\right) h_3\left(q_1, q_2 + \frac{dq_2}{2}, q_3\right) dq_3$$

$$= \left[A_3 + \frac{dq_2}{2}\frac{\partial A_3}{\partial q_2}\right]\left[h_3 + \frac{dq_2}{2}\frac{\partial h_3}{\partial q_2}\right] dq_3$$

$$= A_3 h_3 + \frac{1}{2}\frac{\partial}{\partial q_2}(h_3 A_3)\, dq_2\, dq_3$$

$$(3)\ \mathbf{A}_{\mathrm{B}}\cdot d\mathbf{r}_{\mathrm{B}} = \mathbf{A}_{\mathrm{B}}\cdot(\hat{\mathbf{e}}_2\, d\ell_{\mathrm{B}}) = A_{2\mathrm{B}}\, d\ell_{\mathrm{B}} = A_{2\mathrm{B}} h_{2\mathrm{B}}\, dq_2$$

$$= A_2\left(q_1, q_2, q_3 - \frac{dq_3}{2}\right) h_2\left(q_1, q_2, q_3 - \frac{dq_3}{2}\right) dq_2$$

$$= \left[A_2 - \frac{dq_3}{2}\frac{\partial A_2}{\partial q_3}\right]\left[h_2 - \frac{dq_3}{2}\frac{\partial h_2}{\partial q_3}\right] dq_2$$

$$= A_2 h_2 - \frac{1}{2}\frac{\partial}{\partial q_3}(h_2 A_2)\, dq_2\, dq_3$$

$$(4)\ \mathbf{A}_{\mathrm{T}}\cdot d\mathbf{r}_{\mathrm{T}} = \mathbf{A}_{\mathrm{T}}\cdot(-\hat{\mathbf{e}}_2\, d\ell_{\mathrm{T}}) = -A_{2\mathrm{T}}\, d\ell_{\mathrm{T}} = -A_{2\mathrm{T}} h_{2\mathrm{T}}\, dq_2$$

$$= -A_2\left(q_1, q_2, q_3 + \frac{dq_3}{2}\right) h_2\left(q_1, q_2, q_3 + \frac{dq_3}{2}\right) dq_2$$

$$= -A_2 h_2 - \frac{1}{2}\frac{\partial}{\partial q_3}(h_2 A_2)\, dq_2\, dq_3$$

Summing up all these contributions, we obtain

$$\oint_C \mathbf{A}\cdot d\mathbf{r} = \left[\frac{\partial}{\partial q_2}(h_3 A_3) - \frac{\partial}{\partial q_3}(h_2 A_2)\right] dq_2\, dq_3$$

For the path in Fig. 1.35

$$da = h_2 dq_2 h_3 dq_3$$

By (1.52), then, the first component of the curl of **A** becomes

$$(\nabla \times \mathbf{A})_1 = \frac{1}{h_2 h_3}\left[\frac{\partial}{\partial q_2}(h_3 A_3) - \frac{\partial}{\partial q_3}(h_2 A_2)\right]$$

Corresponding expressions for the other two components of the curl can be found by proceeding as above. We can put all of the components together in a mnemonic determinant form:

$$\nabla \times \mathbf{A} = \frac{1}{h_1 h_2 h_3}\begin{vmatrix} \hat{\mathbf{e}}_1 h_1 & \hat{\mathbf{e}}_2 h_2 & \hat{\mathbf{e}}_3 h_3 \\ \dfrac{\partial}{\partial q_1} & \dfrac{\partial}{\partial q_2} & \dfrac{\partial}{\partial q_3} \\ h_1 A_1 & h_2 A_2 & h_3 A_3 \end{vmatrix} \tag{1.53}$$

If we substitute the appropriate values for h and q in cylindrical coordinates, we obtain

$$\nabla \times \mathbf{A} = \frac{1}{\rho}\begin{vmatrix} \hat{\mathbf{e}}_\rho & \hat{\mathbf{e}}_\varphi \rho & \hat{\mathbf{e}}_z \\ \dfrac{\partial}{\partial \rho} & \dfrac{\partial}{\partial \varphi} & \dfrac{\partial}{\partial z} \\ A_\rho & \rho A_\varphi & A_z \end{vmatrix}$$

whereas in spherical coordinates we have

$$\nabla \times \mathbf{A} = \frac{1}{r^2 \sin\theta}\begin{vmatrix} \hat{\mathbf{e}}_r & \hat{\mathbf{e}}_\theta r & \hat{\mathbf{e}}_\varphi r\sin\theta \\ \dfrac{\partial}{\partial r} & \dfrac{\partial}{\partial \theta} & \dfrac{\partial}{\partial \varphi} \\ A_r & rA_\theta & r\sin\theta\, A_\varphi \end{vmatrix}$$

Example 1.6.3

Consider the vector field **B** described in cylindrical coordinates as

$$\mathbf{B} = \frac{k}{\rho}\hat{\mathbf{e}}_\varphi$$

where k is a constant. The curl of **B** is easily found to be zero:

$$\nabla \times \mathbf{B} = \frac{1}{\rho}\begin{vmatrix} \hat{\mathbf{e}}_\rho & \rho\hat{\mathbf{e}}_\varphi & \hat{\mathbf{e}}_z \\ \dfrac{\partial}{\partial \rho} & \dfrac{\partial}{\partial \varphi} & \dfrac{\partial}{\partial z} \\ 0 & \rho\left(\dfrac{k}{\rho}\right) & 0 \end{vmatrix} = 0$$

However, for any circle (of radius a, for example) centered at the origin and located in the xy-plane, we get

$$\oint_C \mathbf{B}\cdot d\mathbf{r} = \int_0^{2\pi} \frac{k}{a}\hat{\mathbf{e}}_\varphi \cdot \underbrace{(\hat{\mathbf{e}}_\varphi a\,d\varphi)}_{d\mathbf{r}} = k2\pi \neq 0$$

The reason for this surprising result is that the circle is *not* contractable to zero, because at the origin, which is inside the circle and at which $\rho = 0$, $\mathbf{B}$ is not defined.

This vector field should look familiar. It is the magnetic field due to a long straight wire carrying a current along the z-axis. The line integral of $\mathbf{B}$ along any closed curve encircling the wire (in particular, the above circle) gives, essentially, the current in the wire, and this current is not zero. ●

Example 1.6.4

A vector field that can be written as

$$\mathbf{F} = f(r)\mathbf{r}$$

where $\mathbf{r}$ is the displacement vector from the origin, is conservative. It is instructive to show this using both Cartesian and spherical coordinate systems.

First, with Cartesian coordinates

$$\mathbf{F} = \hat{\mathbf{e}}_x[xf(r)] + \hat{\mathbf{e}}_y[yf(r)] + \hat{\mathbf{e}}_z[zf(r)]$$

The curl is

$$\nabla \times \mathbf{F} = \begin{vmatrix} \hat{\mathbf{e}}_x & \hat{\mathbf{e}}_y & \hat{\mathbf{e}}_z \\ \dfrac{\partial}{\partial x} & \dfrac{\partial}{\partial y} & \dfrac{\partial}{\partial z} \\ xf & yf & zf \end{vmatrix}$$

$$= \hat{\mathbf{e}}_x\left[\frac{\partial}{\partial y}(zf) - \frac{\partial}{\partial z}(yf)\right] + \hat{\mathbf{e}}_y\left[\frac{\partial}{\partial z}(xf) - \frac{\partial}{\partial x}(zf)\right] + \hat{\mathbf{e}}_z\left[\frac{\partial}{\partial x}(yf) - \frac{\partial}{\partial y}(xf)\right]$$

Concentrating on the x component first and denoting df/dr by f', we have

$$\frac{\partial}{\partial y}(zf) = z\frac{\partial f}{\partial y} = z\frac{df}{dr}\frac{\partial r}{\partial y} \equiv zf'\frac{\partial r}{\partial y}$$

But
$$\frac{\partial r}{\partial y} = \frac{\partial}{\partial y}[(x^2 + y^2 + z^2)^{1/2}] = \frac{y}{r}$$

Thus,
$$\frac{\partial}{\partial y}(zf) = yzf'$$

Similarly,
$$\frac{\partial}{\partial z}(yf) = zyf'$$

Therefore, the x component of $\nabla \times \mathbf{F}$ is zero. The y and z components are also zero, and we get

$$\nabla \times \mathbf{F} = 0$$

Using spherical coordinates, we easily obtain

$$\nabla \times \mathbf{F} = \frac{1}{r^2 \sin\theta} \begin{vmatrix} \hat{\mathbf{e}}_r & r\hat{\mathbf{e}}_\theta & r\sin\theta\,\hat{\mathbf{e}}_\varphi \\ \dfrac{\partial}{\partial r} & \dfrac{\partial}{\partial \theta} & \dfrac{\partial}{\partial \varphi} \\ rf(r) & 0 & 0 \end{vmatrix} = 0$$

Obviously, the use of spherical coordinates considerably simplifies the calculation.

We have just shown that any *well-behaved* vector field whose magnitude is only a function of radial distance, r, and whose direction is along $\mathbf{r}$ is conservative. Such vector fields are generally known as *central vector fields*. •

1.6.6 The Laplacian

Combining divergence and the gradient gives the Laplacian. Using (1.49) in (1.51), we get

$$\nabla^2 f = \nabla \cdot (\nabla f) = \frac{1}{h_1 h_2 h_3}\left[\frac{\partial}{\partial q_1}\left(\frac{h_2 h_3}{h_1}\frac{\partial f}{\partial q_1}\right) + \frac{\partial}{\partial q_2}\left(\frac{h_1 h_3}{h_2}\frac{\partial f}{\partial q_2}\right) + \frac{\partial}{\partial q_3}\left(\frac{h_1 h_2}{h_3}\frac{\partial f}{\partial q_3}\right)\right] \tag{1.54}$$

For cylindrical coordinates the Laplacian is

$$\nabla^2 f = \frac{1}{\rho}\frac{\partial}{\partial \rho}\left(\rho\frac{\partial f}{\partial \rho}\right) + \frac{1}{\rho^2}\frac{\partial^2 f}{\partial \varphi^2} + \frac{\partial^2 f}{\partial z^2}$$

and for spherical coordinates it is

$$\nabla^2 f = \frac{1}{r^2}\frac{\partial}{\partial r}\left(r^2\frac{\partial f}{\partial r}\right) + \frac{1}{r^2\sin\theta}\left[\frac{\partial}{\partial \theta}\left(\sin\theta\frac{\partial f}{\partial \theta}\right) + \frac{1}{\sin\theta}\frac{\partial^2 f}{\partial \varphi^2}\right]$$

Exercises

1.6.1 Show that if f is a function of only r, then

$$\nabla \cdot (f\mathbf{r}) = r\frac{df}{dr} + 3f$$

1.6.2 Show that

$$\nabla \cdot \hat{\mathbf{e}}_r = \frac{2}{r}$$

Note that although $\|\hat{\mathbf{e}}_r\| = 1$ by definition, its divergence does not vanish, because the direction of $\hat{\mathbf{e}}_r$ changes with position.

1.6.3 Find the velocity of a particle in the general curvilinear coordinate system. Specialize the result to cylindrical and spherical coordinates.

1.7 SUMMARY OF VECTOR IDENTITIES

The most frequently used vector identities are collected in Table 1.2. Most of these identities can be proved easily by expanding in terms of components. Some of them have been proved in the text or in examples. The rest are left as problems for the reader. All the vectors in the table are either in the plane or in space. The vectors are designated by Latin letters and their components by the corresponding Greek letters. The vectors $\hat{\mathbf{e}}_1$, $\hat{\mathbf{e}}_2$, and $\hat{\mathbf{e}}_3$ are the three mutually orthonormal unit vectors in three dimensions, which are related by a right-hand rule: $\hat{\mathbf{e}}_3 = \hat{\mathbf{e}}_1 \times \hat{\mathbf{e}}_2$.

TABLE 1.2 COMMON VECTOR IDENTITIES

(I.1)	$\mathbf{a}\cdot\mathbf{b} = \mathbf{b}\cdot\mathbf{a} = \alpha_1\beta_1 + \alpha_2\beta_2 + \alpha_3\beta_3 = \|\mathbf{a}\|\,\|\mathbf{b}\|\cos\theta,\ \|\mathbf{a}\| = \sqrt{\alpha_1^2 + \alpha_2^2 + \alpha_3^2}$
(I.2)	$\mathbf{a}\times\mathbf{b} = -\,\mathbf{b}\times\mathbf{a} = \begin{vmatrix} \hat{\mathbf{e}}_1 & \hat{\mathbf{e}}_2 & \hat{\mathbf{e}}_3 \\ \alpha_1 & \alpha_2 & \alpha_3 \\ \beta_1 & \beta_2 & \beta_3 \end{vmatrix}$ and $\|\mathbf{a}\times\mathbf{b}\| = \|\mathbf{a}\|\,\|\mathbf{b}\|\sin\theta$
(I.3)	$\mathbf{a}\cdot(\mathbf{b}\times\mathbf{c}) = \mathbf{c}\cdot(\mathbf{a}\times\mathbf{b}) = \mathbf{b}\cdot(\mathbf{c}\times\mathbf{a}) = \begin{vmatrix} \alpha_1 & \alpha_2 & \alpha_3 \\ \beta_1 & \beta_2 & \beta_3 \\ \gamma_1 & \gamma_2 & \gamma_3 \end{vmatrix}$
(I.4)	$\mathbf{a}\times(\mathbf{b}\times\mathbf{c}) = \mathbf{b}(\mathbf{a}\cdot\mathbf{c}) - \mathbf{c}(\mathbf{a}\cdot\mathbf{b})$
(I.5)	$(\mathbf{a}\times\mathbf{b})\cdot(\mathbf{c}\times\mathbf{d}) = (\mathbf{a}\cdot\mathbf{c})(\mathbf{b}\cdot\mathbf{d}) - (\mathbf{a}\cdot\mathbf{d})(\mathbf{b}\cdot\mathbf{c})$
(I.6)	$(\mathbf{a}\times\mathbf{b})\times(\mathbf{c}\times\mathbf{d}) = \mathbf{b}[\mathbf{a}\cdot(\mathbf{c}\times\mathbf{d})] - \mathbf{a}[\mathbf{b}\cdot(\mathbf{c}\times\mathbf{d})] = \mathbf{c}[\mathbf{a}\cdot(\mathbf{b}\times\mathbf{d})] - \mathbf{d}[\mathbf{a}\cdot(\mathbf{b}\times\mathbf{c})]$
(I.7)	$\nabla\cdot(\nabla\times\mathbf{A}) \equiv 0$
(I.8)	$\nabla\times(\nabla f) \equiv 0$
(I.9)	$\nabla\cdot(\mathbf{a}\times\mathbf{b}) = \mathbf{b}\cdot(\nabla\times\mathbf{a}) - \mathbf{a}\cdot(\nabla\times\mathbf{b})$
(I.10)	$\nabla\times(\mathbf{a}\times\mathbf{b}) = (\mathbf{b}\cdot\nabla)\mathbf{a} - (\mathbf{a}\cdot\nabla)\mathbf{b} + \mathbf{a}\times(\nabla\times\mathbf{b}) - \mathbf{b}\times(\nabla\times\mathbf{a})$
(I.11)	$\nabla(\mathbf{a}\cdot\mathbf{b}) = (\mathbf{b}\cdot\nabla)\mathbf{a} + (\mathbf{a}\cdot\nabla)\mathbf{b} + \mathbf{a}\times(\nabla\times\mathbf{b}) + \mathbf{b}\times(\nabla\times\mathbf{a})$
(I.12)	$\nabla\cdot(f\mathbf{a}) = (\nabla f)\cdot\mathbf{a} + f\nabla\cdot\mathbf{a}$
(I.13)	$\nabla\times(f\mathbf{a}) = (\nabla f)\times\mathbf{a} + f(\nabla\times\mathbf{a})$
(I.14)	$\nabla\times(\nabla\times\mathbf{a}) = \nabla(\nabla\cdot\mathbf{a}) - \nabla^2\mathbf{a}$ (only in Cartesian coordinates)
	$\nabla^2\mathbf{a} \equiv \hat{\mathbf{e}}_1\nabla^2\alpha_1 + \hat{\mathbf{e}}_2\nabla^2\alpha_2 + \hat{\mathbf{e}}_3\nabla^2\alpha_3$
(I.15)	$\nabla\cdot[\mathbf{r}f(r)] = r\dfrac{df}{dr} + 3f;\ \nabla\cdot\mathbf{r} = 3;\ \nabla\cdot\hat{\mathbf{e}}_r = \dfrac{2}{r};\ r = \sqrt{x^2 + y^2 + z^2}$

The definition of $\nabla^2\mathbf{a}$ in (I.14) can also be used in any other coordinate system; thus, $\nabla^2\mathbf{a} \equiv \nabla(\nabla\cdot\mathbf{a}) - \nabla\times(\nabla\times\mathbf{a})$.

TABLE 1.2 (Continued)

(I.16) $\nabla \times [\mathbf{r} f(r)] = 0$

(I.17) $(\mathbf{a} \cdot \nabla)\mathbf{r} = \mathbf{a}$

(I.18) $\nabla(fg) = f\nabla g + g\nabla f;\ \nabla\left(\dfrac{f}{g}\right) = \dfrac{1}{g^2}(g\nabla f - f\nabla g)$

(I.19) $\nabla[f(r)] = \dfrac{df}{dr}\hat{\mathbf{e}}_r,\ \nabla r = \hat{\mathbf{e}}_r$

(I.20) $\nabla \cdot [f\nabla g] = f\nabla^2 g + (\nabla f)\cdot(\nabla g)$

(I.21) $\nabla \cdot (f\nabla g - g\nabla f) = f\nabla^2 g - g\nabla^2 f$

(I.22) Divergence theorem: $\displaystyle\iint_S \mathbf{A}\cdot d\mathbf{a} = \iiint_V \nabla\cdot\mathbf{A}\, dV$

(I.23) Stokes' theorem: $\displaystyle\oint_C \mathbf{A}\cdot d\mathbf{r} = \iint_S (\nabla\times\mathbf{A})\cdot\hat{\mathbf{e}}_n\, da$

(I.24) $\displaystyle\iint_S \hat{\mathbf{e}}_n \times \mathbf{A}\, da = \iiint_V \nabla\times\mathbf{A}\, dV$

(I.25) $\displaystyle\iint_S f\hat{\mathbf{e}}_n\, da = \iiint_V (\nabla f)\, dV$

(I.26) $\displaystyle\oint f d\mathbf{r} = \iiint_V \hat{\mathbf{e}}_n \times (\nabla f)\, da$

Curvilinear coordinates

Cartesian: $q_1 = x,\ q_2 = y,\ q_3 = z;\ h_1 = h_2 = h_3 = 1$

Cylindrical: $q_1 = \rho,\ q_2 = \varphi,\ q_3 = z;\ h_1 = h_3 = 1,\ h_2 = \rho$

Spherical: $q_1 = r,\ q_2 = \theta,\ q_3 = \varphi;\ h_1 = 1,\ h_2 = r,\ h_3 = r\sin\theta$

(I.27) Gradient: $\nabla f = \hat{\mathbf{e}}_1 \dfrac{1}{h_1}\dfrac{\partial f}{\partial q_1} + \hat{\mathbf{e}}_2 \dfrac{1}{h_2}\dfrac{\partial f}{\partial q_2} + \hat{\mathbf{e}}_3 \dfrac{1}{h_3}\dfrac{\partial f}{\partial q_3}$

(I.28) Divergence: $\nabla\cdot\mathbf{A} = \dfrac{1}{h_1 h_2 h_3}\left[\dfrac{\partial}{\partial q_1}(h_2 h_3 A_1) + \dfrac{\partial}{\partial q_2}(h_1 h_3 A_2) + \dfrac{\partial}{\partial q_3}(h_1 h_2 A_3)\right]$

(I.29) Curl: $\nabla\times\mathbf{A} = \dfrac{1}{h_1 h_2 h_3}\begin{vmatrix} \hat{\mathbf{e}}_1 h_1 & \hat{\mathbf{e}}_2 h_2 & \hat{\mathbf{e}}_3 h_3 \\ \partial/\partial q_1 & \partial/\partial q_2 & \partial/\partial q_3 \\ h_1 A_1 & h_2 A_2 & h_3 A_3 \end{vmatrix}$

(I.30) Laplacian: $\nabla^2 f = \dfrac{1}{h_1 h_2 h_3}\left[\dfrac{\partial}{\partial q_1}\left(\dfrac{h_2 h_3}{h_1}\dfrac{\partial f}{\partial q_1}\right) + \dfrac{\partial}{\partial q_2}\left(\dfrac{h_1 h_3}{h_2}\dfrac{\partial f}{\partial q_2}\right) + \dfrac{\partial}{\partial q_3}\left(\dfrac{h_1 h_2}{h_3}\dfrac{\partial f}{\partial q_3}\right)\right]$

PROBLEMS

1.1 Show that

$$\mathbf{A} = 9\hat{\mathbf{e}}_x + \hat{\mathbf{e}}_y - 6\hat{\mathbf{e}}_z \qquad \text{and} \qquad \mathbf{B} = 4\hat{\mathbf{e}}_x - 6\hat{\mathbf{e}}_y + 5\hat{\mathbf{e}}_z$$

are perpendicular.

1.2 Show that the vectors

$$\mathbf{A} = 2\hat{\mathbf{e}}_x - \hat{\mathbf{e}}_y + \hat{\mathbf{e}}_z$$

$$\mathbf{B} = \hat{\mathbf{e}}_x - 3\hat{\mathbf{e}}_y - 5\hat{\mathbf{e}}_z$$

and

$$\mathbf{C} = 3\hat{\mathbf{e}}_x - 4\hat{\mathbf{e}}_y - 4\hat{\mathbf{e}}_z$$

form the sides of a right triangle.

1.3 Find the angle between $\mathbf{A} = 2\hat{\mathbf{e}}_x + 3\hat{\mathbf{e}}_y + \hat{\mathbf{e}}_z$ and $\mathbf{B} = \hat{\mathbf{e}}_x - 6\hat{\mathbf{e}}_y + \hat{\mathbf{e}}_z$.

1.4 Take the dot product of $\mathbf{A} = \mathbf{B} - \mathbf{C}$ with itself and prove the law of cosines by interpreting the result geometrically.

1.5 Show that

$$\mathbf{A} = \hat{\mathbf{e}}_x \cos\theta + \hat{\mathbf{e}}_y \sin\theta \qquad \text{and} \qquad \mathbf{B} = \hat{\mathbf{e}}_x \cos\varphi + \hat{\mathbf{e}}_y \sin\varphi$$

are unit vectors in the xy-plane making angles θ and φ with the x-axis. Then take their inner product and obtain a formula for $\cos(\theta - \varphi)$.

1.6 Given **A**, **B**, and **C**, vectors from the origin to the points A, B, and C, show that $(\mathbf{A} \times \mathbf{B}) + (\mathbf{B} \times \mathbf{C}) + (\mathbf{C} \times \mathbf{A})$ is perpendicular to the plane ABC.

1.7 Vectors **A** and **B** are the sides of a parallelogram, **C** and **D** are the diagonals, and θ is the angle between **A** and **B**. Show that

$$(\|\mathbf{C}\|^2 + \|\mathbf{D}\|^2) = 2(\|\mathbf{A}\|^2 + \|\mathbf{B}\|^2)$$

and that

$$(\|\mathbf{C}\|^2 - \|\mathbf{D}\|^2) = 4\|\mathbf{A}\|\,\|\mathbf{B}\|\cos\theta$$

1.8 Show that $\mathbf{A}\cdot(\mathbf{B} \times \mathbf{C}) = \mathbf{C}\cdot(\mathbf{A} \times \mathbf{B}) = \mathbf{B}\cdot(\mathbf{C} \times \mathbf{A})$.

1.9 Show that $\mathbf{A} \times (\mathbf{B} \times \mathbf{C}) = \mathbf{B}(\mathbf{A}\cdot\mathbf{C}) - \mathbf{C}(\mathbf{A}\cdot\mathbf{B})$. (This is called the *bac cab rule.*)

1.10 Prove Identities (I.5) and (I.6) in Table 1.2.

1.11 Show that

$$(\mathbf{a}_1 \times \mathbf{a}_2)\cdot(\mathbf{a}_1 \times \mathbf{a}_2) = \|\mathbf{a}_1\|^2\|\mathbf{a}_2\|^2 - (\mathbf{a}_1\cdot\mathbf{a}_2)^2$$

1.12 Using vector methods, show that the diagonals of a rhombus are orthogonal.

1.13 Determine an equation for the plane passing through the points $P_1(2, -1, 1)$, $P_2(3, 2, -1)$, and $P_3(-1, 3, 2)$.

1.14 Use the Gram-Schmidt process to find an orthonormal basis in three dimensions from each of the following.

(a) $(-1, 1, 1)$, $(1, -1, 1)$, $(1, 1, -1)$ **(b)** $(1, 2, 2)$, $(0, 0, 1)$, $(0, 1, 0)$

1.15 A particle moves in the xy-plane such that its position vector is

$$\mathbf{r} = \hat{\mathbf{e}}_x a\cos\omega t + \hat{\mathbf{e}}_y a\sin\omega t$$

where ω is a constant. Show that (**a**) the velocity, $\mathbf{v}$, is perpendicular to $\mathbf{r}$, (**b**) the acceleration is directed toward the origin, and (**c**) the angular momentum, $\mathbf{L} = \mathbf{r} \times m\mathbf{v}$, is a constant vector (independent of time).

1.16 Show, by the definition of the derivative, that

$$\frac{d}{dt}(\mathbf{a}_1 \cdot \mathbf{a}_2) = \left(\frac{d\mathbf{a}_1}{dt}\right) \cdot \mathbf{a}_2 + \mathbf{a}_1 \cdot \left(\frac{d\mathbf{a}_2}{dt}\right)$$

and

$$\frac{d}{dt}(\mathbf{a}_1 \times \mathbf{a}_2) = \left(\frac{d\mathbf{a}_1}{dt}\right) \times \mathbf{a}_2 + \mathbf{a}_1 \times \left(\frac{d\mathbf{a}_2}{dt}\right)$$

1.17 Find a unit vector normal to the surface defined by the equation $3x^2 z - 2xy^2 - 4y = 8$ at the point $(1, -2, 3)$.

1.18 Find the divergence and the curl of the following vector field:

$$\mathbf{F} = \hat{\mathbf{e}}_x(x^2 + 2yz) + \hat{\mathbf{e}}_y(y^2 + 2xz) + \hat{\mathbf{e}}_z(z^2 + 2xy)$$

1.19 Show that the divergence of $\mathbf{A} \times \mathbf{B}$ is zero if both $\mathbf{A}$ and $\mathbf{B}$ are conservative vector fields.

1.20 Prove Green's theorem:

$$\iiint_V (f\nabla^2 g - g\nabla^2 f)\, dV = \iint_S (f\nabla g - g\nabla f) \cdot d\mathbf{a}$$

1.21 Prove that $\nabla \times (f\nabla f) = 0$.

1.22 Show that

$$(\mathbf{r} \times \nabla) \cdot (\mathbf{r} \times \nabla) f = r^2 \nabla^2 f - r^2 \frac{\partial^2 f}{\partial r^2} - 2r \frac{\partial f}{\partial r}$$

1.23 Use the divergence theorem to prove that $\iint_S d\mathbf{a} = 0$ for a closed surface, S.

1.24 Use the divergence theorem to prove that $\frac{1}{3}\iint_S \mathbf{r} \cdot d\mathbf{a} = V$, where V is the volume enclosed by S.

1.25 Show that for any closed surface S, $\iint_S \mathbf{A} \cdot d\mathbf{a} = 0$ if $\mathbf{A}$ is the curl of a vector field.

1.26 Show that the line integral of a conservative vector field between any two points, P_1 and P_2, is independent of the path from P_1 to P_2. (*Hint*: Two different paths from P_1 to P_2 form a closed loop.)

1.27 Determine the constants a, b, and c if the following vector is conservative:

$$\mathbf{F} = \hat{\mathbf{e}}_x(x + 2y + az) + \hat{\mathbf{e}}_y(bx - 3y - z) + \hat{\mathbf{e}}_z(4x + cy + 2z)$$

Find the potential of $\mathbf{F}$.

1.28 Show that the following vector field is conservative:

$$\mathbf{F} = \hat{\mathbf{e}}_x(3x^2y^2 + 4xz^3) + \hat{\mathbf{e}}_y(2x^3y + 9y^2z^2) + \hat{\mathbf{e}}_z(6x^2z^2 + 6y^3z)$$

Find the potential of $\mathbf{F}$.

1.29 For the vector field of Problem 1.27, evaluate the line integral along the straight line from the origin to the point $(1, 2, 0)$, along the parabola $y = 2x^2$, and along a broken path from $(0, 0, 0)$ to $(1, 0, 0)$ and then to $(1, 2, 0)$.

1.30 Let $P(\mathbf{r})$ be a point at $\mathbf{r} = (x, y, z)$ and $P'(\mathbf{r}')$ another point at $\mathbf{r}' = (x', y', z')$. Show that

$$\nabla\left(\frac{1}{\|\mathbf{r}-\mathbf{r}'\|}\right) = \frac{\mathbf{r}'-\mathbf{r}}{\|\mathbf{r}-\mathbf{r}'\|^3} \quad \text{and} \quad \nabla'\left(\frac{1}{\|\mathbf{r}-\mathbf{r}'\|}\right) = \frac{\mathbf{r}-\mathbf{r}'}{\|\mathbf{r}-\mathbf{r}'\|^3}$$

where ∇ differentiates with respect to (x, y, z) and ∇' differentiates with respect to (x', y', z').

1.31 Show that $\mathbf{F}\cdot d\mathbf{F} = 0$ if and only if $\|\mathbf{F}\|$ is a constant.

1.32 Prove Identity (I.9) in Table 1.2.

1.33 Prove Identity (I.13) in Table 1.2.

1.34 Show that the identity $\nabla \times (\nabla \times \mathbf{A}) = \nabla(\nabla \cdot \mathbf{A}) - \nabla^2\mathbf{A}$ holds in Cartesian coordinates.

1.35 Prove Identities (I.20) and (I.21) in Table 1.2.

1.36 Use tricks similar to that employed in solving Exercise 1.5.8 to prove Identities (I.25) and (I.26) in Table 1.2.

2

Finite-Dimensional Vector Spaces I: Vectors and Operators

Familiarity with two- and three-dimensional vector spaces allows generalization to any number of dimensions. It is worthwhile to introduce the ideas in an abstract way because such abstraction is powerful, elegant, general, and just as easy to study. Instead of N-dimensional Euclidean spaces, which are the most direct generalizations of the ideas developed so far, we will take one further step into abstraction and consider general N-dimensional vector spaces. The discussion in this chapter will maintain a balance between formality and convenience. Although we will be concentrating on finite-dimensional vector spaces, the concepts, ideas, and examples that are introduced may apply to infinite-dimensional cases.

2.1 VECTOR SPACES

Let us begin with the definition of an abstract vector space.

2.1.1 Definition A vector space $\mathcal{V}$ over $\mathbb{C}$ is a set of objects denoted by $|a\rangle$, $|b\rangle$, $|c\rangle$, and so on, and called vectors, with the following properties:[1]

(A) To every pair of vectors, $|a\rangle$ and $|b\rangle$, in $\mathcal{V}$ there corresponds a vector $|a\rangle + |b\rangle$, also in $\mathcal{V}$, called the *sum* of $|a\rangle$ and $|b\rangle$, such that

(i) $|a\rangle + |b\rangle = |b\rangle + |a\rangle$

(ii) $|a\rangle + (|b\rangle + |c\rangle) = (|a\rangle + |b\rangle) + |c\rangle$

[1] The bra, $\langle\,|$, and ket, $|\,\rangle$, notation for vectors, invented by Dirac, is very useful when dealing with complex vector spaces. However, it is somewhat clumsy for discussing norm and metrics and will therefore be abandoned for those discussions.

(iii) there exists a unique vector $|0\rangle \in \mathscr{V}$, called the *zero vector*, such that $|a\rangle + |0\rangle = |a\rangle$ for every vector $|a\rangle$

(iv) to every vector $|a\rangle \in \mathscr{V}$ there corresponds a unique vector $-|a\rangle$ such that

$$|a\rangle + (-|a\rangle) = |0\rangle$$

(B) To every complex number α, called a scalar, and every vector $|a\rangle$ there corresponds a vector $\alpha|a\rangle$ in $\mathscr{V}$, such that[2]

(i) $\alpha(\beta|a\rangle) = (\alpha\beta)|a\rangle$

(ii) $1(|a\rangle) = |a\rangle$

(C) Multiplication involving vectors and scalars is distributive.

(i) Multiplication by complex numbers (scalars) is distributive with respect to vector addition:

$$\alpha(|a\rangle + |b\rangle) = \alpha|a\rangle + \alpha|b\rangle$$

(ii) Multiplication by vectors is distributive with respect to scalar addition:

$$(\alpha + \beta)|a\rangle = \alpha|a\rangle + \beta|a\rangle$$

Example 2.1.1

Consider the following examples of vector spaces.

(a) If, instead of the complex numbers, the real numbers are used as the field of scalars, then the set of all real numbers, $\mathbb{R}$, forms a vector space. In this case the objects $|a\rangle$, $|b\rangle$, and so forth are the same as α, β, and so forth. All the properties of vector spaces can easily be verified for this case.

(b) Let $\mathscr{V} = \mathbb{C}$ and the set of scalars be $\mathbb{R}$. Then $\mathbb{C}$ is a vector space over the field of real numbers.

(c) Let $\mathscr{V} = \mathbb{C}$ and the set of scalars be $\mathbb{C}$. Then $\mathbb{C}$ is a vector space over the field of complex numbers.

(d) Let $\mathscr{V} = \mathbb{R}$ and the field of scalars be $\mathbb{C}$. This is *not* a vector space, because property (B) of Definition 2.1.1 is not satisfied. A complex number times a real number is *not* a real number and, therefore, does not belong to $\mathscr{V}$.

(e) The set of arrows in the plane forms a vector space over $\mathbb{R}$ under the usual addition of vectors.

(f) The set of arrows in space also forms a vector space over $\mathbb{R}$ under the usual addition of vectors.

[2] One can be more abstract and, instead of complex numbers, use an arbitrary *field*. A field is an abstraction of real and complex numbers. For the purposes of this book, the field of real numbers, denoted by $\mathbb{R}$, and the field of complex numbers, denoted by $\mathbb{C}$, are sufficient. There is a self-contained discussion of complex numbers in Section 6.1. There are other fields besides $\mathbb{R}$ and $\mathbb{C}$; however, there will be no occasion to use them in this book.

(g) Let $\mathscr{P}^{c}$ be the set of all polynomials with complex coefficients in a variable, t. Then $\mathscr{P}^{c}$ is a vector space under the ordinary addition of polynomials and the multiplication of a polynomial by a complex number. In this case the zero vector is the zero polynomial.

(h) Let $\mathbb{C}^n$, where $n = 1, 2, 3, \ldots$, be the set of all complex n-tuples. If $|a\rangle = (\alpha_1, \alpha_2, \ldots, \alpha_n)$ and $|b\rangle = (\beta_1, \beta_2, \ldots, \beta_n)$ are elements of $\mathbb{C}^n$, and α is a complex number, then we can define

$$
\begin{aligned}
|a\rangle + |b\rangle &= (\alpha_1 + \beta_1, \alpha_2 + \beta_2, \ldots, \alpha_n + \beta_n) \\
\alpha|a\rangle &= (\alpha\alpha_1, \alpha\alpha_2, \ldots, \alpha\alpha_n) \\
|0\rangle &= (0, 0, \ldots, 0) \\
-|a\rangle &= (-\alpha_1, -\alpha_2, \ldots, -\alpha_n)
\end{aligned}
$$

It is easy to verify that $\mathbb{C}^n$ is a vector space over the complex field. It is called the *n-dimensional complex coordinate space.*

(i) Let $\mathscr{M}^{m \times n}$ be the set of all $m \times n$ matrices with complex entries. Then under the usual addition and multiplication of matrices by scalars, $\mathscr{M}^{m \times n}$ becomes a vector space. The zero vector is the $m \times n$ matrix with all entries equal to zero.

(j) For each positive integer n, let $\mathscr{P}_n^{c}$ be the set of all polynomials with complex coefficients of degree less than or equal to $n - 1$. Again it is easy to verify that $\mathscr{P}_n^{c}$ is a vector space under the usual addition of polynomials and their multiplication by complex scalars. In particular, the sum of two polynomials of degree less than or equal to $n - 1$ is also a polynomial of degree less than or equal to $n - 1$, and multiplying a polynomial having complex coefficients by a complex number gives another polynomial of the same type. Here the zero polynomial is the zero vector.

(k) The set $\mathscr{P}_n^{r}$ of polynomials of degree less than or equal to $n - 1$ with real coefficients is a vector space over the reals, but it is *not* a vector space over the complex numbers, because multiplying a polynomial having real coefficients by a complex number destroys the reality of the coefficients, resulting in a polynomial that does not belong to $\mathscr{P}_n^{r}$.

(l) Let $\mathbb{R}^n$, where n is a positive integer, be the set of all real n-tuples. [This is a special case of (h) using the field of reals instead of the field of complex numbers.] Then $\mathbb{R}^n$ is a vector space *over the field of real numbers*. It is called the *n-dimensional real coordinate space*, or the *Cartesian n-space.*

(m) Let $C(a, b)$ denote the set of all complex-valued functions of a single real variable that are continuous in the real interval $[a, b]$. It is straightforward to show that $C(a, b)$ is a vector space over the complex field.

(n) Let $C^{\infty}(a, b)$ denote the set of all real-valued functions of a single real variable that possess derivatives of all order. Here $C^{\infty}(a, b)$ forms a vector space over the reals. •

It is clear from Example 2.1.1 that the existence of a vector space depends as much on the nature of vectors as on the nature of scalars.

2.1.2 Definition The vectors $|a_1\rangle, |a_2\rangle, \ldots, |a_n\rangle$ are *linearly independent* if $\sum_{i=1}^{n} \alpha_i |a_i\rangle = 0$, where $\alpha_i \in \mathbb{C}$, implies that $\alpha_1 = \alpha_2 = \cdots = \alpha_n = 0$.

2.1.3 Theorem If $\{|a_i\rangle\}_{i=1}^n \equiv \{|a_1\rangle, |a_2\rangle, \ldots, |a_n\rangle\}$ are linearly independent elements of $\mathcal{V}$, then for each $|x\rangle \in \mathcal{V}$, the equation

$$|x\rangle = \sum_{i=1}^{n} \alpha_i |a_i\rangle \tag{2.1}$$

has at most one solution for the set $\{\alpha_i\}_{i=1}^n$. ■

The proof of Theorem 2.1.3 is left as an exercise. The vector $|x\rangle$ is said to be a *linear combination* of $|a_1\rangle, |a_2\rangle, \ldots, |a_n\rangle$.

2.1.4 Definition A *basis* in a vector space $\mathcal{V}$ is a set, B, of linearly independent vectors such that every vector in $\mathcal{V}$ is a linear combination of elements of B. A vector space is *finite-dimensional* if it has a finite basis; otherwise, it is *infinite-dimensional.*

A basis is denoted by an actual enumeration such as $B = \{|a_1\rangle, |a_2\rangle, \ldots, |a_n\rangle\}$ or by a symbolic enumeration such as $B = \{|a_i\rangle\}_{i=1}^n$.

2.1.5 Theorem All bases of a given finite-dimensional vector space have the same number of independent vectors. This number is called the *dimension* of that vector space. A vector space with N dimensions is sometimes denoted as $\mathcal{V}_N$.

Proof. Let $B = \{|a_i\rangle\}_{i=1}^n$ and $B' = \{|a'_j\rangle\}_{j=1}^m$ be two bases of the vector space $\mathcal{V}$. For definiteness, assume that $m > n$. Since B is a basis, we can expand all vectors in B′ in terms of those in B:

$$\begin{aligned}
|a'_1\rangle &= \alpha_1 |a_1\rangle + \alpha_2 |a_2\rangle + \cdots + \alpha_n |a_n\rangle \\
|a'_2\rangle &= \beta_1 |a_1\rangle + \beta_2 |a_2\rangle + \cdots + \beta_n |a_n\rangle \\
&\ \vdots \\
|a'_m\rangle &= \mu_1 |a_1\rangle + \mu_2 |a_2\rangle + \cdots + \mu_n |a_n\rangle
\end{aligned}$$

Since $m > n$, we can eliminate $|a_1\rangle, |a_2\rangle, \ldots, |a_n\rangle$ from the first n equations. This can be done because the above equations are linearly independent as a result of the linear independence of the LHS. Thus, we can express $|a_1\rangle, |a_2\rangle, \ldots, |a_n\rangle$ in terms of just the first n vectors, $|a'_1\rangle, |a'_2\rangle, \ldots, |a'_n\rangle$. Substituting the former in terms of the latter in the remaining $m - n$ equations gives $|a'_{n+1}\rangle, |a'_{n+2}\rangle, \ldots, |a'_m\rangle$ expressed in terms of $|a'_1\rangle, |a'_2\rangle, \ldots, |a'_n\rangle$. This contradicts the assumption that $|a'_1\rangle, |a'_2\rangle, \ldots, |a'_m\rangle$ are linearly independent. Thus, $m \ngtr n$. Similarly, $n \ngtr m$. Therefore, $m = n$ and the two bases have the same number of vectors. ■

If $|a\rangle$ is a vector in an N-dimensional vector space $\mathcal{V}$ and $B = \{|a_i\rangle\}_{i=1}^N$ is a basis in that space, then, by the definition of a basis, there exists a set of scalars, $\{\alpha_1, \alpha_2, \ldots, \alpha_n\}$, such that $|a\rangle = \sum_{i=1}^N \alpha_i |a_i\rangle$. The set $\{\alpha_i\}_{i=1}^N$ is called the *components of* $|a\rangle$ with respect to the basis B.

2.1.6 Definition A *subspace*, $\mathcal{M}$, of a vector space $\mathcal{V}$ is a subset of $\mathcal{V}$ with the property that if $|a\rangle \in \mathcal{M}$ and $|b\rangle \in \mathcal{M}$, then $|c\rangle = \alpha|a\rangle + \beta|b\rangle$ also belongs to $\mathcal{M}$ $\forall \alpha, \beta \in \mathbb{C}$.

It is clear that a subspace is also a vector space in its own right.

Example 2.1.2

The following are subspaces of some of the vector spaces considered in Example 2.1.1.

(b) A subspace of $\mathcal{V} = \mathbb{C}$ *over the reals* is the space of real numbers as given in Example 2.1.1(a).

(c) Here $\mathbb{R}$ is *not* a subspace of $\mathcal{V} = \mathbb{C}$ over the complex numbers, because, as explained in Example 2.1.1(d), $\mathbb{R}$ cannot be a vector space over the complex numbers.

(e) A subspace of arrows in the plane over $\mathbb{R}$ is the set of all vectors along a given line going through the origin.

(f) A subspace of arrows in the space over $\mathbb{R}$ is the set of all vectors along a given line through the origin. Another subspace is the set of all arrows that lie on a given plane going through the origin.

(g) A subspace of $\mathcal{P}^c$ is $\mathcal{P}^c_n$.

(h) A subspace of $\mathbb{C}^n$ is $\mathbb{C}^{n-1}$. In general, a subspace of $\mathbb{C}^n$ is $\mathbb{C}^m$ for $m < n$.

(i) A subspace of $\mathcal{M}^{m \times n}$ is $\mathcal{M}^{r \times s}$ for $r \leqslant m$ and $s \leqslant n$. That is, all 2×2 matrices form a subspace of the 3×3 matrices, and so on.

(j) A subspace of $\mathcal{P}^c_n$ is $\mathcal{P}^c_m$ for $m < n$.

(k) A subspace of $\mathcal{P}^r_n$ is $\mathcal{P}^r_m$ for $m < n$. Note that both $\mathcal{P}^r_n$ and $\mathcal{P}^r_m$ are vector spaces over the reals only.

(l) A subspace of $\mathbb{R}^n$ is $\mathbb{R}^m$ for $m < n$. Therefore, $\mathbb{R}^2$, the plane, is a subspace of $\mathbb{R}^3$, the space. Also, $\mathbb{R}^1 \equiv \mathbb{R}$ is a subspace of both $\mathbb{R}^2$ and $\mathbb{R}^3$. ●

Example 2.1.3

The following are bases for the corresponding vector spaces given in Example 2.1.1.

(a) The number 1 is a basis for this space, which is therefore one-dimensional.

(b) The numbers 1 and $i = \sqrt{-1}$ are basis vectors here. Thus, this space is two-dimensional.

(c) Here again 1 is a basis, and the space is one-dimensional. Note that, although the vectors are the same as in (b), changing the nature of the scalars changes the dimensionality of the space.

(f) The set $\{\hat{\mathbf{i}}, \hat{\mathbf{j}}, \hat{\mathbf{k}}\}$ (or $\{\hat{\mathbf{e}}_1, \hat{\mathbf{e}}_2, \hat{\mathbf{e}}_3\}$) forms a basis. The space is three-dimensional.

(g) A basis can be formed by the monomials $1, t, t^2, \ldots, t^n, \ldots$. It is clear that this space is infinite-dimensional.

(h) A basis is given by $\hat{\mathbf{e}}_1, \hat{\mathbf{e}}_2, \hat{\mathbf{e}}_3, \ldots, \hat{\mathbf{e}}_n$, where

$$\hat{\mathbf{e}}_j = (0, 0, \ldots, \underset{\substack{\uparrow \\ j\text{th position}}}{1}, \ldots, 0, 0)$$

has zeros everywhere except at the jth position. Clearly, the space has n dimensions.

(i) A basis is given by $\hat{e}_{11}, \hat{e}_{12}, \ldots, \hat{e}_{ij}, \ldots, \hat{e}_{mn}$, where

$$\hat{e}_{ij} = \begin{pmatrix} 0 & 0 & \cdots & 0 & \cdots & 0 \\ 0 & 0 & \cdots & 0 & \cdots & 0 \\ \vdots & \vdots & & \vdots & & \vdots \\ 0 & 0 & \cdots & 1 & \cdots & 0 \\ \vdots & \vdots & & \vdots & & \vdots \\ 0 & 0 & \cdots & 0 & \cdots & 0 \end{pmatrix} \begin{matrix} \\ \\ \\ \leftarrow i\text{th row} \\ \\ \\ \end{matrix}$$

$$\uparrow \; j\text{th column}$$

has zeros everywhere except at the intersection of the ith row and the jth column, where it has a one.

(j) A set consisting of the monomials $1, t, t^2, \ldots, t^{n-1}$ forms a basis. Thus, this space is n-dimensional.

(l) The basis is the same as that given in (h).

(n) If we assume that $a = 0$, then the set of monomials $1, x, x^2, \ldots, x^n, \ldots$ forms a basis, because by Taylor's theorem any function belonging to $C^\infty(0, b)$ can be expanded in an infinite series. Thus, this space is infinite-dimensional. ●

2.1.7 Theorem If S is any set of vectors in a vector space $\mathcal{V}$, then the set $\mathcal{M}$ of all linear combinations of vectors in S is a subspace of $\mathcal{V}$. We say that $\mathcal{M}$ is the *span* of S, or that S *spans* $\mathcal{M}$, or that $\mathcal{M}$ is *spanned* by S. ■

The proof of Theorem 2.1.7 is easy and is left as a problem at the end of the chapter.

Given a space $\mathcal{V}$ with a basis $B = \{|a_i\rangle\}_{i=1}^N$, we can construct a subspace of $\mathcal{V}$ by choosing m vectors ($m < n$) from B and forming a basis out of them. The space spanned by these vectors is an m-dimensional subspace of the original space. For instance, in Example 2.1.1 the vector space in (a) is a one-dimensional subspace of that in (l); the space in (e) is a two-dimensional subspace of those in (f) and (l). Similarly, the space in (j) is an n-dimensional subspace of that in (g), and so forth.

2.1.1 Dual Spaces

In Chapter 1 (bilinear) mappings from a vector space to the set of scalars gave rise to metric tensor and scalar (inner) products. We can generalize those concepts to

N dimensions; however, it is more convenient (and more systematic) to investigate (mono)linear mappings from a vector space to the complex numbers (field of scalars). Thus, we begin with the idea of a linear functional.

2.1.8 Definition A *linear functional* on a vector space $\mathscr{V}$ is a scalar-valued function, $\mathbb{f}$, such that

$$\mathbb{f}(\alpha|a\rangle + \beta|b\rangle) = \alpha\mathbb{f}(|a\rangle) + \beta\mathbb{f}(|b\rangle) \tag{2.2}$$

for all $\alpha, \beta \in \mathbb{C}$ and $|a\rangle, |b\rangle \in \mathscr{V}$.

If the addition of two linear functionals, $\mathbb{f}$ and $\mathbb{g}$, is defined as $(\mathbb{f} + \mathbb{g})(|a\rangle) \equiv \mathbb{f}(|a\rangle) + \mathbb{g}(|a\rangle)$ and the product of a linear functional and a scalar is defined as $(\alpha\mathbb{f})(|a\rangle) \equiv \alpha(\mathbb{f}(|a\rangle))$, then it is fairly straightforward to show that the set of all linear functionals forms a vector space. The zero vector of this space is the linear functional defined by $\mathbb{f}(|a\rangle) = 0$ for all $|a\rangle \in \mathscr{V}$. The vector space of all linear functionals is called the *dual space* of $\mathscr{V}$ and denoted by $\mathscr{V}^*$.

Example 2.1.4

A convenient way of visualizing linear functionals geometrically is as parallel surfaces in space. Such a visualization is analogous to representing vectors by arrows. When a linear functional acts on a vector, the corresponding collection of parallel surfaces representing that linear functional is pierced by the arrow representing the vector; see Fig. 2.1. For a good discussion along these lines, see Misner, Thorne, and Wheeler (1973). ●

There are several important linear functionals. Consider an N-dimensional vector space with a basis $B = \{|a_1\rangle, \ldots, |a_N\rangle\}$. If $\{\alpha_1, \ldots, \alpha_n\}$ is any given set of N scalars, the linear functional $\mathbb{f}$ can be defined by $\mathbb{f}(|a_i\rangle) = \alpha_i$. When $\mathbb{f}$ acts on any arbitrary vector $|b\rangle = \sum_{i=1}^{N} \beta_i|a_i\rangle$ in $\mathscr{V}$, the result is

$$\mathbb{f}(|b\rangle) = \mathbb{f}\left(\sum_{i=1}^{N} \beta_i|a_i\rangle\right) = \sum_{i=1}^{N} \beta_i\mathbb{f}(|a_i\rangle) = \sum_{i=1}^{N} \beta_i\alpha_i \tag{2.3}$$

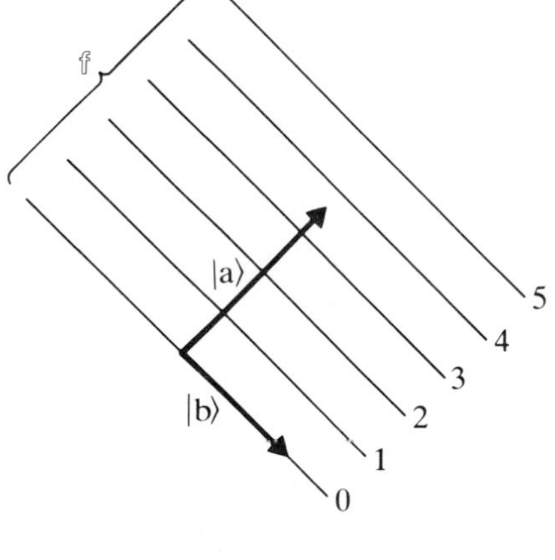

Figure 2.1 A linear functional, $\mathbb{f}$, and two vectors, $|a\rangle$ and $|b\rangle$, such that $\mathbb{f}(|a\rangle) = 3.5$ and $\mathbb{f}(|b\rangle) = 0$. The vector $|a\rangle$ pierces 3.5 surfaces; $|b\rangle$ is parallel to the surfaces and, therefore, does not pierce any of them.

This expression suggests that $|b\rangle$ can be represented as a column vector with entries $\beta_1, \beta_2, \ldots, \beta_N$ and $\mathbb{f}$ as a row vector with entries $\alpha_1, \alpha_2, \ldots, \alpha_N$. Then $\mathbb{f}(|b\rangle)$ is merely the matrix product of the row vector (on the left) and the column vector (on the right). Note that $\mathbb{f}$ is uniquely determined by the set $\{\alpha_1, \alpha_2, \ldots, \alpha_N\}$. In other words, corresponding to every set of scalars $\{\alpha_1, \alpha_2, \ldots, \alpha_N\}$ there exists a unique linear functional. This leads us to a particular set of functionals, $\mathbb{f}_1, \mathbb{f}_2, \ldots, \mathbb{f}_N$, corresponding, respectively, to the sets of scalars $\{1, 0, 0, \ldots, 0\}$, $\{0, 1, 0, \ldots, 0\}, \ldots,$ $\{0, 0, 0, \ldots, 1\}$. This means that

$$\mathbb{f}_1(|a_1\rangle) = 1 \quad \text{and} \quad \mathbb{f}_1(|a_j\rangle) = 0 \quad \text{for } j \neq 1$$

$$\mathbb{f}_2(|a_2\rangle) = 1 \quad \text{and} \quad \mathbb{f}_2(|a_j\rangle) = 0 \quad \text{for } j \neq 2$$

$$\vdots$$

$$\mathbb{f}_N(|a_N\rangle) = 1 \quad \text{and} \quad \mathbb{f}_N(|a_j\rangle) = 0 \quad \text{for } j \neq N$$

or that

$$\mathbb{f}_j(|a_i\rangle) = \delta_{ij} \tag{2.4}$$

where δ_{ij} is the *Kronecker delta*, defined by

$$\delta_{ij} = \begin{cases} 1 & \text{if } i = j \\ 0 & \text{if } i \neq j \end{cases} \tag{2.5}$$

The functionals of (2.4) form a basis in the dual space, $\mathcal{V}^*$. To show this we suppose a $\mathbb{g} \in \mathcal{V}^*$, which is uniquely determined by its action on the vectors in the basis $B = \{|a_1\rangle, \ldots, |a_N\rangle\}$. Assume that $\mathbb{g}(|a_i\rangle) = \gamma_i \in \mathbb{C}$. Then we claim that $\mathbb{g} = \sum_{i=1}^{N} \gamma_i \mathbb{f}_i$. In fact, consider an arbitrary vector $|a\rangle$ in $\mathcal{V}$ with components $\{\alpha_1, \alpha_2, \ldots, \alpha_N\}$ with respect to B. Then

$$\mathbb{g}(|a\rangle) = \mathbb{g}\left(\sum_{i=1}^{N} \alpha_i |a_i\rangle\right) = \sum_{i=1}^{N} \alpha_i \mathbb{g}(|a_i\rangle) = \sum_{i=1}^{N} \alpha_i \gamma_i \tag{2.6}$$

On the other hand,

$$\begin{aligned} \left(\sum_{i=1}^{N} \gamma_i \mathbb{f}_i\right)(|a\rangle) &= \left(\sum_{i=1}^{N} \gamma_i \mathbb{f}_i\right)\left(\sum_{j=1}^{N} \alpha_j |a_j\rangle\right) \\ &= \sum_{i=1}^{N} \gamma_i \sum_{j=1}^{N} \alpha_j \mathbb{f}_i(|a_j\rangle) \\ &= \sum_{i=1}^{N} \gamma_i \sum_{j=1}^{N} \alpha_j \delta_{ij} = \sum_{i=1}^{N} \gamma_i \alpha_i \end{aligned} \tag{2.7}$$

Since (2.6) and (2.7) are equal for arbitrary $|a\rangle$, we conclude that

$$\mathbb{g} = \sum_{i=1}^{N} \gamma_i \mathbb{f}_i$$

Thus, we have the following theorem.

2.1.9 Theorem If $\mathscr{V}$ is an N-dimensional vector space with a basis $\mathrm{B} = \{|\mathrm{a}_1\rangle, \ldots, |\mathrm{a}_N\rangle\}$, then there is a corresponding unique basis $\mathrm{B}^* = \{\mathfrak{f}_1, \ldots, \mathfrak{f}_N\}$ in $\mathscr{V}^*$, with the property that $\mathfrak{f}_j(|\mathrm{a}_i\rangle) = \delta_{ij}$. ■

By this theorem the dual space of an N-dimensional vector space is also N-dimensional. The basis B^* is called the *dual basis* of B. A corollary to Theorem 2.1.9 is that to every vector in $\mathscr{V}$ there corresponds a *unique* linear functional in $\mathscr{V}^*$. This can be seen by noting that every vector $|\mathrm{a}\rangle$ is uniquely determined by its components, $(\alpha_1, \alpha_2, \ldots, \alpha_N)$, in a basis B. The unique linear functional $\mathfrak{f}_\mathrm{a}$ corresponding to $|\mathrm{a}\rangle$ is simply $\mathfrak{f}_\mathrm{a} = \sum_{i=1}^{N} \alpha_i \mathfrak{f}_i$, where $\mathfrak{f}_i \in \mathrm{B}^*$.

2.2 INNER (SCALAR) PRODUCT

We saw in Chapter 1 that a vector space as given by Definition 2.1.1 is too structureless to be of physical interest. The remedy for this lack of structure consists of the ideas of metric and scalar products. Let us therefore now consider the (complex) metric. First let us use the definition in Chapter 1 and define the metric g as a bilinear function with these properties:

$$g(|\mathrm{a}\rangle, |\mathrm{b}\rangle) = g(|\mathrm{b}\rangle, |\mathrm{a}\rangle) \tag{2.8a}$$

$$g(|\mathrm{a}\rangle, \beta|\mathrm{b}\rangle + \gamma|\mathrm{c}\rangle) = \beta g(|\mathrm{a}\rangle, |\mathrm{b}\rangle) + \gamma g(|\mathrm{a}\rangle, |\mathrm{c}\rangle) \tag{2.8b}$$

$$g(\alpha|\mathrm{a}\rangle + \beta|\mathrm{b}\rangle, |\mathrm{c}\rangle) = \alpha g(|\mathrm{a}\rangle, |\mathrm{c}\rangle) + \beta g(|\mathrm{b}\rangle, |\mathrm{c}\rangle) \tag{2.8c}$$

$$g(|\mathrm{a}\rangle, |\mathrm{a}\rangle) \geqslant 0 \qquad \text{and} \qquad g(|\mathrm{a}\rangle, |\mathrm{a}\rangle) = 0 \quad \text{iff } |\mathrm{a}\rangle = |0\rangle \tag{2.8d}$$

However, we immediately run into difficulty. Using (2.8b) and (2.8c), we obtain

$$g(i|\mathrm{a}\rangle, i|\mathrm{a}\rangle) = i^2 g(|\mathrm{a}\rangle, |\mathrm{a}\rangle) = -g(|\mathrm{a}\rangle, |\mathrm{a}\rangle)$$

for any $|\mathrm{a}\rangle$. But this is inconsistent with (2.8d), which requires $g(|\mathrm{a}\rangle, |\mathrm{a}\rangle)$ to be nonnegative for all vectors (including $i|\mathrm{a}\rangle$). The source of the problem is in (2.8a). We have to change it in such a way that $g(|\mathrm{a}\rangle, |\mathrm{a}\rangle)$ will (automatically) be real. This is a necessary requirement for $g(|\mathrm{a}\rangle, |\mathrm{a}\rangle)$ if (2.8d) is to make any sense. Thus, we change (2.8a) to

$$g(|\mathrm{a}\rangle, |\mathrm{b}\rangle) = (g(|\mathrm{b}\rangle, |\mathrm{a}\rangle))^* \tag{2.8a$'$}$$

from which the reality of $g(|\mathrm{a}\rangle, |\mathrm{a}\rangle)$ follows immediately. The change in (2.8a) automatically entails a change in (2.8c). This can be seen by taking the complex conjugate of (2.8b) and using (2.8a′):

$$\begin{aligned}(g(|\mathrm{a}\rangle, \beta|\mathrm{b}\rangle + \gamma|\mathrm{c}\rangle))^* &= g(\beta|\mathrm{b}\rangle + \gamma|\mathrm{c}\rangle, |\mathrm{a}\rangle) \\ &= \beta^*(g(|\mathrm{a}\rangle, |\mathrm{b}\rangle))^* + \gamma^*(g(|\mathrm{a}\rangle, |\mathrm{c}\rangle))^* \qquad (2.8\mathrm{c}') \\ &= \beta^* g(|\mathrm{b}\rangle, |\mathrm{a}\rangle) + \gamma^* g(|\mathrm{c}\rangle, |\mathrm{a}\rangle)\end{aligned}$$

Equation (2.8c′) says that, by definition, the scalars on the RHS of (2.8c) must be complex conjugated.

The question of the existence of a metric on a space is a deep problem in higher analysis, and if the metric exists, there are many ways to introduce it on a vector space. However, it can be shown that for a finite-dimensional vector space at least one metric exists; and if there are several metrics, then all of them are equivalent (Abraham, Marsden, and Ratiu 1983). So, for all practical purposes we can speak of *the* metric on a finite-dimensional vector space. Thus, as with the two- and three-dimensional cases, we can omit the letter g and use a notation that involves only the vectors. There are several such notations in use, but the one that will be employed in this book is the *Dirac bra and ket notation* by which $g(|a\rangle, |b\rangle)$ is denoted as $\langle a|b\rangle$. Using this notation, we have the following definition.

2.2.1 Definition The *inner (scalar) product* of any two vectors, $|a\rangle$ and $|b\rangle$, in a vector space $\mathcal{V}$ is a complex number, $\langle a|b\rangle \in \mathbb{C}$, such that

$$\langle a|b\rangle = \langle b|a\rangle^* \tag{2.9a}$$

$$\langle a|(\beta|b\rangle + \gamma|c\rangle) = \beta\langle a|b\rangle + \gamma\langle a|c\rangle \tag{2.9b}$$

$$\langle a|a\rangle \geqslant 0 \quad \text{and} \quad \langle a|a\rangle = 0 \ \text{ iff } \ |a\rangle = |0\rangle \tag{2.9c}$$

Note that the relation equivalent to (2.8c) is absent, because, as explained earlier, it would be inconsistent with (2.9a). The consistent version, Eq. (2.8c′), follows from (2.9a) and (2.9b). Because of the complex conjugation in (2.8c′) the function g is not truly bilinear; it is commonly named a *sesquilinear map*.

The vector space $\mathcal{V}$ on which an inner product is defined is called an *inner product space*. The above discussion implies that all finite-dimensional vector spaces can be turned into inner product spaces.

It is possible to connect the inner product with linear functionals and dual spaces, by identifying $\langle a|$ with an element of the dual space. One way to do this is to consider a basis $\{|a_1\rangle, \ldots, |a_N\rangle\}$ and define $\alpha_1 = \langle a|a_1\rangle, \ldots, \alpha_N = \langle a|a_N\rangle$. As noted earlier, the set of scalars $\{\alpha_i\}_{i=1}^N$ defines a unique linear functional $\mathfrak{f}_a$ such that $\mathfrak{f}_a(|a_i\rangle) = \alpha_i$. Since $\langle a|a_i\rangle$ is also equal to α_i, it is natural to write $\mathfrak{f}_a \equiv \langle a|$. So we identify the unique dual vector of $|a\rangle$ as $\langle a|$ and write

$$(|a\rangle)^\dagger \equiv \langle a| \tag{2.10}$$

where the symbol † means "dual of."[3]

What about the dual of a linear combination of vectors? Let $|c\rangle = \alpha|a\rangle + \beta|b\rangle$ and take the inner product of $|c\rangle$ with an arbitrary vector $|x\rangle$ using property (2.9b):

$$\langle x|c\rangle = \alpha\langle x|a\rangle + \beta\langle x|b\rangle$$

Now use (2.9a) and write

$$\langle c|x\rangle = \langle x|c\rangle^* = \alpha^*\langle a|x\rangle + \beta^*\langle b|x\rangle$$

[3] The foregoing, although proved only for finite-dimensional vector spaces, actually holds for special infinite-dimensional vector spaces called Hilbert spaces. The proof is the content of an important result in functional analysis called the Riesz lemma. See Reed and Simon (1980, p. 43).

Since this is true for all $|x\rangle$, we must have

$$(|c\rangle)^\dagger \equiv \langle c| = \alpha^* \langle a| + \beta^* \langle b|$$

Therefore, in a duality "operation" the complex scalars must be complex conjugated. So, very generally, we have

$$(\alpha|a\rangle + \beta|b\rangle)^\dagger = \alpha^* \langle a| + \beta^* \langle b| \tag{2.11}$$

Example 2.2.1

For $|a\rangle, |b\rangle \in \mathbb{C}^n$, where $|a\rangle = (\alpha_1, \alpha_2, \ldots, \alpha_n)$ and $|b\rangle = (\beta_1, \beta_2, \ldots, \beta_n)$, we can define

$$\langle a|b\rangle \equiv \alpha_1^* \beta_1 + \alpha_2^* \beta_2 + \cdots + \alpha_n^* \beta_n = \sum_{i=1}^{n} \alpha_i^* \beta_i$$

That this product satisfies all the required properties of an inner product is easily checked. Similarly, for $|a\rangle, |b\rangle \in \mathbb{R}^n$ this definition (without the complex conjugation) satisfies all the properties of an inner product. Thus, if $|a\rangle = (\alpha_1, \alpha_2, \ldots, \alpha_n)$ and $|b\rangle = (\beta_1, \beta_2, \ldots, \beta_n)$ with α_i and β_i real, then

$$\langle a|b\rangle = \sum_{i=1}^{n} \alpha_i \beta_i$$

defines a scalar product in $\mathbb{R}^n$.

It is more customary to represent $|a\rangle$ as defined above as a column vector

$$|a\rangle \equiv \begin{pmatrix} \alpha_1 \\ \alpha_2 \\ \vdots \\ \alpha_n \end{pmatrix}$$

Then the definition of the complex inner product suggests that the dual of $|a\rangle$ must be represented as a row vector with complex conjugate entries:

$$\langle a| = (\alpha_1^* \quad \alpha_2^* \quad \cdots \quad \alpha_n^*)$$

Then the inner product can be written as the (matrix) product

$$\langle a|b\rangle = (\alpha_1^* \quad \alpha_2^* \quad \cdots \quad \alpha_n^*) \begin{pmatrix} \beta_1 \\ \beta_2 \\ \vdots \\ \beta_n \end{pmatrix}$$

In particular, if $|b\rangle = |a\rangle$, we obtain $\langle a|a\rangle = |\alpha_1|^2 + |\alpha_2|^2 + \cdots + |\alpha_n|^2$, which is clearly nonnegative. •

Example 2.2.2

Let $x(t)$, $y(t) \in \mathscr{P}^c$, the space of all polynomials in t. Define

$$\langle x|y\rangle \equiv \int_a^b \rho(t) x^*(t) y(t)\, dt \tag{1}$$

where a and b are any pair of real numbers such that the integral exists (a and b could be infinite) and where $\rho(t)$ is a nonzero, real-valued, continuous function that is always positive in the interval $[a, b]$. Then (1) defines an inner product. We can show this by proving Eqs. (2.9a)–(2.9c):

$$(2.9\text{a})\quad \langle \mathrm{y}|\mathrm{x}\rangle^* = \left[\int_a^b \rho(t)y^*(t)x(t)\,dt\right]^* = \int_a^b \rho(t)[y^*(t)]^*[x(t)]^*\,dt$$

$$= \int_a^b \rho(t)y(t)x^*(t)\,dt = \langle \mathrm{x}|\mathrm{y}\rangle$$

$$(2.9\text{b})\quad \langle \mathrm{x}|(\beta|\mathrm{y}\rangle + \gamma|\mathrm{u}\rangle)\rangle = \int_a^b \rho(t)x^*(t)[\beta y(t) + \gamma u(t)]\,dt$$

$$= \beta\int_a^b \rho(t)x^*(t)\,y(t)\,dt + \gamma\int_a^b \rho(t)x^*(t)u(t)\,dt$$

$$= \beta\langle \mathrm{x}|\mathrm{y}\rangle + \gamma\langle \mathrm{x}|\mathrm{u}\rangle$$

$$(2.9\text{c})\quad \langle \mathrm{x}|\mathrm{x}\rangle = \int_a^b \rho(t)x^*(t)x(t)\,dt = \int_a^b \rho(t)|x(t)|^2\,dt \geqslant 0$$

because both $\rho(t)$ and $|x(t)|^2$ are positive ($|x(t)|^2$ can be zero) in the interval $[a, b]$. The last integral will be zero only if $|x(t)|^2 = 0$ for all t [$\rho(t)$ is not zero by assumption]. But $|x(t)|^2 = 0$ if and only if $x(t) = 0$, that is, $x(t)$ is the zero polynomial.

Thus, (1) does define an inner product. Note that, depending on the so-called weight function, $\rho(t)$, there can be many different inner products defined in $\mathscr{P}^c$. •

Example 2.2.3

Let $f(x)$, $g(x) \in \mathbb{C}(a, b)$ and define their inner product by

$$\langle \mathrm{f}|\mathrm{g}\rangle \equiv \int_a^b f^*(x)g(x)\,dx$$

It is easily shown that $\langle \mathrm{f}|\mathrm{g}\rangle$ satisfies all the requirements of the inner product. Note that the correspondence $|\mathrm{g}\rangle \leftrightarrow g(x)$ demands the "dual" correspondence $\langle \mathrm{f}| \leftrightarrow f^*(x)$. •

2.2.1 Orthogonality

In two and three dimensions we found it convenient to express vectors in terms of orthogonal vectors of unit length. It is important to have a generalization of this concept for N-dimensional vector spaces. We say that vectors $|\mathrm{a}\rangle, |\mathrm{b}\rangle \in \mathcal{V}_N$ are *orthogonal* if

$$\langle \mathrm{a}|\mathrm{b}\rangle = 0 \tag{2.12a}$$

A *normal vector* (or normalized vector) $|\mathrm{e}\rangle$ is one for which

$$\langle \mathrm{e}|\mathrm{e}\rangle = 1 \tag{2.12b}$$

2.2.2 Definition A basis $\mathrm{B} = \{|\mathrm{e}_i\rangle\}_{i=1}^N$ in an N-dimensional vector space $\mathcal{V}_N$ is an *orthonormal basis* if

$$\langle \mathrm{e}_i|\mathrm{e}_j\rangle = \delta_{ij} \qquad \text{for } i, j = 1, 2, \ldots, N \tag{2.13}$$

where δ_{ij} is the Kronecker delta.

This definition implies that all the vectors in the basis are normalized, and that each pair of vectors are mutually orthogonal.

Example 2.2.4

Let

$$|\mathrm{e}_1\rangle = \begin{pmatrix} 1 \\ 0 \\ \vdots \\ 0 \end{pmatrix}, |\mathrm{e}_2\rangle = \begin{pmatrix} 0 \\ 1 \\ \vdots \\ 0 \end{pmatrix}, \ldots, |\mathrm{e}_i\rangle = \begin{pmatrix} 0 \\ 0 \\ \vdots \\ 1 \\ \vdots \\ 0 \end{pmatrix}$$

That is, the only nonzero element of the column vector $|\mathrm{e}_i\rangle$ is 1 and it occurs at the ith row. These *standard basis vectors* of $\mathbb{R}^n$ (or $\mathbb{C}^n$) are orthonormal if the inner product is defined as in Example 2.2.1.

Similarly, let

$$|\mathrm{e}_k\rangle = \frac{e^{ikx}}{\sqrt{2\pi}}$$

and define the inner product as in Example 2.2.3 with $[a, b] = [0, 2\pi]$. Then

$$\langle \mathrm{e}_k|\mathrm{e}_k\rangle = \frac{1}{2\pi}\int_0^{2\pi} e^{-ikx} e^{ikx}\, dx = 1$$

and, for $l \neq k$,

$$\langle \mathrm{e}_l|\mathrm{e}_k\rangle = \frac{1}{2\pi}\int_0^{2\pi} e^{-ilx} e^{ikx}\, dx = \frac{1}{2\pi}\int_0^{2\pi} e^{i(k-l)x}\, dx = 0$$

Thus, $\langle \mathrm{e}_l|\mathrm{e}_k\rangle = \delta_{lk}$. ●

The Gram-Schmidt process and bases. It is always possible to convert any basis in $\mathcal{V}$ into an orthonormal basis. The process by which this is accomplished is Gram-Schmidt orthonormalization, a generalization of what was discussed in

Chapter 1. Consider a basis, $B = \{|a_1\rangle, |a_2\rangle, \ldots, |a_N\rangle\}$. We intend to take linear combinations of $|a_i\rangle$ in such a way that the resulting vectors are orthonormal. First, we let $|e_1\rangle = |a_1\rangle/\sqrt{\langle a_1|a_1\rangle}$. Clearly, $\langle e_1|e_1\rangle = 1$; in fact, since $\langle a_1|a_1\rangle$ is real, the dual of $|e_1\rangle$ is

$$\langle e_1| = \frac{1}{\sqrt{\langle a_1|a_1\rangle}}\langle a_1|$$

and

$$\langle e_1|e_1\rangle = \frac{1}{\sqrt{\langle a_1|a_1\rangle}}\langle a_1|\left(\frac{1}{\sqrt{\langle a_1|a_1\rangle}}|a_1\rangle\right) = \frac{1}{\langle a_1|a_1\rangle}\langle a_1|a_1\rangle = 1$$

For the second vector we consider

$$|e'_2\rangle = |a_2\rangle - \langle e_1|a_2\rangle|e_1\rangle$$

which can be written more symmetrically as

$$|e'_2\rangle = |a_2\rangle - |e_1\rangle\langle e_1|a_2\rangle$$

Clearly, this vector is orthogonal to $|e_1\rangle$; in fact,

$$\langle e_1|e'_2\rangle = \langle e_1|a_2\rangle - \langle e_1|e_1\rangle\langle e_1|a_2\rangle = \langle e_1|a_2\rangle - \langle e_1|a_2\rangle = 0$$

In order to normalize $|e'_2\rangle$ we divide it by $\sqrt{\langle e'_2|e'_2\rangle}$. Then

$$|e_2\rangle = \frac{|e'_2\rangle}{\sqrt{\langle e'_2|e'_2\rangle}}$$

will be a normal vector orthogonal to $|e_1\rangle$. For the third vector we consider

$$\begin{aligned}|e'_3\rangle &= |a_3\rangle - |e_1\rangle\langle e_1|a_3\rangle - |e_2\rangle\langle e_2|a_3\rangle \\ &= |a_3\rangle - \sum_{i=1}^{2} |e_i\rangle\langle e_i|a_3\rangle\end{aligned}$$

We note that $|e'_3\rangle$ is orthogonal to both $|e_1\rangle$ and $|e_2\rangle$:

$$\begin{aligned}\langle e_1|e'_3\rangle &= \langle e_1|a_3\rangle - \langle e_1|e_1\rangle\langle e_1|a_3\rangle - \langle e_1|e_2\rangle\langle e_2|a_3\rangle \\ &= \langle e_1|a_3\rangle - \langle e_1|a_3\rangle - 0 = 0\end{aligned}$$

Similarly, $\langle e_2|e'_3\rangle = 0$.

In general, if we have calculated m orthonormal vectors, $|e_1\rangle, |e_2\rangle, \ldots, |e_m\rangle$, with $m < n$, then we can find the next one using these relations:

$$|e'_{m+1}\rangle = |a_{m+1}\rangle - \sum_{i=1}^{m} |e_i\rangle\langle e_i|a_{m+1}\rangle \qquad (2.14a)$$

$$|e_{m+1}\rangle = \frac{|e'_{m+1}\rangle}{\sqrt{\langle e'_{m+1}|e'_{m+1}\rangle}} \qquad (2.14b)$$

Note that, even though we have been discussing finite-dimensional vector spaces, the process of (2.14) can continue for infinite-dimensional vector spaces as well.

The Schwarz inequality. Let us now consider an inequality due to Cauchy and Schwarz, which is valid in both finite and infinite dimensions and whose restriction to two and three dimensions is equivalent to the fact that $\cos\theta \leqslant 1$.

2.2.3 Theorem For any pair of vectors $|a\rangle, |b\rangle \in \mathscr{V}$ where $\mathscr{V}$ is an inner product space, the following inequality holds

$$\langle a|a\rangle\langle b|b\rangle \geqslant |\langle a|b\rangle|^2$$

This is called the *Schwarz inequality* (or sometimes the Cauchy-Schwarz inequality).

Proof. Using (2.9c) for $|c\rangle = \alpha|a\rangle + |b\rangle$, where α is an arbitrary complex scalar, we have

$$0 \leqslant \langle c|c\rangle = (\alpha^*\langle a| + \langle b|)(\alpha|a\rangle + |b\rangle)$$
$$= \alpha^*\alpha\langle a|a\rangle + \alpha^*\langle a|b\rangle + \alpha\langle b|a\rangle + \langle b|b\rangle \tag{2.15}$$

This inequality is satisfied for *any* complex scalar, α. In particular, we can choose an α that minimizes the expression on the RHS of (2.15). Such an α is obtained by differentiating the RHS, setting it equal to zero, and solving for α. To do so, we substitute $\alpha \equiv x + iy$ and $\langle a|b\rangle \equiv x_0 + iy_0$ in the above expression to get

$$0 \leqslant (x^2 + y^2)\langle a|a\rangle + (x - iy)(x_0 + iy_0) + (x + iy)(x_0 - iy_0) + \langle b|b\rangle$$

or

$$0 \leqslant (x^2 + y^2)\langle a|a\rangle + 2x_0x + 2y_0y + \langle b|b\rangle$$

This is true for any x and y. In particular, it must be true for the pair (x, y) that minimizes the RHS. To obtain this specific pair, we differentiate once with respect to x and again with respect to y and set the results equal to zero to obtain the following two equations:

$$2x\langle a|a\rangle + 2x_0 = 0 \quad \Rightarrow \quad x = -\frac{x_0}{\langle a|a\rangle}$$

$$2y\langle a|a\rangle + 2y_0 = 0 \quad \Rightarrow \quad y = -\frac{y_0}{\langle a|a\rangle}$$

or

$$\alpha = x + iy = -\frac{x_0 + iy_0}{\langle a|a\rangle} = -\frac{\langle a|b\rangle}{\langle a|a\rangle}$$

Now we substitute this expression for α in (2.15) to get

$$0 \leqslant \frac{\langle b|a\rangle}{\langle a|a\rangle}\frac{\langle a|b\rangle}{\langle a|a\rangle}\langle a|a\rangle - \frac{\langle b|a\rangle}{\langle a|a\rangle}\langle a|b\rangle - \frac{\langle a|b\rangle}{\langle a|a\rangle}\langle b|a\rangle + \langle b|b\rangle$$

or

$$0 \leqslant -\langle b|a\rangle\langle a|b\rangle + \langle a|a\rangle\langle b|b\rangle \tag{2.16}$$

But $\langle b|a\rangle = \langle a|b\rangle^*$, so $\langle b|a\rangle\langle a|b\rangle = \langle a|b\rangle^*\langle a|b\rangle = |\langle a|b\rangle|^2$. Substitution in (2.16) yields the Schwarz inequality. ■

It is worthwhile to emphasize the power of abstraction. We have derived the Schwarz inequality solely from the basic assumptions of inner product spaces. This means that *any* vector space that has an inner product defined on it satisfies this inequality. Therefore, we do not have to prove the Schwarz inequality every time we encounter a new vector space.

Length of a vector. Chapter 1 introduced the intuitive idea of the length of a vector, which was used in the definition of the dot product. Sometimes it is more convenient to introduce the inner product first and then define the length. That is what we shall do now.

2.2.4 Definition The *norm* (or length) of a vector $|a\rangle$ is denoted by $\|\mathbf{a}\|$ and defined as

$$\|\mathbf{a}\| \equiv \sqrt{\langle a|a\rangle}$$

A vector space on which a norm is defined is called a *normed linear space*. A normed linear space is automatically a metric space (see the Introduction) because one can define $d(a, b) \equiv \|\mathbf{a} - \mathbf{b}\|$, where the RHS is simply the norm of $|a\rangle - |b\rangle$. Any inner product space is a normed linear space by Definition 2.2.4. However, the converse is not true; there are normed linear spaces that cannot be promoted to inner product spaces. We will encounter examples of such spaces in Chapter 5 (see also Richtmeyer 1978, p. 6).

Example 2.2.5

Let the space be $\mathbb{C}^n$ and define the distance between $a = (\alpha_1, \alpha_2, \ldots, \alpha_n)$ and $b = (\beta_1, \beta_2, \ldots, \beta_n)$ by

$$d(a, b) = \sum_{i=1}^{n} |\alpha_i - \beta_i|$$

We easily verify that $d(a, b) = d(b, a) \geq 0$; that if $a = b$, then $d(a, b) = 0$; and that if $d(a, b) = 0$, $\sum_{i=1}^{n} |\alpha_i - \beta_i| = 0$, or $|\alpha_i - \beta_i| = 0$ for all i, or $\alpha_i = \beta_i$ for all i. Thus,

$$d(a, b) = 0 \quad \Rightarrow \quad a = b$$

and the second property is also satisfied. To verify the last property, the triangle inequality, we note that

$$|\alpha_i - \beta_i| + |\beta_i - \gamma_i| \geq |\gamma_i - \alpha_i| \qquad \forall\ i$$

This follows from the triangle inequality for complex numbers. If we sum this over all i, we obtain

$$\sum_{i=1}^{n} |\alpha_i - \beta_i| + \sum_{i=1}^{n} |\beta_i - \gamma_i| \geq \sum_{i=1}^{n} |\gamma_i - \alpha_i|$$

which is the triangle inequality for the given distance function.

It can be shown that for any positive integer, p,

$$d(a, b) = \left(\sum_{i=1}^{n} |\alpha_i - \beta_i|^p \right)^{1/p}$$

is a distance function. The distance introduced at the beginning of this example is a special case, where $p = 1$. Another commonly used function is that where $p = 2$:

$$d(a, b) = \sqrt{\sum_{i=1}^{n} |\alpha_i - \beta_i|^2}$$ •

Exercises

2.2.1 Given the linearly independent vectors $x(t) = t^n$, for $n = 0, 1, 2, \ldots$, in $\mathscr{P}^c$ and the inner product

$$\langle \mathbf{x} | \mathbf{y} \rangle = \int_{-1}^{1} x^*(t) y(t)\, dt$$

use the Gram-Schmidt process to find the orthonormal polynomials $e_0(t)$, $e_1(t)$, and $e_2(t)$.

2.2.2 Repeat Exercise 2.2.1 with the inner product defined as

$$\langle \mathbf{x} | \mathbf{y} \rangle = \int_{-\infty}^{\infty} e^{-t^2} x^*(t) y(t)\, dt$$

Here $\rho(t) = e^{-t^2}$ and is always positive, as it should be. Hint:

$$\int_{-\infty}^{\infty} e^{-t^2} t^n\, dt = \begin{cases} \sqrt{2\pi} & \text{if } n = 0 \\ 0 & \text{if } n \text{ is odd} \\ \sqrt{2\pi}[1 \cdot 3 \cdot 5 \cdots (n-1)] & \text{if } n \text{ is even} \end{cases}$$

2.2.3 Show that for any set of n complex numbers, $\alpha_1, \alpha_2, \ldots, \alpha_n$, we have

$$|\alpha_1 + \alpha_2 + \cdots + \alpha_n|^2 \leqslant n(|\alpha_1|^2 + |\alpha_2|^2 + \cdots + |\alpha_n|^2)$$

2.2.4 Show that the function d defined by $d(a, b) = \|\mathbf{a} - \mathbf{b}\|$ is a distance function.

2.3 LINEAR OPERATORS

We have made considerable progress in enriching vector spaces. This enrichment, although important, will be of little value if it is imprisoned in a single vector space. We would like to give vector space properties freedom of movement, so they can go from one space to another. The vehicle that carries these properties is a linear operator that is the subject of this section. However, first it is instructive to review the concept of mapping (discussed in the introduction) by considering some examples relevant to the present discussion.

Example 2.3.1

The following are a few familiar examples of mappings.

(a) Let $F: \mathbb{R} \to \mathbb{R}^+$ given by $F(x) = x^2$. Here $\mathbb{R}^+$ is the set of positive real numbers.

(b) Let $F: \mathbb{R}^2 \to \mathbb{R}$ given by $F(x, y) = x^2 + y^2 - 4$.

(c) Let $F: \mathbb{R}^2 \to \mathbb{C}$ given by $F(x, y) = U(x, y) + iV(x, y)$, where $U: \mathbb{R}^2 \to \mathbb{R}$ and $V: \mathbb{R}^2 \to \mathbb{R}$.

(d) Let $\mathscr{V}$ be a vector space. Then a linear functional $\mathbb{f}$ can be considered a mapping, $f: \mathscr{V} \to \mathbb{C}$.

(e) If we define the *Cartesian product*, $\mathrm{A} \times \mathrm{B}$, of two sets A and B as the set of ordered pairs (a, b) such that $a \in \mathrm{A}$ and $b \in \mathrm{B}$, then a metric g on a vector space $\mathscr{V}$ over $\mathbb{C}$ can be considered as a mapping, $g: \mathscr{V} \times \mathscr{V} \to \mathbb{C}$, given by $g(|\mathrm{a}\rangle, |\mathrm{b}\rangle) = \langle \mathrm{a} | \mathrm{b} \rangle$.

(f) Motion of a point particle in space can be considered as a mapping, $M: [a, b] \to \mathbb{R}^3$, where $[a, b]$ is an interval of the real line. For each $t \in [a, b]$ we define

$$M(t) = (x(t), y(t), z(t))$$

where $x(t)$, $y(t)$, and $z(t)$ are real-valued functions of t. If we identify t with time, which is assumed to lie in the interval $[a, b]$, then $M(t)$ describes the path of the particle as a function of time, and a and b are the beginning and the end of the motion, respectively. ●

2.3.1 Linear Transformations

Let us consider an arbitrary mapping, $F: \mathscr{V} \to \mathscr{W}$, from a vector space $\mathscr{V}$ into another vector space $\mathscr{W}$. It is assumed that the two vector spaces are over the same field, say $\mathbb{C}$. Consider $|\mathrm{a}\rangle$ and $|\mathrm{b}\rangle$ in $\mathscr{V}$ and $|\mathrm{x}\rangle$ and $|\mathrm{y}\rangle$ in $\mathscr{W}$ such that $F(|\mathrm{a}\rangle) = |\mathrm{x}\rangle$ and $F(|\mathrm{b}\rangle) = |\mathrm{y}\rangle$. In general, F does not preserve the vector space structure. That is, the image of a linear combination of vectors is *not* the same as the linear combination of the images:

$$F(\alpha|\mathrm{a}\rangle + \beta|\mathrm{b}\rangle) \neq \alpha|\mathrm{x}\rangle + \beta|\mathrm{y}\rangle$$

This is obvious in part (a) of Example 2.3.1 if we consider $\mathbb{R}$ a vector space over reals.

There are many applications in which the preservation of the vector space structure (preservation of the linear combination) is desired. Mappings that have this property are called *linear mappings*, or *linear transformations*. We, therefore, have the following definition.

2.3.1 Definition A linear transformation between vector spaces $\mathscr{V}$ and $\mathscr{W}$ is a mapping $\mathbb{T}: \mathscr{V} \to \mathscr{W}$ such that

$$\mathbb{T}(\alpha|\mathrm{a}\rangle + \beta|\mathrm{b}\rangle) = \alpha\mathbb{T}(|\mathrm{a}\rangle) + \beta\mathbb{T}(|\mathrm{b}\rangle)$$

Linear transformations are often called *linear operators*, and their action on a vector is written without the parentheses: $\mathbb{T}|\mathrm{a}\rangle \equiv \mathbb{T}(|\mathrm{a}\rangle)$.

Example 2.3.2

The following are some examples of linear operators in various vector spaces. The proofs of linearity are simple in all cases and are left as exercises for the reader.

(a) Let $\{|a_1\rangle, |a_2\rangle, \ldots, |a_n\rangle\}$ be an arbitrary finite set of vectors in $\mathscr{V}$ and $\{\mathfrak{f}_1, \mathfrak{f}_2, \ldots, \mathfrak{f}_n\}$ be a corresponding set of linear functionals on $\mathscr{V}$, and write

$$\mathbb{A}|x\rangle = \mathfrak{f}_1(|x\rangle)|a_1\rangle + \mathfrak{f}_2(|x\rangle)|a_2\rangle + \cdots + \mathfrak{f}_n(|x\rangle)|a_n\rangle$$

Then $\mathbb{A}$ is a linear operator on $\mathscr{V}$.

(b) Let π be a permutation (shuffling) of the integers $\{1, 2, \ldots, n\}$.[4] If $|x\rangle = (\eta_1, \eta_2, \ldots, \eta_n)$ is a vector in $\mathbb{C}^n$, we can write

$$\mathbb{A}_\pi|x\rangle = (\eta_{\pi(1)}, \eta_{\pi(2)}, \ldots, \eta_{\pi(n)})$$

Then $\mathbb{A}_\pi$ is a linear operator that reshuffles the components of the vector $|x\rangle$.

(c) Let $|u\rangle$ be a polynomial with complex coefficients, that is, $|u\rangle \in \mathscr{P}^c$. If $|x\rangle \in \mathscr{P}^c$, we can define

$$\mathbb{A}_u|x\rangle = |y\rangle$$

where the polynomial $|y\rangle$ is defined as $y(t) = x(u(t))$. For example, if $u(t) = 1 - t^2$ and $x(t) = 2 + 3t + 5t^2$, then

$$\begin{aligned} y(t) = x(u(t)) &= 2 + 3u(t) + 5(u(t))^2 \\ &= 2 + 3(1 - t^2) + 5(1 - t^2)^2 \\ &= 10 - 13t^2 + 5t^4 \end{aligned}$$

(d) For any $|x\rangle \in \mathscr{P}_n^c$, where $x(t) = \sum_{k=0}^{n-1} \xi_k t^k$, we can write $\mathbb{D}|x\rangle = |y\rangle$, where $|y\rangle \in \mathscr{P}_n^c$ is defined as

$$y(t) = \sum_{k=0}^{n-1} k\xi_k t^{k-1}$$

Then $\mathbb{D}$ is a linear operator. (Note that $\mathbb{D}$ is the derivative operator.)

(e) For every $|x\rangle \in \mathscr{P}^c$, where $x(t) = \sum_{k=0}^{n-1} \xi_k t^k$, we can write $\mathbb{S}|x\rangle = |y\rangle$, where $|y\rangle \in \mathscr{P}^c$ is defined as

$$y(t) = \sum_{k=0}^{n-1} \left(\frac{\xi_k}{k+1}\right) t^{k+1}$$

Then $\mathbb{S}$ is a linear operator. (Note that $\mathbb{S}$ is the integral operator.)

(f) Let $C^{(n)}(a, b)$ denote the space of all real-valued functions that are defined in the interval $[a, b]$ on the real line and have up to the nth derivative ($n \geqslant 1$). Then both $\mathbb{D}$ (the derivative operator) and $\mathbb{S}$ (the integral from a to b) are linear operators on $C^{(n)}(a, b)$.

(g) Let $w(t)$ be a polynomial with complex coefficients. For any $|x\rangle \in \mathscr{P}^c$ we can define $\mathbb{W}|x\rangle = |y\rangle$, where $|y\rangle \in \mathscr{P}^c$ is defined as $y(t) = w(t)x(t)$; that is, $y(t)$ is the polynomial obtained by multiplying the two polynomials $w(t)$ and $x(t)$. Then $\mathbb{W}$ is a linear operator.

[4] A self-contained discussion of permutations is found in Section 2.4.

For the special case when $w(t) = t$, we denote the corresponding linear operator by $\mathbb{T}$. Thus, we have

$$\mathbb{T}|\mathrm{x}\rangle = tx(t)$$

(h) Let $|\mathrm{f}\rangle \in \mathrm{C}^{(n)}(a, b)$ as defined in part (f). Let $g(t)$ be any other function in $\mathrm{C}^{(n)}(a, b)$. Define $\mathbb{G}|\mathrm{f}\rangle = |\mathrm{u}\rangle$, where $u(t) = g(t)f(t)$, $t \in [a, b]$. Then $\mathbb{G}$ is linear. In particular, the operation of multiplying by t, whose operator is denoted by $\mathbb{T}$, is linear. •

An immediate consequence of Definition 2.3.1 is that the image of the zero vector in $\mathscr{V}$ is the zero vector in $\mathscr{W}$. This is *not* true for a general mapping, but it is necessarily true for a linear mapping. To see this, let $\alpha = 0 = \beta$ in Definition 2.3.1 and recall that the number 0 times any vector gives the zero vector. Therefore, for $|\mathrm{a}\rangle$, $|\mathrm{b}\rangle \in \mathscr{V}$,

$$0|\mathrm{a}\rangle + 0|\mathrm{b}\rangle = |0\rangle_{\mathscr{V}} + |0\rangle_{\mathscr{V}} = |0\rangle_{\mathscr{V}}$$

and

$$0(\mathbb{T}|\mathrm{a}\rangle) + 0(\mathbb{T}|\mathrm{b}\rangle) = |0\rangle_{\mathscr{W}} + |0\rangle_{\mathscr{W}} = |0\rangle_{\mathscr{W}}$$

Thus,

$$\mathbb{T}|0\rangle_{\mathscr{V}} = \mathbb{T}(0|\mathrm{a}\rangle + 0|\mathrm{b}\rangle) = 0(\mathbb{T}|\mathrm{a}\rangle) + 0(\mathbb{T}|\mathrm{b}\rangle) = |0\rangle_{\mathscr{W}}$$

where $|0\rangle_{\mathscr{V}}$ is the zero vector in $\mathscr{V}$ and $|0\rangle_{\mathscr{W}}$ is that in $\mathscr{W}$.

The zero vector of $\mathscr{V}$ is necessarily mapped onto the zero vector of $\mathscr{W}$. However, other vectors of $\mathscr{V}$ may also be dragged along onto the zero vector of $\mathscr{W}$. In fact, we have the following theorem.

2.3.2 Theorem The set of vectors in $\mathscr{V}$ that are mapped onto the zero vector of $\mathscr{W}$ under the linear transformation $\mathbb{T}: \mathscr{V} \to \mathscr{W}$ form a *subspace* of $\mathscr{V}$ called the *null space*, or *kernel*, of $\mathbb{T}$ and denoted $\mathscr{N}(\mathbb{T})$. ■

The proof of this theorem is left as an exercise. The dimension of the null space of $\mathbb{T}$ is called the *nullity of* $\mathbb{T}$ and denoted $\mathrm{n}(\mathbb{T})$.

We already know that the range of $\mathbb{T}$, like the range of any mapping, is a subset of $\mathscr{W}$. However, it can easily be shown that it is a subspace of $\mathscr{W}$ as well.

2.3.3 Theorem For a linear transformation $\mathbb{T}: \mathscr{V} \to \mathscr{W}$, where $\mathscr{V}$ and $\mathscr{W}$ are linear vector spaces, $\mathbb{T}(\mathscr{V})$ is a subspace of $\mathscr{W}$. The dimension of $\mathbb{T}(\mathscr{V})$ is called the *rank of* $\mathbb{T}$ and denoted rank $(\mathbb{T})$. ■

Suppose we start with a basis of $\mathscr{N}(\mathbb{T})$ and adjunct enough linearly independent vectors to it to get a basis for $\mathscr{V}$. Without loss of generality, let us assume that the first n vectors in this basis form a basis of $\mathscr{N}(\mathbb{T})$. We claim that the images of the remaining vectors form a basis for $\mathbb{T}(\mathscr{V})$. Let $\mathrm{B} = \{|\mathrm{a}_1\rangle, \ldots, |\mathrm{a}_N\rangle\}$, where $N = \dim \mathscr{V}$, be a basis for $\mathscr{V}$ and $\mathrm{B}' = \{|\mathrm{a}_1\rangle, \ldots, |\mathrm{a}_n\rangle\}$, where $n = \mathrm{n}(\mathbb{T})$, be a basis for $\mathscr{N}(\mathbb{T})$. We want to show that $\mathbb{T}(\mathrm{B}'') \equiv \{\mathbb{T}|\mathrm{a}_{n+1}\rangle, \ldots, \mathbb{T}|\mathrm{a}_N\rangle\}$ is a basis for $\mathbb{T}(\mathscr{V})$. We pick any arbitrary vector in $\mathbb{T}(\mathscr{V})$, say $|\mathrm{x}\rangle$. Then there exists a vector $|\mathrm{a}\rangle \in \mathscr{V}$ such that $\mathbb{T}|\mathrm{a}\rangle = |\mathrm{x}\rangle$. But since B is a basis of $\mathscr{V}$, $|\mathrm{a}\rangle$ can be written as

$$|\mathrm{a}\rangle = \sum_{i=1}^{N} \alpha_i |\mathrm{a}_i\rangle = \sum_{i=1}^{n} \alpha_i |\mathrm{a}_i\rangle + \sum_{i=n+1}^{N} \alpha_i |\mathrm{a}_i\rangle$$

Applying $\mathbb{T}$ on both sides, we obtain

$$\mathbb{T}|\mathrm{a}\rangle = |\mathrm{x}\rangle = \mathbb{T}\left(\sum_{i=1}^{n} \alpha_i|\mathrm{a}_i\rangle\right) + \mathbb{T}\left(\sum_{i=n+1}^{N} \alpha_i|\mathrm{a}_i\rangle\right)$$

$$= \sum_{i=1}^{n} \alpha_i\underbrace{\mathbb{T}|\mathrm{a}_i\rangle}_{0} + \sum_{i=n+1}^{N} \alpha_i\mathbb{T}|\mathrm{a}_i\rangle$$

$$= \sum_{i=n+1}^{N} \alpha_i\mathbb{T}|\mathrm{a}_i\rangle$$

This shows that $\{\mathbb{T}|\mathrm{a}_{n+1}\rangle, \mathbb{T}|\mathrm{a}_{n+2}\rangle, \ldots, \mathbb{T}|\mathrm{a}_N\rangle\}$ spans $\mathbb{T}(\mathscr{V})$. What remains to be shown is that these are linearly independent. Let us assume otherwise, so that there exist $\beta_1, \beta_2, \ldots, \beta_m$ such that

$$\beta_1\mathbb{T}|\mathrm{a}_{n+1}\rangle + \beta_2\mathbb{T}|\mathrm{a}_{n+2}\rangle + \cdots + \beta_m\mathbb{T}|\mathrm{a}_N\rangle = 0$$

or

$$\mathbb{T}(\beta_1|\mathrm{a}_{n+1}\rangle + \beta_2|\mathrm{a}_{n+2}\rangle + \cdots + \beta_m|\mathrm{a}_N\rangle) = 0$$

This indicates that $\beta_1|\mathrm{a}_{n+1}\rangle + \beta_2|\mathrm{a}_{n+2}\rangle + \cdots + \beta_m|\mathrm{a}_N\rangle$ belongs to $\mathscr{N}(\mathbb{T})$ and, thus, can be expressed as a linear combination of $|\mathrm{a}_1\rangle, |\mathrm{a}_2\rangle, \ldots, |\mathrm{a}_n\rangle$. This is not possible because, by assumption, $|\mathrm{a}_1\rangle, |\mathrm{a}_2\rangle, \ldots, |\mathrm{a}_N\rangle$ are all independent. We, therefore, have the following theorem.

2.3.4 Theorem Let $\mathbb{T}: \mathscr{V} \to \mathscr{W}$ be a linear transformation. Then

$$\dim \mathscr{V} = \dim \mathscr{N}(\mathbb{T}) + \dim \mathbb{T}(\mathscr{V})$$ ■

Example 2.3.3

Let $\mathbb{T}: \mathbb{R}^4 \to \mathbb{R}^3$ be given by

$$\mathbb{T}(x_1, x_2, x_3, x_4) = (2x_1 + x_2 + x_3 - x_4, x_1 + x_2 + 2x_3 + 2x_4, x_1 - x_3 - 3x_4)$$

Let us try to find the null space of the operator $\mathbb{T}$. To find $\mathscr{N}(\mathbb{T})$, we must look for (x_1, x_2, x_3, x_4) such that $\mathbb{T}(x_1, x_2, x_3, x_4) = (0, 0, 0)$, or

$$\begin{aligned} 2x_1 + x_2 + x_3 - x_4 &= 0 \\ x_1 + x_2 + 2x_3 + 2x_4 &= 0 \\ x_1 - x_3 - 3x_4 &= 0 \end{aligned} \tag{1}$$

The third equation can be rewritten as

$$x_1 = x_3 + 3x_4$$

Substitution in the first or second equation then yields

$$x_2 + 3x_3 + 5x_4 = 0 \quad \Rightarrow \quad x_2 = -3x_3 - 5x_4$$

Thus, to satisfy (1) the vector in $\mathbb{R}^4$ must be of the form

$$(x_3 + 3x_4, \ -3x_3 - 5x_4, x_3, x_4) = x_3(1, -3, 1, 0) + x_4(3, -5, 0, 1)$$

where x_3 and x_4 are arbitrary real numbers.

We can restate the above discussion as follows. The null space $\mathcal{N}(\mathbb{T})$ consists of vectors that can be written as $\alpha_1(1,-3, 1, 0) + \alpha_2(3,-5, 0, 1)$, that is, vectors that are linear combinations of the two vectors $(1,-3, 1, 0)$ and $(3,-5, 0, 1)$. These two vectors are linearly independent. Therefore, $\dim \mathcal{N}(\mathbb{T}) = 2$. Theorem 2.3.4 then says that $\dim \mathbb{T}(\mathcal{V}) = 2$; that is, the range of $\mathbb{T}$ is two-dimensional. This becomes obvious when one notices that

$$\mathbb{T}(x_1, \ldots, x_4) = (2x_1 + x_2 + x_3 - x_4)(1, 0, 1) + (x_1 + x_2 + 2x_3 + 2x_4)(0, 1, -1)$$

and, therefore, $\mathbb{T}(x_1, x_2, x_3, x_4)$ is a linear combination of *only two* vectors, namely $(1, 0, 1)$ and $(0, 1, -1)$. ●

An important linear mapping is given by the following definition.

2.3.5 Definition A vector space $\mathcal{V}$ is said to be *isomorphic* to another vector space $\mathcal{W}$ if there exists a bijective linear mapping $\mathbb{T}: \mathcal{V} \to \mathcal{W}$. Then $\mathbb{T}$ is called an *isomorphism.*

Isomorphism is essentially the same as identity. This means that, for all practical purposes, two isomorphic vector spaces are really the same, although they may appear to be different. For instance, the set of complex numbers, $\mathbb{C}$, and the set of vectors in a plane, $\mathbb{R}^2$, may appear different; however, over the real numbers they are identical vector spaces. In fact, the correspondence

$$(x, y) \leftrightarrow x + iy$$

establishes an isomorphism between the two vector spaces. It should be emphasized that *only as vector spaces* are $\mathbb{C}$ and $\mathbb{R}^2$ isomorphic. If we go beyond the vector space structures, the two sets are quite different. For example, $\mathbb{C}$ accepts a natural multiplication for its elements, but $\mathbb{R}^2$ does not. The following theorem gives a working criterion for isomorphism.

2.3.6 Theorem A linear surjective mapping $\mathbb{T}: \mathcal{V} \to \mathcal{W}$ is an isomorphism if and only if its nullity is zero.

Proof. Assume that $\mathbb{T}$ is an isomorphism. Since $\mathbb{T}$ is linear, it carries $|0\rangle_{\mathcal{V}}$ onto $|0\rangle_{\mathcal{W}}$. Because $\mathbb{T}$ is bijective, it cannot carry any other vector of $\mathcal{V}$ onto $|0\rangle_{\mathcal{W}}$. Thus, $\dim \mathcal{N}(\mathbb{T}) = 0$. Conversely, if we assume that $\dim \mathcal{N}(\mathbb{T}) = 0$, that is, that $\mathbb{T}$ carries *only* $|0\rangle_{\mathcal{V}}$ onto $|0\rangle_{\mathcal{W}}$, then we can show that $\mathbb{T}$ is bijective. It is surjective by assumption. On the other hand,

$$\mathbb{T}|x\rangle = \mathbb{T}|y\rangle \quad \Rightarrow \quad \mathbb{T}(|x\rangle - |y\rangle) = |0\rangle_{\mathcal{W}}$$

$$\Rightarrow \quad |x\rangle - |y\rangle = |0\rangle_{\mathcal{V}} \quad \Rightarrow \quad |x\rangle = |y\rangle + |0\rangle_{\mathcal{V}} = |y\rangle$$

Thus, $\mathbb{T}$ must be injective also. This shows that $\mathbb{T}$ is bijective and, therefore, an isomorphism. ■

2.3.7 Theorem An isomorphism $\mathbb{T}:\mathcal{V}\to\mathcal{W}$ carries linearly independent sets of vectors onto linearly independent sets of vectors.

Proof. Assume that $\{|a_i\rangle\}_{i=1}^m$ is a set of linearly independent vectors. We want to show that $\{\mathbb{T}|a_i\rangle\}_{i=1}^m$ is also linearly independent. If there exist $\alpha_1, \alpha_2, \ldots, \alpha_m$ such that

$$\sum_{i=1}^{m} \alpha_i \mathbb{T}|a_i\rangle = |0\rangle_{\mathcal{W}}$$

then the linearity of $\mathbb{T}$ gives

$$\mathbb{T}\left(\sum_{i=1}^{m} \alpha_i |a_i\rangle\right) = |0\rangle_{\mathcal{W}}$$

From Theorem 2.3.6 we have

$$\sum_{i=1}^{m} \alpha_i |a_i\rangle = |0\rangle_{\mathcal{V}}$$

and the linear independence of the $|a_i\rangle$ implies that $\alpha_i = 0$ for all i. Therefore, $\mathbb{T}|a_1\rangle, \mathbb{T}|a_2\rangle, \ldots, \mathbb{T}|a_m\rangle$ must be linearly independent. ■

The following theorem shows that the number of different finite-dimensional vector spaces is severely limited.

2.3.8 Theorem Two finite-dimensional vector spaces are isomorphic if and only if they have the same dimensions.

Proof. First we prove that if two vector spaces have the same dimensions, they are isomorphic. Let $B_{\mathcal{V}} = \{|a_i\rangle\}_{i=1}^N$ be a basis for $\mathcal{V}$ and $B_{\mathcal{W}} = \{|b_i\rangle\}_{i=1}^N$ a basis for $\mathcal{W}$. Let $\mathbb{T}:\mathcal{V}\to\mathcal{W}$ be a linear transformation defined by

$$\mathbb{T}|a_i\rangle = |b_i\rangle \qquad i = 1, 2, \ldots, N$$

We claim that $\mathbb{T}$ is an isomorphism.

To show that $\mathbb{T}$ is surjective, let

$$|b\rangle \equiv \sum_{i=1}^{N} \beta_i |b_i\rangle$$

be an arbitrary vector in $\mathcal{W}$. Then

$$|b\rangle = \sum_{i=1}^{N} \beta_i \mathbb{T}|a_i\rangle = \sum_{i=1}^{N} \mathbb{T}(\beta_i|a_i\rangle) = \mathbb{T}\left(\sum_{i=1}^{N} \beta_i |a_i\rangle\right)$$

Thus, $|b\rangle$ is the image of some vector in $\mathcal{V}$, namely $\sum_{i=1}^{N} \beta_i |a_i\rangle$.

To show that $\mathbb{T}$ is injective, we use the preceding result and Theorem 2.3.6. Assume that $\mathbb{T}|a\rangle = |0\rangle_{\mathcal{W}}$ for some $|a\rangle \in \mathcal{V}$. Then

$$|0\rangle_{\mathcal{W}} = \mathbb{T}|a\rangle = \mathbb{T}\left(\sum_{i=1}^{N} \alpha_i |a_i\rangle\right) = \sum_{i=1}^{N} \alpha_i \mathbb{T}|a_i\rangle = \sum_{i=1}^{N} \alpha_i |b_i\rangle$$

But the vectors $|b_i\rangle$ are linearly independent. Thus, $\alpha_i = 0$, for $i = 1, 2, \ldots, N$, and $|a\rangle = |0\rangle_{\mathcal{V}}$. This shows that the only vector in $\mathcal{V}$ that is mapped onto $|0\rangle_{\mathcal{W}}$ is $|0\rangle_{\mathcal{V}}$. Therefore, $\dim \mathcal{N}(\mathbb{T}) = 0$, and by Theorem 2.3.6 $\mathbb{T}$ is an isomorphism.

Conversely, assume that there exists an isomorphism $\mathbb{T}: \mathcal{V} \to \mathcal{W}$. We must show that $\mathcal{V}$ and $\mathcal{W}$ have the same dimensions. Let $B_{\mathcal{V}} = \{|a_i\rangle\}_{i=1}^N$ be a basis for $\mathcal{V}$. It is sufficient to show that $B'_{\mathcal{W}} = \{\mathbb{T}|a_i\rangle\}_{i=1}^N$ is a basis for $\mathcal{W}$. We already know from Theorem 2.3.7 that $\mathbb{T}|a_i\rangle$ are linearly independent. All we have to show is that they span $\mathcal{W}$. Let $|b\rangle \in \mathcal{W}$ be an *arbitrary* vector. Because $\mathbb{T}$ is surjective, there exists $|a\rangle \in \mathcal{V}$ such that $\mathbb{T}|a\rangle = |b\rangle$. On the other hand, since $B_{\mathcal{V}}$ is a basis in $\mathcal{V}$, we have

$$|b\rangle = \mathbb{T}|a\rangle = \mathbb{T}\left(\sum_{i=1}^N \alpha_i |a_i\rangle\right) = \sum_{i=1}^N \alpha_i \mathbb{T}|a_i\rangle$$

We have written $|b\rangle$ as a linear combination of $\mathbb{T}|a_i\rangle$'s. This completes the proof that $B'_{\mathcal{W}}$ is a basis for $\mathcal{W}$. Since the number of $\mathbb{T}|a_i\rangle$'s is precisely N, the equality of $\dim \mathcal{V}$ and $\dim \mathcal{W}$ is established. ■

A consequence of Theorem 2.3.8 is that all N-dimensional vector spaces over $\mathbb{R}$ are isomorphic to $\mathbb{R}^N$, and all complex N-dimensional vector spaces are isomorphic to $\mathbb{C}^N$. So, for all practical purposes, we have only two N-dimensional vector spaces, $\mathbb{R}^N$ and $\mathbb{C}^N$.

2.3.2 Operator Algebra

The set of linear transformations between vector spaces has some important properties that will be investigated in this subsection. First, we note that if we define the sum of two linear transformations, $\mathbb{T}_1$ and $\mathbb{T}_2$, from $\mathcal{V}$ to $\mathcal{W}$ by

$$(\mathbb{T}_1 + \mathbb{T}_2)|a\rangle \equiv \mathbb{T}_1|a\rangle + \mathbb{T}_2|a\rangle$$

and the product of $\mathbb{T}_1$ (or $\mathbb{T}_2$) with a scalar, α, by

$$(\alpha\mathbb{T}_1)|a\rangle = \alpha(\mathbb{T}_1|a\rangle)$$

then the set of such linear transformations forms a vector space denoted by $\mathcal{L}(\mathcal{V}, \mathcal{W})$. The zero vector (operator) is, by definition, that operator which yields the zero vector when acting on any vector. The zero operator is denoted by $\mathbb{O}$.

Second, there is an extra operation defined for linear operators that is not, in general, defined for ordinary vector spaces. This operation is multiplication of two linear transformations. If $\mathbb{T}_1: \mathcal{V} \to \mathcal{W}$ and $\mathbb{T}_2: \mathcal{W} \to \mathcal{U}$ are linear operators, then the composition $\mathbb{T}_2 \circ \mathbb{T}_1: \mathcal{V} \to \mathcal{U}$ is also a linear operator, as can easily be verified. This composition is the *product* of $\mathbb{T}_2$ and $\mathbb{T}_1$. It is customary to omit the circle between the two operators, that is,

$$\mathbb{T}_2\mathbb{T}_1 \equiv \mathbb{T}_2 \circ \mathbb{T}_1$$

This product, however, is not defined on a *single* vector space, but is such that it takes an element in $\mathcal{L}(\mathcal{V}, \mathcal{W})$ and another element in $\mathcal{L}(\mathcal{W}, \mathcal{U})$ and gives a third element in

$\mathcal{L}(\mathcal{V}, \mathcal{U})$. It is desirable to work with a single space and define products of operators belonging to that space. If we let $\mathcal{V} = \mathcal{W} = \mathcal{U}$, then the three spaces collapse to the single space $\mathcal{L}(\mathcal{V}, \mathcal{V})$, which we can abbreviate as $\mathcal{L}(\mathcal{V})$ and to which $\mathbb{T}_1$, $\mathbb{T}_2$, $\mathbb{T}_2\mathbb{T}_1$, and $\mathbb{T}_1\mathbb{T}_2$ belong. The space $\mathcal{L}(\mathcal{V})$ is not only a vector space but an *algebra*, the algebra of the linear operators on $\mathcal{V}$.

This algebra has a zero element, $\mathbb{O}$ (often denoted simply as 0), and an identity element, $\mathbb{1}$, and possesses the following properties:

(i) $\mathbb{T}\mathbb{O} = \mathbb{O}\mathbb{T} = \mathbb{O}$	existence of zero element
(ii) $\mathbb{T}\mathbb{1} = \mathbb{1}\mathbb{T} = \mathbb{T}$	existence of identity element
(iii) $\mathbb{T}_1(\mathbb{T}_2 + \mathbb{T}_3) = \mathbb{T}_1\mathbb{T}_2 + \mathbb{T}_1\mathbb{T}_3$	right distributive property
(iv) $(\mathbb{T}_1 + \mathbb{T}_2)\mathbb{T}_3 = \mathbb{T}_1\mathbb{T}_3 + \mathbb{T}_2\mathbb{T}_3$	left distributive property
(v) $\mathbb{T}_1(\mathbb{T}_2\mathbb{T}_3) = (\mathbb{T}_1\mathbb{T}_2)\mathbb{T}_3$	associative property

The zero transformation has been defined before. The identity transformation is the operator that satisfies

$$\mathbb{1}\,|a\rangle = |a\rangle \qquad \forall\ |a\rangle \in \mathcal{V}$$

Having defined $\mathbb{1}$, we can ask whether it is possible to find an operator $\mathbb{T}^{-1}$ with the property that $\mathbb{T}\mathbb{T}^{-1} = \mathbb{T}^{-1}\mathbb{T} = \mathbb{1}$. First, let us see if $\mathbb{T}^{-1}$ exists at all. The general discussion of mappings noted that inverses are defined only for bijective mappings. Specializing that definition to linear mappings, we conclude that *all isomorphisms (and only isomorphisms) acting on a vector space are invertible* (have inverses).

Example 2.3.4

Let the linear operator $\mathbb{T}: \mathbb{R}^3 \to \mathbb{R}^3$ be defined by

$$\mathbb{T}(x_1, x_2, x_3) = (x_1 + x_2, x_2 + x_3, x_1 + x_3)$$

We want to see if $\mathbb{T}$ is invertible and, if so, find its inverse.

The operator $\mathbb{T}$ has an inverse if and only if $\mathbb{T}$ is bijective, which is true if and only if $\dim \mathcal{N}(\mathbb{T}) = 0$. So we construct $\mathcal{N}(\mathbb{T})$, which is the set of all vectors satisfying

$$\mathbb{T}(x_1, x_2, x_3) = (0, 0, 0)$$

or

$$x_1 + x_2 = 0 \qquad x_2 + x_3 = 0 \qquad x_1 + x_3 = 0$$

This has the unique solution $x_1 = x_2 = x_3 = 0$. Thus, the only vector belonging to $\mathcal{N}(\mathbb{T})$ is the zero vector, and $\dim \mathcal{N}(\mathbb{T}) = 0$. Therefore, $\mathbb{T}$ has an inverse.

To find $\mathbb{T}^{-1}$ we use $\mathbb{T}^{-1}\mathbb{T} = \mathbb{1}$ to obtain

$$\mathbb{T}^{-1}\mathbb{T}(x_1, x_2, x_3) = \mathbb{T}^{-1}(x_1 + x_2, x_2 + x_3, x_1 + x_3) = (x_1, x_2, x_3)$$

The last equality demonstrates how $\mathbb{T}^{-1}$ acts on vectors. To make this more apparent, we let $x_1 + x_2 = x$, $x_2 + x_3 = y$, $x_1 + x_3 = z$, solve for x_1, x_2, and x_3 in terms of x, y, and z, and substitute in the preceding equation to obtain

$$\mathbb{T}^{-1}(x, y, z) = \left(\tfrac{1}{2}(x - y + z), \tfrac{1}{2}(x + y - z), \tfrac{1}{2}(-x + y + z)\right)$$
$$= \tfrac{1}{2}(x - y + z,\ x + y - z,\ -x + y + z)$$

Rewriting this equation in terms of x_1, x_2, x_3 gives

$$\mathbb{T}^{-1}(x_1, x_2, x_3) = \tfrac{1}{2}(x_1 - x_2 + x_3,\ x_1 + x_2 - x_3,\ -x_1 + x_2 + x_3)$$

We can easily verify that $\mathbb{T}^{-1}\mathbb{T} = \mathbb{1}$:

$$\begin{aligned}\mathbb{T}^{-1}\mathbb{T}(x_1, x_2, x_3) &= \mathbb{T}^{-1}(x_1 + x_2,\ x_2 + x_3,\ x_1 + x_3)\\ &= \tfrac{1}{2}((x_1 + x_2) - (x_2 + x_3) + (x_1 + x_3), (x_1 + x_2) + (x_2 + x_3)\\ &\quad - (x_1 + x_3),\ -(x_1 + x_2) + (x_2 + x_3) + (x_1 + x_3))\\ &= \tfrac{1}{2}(2x_1, 2x_2, 2x_3) = (x_1, x_2, x_3)\end{aligned}$$

Thus, the effect of $\mathbb{T}^{-1}\mathbb{T}$ on an arbitrary element of $\mathbb{R}^3$ is to keep it unchanged. Therefore, $\mathbb{T}^{-1}\mathbb{T} = \mathbb{1}$. Similarly, one can show that $\mathbb{T}\mathbb{T}^{-1} = \mathbb{1}$. ●

As indicated above, not all linear operators are invertible. Only linear isomorphisms have inverses. The following theorem shows that if an inverse exists, it must be unique. This is the justification for using the symbol $\mathbb{T}^{-1}$ for the inverse of $\mathbb{T}$.

2.3.9 Theorem The inverse of a linear operator is unique.

Proof. Let $\mathbb{A}$ and $\mathbb{B}$ be two inverses for $\mathbb{T}$. Then by definition

$$\mathbb{T}\mathbb{A} = \mathbb{1}$$

Multiplying both sides by $\mathbb{B}$ on the left gives

$$\mathbb{B}\mathbb{T}\mathbb{A} = \mathbb{B}\mathbb{1} = \mathbb{B}$$

But $\mathbb{B}\mathbb{T} = \mathbb{1}$ because $\mathbb{B}$ is also an inverse. Thus,

$$\mathbb{1}\mathbb{A} = \mathbb{B} \quad \Rightarrow \quad \mathbb{A} = \mathbb{B}$$ ■

The following theorem can be proved easily.

2.3.10 Theorem If $\mathbb{T}_1$ and $\mathbb{T}_2$ are two invertible linear operators, then $\mathbb{T}_1\mathbb{T}_2$ is also invertible and

$$(\mathbb{T}_1\mathbb{T}_2)^{-1} = \mathbb{T}_2^{-1}\mathbb{T}_1^{-1}$$ ■

Theorem 2.3.8 can be used to prove another theorem.

2.3.11 Theorem A linear transformation $\mathbb{T}: \mathcal{V} \to \mathcal{V}$ is invertible if and only if it carries a basis of $\mathcal{V}$ onto another basis of $\mathcal{V}$. ■

Polynomials of operators. Having defined both products and sums of operators, we can construct polynomials of operators. We define powers of $\mathbb{T}$ inductively as

$$\mathbb{T}^m = \mathbb{T} \circ \mathbb{T}^{m-1} = \mathbb{T}^{m-1} \circ \mathbb{T}$$

for all positive integers $m \geqslant 1$. The consistency of this equation (for $m = 1$) demands this relation:

$$\mathbb{T}^0 = \mathbb{1}$$

We can now define a polynomial in $\mathbb{T}$ of degree n as

$$P(\mathbb{T}) = \alpha_0 \mathbb{1} + \alpha_1 \mathbb{T} + \alpha_2 \mathbb{T}^2 + \cdots + \alpha_n \mathbb{T}^n \tag{2.17}$$

Example 2.3.5

Let $\mathbb{T}_\theta : \mathbb{R}^2 \to \mathbb{R}^2$ be the linear operator that rotates vectors in the xy-plane through angle θ, that is,

$$\mathbb{T}_\theta(x, y) = (x \cos\theta - y \sin\theta,\ x \sin\theta + y \cos\theta)$$

We are interested in powers of $\mathbb{T}_\theta$.

First, let us evaluate $\mathbb{T}_\theta^2$:

$$\begin{aligned}
\mathbb{T}_\theta^2(x, y) &= \mathbb{T}_\theta(\underbrace{x \cos\theta - y \sin\theta}_{x'},\ \underbrace{x \sin\theta + y \cos\theta}_{y'}) \\
&= (x' \cos\theta - y' \sin\theta,\ x' \sin\theta + y' \cos\theta) \\
&= ((x \cos\theta - y \sin\theta)\cos\theta - (x \sin\theta + y \cos\theta)\sin\theta, \\
&\qquad (x \cos\theta - y \sin\theta)\sin\theta + (x \sin\theta + y \cos\theta)\cos\theta) \\
&= (x(\cos^2\theta - \sin^2\theta) - y(2 \sin\theta \cos\theta),\ x(2 \sin\theta \cos\theta) + y(\cos^2\theta - \sin^2\theta)) \\
&= (x \cos 2\theta - y \sin 2\theta,\ x \sin 2\theta + y \cos 2\theta)
\end{aligned}$$

Thus, $\mathbb{T}_\theta^2$ rotates (x, y) by 2θ. Similarly,

$$\begin{aligned}
\mathbb{T}_\theta^3(x, y) &= \mathbb{T}_\theta(\mathbb{T}_\theta^2(x, y)) = \mathbb{T}_\theta(\underbrace{x \cos 2\theta - y \sin 2\theta}_{x''},\ \underbrace{x \sin 2\theta + y \cos 2\theta}_{y''}) \\
&= (x'' \cos\theta - y'' \sin\theta,\ x'' \sin\theta + y'' \cos\theta) \\
&= (x(\cos 2\theta \cos\theta - \sin 2\theta \sin\theta) - y(\sin 2\theta \cos\theta + \cos 2\theta \sin\theta), \\
&\qquad x(\sin 2\theta \cos\theta + \cos 2\theta \sin\theta) + y(\cos 2\theta \cos\theta - \sin 2\theta \sin\theta)) \\
&= (x \cos 3\theta - y \sin 3\theta,\ x \sin 3\theta + y \cos 3\theta)
\end{aligned}$$

Using mathematical induction, we can show that

$$\mathbb{T}_\theta^n(x, y) = (x \cos n\theta - y \sin n\theta,\ x \sin n\theta + y \cos n\theta)$$

which shows that $\mathbb{T}_\theta^n$ is a rotation of (x, y) by an angle $n\theta$, that is,

$$\mathbb{T}_\theta^n = \mathbb{T}_{n\theta}$$

This result should have been obvious because $\mathbb{T}_\theta^n$ is equivalent to rotating (x, y) n times by angle θ each time. •

Negative powers of an invertible linear operator $\mathbb{T}$ are defined by

$$\mathbb{T}^{-m} \equiv (\mathbb{T}^{-1})^m$$

Note that $\mathbb{T}^{-m}$ is the inverse of $\mathbb{T}^m$.

The exponents of $\mathbb{T}$ satisfy the usual rules. In particular, for any integers m and n (positive or negative), we have

$$\mathbb{T}^m\mathbb{T}^n = \mathbb{T}^{m+n}$$

$$(\mathbb{T}^m)^n = \mathbb{T}^{mn}$$

We can further generalize m and n to include fractions and ultimately all real numbers.

Example 2.3.6

Let us evaluate $\mathbb{T}_\theta^{-n}$ where $\mathbb{T}_\theta$ is the linear operator of Example 2.3.5.

First, let us find $\mathbb{T}_\theta^{-1}$. We are looking for an operator such that

$$\mathbb{T}_\theta^{-1}\mathbb{T}_\theta(x, y) = (x, y) \tag{1}$$

for all (x, y). We rewrite this as

$$\mathbb{T}_\theta^{-1}(x\cos\theta - y\sin\theta, x\sin\theta + y\cos\theta) = (x, y)$$

We define

$$x' = x\cos\theta - y\sin\theta \qquad \text{and} \qquad y' = x\sin\theta + y\cos\theta$$

and solve x and y in terms of x' and y', to obtain

$$x = x'\cos\theta + y'\sin\theta \qquad \text{and} \qquad y = -x'\sin\theta + y'\cos\theta$$

Substituting for x and y in (1) yields

$$\mathbb{T}_\theta^{-1}(x', y') = (x'\cos\theta + y'\sin\theta, \ -x'\sin\theta + y'\cos\theta)$$

Comparing this with the action of $\mathbb{T}_\theta$ (see Example 2.3.5), we note that the only difference is in the sign of the $\sin\theta$ term. We conclude that $\mathbb{T}_\theta^{-1}$ has the same effect as $\mathbb{T}_{-\theta}$. So we have

$$\mathbb{T}_\theta^{-1} = \mathbb{T}_{-\theta}$$

and

$$\mathbb{T}_\theta^{-n} = (\mathbb{T}_{-\theta})^n = \mathbb{T}_{-n\theta}$$

It is instructive to verify that $\mathbb{T}_\theta^{-n}\mathbb{T}_\theta^n = \mathbb{1}$:

$$\mathbb{T}_\theta^{-n}\mathbb{T}_\theta^n(x, y) = \mathbb{T}_\theta^{-n}(\underbrace{x\cos n\theta - y\sin n\theta}_{x'}, \underbrace{x\sin n\theta + y\cos n\theta}_{y'})$$

$$= (x'\cos n\theta + y'\sin n\theta, \ -x'\sin n\theta + y'\cos n\theta)$$

$$= ((x\cos n\theta - y\sin n\theta)\cos n\theta + (x\sin n\theta + y\cos n\theta)\sin n\theta,$$

$$-(x\cos n\theta - y\sin n\theta)\sin n\theta + (x\sin n\theta + y\cos n\theta)\cos n\theta)$$

$$= (x(\cos^2 n\theta + \sin^2 n\theta), y(\sin^2 n\theta + \cos^2 n\theta))$$

$$= (x, y)$$

Similarly,

$$\mathbb{T}_\theta^n\mathbb{T}_\theta^{-n}(x, y) = (x, y)$$

•

It is important to realize that $P(\mathbb{T})$ is not, in general, invertible, even if $\mathbb{T}$ is. In fact, $\mathbb{T}_1 + \mathbb{T}_2$ is not necessarily invertible, even if both $\mathbb{T}_1$ and $\mathbb{T}_2$ are. A simple illustration of this fact is exhibited by choosing $\mathbb{T}_2$ to be $-\mathbb{T}_1$. Then $\mathbb{T}_1 + \mathbb{T}_2 = \mathbb{0}$, which is obviously not invertible. Therefore, the sum of invertible operators is not necessarily invertible. In particular, polynomials of an invertible operator are not necessarily invertible.

Functions of operators. We can go one step beyond polynomials of operators and, via Taylor expansion, define functions of them. Consider an ordinary function, $f(x)$, which has the Taylor expansion

$$f(x) = \sum_{n=0}^{\infty} \frac{(x - x_0)^n}{n!} \left(\frac{d^n f}{dx^n}\right)_{x_0}$$

in which x_0 is an arbitrary point where $f(x)$ and all its derivatives are defined. To this function, there corresponds a function of the operator $\mathbb{T}$, defined as follows:

$$f(\mathbb{T}) \equiv \sum_{n=0}^{\infty} \frac{(\mathbb{T} - x_0 \mathbb{1})^n}{n!} \left(\frac{d^n f}{dx^n}\right)_{x_0}$$

Because this series is an infinite sum of operators, difficulties may arise concerning its convergence. However, as will be shown in Section 3.6, for finite-dimensional vector spaces $f(\mathbb{T})$ is always defined, and, in fact, it is always a polynomial in $\mathbb{T}$. For the time being, we will think of $f(\mathbb{T})$ as an infinite series.

A simplification results when the function can be expanded around $x_0 = 0$. In this case we obtain

$$f(\mathbb{T}) = \sum_{n=0}^{\infty} \frac{\mathbb{T}^n}{n!} \left(\frac{d^n f}{dx^n}\right)_{x=0}$$

A widely used function is the exponential, whose expansion is easily found to be

$$e^{\mathbb{T}} \equiv \exp(\mathbb{T}) = \sum_{n=0}^{\infty} \frac{\mathbb{T}^n}{n!} \tag{2.18}$$

Example 2.3.7

Let us evaluate $e^{\mathbb{T}}$ when $\mathbb{T}: \mathbb{R}^2 \to \mathbb{R}^2$ is given by

$$\mathbb{T}(x, y) = (-\alpha y, \alpha x)$$

We can find a general formula for the action of $\mathbb{T}^n$ on (x, y). For an arbitrary (x, y),

$$\mathbb{T}^2(x, y) = \mathbb{T}(-\alpha y, \alpha x) = (-\alpha(\alpha x), \alpha(-\alpha y))$$
$$= -\alpha^2(x, y) = -\alpha^2 \mathbb{1}(x, y)$$

Thus,

$$\mathbb{T}^2 = -\alpha^2 \mathbb{1}$$

From $\mathbb{T}$ and $\mathbb{T}^2$ we can easily obtain higher powers of $\mathbb{T}$, for example:

$$\mathbb{T}^3 = \mathbb{T}(\mathbb{T}^2) = -\alpha^2 \mathbb{T}$$

$$\mathbb{T}^4 = \mathbb{T}^2 \mathbb{T}^2 = \alpha^4 \mathbb{1}$$

In general, then

$$\mathbb{T}^{2n} = (-1)^n \alpha^{2n} \mathbb{1} \qquad \text{for } n = 0, 1, 2, \ldots$$

$$\mathbb{T}^{2n+1} = (-1)^n \alpha^{2n} \mathbb{T} \qquad \text{for } n = 0, 1, 2, \ldots$$

Thus,

$$e^{\mathbb{T}} = \sum_{n \text{ odd}} \frac{\mathbb{T}^n}{n!} + \sum_{n \text{ even}} \frac{\mathbb{T}^n}{n!} = \sum_{n=0}^{\infty} \frac{(-1)^n \alpha^{2n}}{(2n+1)!} \mathbb{T} + \sum_{n=0}^{\infty} \frac{(-1)^n \alpha^{2n}}{(2n)!} \mathbb{1}$$

$$= \frac{1}{\alpha} \mathbb{T} \sum_{n=0}^{\infty} \frac{(-1)^n \alpha^{2n+1}}{(2n+1)!} + \mathbb{1} \sum_{n=0}^{\infty} \frac{(-1)^n \alpha^{2n}}{(2n)!}$$

The two series are recognizable as $\sin \alpha$ and $\cos \alpha$, respectively. Therefore, we get

$$e^{\mathbb{T}} = \frac{1}{\alpha} \mathbb{T} \sin \alpha + \mathbb{1} \cos \alpha$$

which shows that $e^{\mathbb{T}}$ is a polynomial (of first degree) in $\mathbb{T}$.

The action of $e^{\mathbb{T}}$ on (x, y) is given by

$$e^{\mathbb{T}}(x, y) = \left(\frac{\sin \alpha}{\alpha} \mathbb{T} + \cos \alpha \mathbb{1} \right)(x, y) = \frac{\sin \alpha}{\alpha} \mathbb{T}(x, y) + \cos \alpha \mathbb{1}(x, y)$$

$$= \frac{\sin \alpha}{\alpha}(-\alpha y, \alpha x) + (x \cos \alpha, y \cos \alpha)$$

$$= (-y \sin \alpha, x \sin \alpha) + (x \cos \alpha, y \cos \alpha)$$

$$= (x \cos \alpha - y \sin \alpha, x \sin \alpha + y \cos \alpha)$$

The reader will recognize the final expression as a rotation in the xy-plane by an angle α. Thus, we can think of $e^{\mathbb{T}}$ as a rotation operator around the z-axis. In this context $\mathbb{T}$ is called the *generator* of rotation. For infinitesimal α the action equation above gives

$$e^{\mathbb{T}}(x, y) \xrightarrow[\alpha \to 0]{} (x - \alpha y, \alpha x + y) = (x, y) + (-\alpha y, \alpha x) = \mathbb{1}(x, y) + \mathbb{T}(x, y) = (\mathbb{1} + \mathbb{T})(x, y)$$

Therefore, for small α the rotation in the xy-plane becomes simply $\mathbb{1} + \mathbb{T}$; that is, $\mathbb{T}$, by itself, determines the infinitesimal rotation. On the other hand, any *finite* rotation can be *generated* by repeated application of infinitesimal rotations. That is why $\mathbb{T}$ is called a generator. •

Exercise 1.3.4 presented a problem similar to Example 2.3.7 but in the language of matrices. This is not a coincidence but a result of a deep connection between linear operators and matrices, which will be developed in Chapter 3.

Commutators. In general, when we multiply two operators, the result depends on the order in which this is done. This means that if $\mathbb{T}, \mathbb{U} \in \mathcal{L}(\mathcal{V})$, then $\mathbb{T}\mathbb{U} \in \mathcal{L}(\mathcal{V})$ and $\mathbb{U}\mathbb{T} \in \mathcal{L}(\mathcal{V})$; however, $\mathbb{T}\mathbb{U}$ is not equal to $\mathbb{U}\mathbb{T}$ in general. The following defines the equality of operators precisely.

2.3.12 Definition Two linear operators $\mathbb{T}_1, \mathbb{T}_2 \in \mathcal{L}(\mathcal{V})$ are equal if and only if

$$\mathbb{T}_1|a\rangle = \mathbb{T}_2|a\rangle \qquad \forall\ |a\rangle \in \mathcal{V}$$

Equivalently,
$$\mathbb{T}_1 = \mathbb{T}_2 \quad \text{iff} \quad \mathbb{T}_1 - \mathbb{T}_2 = \mathbb{O}$$

Thus, when we say that $\mathbb{U}\mathbb{T} \neq \mathbb{T}\mathbb{U}$, we mean that we can find a vector $|a\rangle \in \mathcal{V}$ such that $\mathbb{U}\mathbb{T}|a\rangle \neq \mathbb{T}\mathbb{U}|a\rangle$. When this is the case, we say that $\mathbb{U}$ and $\mathbb{T}$ *do not commute.*

Example 2.3.8

The two linear operators $\mathbb{T}, \mathbb{U}: \mathbb{R}^2 \to \mathbb{R}^2$ defined by

$$\mathbb{T}(x, y) = (x + 2y, 2x + y) \qquad \text{and} \qquad \mathbb{U}(x, y) = (x - y, x + y)$$

do not commute, because

$$\mathbb{T}\mathbb{U}(x, y) = \mathbb{T}(x - y, x + y) = (x - y + 2(x + y), 2(x - y) + x + y) = (3x + y, 3x - y)$$

whereas

$$\begin{aligned}\mathbb{U}\mathbb{T}(x, y) = \mathbb{U}(x + 2y, 2x + y) &= (x + 2y - (2x + y), (x + 2y) + (2x + y) \\ &= (-x + y, 3x + 3y)\end{aligned}$$

Clearly, the two products are not equal for arbitrary (x, y). ●

To see "by how much" two operators do not commute, we take the difference between their products in different orders and call the result the commutator of the two operators. The following definition states this more precisely.

2.3.13 Definition The commutator $[\mathbb{U}, \mathbb{T}]$ of the two operators $\mathbb{U}$ and $\mathbb{T}$ belonging to $\mathcal{L}(\mathcal{V})$ is an operator in $\mathcal{L}(\mathcal{V})$, defined as

$$[\mathbb{U}, \mathbb{T}] \equiv \mathbb{U}\mathbb{T} - \mathbb{T}\mathbb{U}$$

An immediate consequence of Definition 2.3.13 is the following proposition.

2.3.14 Proposition For $\mathbb{S}, \mathbb{T}, \mathbb{U} \in \mathcal{L}(\mathcal{V})$ and $\alpha, \beta \in \mathbb{C}$ (or $\mathbb{R}$), we have

(i)	$[\mathbb{U}, \mathbb{T}] = -[\mathbb{T}, \mathbb{U}]$	antisymmetry
(ii)	$[\alpha\mathbb{U}, \beta\mathbb{T}] = \alpha\beta[\mathbb{U}, \mathbb{T}]$	linearity
(iii)	$[\mathbb{S}, \mathbb{T} + \mathbb{U}] = [\mathbb{S}, \mathbb{T}] + [\mathbb{S}, \mathbb{U}]$	linearity in the right entry
(iv)	$[\mathbb{S} + \mathbb{T}, \mathbb{U}] = [\mathbb{S}, \mathbb{U}] + [\mathbb{T}, \mathbb{U}]$	linearity in the left entry
(v)	$[\mathbb{S}, \mathbb{T}\mathbb{U}] = [\mathbb{S}, \mathbb{T}]\mathbb{U} + \mathbb{T}[\mathbb{S}, \mathbb{U}]$	left derivation property
(vi)	$[\mathbb{S}\mathbb{T}, \mathbb{U}] = \mathbb{S}[\mathbb{T}, \mathbb{U}] + [\mathbb{S}, \mathbb{U}]\mathbb{T}$	right derivation property
(vii)	$[[\mathbb{S}, \mathbb{T}], \mathbb{U}] + [[\mathbb{U}, \mathbb{S}], \mathbb{T}] + [[\mathbb{T}, \mathbb{U}], \mathbb{S}] = 0$	Jacobi identity

Proof. In almost all cases the proof follows immediately from Definition 2.3.13. The only (minor) exceptions are the derivation properties. We prove the left derivation property by rewriting (v) as follows:

$$[\mathbb{S}, \mathbb{TU}] = \mathbb{S}(\mathbb{TU}) - (\mathbb{TU})\mathbb{S} = \mathbb{STU} - \mathbb{TUS} + \underbrace{\mathbb{TSU} - \mathbb{TSU}}_{\equiv \mathbb{0}}$$

$$= (\mathbb{ST} - \mathbb{TS})\mathbb{U} + \mathbb{T}(\mathbb{SU} - \mathbb{US})$$

$$= [\mathbb{S}, \mathbb{T}]\mathbb{U} + \mathbb{T}[\mathbb{S}, \mathbb{U}]$$

The right derivation property is proved in exactly the same way. ■

Useful consequences of property (i) of the foregoing proposition (or, more directly, of the definition of the commutator) are

$$[\mathbb{A}, \mathbb{A}] = \mathbb{0}$$

and

$$[\mathbb{A}, \mathbb{A}^m] = \mathbb{0} \qquad \text{for } m = 0, \pm 1, \pm 2, \ldots$$

In particular,

$$[\mathbb{A}, \mathbb{1}] = \mathbb{0} \qquad \forall\, \mathbb{A}$$

and

$$[\mathbb{A}, \mathbb{A}^{-1}] = \mathbb{0}$$

2.3.3 Operator-Valued Functions and the Derivative of Operators

Up to this point we have been discussing the algebraic properties of operators, static objects that obey certain algebraic rules and fulfill the static needs of some applications. However, physical quantities are dynamic objects, and if we want operators to represent physical quantities, we must allow them to change with time. This dynamism is best illustrated in quantum mechanics, where physical observables, truly dynamic entities, are represented by operators.

Therefore, let us consider a mapping $\mathbb{H}: \mathbb{R} \to \mathcal{L}(\mathcal{V})$, which takes in a real number and gives out a linear operator on the vector space $\mathcal{V}$.[5] We denote the image of $t \in \mathbb{R}$ by $\mathbb{H}(t)$, which acts on the underlying vector space, $\mathcal{V}$. The physical meaning of this is that as t (usually time) varies, its image, $\mathbb{H}(t)$, also varies. Therefore, for different values of t, we have *different* operators. In particular,

$$[\mathbb{H}(t), \mathbb{H}(t')] \neq \mathbb{0} \qquad \text{for } t \neq t'$$

A concrete example is the operator that is a linear combination of operators $\mathbb{D}$ and $\mathbb{T}$ presented in Example 2.3.2, using scalars that are functions of time instead of constant scalars. To be specific, let

$$\mathbb{H}(t) = \mathbb{D}\cos\omega t + \mathbb{T}\sin\omega t$$

[5] Strictly speaking, the domain of $\mathbb{H}$ must be an interval $[a, b]$ of the real line, because $\mathbb{H}$ may not be defined for all $\mathbb{R}$. However, we need not make such fine distinctions.

where ω is a constant. As time goes by, $\mathbb{H}(t)$ changes its identity from $\mathbb{D}$ to $\mathbb{T}$ and back to $\mathbb{D}$. Sometimes it even has a hybrid identity! Since $\mathbb{T}$ and $\mathbb{D}$ do not commute, values of $\mathbb{H}(t)$ for different t's do not necessarily commute.

Of particular interest are operators that can be written as $e^{\mathbb{H}(t)}$, where $\mathbb{H}(t)$ are simple operators. That is, the dependence of $\mathbb{H}(t)$ on t is simpler than the corresponding dependence of $e^{\mathbb{H}(t)}$. We have already encountered such a situation in Example 2.3.7, where it was shown that the operation of rotation around the z-axis could be written as $e^{\mathbb{T}}$, and the action of $\mathbb{T}$ on (x, y) was a great deal simpler than the corresponding action of $e^{\mathbb{T}}$.

Such a state of affairs is very common in physics. In fact, it can be shown that all operators of physical interest can be written as a product of simpler operators, each being of the form $e^{\mathbb{T}}$ (see Varadarajan 1984). For example, we saw in Example 1.4.2 that an arbitrary rotation in three dimensions can be written as a product of three simpler rotations, each one being a rotation about an axis.

Operators of the form $\exp\{\mathbb{H}(t)\}$ are important enough to warrant further study. However, we will restrict ourselves to the evaluation of some derivatives that are important in physical applications. Let us start with a definition.

2.3.15 Definition For the mapping $\mathbb{H} : \mathbb{R} \to \mathcal{L}(\mathcal{V})$ we define the derivative as

$$\frac{d\mathbb{H}}{dt} = \lim_{\Delta t \to 0} \frac{\mathbb{H}(t + \Delta t) - \mathbb{H}(t)}{\Delta t}$$

This derivative also belongs to $\mathcal{L}(\mathcal{V})$.

The derivative of an operator is exactly the same as that of an ordinary function. In fact, as long as we keep track of the order, all the rules of differentiation apply to operators. For example,

$$\frac{d}{dt}(\mathbb{U}\mathbb{T}) = \left(\frac{d\mathbb{U}}{dt}\right)\mathbb{T} + \mathbb{U}\left(\frac{d\mathbb{T}}{dt}\right)$$

We are not allowed to change the order of multiplication, not even when both operators being multiplied are the same. For instance, if we let $\mathbb{U} = \mathbb{T} = \mathbb{H}$ in the preceding equation, we obtain

$$\frac{d}{dt}(\mathbb{H}^2) = \left(\frac{d\mathbb{H}}{dt}\right)\mathbb{H} + \mathbb{H}\left(\frac{d\mathbb{H}}{dt}\right)$$

This is *not*, in general, equal to $2\mathbb{H}(d\mathbb{H}/dt)$.

Example 2.3.9

Let us find the derivative of $e^{t\mathbb{H}}$, where $\mathbb{H}$ is independent of t. Using Definition 2.3.15, we have

$$\frac{d}{dt}(e^{t\mathbb{H}}) = \lim_{\Delta t \to 0} \frac{e^{(t+\Delta t)\mathbb{H}} - e^{t\mathbb{H}}}{\Delta t}$$

However, for infinitesimal Δt we have

$$e^{(t+\Delta t)\mathbb{H}} - e^{t\mathbb{H}} = e^{t\mathbb{H}}e^{(\Delta t)\mathbb{H}} - e^{t\mathbb{H}} = e^{t\mathbb{H}}(1 + \mathbb{H}\Delta t) - e^{t\mathbb{H}}$$
$$= e^{t\mathbb{H}}\mathbb{H}\Delta t$$

Therefore,

$$\frac{d}{dt}(e^{t\mathbb{H}}) = \lim_{\Delta t \to 0} \frac{e^{t\mathbb{H}}\mathbb{H}\Delta t}{\Delta t} = e^{t\mathbb{H}}\mathbb{H}$$

Since $\mathbb{H}$ and $e^{t\mathbb{H}}$ commute, we also have

$$\frac{d}{dt}(e^{t\mathbb{H}}) = \mathbb{H}e^{t\mathbb{H}} = e^{t\mathbb{H}}\mathbb{H}$$

Note that in deriving the equation for the derivative of $e^{t\mathbb{H}}$, we have used the relation

$$e^{t\mathbb{H}}e^{(\Delta t)\mathbb{H}} = e^{(t+\Delta t)\mathbb{H}}$$

This may seem trivial, but it will be shown later that, in general, $e^{\mathbb{H}_1 + \mathbb{H}_2} \neq e^{\mathbb{H}_1}e^{\mathbb{H}_2}$ (see Exercise 2.3.8, however). ●

Now let us consider the exponential of a more general time-dependent operator, $e^{\mathbb{H}(t)}$. We use Definition 2.3.15 to evaluate the derivative of such an operator:

$$\frac{d}{dt}(e^{\mathbb{H}(t)}) = \lim_{\Delta t \to 0} \frac{e^{\mathbb{H}(t+\Delta t)} - e^{\mathbb{H}(t)}}{\Delta t}$$

If $\mathbb{H}(t)$ possesses a derivative, we have, to the first order in Δt,

$$\mathbb{H}(t + \Delta t) = \mathbb{H}(t) + \Delta t\left(\frac{d\mathbb{H}}{dt}\right)$$

and we can write

$$e^{\mathbb{H}(t+\Delta t)} = e^{\mathbb{H}(t) + \Delta t(d\mathbb{H}/dt)}$$

It is very tempting to factor out the $e^{\mathbb{H}(t)}$ and expand the remaining part. However, as we will see presently, this is not possible in general. As preparation, consider the following example, which concerns the integration of an operator.

Example 2.3.10

In physics we often encounter *operator differential equations* of the form

$$\frac{d\mathbb{U}}{dt} = \mathbb{H}\mathbb{U}(t)$$

where $\mathbb{H}$ is *not* dependent on t. We can find a solution to such an equation by repeated differentiation followed by Taylor-series expansion. Thus,

$$\frac{d^2\mathbb{U}}{dt^2} = \mathbb{H}\frac{d\mathbb{U}}{dt} = \mathbb{H}[\mathbb{H}\mathbb{U}(t)] = \mathbb{H}^2\mathbb{U}(t)$$

$$\frac{d^3\mathbb{U}}{dt^3} = \frac{d}{dt}[\mathbb{H}^2\mathbb{U}(t)] = \mathbb{H}^2\frac{d\mathbb{U}}{dt} = \mathbb{H}^3\mathbb{U}(t)$$

In general,

$$\frac{d^n\mathbb{U}}{dt^n} = \mathbb{H}^n\mathbb{U}(t)$$

Assuming that $\mathbb{U}(t)$ is well-defined at $t = 0$, the above relations say that all derivatives of $\mathbb{U}(t)$ are also well-defined at $t = 0$. Therefore, we can expand $\mathbb{U}(t)$ around $t = 0$ to obtain

$$\mathbb{U}(t) = \sum_{n=0}^{\infty} \frac{1}{n!} t^n \left(\frac{d^n\mathbb{U}}{dt^n}\right)_{t=0} = \sum_{n=0}^{\infty} \frac{t^n}{n!}\mathbb{H}^n\mathbb{U}(0)$$

$$= \left(\sum_{n=0}^{\infty} \frac{(t\mathbb{H})^n}{n!}\right)\mathbb{U}(0) = e^{t\mathbb{H}}\mathbb{U}(0) \qquad \bullet$$

Let us now consider the feasibility of $e^{\mathbb{H}_1 + \mathbb{H}_2} = e^{\mathbb{H}_1}e^{\mathbb{H}_2}$. We define $[\mathbb{H}_1, \mathbb{H}_2] = \mathbb{T}$ and assume that $[\mathbb{T}, \mathbb{H}_1] = 0 = [\mathbb{T}, \mathbb{H}_2]$. Then we consider the operator $\mathbb{U}(t)$, defined by

$$\mathbb{U}(t) = e^{t\mathbb{H}_1}e^{t\mathbb{H}_2}e^{-t(\mathbb{H}_1 + \mathbb{H}_2)}$$

We differentiate both sides, using the result of Example 2.3.9 and the product rule for differentiation:

$$\frac{d\mathbb{U}}{dt} = \mathbb{H}_1 e^{t\mathbb{H}_1}e^{t\mathbb{H}_2}e^{-t(\mathbb{H}_1 + \mathbb{H}_2)} + e^{t\mathbb{H}_1}\mathbb{H}_2 e^{t\mathbb{H}_2}e^{-t(\mathbb{H}_1 + \mathbb{H}_2)} - e^{t\mathbb{H}_1}e^{t\mathbb{H}_2}(\mathbb{H}_1 + \mathbb{H}_2)e^{-t(\mathbb{H}_1 + \mathbb{H}_2)} \tag{2.19}$$

The three factors of $\mathbb{U}(t)$ are present in all terms; however, in some terms there is an operator that cannot be pushed to the left without resistance. We can partially overcome this resistance by using the commutation relation between $\mathbb{H}_1$ and $\mathbb{H}_2$. For instance,

$$e^{t\mathbb{H}_1}\mathbb{H}_2 = \mathbb{H}_2 e^{t\mathbb{H}_1} + [e^{t\mathbb{H}_1}, \mathbb{H}_2]$$

It is left as a problem for the reader to show that if $[\mathbb{H}_1, \mathbb{H}_2] = \mathbb{T}$ and $\mathbb{T}$ commutes with $\mathbb{H}_1$ and $\mathbb{H}_2$, then

$$[e^{t\mathbb{H}_1}, \mathbb{H}_2] = t\mathbb{T}e^{t\mathbb{H}_1}$$

and, therefore,

$$e^{t\mathbb{H}_1}\mathbb{H}_2 = \mathbb{H}_2 e^{t\mathbb{H}_1} + t\mathbb{T}e^{t\mathbb{H}_1}$$

Similarly, it is easily shown that

$$e^{t\mathbb{H}_1}e^{t\mathbb{H}_2}(\mathbb{H}_1 + \mathbb{H}_2) = (\mathbb{H}_1 + \mathbb{H}_2)e^{t\mathbb{H}_1}e^{t\mathbb{H}_2}$$

Substituting from these last two equations in (2.19) yields

$$\frac{d\mathbb{U}}{dt} = \mathbb{H}_1\mathbb{U}(t) + \mathbb{H}_2\mathbb{U}(t) + t\mathbb{T}\mathbb{U}(t) - (\mathbb{H}_1 + \mathbb{H}_2)\mathbb{U}(t) = t\mathbb{T}\mathbb{U}(t)$$

The solution to this equation is (see Exercise 2.3.9)

$$\mathbb{U}(t) = e^{(1/2)t^2\mathbb{T}}\,\mathbb{U}(0) = e^{(1/2)t^2[\mathbb{H}_1, \mathbb{H}_2]}$$

because $\mathbb{U}(0) = \mathbb{1}$, from the definition of $\mathbb{U}(t)$, and $\mathbb{T} = [\mathbb{H}_1, \mathbb{H}_2]$. Recalling what $\mathbb{U}(t)$ was, we have

$$e^{t\mathbb{H}_1} e^{t\mathbb{H}_2} e^{-t(\mathbb{H}_1 + \mathbb{H}_2)} = e^{(1/2)t^2[\mathbb{H}_1, \mathbb{H}_2]}$$

which can be rewritten as

$$e^{t\mathbb{H}_1} e^{t\mathbb{H}_2} e^{-(1/2)t^2[\mathbb{H}_1, \mathbb{H}_2]} = e^{t(\mathbb{H}_1 + \mathbb{H}_2)} \tag{2.20}$$

For $t = 1$ Eq. (2.20) yields

$$e^{\mathbb{H}_1 + \mathbb{H}_2} = e^{\mathbb{H}_1} e^{\mathbb{H}_2} e^{-(1/2)[\mathbb{H}_1, \mathbb{H}_2]}$$

This result is summarized in the following proposition.[6]

2.3.16 Proposition Let $\mathbb{H}_1, \mathbb{H}_2 \in \mathscr{L}(\mathscr{V})$. If $[\mathbb{H}_1, [\mathbb{H}_1, \mathbb{H}_2]] = 0 = [\mathbb{H}_2, [\mathbb{H}_1, \mathbb{H}_2]]$, then

$$e^{\mathbb{H}_1 + \mathbb{H}_2} = e^{\mathbb{H}_1} e^{\mathbb{H}_2} e^{-(1/2)[\mathbb{H}_1, \mathbb{H}_2]}$$

In particular,

$$e^{\mathbb{H}_1 + \mathbb{H}_2} = e^{\mathbb{H}_1} e^{\mathbb{H}_2} \quad \text{iff} \quad [\mathbb{H}_1, \mathbb{H}_2] = 0$$ ■

We can use Proposition 2.3.16 to finish the job of finding the derivative of $e^{\mathbb{H}(t)}$. We assume (and this is true for almost all cases of physical interest) that *both* $\mathbb{H}(t)$ *and* $d\mathbb{H}/dt$ *commute with* $[\mathbb{H}, d\mathbb{H}/dt]$. However, we do not assume that $\mathbb{H}$ and $d\mathbb{H}/dt$ commute. Letting $\mathbb{H}_1 = \mathbb{H}(t)$ and $\mathbb{H}_2 = \Delta t(d\mathbb{H}/dt)$, we obtain

$$e^{\mathbb{H}(t+\Delta t)} = e^{\mathbb{H}(t)} e^{\Delta t(d\mathbb{H}/dt)} e^{-(1/2)[\mathbb{H}(t), \Delta t(d\mathbb{H}/dt)]}$$

For infinitesimal Δt, this yields

$$\begin{aligned} e^{\mathbb{H}(t+\Delta t)} &= e^{\mathbb{H}(t)}\left(\mathbb{1} + \Delta t\left(\frac{d\mathbb{H}}{dt}\right)\right)\left(\mathbb{1} - \frac{1}{2}\Delta t\left[\mathbb{H}(t), \frac{d\mathbb{H}}{dt}\right]\right) \\ &= e^{\mathbb{H}(t)}\left\{\mathbb{1} - \frac{1}{2}\Delta t\left[\mathbb{H}, \frac{d\mathbb{H}}{dt}\right] + \Delta t\left(\frac{d\mathbb{H}}{dt}\right)\right\} \end{aligned}$$

and we have

$$\frac{d}{dt}(e^{\mathbb{H}(t)}) = e^{\mathbb{H}}\frac{d\mathbb{H}}{dt} - \frac{1}{2}e^{\mathbb{H}}\left[\mathbb{H}, \frac{d\mathbb{H}}{dt}\right]$$

We can also write

$$\begin{aligned} e^{\mathbb{H}(t+\Delta t)} &= e^{\mathbb{H}(t) + \Delta t(d\mathbb{H}/dt)} = e^{\Delta t(d\mathbb{H}/dt) + \mathbb{H}(t)} \\ &= e^{\Delta t(d\mathbb{H}/dt)} e^{\mathbb{H}(t)} e^{-(1/2)[\Delta t(d\mathbb{H}/dt), \mathbb{H}(t)]} \end{aligned}$$

[6] The content of this proposition is a special case of a more general result known as the Baker-Campbell-Hausdorff formula and is proved in Lie group theory. For a discussion, see Varadarajan (1984).

which yields

$$\frac{d}{dt}(e^{\mathbb{H}(t)}) = \frac{d\mathbb{H}}{dt}e^{\mathbb{H}} + \frac{1}{2}e^{\mathbb{H}}\left[\mathbb{H}(t), \frac{d\mathbb{H}}{dt}\right]$$

Adding the above two expressions for the derivative of $e^{\mathbb{H}(t)}$ and dividing by 2 yields the symmetric expression for the derivative:

$$\frac{d}{dt}(e^{\mathbb{H}(t)}) = \frac{1}{2}\left(e^{\mathbb{H}}\frac{d\mathbb{H}}{dt} + \frac{d\mathbb{H}}{dt}e^{\mathbb{H}}\right) = \frac{1}{2}\left\{e^{\mathbb{H}}, \frac{d\mathbb{H}}{dt}\right\}$$

where $$\{\mathbb{T}_1, \mathbb{T}_2\} \equiv \mathbb{T}_1\mathbb{T}_2 + \mathbb{T}_2\mathbb{T}_1$$

is called the *anticommutator* of the operators $\mathbb{T}_1$ and $\mathbb{T}_2$.

We, therefore, have the following proposition.

2.3.17 Proposition Let $\mathbb{H}: \mathbb{R} \to \mathscr{L}(\mathscr{V})$ and $[\mathbb{H}, [\mathbb{H}, d\mathbb{H}/dt]] = 0 = [d\mathbb{H}/dt, [\mathbb{H}, d\mathbb{H}/dt]]$. Then

$$\frac{d}{dt}(e^{\mathbb{H}(t)}) = \frac{1}{2}\left\{e^{\mathbb{H}(t)}, \frac{d\mathbb{H}}{dt}\right\}$$

In particular, if $[\mathbb{H}, d\mathbb{H}/dt] = 0$, then

$$\frac{d}{dt}(e^{\mathbb{H}(t)}) = \left(\frac{d\mathbb{H}}{dt}\right)e^{\mathbb{H}(t)} = e^{\mathbb{H}(t)}\left(\frac{d\mathbb{H}}{dt}\right)$$ ■

This subsection concludes with some further examples of the application of derivatives.

Example 2.3.11

Consider the operator

$$\mathbb{F}(t) = e^{t\mathbb{A}}\mathbb{B}e^{-t\mathbb{A}}$$

where $\mathbb{A}$ and $\mathbb{B}$ are t-independent operators on $\mathscr{V}$. It is straightforward (and left as a problem for the reader) to show that

$$\frac{d\mathbb{F}}{dt} = [\mathbb{A}, \mathbb{F}(t)] \qquad \text{and} \qquad \frac{d}{dt}[\mathbb{A}, \mathbb{F}(t)] = \left[\mathbb{A}, \frac{d\mathbb{F}}{dt}\right]$$

Using these results, we can write

$$\frac{d^2\mathbb{F}}{dt^2} = \frac{d}{dt}[\mathbb{A}, \mathbb{F}(t)] = [\mathbb{A}, [\mathbb{A}, \mathbb{F}(t)]] \equiv \mathbb{A}^2[\mathbb{F}(t)]$$

and, in general,

$$\frac{d^n\mathbb{F}}{dt^n} = \mathbb{A}^n[\mathbb{F}(t)]$$

where $\mathbb{A}^n[\mathbb{F}(t)]$ is defined inductively as

$$\mathbb{A}^n[\mathbb{F}(t)] \equiv [\mathbb{A}, \mathbb{A}^{n-1}[\mathbb{F}(t)]]$$

with $\mathbb{A}^0[\mathbb{F}(t)] \equiv \mathbb{F}(t)$. For example,

$$\mathbb{A}^3[\mathbb{F}(t)] \equiv [\mathbb{A}, \mathbb{A}^2[\mathbb{F}(t)]] = [\mathbb{A}, [\mathbb{A}, \mathbb{A}[\mathbb{F}(t)]]]$$
$$= [\mathbb{A}, [\mathbb{A}, [\mathbb{A}, \mathbb{F}(t)]]]$$

Evaluating $\mathbb{F}(t)$ and all its derivatives at $t = 0$ and substituting in the Taylor expansion about $t = 0$, we get

$$\mathbb{F}(t) = \sum_{n=0}^{\infty} \frac{t^n}{n!}\left(\frac{d^n\mathbb{F}}{dt^n}\right)_{t=0} = \sum_{n=0}^{\infty} \frac{t^n}{n!}\mathbb{A}^n[\mathbb{F}(0)]$$
$$= \sum_{n=0}^{\infty} \frac{t^n}{n!}\mathbb{A}^n[\mathbb{B}]$$

That is,

$$e^{t\mathbb{A}}\mathbb{B}e^{-t\mathbb{A}} = \mathbb{B} + t[\mathbb{A}, \mathbb{B}] + \frac{t^2}{2!}[\mathbb{A}, [\mathbb{A}, \mathbb{B}]] + \cdots$$
$$\equiv \sum_{n=0}^{\infty} \frac{t^n}{n!}\mathbb{A}^n[\mathbb{B}]$$

Sometimes this is written symbolically as

$$e^{t\mathbb{A}}\mathbb{B}e^{-t\mathbb{A}} \equiv \left(\sum_{n=0}^{\infty} \frac{t^n}{n!}\mathbb{A}^n\right)[\mathbb{B}] \equiv (e^{t\mathbb{A}})[\mathbb{B}]$$

where the RHS is merely a definition of the infinite sum to its left.

For $t = 1$ we obtain a widely used formula:

$$e^{\mathbb{A}}\mathbb{B}e^{-\mathbb{A}} = e^{\mathbb{A}}[\mathbb{B}] \equiv \sum_{n=0}^{\infty} \frac{1}{n!}\mathbb{A}^n[\mathbb{B}] = \mathbb{B} + [\mathbb{A}, \mathbb{B}] + \frac{1}{2!}[\mathbb{A}, [\mathbb{A}, \mathbb{B}]] + \cdots$$ ●

Example 2.3.12

If $\mathbb{A}$ in Example 2.3.11 commutes with $[\mathbb{A}, \mathbb{B}]$, then the infinite series truncates at the second term, and we have

$$e^{t\mathbb{A}}\mathbb{B}e^{-t\mathbb{A}} = \mathbb{B} + t[\mathbb{A}, \mathbb{B}]$$

For instance, if $\mathbb{A}$ and $\mathbb{B}$ are replaced by $\mathbb{D}$ and $\mathbb{T}$ of Example 2.3.2, we get

$$e^{t\mathbb{D}}\mathbb{T}e^{-t\mathbb{D}} = \mathbb{T} + t[\mathbb{D}, \mathbb{T}] = \mathbb{T} + t\mathbb{1} \equiv \mathbb{T} + t$$

This shows that the operator $\mathbb{T}$ has been *translated* by an amount t. We, therefore, call $e^{t\mathbb{D}}$ the translation operator of $\mathbb{T}$ by t, and we call $\mathbb{D}$ the *generator* of translation.

With a little modification $\mathbb{T}$ and $\mathbb{D}$ become, respectively, the position and momentum operators in quantum mechanics. Thus, momentum is thought of as the generator of translation in quantum mechanics. ●

2.3.4 Conjugation of Operators

We have discussed the notion of the dual of a vector in conjunction with inner products. We now incorporate linear operators into this notion. Let $|b\rangle, |c\rangle \in \mathcal{V}$ and let $|c\rangle = \mathbb{T}|b\rangle$. We know that there are linear functionals in the dual space $\mathcal{V}^*$ that are associated with $|b\rangle$ and $|c\rangle$; these are $(|b\rangle)^\dagger = \langle b|$ and $(|c\rangle)^\dagger = \langle c|$. The natural question is whether there is a linear operator belonging to $\mathcal{L}(\mathcal{V}^*)$ that, somehow, corresponds to $\mathbb{T}$. In other words, can we find a linear operator that relates $\langle b|$ and $\langle c|$ just as $\mathbb{T}$ relates $|b\rangle$ and $|c\rangle$? The answer comes in the form of a definition.

2.3.18 Definition Let $\mathbb{T} \in \mathcal{L}(\mathcal{V})$ and $|b\rangle, |c\rangle \in \mathcal{V}$ such that $|c\rangle = \mathbb{T}|b\rangle$. An operator $\mathbb{T}^\dagger \in \mathcal{L}(\mathcal{V}^*)$, called the *adjoint*, or *hermitian conjugate*, of $\mathbb{T}$, is defined by

$$\langle c| = \langle b|\mathbb{T}^\dagger \tag{2.21}$$

where $\langle b|$ and $\langle c|$ are duals of $|b\rangle$ and $|c\rangle$, respectively.

Note that $\mathbb{T}^\dagger$ (read as "tee dagger") stands to the right of a (bra) vector in the dual space $\mathcal{V}^*$. Thus, we say that $\mathbb{T}^\dagger$ "acts to the left." However, we can also think of $\mathbb{T}^\dagger$ as an operator *in* $\mathcal{L}(\mathcal{V})$ so that it can act on kets to the right. Equation (2.21) is dying for such an interpretation! On the RHS of the equation, $\mathbb{T}^\dagger$ is waiting for a ket on which to act just as much as $\langle c|$ on the LHS is waiting for a ket with which to contract. Similar arguments lead us to interpret $\mathbb{T}$ not only as an operator in $\mathcal{L}(\mathcal{V})$, but also as an operator in $\mathcal{L}(\mathcal{V}^*)$.

The above discussion can be clarified if we "multiply" both sides of Eq. (2.21) by a ket, such as $|a\rangle$. We then get

$$\langle c|a\rangle = (\langle b|\mathbb{T}^\dagger)|a\rangle = \langle b|(\mathbb{T}^\dagger|a\rangle) = \langle b|\mathbb{T}^\dagger|a\rangle$$

$$\underset{\text{by definition}}{\uparrow} \qquad \underset{\text{by reinterpretation}}{\uparrow} \qquad \underset{\text{by economy in writing}}{\uparrow}$$

Similarly, if we contract $|c\rangle = \mathbb{T}|b\rangle$ with a bra, such as $\langle a|$, we get

$$\langle a|c\rangle = \langle a|(\mathbb{T}|b\rangle) = (\langle a|\mathbb{T})|b\rangle = \langle a|\mathbb{T}|b\rangle$$

Taking the complex conjugate of the latter equation and identifying $\langle a|c\rangle^*$ with $\langle c|a\rangle$, we obtain

$$\langle a|\mathbb{T}|b\rangle^* = \langle b|\mathbb{T}^\dagger|a\rangle \tag{2.22}$$

This equation is sometimes used as the definition of the hermitian conjugate.

Some of the properties of conjugation are listed in the following theorem.

2.3.19 Theorem Let $\mathbb{U}, \mathbb{T} \in \mathcal{L}(\mathcal{V})$ and $\alpha \in \mathbb{C}$, then

(i) $(\mathbb{U} + \mathbb{T})^\dagger = \mathbb{U}^\dagger + \mathbb{T}^\dagger$

(ii) $(\mathbb{U}\mathbb{T})^\dagger = \mathbb{T}^\dagger\mathbb{U}^\dagger$ (Note the change in order!)

(iii) $(\alpha\mathbb{T})^\dagger = \alpha^*\mathbb{T}^\dagger$

(iv) $(\mathbb{T}^\dagger)^\dagger = \mathbb{T}$

Proof. (i) Let $|c\rangle = (\mathbb{U} + \mathbb{T})|b\rangle = |b_1\rangle + |b_2\rangle$, where $|b_1\rangle = \mathbb{U}|b\rangle$ and $|b_2\rangle = \mathbb{T}|b\rangle$. From the definition of conjugation, we have

$$\langle c| = \langle b|(\mathbb{U} + \mathbb{T})^\dagger = \langle b_1| + \langle b_2|$$

However, $$|b_1\rangle = \mathbb{U}|b\rangle \Rightarrow \langle b_1| = \langle b|\mathbb{U}^\dagger$$

and $$|b_2\rangle = \mathbb{T}|b\rangle \Rightarrow \langle b_2| = \langle b|\mathbb{T}^\dagger$$

Thus, $$\langle b|(\mathbb{U} + \mathbb{T})^\dagger = \langle b|\mathbb{U}^\dagger + \langle b|\mathbb{T}^\dagger = \langle b|(\mathbb{U}^\dagger + \mathbb{T}^\dagger)$$

This is true for all $\langle b|$. Therefore,

$$(\mathbb{U} + \mathbb{T})^\dagger = \mathbb{U}^\dagger + \mathbb{T}^\dagger$$

(ii) Let $|c\rangle = \mathbb{U}\mathbb{T}|b\rangle = \mathbb{U}|b_2\rangle$. Then

$$\langle c| = \langle b_2|\mathbb{U}^\dagger = \langle b|\mathbb{T}^\dagger\mathbb{U}^\dagger$$

On the other hand, by definition

$$\langle c| = \langle b|(\mathbb{U}\mathbb{T})^\dagger$$

Comparing these two expressions for $\langle c|$ yields the desired result.

(iii) Let $|c\rangle = (\alpha\mathbb{T})|b\rangle = \mathbb{T}(\alpha|b\rangle) = \mathbb{T}|b'\rangle$. Then

$$\langle c| = \langle b'|\mathbb{T}^\dagger$$

From Eq. (2.11) we get $\langle b'| = \alpha^*\langle b|$; therefore, the preceding equation becomes

$$\langle c| = \alpha^*\langle b|\mathbb{T}^\dagger = \langle b|(\alpha^*\mathbb{T}^\dagger)$$

On the other hand,

$$\langle c| = \langle b|(\alpha\mathbb{T})^\dagger$$

Comparison of these last two equations yields the desired identity.

(iv) We use Eq. (2.22) twice, as follows:

$$\langle a|\mathbb{T}|b\rangle = (\langle a|\mathbb{T}|b\rangle^*)^* = (\langle b|\mathbb{T}^\dagger|a\rangle)^* = \langle a|(\mathbb{T}^\dagger)^\dagger|b\rangle$$

Since this is true for all $|a\rangle$ and $|b\rangle$, the identity of the two operators must hold.[7] ∎

The foregoing concepts can be illustrated with some simple examples.

Example 2.3.13

In previous examples dealing with linear operators $\mathbb{T}: \mathbb{R}^n \to \mathbb{R}^n$, an element of $\mathbb{R}^n$ was denoted by a row vector, such as (x, y) for $\mathbb{R}^2$ and (x, y, z) for $\mathbb{R}^3$. There was no confusion

[7] The identity does not hold for infinite-dimensional vector spaces in general.

because we were operating only in $\mathcal{V}$. However, since elements of both $\mathcal{V}$ and $\mathcal{V}^*$ are required when discussing $\mathbb{T}$ and $\mathbb{T}^\dagger$, it is helpful to make a distinction between them. We, therefore, resort to the convention introduced in Example 2.2.1 by which kets are represented as column vectors and bras as row vectors.

Let us, therefore, consider $\mathbb{T}: \mathbb{R}^2 \to \mathbb{R}^2$ given by

$$\mathbb{T}\begin{pmatrix} x \\ y \end{pmatrix} = \begin{pmatrix} x + y \\ x - y \end{pmatrix}$$

We attempt to find a similar expression for $\mathbb{T}^\dagger$ *as it acts on a vector in* $\mathcal{V}$. The action of $\mathbb{T}^\dagger$ on $\mathcal{V}^*$ is trivially obtained from the above:

$$(x \quad y)\mathbb{T}^\dagger = (x + y \quad x - y)$$

which contains no information about $\mathbb{T}^\dagger$ as a member of $\mathcal{L}(\mathcal{V})$.

To find such an expression, we use Eq. (2.22), noting that the scalars here are real. So, let

$$|\mathrm{a}\rangle = \begin{pmatrix} x_1 \\ y_1 \end{pmatrix} \qquad \text{and} \qquad |\mathrm{b}\rangle = \begin{pmatrix} x_2 \\ y_2 \end{pmatrix}$$

Then

$$\langle \mathrm{a}|\mathbb{T}|\mathrm{b}\rangle = \langle \mathrm{a}|\mathbb{T}|\mathrm{b}\rangle^* = (x_1 \quad y_1)\mathbb{T}\begin{pmatrix} x_2 \\ y_2 \end{pmatrix} = (x_1 \quad y_1)\begin{pmatrix} x_2 + y_2 \\ x_2 - y_2 \end{pmatrix}$$

↑ because everything is real ↑ by definition of $\mathbb{T}$

$$= x_1x_2 + x_1y_2 + y_1x_2 - y_1y_2 = x_2(x_1 + y_1) + y_2(x_1 - y_1)$$

$$= (x_2 \quad y_2)\begin{pmatrix} x_1 + y_1 \\ x_1 - y_1 \end{pmatrix} = \langle \mathrm{b}|\mathbb{T}^\dagger|\mathrm{a}\rangle$$

↑ by Eq. (2.22)

$$= (x_2 \quad y_2)\mathbb{T}^\dagger\begin{pmatrix} x_1 \\ y_1 \end{pmatrix}$$

The last equality holds for all $(x_2 \quad y_2)$. Therefore, we have

$$\mathbb{T}^\dagger\begin{pmatrix} x_1 \\ y_1 \end{pmatrix} = \begin{pmatrix} x_1 + y_1 \\ x_1 - y_1 \end{pmatrix}$$

or, in general,

$$\mathbb{T}^\dagger\begin{pmatrix} x \\ y \end{pmatrix} = \begin{pmatrix} x + y \\ x - y \end{pmatrix}$$

We note that in this case

$$\mathbb{T}^\dagger = \mathbb{T}$$

However, this is not always true, as the next example shows. •

Example 2.3.14

Now let $\mathbb{T}:\mathbb{R}^2 \to \mathbb{R}^2$ be given by

$$\mathbb{T}\begin{pmatrix} x \\ y \end{pmatrix} = \begin{pmatrix} 2x - y \\ x + 2y \end{pmatrix}$$

To find $\mathbb{T}^\dagger$ we go through the same steps as in the preceding example:

$$\begin{aligned} \langle \mathrm{a}|\mathbb{T}|\mathrm{b}\rangle &= (x_1 \quad y_1)\mathbb{T}\begin{pmatrix} x_2 \\ y_2 \end{pmatrix} = (x_1 \quad y_1)\begin{pmatrix} 2x_2 - y_2 \\ x_2 + 2y_2 \end{pmatrix} \\ &= 2x_1x_2 - x_1y_2 + y_1x_2 + 2y_1y_2 = x_2(2x_1 + y_1) + y_2(-x_1 + 2y_1) \\ &= (x_2 \quad y_2)\begin{pmatrix} 2x_1 + y_1 \\ -x_1 + 2y_1 \end{pmatrix} \equiv (x_2 \quad y_2)\mathbb{T}^\dagger\begin{pmatrix} x_1 \\ y_1 \end{pmatrix} \end{aligned}$$

Therefore, in this case

$$\mathbb{T}^\dagger\begin{pmatrix} x \\ y \end{pmatrix} = \begin{pmatrix} 2x + y \\ -x + 2y \end{pmatrix}$$

which clearly differs from $\mathbb{T}$. •

Example 2.3.15

Let us now consider a slightly more complicated operator, $\mathbb{T}:\mathbb{C}^3 \to \mathbb{C}^3$, given by

$$\mathbb{T}\begin{pmatrix} \alpha_1 \\ \alpha_2 \\ \alpha_3 \end{pmatrix} = \begin{pmatrix} \alpha_1 - i\alpha_2 + \alpha_3 \\ i\alpha_1 - \alpha_3 \\ \alpha_1 - \alpha_2 + i\alpha_3 \end{pmatrix}$$

Let
$$|\mathrm{a}\rangle = \begin{pmatrix} \alpha_1 \\ \alpha_2 \\ \alpha_3 \end{pmatrix} \quad \text{and} \quad |\mathrm{b}\rangle = \begin{pmatrix} \beta_1 \\ \beta_2 \\ \beta_3 \end{pmatrix}$$

with dual vectors $\langle \mathrm{a}| = (\alpha_1^* \quad \alpha_2^* \quad \alpha_3^*)$ and $\langle \mathrm{b}| = (\beta_1^* \quad \beta_2^* \quad \beta_3^*)$, respectively.

We use Eq. (2.22) to find $\mathbb{T}^\dagger$:

$$\begin{aligned} \langle \mathrm{a}|\mathbb{T}|\mathrm{b}\rangle^* &= \left[(\alpha_1^* \quad \alpha_2^* \quad \alpha_3^*)\mathbb{T}\begin{pmatrix} \beta_1 \\ \beta_2 \\ \beta_3 \end{pmatrix}\right]^* = \left[(\alpha_1^* \quad \alpha_2^* \quad \alpha_3^*)\begin{pmatrix} \beta_1 - i\beta_2 + \beta_3 \\ i\beta_1 - \beta_3 \\ \beta_1 - \beta_2 + i\beta_3 \end{pmatrix}\right]^* \\ &= [\alpha_1^*\beta_1 - i\alpha_1^*\beta_2 + \alpha_1^*\beta_3 + i\alpha_2^*\beta_1 - \alpha_2^*\beta_3 + \alpha_3^*\beta_1 - \alpha_3^*\beta_2 + i\alpha_3^*\beta_3]^* \\ &= \alpha_1\beta_1^* + i\alpha_1\beta_2^* + \alpha_1\beta_3^* - i\alpha_2\beta_1^* - \alpha_2\beta_3^* + \alpha_3\beta_1^* - \alpha_3\beta_2^* - i\alpha_3\beta_3^* \\ &= \beta_1^*(\alpha_1 - i\alpha_2 + \alpha_3) + \beta_2^*(i\alpha_1 - \alpha_3) + \beta_3^*(\alpha_1 - \alpha_2 - i\alpha_3) \\ &= (\beta_1^* \quad \beta_2^* \quad \beta_3^*)\begin{pmatrix} \alpha_1 - i\alpha_2 + \alpha_3 \\ i\alpha_1 - \alpha_3 \\ \alpha_1 - \alpha_2 - i\alpha_3 \end{pmatrix} \underset{\substack{\uparrow \\ \text{by Eq. (2.22)}}}{=} \langle \mathrm{b}|\mathbb{T}^\dagger|\mathrm{a}\rangle \end{aligned}$$

Therefore, we obtain

$$\mathbb{T}^{\dagger}\begin{pmatrix}\alpha_1\\ \alpha_2\\ \alpha_3\end{pmatrix} = \begin{pmatrix}\alpha_1 - i\alpha_2 + \alpha_3\\ i\alpha_1 - \alpha_3\\ \alpha_1 - \alpha_2 - i\alpha_3\end{pmatrix}$$

Note that $\mathbb{T}^{\dagger} \neq \mathbb{T}$. •

Hermitian and unitary operators. The process of conjugation for linear operators looks very much like complex conjugation of complex numbers. This is clear from Eq. (2.22) and also from (iii) and (iv) of Theorem 2.3.19. It is, therefore, natural to look for real operators. The following definition is in complete analogy with the fact that the real numbers, as a subset of the complex numbers, have the property that $z^* = z$.

2.3.20 Definition A linear operator $\mathbb{H} \in \mathscr{L}(\mathscr{V})$ is called *hermitian* if $\mathbb{H}^{\dagger} = \mathbb{H}$. Similarly, $\mathbb{A} \in \mathscr{L}(\mathscr{V})$ is called *antihermitian* if $\mathbb{A}^{\dagger} = -\mathbb{A}$.

Hermitian operators correspond to the reals if the correspondence rule is $\dagger \leftrightarrow *$. What do antihermitian operators correspond to? Example 2.3.17 and Exercise 2.3.10 give the answer.

Example 2.3.16

The *expectation value* of an operator $\mathbb{A}$ in (state) $|a\rangle$ is defined by

$$\langle \mathbb{A} \rangle \equiv \langle a|\mathbb{A}|a\rangle$$

In general, the complex conjugate of the expectation value is

$$\langle \mathbb{A} \rangle^* = \langle a|\mathbb{A}|a\rangle^* = \langle a|\mathbb{A}^{\dagger}|a\rangle$$

In particular, if $\mathbb{H}$ is a hermitian operator, then

$$\langle \mathbb{H} \rangle^* = \langle a|\mathbb{H}^{\dagger}|a\rangle = \langle a|\mathbb{H}|a\rangle = \langle \mathbb{H} \rangle$$

which implies that $\langle \mathbb{H} \rangle$ *is real.* •

Example 2.3.17

Any antihermitian operator can be written as

$$\mathbb{A} = i\mathbb{H}$$

where $\mathbb{H}$ is some hermitian operator. In fact, taking the adjoint of both sides and using Theorem 2.3.19, we obtain

$$\mathbb{A}^{\dagger} = (i\mathbb{H})^{\dagger} = (i)^* \mathbb{H}^{\dagger} = -i\mathbb{H} = -\mathbb{A}$$

This relation says that if $\mathbb{A}$ is antihermitian, then $i\mathbb{A}$ is hermitian. •

Example 2.3.18

If $\mathbb{A}$ and $\mathbb{B}$ are both hermitian, then $\mathbb{AB}$ is *not*, in general, hermitian. This is because

$$(\mathbb{AB})^\dagger = \mathbb{B}^\dagger\mathbb{A}^\dagger = \mathbb{BA} \neq \mathbb{AB}$$

However, *if* $\mathbb{A}$ *and* $\mathbb{B}$ *commute*, then their hermiticity implies that their product, $\mathbb{AB}$, is hermitian. ●

In Chapter 1 we saw that rigid rotations preserve the scalar product, and we discovered that such operations were represented by orthogonal matrices. What sort of generalization of those results, if any, is possible for arbitrary vector spaces?

Let $|a\rangle, |b\rangle \in \mathscr{V}$, and let $\mathbb{U} \in \mathscr{L}(\mathscr{V})$ be an operator that preserves the scalar product; that is, given $|b'\rangle \equiv \mathbb{U}|b\rangle$ and $|a'\rangle \equiv \mathbb{U}|a\rangle$, then $\langle a'|b'\rangle = \langle a|b\rangle$. Using Eq. (2.21) for $|a'\rangle$, we obtain

$$\langle a'|b'\rangle = (\langle a|\mathbb{U}^\dagger)(\mathbb{U}|b\rangle) = \langle a|\mathbb{U}^\dagger\mathbb{U}|b\rangle = \langle a|b\rangle$$

Since this is true for arbitrary $|a\rangle$ and $|b\rangle$, we obtain

$$\mathbb{U}^\dagger\mathbb{U} = \mathbb{1}$$

Employing the result of Problem 2.23, we conclude that *if* dim $\mathscr{V}$ *is finite*, $\mathbb{U}$ is invertible, and, therefore, its *unique* inverse is $\mathbb{U}^\dagger$. We have just shown the following.

2.3.21 Theorem The necessary and sufficient condition for an operator $\mathbb{U} \in \mathscr{L}(\mathscr{V})$, where dim $\mathscr{V}$ is finite, to preserve the inner product is

$$\mathbb{U}^\dagger = \mathbb{U}^{-1} \tag{2.23}$$

■

Linear operators for which Eq. (2.23) is satisfied are called *unitary transformations*, or, more generally, *isometries.*

Example 2.3.19

Consider the linear transformation $\mathbb{T}: \mathbb{C}^3 \to \mathbb{C}^3$, given by

$$\mathbb{T}\begin{pmatrix}\alpha_1\\ \alpha_2\\ \alpha_3\end{pmatrix} = \begin{pmatrix}(\alpha_1 - i\alpha_2)/\sqrt{2}\\ (\alpha_1 + i\alpha_2 - 2\alpha_3)/\sqrt{6}\\ [\alpha_1 - \alpha_2 + \alpha_3 + i(\alpha_1 + \alpha_2 + \alpha_3)]/\sqrt{6}\end{pmatrix}$$

Let us show that $\mathbb{T}$ is unitary.

Let

$$|a\rangle = \begin{pmatrix}\alpha_1\\ \alpha_2\\ \alpha_3\end{pmatrix} \quad \text{and} \quad |b\rangle = \begin{pmatrix}\beta_1\\ \beta_2\\ \beta_3\end{pmatrix}$$

with duals $\langle \mathrm{a}| = (\alpha_1^* \quad \alpha_2^* \quad \alpha_3^*)$ and $\langle \mathrm{b}| = (\beta_1^* \quad \beta_2^* \quad \beta_3^*)$. Using Eq. (2.22), we get

$$\langle \mathrm{a}|\mathbb{T}|\mathrm{b}\rangle^* = \left[(\alpha_1^* \quad \alpha_2^* \quad \alpha_3^*)\mathbb{T}\begin{pmatrix}\beta_1\\ \beta_2\\ \beta_3\end{pmatrix}\right]^*$$

$$= \left[(\alpha_1^* \quad \alpha_2^* \quad \alpha_3^*)\begin{pmatrix}(\beta_1 - i\beta_2)/\sqrt{2}\\ (\beta_1 + i\beta_2 - 2\beta_3)/\sqrt{6}\\ [\beta_1 - \beta_2 + \beta_3 + i(\beta_1 + \beta_2 + \beta_3)]/\sqrt{6}\end{pmatrix}\right]^*$$

$$= \left[\frac{\alpha_1^*\beta_1 - i\alpha_1^*\beta_2}{\sqrt{2}} + \frac{\alpha_2^*\beta_1 + i\alpha_2^*\beta_2 - 2\alpha_2^*\beta_3}{\sqrt{6}} + \frac{\alpha_3^*(\beta_1 - \beta_2 + \beta_3) + i\alpha_3^*(\beta_1 + \beta_2 + \beta_3)}{\sqrt{6}}\right]^*$$

$$= \beta_1^*\left(\frac{\alpha_1}{\sqrt{2}} + \frac{\alpha_2}{\sqrt{6}} + \frac{\alpha_3(1-i)}{\sqrt{6}}\right) + \beta_2^*\left(\frac{i\alpha_1}{\sqrt{2}} - \frac{i\alpha_2}{\sqrt{6}} - \frac{\alpha_3(1+i)}{\sqrt{6}}\right) + \beta_3^*\left(-\frac{2\alpha_2}{\sqrt{6}} + \frac{\alpha_3(1-i)}{\sqrt{6}}\right)$$

$$= (\beta_1^* \quad \beta_2^* \quad \beta_3^*)\begin{pmatrix}\dfrac{\alpha_1}{\sqrt{2}} + \dfrac{\alpha_2}{\sqrt{6}} + \dfrac{\alpha_3(1-i)}{\sqrt{6}}\\ \dfrac{i\alpha_1}{\sqrt{2}} - \dfrac{i\alpha_2}{\sqrt{6}} - \dfrac{\alpha_3(1+i)}{\sqrt{6}}\\ -\dfrac{2\alpha_2}{\sqrt{6}} + \dfrac{\alpha_3(1-i)}{\sqrt{6}}\end{pmatrix} = \langle \mathrm{b}|\mathbb{T}^\dagger|\mathrm{a}\rangle$$

Therefore,

$$\mathbb{T}^\dagger\begin{pmatrix}\alpha_1\\ \alpha_2\\ \alpha_3\end{pmatrix} = \begin{pmatrix}\dfrac{\alpha_1}{\sqrt{2}} + \dfrac{\alpha_2}{\sqrt{6}} + \dfrac{\alpha_3(1-i)}{\sqrt{6}}\\ \dfrac{i\alpha_1}{\sqrt{2}} - \dfrac{i\alpha_2}{\sqrt{6}} - \dfrac{\alpha_3(1+i)}{\sqrt{6}}\\ -\dfrac{2\alpha_2}{\sqrt{6}} + \dfrac{\alpha_3(1-i)}{\sqrt{6}}\end{pmatrix}$$

and we can verify that

$$\mathbb{T}\mathbb{T}^\dagger\begin{pmatrix}\alpha_1\\ \alpha_2\\ \alpha_3\end{pmatrix} = \begin{pmatrix}\alpha_1\\ \alpha_2\\ \alpha_3\end{pmatrix}$$

Thus $\mathbb{T}\mathbb{T}^\dagger = \mathbb{1}$. Similarly, we can show that $\mathbb{T}^\dagger\mathbb{T} = \mathbb{1}$ and, therefore, that $\mathbb{T}$ is unitary. ●

Projection operators. We have already considered subspaces briefly. The significance of subspaces is that physics frequently takes place not in all of the vector space on hand but in one of its subspaces. For instance, although motion, in general, takes place in a three-dimensional space, it may restrict itself to a plane either because constraints are imposed or because of the nature of the force responsible for the motion. An example is planetary motion.

It is, therefore, appropriate to ask how we can go from a full space to one of its subspaces in the context of linear operators. Let us first consider a simple example. A point in the plane is designated by the coordinates (x, y). A subspace of the plane is the x-axis. Is there an operator, $\mathbb{P}$, that acts on such a point and somehow sends it into that subspace? Of course, there are many operators from $\mathbb{R}^2$ to $\mathbb{R}$. However, we are looking for a specific one. We want $\mathbb{P}$ to *project* the point onto the x-axis. Such an operator has to act on (x, y) and produce $(x, 0)$:

$$\mathbb{P}(x, y) = (x, 0)$$

This operator is called a *projection operator*. Note that if the point already lies on the x-axis, $\mathbb{P}$ does not change it; that is, $\mathbb{P}$ acts as the unit operator on it. In particular, if we apply $\mathbb{P}$ twice, we get the same result as if we apply it once. And this is true for any point in the plane. Therefore, for a projection operator we have

$$\mathbb{P}^2 = \mathbb{P}$$

We can generalize the above discussion in a definition.

2.3.22 Definition A hermitian operator, $\mathbb{P} \in \mathscr{L}(\mathscr{V})$, is called a projection operator if

$$\mathbb{P}^2 = \mathbb{P} \tag{2.24}$$

From this definition it immediately follows that the only projection operator with an inverse is the identity operator. Multiplying both sides of (2.24) by $\mathbb{P}^{-1}$ gives

$$\mathbb{P}^2\mathbb{P}^{-1} = \mathbb{P}\mathbb{P}^{-1} \Rightarrow \mathbb{P}(\mathbb{P}\mathbb{P}^{-1}) = \mathbb{P}\mathbb{P}^{-1}$$

or

$$\mathbb{P}\mathbb{1} = \mathbb{1} \Rightarrow \mathbb{P} = \mathbb{1}$$

We assume that projection operators are not invertible, so they are different from $\mathbb{1}$. Consider two projection operators, $\mathbb{P}_1$ and $\mathbb{P}_2$. We want to investigate the possibility of $\mathbb{P}_1 + \mathbb{P}_2$ being a projection operator. By definition,

$$(\mathbb{P}_1 + \mathbb{P}_2)^2 = \mathbb{P}_1^2 + \mathbb{P}_1\mathbb{P}_2 + \mathbb{P}_2\mathbb{P}_1 + \mathbb{P}_2^2 = \mathbb{P}_1 + \mathbb{P}_2 + \mathbb{P}_1\mathbb{P}_2 + \mathbb{P}_2\mathbb{P}_1$$

So $\mathbb{P}_1 + \mathbb{P}_2$ is a projection operator if and only if

$$\mathbb{P}_1\mathbb{P}_2 + \mathbb{P}_2\mathbb{P}_1 = 0 \tag{2.25}$$

Multiply on the left by $\mathbb{P}_1$ to get

$$\mathbb{P}_1^2\mathbb{P}_2 + \mathbb{P}_1\mathbb{P}_2\mathbb{P}_1 = \mathbb{0} \Rightarrow \mathbb{P}_1\mathbb{P}_2 + \mathbb{P}_1\mathbb{P}_2\mathbb{P}_1 = 0$$

and on the right by $\mathbb{P}_1$ to get

$$\mathbb{P}_1\mathbb{P}_2\mathbb{P}_1 + \mathbb{P}_2\mathbb{P}_1^2 = 0 \Rightarrow \mathbb{P}_1\mathbb{P}_2\mathbb{P}_1 + \mathbb{P}_2\mathbb{P}_1 = 0$$

These last two equations yield

$$\mathbb{P}_1\mathbb{P}_2 - \mathbb{P}_2\mathbb{P}_1 = 0 \tag{2.26}$$

The solution to (2.25) and (2.26) is

$$\mathbb{P}_1\mathbb{P}_2 = \mathbb{P}_2\mathbb{P}_1 = 0$$

We, therefore, have the following proposition.

2.3.23 Proposition Let $\mathbb{P}_1, \mathbb{P}_2 \in \mathcal{L}(\mathcal{V})$ be projection operators. Then $\mathbb{P}_1 + \mathbb{P}_2$ is a projection operator if and only if $\mathbb{P}_1\mathbb{P}_2 = \mathbb{P}_2\mathbb{P}_1 = 0$. Operators satisfying this condition are called *orthogonal operators.* ■

More generally, if there is a set $\{\mathbb{P}_i\}_{i=1}^N$ of projection operators satisfying

$$\mathbb{P}_i\mathbb{P}_j = \begin{cases} \mathbb{P}_i & \text{if } i = j \\ \mathbb{0} & \text{if } i \neq j \end{cases}$$

then $\mathbb{P} = \sum_{i=1}^N \mathbb{P}_i$ is also a projection operator.

A prototypical projection operator is given in terms of a normal vector, $|e_1\rangle$, as follows:

$$\mathbb{P} = |e_1\rangle\langle e_1|$$

Clearly, $\mathbb{P}$ is hermitian

$$\mathbb{P}^\dagger = (|e_1\rangle\langle e_1|)^\dagger = |e_1\rangle\langle e_1|$$

and

$$\mathbb{P}^2 = (|e_1\rangle\langle e_1|)(|e_1\rangle\langle e_1|) = |e_1\rangle\underbrace{\langle e_1|e_1\rangle}_{=1}\langle e_1|$$
$$= |e_1\rangle\langle e_1| = \mathbb{P}$$

In fact, we can take an orthonormal basis, $B = \{|e_i\rangle\}_{i=1}^N$, and construct a set of projection operators, $\{\mathbb{P}_i = |e_i\rangle\langle e_i|\}_{i=1}^N$. The operators $\mathbb{P}_i$ are mutually orthogonal. Thus, their sum, $\sum_{i=1}^N |e_i\rangle\langle e_i|$, is also a projection operator, one that is so special that it is introduced in a proposition.

2.3.24 Proposition Let $B = \{|e_i\rangle\}_{i=1}^N$ be an orthonormal basis for $\mathcal{V}_N$. Then the set $\{\mathbb{P}_i = |e_i\rangle\langle e_i|\}_{i=1}^N$ consists of mutually orthogonal projection operators and

$$\sum_{i=1}^N \mathbb{P}_i = \sum_{i=1}^N |e_i\rangle\langle e_i| = \mathbb{1}$$

This relation is called the *completeness relation.*

Proof. The mutual orthogonality of the $\mathbb{P}_i$ is an immediate consequence of the orthonormality of $|e_i\rangle$. To prove the proposition we show that both sides of the equality give the same vector when applied to an arbitrary vector. Consider an arbitrary vector $|a\rangle$, written in terms of $|e_i\rangle$:

$$|a\rangle = \sum_{j=1}^{N} \alpha_j |e_j\rangle$$

Apply $\mathbb{P}_i$ to $|a\rangle$ to obtain

$$\mathbb{P}_i|a\rangle = \mathbb{P}_i\left(\sum_{j=1}^{N} \alpha_j|e_j\rangle\right) = \sum_{j=1}^{N} \alpha_j \mathbb{P}_i|e_j\rangle$$

$$= \sum_{j=1}^{N} \alpha_j|e_i\rangle \underbrace{\langle e_i|e_j\rangle}_{\delta_{ij}} = \alpha_i|e_i\rangle$$

Therefore, we have

$$|a\rangle = \sum_{i=1}^{N} \alpha_i|e_i\rangle = \sum_{i=1}^{N} \mathbb{P}_i|a\rangle = \left(\sum_{i=1}^{N} \mathbb{P}_i\right)|a\rangle$$

for an arbitrary $|a\rangle$. ■

If we choose only the first $m < N$ vectors instead of the entire basis, then the projection operator $\mathbb{P}_m \equiv \sum_{i=1}^{m} |e_i\rangle\langle e_i|$ projects arbitrary vectors into the subspace spanned by the m vectors, $\{|e_i\rangle\}_{i=1}^{m}$. In other words, when $\mathbb{P}_m$ acts on any vector $|a\rangle \in \mathcal{V}$, the result will be a linear combination of only the first m vectors. The easy proof of this is left as an exercise.

These points are illustrated in the following example.

Example 2.3.20

Consider three orthonormal vectors, $\{|e_i\rangle\}_{i=1}^{3} \in \mathbb{R}^3$, given by

$$|e_1\rangle = \frac{1}{\sqrt{2}}\begin{pmatrix}1\\1\\0\end{pmatrix} \qquad |e_2\rangle = \frac{1}{\sqrt{6}}\begin{pmatrix}1\\-1\\2\end{pmatrix} \qquad |e_3\rangle = \frac{1}{\sqrt{3}}\begin{pmatrix}-1\\1\\1\end{pmatrix}$$

The projection operators associated with each of these can be obtained by noting that $\langle e_i|$ is a row vector. Therefore,

$$\mathbb{P}_1 = |e_1\rangle\langle e_1| = \frac{1}{\sqrt{2}}\begin{pmatrix}1\\1\\0\end{pmatrix}\frac{1}{\sqrt{2}}(1 \quad 1 \quad 0) = \frac{1}{2}\underbrace{\begin{pmatrix}1\\1\\0\end{pmatrix}}_{\text{a } 3\times 1 \text{ matrix}}\underbrace{(1 \quad 1 \quad 0)}_{\text{a } 1\times 3 \text{ matrix}} = \frac{1}{2}\begin{pmatrix}1 & 1 & 0\\1 & 1 & 0\\0 & 0 & 0\end{pmatrix}$$

Similarly, $$\mathbb{P}_2 = \frac{1}{\sqrt{6}}\begin{pmatrix}1\\-1\\2\end{pmatrix}\frac{1}{\sqrt{6}}(1 \quad -1 \quad 2) = \frac{1}{6}\begin{pmatrix}1 & -1 & 2\\-1 & 1 & -2\\2 & -2 & 4\end{pmatrix}$$

and
$$\mathbb{P}_3 = \frac{1}{3}\begin{pmatrix} 1 & -1 & -1 \\ -1 & 1 & 1 \\ -1 & 1 & 1 \end{pmatrix}$$

Note that $\mathbb{P}_i$ projects onto the line along $|e_i\rangle$. This can be tested by letting $\mathbb{P}_i$ act on an arbitrary vector and showing that the resulting vector is perpendicular to the other two vectors. For example, let $\mathbb{P}_2$ act on an arbitrary column vector, (x, y, z):

$$|a\rangle \equiv \mathbb{P}_2\begin{pmatrix} x \\ y \\ z \end{pmatrix} = \frac{1}{6}\begin{pmatrix} 1 & -1 & 2 \\ -1 & 1 & -2 \\ 2 & -2 & 4 \end{pmatrix}\begin{pmatrix} x \\ y \\ z \end{pmatrix} = \frac{1}{6}\begin{pmatrix} x - y + 2z \\ -x + y - 2z \\ 2x - 2y + 4z \end{pmatrix}$$

We verify that $|a\rangle$ is perpendicular to both $|e_1\rangle$ and $|e_3\rangle$:

$$\langle e_1|a\rangle = \frac{1}{\sqrt{2}}(1 \quad 1 \quad 0)\frac{1}{6}\begin{pmatrix} x - y + 2z \\ -x + y - 2z \\ 2x - 2y + 4z \end{pmatrix} = 0$$

Similarly,
$$\langle e_3|a\rangle = \frac{1}{6\sqrt{3}}(-1 \quad 1 \quad 1)\begin{pmatrix} x - y + 2z \\ -x + y - 2z \\ 2x - 2y + 4z \end{pmatrix} = 0$$

So indeed $|a\rangle$ is along $|e_3\rangle$.

We can find the operator that projects onto the plane formed by $|e_1\rangle$ and $|e_2\rangle$. This is

$$\mathbb{P}_1 + \mathbb{P}_2 = \frac{1}{3}\begin{pmatrix} 2 & 1 & 1 \\ 1 & 2 & -1 \\ 1 & -1 & 2 \end{pmatrix}$$

When this operator acts on the column vector (x, y, z), it produces a vector lying in the plane of $|e_1\rangle$ and $|e_2\rangle$, or perpendicular to $|e_3\rangle$:

$$|b\rangle \equiv (\mathbb{P}_1 + \mathbb{P}_2)\begin{pmatrix} x \\ y \\ z \end{pmatrix} = \frac{1}{3}\begin{pmatrix} 2 & 1 & 1 \\ 1 & 2 & -1 \\ 1 & -1 & 2 \end{pmatrix}\begin{pmatrix} x \\ y \\ z \end{pmatrix} = \frac{1}{3}\begin{pmatrix} 2x + y + z \\ x + 2y - z \\ x - y + 2z \end{pmatrix}$$

and
$$\langle e_3|b\rangle = \frac{1}{3\sqrt{3}}(-1 \quad 1 \quad 1)\begin{pmatrix} 2x + y + z \\ x + 2y - z \\ x - y + 2z \end{pmatrix} = 0$$

The operators that project onto planes formed by $|e_1\rangle$ and $|e_3\rangle$ and by $|e_2\rangle$ and $|e_3\rangle$ are obtained similarly.

Finally, we verify easily that

$$\mathbb{P}_1 + \mathbb{P}_2 + \mathbb{P}_3 = \begin{pmatrix} 1 & 0 & 0 \\ 0 & 1 & 0 \\ 0 & 0 & 1 \end{pmatrix} = \mathbb{1}_3$$ •

Exercises

2.3.1 Prove Theorem 2.3.2.

2.3.2 Let $\mathscr{V}$ be the vector space of all 2×2 matrices and let $\mathscr{W} = \mathbb{R}$. Show that the mapping $F: \mathscr{V} \to \mathbb{R}$ given by $F(\mathsf{A}) = \det \mathsf{A}$, where A is any 2×2 matrix, is *not* linear.

2.3.3 Use mathematical induction to show that $[\mathbb{A}, \mathbb{A}^m] = 0$.

2.3.4 For $\mathbb{D}$ and $\mathbb{T}$ as defined in Example 2.3.2, show that $[\mathbb{D}, \mathbb{T}] = \mathbb{1}$.

2.3.5 Show that if $[\mathbb{A}, \mathbb{B}]$ commutes with $\mathbb{A}$, then, for every positive integer k,

$$[\mathbb{A}^k, \mathbb{B}] = k\mathbb{A}^{k-1}[\mathbb{A}, \mathbb{B}]$$

Use mathematical induction.

2.3.6 Show that for $\mathbb{D}$ and $\mathbb{T}$ as defined in Example 2.3.2,

$$[\mathbb{D}^k, \mathbb{T}] = k\mathbb{D}^{k-1} \qquad \text{and} \qquad [\mathbb{T}^k, \mathbb{D}] = -k\mathbb{T}^{k-1}$$

2.3.7 Evaluate the derivative of $\mathbb{H}^{-1}(t)$.

2.3.8 Show that for any $\alpha, \beta \in \mathbb{R}$ and any $\mathbb{H} \in \mathscr{L}(\mathscr{V})$, we have

$$e^{\alpha\mathbb{H}} e^{\beta\mathbb{H}} = e^{(\alpha+\beta)\mathbb{H}}$$

2.3.9 Find the solution to the following operator differential equation:

$$\frac{d\mathbb{U}}{dt} = t\mathbb{H}\mathbb{U}(t)$$

2.3.10 Show that any operator can be written as the sum of a hermitian and an antihermitian operator.

2.3.11 Let $|\mathrm{f}\rangle, |\mathrm{g}\rangle \in \mathrm{C}^{(n)}(a, b)$ with the additional property that $f(a) = g(a) = f(b) = g(b) = 0$. Show that for such functions the derivative operator, $\mathbb{D}$, is antihermitian. The inner product is defined as usual:

$$\langle \mathrm{f}|\mathrm{g}\rangle \equiv \int_a^b f^*(t)g(t)\,dt$$

2.3.12 Show that $\mathbb{U} = e^{\mathbb{A}}$ is unitary if and only if $\mathbb{A}$ is antihermitian.[8]

2.3.13 Using the notation of Example 2.2.1, find the projection operator $\mathbb{P}_{\mathrm{a}}$ associated with the vector

$$|\mathrm{a}\rangle = \frac{1}{\sqrt{2}}\begin{pmatrix} 0 \\ 1 \\ -1 \\ 0 \end{pmatrix}$$

Also, show that $\mathbb{P}_{\mathrm{a}}$ does project an arbitrary vector in $\mathbb{C}^4$ along $|\mathrm{a}\rangle$.

[8] A more rigorous proof than that given in the solution to this exercise requires the machinery of the spectral decomposition theorem given in Chapter 3.

2.3.14 Let $\mathbb{P}_m = \sum_{i=1}^{m} |e_i\rangle\langle e_i|$ be a projection operator constructed out of the first m orthonormal vectors of the basis $\mathbf{B} = \{|e_i\rangle\}_{i=1}^{N}$ of $\mathcal{V}$. Show that $\mathbb{P}_m$ projects into the subspace spanned by the first m vectors in B.

2.3.15 What is the length of the projection of the vector $(3, 4, -4)$ onto the line $x = 2t + 1$, $y = -t + 3$, $z = t - 1$?

2.3.16 Let the operator $\mathbb{U}: \mathbb{C}^2 \to \mathbb{C}^2$ be given by

$$\mathbb{U}\begin{pmatrix}\alpha_1\\ \alpha_2\end{pmatrix} = \begin{pmatrix}\dfrac{i}{\sqrt{2}}\alpha_1 - \dfrac{i}{\sqrt{2}}\alpha_2\\ \dfrac{\alpha_1}{\sqrt{2}} + \dfrac{\alpha_2}{\sqrt{2}}\end{pmatrix}$$

Show that $\mathbb{U}$ is unitary. Find its inverse, $\mathbb{U}^\dagger$.

2.4 PERMUTATIONS

Some elementary notions concerning permutations will be required for the discussion of determinants and tensors in the next two chapters. Therefore, this chapter concludes with an elementary introduction to permutations.

In the context of group theory (the abstract study of symmetry), permutations are probably the oldest and best understood objects. They have a very rich group-theoretic structure and are used extensively in those areas of physics where group theory plays a dominant role.

Here we will study permutations only as much as is needed to discuss determinants and tensors. For this purpose we need only the mapping (not the group-theoretic) structure of permutations. Thus, we start with the following definition.

2.4.1 Definition Let $\mathrm{A}_n \equiv \{a_i\}_{i=1}^{n}$ denote any set of n objects. A *permutation* of A_n is a bijective map $\pi: \mathrm{A}_n \to \mathrm{A}_n$ on the set A_n.

Like all bijective maps, π has an inverse, π^{-1}, which is also a permutation.

Because there is a one-to-one correspondence between A^n and the set $\{1, 2, \ldots, n\}$, permutations are usually studied as actions on those integers. Thus, $\pi(i)$ is the image of i under the permutation π. Clearly, $\pi(i)$ is just another number among $\{1, 2, \ldots, n\}$.

It is customary to exhibit the mapping π by actual enumeration since it is a finite set. Thus, a generic permutation is shown as

$$\pi = \begin{pmatrix} 1 & 2 & 3 & \cdots & i & \cdots & n\\ \pi(1) & \pi(2) & \pi(3) & \cdots & \pi(i) & \cdots & \pi(n)\end{pmatrix} \tag{2.27}$$

That is, the element 1 goes to $\pi(1)$, 2 goes to $\pi(2)$, and so forth under the mapping π. Because the mapping is bijective, no two elements can have the same image, and $\pi(1), \pi(2), \ldots, \pi(n)$ exhaust all the elements in $\{1, 2, \ldots, n\}$. So, a permutation is merely a reshuffling of the elements $\{i\}_{i=1}^{n}$. For instance, the permutation

$$\pi = \begin{pmatrix} 1 & 2 & 3 & 4 \\ 3 & 4 & 1 & 2 \end{pmatrix}$$

takes the number 1 to the number 3, and we write $\pi(1) = 3$. Similarly, $\pi(2) = 4$, $\pi(3) = 1$, and $\pi(4) = 2$. It is easily verified that

$$\pi^{-1} = \begin{pmatrix} 1 & 2 & 3 & 4 \\ 3 & 4 & 1 & 2 \end{pmatrix} = \pi$$

The set of all permutations on n objects is denoted S_n. It is clear that if $\pi_1 \in S_n$ and $\pi_2 \in S_n$, then the composition (product) $\pi_2 \circ \pi_1$ is also in S_n. Its action on any integer is simply the succession of actions of π_1 and π_2:

$$\pi_2 \circ \pi_1(i) = \pi_2(\pi_1(i))$$

We can display this product using Eq. (2.27). For instance, if

$$\pi_1 = \begin{pmatrix} 1 & 2 & 3 & 4 \\ 3 & 4 & 1 & 2 \end{pmatrix} \quad \text{and} \quad \pi_2 = \begin{pmatrix} 1 & 2 & 3 & 4 \\ 2 & 4 & 3 & 1 \end{pmatrix}$$

then the product is obtained by following the appropriate paths:

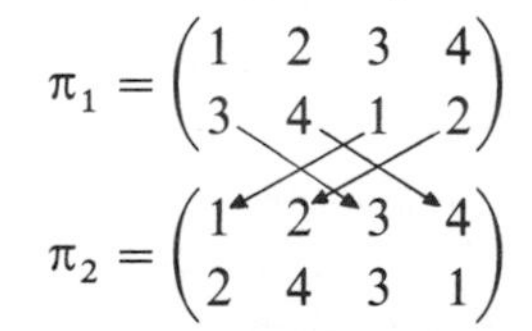

$$\pi_1 = \begin{pmatrix} 1 & 2 & 3 & 4 \\ 3 & 4 & 1 & 2 \end{pmatrix}$$
$$\pi_2 = \begin{pmatrix} 1 & 2 & 3 & 4 \\ 2 & 4 & 3 & 1 \end{pmatrix} \quad \Rightarrow \quad \pi_2 \circ \pi_1 = \begin{pmatrix} 1 & 2 & 3 & 4 \\ 3 & 1 & 2 & 4 \end{pmatrix}$$

Because we are dealing with finite numbers, repeated application of a permutation to any integer in the set $\{1, 2, \ldots, n\}$ eventually produces the initial integer. This leads us to the following definition.

2.4.2 Definition Let $\pi \in S_n$, $i \in \{1, 2, \ldots, n\}$, and let r be the smallest positive integer such that $\pi^r(i) = (i)$. Then the set of r distinct elements, $\{i, \pi(i), \pi^2(i), \ldots, \pi^{r-1}(i)\}$, is called an *orbit* of π of length r.

It is clear that any permutation can be broken up into *disjoint orbits*. Starting with 1 and applying π to it repeatedly until 1 is obtained again forms an orbit in which 1 (plus other elements) is contained. Then a second number that is not in this orbit is selected and π is applied to it repeatedly until it is obtained again. Continuing in this way produces a set of disjoint orbits that exhausts all elements of $\{1, 2, \ldots, n\}$. We thus have a proposition.

2.4.3 Proposition Any permutation can be broken up into disjoint orbits. ■

It is customary to write elements of each orbit in some specific order within parentheses starting with the first element, say i, on the left, then $\pi(i)$ immediately to its right, followed by $\pi^2(i)$, and so on. For example, the permutation π_1 discussed above has the orbit structure $\pi_1 = (13)(24)$, because $\pi_1(1) = 3$, $\pi_1^2(1) = \pi_1(3) = 1$, and $\pi_1(2) = 4$, $\pi_1^2(2) = \pi_1(4) = 2$. Similarly, $\pi_2 = (124)(3)$, and $\pi_2 \circ \pi_1 = (132)(4)$.

Example 2.4.1

Let $\pi_1, \pi_2 \in S_8$ be given as follows:

$$\pi_1 = \begin{pmatrix} 1 & 2 & 3 & 4 & 5 & 6 & 7 & 8 \\ 3 & 5 & 7 & 1 & 2 & 8 & 4 & 6 \end{pmatrix}$$

$$\pi_2 = \begin{pmatrix} 1 & 2 & 3 & 4 & 5 & 6 & 7 & 8 \\ 2 & 5 & 6 & 8 & 1 & 7 & 4 & 3 \end{pmatrix}$$

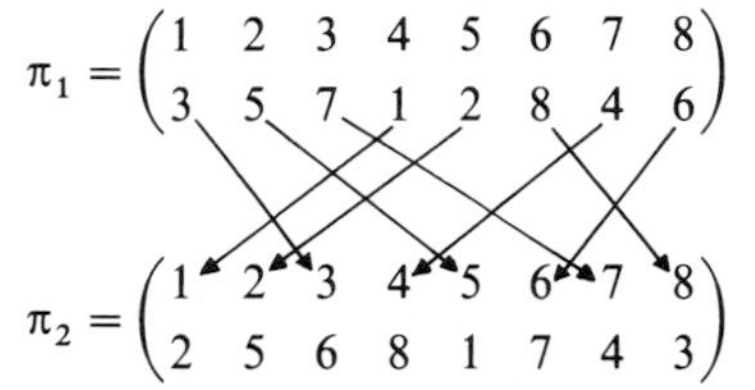

Then $\pi_2 \circ \pi_1$ (note the ordering!) can be obtained by keeping track of where each element goes when π_1 and π_2 are applied in succession, as indicated by the arrows. The product is

$$\pi_2 \circ \pi_1 = \begin{pmatrix} 1 & 2 & 3 & 4 & 5 & 6 & 7 & 8 \\ 6 & 1 & 4 & 2 & 5 & 3 & 8 & 7 \end{pmatrix}$$

Let us find the orbit structures of the two permutations and their product. We start with 1 within each permutation and follow the orbit as shown for π_1:

$$\pi_1 = \begin{pmatrix} 1 & 2 & 3 & 4 & 5 & 6 & 7 & 8 \\ 3 & 5 & 7 & 1 & 2 & 8 & 4 & 6 \end{pmatrix}$$

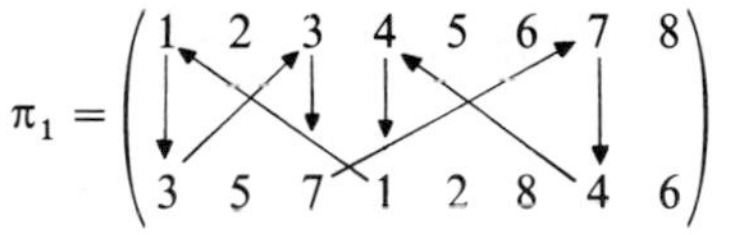

This leads to

$$\pi_1 = (1374)(25)(68)$$

Similarly,

$$\pi_2 = (125)(36748)$$

and

$$\pi_2 \circ \pi_1 = (16342)(5)(78)$$

Note that, in all cases, the orbits have no element in common. This follows from the way orbits are constructed.

In general, permutations do not commute. To see this, let us consider $\pi_1 \circ \pi_2$, which is easily seen to be

$$\pi_1 \circ \pi_2 = \begin{pmatrix} 1 & 2 & 3 & 4 & 5 & 6 & 7 & 8 \\ 5 & 2 & 8 & 6 & 3 & 4 & 1 & 7 \end{pmatrix} = (15387)(2)(46)$$

This product clearly differs from $\pi_2 \circ \pi_1$. However, note that $\pi_1 \circ \pi_2$ and $\pi_2 \circ \pi_1$ have the same orbit structure in that orbits of equal length appear in both. This is a general property of all permutations and follows from their group-theoretic structure.   ●

2.4.4 Definition   Let $\pi \in S_n$. If π has an orbit of length r and all other orbits of π have only one element, then π is called a *cyclic permutation* of length r.

Thus, $\pi_2 \in S_4$ as defined earlier is a cyclic permutation of length 3. Similarly,

$$\pi = \begin{pmatrix} 1 & 2 & 3 & 4 & 5 & 6 \\ 6 & 2 & 1 & 3 & 5 & 4 \end{pmatrix}$$

is a cyclic permutation of length 4 (verify this).

An important permutation is defined as follows.

2.4.5 Definition A cyclic permutation of length 2 is called a *transposition.*

A transposition (ij) merely switches the two integers i and j. The following example shows how products of cycles (orbits) act on $\{1, 2, \ldots, n\}$.

Example 2.4.2

Consider the product of cycles $P \in S_6$:

$$P \equiv (1435)(26)$$

Let us see what permutation is associated with this product. Starting with 1, we let the above product act on all the integers in $\{1, 2, \ldots, 6\}$ and record the results:

$$P(1) = 4 \qquad P(2) = 6 \qquad P(3) = 5 \qquad P(4) = 3 \qquad P(5) = 1 \qquad P(6) = 2$$

This is evident from the definition of an orbit and the fact that the image of the last member of a cycle is its first member. Therefore, the permutation associated with the product P is

$$\begin{pmatrix} 1 & 2 & 3 & 4 & 5 & 6 \\ 4 & 6 & 5 & 3 & 1 & 2 \end{pmatrix}$$

Products of cycles may sometimes contain a common element. In that case, ordering of the cycles is important. Of course, in a single permutation the cycles are disjoint. However, in products of permutations in which the permutations are written as products of cycles, there may be several cycles (one from each permutation) having common elements. For example, let $\pi_1 \in S_6$ be given as a product of cycles by

$$\pi_1 = (143)(24)(456)$$

We want to exhibit π_1 in the usual manner. We start with 1 and follow the action of the cycles on it, starting from the right. The first and second cycles leave 1 alone, and the last cycle takes 1 to 4. Thus, $\pi_1(1) = 4$. For 2 we note that the first cycle leaves it alone, the second cycle takes it to 4, and the last cycle takes 4 to 3. Thus, $\pi(2) = 3$. Similarly, $\pi_1(3) = 1$, $\pi_1(4) = 5$, $\pi_1(5) = 6$, $\pi_1(6) = 2$. Therefore,

$$\pi_1 = \begin{pmatrix} 1 & 2 & 3 & 4 & 5 & 6 \\ 4 & 3 & 1 & 5 & 6 & 2 \end{pmatrix}$$

We note that π_1 is a cyclic permutation of length 6. (Follow the orbit starting with 1!)

It is left to the reader to show that the permutation $\pi_2 \in S_5$ given by the product

$$\pi_2 = (13)(15)(12)(14)$$

is
$$\pi_2 = \begin{pmatrix} 1 & 2 & 3 & 4 & 5 \\ 4 & 5 & 1 & 2 & 3 \end{pmatrix}$$

whose cycle structure is
$$\pi_2 = (14253)$$

That is, the permutation is cyclic. •

A cycle of length 1, such as (i), can be written as the product of transpositions as follows
$$(i) = (ij)(ij) \equiv [(ij)]^2$$
where j is an arbitrary element of $\{1, 2, \ldots, n\}$. Obviously, this decomposition is not unique. In fact, the product of any transposition with itself is identity. Therefore, we can include such a product in any product of permutations without changing anything.

2.4.6 Proposition An orbit of length r, $(i_1 i_2 \cdots i_r)$, can be decomposed into the product of $r - 1$ transpositions as follows:
$$(i_1 i_2 \cdots i_r) = (i_1 i_r)(i_1 i_{r-1}) \cdots (i_1 i_3)(i_1 i_2)$$ ■

The proof of this proposition is left as a problem for the reader.

Although a decomposition such as that in Proposition 2.4.6 is not unique, it can be shown that if $r - 1$, the number of transpositions in the decomposition, is even (odd), then any other decomposition must also be even (odd). For instance, it is easy to verify that
$$\begin{aligned}(1234) &= (14)(13)(12) = (21)(24)(23) \\ &= (14)(13)(12)\underbrace{(34)(34)}_{1}\underbrace{(23)\underbrace{(12)(12)}_{1}(23)}_{1}\end{aligned}$$

That is, (1234) is written as a product of 3 or 9 transpositions, both of which are odd.

We have already seen that any permutation can be written as a product of cycles. In addition, Proposition 2.4.6 says that these cycles can be further broken down into products of transpositions. This implies the following (see Paley and Weichsel 1966).

2.4.7 Proposition Any permutation can be written as products of transpositions. Such a product is not unique. However, the evenness or oddness of the number of transpositions *is* unique. ■

2.4.8 Definition A permutation is even (odd) if it can be expressed as a product of an even (odd) number of transpositions.

The evenness or oddness of a permutation can be determined from its cycle structure. Assume that there are k cycles in π and that the jth cycle has length r_j. Then Proposition 2.4.6 tells us directly that the jth cycle can be expressed as the product of $r_j - 1$ transpositions. Summing all these numbers, we obtain

$$\text{number of transpositions in } \pi = \sum_{j=1}^{k} (r_j - 1) = \sum_{j=1}^{k} r_j - k$$

It should be clear that the product of two even (odd) permutations is always even. On the other hand, the product of an odd and an even permutation is always odd. It is, therefore, natural to assign the number $+1$ to even permutations and the number -1 to odd ones. We, therefore, define

$$\delta_\pi \equiv \begin{cases} +1 & \text{if } \pi \text{ is even} \\ -1 & \text{if } \pi \text{ is odd} \end{cases} \tag{2.28}$$

The quantity δ_π is sometimes called the *signature*, or the *sign*, of the permutation and denoted by $\operatorname{sgn}(\pi)$.

Another quantity of interest is the so-called Levi-Civita tensor, $\varepsilon_{i_1 i_2 \cdots i_n}$ (see Chapter 4 for details). This quantity has n indices and is defined by

$$\varepsilon_{\pi(1), \pi(2), \ldots, \pi(n)} = \delta_\pi$$

We note that $\varepsilon_{12\cdots n} = +1$ because in this case π is the identity permutation, which takes i to itself; that is, $\pi(i) = i$, for all i. Thus, all cycles of the identity permutation are of length 1, and since $(i) = [(ij)]^2$, there are $2n$ (or some other even number of) transpositions in its decomposition.

We will return to some of the results obtained here in Chapters 3 and 4.

Exercises

2.4.1 Find the inverses of the permutations

$$\pi_1 = \begin{pmatrix} 1 & 2 & 3 & 4 & 5 & 6 & 7 & 8 \\ 3 & 5 & 7 & 1 & 2 & 8 & 4 & 6 \end{pmatrix} \quad \text{and} \quad \pi_2 = \begin{pmatrix} 1 & 2 & 3 & 4 & 5 & 6 & 7 & 8 \\ 2 & 5 & 6 & 8 & 1 & 7 & 4 & 3 \end{pmatrix}$$

and show directly that $(\pi_1 \circ \pi_2)^{-1} = \pi_2^{-1} \circ \pi_1^{-1}$.

2.4.2 Express each of the following products in terms of disjoint cycles. Assume that all permutations are in S_7.

(a) (123) (347) (456) (145) **(b)** (34) (562) (273) **(c)** (1345) (134) (13)

2.4.3 Express the permutation

$$\pi = \begin{pmatrix} 1 & 2 & 3 & 4 & 5 & 6 & 7 & 8 \\ 2 & 4 & 1 & 3 & 6 & 8 & 7 & 5 \end{pmatrix}$$

as a product of transpositions. Is the permutation even or odd?

PROBLEMS

2.1 Let $\mathbb{R}^+$ denote the set of positive real numbers. Define the "sum" of two elements of $\mathbb{R}^+$ to be their product in the usual sense, and define scalar multiplication by elements of $\mathbb{R}$ as being given by $r \cdot p \equiv p^r$ where $r \in \mathbb{R}$ (the scalar) and $p \in \mathbb{R}^+$. With these operations, show that $\mathbb{R}^+$ is a vector space over $\mathbb{R}$.

2.2 Prove Theorem 2.1.3.

2.3 Show that the following n vectors form a basis in $\mathbb{C}^n$ (or $\mathbb{R}^n$).

$$|a_1\rangle = \begin{pmatrix} 1 \\ 1 \\ 1 \\ \vdots \\ 1 \\ 1 \end{pmatrix}, |a_2\rangle = \begin{pmatrix} 1 \\ 1 \\ 1 \\ \vdots \\ 1 \\ 0 \end{pmatrix}, |a_3\rangle = \begin{pmatrix} 1 \\ 1 \\ 1 \\ \vdots \\ 0 \\ 0 \end{pmatrix}, \ldots, |a_n\rangle = \begin{pmatrix} 1 \\ 0 \\ 0 \\ \vdots \\ 0 \\ 0 \end{pmatrix}$$

2.4 Prove Theorem 2.1.7.

2.5 Show that in Definition 0.4.1 in the Introduction, (iii) implies (ii).

2.6 Find a_0, a_1, a_2, b_1, b_2, and b_3 such that the polynomials $a_0, a_1x + a_2$, and $b_1x^2 + b_2x + b_3$ are mutually orthonormal in the interval $[0, 1]$. The inner product is defined as in Example 2.3.2.

2.7 Show that $\mathbb{T}: \mathbb{R}^2 \to \mathbb{R}^3$ given by $\mathbb{T}(x, y) = (x^2 + y^2, x + y, 2x - y)$ is not a linear mapping.

2.8 Show that all the transformations in Example 2.3.2 are linear.

2.9 Show that the following operators are linear.
(a) $\mathcal{V}$ is $\mathbb{C}$, over the reals, and $\mathbb{A}|z\rangle = z^*$
(b) $\mathcal{V}$ is $\mathcal{P}^c$ and $\mathbb{A}|x\rangle = |y\rangle$, where $y(t) = x(t + 1) - x(t)$

2.10 Consider the three linear operators $\mathbb{A}_1$, $\mathbb{A}_2$, and $\mathbb{A}_3$ such that $[\mathbb{A}_1, \mathbb{A}_2] = \mathbb{A}_3$, $[\mathbb{A}_3, \mathbb{A}_1] = \mathbb{A}_2$, and $[\mathbb{A}_2, \mathbb{A}_3] = \mathbb{A}_1$. Show that the operator $\mathbb{A}$, defined by $\mathbb{A} \equiv \mathbb{A}_1^2 + \mathbb{A}_2^2 + \mathbb{A}_3^2$, commutes with $\mathbb{A}_1, \mathbb{A}_2$, and $\mathbb{A}_3$.

2.11 Calculate the linear transformation $\mathbb{D}^n\mathbb{T}^n$ and $\mathbb{T}^n\mathbb{D}^n$ for $n = 1, 2, 3$ on polynomials as defined in Example 2.3.2.

2.12 Consider a linear operator $\mathbb{A}$ on a finite-dimensional vector space $\mathcal{V}$. Show that there exists a polynomial P such that $P(\mathbb{A}) = 0$. (Hint: Take an arbitrary vector $|x\rangle \in \mathcal{V}$ and consider the vectors $\mathbb{A}^k|x\rangle$ for various values of k.)

2.13 Prove the rest of the properties in Proposition 2.3.14.

2.14 Show that if $\mathbb{A}$ and $\mathbb{B}$ are hermitian, then $i[\mathbb{A}, \mathbb{B}]$ is also hermitian.

2.15 Show that

$$(\mathbb{U} + \mathbb{T})(\mathbb{U} - \mathbb{T}) = \mathbb{U}^2 - \mathbb{T}^2 \quad \text{iff} \quad [\mathbb{U}, \mathbb{T}] = 0$$

2.16 Show that

$$\frac{d}{dt}\mathbb{H}^3 = \left(\frac{d\mathbb{H}}{dt}\right)\mathbb{H}^2 + \mathbb{H}\left(\frac{d\mathbb{H}}{dt}\right)\mathbb{H} + \mathbb{H}^2\left(\frac{d\mathbb{H}}{dt}\right)$$

2.17 Show that if $[\mathbb{H}_1, \mathbb{H}_2] = \mathbb{T}$ and $[\mathbb{T}, \mathbb{H}_1] = 0 = [\mathbb{T}, \mathbb{H}_2]$, then
(a) $[\mathbb{H}_1, \mathbb{H}_2^n] = n\mathbb{T}\mathbb{H}_2^{n-1}$
(b) $[\mathbb{H}_1, e^{t\mathbb{H}_2}] = t\mathbb{T}e^{t\mathbb{H}_2}$

2.18 Prove that

$$e^{\mathbb{H}_1 + \mathbb{H}_2 + \mathbb{H}_3} = e^{\mathbb{H}_1}e^{\mathbb{H}_2}e^{\mathbb{H}_3}e^{-1/2([\mathbb{H}_1, \mathbb{H}_2] + [\mathbb{H}_1, \mathbb{H}_3] + [\mathbb{H}_2, \mathbb{H}_3])}$$

provided that $\mathbb{H}_1$, $\mathbb{H}_2$, and $\mathbb{H}_3$ commute with all the commutators. What is the generalization to $\mathbb{H}_1 + \mathbb{H}_2 + \cdots + \mathbb{H}_n$?

2.19 Prove that if $\mathbb{H}$ is independent of t, then

$$\frac{d}{dt}([\mathbb{H}, \mathbb{U}(t)]) = \left[\mathbb{H}, \frac{d\mathbb{U}}{dt}\right]$$

2.20 Show that if $\mathbb{A}(t) \equiv e^{t\mathbb{H}}\mathbb{A}_0 e^{-t\mathbb{H}}$, where $\mathbb{H}$ and $\mathbb{A}_0$ belong to $\mathscr{L}(\mathscr{V})$ and are independent of t, then

$$\frac{d\mathbb{A}(t)}{dt} = [\mathbb{H}, \mathbb{A}(t)]$$

What happens when $\mathbb{H}$ commutes with $\mathbb{A}(t)$?

2.21 Find $\mathbb{T}^\dagger$ for $\mathbb{T}: \mathbb{R}^3 \to \mathbb{R}^3$ given by

$$\mathbb{T}\begin{pmatrix} x \\ y \\ z \end{pmatrix} = \begin{pmatrix} x + 2y - z \\ 3x - y + 2z \\ -x + 2y + 3z \end{pmatrix}$$

2.22 Find $\mathbb{T}^\dagger$ for each of the following linear operators.
(a) $\mathbb{T}: \mathbb{R}^2 \to \mathbb{R}^2$ given by

$$\mathbb{T}\begin{pmatrix} x \\ y \end{pmatrix} = \begin{pmatrix} x\cos\theta - y\sin\theta \\ x\sin\theta + y\cos\theta \end{pmatrix}$$

where θ is a real number (what is $\mathbb{T}^\dagger\mathbb{T}$?)
(b) $\mathbb{T}: \mathbb{C}^2 \to \mathbb{C}^2$ given by

$$\mathbb{T}\begin{pmatrix} \alpha_1 \\ \alpha_2 \end{pmatrix} = \begin{pmatrix} \alpha_1 - i\alpha_2 \\ i\alpha_1 + \alpha_2 \end{pmatrix}$$

(c) $\mathbb{T}: \mathbb{C}^3 \to \mathbb{C}^3$ given by

$$\mathbb{T}\begin{pmatrix} \alpha_1 \\ \alpha_2 \\ \alpha_3 \end{pmatrix} = \begin{pmatrix} \alpha_1 + i\alpha_2 - 2i\alpha_3 \\ -2i\alpha_1 + \alpha_2 + i\alpha_3 \\ i\alpha_1 - 2i\alpha_2 + \alpha_3 \end{pmatrix}$$

2.23 Let $\mathbb{U}, \mathbb{T} \in \mathscr{L}(\mathscr{V})$, where $\mathscr{V}$ is finite-dimensional. Assume that $\mathbb{U}\mathbb{T} = \mathbb{1}$. Show that $\mathbb{U}$ and $\mathbb{T}$ are both invertible.

2.24 Show that if $\mathbb{P}$ is a projection operator, so is each of the following.
(a) $\mathbb{U}^\dagger\mathbb{P}\mathbb{U}$, for any unitary operator $\mathbb{U}$ **(b)** $\mathbb{1} - \mathbb{P}$

2.25 Show that the product of two unitary (orthogonal) operators is always unitary (orthogonal), but the product of two hermitian (symmetric) operators is hermitian (symmetric) if and only if the two operators commute.

2.26 Find the products $\pi_1 \circ \pi_2$ and $\pi_2 \circ \pi_1$ of the two permutations

$$\pi_1 = \begin{pmatrix} 1 & 2 & 3 & 4 & 5 & 6 \\ 3 & 4 & 6 & 5 & 1 & 2 \end{pmatrix} \quad \text{and} \quad \pi_2 = \begin{pmatrix} 1 & 2 & 3 & 4 & 5 & 6 \\ 2 & 1 & 3 & 6 & 5 & 4 \end{pmatrix}$$

2.27 Find the inverse of each of the following permutations.

(a) $\pi_1 = \begin{pmatrix} 1 & 2 & 3 & 4 \\ 3 & 2 & 4 & 1 \end{pmatrix}$ **(b)** $\pi_2 = \begin{pmatrix} 1 & 2 & 3 & 4 & 5 \\ 1 & 4 & 2 & 5 & 3 \end{pmatrix}$

(c) $\pi_3 = \begin{pmatrix} 1 & 2 & 3 & 4 & 5 & 6 \\ 6 & 5 & 4 & 3 & 2 & 1 \end{pmatrix}$ **(d)** $\pi_4 = \begin{pmatrix} 1 & 2 & 3 & 4 & 5 \\ 3 & 4 & 5 & 1 & 2 \end{pmatrix}$

2.28 Express the following permutations as products of disjoint cycles, and determine which are cyclic.

(a) $\begin{pmatrix} 1 & 2 & 3 & 4 & 5 & 6 \\ 1 & 3 & 4 & 5 & 6 & 2 \end{pmatrix}$ **(b)** $\begin{pmatrix} 1 & 2 & 3 & 4 & 5 & 6 \\ 2 & 1 & 4 & 5 & 6 & 3 \end{pmatrix}$ **(c)** $\begin{pmatrix} 1 & 2 & 3 & 4 & 5 \\ 1 & 3 & 5 & 4 & 2 \end{pmatrix}$

2.29 Express the following permutations as products of transpositions, and determine whether they are even or odd.

(a) $\begin{pmatrix} 1 & 2 & 3 & 4 & 5 \\ 3 & 4 & 2 & 1 & 5 \end{pmatrix}$ **(b)** $\begin{pmatrix} 1 & 2 & 3 & 4 & 5 & 6 & 7 & 8 \\ 4 & 1 & 7 & 8 & 3 & 6 & 5 & 2 \end{pmatrix}$

(c) $\begin{pmatrix} 1 & 2 & 3 & 4 & 5 & 6 \\ 6 & 4 & 5 & 3 & 2 & 1 \end{pmatrix}$ **(d)** $\begin{pmatrix} 1 & 2 & 3 & 4 & 5 & 6 & 7 \\ 6 & 7 & 2 & 4 & 1 & 5 & 3 \end{pmatrix}$

2.30 Show that the product of two even (odd) permutations is always even, and the product of an even and an odd permutation is always odd.

2.31 Show that π and π^{-1} have the same parity (both even or both odd).

3

Finite-Dimensional Vector Spaces II: Matrices and Spectral Decomposition

So far our theoretical investigations have been dealing mostly with abstract vectors and abstract operators. As we have seen in examples and exercises, concrete representations of vectors and operators are necessary in most applications. Such representations are obtained by choosing a basis and expressing all operations in terms of components of vectors and matrix representations of operators.

3.1 MATRICES

Let us choose a basis $\mathrm{B} = \{|\mathrm{a}_i\rangle\}_{i=1}^{N}$ of a vector space $\mathscr{V}_N$, and write an arbitrary vector $|\mathrm{x}\rangle$ in this basis:

$$|\mathrm{x}\rangle = \sum_{i=1}^{N} \eta_i |\mathrm{a}_i\rangle$$

where $\eta_1, \eta_2, \ldots, \eta_N$ are the components of $|\mathrm{x}\rangle$ in B. Now consider the vector

$$|\mathrm{y}\rangle = \mathbb{A}|\mathrm{x}\rangle$$

where $\mathbb{A} \in \mathscr{L}(\mathscr{V})$ is a linear operator. The vector $|\mathrm{y}\rangle$ can also be written as a linear combination of $|\mathrm{a}_i\rangle$:

$$|\mathrm{y}\rangle = \sum_{j=1}^{N} \rho_j |\mathrm{a}_j\rangle$$

It is clear that the ρ_j's depend, somehow, on $\mathbb{A}$. To see how ρ_j, $\mathbb{A}$, and η_i are related, we let $\mathbb{A}$ act on $|x\rangle$:

$$\mathbb{A}|x\rangle = \mathbb{A}\left(\sum_{i=1}^{N} \eta_i|a_i\rangle\right) \underset{\uparrow}{=} \sum_{i=1}^{N} \eta_i(\mathbb{A}|a_i\rangle)$$

by the linearity of $\mathbb{A}$

$$= \sum_{i=1}^{N} \eta_i\left(\sum_{j=1}^{N} \alpha_{ji}|a_j\rangle\right)$$

jth component of the vector $\mathbb{A}|a_i\rangle$

$$= \sum_{j=1}^{N}\left(\sum_{i=1}^{N} \eta_i\alpha_{ji}\right)|a_j\rangle$$

Since the LHS's of the two preceding equations are equal, we obtain

$$\sum_{j=1}^{N} \rho_j|a_j\rangle = \sum_{j=1}^{N}\left(\sum_{i=1}^{N} \eta_i\alpha_{ji}\right)|a_j\rangle$$

or

$$\sum_{j=1}^{N}\left[\rho_j - \sum_{i=1}^{N} \alpha_{ji}\eta_i\right]|a_j\rangle = 0$$

The linear independence of the $|a_j\rangle$'s gives

$$\rho_j = \sum_{i=1}^{N} \alpha_{ji}\eta_i, \qquad \text{for } j = 1, 2, \ldots, N \tag{3.1}$$

We can now think of α_{ji} as elements of a matrix A:

$$\mathsf{A} = \begin{pmatrix} \alpha_{11} & \alpha_{12} & \cdots & \alpha_{1N} \\ \alpha_{21} & \alpha_{22} & \cdots & \alpha_{2N} \\ \vdots & \vdots & & \vdots \\ \alpha_{N1} & \alpha_{N2} & \cdots & \alpha_{NN} \end{pmatrix}$$

This matrix is the representation of the operator $\mathbb{A}$ in the basis $B = \{|a_i\rangle\}_{i=1}^{N}$. If we also let **x** and **y** denote, respectively, the column vectors

$$\begin{pmatrix} \eta_1 \\ \eta_2 \\ \vdots \\ \eta_N \end{pmatrix} \quad \text{and} \quad \begin{pmatrix} \rho_1 \\ \rho_2 \\ \vdots \\ \rho_N \end{pmatrix}$$

then (3.1) can be written in matrix form as

$$\mathbf{y} = \mathsf{A}\mathbf{x}$$

This equation is the *representation* of the operator equation $|y\rangle = \mathbb{A}|x\rangle$ *in the basis* **B**.

It is trivial to show that if the ith components of $|x\rangle$ and $|y\rangle$ in **B** are, respectively, η_i and ρ_i, then the ith component of $|z\rangle = |x\rangle + |y\rangle$ is $\mu_i = \eta_i + \rho_i$.

Similarly, if α_{ij} and β_{ij} are, respectively, the ijth elements of the matrix representations of the operators $\mathbb{A}$ and $\mathbb{B}$ in B, then the ijth element of their sum, $\mathbb{S} = \mathbb{A} + \mathbb{B}$, in **B** will be $\sigma_{ij} = \alpha_{ij} + \beta_{ij}$.

The elements of the product of two operators can be found by letting such a product, for example, $\mathbb{C} = \mathbb{B}\mathbb{A}$ (note the order in this definition), act on an arbitrary vector $|x\rangle$ to give the vector $|z\rangle = \mathbb{C}|x\rangle = \mathbb{B}\mathbb{A}|x\rangle = \mathbb{B}(\mathbb{A}|x\rangle) = \mathbb{B}|y\rangle$, where $|y\rangle$ is defined as before. If ρ_i, μ_i, and β_{ij} are, respectively, the ith component of $|y\rangle$, the ith component of $|z\rangle$, and the ijth element of B in the basis B, then (3.1) gives

$$\mu_i = \sum_{i=1}^{N} \beta_{ij}\rho_j \qquad \text{for } i = 1, 2, \ldots, N$$

A second use of (3.1), relating $|y\rangle$, $\mathbb{A}$, and $|x\rangle$ in the preceding equation, results in

$$\mu_i = \sum_{j=1}^{N} \beta_{ij} \sum_{k=1}^{N} \alpha_{jk}\eta_k = \sum_{j=1}^{N}\sum_{k}^{N} \beta_{ij}\alpha_{jk}\eta_k .$$

$$\equiv \sum_{i,j=1}^{N} \beta_{ij}\alpha_{jk}\eta_k \equiv \sum_{i,j} \beta_{ij}\alpha_{jk}\eta_k \tag{3.2}$$

Defining $\gamma_{ik} \equiv \sum_{j=1}^{N} \beta_{ij}\alpha_{jk}$ and comparing with the definition of products of matrices in Chapter 1 shows that this is the product of the two matrices B (with elements β_{ij}) and A (with elements α_{ij}). If we define a third matrix C (with elements γ_{ij}), then the above equation can be written in matrix form as

$$\mathsf{C} = \mathsf{BA}$$

which says that the matrix representation of an *operator* product is the *matrix* product of the matrix representations of the two operators. So Eq. (3.2) can be written in matrix form as

$$\mathbf{z} = \mathsf{BA}\mathbf{x}$$

where $\mathbf{x}$ and $\mathbf{z}$ are the column vectors representing $|x\rangle$ and $|z\rangle$ in the basis **B**.

It is not hard to show that all operations performed on operators carry over to their matrix representations as well. For instance, the matrix representation of $\alpha\mathbb{A}$ has $\alpha\alpha_{ij}$ as its ijth element; $-\alpha_{ij}$ is the ijth element of the matrix associated with the operator $-\mathbb{A}$; and the zero operator has all zeros in its matrix representation. It is straightforward to show that the ijth element of the matrix representing the unit operator $\mathbb{1}$ in *any* basis is δ_{ij}, the Kronecker delta, defined by

$$\delta_{ij} = \begin{cases} 1 & \text{if } i = j \\ 0 & \text{if } i \neq j \end{cases}$$

The above discussion is summarized in the following proposition.

3.1.1 Proposition Let $\mathbb{A}, \mathbb{B} \in \mathcal{L}(\mathcal{V})$ be operators acting in an N-dimensional vector space $\mathcal{V}$. Then for any basis $\mathbf{B} = \{|a_i\rangle\}_{i=1}^{N}$ of $\mathcal{V}$, there exists a mapping, $\text{Mat}: \mathcal{L}(\mathcal{V}) \to \mathcal{M}_{\mathbf{B}}^{(N)}$, from the space of linear operators on $\mathcal{V}$ onto the space of

$N \times N$ matrices given by $(\text{Mat}(\mathbb{A}))_{ji} = a_{ji}$, where $\{a_{ji}\}_{j=1}^N$ are the components of $\mathbb{A}|\text{a}_i\rangle$ in B.[1] Furthermore, this mapping has the following properties:

(i) $(\text{Mat}(\mathbb{A} + \mathbb{B}))_{ij} = (\text{Mat}(\mathbb{A}))_{ij} + (\text{Mat}(\mathbb{B}))_{ij} = a_{ij} + b_{ij}$

(ii) $(\text{Mat}(\alpha\mathbb{A}))_{ij} = \alpha(\text{Mat}(\mathbb{A}))_{ij} = \alpha a_{ij} \qquad \forall\, \alpha \in \mathbb{C}$ (or $\mathbb{R}$)

(iii) $(\text{Mat}(\mathbb{A}\mathbb{B}))_{ij} = \sum_{k=1}^{N} a_{ik} b_{kj}$

(iv) $(\text{Mat}(\mathbb{O}))_{ij} = 0 \qquad \forall\, i, j = 1, \ldots, N$

(v) $(\text{Mat}(\mathbb{1}))_{ij} = \delta_{ij}$

From (i) and (ii) we conclude that this mapping is linear. ■

It must be emphasized that the mapping of Proposition 3.1.1 is defined *for a given basis*. That is why the subscript B is used for the $N \times N$ matrix space in the definition of Mat.

Proposition 3.1.1 associates with every linear operator a matrix representation of that operator in a given basis. Is this matrix representation unique in that given basis? The following proposition answers this.

3.1.2 Proposition The mapping $\text{Mat}: \mathscr{L}(\mathscr{V}) \to \mathscr{M}_{\text{B}}^{(N)}$ is a *linear isomorphism*.

Proof. To show this we have to prove that Mat is bijective, that is, both surjective and injective.

First, we prove that Mat is surjective. Given $\text{B} = \{|\text{a}_i\rangle\}_{i=1}^N$ and any matrix $\mathsf{A} \in \mathscr{M}_{\text{B}}^{(N)}$ with elements a_{ij}, we can define a linear operator $\mathbb{T}_{\mathsf{A}}$ as follows. Let $|\text{a}\rangle = \sum_{i=1}^N \alpha_i |\text{a}_i\rangle$ be an arbitrary vector. Then

$$\mathbb{T}_{\mathsf{A}}|\text{a}\rangle = \mathbb{T}_{\mathsf{A}}\left(\sum_{i=1}^{N} \alpha_i |\text{a}_i\rangle\right) \equiv \sum_{i=1}^{N} \alpha_i \mathbb{T}|\text{a}_i\rangle$$

$$= \sum_{i=1}^{N} \alpha_i \sum_{j=1}^{N} a_{ji} |\text{a}_j\rangle$$

It is easily shown that $\mathbb{T}_{\mathsf{A}}$ is linear. We have just shown that for any A there is a linear transformation $\mathbb{T}_{\mathsf{A}}$. Therefore, Mat is surjective.

Now we prove that Mat is injective. If $\text{Mat}(\mathbb{A}) = \text{Mat}(\mathbb{B})$, then $a_{ij} = b_{ij}$, which implies that all elements of the two matrices are equal, which implies that the two matrices are equal. Thus, no two distinct operators are mapped into the same matrix. Therefore, Mat is injective. ■

Because of the isomorphism all practical distinctions between operators acting on an N-dimensional vector space and the set of $N \times N$ matrices vanishes. The

[1] Letters of the Latin alphabet are often used for elements of a matrix or components of a vector. Note that the notation for the space of $N \times N$ matrices differs from that used in Chapter 1.

algebraic structures of $\mathcal{L}(\mathcal{V})$ and $\mathcal{M}_{\mathrm{B}}^{(N)}$ are identical. The advantage of using the latter is that one deals with arrays of *numbers* and can therefore compute with them.

Example 3.1.1

Let us find the matrix representation of the linear operator $\mathbb{A} \in \mathcal{L}(\mathbb{R}^3)$, given by

$$\mathbb{A}\begin{pmatrix} x \\ y \\ z \end{pmatrix} = \begin{pmatrix} x - y + 2z \\ 3x - z \\ 2y + z \end{pmatrix} \tag{1}$$

in the basis $\mathrm{B} = \left\{ \begin{pmatrix} 1 \\ 1 \\ 0 \end{pmatrix}, \begin{pmatrix} 1 \\ 0 \\ 1 \end{pmatrix}, \begin{pmatrix} 0 \\ 1 \\ 1 \end{pmatrix} \right\}$.

Recall that the jith element α_{ji} of $\mathrm{Mat}(\mathbb{A})$ is obtained by applying $\mathbb{A}$ to $|\mathrm{a}_i\rangle$. Since the first index is constant over a row and the second over a column, the coefficients of $\mathbb{A}|\mathrm{a}_i\rangle$ yield the *ith column* of $\mathrm{Mat}(\mathbb{A})$. Thus, to obtain the first column, we consider

$$\mathbb{A}|\mathrm{a}_1\rangle \equiv \mathbb{A}\begin{pmatrix} 1 \\ 1 \\ 0 \end{pmatrix} = \begin{pmatrix} 0 \\ 3 \\ 2 \end{pmatrix} = \frac{1}{2}\begin{pmatrix} 1 \\ 1 \\ 0 \end{pmatrix} - \frac{1}{2}\begin{pmatrix} 1 \\ 0 \\ 1 \end{pmatrix} + \frac{5}{2}\begin{pmatrix} 0 \\ 1 \\ 1 \end{pmatrix}$$

So the first column of the matrix is $\begin{pmatrix} \frac{1}{2} \\ -\frac{1}{2} \\ \frac{5}{2} \end{pmatrix}$. To get the second column, we consider

$$\mathbb{A}|\mathrm{a}_2\rangle \equiv \mathbb{A}\begin{pmatrix} 1 \\ 0 \\ 1 \end{pmatrix} = \begin{pmatrix} 3 \\ 2 \\ 1 \end{pmatrix} = 2\begin{pmatrix} 1 \\ 1 \\ 0 \end{pmatrix} + \begin{pmatrix} 1 \\ 0 \\ 1 \end{pmatrix} + 0\begin{pmatrix} 0 \\ 1 \\ 1 \end{pmatrix}$$

which gives $\begin{pmatrix} 2 \\ 1 \\ 0 \end{pmatrix}$ as the second column. Similarly,

$$\mathbb{A}|\mathrm{a}_3\rangle \equiv \mathbb{A}\begin{pmatrix} 0 \\ 1 \\ 1 \end{pmatrix} = \begin{pmatrix} 1 \\ -1 \\ 3 \end{pmatrix} = -\frac{3}{2}\begin{pmatrix} 1 \\ 1 \\ 0 \end{pmatrix} + \frac{5}{2}\begin{pmatrix} 1 \\ 0 \\ 1 \end{pmatrix} + \frac{1}{2}\begin{pmatrix} 0 \\ 1 \\ 1 \end{pmatrix}$$

gives $\begin{pmatrix} -\frac{3}{2} \\ \frac{5}{2} \\ \frac{1}{2} \end{pmatrix}$ as the third column. The whole matrix, then, is

$$\mathsf{A} = \begin{pmatrix} \frac{1}{2} & 2 & -\frac{3}{2} \\ -\frac{1}{2} & 1 & \frac{5}{2} \\ \frac{5}{2} & 0 & \frac{1}{2} \end{pmatrix} \tag{2}$$

As long as all vectors are represented by column vectors whose entries are expansion coefficients *of the vectors in* B, (1) and (2) are indistinguishable. However, the action of (2) on the column vector (x, y, z) will not yield (1)! The column vector on the LHS of (1) is really (although this is not usually emphasized) the vector

$$x\begin{pmatrix}1\\0\\0\end{pmatrix}+y\begin{pmatrix}0\\1\\0\end{pmatrix}+z\begin{pmatrix}0\\0\\1\end{pmatrix}$$

which is an expansion in terms of the so-called *standard basis* of $\mathbb{R}^3$ rather than in terms of **B**.

We can expand $\mathbb{A}\begin{pmatrix}x\\y\\z\end{pmatrix}$ in terms of **B**, yielding

$$\mathbb{A}\begin{pmatrix}x\\y\\z\end{pmatrix}=\begin{pmatrix}x-y+2z\\3x-z\\2y+z\end{pmatrix}$$

$$=(2x-\tfrac{3}{2}y)\begin{pmatrix}1\\1\\0\end{pmatrix}+\left(-x+\frac{y}{2}+2z\right)\begin{pmatrix}1\\0\\1\end{pmatrix}+(x+\tfrac{3}{2}y-z)\begin{pmatrix}0\\1\\1\end{pmatrix}$$

This says that *in the basis* **B** this vector has the representation

$$\left(\mathbb{A}\begin{pmatrix}x\\y\\z\end{pmatrix}\right)_{\mathbf{B}}=\begin{pmatrix}2x-\frac{3}{2}y\\-x+\frac{y}{2}+2z\\x+\frac{3}{2}y-z\end{pmatrix}\tag{3}$$

Similarly, $\begin{pmatrix}x\\y\\z\end{pmatrix}$ is represented by

$$\begin{pmatrix}x\\y\\z\end{pmatrix}_{\mathbf{B}}=\begin{pmatrix}\frac{x}{2}+\frac{y}{2}-\frac{z}{2}\\\frac{x}{2}-\frac{y}{2}+\frac{z}{2}\\-\frac{x}{2}+\frac{y}{2}+\frac{z}{2}\end{pmatrix}\tag{4}$$

Applying the matrix in (2) to (4) yields (3), as it should. •

It is sometimes useful to consider matrix representations of operators that map $\mathscr{V}$ into $\mathscr{W}$, where $N = \dim \mathscr{V} \neq \dim \mathscr{W} = N'$. To construct such representations we let $\mathrm{B} = \{|\mathrm{a}_i\rangle\}_{i=1}^{N}$ be a basis in $\mathscr{V}$ and $\mathrm{B}' = \{|\mathrm{b}_j\rangle\}_{j=1}^{N'}$ be a basis in $\mathscr{W}$. We define the mapping $\mathrm{Mat}: \mathscr{L}(\mathscr{V}, \mathscr{W}) \to \mathscr{M}^{N' \times N}$ by $(\mathrm{Mat}(\mathbb{A}))_{ji} = \alpha_{ji}$, where $\{\alpha_{ji}\}_{j=1}^{N'}$ are the components of $\mathbb{A}|\mathrm{a}_i\rangle$ in B′. Since j runs from 1 to N' and i runs from 1 to N, the matrix representing $\mathbb{A}$ will be $N' \times N$.

Example 3.1.2

Consider the operator $\mathbb{A}: \mathbb{R}^3 \to \mathbb{R}^2$ given by $\mathbb{A}(x, y, z) = (2x + y - 3z, x + y - z)$. Let us construct the matrix representing $\mathbb{A}$ in the standard basis of $\mathbb{R}^3$ and $\mathbb{R}^2$. If we write the vectors as columns, the definition of $\mathbb{A}$ gives

$$\mathbb{A}\begin{pmatrix} x \\ y \\ z \end{pmatrix} = \begin{pmatrix} 2x + y - 3z \\ x + y - z \end{pmatrix} \quad \Rightarrow \quad \mathbb{A}\begin{pmatrix} 1 \\ 0 \\ 0 \end{pmatrix} = \begin{pmatrix} 2 \\ 1 \end{pmatrix} = 2\begin{pmatrix} 1 \\ 0 \end{pmatrix} + 1\begin{pmatrix} 0 \\ 1 \end{pmatrix}$$

So the first column of A is $\begin{pmatrix} 2 \\ 1 \end{pmatrix}$. Similarly,

$$\mathbb{A}\begin{pmatrix} 0 \\ 1 \\ 0 \end{pmatrix} = \begin{pmatrix} 1 \\ 1 \end{pmatrix} = 1\begin{pmatrix} 1 \\ 0 \end{pmatrix} + 1\begin{pmatrix} 0 \\ 1 \end{pmatrix} \quad \Rightarrow \quad \text{second column is } \begin{pmatrix} 1 \\ 1 \end{pmatrix}$$

$$\mathbb{A}\begin{pmatrix} 0 \\ 0 \\ 1 \end{pmatrix} = \begin{pmatrix} -3 \\ -1 \end{pmatrix} = -3\begin{pmatrix} 1 \\ 0 \end{pmatrix} - 1\begin{pmatrix} 0 \\ 1 \end{pmatrix} \quad \Rightarrow \quad \text{third column is } \begin{pmatrix} -3 \\ -1 \end{pmatrix}$$

Therefore, the matrix representing $\mathbb{A}$ is this 2×3 matrix:

$$\mathsf{A} = \begin{pmatrix} 2 & 1 & -3 \\ 1 & 1 & -1 \end{pmatrix}$$

Such a matrix acts on a three-dimensional vector to give a two-dimensional vector, as it should. ●

3.1.1 Operations on Matrices

Given a matrix, we can obtain new matrices by performing certain operations on it. These operations include transposition and complex conjugation.

Transposition is an operation under which the rows and columns of the matrix are interchanged. If we denote the transpose of a matrix A by $\tilde{\mathsf{A}}$, then

$$(\tilde{\mathsf{A}})_{ij} = a_{ji} \tag{3.3}$$

where a_{ij} are the elements of A. The following theorem gives the properties of transposition.

3.1.3 Theorem

(i) $\widetilde{(\tilde{\mathsf{A}})} = \mathsf{A}$

(ii) $\widetilde{(\mathsf{A} + \mathsf{B})} = \tilde{\mathsf{A}} + \tilde{\mathsf{B}}$

(iii) $\widetilde{\mathsf{AB}} = \tilde{\mathsf{B}}\tilde{\mathsf{A}}$ ■

The proof follows immediately from the definition of transposition and is left to the reader.

Of special interest is a matrix that is identical to its transpose, or one satisfying $\mathsf{A} = \tilde{\mathsf{A}}$. Such matrices occur frequently in physics and are called *symmetric matrices.* Similarly, *antisymmetric matrices* are those satisfying $\mathsf{A} = -\tilde{\mathsf{A}}$. Note that any matrix can be written as

$$\mathsf{A} = \tfrac{1}{2}(\mathsf{A} + \tilde{\mathsf{A}}) + \tfrac{1}{2}(\mathsf{A} - \tilde{\mathsf{A}})$$

where the first term is symmetric and the second is antisymmetric. That is, any matrix can be decomposed into a symmetric matrix plus an antisymmetric matrix.

The elements of a symmetric matrix are related. This can be seen from the following:

$$(\tilde{\mathsf{A}})_{ij} = (\mathsf{A}_{ij}) \equiv a_{ij}$$

or

$$a_{ji} = a_{ij}$$

Therefore, symmetry means symmetry under reflection through the main diagonal. On the other hand, for an antisymmetric matrix we have

$$a_{ij} = -a_{ji}$$

A matrix satisfying $\mathsf{A}\tilde{\mathsf{A}} = \tilde{\mathsf{A}}\mathsf{A} = 1$ is called *orthogonal.*

Complex conjugation is an operation under which all elements of a matrix are complex conjugated. Thus, if we denote complex conjugation of a matrix by $*$, then

$$(\mathsf{A}^*)_{ij} = a_{ij}^* \tag{3.4}$$

A matrix is called *real* if $\mathsf{A}^* = \mathsf{A}$. Clearly, $(\mathsf{A}^*)^* = \mathsf{A}$.

Under the combined operation of complex conjugation and transposition, the rows and columns of a matrix are interchanged and all of its elements are complex conjugated. This combined operation is called *hermitian conjugation* and is denoted by †, as with operators. Thus, we have

$$\mathsf{A}^\dagger = \widetilde{(\mathsf{A}^*)} = (\tilde{\mathsf{A}})^* \tag{3.5a}$$

and, therefore,

$$(\mathsf{A}^\dagger)_{ij} = a_{ji}^* \tag{3.5b}$$

Table 3.1 shows examples of these operations on a matrix.

Two types of matrices are so important that they are defined separately.

TABLE 3.1 SOME COMMON OPERATIONS PERFORMED ON A TYPICAL MATRIX, A

Matrix	Components	Example
A	a_{ij}	$\begin{pmatrix} i & 1 & 1-i \\ 2 & e^{i\varphi} & -i \\ 0 & 2+3i & 1 \end{pmatrix}$
$\tilde{A}$ (transpose)	$(\tilde{A})_{ij} = a_{ji}$	$\begin{pmatrix} i & 2 & 0 \\ 1 & e^{i\varphi} & 2+3i \\ 1-i & -i & 1 \end{pmatrix}$
A^* (complex conjugate)	$(A^*)_{ij} = a_{ij}^*$	$\begin{pmatrix} -i & 1 & 1+i \\ 2 & e^{-i\varphi} & i \\ 0 & 2-3i & 1 \end{pmatrix}$
$A^\dagger$ (adjoint)	$(A^\dagger)_{ij} = a_{ji}^*$	$\begin{pmatrix} -i & 2 & 0 \\ 1 & e^{-i\varphi} & 2-3i \\ 1+i & i & 1 \end{pmatrix}$

3.1.4 Definition A *hermitian matrix* H satisfies $H^\dagger = H$ or, in terms of elements, $h_{ij}^* = h_{ji}$. A *unitary matrix* U satisfies $UU^\dagger = U^\dagger U = 1$ or, in terms of elements, $\sum_{k=1}^{N} u_{ik} u_{jk}^* = \sum_{k=1}^{N} u_{ki}^* u_{kj} = \delta_{ij}$.

It follows immediately from Definition 3.1.4 that

(1) The diagonal elements of a hermitian matrix are real.
(2) The kth column of a hermitian matrix is the complex conjugate of its kth row, and vice versa.
(3) A real hermitian matrix is necessarily symmetric.
(4) The rows of an $N \times N$ unitary matrix, when considered as vectors in $\mathbb{C}^N$, are orthonormal to each other, as are the columns.
(5) A real unitary matrix is necessarily orthogonal.

It is sometimes possible (and desirable) to bring a matrix into a form in which all of its off-diagonal elements are zeros. Such a matrix is called a *diagonal matrix*.

Table 3.2 summarizes the properties of the types of matrices discussed so far.

TABLE 3.2 SOME SPECIAL MATRICES AND THEIR PROPERTIES

	Property		
Matrix	Component Form	Matrix Form	Example
S (symmetric)	$s_{ij} = s_{ji}$	$\tilde{S} = S$	$\begin{pmatrix} 2 & 3i & 4 \\ 3i & -2 & -5i \\ 4 & -5i & 0 \end{pmatrix}$
A (antisymmetric)	$a_{ij} = -a_{ji}$	$\tilde{A} = -A$	$\begin{pmatrix} 0 & 2 & 3 \\ -2 & 0 & 4i+2 \\ -3 & -4i-2 & 0 \end{pmatrix}$
H (hermitian)	$h_{ij} = h^*_{ji}$	$H^\dagger = H$	$\begin{pmatrix} 1 & 2i & 1-i \\ -2i & 3 & e^{i\varphi} \\ 1+i & e^{-i\varphi} & 4 \end{pmatrix}$
O (orthogonal)	$\sum_k o_{ik} o_{jk} = \delta_{ij}$	$O\tilde{O} = \tilde{O}O = 1$	$\begin{pmatrix} \cos\theta & -\sin\theta \\ \sin\theta & \cos\theta \end{pmatrix}$
U (unitary)	$\sum_k u_{ik} u^*_{jk} = \delta_{ij}$	$U^\dagger U = UU^\dagger = 1$	$\begin{pmatrix} \frac{1}{\sqrt{2}} & \frac{-i}{\sqrt{2}} & 0 \\ \frac{1}{\sqrt{6}} & \frac{i}{\sqrt{6}} & \frac{-2}{\sqrt{6}} \\ \frac{1+i}{\sqrt{6}} & \frac{-1+i}{\sqrt{6}} & \frac{1+i}{\sqrt{6}} \end{pmatrix}$
D (diagonal)	$d_{ij} = 0 \quad \text{if } i \neq j$		$\begin{pmatrix} 1+i & 0 & 0 \\ 0 & 2 & 0 \\ 0 & 0 & e^{i\varphi} \end{pmatrix}$

It can easily be shown that the number of *independent* real parameters in an $N \times N$

symmetric matrix is $\frac{1}{2} N(N+1)$
antisymmetric matrix is $\frac{1}{2} N(N-1)$
(complex) unitary matrix is N^2
orthogonal matrix is $\frac{1}{2} N(N-1)$
(complex) hermitian matrix is N^2

Example 3.1.3

(a) A prototypical symmetric matrix is the moment of inertia matrix encountered in mechanics. The ijth element of this matrix is defined as

$$\mathsf{I}_{ij} \equiv \iiint \rho(x_1, x_2, x_3) x_i x_j \, dV$$

where x_i is the ith (Cartesian) coordinate of a point in the distribution of mass described by the volume density $\rho(x_1, x_2, x_3)$. It is clear that $\mathsf{I}_{ij} = \mathsf{I}_{ji}$, or $\mathsf{I} = \tilde{\mathsf{I}}$. The moment of inertia matrix can be represented as

$$\mathsf{I} = \begin{pmatrix} I_{11} & I_{12} & I_{13} \\ I_{12} & I_{22} & I_{23} \\ I_{13} & I_{23} & I_{33} \end{pmatrix}$$

and has six independent elements.

(b) An important antisymmetric matrix occurs in electromagnetism. In the context of Einstein's special theory of relativity, the three components of the vector potential, **A**, are combined with the scalar potential, ϕ, to form a four-vector $(A_1, A_2, A_3, i\phi)$, where $i = \sqrt{-1}$.[2] Similarly, with c representing the velocity of light, ict is the fourth component of the displacement four-vector, (x_1, x_2, x_3, ict). These four-vectors are written as (A_1, A_2, A_3, A_4) and (x_1, x_2, x_3, x_4), and a component is subscripted with a Greek letter. Thus, A_μ and x_ν are typical components of these four-vectors.

The electromagnetic field tensor is defined as

$$F_{\mu\nu} = \frac{\partial A_\mu}{\partial x_\nu} - \frac{\partial A_\nu}{\partial x_\mu} \qquad \text{for } \mu, \nu = 1, 2, 3, 4$$

This definition yields

$$F_{j4} = \frac{\partial A_j}{\partial x_4} - \frac{\partial A_4}{\partial x_j} = -\frac{i}{c}\frac{\partial A_j}{\partial t} - i\frac{\partial \phi}{\partial x_j} \tag{1}$$

Comparing (1) with the relation between the electric field, **E**, and the potentials, given by

$$\mathbf{E} = -\frac{1}{c}\frac{\partial \mathbf{A}}{\partial t} - \nabla\phi$$

suggests

$$F_{j4} = iE_j$$

[2]This formalism, in which the time component of a four-vector is multiplied by $i = \sqrt{-1}$, is completely outmoded. It is now common to use real-time components and employ the Minkowski metric, which is discussed in Chapter 4. However, the use of imaginary time components facilitates this discussion.

Similarly,

$$F_{12} = \frac{\partial A_1}{\partial x_2} - \frac{\partial A_2}{\partial x_1} = -(\nabla \times \mathbf{A})_3 = -B_3$$

$$F_{13} = \frac{\partial A_1}{\partial x_3} - \frac{\partial A_3}{\partial x_1} = (\nabla \times \mathbf{A})_2 = B_2$$

$$F_{23} = \frac{\partial A_2}{\partial x_3} - \frac{\partial A_3}{\partial x_2} = -(\nabla \times \mathbf{A})_1 = -B_1$$

give the components of the magnetic field, **B**. Clearly, all the diagonal elements vanish, and we can write the electromagnetic field tensor as an antisymmetric matrix:

$$\mathsf{F} = \begin{pmatrix} 0 & -B_3 & B_2 & iE_1 \\ B_3 & 0 & -B_1 & iE_2 \\ -B_2 & B_1 & 0 & iE_3 \\ -iE_1 & -iE_2 & -iE_3 & 0 \end{pmatrix}$$

This demonstrates the unity of electric and magnetic phenomena.

(c) Examples of hermitian matrices are the 2×2 Pauli spin matrices:

$$\begin{pmatrix} 0 & 1 \\ 1 & 0 \end{pmatrix} \quad \begin{pmatrix} 0 & -i \\ i & 0 \end{pmatrix} \quad \begin{pmatrix} 1 & 0 \\ 0 & -1 \end{pmatrix}$$

(d) We encountered an orthogonal matrix in Chapter 1 in conjunction with rotations of rigid objects in terms of Euler angles. ●

Exercises

3.1.1 (**a**) Construct the matrix representations of $\mathbb{D}$ and $\mathbb{T}$, the derivative and multiplication-by-t operators, acting on the space of polynomials of degree 2. (**b**) What is the commutation relation for these matrices?

3.1.2 Show that the number of independent components of a symmetric $N \times N$ matrix is $\frac{1}{2}N(N+1)$.

3.1.3 The linear transformation $\mathbb{T}: \mathbb{R}^3 \to \mathbb{R}^3$ is defined as

$$\mathbb{T}(x_1, x_2, x_3) = (x_1 + x_2 - x_3, \; -x_3 + 2x_1, 2x_2 + x_1)$$

Find the matrix representation of $\mathbb{T}$ in (**a**) the standard basis in $\mathbb{R}^3$, that is, $|e_1\rangle = (1, 0, 0)$, $|e_2\rangle = (0, 1, 0)$, and $|e_3\rangle = (0, 0, 1)$, and (**b**) the basis consisting of $|a_1\rangle = (1, 1, 0)$, $|a_2\rangle = (1, 0, -1)$, and $|a_3\rangle = (0, 2, 3)$.

3.1.4 Find the matrix representing the derivative operator d/dx in the basis $\{1, x, x^2, x^3\}$ of the space of polynomials of degree 3. Use this matrix to obtain the first and second derivatives of a polynomial of degree 3.

3.2 ORTHONORMAL BASES

We have been dealing with general bases, but in practical situations calculations are facilitated if we choose an *orthonormal* basis, $B = \{|e_i\rangle\}_{i=1}^N$. The matrix elements of an operator $\mathbb{A}$ can be found in such a basis by the following procedure. Apply $\mathbb{A}$ to $|e_i\rangle$, and write this (a vector in the vector space $\mathcal{V}$ spanned by B) as a linear combination of $|e_1\rangle, |e_2\rangle, \ldots, |e_N\rangle$:

$$\mathbb{A}|e_i\rangle = \sum_{k=1}^{N} a_{ki}|e_k\rangle$$

Now "multiply" both sides on the left by $\langle e_j|$ and use the orthonormality condition, $\langle e_j|e_i\rangle = \delta_{ij}$, and the linearity of the scalar product, $\langle c|(\alpha|a\rangle + \beta|b\rangle) = \alpha\langle c|a\rangle + \beta\langle c|b\rangle$:

$$\langle e_j|\mathbb{A}|e_i\rangle = \langle e_j|\left(\sum_{k=1}^{N} a_{ki}|e_k\rangle\right) = \sum_{k=1}^{N} a_{ki}\langle e_j|e_k\rangle$$

$$= \sum_{k=1}^{N} a_{ki}\delta_{jk} = a_{ji}$$

This equation tells us that if we want the ijth element of the matrix representing an operator *in an orthonormal basis*, we have to "sandwich" the operator between the ith (on the left) and the jth (on the right) vectors of this basis. We rewrite this as

$$a_{ij} = \langle e_i|\mathbb{A}|e_j\rangle \tag{3.6}$$

We can also find the components of a vector in an analogous manner. Consider an arbitrary vector $|a\rangle$, which can be written in an orthonormal basis B as

$$|a\rangle = \sum_{j=1}^{N} \alpha_j|e_j\rangle$$

Now "multiply" both sides on the left by $\langle e_i|$:

$$\langle e_i|a\rangle = \langle e_i|\left(\sum_{j=1}^{N} \alpha_j|e_j\rangle\right) = \sum_{j=1}^{N} \alpha_j\langle e_i|e_j\rangle$$

$$= \sum_{j=1}^{N} \alpha_j\delta_{ij} = \alpha_i$$

Thus, *in an orthonormal basis*, the ith component of a vector is found by multiplying it by $\langle e_i|$. This expression for α_i allows us to write the expansion of $|a\rangle$ as

$$|a\rangle = \sum_{j=1}^{N} \langle e_j|a\rangle|e_j\rangle = \sum_{j=1}^{N} |e_j\rangle\langle e_j|a\rangle$$

which is the same as in Proposition 2.3.24 but with $\mathbb{P}_j = |e_j\rangle\langle e_j|$.

The identity

$$\mathbb{1} = \sum_{i=1}^{N} |e_i\rangle\langle e_i| \tag{3.7}$$

holds for all (finite) orthonormal bases and is very handy for calculations. For instance, we can write any linear operator $\mathbb{A}$ as a sum over a complete set of more elementary operators. We do this by introducing a factor of $\mathbb{1}$ on either side of $\mathbb{A}$ and using (3.7) (note the use of *different* dummy indices for the two $\mathbb{1}$'s) and then (3.6):

$$\begin{aligned}\mathbb{A} = \mathbb{1}\mathbb{A}\mathbb{1} &= \left(\sum_i |e_i\rangle\langle e_i|\right)\mathbb{A}\left(\sum_j |e_j\rangle\langle e_j|\right)\\ &= \sum_{i,j} |e_i\rangle\langle e_i|\mathbb{A}|e_j\rangle\langle e_j| = \sum_{i,j} |e_i\rangle a_{ij}\langle e_j|\\ &= \sum_{i,j} a_{ij}|e_i\rangle\langle e_j| \equiv \sum_{i,j} a_{ij}\mathbb{P}_{ij}\end{aligned}$$

This equation gives $\mathbb{A}$ in terms of operators $\mathbb{P}_{ij}$, defined by

$$\mathbb{P}_{ij} = |e_i\rangle\langle e_j|$$

Let us now investigate the representation of the special operators discussed in Section 2.3.4 and find the connection between those operators and the matrices encountered in Section 3.1.

Let us begin by calculating the matrix representing the hermitian conjugate of an operator $\mathbb{A}$. In an orthonormal basis the elements of this matrix are given by (3.6), $a_{ij} = \langle e_i|\mathbb{A}|e_j\rangle$. Taking the complex conjugate of this equation and using the definition of $\mathbb{A}^\dagger$ given in Eq. (2.22), we obtain

$$a_{ij}^* = \langle e_i|\mathbb{A}|e_j\rangle^* = \langle e_j|\mathbb{A}^\dagger|e_i\rangle$$

Using $(\mathbb{A}^\dagger)_{ji}$ to denote the jith element of the operator $\mathbb{A}^\dagger$ in the orthonormal basis $\{|e_1\rangle, \ldots, |e_N\rangle\} = \mathrm{B}$, we obtain

$$(\mathbb{A}^\dagger)_{ji} = a_{ij}^*$$

or

$$(\mathbb{A}^\dagger)_{ij} = a_{ji}^*$$

This is precisely how the adjoint of a matrix was defined [see Eq. (3.5b)].

Note how crucially this conclusion depends on the orthonormality of the basis vectors. If the basis were not orthonormal, we could not use (3.6), on which the conclusion is based. Therefore, *only in an orthonormal basis is the adjoint of an operator represented by the adjoint of the matrix representing that operator.* In particular, a hermitian operator is represented by a hermitian matrix only if an orthonormal basis is used. Example 3.2.1 illustrates this point.

Example 3.2.1

Consider the matrix representation of the hermitian operator $\mathbb{H}$ in a general (not orthonormal) basis, $\mathrm{B} = \{|a_i\rangle\}_{i=1}^N$. The elements of the matrix corresponding to $\mathbb{H}$ are given by

$$\mathbb{H}|a_k\rangle = \sum_{j=1}^N h_{jk}|a_j\rangle$$

and also by

$$\mathbb{H}|a_i\rangle = \sum_{j=1}^{N} h_{ji}|a_j\rangle \tag{1}$$

Taking the product of the first equation with $\langle a_i|$ and complex-conjugating the result gives

$$\langle a_i|\mathbb{H}|a_k\rangle^* = \left(\sum_{j=1}^{N} h_{jk}\langle a_i|a_j\rangle\right)^* = \sum_{j=1}^{N} h_{jk}^*\langle a_j|a_i\rangle$$

But, by the definition of a hermitian operator, the LHS is simply

$$\langle a_i|\mathbb{H}|a_k\rangle^* = \langle a_k|\mathbb{H}^\dagger|a_i\rangle = \langle a_k|\mathbb{H}|a_i\rangle$$

So we have

$$\langle a_k|\mathbb{H}|a_i\rangle = \sum_{j=1}^{N} h_{jk}^*\langle a_j|a_i\rangle$$

On the other hand, multiplying (1) by $\langle a_k|$ gives

$$\langle a_k|\mathbb{H}|a_i\rangle = \sum_{j=1}^{N} h_{ji}\langle a_k|a_j\rangle$$

The LHS's of the last two equations are equal, so

$$\sum_{j=1}^{N} h_{jk}^*\langle a_j|a_i\rangle = \sum_{j=1}^{N} h_{ji}\langle a_k|a_j\rangle$$

Because this equation does not say anything about each individual h_{ij}, we cannot conclude, in general, that $h_{ij}^* = h_{ji}$. However, if the $|a_i\rangle$'s are orthonormal, then

$$\langle a_j|a_i\rangle = \delta_{ji} \qquad \text{and} \qquad \langle a_k|a_j\rangle = \delta_{kj}$$

and we obtain

$$\sum_{j=1}^{N} h_{jk}^*\delta_{ij} = \sum_{j=1}^{N} h_{ji}\delta_{kj}$$

or

$$h_{ik}^* = h_{ki}$$

as expected for an orthonormal basis. •

Similarly, we expect the matrices representing unitary operators to be unitary only if the basis is orthonormal. This is an immediate consequence of Eq. (3.5b), but we can prove it to provide another example of how Eq. (3.7) is used. Since $\mathbb{U}\mathbb{U}^\dagger = \mathbb{1}$, we have

$$\langle e_i|\mathbb{U}\mathbb{U}^\dagger|e_j\rangle = \langle e_i|\mathbb{1}|e_j\rangle = \delta_{ij}$$

We insert $\mathbb{1} = \sum_{k=1}^{N}|e_k\rangle\langle e_k|$ between $\mathbb{U}$ and $\mathbb{U}^\dagger$ on the LHS:

$$\begin{aligned}\langle e_i|\mathbb{U}\left(\sum_{k=1}^{N}|e_k\rangle\langle e_k|\right)\mathbb{U}^\dagger|e_j\rangle &= \sum_{k=1}^{N}\langle e_i|\mathbb{U}|e_k\rangle\langle e_k|\mathbb{U}^\dagger|e_j\rangle \\ &\equiv \sum_{k=1}^{N} u_{ik}(\mathbb{U}^\dagger)_{kj} = \delta_{ij}\end{aligned}$$

where u_{ik} is the ikth element of the matrix representing $\mathbb{U}$ in the basis B and $(\mathbb{U}^\dagger)_{kj}$ is similarly interpreted. Using (3.5b), we obtain

$$\sum_{k=1}^{N} u_{ik} u_{jk}^* = \delta_{ij}$$

which is the first half of the requirement for a unitary matrix given in Definition 3.1.4. By redoing the calculation for $\mathbb{U}^\dagger\mathbb{U}$, we could obtain the second half of that requirement.

3.3 CHANGE OF BASES AND SIMILARITY TRANSFORMATIONS

A physical problem must often be described in a particular basis because it takes a simpler form there, but the general form of the result may still be of importance. In such cases the problem is solved in one basis, and the result is transformed to other bases. Let us investigate this point in some detail.

Given a basis $B = \{|a_i\rangle\}_{i=1}^N$, we can write an arbitrary vector $|a\rangle$ with components $\{\alpha_1, \alpha_2, \ldots, \alpha_N\}$ in B as $|a\rangle = \sum_{i=1}^N \alpha_i |a_i\rangle$. Now suppose that we change the basis to $B' = \{|a'_1\rangle, \ldots, |a'_N\rangle\}$. What are the components of $|a\rangle$ in B'? To answer this question, we write $|a_i\rangle$ in terms of B' vectors:

$$|a_i\rangle = \sum_{j=1}^{N} r_{ji} |a'_j\rangle$$

where r_{ji} are components of $|a_i\rangle$ in B' (note the use of Latin instead of Greek symbols for scalars!). Now we substitute for $|a_i\rangle$ in the above expansion of $|a\rangle$, obtaining

$$|a\rangle = \sum_{i=1}^{N} \alpha_i \sum_{j=1}^{N} r_{ji} |a'_j\rangle = \sum_{i,j} \alpha_i r_{ji} |a'_j\rangle$$

If we denote the jth component of $|a\rangle$ in B' by α'_j then this equation tells us that

$$\alpha'_j = \sum_{i=1}^{N} r_{ji}\alpha_i \qquad \text{for } j = 1, 2, \ldots, N \tag{3.8a}$$

If we use $\boldsymbol{a}'$, R, and $\boldsymbol{a}$, respectively, to designate a column vector with elements α'_j, an $N \times N$ matrix with elements r_{ji}, and a column vector with elements α_i, then (3.8a) can be written in matrix form as

$$\boldsymbol{a}' = \mathsf{R}\boldsymbol{a}$$

or

$$\begin{pmatrix} \alpha'_1 \\ \alpha'_2 \\ \vdots \\ \alpha'_N \end{pmatrix} = \begin{pmatrix} r_{11} & r_{12} & \cdots & r_{1N} \\ r_{21} & r_{22} & \cdots & r_{2N} \\ \vdots & \vdots & & \vdots \\ r_{N1} & r_{N2} & \cdots & r_{NN} \end{pmatrix} \begin{pmatrix} \alpha_1 \\ \alpha_2 \\ \vdots \\ \alpha_N \end{pmatrix} \tag{3.8b}$$

The matrix R is called the *basis transformation matrix*. It is invertible because it is a linear transformation that maps one basis onto another (see Theorem 2.3.11).

As usual, (3.8) can also be thought of as an active transformation in which the vector $|a\rangle$ is transformed into another vector, $|a'\rangle$, with components $\{\alpha'_1, \alpha'_2, \ldots, \alpha'_N\}$ *in the basis* B. The two interpretations are, of course, equivalent.

What happens to a matrix when we transform the basis? Consider the equation

$$|b\rangle = \mathbb{A}|a\rangle$$

where $|a\rangle$ and $|b\rangle$ have components $\{\alpha_i\}_{i=1}^N$ and $\{\beta_i\}_{i=1}^N$, respectively, in B. This equation has a corresponding matrix equation:

$$\boldsymbol{\beta} = \mathsf{A}\boldsymbol{\alpha} \tag{3.9a}$$

Now, if we change the basis (or the vectors), the components of $|a\rangle$ and $|b\rangle$ will change to those of $\boldsymbol{\alpha}'$ and $\boldsymbol{\beta}'$, respectively. We seek a matrix A′ such that

$$\boldsymbol{\beta}' = \mathsf{A}'\boldsymbol{\alpha}'$$

This matrix will clearly be the transform of A. Using (3.8), we write $\mathsf{R}\boldsymbol{\beta} = \mathsf{A}'\mathsf{R}\boldsymbol{\alpha}$, or

$$\boldsymbol{\beta} = (\mathsf{R}^{-1}\mathsf{A}'\mathsf{R})\boldsymbol{\alpha} \tag{3.9b}$$

Comparing this with (3.9a) and applying the fact that both equations hold for arbitrary $\boldsymbol{\alpha}$ and $\boldsymbol{\beta}$, we conclude that

$$\mathsf{R}^{-1}\mathsf{A}'\mathsf{R} = \mathsf{A}$$

or

$$\mathsf{A}' = \mathsf{R}\mathsf{A}\mathsf{R}^{-1} \tag{3.10}$$

This is called a *similarity transformation* on A, and A′ is said to be *similar to* A.

The transformation matrix R can easily be found for orthonormal bases. We have

$$|e_i\rangle = \sum_{k=1}^N r_{ki}|e'_k\rangle$$

where $\mathrm{B} = \{|e_i\rangle\}_{i=1}^N$ and $\mathrm{B}' = \{|e'_i\rangle\}_{i=1}^N$ are orthonormal bases. We multiply this equation by $\langle e'_j|$ to obtain

$$\langle e'_j|e_i\rangle = \sum_{k=1}^N r_{ki}\langle e'_j|e'_k\rangle = \sum_k r_{ki}\delta_{jk} = r_{ji} \tag{3.11}$$

That is, to find the ijth element of the matrix that changes the components of a vector in the orthonormal basis B into those of the same vector in the orthonormal basis B′ [see (3.8)], we take the jth ket in B and multiply it by the ith bra in B′.

To find the ijth element of the matrix that changes B′ into B, we take the jth ket in B′, which is $|e'_j\rangle$, and multiply it by the ith bra in B, which is $\langle e_i|$:

$$r'_{ij} = \langle e_i|e'_j\rangle$$

However, the matrix R' must be R^{-1}, as can be seen from (3.8). On the other hand,

$$(r'_{ij})^* = \langle e_i | e'_j \rangle^* = \langle e'_j | e_i \rangle = r_{ji}$$

or

$$(\mathsf{R}^{-1})^*_{ij} = r_{ji}$$

or

$$(\mathsf{R}^{-1})_{ij} = r^*_{ji} = (\mathsf{R}^{+})_{ij} \tag{3.12}$$

This shows that R is a unitary matrix and yields an important theorem.

3.3.1 Theorem The matrix that transforms one orthonormal basis into another is necessarily unitary. ■

From Eqs. (3.11) and (3.12) we have $(\mathsf{R}^\dagger)_{ij} = \langle e_i | e'_j \rangle$. Thus, to obtain the jth column of $\mathsf{R}^\dagger$, we take the jth vector in the *new* basis and successively "multiply" it by $\langle e_i |$ for $i = 1, 2, \ldots, N$. The numbers we obtain give the jth column of $\mathsf{R}^\dagger$. In particular, if the *original* basis is the standard basis of $\mathbb{C}^N$ and $|e'_j\rangle$ is represented by a column vector in that basis, then the jth column of $\mathsf{R}^\dagger$ is simply the vector $|e'_j\rangle$.

Example 3.3.1

The vectors

$$|a_1\rangle = \begin{pmatrix} \frac{1}{\sqrt{2}} & \frac{1}{\sqrt{6}} & \frac{1+i}{\sqrt{6}} \end{pmatrix}$$

$$|a_2\rangle = \begin{pmatrix} \frac{-i}{\sqrt{2}} & \frac{i}{\sqrt{6}} & \frac{-1+i}{\sqrt{6}} \end{pmatrix}$$

$$|a_3\rangle = \begin{pmatrix} 0 & \frac{-2}{\sqrt{6}} & \frac{1+i}{\sqrt{6}} \end{pmatrix}$$

are orthonormal vectors in $\mathbb{C}^3$. Let us find the transformation matrix connecting the standard basis to this basis and show that it is unitary.

The elements of the desired matrix are given by

$$r_{ij} = \langle a_i | e_j \rangle$$

where $|e_1\rangle = (1, 0, 0)$, $|e_2\rangle = (0, 1, 0)$, and $|e_3\rangle = (0, 0, 1)$. Therefore,

$$\mathsf{R} = \begin{pmatrix} \frac{1}{\sqrt{2}} & \frac{1}{\sqrt{6}} & \frac{1-i}{\sqrt{6}} \\ \frac{i}{\sqrt{2}} & \frac{-i}{\sqrt{6}} & \frac{-1-i}{\sqrt{6}} \\ 0 & -\frac{2}{\sqrt{6}} & \frac{1-i}{\sqrt{6}} \end{pmatrix}$$

and $\mathsf{R}^\dagger$ is obtained by complex-conjugating R and then interchanging its rows and columns:

$$\mathsf{R}^\dagger = \begin{pmatrix} \frac{1}{\sqrt{2}} & \frac{-i}{\sqrt{2}} & 0 \\ \frac{1}{\sqrt{6}} & \frac{i}{\sqrt{6}} & -\frac{2}{\sqrt{6}} \\ \frac{1+i}{\sqrt{6}} & \frac{-1+i}{\sqrt{6}} & \frac{1+i}{\sqrt{6}} \end{pmatrix}$$

It is easily verified that

$$\mathsf{R}\mathsf{R}^\dagger = \mathsf{R}^\dagger\mathsf{R} = 1$$

●

Exercises

3.3.1 Find the transformation matrix R that relates the orthonormal basis

$$\left\{ |e_1\rangle = \begin{pmatrix} 1 \\ 0 \\ 0 \end{pmatrix}, |e_2\rangle = \begin{pmatrix} 0 \\ 1 \\ 0 \end{pmatrix}, |e_3\rangle = \begin{pmatrix} 0 \\ 0 \\ 1 \end{pmatrix} \right\}$$

to the orthonormal basis obtained via the Gram-Schmidt process from these vectors:

$$|a_1\rangle = \begin{pmatrix} 1 \\ i \\ 0 \end{pmatrix} \qquad |a_2\rangle = \begin{pmatrix} 0 \\ 1 \\ -i \end{pmatrix} \qquad |a_3\rangle = \begin{pmatrix} i \\ 0 \\ -1 \end{pmatrix}$$

3.3.2 Verify that the matrix R obtained in Exercise 3.3.1 is unitary, as expected from Theorem 3.3.1.

3.3.3 If the matrix representation of a linear transformation $\mathbb{A}$ on $\mathbb{C}^2$ with respect to the basis $\{(1, 0), (0, 1)\}$ is $\begin{pmatrix} 1 & 1 \\ 1 & 1 \end{pmatrix}$, what is the matrix representation of $\mathbb{A}$ with respect to the basis $\{(1, 1), (1, -1)\}$?

3.3.4 If the matrix representation of a linear transformation $\mathbb{A}$ on $\mathbb{C}^3$ with respect to the basis $\{(1, 0, 0), (0, 1, 0), (0, 0, 1)\}$ is

$$\begin{pmatrix} 0 & 1 & 1 \\ 1 & 0 & -1 \\ -1 & -1 & 0 \end{pmatrix}$$

what is the matrix representation of $\mathbb{A}$ with respect to the basis $\{(0, 1, -1), (1, -1, 1), (-1, 1, 0)\}$?

3.4 DETERMINANTS AND TRACES

An important concept associated with linear operators and their matrix representations is the determinant. It is possible (and will be done in Chapter 4) to define determinants of operators without resort to a specific representation of the operator in terms of matrices. Here we will define determinants in terms of matrices.

In the discussion of permutations in Section 2.4, the permutation symbol $\varepsilon_{i_1 i_2 \ldots i_N}$ was defined by

$$\varepsilon_{12\ldots N} = +1 \tag{3.13a}$$

and

$$\varepsilon_{i_1 i_2 \ldots i_j \ldots i_k \ldots i_N} = -\varepsilon_{i_1 i_2 \ldots i_k \ldots i_j \ldots i_N} \tag{3.13b}$$

In other words, $\varepsilon_{i_1 i_2 \ldots i_N}$ is completely antisymmetric (or skew-symmetric) under interchange of any pair of its indices. We will use this permutation symbol in discussing determinants.

Example 3.4.1

A consequence of Eq. (3.13b) is that $\varepsilon_{i_1 i_2 \ldots i_N}$ will be zero if any two of its indices are equal. Suppose that the two indices i_j and i_k are equal. Call both of them l. Then (3.13b) becomes

$$\varepsilon_{i_1 i_2 \ldots l \ldots l \ldots i_N} - \quad \varepsilon_{i_1 i_2 \ldots l \ldots l \ldots i_N}$$

Transferring the RHS to the LHS yields

$$2\varepsilon_{i_1 i_2 \ldots l \ldots l \ldots i_N} = 0$$

or

$$\varepsilon_{i_1 i_2 \ldots l \ldots l \ldots i_N} = 0$$

•

3.4.1 Determinant of a Matrix

Let us define the determinant of the matrix A, which has elements a_{ij}.

3.4.1 Definition The determinant is a mapping, $\det : \mathscr{M}^{(N)} \rightarrow \mathbb{R}$, given, in term of elements a_{ij} of a matrix $\mathsf{A} \in \mathscr{M}^{(N)}$, by

$$\det \mathsf{A} \equiv \sum_{i_1 \ldots i_N}^{N} \varepsilon_{i_1 \ldots i_N} a_{1i_1} \cdots a_{Ni_N} \tag{3.14a}$$

Note that a_{1i_1} is always one of the entries in the first row of the matrix A, a_{2i_2} is one entry of the second row, and so forth. Also note that $\varepsilon_{i_1 i_2 \ldots i_N}$ is $+1$ if $(i_1 i_2 \ldots i_N)$ is an even permutation of $(1\ 2 \ldots N)$, and -1 if it is an odd permutation of $(1\ 2 \ldots N)$.

Example 3.4.2

As practice for handling multiple sums, let us evaluate (3.14a) for $N = 3$:

$$\det \mathsf{A} = \sum_{i_3=1}^{3} \sum_{i_2=1}^{3} \sum_{i_1=1}^{3} \varepsilon_{i_1 i_2 i_3} a_{1i_1} a_{2i_2} a_{3i_3}$$

Let us sum over i_1 first:

$$\det \mathsf{A} = \sum_{i_3=1}^{3}\sum_{i_2=1}^{3} \varepsilon_{1i_2i_3}a_{11}a_{2i_2}a_{3i_3} + \sum_{i_3=1}^{3}\sum_{i_2=1}^{3} \varepsilon_{2i_2i_3}a_{12}a_{2i_2}a_{3i_3} + \sum_{i_3=1}^{3}\sum_{i_2=1}^{3} \varepsilon_{3i_2i_3}a_{13}a_{2i_2}a_{3i_3}$$

Next, we sum over i_2. In the first sum on the RHS, i_2 can take the values 2 and 3 (if $i_2 = 1$, then ε will have two equal indices and must vanish, as demonstrated in Example 3.4.1). In the second sum, i_2 can be 1 and 3, and in the third sum, 1 and 2. Thus, summing over i_2 gives

$$\det \mathsf{A} = \sum_{i_3=1}^{3} \varepsilon_{12i_3}a_{11}a_{22}a_{3i_3} + \sum_{i_3=1}^{3} \varepsilon_{13i_3}a_{11}a_{23}a_{3i_3} + \sum_{i_3=1}^{3} \varepsilon_{21i_3}a_{12}a_{21}a_{3i_3} + \sum_{i_3=1}^{3} \varepsilon_{23i_3}a_{12}a_{23}a_{3i_3} + \sum_{i_3=1}^{3} \varepsilon_{31i_3}a_{13}a_{21}a_{3i_3} + \sum_{i_3=1}^{3} \varepsilon_{32i_3}a_{13}a_{22}a_{3i_3}$$

Now we have to sum over i_3. But note that in each of the six sums above i_3 can take only one value. For instance, in the first sum i_3 can only be 3, in the second sum it can only be 2, and so on. Using this fact, we can easily write

$$\det \mathsf{A} = \varepsilon_{123}a_{11}a_{22}a_{33} + \varepsilon_{132}a_{11}a_{23}a_{32} + \varepsilon_{213}a_{12}a_{21}a_{33} + \varepsilon_{231}a_{12}a_{23}a_{31} + \varepsilon_{312}a_{13}a_{21}a_{32} + \varepsilon_{321}a_{13}a_{22}a_{31}$$

Finally, putting in the values for the permutation symbol, we obtain

$$\det \mathsf{A} = a_{11}(a_{22}a_{33} - a_{23}a_{32}) - a_{12}(a_{21}a_{33} - a_{23}a_{31}) + a_{13}(a_{21}a_{32} - a_{22}a_{31}) \tag{1}$$

Similarly, for a 2×2 matrix,

$$\det\begin{pmatrix} a & b \\ c & d \end{pmatrix} = ad - bc$$

•

Example 3.4.3

The following trick is useful for evaluating determinants of 3×3 (and only 3×3) matrices. Write the first and the second columns next to the third column, take products along the diagonal lines and give them the signs shown, and then add all the products:

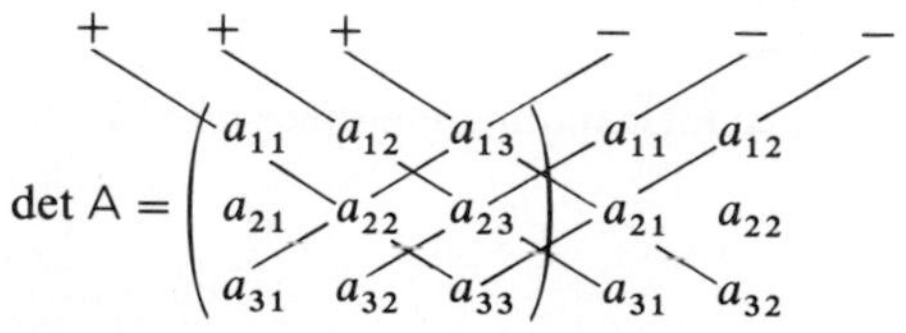

•

Definition 3.4.1 gives det A in terms of an expansion in rows, so the first entry is from the first row, the second from the second row, and so on. It is also possible to expand in terms of columns, as the following theorem shows.

3.4.2 Theorem The determinant of a matrix A can be written as

$$\det \mathsf{A} = \sum_{i_1 \ldots i_N}^{N} \varepsilon_{i_1 i_2 \ldots i_N} a_{i_1 1} a_{i_2 2} \cdots a_{i_N N} \tag{3.14b}$$

Therefore, $\det \mathsf{A} = \det \tilde{\mathsf{A}}$.

Proof. In (3.14a) $i_1, \ldots, i_N$ are all different and form a permutation of $(1\ 2 \ldots N)$. So one of the a's must have 1 as its second index. We assume that it is the j_1th term; that is, $i_{j_1} = 1$, and

$$a_{1i_1} \cdots a_{j_1 i_{j_1}} \cdots a_{Ni_N} = a_{1i_1} \cdots a_{j_1 1} \cdots a_{Ni_N}$$

We move this term all the way to the left to get $a_{j_1 1} a_{1i_1} a_{2i_2} \cdots a_{Ni_N}$. Now we look for the entry with 2 as the second index and assume that it occurs at j_2th position; that is, $i_{j_2} = 2$. We move this to the left, next to $a_{j_1 1}$, and write $a_{j_1 1} a_{j_2 2} a_{1i_1} a_{2i_2} \cdots a_{Ni_N}$. We continue in this fashion until we get $a_{j_1 1} a_{j_2 2} \cdots a_{j_N N}$. Since $j_1 j_2 \ldots j_N$ is really a reshuffling of $i_1 i_2 \ldots i_N$, the summation indices can be changed to $j_1 j_2 \ldots j_N$, and we can write

$$\det \mathsf{A} = \sum_{j_1 \ldots j_N} \varepsilon_{i_1 i_2 \ldots i_N} a_{j_1 1} a_{j_2 2} \cdots a_{j_N N}$$

If we can show that $\varepsilon_{i_1 \ldots i_N} = \varepsilon_{j_1 \ldots j_N}$, we are done. In (3.14a) the sequence of integers $(i_1\ i_2 \ldots i_N)$ is obtained by some shuffling of $(1\ 2 \ldots N)$. What we have done above is to reshuffle $(i_1\ i_2 \ldots i_N)$ *in reverse order* so that we get back to $(1\ 2 \ldots N)$. Thus, if the shuffling in (3.14a) is even (odd), the reshuffling will also be even (odd). Thus, $\varepsilon_{i_1 i_2 \ldots i_N} = \varepsilon_{j_1 j_2 \ldots j_N}$, and we obtain

$$\det \mathsf{A} = \sum_{j_1 \ldots j_N} \varepsilon_{j_1 j_2 \ldots j_N} a_{j_1 1} a_{j_2 2} \cdots a_{j_N N}$$

The theorem is established. ■

3.4.3 Theorem Interchanging two rows (or two columns) of a matrix changes the sign of its determinant.

Proof. We will prove the theorem for the interchange of two rows using (3.14a). Similar arguments can be applied using (3.14b) to prove the same result for two columns. Interchanging the two rows j and k, for example, in (3.14a) means writing $a_{1i_1} \cdots a_{ki_j} \cdots a_{ji_k} \cdots a_{Ni_N}$ instead of $a_{1i_1} \cdots a_{ji_j} \cdots a_{ki_k} \cdots a_{Ni_N}$. So the RHS of (3.14a) becomes

$$\begin{aligned}
&\sum \varepsilon_{i_1 \ldots i_j \ldots i_k \ldots i_N} a_{1i_1} \cdots a_{ki_j} \cdots a_{ji_k} \cdots a_{Ni_N} \\
&\qquad = -\sum \varepsilon_{i_1 \ldots i_k \ldots i_j \ldots i_N} a_{1i_1} \cdots a_{ki_j} \cdots a_{ji_k} \cdots a_{Ni_N} \\
&\qquad = -\sum_{i_1 \ldots i_j \ldots i_k \ldots i_N} \varepsilon_{i_1 \ldots i_k \ldots i_j \ldots i_N} a_{1i_1} \cdots a_{ji_k} \cdots a_{ki_j} \cdots a_{Ni_N}
\end{aligned}$$

Recalling that $i_1 \ldots i_N$ are dummy variables, we can conclude that the above sum is exactly the same as the one in (3.14a). In fact, if we explicitly sum over i_j and i_k in (3.14a), we obtain

$$\det \mathsf{A} = \sum_{i_1 \ldots i_j \ldots i_k \ldots i_N} \varepsilon_{i_1 \ldots i_j \ldots i_k \ldots i_N} a_{1i_1} \cdots a_{ji_j} \cdots a_{ki_k} \cdots a_{Ni_N}$$

which is exactly the same as the sum above except that there we have interchanged the dummy indices i_k and i_j. But the choice of dummy index does not affect the sum, so the theorem is proved. ■

An immediate consequence of Theorem 3.4.3 is the following corollary.

3.4.4 Corollary The determinant of a matrix with two equal rows (or two equal columns) is zero. ■

Since every term of (3.14) contains *one and only one* element from each row, we can collect together all the terms of the ith row containing $a_{i1}, a_{i2}, \ldots, a_{iN}$ and write

$$\det \mathsf{A} = a_{i1}A_{i1} + a_{i2}A_{i2} + \cdots + a_{iN}A_{iN} = \sum_{j=1}^{N} a_{ij}A_{ij} \tag{3.15}$$

where A_{ij} contains products of elements of the matrix A other than the element a_{ij}. Since each element of a row or column occurs at most once in each term of the expansion, A_{ij} cannot contain *any* element from the ith row *or* the jth column. The quantity A_{ij} is called the *cofactor of* a_{ij}, and (3.15) is known as the (Laplace) *expansion of det* A *by its ith row*. Clearly, there is a similar expansion by the ith column of det A, which is obtained by a similar argument using (3.14b).

3.4.5 Proposition If $i \neq k$, then

$$\sum_{j=1}^{N} a_{ij}A_{kj} = 0 \qquad \text{and} \qquad \sum_{j=1}^{N} a_{ji}A_{jk} = 0$$

Proof. Consider the matrix B obtained from A by replacing row k by row i, where $k \neq i$ (row i, of course, remains unchanged). The matrix B has two rows equal, and its determinant is therefore zero. Now, if we expand det B by its kth row according to (3.15), we obtain

$$0 = \det \mathsf{B} = \sum_{j=1}^{N} b_{kj}B_{kj}$$

But the elements of the kth row of B are the same as the elements of the ith row of A, that is, $b_{kj} = a_{ij}$; and the cofactors of the kth row of B are the same as those of A, that is, $B_{kj} = A_{kj}$. Thus, the first equation of the proposition is established. The second equation can be established by using expansion by columns. ■

Example 3.4.4

Let us demonstrate Proposition 3.4.5 using a 3 × 3 matrix. Taking $i = 1$ and $k = 2$ and summing over j, we obtain

$$\sum_{j=1}^{3} a_{1j}A_{2j} = a_{11}A_{21} + a_{12}A_{22} + a_{13}A_{23}$$

We can read off A_{21}, A_{22}, and A_{23} from Eq. (1) in Example 3.4.2, using the definition of A_{ij} as the coefficient of a_{ij} in the expansion of det A:

$$A_{21} = a_{13}a_{32} - a_{12}a_{33} \qquad A_{22} = a_{11}a_{33} - a_{13}a_{31} \qquad A_{23} = a_{12}a_{31} - a_{11}a_{32}$$

Substituting these in the preceding equation, we obtain

$$\begin{aligned}\sum_{j=1}^{3} a_{1j}A_{2j} &= a_{11}(a_{13}a_{32} - a_{12}a_{33}) + a_{12}(a_{11}a_{33} - a_{13}a_{31}) + a_{13}(a_{12}a_{31} - a_{11}a_{32}) \\ &= 0\end{aligned}$$

●

The *minor of order* $N - 1$ of an $N \times N$ matrix A is the determinant of a matrix obtained by striking out one row and one column of A. If we strike out the ith row and jth column of A, then the minor is denoted by M_{ij}.

3.4.6 Theorem $A_{ij} = (-1)^{i+j}M_{ij}$

Proof. Let us split the sum over i_1 in (3.14) into two parts, one in which $i_1 = 1$ and the other in which i_1 takes values other than 1:

$$\begin{aligned}\det A &= \sum_{i_2 \ldots i_N} \varepsilon_{1i_2\ldots i_N} a_{11}a_{2i_2} \cdots a_{Ni_N} + \sum_{\substack{i_1 \ldots i_N \\ i_1 \neq 1}} \varepsilon_{i_1\ldots i_N} a_{1i_1} \cdots a_{Ni_N} \\ &= a_{11} \sum_{i_2 \ldots i_N} \varepsilon_{i_2\ldots i_N} a_{2i_2} \cdots a_{Ni_N} + \text{terms not involving } a_{11} \qquad (3.16) \\ &= a_{11}M_{11} + \text{terms not involving } a_{11}\end{aligned}$$

Here we used the definition of M_{11} and the fact that the first sum in (3.16) is a determinant involving all a's except those in the first row and the first column. Note that by its definition $\varepsilon_{i_2 i_3\ldots i_N}$ has $N - 1$ indices, none of which can be 1; thus, the sum on the second line of (3.16) is the determinant of an $(N - 1) \times (N - 1)$ matrix, namely M_{11}. But that sum is also A_{11}, or what multiplies a_{11} in the expansion of the determinant [see (3.15)]:

$$A_{11} = \sum_{i_2 \ldots i_N} \varepsilon_{1i_2\ldots i_N} a_{2i_2} \cdots a_{Ni_N}$$

Therefore,

$$M_{11} = A_{11}$$

For a general a_{ij}, we shift it to the (1, 1) position by $i - 1$ successive interchanges of adjacent rows followed by $j-1$ interchanges of adjacent columns. The new matrix B obtained in this way has the determinant

$$\det \mathsf{B} = (-1)^{i-1}(-1)^{j-1}\det \mathsf{A} = (-1)^{i+j}\det \mathsf{A}$$

because each interchange of two rows or two columns introduces one minus sign. Applying (3.16) to det B, we can write (with obvious notation)

$$\begin{aligned}\det \mathsf{B} &= b_{11}M_{11}^{(\mathsf{B})} + \text{terms not involving } b_{11}\\ &= a_{ij}M_{ij} + \text{terms not involving } a_{ij}\end{aligned}$$

because $b_{11} = a_{ij}$ and $M_{11} = M_{ij}$ by the way B was constructed. On the other hand, using (3.15), we can write

$$\det \mathsf{A} = a_{ij}A_{ij} + \text{terms not involving } a_{ij}$$

Comparing the preceding three equations, we obtain

$$a_{ij}M_{ij} = (-1)^{i+j}a_{ij}A_{ij}$$

or

$$A_{ij} = (-1)^{i+j}M_{ij}$$ ■

The combination of (3.15) and Theorem 3.4.6 gives a practical (and probably familiar) way of evaluating the determinant of a matrix.

Example 3.4.5

Let us find the cofactors of this matrix:

$$\mathsf{A} = \begin{pmatrix} 1 & 2 & -1 \\ 0 & 1 & -2 \\ 2 & 1 & -1 \end{pmatrix}$$

By Theorem 3.4.6 we have

$$A_{11} = (-1)^{1+1}\det\begin{pmatrix} 1 & -2 \\ 1 & -1 \end{pmatrix} = 1 \qquad A_{12} = (-1)^{1+2}\det\begin{pmatrix} 0 & -2 \\ 2 & -1 \end{pmatrix} = -4$$

$$A_{13} = (-1)^{1+3}\det\begin{pmatrix} 0 & 1 \\ 2 & 1 \end{pmatrix} = -2 \qquad A_{21} = (-1)^{2+1}\det\begin{pmatrix} 2 & -1 \\ 1 & -1 \end{pmatrix} = 1$$

$$A_{22} = (-1)^{2+2}\det\begin{pmatrix} 1 & -1 \\ 2 & -1 \end{pmatrix} = 1 \qquad A_{23} = (-1)^{2+3}\det\begin{pmatrix} 1 & 2 \\ 2 & 1 \end{pmatrix} = 3$$

$$A_{31} = -3 \qquad A_{32} = 2 \qquad A_{33} = 1$$ •

3.4.2 Inverse of a Matrix

We are ready to investigate the conditions under which the inverse of a matrix exists. First, let us combine (3.15) and the content of Proposition 3.4.5 into a single equation:

$$\sum_{j=1}^{N} a_{ij}A_{kj} = (\det \mathsf{A})\delta_{ik} \tag{3.17}$$

We construct a matrix $\mathsf{C}(\mathsf{A})$, whose elements c_{ij} are the cofactors of the elements of the matrix A. That is, $c_{ij} = A_{ij}$ or

$$\mathsf{C}(\mathsf{A}) = \begin{pmatrix} A_{11} & A_{12} & \cdots & A_{1N} \\ A_{21} & A_{22} & \cdots & A_{2N} \\ \vdots & \vdots & & \vdots \\ A_{N1} & A_{N2} & \cdots & A_{NN} \end{pmatrix} \tag{3.18}$$

Then (3.17) can be written as

$$\sum_{j=1}^{N} a_{ij} c_{kj} = \sum_{j=1}^{N} a_{ij} (\widetilde{\mathsf{C}(\mathsf{A})})_{jk} = (\det \mathsf{A}) \delta_{ij}$$

or, in matrix form, as

$$\mathsf{A}(\widetilde{\mathsf{C}(\mathsf{A})}) = (\det \mathsf{A})1 \qquad \text{or} \qquad \mathsf{A}\left(\frac{\widetilde{\mathsf{C}(\mathsf{A})}}{\det \mathsf{A}}\right) = 1$$

We thus have the following theorem.

3.4.7 Theorem The matrix A has an inverse if and only if $\det \mathsf{A} \neq 0$. Furthermore,

$$\mathsf{A}^{-1} = \frac{\widetilde{\mathsf{C}(\mathsf{A})}}{\det \mathsf{A}} \tag{3.19}$$

where $\mathsf{C}(\mathsf{A})$ is the matrix of the cofactors of A. ■

In particular, the inverse of a 2×2 matrix is easily found:

$$\begin{pmatrix} a & b \\ c & d \end{pmatrix}^{-1} = \frac{1}{ad - bc} \begin{pmatrix} d & -b \\ -c & a \end{pmatrix} \tag{3.20}$$

Example 3.4.6

Let us find the inverse of the matrix A of Example 3.4.5. First, we calculate the determinant:

$$\begin{aligned} \det \mathsf{A} \\ &= \det \begin{pmatrix} 1 & 2 & -1 \\ 0 & 1 & -2 \\ 2 & 1 & -1 \end{pmatrix} \\ &= (-1)^{1+1} \det \begin{pmatrix} 1 & -2 \\ 1 & -1 \end{pmatrix} + 2(-1)^{1+2} \det \begin{pmatrix} 0 & -2 \\ 2 & -1 \end{pmatrix} + (-1)(-1)^{1+3} \det \begin{pmatrix} 0 & 1 \\ 2 & 1 \end{pmatrix} \\ &= -5 \end{aligned}$$

Thus, using Eq. (3.18) in conjunction with the results of Example 3.4.5 and Eq. (3.19), we obtain

$$\mathsf{A}^{-1} = \frac{1}{-5}\begin{pmatrix} 1 & 1 & -3 \\ -4 & 1 & 2 \\ -2 & 3 & 1 \end{pmatrix} = \begin{pmatrix} -\frac{1}{5} & -\frac{1}{5} & \frac{3}{5} \\ \frac{4}{5} & -\frac{1}{5} & -\frac{2}{5} \\ \frac{2}{5} & -\frac{3}{5} & -\frac{1}{5} \end{pmatrix}$$

This result can easily be checked by multiplying A and A^{-1}. ●

There is a much more practical way of calculating the inverse of matrices. The following discussion of this method will omit most proofs because those would carry us too far into matrix theory. We start with some definitions.

3.4.8 Definition An *elementary row operation* on a matrix is one of the following:

I. Interchange of two rows of the matrix
II. Multiplication of a row by a nonzero number
III. Addition of a multiple of one row to another

Elementary column operations are defined analogously.

3.4.9 Definition A matrix is in *triangular*, or *row-echelon, form* if it satisfies the following three conditions:

(i) Any row consisting of only zeros is below any row that contains at least one nonzero element.
(ii) Going from left to right, the first nonzero entry of any row is to the left of the first nonzero entry of any lower row.
(iii) The first nonzero entry of each row is 1.

The matrix

$$\begin{pmatrix} 1 & 0 & -1 & 3 \\ 0 & 0 & 1 & -1 \\ 0 & 0 & 0 & 1 \\ 0 & 0 & 0 & 0 \end{pmatrix}$$

is in triangular form, but the three matrices

$$\mathsf{A} = \begin{pmatrix} 0 & -1 & 2 & 3 & 0 \\ 0 & 0 & 2 & 0 & 1 \\ 0 & 0 & 0 & 1 & 0 \end{pmatrix} \qquad \mathsf{B} = \begin{pmatrix} 1 & 0 & 1 \\ 1 & 2 & 0 \\ 0 & 0 & 0 \end{pmatrix} \qquad \mathsf{C} = \begin{pmatrix} 0 & 0 & 0 \\ 0 & 1 & 2 \\ 0 & 0 & 1 \end{pmatrix}$$

are not. Matrix A violates (iii), B violates (ii), and C violates (i).

3.4.10 Theorem An invertible matrix can be transformed to 1 by means of a finite number of elementary row operations. ■

Theorem 3.4.10 is the crucial theorem on which the method for finding inverses hinges. (For a proof, see Birkhoff and MacLane 1977, p. 246.)

3.4.11 Definition An $n \times n$ matrix E is called an *elementary matrix* if it can be obtained from $\mathsf{1}$ by a single elementary row or column operation.

For example,

$$\mathsf{E} = \begin{pmatrix} 1 & 0 & 0 \\ -3 & 1 & 0 \\ 0 & 0 & 1 \end{pmatrix}$$

is an elementary matrix because it is obtained from $\mathsf{1}$ by multiplying its first row by -3 and adding it to the second row.

The next theorem shows that elementary row and column operations are equivalent. (Again a proof is found in Birkhoff and MacLane 1977.)

3.4.12 Theorem Any elementary matrix can be obtained from $\mathsf{1}$ by either an elementary row operation or an elementary column operation. ■

The following theorem paves the way for the final result (again, see Birkhoff and MacLane 1977 for a proof).

3.4.13 Theorem Let A and B be matrices.

(i) If B can be obtained from A by an elementary row operation, then $\mathsf{B} = \mathsf{EA}$, where E is the elementary matrix obtained from $\mathsf{1}$ by the same row operation.

(ii) If B can be obtained from A by an elementary column operation, then $\mathsf{B} = \mathsf{AE}$, where E is the elementary matrix obtained from $\mathsf{1}$ by the same column operation. ■

Example 3.4.7

Consider the matrices

$$\mathsf{A} = \begin{pmatrix} 2 & 0 & -1 \\ 3 & 1 & -2 \\ 4 & 2 & -3 \end{pmatrix} \qquad \mathsf{B}_1 = \begin{pmatrix} 2 & 0 & -1 \\ -1 & 1 & 0 \\ 4 & 2 & -3 \end{pmatrix} \qquad \mathsf{B}_2 = \begin{pmatrix} 2 & 0 & -1 \\ 3 & 1 & 1 \\ 4 & 2 & 3 \end{pmatrix}$$

where B_1 is obtained from A by multiplying A's first row by -2 and adding it to the second row. The corresponding elementary matrix is

$$\mathsf{E}_1 = \begin{pmatrix} 1 & 0 & 0 \\ -2 & 1 & 0 \\ 0 & 0 & 1 \end{pmatrix}$$

and

$$\mathsf{E}_1\mathsf{A} = \begin{pmatrix} 1 & 0 & 0 \\ -2 & 1 & 0 \\ 0 & 0 & 1 \end{pmatrix}\begin{pmatrix} 2 & 0 & -1 \\ 3 & 1 & -2 \\ 4 & 2 & -3 \end{pmatrix} = \begin{pmatrix} 2 & 0 & -1 \\ -1 & 1 & 0 \\ 4 & 2 & -3 \end{pmatrix} = \mathsf{B}_1$$

Similarly, B_2 is obtained from A by multiplying A's second column by 3 and adding it to the third column. The corresponding elementary matrix is

$$\mathsf{E}_2 = \begin{pmatrix} 1 & 0 & 0 \\ 0 & 1 & 3 \\ 0 & 0 & 1 \end{pmatrix}$$

and

$$\mathsf{A}\mathsf{E}_2 = \begin{pmatrix} 2 & 0 & -1 \\ 3 & 1 & -2 \\ 4 & 2 & -3 \end{pmatrix}\begin{pmatrix} 1 & 0 & 0 \\ 0 & 1 & 3 \\ 0 & 0 & 1 \end{pmatrix} = \begin{pmatrix} 2 & 0 & -1 \\ 3 & 1 & 1 \\ 4 & 2 & 3 \end{pmatrix} = \mathsf{B}_2$$ •

3.4.14 Proposition Every elementary matrix E is invertible. Furthermore, E^{-1} is also elementary.

Proof. Every elementary row operation can be "undone" by a similar operation. Therefore, if we obtain E from $\mathsf{1}$ by some elementary operation, we can perform the undoing operation on E to get $\mathsf{1}$. But Theorem 3.4.13 says that this undoing can be performed by multiplying E by the corresponding elementary matrix, which we can denote by E'. We can, therefore, write

$$\mathsf{E}'\mathsf{E} = \mathsf{1}$$

and E is invertible. ∎

Finally, we come to the final step of the development of the method of finding inverses.

3.4.15 Proposition Every invertible matrix is a product of elementary matrices.

Proof. From Theorems 3.4.10 and 3.4.13 we can write $\mathsf{E}_m\mathsf{E}_{m-1}\cdots\mathsf{E}_1\mathsf{A} = \mathsf{1}$, or

$$\mathsf{A} = (\mathsf{E}_m\mathsf{E}_{m-1}\cdots\mathsf{E}_1)^{-1} = \mathsf{E}_1^{-1}\mathsf{E}_2^{-1}\cdots\mathsf{E}_m^{-1}$$

By Proposition 3.4.14 $\{\mathsf{E}_i^{-1}\}_{i=1}^m$ are elementary. ∎

We can now state the main theorem.

3.4.16 Theorem For any invertible $n \times n$ matrix A,

(i) The matrix $(\mathsf{A}|\mathsf{1})$ can be transformed into the $n \times 2n$ matrix $(\mathsf{1}|\mathsf{A}^{-1})$ by means of a finite number of elementary row operations.[3]

[3] The matrix $(\mathsf{A}|\mathsf{1})$ means the $n \times 2n$ matrix obtained by juxtaposing the $n \times n$ unit matrix to the right of A. It can easily be shown that if A, B, and C are $n \times n$ matrices, then $\mathsf{A}(\mathsf{B}|\mathsf{C}) = (\mathsf{AB}|\mathsf{AC})$.

(ii) If $(\mathsf{A}|1)$ is transformed into $(1|\mathsf{B})$ by means of elementary row operations, then $\mathsf{B} = \mathsf{A}^{-1}$.

Proof. (i) We can obtain $(1|\mathsf{A}^{-1})$ from $(\mathsf{A}|1)$ by multiplying the latter by A^{-1}:

$$\mathsf{A}^{-1}(\mathsf{A}|1) = (\mathsf{A}^{-1}\mathsf{A}|\mathsf{A}^{-1}1) = (1|\mathsf{A}^{-1})$$

But, from Proposition 3.4.15, A^{-1} is a product of elementary matrices. Therefore,

$$\mathsf{E}_1\mathsf{E}_2\cdots\mathsf{E}_m(\mathsf{A}|1) = (1|\mathsf{A}^{-1})$$

However, Theorem 3.4.13 translates the above product of elementary matrices into elementary row operations.

(ii) Again, by Theorem 3.4.13 elementary row operations correspond to multiplication by elementary matrices. So we let

$$(1|\mathsf{B}) = \mathsf{E}_1\mathsf{E}_2\cdots\mathsf{E}_m(\mathsf{A}|1) \equiv \mathsf{C}(\mathsf{A}|1) = (\mathsf{CA}|\mathsf{C})$$

This shows that $\mathsf{CA} = 1$ and $\mathsf{C} = \mathsf{B}$. Thus, $\mathsf{C} = \mathsf{B} = \mathsf{A}^{-1}$. ■

A systematic way of transforming $(\mathsf{A}|1)$ into $(1|\mathsf{A}^{-1})$ is to first bring A into triangular form and then eliminate all nonzero elements of each column by elementary row operations.

Example 3.4.8

Let us evaluate the inverse of

$$\mathsf{A} = \begin{pmatrix} 1 & 2 & -1 \\ 0 & 1 & -2 \\ 2 & 1 & -1 \end{pmatrix}$$

We start with

$$\left(\begin{array}{ccc|ccc} 1 & 2 & -1 & 1 & 0 & 0 \\ 0 & 1 & -2 & 0 & 1 & 0 \\ 2 & 1 & -1 & 0 & 0 & 1 \end{array}\right) \equiv \mathsf{M}$$

and apply elementary row operations to M to bring the left half of it into row-echelon form. If we denote the kth row as (k) and the three operations of Definition 3.4.8, respectively, as $(k) \leftrightarrow (j)$, $\alpha(k)$, and $\alpha(k) + (j)$, we get

$$\mathsf{M} \xrightarrow[-2(1)+(3)]{} \left(\begin{array}{ccc|ccc} 1 & 2 & -1 & 1 & 0 & 0 \\ 0 & 1 & -2 & 0 & 1 & 0 \\ 0 & -3 & 1 & -2 & 0 & 1 \end{array}\right) \xrightarrow[3(2)+(3)]{} \left(\begin{array}{ccc|ccc} 1 & 2 & -1 & 1 & 0 & 0 \\ 0 & 1 & -2 & 0 & 1 & 0 \\ 0 & 0 & -5 & -2 & 3 & 1 \end{array}\right)$$

$$\xrightarrow[-\frac{1}{5}(3)]{} \left(\begin{array}{ccc|ccc} 1 & 2 & -1 & 1 & 0 & 0 \\ 0 & 1 & -2 & 0 & 1 & 0 \\ 0 & 0 & 1 & \frac{2}{5} & -\frac{3}{5} & \frac{1}{5} \end{array}\right) \equiv \mathsf{M}'$$

It is easy to make all column entries above a one on the left side of M′ equal to zero by appropriate use of type III elementary row operations:

$$\mathsf{M}' \xrightarrow[-2(2)+(1)]{} \left(\begin{array}{ccc|ccc} 1 & 0 & 3 & 1 & -2 & 0 \\ 0 & 1 & -2 & 0 & 1 & 0 \\ 0 & 0 & 1 & \frac{2}{5} & -\frac{3}{5} & -\frac{1}{5} \end{array}\right)$$

$$\xrightarrow[-3(3)+1]{} \left(\begin{array}{ccc|ccc} 1 & 0 & 0 & -\frac{1}{5} & -\frac{1}{5} & \frac{3}{5} \\ 0 & 1 & -2 & 0 & 1 & 0 \\ 0 & 0 & 1 & \frac{2}{5} & -\frac{3}{5} & -\frac{1}{5} \end{array}\right)$$

$$\xrightarrow[2(3)+(2)]{} \left(\begin{array}{ccc|ccc} 1 & 0 & 0 & -\frac{1}{5} & -\frac{1}{5} & \frac{3}{5} \\ 0 & 1 & 0 & \frac{4}{5} & -\frac{1}{5} & -\frac{2}{5} \\ 0 & 0 & 1 & \frac{2}{5} & -\frac{3}{5} & -\frac{1}{5} \end{array}\right)$$

The right half of the resulting matrix is A^{-1}. That is,

$$\mathsf{A}^{-1} = \begin{pmatrix} -\frac{1}{5} & -\frac{1}{5} & \frac{3}{5} \\ \frac{4}{5} & -\frac{1}{5} & -\frac{2}{5} \\ \frac{2}{5} & -\frac{3}{5} & -\frac{1}{5} \end{pmatrix}$$

This is the same as the matrix we found in Example 3.4.6. •

Example 3.4.9

It is instructive to start with a matrix that is *not* invertible and show that it is impossible to turn it into 1 by elementary row operations. Consider the matrix

$$\mathsf{B} = \begin{pmatrix} 2 & -1 & 3 \\ 1 & -2 & 1 \\ -1 & 5 & 0 \end{pmatrix}$$

Let us systematically bring it into row-echelon form:

$$\mathsf{M} = \left(\begin{array}{ccc|ccc} 2 & -1 & 3 & 1 & 0 & 0 \\ 1 & -2 & 1 & 0 & 1 & 0 \\ -1 & 5 & 0 & 0 & 0 & 1 \end{array}\right) \xrightarrow[(1)\leftrightarrow(2)]{} \left(\begin{array}{ccc|ccc} 1 & -2 & 1 & 0 & 1 & 0 \\ 2 & -1 & 3 & 1 & 0 & 0 \\ -1 & 5 & 0 & 0 & 0 & 1 \end{array}\right)$$

$$\xrightarrow[-2(1)+(2)]{} \left(\begin{array}{ccc|ccc} 1 & -2 & 1 & 0 & 1 & 0 \\ 0 & 3 & 1 & 1 & -2 & 0 \\ -1 & 5 & 0 & 0 & 0 & 1 \end{array}\right) \xrightarrow[(1)+(3)]{} \left(\begin{array}{ccc|ccc} 1 & -2 & 1 & 0 & 1 & 0 \\ 0 & 3 & 1 & 1 & -2 & 0 \\ 0 & 3 & 1 & 0 & 1 & 1 \end{array}\right)$$

$$\xrightarrow[-(2)+(3)]{} \left(\begin{array}{ccc|ccc} 1 & -2 & 1 & 0 & 1 & 0 \\ 0 & 3 & 1 & 1 & -2 & 0 \\ 0 & 0 & 0 & -1 & 3 & 1 \end{array}\right) \xrightarrow[\frac{1}{3}(2)]{} \left(\begin{array}{ccc|ccc} 1 & -2 & 1 & 0 & 1 & 0 \\ 0 & 1 & \frac{1}{3} & \frac{1}{3} & -\frac{2}{3} & 0 \\ 0 & 0 & 0 & -1 & 3 & 1 \end{array}\right)$$

The matrix B is now in row-echelon form, but its third row contains all zeros. There is no way we can bring this into the form of a unit matrix. We, therefore, conclude that B is not invertible. This is, of course, obvious, since it can easily be verified that $\det \mathsf{B} = 0$. •

3.4.3 Determinants of Products of Matrices

Chapter 4 will discuss determinants using the elegant tools of exterior algebra. The ease and power of exterior algebra can be better appreciated by proving the following theorem "the hard way."

3.4.17 Theorem $\det(\mathsf{AB}) = (\det \mathsf{A})(\det \mathsf{B})$

Proof. Let $\mathsf{P} = \mathsf{AB}$. Then we can use (3.14a) to obtain

$$\begin{aligned}\det \mathsf{P} &= \sum \varepsilon_{i_1 i_2 \ldots i_N} p_{1i_1} p_{2i_2} \cdots p_{Ni_N}\\ &= \sum \varepsilon_{i_1 i_2 \ldots i_N}\left(\sum_{j_1} a_{1j_1} b_{j_1 i_1}\right)\cdots\left(\sum_{j_N} a_{Ni_N} b_{j_N i_N}\right)\\ &= \sum_{i,j} \varepsilon_{i_1 \ldots i_N} a_{1j_1} a_{2j_2} \cdots a_{Nj_N} b_{j_1 i_1} b_{j_2 i_2} \cdots b_{j_N i_N}\\ &= \sum_j a_{1j_1} \cdots a_{Nj_N} \sum_i \varepsilon_{i_1 \ldots i_N} b_{j_1 i_1} \cdots b_{j_N i_N}\end{aligned} \tag{3.21}$$

Let us concentrate on the terms in the second sum. We rearrange the b's so that the first index of the first b is 1, the first index of the second b is 2, and so forth. We then have $b_{1i_{k_1}} b_{2i_{k_2}} \cdots b_{Ni_{k_N}}$. It is clear that $(k_1\, k_2 \ldots k_N)$ is a permutation of $(1\, 2 \ldots N)$ that brings $(j_1\, j_2 \ldots j_N)$ into $(1\, 2 \ldots N)$. It is also obvious that $(i_{k_1}\, i_{k_2} \ldots i_{k_N})$ is the *same* permutation of $(i_1\, i_2 \ldots i_N)$ as $(j_1\, j_2 \ldots j_N)$ is of $(1\, 2 \ldots N)$. We now claim that

$$\varepsilon_{i_1 i_2 \ldots i_N} = \varepsilon_{i_{k_1} i_{k_2} \ldots i_{k_N}} \varepsilon_{j_1 j_2 \ldots j_N}$$

This can be seen by noting that $\varepsilon_{i_{k_1} i_{k_2} \ldots i_{k_N}} = \varepsilon_{i_1 i_2 \ldots i_N}$ if $(i_{k_1}\, i_{k_2} \ldots i_{k_N})$ is an even permutation of $(i_1\, i_2 \ldots i_N)$ [in which case $(j_1\, j_2 \ldots j_N)$ is also an even permutation of $(1\, 2 \ldots N)$ and $\varepsilon_{j_1 j_2 \ldots j_N} = +1$] and that $\varepsilon_{i_{k_1} i_{k_2} \ldots i_{k_N}} = -\varepsilon_{i_1 i_2 \ldots i_N}$ if $(i_{k_1}\, i_{k_2} \ldots i_{k_N})$ is an odd permutation of $(i_1\, i_2 \ldots i_N)$ [in which case $(j_1\, j_2 \ldots j_N)$ is also an odd permutation of $(1\, 2 \ldots N)$ and $\varepsilon_{j_1 j_2 \ldots j_N} = -1$]. In either case our claim is established. Thus, we can write (3.21) as

$$\begin{aligned}\det \mathsf{P} &= \sum_j a_{1j_1} \cdots a_{Nj_N} \sum_i \varepsilon_{j_1 \ldots j_N} \varepsilon_{i_{k_1} \ldots i_{k_N}} b_{1i_{k_1}} \cdots b_{Ni_{k_N}}\\ &= \sum_j a_{1j_1} \cdots a_{Nj_N} \varepsilon_{j_1 \ldots j_N} \sum_i \varepsilon_{k_1 \ldots k_N} b_{1k_1} \cdots b_{Nk_N}\\ &= (\det \mathsf{A})(\det \mathsf{B})\end{aligned}$$

where we rearrange the dummy variables in the second sum and use $k_1, k_2, \ldots, k_N$ instead of the awkward set $i_{k_1}, i_{k_2}, \ldots, i_{k_N}$. ■

Example 3.4.10

Let O and U denote, respectively, an orthogonal and a unitary $n \times n$ matrix; that is,

$$\mathsf{O}\tilde{\mathsf{O}} = \tilde{\mathsf{O}}\mathsf{O} = 1$$

$$\mathsf{U}\mathsf{U}^\dagger = \mathsf{U}^\dagger\mathsf{U} = 1$$

Taking the determinant of the first equation and using Theorems 3.4.2 and 3.4.17, we obtain

$$(\det \mathsf{O})(\det \tilde{\mathsf{O}}) = (\det \mathsf{O})^2 = \det \mathsf{1} = 1$$

Therefore, for an orthogonal matrix, we get

$$\det \mathsf{O} = \pm 1$$

Orthogonal transformations preserve a real inner product. Among such transformations are the so-called *inversions*, which, in their simplest form, multiply a vector by -1. In three dimensions this corresponds to a reflection through the origin. The matrix associated with this operation is $-\mathsf{1}_3$:

$$\begin{pmatrix} x \\ y \\ z \end{pmatrix} \to \begin{pmatrix} -x \\ -y \\ -z \end{pmatrix} = \begin{pmatrix} -1 & 0 & 0 \\ 0 & -1 & 0 \\ 0 & 0 & -1 \end{pmatrix} \begin{pmatrix} x \\ y \\ z \end{pmatrix}$$

which has a determinant of -1. This is a prototype of other, more complicated, orthogonal transformations whose determinants are -1.

The other group of orthogonal transformations, whose determinants are $+1$, are of special interest because (in three dimensions) they correspond to rotations. The set of orthogonal transformations in n dimensions having the determinant $+1$ is denoted as $\mathsf{SO}(n)$. These transformations are special because the identity operator belongs to them. Furthermore, they have the mathematical structure of a (continuous) group, which finds application in all areas of advanced physics.[4]

We can obtain a similar result for unitary transformations. We take the determinant of both sides of $\mathsf{U}\mathsf{U}^\dagger = \mathsf{1}$:

$$\det(\mathsf{U}\mathsf{U}^\dagger) = (\det \mathsf{U})(\det \widetilde{\mathsf{U}^*}) = (\det \mathsf{U})(\det \mathsf{U}^*)$$
$$= (\det \mathsf{U})(\det \mathsf{U})^* = |\det \mathsf{U}|^2 = 1$$

Thus, we can generally write

$$\det \mathsf{U} = e^{ia} \qquad \text{where } a \in \mathbb{R}$$

The set of those transformations with $a = 0$ form a group to which $\mathsf{1}$ belongs and which is denoted as $\mathsf{SU}(n)$. This group has recently found many applications in high-energy physics, especially in attempts at unifying the fundamental interactions. •

3.4.4 The Trace

As is shown in the solution of Problem 3.10, a similarity transformation does not change the determinant of a matrix. Consider an operator whose matrix representation is given in one basis. In another basis the operator will have a different matrix representation, which is related to the original one by a similarity transformation. This implies that the determinant of the matrix representation of an operator is

[4] Lack of space unfortunately precludes treatment of this beautiful aspect of mathematical physics, but see Weyl (1946).

independent of the basis chosen to represent the operator. Thus, we can think of the determinant as an *intrinsic property of an operator*. (An elegant proof of this is presented in Chapter 4.)

Another quantity that is associated with a matrix and does not change with similarity transformations is called the trace of the matrix and is defined as follows.

3.4.18 Definition Let A be an $N \times N$ matrix. The mapping $\mathrm{tr}: \mathcal{M}^{(N)} \to \mathbb{C}$ (or $\mathbb{R}$) given by

$$\mathrm{tr}\,\mathsf{A} = \sum_{i=1}^{N} a_{ii}$$

is called the *trace of* A.

3.4.19 Theorem The trace is a linear mapping. Furthermore,

$$\mathrm{tr}(\mathsf{AB}) = \mathrm{tr}(\mathsf{BA})$$

Proof. The linearity of the trace follows directly from its definition. To prove the identity of the theorem, we use the definitions of the trace and the matrix product:

$$\begin{aligned}\mathrm{tr}(\mathsf{AB}) &= \sum_{i=1}^{N} (\mathsf{AB})_{ii} = \sum A_{ij}B_{ji} = \sum_{i,j} B_{ji}A_{ij} \\ &= \sum_{j=1}^{N}\left(\sum_{i=1}^{N} B_{ji}A_{ij}\right) = \sum_{j=1}^{N} (\mathsf{BA})_{jj} = \mathrm{tr}(\mathsf{BA})\end{aligned}$$

■

3.4.20 Proposition Similar matrices have the same trace.

Proof. Let $\mathsf{A}' = \mathsf{RAR}^{-1}$. Then

$$\begin{aligned}\mathrm{tr}\,\mathsf{A}' &= \mathrm{tr}(\mathsf{RAR}^{-1}) = \mathrm{tr}[\mathsf{R}(\mathsf{AR}^{-1})] = \mathrm{tr}[(\mathsf{AR}^{-1})\mathsf{R}] \\ &= \mathrm{tr}(\mathsf{AR}^{-1}\mathsf{R}) = \mathrm{tr}(\mathsf{A1}) = \mathrm{tr}\,\mathsf{A}\end{aligned}$$

■

Example 3.4.11

As an example of trace calculations and to gain familiarity with the summation notation, let us prove that

$$\mathrm{tr}(\mathsf{AS}) = 0$$

where A is an antisymmetric matrix and S a symmetric one.

Recall that a symmetric matrix is one that is equal to its transpose:

$$\tilde{\mathsf{S}} = \mathsf{S} \quad \Leftrightarrow \quad s_{ij} = s_{ji}$$

On the other hand, an antisymmetric matrix is the negative of its transpose:

$$\tilde{\mathsf{A}} = -\mathsf{A} \quad \Leftrightarrow \quad a_{ij} = -a_{ji}$$

Using these two facts, we can write the first equation of this example as

$$\begin{aligned}\mathrm{tr}(\mathsf{AS}) &= \sum_{i=1}^{N} (\mathsf{AS})_{ii} = \sum_{i=1}^{N}\sum_{j=1}^{N} a_{ij}s_{ji}\\ &= \sum_{i=1}^{N}\sum_{j=1}^{N} (-a_{ji})(s_{ij})\\ &= -\sum_{j=1}^{N}\left(\sum_{i=1}^{N} a_{ji}s_{ij}\right) = -\sum_{j=1}^{N} (\mathsf{AS})_{jj}\\ &= -\mathrm{tr}(\mathsf{AS})\end{aligned}$$

Transferring the RHS to the LHS, we get

$$2\mathrm{tr}(\mathsf{AS}) = 0 \quad \Rightarrow \quad \mathrm{tr}(\mathsf{AS}) = 0$$ •

Example 3.4.12

Let us show that

$$\det \mathsf{A} = e^{\mathrm{tr}(\ln \mathsf{A})}$$

for any matrix A that can be brought to a diagonal form via a similarity transformation. That is, for some R

$$\mathsf{D} = \mathsf{RAR}^{-1}$$

where D is a diagonal matrix, usually denoted by diag $(\lambda_1, \lambda_2, \ldots, \lambda_n)$, where the lambdas are the diagonal entries.

From Exercise 3.4.8 we have

$$\det \mathsf{D} = e^{\mathrm{tr}(\ln \mathsf{D})}$$

Thus,
$$\det(\mathsf{RAR}^{-1}) = e^{\mathrm{tr}[\ln(\mathsf{RAR}^{-1})]}$$

or, from Theorem 3.4.17 and the fact that $\det \mathsf{R}^{-1} = 1/\det \mathsf{R}$,

$$\det \mathsf{A} = e^{\mathrm{tr}[\ln(\mathsf{RAR}^{-1})]}$$

If we can show that

$$\mathrm{tr}[\ln(\mathsf{RAR}^{-1})] = \mathrm{tr}(\ln \mathsf{A})$$

we are done. Let us consider a general function of A that can be expanded in Taylor series:

$$f(\mathsf{A}) = \sum_{k=0}^{\infty} a_k \mathsf{A}^k$$

Applying a similarity transformation on both sides gives

$$\begin{aligned}\mathsf{R}f(\mathsf{A})\mathsf{R}^{-1} &= \sum_{k=0}^{\infty} a_k \mathsf{RA}^k\mathsf{R}^{-1} = \sum_{k=0}^{\infty} a_k \mathsf{R}\,\underbrace{\mathsf{AA}\cdots\mathsf{A}}_{k \text{ times}}\mathsf{R}^{-1}\\ &= \sum_{k=0}^{\infty} a_k \underbrace{\mathsf{RAR}^{-1}\mathsf{RAR}^{-1}\mathsf{R}\cdots\mathsf{RAR}^{-1}}_{k \text{ times}}\end{aligned}$$

where a factor of $R^{-1}R = 1$ has been introduced between the A's. So we have

$$Rf(A)R^{-1} = \sum_{k=0}^{\infty} a_k(RAR^{-1})^k = f(RAR^{-1})$$

This tells us very generally that the similarity transform of a function of a matrix is the function of the transform of that matrix. In particular, we have

$$\ln(RAR^{-1}) = R(\ln A)R^{-1}$$

Taking the trace of this equation and using Theorem 3.4.19, we obtain

$$\mathrm{tr}[\ln(RAR^{-1})] = \mathrm{tr}[R(\ln A)R^{-1}] = \mathrm{tr}(\ln A)$$

which is what we set out to show. ●

Both the determinant and the trace are mappings from $\mathcal{M}^{(N)}$ to $\mathbb{R}$. Counterexamples show that the determinant is not a linear mapping (see Exercise 3.4.3). However, the trace is, and this opens up the possibility of defining an inner product in the vector space of $N \times N$ matrices. In fact, we have the following proposition.

3.4.21 Proposition For any two matrices, $A, B \in \mathcal{M}^{(N)}$, the mapping $g:(\mathcal{M}^{(N)}) \times (\mathcal{M}^{(N)}) \to \mathbb{R}$ can be defined by

$$g(A, B) = \mathrm{tr}(A^\dagger B)$$

Then g is a sesquilinear inner product.

Proof. We have to show that g satisfies Eqs. (2.9). For (2.9a) we have

$$\begin{aligned} g(B,A) &= \mathrm{tr}(B^\dagger A) = \sum_{i,j} (B^\dagger)_{ij}(A)_{ji} \\ &= \sum_{i,j} b_{ji}^* a_{ji} = \left(\sum_{i,j} a_{ji}^* b_{ji}\right)^* = \left(\sum_{i,j} (A^\dagger)_{ij}(B)_{ji}\right)^* \\ &= (\mathrm{tr}(A^\dagger B))^* = (g(A,B))^* \end{aligned}$$

For (2.9b) we have

$$\begin{aligned} g(A, \beta B + \gamma C) &= \mathrm{tr}(A^\dagger(\beta B + \gamma C)) = \mathrm{tr}(\beta A^\dagger B + \gamma A^\dagger C) \\ &\underset{\uparrow}{=} \mathrm{tr}(\beta A^\dagger B) + \mathrm{tr}(\gamma A^\dagger C) = \beta g(A,B) + \gamma g(A,C) \end{aligned}$$

by the linearity of the trace

For (2.9c) we have

$$g(A,A) = \mathrm{tr}(A^\dagger A) = \sum_{i,j} a_{ji}^* a_{ji} = \sum_{i,j} |a_{ji}|^2$$

Clearly, $g(A,A) \geqslant 0$ and

$$g(A,A) = 0 \quad \Leftrightarrow \quad \sum_{i,j} |a_{ji}|^2 = 0 \quad \Leftrightarrow \quad a_{ji} = 0 \quad \Leftrightarrow \quad A = 0$$ ■

Exercises

3.4.1 Show that $\det(\alpha\mathsf{A}) = \alpha^N \det\mathsf{A}$ for an $N \times N$ matrix A and a complex number α.

3.4.2 Show that $\det \mathsf{1} = 1$ for any unit matrix.

3.4.3 Show that, in general, $\det(\mathsf{A} + \mathsf{B}) \neq \det\mathsf{A} + \det\mathsf{B}$. Therefore, the determinant is *not* a linear mapping.

3.4.4 Find the inverse of this matrix:

$$\mathsf{A} = \begin{pmatrix} 3 & -1 & 2 \\ 1 & 0 & -3 \\ -2 & 1 & -1 \end{pmatrix}$$

3.4.5 Show explicitly that $\det(\mathsf{AB}) = (\det\mathsf{A})(\det\mathsf{B})$ for 2×2 matrices.

3.4.6 Given two $N \times N$ matrices A and B such that $\mathsf{AB} = \mathsf{1}$, show that both A and B must be invertible.

3.4.7 Consider the three $N \times N$ matrices $\mathsf{L}_1, \mathsf{L}_2$, and L_3 such that $[\mathsf{L}_1, \mathsf{L}_2] = i\mathsf{L}_3$, $[\mathsf{L}_3, \mathsf{L}_1] = i\mathsf{L}_2$, and $[\mathsf{L}_2, \mathsf{L}_3] = i\mathsf{L}_1$. Show that

$$\operatorname{tr} \mathsf{L}_k = 0 \qquad \text{for } k = 1, 2, 3$$

3.4.8 Show that $\det \mathsf{D} = e^{\operatorname{tr}(\ln \mathsf{D})}$ where $\mathsf{D} \equiv \operatorname{diag}(\lambda_1, \lambda_2, \ldots, \lambda_n)$, a diagonal matrix whose nonzero entries are $\lambda_1, \lambda_2, \ldots, \lambda_n$.

3.4.9 Show directly that the similarity transformation induced by

$$\mathsf{R} = \begin{pmatrix} 1 & 2 & -1 \\ 0 & 1 & -2 \\ 2 & 1 & -1 \end{pmatrix}$$

does not change the determinant or the trace of

$$\mathsf{A} = \begin{pmatrix} 3 & -1 & 2 \\ 0 & 1 & -2 \\ 1 & -3 & -1 \end{pmatrix}$$

3.5 DIRECT SUMS AND INVARIANT SUBSPACES

This section presents some of the formal results needed for the remaining parts of this chapter.

3.5.1 Direct Sums

Sometimes it is possible, and convenient, to break up a vector space into special (disjoint) subspaces. For instance, for a plane in $\mathbb{R}^3$ going through the origin, it may be convenient to decompose any vector into its projection on that plane and its remaining component perpendicular to that plane. Such a decomposition breaks down an arbitrary vector in $\mathbb{R}^3$ into a vector in $\mathbb{R}^2$ and one in $\mathbb{R}$. We can generalize this as a definition.

3.5.1 Definition Let $\mathcal{U}$ and $\mathcal{W}$ be subspaces of a vector space $\mathcal{V}$ such that $\mathcal{V} = \mathcal{U} + \mathcal{W}$ and $\mathcal{U} \cap \mathcal{W} = \mathbf{0}$ (this means that the only vector common to both $\mathcal{U}$ and $\mathcal{W}$ is the zero vector). Then we say that $\mathcal{V}$ is the *direct sum* of $\mathcal{U}$ and $\mathcal{W}$ and write

$$\mathcal{V} = \mathcal{U} \oplus \mathcal{W}$$

Note that $\mathcal{V} = \mathcal{U} \oplus \mathcal{W}$ means that any vector in $\mathcal{V}$ can be decomposed into a vector in $\mathcal{U}$ plus a vector in $\mathcal{W}$. The following proposition shows that this decomposition is unique.

3.5.2 Proposition Let $\mathcal{U}$ and $\mathcal{W}$ be subspaces of $\mathcal{V}$. Then $\mathcal{V} = \mathcal{U} \oplus \mathcal{W}$ if and only if any vector in $\mathcal{V}$ can be written as a vector in $\mathcal{U}$ plus a vector in $\mathcal{W}$ *in a unique way.*

Proof. Let $\mathcal{V} = \mathcal{U} \oplus \mathcal{W}$, $|\mathrm{v}\rangle \in \mathcal{V}$, $|\mathrm{u}\rangle \in \mathcal{U}$, $|\mathrm{w}\rangle \in \mathcal{W}$, and $|\mathrm{v}\rangle = |\mathrm{u}\rangle + |\mathrm{w}\rangle$. If there are $|\mathrm{u}'\rangle \in \mathcal{U}$ and $|\mathrm{w}'\rangle \in \mathcal{W}$ such that

$$|\mathrm{v}\rangle = |\mathrm{u}'\rangle + |\mathrm{w}'\rangle$$

then

$$|\mathrm{u}\rangle + |\mathrm{w}\rangle = |\mathrm{u}'\rangle + |\mathrm{w}'\rangle \qquad \text{or} \qquad |\mathrm{u}\rangle - |\mathrm{u}'\rangle = |\mathrm{w}'\rangle - |\mathrm{w}\rangle$$

However, $(|\mathrm{u}\rangle - |\mathrm{u}'\rangle) \in \mathcal{U}$ and $(|\mathrm{w}'\rangle - |\mathrm{w}\rangle) \in \mathcal{W}$. Because of the definition of the direct sum, we must have

$$|\mathrm{u}\rangle - |\mathrm{u}'\rangle = \mathbf{0} = |\mathrm{w}\rangle - |\mathrm{w}'\rangle$$

and the uniqueness of the decomposition is established.

Conversely, if $|\mathrm{a}\rangle \in \mathcal{U}$ and also $|\mathrm{a}\rangle \in \mathcal{W}$, then

$$|\mathrm{a}\rangle = \underset{\text{in } \mathcal{U}}{\underset{\uparrow}{|\mathrm{a}\rangle}} + \underset{\text{in } \mathcal{W}}{\underset{\uparrow}{\mathbf{0}}}$$

and

$$|\mathrm{a}\rangle = \underset{\text{in } \mathcal{U}}{\underset{\uparrow}{\mathbf{0}}} + \underset{\text{in } \mathcal{W}}{\underset{\uparrow}{|\mathbf{a}\rangle}}$$

This decomposition is clearly *not* unique unless $|\mathrm{a}\rangle = \mathbf{0}$. Therefore, the only vector common to both $\mathcal{U}$ and $\mathcal{W}$ is the zero vector. This implies that $\mathcal{V} = \mathcal{U} \oplus \mathcal{W}$. ■

3.5.3 Proposition The dimension of a direct sum is the sum of the dimensions of its summands. In other words, if $\mathcal{V} = \mathcal{U} \oplus \mathcal{W}$, then

$$\dim \mathcal{V} = \dim \mathcal{U} + \dim \mathcal{W}$$

Proof. Let $\{|\mathrm{u}_i\rangle\}_{i=1}^{m}$ be a basis for $\mathcal{U}$ and $\{|\mathrm{w}_i\rangle\}_{i=1}^{k}$ be a basis for $\mathcal{W}$. Then it is easily verified that $\{|\mathrm{u}_1\rangle, |\mathrm{u}_2\rangle, \ldots, |\mathrm{u}_m\rangle, |\mathrm{w}_1\rangle, |\mathrm{w}_2\rangle, \ldots, |\mathrm{w}_k\rangle\}$ is a basis for $\mathcal{V}$. The rest of the proof is left as an easy problem. ■

Example 3.5.1

Let $\mathcal{V}$ be an inner product space. Let $\mathcal{M}$ be any subspace of $\mathcal{V}$. Denote by $\mathcal{M}^{\perp}$ the set of all vectors in $\mathcal{V}$ orthogonal to all the vectors in $\mathcal{M}$. It is easily shown that $\mathcal{M}^{\perp}$ is also a subspace of $\mathcal{V}$. In fact, if $|a\rangle, |b\rangle \in \mathcal{M}^\perp$, then for any vector $|c\rangle \in \mathcal{M}$, we have

$$\langle c|(\alpha|a\rangle + \beta|b\rangle) = \underbrace{\alpha\langle c|a\rangle}_{\substack{=0,\\ \text{because}\\ |a\rangle\in\mathcal{M}^\perp}} + \underbrace{\beta\langle c|b\rangle}_{\substack{=0,\\ \text{because}\\ |b\rangle\in\mathcal{M}^\perp}} = 0$$

$\mathcal{M}^\perp$ (pronounced "em perp") is called the *orthogonal complement of* $\mathcal{M}$.

Now consider an orthonormal basis $B_1 = \{|e_i\rangle\}_{i=1}^m$ for $\mathcal{M}$, where $m = \dim \mathcal{M}$. The set of all vectors $|a_k\rangle \in \mathcal{V}$ such that $\langle e_i|a_k\rangle = 0$, for all i, clearly spans $\mathcal{M}^\perp$. The largest set of linearly independent vectors among $|a_k\rangle$ forms a basis $B_2 = \{|a_k\rangle_{k=1}^l$ for $\mathcal{M}^\perp$ with $l = \dim \mathcal{M}^\perp$. Now we construct a projection operator,

$$\mathbb{P}_{\mathcal{M}} \equiv \sum_{i=1}^{m} |e_i\rangle\langle e_i|$$

This is the operator that projects an arbitrary vector $|a\rangle \in \mathcal{V}$ onto the subspace $\mathcal{M}$. In fact, for any $|a\rangle \in \mathcal{V}$,

$$\mathbb{P}_{\mathcal{M}}|a\rangle = \sum_{i=1}^{m} |e_i\rangle\langle e_i|a\rangle \equiv \sum_{i=1}^{m} \alpha_i|e_i\rangle \in \mathcal{M}$$

We now claim that $\mathbb{1} - \mathbb{P}_{\mathcal{M}}$ is the projection operator that projects onto $\mathcal{M}^\perp$. Clearly, $\mathbb{1} - \mathbb{P}_{\mathcal{M}}$ is hermitian. Furthermore,

$$(\mathbb{1} - \mathbb{P}_{\mathcal{M}})^2 = (\mathbb{1} - \mathbb{P}_{\mathcal{M}})(\mathbb{1} - \mathbb{P}_{\mathcal{M}}) = \mathbb{1} - 2\mathbb{P}_{\mathcal{M}} + \mathbb{P}_{\mathcal{M}}^2$$
$$= \mathbb{1} - 2\mathbb{P}_{\mathcal{M}} + \mathbb{P}_{\mathcal{M}} = \mathbb{1} - \mathbb{P}_{\mathcal{M}}$$

Thus, $\mathbb{1} - \mathbb{P}_{\mathcal{M}}$ is a projection operator. For any $|a\rangle \in \mathcal{V}$, let $|b\rangle = (\mathbb{1} - \mathbb{P}_{\mathcal{M}})|a\rangle$. Then

$$\langle e_i|b\rangle = \langle e_i|(\mathbb{1} - \mathbb{P}_{\mathcal{M}})|a\rangle = \langle a|(\mathbb{1} - \mathbb{P}_{\mathcal{M}})^\dagger|e_i\rangle^* = \langle a|(\mathbb{1} - \mathbb{P}_{\mathcal{M}})|e_i\rangle^*$$
$$= \langle a|e_i\rangle^* - \langle a|\mathbb{P}_{\mathcal{M}}|e_i\rangle^* = \langle a|e_i\rangle^* - \langle a|e_i\rangle^* = 0$$

because $\mathbb{P}_{\mathcal{M}}|e_i\rangle = |e_i\rangle$. Therefore, $|b\rangle = (\mathbb{1} - \mathbb{P}_{\mathcal{M}})|a\rangle$ belongs to $\mathcal{M}^\perp$ for arbitrary $|a\rangle$. This shows that $\mathbb{1} - \mathbb{P}_{\mathcal{M}}$ projects onto $\mathcal{M}^\perp$.

An arbitrary vector $|a\rangle \in \mathcal{V}$ can be written as

$$|a\rangle = (\mathbb{P}_{\mathcal{M}} + \mathbb{1} - \mathbb{P}_{\mathcal{M}})|a\rangle = \underbrace{\mathbb{P}_{\mathcal{M}}|a\rangle}_{\in\mathcal{M}} + \underbrace{(\mathbb{1} - \mathbb{P}_{\mathcal{M}})|a\rangle}_{\in\mathcal{M}^\perp}$$

Furthermore, the only vector that can be in both $\mathcal{M}$ and $\mathcal{M}^\perp$ is the zero vector, which is the only vector orthogonal to itself. We, therefore, conclude that $\mathcal{V} = \mathcal{M} \oplus \mathcal{M}^\perp$. •

3.5.2 Invariant Subspaces

This subsection explores the possibility of obtaining subspaces by means of the action of a linear operator on vectors of $\mathcal{V}$. Let $|a\rangle$ be any vector in $\mathcal{V}$ and $\mathbb{A}$ be a linear operator on $\mathcal{V}$. Then the sequence of vectors

$$|a\rangle, \mathbb{A}|a\rangle, \mathbb{A}^2|a\rangle, \ldots, \mathbb{A}^k|a\rangle, \ldots$$

all belong to $\mathscr{V}$. Let m be the maximum number of linearly independent vectors among that sequence. Clearly, $m \leqslant N$. Now let $\mathscr{M}_{\mathbb{A}}^{a}$ be the m-dimensional subspace of $\mathscr{V}$ spanned by the above vectors. The superscript on $\mathscr{M}$ indicates that the subspace is obtained by repeated action of $\mathbb{A}$ on $|a\rangle$. The subspace has the property that for any vector $|x\rangle \in \mathscr{M}_{\mathbb{A}}^{a}$ the vector $\mathbb{A}|x\rangle$ also belongs to $\mathscr{M}_{\mathbb{A}}^{a}$. In other words, no vector in $\mathscr{M}_{\mathbb{A}}^{a}$ "leaves" the subspace when acted on by $\mathbb{A}$. Of course, if another operator acts on $|x\rangle$, in general, the result will *not* be in $\mathscr{M}_{\mathbb{A}}^{a}$. We say that $\mathscr{M}_{\mathbb{A}}^{a}$ is an *invariant subspace* of the operator $\mathbb{A}$.

Now consider a basis $B = \{|a_i\rangle\}_{i=1}^{N}$ of $\mathscr{V}$, whose first m vectors span $\mathscr{M}_{\mathbb{A}}^{a}$. This can always be done. In fact, given any set of $k < N$ linearly independent vectors in $\mathscr{V}$, we can add other linearly independent vectors to form a basis in $\mathscr{V}$. Let us look at the matrix representation of $\mathbb{A}$ in such a basis. This is given by the relation

$$\mathbb{A}|a_i\rangle = \sum_{j=1}^{N} a_{ji}|a_j\rangle \qquad \text{for } i = 1, 2, \ldots, N$$

If $i \leqslant m$, then $a_{ji} = 0$ for $j > m$, because $\mathbb{A}|a_i\rangle$ belongs to $\mathscr{M}_{\mathbb{A}}^{a}$ when $i \leqslant m$ and, therefore, can be written as a linear combination of *only* $\{|a_1\rangle, \ldots, |a_m\rangle\}$. Thus, the matrix representation of $\mathbb{A}$ in B will have the form

$$\mathsf{A} = \begin{pmatrix} \boxed{\quad \mathsf{A}_1 \quad} & \\ 0 \quad 0 \cdots 0 & \\ 0 \quad 0 \cdots 0 & \mathsf{A}_2 \\ \vdots \quad \vdots \quad\ \ \vdots & \\ 0 \quad 0 \cdots 0 & \end{pmatrix}$$

where A_1 is an $m \times m$ matrix and A_2 is an $N \times (N - m)$ matrix. We say that

$$\mathsf{A}_1 = \begin{pmatrix} a_{11} & a_{12} & \cdots & a_{1m} \\ a_{21} & a_{22} & \cdots & a_{2m} \\ \vdots & \vdots & & \vdots \\ a_{m1} & a_{m2} & \cdots & a_{mm} \end{pmatrix}$$

represents the operator $\mathbb{A}$ in the m-dimensional subspace $\mathscr{M}_{\mathbb{A}}^{a}$.

It may also be possible to choose the remaining basis vectors in B, namely $|a_{m+1}\rangle, |a_{m+2}\rangle, \ldots, |a_N\rangle$, in such a way that the $N \times (N - m)$ matrix

$$\mathsf{A}_2 = \begin{pmatrix} a_{1,m+1} & a_{1,m+2} & \cdots & a_{1N} \\ a_{2,m+1} & a_{2,m+2} & \cdots & a_{2N} \\ \vdots & \vdots & & \vdots \\ a_{N,m+1} & a_{N,m+2} & \cdots & a_{NN} \end{pmatrix}$$

takes the form

$$\mathsf{A}_2 = \begin{pmatrix} 0 & 0 & \dots & 0 \\ 0 & 0 & \dots & 0 \\ \vdots & \vdots & & \\ 0 & 0 & \dots & 0 \\ \multicolumn{4}{|c|}{\mathsf{A}_3} \end{pmatrix}$$

where A_3 is an $(N-m) \times (N-m)$ matrix. This is possible if the remaining vectors, $|\mathsf{a}_{m+1}\rangle, \dots, |\mathsf{a}_N\rangle$, also form a (different) invariant subspace. In that case the matrix A will be decomposed into

$$\mathsf{A} = \begin{pmatrix} \mathsf{A}_1 & \mathsf{0}_1 \\ \mathsf{0}_2 & \mathsf{A}_3 \end{pmatrix}$$

where A_1 is an $m \times m$ matrix and A_3 is an $(N-m) \times (N-m)$ matrix. $\mathsf{0}_1$ and $\mathsf{0}_2$ are, respectively, the $m \times (N-m)$ and $(N-m) \times m$ zero matrices. A matrix that can be brought into this form by a suitable choice of a basis is said to be *reducible*. If it is impossible to bring a matrix into this above form by a change of basis, then the matrix is said to be *irreducible*. If a matrix A is reducible into A_1 and A_2 such that

$$\mathsf{A} = \begin{pmatrix} \mathsf{A}_1 & \mathsf{0} \\ \mathsf{0} & \mathsf{A}_2 \end{pmatrix}$$

we write

$$\mathsf{A} = \mathsf{A}_1 \oplus \mathsf{A}_2 \tag{3.22}$$

3.5.4 Proposition For an $N \times N$ matrix A having the block form

$$\mathsf{A} = \begin{pmatrix} \boxed{\mathsf{A}_1} & \mathsf{0} \\ \boxed{\mathsf{B}} & \boxed{\mathsf{A}_2} \end{pmatrix} \quad \text{or} \quad \mathsf{A} = \begin{pmatrix} \boxed{\mathsf{A}_1} & \boxed{\mathsf{B}} \\ \mathsf{0} & \boxed{\mathsf{A}_2} \end{pmatrix}$$

we have

$$\det \mathsf{A} = (\det \mathsf{A}_1)(\det \mathsf{A}_2)$$

Proof. Let us prove the theorem for the first block form shown. The proof for the second block form follows from taking the transpose of A and noting that the determinant does not change in this transposition.

Let us assume that A_1 is $m \times m$, A_2 is $(N-m) \times (N-m)$, and B is $m \times (N-m)$. From the definition of the determinant, we have

$$\begin{aligned}\det \mathsf{A} &= \sum_i \varepsilon_{i_1 i_2 \dots i_N} a_{1i_1} a_{2i_2} \cdots a_{Ni_N} \\ &= \sum_i \varepsilon_{i_1 i_2 \cdots i_N} \underbrace{a_{1i_1} \cdots a_{mi_m}}_{\substack{\text{all } = 0 \text{ unless} \\ i_1, i_2, \dots, i_m \\ \text{are all } \leqslant m}} \underbrace{a_{m+1, i_{m+1}} \cdots a_{Ni_N}}_{\substack{\text{all } = 0 \text{ unless} \\ i_{m+1}, \dots, i_N \\ \text{are all } \geqslant m}} \\ &= \sum_{i_1 i_2 \dots i_m = 1}^{m} \; \sum_{i_{m+1} \dots i_N = m+1}^{N} \varepsilon_{i_1 i_2 \dots i_N} a_{1i_1} \cdots a_{mi_m} a_{m+1, i_{m+1}} \cdots a_{Ni_N}\end{aligned}$$

Since the first m indices of $\varepsilon_{i_1 i_2 \cdots i_N}$ are never exchanged with any of the remaining $N - m$ indices, we can write

$$\varepsilon_{i_1 \ldots i_m i_{m+1} \ldots i_N} = \varepsilon_{i_1 \ldots i_m} \varepsilon_{i_{m+1} \ldots i_N}$$

Thus, the above expansion becomes

$$\det \mathsf{A} = \sum_{i_1 \ldots i_m}^{m} \varepsilon_{i_1 \ldots i_m} a_{1 i_1} \cdots a_{m i_m} \sum_{i_{m+1} \ldots i_N}^{N} \varepsilon_{i_{m+1} \ldots i_N} a_{m+1, i_{m+1}} \cdots a_{N i_N}$$

The first sum is $\det \mathsf{A}_1$, and the second is $\det \mathsf{A}_2$. Thus, we have $\det \mathsf{A} = (\det \mathsf{A}_1)(\det \mathsf{A}_2)$. ■

Note that the block B plays no role here. In particular, if A is reducible, so

$$\mathsf{A} = \mathsf{A}_1 \oplus \mathsf{A}_2$$

then, again,

$$\det \mathsf{A} = (\det \mathsf{A}_1)(\det \mathsf{A}_2)$$

3.6 SPECTRAL DECOMPOSITION AND DIAGONALIZATION

The main goal of this section is to prove that hermitian operators are diagonalizable, that is, that we can always find an orthonormal basis in which a hermitian operator is represented by a diagonal matrix. Furthermore, if a hermitian operator is represented in any orthonormal basis to begin with, there exists a unitary transformation matrix that brings that matrix representation to a diagonal form. In addition, we will prove the so-called spectral theorem, by which the hermitian operator is written as a linear combination of certain projection operators.

This section is admittedly a formal one. However, the mathematical enrichment involved makes the formalism worthwhile.

3.6.1 Eigenvalues and Eigenvectors

Let us begin by considering eigenvalues and eigenvectors, which are generalizations of familiar concepts in two and three dimensions. In particular, consider the operation of rotation about the z-axis by an angle θ, denoted by $\mathbb{R}_z(\theta)$. Such a rotation takes any vector (x, y) in the xy-plane to a new vector $(x \cos\theta - y \sin\theta, x \sin\theta + y \cos\theta)$. Thus, unless $(x, y) = (0, 0)$ or $\theta = 0$, the vector will change. Is there a vector that is so special that it does not change when acted on by $\mathbb{R}_z(\theta)$? If we lift ourselves up from the two-dimensional xy-plane, we immediately encounter many such vectors, all of which lie along the z-axis.

The foregoing example can be generalized to any rotation (normally specified by Euler angles). In fact, the methods developed in this section can be used to show that a general rotation, given by Euler angles, always has an unchanged vector lying along the axis around which the rotation takes place. This concept is further generalized in the following definition.

3.6.1 Definition A scalar λ is an *eigenvalue* and a nonzero vector $|a\rangle \in \mathcal{V}$ is an *eigenvector* of the linear transformation $\mathbb{A} \in \mathcal{L}(\mathcal{V})$ if

$$\mathbb{A}|a\rangle = \lambda|a\rangle \tag{3.23}$$

3.6.2 Proposition If the zero vector is added to the set of all eigenvectors of $\mathbb{A}$ belonging to the *same* eigenvalue λ and the resulting set is called $\mathcal{M}_\lambda$, then $\mathcal{M}_\lambda$ is a subspace of $\mathcal{V}$.

Proof. The proof follows immediately from Definition 3.6.1 and the definition of a subspace. ■

The dimension of $\mathcal{M}$ is referred to as the *geometric multiplicity* of λ. An eigenvalue is called *simple* if its geometric multiplicity is 1. The set of eigenvalues of $\mathbb{A}$ is called the *spectrum* of $\mathbb{A}$.

Let us rewrite (3.23) as

$$(\mathbb{A} - \lambda\mathbb{1})|a\rangle = 0$$

This equation says that $|a\rangle$ is an eigenvector of $\mathbb{A}$ if and only if $|a\rangle$ belongs to the null space of $\mathbb{A} - \lambda\mathbb{1}$. If $\mathbb{A} - \lambda\mathbb{1}$ is invertible, then the null space will consist of only the zero vector, which is not acceptable as a solution of (3.23). Thus, if we are to obtain nontrivial solutions, $\mathbb{A} - \lambda\mathbb{1}$ *must not have an inverse*. The isomorphism between linear operators and the $N \times N$ matrices representing them implies that the matrix representing $\mathbb{A} - \lambda\mathbb{1}$ must not be invertible. This is true if and only if

$$\det(\mathbb{A} - \lambda\mathbb{1}) = 0 \tag{3.24}$$

The invariance of the determinant under a basis transformation assures us that if (3.24) holds in one basis it will hold in all bases.

The determinant in (3.24) is a polynomial in λ, called the *characteristic polynomial of* $\mathbb{A}$. The roots of this polynomial are called *characteristic roots* and are simply the eigenvalues of $\mathbb{A}$. The coefficient of λ^n in the characteristic polynomial is clearly $(-1)^n$. Thus, we can write

$$\det(\mathbb{A} - \lambda\mathbb{1}) = \alpha_0 + \alpha_1\lambda + \cdots + \alpha_{n-1}\lambda^{n-1} + (-1)^n\lambda^n \tag{3.25}$$

Setting $\lambda = 0$ on both sides gives

$$\det \mathbb{A} = \alpha_0$$

On the other hand, any polynomial in $\mathbb{C}$ can be completely factored. Let $\lambda_1, \lambda_2, \ldots, \lambda_p$ be the distinct roots of the characteristic polynomial of $\mathbb{A}$, and let λ_j occur m_j times. Then (3.25) can be written as

$$\det(\mathbb{A} - \lambda\mathbb{1}) = (\lambda_1 - \lambda)^{m_1} \cdots (\lambda_p - \lambda)^{m_p} \equiv \prod_{j=1}^{p} (\lambda_j - \lambda)^{m_j} \tag{3.26}$$

which for $\lambda = 0$ gives

$$\det \mathbb{A} = \lambda_1^{m_1}\lambda_2^{m_2}\cdots\lambda_p^{m_p} = \prod_{j=1}^{p}\lambda_j^{m_j} \tag{3.27}$$

Equation (3.27) states that the determinant of an operator is the product of all its eigenvalues. In particular, if one of the eigenvalues is zero, then the operator is not invertible.

Example 3.6.1

Let us find the eigenvalues of a projection operator, $\mathbb{P}$. If $|a\rangle$ is an eigenvector, then

$$\mathbb{P}|a\rangle = \lambda|a\rangle$$

Applying $\mathbb{P}$ on both sides again, we obtain

$$\mathbb{P}^2|a\rangle = \lambda\mathbb{P}|a\rangle = \lambda(\lambda|a\rangle) = \lambda^2|a\rangle$$

But $\mathbb{P}^2 = \mathbb{P}$; thus,

$$\mathbb{P}|a\rangle = \lambda^2|a\rangle$$

This implies that $\lambda^2|a\rangle = \lambda|a\rangle$, or $(\lambda^2 - \lambda)|a\rangle = 0$. Since $|a\rangle \neq 0$, we get $\lambda(\lambda - 1) = 0$, or $\lambda = 0, 1$. Thus, the only eigenvalues of $\mathbb{P}$ are 0 and 1. The presence of zero as an eigenvalue of $\mathbb{P}$ is another indication that $\mathbb{P}$ cannot be invertible. ●

In Eqs. (3.26) and (3.27) m_j is called the *algebraic multiplicity* of λ_j. We have already defined the geometric multiplicity of λ_j as the dimension of the subspace spanned by the eigenvectors corresponding to λ_j. If we denote the geometric multiplicity of λ_j as g_j, we have a proposition.

3.6.3 Proposition For any eigenvalue λ_j of $\mathbb{A}$, $g_j \leqslant m_j$.

Proof. We note that $\mathcal{M}_{\lambda_j}$ is a g_j-dimensional invariant subspace, because if $|a\rangle \in \mathcal{M}_{\lambda_j}$, then $\mathbb{A}|a\rangle = \lambda_j|a\rangle$. Thus, in a basis of $\mathcal{V}$ whose first g_j vectors form a basis of $\mathcal{M}_{\lambda_j}$, the matrix of $\mathbb{A}$ will look like

$$\mathsf{A} = \begin{pmatrix} \mathsf{A}_j & \mathsf{B} \\ 0 & \mathsf{C} \end{pmatrix}$$

and

$$\mathsf{A} - \lambda\mathsf{1} = \begin{pmatrix} \mathsf{A}_j - \lambda\mathsf{1}_{g_j} & \mathsf{B} \\ 0 & \mathsf{C} - \lambda_{N-g_j} \end{pmatrix}$$

where A_j and $\mathsf{1}_{g_j}$ are $g_j \times g_j$ matrices and C and $\mathsf{1}_{N-g_j}$ are $(N - g_j) \times (N - g_j)$ matrices. From Proposition 3.5.4 we have

$$\det(\mathsf{A} - \lambda\mathsf{1}) = [\det(\mathsf{A}_j - \lambda\mathsf{1}_{g_j})][\det(\mathsf{C} - \lambda\mathsf{1}_{N-g_j})]$$

The first factor on the right is simply the characteristic polynomial of A_j. However, by the definition of $\mathcal{M}_{\lambda_j}$, the only root of this polynomial is λ_j and it occurs g_j times. Thus, we may write

$$\det(\mathsf{A} - \lambda\mathsf{1}) = (\lambda_j - \lambda)^{g_j}\det(\mathsf{C} - \lambda\mathsf{1}_{N-g_j})$$

On the other hand, we already know from (3.26) that the algebraic multiplicity of λ_j cannot exceed m_j. We, therefore, conclude that

$$g_j \leqslant m_j \qquad \blacksquare$$

From Proposition 3.6.3 we get a useful corollary.

3.6.4 Corollary If $\lambda_1, \lambda_2, \ldots, \lambda_p$ are the distinct eigenvalues of $\mathbb{A}$ with respective geometric multiplicities $g_1, g_2, \ldots, g_p$, and if $\sum_{j=1}^{p} g_j = N$, then $g_j = m_j$ for all $j = 1, \ldots, p$, where m_j is the algebraic multiplicity of λ_j. ■

3.6.2 Eigenvalues of Hermitian Operators

This subsection focuses on hermitian operators with occasional reference to unitary operators. Let us start by proving some general results.

3.6.5 Theorem A linear operator $\mathbb{A}$ on an inner product space is $\mathbb{O}$ if and only if $\langle \mathbb{A}a|b\rangle = 0$ for all $|a\rangle$ and $|b\rangle$.[5]

Proof.

$$\mathbb{A} = \mathbb{O} \quad \Rightarrow \quad \mathbb{A}|a\rangle \equiv |\mathbb{A}a\rangle = 0$$

Conversely, if

$$\langle \mathbb{A}a|b\rangle = 0 \qquad \forall\ |a\rangle, |b\rangle$$

then for $|b\rangle = \mathbb{A}|a\rangle \equiv |\mathbb{A}a\rangle$ in particular, we must have

$$\langle \mathbb{A}a|\mathbb{A}a\rangle = 0 \quad \forall\ |a\rangle \quad \Rightarrow \quad \mathbb{A}|a\rangle = 0 \quad \forall\ |a\rangle \quad \Rightarrow \quad \mathbb{A} = \mathbb{O} \qquad \blacksquare$$

3.6.6 Theorem A linear operator $\mathbb{A}$ on a complex inner product space (also called a *unitary space*) is $\mathbb{O}$ if and only if $\langle \mathbb{A}a|a\rangle = 0$ for all $|a\rangle$.

Proof. Clearly, if $\mathbb{A} = \mathbb{O}$, then $\langle \mathbb{A}a|a\rangle = 0$. The converse can be shown by considering

$$\begin{aligned}\langle \mathbb{A}(\alpha a + \beta b)|(\alpha a + \beta b)\rangle &\equiv [\mathbb{A}(\alpha|a\rangle + \beta|b\rangle)]^\dagger(\alpha|a\rangle + \beta|b\rangle)\\ &= (\alpha^*\langle a| + \beta^*\langle b|)\mathbb{A}^\dagger(\alpha|a\rangle + \beta|b\rangle) = |\alpha|^2\langle a|\mathbb{A}^\dagger|a\rangle\\ &\quad + \alpha^*\beta\langle a|\mathbb{A}^\dagger|b\rangle + \beta^*\alpha\langle b|\mathbb{A}^\dagger|a\rangle + |\beta|^2\langle b|\mathbb{A}^\dagger|b\rangle\\ &\equiv |\alpha|^2\langle \mathbb{A}a|a\rangle + \alpha^*\beta\langle \mathbb{A}a|b\rangle + \beta^*\alpha\langle \mathbb{A}b|a\rangle + |\beta|^2\langle \mathbb{A}b|b\rangle\end{aligned}$$

We rewrite this in the form known as the *polarization identity*:

$$\alpha\beta^*\langle \mathbb{A}b|a\rangle + \alpha^*\beta\langle \mathbb{A}a|b\rangle = \langle \mathbb{A}(\alpha a + \beta b)|\alpha a + \beta b\rangle - |\alpha|^2\langle \mathbb{A}a|a\rangle - |\beta|^2\langle \mathbb{A}b|b\rangle$$

[5] It is very convenient to use the notation $|\mathbb{A}a\rangle$ for $\mathbb{A}|a\rangle$, because the hermitian conjugate $\langle a|\mathbb{A}^\dagger$ is then simply written as $\langle \mathbb{A}a|$. This should be kept in mind. For example, $\langle \lambda a| = \langle a|\lambda^* = \lambda^*\langle a|$, by the definition of hermitian conjugate of a ket.

According to the assumption of the theorem, the RHS is zero. Thus,

$$\alpha\beta^*\langle\mathbb{A}b|a\rangle + \alpha^*\beta\langle\mathbb{A}a|b\rangle = 0$$

This is true for all α and β. We let $\alpha = \beta = 1$ to obtain

$$\langle\mathbb{A}b|a\rangle + \langle\mathbb{A}a|b\rangle = 0$$

Now we let $\alpha = i$ and $\beta = 1$ to get

$$i\langle\mathbb{A}b|a\rangle - i\langle\mathbb{A}a|b\rangle = 0$$

These two equations give

$$\langle\mathbb{A}a|b\rangle = 0 \qquad \forall\ |a\rangle, |b\rangle$$

By Theorem 3.6.5 $\mathbb{A} = \mathbb{O}$. ■

We are now ready to prove a fundamental theorem concerning hermitian operators.

3.6.7 Theorem A linear transformation $\mathbb{H}$ on a unitary space is hermitian if and only if $\langle\mathbb{H}a|a\rangle$ is *real* for all $|a\rangle$.

Proof. If $\mathbb{H} = \mathbb{H}^\dagger$, then

$$\langle\mathbb{H}a|a\rangle^* = ((\mathbb{H}|a\rangle)^\dagger|a\rangle)^* = \langle a|\mathbb{H}^\dagger|a\rangle^* = \langle a|(\mathbb{H}^\dagger)^\dagger|a\rangle = \langle a|\mathbb{H}|a\rangle$$
$$= (\mathbb{H}^\dagger|a\rangle)^\dagger|a\rangle = \langle\mathbb{H}^\dagger a|a\rangle = \langle\mathbb{H}a|a\rangle$$

and $\langle\mathbb{H}a|a\rangle$ must be real.

Conversely, assume that $\langle\mathbb{H}a|a\rangle$ is real for all $|a\rangle$. Then

$$\langle\mathbb{H}a|a\rangle = \langle\mathbb{H}a|a\rangle^* = \langle a|\mathbb{H}^\dagger|a\rangle^* = \langle a|\mathbb{H}|a\rangle = \langle\mathbb{H}^\dagger a|a\rangle \qquad \forall\ |a\rangle$$

Thus, we have

$$\langle(\mathbb{H} - \mathbb{H}^\dagger)a|a\rangle = 0 \qquad \forall\ |a\rangle$$

By Theorem 3.6.6 we must have $\mathbb{H} = \mathbb{H}^\dagger$. ■

Example 3.6.2

The matrix $\mathsf{H} = \begin{pmatrix} 0 & -i \\ i & 0 \end{pmatrix}$ is hermitian and acts on $\mathbb{C}^2$. Let us take an arbitrary vector $|a\rangle = \begin{pmatrix} \alpha_1 \\ \alpha_2 \end{pmatrix}$ and evaluate $\langle\mathbb{H}a|a\rangle$. We have

$$|\mathbb{H}a\rangle \equiv \mathbb{H}|a\rangle = \begin{pmatrix} 0 & -i \\ i & 0 \end{pmatrix}\begin{pmatrix} \alpha_1 \\ \alpha_2 \end{pmatrix} = \begin{pmatrix} -i\alpha_2 \\ i\alpha_1 \end{pmatrix}$$

Therefore,

$$\langle \mathbb{H}a|a\rangle = (i\alpha_2^* \quad -i\alpha_1^*)\begin{pmatrix}\alpha_1\\ \alpha_2\end{pmatrix} = i\alpha_2^*\alpha_1 - i\alpha_1^*\alpha_2$$

$$= i\alpha_2^*\alpha_1 + (i\alpha_2^*\alpha_1)^* = 2\mathrm{Re}(i\alpha_2^*\alpha_1)$$

and $\langle \mathbb{H}a|a\rangle$ is real.

Let us now consider the most general 2×2 matrix, $\mathsf{H} = \begin{pmatrix} a & b\\ b^* & c\end{pmatrix}$, where a and c are real. Then

$$|\mathbb{H}a\rangle = \begin{pmatrix} a & b\\ b^* & c\end{pmatrix}\begin{pmatrix}\alpha_1\\ \alpha_2\end{pmatrix} = \begin{pmatrix} a\alpha_1 + b\alpha_2\\ b^*\alpha_1 + c\alpha_2\end{pmatrix}$$

and

$$\langle \mathbb{H}a|a\rangle = (a\alpha_1^* + b^*\alpha_2^* \quad b\alpha_1^* + c\alpha_2^*)\begin{pmatrix}\alpha_1\\ \alpha_2\end{pmatrix} = a|\alpha_1|^2 + b^*\alpha_2^*\alpha_1 + b\alpha_1^*\alpha_2 + c|\alpha_2|^2$$

$$= a|\alpha_1|^2 + c|\alpha_2|^2 + b^*\alpha_2^*\alpha_1 + (b^*\alpha_2^*\alpha_1)^*$$

$$= a|\alpha_1|^2 + c|\alpha_2|^2 + 2\mathrm{Re}(b^*\alpha_2^*\alpha_1)$$

Again $\langle \mathbb{H}a|a\rangle$ is real.

These are concrete examples of Theorem 3.6.7, which asserts the above results for all hermitian operators. ●

An operator $\mathbb{A}$ on an inner product space is called *positive* (written $\mathbb{A} \geqslant 0$) if $\mathbb{A}$ is hermitian and

$$\langle \mathbb{A}a|a\rangle \geqslant 0 \qquad \forall\, |a\rangle \in \mathcal{V}$$

Example 3.6.3

An example of a positive operator is the square of a hermitian operator. We note that for any hermitian operator $\mathbb{H}$

$$\mathbb{H}^2 = (\mathbb{H}^\dagger)^2 = (\mathbb{H}^2)^\dagger$$

Also, $$\langle \mathbb{H}^2 a|a\rangle = \langle a|(\mathbb{H}^2)^\dagger|a\rangle = \langle a|\mathbb{H}^\dagger\mathbb{H}|a\rangle = \langle \mathbb{H}a|\mathbb{H}a\rangle \geqslant 0$$

because of the properties of the inner product. Furthermore, if $\langle \mathbb{H}a|\mathbb{H}a\rangle = 0$ for an invertible $\mathbb{H}$, we must have

$$|\mathbb{H}a\rangle = \mathbb{H}|a\rangle = 0 \quad \Rightarrow \quad |a\rangle = \mathbf{0}$$

An operator satisfying the extra condition that $\langle \mathbb{A}a|a\rangle = 0$ implies $|a\rangle = \mathbf{0}$, is called a *positive definite operator*. From the above argument, we conclude that the square of an invertible hermitian operator is positive definite. ●

3.6.8 Theorem If $\mathbb{H}$ is a hermitian operator on a unitary space, then every eigenvalue of $\mathbb{H}$ is real. Furthermore, if $\mathbb{H}$ is positivc or positive definite, then so is every eigenvalue of $\mathbb{H}$.

Proof. Let $|\mathrm{h}\rangle$ be an eigenvector of $\mathbb{H}$ with eigenvalue λ. Then $\mathbb{H}|\mathrm{h}\rangle = \lambda|\mathrm{h}\rangle$ with $|\mathrm{h}\rangle \neq \mathbf{0}$, and we can write

$$\frac{\langle \mathbb{H}\mathrm{h}|\mathrm{h}\rangle}{\langle \mathrm{h}|\mathrm{h}\rangle} = \frac{\langle \lambda\mathrm{h}|\mathrm{h}\rangle}{\langle \mathrm{h}|\mathrm{h}\rangle} = \frac{\lambda^*\langle \mathrm{h}|\mathrm{h}\rangle}{\langle \mathrm{h}|\mathrm{h}\rangle} = \lambda^*$$

However, Theorem 3.6.7 implies that the LHS is real. Thus, λ^* is real. This implies that λ is real. Furthermore, $\langle \mathbb{H}\mathrm{h}|\mathrm{h}\rangle \geqslant 0$ implies that $\lambda \geqslant 0$, and so forth. ■

Example 3.6.4

Let us find the eigenvalues and eigenvectors of $\mathbb{H} = \begin{pmatrix} 0 & -i \\ i & 0 \end{pmatrix}$. We have

$$\det(\mathbb{H} - \lambda\mathbb{1}) = \det\begin{pmatrix} -\lambda & -i \\ i & -\lambda \end{pmatrix} = \lambda^2 - 1 = (\lambda - 1)(\lambda + 1)$$

Thus, the eigenvalues are $\lambda = \pm 1$, which are real, as expected.

To find the eigenvectors, we write

$$0 = (\mathbb{H} - \lambda_1\mathbb{1})|\mathrm{a}_1\rangle = (\mathbb{H} - \mathbb{1})|\mathrm{a}_1\rangle = \begin{pmatrix} -1 & -i \\ i & -1 \end{pmatrix}\begin{pmatrix} \alpha_1 \\ \alpha_2 \end{pmatrix} = \begin{pmatrix} -\alpha_1 - i\alpha_2 \\ i\alpha_1 - \alpha_2 \end{pmatrix}$$

or $\alpha_1 = -i\alpha_2$, which gives

$$|\mathrm{a}_1\rangle = \begin{pmatrix} -i\alpha_2 \\ \alpha_2 \end{pmatrix} = \alpha_2\begin{pmatrix} -i \\ 1 \end{pmatrix} \equiv \alpha\begin{pmatrix} -i \\ 1 \end{pmatrix}$$

where α is an arbitrary complex number. Also,

$$0 = (\mathbb{H} - \lambda_2\mathbb{1})|\mathrm{a}_2\rangle = (\mathbb{H} + \mathbb{1})|\mathrm{a}_2\rangle = \begin{pmatrix} 1 & -i \\ i & 1 \end{pmatrix}\begin{pmatrix} \beta_1 \\ \beta_2 \end{pmatrix} = \begin{pmatrix} \beta_1 - i\beta_2 \\ i\beta_1 + \beta_2 \end{pmatrix}$$

or $\beta_1 = i\beta_2$, which gives

$$|\mathrm{a}_2\rangle = \beta_2\begin{pmatrix} i \\ 1 \end{pmatrix} \equiv \beta\begin{pmatrix} i \\ 1 \end{pmatrix}$$

where β is an arbitrary complex number.

It is desirable, in most situations, to choose α and β such that $|\mathrm{a}_1\rangle$ and $|\mathrm{a}_2\rangle$ become normalized vectors. In such a case we demand that

$$1 = \langle \mathrm{a}_1|\mathrm{a}_1\rangle = \alpha^*(i \quad 1)\alpha\begin{pmatrix} -i \\ 1 \end{pmatrix} = |\alpha|^2(1 + 1) = 2|\alpha|^2$$

or that

$$|\alpha| = \frac{1}{\sqrt{2}} \quad \Rightarrow \quad \alpha = \frac{1}{\sqrt{2}}e^{i\varphi} \qquad \text{for some } \varphi \in \mathbb{R}$$

A common choice is $\varphi = 0$, for which we obtain

$$|\mathrm{a}_1\rangle = \frac{1}{\sqrt{2}}\begin{pmatrix} -i \\ 1 \end{pmatrix}$$

We look at the matrix equation $(\mathsf{H} + 2\mathsf{1})|\mathrm{a}\rangle = 0$, or

$$\begin{pmatrix} 2 & 0 & -1+i & -1-i \\ 0 & 2 & -1+i & 1+i \\ -1-i & -1-i & 2 & 0 \\ -1+i & 1-i & 0 & 2 \end{pmatrix} \begin{pmatrix} a_1 \\ a_2 \\ a_3 \\ a_4 \end{pmatrix} = 0$$

This is a system of linear equations whose solution is

$$a_3 = \tfrac{1}{2}(1+i)(a_1 + a_2)$$
$$a_4 = \tfrac{1}{2}(1-i)(a_1 - a_2)$$

We have two arbitrary parameters, so we expect two linearly independent solutions. For the two choices $a_1 = 1$, $a_2 = 0$, and $a_1 = 0$, $a_2 = 1$, we obtain, respectively,

$$|\mathrm{a}_1\rangle = \begin{pmatrix} 1 \\ 0 \\ \tfrac{1}{2}(1+i) \\ \tfrac{1}{2}(1-i) \end{pmatrix} \quad \text{and} \quad |\mathrm{a}_2\rangle = \begin{pmatrix} 0 \\ 1 \\ \tfrac{1}{2}(1+i) \\ -\tfrac{1}{2}(1-i) \end{pmatrix}$$

which happen to be orthogonal. Thus, we simply normalize them, obtaining

$$|\mathrm{e}'_1\rangle = \frac{1}{\sqrt{2}}\begin{pmatrix} 1 \\ 0 \\ \tfrac{1}{2}(1+i) \\ \tfrac{1}{2}(1-i) \end{pmatrix} \quad \text{and} \quad |\mathrm{e}'_2\rangle = \frac{1}{\sqrt{2}}\begin{pmatrix} 0 \\ 1 \\ \tfrac{1}{2}(1+i) \\ -\tfrac{1}{2}(1-i) \end{pmatrix}$$

Similarly, the second eigenvalue equation $(\mathsf{H} - 2\mathsf{1})|\mathrm{a}\rangle = 0$ gives

$$\begin{pmatrix} -2 & 0 & -1+i & -1-i \\ 0 & -2 & -1+i & 1+i \\ -1-i & -1-i & -2 & 0 \\ -1+i & 1-i & 0 & -2 \end{pmatrix} \begin{pmatrix} a_1 \\ a_2 \\ a_3 \\ a_4 \end{pmatrix} = 0$$

which gives rise to the conditions

$$a_3 = -\tfrac{1}{2}(1+i)(a_1 + a_2)$$
$$a_4 = \tfrac{1}{2}(1-i)(a_2 - a_1)$$

which produce the orthonormal vectors

$$|\mathrm{e}'_3\rangle = \frac{1}{\sqrt{2}}\begin{pmatrix} 1 \\ 0 \\ -\tfrac{1}{2}(1+i) \\ -\tfrac{1}{2}(1-i) \end{pmatrix} \quad \text{and} \quad |\mathrm{e}'_4\rangle = \frac{1}{\sqrt{2}}\begin{pmatrix} 0 \\ 1 \\ -\tfrac{1}{2}(1+i) \\ \tfrac{1}{2}(1-i) \end{pmatrix}$$

The unitary matrix that diagonalizes H can be constructed from these column vectors using the remarks at the end of Section 3.2, which imply that if we simply put the vectors $|e'_j\rangle$ together as columns, the resulting matrix is $R^\dagger$ (or $U^\dagger$):

$$U^\dagger = \frac{1}{\sqrt{2}} \begin{pmatrix} 1 & 0 & 1 & 0 \\ 0 & 1 & 0 & 1 \\ \frac{1}{2}(1+i) & \frac{1}{2}(1+i) & -\frac{1}{2}(1+i) & -\frac{1}{2}(1+i) \\ \frac{1}{2}(1-i) & -\frac{1}{2}(1-i) & -\frac{1}{2}(1-i) & \frac{1}{2}(1-i) \end{pmatrix}$$

and the unitary matrix will be

$$U = (U^\dagger)^\dagger = \frac{1}{\sqrt{2}} \begin{pmatrix} 1 & 0 & \frac{1}{2}(1-i) & \frac{1}{2}(1+i) \\ 0 & 1 & \frac{1}{2}(1-i) & -\frac{1}{2}(1+i) \\ 1 & 0 & -\frac{1}{2}(1-i) & -\frac{1}{2}(1+i) \\ 0 & 1 & -\frac{1}{2}(1-i) & \frac{1}{2}(1+i) \end{pmatrix}$$

We can easily check that U diagonalizes H (that $UHU^\dagger$ is diagonal). •

Example 3.6.6

In many physical applications diagonalization of matrices comes in very handy. As a simple but illustrative example, let us consider the motion of a charged particle in a constant magnetic field pointing in the z direction. The equation of motion for such a particle is

$$m\frac{d\mathbf{v}}{dt} = q\mathbf{v} \times \mathbf{B} = q \det \begin{pmatrix} \hat{\mathbf{e}}_x & \hat{\mathbf{e}}_y & \hat{\mathbf{e}}_z \\ v_x & v_y & v_z \\ 0 & 0 & B \end{pmatrix}$$

which, in component form, becomes

$$\frac{dv_x}{dt} = \frac{qB}{m}v_y \qquad \frac{dv_y}{dt} = -\frac{qB}{m}v_x \qquad \frac{dv_z}{dt} = 0$$

Ignoring the uniform motion in the z direction, we need to solve the first two *coupled* equations.

We can write those two equations in matrix form:

$$\frac{d}{dt}\begin{pmatrix} v_x \\ v_y \end{pmatrix} = \frac{qB}{m}\begin{pmatrix} 0 & 1 \\ -1 & 0 \end{pmatrix}\begin{pmatrix} v_x \\ v_y \end{pmatrix} \tag{1}$$

If the 2×2 matrix were diagonal, we would get two *uncoupled* equations, which we could solve easily. Diagonalizing it involves finding a matrix R such that

$$D \equiv R\begin{pmatrix} 0 & 1 \\ -1 & 0 \end{pmatrix}R^{-1} \equiv \begin{pmatrix} \lambda_1 & 0 \\ 0 & \lambda_2 \end{pmatrix}$$

is diagonal. The utility of such a matrix (which is time-independent) manifests itself when we multiply (1) by it. This gives

$$\frac{d}{dt}\mathsf{R}\begin{pmatrix} v_x \\ v_y \end{pmatrix} = \frac{qB}{m}\mathsf{R}\begin{pmatrix} 0 & 1 \\ -1 & 0 \end{pmatrix}\mathsf{R}^{-1}\mathsf{R}\begin{pmatrix} v_x \\ v_y \end{pmatrix}$$

which can be written as

$$\frac{d}{dt}\begin{pmatrix} v'_x \\ v'_y \end{pmatrix} = \omega\begin{pmatrix} \lambda_1 & 0 \\ 0 & \lambda_2 \end{pmatrix}\begin{pmatrix} v'_x \\ v'_y \end{pmatrix}$$

where $\omega = qB/m$ and

$$\begin{pmatrix} v'_x \\ v'_y \end{pmatrix} \equiv \mathsf{R}\begin{pmatrix} v_x \\ v_y \end{pmatrix}$$

We thus have

$$\frac{dv'_x}{dt} = \lambda_1 \omega v'_x$$

which has $v'_x = v'_{0x}e^{\lambda_1 \omega t}$ as a solution, and

$$\frac{dv'_y}{dt} = \lambda_2 \omega v'_y$$

which has $v'_y = v'_{0y}e^{\lambda_2 \omega t}$ as a solution, where v'_{0x} and v'_{0y} are integration constants.

To find R, we look at the characteristic equation

$$\det\begin{pmatrix} -\lambda & 1 \\ -1 & -\lambda \end{pmatrix} = \lambda^2 + 1$$

with roots $\lambda_1 = i$ and $\lambda_2 = -i$. Next, we find the normalized eigenvectors. For $\lambda_1 = i$ we have

$$\begin{pmatrix} -i & 1 \\ -1 & -i \end{pmatrix}\begin{pmatrix} a_1 \\ a_2 \end{pmatrix} = 0 \quad \Rightarrow \quad a_2 = ia_1$$

giving the normalized eigenvector

$$|\mathrm{a}_1\rangle = \frac{1}{\sqrt{2}}\begin{pmatrix} 1 \\ i \end{pmatrix}$$

Similarly, we find

$$|\mathrm{a}_2\rangle = \frac{1}{\sqrt{2}}\begin{pmatrix} 1 \\ -i \end{pmatrix}$$

From comments at the end of Section 3.2, we get

$$\mathsf{R}^{-1} = \mathsf{R}^\dagger = \begin{pmatrix} 1/\sqrt{2} & 1/\sqrt{2} \\ i/\sqrt{2} & -i/\sqrt{2} \end{pmatrix} \quad \Rightarrow \quad \mathsf{R} = (\mathsf{R}^\dagger)^\dagger = \begin{pmatrix} 1/\sqrt{2} & -i/\sqrt{2} \\ 1/\sqrt{2} & i/\sqrt{2} \end{pmatrix}$$

Having found R^{-1}, we can write

$$\begin{pmatrix} v_x \\ v_y \end{pmatrix} = \mathsf{R}^\dagger \begin{pmatrix} v'_x \\ v'_y \end{pmatrix} = \begin{pmatrix} 1/\sqrt{2} & 1/\sqrt{2} \\ i/\sqrt{2} & -i/\sqrt{2} \end{pmatrix} \begin{pmatrix} v'_{0x} e^{i\omega t} \\ v'_{0y} e^{-i\omega t} \end{pmatrix} \tag{2}$$

If the x and y components of velocity at $t = 0$ are v_{0x} and v_{0y}, respectively, then

$$\begin{pmatrix} v_{0x} \\ v_{0y} \end{pmatrix} = \begin{pmatrix} 1/\sqrt{2} & 1/\sqrt{2} \\ i/\sqrt{2} & -i/\sqrt{2} \end{pmatrix} \begin{pmatrix} v'_{0x} \\ v'_{0y} \end{pmatrix}$$

or

$$\begin{pmatrix} v'_{0x} \\ v'_{0y} \end{pmatrix} = \mathsf{R} \begin{pmatrix} v_{0x} \\ v_{0y} \end{pmatrix} = \frac{1}{\sqrt{2}} \begin{pmatrix} v_{0x} - iv_{0y} \\ v_{0x} + iv_{0y} \end{pmatrix}$$

Substituting in (2), we obtain

$$\begin{pmatrix} v_x \\ v_y \end{pmatrix} = \begin{pmatrix} 1/\sqrt{2} & 1/\sqrt{2} \\ i/\sqrt{2} & -i/\sqrt{2} \end{pmatrix} \begin{pmatrix} (1/\sqrt{2})(v_{0x} - iv_{0y})e^{i\omega t} \\ (1/\sqrt{2})(v_{0x} + iv_{0y})e^{-i\omega t} \end{pmatrix}$$

$$= \begin{pmatrix} v_{0x} \cos \omega t + v_{0y} \sin \omega t \\ -v_{0x} \sin \omega t + v_{0y} \cos \omega t \end{pmatrix}$$

This gives the velocity as a function of time. Antidifferentiating once with respect to time yields the position vector. ●

3.6.3 Normal Transformations

There is no doubt that diagonal matrices are extremely useful, as Example 3.6.6 illustrates. Since hermitian operators are so important in applications, it is reassuring to have Theorem 3.6.12, which guarantees the diagonalizability of such operators. However, hermitian operators are not the only ones used in applications. For example, unitary operators are clearly important in the preservation of inner products. Therefore, it is natural to seek the most general operators that have a spectral decomposition.

It might seem at first glance that Theorem 3.6.12 exhausts all possibilities. After all, Exercise 2.3.10 tells us that any arbitrary operator $\mathbb{A}$ can be written in terms of its so-called Cartesian components as

$$\mathbb{A} = \mathbb{H} + i\mathbb{H}' \tag{3.29}$$

where both $\mathbb{H}$ and $\mathbb{H}'$ are hermitian and can therefore be decomposed according to Theorem 3.6.12. What is wrong with concluding that $\mathbb{A}$ is also decomposable? The answer is that the projection operators used in the decomposition of $\mathbb{H}$ may not be the same as those used for $\mathbb{H}'$.

Let us investigate this idea systematically starting with a definition.

3.6.14 Definition Two operators are said to be *simultaneously diagonalizable* if they can be written in terms of the same set of projection operators, as in Theorem 3.6.12.

This definition is consistent with the matrix representation of the two operators, because if we take the orthonormal basis $B = \{|e_j^{(i)}\rangle\}$ discussed in the preceding section, we obtain diagonal matrices for both operators. If both $\mathbb{H}$ and $\mathbb{H}'$ in Eq. (3.29) are simultaneously diagonalizable such that

$$\mathbb{H} = \sum_{i=1}^{r} \lambda_i \mathbb{P}_i \qquad \text{and} \qquad \mathbb{H}' = \sum_{i=1}^{r} \lambda'_i \mathbb{P}_i \tag{3.30a}$$

then

$$\mathbb{A} = \sum_{j=1}^{r} (\lambda_j + i\lambda'_j)\mathbb{P}_j \tag{3.30b}$$

and $\mathbb{A}$ has a spectral decomposition. What are the conditions under which (3.30a) is satisfied? The answer lies in the following theorem.

3.6.15 Theorem A necessary and sufficient condition for two hermitian operators $\mathbb{H}$ and $\mathbb{H}'$ to be simultaneously diagonalizable is $[\mathbb{H}, \mathbb{H}'] = \mathbb{O}$.

Proof. If $\mathbb{H}$ and $\mathbb{H}'$ are simultaneously diagonalizable, then (3.30a) gives

$$\mathbb{H}\mathbb{H}' = \sum_{i=1}^{r} \lambda_i \mathbb{P}_i \sum_{j=1}^{r} \lambda'_j \mathbb{P}_j = \sum_{i,j} \lambda_i \lambda'_j \mathbb{P}_i \mathbb{P}_j$$

But $\mathbb{P}_i\mathbb{P}_j = \mathbb{P}_j\mathbb{P}_i$ for all i and j because $\mathbb{P}_i$ and $\mathbb{P}_j$ are mutually orthogonal ($\mathbb{P}_i\mathbb{P}_j = \mathbb{O}$ for $i \neq j$). Thus, we have

$$\mathbb{H}\mathbb{H}' = \sum_{i,j} \lambda_i \lambda'_j \mathbb{P}_j \mathbb{P}_i = \sum_j \lambda'_j \mathbb{P}_j \sum_i \lambda_i \mathbb{P}_i = \mathbb{H}'\mathbb{H}$$

On the other hand, if $[\mathbb{H}, \mathbb{H}'] = \mathbb{O}$ and if $|a\rangle$ is an eigenvector of $\mathbb{H}$ with eigenvalue λ (that is, $\mathbb{H}|a\rangle = \lambda|a\rangle$), then

$$\mathbb{H}(\mathbb{H}'|a\rangle) = \mathbb{H}\mathbb{H}'|a\rangle = \mathbb{H}'(\mathbb{H}|a\rangle) = \mathbb{H}'(\lambda|a\rangle) = \lambda(\mathbb{H}'|a\rangle)$$

That is, $\mathbb{H}'|a\rangle$ is also an eigenvector of $\mathbb{H}$ with the *same* eigenvalue as $|a\rangle$. In particular, the subspaces $\mathcal{M}_{\lambda_j}$ corresponding to the eigenvalues λ_j of $\mathbb{H}$ are all invariant under the action of $\mathbb{H}'$. This means that for any particular λ_j, we can think of $\mathcal{M}_{\lambda_j}$ as a vector space in its own right and apply Theorem 3.6.12 to it. The only difference is that the sum of the projection operators in (ii) will be $\mathbb{P}_j$ rather than $\mathbb{1}$.

Specifically, let $\lambda_1'^{(j)}, \lambda_2'^{(j)}, \ldots, \lambda_{s_j}'^{(j)}$ be the eigenvalues of $\mathbb{H}'$ corresponding to its eigenvectors in $\mathcal{M}_{\lambda_j}$. Then Theorem 3.6.12 says that there exist projection operators $\mathbb{P}_1^{(j)}, \mathbb{P}_2^{(j)}, \ldots, \mathbb{P}_{s_j}^{(j)}$ such that

(i) $\mathbb{P}_k^{(j)}\mathbb{P}_l^{(j)} = 0 \qquad$ for $k \neq l$

(ii) $\sum_{k=1}^{s_j} \mathbb{P}_k^{(j)} = \mathbb{P}_j$

(iii) $\sum_{k=1}^{s_j} \lambda_k'^{(j)}\mathbb{P}_k^{(j)} = \mathbb{H}'_j$ where $\mathbb{H}'_j$ is the restriction of $\mathbb{H}'$ to $\mathcal{M}_{\lambda_j}$

Invariance of $\mathcal{M}_{\lambda_j}$, on the other hand, reduces $\mathbb{H}'$ to

$$\mathbb{H}' = \mathbb{H}'_1 \oplus \mathbb{H}'_2 \oplus \cdots \oplus \mathbb{H}'_r$$

Putting all this together, we obtain

$$\mathbb{H}' = \left(\sum_{k_1=1}^{s_1} \lambda'^{(1)}_{k_1}\mathbb{P}^{(1)}_{k_1}\right) \oplus \left(\sum_{k_2=1}^{s_2} \lambda'^{(2)}_{k_2}\mathbb{P}^{(2)}_{k_2}\right) \oplus \cdots \oplus \left(\sum_{k_r=1}^{s_r} \lambda'^{(r)}_{k_r}\mathbb{P}^{(r)}_{k_r}\right)$$

$$= \sum_{j=1}^{r}\sum_{k=1}^{s_j} \lambda'^{(j)}_{k}\mathbb{P}^{(j)}_{k}$$

On the other hand,

$$\mathbb{H} = \sum_{j=1}^{r} \lambda_j \mathbb{P}_j = \sum_{j=1}^{r} \lambda_j \sum_{k=1}^{s_j} \mathbb{P}^{(j)}_k = \sum_{j=1}^{r}\sum_{k=1}^{s_j} \lambda_j \mathbb{P}^{(j)}_k$$

Comparison of these two expansions shows that $\mathbb{H}$ and $\mathbb{H}'$ are simultaneously diagonalizable. ■

Applying this result to the real and imaginary parts of a general operator $\mathbb{A}$, we conclude that $\mathbb{A}$ has a spectral decomposition if and only if

$$\left[\frac{1}{2}(\mathbb{A} + \mathbb{A}^\dagger), \frac{i}{2}(\mathbb{A} - \mathbb{A}^\dagger)\right] = 0$$

or

$$\frac{i}{4}\{-[\mathbb{A}, \mathbb{A}^\dagger] + [\mathbb{A}^\dagger, \mathbb{A}]\} = 0$$

or

$$[\mathbb{A}, \mathbb{A}^\dagger] = 0 \quad \Leftrightarrow \quad \mathbb{A}\mathbb{A}^\dagger = \mathbb{A}^\dagger\mathbb{A} \tag{3.31}$$

A transformation satisfying (3.31) is called a *normal transformation.* We have just established the following theorem.

3.6.16 Theorem An operator has a spectral decomposition in a unitary (complex) space if and only if it is normal. ■

Clearly, hermitian and unitary operators are normal. It should also be evident that, when speaking of the spectral decomposition of general normal operators, we must lift the restriction as to the reality of the eigenvalues, as stated in Theorem 3.6.12.

Example 3.6.7

Let us find the spectral decomposition for this Pauli spin matrix:

$$\boldsymbol{\sigma}_2 = \begin{pmatrix} 0 & -i \\ i & 0 \end{pmatrix}$$

First, we find its eigenvalues:

$$\det(\boldsymbol{\sigma}_2 - \lambda 1) = \det\begin{pmatrix} -\lambda & -i \\ i & -\lambda \end{pmatrix} = \lambda^2 - 1 = 0 \quad \Rightarrow \quad \lambda = \pm 1$$

Let $\lambda_1 = 1$ and $\lambda_2 = -1$. Then for $|a_1\rangle = \begin{pmatrix}\alpha_1\\ \alpha_2\end{pmatrix}$ we get

$$\begin{pmatrix} -1 & -i \\ i & -1 \end{pmatrix}\begin{pmatrix}\alpha_1\\ \alpha_2\end{pmatrix} = 0 \quad \Rightarrow \quad \alpha_2 = i\alpha_1$$

and the normalized eigenvector is

$$|e_1\rangle = \frac{1}{\sqrt{2}}\begin{pmatrix}1\\ i\end{pmatrix}$$

Similarly, for λ_2 we get

$$|e_2\rangle = \frac{1}{\sqrt{2}}\begin{pmatrix}1\\ -i\end{pmatrix}$$

The subspaces $\mathcal{M}_{\lambda_i}$ are one-dimensional; therefore,

$$\mathbb{P}_1 = |e_1\rangle\langle e_1| = \frac{1}{\sqrt{2}}\begin{pmatrix}1\\ i\end{pmatrix}\frac{1}{\sqrt{2}}(1 \quad -i) = \frac{1}{2}\begin{pmatrix}1 & -i\\ i & 1\end{pmatrix}$$

$$\mathbb{P}_2 = |e_2\rangle\langle e_2| = \frac{1}{2}\begin{pmatrix}1 & i\\ -i & 1\end{pmatrix}$$

We check:

$$\mathbb{P}_1 + \mathbb{P}_2 = \begin{pmatrix}1 & 0\\ 0 & 1\end{pmatrix}$$

and

$$\lambda_1\mathbb{P}_1 + \lambda_2\mathbb{P}_2 = \frac{1}{2}\begin{pmatrix}1 & -i\\ i & 1\end{pmatrix} - \frac{1}{2}\begin{pmatrix}1 & i\\ -i & 1\end{pmatrix} = \begin{pmatrix}0 & -i\\ i & 0\end{pmatrix} = \sigma_2$$ •

3.6.4 Functions of Transformations

Functions of transformations were discussed in Chapter 2. With the power of spectral decomposition at our disposal, we can draw many important conclusions about them.

First, we note that if

$$\mathbb{A} = \sum_{i=1}^{r} \lambda_i \mathbb{P}_i$$

then

$$\mathbb{A}^2 = \sum_{i=1}^{r} \lambda_i^2 \mathbb{P}_i$$

and, in general,

$$\mathbb{A}^n = \sum_{i=1}^{r} \lambda_i^n \mathbb{P}_i$$

Thus, any polynomial p in $\mathbb{A}$ has a spectral decomposition,

$$p(\mathbb{A}) = \sum_{i=1}^{r} p(\lambda_i)\mathbb{P}_i \tag{3.32a}$$

Generalizing this to functions expandable in power series gives

$$f(\mathbb{A}) = \sum_{i=1}^{r} f(\lambda_i)\mathbb{P}_i \tag{3.32b}$$

Example 3.6.8

Let us investigate the spectral decomposition of the following unitary (actually orthogonal) matrix:

$$\mathsf{U} = \begin{pmatrix} \cos\theta & -\sin\theta \\ \sin\theta & \cos\theta \end{pmatrix}$$

We find the eigenvalues:

$$\det\begin{pmatrix} \cos\theta - \lambda & -\sin\theta \\ \sin\theta & \cos\theta - \lambda \end{pmatrix} = \lambda^2 - 2\cos\theta\lambda + 1 = 0$$

$$\lambda = \cos\theta \pm \sqrt{\cos^2\theta - 1} = \cos\theta \pm i\sin\theta$$

$$\lambda_1 = e^{-i\theta} \text{ and } \lambda_2 = e^{i\theta}$$

Now we find the eigenvectors. For λ_1 we have

$$\begin{pmatrix} \cos\theta - e^{-i\theta} & -\sin\theta \\ \sin\theta & \cos\theta - e^{-i\theta} \end{pmatrix}\begin{pmatrix} \alpha_1 \\ \alpha_2 \end{pmatrix} = 0 \quad\Rightarrow\quad \alpha_2 = i\alpha_1 \quad\Rightarrow\quad |\mathsf{e}_1\rangle = \frac{1}{\sqrt{2}}\begin{pmatrix} 1 \\ i \end{pmatrix}$$

and for λ_2 we have

$$\begin{pmatrix} \cos\theta - e^{i\theta} & -\sin\theta \\ \sin\theta & \cos\theta - e^{i\theta} \end{pmatrix}\begin{pmatrix} \alpha_1 \\ \alpha_2 \end{pmatrix} = 0 \quad\Rightarrow\quad \alpha_2 = -i\alpha_1 \quad\Rightarrow\quad |\mathsf{e}_2\rangle = \frac{1}{\sqrt{2}}\begin{pmatrix} 1 \\ -i \end{pmatrix}$$

We note that the $\mathscr{M}_{\lambda_i}$ are one-dimensional and spanned by $|\mathsf{e}_i\rangle$. Thus,

$$\mathsf{P}_1 = |\mathsf{e}_1\rangle\langle\mathsf{e}_1| = \frac{1}{2}\begin{pmatrix} 1 \\ i \end{pmatrix}(1 \quad -i) = \frac{1}{2}\begin{pmatrix} 1 & -i \\ i & 1 \end{pmatrix}$$

$$\mathsf{P}_2 = |\mathsf{e}_2\rangle\langle\mathsf{e}_2| = \frac{1}{2}\begin{pmatrix} 1 & i \\ -i & 1 \end{pmatrix}$$

Clearly, $\mathsf{P}_1 + \mathsf{P}_2 = 1$, and

$$\lambda_1\mathsf{P}_1 + \lambda_2\mathsf{P}_2 = \frac{1}{2}\begin{pmatrix} e^{-i\theta} & -ie^{-i\theta} \\ ie^{-i\theta} & e^{-i\theta} \end{pmatrix} + \frac{1}{2}\begin{pmatrix} e^{i\theta} & ie^{i\theta} \\ -ie^{i\theta} & e^{i\theta} \end{pmatrix} = \mathsf{U}$$

If we take the natural log of

$$\mathsf{U} = e^{-i\theta}\mathsf{P}_1 + e^{i\theta}\mathsf{P}_2$$

and use (3.32b), we obtain

$$\begin{aligned} \ln\mathsf{U} &= \ln(e^{-i\theta})\mathsf{P}_1 + \ln(e^{i\theta})\mathsf{P}_2 = -i\theta\mathsf{P}_1 + i\theta\mathsf{P}_2 \\ &= i(-\theta\mathsf{P}_1 + \theta\mathsf{P}_2) \equiv i\mathsf{H} \end{aligned} \tag{1}$$

where $\mathsf{H} \equiv -\theta \mathsf{P}_1 + \theta \mathsf{P}_2$ is a hermitian operator because θ is real and P_1 and P_2 are hermitian. Inverting Eq. (1) gives

$$\mathsf{U} = e^{i\mathsf{H}}$$

where

$$\mathsf{H} = \theta[-\mathsf{P}_1 + \mathsf{P}_2] = \theta\begin{pmatrix} 0 & i \\ -i & 0 \end{pmatrix}$$

We have shown that the unitary matrix U can be written as an exponential of an antihermitian operator. This is a general result which is proved in Exercise 3.6.7. ●

An important function of an operator is its square root. A natural way of defining the square root of an operator $\mathbb{A}$ is

$$\sqrt{\mathbb{A}} = \sum_{i=1}^{r} (\pm\sqrt{\lambda_i})\mathbb{P}_i \tag{3.33}$$

This clearly gives many candidates for the root, because each term in (3.33) can have either the plus sign or the minus sign. To make $\sqrt{\mathbb{A}}$ unique, it is customary to define it only for positive operators, which are hermitian operators, $\mathbb{A}$, with the property that $\langle \mathbb{A}a|a\rangle \geqslant 0$, for all $|a\rangle \in \mathscr{V}$. In particular, all eigenvalues of a positive operator are nonnegative.

3.6.17 Definition The *positive square root* of a positive operator $\mathbb{A} = \sum_{i=1}^{r} \lambda_i \mathbb{P}_i$ is $\sqrt{\mathbb{A}} = \sum_{i=1}^{r} \sqrt{\lambda_i}\mathbb{P}_i$.

The uniqueness of the spectral decomposition implies that the positive square root of a positive operator is unique.

Example 3.6.9

Let us evaluate $\sqrt{\mathbb{A}}$ where

$$\mathsf{A} = \begin{pmatrix} 5 & 3i \\ -3i & 5 \end{pmatrix}$$

First, we have to spectrally decompose A. Its characteristic equation is

$$\lambda^2 - 10\lambda + 16 = 0$$

with roots $\lambda_1 = 8$ and $\lambda_2 = 2$. Since both eigenvalues are positive and A is hermitian, we conclude that A is indeed positive. We can also easily find its normalized eigenvectors:

$$|\mathsf{e}_1\rangle = \frac{1}{\sqrt{2}}\begin{pmatrix} i \\ 1 \end{pmatrix} \qquad \text{and} \qquad |\mathsf{e}_2\rangle = \frac{1}{\sqrt{2}}\begin{pmatrix} -i \\ 1 \end{pmatrix}$$

Thus,

$$\mathsf{P}_1 = |\mathsf{e}_1\rangle\langle \mathsf{e}_1| = \frac{1}{2}\begin{pmatrix} 1 & i \\ -i & 1 \end{pmatrix}$$

$$\mathsf{P}_2 = |\mathsf{e}_2\rangle\langle \mathsf{e}_2| = \frac{1}{2}\begin{pmatrix} 1 & -i \\ i & 1 \end{pmatrix}$$

and

$$\sqrt{\mathsf{A}} = \sqrt{\lambda_1}\mathsf{P}_1 + \sqrt{\lambda_2}\mathsf{P}_2 = \sqrt{8}\frac{1}{2}\begin{pmatrix} 1 & i \\ -i & 1 \end{pmatrix} + \sqrt{2}\frac{1}{2}\begin{pmatrix} 1 & -i \\ i & 1 \end{pmatrix} = \begin{pmatrix} 3/\sqrt{2} & i/\sqrt{2} \\ -i/\sqrt{2} & 3/\sqrt{2} \end{pmatrix}$$

We can easily check that $(\sqrt{\mathsf{A}})^2 = \mathsf{A}$. ●

Exercise 3.6.6 shows that $\mathbb{P}_i$ are polynomials in $\mathbb{A}$. We, therefore, conclude that $\sqrt{\mathbb{A}}$ is also a polynomial in $\mathbb{A}$. In fact, from Exercise 3.6.6 we have

$$\sqrt{\mathbb{A}} = \sum_{i=1}^{r} \sqrt{\lambda_i}\, p_i(\mathbb{A}) = \sum_{i=1}^{r} \sqrt{\lambda_i} \prod_{k \neq i} \frac{\mathsf{A} - \lambda_k}{\lambda_i - \lambda_k} \tag{3.34}$$

Example 3.6.10

Let us write $\sqrt{\mathsf{A}}$ of Example 3.6.9 as a polynomial in A. We have

$$p_1(\mathsf{A}) = \prod_{k \neq 1} \frac{\mathsf{A} - \lambda_k}{\lambda_1 - \lambda_k} = \frac{\mathsf{A} - \lambda_2}{\lambda_1 - \lambda_2} = \frac{1}{6}(\mathsf{A} - 2)$$

$$p_2(\mathsf{A}) = \prod_{k \neq 2} \frac{\mathsf{A} - \lambda_k}{\lambda_2 - \lambda_k} = \frac{\mathsf{A} - \lambda_1}{\lambda_2 - \lambda_1} = -\frac{1}{6}(\mathsf{A} - 8)$$

Substituting in the expansion of $\sqrt{\mathsf{A}}$ given in (3.34), we obtain

$$\sqrt{\mathsf{A}} = \sqrt{\lambda_1}p_1(\mathsf{A}) + \sqrt{\lambda_2}p_2(\mathsf{A}) = \frac{2\sqrt{2}}{6}(\mathsf{A} - 2) - \frac{\sqrt{2}}{6}(\mathsf{A} - 8) = \frac{\sqrt{2}}{6}\mathsf{A} + \frac{2\sqrt{2}}{3}$$

The RHS is clearly a (first-degree) polynomial in A, and it is easy to verify that it is the matrix of $\sqrt{\mathsf{A}}$. ●

3.6.5 Polar Decomposition

There are clearly many similarities between operators and complex numbers. For instance, hermitian operators behave very much like the real numbers: they have real eigenvalues; their squares are positive; every operator can be written as $\mathbb{H} + i\mathbb{H}'$, where both $\mathbb{H}$ and $\mathbb{H}'$ are hermitian; and so forth. Also, unitary operators can be written as $e^{i\mathbb{H}}$, where $\mathbb{H}$ is hermitian. So unitary operators are the analog of complex numbers of unit magnitude, such as $e^{i\theta}$. A complex number can also be written as $re^{i\theta}$. Can we write an arbitrary operator in an analogous way? The following theorem provides the answer.

3.6.18 Theorem If $\mathbb{A}$ is an arbitrary operator on a finite-dimensional unitary space, then there exists a unique positive operator, $\mathbb{P}$, and a unitary operator, $\mathbb{U}$, such that $\mathbb{A} = \mathbb{U}\mathbb{P}$. If $\mathbb{A}$ is invertible, then $\mathbb{U}$ is also unique.

Proof. We will prove the theorem for the case where $\mathbb{A}$ is invertible. The proof of the general case can be found in books on linear algebra (such as Halmos 1958).

The operator $\mathbb{A}^\dagger\mathbb{A}$ is positive because

$$\langle \mathbb{A}^\dagger\mathbb{A}a|a\rangle = \langle \mathbb{A}a|(\mathbb{A}^\dagger)^\dagger|a\rangle = \langle \mathbb{A}a|\mathbb{A}a\rangle \geqslant 0$$

Therefore, it has a unique positive square root, $\mathbb{P} = \sqrt{\mathbb{A}^\dagger\mathbb{A}}$. We let $\mathbb{V} = \mathbb{P}\mathbb{A}^{-1}$, or $\mathbb{V}\mathbb{A} = \mathbb{P}$, and show that $\mathbb{V}$ is unitary.

$$\mathbb{V}^\dagger = (\mathbb{P}\mathbb{A}^{-1})^\dagger = (\mathbb{A}^{-1})^\dagger\mathbb{P}^\dagger = (\mathbb{A}^\dagger)^{-1}\mathbb{P}$$

and $$\mathbb{V}^\dagger\mathbb{V} = (\mathbb{A}^\dagger)^{-1}\mathbb{P}\mathbb{P}\mathbb{A}^{-1} = (\mathbb{A}^\dagger)^{-1}\mathbb{P}^2\mathbb{A}^{-1} = (\mathbb{A}^\dagger)^{-1}\mathbb{A}^\dagger\mathbb{A}\mathbb{A}^{-1} = \mathbb{1}$$

Now we let $\mathbb{U} = \mathbb{V}^\dagger$. Then $\mathbb{V}\mathbb{A} = \mathbb{P}$ gives $\mathbb{U}^\dagger\mathbb{A} = \mathbb{P}$, or $\mathbb{A} = \mathbb{U}\mathbb{P}$. To prove uniqueness we note that $\mathbb{U}\mathbb{P} = \mathbb{U}'\mathbb{P}'$ implies that $\mathbb{P} = \mathbb{U}^\dagger\mathbb{U}'\mathbb{P}'$ and

$$\begin{aligned}\mathbb{P}^2 = \mathbb{P}^\dagger\mathbb{P} &= (\mathbb{U}^\dagger\mathbb{U}'\mathbb{P}')^\dagger(\mathbb{U}^\dagger\mathbb{U}'\mathbb{P}') = \mathbb{P}'^\dagger\mathbb{U}'^\dagger\mathbb{U}\mathbb{U}^\dagger\mathbb{U}'\mathbb{P}'\\ &= \mathbb{P}'^\dagger\mathbb{P}' = \mathbb{P}'^2\end{aligned}$$

Since the positive transformation $\mathbb{P}^2$ (or $\mathbb{P}'^2$) has only one square root, it follows that $\mathbb{P} = \mathbb{P}'$.

If $\mathbb{A}$ is invertible, then so is $\mathbb{P} = \mathbb{U}^\dagger\mathbb{A}$. Therefore, $\mathbb{U}\mathbb{P} = \mathbb{U}'\mathbb{P}' = \mathbb{U}'\mathbb{P}$ gives $\mathbb{U}\mathbb{P}\mathbb{P}^{-1} = \mathbb{U}'\mathbb{P}\mathbb{P}^{-1}$, or $\mathbb{U} = \mathbb{U}'$, and $\mathbb{U}$ is also unique. ■

In practice, $\mathbb{P}$ and $\mathbb{U}$ are found by noting that

$$\mathbb{A}\mathbb{A}^\dagger = \mathbb{P}\mathbb{U}(\mathbb{P}\mathbb{U})^\dagger = \mathbb{P}\mathbb{U}\mathbb{U}^\dagger\mathbb{P} = \mathbb{P}^2$$

Thus, finding $\mathbb{P}$ involves spectrally decomposing $\mathbb{A}\mathbb{A}^\dagger$ and taking its positive square root. Once $\mathbb{P}$ is found, $\mathbb{U}$ can be calculated from the definition $\mathbb{A} = \mathbb{P}\mathbb{U}$. In general, $\mathbb{U}$ is not unique. However, if $\mathbb{P}$ (and $\mathbb{A}$) is invertible, then $\mathbb{U}$ is uniquely determined by $\mathbb{U} = \mathbb{P}^{-1}\mathbb{A}$.

3.6.6 Real Vector Spaces

The treatment so far in this chapter has focused on complex (unitary) inner product spaces. The complex number system is far richer than the real number system. For example, in preparation for the proof of the spectral decomposition theorem, we used the existence of n roots of a polynomial of degree n on the complex field (this is the fundamental theorem of algebra). A polynomial on reals, on the other hand, *does not* necessarily have all its roots in the real number system.

It may, therefore, seem that vector spaces on reals will not satisfy the useful theorems and results developed for complex spaces. However, through a process called *complexification* of a real vector space, in which an imaginary part is added to such a space, it is possible to prove practically all the results obtained here for complex vector spaces. This task is accomplished in the excellent book by Halmos (1958). Only the results are given here.

3.6.19 Theorem A real symmetric operator has a spectral decomposition as stated in Theorem 3.6.12. ■

This theorem is especially useful in applications of classical physics, which deal mostly with real vector spaces. A typical situation involves a vector that is related to another vector by a symmetric matrix. It is then necessary to find a coordinate system in which the two vectors are related in a simple manner. This involves diagonalizing the symmetric matrix by a rotation (a real orthogonal matrix). Theorem 3.6.19 reassures us that such a diagonalization is possible.

Example 3.6.11

For a system of N point particles, the total angular momentum is defined as

$$\mathbf{L} = \sum_{i=1}^{N} m_i(\mathbf{r}_i \times \mathbf{v}_i)$$

If the particles constitute a rigid body rotating with angular velocity ω, then $\mathbf{v}_i = \omega \times \mathbf{r}_i$, and we have

$$\begin{aligned}\mathbf{L} &= \sum_{i=1}^{N} m_i[\mathbf{r}_i \times (\omega \times \mathbf{r}_i)] \\ &= \sum_{i=1}^{N} m_i[\omega \mathbf{r}_i \cdot \mathbf{r}_i - \mathbf{r}_i(\mathbf{r}_i \cdot \omega)]\end{aligned}$$

Let $\mathbf{r}_i = (x_i, y_i, z_i)$ and $\omega = (\omega_x, \omega_y, \omega_z)$. Then the angular momentum is given by

$$\mathbf{L} = \sum_{i=1}^{N} m_i[\omega r_i^2 - \mathbf{r}_i(x_i\omega_x + y_i\omega_y + z_i\omega_z)] \tag{1}$$

Writing each component of this equation gives

$$\begin{aligned}L_x &= \sum_{i=1}^{N} m_i[\omega_x(r_i^2 - x_i^2) - x_i y_i \omega_y - x_i z_i \omega_z] \\ L_y &= \sum_{i=1}^{N} m_i[\omega_y(r_i^2 - y_i^2) - x_i y_i \omega_x - y_i z_i \omega_z] \\ L_z &= \sum_{i=1}^{N} m_i[\omega_z(r_i^2 - z_i^2) - x_i z_i \omega_x - y_i z_i \omega_y]\end{aligned}$$

If we define

$$I_{xx} = \sum_{i=1}^{N} m_i(r_i^2 - x_i^2) \qquad I_{yy} = \sum_{i=1}^{N} m_i(r_i^2 - y_i^2) \qquad I_{zz} = \sum_{i=1}^{N} m_i(r_i^2 - z_i^2)$$

$$\begin{aligned}I_{xy} &= I_{yx} = -\sum m_i x_i y_i \\ I_{xz} &= I_{zx} = -\sum m_i x_i z_i \\ I_{yz} &= I_{zy} = -\sum m_i y_i z_i\end{aligned}$$

then Eq. (1) can be written in matrix form as

$$\begin{pmatrix} L_x \\ L_y \\ L_z \end{pmatrix} = \begin{pmatrix} I_{xx} & I_{xy} & I_{xz} \\ I_{yx} & I_{yy} & I_{yz} \\ I_{zx} & I_{zy} & I_{zz} \end{pmatrix} \begin{pmatrix} \omega_x \\ \omega_y \\ \omega_z \end{pmatrix}$$

The 3×3 matrix is denoted by I and called the moment of inertia matrix. It is symmetric, and Theorem 3.6.19 permits its diagonalization by an orthogonal transformation (the counterpart of a unitary transformation in a real vector space). But an orthogonal transformation in three dimensions is merely a rotation of coordinates. Thus, Theorem 3.6.19 says that it is always possible to choose coordinate systems in which the moment of inertia matrix is diagonal. In such a coordinate system we have

$$L_x = I_{xx}\omega_x \qquad L_y = I_{yy}\omega_y \qquad L_z = I_{zz}\omega_z$$

simplifying the equations considerably.

Similarly, the kinetic energy of the rigid rotating body,

$$T = \sum_{i=1}^{N} \tfrac{1}{2} m_i v_i^2 = \sum \tfrac{1}{2} m_i \mathbf{v}_i \cdot (\boldsymbol{\omega} \times \mathbf{r}_i)$$

$$= \sum \tfrac{1}{2} m_i \boldsymbol{\omega} \cdot (\mathbf{r}_i \times \mathbf{v}_i) = \tfrac{1}{2} \boldsymbol{\omega} \cdot \mathbf{L}$$

which, in general, has off-diagonal terms involving I_{xy}, and so forth, reduces to a simple form:

$$T = \tfrac{1}{2} I_{xx}\omega_x^2 + \tfrac{1}{2} I_{yy}\omega_y^2 + \tfrac{1}{2} I_{zz}\omega_z^2$$ •

Another application of Theorem 3.6.19 is in the study of conic sections. The following example illustrates such an application (see Exercise 3.6.9 also).

Example 3.6.12

The most general form of the equation of a conic section is

$$a_1x^2 + a_2y^2 + a_3xy + a_4x + a_5y + a_6 = 0$$

where $a_1, \ldots, a_6$ are all constants. If the coordinate axes coincide with the principal axes of the conic section, the xy term will be absent, and the conic section can easily be graphed. Thus, on geometrical grounds we have to be able to rotate xy-coordinates into coincidence with the principal axes. Let us do this using the ideas discussed in this section.

First, we note that the general equation for a conic section can be written in matrix form as

$$(x \quad y)\begin{pmatrix} a_1 & a_3/2 \\ a_3/2 & a_2 \end{pmatrix}\begin{pmatrix} x \\ y \end{pmatrix} + (a_4 \quad a_5)\begin{pmatrix} x \\ y \end{pmatrix} + a_6 = 0$$

The 2×2 matrix is symmetric and can therefore be diagonalized by means of an orthogonal matrix. We call this orthogonal matrix R. Then $\tilde{\mathsf{R}}\mathsf{R} = 1$, and we can write

$$(x \quad y)\tilde{\mathsf{R}}\mathsf{R}\begin{pmatrix} a_1 & a_3/2 \\ a_3/2 & a_2 \end{pmatrix}\tilde{\mathsf{R}}\mathsf{R}\begin{pmatrix} x \\ y \end{pmatrix} + (a_4 \quad a_5)\tilde{\mathsf{R}}\mathsf{R}\begin{pmatrix} x \\ y \end{pmatrix} + a_6 = 0$$

Let

$$\mathsf{R}\begin{pmatrix} x \\ y \end{pmatrix} = \begin{pmatrix} x' \\ y' \end{pmatrix} \qquad \mathsf{R}\begin{pmatrix} a_1 & a_3/2 \\ a_3/2 & a_2 \end{pmatrix}\tilde{\mathsf{R}} = \begin{pmatrix} a'_1 & 0 \\ 0 & a'_2 \end{pmatrix} \qquad \mathsf{R}\begin{pmatrix} a_4 \\ a_5 \end{pmatrix} = \begin{pmatrix} a'_4 \\ a'_5 \end{pmatrix}$$

Then we get

$$(x' \quad y')\begin{pmatrix} a'_1 & 0 \\ 0 & a'_2 \end{pmatrix}\begin{pmatrix} x' \\ y' \end{pmatrix} + (a'_4 \quad a'_5)\begin{pmatrix} x' \\ y' \end{pmatrix} + a_6 = 0$$

or

$$a'_1(x')^2 + a'_2(y')^2 + a'_4 x' + a'_5 y' + a_6 = 0$$

The cross term has disappeared. The orthogonal matrix R is simply a rotation. In fact, it is exactly the rotation that takes the original coordinate system into coincidence with the principal axes. ●

Although the case of symmetric operators is fairly straightforward, the case of orthogonal operators (the counterpart of unitary operators in a real vector space) is more complicated. In fact, we have already seen in Example 3.6.8 that the eigenvalues of an orthogonal transformation in two dimensions are, in general, complex. This is in contrast to symmetric transformations.

We can show that if an eigenvalue of an orthogonal operator is not ± 1, then it *must* be complex. Think of the othogonal operator $\mathbb{O}$ as a unitary operator on a complex vector space. Then by Exercise 3.6.4 the absolute value of each of its eigenvalues is 1. The only real solutions are, therefore, ± 1. To find the other eigenvalues we note that, as a unitary operator *in a unitary space*, $\mathbb{O}$ can be written as $e^{\mathbb{A}}$, where $\mathbb{A}$ is antihermitian (see Exercise 3.6.7). It is left as a problem for the reader to show that if $[\mathbb{A},\mathbb{B}] = 0$ for any two operators $\mathbb{A}$ and $\mathbb{B}$, then $[f(\mathbb{A}), g(\mathbb{B})] = 0$ for any two functions f and g. In particular, since $[\mathbb{O},\tilde{\mathbb{O}}] = 0$ $(\mathbb{O}\tilde{\mathbb{O}} = \tilde{\mathbb{O}}\mathbb{O} = \mathbb{1})$, we have $[\ln \mathbb{O}, \ln \tilde{\mathbb{O}}] = 0$, or $[\mathbb{A},\tilde{\mathbb{A}}] = 0$. Thus, by Proposition 2.3.16 we have

$$\mathbb{O}\tilde{\mathbb{O}} = e^{\mathbb{A}}e^{\tilde{\mathbb{A}}} = e^{\mathbb{A}+\tilde{\mathbb{A}}} = \mathbb{1}$$

or

$$\mathbb{A} + \tilde{\mathbb{A}} = \mathbb{0} \quad \Rightarrow \quad \mathbb{A} = -\tilde{\mathbb{A}}$$

and $\mathbb{A}$ is antisymmetric. Because $\mathbb{O}$ is real, so is $\mathbb{A}$.

Let us now consider the eigenvalues of $\mathbb{A}$. If λ is an eigenvalue of $\mathbb{A}$ corresponding to eigenvector $|\mathrm{a}\rangle$, then

$$\langle \mathrm{a}|\mathbb{A}|\mathrm{a}\rangle = \lambda\langle \mathrm{a}|\mathrm{a}\rangle$$

Taking the complex conjugate of both sides gives

$$\langle \mathrm{a}|\mathbb{A}^\dagger|\mathrm{a}\rangle = \lambda^*\langle \mathrm{a}|\mathrm{a}\rangle$$

but $\mathbb{A}^\dagger = \widetilde{\mathbb{A}^*} = \tilde{\mathbb{A}} = -\mathbb{A}$ because $\mathbb{A}$ is real and antisymmetric. We, therefore, have

$$\langle \mathrm{a}|\mathbb{A}|\mathrm{a}\rangle = -\lambda^*\langle \mathrm{a}|\mathrm{a}\rangle$$

which gives

$$\lambda^* = -\lambda$$

If λ is real, then its only possibility is zero. If λ is not real, then it must be *purely* imaginary. Therefore, the diagonal form of $\mathbb{A}$ looks like this:

$$\mathsf{A}_{\text{diag}} = \begin{pmatrix} 0 & & & & & & & \\ & 0 & & & & & \bigcirc & \\ & & \ddots & & & & & \\ & & & 0 & & & & \\ & & & & i\theta_1 & & & \\ & & \bigcirc & & & i\theta_2 & & \\ & & & & & & \ddots & \\ & & & & & & & i\theta_k \end{pmatrix}$$

which gives $\mathbb{O}$ the following diagonal form

$$\mathsf{O}_{\text{diag}} = e^{\mathbb{A}_{\text{diag}}} = \begin{pmatrix} e^0 & & & & & & & \\ & e^0 & & & & & \bigcirc & \\ & & \ddots & & & & & \\ & & & e^0 & & & & \\ & & & & e^{i\theta_1} & & & \\ & & \bigcirc & & & e^{i\theta_2} & & \\ & & & & & & \ddots & \\ & & & & & & & e^{i\theta_k} \end{pmatrix}$$

with $\theta_1, \theta_2, \ldots, \theta_k$ all real. It is clear that if $\mathbb{O}$ has -1 as its eigenvalue, then some of the θ's must equal $\pm\pi$. Separating the π's from the rest of θ's and putting all of the above arguments together, we get

$$\mathsf{O}_{\text{diag}} = \begin{pmatrix} 1_{N_+} & & & & & \\ & -1_{N_-} & & & \bigcirc & \\ & & e^{i\theta_1} & & & \\ & & & e^{i\theta_2} & & \\ & \bigcirc & & & \ddots & \\ & & & & & e^{i\theta_l} \end{pmatrix}$$

where $N_+ + N_- + l = \dim \mathsf{O}$.

Getting insight from Example 3.6.8, we can argue, admittedly in a nonrigorous way, that corresponding to each pair of $e^{i\theta_i}$ eigenvalues is a 2×2 matrix of the form

$$\begin{pmatrix} \cos\theta_i & -\sin\theta_i \\ \sin\theta_i & \cos\theta_i \end{pmatrix} \equiv \mathsf{R}_2(\theta_i) \tag{3.35}$$

We, therefore, have the following theorem (refer to Halmos 1958 for a rigorous treatment).

3.6.20 Theorem A real orthogonal operator on a real inner product space $\mathcal{V}$ cannot, in general, be completely diagonalized. The closest it can get to a diagonal form is

$$\begin{pmatrix} 1_{N_+} & & & & & \bigcirc \\ & -1_{N_-} & & & & \\ & & \mathsf{R}_2(\theta_1) & & & \\ & & & \mathsf{R}_2(\theta_2) & & \\ & & & & \ddots & \\ & \bigcirc & & & & \mathsf{R}_2(\theta_j) \end{pmatrix}$$

where $N_+ + N_- + 2j = \dim \mathcal{V}$, $1_{N_\pm}$ is the unit $N_\pm \times N_\pm$ matrix, and $\mathsf{R}_2(\theta_i)$ is as given in (3.35). ■

Finally, we arrive at the analogue of polar decomposition for real inner product spaces.

3.6.21 Theorem Any real operator (matrix) $\mathbb{A}$ can be written as $\mathbb{A} = \mathbb{P}\mathbb{O}$, where $\mathbb{P}$ is symmetric positive and $\mathbb{O}$ orthogonal. ■

Example 3.6.13

Let us decompose the following matrix in its polar form:

$$\mathsf{A} = \begin{pmatrix} 2 & 3 \\ 0 & -2 \end{pmatrix}$$

We have

$$\mathsf{P}^2 = \mathsf{A}\tilde{\mathsf{A}} = \begin{pmatrix} 2 & 3 \\ 0 & -2 \end{pmatrix}\begin{pmatrix} 2 & 0 \\ 3 & -2 \end{pmatrix} = \begin{pmatrix} 13 & -6 \\ -6 & 4 \end{pmatrix}$$

with eigenvalues $\lambda_1 = 1$ and $\lambda_2 = 16$ and eigenvectors

$$|\mathsf{e}_1\rangle = \frac{1}{\sqrt{5}}\begin{pmatrix} 1 \\ 2 \end{pmatrix} \quad \text{and} \quad |\mathsf{e}_2\rangle = \frac{1}{\sqrt{5}}\begin{pmatrix} 2 \\ -1 \end{pmatrix}$$

The projection operators are

$$\mathbb{P}_1 = |\mathsf{e}_1\rangle\langle\mathsf{e}_1| = \frac{1}{5}\begin{pmatrix} 1 & 2 \\ 2 & 4 \end{pmatrix}$$

$$\mathbb{P}_2 = |\mathsf{e}_2\rangle\langle\mathsf{e}_2| = \frac{1}{5}\begin{pmatrix} 4 & -2 \\ -2 & 1 \end{pmatrix}$$

Thus, we have

$$\mathsf{P} = \sqrt{\mathsf{P}^2} = \sqrt{\lambda_1}\mathbb{P}_1 + \sqrt{\lambda_2}\mathbb{P}_2 = \frac{1}{5}\begin{pmatrix} 1 & 2 \\ 2 & 4 \end{pmatrix} + \frac{4}{5}\begin{pmatrix} 4 & -2 \\ -2 & 1 \end{pmatrix}$$

$$= \frac{1}{5}\begin{pmatrix} 17 & -6 \\ -6 & 8 \end{pmatrix}$$

We note that A is invertible. Thus, P is also invertible, and

$$\mathsf{P}^{-1} = \frac{1}{20}\begin{pmatrix} 8 & 6 \\ 6 & 17 \end{pmatrix}$$

Now, if A = PO, then $\mathsf{O} = \mathsf{P}^{-1}\mathsf{A}$, or

$$\mathsf{O} = \frac{1}{5}\begin{pmatrix} 4 & 3 \\ 3 & -4 \end{pmatrix}$$

It is easily verified that O is indeed orthogonal. ●

Our rather long excursion through operator algebra and matrix theory has revealed that there are many different kinds of operators that are diagonalizable. Could it be perhaps that *all* operators are diagonalizable? In other words, given any operator, can we find a basis in which the matrix representing that operator is diagonal? The answer is, in general, no! Discussion of this topic would take us into the Hamilton-Cayly theorem and the Jordan canonical form of a matrix—a digression we cannot afford. The interested reader can find such discussion in books on linear algebra and matrix theory. One result is worth mentioning. If the roots of the characteristic polynomial of a matrix are all simple, then it can be brought to a diagonal form by a similarity (not necessarily a unitary) transformation.

Example 3.6.14

As a further example of the application of the results of this section, let us evaluate the n-tuple integral

$$I_n = \int_0^\infty dx_1 \int_0^\infty dx_2 \cdots \int_0^\infty dx_n e^{-\sum_{i,j=1}^n m_{ij}x_ix_j} \tag{1}$$

where m_{ij} are elements of a real, symmetric, positive definite matrix.

Let M be the matrix with elements m_{ij}. Because it is symmetric, it can be diagonalized by an orthogonal matrix R so that $\mathsf{R}\mathsf{M}\tilde{\mathsf{R}} = \mathsf{D}$ is a diagonal matrix whose diagonal entries are the eigenvalues, $\lambda_1, \lambda_2, \ldots, \lambda_n$, of M. Because M is positive definite, *none of these eigenvalues is zero or negative.*

The exponent in (1) can be written as

$$\sum_{i,j=1}^n m_{ij}x_ix_j = \tilde{\mathbf{x}}\mathsf{M}\mathbf{x} = \tilde{\mathbf{x}}\tilde{\mathsf{R}}\mathsf{R}\mathsf{M}\tilde{\mathsf{R}}\mathsf{R}\mathbf{x} = \tilde{\mathbf{x}}\tilde{\mathsf{R}}\mathsf{D}\mathsf{R}\mathbf{x}$$

$$= \widetilde{\mathbf{x}'}\,\mathsf{D}\mathbf{x}' = \lambda_1(x'_1)^2 + \lambda_2(x'_2)^2 + \cdots + \lambda_n(x'_n)^2$$

where

$$\mathbf{x}' = \begin{pmatrix} x'_1 \\ x'_2 \\ \vdots \\ x'_n \end{pmatrix} = \mathsf{R}\mathbf{x} = \mathsf{R}\begin{pmatrix} x_1 \\ x_2 \\ \vdots \\ x_n \end{pmatrix}$$

or, in component form,

$$x'_i = \sum_{j=1}^{n} r_{ij} x_j \qquad \text{for } i = 1, 2, \ldots, n$$

Similarly, since $\mathbf{x} = \tilde{\mathsf{R}}\mathbf{x}'$,

$$x_i = \sum_{j=1}^{n} r_{ji} x'_j \qquad \text{for } i = 1, 2, \ldots, n$$

On the other hand, the "volume element" $dx_1\, dx_2 \cdots dx_n$ is related to the "volume element" $dx'_1\, dx'_2 \cdots dx'_n$ as follows:

$$dx_1\, dx_2 \cdots dx_n = \left|\frac{\partial(x_1, x_2, \ldots, x_n)}{\partial(x'_1, x'_2, \ldots, x'_n)}\right| dx'_1\, dx'_2 \cdots dx'_n$$

$$\equiv |\det \mathsf{J}|\, dx'_1\, dx'_2 \cdots dx'_n$$

where J is the Jacobian matrix whose ijth element is $\partial x_i/\partial x'_j$. But

$$\partial x_i/\partial x'_j = r_{ji} \quad \Rightarrow \quad \mathsf{J} = \tilde{\mathsf{R}} \quad \Rightarrow \quad |\det \mathsf{J}| = |\det \tilde{\mathsf{R}}| = 1$$

Therefore, in terms of x', the integral I_n becomes

$$I_n = \int_0^\infty dx'_1 \int_0^\infty dx'_2 \cdots \int_0^\infty dx'_n e^{-\lambda_1 (x'_1)^2 - \lambda_2 (x'_2)^2 - \cdots - \lambda_n (x'_n)^2}$$

$$= \left(\int_0^\infty dx'_1 e^{-\lambda_1 (x'_1)^2}\right)\left(\int_0^\infty dx'_2 e^{-\lambda_2 (x'_2)^2}\right) \cdots \left(\int_0^\infty dx'_n e^{-\lambda_n (x'_n)^2}\right)$$

$$= \left(\frac{1}{2}\sqrt{\frac{\pi}{\lambda_1}}\right)\left(\frac{1}{2}\sqrt{\frac{\pi}{\lambda_2}}\right) \cdots \left(\frac{1}{2}\sqrt{\frac{\pi}{\lambda_n}}\right)$$

$$= \left(\frac{\sqrt{\pi}}{2}\right)^n \frac{1}{\sqrt{\lambda_1 \lambda_2 \cdots \lambda_n}} = \left(\frac{\sqrt{\pi}}{2}\right)^n (\det \mathsf{M})^{-1/2}$$

where the fact that the determinant of a matrix is the product of its eigenvalues has been used. •

Exercises

3.6.1 Find the eigenvalues and eigenvectors of the following matrix:

$$\mathsf{A} = \begin{pmatrix} 0 & 1 & 1 \\ 1 & 0 & -1 \\ -1 & -1 & 0 \end{pmatrix}$$

3.6.2 Consider the operator π_{ij}, which interchanges α_i and α_j in $(\alpha_1\ \alpha_2\ \ldots\ \alpha_n) \in \mathbb{C}^n$. Show that the eigenvalues of π_{ij} can only be ± 1.

3.6.3 Find the eigenvalues and eigenvectors of the operator $-i\,d/dx$ acting in $\mathrm{C}^{(1)}(-\infty, +\infty)$.

3.6.4 Show that the eigenvalues of a unitary operator $\mathbb{U}$ have absolute value 1.

3.6.5 What are the spectral decomposition of (**a**) $\mathbb{A}^{-1}$ and (**b**) $\mathbb{A}^\dagger\mathbb{A}$ for a normal operator $\mathbb{A}$?

3.6.6 Show that $\mathbb{P}_i$ in (3.32) could be written as a polynomial in $\mathbb{A}$.

3.6.7 Prove that, corresponding to every unitary operator $\mathbb{U}$, acting on a finite-dimensional vector space, there is a hermitian operator $\mathbb{H}$ such that $\mathbb{U} = e^{i\mathbb{H}}$.

3.6.8 Find the polar decomposition of this matrix:

$$\mathsf{A} = \begin{pmatrix} 2i & 0 \\ \sqrt{7} & 3 \end{pmatrix}$$

3.6.9 What is the angle between the principal axes of the conic section $2x^2 - 4xy + 5y^2 - 36 = 0$ and the xy coordinate system? What sort of conic section is this?

3.6.10 A point, $\mathbf{a} = (a_1, a_2, \ldots, a_n) \in \mathbb{R}^n$, is a maximum (minimum) of a function, $f(x_1, x_2, \ldots, x_n) \equiv f(\mathbf{r})$, if $\nabla f|_{\mathbf{r}=\mathbf{a}} \equiv (\partial f/\partial x_1, \partial f/\partial x_2, \ldots, \partial f/\partial x_n)_{\mathbf{r}=\mathbf{a}} = 0$ and, for *small* $\boldsymbol{\delta} = \mathbf{r} - \mathbf{a}$, the difference $f(\mathbf{r}) - f(\mathbf{a})$ is negative (positive). Show that $f(\mathbf{r})$ has a minimum (maximum) if and only if the $n \times n$ matrix $(\mathsf{D})_{ij} \equiv (\partial^2 f/\partial x_i \partial x_j)_{\mathbf{a}}$ has only nonnegative (nonpositive) eigenvalues.

PROBLEMS

3.1 Show that if $|z\rangle = |x\rangle + |y\rangle$, then in any basis the components of $|z\rangle$ are equal to the sum of the corresponding components of $|x\rangle$ and $|y\rangle$. Also show that the elements of the matrix representing the sum of two operators are the sums of the elements of the matrices representing those two operators.

3.2 Show that the unit operator, $\mathbb{1}$, is represented by the unit matrix in any basis.

3.3 Show that the diagonal elements of an antisymmetric matrix are all zero.

3.4 Show that the number of independent real parameters
(**a**) for an $N \times N$ antisymmetric matrix is $N(N-1)/2$
(**b**) for an $N \times N$ (complex) unitary matrix is N^2
(**c**) for an $N \times N$ (complex) hermitian matrix is N^2
(**d**) for an $N \times N$ (real) orthogonal matrix is $N(N-1)/2$

3.5 Prove that if a matrix M satisfies $\mathsf{M}\mathsf{M}^\dagger = 0$, then $\mathsf{M} = 0$. Note that, in general, $\mathsf{AB} = 0$ does not imply that either matrix is zero. In fact, the nonzero 2×2 matrix $\begin{pmatrix} 2 & \frac{1}{2} \\ -8 & -2 \end{pmatrix}$ has a zero square.

3.6 Show that in the expansion of the determinant given by Eqs. (3.14), no two elements of the same row or the same column can appear in each term of the sum.

3.7 Find inverses for the following matrices using both methods discussed in this chapter.

$$\mathsf{A} = \begin{pmatrix} 2 & 1 & -1 \\ 2 & 1 & 2 \\ -1 & -2 & -2 \end{pmatrix} \quad \mathsf{B} = \begin{pmatrix} 1 & 2 & -1 \\ 0 & 1 & -2 \\ 2 & 1 & -1 \end{pmatrix} \quad \mathsf{C} = \begin{pmatrix} 1 & -1 & 1 \\ -1 & 1 & 1 \\ 1 & -1 & -2 \end{pmatrix}$$

$$\mathsf{D} = \begin{pmatrix} 1/\sqrt{2} & 0 & (1-i)/2\sqrt{2} & (1+i)/2\sqrt{2} \\ 0 & 1/\sqrt{2} & (1-i)/2\sqrt{2} & -(1+i)/2\sqrt{2} \\ 1/\sqrt{2} & 0 & -(1-i)/2\sqrt{2} & -(1+i)/2\sqrt{2} \\ 0 & 1/\sqrt{2} & -(1-i)/2\sqrt{2} & (1+i)/2\sqrt{2} \end{pmatrix}$$

3.8 Use Theorem 3.4.3 to prove its corollary.

3.9 For which values of α are the following matrices invertible? Find the inverses whenever possible.

$$\mathsf{A} = \begin{pmatrix} 1 & \alpha & 0 \\ \alpha & 1 & \alpha \\ 0 & \alpha & 1 \end{pmatrix} \quad \mathsf{B} = \begin{pmatrix} \alpha & 1 & 0 \\ 1 & \alpha & 1 \\ 0 & 1 & \alpha \end{pmatrix} \quad \mathsf{C} = \begin{pmatrix} 0 & 1 & \alpha \\ 1 & \alpha & 0 \\ \alpha & 0 & 1 \end{pmatrix} \quad \mathsf{D} = \begin{pmatrix} 1 & 1 & 1 \\ 1 & 1 & \alpha \\ 1 & \alpha & 1 \end{pmatrix}$$

3.10 Show that $\det \mathsf{A}^{-1} = (\det \mathsf{A})^{-1}$. Use this fact to show that a similarity transformation does not change the determinant of a matrix.

3.11 Verify that adding a multiple of a row (column) to another row (column) does not affect the determinant of a matrix.

3.12 Let $\{\mathbf{a}_i\}_{i=1}^N$ be the set consisting of the N rows of an $N \times N$ matrix A and assume that the $\mathbf{a}_i$ are orthogonal to each other. Show that

$$|\det \mathsf{A}| = \|\mathbf{a}_1\| \, \|\mathbf{a}_2\| \cdots \|\mathbf{a}_N\|$$

(Hint: Consider $\mathsf{A}^\dagger \mathsf{A}$.)

3.13 Assume that A and A' are similar matrices. Show that they have the same eigenvalues.

3.14 Prove that a set of n homogeneous linear equations in n unknowns has a nontrivial solution if and only if the determinant of the matrix of coefficients is zero.

3.15 Use determinants to show that an antisymmetric matrix whose dimension is odd cannot have an inverse.

3.16 For $\mathsf{A} = \begin{pmatrix} 1 & x \\ 0 & 1 \end{pmatrix}$, where $x \neq 0$, show that it is impossible to find an invertible 2×2 matrix R such that RAR^{-1} is diagonal. (This shows that not all operators are diagonalizable.)

3.17 Complete the proof of Proposition 3.5.3.

3.18 Let π be a permutation of the integers $\{1, 2, \ldots, n\}$. If $|x\rangle = (\alpha_1, \alpha_2, \ldots, \alpha_n)$ is a vector in $\mathbb{C}^n$, write $\mathbb{A}|x\rangle = (\alpha_{\pi(1)}, \ldots, \alpha_{\pi(n)})$. What is the spectrum of $\mathbb{A}$?

3.19 Show that for any involutive operator (that is, an operator $\mathbb{A}$ with the property $\mathbb{A}^2 = \mathbb{1}$), the possible eigenvalues are ± 1.

3.20 Show that for a normal operator $\mathbb{A}$ and any other operator $\mathbb{B}$,

$$[\mathbb{A}, \mathbb{B}] = 0 \quad \Leftrightarrow \quad [\mathbb{P}_i, \mathbb{B}] = 0$$

where $\mathbb{P}_i$ are the projection operators in the spectral decomposition of $\mathbb{A}$. (Hint: Use Exercise 3.6.6.)

3.21 Show that an arbitrary orthogonal 2×2 matrix can be written in one of the following two forms:

$$\begin{pmatrix} \cos\theta & -\sin\theta \\ \sin\theta & \cos\theta \end{pmatrix} \quad \text{or} \quad \begin{pmatrix} \cos\theta & \sin\theta \\ \sin\theta & -\cos\theta \end{pmatrix}$$

The first is a pure rotation (its determinant is $+1$), and the second involves a rotation and an inversion of coordinates (its determinant is -1). The two choices are such that the first entry of the matrix reduces to 1 when $\theta = 0$.

3.22 In each case, determine the counterclockwise rotation of the xy-axes that brings the conic section into the standard form and determine the conic section.
(a) $11x^2 + 3y^2 + 6xy - 12 = 0$
(b) $5x^2 - 3y^2 - 6xy + 6 = 0$
(c) $2x^2 - y^2 - 4xy - 3 = 0$
(d) $6x^2 + 3y^2 - 4xy - 7 = 0$

3.23 Let $|\mathrm{a}\rangle, |\mathrm{b}\rangle \in \mathbb{C}^n$. Show that $\mathrm{tr}(|\mathrm{a}\rangle\langle \mathrm{b}|) = \langle \mathrm{b}|\mathrm{a}\rangle$.

3.24 Show that if two invertible $N \times N$ matrices A and B anticommute (that is, $\mathsf{AB} + \mathsf{BA} = 0$), then **(a)** N must be even, and **(b)** $\mathrm{tr}(\mathsf{A}) = \mathrm{tr}(\mathsf{B}) = 0$.

3.25 Let $\mathbb{M}$, $\mathbb{S}$, and $\mathbb{A}$ be an arbitrary, a symmetric, and an antisymmetric operator, respectively. Show that
(a) $\mathrm{tr}(\mathbb{M}) = \mathrm{tr}(\tilde{\mathbb{M}})$
(b) $\mathrm{tr}(\mathbb{S}\mathbb{A}) = 0$; in particular, $\mathrm{tr}(\mathbb{A}) = 0$
(c) $\mathbb{A}\mathbb{S}$ is antisymmetric if and only if $[\mathbb{A}, \mathbb{S}] = 0$
(d) $\mathsf{MS\tilde{M}}$ is symmetric and $\mathsf{MA\tilde{M}}$ is antisymmetric
(e) $\mathsf{MHM}^\dagger$ is hermitian if H is

3.26 Express the sum of the squares of elements of a matrix as a trace. Show that this sum is invariant under an orthogonal transformation.

3.27 Derive the formulas

$$\cos(\theta_1 + \theta_2) = \cos\theta_1 \cos\theta_2 - \sin\theta_1 \sin\theta_2$$

and

$$\sin(\theta_1 + \theta_2) = \sin\theta_1 \cos\theta_2 + \cos\theta_1 \sin\theta_2$$

by noting that the rotation of the angle $\theta_1 + \theta_2$ in the xy-plane is the product of two rotations.

3.28 Show that for a rotation $\mathsf{R}_{\hat{n}}(\theta)$ of an angle θ about an arbitrary axis $\hat{n}$, $\mathrm{tr}[\mathsf{R}_{\hat{n}}(\theta)] = 1 + 2\cos\theta$.

3.29 Show that if A is invertible, then the eigenvectors of A^{-1} are the same as those of A and the eigenvalues of A^{-1} are the reciprocals of those of A.

3.30 Find all eigenvalues and eigenvectors of the following matrices:

$$\mathsf{A}_1 = \begin{pmatrix} 1 & 1 \\ 0 & i \end{pmatrix} \qquad \mathsf{B}_1 = \begin{pmatrix} 0 & 1 \\ 0 & 0 \end{pmatrix} \qquad \mathsf{C}_1 = \begin{pmatrix} 1 & 1 & 1 \\ 1 & 1 & 1 \\ 1 & 1 & 1 \end{pmatrix}$$

$$\mathsf{A}_2 = \begin{pmatrix} 1 & 0 & 1 \\ 0 & 1 & 0 \\ 1 & 0 & 1 \end{pmatrix} \qquad \mathsf{B}_2 = \begin{pmatrix} 1 & 1 & 0 \\ 1 & 0 & 1 \\ 0 & 1 & 1 \end{pmatrix} \qquad \mathsf{C}_2 = \begin{pmatrix} 0 & 1 & 1 \\ 1 & 0 & 1 \\ 1 & 1 & 0 \end{pmatrix}$$

$$\mathsf{A}_3 = \begin{pmatrix} 1 & 1 & 1 \\ 0 & 1 & 1 \\ 0 & 0 & 1 \end{pmatrix} \qquad \mathsf{B}_3 = \begin{pmatrix} -1 & 1 & 1 \\ 1 & -1 & 1 \\ 1 & 1 & -1 \end{pmatrix} \qquad \mathsf{C}_3 = \begin{pmatrix} 2 & -2 & -1 \\ -1 & 3 & 1 \\ 2 & -4 & -1 \end{pmatrix}$$

3.31 Three equal point masses are located at $(a, a, 0)$, $(a, 0, a)$, and $(0, a, a)$.

(a) Find the moment of inertia matrix.

(b) Find its eigenvalues and the corresponding eigenvectors.

3.32 Use the spectral decomposition theorem to show that if two hermitian matrices have the same set of eigenvalues, then they are unitarily related.

3.33 Prove that in a finite-dimensional vector space any function of an operator is simply a polynomial in that operator. (Hint: See Exercise 3.6.6.)

3.34 Show that if two operators commute, any functions of those two operators also commute.

3.35 Show that an arbitrary matrix A can be "diagonalized" as $\mathsf{D} = \mathsf{U}\mathsf{A}\mathsf{V}^\dagger$, where U and V are unitary matrices and D is a real diagonal matrix with only nonnegative eigenvalues.

3.36 Show that if λ is an eigenvalue of an antisymmetric operator $\mathbb{A}$, then so is $-\lambda$. Now conclude that antisymmetric operators (matrices) of odd dimension cannot be inverted.

3.37 Find the unitary matrices that diagonalize the following hermitian matrices:

$$\begin{pmatrix} 3 & i \\ -i & 3 \end{pmatrix} \qquad \begin{pmatrix} 2 & -1+i \\ -1-i & -1 \end{pmatrix} \qquad \begin{pmatrix} 1 & -i \\ i & 0 \end{pmatrix}$$

$$\begin{pmatrix} 1 & -1 & -i \\ -1 & 0 & i \\ i & -i & -1 \end{pmatrix} \qquad \begin{pmatrix} 2 & 0 & i \\ 0 & -1 & -i \\ -i & i & 0 \end{pmatrix}$$

(Warning! You may have to resort to numerical methods for some of these.)

4

Differential Geometry and Tensor Analysis

It used to be that tensors were almost completely synonymous with relativity (except for minor use in hydrodynamics). Students of physics did not need to study tensors until they took a course in the general theory of relativity. Then they would read the introductory chapter on tensor algebra and analysis, solve a few problems to condition themselves for index "gymnastics," read through the book, learn some basic facts about relativity, and finally abandon it (unless they became relativists).

Now, with the advent of gauge theories of fundamental particles, the realization that gauge fields are to be thought of as geometrical objects, and the widespread belief that all fundamental interactions (including gravity) are different manifestations of the same superforce, the picture has changed drastically.

Two important developments have taken place as a consequence: tensors have crept into other interactions besides gravity (such as the weak and strong nuclear interactions); and the geometrical (coordinate-independent) aspects of tensors have become more and more significant in the study of all interactions. The coordinate-independent study of tensors is the focus of the fascinating field of differential geometry, on which this chapter will briefly touch.

This chapter is divided into two major parts. Section 4.1 covers tensor algebra, and the remainder of the chapter is devoted to tensor analysis. Since this will be merely a brief introduction to the subject, some proofs will be omitted.

As is customary, we will consider only real vector spaces and abandon—temporarily—the Dirac bra and ket notation, whose implementation is most advantageous in unitary (complex) spaces. In this chapter the basis vectors of a vector space $\mathcal{V}$ will be denoted using a subscript and those of its dual space with a superscript. That is, $\{\mathbf{e}_i\}_{i=1}^N$ is a basis in $\mathcal{V}$, and $\{\boldsymbol{\varepsilon}^j\}_{j=1}^N$ is a basis in $\mathcal{V}^*$. Also, Einstein's summation convention will be used. Repeated indices, of which one is an upper and the other a lower index, are assumed to be summed over. That is, $a_i^k b_j^i$ means $\sum_{i=1}^{N} a_i^k b_j^i$. This makes it more natural to label elements of the matrix representation of an operator $\mathbb{A}$ by α_i^j (rather than α_{ji}) because then $\mathsf{A}\mathbf{e}_i = \alpha_i^j \mathbf{e}_j$.

4.1 TENSOR ALGEBRA

Since tensors are special kinds of linear operators on vector spaces, let us reconsider $\mathcal{L}(\mathcal{V},\mathcal{W})$, the space of all linear mappings from the (real) vector space $\mathcal{V}$ to the (real) vector space $\mathcal{W}$. We noted before that $\mathcal{L}(\mathcal{V},\mathcal{W})$ is isomorphic to the space of $N_2 \times N_1$ matrices if $\dim \mathcal{V} = N_1$ and $\dim \mathcal{W} = N_2$. In particular, this says that $\dim \mathcal{L}(\mathcal{V},\mathcal{W}) = N_2 N_1$. The following proposition shows this directly.

4.1.1 Proposition Let $\{\mathbf{e}_i\}_{i=1}^{N_1}$ be a basis for $\mathcal{V}$ and $\{\mathbf{e}'_\alpha\}_{\alpha=1}^{N_2}$ a basis for $\mathcal{W}$.

(i) Then the linear operators $\mathbb{T}_\beta^j : \mathcal{V} \to \mathcal{W}$ defined by (note the new way of writing the Kronecker delta)

$$\mathbb{T}_\beta^j \mathbf{e}_i = \delta_i^j \mathbf{e}'_\beta \qquad \text{for } j = 1, \ldots, N_1;\ \beta = 1, \ldots, N_2$$

form a basis in $\mathcal{L}(\mathcal{V},\mathcal{W})$. In particular, $\dim \mathcal{L}(\mathcal{V},\mathcal{W}) = N_1 N_2$.

(ii) If τ_j^α are the elements of the matrix of an operator $\mathbb{T}$, then

$$\mathbb{T} = \tau_j^\alpha \mathbb{T}_\alpha^j$$

Proof. The $\mathbb{T}_\beta^j$ are linearly independent because if there are real numbers α_j^β such that the sum $\alpha_j^\beta \mathbb{T}_\beta^j$ is zero, then for all basis vectors $\mathbf{e}_i$

$$0 = \alpha_j^\beta \mathbb{T}_\beta^j \mathbf{e}_i = \alpha_j^\beta \delta_i^j \mathbf{e}'_\beta = \alpha_i^\beta \mathbf{e}'_\beta$$

The linear independence of $\mathbf{e}'_\beta$ implies that $\alpha_i^\beta = 0$ for $\beta = 1, 2, \ldots, N_2$. That the above equation is true for all $\mathbf{e}_i$ implies that $\alpha_i^\beta = 0$ for every i and β.

For any $\mathbb{T} \in \mathcal{L}(\mathcal{V},\mathcal{W})$ whose matrix elements in the above basis are τ_j^α, we have

$$\mathbb{T}\mathbf{e}_i = \tau_i^\alpha \mathbf{e}'_\alpha = \delta_i^j \tau_j^\alpha \mathbf{e}'_\alpha = \tau_j^\alpha(\delta_i^j \mathbf{e}'_\alpha) = \tau_j^\alpha(\mathbb{T}_\alpha^j \mathbf{e}_i) = (\tau_j^\alpha \mathbb{T}_\alpha^j)\mathbf{e}_i$$

Therefore, $\mathbb{T} = \tau_j^\alpha \mathbb{T}_\alpha^j$ implies that the $\mathbb{T}_\alpha^j$ span $\mathcal{L}(\mathcal{V},\mathcal{W})$. Thus, $\{\mathbb{T}_\alpha^j\}_{j,\alpha=1}^{N_1,N_2}$ is a basis for $\mathcal{L}(\mathcal{V},\mathcal{W})$. Note that we have also proved part (ii). ∎

The dual space $\mathcal{V}^*$ is simply the space $\mathcal{L}(\mathcal{V},\mathbb{R})$. Proposition 4.1.1 (with $N_2 = 1$) then implies that $\dim \mathcal{V}^* = \dim \mathcal{V}$, which was shown in Chapter 2. The dual space is important in the discussion of tensors, so we consider some of its properties below.

The basis $\{\mathbb{T}_\beta^j\}$ of Proposition 4.1.1 reduces to $\{\mathbb{T}_1^j\}$ when $\mathcal{W} = \mathbb{R}$ and is denoted as $\{\varepsilon^j\}_{i=1}^{N}$, with $N = \dim \mathcal{V}^* = \dim \mathcal{V}$. The ε^j have the property that

$$\varepsilon^j \mathbf{e}_i = \delta_i^j \mathbf{1} = \delta_i^j \tag{4.1}$$

where $\mathbf{1}$ is a basis for $\mathbb{R}$. This relation was established in Chapter 2. The basis

$B^* \equiv \{\varepsilon^j\}_{j=1}^N$ is simply the dual of the basis $B = \{\mathbf{e}_i\}_{i=1}^N$. Note the "natural" position of the index for B and B*.

Now suppose that $\{\mathbf{f}_i\}_{i=1}^N \equiv B'$ is another basis of $\mathscr{V}$ and R is the (invertible) matrix carrying B onto B′. Let $B'^* = \{\varphi^j\}_{j=1}^N$ be the dual of B′. We want to find the matrix that carries B* onto B′*. If we denote this matrix by A and its elements by a_i^j, we then have

$$\underset{\substack{\uparrow\\ \text{because } \varphi^k \\ \text{and } \mathbf{f}_i \text{ are} \\ \text{dual}}}{\delta_i^k =} \varphi^k \mathbf{f}_i \underset{\substack{\uparrow\\ \text{by the definition}\\ \text{of A and R}}}{=} (a_l^k \varepsilon^l)(r_i^j \mathbf{e}_j) = a_l^k r_i^j (\varepsilon^l \mathbf{e}_j) \underset{\substack{\uparrow\\ \text{because } \varepsilon^l \\ \text{and } \mathbf{e}_j \text{ are dual}}}{=} a_l^k r_i^j \delta_j^l$$

$$= a_l^k r_i^l \underset{\substack{\uparrow\\ \text{by the definition of}\\ \text{matrix multiplication}}}{=} (\mathsf{AR})_i^k \quad \Rightarrow \quad \mathsf{AR} = 1 \quad \Rightarrow \quad \mathsf{A} = \mathsf{R}^{-1}$$

Thus, the matrix that transforms bases of $\mathscr{V}^*$ is the inverse of the matrix that transforms the corresponding bases of $\mathscr{V}$.

It is important to emphasize that in the above equations the upper index in a_l^k or r_i^j labels *rows* and the lower index labels *columns*. This can be remembered by noting that the column vectors $\mathbf{e}_i$ can be thought of as columns of a matrix, and the lower index i then labels those columns. Similarly, ε^j can be thought of as rows of a matrix.

4.1.1 Multilinear Mappings

We can now generalize the concept of linear functionals.

4.1.2 Definition Let $\mathscr{V}_1$, $\mathscr{V}_2$, and $\mathscr{W}$ be vector spaces. A map $\mathsf{T}: \mathscr{V}_1 \times \mathscr{V}_2 \to \mathscr{W}$ is called *bilinear* if it is linear in each variable, that is, if

$$\mathsf{T}(\alpha_1 \mathbf{v}_1 + \beta_1 \mathbf{v}_1', \mathbf{v}_2) = \alpha_1 \mathsf{T}(\mathbf{v}_1, \mathbf{v}_2) + \beta_1 \mathsf{T}(\mathbf{v}_1', \mathbf{v}_2)$$

$$\mathsf{T}(\mathbf{v}_1, \alpha_2 \mathbf{v}_2 + \beta_2 \mathbf{v}_2') = \alpha_2 \mathsf{T}(\mathbf{v}_1, \mathbf{v}_2) + \beta_2 \mathsf{T}(\mathbf{v}_1, \mathbf{v}_2')$$

for all $\mathbf{v}_i, \mathbf{v}_i' \in \mathscr{V}_i$, where $i = 1, 2$, and $\alpha_i, \beta_i \in \mathbb{R}$.

Recall that $\mathscr{V}_1 \times \mathscr{V}_2$ is the Cartesian product of $\mathscr{V}_1$ and $\mathscr{V}_2$, or the set of ordered pairs $(\mathbf{v}_1, \mathbf{v}_2)$ with $\mathbf{v}_1 \in \mathscr{V}_1$ and $\mathbf{v}_2 \in \mathscr{V}_2$. We can extend Definition 4.1.2 to multilinear mappings. In particular, an r-linear mapping, $\mathsf{T}: \mathscr{V}_1 \times \mathscr{V}_2 \times \cdots \times \mathscr{V}_r \to \mathscr{W}$, is linear in all its variables; that is,

$$\mathsf{T}(\mathbf{v}_1, \ldots, \alpha \mathbf{v}_i + \alpha' \mathbf{v}_i', \ldots, \mathbf{v}_r) = \alpha \mathsf{T}(\mathbf{v}_1, \ldots, \mathbf{v}_i, \ldots, \mathbf{v}_r) + \alpha' \mathsf{T}(\mathbf{v}_1, \ldots, \mathbf{v}_i', \ldots, \mathbf{v}_r)$$

We can easily construct a bilinear mapping. Let $\tau_1 \in \mathscr{V}_1^*$ and $\tau_2 \in \mathscr{V}_2^*$. We define the mapping $\tau_1 \otimes \tau_2 : \mathscr{V}_1 \times \mathscr{V}_2 \to \mathbb{R}$ by

$$\tau_1 \otimes \tau_2(\mathbf{v}_1, \mathbf{v}_2) = (\tau_1(\mathbf{v}_1))(\tau_2(\mathbf{v}_2)) \tag{4.2a}$$

The expression $\tau_1 \otimes \tau_2$ is called the *tensor product* of τ_1 and τ_2. Clearly, since τ_1 and τ_2 are separately linear, so is $\tau_1 \otimes \tau_2$.

Multilinear mappings can be multiplied by scalars, and two multilinear mappings of the same kind can be added; in each case the result is a multilinear mapping of the same kind. Thus, the set of r-linear mappings from $\mathcal{V}_1 \times \mathcal{V}_2 \times \cdots \times \mathcal{V}_r$ into $\mathcal{W}$ forms a vector space that is denoted by $\mathcal{L}(\mathcal{V}_1, \ldots, \mathcal{V}_r; \mathcal{W})$.

We can also construct multilinear mappings on the dual space. First, we note that we can define a *natural* linear functional on $\mathcal{V}^*$ as follows. We let $\tau \in \mathcal{V}^*$ and $\mathbf{v} \in \mathcal{V}$; then $\tau(\mathbf{v}) \in \mathbb{R}$. Now we twist this around and write $\mathbf{v}(\tau) \equiv \tau(\mathbf{v})$; that is, we *define* a mapping $\mathbf{v}: \mathcal{V}^* \to \mathbb{R}$ given by $\mathbf{v}(\boldsymbol{\tau}) = \boldsymbol{\tau}(\mathbf{v})$.[1] It is easily shown that this mapping is linear. Thus, we have naturally constructed a linear functional on $\mathcal{V}^*$ by identifying $(\mathcal{V}^*)^*$ with $\mathcal{V}$.

Construction of multilinear mappings on $\mathcal{V}^*$ is now trivial. For example, we can construct a bilinear map analogously to the way we did the bilinear construction on $\mathcal{V}_1 \times \mathcal{V}_2$. To be specific, let $\mathbf{v}_1 \in \mathcal{V}_1$ and $\mathbf{v}_2 \in \mathcal{V}_2$ and define the tensor product $\mathbf{v}_1 \otimes \mathbf{v}_2 : \mathcal{V}_1^* \times \mathcal{V}_2^* \to \mathbb{R}$ by

$$\mathbf{v}_1 \otimes \mathbf{v}_2(\boldsymbol{\tau}_1, \boldsymbol{\tau}_2) = (\mathbf{v}_1(\boldsymbol{\tau}_1))(\mathbf{v}_2(\boldsymbol{\tau}_2)) = (\boldsymbol{\tau}_1(\mathbf{v}_1))(\boldsymbol{\tau}_2(\mathbf{v}_2)) \tag{4.2b}$$

We can also construct mixed multilinear mappings such as $\mathbf{v} \otimes \boldsymbol{\tau} : \mathcal{V}^* \times \mathcal{V} \to \mathbb{R}$ given by

$$\mathbf{v} \otimes \boldsymbol{\tau}(\boldsymbol{\theta}, \mathbf{u}) = \mathbf{v}(\boldsymbol{\theta})\boldsymbol{\tau}(\mathbf{u}) = \boldsymbol{\theta}(\mathbf{v})\boldsymbol{\tau}(\mathbf{u}) \tag{4.2c}$$

There is a bilinear map $\mathbf{h}: \mathcal{V}^* \times \mathcal{V} \to \mathbb{R}$ that naturally pairs $\mathcal{V}$ and $\mathcal{V}^*$; it is given by $\mathbf{h}(\boldsymbol{\theta}, \mathbf{v}) \equiv \boldsymbol{\theta}(\mathbf{v})$. This mapping is called the *natural pairing* of $\mathcal{V}$ and $\mathcal{V}^*$ into $\mathbb{R}$ and is denoted using angle brackets:

$$\mathbf{h}(\boldsymbol{\theta}, \mathbf{v}) \equiv \langle \boldsymbol{\theta}, \mathbf{v} \rangle \equiv \boldsymbol{\theta}(\mathbf{v})$$

Tensors. For the general case we have the following definition.

4.1.3 Definition Let $\mathcal{V}$ be a vector space with dual space $\mathcal{V}^*$. Then a *tensor of type* (r, s) is a multilinear mapping $\mathbf{T}_s^r : \mathcal{V}^* \times \mathcal{V}^* \times \cdots \times \mathcal{V}^* \times \mathcal{V} \times \mathcal{V} \times \cdots \times \mathcal{V} \to \mathbb{R}$ (where $\mathcal{V}^*$ appears r times and $\mathcal{V}$ appears s times). The set of all such mappings, for fixed r and s, forms a vector space denoted $\mathcal{T}_s^r(\mathcal{V})$. The quantity r is called the *contravariant degree* of the tensor, and s is called the *covariant degree* of the tensor.

Thus, $\mathbf{T}_s^r$ takes r elements of $\mathcal{V}^*$ and s elements of $\mathcal{V}$ and produces a real number from them. As an example, let $\mathbf{v}_1, \mathbf{v}_2, \ldots, \mathbf{v}_r \in \mathcal{V}$ and $\boldsymbol{\tau}^1, \boldsymbol{\tau}^2, \ldots, \boldsymbol{\tau}^s \in \mathcal{V}^*$ and define the tensor product

$$\mathbf{T}_s^r \equiv \mathbf{v}_1 \otimes \cdots \otimes \mathbf{v}_r \otimes \boldsymbol{\tau}^1 \otimes \cdots \otimes \boldsymbol{\tau}^s : \underbrace{\mathcal{V}^* \times \cdots \times \mathcal{V}^*}_{r \text{ times}} \times \underbrace{\mathcal{V} \times \cdots \times \mathcal{V}}_{s \text{ times}} \to \mathbb{R}$$

[1] From now on the same notation will be used for vectors and linear functionals and for multilinear maps.

by

$$\mathbf{v}_1 \otimes \cdots \otimes \mathbf{v}_r \otimes \boldsymbol{\tau}^1 \otimes \cdots \otimes \boldsymbol{\tau}^s(\boldsymbol{\theta}^1, \ldots, \boldsymbol{\theta}^r, \mathbf{u}_1, \ldots, \mathbf{u}_s)$$
$$= \mathbf{v}_1(\boldsymbol{\theta}^1) \cdots \mathbf{v}_r(\boldsymbol{\theta}^r)\boldsymbol{\tau}^1(\mathbf{u}_1) \cdots \boldsymbol{\tau}^s(\mathbf{u}_s)$$
$$= \prod_{i=1}^{r} \prod_{j=1}^{s} \boldsymbol{\theta}^i(\mathbf{v}_i)\boldsymbol{\tau}^j(\mathbf{u}_j)$$

The reader should not confuse the tensor product $\otimes$, which is an algebraic operation, with the Cartesian product $\times$, which is merely a set-theoretic operation. Each $\mathbf{v}$ in the tensor product requires an element of $\mathscr{V}^*$; that is why the number of factors of $\mathscr{V}^*$ in the Cartesian product equals the number of $\mathbf{v}$'s in the tensor product. The Cartesian product $\mathscr{V} \times \mathscr{V} \times \cdots \times \mathscr{V}$ with s factors is sometimes denoted $\mathscr{V}^s$ (similarly for $\mathscr{V}^*$).

A tensor of type (0, 0) is defined to be a scalar, so $\mathscr{T}_0^0(\mathscr{V}) = \mathbb{R}$. A tensor of type (1, 0) is called a *contravariant vector*, and one of type (0, 1) a *covariant vector*. A tensor of type $(r, 0)$ is called a *contravariant tensor of rank r*, and one of type $(0, s)$ is called a *covariant tensor of rank s*.

Algebra of tensors. The set of tensors of type (r, s) on a vector space $\mathscr{V}$ can be given a vector space structure, denoted by $\mathscr{T}_s^r(\mathscr{V})$. If we collect all these spaces together, we can define a product on that collection. This turns the union of all $\mathscr{T}_s^r(\mathscr{V})$ into an algebra, called the *algebra of tensors*.

4.1.4 Definition The *tensor product* of a tensor $\mathbf{T}$ of type (r, s) and a tensor $\mathbf{U}$ of type (k, l) is a tensor $\mathbf{T} \otimes \mathbf{U}$ of type $(r + k, s + l)$, defined, as an operator on $(\mathscr{V}^*)^{r+k} \times \mathscr{V}^{s+l}$, by

$$\mathbf{T} \otimes \mathbf{U}(\boldsymbol{\tau}^1, \ldots, \boldsymbol{\tau}^{r+k}, \mathbf{v}_1, \ldots, \mathbf{v}_{s+l})$$
$$= \mathbf{T}(\boldsymbol{\tau}^1, \ldots, \boldsymbol{\tau}^r, \mathbf{v}_1, \ldots, \mathbf{v}_s)\mathbf{U}(\boldsymbol{\tau}^{r+1}, \ldots, \boldsymbol{\tau}^{r+k}, \mathbf{v}_{s+1}, \ldots, \mathbf{v}_{s+l})$$

This definition is a generalization of Eqs. (4.2). It is easily verified that if $\mathbf{S}$, $\mathbf{T}$, and $\mathbf{U}$ are tensors, then

$$(\mathbf{S} \otimes \mathbf{T}) \otimes \mathbf{U} = \mathbf{S} \otimes (\mathbf{T} \otimes \mathbf{U})$$
$$\mathbf{S} \otimes (\mathbf{T} + \mathbf{U}) = \mathbf{S} \otimes \mathbf{T} + \mathbf{S} \otimes \mathbf{U}$$
$$(\mathbf{S} + \mathbf{T}) \otimes \mathbf{U} = \mathbf{S} \otimes \mathbf{U} + \mathbf{T} \otimes \mathbf{U}$$

However, in general, $\mathbf{T} \otimes \mathbf{U} \neq \mathbf{U} \otimes \mathbf{T}$; that is, the tensor product is not commutative.

Components of tensors. Making computations with tensors requires choosing a basis for $\mathscr{V}$ and one for $\mathscr{V}^*$ and representing the tensors in terms of numbers (components). This process is not, of course, new. Linear operators are represented by arrays of numbers in the form of matrices. The case of tensors is merely a generalization of that of linear operators and can be stated in the form of a theorem.

4.1.5 Theorem Let $\{\mathbf{e}_i\}_{i=1}^N = \mathrm{B}$ and $\mathrm{B}^* = \{\boldsymbol{\epsilon}^j\}_{j=1}^N$ be bases for $\mathscr{V}$ and $\mathscr{V}^*$, respectively. Then the set of all tensor products $\mathbf{e}_{i_1} \otimes \cdots \otimes \mathbf{e}_{i_r} \otimes \boldsymbol{\epsilon}^{j_1} \otimes \cdots \otimes \boldsymbol{\epsilon}^{j_s}$ forms a basis for $\mathscr{T}^r_s(\mathscr{V})$. Furthermore, the components of any tensor, $\mathbf{A} \in \mathscr{T}^r_s(\mathscr{V})$, are

$$A^{j_1 \dots j_r}_{i_1 \dots i_s} = \mathbf{A}(\boldsymbol{\epsilon}^{j_1}, \dots, \boldsymbol{\epsilon}^{j_r}, \mathbf{e}_{i_1}, \dots, \mathbf{e}_{i_s})$$

Proof. For any $\boldsymbol{\tau}^1, \boldsymbol{\tau}^2, \dots, \boldsymbol{\tau}^r \in \mathscr{V}^*$ and $\mathbf{v}_1, \mathbf{v}_2, \dots, \mathbf{v}_s \in \mathscr{V}$, we have $\boldsymbol{\tau}^p = a^p_j \boldsymbol{\epsilon}^j$ and $\mathbf{v}_q = b^i_q \mathbf{e}_i$, for $p = 1, \dots, r$ and $q = 1, \dots, s$, and

$$\begin{aligned}
\mathbf{A}(\boldsymbol{\tau}^1, \dots, \boldsymbol{\tau}^r, \mathbf{v}_1, \dots, \mathbf{v}_s) &= \mathbf{A}(a^1_{j_1} \boldsymbol{\epsilon}^{j_1}, \dots, a^r_{j_r} \boldsymbol{\epsilon}^{j_r}, b^{i_1}_1 \mathbf{e}_{i_1}, \dots, b^{i_s}_s \mathbf{e}_{i_s}) \\
&\underset{\uparrow}{=} a^1_{j_1} \cdots a^r_{j_r} b^{i_1}_1 \cdots b^{i_s}_s \mathbf{A}(\boldsymbol{\epsilon}^{j_1}, \dots, \boldsymbol{\epsilon}^{j_r}, \mathbf{e}_{i_1}, \dots, \mathbf{e}_{i_s})
\end{aligned}$$

by the multilinearity of $\mathbf{A}$ (4.3)

$$\begin{aligned}
&= A^{j_1 \dots j_r}_{i_1 \dots i_s} \mathbf{e}_{j_1}(\boldsymbol{\tau}^1) \cdots \mathbf{e}_{j_r}(\boldsymbol{\tau}^r) \boldsymbol{\epsilon}^{i_1}(\mathbf{v}_1) \cdots \boldsymbol{\epsilon}^{i_s}(\mathbf{v}_s) \\
&= A^{j_1 \dots j_r}_{i_1 \dots i_s} \mathbf{e}_{j_1} \otimes \cdots \otimes \mathbf{e}_{j_r} \otimes \boldsymbol{\epsilon}^{i_1} \otimes \cdots \otimes \boldsymbol{\epsilon}^{i_s} \\
&\quad \times (\boldsymbol{\tau}^1, \dots, \boldsymbol{\tau}^r, \mathbf{v}_1, \dots, \mathbf{v}_s)
\end{aligned}$$

In thc last step we used the relations

$$\mathbf{e}_j(\boldsymbol{\tau}^p) = \mathbf{e}_j(a^p_i \boldsymbol{\epsilon}^i) = a^p_i \mathbf{e}_j(\boldsymbol{\epsilon}^i) = a^p_i \delta^i_j = a^p_j$$

and

$$\boldsymbol{\epsilon}^i(\mathbf{v}_q) = \boldsymbol{\epsilon}^i(b^j_q \mathbf{e}_j) = b^j_q \boldsymbol{\epsilon}^i(\mathbf{e}_j) = b^j_q \delta^i_j = b^i_q$$

which are consequences of the fact that $\mathbf{e}_j$ is a linear functional on $\mathscr{V}^*$ and $\boldsymbol{\epsilon}^i$ is a linear functional on $\mathscr{V}$. Equation (4.3) holds for arbitrary $\boldsymbol{\tau}^p$ and $\mathbf{v}_q$; therefore, the operators on both sides must be equal. Thus,

$$\mathbf{A} = A^{j_1 \dots j_r}_{i_1 \dots i_s} \mathbf{e}_{j_1} \otimes \cdots \otimes \mathbf{e}_{j_r} \otimes \boldsymbol{\epsilon}^{i_1} \otimes \cdots \otimes \boldsymbol{\epsilon}^{i_s} \tag{4.4}$$

This implies that the products $\mathbf{e}_{j_1} \otimes \cdots \otimes \mathbf{e}_{j_r} \otimes \boldsymbol{\epsilon}^{i_1} \otimes \cdots \otimes \boldsymbol{\epsilon}^{i_s}$ span $\mathscr{T}^r_s(\mathscr{V})$. Exercise 4.1.1 shows that they are also linearly independent. Thus, they form a basis of $\mathscr{T}^r_s(\mathscr{V})$. ∎

Note that for every factor in the tensor product $\mathbf{e}_{j_1} \otimes \cdots \otimes \mathbf{e}_{j_r} \otimes \boldsymbol{\epsilon}^{i_1} \otimes \cdots \otimes \boldsymbol{\epsilon}^{i_s}$ there are N possibilities. Thus, the number of possible tensor products is N^{r+s}, and we have a corollary to the preceding theorem.

4.1.6 Corollary The dimension of $\mathscr{T}^r_s(\mathscr{V})$ is N^{r+s}, where $N = \dim \mathscr{V}$. ∎

Example 4.1.1

Let us consider the special case of $\mathscr{T}^1_1(\mathscr{V})$ as an illustration. We can write $\mathbf{A} \in \mathscr{T}^1_1(\mathscr{V})$ as $\mathbf{A} = A^i_j \mathbf{e}_i \otimes \boldsymbol{\epsilon}^j$. Given any $\mathbf{v} \in \mathscr{V}$, we can define the action of $\mathbf{A}$ on $\mathbf{v}$ in a natural way as

$$\mathbf{A}(\mathbf{v}) = (A^i_j \mathbf{e}_i \otimes \boldsymbol{\epsilon}^j)(\mathbf{v}) \equiv \underset{\substack{\uparrow \\ \in \mathbb{R}}}{A^i_j} \mathbf{e}_i (\underset{\substack{\uparrow \\ \in \mathbb{R}}}{\boldsymbol{\epsilon}^j(\mathbf{v})})$$

This shows that $\mathbf{A}(\mathbf{v}) \in \mathcal{V}$. That is, $\mathbf{A}$ can be interpreted as an operator that takes a vector in $\mathcal{V}$ and gives another vector, also in $\mathcal{V}$. Thus, $\mathbf{A}$ can be thought of as a linear operator on $\mathcal{V}$, and $\mathbf{A} \in \mathcal{L}(\mathcal{V})$. Similarly, for $\boldsymbol{\tau} \in \mathcal{V}^*$ we can define

$$\mathbf{A}(\boldsymbol{\tau}) = (A^i_j \mathbf{e}_i \otimes \boldsymbol{\epsilon}^j)(\boldsymbol{\tau}) = \underset{\in\mathbb{R}}{\underset{\uparrow}{A^i_j}} \underset{\in\mathbb{R}}{\underset{\uparrow}{[\mathbf{e}_i(\boldsymbol{\tau})]}} \boldsymbol{\epsilon}^j$$

which shows that $\mathbf{A}(\boldsymbol{\tau}) \in \mathcal{V}^*$. Thus, $\mathbf{A} \in \mathcal{L}(\mathcal{V}^*)$. We have shown that the three spaces $\mathcal{T}^1_1(\mathcal{V})$, $\mathcal{L}(\mathcal{V})$, and $\mathcal{L}(\mathcal{V}^*)$ are the same in a way that requires qualification. What we have shown is that *given* $\mathbf{A} \in \mathcal{T}^1_1(\mathcal{V})$, there *corresponds* a linear operator belonging to $\mathcal{L}(\mathcal{V})$ [or $\mathcal{L}(\mathcal{V}^*)$] and having a natural relation to $\mathbf{A}$. Similarly, given any $\mathbb{A} \in \mathcal{L}(\mathcal{V})$ [or $\mathcal{L}(\mathcal{V}^*)$] with a matrix representation in some basis of $\mathcal{V}$ (or $\mathcal{V}^*$) given by A^i_j, then corresponding to it in a natural way is a tensor in $\mathcal{T}^1_1(\mathcal{V})$, namely $A^i_j \mathbf{e}_i \otimes \boldsymbol{\epsilon}^j$. Therefore, there is a *natural* one-to-one correspondence among $\mathcal{T}^1_1(\mathcal{V})$, $\mathcal{L}(\mathcal{V})$, and $\mathcal{L}(\mathcal{V}^*)$. This natural correspondence is called an *isomorphism* and is what is meant by saying the spaces are the same. ●

We have defined tensors as multilinear functions that act like a multilinear machine that takes in a vector from the Cartesian product space of $\mathcal{V}$'s and $\mathcal{V}^*$'s and manufactures a real number. Given the representation in Eq. (4.4), however, we can interpret a tensor as a linear machine that takes a vector belonging to a Cartesian product space and manufactures a tensor. This corresponds to a situation in which not all factors of (4.4) find "partners." An illustration of this situation was presented in Example 4.1.1.

To clarify this let us consider $\mathbf{A} \in \mathcal{T}^1_2(\mathcal{V})$, given by

$$\mathbf{A} = A^i_{jk} \mathbf{e}_i \otimes \boldsymbol{\epsilon}^j \otimes \boldsymbol{\epsilon}^k$$

This machine needs a Cartesian-product vector of the form $(\boldsymbol{\tau}, \mathbf{v}_1, \mathbf{v}_2)$, with $\boldsymbol{\tau} \in \mathcal{V}^*$ and $\mathbf{v}_1, \mathbf{v}_2 \in \mathcal{V}$, to give a real number. However, if it is not fed enough, it will not complete its job. For instance, if we feed it only a vector $\boldsymbol{\tau}$ belonging to $\mathcal{V}^*$, it will give a tensor belonging to $\mathcal{T}^0_2(\mathcal{V})$:

$$\mathbf{A}(\boldsymbol{\tau}) = (A^i_{jk} \mathbf{e}_i \otimes \boldsymbol{\epsilon}^j \otimes \boldsymbol{\epsilon}^k)(\boldsymbol{\tau}) = A^i_{jk} \mathbf{e}_i(\boldsymbol{\tau}) \boldsymbol{\epsilon}^j \otimes \boldsymbol{\epsilon}^k$$

If we feed it a double vector $(\mathbf{v}_1, \mathbf{v}_2) \in \mathcal{V} \times \mathcal{V}$, it will manufacture a vector in $\mathcal{V}$:

$$\begin{aligned}\mathbf{A}(\mathbf{v}_1, \mathbf{v}_2) &= (A^i_{jk} \mathbf{e}_i \otimes \boldsymbol{\epsilon}^j \otimes \boldsymbol{\epsilon}^k)(\mathbf{v}_1, \mathbf{v}_2) \\ &= A^i_{jk} \mathbf{e}_i \boldsymbol{\epsilon}^j(\mathbf{v}_1) \boldsymbol{\epsilon}^k(\mathbf{v}_2) \in \mathcal{V}\end{aligned}$$

What if we feed it simply a vector $\mathbf{v} \in \mathcal{V}$? It will go haywire, because it does not know whether to give $\mathbf{v}$ to $\boldsymbol{\epsilon}^j$ or $\boldsymbol{\epsilon}^k$ (it is smart enough to know that it cannot give $\mathbf{v}$ to $\mathbf{e}_i$). That is why we have to inform the machine as to which factor of $\boldsymbol{\epsilon}$ $\mathbf{v}$ has to go. This is done by positioning $\mathbf{v}$ in a pair of parentheses divided into regions separated by commas. Thus, if we write $(.\,, \mathbf{v}, .)$, the machine will know that $\mathbf{v}$ belongs to $\boldsymbol{\epsilon}^j$, and $(.\,, .\,, \mathbf{v})$ tells the machine to contract $\mathbf{v}$ with $\boldsymbol{\epsilon}^k$. If we write $(\mathbf{v}, .\,, .)$, the machine will give us an "error message" because it cannot contract $\mathbf{v}$ with $\mathbf{e}_i$!

Transformation laws. The components of a tensor $\mathbf{A}$, as given in Eq. (4.4), depend on the basis in which they are described. If the basis is changed, the components change. The relation between components of a tensor in different bases is called the *transformation law* for that particular tensor. Let us investigate this concept.

We can use overbars to distinguish among various bases. For instance, $B = \{\mathbf{e}_i\}_{i=1}^N$, $\bar{B} = \{\bar{\mathbf{e}}_j\}_{j=1}^N$, and $\bar{\bar{B}} = \{\bar{\bar{\mathbf{e}}}_k\}_{k=1}^N$ are three different bases of $\mathscr{V}$. Similarly, $B^* = \{\boldsymbol{\epsilon}^i\}_{i=1}^N$, $\bar{B}^* = \{\bar{\boldsymbol{\epsilon}}^j\}_{j=1}^N$, and $\bar{\bar{B}}^* = \{\bar{\bar{\boldsymbol{\epsilon}}}^k\}_{k=1}^N$ are bases for $\mathscr{V}^*$. The components are also distinguished with overbars. Recall that if R is the matrix connecting B and $\bar{B}$, then $\mathsf{S} = \mathsf{R}^{-1}$ connects B^* and $\bar{B}^*$. For a tensor $\mathbf{A}$ of type (1,2), from Theorem 4.1.5 we have

$$\begin{aligned} \bar{A}^i_{jk} &= \mathbf{A}(\bar{\boldsymbol{\epsilon}}^i, \bar{\mathbf{e}}_j, \bar{\mathbf{e}}_k) = \mathbf{A}(s^i_m \boldsymbol{\epsilon}^m, r^n_j \mathbf{e}_n, r^p_k \mathbf{e}_p) \\ &= s^i_m r^n_j r^p_k \mathbf{A}(\boldsymbol{\epsilon}^m, \mathbf{e}_n, \mathbf{e}_p) \\ &= s^i_m r^n_j r^p_k A^m_{np} \end{aligned} \tag{4.5}$$

This is the law that transforms the components of a tensor from one basis to another. In the coordinate-dependent treatment of tensors, Eq. (4.5) was the *defining* relation for a tensor of type (1,2). In other words, a tensor of type (1,2) was defined to be a "bunch" of numbers, A^m_{np}, which transformed to another "bunch" of numbers, $\bar{A}^i_{jk}$, according to the rule in (4.5) when the basis was changed. In the modern treatment of tensors it is not necessary to introduce any basis to define tensors. Only when the components of a tensor are needed must bases be introduced. The advantage of the modern treatment is obvious, since a (1,2)-type tensor has 27 components in three dimensions and 64 components in four dimensions, all of which are represented by the symbol $\mathbf{A}$. However, the role of components should not be downplayed. After all, when it comes to actual calculations, we are forced to choose a basis and start manipulating the components.

Equation (4.5) is the transformation law for a (1,2)-type tensor. The alterations necessary for obtaining the laws for other types can easily be made and will not be given here.

Functions of tensors. Since $\mathscr{T}^r_s(\mathscr{V})$ are vector spaces, it is possible to construct mappings $h: \mathscr{T}^r_s(\mathscr{V}) \to \mathscr{T}^k_l(\mathscr{V})$. We will be interested in linear mappings as usual. For example, $\mathfrak{f}: \mathscr{T}^1_0(\mathscr{V}) \to \mathscr{T}^0_0(\mathscr{V}) = \mathbb{R}$ is what was called a linear functional before. Similarly, $\mathfrak{t}: \mathscr{T}^1_0(\mathscr{V}) \to \mathscr{T}^1_0(\mathscr{V})$ is a linear transformation on $\mathscr{V}$.

A special linear transformation is $\operatorname{tr}: \mathscr{T}^1_1(\mathscr{V}) \to \mathscr{T}^0_0(\mathscr{V}) = \mathbb{R}$, given by

$$\operatorname{tr} \mathbf{A} = \operatorname{tr}(A^i_j \mathbf{e}_i \otimes \boldsymbol{\epsilon}^i) \equiv A^i_i \equiv \sum_{i=1}^N A^i_i$$

This is very similar to the trace function encountered in the study of linear transformations in Chapter 2. In fact, the isomorphism of $\mathscr{T}^1_1(\mathscr{V})$ and $\mathscr{L}(\mathscr{V})$ completely identifies this trace with that of linear transformations.

The above definition makes explicit use of components with respect to a basis. So it may be thought that the trace is a basis-dependent function. However, it is easily shown (see Exercise 4.1.3) that it is in fact basis-independent. Functions of tensors that do not depend on bases are called *invariants*. Another example of an invariant is a linear functional. The proof of this fact is easy and is left as a problem.

Example 4.1.2

Let us show that not every expression in terms of components need be an invariant. Consider the tensor $\mathbf{A} \in \mathscr{T}_0^2(\mathbb{R}^2)$ given by

$$\mathbf{A} = \mathbf{e}_1 \otimes \mathbf{e}_1 + \mathbf{e}_2 \otimes \mathbf{e}_1$$

We calculate the analogue of the trace for $\mathbf{A}$:

$$A_{ii} \equiv \sum_{i=1}^{2} A_{ii} = A_{11} + A_{22} = 1 + 0 = 1$$

Now we change to a new basis, $\{\bar{\mathbf{e}}_1, \bar{\mathbf{e}}_2\}$, given by

$$\mathbf{e}_1 = \bar{\mathbf{e}}_1 + 2\bar{\mathbf{e}}_2 \qquad \text{and} \qquad \mathbf{e}_2 = -\bar{\mathbf{e}}_1 + \bar{\mathbf{e}}_2$$

In terms of the new basis vectors, $\mathbf{A}$ is given by

$$\begin{aligned}\mathbf{A} &= (\bar{\mathbf{e}}_1 + 2\bar{\mathbf{e}}_2) \otimes (\bar{\mathbf{e}}_1 + 2\bar{\mathbf{e}}_2) + (-\bar{\mathbf{e}}_1 + \bar{\mathbf{e}}_2) \otimes (\bar{\mathbf{e}}_1 + 2\bar{\mathbf{e}}_2) \\ &= 3\bar{\mathbf{e}}_2 \otimes \bar{\mathbf{e}}_1 + 6\bar{\mathbf{e}}_2 \otimes \bar{\mathbf{e}}_2\end{aligned}$$

with $\bar{A}_{ii} = 0 + 6 = 6 \neq A_{ii}$. ●

Whenever a quantity is defined without reference to a basis, it is clearly invariant. The trace is an example. Another example is the determinant of a linear operator [or a member of $\mathscr{T}_1^1(\mathscr{V})$], which was defined earlier in terms of a matrix representation of that operator, which is clearly basis-dependent. However, a basis-independent definition for the determinant of a tensor is given later in this section.

Besides mappings of the form $h: \mathscr{T}_s^r \to \mathscr{T}_l^k$, which are single-variable functions, we can define mappings that depend on several variables, in other words, that take several elements of $\mathscr{T}_s^r$ and give an element of $\mathscr{T}_l^k$. We can write

$$h: \underbrace{\mathscr{T}_s^r(\mathscr{V}) \times \cdots \times \mathscr{T}_s^r(\mathscr{V})}_{m \text{ times}} \to \mathscr{T}_l^k(\mathscr{V})$$

or simply

$$h: (\mathscr{T}_s^r(\mathscr{V}))^m \to \mathscr{T}_l^k(\mathscr{V})$$

It is then understood that $h(\mathbf{t}_1, \mathbf{t}_2, \ldots, \mathbf{t}_m)$, in which $\mathbf{t}_1, \mathbf{t}_2, \ldots, \mathbf{t}_m \in \mathscr{T}_s^r(\mathscr{V})$, is a tensor of type (k, l). If h is linear in all of its variables, it is called a multilinear function. Furthermore, if $h(\mathbf{t}_1, \mathbf{t}_2, \ldots, \mathbf{t}_m)$ does not depend on the choice of a basis of $\mathscr{T}_s^r$, it is called a *multilinear invariant*. In most cases $k = 0 = l$, and we speak of *scalar-valued invariants*, or simply *invariants*. An example of a multilinear invariant is the determinant considered as a function of the rows of a matrix.

An important class of multilinear invariants is composed of the contractions, which are defined as follows.

4.1.7 Definition A *contraction* of a tensor $\mathbf{A} \in \mathcal{T}^r_s(\mathcal{V})$ with respect to a contravariant index p and a covariant index q is a linear mapping $\mathbb{C}: \mathcal{T}^r_s(\mathcal{V}) \to \mathcal{T}^{r-1}_{s-1}(\mathcal{V})$ given *in components* by

$$(\mathbb{C}(\mathbf{A}))^{i_1 \cdots i_{r-1}}_{j_1 \cdots j_{s-1}} = A^{i_1 \cdots i_{p-1} k i_p \cdots i_{r-1}}_{j_1 \cdots j_{q-1} k j_q \cdots j_{s-1}}$$

It can easily be shown that contractions are invariants. The proof is exactly the same as that for the invariance of the trace. In fact, the trace is a special case of a contraction, in which $r = s = 1$.

4.1.2 Symmetries of Tensors

Many applications deal with tensors that have a symmetry property of some kind. We have already encountered a symmetric tensor—the metric tensor. If $\mathcal{V}$ is a real vector space, as is assumed in this chapter, and $\mathbf{v}_1, \mathbf{v}_2 \in \mathcal{V}$, then $g(\mathbf{v}_1, \mathbf{v}_2) = g(\mathbf{v}_2, \mathbf{v}_1)$. This shows that interchanging $\mathbf{v}_1$ and $\mathbf{v}_2$ does not change the value of g. The following generalizes this.

4.1.8 Definition A tensor $\mathbf{A}$ is symmetric in the ith and jth variables if its values as a multilinear function are unchanged when these variables are interchanged. Clearly, the two variables must be of the same kind.

An immediate consequence of Definition 4.1.8 is that, in any basis, the components of a tensor do not change when the ith and jth indices are interchanged. To illustrate this point, let us consider a tensor $\mathbf{A}$ of type (3,1) and assume that it is symmetric in the second and third contravariant variables. Then

$$A^{ijk}_l = \mathbf{A}(\boldsymbol{\epsilon}^i, \boldsymbol{\epsilon}^j, \boldsymbol{\epsilon}^k, \mathbf{e}_l) \underset{\uparrow}{=} \mathbf{A}(\boldsymbol{\epsilon}^i, \boldsymbol{\epsilon}^k, \boldsymbol{\epsilon}^j, \mathbf{e}_l) = A^{ikj}_l$$

by the definition
of a symmetric tensor

and the interchange of indices does not change the components.

4.1.9 Definition A tensor is *contravariant-symmetric* if it is symmetric in every pair of its contravariant indices and *covariant-symmetric* if it is symmetric in every pair of its covariant indices. A tensor is *symmetric* if it is both contravariant-symmetric and covariant-symmetric.

It is clear that there cannot be symmetry under interchange of a covariant index with a contravariant index.

Algebra of symmetric tensors. The set of all symmetric tensors $\mathcal{S}^r$ of type $(r, 0)$ forms a subspace of the vector space $\mathcal{T}^r_0$; similarly, the set of those of type

(0, s) forms a subspace $\mathcal{S}_s$ of $\mathcal{T}^0_s$. The components of a symmetric tensor $\mathbf{A} \in \mathcal{S}^r$ are $A^{i_1 i_2 \cdots i_r}$, where $i_1 \leqslant i_2 \leqslant \cdots \leqslant i_r$; the other components are given by symmetry.

Although a set of symmetric tensors forms a vector space, it does *not* form an algebra under the usual multiplication of tensors. In fact, even if $\mathbf{A} = A^{ij}\mathbf{e}_i \otimes \mathbf{e}_j$ and $\mathbf{B} = B^{kl}\mathbf{e}_k \otimes \mathbf{e}_l$ are symmetric tensors of type (2, 0), the tensor product

$$\mathbf{A} \otimes \mathbf{B} = A^{ij}B^{kl}\mathbf{e}_i \otimes \mathbf{e}_j \otimes \mathbf{e}_k \otimes \mathbf{e}_l$$

need not be a type (4, 0) symmetric tensor. For instance, $A^{ik}B^{jl}$ may not equal $A^{ij}B^{kl}$. However, we can modify the definition of the tensor product to give a symmetric product out of symmetric factors. Let us first consider the following definition.

4.1.10 Definition A *symmetrizer* is an operator, $\mathbb{S}: \mathcal{T}^r_0 \to \mathcal{S}^r$, given by

$$(\mathbb{S}(\mathbf{A}))(\boldsymbol{\tau}^1, \ldots, \boldsymbol{\tau}^r) = \frac{1}{r!}\sum_{\pi} \mathbf{A}(\boldsymbol{\tau}^{\pi(1)}, \ldots, \boldsymbol{\tau}^{\pi(r)}) \tag{4.6}$$

where the sum is taken over the $r!$ permutations, π, of the integers $1, 2, \ldots, r$, and $\boldsymbol{\tau}^1, \ldots, \boldsymbol{\tau}^r$ are any elements in $\mathcal{V}^*$, and $\pi(j)$ is the number to which j goes under permutation. Sometimes $\mathbb{S}(\mathbf{A})$ is denoted $\mathbf{A}_s$. Clearly, $\mathbf{A}_s$ is a symmetric tensor.

Example 4.1.3

Let us write out Eq. (4.6) explicitly for $r = 2$ and $r = 3$.

(a) For $r = 2$, we have only two permutations, and

$$\mathbf{A}_s(\boldsymbol{\tau}^1, \boldsymbol{\tau}^2) = \tfrac{1}{2}[\mathbf{A}(\boldsymbol{\tau}^1, \boldsymbol{\tau}^2) + \mathbf{A}(\boldsymbol{\tau}^2, \boldsymbol{\tau}^1)] \tag{1}$$

(b) For $r = 3$, we have six permutations of 1, 2, 3. In this case (4.6) gives

$$\begin{aligned}\mathbf{A}_s(\boldsymbol{\tau}^1, \boldsymbol{\tau}^2, \boldsymbol{\tau}^3) = \tfrac{1}{6}[&\mathbf{A}(\boldsymbol{\tau}^1, \boldsymbol{\tau}^2, \boldsymbol{\tau}^3) + \mathbf{A}(\boldsymbol{\tau}^1, \boldsymbol{\tau}^3, \boldsymbol{\tau}^2) + \mathbf{A}(\boldsymbol{\tau}^2, \boldsymbol{\tau}^1, \boldsymbol{\tau}^3) \\ &+ \mathbf{A}(\boldsymbol{\tau}^2, \boldsymbol{\tau}^3, \boldsymbol{\tau}^1) + \mathbf{A}(\boldsymbol{\tau}^3, \boldsymbol{\tau}^1, \boldsymbol{\tau}^2) + \mathbf{A}(\boldsymbol{\tau}^3, \boldsymbol{\tau}^2, \boldsymbol{\tau}^1)]\end{aligned} \tag{2}$$

It is clear that interchanging any pair of $\boldsymbol{\tau}$'s on the RHS of (1) or (2) does not change the sum. Thus, $\mathbf{A}_s$ is indeed a symmetric tensor. ●

A quantity that is of interest is the dimension of $\mathcal{S}^r(\mathcal{V})$. If $\dim \mathcal{V} = N$, then it can be shown that

$$\dim \mathcal{S}^r(\mathcal{V}) = \binom{N + r - 1}{r} \equiv \frac{(N + r - 1)!}{r!(N - 1)!}$$

The proof involves counting the number of different integers $i_1, \ldots, i_r$ for which $1 \leqslant i_m \leqslant i_{m+1} \leqslant N$ for each m.

A similar definition gives the symmetrizer $\mathbb{S}: \mathcal{T}^0_s \to \mathcal{S}_s$. Instead of $\boldsymbol{\tau}^1, \boldsymbol{\tau}^2, \ldots, \boldsymbol{\tau}^r$ in (4.6), we would have $\mathbf{v}_1, \mathbf{v}_2, \ldots, \mathbf{v}_s$.

We are now ready to define a product on $\mathcal{S}^r(\mathcal{V})$ and make it an algebra called the *symmetric algebra*.

4.1.11 Definition The *symmetric product* of symmetric tensors $\mathbf{A} \in \mathscr{S}^r(\mathscr{V})$ and $\mathbf{B} \in \mathscr{S}^t(\mathscr{V})$ is the symmetric tensor $\mathbb{S}(\mathbf{A} \otimes \mathbf{B}) \in \mathscr{S}^{r+t}(\mathscr{V})$. We denote this product simply by $\mathbf{AB}$. Thus, in detail

$$\mathbf{AB}(\boldsymbol{\tau}^1, \ldots, \boldsymbol{\tau}^{r+t}) = \frac{1}{(r+t)!} \sum_{\pi} (\mathbf{A} \otimes \mathbf{B})(\boldsymbol{\tau}^{\pi(1)}, \ldots, \boldsymbol{\tau}^{\pi(r+t)})$$

$$= \frac{1}{(r+t)!} \sum_{\pi} \mathbf{A}(\boldsymbol{\tau}^{\pi(1)}, \ldots, \boldsymbol{\tau}^{\pi(r)}) \mathbf{B}(\boldsymbol{\tau}^{\pi(r+1)}, \ldots, \boldsymbol{\tau}^{\pi(r+t)})$$

where again the sum is over all permutations of $1, 2, \ldots, r + t$. The symmetric product of $\mathbf{A} \in \mathscr{S}_s(\mathscr{V})$ and $\mathbf{B} \in \mathscr{S}_u(\mathscr{V})$ is defined similarly.

Example 4.1.4

Let us construct the symmetric tensor products of ranks 2 and 3. First consider $\mathbf{v}_1, \mathbf{v}_2 \in \mathscr{V}$. We use the definition of the symmetrizer to find the symmetric product of $\mathbf{v}_1$ and $\mathbf{v}_2$:

$$(\mathbf{v}_1\mathbf{v}_2)(\boldsymbol{\tau}^1, \boldsymbol{\tau}^2) \equiv [\mathbb{S}(\mathbf{v}_1 \otimes \mathbf{v}_2)](\boldsymbol{\tau}^1, \boldsymbol{\tau}^2) = \tfrac{1}{2} \sum_{\pi} (\mathbf{v}_1 \otimes \mathbf{v}_2)(\boldsymbol{\tau}^1, \boldsymbol{\tau}^2)$$

$$= \tfrac{1}{2}[(\mathbf{v}_1 \otimes \mathbf{v}_2)(\boldsymbol{\tau}^1, \boldsymbol{\tau}^2) + (\mathbf{v}_1 \otimes \mathbf{v}_2)(\boldsymbol{\tau}^2, \boldsymbol{\tau}^1)]$$

$$= \tfrac{1}{2}[\mathbf{v}_1(\boldsymbol{\tau}^1)\mathbf{v}_2(\boldsymbol{\tau}^2) + \mathbf{v}_1(\boldsymbol{\tau}^2)\mathbf{v}_2(\boldsymbol{\tau}^1)]$$

$$= \tfrac{1}{2}[\mathbf{v}_1(\boldsymbol{\tau}^1)\mathbf{v}_2(\boldsymbol{\tau}^2) + \mathbf{v}_2(\boldsymbol{\tau}^1)\mathbf{v}_1(\boldsymbol{\tau}^2)]$$

$$= \tfrac{1}{2}[(\mathbf{v}_1 \otimes \mathbf{v}_2)(\boldsymbol{\tau}^1, \boldsymbol{\tau}^2) + (\mathbf{v}_2 \otimes \mathbf{v}_1)(\boldsymbol{\tau}^1, \boldsymbol{\tau}^2)]$$

$$= \tfrac{1}{2}(\mathbf{v}_1 \otimes \mathbf{v}_2 + \mathbf{v}_2 \otimes \mathbf{v}_1)(\boldsymbol{\tau}^1, \boldsymbol{\tau}^2)$$

Since this is true for any pair of $\boldsymbol{\tau}^1$ and $\boldsymbol{\tau}^2$, we have

$$\mathbf{v}_1\mathbf{v}_2 = \tfrac{1}{2}(\mathbf{v}_1 \otimes \mathbf{v}_2 + \mathbf{v}_2 \otimes \mathbf{v}_1)$$

Similarly, we can show that

$$(\mathbf{v}_1\mathbf{v}_2)\mathbf{v}_3 = \mathbf{v}_1(\mathbf{v}_2\mathbf{v}_3) = \mathbf{v}_1(\mathbf{v}_3\mathbf{v}_2) = \cdots$$

$$= \tfrac{1}{6}(\mathbf{v}_1 \otimes \mathbf{v}_2 \otimes \mathbf{v}_3 + \mathbf{v}_1 \otimes \mathbf{v}_3 \otimes \mathbf{v}_2 + \mathbf{v}_2 \otimes \mathbf{v}_1 \otimes \mathbf{v}_3$$

$$+ \mathbf{v}_2 \otimes \mathbf{v}_3 \otimes \mathbf{v}_1 + \mathbf{v}_3 \otimes \mathbf{v}_1 \otimes \mathbf{v}_2 + \mathbf{v}_3 \otimes \mathbf{v}_2 \otimes \mathbf{v}_1)$$

●

It is clear from the definition that symmetric multiplication is commutative, associative, and distributive.

Commutative: $\mathbf{AB} = \mathbf{BA}$

Associative: $(\mathbf{AB})\mathbf{C} = \mathbf{A}(\mathbf{BC})$

Distributive: $(\mathbf{A} + \mathbf{B})\mathbf{C} = \mathbf{AC} + \mathbf{BC}$

$\mathbf{A}(\mathbf{B} + \mathbf{C}) = \mathbf{AB} + \mathbf{AC}$

If we choose a basis $\{\mathbf{e}_i\}_{i=1}^N$ for $\mathscr{V}$ and express all symmetric tensors in terms of symmetric products of $\mathbf{e}_i$ using the above properties, then any symmetric tensor can be expressed as a sum of terms of the form $\alpha(\mathbf{e}_1)^{n_1}(\mathbf{e}_2)^{n_2} \ldots (\mathbf{e}_N)^{n_N}$, where $\alpha \in \mathbb{R}$.

Skew-symmetric tensors. Skew-symmetry is the same as symmetry except that in the interchange of variables the tensor changes sign. The following theorem further characterizes skew-symmetric tensors.

4.1.12 Theorem A tensor $\mathbf{A}$ is skew-symmetric in contravariant indices i and j if and only if, for all $\boldsymbol{\tau} \in \mathcal{V}^*$, insertion of $\boldsymbol{\tau}$ for both the ith and jth variables of $\mathbf{A}$ yields zero, regardless of the values of the remaining variables.

Proof. Let $\mathbf{A} \in \mathcal{T}^r_s(\mathcal{V})$ be skew-symmetric; then for arbitrary $\boldsymbol{\tau}^p$ and $\mathbf{v}_q$ we have

$$\mathbf{A}(\boldsymbol{\tau}^1, \ldots, \boldsymbol{\tau}^{i-1}, \boldsymbol{\tau}, \boldsymbol{\tau}^{i+1}, \ldots, \boldsymbol{\tau}^{j-1}, \boldsymbol{\tau}, \boldsymbol{\tau}^{j+1}, \ldots, \boldsymbol{\tau}^r, \mathbf{v}_1, \ldots, \mathbf{v}_s) = 0$$

Let $\boldsymbol{\tau} = \boldsymbol{\alpha} + \boldsymbol{\beta}$ for arbitrary $\boldsymbol{\alpha}, \boldsymbol{\beta} \in \mathcal{V}^*$. Then, after expanding $\mathbf{A}$ using the linearity of $\mathbf{A}$, the above relation gives

$$0 = \mathbf{A}(\ldots, \boldsymbol{\alpha}, \ldots, \boldsymbol{\alpha}, \ldots) + \mathbf{A}(\ldots, \boldsymbol{\alpha}, \ldots, \boldsymbol{\beta}, \ldots)$$
$$+ \mathbf{A}(\ldots, \boldsymbol{\beta}, \ldots, \boldsymbol{\alpha}, \ldots) + \mathbf{A}(\ldots, \boldsymbol{\beta}, \ldots, \boldsymbol{\beta}, \ldots)$$

where dots replace variables except those in the ith and the jth positions. In the above sum the first and last terms vanish by the definition of $\mathbf{A}$. Thus, for arbitrary $\boldsymbol{\alpha}$ and $\boldsymbol{\beta}$ we have

$$\mathbf{A}(\ldots, \boldsymbol{\alpha}, \ldots, \boldsymbol{\beta}, \ldots) = -\mathbf{A}(\ldots, \boldsymbol{\beta}, \ldots, \boldsymbol{\alpha}, \ldots)$$

This shows that $\mathbf{A}$ is skew-symmetric. The other half of the proof is obvious. ■

4.1.13 Definition A *covariant* (*contravariant*) *skew-symmetric tensor* is one that is skew-symmetric in all pairs of covariant (contravariant) variables. A tensor is skew-symmetric if it is *both* covariant and contravariant skew-symmetric.

4.1.3 Exterior Algebra and Its Application to Determinants

The following discussion of exterior algebra will concentrate on tensors of the type $(r, 0)$. However, it is important to keep in mind that simply interchanging the roles of $\mathcal{V}$ and $\mathcal{V}^*$ makes all definitions, theorems, propositions, and conclusions valid for tensors of type $(0, s)$.

The set of all skew-symmetric tensors of type $(p, 0)$ forms a subspace of $\mathcal{T}^p_0(\mathcal{V})$. This subspace is denoted by $\Lambda^p(\mathcal{V})$ (this strange notation will appear more reasonable once we discuss the skew-symmetric product of tensors). It is not, however, an algebra unless we define a skew-symmetric product analogous to that for the symmetric case. First, we need the following definition.

4.1.14 Definition An *anti-(skew-) symmetrizer* is a linear operator $\mathbb{A}: \mathcal{T}^p_0(\mathcal{V}) \to \Lambda^p(\mathcal{V})$ given by

$$[\mathbb{A}(\mathbf{A})](\boldsymbol{\tau}^1, \ldots, \boldsymbol{\tau}^p) = \frac{1}{p!}\sum_{\pi} \delta_\pi \mathbf{A}(\boldsymbol{\tau}^{\pi(1)}, \ldots, \boldsymbol{\tau}^{\pi(p)}) \qquad \text{for } \mathbf{A} \in \mathcal{T}^p_0(\mathcal{V}) \qquad (4.7)$$

The sum is on all permutations of $(1\ 2 \ldots p)$, and $\delta_\pi = \pm 1$, depending on whether the permutation is even or odd. Sometimes $\mathbb{A}(\mathbf{A})$ is denoted by $\mathbf{A}_a$.

Example 4.1.5

Let us write out Eq. (4.7) for $p = 2$ and $p = 3$.

(a) For $p = 2$, using an obvious notation for permutations, we have

$$\begin{aligned}(\mathbb{A}(\mathbf{A}))(\boldsymbol{\tau}^1,\boldsymbol{\tau}^2) &= \tfrac{1}{2}\{\delta_{\pi_{12}}\mathbf{A}(\boldsymbol{\tau}^1,\boldsymbol{\tau}^2) + \delta_{\pi_{21}}\mathbf{A}(\boldsymbol{\tau}^2,\boldsymbol{\tau}^1)\}\\ &= \tfrac{1}{2}\{\mathbf{A}(\boldsymbol{\tau}^1,\boldsymbol{\tau}^2) - \mathbf{A}(\boldsymbol{\tau}^2,\boldsymbol{\tau}^1)\}\end{aligned}$$

Here $\delta_{\pi_{21}} = -1$ because to get from (2,1) to (1,2) we need to make only one (an odd number) interchange of symbols.

(b) For $p = 3$ we have

$$\begin{aligned}(\mathbb{A}(\mathbf{A}))(\boldsymbol{\tau}^1, \boldsymbol{\tau}^2, \boldsymbol{\tau}^3) &= \tfrac{1}{6}\{\delta_{\pi_{123}}\mathbf{A}(\boldsymbol{\tau}^1, \boldsymbol{\tau}^2, \boldsymbol{\tau}^3) + \delta_{\pi_{132}}\mathbf{A}(\boldsymbol{\tau}^1, \boldsymbol{\tau}^3, \boldsymbol{\tau}^2) + \delta_{\pi_{213}}\mathbf{A}(\boldsymbol{\tau}^2, \boldsymbol{\tau}^1, \boldsymbol{\tau}^3)\\ &\quad + \delta_{\pi_{231}}\mathbf{A}(\boldsymbol{\tau}^2, \boldsymbol{\tau}^3, \boldsymbol{\tau}^1) + \delta_{\pi_{312}}\mathbf{A}(\boldsymbol{\tau}^3, \boldsymbol{\tau}^1, \boldsymbol{\tau}^2) + \delta_{\pi_{321}}\mathbf{A}(\boldsymbol{\tau}^3, \boldsymbol{\tau}^2, \boldsymbol{\tau}^1)\}\\ &= \tfrac{1}{6}\{\mathbf{A}(\boldsymbol{\tau}^1, \boldsymbol{\tau}^2, \boldsymbol{\tau}^3) - \mathbf{A}(\boldsymbol{\tau}^1, \boldsymbol{\tau}^3, \boldsymbol{\tau}^2) - \mathbf{A}(\boldsymbol{\tau}^2, \boldsymbol{\tau}^1, \boldsymbol{\tau}^3) + \mathbf{A}(\boldsymbol{\tau}^2, \boldsymbol{\tau}^3, \boldsymbol{\tau}^1)\\ &\quad + \mathbf{A}(\boldsymbol{\tau}^3, \boldsymbol{\tau}^1, \boldsymbol{\tau}^2) - \mathbf{A}(\boldsymbol{\tau}^3, \boldsymbol{\tau}^2, \boldsymbol{\tau}^1)\}\end{aligned}$$

The reader should verify that all terms with a plus sign are obtained from (1 2 3) by an even number of interchanges of symbols, and those with a minus sign by an odd number. ●

We can now define the important product that turns the vector space $\Lambda^p(\mathcal{V})$ into an algebra.

4.1.15 Definition The *exterior product* (also called the wedge, Grassmann, alternating, or veck product) of two skew-symmetric tensors $\mathbf{A} \in \Lambda^p(\mathcal{V})$ and $\mathbf{B} \in \Lambda^q(\mathcal{V})$ is another skew-symmetric tensor belonging to $\Lambda^{p+q}(\mathcal{V})$ and given by

$$\mathbf{A} \wedge \mathbf{B} \equiv \mathbb{A}(\mathbf{A} \otimes \mathbf{B})$$

Example 4.1.6

Let us find the wedge products $\mathbf{v}_1 \wedge \mathbf{v}_2$ and $(\mathbf{v}_1 \wedge \mathbf{v}_2) \wedge \mathbf{v}_3$ for $\mathbf{v}_1, \mathbf{v}_2, \mathbf{v}_3 \in \mathcal{V}$. For arbitrary $\boldsymbol{\tau}^1, \boldsymbol{\tau}^2 \in \mathcal{V}^*$ we have

$$\begin{aligned}(\mathbf{v}_1 \wedge \mathbf{v}_2)(\boldsymbol{\tau}^1, \boldsymbol{\tau}^2) &= [\mathbb{A}(\mathbf{v}_1 \otimes \mathbf{v}_2)](\boldsymbol{\tau}^1, \boldsymbol{\tau}^2) = \frac{1}{2}\sum_\pi \delta_\pi(\mathbf{v}_1 \otimes \mathbf{v}_2)(\boldsymbol{\tau}^{\pi(1)}, \boldsymbol{\tau}^{\pi(2)})\\ &= \tfrac{1}{2}[\delta_{\pi_{12}}(\mathbf{v}_1 \otimes \mathbf{v}_2)(\boldsymbol{\tau}^1, \boldsymbol{\tau}^2) + \delta_{\pi_{21}}(\mathbf{v}_1 \otimes \mathbf{v}_2)(\boldsymbol{\tau}^2, \boldsymbol{\tau}^1)]\\ &= \tfrac{1}{2}[(\mathbf{v}_1 \otimes \mathbf{v}_2)(\boldsymbol{\tau}^1, \boldsymbol{\tau}^2) - \mathbf{v}_1(\boldsymbol{\tau}^2)\mathbf{v}_2(\boldsymbol{\tau}^1)]\\ &= \tfrac{1}{2}[(\mathbf{v}_1 \otimes \mathbf{v}_2)(\boldsymbol{\tau}^1, \boldsymbol{\tau}^2) - (\mathbf{v}_2 \otimes \mathbf{v}_1)(\boldsymbol{\tau}^1, \boldsymbol{\tau}^2)]\\ &= \tfrac{1}{2}(\mathbf{v}_1 \otimes \mathbf{v}_2 - \mathbf{v}_2 \otimes \mathbf{v}_1)(\boldsymbol{\tau}^1, \boldsymbol{\tau}^2)\end{aligned}$$

Since this is true for all $\mathbf{\tau}^1$ and $\mathbf{\tau}^2$, we have

$$\mathbf{v}_1 \wedge \mathbf{v}_2 = \tfrac{1}{2}(\mathbf{v}_1 \otimes \mathbf{v}_2 - \mathbf{v}_2 \otimes \mathbf{v}_1)$$

Similarly, we can show that the definition

$$[(\mathbf{v}_1 \wedge \mathbf{v}_2) \wedge \mathbf{v}_3](\mathbf{\tau}^1, \mathbf{\tau}^2, \mathbf{\tau}^3) = \frac{1}{6}\sum_\pi \delta_\pi(\mathbf{v}_1 \otimes \mathbf{v}_2 \otimes \mathbf{v}_3)(\mathbf{\tau}^{\pi(1)}, \mathbf{\tau}^{\pi(2)}, \mathbf{\tau}^{\pi(3)})$$

leads to

$$(\mathbf{v}_1 \wedge \mathbf{v}_2) \wedge \mathbf{v}_3 = \mathbf{v}_1 \wedge (\mathbf{v}_2 \wedge \mathbf{v}_3) = \tfrac{1}{6}(\mathbf{v}_1 \otimes \mathbf{v}_2 \otimes \mathbf{v}_3 - \mathbf{v}_1 \otimes \mathbf{v}_3 \otimes \mathbf{v}_2 - \mathbf{v}_2 \otimes \mathbf{v}_1 \otimes \mathbf{v}_3$$
$$+ \mathbf{v}_2 \otimes \mathbf{v}_3 \otimes \mathbf{v}_1 - \mathbf{v}_3 \otimes \mathbf{v}_2 \otimes \mathbf{v}_1 + \mathbf{v}_3 \otimes \mathbf{v}_1 \otimes \mathbf{v}_2)$$

This shows that the exterior product is associative, at least for the case at hand.　　●

The following theorem contains the properties of the exterior product. (For a proof, see Abraham, Marsden, and Ratiu 1983, p. 326.)

4.1.16 Theorem The exterior product has the following properties for skew-symmetric tensors **A**, **B**, and **C**.

(i) Associativity:

$$(\mathbf{A} \wedge \mathbf{B}) \wedge \mathbf{C} = \mathbf{A} \wedge (\mathbf{B} \wedge \mathbf{C})$$

(ii) Anticommutativity: If $\mathbf{A} \in \Lambda^p(\mathscr{V})$ and $\mathbf{B} \in \Lambda^q(\mathscr{V})$, then $\mathbf{A} \wedge \mathbf{B} = (-1)^{pq}\mathbf{B} \wedge \mathbf{A}$. In particular, $\mathbf{v}_1 \wedge \mathbf{v}_2 = -\mathbf{v}_2 \wedge \mathbf{v}_1$, for all $\mathbf{v}_1, \mathbf{v}_2 \in \mathscr{V}$.

(iii) Distributivity:

$$(\mathbf{A} + \mathbf{B}) \wedge \mathbf{C} = \mathbf{A} \wedge \mathbf{C} + \mathbf{B} \wedge \mathbf{C}$$

$$\mathbf{A} \wedge (\mathbf{B} + \mathbf{C}) = \mathbf{A} \wedge \mathbf{B} + \mathbf{A} \wedge \mathbf{C}$$　　　　■

The components of $\mathbf{A} \in \Lambda^p(\mathscr{V})$ are given by $A^{i_1, \cdots, i_p}$, where $i_1 < i_2 < \cdots < i_p$. All other components are related to these by skew-symmetry. The number of independent components, which is the dimension of $\Lambda^p(\mathscr{V})$, is equal to the number of ways p numbers can be chosen from among N distinct numbers in such a way that no two of the p numbers are equal. This is simply the combination of N objects taken p at a time. Thus, we have

$$\dim \Lambda^p(\mathscr{V}) = \binom{N}{p} = \frac{N!}{p!(N-p)!} \tag{4.8}$$

In particular, $\dim \Lambda^N(\mathscr{V}) = 1$. This should come as no surprise because from a basis $\{\mathbf{e}_i\}_{i=1}^N$ of $\mathscr{V}$ we can form a basis for $\Lambda^p(\mathscr{V})$ by constructing all products $\mathbf{e}_{i_1} \wedge \mathbf{e}_{i_2} \wedge \cdots \wedge \mathbf{e}_{i_p}$, of which there are $\binom{N}{p}$. However, when $p = N$, there is only one such product, within a multiplicative constant, and it is $\mathbf{e}_1 \wedge \mathbf{e}_2 \wedge \cdots \wedge \mathbf{e}_N$.

An elegant way of determining the linear independence of vectors using the formalism developed so far is given in the following proposition.

4.1.17 Proposition A set of vectors, $\mathbf{v}_1, \mathbf{v}_2, \ldots, \mathbf{v}_p \in \mathcal{V}$, is linearly independent if and only if $\mathbf{v}_1 \wedge \mathbf{v}_2 \wedge \cdots \wedge \mathbf{v}_p \neq 0$.

Proof. If $\{\mathbf{v}_i\}_{i=1}^p$ are independent, then they span a p-dimensional subspace $\mathcal{M}$ of $\mathcal{V}$. Considering $\mathcal{M}$ as a vector space in its own right, we have $\dim \Lambda^p(\mathcal{M}) = 1$. A basis for $\Lambda^p(\mathcal{M})$ is simply $\mathbf{v}_1 \wedge \mathbf{v}_2 \wedge \cdots \wedge \mathbf{v}_p$, which cannot be zero because $\Lambda^p(\mathcal{M})$ is *one*-dimensional.

Conversely, we assume that $\alpha_1\mathbf{v}_1 + \alpha_2\mathbf{v}_2 + \cdots + \alpha_p\mathbf{v}_p = 0$. Then taking the exterior product of the LHS with $\mathbf{v}_2 \wedge \mathbf{v}_3 \wedge \cdots \wedge \mathbf{v}_p$ makes all terms vanish except the first one. Thus, we have $\alpha_1\mathbf{v}_1 \wedge \mathbf{v}_2 \wedge \cdots \wedge \mathbf{v}_p = 0$. The fact that $\mathbf{v}_1 \wedge \mathbf{v}_2 \wedge \cdots \wedge \mathbf{v}_p \neq 0$ forces α_1 to be zero. Similarly, multiplying by $\mathbf{v}_1 \wedge \mathbf{v}_3 \wedge \cdots \wedge \mathbf{v}_p$ shows that $\alpha_2 = 0$, and so on. ■

Example 4.1.7

Let $\{\mathbf{e}_i\}_{i=1}^N$ form a basis for $\mathcal{V}$. Let

$$\mathbf{v}_1 = \mathbf{e}_1 + 2\mathbf{e}_2 - \mathbf{e}_3$$

$$\mathbf{v}_2 = 3\mathbf{e}_1 + \mathbf{e}_2 + 2\mathbf{e}_3$$

$$\mathbf{v}_3 = -\mathbf{e}_1 - 3\mathbf{e}_2 + 2\mathbf{e}_3$$

To see if $\mathbf{v}_1, \mathbf{v}_2$, and $\mathbf{v}_3$ are linearly independent, we take their triple wedge product. First, we take the product of the first two:

$$\mathbf{v}_1 \wedge \mathbf{v}_2 = (\mathbf{e}_1 + 2\mathbf{e}_2 - \mathbf{e}_3) \wedge (3\mathbf{e}_1 + \mathbf{e}_2 + 2\mathbf{e}_3) = -5\mathbf{e}_1 \wedge \mathbf{e}_2 + 5\mathbf{e}_1 \wedge \mathbf{e}_3 + 5\mathbf{e}_2 \wedge \mathbf{e}_3$$

All the wedge products that have repeated factors vanish. Now we multiply by $\mathbf{v}_3$:

$$\begin{aligned}\mathbf{v}_1 \wedge \mathbf{v}_2 \wedge \mathbf{v}_3 &= -5\mathbf{e}_1 \wedge \mathbf{e}_2 \wedge (-\mathbf{e}_1 - 3\mathbf{e}_2 + 2\mathbf{e}_3) + 5\mathbf{e}_1 \wedge \mathbf{e}_3 \wedge (-\mathbf{e}_1 - 3\mathbf{e}_2 + 2\mathbf{e}_3) \\ &\quad + 5\mathbf{e}_2 \wedge \mathbf{e}_3 \wedge (-\mathbf{e}_1 - 3\mathbf{e}_2 + 2\mathbf{e}_3) \\ &= -10\mathbf{e}_1 \wedge \mathbf{e}_2 \wedge \mathbf{e}_3 - 15\mathbf{e}_1 \wedge \mathbf{e}_3 \wedge \mathbf{e}_2 - 5\mathbf{e}_2 \wedge \mathbf{e}_3 \wedge \mathbf{e}_1 \\ &= 0\end{aligned}$$

Thus, the three vectors are linearly dependent. •

Example 4.1.8

As another example of the application of Proposition 4.1.17, let us show Cartan's lemma, which states that if $\{\mathbf{e}_i\}_{i=1}^p$ form a linearly independent set of vectors in $\mathcal{V}$ and $\{\mathbf{v}_i\}_{i=1}^p$ are also vectors in $\mathcal{V}$ such that $\sum_{i=1}^p \mathbf{e}_i \wedge \mathbf{v}_i = 0$, then $\mathbf{v}_i$ are linear combinations of (*only*) the set $\{\mathbf{e}_i\}_{i=1}^p$. Furthermore, if $\mathbf{v}_i = \sum_{j=1}^p A_{ij}\mathbf{e}_j$, then $A_{ij} = A_{ji}$.

To show this we write

$$0 = \mathbf{e}_1 \wedge \mathbf{v}_1 + \mathbf{e}_2 \wedge \mathbf{v}_2 + \cdots + \mathbf{e}_p \wedge \mathbf{v}_p \tag{1}$$

Multiplying both sides by $\mathbf{e}_2 \wedge \mathbf{e}_3 \wedge \cdots \wedge \mathbf{e}_p$ gives

$$0 = -\mathbf{v}_1 \wedge \mathbf{e}_1 \wedge \mathbf{e}_2 \wedge \cdots \wedge \mathbf{e}_p \equiv -\mathbf{v}_1 \wedge \mathbf{\Omega}$$

where $\mathbf{\Omega} = \mathbf{e}_1 \wedge \mathbf{e}_2 \wedge \cdots \wedge \mathbf{e}_p$. Similarly, if we multiply both sides by the wedge product $\mathbf{e}_1 \wedge \mathbf{e}_3 \wedge \cdots \wedge \mathbf{e}_p$, we obtain $-\mathbf{v}_2 \wedge \mathbf{\Omega} = 0$. In general,

$$\mathbf{v}_k \wedge \mathbf{e}_1 \wedge \mathbf{e}_2 \wedge \cdots \wedge \mathbf{e}_p = 0 \qquad \forall k = 1, 2, \ldots, p$$

Proposition 4.1.17 says that $\mathbf{v}_k$ and $\mathbf{e}_1, \ldots, \mathbf{e}_p$ are linearly dependent. Thus,

$$\mathbf{v}_k = \sum_{i=1}^{p} A_{ki}\mathbf{e}_i$$

Equation (1) now becomes

$$0 = \sum_{k=1}^{p} \mathbf{e}_k \wedge \mathbf{v}_k = \sum_{k=1}^{p}\sum_{i=1}^{p} \mathbf{e}_k \wedge (A_{ki}\mathbf{e}_i) = \sum_{k,i} A_{ki}\mathbf{e}_k \wedge \mathbf{e}_i$$

$$= \sum_{k<i} (A_{ki} - A_{ik})\mathbf{e}_k \wedge \mathbf{e}_i \tag{2}$$

But $\{\mathbf{e}_k \wedge \mathbf{e}_i\}$ for $k < i$ forms a basis in $\Lambda^2(\mathscr{M})$ with $\mathscr{M}$ the subspace spanned by $\{\mathbf{e}_i\}_{i=1}^{p}$. Therefore, the coefficients in the RHS of (2) must vanish, and $A_{ki} = A_{ik}$. ●

We have noticed that if a tensor $\mathbf{A}$ belongs to $\Lambda^p(\mathscr{V})$, then it can be expressed as a linear combination of exterior products. Sometimes it is useful to know whether a tensor can be expressed as a *single* exterior product of p vectors. Such tensors are called *decomposable*. That is, $\mathbf{A} \in \Lambda^p(\mathscr{V})$ is decomposable if there exist linearly independent vectors $\mathbf{v}_1, \mathbf{v}_2, \cdots, \mathbf{v}_p \in \mathscr{V}$ such that $\mathbf{A} = \mathbf{v}_1 \wedge \mathbf{v}_2 \wedge \ldots \wedge \mathbf{v}_p$. For $p = 2$ we have a compact criterion.

4.1.18 Proposition A tensor $\mathbf{A} \in \Lambda^2(\mathscr{V})$ is decomposable if and only if $\mathbf{A} \wedge \mathbf{A} = 0$.

Proof. If $\mathbf{A} = \mathbf{v}_1 \wedge \mathbf{v}_2$, then $\mathbf{A} \wedge \mathbf{A} = 0$ trivially. Conversely, assuming that $\mathbf{A} \wedge \mathbf{A} = 0$, we claim that there exists $\mathbf{v} \in \mathscr{V}$ such that $\mathbf{v} \wedge \mathbf{A} = 0$. For any i, the vector $A^{ij}\mathbf{e}_j$ (note the sum over j) is such a vector. We can show this:

$$\begin{aligned}(A^{ij}\mathbf{e}_j) \wedge \mathbf{A} &= (A^{ij}\mathbf{e}_j) \wedge (A^{kl}\mathbf{e}_k \wedge \mathbf{e}_l) = A^{ij}A^{kl}\mathbf{e}_j \wedge \mathbf{e}_k \wedge \mathbf{e}_l \\ &= \tfrac{1}{3}A^{ij}A^{kl}\mathbf{e}_j \wedge \mathbf{e}_k \wedge \mathbf{e}_l + \tfrac{1}{3}A^{ik}A^{jl}\mathbf{e}_k \wedge \mathbf{e}_j \wedge \mathbf{e}_l \\ &\quad + \tfrac{1}{3}A^{il}A^{jk}\mathbf{e}_l \wedge \mathbf{e}_j \wedge \mathbf{e}_k\end{aligned}$$

We have simply used different dummy indices to express the same thing three times. The above expression can be rewritten as

$$(A^{ij}\mathbf{e}_j) \wedge \mathbf{A} = \tfrac{1}{3}(A^{ij}A^{kl} - A^{ik}A^{jl} + A^{il}A^{jk})\mathbf{e}_j \wedge \mathbf{e}_k \wedge \mathbf{e}_l$$

But if $\mathbf{A} \wedge \mathbf{A} = 0$, the expression in parentheses vanishes, by Exercise 4.1.6. Thus, $\mathbf{v} = A^{ij}\mathbf{e}_j$ has the property that $\mathbf{v} \neq 0$ and $\mathbf{v} \wedge \mathbf{A} = 0$. By Exercise 4.1.5 there must be a member of $\Lambda^1(\mathscr{V}) \equiv \mathscr{V}$, that is, a vector $\mathbf{u} \in \mathscr{V}$ such that $\mathbf{A} = \mathbf{v} \wedge \mathbf{u}$. Thus, $\mathbf{A}$ is decomposable. ■

Example 4.1.9

(a) In three dimensions all $\mathbf{A} \in \Lambda^2 \mathscr{V}$ are decomposable. This is because $\mathbf{A} = A^{ij}\mathbf{e}_i \wedge \mathbf{e}_j$ and $\mathbf{A} \wedge \mathbf{A} = A^{ij}A^{kl}\mathbf{e}_i \wedge \mathbf{e}_j \wedge \mathbf{e}_k \wedge \mathbf{e}_l$. Since there are only *three* linearly independent vectors in three dimensions, the exterior products of degree 4 vanish. Thus, $\mathbf{A} \wedge \mathbf{A} = 0$.

(b) In four dimensions $\mathbf{A} = \mathbf{e}_1 \wedge \mathbf{e}_2 + \mathbf{e}_3 \wedge \mathbf{e}_4$ is *not* decomposable. In fact,

$$\mathbf{A} \wedge \mathbf{A} = (\mathbf{e}_1 \wedge \mathbf{e}_2 + \mathbf{e}_3 \wedge \mathbf{e}_4) \wedge (\mathbf{e}_1 \wedge \mathbf{e}_2 + \mathbf{e}_3 \wedge \mathbf{e}_4) = 2\mathbf{e}_1 \wedge \mathbf{e}_2 \wedge \mathbf{e}_3 \wedge \mathbf{e}_4 \neq 0$$ •

One of the most beautiful applications of exterior algebra is in the theory of determinants. We have already considered determinants in detail in Chapter 3, where we noted how messy it was to prove some of the theorems concerning them. With the machinery of exterior algebra at our disposal, we will see how elegant this theory becomes and how trivial some of the proofs will turn out to be.

First, let us recall that $\dim \Lambda^N(\mathscr{V}) = 1$ when $\mathscr{V}$ is N-dimensional. This means that if $\{\mathbf{e}_i\}_{i=1}^N$ is a basis of $\mathscr{V}$, then $\mathbf{e}_1 \wedge \mathbf{e}_2 \wedge \cdots \wedge \mathbf{e}_N$ is the only vector in the corresponding basis of $\Lambda^N(\mathscr{V})$. On the other hand, if $\{\mathbf{v}_i\}_{i=1}^N$ is any set of N vectors, the product $\mathbf{v}_1 \wedge \mathbf{v}_2 \wedge \cdots \wedge \mathbf{v}_N$ is either zero (if the $\mathbf{v}_i$ are linearly dependent) or a nonzero product belonging to $\Lambda^N(\mathscr{V})$. Since $\mathbf{e}_1 \wedge \mathbf{e}_2 \wedge \cdots \wedge \mathbf{e}_N$ is a basis of $\Lambda^N(\mathscr{V})$, we can conclude that for any set of N vectors, $\mathbf{v}_1, \mathbf{v}_2, \ldots, \mathbf{v}_N \in \mathscr{V}$, the product $\mathbf{v}_1 \wedge \mathbf{v}_2 \wedge \cdots \wedge \mathbf{v}_N$ is a multiple (possibly zero) of $\mathbf{e}_1 \wedge \mathbf{e}_2 \wedge \cdots \wedge \mathbf{e}_N$.

Now let $\mathbb{A} \in \mathscr{L}(\mathscr{V})$ be a linear operator on $\mathscr{V}$. Then the set of vectors $\mathbb{A}\mathbf{e}_1, \mathbb{A}\mathbf{e}_2, \ldots, \mathbb{A}\mathbf{e}_N$ all belong to $\mathscr{V}$. By the above remarks $(\mathbb{A}\mathbf{e}_1) \wedge (\mathbb{A}\mathbf{e}_2) \wedge \cdots \wedge (\mathbb{A}\mathbf{e}_N)$ is proportional to $\mathbf{e}_1 \wedge \mathbf{e}_2 \wedge \cdots \wedge \mathbf{e}_N$. We now show that the proportionality constant is simply $\det \mathbb{A}$.

4.1.19 Theorem Let $\mathbb{A}: \mathscr{V} \to \mathscr{V}$ be linear. Let $\{\mathbf{e}_i\}_{i=1}^N$ be a basis for $\mathscr{V}$. Then

$$(\mathbb{A}\mathbf{e}_1) \wedge (\mathbb{A}\mathbf{e}_2) \wedge \cdots \wedge (\mathbb{A}\mathbf{e}_N) = (\det \mathbb{A})\mathbf{e}_1 \wedge \mathbf{e}_2 \wedge \cdots \wedge \mathbf{e}_N \tag{4.9a}$$

Furthermore, $\det \mathbb{A}$ is basis-independent.

Proof. Let $\mathbb{A}\mathbf{e}_r = A_r^{i_r}\mathbf{e}_{i_r}$, for $r = 1, 2, \ldots, N$. Then

$$\begin{aligned}(\mathbb{A}\mathbf{e}_1) \wedge \cdots \wedge (\mathbb{A}\mathbf{e}_N) &= (A_1^{i_1}\mathbf{e}_{i_1}) \wedge \cdots \wedge (A_N^{i_N}\mathbf{e}_{i_N}) \\ &= A_1^{i_1}A_2^{i_2} \cdots A_N^{i_N}\mathbf{e}_{i_1} \wedge \cdots \wedge \mathbf{e}_{i_N}\end{aligned} \tag{4.9b}$$

The exterior products have the following properties:

$$\mathbf{e}_{i_1} \wedge \cdots \wedge \mathbf{e}_{i_N} = \begin{cases} 0 & \text{if any two indices are equal} \\ +\mathbf{e}_1 \wedge \cdots \wedge \mathbf{e}_N & \text{if } (i_1 \ldots i_N) \text{ is an even permutation of } (1\,2 \ldots N) \\ -\mathbf{e}_1 \wedge \cdots \wedge \mathbf{e}_N & \text{if } (i_1 \ldots i_N) \text{ is an odd permutation of } (1\,2 \ldots N) \end{cases}$$

These can be summarized as

$$\mathbf{e}_{i_1} \wedge \cdots \wedge \mathbf{e}_{i_N} = \varepsilon_{i_1 i_2 \ldots i_N} \mathbf{e}_1 \wedge \cdots \wedge \mathbf{e}_N$$

with $\varepsilon_{i_1 \ldots i_N}$ defined as in Chapter 2. We can, therefore, write (4.9b) as

$$(\mathbb{A}\mathbf{e}_1) \wedge \cdots \wedge (\mathbb{A}\mathbf{e}_N) = (A_1^{i_1} A_2^{i_2} \cdots A_N^{i_N} \varepsilon_{i_1 i_2 \ldots i_N}) \mathbf{e}_1 \wedge \cdots \wedge \mathbf{e}_N$$

The expression in parentheses on the RHS is simply the determinant as defined before (recall the summation convention).

Equation (4.9a) can be thought of as a mapping, $\mathbb{A}^{(N)} : \Lambda^N(\mathcal{V}) \to \Lambda^N(\mathcal{V})$ given by $\mathbb{A}^{(N)}(\mathbf{e}_1 \wedge \mathbf{e}_2 \wedge \cdots \wedge \mathbf{e}_N) \equiv (\mathbb{A}\mathbf{e}_1) \wedge (\mathbb{A}\mathbf{e}_2) \wedge \cdots \wedge (\mathbb{A}\mathbf{e}_N)$. Sometimes $\mathbb{A}^{(N)}$ is called the *homomorphic extension of* $\mathbb{A}$. Because $\mathbb{A}^{(N)}$ is an operator in a one-dimensional space, it is entirely determined by the constant $\det \mathbb{A}$. Furthermore, as Problem 4.8 shows, this is independent of the basis chosen for $\Lambda^N(\mathcal{V})$. ■

Example 4.1.10

The symbol $\varepsilon_{i_1 i_2 \ldots i_N}$ is called the *Levi-Civita tensor*. It can be defined by

$$\mathbf{e}_{i_1} \wedge \mathbf{e}_{i_2} \wedge \cdots \wedge \mathbf{e}_{i_N} = \varepsilon_{i_1 i_2 \ldots i_N} \mathbf{e}_1 \wedge \mathbf{e}_2 \wedge \cdots \wedge \mathbf{e}_N \tag{1a}$$

which is equivalent to

$$\mathbf{e}_{j_{i_1}} \wedge \mathbf{e}_{j_{i_2}} \wedge \cdots \wedge \mathbf{e}_{j_{i_N}} = \varepsilon_{i_1 i_2 \ldots i_N} \mathbf{e}_{j_1} \wedge \mathbf{e}_{j_2} \wedge \cdots \wedge \mathbf{e}_{j_N} \tag{1b}$$

Let us see how much $\varepsilon_{i_1 \ldots i_N}$ depends on the choice of a basis. Let $\{\bar{\mathbf{e}}_j\}_{j=1}^N$ be another basis. Then

$$\bar{\mathbf{e}}_{i_1} \wedge \cdots \wedge \bar{\mathbf{e}}_{i_N} = \bar{\varepsilon}_{i_1 \ldots i_N} \bar{\mathbf{e}}_1 \wedge \bar{\mathbf{e}}_2 \wedge \cdots \wedge \bar{\mathbf{e}}_N \tag{2}$$

However, $\bar{\mathbf{e}}_i = r_i^j \mathbf{e}_j$ where r_i^j are elements of an invertible matrix (the transformation matrix). Let $\bar{\mathbf{e}}_k = \mathsf{R}\mathbf{e}_{j_k}$. Then

$$\bar{\mathbf{e}}_1 \wedge \cdots \wedge \bar{\mathbf{e}}_N = (\mathsf{R}\mathbf{e}_{j_1}) \wedge \cdots \wedge (\mathsf{R}\mathbf{e}_{j_N}) = (\det \mathsf{R})\mathbf{e}_{j_1} \wedge \cdots \wedge \mathbf{e}_{j_N}$$

and

$$\bar{\mathbf{e}}_{i_1} \wedge \cdots \wedge \bar{\mathbf{e}}_{i_N} = (\mathsf{R}\mathbf{e}_{j_{i_1}}) \wedge \cdots \wedge (\mathsf{R}\mathbf{e}_{j_{i_N}}) = (\det \mathsf{R})\mathbf{e}_{j_{i_1}} \wedge \cdots \wedge \mathbf{e}_{j_{i_N}}$$

Substituting these in (2), we have

$$(\det \mathsf{R})\mathbf{e}_{j_{i_1}} \wedge \cdots \wedge \mathbf{e}_{j_{i_N}} = \bar{\varepsilon}_{i_1 \ldots i_N} (\det \mathsf{R})\mathbf{e}_{j_1} \wedge \ldots \wedge \mathbf{e}_{j_N}$$

Since R is invertible, $\det \mathsf{R} \neq 0$ and can be cancelled on both sides, yielding

$$\mathbf{e}_{j_{i_1}} \wedge \cdots \wedge \mathbf{e}_{j_{i_N}} = \bar{\varepsilon}_{i_1 \cdots i_N} \mathbf{e}_{j_1} \wedge \cdots \wedge \mathbf{e}_{j_N}$$

Now we use (1b) on the LHS to get

$$\varepsilon_{i_1 \ldots i_N} \mathbf{e}_{j_1} \wedge \cdots \wedge e_{j_N} = \bar{\varepsilon}_{i_1 \ldots i_N} \mathbf{e}_{j_1} \wedge \cdots \wedge \mathbf{e}_{j_N}$$

or

$$\varepsilon_{i_1 \ldots i_N} = \bar{\varepsilon}_{i_1 \ldots i_N}$$

Thus, the Levi-Civita tensor takes the same value in all coordinate systems. This can also be seen by interpreting (1a) as the homomorphic extension of a permutation "operator" whose only action on the basis vectors is to reshuffle them. Equation (1a) is then identical to (4.9a) with $\varepsilon_{i_1 \ldots i_N}$ as the determinant. Since the determinant is basis-independent, so is $\varepsilon_{i_1 i_2 \ldots i_N}$. ●

The definition of $\varepsilon_{i_1 i_2 \ldots i_N}$ in Eq. (1a) of the preceding example assumes an ordering for the basis $\{\mathbf{e}_i\}_{i=1}^N$. If we switch $\mathbf{e}_1$ and $\mathbf{e}_2$, the RHS becomes negative, which changes the sign of $\varepsilon_{i_1 \ldots i_N}$. We will investigate this point later on.

Let us now take two operators, $\mathbb{A}, \mathbb{B} \in \mathcal{L}(\mathcal{V})$. Clearly, $\mathbb{AB} \in \mathcal{L}(\mathcal{V})$. Let us look at the determinant. By definition,

$$(\mathbb{AB})^{(N)}\mathbf{e}_1 \wedge \cdots \wedge \mathbf{e}_N = [\det(\mathbb{AB})]\mathbf{e}_1 \wedge \mathbf{e}_2 \wedge \cdots \wedge \mathbf{e}_N$$

However,

$$\begin{aligned}(\mathbb{AB})^{(N)}\mathbf{e}_1 \wedge \cdots \wedge \mathbf{e}_N &= (\mathbb{AB}\mathbf{e}_1) \wedge (\mathbb{AB}\mathbf{e}_2) \wedge \cdots \wedge (\mathbb{AB}\mathbf{e}_N) \\ &= \mathbb{A}^{(N)}[(\mathbb{B}\mathbf{e}_1) \wedge \cdots \wedge (\mathbb{B}\mathbf{e}_N)] = \mathbb{A}^{(N)}[(\det \mathbb{B})\mathbf{e}_1 \wedge \cdots \wedge \mathbf{e}_N] \\ &= (\det \mathbb{B})\mathbb{A}^{(N)}(\mathbf{e}_1 \wedge \cdots \wedge \mathbf{e}_N) = (\det \mathbb{B})(\det \mathbb{A})\mathbf{e}_1 \wedge \cdots \wedge \mathbf{e}_N\end{aligned}$$

Comparison of this equation with the one above it shows that

$$\det(\mathbb{AB}) = (\det \mathbb{A})\,(\det \mathbb{B}) \tag{4.10}$$

Here the power and elegance of exterior algebra can be truly appreciated—to prove (4.10) in Chapter 3, we had to go through a maze of index shuffling and reshuffling.

4.1.4 The Inner Product and Orthonormal Bases Revisited

In Chapters 1 and 2 the inner product was defined in terms of a metric function that took two vectors as input and manufactured a real number. We now know what kind of machine this is in the language of tensors.

4.1.20 Definition A symmetric *bilinear form* on $\mathcal{V}$ is a symmetric tensor of type (0, 2), that is, a symmetric bilinear function $\mathbf{g}: \mathcal{V} \times \mathcal{V} \to \mathbb{R}$.

If $\{\mathbf{e}_j\}_{j=1}^N$ is a basis of $\mathcal{V}$ and $\{\boldsymbol{\epsilon}^i\}_{i=1}^N$ is its dual basis, then $\mathbf{g} = g_{ij}\boldsymbol{\epsilon}^i\boldsymbol{\epsilon}^j$ (recall Einstein's summation convention), because $\boldsymbol{\epsilon}^i\boldsymbol{\epsilon}^j = \frac{1}{2}(\boldsymbol{\epsilon}^i \otimes \boldsymbol{\epsilon}^j + \boldsymbol{\epsilon}^j \otimes \boldsymbol{\epsilon}^i)$ form a basis of $\mathcal{S}^2(\mathcal{V}^*)$. For any vector $\mathbf{v} \in \mathcal{V}$, we can write

$$\mathbf{g}(\mathbf{v}) = g_{ij}\boldsymbol{\epsilon}^i\boldsymbol{\epsilon}^j(\mathbf{v}) = g_{ij}\boldsymbol{\epsilon}^i\boldsymbol{\epsilon}^j(v^k\mathbf{e}_k) = g_{ij}v^k\boldsymbol{\epsilon}^i\boldsymbol{\epsilon}^j(\mathbf{e}_k) = g_{ij}v^j\boldsymbol{\epsilon}^i \tag{4.11}$$

Thus, $\mathbf{g}(\mathbf{v}) \in \mathcal{V}^*$. This shows that $\mathbf{g}$ can be thought of as a mapping, $\mathbf{g}: \mathcal{V} \to \mathcal{V}^*$, given by Eq. (4.11). For this equation to make sense, it should not matter which factor in the symmetric product $\mathbf{v}$ contracts with. But this is a trivial consequence of the symmetries $g_{ij} = g_{ji}$ and $\boldsymbol{\epsilon}^i\boldsymbol{\epsilon}^j = \boldsymbol{\epsilon}^j\boldsymbol{\epsilon}^i$, so $\mathbf{g}(\mathbf{v}) = g_{ij}\boldsymbol{\epsilon}^i\boldsymbol{\epsilon}^j(\mathbf{v}) = g_{ij}\boldsymbol{\epsilon}^i(\mathbf{v})\boldsymbol{\epsilon}^j$ is uniquely defined. The components $g_{ij}v^j$ of $\mathbf{g}(\mathbf{v})$ in the basis $\{\boldsymbol{\epsilon}^i\}_{i=1}^N$ of $\mathcal{V}^*$ are denoted by v_i, so

$$\mathbf{g}(\mathbf{v}) = v_i\boldsymbol{\epsilon}^i \qquad \text{where } v_i \equiv g_{ij}v^j \tag{4.12a}$$

We have thus *lowered* the index of v^j by the use of the symmetric bilinear form $\mathbf{g}$. In applications v_i is uniquely defined; furthermore, there is a one-to-one correspondence between v^i and v_i. This can happen if and only if the mapping $\mathbf{g}: \mathcal{V} \to \mathcal{V}^*$ is *invertible*;

in which case there must exist a unique $\mathbf{g}^{-1}: \mathscr{V}^* \to \mathscr{V}$, or $\mathbf{g}^{-1} \in \mathscr{S}^2(\mathscr{V})$, such that

$$\mathbf{g}^{-1}\mathbf{g}(\mathbf{v}) = \mathbf{v} = \mathbf{g}^{-1}(v_i \boldsymbol{\epsilon}^i) = v_i \mathbf{g}^{-1}(\boldsymbol{\epsilon}^i) = v_i[(g^{-1})^{jk}\mathbf{e}_j\mathbf{e}_k](\boldsymbol{\epsilon}^i)$$
$$= v_i(g^{-1})^{jk}\, \mathbf{e}_j\mathbf{e}_k(\boldsymbol{\epsilon}^i) = v_i(g^{-1})^{ji}\mathbf{e}_j \underset{\uparrow}{=} v^j\mathbf{e}_j$$

by the definition
of $\mathbf{v}$

or $\qquad v^j = v_i(g^{-1})^{ji}$

It is customary to omit the -1 and simply write

$$v^j = g^{ji}v_i \tag{4.12b}$$

where it is understood that $\mathbf{g}$ with upper indices is the inverse of $\mathbf{g}$ (with lower indices). An invertible bilinear form is called *nondegenerate*. A symmetric bilinear form that is nondegenerate is called an *inner product*.

We, therefore, see that the presence of a symmetric bilinear form naturally connects the vectors in $\mathscr{V}$ and $\mathscr{V}^*$ in a unique way. Going from a vector in $\mathscr{V}$ to its unique image in $\mathscr{V}^*$ is done by simply lowering the index using (4.12a), and going the other way involves using (4.12b) to raise the index. This process can be generalized to all tensors. For instance, although there is no connection among $\mathscr{T}^2_0(\mathscr{V})$, $\mathscr{T}^1_1(\mathscr{V})$, and $\mathscr{T}^0_2(\mathscr{V})$, the introduction of a symmetric bilinear form connects all these spaces in a natural way and establishes a one-to-one correspondence among them. For example, to a tensor in $\mathscr{T}^2_0(\mathscr{V})$ with components t^{ij}, there corresponds a unique tensor in $\mathscr{T}^1_1(\mathscr{V})$, given, in component form, by $t^i_j = g_{jk}t^{ik}$, and another unique tensor in $\mathscr{T}^0_2(\mathscr{V})$, given by $t_{ij} = g_{il}t^l_j = g_{il}g_{jk}t^{lk}$.

Note that raising or lowering an index involves multiplying by g_{ij} or g^{ij} and summing over the index being lowered or raised. Let us try to lower one of the indices of g^{ij}, which is also a tensor and for which the lowering process is defined. Lowering an index of g^{ij} gives g^j_i, which is a tensor belonging to $\mathscr{T}^1_1(\mathscr{V})$. We know that $\mathscr{T}^1_1(\mathscr{V})$ is isomorphic to (the same as) $\mathscr{L}(\mathscr{V})$. So g^j_i can be thought of as the matrix representation of an operator on $\mathscr{V}$. Let us see what this operator is. We have

$$v^i = g^{ij}v_j$$

Multiplying both sides by g_{ik} and summing over i, we get

$$g_{ik}v^i = v_k = g_{ik}g^{ij}v_j = g^j_k v_j$$

This must be true for all $\mathbf{v}$'s. Thus we obtain the result:

$$g^j_k = g_{ik}g^{ij} = \delta^j_k \tag{4.13}$$

That is, the operator corresponding to g^i_j is simply the unit operator. This is, of course, true in all bases.

The inner product has been defined as a nondegenerate symmetric bilinear form. The important criterion of nondegeneracy has equivalences that can be given in a proposition.

4.1.21 Proposition A symmetric bilinear form $\mathbf{g}$ is nondegenerate if and only if

(i) for every nonzero $\mathbf{v} \in \mathscr{V}$, there exists some $\mathbf{w} \in \mathscr{V}$ such that $\mathbf{g}(\mathbf{v},\mathbf{w}) \neq 0$, or
(ii) the matrix of components g_{ij} has a nonvanishing determinant.

Proof. The symmetric bilinear form $\mathbf{g} \in \mathscr{T}^0_2(\mathscr{V})$ is nondegenerate iff $\mathbf{g}$, as the linear operator $\mathbb{g} : \mathscr{V} \to \mathscr{V}^*$, has an inverse, which is true iff det $\mathbb{g}$ is nonzero [this is part (ii)]. If det $\mathbb{g}$ is nonzero, then the nullity of $\mathbb{g}$ is zero iff for nonzero $\mathbf{v} \in \mathscr{V}$,

$$\mathbb{g}(\mathbf{v}) \neq 0 \quad \Leftrightarrow \quad \exists \text{ a } \mathbf{w} \in \mathscr{V} \ni [\mathbf{g}(\mathbf{v})](\mathbf{w}) = \mathbf{g}(\mathbf{v},\mathbf{w}) = \mathbf{g}(\mathbf{w},\mathbf{v}) \neq 0 \qquad \blacksquare$$

Note that $\mathbf{g}(\mathbf{v}) \in \mathscr{V}^*$, so $[\mathbf{g}(\mathbf{v})](\mathbf{w})$ is well-defined. In fact,

$$[\mathbf{g}(\mathbf{v})](\mathbf{w}) = g_{ij} v^j \boldsymbol{\epsilon}^i(\mathbf{w}) = g_{ij} \boldsymbol{\epsilon}^j(\mathbf{v}) \boldsymbol{\epsilon}^i(\mathbf{w}) = \mathbf{g}(\mathbf{v},\mathbf{w}) = \mathbf{g}(\mathbf{w},\mathbf{v})$$

We have already seen part (ii) of Proposition 4.1.21 in action in Eq. (4.13). The following is a direct consequence of the proposition; however, it is so important that it is stated as a corollary.

4.1.22 Corollary The matrix $g_{ij} = \mathbf{g}(\mathbf{e}_i,\mathbf{e}_j)$ of an inner product must necessarily have an inverse. $\blacksquare$

4.1.23 Definition A general (not necessarily nondegenerate) symmetric bilinear form $\mathbf{g}$ can be categorized as follows:

(i) *positive definite* if $\mathbf{g}(\mathbf{v},\mathbf{v}) > 0$ for every $\mathbf{v} \neq 0$
(ii) *negative definite* if $\mathbf{g}(\mathbf{v},\mathbf{v}) < 0$ for every $\mathbf{v} \neq 0$
(iii) *definite* if $\mathbf{g}$ is either positive definite or negative definite
(iv) *positive semidefinite* if $\mathbf{g}(\mathbf{v},\mathbf{v}) \geqslant 0$ for every $\mathbf{v}$
(v) *negative semidefinite* if $\mathbf{g}(\mathbf{v},\mathbf{v}) \leqslant 0$ for every $\mathbf{v}$
(vi) *semidefinite* if $\mathbf{g}$ is either positive semidefinite or negative semidefinite
(vii) *indefinite* if $\mathbf{g}$ is not definite

Example 4.1.11

Some of the categories of Definition 4.1.23 can be illustrated in $\mathbb{R}^2$ with $\mathbf{v}_1 = (x_1,y_1)$, $\mathbf{v}_2 = (x_2,y_2)$, and $\mathbf{v} = (x,y)$.

(a) Positive definite:

$$\mathbf{g}(\mathbf{v}_1,\mathbf{v}_2) = x_1 x_2 + y_1 y_2$$

because if $\mathbf{v} \neq 0$, then either $x \neq 0$ or $y \neq 0$, and $\mathbf{g}(\mathbf{v},\mathbf{v}) = x^2 + y^2 > 0$.

(b) Negative definite:

$$\mathbf{g}(\mathbf{v}_1,\mathbf{v}_2) = \tfrac{1}{2}(x_1 y_2 + x_2 y_1) - x_1 x_2 - y_1 y_2$$

because $$\mathbf{g}(\mathbf{v},\mathbf{v}) = xy - x^2 - y^2 = -\tfrac{1}{2}(x - y)^2 - \tfrac{1}{2}x^2 - \tfrac{1}{2}y^2$$

which is definitely negative for $\mathbf{v} \neq 0$.

(c) Indefinite:

$$\mathbf{g}(\mathbf{v}_1, \mathbf{v}_2) = x_1x_2 - y_1y_2$$

For $x = y$, $\mathbf{g}(\mathbf{v},\mathbf{v}) = 0$. However, $\mathbf{g}$ is nondegenerate, because, in the standard basis of $\mathbb{R}^2$, $\mathbf{g}$ has the matrix

$$\mathsf{g} = \begin{pmatrix} 1 & 0 \\ 0 & -1 \end{pmatrix}$$

This matrix is invertible; thus, by Proposition 4.1.21 $\mathbf{g}$ is nondegenerate.

(d) Positive semidefinite:

$$\mathbf{g}(\mathbf{v}_1,\mathbf{v}_2) = x_1x_2 \quad \Rightarrow \quad \mathbf{g}(\mathbf{v},\mathbf{v}) = x^2$$

For $\mathbf{v} = (0,y)$, this gives zero. However, $\mathbf{g}(\mathbf{v},\mathbf{v})$ is never negative. Therefore,

$$\mathbf{g}(\mathbf{v},\mathbf{v}) \geqslant 0 \qquad \forall\, \mathbf{v} \in \mathscr{V}$$

However, $\mathbf{g}$ is degenerate because its matrix in the standard basis of $\mathbb{R}^2$ is

$$\mathsf{g} = \begin{pmatrix} 1 & 0 \\ 0 & 0 \end{pmatrix}$$

which is *not* invertible. •

As we saw in Chapters 1 and 2, two vectors $\mathbf{u}, \mathbf{v} \in \mathscr{V}$ are orthogonal if $\mathbf{g}(\mathbf{u},\mathbf{v}) = 0$. A *null vector* of $\mathbf{g}$ is a vector that is orthogonal to itself. If $\mathbf{g}$ is definite, then the only null vector is the zero vector. The converse is also true and is stated in the following proposition.

4.1.24 Proposition If $\mathbf{g}$ is not definite, then there exists a nonzero null vector.

Proof. That $\mathbf{g}$ is not positive definite implies that there exists a nonzero vector, $\mathbf{v} \in \mathscr{V}$, such that $\mathbf{g}(\mathbf{v},\mathbf{v}) \leqslant 0$. Similarly, that $\mathbf{g}$ is not negative definite implies that there exists a nonzero vector, $\mathbf{w} \in \mathscr{V}$, such that $\mathbf{g}(\mathbf{w},\mathbf{w}) \geqslant 0$. For various values of α, $0 \leqslant \alpha \leqslant 1$, consider the vectors

$$\mathbf{u} = \alpha\mathbf{v} + (1 - \alpha)\mathbf{w}$$

All these vectors are nonzero unless $\mathbf{v}$ and $\mathbf{w}$ are linearly dependent, in which case

$$\mathbf{v} = \beta\mathbf{w}$$

$$\mathbf{g}(\mathbf{v},\mathbf{v}) = \mathbf{g}(\beta\mathbf{w},\beta\mathbf{w}) = \beta^2\mathbf{g}(\mathbf{w},\mathbf{w}) \geqslant 0$$

But the assumption was that $\mathbf{g}(\mathbf{v},\mathbf{v}) \leqslant 0$. Thus, $\mathbf{g}(\mathbf{v},\mathbf{v}) = 0$, and we are done. If we assume that none of the $\mathbf{u}$'s is zero, then

$$\mathbf{g}(\mathbf{u},\mathbf{u}) = \mathbf{g}[\alpha\mathbf{v} + (1-\alpha)\mathbf{w}, \alpha\mathbf{v} + (1-\alpha)\mathbf{w}] = \alpha^2\mathbf{g}(\mathbf{v},\mathbf{v}) + 2\alpha(1-\alpha)\mathbf{g}(\mathbf{v},\mathbf{w}) + (1-\alpha)^2\mathbf{g}(\mathbf{w},\mathbf{w})$$

is a continuous function of α. For $\alpha = 0$ this function has the value $\mathbf{g}(\mathbf{w},\mathbf{w}) \geqslant 0$, and for $\alpha = 1$ it has the value $\mathbf{g}(\mathbf{v},\mathbf{v}) \leqslant 0$. Thus, there must be some α for which $\mathbf{g}(\mathbf{u},\mathbf{u}) = 0$. ∎

Example 4.1.12

In the special theory of relativity, the inner product of two "displacement" four-vectors, $\mathbf{r}_1 = (ct_1, x_1, y_1, z_1)$ and $\mathbf{r}_2 = (ct_2, x_2, y_2, z_2)$, where c is the velocity of light, is defined as

$$\mathbf{g}(\mathbf{r}_1, \mathbf{r}_2) = c^2 t_1 t_2 - x_1 x_2 - y_1 y_2 - z_1 z_2$$

This is clearly an indefinite symmetric bilinear form. Proposition 4.1.24 tells us that there must exist a nonzero null vector. Such a vector, $\mathbf{r}$, satisfies

$$\mathbf{g}(\mathbf{r}, \mathbf{r}) = c^2 t^2 - x^2 - y^2 - z^2 = 0$$

or

$$c^2 = \frac{x^2 + y^2 + z^2}{t^2} \quad \Rightarrow \quad c = \pm \frac{\sqrt{x^2 + y^2 + z^2}}{t} = \pm \frac{\text{distance}}{\text{time}}$$

This corresponds to a particle moving with the speed of light. Thus, light rays are the null vectors in the special theory of relativity. •

Whenever there is an inner product on a vector space, there is the possibility of orthonormal basis vectors. However, since $\mathbf{g}(\mathbf{v}, \mathbf{v})$ is allowed to be negative or zero, it is impossible to demand normality (as defined in Chapters 1 and 2) for some vectors. We, therefore, define an orthonormal basis as follows.

4.1.25 Definition A basis $\{\mathbf{e}_i\}_{i=1}^N$ of $\mathcal{V}$ is *orthonormal with respect to* $\mathbf{g}$ if $\mathbf{g}(\mathbf{e}_i, \mathbf{e}_j) = 0$ for $i \neq j$ and each $\mathbf{g}(\mathbf{e}_i, \mathbf{e}_i)$ (no sum!) is one of the three values $+1$, -1, and 0. The $\mathbf{g}(\mathbf{e}_i, \mathbf{e}_i)$ are called the *diagonal components* of $\mathbf{g}$. We use n_+, n_-, and n_0 to denote the number of the vectors $\mathbf{e}_i$ for which $\mathbf{g}(\mathbf{e}_i, \mathbf{e}_i)$ is, respectively, $+1$, -1, or 0.

The existence of orthonormal bases was established for positive definite $\mathbf{g}$ by the Gram-Schmidt orthonormalization process in Chapters 1 and 2. One of the steps in this process is division by $\mathbf{g}(\mathbf{v}, \mathbf{v})$, which is legitimate as long as $\mathbf{g}$ is positive (or negative) definite. However, for the general case $\mathbf{g}(\mathbf{v}, \mathbf{v})$ could be zero, and the Gram-Schmidt process would break down. This does not mean that orthonormal bases do not exist for a general (possibly indefinite) $\mathbf{g}$. In fact, they do; however, the proof of their existence is a little harder and is the content of the following theorem (for a proof, see Bishop and Goldberg 1980, p. 104).

4.1.26 Theorem For every symmetric bilinear form $\mathbf{g}$ on $\mathcal{V}$, there is an orthonormal basis. Furthermore, n_+, n_-, and n_0 are the same in all orthonormal bases. ■

The quantity n_- is called the *index* of $\mathbf{g}$, and $s \equiv n_+ - n_-$ is called the *signature* of $\mathbf{g}$.

Example 4.1.13

Let $\mathcal{V} = \mathbb{R}^3$ and $\mathbf{v}_1 = (x_1, y_1, z_1)$, $\mathbf{v}_2 = (x_2, y_2, z_2)$, and $\mathbf{v} = (x, y, z)$. Define the symmetric bilinear form

$$\mathbf{g}(\mathbf{v}_1, \mathbf{v}_2) = \tfrac{1}{2}(x_1 y_2 + x_2 y_1 + y_1 z_2 + y_2 z_1 + x_1 z_2 + x_2 z_1)$$

so $\mathbf{g}(\mathbf{v},\mathbf{v}) = xy + yz + xz$. We wish to find a set of vectors in $\mathbb{R}^3$ that are orthonormal with respect to $\mathbf{g}$.

Clearly, $\mathbf{e}_1 = (1,1,0)$ is such that $\mathbf{g}(\mathbf{e}_1,\mathbf{e}_1) = 1$. So $\mathbf{e}_1$ is one of our vectors. Let us consider $\mathbf{v} = (1,0,1)$. We note that

$$\mathbf{f}_2 = \mathbf{v} - \frac{\mathbf{g}(\mathbf{v},\mathbf{e}_1)}{\mathbf{g}(\mathbf{e}_1,\mathbf{e}_1)}\mathbf{e}_1$$

is orthogonal to $\mathbf{e}_1$; that is, $\mathbf{g}(\mathbf{f}_2,\mathbf{e}_1) = 0$. Furthermore,

$$\mathbf{f}_2 = (1,0,1) - \frac{\frac{3}{2}}{1}(1,1,0) = (-\tfrac{1}{2}, -\tfrac{3}{2}, 1)$$

so $\mathbf{g}(\mathbf{f}_2,\mathbf{f}_2) = \frac{3}{4} - \frac{3}{2} - \frac{1}{2} = -\frac{5}{4}$, and

$$\mathbf{e}_2 = \frac{\mathbf{f}_2}{\sqrt{|\mathbf{g}(\mathbf{f}_2,\mathbf{f}_2)|}} = \frac{\mathbf{f}_2}{\sqrt{\frac{5}{4}}} = \left(-\frac{1}{\sqrt{5}}, -\frac{3}{\sqrt{5}}, \frac{2}{\sqrt{5}}\right)$$

is the second vector, with $\mathbf{g}(\mathbf{e}_2,\mathbf{e}_2) = -1$. Finally, we take $\mathbf{w} = (0,1,1)$. Then

$$\mathbf{f}_3 = \mathbf{w} - \frac{\mathbf{g}(\mathbf{w},\mathbf{e}_1)}{\mathbf{g}(\mathbf{e}_1,\mathbf{e}_1)}\mathbf{e}_1 - \frac{\mathbf{g}(\mathbf{w},\mathbf{e}_2)}{\mathbf{g}(\mathbf{e}_2,\mathbf{e}_2)}\mathbf{e}_2$$

is orthogonal to both $\mathbf{e}_1$ and $\mathbf{e}_2$ and can easily be shown to be

$$\mathbf{f}_3 = \tfrac{4}{10}(-3,1,1)$$

with $\mathbf{g}(\mathbf{f}_3,\mathbf{f}_3) = -\frac{4}{5}$. Thus, the third vector can be chosen to be

$$\mathbf{e}_3 = \frac{\mathbf{f}_3}{\sqrt{\frac{4}{5}}} = \left(-\frac{3}{\sqrt{5}}, \frac{1}{\sqrt{5}}, \frac{1}{\sqrt{5}}\right)$$

We thus have

$$\mathbf{g}(\mathbf{e}_1,\mathbf{e}_1) = 1$$

$$\mathbf{g}(\mathbf{e}_2,\mathbf{e}_2) = -1$$

$$\mathbf{g}(\mathbf{e}_3,\mathbf{e}_3) = -1$$

$$\mathbf{g}(\mathbf{e}_i,\mathbf{e}_j) = 0 \qquad \text{for } i \neq j$$

and $$n_+ = 1 \qquad n_- = 2 \qquad n_0 = 0$$

That is, the index of $\mathbf{g}$ is 2, and its signature is -1. Although we have worked in a particular basis, Theorem 4.1.26 guarantees that n_+, n_-, and n_0 are basis-independent. ●

It is important to emphasize that the invariance of n_+, n_-, and n_0 with respect to bases is true for *orthonormal bases with respect to* $\mathbf{g}$. As a counterexample, consider $\mathbf{g}$ of Example 4.1.13 applied to the standard basis of $\mathbb{R}^3$, $\{\mathbf{e}'_1 = (1,0,0),\ \mathbf{e}'_2 = (0,1,0),\ \mathbf{e}'_3 = (0,0,1)\}$. It is easily verified that

$$\mathbf{g}(\mathbf{e}'_i,\mathbf{e}'_i) = 0 \qquad \text{for } i = 1, 2, 3$$

So it might appear that $n_0 = 3$ for this basis. However, the standard basis is *not* orthonormal with respect to **g**. In fact,

$$g(\mathbf{e}'_1,\mathbf{e}'_2) = \tfrac{1}{2} = g(\mathbf{e}'_1,\mathbf{e}'_3) = g(\mathbf{e}'_2,\mathbf{e}'_3)$$

That is why the nonstandard vectors $\mathbf{e}_1$, **v**, and **w** were chosen in Example 4.1.13.

In an orthonormal basis the matrix of **g** is diagonal, with entries $+1$, -1, and 0. In particular, if **g** is to be nondegenerate, that is, to be an inner product, then n_0 must be zero. Thus, a general inner product on an N-dimensional vector space $\mathcal{V}$ satisfies the conditions

$$n_+ + n_- = N \qquad \text{and} \qquad n_+ - n_- = s$$

which give

$$s = N - 2n_-$$

An inner product space with $n_- = 1$ or $n_- = N - 1$ is called a *Minkowski space.* For $N = 4$ this is the space of the special theory of relativity. An inner product space with $n_- = 0$ is called a *Euclidean space.* This is the space in which all classical physics takes place.

Example 4.1.14

Let $\{\mathbf{e}_i\}_{i=1}^N$ be a basis of $\mathcal{V}$ and $\{\boldsymbol{\epsilon}^j\}_{j=1}^N$ its dual basis. We can define the *permutation tensor*:

$$\delta^{i_1 i_2 \ldots i_N}_{j_1 j_2 \ldots j_N} \equiv N!\, \boldsymbol{\epsilon}^{i_1} \wedge \boldsymbol{\epsilon}^{i_2} \wedge \cdots \wedge \boldsymbol{\epsilon}^{i_N}(\mathbf{e}_{j_1},\mathbf{e}_{j_2}, \ldots, \mathbf{e}_{j_N}) \tag{1}$$

It is clear from this definition that $\delta^{i_1 \ldots i_N}_{j_1 \ldots j_N}$ is completely skew-symmetric in all upper indices. It is not clear, but is also true, that it is also skew-symmetric in the lower indices. This can be seen as follows. Assume that two of the lower indices are equal. This means having two $\mathbf{e}_j$'s equal in (1). These two $\mathbf{e}_j$'s will contract with two $\boldsymbol{\epsilon}^i$'s, say $\boldsymbol{\epsilon}^k$ and $\boldsymbol{\epsilon}^l$. Thus, in the expansion there will be a term $C\boldsymbol{\epsilon}^k(\mathbf{e}_j)\boldsymbol{\epsilon}^l(\mathbf{e}_j)$, where C is the product of all the other factors. Since the product is completely skew-symmetric in the upper indices, there must also exist another term, with a minus sign and in which the upper indices k and l are interchanged: $-C\boldsymbol{\epsilon}^l(\mathbf{e}_j)\boldsymbol{\epsilon}^k(\mathbf{e}_j)$. This makes the sum zero. Theorem 4.1.12 implies that (1) is antisymmetric in the lower indices as well.

This suggests that

$$\delta^{i_1 \ldots i_N}_{j_1 \ldots j_N} \propto \varepsilon^{i_1 \ldots i_N} \varepsilon_{j_1 \ldots j_N}$$

To find the proportionality constant, we note that

$$\begin{aligned}
\delta^{12 \ldots N}_{12 \ldots N} &= N!\, \boldsymbol{\epsilon}^1 \wedge \boldsymbol{\epsilon}^2 \wedge \cdots \wedge \boldsymbol{\epsilon}^N(\mathbf{e}_1,\mathbf{e}_2, \ldots, \mathbf{e}_N) \\
&= N!\, [\mathbb{A}(\boldsymbol{\epsilon}^1 \otimes \boldsymbol{\epsilon}^2 \otimes \cdots \otimes \boldsymbol{\epsilon}^N)](\mathbf{e}_1, \ldots, \mathbf{e}_N) \\
&= \sum_\pi \delta_\pi (\boldsymbol{\epsilon}^1 \otimes \boldsymbol{\epsilon}^2 \otimes \cdots \otimes \boldsymbol{\epsilon}^N)(\mathbf{e}_{\pi(1)}, \ldots, \mathbf{e}_{\pi(N)}) \\
&= \sum_\pi \delta_\pi \boldsymbol{\epsilon}^1(\mathbf{e}_{\pi(1)})\, \boldsymbol{\epsilon}^2(\mathbf{e}_{\pi(2)}) \cdots \boldsymbol{\epsilon}^N(\mathbf{e}_{\pi(N)}) \\
&= \sum_\pi \delta_\pi \delta^1_{\pi(1)}\, \delta^2_{\pi(2)} \cdots \delta^N_{\pi(N)}
\end{aligned}$$

The only contribution to the sum comes from the permutation with the property $\pi(i) = i$. This is the identity permutation for which $\delta_\pi = 1$. Thus, we have

$$\delta^{12\ldots N}_{12\ldots N} = 1$$

On the other hand [see part (b) of Exercise 4.1.8],

$$\varepsilon^{12\ldots N}\varepsilon_{12\ldots N} = \varepsilon^{12\ldots N} = (-1)^{n-}$$

Therefore, the proportionality constant is $(-1)^{n-}$. Thus

$$\varepsilon^{i_1 i_2 \ldots i_N}\varepsilon_{j_1 j_2 \ldots j_N} = (-1)^{n-}\,\delta^{i_1 i_2 \ldots i_N}_{j_1 j_2 \ldots j_N}$$ •

We can find an explicit expression for the permutation tensor of Example 4.1.14. As in that example, we write

$$\begin{aligned}\boldsymbol{\epsilon}^{i_1}\wedge\cdots\wedge\boldsymbol{\epsilon}^{i_N}(\mathbf{e}_{j_1},\ldots,\mathbf{e}_{j_N}) &= \frac{1}{N!}\sum_\pi \delta_\pi \boldsymbol{\epsilon}^{i_1}\otimes\cdots\otimes\boldsymbol{\epsilon}^{i_N}(\mathbf{e}_{\pi(j_1)},\ldots,\mathbf{e}_{\pi(j_N)})\\ &= \frac{1}{N!}\sum_\pi \delta_\pi \boldsymbol{\epsilon}^{i_1}(\mathbf{e}_{\pi(j_1)})\cdots\boldsymbol{\epsilon}^{i_N}(\mathbf{e}_{\pi(j_N)})\\ &= \frac{1}{N!}\sum_\pi \delta_\pi \delta^{i_1}_{\pi(j_1)}\delta^{i_2}_{\pi(j_2)}\cdots\delta^{i_N}_{\pi(j_N)}\end{aligned}$$

The last equality follows because $\{\boldsymbol{\epsilon}^j\}$ and $\{\mathbf{e}_i\}$ are dual bases and so $\boldsymbol{\epsilon}^j(\mathbf{e}_i) = \delta^j_i$. We can now write

$$\delta^{i_1 i_2\ldots i_N}_{j_1 j_2\ldots j_N} = \sum_\pi \delta_\pi \delta^{i_1}_{\pi(j_1)}\delta^{i_2}_{\pi(j_2)}\cdots\delta^{i_N}_{\pi(j_N)} \tag{4.14a}$$

and

$$\varepsilon^{i_1 i_2\ldots i_N}\varepsilon_{j_1 j_2\ldots j_N} = (-1)^{n-}\sum_\pi \delta_\pi \delta^{i_1}_{\pi(j_1)}\delta^{i_2}_{\pi(j_2)}\cdots\delta^{i_N}_{\pi(j_N)} \tag{4.14b}$$

Recall that $\delta_\pi = \pm 1$, depending on whether π is an even or an odd permutation of $(1\ 2\ \ldots\ N)$.

Example 4.1.15

We can write Eq. (4.14a) concisely as a determinant. First, we note that $\delta_\pi = \varepsilon_{\pi(1),\ldots,\pi(N)}$. Now we consider

$$\begin{aligned}\delta^{i_1 i_2\ldots i_N}_{12\ldots N} &= \sum_\pi \varepsilon_{\pi(1),\pi(2),\ldots,\pi(N)}\delta^{i_1}_{\pi(1)}\delta^{i_2}_{\pi(2)}\cdots\delta^{i_N}_{\pi(N)}\\ &\equiv \sum_{k_1\ldots k_N}\varepsilon_{k_1 k_2\ldots k_N}\delta^{i_1}_{k_1}\delta^{i_2}_{k_2}\cdots\delta^{i_N}_{k_N}\end{aligned}$$

The RHS is clearly the determinant of a matrix whose elements are $\delta^{i_1}_{k_1}$, where $i_1, k_1 = 1, \ldots, N$. The same holds true if $1, 2, \ldots, N$ is replaced by $j_1, j_2, \ldots, j_N$; thus,

$$\delta^{i_1 i_2\ldots i_N}_{j_1 j_2\ldots j_N} = \det\begin{pmatrix}\delta^{i_1}_{j_1} & \delta^{i_1}_{j_2} & \cdots & \delta^{i_1}_{j_N}\\ \delta^{i_2}_{j_1} & \delta^{i_2}_{j_2} & \cdots & \delta^{i_2}_{j_N}\\ \vdots & \vdots & & \vdots\\ \delta^{i_N}_{j_1} & \delta^{i_N}_{j_2} & \cdots & \delta^{i_N}_{j_N}\end{pmatrix}$$ •

Example 4.1.16

As an application of the foregoing formalism, we can express the determinant of a 2×2 matrix in terms of traces. Let A be such a matrix with elements A^i_j, then

$$\begin{aligned}\det \mathsf{A} &= \varepsilon_{ij} A^i_1 A^j_2 = \tfrac{1}{2}\{\varepsilon_{ij} A^i_1 A^j_2 - \varepsilon_{ij} A^i_2 A^j_1\} \\ &= \tfrac{1}{2}\{\varepsilon_{ij}\varepsilon^{kl} A^i_k A^j_l\} = \tfrac{1}{2} A^i_k A^j_l (\delta^k_i \delta^l_j - \delta^k_j \delta^l_i) \\ &= \tfrac{1}{2}(A^i_i A^j_j - A^i_j A^j_i) = \tfrac{1}{2}[(\mathrm{tr}\,\mathsf{A})(\mathrm{tr}\,\mathsf{A}) - (\mathsf{A}^2)^i_i] \\ &= \tfrac{1}{2}[(\mathrm{tr}\,\mathsf{A})^2 - \mathrm{tr}(\mathsf{A}^2)]\end{aligned}$$

●

We can generalize the result of Example 4.1.16 and express the determinant of an $N \times N$ matrix as

$$\det \mathsf{A} = \frac{1}{N!}\, \varepsilon^{i_1 i_2 \ldots i_N} \varepsilon_{j_1 j_2 \ldots j_N} A^{j_1}_{i_1} A^{j_2}_{i_2} \cdots A^{j_N}_{i_N}$$

where the underlying space is assumed to be Euclidean with $n_- = 0$. This can be rewritten as

$$\det \mathsf{A} = \frac{1}{N!} \sum_{\pi} \delta_\pi \delta^{i_1}_{\pi(j_1)} \cdots \delta^{i_N}_{\pi(j_N)} A^{j_1}_{i_1} \cdots A^{j_N}_{i_N} \tag{4.15}$$

4.1.5 The Hodge Star Operator

In Chapter 2 it was established that all vector spaces of the same dimensions are isomorphic (identical). On the other hand, the two vector spaces $\Lambda^p(\mathcal{V})$ and $\Lambda^{N-p}(\mathcal{V})$ have the same dimension:

$$\binom{N}{p} = \binom{N}{N-p}$$

Thus, there is an isomorphism between the two spaces. Hodge established a natural isomorphism between the two spaces, which we will investigate shortly.

4.1.27 Definition An *oriented basis* of an N-dimensional vector space $\mathcal{V}$ is an ordered collection of N linearly independent vectors.

If $\{\mathbf{v}_1, \mathbf{v}_2, \ldots, \mathbf{v}_N\}$ is one oriented basis and $\{\mathbf{u}_1, \mathbf{u}_2, \ldots, \mathbf{u}_N\}$ is a second one, then

$$\mathbf{u}_1 \wedge \mathbf{u}_2 \wedge \cdots \wedge \mathbf{u}_N = (\det \mathsf{R})\mathbf{v}_1 \wedge \mathbf{v}_2 \wedge \cdots \wedge \mathbf{v}_N$$

where R is the transformation matrix and $\det \mathsf{R}$ is a nonzero number (R is invertible) which can be positive or negative. Accordingly, we have the following definition.

4.1.28 Definition An *orientation* is the collection of all oriented bases related by a transformation matrix having a positive determinant.

Clearly, there are *only two* orientations in any vector space. Each oriented basis is positively related to any oriented basis belonging to the same orientation and negatively related to any oriented basis belonging to the other orientation.

In $\mathbb{R}^3$ the bases $\{\hat{\mathbf{e}}_x, \hat{\mathbf{e}}_y, \hat{\mathbf{e}}_z\}$ and $\{\hat{\mathbf{e}}_y, \hat{\mathbf{e}}_x, \hat{\mathbf{e}}_z\}$ belong to different orientations because

$$\hat{\mathbf{e}}_x \wedge \hat{\mathbf{e}}_y \wedge \hat{\mathbf{e}}_z = -\hat{\mathbf{e}}_y \wedge \hat{\mathbf{e}}_x \wedge \hat{\mathbf{e}}_z$$

The first basis is called a right-handed coordinate system, and the second is called a left-handed coordinate system. Any other basis is either right-handed or left-handed. There is no third alternative!

An inner product space for which an orientation is specified is called an *oriented vector space.*

4.1.29 Definition Let $\mathcal{V}$ be a space with an inner product $\mathbf{g}$. Let $\mathcal{V}^*$ have the oriented orthonormal (with respect to $\mathbf{g}$) basis $\{\boldsymbol{\epsilon}^i\}_{i=1}^N$. The oriented *volume element* of $\mathcal{V}$ is defined as $\boldsymbol{\mu} \in \Lambda^N(\mathcal{V}^*)$ given by $\boldsymbol{\mu} \equiv \boldsymbol{\epsilon}^1 \wedge \boldsymbol{\epsilon}^2 \wedge \cdots \wedge \boldsymbol{\epsilon}^N$.

Note that if $\{\mathbf{e}_i\}$ is ordered the same way as $\{\boldsymbol{\epsilon}^j\}$, then $\boldsymbol{\mu}(\mathbf{e}_1, \mathbf{e}_2, \ldots, \mathbf{e}_N) = +1$. We say that $\{\mathbf{e}_i\}$ is *positively oriented.* In general, $\{\mathbf{v}_i\}$ is positively oriented if $\boldsymbol{\mu}(\mathbf{v}_1, \mathbf{v}_2, \ldots, \mathbf{v}_N) > 0$. The justification for writing *the* oriented volume element follows from the fact that if $\{\boldsymbol{\varphi}^k\}_{k=1}^N$ is another orthonormal basis *in the same orientation* and related to $\{\boldsymbol{\epsilon}^j\}$ by a matrix R, then

$$\boldsymbol{\varphi}^1 \wedge \boldsymbol{\varphi}^2 \wedge \cdots \wedge \boldsymbol{\varphi}^N = (\det \mathsf{R})\boldsymbol{\epsilon}^1 \wedge \boldsymbol{\epsilon}^2 \wedge \cdots \wedge \boldsymbol{\epsilon}^N$$

Since $\{\boldsymbol{\varphi}^k\}$ and $\{\boldsymbol{\epsilon}^j\}$ are orthonormal, the determinant of $\mathbf{g}$ is $(-1)^{n_-}$ in both of them. Exercise 4.1.11 then implies that $(\det \mathsf{R})^2 = 1$ or $\det \mathsf{R} = \pm 1$. However, $\{\boldsymbol{\varphi}^k\}$ and $\{\boldsymbol{\epsilon}^j\}$ belong to the same orientation. Thus, $\det \mathsf{R} = +1$, and $\{\boldsymbol{\varphi}^k\}$ and $\{\boldsymbol{\epsilon}^j\}$ give the same volume element.

The volume element of $\mathcal{V}$ is defined in terms of a basis for $\mathcal{V}^*$. The reason for this will become apparent later, when we see that dx, dy, and dz form a basis for $(\mathbb{R}^3)^*$ and a volume element of $\mathbb{R}^3$ is $dx\,dy\,dz \equiv dx \wedge dy \wedge dz$. Now we are ready for the definition of the Hodge star operator.

4.1.30 Definition Let $\{\mathbf{e}_i\}_{i=1}^N$ be an ordered orthonormal basis of $\mathcal{V}$ relative to the inner product $\mathbf{g}$. The *Hodge star operator* is a linear mapping, $*: \Lambda^p(\mathcal{V}) \to \Lambda^{N-p}(\mathcal{V})$, given by (remember Einstein's summation convention!)

$$*(\mathbf{e}_{i_1} \wedge \cdots \wedge \mathbf{e}_{i_p}) = \frac{1}{(N-p)!}\varepsilon_{i_1 \ldots i_p}^{\;\;\;\;\;\;\;\;\;i_{p+1} \ldots i_N}\,\mathbf{e}_{i_{p+1}} \wedge \cdots \wedge \mathbf{e}_{i_N} \tag{4.16a}$$

where

$$\varepsilon_{i_1 i_2 \ldots i_p}^{\;\;\;\;\;\;\;\;\;\;\;\;i_{p+1} i_{p+2} \ldots i_N} = g^{j_1 i_{p+1}} g^{j_2 i_{p+2}} \cdots g^{j_{N-p} i_N}\,\varepsilon_{i_1 i_2 \ldots i_p j_1 j_2 \ldots j_{N-p}} \tag{4.16b}$$

Although this definition is based on a choice of basis, it can be shown that the operator is in fact basis-independent. We note that the product of the g's on the RHS of (4.16b) is ± 1. In particular, for Euclidean spaces in which $n_- = 0$, the product is absent.

Example 4.1.17

Let us apply Definition 4.1.30 to $\Lambda^p(\mathbb{R}^3)$ for $p = 0, 1, 2, 3$. Let $\{\mathbf{e}_1, \mathbf{e}_2, \mathbf{e}_3\}$ be an oriented orthonormal basis of $\mathbb{R}^3$.

(a) For $\Lambda^0(\mathbb{R}^3) = \mathbb{R}$ a basis is 1, and (4.16a) gives

$$*1 = \frac{1}{3!}\varepsilon^{ijk}\mathbf{e}_i \wedge \mathbf{e}_j \wedge \mathbf{e}_k = \mathbf{e}_1 \wedge \mathbf{e}_2 \wedge \mathbf{e}_3$$

(b) For $\Lambda^1(\mathbb{R}^3) = \mathbb{R}^3$ a basis is $\{\mathbf{e}_1, \mathbf{e}_2, \mathbf{e}_3\}$, and (4.16a) gives

$$*\mathbf{e}_i = \tfrac{1}{2}\varepsilon_i^{jk}\mathbf{e}_j \wedge \mathbf{e}_k \quad \Rightarrow \quad *\mathbf{e}_1 = \mathbf{e}_2 \wedge \mathbf{e}_3$$

$$*\mathbf{e}_2 = \mathbf{e}_3 \wedge \mathbf{e}_1$$

$$*\mathbf{e}_3 = \mathbf{e}_1 \wedge \mathbf{e}_2$$

(c) For $\Lambda^2(\mathbb{R}^3)$ a basis is $\{\mathbf{e}_1 \wedge \mathbf{e}_2, \mathbf{e}_1 \wedge \mathbf{e}_3, \mathbf{e}_2 \wedge \mathbf{e}_3\}$, and (4.16a) gives

$$*\mathbf{e}_i \wedge \mathbf{e}_j = \varepsilon_{ij}^k\mathbf{e}_k \quad \Rightarrow \quad *\mathbf{e}_1 \wedge \mathbf{e}_2 = \varepsilon_{12}^k\mathbf{e}_k = \varepsilon_{12}^3\mathbf{e}_3 = \varepsilon_{123}\mathbf{e}_3 = \mathbf{e}_3$$

$$*\mathbf{e}_1 \wedge \mathbf{e}_3 = \varepsilon_{13}^k\mathbf{e}_k = \varepsilon_{13}^2\mathbf{e}_2 = \varepsilon_{132}\mathbf{e}_2 = -\mathbf{e}_2$$

$$*\mathbf{e}_2 \wedge \mathbf{e}_3 = \varepsilon_{23}^k\mathbf{e}_k = \mathbf{e}_1$$

(d) For $\Lambda^3(\mathbb{R}^3)$ a basis is $\{\mathbf{e}_1 \wedge \mathbf{e}_2 \wedge \mathbf{e}_3\}$, and (4.16a) yields

$$*\mathbf{e}_1 \wedge \mathbf{e}_2 \wedge \mathbf{e}_3 = \varepsilon_{123} = 1$$

●

Example 4.1.17 may suggest that applying the Hodge star operator twice (composition of $*$ with itself, or $* \circ *$) is the same as applying the identity operator. This is partially true. The following theorem is a precise statement of this conjecture. (For a proof, see Bishop and Goldberg 1980, p. 111.)

4.1.31 Theorem For $\mathbf{A} \in \Lambda^p(\mathcal{V})$, where $\mathcal{V}$ is an oriented space with respect to the inner product $\mathbf{g}$, we have

$$* \circ * \, \mathbf{A} \equiv ** \mathbf{A} = (-1)^{n_-}(-1)^{p(N-p)}\mathbf{A} \tag{4.17}$$

where n_- is the index of $\mathbf{g}$ and $N = \dim \mathcal{V}$. ■

In particular, for Euclidean spaces with an odd number of dimensions, $**\mathbf{A} = \mathbf{A}$. This is the case for $\mathbb{R}^3$.

For an arbitrary $\mathbf{A} \in \Lambda^p(\mathcal{V})$ and an arbitrary (not necessarily orthonormal) basis $\{\mathbf{v}_i\}_{i=1}^N$ of $\mathcal{V}$, which is positively related to $\{\mathbf{e}_i\}$ and in terms of which $\mathbf{A}$ is given by $\mathbf{A} = A^{i_1 i_2 \cdots i_p}\mathbf{v}_{i_1} \wedge \mathbf{v}_{i_2} \wedge \cdots \wedge \mathbf{v}_{i_p}$, we have

$$*\mathbf{A} = (\det \mathsf{R})^{-1}\frac{1}{(N-p)!}A_{j_1 \ldots j_p}\varepsilon^{j_1 \cdots j_N}\mathbf{e}_{j_{p+1}} \wedge \cdots \wedge \mathbf{e}_{j_N}$$

where the raising and lowering is done by the inner product tensor $g_{ij} = \mathbf{g}(\mathbf{v}_i, \mathbf{v}_j)$ as expressed in the basis $\{\mathbf{v}_i\}$. From Exercise 4.1.11, $(\det \mathsf{R})^2 = (-1)^{n_-} \det \mathsf{G}$, where G denotes the matrix of $\mathbf{g}$ in $\{\mathbf{v}_i\}$. Since $\det \mathsf{R} > 0$ in this case, we have $\det \mathsf{R} =$

$|\det G|^{1/2}$. Thus, the above expression for $*\mathbf{A}$ can also be written as follows:

$$*\mathbf{A} = |\det G|^{-1/2} \frac{1}{(N-p)!} A_{j_1 j_2 \ldots j_p} \varepsilon^{j_1 j_2 \ldots j_N} \mathbf{e}_{j_{p+1}} \wedge \mathbf{e}_{j_{p+2}} \wedge \cdots \wedge \mathbf{e}_{j_N} \tag{4.18}$$

It is important to emphasize that in the sum $A^{i_1 \ldots i_p} \mathbf{v}_{i_1} \wedge \cdots \wedge \mathbf{v}_{i_p}$ there is no restriction on the range of summation except for the fact that $A^{i_1 i_2 \ldots i_p}$ is skew-symmetric. This means that if $\mathbf{A}$ is given as a linear combination of only a few of the products $\mathbf{v}_{i_1} \wedge \mathbf{v}_{i_2} \wedge \cdots \wedge \mathbf{v}_{i_p}$, in which not all possible basis vectors are present, then the components of $\mathbf{A}$ must be split among all possible $\mathbf{v}_{i_1} \wedge \cdots \wedge \mathbf{v}_{i_p}$. For example, if $\mathbf{A} = \mathbf{e}_1 \wedge \mathbf{e}_2$, then in the sum $A^{ij}\mathbf{e}_i \wedge \mathbf{e}_j$ the nonzero components consist of $A^{12} = \frac{1}{2}$ and $A^{21} = -\frac{1}{2}$, and all other components are zero. Similarly, when $\mathbf{B} = \mathbf{e}_1 \wedge \mathbf{e}_2 \wedge \mathbf{e}_3$ is written in the form $\mathbf{B} = B^{ijk} e_i \wedge e_j \wedge e_k$, it is understood that the nonzero components of $\mathbf{B}$ are not restricted to $\mathbf{B}^{123}$. Other components, such as B^{132}, B^{231}, and so on, are also nonzero. In fact, we have

$$B^{123} = -B^{132} = -B^{213} = B^{231} = B^{312} = -B^{321} = \tfrac{1}{6}$$

This should be kept in mind when sums over exterior products are encountered. It is, of cou se, obvious that the "different" exterior products in the sum are all linearly dt pendent.

Example 4.1.18

Let $\mathbf{a}, \mathbf{b} \in \mathbb{R}^3$ and $\{\mathbf{e}_1, \mathbf{e}_2, \mathbf{e}_3\}$ be an orthonormal oriented basis of $\mathbb{R}^3$. Then $\mathbf{a} = a^i \mathbf{e}_i$ and $\mathbf{b} = b^j \mathbf{e}_j$. Let us calculate $\mathbf{a} \wedge \mathbf{b}$ and $*(\mathbf{a} \wedge \mathbf{b})$. We assume a Euclidean $\mathbf{g}$ on $\mathbb{R}^3$. Clearly,

$$\mathbf{a} \wedge \mathbf{b} = (a^i \mathbf{e}_i) \wedge (b^j \mathbf{e}_j) = a^i b^j \mathbf{e}_i \wedge \mathbf{e}_j$$

and

$$*(\mathbf{a} \wedge \mathbf{b}) = *(a^i b^j \mathbf{e}_i \wedge \mathbf{e}_j) = a^i b^j (*\mathbf{e}_i \wedge \mathbf{e}_j)$$
$$= a^i b^j (\varepsilon_{ij}^k \mathbf{e}_k) = (a^i b^j \varepsilon_{ijk}) \mathbf{e}_k$$

We see that $*(\mathbf{a} \wedge \mathbf{b})$ is a vector with components

$$[*(\mathbf{a} \wedge \mathbf{b})]^k = a^i b^j \varepsilon_{ij}^k$$

In particular,

$$[*(\mathbf{a} \wedge \mathbf{b})]^1 = a^i b^j \varepsilon_{ij}^1 = a^i b^j \varepsilon_{ij1} = a^i b^2 \varepsilon_{i21} + a^i b^3 \varepsilon_{i31}$$
$$= a^3 b^2 \varepsilon_{321} + a^2 b^3 \varepsilon_{231} = \varepsilon_{231}(a^2 b^3 - a^3 b^2)$$
$$= a^2 b^3 - a^3 b^2$$

Similarly,

$$[*(\mathbf{a} \wedge \mathbf{b})]^2 = a^3 b^1 - a^1 b^3$$
$$[*(\mathbf{a} \wedge \mathbf{b})]^3 = a^1 b^2 - a^2 b^1$$

The last three equations can be written as

$$*(\mathbf{a} \wedge \mathbf{b}) = \mathbf{a} \times \mathbf{b}$$

We thus have a convenient way of writing the cross product of two vectors in $\mathbb{R}^3$ in terms of the Levi-Civita tensor:

$$(\mathbf{a} \times \mathbf{b})^k = a^i b^j \varepsilon_{ij}^k$$

This relation (together with the results of Exercise 4.1.9) comes in very handy when working with cross products.

The correspondence between $\mathbf{a} \wedge \mathbf{b}$ and $\mathbf{a} \times \mathbf{b}$ holds *only* in three dimensions because $\dim[\Lambda^1(\mathcal{V})] = \dim[\Lambda^2(\mathcal{V})]$ only if $\dim \mathcal{V} = 3$. In fact, if $\dim \mathcal{V} = N$, then

$$\frac{N}{2!(N-2)!} = \frac{N}{1!(N-1)!} \quad \Leftrightarrow \quad (N-1)! = 2[(N-2)!] \quad \Leftrightarrow \quad N-1=2 \quad \Leftrightarrow \quad N=3$$

That is why the cross product can be defined as a machine that takes two vectors in $\mathcal{V}$ and manufactures a *vector* in $\mathcal{V}$ only in three dimensions. •

Example 4.1.19

We can use the results of Example 4.1.18 and Exercise 4.1.9 to establish certain vector identities componentwise.

(a) For $\mathbf{a} \times (\mathbf{b} \times \mathbf{c}) = \mathbf{b}(\mathbf{a}\cdot\mathbf{c}) - \mathbf{c}(\mathbf{a}\cdot\mathbf{b})$, we have

$$[\mathbf{a} \times (\mathbf{b} \times \mathbf{c})]^k = a^i (b \times c)^j \varepsilon^k_{ij} = a^i \varepsilon^k_{ij} (b^l c^m \varepsilon^j_{lm})$$

$$= a^i b^l c^m \varepsilon^k_{ij} \varepsilon^j_{lm} = a_i b^l c^m \varepsilon^{ijk} \varepsilon_{lmj}$$

↑

by Problem 4.16

$$= -a_i b^l c^m \varepsilon^{ikj} \varepsilon_{lmj} = -a_i b^l c^m (\delta^i_l \delta^k_m - \delta^i_m \delta^k_l)$$

$$= -a_i b^i c^k + a_i b^k c^i = -\mathbf{a}\cdot\mathbf{b} c^k + \mathbf{a}\cdot\mathbf{c} b^k$$

which is the kth component of $\mathbf{b}(\mathbf{a}\cdot\mathbf{c}) - \mathbf{c}(\mathbf{a}\cdot\mathbf{b})$.

(b) For $\mathbf{a}\cdot(\mathbf{a} \times \mathbf{b}) = \mathbf{b}\cdot(\mathbf{a} \times \mathbf{b}) = 0$, we have

$$\mathbf{a}\cdot(\mathbf{a} \times \mathbf{b}) = a_k (\mathbf{a} \times \mathbf{b})^k = a_k (a^i b^j \varepsilon^k_{ij})$$

$$= a^k a^i b^j \varepsilon_{ijk} = 0$$

↑

by the symmetry of $a^k a^i$, the antisymmetry of ε_{ijk}, and Exercise 4.1.13

(c) For $\nabla \times (\nabla \mathbf{f}) = 0$, we have

$$\nabla \times (\nabla f) = \nabla_i (\nabla f)_j \varepsilon^{ijk} = \nabla_i (\nabla_j f) \varepsilon^{ijk} = (\nabla_i \nabla_j f) \varepsilon^{ijk}$$

$$= 0$$

↑

by the symmetry of $\nabla_i \nabla_j$, the antisymmetry of ε_{ijk}, and Exercise 4.1.13 •

Exercises

4.1.1 Show that $\mathbf{e}_{j_1} \otimes \cdots \otimes \mathbf{e}_{j_r} \otimes \boldsymbol{\epsilon}^{i_1} \otimes \cdots \otimes \boldsymbol{\epsilon}^{i_s}$ are linearly independent.

4.1.2 What is the tensor product of $\mathbf{A} = 2\hat{\mathbf{e}}_x - \hat{\mathbf{e}}_y + 3\hat{\mathbf{e}}_z$ with itself?

4.1.3 Show that $\text{tr}: \mathcal{T}^1_1 \to \mathbb{R}$ is an invariant linear function.

4.1.4 Let $\mathbf{A} \in \mathcal{T}^r_0(\mathcal{V})$ be skew-symmetric. Show that if $\boldsymbol{\tau}^1, \ldots, \boldsymbol{\tau}^r \in \mathcal{V}^*$ are linearly dependent, then $\mathbf{A}(\boldsymbol{\tau}^1, \ldots, \boldsymbol{\tau}^r) = 0$.

4.1.5 Let $\mathbf{v} \in \mathcal{V}$, where $\mathbf{v} \neq 0$, and let $\mathbf{A} \in \Lambda^p(\mathcal{V})$, and show that $\mathbf{v} \wedge \mathbf{A} = 0$ if and only if there exists $\mathbf{B} \in \Lambda^{p-1}(\mathcal{V})$ such that $\mathbf{A} = \mathbf{v} \wedge \mathbf{B}$.

4.1.6 Let $\mathbf{A} \in \Lambda^2(\mathcal{V})$ with components A^{ij}. Show that $\mathbf{A} \wedge \mathbf{A} = 0$ if and only if

$$A^{ij}A^{kl} - A^{ik}A^{jl} + A^{il}A^{jk} = 0$$

for all i, j, k, l in any basis.

4.1.7 Let $\{\mathbf{e}_1, \mathbf{e}_2, \mathbf{e}_3\}$ be *any* basis in $\mathbb{R}^3$. Define an operator $\mathbb{P}_{ijk}: \mathbb{R}^3 \to \mathbb{R}^3$ that permutes any set of three vectors, $\{\mathbf{v}_1, \mathbf{v}_2, \mathbf{v}_3\}$, into $\{\mathbf{v}_i, \mathbf{v}_j, \mathbf{v}_k\}$. Find the matrix representation of this operator and show that $\varepsilon_{ijk} = \det \mathbb{P}_{ijk}$.

4.1.8 Let $\{\mathbf{e}_i\}_{i=1}^N$ be an orthonormal basis with respect to the inner product $\mathbf{g}$.
(a) Show that the matrix g^{ij} is indentical to g_{ij}.
(b) If $\varepsilon^{i_1 \cdots i_N}$ is obtained from $\varepsilon_{i_1 i_2 \ldots i_N}$ (the Levi-Civita tensor) by raising the indices, then show that

$$\varepsilon^{12\ldots N} = (-1)^{n_-} \varepsilon_{12\ldots N} = (-1)^{n_-}$$

4.1.9 For a Euclidean three-dimensional space, find $\varepsilon^{ijk}\varepsilon_{lmn}$, $\varepsilon^{ijk}\varepsilon_{lmk}$, $\varepsilon^{ijk}\varepsilon_{ljk}$, and $\varepsilon^{ijk}\varepsilon_{ijk}$, where the sum over repeated indices is understood.

4.1.10 Use the result of Example 4.1.15 to find the sums specified in Exercise 4.1.9.

4.1.11 Let $\{\mathbf{e}_i\}_{i=1}^N$ be an orthonormal basis of $\mathcal{V}$ with respect to the inner product $\mathbf{g}$. Let $\eta_{ij} = \pm\delta^i_j$ be the matrix of $\mathbf{g}$ in this orthonormal basis. Let $\{\mathbf{v}_j\}_{j=1}^N$ be another (not necessarily orthonormal) basis of $\mathcal{V}$ with a transformation matrix R. Using G to denote the matrix of $\mathbf{g}$ in $\{\mathbf{v}_j\}$, show that

$$\det \mathsf{G} = \det \boldsymbol{\eta}\, (\det \mathsf{R})^2 = (-1)^{n_-} (\det \mathsf{R})^2$$

In particular, the *sign* of this determinant is invariant. Why is $\det \mathsf{G}$ not equal to $\det \boldsymbol{\eta}$?

4.1.12 In relativistic electromagnetic theory the current, $\mathbf{J}$, and the electromagnetic field tensor, $\mathbf{F}$, are, respectively, a four-vector and an antisymmetric tensor of rank 2. That is, $\mathbf{J} = J^i\mathbf{e}_i$ and $\mathbf{F} = F^{ij}\mathbf{e}_i \wedge \mathbf{e}_j$. Find the components of $*\mathbf{J}$ and $*\mathbf{F}$. Recall that $N = 4$ for this case.

4.1.13 For a tensor $\mathbf{S}^{\cdots i \cdots j \cdots}$ that is symmetric under interchange of indices i and j and a second tensor $\mathbf{A}_{\ldots k \ldots l \ldots}$ that is skew-symmetric under interchange of indices k and l, show that

$$S^{\ldots i \ldots j \ldots} A_{\ldots i \ldots j \ldots} = 0$$

4.1.14 The three components of $\mathbb{J}$, the angular momentum operator in quantum mechanics, satisfy these commutation relations (see Chapter 2):

$$[\mathbb{J}_i, \mathbb{J}_j] = i\varepsilon^k_{ij}\,\mathbb{J}_k$$

Show that $|\mathbf{J}|^2 = \mathbb{J}^k\mathbb{J}_k$ commutes with all three components.

4.1.15 A vector operator $\vec{\mathbb{V}}$ is defined as a set of three operators, $\{\mathbb{V}^1, \mathbb{V}^2, \mathbb{V}^3\}$, satisfying the following commutation relations with angular momentum:

$$[\mathbb{V}^i, \mathbb{J}^j] = i\,\varepsilon^{ijk}\,\mathbb{V}_k$$

Show that $|\mathbf{V}|^2 \equiv \mathbb{V}^k\mathbb{V}_k$ commutes with all components of angular momentum.

4.1.16 The Pauli spin matrices

$$\sigma^1 = \begin{pmatrix} 0 & 1 \\ 1 & 0 \end{pmatrix} \qquad \sigma^2 = \begin{pmatrix} 0 & -i \\ i & 0 \end{pmatrix} \qquad \sigma^3 = \begin{pmatrix} 1 & 0 \\ 0 & -1 \end{pmatrix}$$

describe a particle with spin $\frac{1}{2}$ in nonrelativistic quantum mechanics. It is easily verified that these matrices satisfy

$$[\sigma_i, \sigma_j] \equiv \sigma_i\sigma_j - \sigma_j\sigma_i = 2i\varepsilon_{ij}^k\sigma_k \tag{1}$$

and

$$\{\sigma_i, \sigma_j\} \equiv \sigma_i\sigma_j + \sigma_j\sigma_i = 2\delta_{ij}1_2 \tag{2}$$

where 1_2 is the unit 2×2 matrix. Show that

(a) $\sigma_i\sigma_j = i\varepsilon_{ij}^k\sigma_k + \delta_{ij}1_2$, and

(b) for any two vectors **a** and **b**, $(\boldsymbol{\sigma}\cdot\mathbf{a})(\boldsymbol{\sigma}\cdot\mathbf{b}) = \mathbf{a}\cdot\mathbf{b}\,1 + i\boldsymbol{\sigma}\cdot(\mathbf{a}\times\mathbf{b})$.

4.2 VECTORS ON MANIFOLDS

Tensor algebra deals with lifeless vectors and tensors—objects that do not move, do not change, and possess no dynamics. Whenever there is a need for tensors in physics, there is also a need to know the way those tensors change with position and time. Tensors that depend on position and time are called tensor fields and are the subject of this section.

In studying the algebra of tensors, we learned that tensors are generalizations of vectors. Once we have a vector space $\mathcal{V}$ and its dual space $\mathcal{V}^*$, we can define multilinear functions on their Cartesian product space and obtain tensors of various types. We can take the tensor products of various factors of $\mathcal{V}$ and $\mathcal{V}^*$ and create all sorts of tensors. Thus, once we know what a vector is, we can make up tensors from it.

In Section 4.1 we did not concern ourselves with what a vector is; we simply assumed that it exists. Because all the vectors considered there were stationary, their mere existence was enough. However, in tensor analysis, where things keep changing from point to point (and over time), the existence of vectors at one point does not guarantee their existence at all points. Therefore, we now have to demand more from vectors than their mere existence.

Another concept that requires clarification at this point is the notion of space, or space-time. Let us consider this first.

4.2.1 Differentiable Manifolds

Space is one of the undefinables in elementary physics. Length and time intervals are concepts that are "given," and any definitions of these concepts are circular. This is true as long as we are confined within a single space. In classical physics, this space is the three-dimensional Euclidean one in which all kinds of motions take place. In special relativity, space is changed to space-time, and Minkowski replaces Euclid. In nonrelativistic quantum mechanics, the underlying space is the (infinite-dimensional) Hilbert space, and time is the only dynamical parameter. In relativistic quantum field theory, the Hilbert space becomes a function of space and time. In the general theory of relativity, gravitation causes space to be curved.

Obviously, there are many types of spaces, which may look completely different. To assume that they are all undefinable would create a feudalistic environment in which each discipline has its own "space," defined within its own "territory." This is fine, except if one feudal lord tries to find out something about other territories; then there will be clashes and disruptions.

To prevent such conflicts mathematicians have invented a unifying theme that brings all the common factors of all the spaces together. This unifying theme is the theory of differentiable manifolds. A thorough understanding of differentiable manifolds requires familiarity with a multitude of advanced topics, including topological spaces, Banach spaces, algebraic topology, Lie groups, analysis, and so on. These are beyond the scope of this book. However, a *working* understanding of manifold theory is surprisingly simple. Here the word "working" is synonymous with "imprecise," "incomplete," and "intuitive." The following discussion aspires to neither completeness nor precision.

Let us begin with an intuitive definition of a differentiable manifold.

4.2.1 Definition A *differentiable manifold* is a collection of objects called *points* that are related to each other in a smooth fashion such that the neighborhood of each point looks like the neighborhood of an m-dimensional (Cartesian) space; m is called the dimension of the manifold.

As is customary in the literature, the word "manifold" is synonymous with "differentiable manifold" in this book.

Example 4.2.1

The following are examples of differentiable manifolds.

(a) The space $\mathbb{R}^n$ is an n-dimensional manifold.

(b) A sphere is a two-dimensional manifold.

(c) A torus is a two-dimensional manifold.

(d) The collection of all $n \times n$ real matrices whose elements are real functions having derivatives of all orders is an n^2-dimensional manifold. (Here a point is an $n \times n$ matrix!)

(e) The collection of all rotations in $\mathbb{R}^3$ is a three-dimensional manifold. (Here a point is a rotation.)

(f) Any smooth surface in $\mathbb{R}^3$ is a two-dimensional manifold.

(g) The unit n-sphere S^n, which is a collection of points in $\mathbb{R}^{n+1}$ satisfying $x_1^2 + \cdots + x_{n+1}^2 = 1$, is a manifold.

Any surface with sharp kinks, edges, or points cannot be a manifold. Thus, a cone is not a two-dimensional manifold because it is not smooth; it has a sharp point at the vertex. A closed cylinder is not a manifold because it has sharp edges. However, an infinitely long cylinder is a manifold. ●

The terminology used in the definition of a differentiable manifold may need explication. The points themselves are undefinable. Their collection forms a topological space having certain intersection and union properties on which we will not elaborate. For us the concept of neighborhood is important. A neighborhood of a point P is simply a collection of points that are "close enough" to P.

Let U_P denote a neighborhood of P. When we say that this neighborhood looks like an m-dimensional (Cartesian) space, we mean there exists a bijective map, $\varphi: U_P \to \mathbb{R}^m$, from the neighborhood U_P to a neighborhood $\varphi(U_P)$ in $\mathbb{R}^m$, such that as we move the point P continuously in U_P, its image moves *continuously* in $\varphi(U_P)$. Since $\varphi(P) \in \mathbb{R}^m$, we can define functions, $x^i: U_P \to \mathbb{R}$, such that $\varphi(P) = (x^1(P), x^2(P), \ldots, x^m(P)) \in \mathbb{R}^m$. These functions are called *coordinate functions* of φ. The $x^i(P)$ are called *coordinates* of P. The neighborhood U_P together with its mapping φ form a *chart*, denoted by (U_P, φ).

Let (V_P, μ) be another chart at P with coordinate functions $\mu(P) = (y^1(P), y^2(P), \ldots, y^m(P)) \in \mathbb{R}^m$ (see Fig. 4.1). Whenever V_P and U_P overlap, there must be a function $F: \varphi(U_P) \to \mu(V_P)$ with derivatives of *all orders*. We note that $\varphi(U_P)$ and $\mu(V_P)$ are subsets of $\mathbb{R}^m$, so derivatives of F are defined. The set of functions that have derivatives of all orders is denoted by C^∞. When a function such as F exists between two coordinate functions such as $\varphi(U_P)$ and $\mu(V_P)$, we say that the two charts are *C^∞-related*. Such a C^∞ relation defines what is meant by "smooth" in the definition of a manifold. A collection of charts that *cover the manifold* and of which each pair is C^∞-related is called a *C^∞ atlas*.

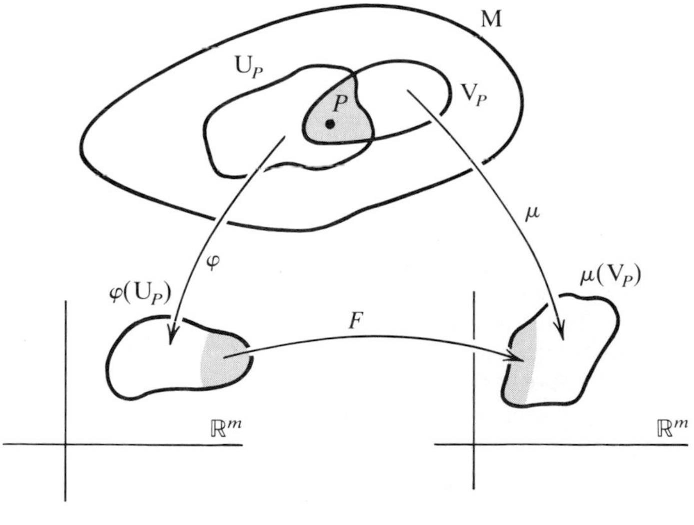

Figure 4.1 Two charts containing P, (U_P, φ) and (V_P, μ), are mapped into the Cartesian $\mathbb{R}^m$. The function F is simply an ordinary function from $\mathbb{R}^m$ to $\mathbb{R}^m$.

Example 4.2.2

For a two-dimensional unit sphere, S^2, we can construct a chart as follows. Let $P = (x_1, x_2, x_3)$, a point in $\mathbb{R}^3$, be a point in S^2. Then $x_1^2 + x_2^2 + x_3^2 = 1$, or

$$x_3 = \pm\sqrt{1 - x_1^2 - x_2^2}$$

The plus sign corresponds to the upper hemisphere, and the minus sign to the lower hemisphere. Let U^+ be the upper hemisphere. Then a chart (U^+, φ) with $\varphi : U^+ \to \mathbb{R}^2$ can be constructed by projection on the x^1x^2-plane:

$$\varphi(x^1, x^2, x^3) = (x^1, x^2)$$

Similarly, (U^-, μ) with $\mu : U^- \to \mathbb{R}^2$ given by $\mu(x^1, x^2, x^3) = (x^1, x^2)$ is a chart for the lower hemisphere.

In manifold theory the neighborhoods, on which mappings of charts are defined, usually have no boundaries. This is because it is more convenient to define limits on boundless (open) neighborhoods. Thus, in the above two charts the equator, which is the boundary for both hemispheres, must be excluded. With this exclusion U^+ and U^- cannot cover S^2; thus, they do not form an atlas. More charts are needed to cover S^2. Two such charts are the right and left hemispheres, for which $x_2 > 0$ and $x_2 < 0$, respectively. However, these two neighborhoods leave two points uncovered, the points (1,0,0) and $(-1,0,0)$. Again this is because boundaries of the right and left hemispheres must be excluded. Adding the front and back hemispheres to the collection covers these two points. Then S^2 is completely covered and we have an atlas.

There is, of course, a lot of overlap among charts, and it can be shown (but will not be here) that these overlaps are C^∞-related. •

Example 4.2.3

For S^2 of the preceding example, we can find a new chart in terms of parameters. If $x_1^2 + x_2^2 + x_3^2 = 1$, then we can write

$$x_1 = \sin\theta\cos\varphi \qquad x_2 = \sin\theta\sin\varphi \qquad x_3 = \cos\theta$$

A chart, then, is given by (S^2, μ) where

$$\mu(\sin\theta\cos\varphi,\ \sin\theta\sin\varphi,\ \cos\theta) = (\theta, \varphi)$$

maps a point of S^2 onto a region in $\mathbb{R}^2$. This is schematically shown in Fig. 4.2.

This chart cannot cover all of S^2, however, because when $\theta = 0$ (or π), the value of φ is not determined. In other words, $\theta = 0$ (or π) determines *one* point of the sphere (the north pole or the south pole), but its image in $\mathbb{R}^2$ is the whole φ-axis. Therefore, we must exclude $\theta = 0$ (or π) from the chart (S^2, μ). To cover these two points we need more charts. •

Example 4.2.4

A third chart for S^2 is the so-called stereographic projection shown in Figure 4.3. In such a mapping the image of a point is obtained by drawing a line from the north pole to that

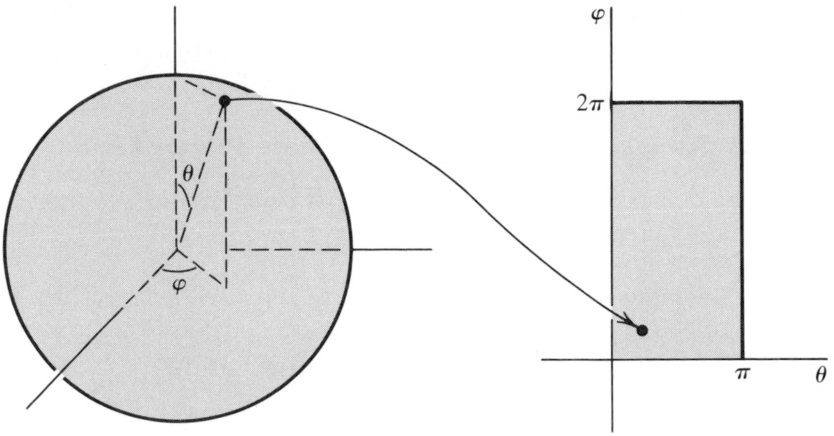

Figure 4.2 A chart mapping points of the sphere S^2 into the Cartesian $\mathbb{R}^2$. Note that the mapping is not defined for $\theta = 0$ and $\theta = \pi$, and therefore at least another chart is required to cover the sphere.

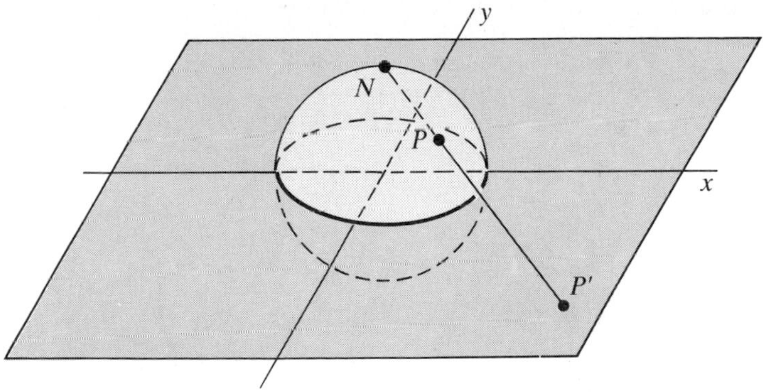

Figure 4.3 Stereographic projection of S^2 into $\mathbb{R}^2$. Note that the north pole does not have an image under this map, so another chart is needed to cover the whole sphere.

point and extending it, if necessary, until it intersects the x_1x_2-plane. It can be verified that the mapping $\varphi: S^2 \to \mathbb{R}^2$ is given by

$$\varphi(x_1, x_2, x_3) = \left(\frac{x_1}{1 - x_3}, \frac{x_2}{1 - x_3}\right)$$

We see that this mapping fails for $x_3 = 1$, that is, the north pole. Therefore, the north pole must be excluded. To cover the north pole we need another stereographic projection—this time from the south pole. Then the two mappings will cover all of S^2.

It is a well-known fact of differential geometry that it is impossible to cover the whole of S^2 with just one chart. ●

We will not dwell any further on the subject of manifolds. The interested reader can consult works such as those listed in the References (in particular, Abraham, Margden, and Ratiu 1983 and Bishop and Goldberg 1980).

4.2.2 Curves and Tangents; Coordinate Vector Fields

Having defined a manifold, we are almost ready for the definitions of vectors and tensors. As preparation, let us make the following definition.

4.2.2 Definition Let M and N be manifolds of m and n dimensions, respectively. Let $\Psi: M \to N$ be a mapping from M to N. We say that Ψ is C^∞ if for every chart (U, φ) in M and every chart (V, μ) in N, the composite function $\mu \circ \Psi \circ \varphi^{-1}: \mathbb{R}^m \to \mathbb{R}^n$ is C^∞ wherever it is defined.

The content of this definition is illustrated in Fig. 4.4. A particularly important special case occurs when $N = \mathbb{R}$; then we call Ψ a (real-valued) function. The collection of all C^∞ functions at a point $P \in M$ is denoted by $F^\infty(P)$. This means that if $f \in F^\infty(P)$, then $f: U_P \to \mathbb{R}$ is C^∞ for some neighborhood U_P of P.

Another special case of Definition 4.2.2 occurs when $M = \mathbb{R}$. This is so important that a separate definition is warranted.

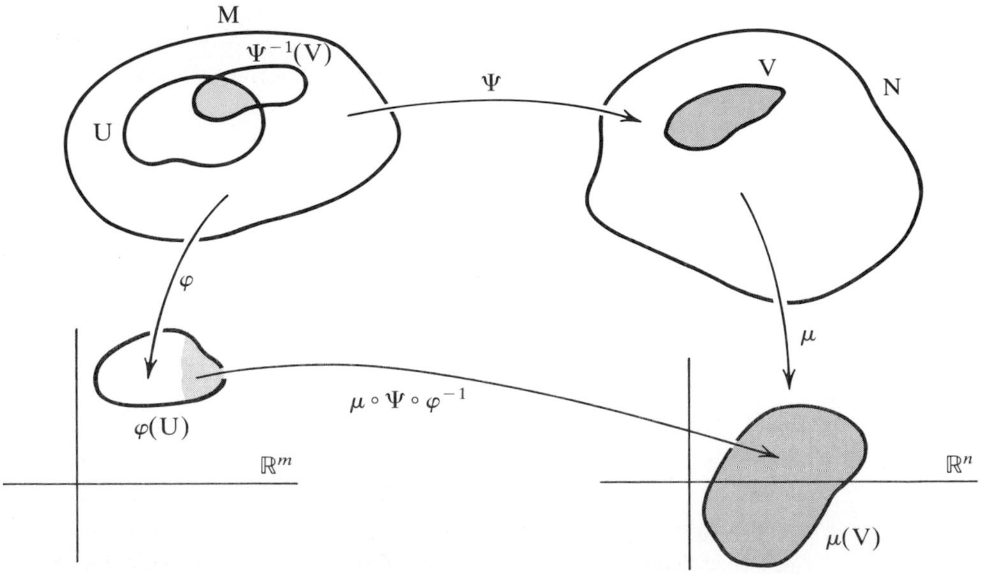

Figure 4.4 Corresponding to every map Ψ: $M \to N$, there exists a coordinate map, $\mu \circ \Psi \circ \varphi^{-1}: \mathbb{R}^m \to \mathbb{R}^n$. The latter is called the *coordinate expression for* Ψ.

4.2.3 Definition A *differentiable curve* is a C^∞ map of an interval of $\mathbb{R}$ to M.

Thus, if $\gamma : [a, b] \to M$ is C^∞, then γ is called a differentiable curve. This should be familiar from calculus where $M = \mathbb{R}^3$ and a curve is given by $(f_1(t), f_2(t), f_3(t))$, or simply by $\mathbf{r}(t)$. The point $\gamma(a) \in M$ is called the *initial point*, and $\gamma(b) \in M$ is called the *final point* of the curve γ. A curve is closed if $\gamma(a) = \gamma(b)$.

We are now ready to consider what a vector at a point is. All the familiar vectors in classical physics, such as position, velocity, momentum, angular momentum, and so forth, are based on the position vector. Let us see how we can generalize such a vector so that it is compatible with the concept of a manifold.

In $\mathbb{R}^2$ we define the position vector from P to Q as a directed straight line that starts at P and ends at Q. Furthermore, the direction of the vector remains the same if we connect P to any other final point on the line segment PQ. This is because $\mathbb{R}^2$ is a flat space, a straight line is well-defined, and there is no ambiguity in the direction of the vector from P to Q.

Things change, however, if we move to a two-dimensional spherical surface such as the globe. How do we define the straight line from New York to Beijing? There is no satisfactory definition of straightness on a curved surface. Furthermore, as we move from New York to Beijing, going westward, the tip of the arrow keeps changing direction. Its direction in New York is a little bit different from its direction in Chicago. In San Francisco the direction is changed even more, and when we reach Beijing, the tip of the arrow is almost opposite to its original direction.

The reason for such a changing arrow is, of course, the curvature of the manifold. We can minimize this curvature effect if we do not go too far from New York. If we stay close to New York, the surface of the earth appears flat, and we can draw arrows between points. The closer the two points, the better the approximation to flatness. Clearly, the concept of a vector is a *local* concept, and the process of constructing a vector is a *limiting* process.

The limiting process in the globe example entailed the notions of "closeness" and "farness." Such notions require the concept of distance, which is natural for a globe but not necessary for a general manifold. For most manifolds it is possible to define a distance function (metric), which gives the "distance" between two points of the manifold, even though the points happen to be matrices or rotation angles! However, the concept of a vector is too general to require such an elaborate structure as a metric. The abstract usefulness of a metric is a result of its real-valuedness: given two points, P_1 and P_2, the distance between them, $d(P_1, P_2)$, is a nonnegative real number. Thus, distances between different points can be compared, and decisions as to the largeness or smallness of these distances can be made.

We have already defined two concepts for manifolds (more basic than the concept of a metric) that, together, can replace the concept of a metric in defining a vector as a limit. These are the concepts of (real-valued) functions and curves. Let us see how functions and curves can replace metrics.

Let $\gamma : [a, b] \to M$ be a curve in the manifold M. Let $P \in M$ be a point of M that lies on the curve γ such that, for some $c \in [a, b]$, $\gamma(c) = P$. Let $f \in F^\infty(P)$ be a C^∞

function at P. Restrict f to the neighboring points of P that lie on the curve γ. Then the composite function $f \circ \gamma : \mathbb{R} \to \mathbb{R}$ is a real-valued function on $\mathbb{R}$.

We can compare values of $f \circ \gamma$ for various real numbers close to c, as in calculus. If $u \in [a, b]$ denotes the variable, then $f \circ \gamma(u) = f(\gamma(u))$ gives the value of $f \circ \gamma$ at various u's. In particular, the difference

$$\Delta(f \circ \gamma) \equiv f(\gamma(u)) - f(\gamma(c))$$

is a measure of how close the point $\gamma(u) \in \mathrm{M}$ is to P. Going one step further, we define

$$\left.\frac{d(f \circ \gamma)}{du}\right|_{u=c} \equiv \lim_{u \to c} \frac{f(\gamma(u)) - f(\gamma(c))}{u - c} \tag{4.19}$$

which is the ordinary derivative of an ordinary function of one variable. However, this derivative depends crucially on γ and on the point P. The function f is merely a *test function*. We could choose any other function to test how things change with movement along γ. What is important is not which function we choose, but how the curve γ causes it to change with movement along γ away from P. This change is determined by the *directional derivative along γ at P*, as given by (4.19). That is why the tangent vector at P along γ is defined to be the directional derivative itself!

The use of a derivative as tangent vector, although new in the physics community, has been familiar to mathematicians for a long time. It is hard for physicists to imagine vectors being charged with the responsibility of measuring the rate of change of functions. It takes some mental reconstruction to get used to this idea. The following simple illustration may help with establishing the vector-derivative connection.

Let us take the familiar case of a plane and consider the vector $\mathbf{a} = a_x \hat{\mathbf{e}}_x + a_y \hat{\mathbf{e}}_y$. How can we associate a directional derivative with $\mathbf{a}$? First we need a curve, $\gamma : \mathbb{R} \to \mathbb{R}^2$, that is somehow related to $\mathbf{a}$. This is easy to construct. We take the straight line along $\mathbf{a}$; that is, we let $\gamma(u) = (a_x u, a_y u)$. This is the curve that passes through the origin of $\mathbb{R}^2$ (the tail of $\mathbf{a}$) for $u = 0$ and through the point (a_x, a_y) (the tip of $\mathbf{a}$) for $u = 1$. The directional derivative at the origin for an arbitrary function $f : \mathbb{R}^2 \to \mathbb{R}$ is given by

$$\begin{aligned} \left.\frac{d(f \circ \gamma)}{du}\right|_{u=0} &= \lim_{u \to 0} \frac{f(\gamma(u)) - f(\gamma(0))}{u} \\ &= \lim_{u \to 0} \frac{f(a_x u, a_y u) - f(0, 0)}{u} \end{aligned} \tag{4.20}$$

Taylor expansion in two dimensions yields

$$f(x, y) = f(0, 0) + x \left.\frac{\partial f}{\partial x}\right|_{\substack{x=0 \\ y=0}} + y \left.\frac{\partial f}{\partial y}\right|_{\substack{x=0 \\ y=0}} + \cdots$$

which for $x = a_x u$ and $y = a_y u$ gives

$$f(a_x u, a_y u) = f(0, 0) + a_x u \left.\frac{\partial f}{\partial x}\right|_{u=0} + a_y u \left.\frac{\partial f}{\partial y}\right|_{u=0} + \cdots$$

Substituting in (4.20), we obtain

$$\left.\frac{d(f\circ\gamma)}{du}\right|_{u=0} = \lim_{u\to 0}\frac{a_x u(\partial f/\partial x)_{u=0} + a_y u(\partial f/\partial y)_{u=0} + \cdots}{u}$$

$$= a_x\frac{\partial f}{\partial x} + a_y\frac{\partial f}{\partial y} + 0$$

$$= \left(a_x\frac{\partial}{\partial x} + a_y\frac{\partial}{\partial y}\right) f$$

This clearly shows the connection between directional derivatives and vectors. In fact, the correspondences $\partial/\partial x \leftrightarrow \hat{\mathbf{e}}_x$ and $\partial/\partial y \leftrightarrow \hat{\mathbf{e}}_y$ establish this connection very naturally.

Note that the curve γ chosen above is by no means unique. In fact, there are infinitely many curves that have the same tangent at the origin and give the same directional derivative.

Now we are ready to define a tangent.

4.2.4 Definition Let M be a differentiable manifold. A *tangent* at $P \in$ M is an operator, $\mathfrak{t}:\mathrm{F}^\infty(P) \to \mathbb{R}$, such that for every $f, g \in \mathrm{F}^\infty(P)$ and $a, b \in \mathbb{R}$

(i) $\mathfrak{t}$ is linear:

$$\mathfrak{t}(af + bg) = a\mathfrak{t}(f) + b\mathfrak{t}(g)$$

and

(ii) $\mathfrak{t}$ satisfies the *derivation property*:

$$\mathfrak{t}(fg) = (\mathfrak{t}(f))g(P) + f(P)(\mathfrak{t}(g))$$

The operator $\mathfrak{t}$ is an abstraction of the derivative operator. Note that $\mathfrak{t}(f)$, $g(P)$, $f(P)$, and $\mathfrak{t}(g)$ are all real numbers.

If addition and scalar multiplication of directional derivatives are defined in an obvious way, the set of all tangents at $P \in$ M becomes a vector space called the *tangent space* at P and denoted by $\mathcal{T}_P(\mathrm{M})$.

Definition 4.2.4 was motivated by Eqs. (4.19) and (4.20). Let us go backwards and see if (4.19) is indeed a tangent, that is, if it satisfies the two conditions of the definition.

4.2.5 Proposition Let γ be a C^∞ curve in M such that $\gamma(c) = P$. Define $\boldsymbol{\gamma}_*(c)$ by requiring, for every $f \in \mathrm{F}^\infty(\mathrm{P})$, that

$$(\boldsymbol{\gamma}_*(c))(f) \equiv \left.\frac{d}{du} f\circ\gamma\right|_{u=c}$$

Then $\boldsymbol{\gamma}_*(c) \in \mathcal{T}_P(M)$. That is, $\boldsymbol{\gamma}_*(c)$ is a tangent at P called the *tangent to* γ *at* c.

Proof. We have to show that conditions (i) and (ii) of Definition 4.2.4 are satisfied for $f, g \in \mathrm{F}^\infty(P)$ and $a, b \in \mathbb{R}$.

$$\text{(i)}\ (\boldsymbol{\gamma}_*(c))(af + bg) = \frac{d}{du}(af + bg)\circ\gamma\bigg|_{u=c} = \frac{d}{du}(af\circ\gamma + bg\circ\gamma)\bigg|_{u=c}$$

$$= a\frac{d}{du}f\circ\gamma\bigg|_{u=c} + b\frac{d}{du}g\circ\gamma\bigg|_{u=c}$$

$$= a(\boldsymbol{\gamma}_*(c))(f) + b(\boldsymbol{\gamma}_*(c))(g)$$

$$\text{(ii)}\ (\boldsymbol{\gamma}_*(c))(fg) = \frac{d}{du}(fg)\circ\gamma\bigg|_{u=c} \equiv \frac{d}{du}[(f\circ\gamma)(g\circ\gamma)]\bigg|_{u=c}$$

$$= \left[\frac{d}{du}(f\circ\gamma)\bigg|_{u=c}\right](g\circ\gamma)_{u=c} + (f\circ\gamma)_{u=c}\left[\frac{d}{du}(g\circ\gamma)\bigg|_{u=c}\right]$$

$$= [(\boldsymbol{\gamma}_*(c))(f)]g(\gamma(c)) + f(\gamma(c))[(\boldsymbol{\gamma}_*(c))(f)]$$

$$= [(\boldsymbol{\gamma}_*(c))(f)]g(P) + f(P)[(\boldsymbol{\gamma}_*(c))(f)]$$

Note that in going from the second equality to the third in (ii), we used the chain rule for the ordinary derivative. Also, to get from the first equality to the second, we used the fact that, by definition, the product of two functions evaluated at a point is the product of the *values* of the two functions at that point. Thus, conditions (i) and (ii) are satisfied, and $\boldsymbol{\gamma}_*(c)$ is indeed a tangent at P. ■

Let us now consider a special curve and corresponding tangent vector that are of extreme importance in applications.

Let $\varphi = (x^1, x^2, \ldots, x^m)$ be a coordinate system at P, where $x^i : \mathrm{M} \to \mathbb{R}$ is the ith coordinate function; φ is a bijective C^∞ mapping from the manifold M onto $\mathbb{R}^m$. Its inverse, φ^{-1}, then, is also a C^∞ mapping from $\mathbb{R}^m$ to M. We, therefore, write $\varphi^{-1} : \mathbb{R}^m \to \mathrm{M}$. Now, for a point $P \in \mathrm{M}$ with a coordinate system in a neighborhood of P, the jth coordinate of P is the real number $x^j(P)$. Suppose that all coordinates of P are fixed except the ith one, which is allowed to vary. Let u describe this variation; thus $x^i(P) = u \in \mathbb{R}$. We now have the following definition.

4.2.6 Definition Let $P \in \mathrm{M}$ and (U_P, φ) be a chart at P. Then the curve $\gamma_i : \mathbb{R} \to \mathrm{M}$, defined by

$$\gamma_i(u) = \varphi^{-1}(x^1(P), \ldots, x^{i-1}(P), u, x^{i+1}(P), \ldots, x^m(P))$$

is called the ith *coordinate curve through* P. The tangent to this curve at P is denoted $\partial_i(P)$ and is called the ith *coordinate vector field.*

Note that $\partial_i(P)$ does not imply any operation on P; it means *evaluation* at P. That is, $\partial_i(P)$ means operate on functions first, then evaluate the result at P.

Let $c = x^i(P)$. Then for $f \in \mathrm{F}^\infty(P)$, we have

$$(\partial_i(P))f = (\mathbf{\gamma}_{i*}(c))f = \frac{d}{du} f \circ \gamma_i \bigg|_{u=c} = \frac{d}{du} f(\gamma_i(u)) \bigg|_{u=c}$$

$$= \frac{d}{du} f(\varphi^{-1}(x^1(P), \ldots, u, \ldots, x^m(P))) \bigg|_{u=c}$$

$$\equiv \frac{\partial f}{\partial x^i}\bigg|_P$$

where the last equality is a (natural) definition of the partial derivative of f with respect to the ith coordinate and evaluated at the point P. This partial derivative is again a C^∞ function on M. We, therefore, have the following proposition.

4.2.7 Proposition The coordinate vector fields at P, denoted $\partial_i(P)$, where $i = 1, 2, \ldots, m$, are operators, $\partial_i(P): \mathrm{F}^\infty(P) \to \mathrm{F}^\infty(P)$ given by

$$(\partial_i(P))f = \frac{\partial f}{\partial x^i}\bigg|_P \qquad \text{for } i = 1, 2, \ldots, m;\ f \in \mathrm{F}^\infty(P) \tag{4.21}$$

which are the partial derivatives at P. A common notation for $\partial f/\partial x^i$ is $f_{,i}$. ■

Note that this proposition does not contradict Definition 4.2.4, in which a tangent $\mathfrak{t}$ maps $\mathrm{F}^\infty(P)$ onto real numbers. In Eq. (4.21), $\partial f/\partial x^i|_P$ *is* a real number, but the equation goes beyond it and gives a real-valued function. The reason for the use of the word "field" in the above discussion will be explained shortly.

Example 4.2.5

Let us pick a point $P = (\sin\theta\cos\varphi, \sin\theta\sin\varphi, \cos\theta)$ on the sphere S^2 in a chart (U_P, μ) given by

$$\mu(\sin\theta\cos\varphi, \sin\theta\sin\varphi, \cos\theta) = (\theta, \varphi)$$

If θ is kept constant and φ is allowed to vary over values given by u, then the coordinate curve associated with φ is given by

$$\gamma_\varphi(u) = \mu^{-1}(\theta, u) = (\sin\theta\cos u, \sin\theta\sin u, \cos\theta)$$

As u varies, $\gamma_\varphi(u)$ describes a curve on S^2. This curve is simply a circle of radius $\sin\theta$.

The tangent to this curve at any point is given by $\partial/\partial\varphi$, or simply ∂_φ, the derivative with respect to the coordinate φ.

Similarly, the curve $\gamma_\theta(u)$ describes a great circle on S^2 with tangent $\partial_\theta \equiv \partial/\partial\theta$. ●

The vector space $\mathcal{T}_P(\mathrm{M})$ of all tangents at P was mentioned earlier. In the case of S^2 this tangent space is simply a plane tangent to the sphere at a point. Also, the two vectors encountered above, ∂_θ and ∂_φ in this plane, are clearly linearly independent. Thus, they form a basis for the tangent plane. This argument can be generalized to any manifold. The following theorem is such a generalization (for a proof, see Bishop and Goldberg 1980).

4.2.8 Theorem Let M be a manifold and $P \in \mathrm{M}$. Then the set $\{\partial_i(P)\}_{i=1}^m$ forms a basis of $\mathscr{T}_P(\mathrm{M})$. In particular, $\mathscr{T}_P(\mathrm{M})$ is m-dimensional. An arbitrary vector, $\mathfrak{t} \in \mathscr{T}_P(\mathrm{M})$, can be written (using Einstein's summation convention) as

$$\mathfrak{t} = a^i \partial_i(P) \qquad \text{where } a^i = \mathfrak{t}(x^i)$$ ■

See Exercise 4.2.1 for the last statement of this theorem.

Suppose we have two coordinate systems at P: (x^i) with tangents $\partial_i(P)$ and (y^j) with tangents $\nabla_j(P)$. Let $\mathfrak{t} \in \mathscr{T}_P(\mathrm{M})$. Then $\mathfrak{t}$ can be expressed either in terms of ∂_i or in terms of ∇_j:

$$\mathfrak{t} = a^i \partial_i(P) = b^j \nabla_j(P)$$

We can use this relation to obtain a^i in terms of b^j. From Theorem 4.2.8 we have

$$\begin{aligned} a^i &= \mathfrak{t}(x^i) = (b^j \nabla_j(P))(x^i) = \left[b^j \frac{\partial}{\partial y^j}(P) \right](x^i) \\ &= b^j \left. \frac{\partial x^i}{\partial y^j} \right|_P \end{aligned} \tag{4.22}$$

In particular, if $\mathfrak{t} = \nabla_k(P)$, then $b^j = \mathfrak{t}(y^j) = [\nabla_k(P)](y^j) = \delta^j_k$, and (4.22) gives

$$a^i = \delta^j_k \frac{\partial x^i}{\partial y^j} = \frac{\partial x^i}{\partial y^k}$$

Thus,

$$\frac{\partial}{\partial y^k}(P) = a^i \partial_i(P) = \frac{\partial x^i}{\partial y^k} \frac{\partial}{\partial x^i}(P) \tag{4.23}$$

This is the chain rule for manifolds. Remember that $\partial/\partial x^i(P)$ means evaluate the derivative at P. Also note that the index i is summed over in (4.23). For any function $f \in \mathrm{F}^\infty(P)$, Eq. (4.23) yields

$$\left[\frac{\partial}{\partial y^k}(P) \right] f = \left. \frac{\partial f}{\partial y^k} \right|_P = \left. \frac{\partial x^i}{\partial y^k} \right|_P \left[\frac{\partial}{\partial x^i}(P) \right] f = \left. \frac{\partial x^i}{\partial y^k} \right|_P \left. \frac{\partial f}{\partial x^i} \right|_P$$

This clearly shows the chain rule in differentiation.

4.2.3 Differential of a Map

Now that we have constructed tangent spaces and defined bases for them, we are ready to consider the notion of the differential (derivative) of a map between manifolds. This notion will be used to construct differential forms, which are important in tensor analysis.

4.2.9 Definition Let M and N be manifolds of dimensions m and n, respectively, and let $\Psi: \mathrm{M} \to \mathrm{N}$ be a C^∞ map. Let $P \in \mathrm{M}$, and let $Q = \Psi(P) \in \mathrm{N}$ be the image of P. Let $\mathcal{T}_P(\mathrm{M})$ and $\mathcal{T}_Q(\mathrm{N})$ be the tangent spaces at P and Q, respectively. Then there is induced a map, $\Psi_{*P}: \mathcal{T}_P(\mathrm{M}) \to \mathcal{T}_Q(\mathrm{N})$, called the *differential of* Ψ *at* P and given as follows. Let $\mathfrak{t} \in \mathcal{T}_P(\mathrm{M})$ and $f \in F^\infty(Q)$; that is, f is a C^∞ function defined in a neighborhood of Q in N. It is clear that $f \circ \Psi$ is a C^∞ function on M. Now the action of $\Psi_{*P}(\mathfrak{t}) \in \mathcal{T}_Q(\mathrm{N})$ on f is defined as

$$(\Psi_{*P}(\mathfrak{t}))(f) \equiv \mathfrak{t}(f \circ \Psi) \tag{4.24}$$

Since $f \circ \Psi \in F^\infty(P)$, we know how $\mathfrak{t}$ acts on it to give a real number. Thus, the RHS of (4.24) is a well-defined real number. On the LHS f is a real-valued function on N and $\Psi_{*P}(\mathfrak{t})$ is a tangent vector at the image point of P, whose action on f is uniquely given by the RHS.

Let us see how (4.24) looks in terms of coordinate functions. Suppose that x^i, for $i = 1, 2, \ldots, m$, are coordinates at P and y^α, for $\alpha = 1, 2, \ldots, n$, are coordinates at $Q = \Psi(P)$. We note that $y^\alpha \circ \Psi$ is a real-valued C^∞ function on M. Thus, we may write, with the function expressed in terms of coordinates,

$$y^\alpha \circ \Psi = f^\alpha(x^1, x^2, \ldots, x^m)$$

Let us write $\mathfrak{t} = \sum_{i=1}^m a^i \partial_i(P)$. Similarly,

$$\Psi_{*P}(\mathfrak{t}) = \sum_{\alpha=1}^n b^\alpha \frac{\partial}{\partial y^\alpha}(Q)$$

because $\Psi_{*P}(\mathfrak{t}) \in \mathcal{T}_Q(N)$ and $\{\partial/\partial y^\alpha\}_{\alpha=1}^n$ is a basis of $\mathcal{T}_Q(\mathrm{N})$. Theorem 4.2.8 and Definition 4.2.9 now give

$$\begin{aligned} b^\alpha &= [\Psi_{*P}(\mathfrak{t})](y^\alpha) = \mathfrak{t}(y^\alpha \circ \Psi) = \mathfrak{t}(f^\alpha) \\ &= \left[\sum_{i=1}^m a^i \partial_i(P)\right](f^\alpha) = \sum_{i=1}^m a^i \left.\frac{\partial f^\alpha}{\partial x^i}\right|_P \end{aligned}$$

This can be written in matrix form as follows:

$$\begin{pmatrix} b^1 \\ b^2 \\ \vdots \\ b^n \end{pmatrix} = \begin{pmatrix} \partial f^1/\partial x^1 & \partial f^1/\partial x^2 & \cdots & \partial f^1/\partial x^m \\ \partial f^2/\partial x^1 & \partial f^2/\partial x^2 & \cdots & \partial f^2/\partial x^m \\ \vdots & \vdots & & \vdots \\ \partial f^n/\partial x^1 & \partial f^n/\partial x^2 & \cdots & \partial f^n/\partial x^m \end{pmatrix} \begin{pmatrix} a^1 \\ a^2 \\ \vdots \\ a^m \end{pmatrix} \tag{4.25}$$

The $n \times m$ matrix is denoted by J and is called the *Jacobian matrix of* Ψ with respect to the coordinates x^i and y^α.

Two special cases that merit closer attention are those where $\mathrm{M} = \mathbb{R}$ and where $\mathrm{N} = \mathbb{R}$. In either case $\mathcal{T}_c(\mathbb{R})$ is one-dimensional with the basis vector $\mathbf{d}/\mathbf{d}u(c)$.

When M $=\mathbb{R}$, the mapping becomes a curve, $\gamma:\mathbb{R}\to$ N. The only vector whose image we are interested in is $\mathbf{t}=\mathbf{d}/\mathbf{d}u(c)$, where $\gamma(c)=P$. From (4.24) and using Proposition 4.2.5 in the last step, we have

$$\left[\gamma_{*c}\frac{\mathbf{d}}{\mathbf{d}u}(c)\right]f=\frac{d}{du}f\circ\gamma\bigg|_{u=c}=[\boldsymbol{\gamma}_*(c)](f)$$

This tells us that the differential of a curve at c is simply its tangent vector at $\gamma(c)$.

When N $=\mathbb{R}$, we are dealing with a real-valued function, f: M $\to\mathbb{R}$. The differential of f at P is $f_*:\mathcal{T}_P(\mathrm{M})\to\mathcal{T}_c(\mathbb{R})$, where $c=f(p)$. Since $\mathcal{T}_c(\mathbb{R})$ is one-dimensional, for a tangent, $\mathbf{t}\in\mathcal{T}_P(\mathrm{M})$, we have

$$f_*(\mathbf{t})=a\frac{\mathbf{d}}{\mathbf{d}u}(c) \tag{4.26a}$$

Let $g:\mathbb{R}\to\mathbb{R}$ be an arbitrary function on $\mathbb{R}$. Then

$$[f_*(\mathbf{t})](g)=a\frac{dg}{du}\bigg|_c$$

The LHS gives $[f_*(\mathbf{t})](g)=\mathbf{t}(f\circ g)$. Thus, we have

$$\mathbf{t}(f\circ g)=a\frac{dg}{du}$$

To find a we set $g(u)=u$, which is the identity function; then $dg/du=1$ and $\mathbf{t}(f\circ g)=\mathbf{t}(f)=a$. Substituting in (4.26a), we obtain

$$f_*(\mathbf{t})=\mathbf{t}(f)\frac{\mathbf{d}}{\mathbf{d}u}(c) \tag{4.26b}$$

Since $\mathcal{T}_c(\mathbb{R})$ is one-dimensional, there is no need to write $\mathbf{d}/\mathbf{d}u(c)$. Thus, we define the *differential of f*, denoted by $\mathbf{d}f\equiv f_*$, as a map, $\mathbf{d}f:\mathcal{T}_P(\mathrm{M})\to\mathbb{R}$ given by

$$(\mathbf{d}f)(\mathbf{t})=\mathbf{t}(f) \tag{4.27}$$

In particular, if f is the coordinate function x^i and $\mathbf{t}$ is the tangent to the jth coordinate curve $\partial_i(P)$, we obtain

$$\mathbf{d}x^i(\partial_j(P))=\partial_j x^i(P)=\delta^i_j \tag{4.28}$$

This shows that $\{\mathbf{d}x^i\}_{i=1}^m$ is dual to the basis $\{\partial_j(P)\}_{j=1}^m$ of $\mathcal{T}_P(\mathrm{M})$.

Example 4.2.6

Let f: M $\to\mathbb{R}$ be a real-valued function on M. Let x^i be coordinates at P. Let us express $\mathbf{d}f$ in terms of coordinate functions. For $\mathbf{t}\in\mathcal{T}_P(\mathrm{M})$ we can write

$$\mathbf{t}=\sum_{i=1}^m a^i\partial_i(P)$$

and
$$\mathbf{d}f(\mathfrak{t}) = \mathfrak{t}(f) = \sum_{i=1}^{m} a^i[\boldsymbol{\partial}_i(P)](f) = \sum_{i=1}^{m} a^i \partial_i f$$

But from Theorem 4.2.8 $a^i = \mathfrak{t}(x^i)$. Thus, Eq. (4.27) gives

$$a^i = \mathfrak{t}(x^i) = (\mathbf{d}x^i)(\mathfrak{t})$$

We thus have

$$(\mathbf{d}f)(\mathfrak{t}) = \sum_{i=1}^{m} (\partial_i f)[(\mathbf{d}x^i)(\mathfrak{t})]$$

Since this is true for all $\mathfrak{t} \in \mathcal{T}_P(\mathrm{M})$, we get

$$\mathbf{d}f = \sum_{i=1}^{m} (\partial_i f)\mathbf{d}x^i$$

This is the classical formula for the differential of a function f. •

Exercise

4.2.1 Evaluate $\partial_i x^j$ and show that if $\mathfrak{t} = \sum a^i \boldsymbol{\partial}_i(P)$ is a tangent at P, then

$$a^i = \mathfrak{t}(x^i) \qquad \text{and} \qquad \mathfrak{t} = \sum_{i=1}^{m} (\mathfrak{t}x^i)\boldsymbol{\partial}_i(P)$$

4.3 TENSOR ANALYSIS ON MANIFOLDS

So far we have studied vector spaces, learned how to construct tensors out of vectors, touched on manifolds (the abstraction of spaces), seen how to construct vectors *at a single point* in a manifold by the use of the tangent-at-a-curve idea, and even found the dual vectors $\mathbf{d}x^i$ to the coordinate vectors $\boldsymbol{\partial}_i(P)$ at the point P of a manifold. We have everything we need to study the analysis of tensors.

4.3.1 Vector Fields

We want to find out what a vector field is in a manifold. To begin with, let us consider the following definition.

4.3.1 Definition The collection (union) of all tangents at different points of a manifold M is denoted by T(M) and called the *tangent bundle* of M:

$$\mathrm{T(M)} = \bigcup_{P \in \mathrm{M}} \mathcal{T}_P(\mathrm{M})$$

It can be shown that T(M) is a *manifold* of dimension $2m$, with $m = \dim \mathrm{M}$.

A vector field can now be defined.

4.3.2 Definition A *vector field* $\mathbf{X}$ on a subset U of a manifold M is a mapping, $\mathbf{X}:\mathrm{U} \to \mathrm{T(M)}$, such that $\mathbf{X}(P) \in \mathcal{T}_P(\mathrm{M})$. The domain of $\mathbf{X}$ is U, and its range space is T(M).

Simply stated, a vector field on U is a machine on U that gives a tangent vector when you feed it a point of U.

In terms of the coordinates x^i, at each point $P \in \mathrm{M}$ we have

$$\mathbf{X}(P) = X_P^i \partial_i(P)$$

where the real numbers X_P^i are components of $\mathbf{X}(P)$ in the basis $\{\partial_i(P)\}$. As P moves around in U, the real number X_P^i keeps changing. Thus, we can think of X_P^i as a function of P and define the real-valued function $X^i:\mathrm{M} \to \mathbb{R}$ by $X^i(P) \equiv X_P^i$. Thus, the *components of a vector field are real-valued functions on M.*

Example 4.3.1

Let $\mathrm{M} = \mathbb{R}^3$. At each point $P = (x, y, z) \in \mathbb{R}^3$, let $\{\hat{\mathbf{e}}_x, \hat{\mathbf{e}}_y, \hat{\mathbf{e}}_z\}$ be a basis for $\mathbb{R}^3$. Let $\mathcal{V}_P$ be the vector space at P. Then $\mathrm{T}(\mathbb{R}^3)$ is the collection of all vector spaces $\mathcal{V}_P$ for all $P \in \mathbb{R}^3$.

We can determine the value of an electric field at a point in $\mathbb{R}^3$ by first specifying the point, as (x_0, y_0, z_0), for example. This uniquely determines the tangent space $\mathcal{T}_{(x_0, y_0, z_0)}(\mathbb{R}^3)$. Once we have the vector space, we can ask what the components of the electric field are in that space. These components are given by three numbers: $E_x(x_0, y_0, z_0)$, $E_y(x_0, y_0, z_0)$, and $E_z(x_0, y_0, z_0)$. The argument is the same for any other vector field.

To specify a "point" in $\mathrm{T}(\mathbb{R}^3)$, we need three numbers to determine the location in $\mathbb{R}^3$ and another three numbers to determine the components of a vector field at that point. Thus, a "point" in $\mathrm{T}(\mathbb{R}^3)$ is given by six "coordinates" (x, y, z, E_x, E_y, E_z), and $\mathrm{T}(\mathbb{R}^3)$ is a six-dimensional manifold. ●

We know how a vector (tangent), $\mathbf{t}$, at a point $P \in \mathrm{M}$ acts on a function, $f \in \mathrm{F}^\infty(P)$, to give a real number, $\mathbf{t}(f)$. We can extend this, point by point, for a vector field $\mathbf{X}$ and define a function $\mathbf{X}(f)$ by

$$[\mathbf{X}(f)](P) \equiv [\mathbf{X}(P)](f) \qquad \forall P \in \mathrm{U} \tag{4.29}$$

where U is a subset of M on which both $\mathbf{X}$ and f are defined. The RHS is well-defined because we know how $\mathbf{X}(P)$, the *vector at P*, acts on functions at P to give the real number $[\mathbf{X}(P)](f)$. On the LHS, on the other hand, we have $\mathbf{X}(f)$, which maps the point P onto a real *number*. Thus, $\mathbf{X}(f)$ is indeed a real-valued function on M. We can, therefore, define vector fields directly as operators on C^∞ functions satisfying both

$$\mathbf{X}(af + bg) = a\mathbf{X}(f) + b\mathbf{X}(g)$$

and

$$\mathbf{X}(fg) = [\mathbf{X}(f)]g + f\mathbf{X}(g)$$

A prototypical vector field is the coordinate vector field ∂_i.

In general, $\mathbf{X}(f)$ is not a C^∞ function even if f is. A vector field that produces a C^∞ function $\mathbf{X}(f)$ for every C^∞ function f is called a C^∞ *vector field*. Such a vector field has components that are C^∞ functions on M.

4.3.2 One-Forms and Tensor Fields

We have defined vector spaces, $\mathcal{T}_P(\mathrm{M})$, at each point of M. We have also constructed coordinate bases, $\{\partial_i(P)\}_{i=1}^m$, for these vector spaces. At the end of Section 4.2, we showed that the differentials $\{\mathbf{d}x^i\}_{i=1}^m$ form a basis that is dual to $\{\partial_i(P)\}$. Let us concentrate on this dual space, which we will denote by $\mathcal{T}_P^*(\mathrm{M})$.

Taking the union of all $\mathcal{T}_P^*(\mathrm{M})$ at all points of M, we obtain the *cotangent bundle* of M:

$$\mathrm{T}^*(\mathrm{M}) = \bigcup_{P \in \mathrm{M}} \mathcal{T}_P^*(\mathrm{M}) \tag{4.30}$$

This is the dual space of T(M) at each point of M. We can now define the analogue of the vector field for the cotangent bundle.

4.3.3 Definition A *one-form* $\boldsymbol{\omega}$ on a subset U of a manifold M is a mapping $\boldsymbol{\omega}: \mathrm{U} \to \mathrm{T}^*(\mathrm{M})$ such that $\boldsymbol{\omega}(P) \in \mathcal{T}_P^*(\mathrm{M})$.

Thus, a one-form is a "function" that, when given a point $P \in \mathrm{M}$, produces a dual vector $\boldsymbol{\omega}(P) \in \mathcal{T}_P^*(\mathrm{M})$.

If $\boldsymbol{\omega}$ is a one-form and $\mathbf{X}$ is a vector field on M, then $\boldsymbol{\omega}(\mathbf{X})$ is a real-valued function on M defined, naturally, by

$$[\boldsymbol{\omega}(\mathbf{X})](P) = \underbrace{\underset{\substack{\uparrow \\ \text{a linear functional} \\ \text{at } P}}{[\boldsymbol{\omega}(P)]}\underset{\substack{\uparrow \\ \text{a vector} \\ \text{at } P}}{[\mathbf{X}(P)]}}_{\text{a real number at } P}$$

A prototypical one-form is the coordinate differential, $\mathbf{d}x^i$.

In complete analogy to the case of vector fields, $\boldsymbol{\omega}$ can be written in terms of the basis $\{\mathbf{d}x^i\}$:

$$\boldsymbol{\omega} = \omega_i \mathbf{d}x^i$$

Here ω_i, the components of $\boldsymbol{\omega}$, are real-valued functions on M.

With the vector spaces $\mathcal{T}_P(\mathrm{M})$ and $\mathcal{T}_P^*(\mathrm{M})$ at our disposal, we can construct all kinds of tensors at each point P. The union of all these tensors is called the bundle of tensors, and a tensor field can be defined as usual. Thus, we have the following definition.

4.3.4 Definition Let $\mathcal{T}_P(\mathrm{M})$ and $\mathcal{T}_P^*(\mathrm{M})$ be the tangent and cotangent spaces at $P \in \mathrm{M}$. Then the set of tensors of type (r, s) on $\mathcal{T}_P(\mathrm{M})$ is denoted by $\mathcal{T}^r_{P,s}(\mathrm{M})$. The *bundle of tensors of type* (r, s) *over* M, denoted by $\mathrm{T}^r_s(\mathrm{M})$, is

$$\mathrm{T}^r_s(\mathrm{M}) = \bigcup_{P \in \mathrm{M}} \mathcal{T}^r_{P,s}(\mathrm{M})$$

A *tensor field* $\mathbf{T}^r_s$ *of type* (r, s) over a subset U of M is a mapping $\mathbf{T}^r_s: \mathrm{U} \to \mathrm{T}^r_s(\mathrm{M})$ such that $\mathbf{T}^r_s(P) \in \mathcal{T}^r_{P,s}(\mathrm{M})$.

In particular, if $r = s = 0$, we are dealing with real-valued functions. For $r = 1$ and $s = 0$, we have $\mathrm{T}^1_0(\mathrm{M}) = \mathrm{T}(\mathrm{M})$, and the tensor field is simply a vector field. For $r = 0$ and $s = 1$, $\mathrm{T}^0_1(\mathrm{M}) = \mathrm{T}^*(\mathrm{M})$, and the tensor field is simply a one-form.

The *components of* $\mathbf{T}^r_s$ *with respect to coordinates* x^i are the m^{r+s} real-valued functions

$$T^{i_1 i_2 \ldots i_r}_{j_1 j_2 \ldots j_s} = \mathbf{T}^r_s(\mathbf{d}x^{i_1}, \mathbf{d}x^{i_2}, \ldots, \mathbf{d}x^{i_r}, \boldsymbol{\partial}_{j_1}, \boldsymbol{\partial}_{j_2}, \ldots, \boldsymbol{\partial}_{j_s})$$

Exercise

4.3.1 Suppose that x^i are coordinate functions on a subset of M. Express $\boldsymbol{\omega}(\mathbf{X})$ in terms of component functions of $\boldsymbol{\omega}$ and $\mathbf{X}$.

4.4 EXTERIOR CALCULUS

Skew-symmetric tensors are of special importance to applications. We studied these tensors in their algebraic format in Section 4.1. Let us now investigate them as functions over manifolds.

4.4.1 Differential Forms

4.4.1 Definition Let M be a manifold and P a point of M. Let $\Lambda^p_P(\mathrm{M})$ denote the space of all antisymmetric tensors of rank p over the tangent space at P. Let $\Lambda^p(\mathrm{M})$ be the union of all $\Lambda^p_P(\mathrm{M})$ for all $P \in \mathrm{M}$. A (differential) *p-form* $\boldsymbol{\omega}$ is a mapping, $\boldsymbol{\omega}: \mathrm{U} \to \Lambda^p(\mathrm{M})$, such that $\boldsymbol{\omega}(P) \in \Lambda^p_P(\mathrm{M})$, where U is, as usual, a subset of M.

Since $\{\mathbf{d}x^i\}_{i=1}^m$ is a basis for $\mathrm{T}^*_P(\mathrm{M})$ at every $P \in \mathrm{M}$, $\{\mathbf{d}x^{i_1} \wedge \mathbf{d}x^{i_2} \wedge \cdots \wedge \mathbf{d}x^{i_p}\}$ is a basis for the p-forms. All the algebraic properties established in Section 4.1 apply to these p-forms at every point $P \in \mathrm{M}$.

Since $\boldsymbol{\omega}$ varies from point to point, we can define derivatives of $\boldsymbol{\omega}$. Recall that $\Lambda^0(\mathrm{M})$, the collection of zero-forms, are simply real-valued functions on M. Also recall that if f is a zero-form, then $\mathbf{d}f$, the differential of f, is a one-form. Thus, the differential operator $\mathbf{d}$ creates a one-form from a zero-form. The fact that this can be generalized to p-forms is the subject of the next theorem.

The exterior derivative. Let us consider the following theorem.

4.4.2 Theorem For some subset U of M, there exists a unique operator

$$\mathbf{d}: \Lambda^p(\mathrm{U}) \to \Lambda^{p+1}(\mathrm{U})$$

such that, for any $\boldsymbol{\omega} \in \Lambda^p(\mathrm{M})$ and $\boldsymbol{\eta} \in \Lambda^q(\mathrm{M})$,

(i) $\mathbf{d}(\boldsymbol{\omega} + \boldsymbol{\eta}) = \mathbf{d}\boldsymbol{\omega} + \mathbf{d}\boldsymbol{\eta}$ if $q = p$; otherwise, the sum is not defined

(ii) $\mathbf{d}(\boldsymbol{\omega} \wedge \boldsymbol{\eta}) = (\mathbf{d}\boldsymbol{\omega}) \wedge \boldsymbol{\eta} + (-1)^p \boldsymbol{\omega} \wedge (\mathbf{d}\boldsymbol{\eta})$

(iii) for each $\boldsymbol{\omega}$, $\mathbf{d}(\mathbf{d}\boldsymbol{\omega}) = 0$

(iv) for each real-valued function f, $\mathbf{d}f = (\partial_i f)\mathbf{d}x^i$ ■

The operator $\mathbf{d}$ is called the *exterior derivative operator*. (For a proof of this theorem, see Abraham, Marsden, and Ratiu 1983, p. 357.)

Example 4.4.1

In relativistic electromagnetic theory the electric and magnetic fields are combined to form the electromagnetic field tensor. This is a skew-symmetric tensor field of rank 2, which can be written as

$$\begin{aligned}\mathbf{F} = & -E_x \mathbf{d}t \wedge \mathbf{d}x - E_y \mathbf{d}t \wedge \mathbf{d}y - E_z \mathbf{d}t \wedge \mathbf{d}z + B_x \mathbf{d}y \wedge \mathbf{d}z \\ & + B_y \mathbf{d}z \wedge \mathbf{d}x + B_z \mathbf{d}x \wedge \mathbf{d}y \end{aligned} \tag{1}$$

where t is the time coordinate and the units are such that c, the velocity of light, is equal to 1.

Let us take the exterior derivative of $\mathbf{F}$. In the process we use the facts that $\mathbf{d}f = \partial_i f \mathbf{d}x^i$ and that $\mathbf{d}(\mathbf{d}x^i \wedge \mathbf{d}x^j) = 0$, and, in taking $\mathbf{d}E_i$ or $\mathbf{d}B_j$, we include only the terms that give a nonzero contribution.

$$\begin{aligned}\mathbf{dF} = & -\left(\frac{\partial E_x}{\partial y}\mathbf{d}y + \frac{\partial E_x}{\partial z}\mathbf{d}z\right) \wedge \mathbf{d}t \wedge \mathbf{d}x - \left(\frac{\partial E_y}{\partial x}\mathbf{d}x + \frac{\partial E_y}{\partial z}\mathbf{d}z\right) \wedge \mathbf{d}t \wedge \mathbf{d}y \\ & -\left(\frac{\partial E_z}{\partial x}\mathbf{d}x + \frac{\partial E_z}{\partial y}\mathbf{d}y\right) \wedge \mathbf{d}t \wedge \mathbf{d}z + \left(\frac{\partial B_x}{\partial t}\mathbf{d}t + \frac{\partial B_x}{\partial x}\mathbf{d}x\right) \wedge \mathbf{d}y \wedge \mathbf{d}z \\ & + \left(\frac{\partial B_y}{\partial t}\mathbf{d}t + \frac{\partial B_y}{\partial y}\mathbf{d}y\right) \wedge \mathbf{d}z \wedge \mathbf{d}x + \left(\frac{\partial B_z}{\partial t}\mathbf{d}t + \frac{\partial B_z}{\partial z}\mathbf{d}z\right) \wedge \mathbf{d}x \wedge \mathbf{d}y\end{aligned}$$

Collecting all similar terms, taking into account changes of sign due to the antisymmetry of the exterior products, gives

$$\begin{aligned}\mathbf{dF} = & \left(-\frac{\partial E_x}{\partial y} + \frac{\partial E_y}{\partial x} + \frac{\partial B_z}{\partial t}\right)\mathbf{d}t \wedge \mathbf{d}x \wedge \mathbf{d}y + \left(-\frac{\partial E_x}{\partial z} + \frac{\partial E_z}{\partial x} - \frac{\partial B_y}{\partial t}\right)\mathbf{d}t \wedge \mathbf{d}x \wedge \mathbf{d}z \\ & + \left(-\frac{\partial E_y}{\partial z} + \frac{\partial E_z}{\partial y} + \frac{\partial B_x}{\partial t}\right)\mathbf{d}t \wedge \mathbf{d}y \wedge \mathbf{d}z + \left(\frac{\partial B_x}{\partial x} + \frac{\partial B_y}{\partial y} + \frac{\partial B_z}{\partial z}\right)\mathbf{d}x \wedge \mathbf{d}y \wedge \mathbf{d}z \\ = & \left[\left(\nabla \times \mathbf{E} + \frac{\partial \mathbf{B}}{\partial t}\right)_z\right]\mathbf{d}t \wedge \mathbf{d}x \wedge \mathbf{d}y + \left[\left(\nabla \times \mathbf{E} + \frac{\partial \mathbf{B}}{\partial t}\right)_y\right]\mathbf{d}t \wedge \mathbf{d}z \wedge \mathbf{d}x \\ & + \left[\left(\nabla \times \mathbf{E} + \frac{\partial \mathbf{B}}{\partial t}\right)_x\right]\mathbf{d}t \wedge \mathbf{d}y \wedge \mathbf{d}z + (\nabla \cdot \mathbf{B})\mathbf{d}x \wedge \mathbf{d}y \wedge \mathbf{d}z\end{aligned}$$

Each component of $\mathbf{dF}$ vanishes because of Maxwell's equations. Thus, when we define $\mathbf{F}$ as in (1), we can combine two of Maxwell's equations, the so-called homogeneous ones, into the deceptively simple equation

$$\mathbf{dF} = 0$$

●

Example 4.4.2

Let us write the electromagnetic field tensor as

$$\mathbf{F} = F_{\alpha\beta}\mathbf{d}x^\alpha \wedge \mathbf{d}x^\beta$$

where α and β run over the values 0, 1, 2, and 3 with 0 as the time index.

Let $\mathbf{p} = p_\alpha \mathbf{d}x^\alpha$ be the momentum one-form and

$$\frac{d\mathbf{p}}{d\tau} \equiv \left(\frac{dp_\alpha}{d\tau}\right)\mathbf{d}x^\alpha$$

its derivative with respect to proper time, τ. Also let $\mathbf{u} = u^\beta \boldsymbol{\partial}_\beta$ be the velocity four-vector of a particle. Then the Lorentz force law can be written simply as

$$\frac{d\mathbf{p}}{d\tau} = q\mathbf{F}(\mathbf{u})$$

where q is the electric charge of the particle whose velocity is $\mathbf{u}$. Note that $\mathbf{F}$, a two-form, contracts with $\mathbf{u}$, a vector, to give a one-form on the RHS. Thus, both sides are of the same type. This claim can be verified by

$$\begin{aligned}
\frac{dp_\alpha}{d\tau}\mathbf{d}x^\alpha &= q(F_{\alpha\beta}\mathbf{d}x^\alpha \wedge \mathbf{d}x^\beta)(u^\gamma\boldsymbol{\partial}_\gamma) = qF_{\alpha\beta}u^\gamma[\tfrac{1}{2}(\mathbf{d}x^\alpha \otimes \mathbf{d}x^\beta - \mathbf{d}x^\beta \otimes \mathbf{d}x^\alpha)(\boldsymbol{\partial}_\gamma)] \\
&= qF_{\alpha\beta}u^\gamma[\tfrac{1}{2}(\mathbf{d}x^\alpha)\mathbf{d}x^\beta(\boldsymbol{\partial}_\gamma) - \tfrac{1}{2}(\mathbf{d}x^\beta)\mathbf{d}x^\alpha(\boldsymbol{\partial}_\gamma)] \\
&= qF_{\alpha\beta}u^\gamma[\tfrac{1}{2}\mathbf{d}x^\alpha\delta^\beta_\gamma - \tfrac{1}{2}\mathbf{d}x^\beta\delta^\alpha_\gamma] = \tfrac{1}{2}qF_{\alpha\beta}(u^\beta\mathbf{d}x^\alpha - u^\alpha\mathbf{d}x^\beta) \\
&= \tfrac{1}{2}q(F_{\alpha\beta} - F_{\beta\alpha})u^\beta\,\mathbf{d}x^\alpha = (qF_{\alpha\beta}u^\beta)\mathbf{d}x^\alpha
\end{aligned}$$

Equating the components on both sides, we get

$$\frac{dp_\alpha}{d\tau} = qF_{\alpha\beta}u^\beta$$

For $\alpha = 1$, we have

$$\frac{dp_1}{d\tau} = qF_{1\beta}u^\beta = q[F_{10}u^0 + F_{12}u^2 + F_{13}u^3] \tag{1}$$

Recall that $u^\alpha = dx^\alpha/d\tau$, where

$$\begin{aligned}
(d\tau)^2 \equiv d\tau^2 &= (dt)^2 - (dx^1)^2 - (dx^2)^2 - (dx^3)^2 \\
&= (dt)^2\left\{1 - \left[\left(\frac{dx^1}{dt}\right)^2 + \left(\frac{dx^2}{dt}\right)^2 + \left(\frac{dx^3}{dt}\right)^2\right]\right\} \\
&= (dt)^2(1 - v^2)
\end{aligned}$$

and $\mathbf{v} = (dx^1/dt, dx^2/dt, dx^3/dt)$ is the velocity of the particle. Since $x^0 = t$, we get

$$u^0 = \frac{dt}{d\tau} = \frac{1}{\sqrt{1 - v^2}} \qquad \mathbf{u} = \frac{\mathbf{v}}{\sqrt{1 - v^2}}$$

Substituting this in (1) and remembering that $F_{10} = -F_{01} = +E_1$, $F_{12} = B_3$, and $F_{13} = -F_{31} = -B_2$, we obtain

$$\frac{dp_1}{dt\sqrt{1-v^2}} = q\left[E_1\frac{1}{\sqrt{1-v^2}} + B_3\frac{v_2}{\sqrt{1-v^2}} - B_2\frac{v_3}{\sqrt{1-v^2}}\right]$$

or

$$\frac{dp_1}{dt} = q[E_1 + (v_2B_3 - v_3B_2)] = [q(\mathbf{E} + \mathbf{v}\times\mathbf{B})]_1$$

The other components are obtained similarly. Thus, in vector form we have

$$\frac{d\mathbf{p}}{dt} = q(\mathbf{E} + \mathbf{v}\times\mathbf{B})$$

This is the familiar form of the Lorentz force law for electromagnetism. Again, note the simplification offered by the language of forms. ●

A combination of operators that is extremely useful is that of the exterior derivative and the Hodge star operator. Recall that the latter is defined by

$$*(\mathrm{d}x^{i_1}\wedge\ldots\wedge\mathrm{d}x^{i_p}) = \frac{1}{(N-p)!}\varepsilon^{i_1\ldots i_p}_{i_{p+1}\ldots i_m}\mathrm{d}x^{i_{p+1}}\wedge\cdots\wedge\mathrm{d}x^{i_m} \tag{4.31}$$

where m is the dimension of the manifold.

Example 4.4.3

Let us calculate $*\mathsf{F}$ and $\mathsf{d}(*\mathsf{F})$ where $\mathsf{F} = F_{\alpha\beta}\,\mathrm{d}x^\alpha\wedge\mathrm{d}x^\beta$ is the electromagnetic field tensor.

$$*\mathsf{F} = *(F_{\alpha\beta}\,\mathrm{d}x^\alpha\wedge\mathrm{d}x^\beta) = F_{\alpha\beta}*(\mathrm{d}x^\alpha\wedge\mathrm{d}x^\beta) = F_{\alpha\beta}\frac{1}{2!}\varepsilon^{\alpha\beta}_{\mu\nu}\,\mathrm{d}x^\mu\wedge\mathrm{d}x^\nu$$

Now we apply d to this two-form to get

$$\begin{aligned}\mathsf{d}(*\mathsf{F}) &= \mathsf{d}(\tfrac{1}{2}F_{\alpha\beta}\,\varepsilon^{\alpha\beta}_{\mu\nu}\,\mathrm{d}x^\mu\wedge\mathrm{d}x^\nu)\\ &= \tfrac{1}{2}\varepsilon^{\alpha\beta}_{\mu\nu}F_{\alpha\beta,\gamma}\,\mathrm{d}x^\gamma\wedge\mathrm{d}x^\mu\wedge\mathrm{d}x^\nu\end{aligned}$$

where $F_{\alpha\beta,\gamma} \equiv \partial F_{\alpha\beta}/\partial x^\gamma$. Let us write F in terms of $\mathbf{E}$ and $\mathbf{B}$:

$$\mathsf{F} = -E_x\mathrm{d}t\wedge\mathrm{d}x - E_y\mathrm{d}t\wedge\mathrm{d}y - E_z\mathrm{d}t\wedge\mathrm{d}z + B_x\mathrm{d}y\wedge\mathrm{d}z + B_y\mathrm{d}z\wedge\mathrm{d}x + B_z\mathrm{d}x\wedge\mathrm{d}y \tag{1}$$

This expression says, in particular, that if $\mathsf{F} = F_{\alpha\beta}\,\mathrm{d}x^\alpha\wedge\mathrm{d}x^\beta$, then $F_{01} = -F_{10} = -\frac{1}{2}E$, and so on. The factor $\frac{1}{2}$ is important in subsequent calculations.

We can now expand $\mathsf{d}(*\mathsf{F})$ and use Eq. (1) to write it in terms of $\mathbf{E}$ and $\mathbf{B}$. After a long but straightforward calculation, we obtain

$$\begin{aligned}\mathsf{d}(*\mathsf{F}) = &\left[\left(\frac{\partial\mathbf{E}}{\partial t} - \nabla\times\mathbf{B}\right)_z\right]\mathrm{d}t\wedge\mathrm{d}x\wedge\mathrm{d}y + \left[\left(\frac{\partial\mathbf{E}}{\partial t} - \nabla\times\mathbf{B}\right)_y\right]\mathrm{d}t\wedge\mathrm{d}z\wedge\mathrm{d}x\\ &+\left[\left(\frac{\partial\mathbf{E}}{\partial t} - \nabla\times\mathbf{B}\right)_x\right]\mathrm{d}t\wedge\mathrm{d}y\wedge\mathrm{d}z + (\nabla\cdot\mathbf{E})\mathrm{d}x\wedge\mathrm{d}y\wedge\mathrm{d}z\end{aligned} \tag{2}$$

The inhomogeneous pair of Maxwell's equations is

$$\nabla \times \mathbf{B} = \frac{\partial \mathbf{E}}{\partial t} + 4\pi \mathbf{J} \qquad \nabla \cdot \mathbf{E} = 4\pi\rho \tag{3}$$

where ρ and $\mathbf{J}$ are charge and current densities, respectively. We can put these two densities together and form a four-current one-form:

$$\mathsf{J} = J_\alpha \mathsf{d}x^\alpha$$

Thus,

$$\begin{aligned} *\mathsf{J} = J_\alpha(*\mathsf{d}x^\alpha) &= J_\alpha \frac{1}{3!} \varepsilon^\alpha_{\mu\nu\rho}\, \mathsf{d}x^\mu \wedge \mathsf{d}x^\nu \wedge \mathsf{d}x^\rho \\ &= -J_0 \mathsf{d}x \wedge \mathsf{d}y \wedge \mathsf{d}z - J_x \mathsf{d}t \wedge \mathsf{d}y \wedge \mathsf{d}z - J_y \mathsf{d}t \wedge \mathsf{d}z \wedge \mathsf{d}x \\ &\quad - J_z \mathsf{d}t \wedge \mathsf{d}x \wedge \mathsf{d}y \\ &= \rho \mathsf{d}x \wedge \mathsf{d}y \wedge \mathsf{d}z - J_x \mathsf{d}t \wedge \mathsf{d}y \wedge \mathsf{d}z - J_y \mathsf{d}t \wedge \mathsf{d}z \wedge \mathsf{d}x \\ &\quad - J_z \mathsf{d}t \wedge \mathsf{d}x \wedge \mathsf{d}y \end{aligned} \tag{4}$$

where we have used the facts that $\rho = J^0 = -J_0$ and $\mathbf{J} = (J^x, J^y, J^z) = (J_x, J_y, J_z)$.

Comparing Eqs. (2), (3), and (4), we note that in the language of forms the inhomogeneous pair of Maxwell's equations has the simple appearance

$$\mathsf{d}(*\mathsf{F}) = 4\pi(*\mathsf{J})$$

●

Closed and exact forms. The solution to Exercise 4.4.2 shows that the relation $\mathsf{d}^2\boldsymbol{\omega} = 0$ is equivalent, at least in $\mathbb{R}^3$, to $\nabla \times (\nabla f) = 0$ and $\nabla \cdot (\nabla \times \mathbf{A}) = 0$. It is customary in physics to go backwards as well, that is, given that $\nabla \times \mathbf{E} = 0$, to assume that $\mathbf{E} = \nabla f$ for some function f. Similarly, $\nabla \cdot \mathbf{B} = 0$ implies that $\mathbf{B} = \nabla \times \mathbf{A}$.

What is the analogue of the above statement for a general p-form? A form $\boldsymbol{\omega}$ that satisfies $\mathsf{d}\boldsymbol{\omega} = 0$ is called a *closed form.* An *exact form* is one that can be written as $\boldsymbol{\omega} = \mathsf{d}\boldsymbol{\eta}$. Thus, *every exact form is automatically closed.* This is the Poincaré lemma. The converse of this lemma is true only if the region of definition of the form is topologically simple, as explained in the following.

Consider a p-form $\boldsymbol{\omega}$ defined on a region U of a manifold M. If all closed curves in U can be shrunk to a point *in U*, we say that U is *contractable to a point.* If $\boldsymbol{\omega}$ is not defined for a point P on M, then any U that contains P is *not* contractable to a point. This is because the *domain of definition of* $\boldsymbol{\omega}$ would be U′ = {all points of U except P}, and a closed curve around P cannot be shrunk to a point *in U′*. We encountered an example of this in Chapter 1.

We can now state the converse of the Poincaré lemma as a theorem.

4.4.3 Theorem (Converse of the Poincaré Lemma) Let U be a region in a manifold M, such that U is contractable to a point. Let $\boldsymbol{\omega}$ be a p-form on U such that $\mathsf{d}\boldsymbol{\omega} = 0$. Then there is a $(p-1)$-form $\boldsymbol{\eta}$ on U such that

$$\boldsymbol{\omega} = \mathsf{d}\boldsymbol{\eta}$$

■

This theorem is illustrated by the following example (for a proof, see Bishop and Goldberg 1980, p. 175).

Example 4.4.4

The electromagnetic field tensor $\mathbf{F} = F_{\alpha\beta}\,\mathbf{d}x^\alpha \wedge \mathbf{d}x^\beta$ is a two-form that satisfies $\mathbf{dF} = 0$. Therefore, it is a closed two-form. The converse of the Poincaré lemma says that if $\mathbf{F}$ is well-behaved in a region U of $\mathbb{R}^4$, then there must exist a one-form $\boldsymbol{\eta}$ such that $\mathbf{F} = \mathbf{d}\boldsymbol{\eta}$.

Let us write this one-form as

$$\boldsymbol{\eta} = A_\alpha\,\mathbf{d}x^\alpha$$

Then $\mathbf{d}\boldsymbol{\eta} = A_{\alpha,\beta}\,\mathbf{d}x^\beta \wedge \mathbf{d}x^\alpha$ and we have

$$F_{\alpha\beta}\,\mathbf{d}x^\alpha \wedge \mathbf{d}x^\beta = A_{\alpha,\beta}\,\mathbf{d}x^\beta \wedge \mathbf{d}x^\alpha$$

or

$$(F_{\alpha\beta} - A_{\beta,\alpha})\,\mathbf{d}x^\alpha \wedge \mathbf{d}x^\beta = 0$$

To equate the coefficient to zero, we must make sure that $i < j$ in $\mathbf{d}x^i \wedge \mathbf{d}x^j$. This gives

$$\sum_{\alpha<\beta} (F_{\alpha\beta} - A_{\beta,\alpha} + A_{\alpha,\beta})\mathbf{d}x^\alpha \wedge \mathbf{d}x^\beta = 0$$

Thus,

$$F_{\alpha\beta} = \frac{\partial A_\beta}{\partial x^\alpha} - \frac{\partial A_\alpha}{\partial x^\beta}$$

The four-vector A^α is simply the four-potential of relativistic electromagnetic theory. ●

Note that the $(p-1)$-form of Theorem 4.4.3 is *not* unique. In fact, if $\boldsymbol{\alpha}$ is any $(p-2)$-form, then $\boldsymbol{\omega}$ can be written as

$$\boldsymbol{\omega} = \mathbf{d}(\boldsymbol{\eta} + \mathbf{d}\boldsymbol{\alpha})$$

because $\mathbf{d}(\mathbf{d}\boldsymbol{\alpha})$ is identical to zero. This freedom of choice in selecting $\boldsymbol{\eta}$ is called *gauge invariance*, and its generalization plays an important role in the physics of fundamental interactions.

4.4.2 Riemannian Geometry and Vector-Valued Forms

It used to be that only relativity theory was concerned with geometry. However, recently, with the advent of gauge field theory, a deep connection has been made between the four fundamental interactions (gravitation, electromagnetism, weak force, and strong force) and geometrical ideas. Thus, geometrical ideas and methods are becoming increasingly important for physicists.

Aside from its applications, (differential) geometry has such an inner simplicity and elegance that its study gives an enormous amount of intellectual satisfaction. So let us indulge ourselves, starting with a definition.

4.4.4 Definition A Riemannian manifold is a differentiable manifold M with a symmetric C^∞ tensor field, $\mathbf{g} \in \mathscr{T}^0_2(\mathrm{M})$, such that at each point $P \in \mathrm{M}$, $\mathbf{g}(P)$ is an inner product.

Thus, if $\mathbf{u}$ and $\mathbf{v}$ are C^∞ vector fields on M, then $\mathbf{g}(\mathbf{u},\mathbf{v})$ is a C^∞ function on M.

With $\mathbf{g}$ defined on M, we can obtain orthonormal vectors at each point of M. That is, we can construct *orthonormal vector fields* $\{\mathbf{e}_i\}$ such that

$$\mathbf{e}_i \cdot \mathbf{e}_j \equiv \mathbf{g}(\mathbf{e}_i, \mathbf{e}_j) = \eta_{ij} \equiv \pm \delta_{ij}$$

at each point $P \in$ M.

A Riemannian manifold is, in general, not flat (flatness will be defined later). One way to obtain the flatness or curvature of a space *intrinsically* is to translate a vector parallel to itself by the same amount in two perpendicular directions, find the change in the vector, then reverse the order of translation, find the change again, and compare the two changes. In a flat space the changes will be the same, but they will not be in general. An illustration is provided by the surface of a sphere. Assume that we have a vector perpendicular to the equator. To exaggerate the effect of curvature, we move the vector parallel to itself on the equator a quarter of the way around the sphere, then all the way to the north pole. We measure how much it has changed, We start with the vector again perpendicular to the equator, but this time we move it parallel to itself first to the north pole and then perpendicular to this path a quarter of the way around the globe. We again measure the change. Clearly, the two changes will not be equal.

The above intuitive discussion should help to make it clear that the curvature of a manifold (space) is related naturally to changes (double derivatives) in vectors. Thus, to find the curvature of space, we look at how vectors change. Since a change is calculated by applying the exterior derivative to objects, the change in a vector $\mathbf{X}$ is $\mathbf{dX}$, which is both a vector and a one-form. That is why we study vector-valued one-forms now. Our goal is to find the curvature of space by looking at changes in vectors. We apply the exterior derivative to vectors enough times to reach what we can call the curvature of the manifold.

In differential geometry everything takes place locally, and translations and movements are all infinitesimal. We have already encountered an operator suitable for such infinitesimal changes—the exterior derivative operator. Let us take a closer look at it. For a function on M, we have

$$\mathbf{d}f = \frac{\partial f}{\partial x^i}\mathbf{d}x^i = \mathbf{d}x^i \frac{\partial f}{\partial x^i} \tag{4.32}$$

where x^i are coordinate functions. We note that

$$\mathbf{d}(\mathbf{d}f) = \mathbf{d}\left(\frac{\partial f}{\partial x^i}\right) \wedge \mathbf{d}x^i = \frac{\partial^2 f}{\partial x^j \partial x^i}\mathbf{d}x^j \wedge \mathbf{d}x^i$$

$$= \sum_{j<i}\left(\frac{\partial^2 f}{\partial x^j \partial x^i} - \frac{\partial^2 f}{\partial x^i \partial x^j}\right)\mathbf{d}x^j \wedge \mathbf{d}x^i$$

Thus, $\mathbf{d}^2 f = 0$ means that the mixed partial derivatives are order-independent. Geometrically, this means that for *small displacements*, $\mathbf{d}x^i$ and $\mathbf{d}x^j$, the value of a function is the same if there is movement in two perpendicular directions, once in a given order and then in reverse order. This is true even if the space is curved.

The function f is arbitrary. Thus, we can think of Eq. (4.32) as representing a general operator $\mathbb{d}$ defined by

$$\mathbb{d} = \mathbf{d}x^i \frac{\partial}{\partial x^i} \tag{4.33a}$$

such that $\mathbb{d}f = \mathbf{d}f$ and $\mathbb{d}^2 f = \mathbf{d}^2 f = 0$. It is customary to attach a point P to $\mathbb{d}$ to indicate evaluation at the point or infinitesimal motions starting from the point. Thus, we write

$$\mathbb{d}P = \mathbf{d}x^i \frac{\partial}{\partial x^i} \tag{4.33b}$$

Note that $\mathbb{d}P$ is a *vector-valued one-form*, a vector whose components are the one-forms $\mathbf{d}x^i$. Although $\mathbb{d}P$ is defined in terms of the coordinate system (x^i), it is independent of coordinates (see Exercise 4.4.6).

In general, $\partial/\partial x^i$ are not orthonormal. However, it is convenient to use orthonormal bases at each point of a manifold. Therefore, at each point $P \in \mathrm{M}$, we pick orthonormal vector fields $\{\mathbf{e}_i\}$ and their corresponding one-forms $\{\boldsymbol{\epsilon}^j\}$. We then have

$$\mathbb{d}P = \boldsymbol{\epsilon}^j \mathbf{e}_j \tag{4.33c}$$

Keep in mind that $\boldsymbol{\epsilon}^j$ and $\mathbf{e}_j$ are *not* components, but the jth members of $\{\boldsymbol{\epsilon}^i\}$ and $\{\mathbf{e}_i\}$.

Let us now apply $\mathbb{d}$ to (4.33c) using the fact that $\mathbb{d}^2 P = 0$ (because $\mathbb{d}^2 f = 0$ for all functions). We then have

$$0 = \mathbb{d}^2 P = \mathbb{d}(\boldsymbol{\epsilon}^j \mathbf{e}_j) = \mathbb{d}\boldsymbol{\epsilon}^j \mathbf{e}_j + (-1)^1 \boldsymbol{\epsilon}^j \wedge \mathbb{d}\mathbf{e}_j \tag{4.34a}$$

where we have used property (ii) of Theorem 4.4.2 concerning the exterior derivative operator, of which $\mathbb{d}$ is a generalization. We know how $\mathbb{d}$ acts on one-forms: it is simply the exterior derivative. On the other hand, since $\mathbf{e}_j$ is a vector, we expect $\mathbb{d}\mathbf{e}_j$ to be a vector-valued one-form. Thus, we can write it as

$$\mathbb{d}\mathbf{e}_j = \boldsymbol{\omega}_j^k \mathbf{e}_k \tag{4.34b}$$

where $\boldsymbol{\omega}_j^k$ are one-forms. Substituting (4.34b) in (4.34a) gives

$$0 = \mathbf{d}\boldsymbol{\epsilon}^j \mathbf{e}_j - \boldsymbol{\epsilon}^j \wedge \boldsymbol{\omega}_j^k \mathbf{e}_k = (\mathbf{d}\boldsymbol{\epsilon}^k - \boldsymbol{\epsilon}^j \wedge \boldsymbol{\omega}_j^k)\mathbf{e}_k$$

The linear independence of $\mathbf{e}_k$ gives

$$\mathbf{d}\boldsymbol{\epsilon}^k = \boldsymbol{\epsilon}^j \wedge \boldsymbol{\omega}_j^k \tag{4.35a}$$

The orthonormality relation $\mathbf{e}_i \cdot \mathbf{e}_j = \pm\delta_{ij}$ puts the following antisymmetry restriction on the one-forms $\boldsymbol{\omega}_i^j$:

$$\boldsymbol{\omega}_i^j + \boldsymbol{\omega}_j^i = 0 \tag{4.35b}$$

It is clear that $\{\boldsymbol{\omega}_i^j\}$ gives all the information about how the bases $\{\mathbf{e}_i\}$ and $\{\boldsymbol{\epsilon}^i\}$ change with infinitesimal movement away from a point P. If we can find the $\boldsymbol{\omega}_i^j$, we will know the geometry of the space (at least locally). But these $\boldsymbol{\omega}_i^j$ must satisfy Eqs. (4.35), and, as we will see shortly, *Eqs.* (4.35) *determine the* $\boldsymbol{\omega}_i^j$ *uniquely*.

Since the $\boldsymbol{\epsilon}^j$ form a basis, we can express $\boldsymbol{\omega}_i^j$ in terms of them:

$$\boldsymbol{\omega}_{ij} = \Gamma_{ijk}\boldsymbol{\epsilon}^k$$

where $\boldsymbol{\omega}_{ij} \equiv g_{ik}\boldsymbol{\omega}_i^k$ is the lowered-index version of $\boldsymbol{\omega}_i^j$. The real-valued functions Γ_{ijk} are called the *connection coefficients.* Because of (4.35b), these coefficients satisfy

$$\Gamma_{ijk} + \Gamma_{jik} = 0 \tag{4.36a}$$

On the other hand, $\mathbf{d}\boldsymbol{\epsilon}^k$, being a two-form, can be expressed as

$$\mathbf{d}\boldsymbol{\epsilon}^k = \tfrac{1}{2} C_{ij}^k \boldsymbol{\epsilon}^i \wedge \boldsymbol{\epsilon}^j$$

with

$$C_{ij}^k + C_{ji}^k = 0$$

where C_{ij}^k are simply the coefficients (functions) of expansion. We can also write $\mathbf{d}\boldsymbol{\epsilon}^k$ directly in terms of Γ_{ijk}:

$$\begin{aligned}\mathbf{d}\boldsymbol{\epsilon}^k &= g^{ik}\boldsymbol{\epsilon}^j \wedge \boldsymbol{\omega}_{ji} = g^{ik}\boldsymbol{\epsilon}^j \wedge (\Gamma_{jil}\boldsymbol{\epsilon}^l)\\ &= g^{ik}\Gamma_{jil}\boldsymbol{\epsilon}^j \wedge \boldsymbol{\epsilon}^l = \tfrac{1}{2} g^{ik}(\Gamma_{jil} - \Gamma_{eij})\boldsymbol{\epsilon}^j \wedge \boldsymbol{\epsilon}^l\end{aligned}$$

Comparing the two expressions for $\mathbf{d}\boldsymbol{\epsilon}^k$, we conclude that

$$C_{jl}^k = g^{ik}(\Gamma_{jil} - \Gamma_{lij})$$

Lowering the indices on both sides gives

$$C_{ijl} = \Gamma_{jil} - \Gamma_{lij} \tag{4.36b}$$

Equations (4.36) are exactly equivalent to (4.35). If we can show that there are *unique* Γ's satisfying (4.36a) and (4.36b), then the uniqueness of the $\boldsymbol{\omega}_{ij}$ follows. The uniqueness of the Γ's is shown in Exercise 4.4.8. The result is

$$\Gamma_{ijk} = \tfrac{1}{2}(-C_{ijk} - C_{jki} + C_{kij}) \tag{4.37}$$

Let us introduce the matrices

$$\boldsymbol{\varepsilon} \equiv (\boldsymbol{\epsilon}^1, \boldsymbol{\epsilon}^2, \ldots, \boldsymbol{\epsilon}^m) \qquad \mathsf{e} \equiv \begin{pmatrix} \mathbf{e}_1 \\ \mathbf{e}_2 \\ \vdots \\ \mathbf{e}_m \end{pmatrix} \qquad \boldsymbol{\Omega} \equiv \begin{pmatrix} 0 & \boldsymbol{\omega}_1^2 & \cdots & \boldsymbol{\omega}_1^m \\ -\boldsymbol{\omega}_1^2 & 0 & \cdots & \boldsymbol{\omega}_2^m \\ \vdots & \vdots & & \vdots \\ -\boldsymbol{\omega}_1^m & -\boldsymbol{\omega}_2^m & \cdots & 0 \end{pmatrix}$$

whose *elements are one-forms or vectors.* We then write Eqs. (4.33c), (4.34b), and (4.35b) as

$$\begin{aligned}\mathbb{d}P &= \boldsymbol{\varepsilon}\mathsf{e},\\ \mathbb{d}\mathsf{e} &= \boldsymbol{\Omega}\mathsf{e}\\ \boldsymbol{\Omega} + \tilde{\boldsymbol{\Omega}} &= 0\end{aligned} \tag{4.38}$$

These are called the *structure equations.* We can also write Eq. (4.35a) in matrix form:

$$\mathbf{d}\boldsymbol{\varepsilon} = \boldsymbol{\varepsilon} \wedge \boldsymbol{\Omega} \tag{4.39}$$

This is called the *integrability condition.*

We have seen that $\mathbb{d}^2 P = 0$ is equivalent to $\partial^2 f/\partial x^i \partial x^j = \partial^2 f/\partial x^j \partial x^i$, which in turn was the result of the fact that infinitesimal changes in functions with movement in two perpendicular directions are the same in either order. We also noted that for vectors this may not be the case. Thus, we should not be surprised if $\mathbb{d}^2\mathbf{e}$ is not zero. In fact, it must be related to the curvature of the manifold. Let us calculate it

$$\begin{aligned} \mathbb{d}^2 \mathsf{e} &= \mathbb{d}(\mathbb{d}\mathsf{e}) = \mathbb{d}(\Omega \mathsf{e}) = \mathbf{d}\Omega \mathsf{e} + (-1)^1 \Omega \wedge \mathbb{d}\mathsf{e} \\ &= \mathbf{d}\Omega \mathsf{e} - \Omega \wedge (\Omega \mathsf{e}) \\ &= (\mathbf{d}\Omega - \Omega \wedge \Omega)\mathsf{e} \\ &\equiv \Theta \mathsf{e} \end{aligned} \tag{4.40}$$

where

$$\Theta \equiv \mathbf{d}\Omega - \Omega \wedge \Omega \tag{4.41}$$

is called the *curvature matrix*.

We can derive further integrability conditions. Applying $\mathbf{d}$ to (4.39) gives

$$\begin{aligned} \mathbf{d}(\mathbf{d}\boldsymbol{\varepsilon}) = 0 &= \mathbf{d}\boldsymbol{\varepsilon} \wedge \Omega + (-1)^1 \boldsymbol{\varepsilon} \wedge \mathbf{d}\Omega \\ &= \boldsymbol{\varepsilon} \wedge \Omega \wedge \Omega - \boldsymbol{\varepsilon} \wedge \mathbf{d}\Omega \\ &= \boldsymbol{\varepsilon} \wedge (-\Theta) \quad \Rightarrow \quad \boldsymbol{\varepsilon} \wedge \Theta = 0 \end{aligned} \tag{4.42}$$

Similarly, applying $\mathbf{d}$ to (4.41) gives

$$\mathbf{d}\Theta = \Omega \wedge \Theta - \Theta \wedge \Omega \tag{4.43}$$

as shown in Exercise 4.4.9. This is called the *Bianchi identity*.

It is clear from its definition that Θ is a matrix whose elements are two-forms. Thus, we can write

$$\boldsymbol{\theta}_{ij} = R_{ijkl} \boldsymbol{\epsilon}^k \wedge \boldsymbol{\epsilon}^l \tag{4.44}$$

which defines the *Riemann curvature tensor*, R_{ijkl}. The antisymmetry of Θ (showing this is left as a problem for the reader) and Eq. (4.44) give

$$R_{ijkl} + R_{jikl} = 0 \tag{4.45a}$$

$$R_{ijkl} + R_{ijlk} = 0 \tag{4.45b}$$

Similarly, $\boldsymbol{\varepsilon} \wedge \Theta = 0$ is equivalent to

$$R_{ijkl} + R_{iklj} + R_{iljk} = 0 \tag{4.45c}$$

as shown in Exercise 4.4.10. Another symmetry of R_{ijkl}, which can be derived using Eqs. (4.45), is

$$R_{ijkl} = R_{klij} \tag{4.45d}$$

Deriving this is left as a problem for the reader.

In applications it is usual to start with the metric tensor $\mathbf{g}$ given in terms of coordinate vector fields:

$$\mathbf{g} = g_{ij}\mathbf{d}x^i \otimes \mathbf{d}x^j \qquad \text{for } g_{ij} = g_{ji} \tag{4.46}$$

with

$$g_{ij} = g(\partial_i, \partial_j) \equiv \partial_i \cdot \partial_j$$

Then the orthonormal bases $\{\mathbf{e}_i\}$ and $\{\boldsymbol{\epsilon}^i\}$ are constructed in terms of $\{\partial_i\}$ and $\{\mathbf{d}x^i\}$, respectively, and are utilized as outlined above. We will look at examples of such a procedure shortly. However, it is also possible to work directly in terms of $\{\partial_i\}$ and $\{\mathbf{d}x^i\}$ to construct the essential connection coefficients. The following example demonstrates how this is done.

Example 4.4.5

Let $\{\mathbf{d}x^i\}$ and $\{\partial_i\}$ be coordinate fields. We have

$$\mathbb{d}P = \mathbf{d}x^i\partial_i$$

and

$$\mathbb{d}(\partial_i) = \mathfrak{a}_i^j\partial_j$$

where $\mathfrak{a}_i^j$ are one-forms and can, therefore, be written as linear combinations of $\mathbf{d}x^i$:

$$\mathfrak{a}_i^j = \begin{Bmatrix} j \\ ik \end{Bmatrix} \mathbf{d}x^k$$

The coefficients $\begin{Bmatrix} j \\ ik \end{Bmatrix}$ are connection coefficients, or *Christoffel symbols.*

We also have

$$g_{ij} = g(\partial_i, \partial_j) = \partial_i \cdot \partial_j$$

Thus,

$$\begin{aligned} \mathbf{d}g_{ij} &= \mathbb{d}(\partial_i)\cdot\partial_j + \partial_i\cdot\mathbb{d}(\partial_j) = (\mathfrak{a}_i^k\partial_k)\cdot\partial_j + \partial_i\cdot(\mathfrak{a}_j^k\partial_k) \\ &= \mathfrak{a}_i^k g_{kj} + \mathfrak{a}_j^k g_{ik} \\ &= \begin{Bmatrix} k \\ il \end{Bmatrix} g_{kj}\,\mathbf{d}x^l + \begin{Bmatrix} k \\ jl \end{Bmatrix} g_{ik}\,\mathbf{d}x^l \end{aligned}$$

On the other hand, since g_{ij} is a real-valued *function*, we have

$$\mathbf{d}g_{ij} = \frac{\partial g_{ij}}{\partial x^l}\mathbf{d}x^l$$

The last two equations yield

$$\frac{\partial g_{ij}}{\partial x^l} = \begin{Bmatrix} k \\ il \end{Bmatrix} g_{kj} + \begin{Bmatrix} k \\ jl \end{Bmatrix} g_{ik}$$

which, when the index is lowered so that

$$[j, il] \equiv \begin{Bmatrix} k \\ il \end{Bmatrix} g_{kj}$$

becomes

$$[j, il] + [i, jl] = \frac{\partial g_{ij}}{\partial x^l} \tag{1}$$

Furthermore, the relation $\mathbb{d}^2 P = 0$ gives

$$[i, jk] + [i, kj] = 0 \tag{2}$$

Equations (1) and (2) are similar to Eqs. (4.36). The unique solution is

$$[i, jk] = \frac{1}{2}\left(\frac{\partial g_{ij}}{\partial x^k} + \frac{\partial g_{jk}}{\partial x^i} - \frac{\partial g_{ik}}{\partial x_j}\right)$$

This is the Christoffel symbol used in classical (index-oriented) tensor analysis. •

Now we can consider what an infinitesimal displacement means in a (curved) manifold. Let it have the most natural meaning that can be given to such a notion. Let P be a point of M. Let γ be a curve through P such that $\gamma(c) = P$. For an infinitesimal δu, let $P' = \gamma(c + \delta u)$ be a point on γ close to P. The coordinates of P are $\{x^i(P)\}$ and those of P' are $\{x^i(P')\}$. Since the x^i are well-behaved functions, $x^i(P') - x^i(P)$ are infinitesimal real numbers. Let $\xi^i = x^i(P') - x^i(P)$, and construct the vector

$$\mathbf{v} = \xi^i \boldsymbol{\partial}_i$$

where $\{\boldsymbol{\partial}_i\}$ consists of tangent vectors at P. We call $\mathbf{v}$ *the infinitesimal displacement at* P. The length of this vector, $\mathbf{g}(\mathbf{v},\mathbf{v})$, is shown to be (see Exercise 4.4.11)

$$\mathbf{g}(\mathbf{v}, \mathbf{v}) = g_{ij}\xi^i \xi^j$$

This is called the *arc length* from P to P' and is naturally written as

$$ds^2 = g_{ij}\xi^i \xi^j$$

It is customary to write dx^i (*not* a one-form!) for ξ^i, yielding

$$ds^2 = g_{ij} dx^i dx^j \tag{4.47}$$

where dx^i are infinitesimal real numbers.

The equivalence of the arc length [Eq. (4.47)] and the metric [Eq. (4.46)] is the reason why it is the arc length that is given in most practical problems, rather than the metric. Once the arc length is known, the metric g_{ij} can be read off. All the relevant geometric quantities can then be calculated from g_{ij}.

Example 4.4.6

Let us look at a few examples of arc lengths and the corresponding metrics.

(a) For $ds^2 = dx^2 + dy^2 + dz^2$, the metric is the Euclidean $\mathbb{R}^3$, with $g_{ij} = \delta_{ij}$.

(b) For $ds^2 = dx^2 + dy^2 + dz^2 - dt^2$, the metric is the Minkowskian $\mathbb{R}^4$, with $g_{ij} = \eta_{ij}$, where $\eta_{xx} = \eta_{yy} = \eta_{zz} = -\eta_{tt} = 1$ and $\eta_{ij} = 0$ for $i \neq j$.

(c) For $ds^2 = dr^2 + r^2(d\theta^2 + \sin^2\theta\, d\varphi^2)$, the metric consists of the spherical coordinates in $\mathbb{R}^3$ with $g_{rr} = 1$, $g_{\theta\theta} = r$, $g_{\varphi\varphi} = r\sin\theta$, and all other components zero.

(d) For $ds^2 = a^2\, d\theta^2 + a^2 \sin^2\theta\, d\varphi^2$, the metric is a two-dimensional spherical surface, with $g_{\theta\theta} = a$, $g_{\varphi\varphi} = a\sin\theta$, and all other components zero.

(e) For $ds^2 = -dt^2 + a^2(t)[d\chi^2 + \sin^2\chi(d\theta^2 + \sin^2\theta\, d\varphi^2)]$, the metric is the *Friedmann metric* used in cosmology. Here $g_{tt} = -1$, $g_{\chi\chi} = a(t)$, $g_{\theta\theta} = a(t)\sin\chi$, $g_{\varphi\varphi} = a(t)\sin\chi\sin\theta$, and all other components are zero.

(f) For $ds^2 = -(1 - 2M/r)\,dt^2 + dr^2/(1 - 2M/r) + r^2(d\theta^2 + \sin^2\theta\, d\varphi^2)$, the metric is the *Schwarzschild metric* with $g_{tt} = -(1 - 2M/r)^{1/2}$, $g_{rr} = (1 - 2M/r)^{-1/2}$, $g_{\theta\theta} = r$, $g_{\varphi\varphi} = r\sin\theta$, and all other components zero.

Note that in all of the above the differentials are infinitesimal *numbers*. For each of the arc lengths, we have an orthonormal basis of one-forms:

(a) $\mathbf{g} = \boldsymbol{\epsilon}^1 \otimes \boldsymbol{\epsilon}^1 + \boldsymbol{\epsilon}^2 \otimes \boldsymbol{\epsilon}^2 + \boldsymbol{\epsilon}^3 \otimes \boldsymbol{\epsilon}^3$ with $\boldsymbol{\epsilon}^1 = \mathbf{d}x$, $\boldsymbol{\epsilon}^2 = \mathbf{d}y$, $\boldsymbol{\epsilon}^3 = \mathbf{d}z$

(b) $\mathbf{g} = \boldsymbol{\epsilon}^1 \otimes \boldsymbol{\epsilon}^1 + \boldsymbol{\epsilon}^2 \otimes \boldsymbol{\epsilon}^2 + \boldsymbol{\epsilon}^3 \otimes \boldsymbol{\epsilon}^3 - \boldsymbol{\epsilon}^0 \otimes \boldsymbol{\epsilon}^0$ with $\boldsymbol{\epsilon}^1 = \mathbf{d}x$, $\boldsymbol{\epsilon}^2 = \mathbf{d}y$, $\boldsymbol{\epsilon}^3 = \mathbf{d}z$, $\boldsymbol{\epsilon}^0 = \mathbf{d}t$

(c) $\mathbf{g} = \boldsymbol{\epsilon}^r \otimes \boldsymbol{\epsilon}^r + \boldsymbol{\epsilon}^\theta \otimes \boldsymbol{\epsilon}^\theta + \boldsymbol{\epsilon}^\varphi \otimes \boldsymbol{\epsilon}^\varphi$ with $\boldsymbol{\epsilon}^r = \mathbf{d}r$, $\boldsymbol{\epsilon}^\theta = r\mathbf{d}\theta$, $\boldsymbol{\epsilon}^\varphi = r\sin\theta\,\mathbf{d}\varphi$

(d) $\mathbf{g} = \boldsymbol{\epsilon}^\theta \otimes \boldsymbol{\epsilon}^\theta + \boldsymbol{\epsilon}^\varphi \otimes \boldsymbol{\epsilon}^\varphi$ with $\boldsymbol{\epsilon}^\theta = a\mathbf{d}\theta$, $\boldsymbol{\epsilon}^\varphi = a\sin\theta\,\mathbf{d}\varphi$

(e) $\mathbf{g} = -\boldsymbol{\epsilon}^t \otimes \boldsymbol{\epsilon}^t + \boldsymbol{\epsilon}^\chi \otimes \boldsymbol{\epsilon}^\chi + \boldsymbol{\epsilon}^\theta \otimes \boldsymbol{\epsilon}^\theta + \boldsymbol{\epsilon}^\varphi \otimes \boldsymbol{\epsilon}^\varphi$ with $\boldsymbol{\epsilon}^t = \mathbf{d}t$, $\boldsymbol{\epsilon}^\chi = a(t)\,\mathbf{d}\chi$, $\boldsymbol{\epsilon}^\theta = a(t)\sin\chi\,\mathbf{d}\theta$, $\boldsymbol{\epsilon}^\varphi = a(t)\sin\chi\sin\theta\,\mathbf{d}\varphi$

(f) $\mathbf{g} = -\boldsymbol{\epsilon}^t \otimes \boldsymbol{\epsilon}^t + \boldsymbol{\epsilon}^r \otimes \boldsymbol{\epsilon}^r + \boldsymbol{\epsilon}^\theta \otimes \boldsymbol{\epsilon}^\theta + \boldsymbol{\epsilon}^\varphi \otimes \boldsymbol{\epsilon}^\varphi$ with $\boldsymbol{\epsilon}^t = \left(1 - \frac{2M}{r}\right)^{1/2}\mathbf{d}t$, $\boldsymbol{\epsilon}^r = \left(1 - \frac{2M}{r}\right)^{-1/2}\mathbf{d}r$, $\boldsymbol{\epsilon}^\theta = r\,\mathbf{d}\theta$, $\boldsymbol{\epsilon}^\varphi = r\sin\theta\,\mathbf{d}\varphi$

Note that in all of these the differentials are *one-forms*! •

A vector field $\mathbf{X} \in T(M)$ that does not change under the action of $\mathbb{d}$ is said to be *parallel-transported*. Thus, for a parallel-transported $\mathbf{X}$, we have

$$\mathbb{d}\mathbf{X} = 0$$

If $\mathbf{X}$ and $\mathbf{Y}$ are vector fields, then

$$\mathbb{d}(\mathbf{X}\cdot\mathbf{Y}) = (\mathbb{d}\mathbf{X})\cdot\mathbf{Y} + \mathbf{X}\cdot(\mathbb{d}\mathbf{Y})$$

In particular, if *both* $\mathbf{X}$ and $\mathbf{Y}$ are parallel-transported, then their inner product remains constant:

$$\mathbb{d}(\mathbf{X}\cdot\mathbf{Y}) = 0$$

A curve $\gamma: [a, b] \to M$ is called a *geodesic* if its tangent gets parallel-transported along the curve *when the curve is parametrized by the arc length s.*

Example 4.4.7

Let $M = \mathbb{R}^2$. The arc length given by

$$ds^2 = \frac{dx^2 + dy^2}{y^2}$$

can be written as $\mathbf{d}s^2 = \boldsymbol{\epsilon}^1 \otimes \boldsymbol{\epsilon}^1 + \boldsymbol{\epsilon}^2 \otimes \boldsymbol{\epsilon}^2$ if we define

$$\boldsymbol{\epsilon}^1 \equiv \frac{\mathbf{d}x}{y} \qquad \text{and} \qquad \boldsymbol{\epsilon}^2 \equiv \frac{\mathbf{d}y}{y}$$

Thus, $\{\boldsymbol{\epsilon}^1, \boldsymbol{\epsilon}^2\}$ are orthonormal one-forms. Inspection of $\boldsymbol{\epsilon}^1$ and $\boldsymbol{\epsilon}^2$, along with the facts that $\mathbf{d}x(\partial_x) = \mathbf{d}y(\partial_y) = 1$ and $\mathbf{d}x(\partial_y) = \mathbf{d}y(\partial_x) = 0$, immediately gives $\mathbf{e}_1 = y\partial_x$ and $\mathbf{e}_2 = y\partial_y$.

To find the curvature tensor, we have to evaluate $\mathbf{d}\mathrm{e}$. This is immediately obtained by taking the exterior derivatives of the $\boldsymbol{\epsilon}$'s:

$$\begin{aligned} \mathbf{d}\boldsymbol{\epsilon}^1 &= \mathbf{d}\left(\frac{1}{y}\mathbf{d}x\right) = \frac{1}{y^2}\mathbf{d}x \wedge \mathbf{d}y = \boldsymbol{\epsilon}^1 \wedge \boldsymbol{\epsilon}^2 \\ \mathbf{d}\boldsymbol{\epsilon}^2 &= \mathbf{d}\left(\frac{1}{y}\mathbf{d}y\right) = -\frac{1}{y^2}\mathbf{d}y \wedge \mathbf{d}y = 0 \end{aligned} \tag{1}$$

Thus, we use (4.35) to find $\boldsymbol{\omega}^i_j$ such that

$$\boldsymbol{\omega}^1_1 = \boldsymbol{\omega}^2_2 = 0$$

$$\boldsymbol{\omega}^2_1 = -\boldsymbol{\omega}^1_2$$

and

$$\mathbf{d}\boldsymbol{\epsilon}^k = \boldsymbol{\epsilon}^i \wedge \boldsymbol{\omega}^k_i$$

For $k = 1$, we have

$$\mathbf{d}\boldsymbol{\epsilon}^1 = \boldsymbol{\epsilon}^1 \wedge \boldsymbol{\omega}^1_1 + \boldsymbol{\epsilon}^2 \wedge \boldsymbol{\omega}^1_2 = \boldsymbol{\epsilon}^2 \wedge \boldsymbol{\omega}^1_2 = -\boldsymbol{\omega}^1_2 \wedge \boldsymbol{\epsilon}^2$$

Comparing this with (1) yields

$$\boldsymbol{\omega}^1_2 = -\boldsymbol{\epsilon}^1 = -\boldsymbol{\omega}^2_1$$

Thus, the matrix $\boldsymbol{\Omega}$ is

$$\boldsymbol{\Omega} = \begin{pmatrix} 0 & \boldsymbol{\epsilon}^1 \\ -\boldsymbol{\epsilon}^1 & 0 \end{pmatrix}$$

which gives

$$\begin{aligned} \mathbf{d}\boldsymbol{\Omega} &= \begin{pmatrix} 0 & \mathbf{d}\boldsymbol{\epsilon}^1 \\ -\mathbf{d}\boldsymbol{\epsilon}^1 & 0 \end{pmatrix} = \begin{pmatrix} 0 & \boldsymbol{\epsilon}^1 \wedge \boldsymbol{\epsilon}^2 \\ -\boldsymbol{\epsilon}^1 \wedge \boldsymbol{\epsilon}^2 & 0 \end{pmatrix} \\ &= \begin{pmatrix} 0 & 1 \\ -1 & 0 \end{pmatrix} \boldsymbol{\epsilon}^1 \wedge \boldsymbol{\epsilon}^2 \end{aligned}$$

and

$$\boldsymbol{\Omega} \wedge \boldsymbol{\Omega} = \begin{pmatrix} 0 & \boldsymbol{\epsilon}^1 \\ -\boldsymbol{\epsilon}^1 & 0 \end{pmatrix} \wedge \begin{pmatrix} 0 & \boldsymbol{\epsilon}^1 \\ -\boldsymbol{\epsilon}^1 & 0 \end{pmatrix} = 0$$

Therefore, the curvature matrix is

$$\boldsymbol{\Theta} = \mathbf{d}\boldsymbol{\Omega} = \begin{pmatrix} 0 & 1 \\ -1 & 0 \end{pmatrix} \boldsymbol{\epsilon}^1 \wedge \boldsymbol{\epsilon}^2$$

This shows that the only nonzero independent component of the Riemann curvature tensor is $R_{1212} = 1$. ●

Example 4.4.8

For a spherical surface of radius a, the element of length is

$$ds^2 = a^2\, d\theta^2 + a^2 \sin^2\theta\, d\varphi^2$$

Thus, the orthonormal forms are

$$\boldsymbol{\epsilon}^\theta = a\mathbf{d}\theta \qquad \text{and} \qquad \boldsymbol{\epsilon}^\varphi = a\sin\theta\,\mathbf{d}\varphi$$

The corresponding vectors are

$$\mathbf{e}_\theta = \frac{1}{a}\partial_\theta \qquad \text{and} \qquad \mathbf{e}_\varphi = \frac{1}{a\sin\theta}\partial_\varphi$$

This is easily verified, because

$$\boldsymbol{\epsilon}^\theta(\mathbf{e}_\theta) = (a\,\mathbf{d}\theta)\left(\frac{1}{a}\partial_\theta\right) = \mathbf{d}\theta(\partial\theta) = 1$$

and so on. We therefore have

$$\mathbf{d}\boldsymbol{\epsilon}^\theta = a\mathbf{d}^2\theta = 0$$

$$\mathbf{d}\boldsymbol{\epsilon}^\varphi = a(\cos\theta)\,\mathbf{d}\theta \wedge \mathbf{d}\varphi = \frac{1}{a}\cot\theta\,\boldsymbol{\epsilon}^\theta \wedge \boldsymbol{\epsilon}^\varphi$$

This gives

$$(\mathbf{d}\boldsymbol{\epsilon}^\theta \quad \mathbf{d}\boldsymbol{\epsilon}^\varphi) = (\boldsymbol{\epsilon}^\theta \quad \boldsymbol{\epsilon}^\varphi) \wedge \begin{pmatrix} 0 & \dfrac{\cot\theta}{a}\boldsymbol{\epsilon}^\varphi \\ -\dfrac{\cot\theta}{a}\boldsymbol{\epsilon}^\varphi & 0 \end{pmatrix}$$

Thus,

$$\boldsymbol{\Omega} = \begin{pmatrix} 0 & \dfrac{\cot\theta}{a}\boldsymbol{\epsilon}^\varphi \\ -\dfrac{\cot\theta}{a}\boldsymbol{\epsilon}^\varphi & 0 \end{pmatrix}$$

A straightforward exterior differentiation yields

$$\mathbf{d}\boldsymbol{\Omega} = \begin{pmatrix} 0 & -\dfrac{1}{a^2}\boldsymbol{\epsilon}^\theta \wedge \boldsymbol{\epsilon}^\varphi \\ \dfrac{1}{a^2}\boldsymbol{\epsilon}^\theta \wedge \boldsymbol{\epsilon}^\varphi & 0 \end{pmatrix}$$

Similarly, $\boldsymbol{\Omega} \wedge \boldsymbol{\Omega} = 0$. Therefore, the curvature matrix is

$$\boldsymbol{\Theta} = \mathbf{d}\boldsymbol{\Omega} = -\frac{1}{a^2}\begin{pmatrix} 0 & 1 \\ -1 & 0 \end{pmatrix}\boldsymbol{\epsilon}^\theta \wedge \boldsymbol{\epsilon}^\varphi$$

The only independent component of the Riemann curvature tensor is

$$R_{\theta\varphi\theta\varphi} = -\frac{1}{a^2}$$

which is constant, as expected for a spherical surface. •

It is clear that when the g_{ij} in the expression for the line element are all constants for all points in the manifold, then $\boldsymbol{\epsilon}^i$ will be proportional to $\mathbf{d}x^i$ and $\mathbf{d}\boldsymbol{\epsilon}^i = 0$, for all i. This immediately tells us that $\boldsymbol{\Omega} = 0$ and, therefore, $\boldsymbol{\Theta} = 0$; that is, the manifold has no curvature. We call such a manifold flat. Thus, for $ds^2 = dx^2 + dy^2 + dz^2$, $\boldsymbol{\epsilon}^x = \mathbf{d}x$, $\boldsymbol{\epsilon}^y = \mathbf{d}y$, $\boldsymbol{\epsilon}^z = \mathbf{d}z$, and $\mathbf{d}\boldsymbol{\epsilon}^x = \mathbf{d}\boldsymbol{\epsilon}^y = \mathbf{d}\boldsymbol{\epsilon}^z = 0$. The space is flat.

However, arc lengths of a flat space come in various disguises with nontrivial coefficients. Does the curvature matrix $\boldsymbol{\Theta}$ recognize the flat arc length, or is it possible to fool $\boldsymbol{\Theta}$ into believing that it is privileged with a curvature when in reality it is still zero? The following example shows that the curvature matrix can detect flatness no matter how disguised the line element is!

Example 4.4.9

In spherical coordinates the line element (arc length) of the *flat* Euclidean space $\mathbb{R}^3$ is

$$ds^2 = dr^2 + r^2\, d\theta^2 + r^2 \sin^2\theta\, d\varphi^2$$

Let us calculate the curvature matrix $\boldsymbol{\Theta}$. We first need an orthonormal set of one-forms, which are immediately obtainable from the expression above:

$$\boldsymbol{\epsilon}^r = \mathbf{d}r \qquad \boldsymbol{\epsilon}^\theta = r\,\mathbf{d}\theta \qquad \boldsymbol{\epsilon}^\varphi = r\sin\theta\,\mathbf{d}\varphi$$

Equation (4.35a) tells us that to find $\boldsymbol{\Omega}$ we must take the exterior derivatives of these one-forms:

$$\mathbf{d}\boldsymbol{\epsilon}^r = \mathbf{d}^2 r = 0$$

$$\mathbf{d}\boldsymbol{\epsilon}^\theta = \mathbf{d}r \wedge \mathbf{d}\theta + r\underbrace{\mathbf{d}^2\theta}_{0} = \boldsymbol{\epsilon}^r \wedge \left(\frac{\boldsymbol{\epsilon}^\theta}{r}\right) = \frac{1}{r}\boldsymbol{\epsilon}^r \wedge \boldsymbol{\epsilon}^\theta$$

$$\begin{aligned}\mathbf{d}\boldsymbol{\epsilon}^\varphi &= \mathbf{d}(r\sin\theta) \wedge \mathbf{d}\varphi = \sin\theta\,\mathbf{d}r \wedge \mathbf{d}\varphi + r\cos\theta\,\mathbf{d}\theta \wedge \mathbf{d}\varphi \\ &= \sin\theta\,\boldsymbol{\epsilon}^r \wedge \left(\frac{\boldsymbol{\epsilon}^\varphi}{r\sin\theta}\right) + r\cos\theta\left(\frac{\boldsymbol{\epsilon}^\theta}{r}\right) \wedge \left(\frac{\boldsymbol{\epsilon}^\varphi}{r\sin\theta}\right) \\ &= \frac{1}{r}\boldsymbol{\epsilon}^r \wedge \boldsymbol{\epsilon}^\varphi + \frac{\cot\theta}{r}\boldsymbol{\epsilon}^\theta \wedge \boldsymbol{\epsilon}^\varphi\end{aligned}$$

We can now use the definition of $\boldsymbol{\omega}^i_j$ [Eq. (4.35a)] to find the matrix of one-forms, $\boldsymbol{\Omega}$. In calculating the elements of $\boldsymbol{\Omega}$, we remember that it is a skew-symmetric matrix, so $\boldsymbol{\omega}^i_i$ (no sum) is zero and $\boldsymbol{\omega}^j_i + \boldsymbol{\omega}^i_j = 0$. We also note that $\mathbf{d}\boldsymbol{\epsilon}^k = 0$ does *not* imply that $\boldsymbol{\omega}^k_j = 0$. Keeping these facts in mind, we can easily obtain $\boldsymbol{\Omega}$ (the calculation is left as a problem for the reader):

$$\boldsymbol{\Omega} = \begin{pmatrix} 0 & \frac{1}{r}\boldsymbol{\epsilon}^\theta & \frac{1}{r}\boldsymbol{\epsilon}^\varphi \\ -\frac{1}{r}\boldsymbol{\epsilon}^\theta & 0 & \frac{\cot\theta}{r}\boldsymbol{\epsilon}^\varphi \\ -\frac{1}{r}\boldsymbol{\epsilon}^\varphi & -\frac{\cot\theta}{r}\boldsymbol{\epsilon}^\varphi & 0 \end{pmatrix}$$

The exterior derivative of this matrix is found to be

$$\mathbf{d}\boldsymbol{\Omega} = \begin{pmatrix} 0 & 0 & \dfrac{\cot\theta}{r^2}\boldsymbol{\epsilon}^\theta \wedge \boldsymbol{\epsilon}^\varphi \\ 0 & 0 & -\dfrac{1}{r^2}\boldsymbol{\epsilon}^\theta \wedge \boldsymbol{\epsilon}^\varphi \\ -\dfrac{\cot\theta}{r^2}\boldsymbol{\epsilon}^\theta \wedge \boldsymbol{\epsilon}^\varphi & \dfrac{1}{r^2}\boldsymbol{\epsilon}^\theta \wedge \boldsymbol{\epsilon}^\varphi & 0 \end{pmatrix}$$

which is precisely the exterior product $\boldsymbol{\Omega} \wedge \boldsymbol{\Omega}$. Thus,

$$\boldsymbol{\Theta} = \mathbf{d}\boldsymbol{\Omega} - \boldsymbol{\Omega} \wedge \boldsymbol{\Omega} = 0$$

and the space is indeed flat! •

It is important to note that the curvature is calculated *intrinsically*, as seen in the foregoing examples. We never have to leave the space and go to a higher dimension to "see" the curvature. For example, in the case of the sphere, the only information we had was the line element in terms of the coordinates *on the sphere*. We never had to resort to any three-dimensional analysis to see a globe embedded in the Euclidean $\mathbb{R}^3$.

Example 4.4.10

Let us determine the geodesics of the space whose arc length is given in Example 4.4.7.

Let $\gamma: [a, b] \to \mathbb{R}^2$ be such a geodesic. We have to parametrize γ by s. Thus, we write

$$\gamma(s) = (x(s), y(s)) \in \mathbb{R}^2$$

The vector field tangent to this curve (see Proposition 4.2.5), for an arbitrary real-valued function $f(x, y)$, is given by

$$\begin{aligned} \mathbf{v}_* f &= \frac{d}{ds} f \circ \gamma(s) = \frac{d}{ds} f(\gamma(s)) = \frac{d}{ds} f(x(s), y(s)) \\ &= \frac{dx}{ds}\frac{\partial f}{\partial x} + \frac{dy}{ds}\frac{\partial f}{\partial y} = \left(\frac{dx}{ds}\frac{\boldsymbol{\partial}}{\boldsymbol{\partial} x} + \frac{dy}{ds}\frac{\boldsymbol{\partial}}{\boldsymbol{\partial} y}\right) f \end{aligned}$$

Thus, the components of $\mathbf{v}_{*s}$ in the coordinate basis $(\boldsymbol{\partial}/\boldsymbol{\partial} x, \boldsymbol{\partial}/\boldsymbol{\partial} y)$ are simply dx/ds and dy/ds. In terms of $\mathbf{e}_1$ and $\mathbf{e}_2$ this becomes

$$\mathbf{v}_{*s} = \frac{dx}{ds}\frac{1}{y}\mathbf{e}_1 + \frac{dy}{ds}\frac{1}{y}\mathbf{e}_2$$

Now we take the exterior derivative of both sides and set them equal to zero (by the definition of a geodesic):

$$0 = \mathbb{d}\mathbf{v}_{*s} = \mathbf{d}\left(\frac{dx}{ds}\frac{1}{y}\right)\mathbf{e}_1 + \frac{dx}{ds}\frac{1}{y}\mathbb{d}\mathbf{e}_1 + \mathbf{d}\left(\frac{dy}{ds}\frac{1}{y}\right)\mathbf{e}_2 + \frac{dy}{ds}\frac{1}{y}\mathbb{d}\mathbf{e}_2 \tag{1}$$

From $\mathbb{d}\mathbf{e}_i = \boldsymbol{\omega}_i^j \mathbf{e}_j$ and the result of Example 4.4.7, we obtain

$$\mathbb{d}\mathbf{e}_1 = \boldsymbol{\epsilon}^1 \mathbf{e}_2 \qquad \mathbb{d}\mathbf{e}_2 = -\boldsymbol{\epsilon}^1 \mathbf{e}_1 \qquad \boldsymbol{\epsilon}^1 = \frac{\mathbf{d}x}{y}$$

So Eq. (1) yields

$$0 = \left[\mathbf{d}\left(\frac{1}{y}\frac{dx}{ds}\right) - \left(\frac{1}{y}\frac{dy}{ds}\right)\frac{\mathbf{d}x}{y}\right]\mathbf{e}_1 + \left[\mathbf{d}\left(\frac{1}{y}\frac{dy}{ds}\right) + \left(\frac{1}{y}\frac{dx}{ds}\right)\frac{\mathbf{d}x}{y}\right]\mathbf{e}_2$$

The linear independence of $\mathbf{e}_1$ and $\mathbf{e}_2$ gives two equations:

$$d\left(\frac{1}{y}\frac{dx}{ds}\right) - \left(\frac{1}{y}\frac{dy}{ds}\right)\frac{dx}{y} = 0 \tag{2a}$$

$$d\left(\frac{1}{y}\frac{dy}{ds}\right) + \left(\frac{1}{y}\frac{dx}{ds}\right)\frac{dx}{y} = 0 \tag{2b}$$

It can easily be verified that these two equations are the same.

With $y' \equiv dy/dx$, it can be shown that Eqs. (2a) and (2b) simplify to

$$y\,dy' + [1 + (y')^2]\,dx = 0 \tag{3}$$

Let $v \equiv y' = dy/dx$ so that $dx = dy/v$, and substitute this in (3). This gives

$$y\,dv + (1 + v^2)\frac{dy}{v} = 0$$

whose solution is simply $1 + v^2 = r^2/y^2$, where r^2 is the constant of integration. Solving for v, we obtain

$$v = \frac{dy}{dx} = \sqrt{\frac{r^2}{y^2} - 1}$$

or

$$\frac{y\,dy}{\sqrt{r^2 - y^2}} = dx$$

which gives the solution

$$(x + a)^2 + y^2 = r^2$$

where a is another arbitrary constant.

Thus, the geodesics are circles with arbitrary radii and centers lying on the x-axis. ●

Example 4.4.11

For the metric $ds^2 = dx^2 + dy^2 + dz^2$, we can calculate the geodesics as follows. We first note that $\gamma: [a, b] \to \mathbb{R}^3$ can be written as

$$\gamma(s) = (x(s), y(s), z(s))$$

So, for $f \in F^\infty(\mathbb{R}^3)$, we have

$$\mathbf{V}_{*s} f = \frac{d}{ds} f \circ \gamma(s) = \frac{d}{ds}[f(\gamma(s))] = \frac{d}{ds} f(x(s), y(s), z(s))$$

$$= \frac{dx}{ds}\frac{\partial f}{\partial x} + \frac{dy}{ds}\frac{\partial f}{\partial y} + \frac{dz}{ds}\frac{\partial f}{\partial z}$$

Thus,

$$\mathbf{V}_{*s} = \frac{dx}{ds}\frac{\partial}{\partial_x} + \frac{dy}{ds}\frac{\partial}{\partial_y} + \frac{dz}{ds}\frac{\partial}{\partial_z} \tag{1}$$

The orthonormal bases are $\{\mathbf{d}x, \mathbf{d}y, \mathbf{d}z\}$ and $\{\partial_x, \partial_y, \partial_z\}$. Since $\mathbf{d}(\mathbf{d}x) = \mathbf{d}(\mathbf{d}y) = \mathbf{d}(\mathbf{d}z) = 0$, all of the $\boldsymbol{\omega}_{ij}$ are zero. Thus, $\mathbb{d}(\partial_x) = \mathbb{d}(\partial_y) = \mathbb{d}(\partial_z) = 0$, and (1) yields

$$0 = \mathbb{d}\mathbf{Y}_{*s} = \mathbf{d}\left(\frac{dx}{ds}\right)\partial_x + \mathbf{d}\left(\frac{dy}{ds}\right)\partial_y + \mathbf{d}\left(\frac{dz}{ds}\right)\partial_z$$

Therefore,

$$d\left(\frac{dx}{ds}\right) = 0 \quad \Rightarrow \quad \frac{dx}{ds} = \text{constant} \equiv a_x \Rightarrow x = a_x s + b_x$$

$$d\left(\frac{dy}{ds}\right) = 0 \quad \Rightarrow \quad y = a_y s + b_y$$

$$d\left(\frac{dz}{ds}\right) = 0 \quad \Rightarrow \quad z = a_z s + b_z$$

The geodesics are straight lines. •

Example 4.4.12

As another illustration of the use of the foregoing formalism, let us reconsider the curvilinear coordinates of Chapter 1. Recall that in terms of these coordinates the displacement is given as

$$ds^2 = h_1^2(dq_1)^2 + h_2^2(dq_2)^2 + h_3^2(dq_3)^2$$

Therefore, the orthonormal one-forms are

$$\boldsymbol{\epsilon}^1 = h_1\,\mathbf{d}q_1 \qquad \boldsymbol{\epsilon}^2 = h_2\,\mathbf{d}q_2 \qquad \boldsymbol{\epsilon}^3 = h_3\,\mathbf{d}q_3$$

We also note (from Exercise 4.4.4) that

$$*\mathbf{d}*\mathbf{d}f = \left(\frac{\partial^2 f}{\partial x^2} + \frac{\partial^2 f}{\partial y^2} + \frac{\partial^2 f}{\partial z^2}\right)\mathbf{d}x \wedge \mathbf{d}y \wedge \mathbf{d}z \equiv \nabla^2 f\,\mathbf{d}x \wedge \mathbf{d}y \wedge \mathbf{d}z \tag{1}$$

We use this definition to find the Laplacian in terms of q_1, q_2, and q_3:

$$\begin{aligned}\mathbf{d}f &= \frac{\partial f}{\partial q_1}\mathbf{d}q_1 + \frac{\partial f}{\partial q_2}\mathbf{d}q_2 + \frac{\partial f}{\partial q_3}\mathbf{d}q_3 \\ &= \left(\frac{1}{h_1}\frac{\partial f}{\partial q_1}\right)\boldsymbol{\epsilon}^1 + \left(\frac{1}{h_2}\frac{\partial f}{\partial q_2}\right)\boldsymbol{\epsilon}^2 + \left(\frac{1}{h_3}\frac{\partial f}{\partial q_3}\right)\boldsymbol{\epsilon}^3\end{aligned}$$

Now we apply the Hodge star operator:

$$\begin{aligned}*\mathbf{d}f &= \left(\frac{1}{h_1}\frac{\partial f}{\partial q_1}\right)*\boldsymbol{\epsilon}^1 + \left(\frac{1}{h_2}\frac{\partial f}{\partial q_2}\right)*\boldsymbol{\epsilon}^2 + \left(\frac{1}{h_3}\frac{\partial f}{\partial q_3}\right)*\boldsymbol{\epsilon}^3 \\ &= \left(\frac{1}{h_1}\frac{\partial f}{\partial q_1}\right)\boldsymbol{\epsilon}^2 \wedge \boldsymbol{\epsilon}^3 + \left(\frac{1}{h_2}\frac{\partial f}{\partial q_2}\right)\boldsymbol{\epsilon}^3 \wedge \boldsymbol{\epsilon}^1 + \left(\frac{1}{h_3}\frac{\partial f}{\partial q_3}\right)\boldsymbol{\epsilon}^1 \wedge \boldsymbol{\epsilon}^2 \\ &= \left(\frac{h_2 h_3}{h_1}\frac{\partial f}{\partial q_1}\right)\mathbf{d}q_2 \wedge \mathbf{d}q_3 + \left(\frac{h_1 h_3}{h_2}\frac{\partial f}{\partial q_2}\right)\mathbf{d}q_3 \wedge \mathbf{d}q_1 + \left(\frac{h_1 h_2}{h_3}\frac{\partial f}{\partial q_3}\right)\mathbf{d}q_1 \wedge \mathbf{d}q_2\end{aligned}$$

Differentiating once more, we get

$$\begin{aligned}\mathbf{d}*\mathbf{d}f &= \frac{\partial}{\partial q_1}\left(\frac{h_2 h_3}{h_1}\frac{\partial f}{\partial q_1}\right)\mathbf{d}q_1\wedge\mathbf{d}q_2\wedge\mathbf{d}q_3 + \frac{\partial}{\partial q_2}\left(\frac{h_1 h_3}{h_2}\frac{\partial f}{\partial q_2}\right)\mathbf{d}q_2\wedge\mathbf{d}q_3\wedge\mathbf{d}q_1\\ &+ \frac{\partial}{\partial q_3}\left(\frac{h_1 h_2}{h_3}\frac{\partial f}{\partial q_3}\right)\mathbf{d}q_3\wedge\mathbf{d}q_1\wedge\mathbf{d}q_2\\ &= \left\{\frac{1}{h_1 h_2 h_3}\left[\frac{\partial}{\partial q_1}\left(\frac{h_2 h_3}{h_1}\frac{\partial f}{\partial q_1}\right) + \frac{\partial}{\partial q_2}\left(\frac{h_1 h_3}{h_2}\frac{\partial f}{\partial q_2}\right)\right.\right.\\ &\left.\left.+ \frac{\partial}{\partial q_3}\left(\frac{h_1 h_2}{h_3}\frac{\partial f}{\partial q_3}\right)\right]\right\}\boldsymbol{\epsilon}^1\wedge\boldsymbol{\epsilon}^2\wedge\boldsymbol{\epsilon}^3\end{aligned} \tag{2}$$

Since $\{\boldsymbol{\epsilon}^1, \boldsymbol{\epsilon}^2, \boldsymbol{\epsilon}^3\}$ are orthonormal one-forms (as are $\{\mathbf{d}x, \mathbf{d}y, \mathbf{d}z\}$), the volume elements $\boldsymbol{\epsilon}^1\wedge\boldsymbol{\epsilon}^2\wedge\boldsymbol{\epsilon}^3$ and $\mathbf{d}x\wedge\mathbf{d}y\wedge\mathbf{d}z$ are equal. Thus, we substitute $\mathbf{d}x\wedge\mathbf{d}y\wedge\mathbf{d}z$ for $\boldsymbol{\epsilon}^1\wedge\boldsymbol{\epsilon}^2\wedge\boldsymbol{\epsilon}^3$ in (2), compare with (1), and conclude that

$$\nabla^2 f = \frac{1}{h_1 h_2 h_3}\left[\frac{\partial}{\partial q_1}\left(\frac{h_2 h_3}{h_1}\frac{\partial f}{\partial q_1}\right) + \frac{\partial}{\partial q_2}\left(\frac{h_1 h_3}{h_2}\frac{\partial f}{\partial q_2}\right) + \frac{\partial}{\partial q_3}\left(\frac{h_1 h_2}{h_3}\frac{\partial f}{\partial q_3}\right)\right]$$

which is the result obtained in Chapter 1. ●

Example 4.4.12 also shows why in Definition 4.1.29 the product of one-forms is called a volume element. In the Cartesian coordinate system this is too obvious to require any discussion.

As mentioned earlier, if a space has line elements with $g_{ij} = \pm\delta_{ij}$, then the Riemann curvature vanishes trivially. We have also seen examples in which the components of a metric tensor were by no means trivial, but $\boldsymbol{\Theta}$ was smart enough to detect the flatness in disguise. Is it possible in flat spaces, namely, those in which $\boldsymbol{\Theta} = 0$, to choose coordinate systems in terms of which the line elements have $g_{ij} = \pm\delta_{ij}$? This question is answered by a theorem that we will partially prove, using the following result (proved in Flanders 1963). If $\boldsymbol{\Omega}$ is a matrix of one-forms such that

$$\mathbf{d}\boldsymbol{\Omega} = \boldsymbol{\Omega}\wedge\boldsymbol{\Omega}$$

then there exists an orthogonal matrix A such that

$$\boldsymbol{\Omega} = (\mathbf{d}\mathsf{A})\mathsf{A}^{-1} \tag{4.48}$$

Now we are ready for the theorem.

4.4.5 Theorem Let M be a Riemannian manifold with $\boldsymbol{\Theta} = 0$. Then M is flat; that is, there exists a local coordinate system (x^i) for which $\{\partial_i\}$ is an orthonormal basis.

Proof. Since $\boldsymbol{\Theta} = 0$, we have $\mathbf{d}\boldsymbol{\Omega} = \boldsymbol{\Omega}\wedge\boldsymbol{\Omega}$. Thus, by (4.48) there exists an orthogonal matrix A such that

$$\boldsymbol{\Omega} = (\mathbf{d}\mathsf{A})\mathsf{A}^{-1} \quad\Rightarrow\quad \mathbf{d}\mathsf{A} = \boldsymbol{\Omega}\mathsf{A}$$

Now we define the one-form row matrix $\mathsf{T} = (\mathsf{T}^1, \mathsf{T}^2, \ldots, \mathsf{T}^m)$ by $\mathsf{T} = \boldsymbol{\varepsilon}\mathsf{A}$, where $\boldsymbol{\varepsilon}$ is the one-form row matrix in terms of which $\boldsymbol{\Omega}$ is given. Then, using (4.39), we have

$$\begin{aligned} \mathbf{d}\mathsf{T} &= \mathbf{d}(\boldsymbol{\varepsilon}\mathsf{A}) = (\mathbf{d}\boldsymbol{\varepsilon})\mathsf{A} - \boldsymbol{\varepsilon} \wedge \mathbf{d}\mathsf{A} \\ &= (\boldsymbol{\varepsilon} \wedge \boldsymbol{\Omega})\mathsf{A} - \boldsymbol{\varepsilon} \wedge (\boldsymbol{\Omega}\mathsf{A}) \end{aligned}$$

Thus,

$$\mathbf{d}\mathsf{T}^i = 0 \qquad \forall\, i$$

By Theorem 4.4.3 there must exist zero-forms (functions) x^i such that

$$\mathsf{T}^i = \mathbf{d}x^i \qquad \forall\, i$$

These x^i are the coordinates we are after. The basis $\{\partial_i\}$ is obtained using the inverse of A (see the discussion following Proposition 4.1.1):

$$\begin{pmatrix} \partial_1 \\ \partial_2 \\ \vdots \\ \partial_m \end{pmatrix} = \mathsf{A}^{-1} \begin{pmatrix} \mathbf{e}_1 \\ \mathbf{e}_2 \\ \vdots \\ \mathbf{e}_m \end{pmatrix}$$

Since A is orthogonal, both $\{\mathbf{d}x^i\}$ and $\{\partial_i\}$ are orthonormal bases. ■

This chapter has covered the essentials of tensor algebra and differential tensor analysis. Unfortunately, space limitations preclude discussion of *integral* tensor analysis, which would cover the generalization of such topics as line and surface integrals, Stokes' theorem, the divergence theorem, and so forth. Also, a complete undertaking of these topics on a general manifold requires a thorough understanding of certain topological concepts, which are beyond the scope of this book.

Exercises

4.4.1 Let $\mathrm{M} = \mathbb{R}^3$ and let f be a real-valued function. Let $\boldsymbol{\omega} = a_i \mathbf{d}x^i$ be a one-form and $\boldsymbol{\eta} = b_1 \mathbf{d}x^2 \wedge \mathbf{d}x^3 + b_2 \mathbf{d}x^3 \wedge \mathbf{d}x^1 + b_3 \mathbf{d}x^1 \wedge \mathbf{d}x^2$ be a two-form on $\mathbb{R}^3$. Show that **(a)** $\mathbf{d}f$ gives the gradient of f, **(b)** $\mathbf{d}\boldsymbol{\omega}$ gives the curl of the vector $\mathbf{A} = (a_1, a_2, a_3)$, and **(c)** $\mathbf{d}\boldsymbol{\eta}$ gives the divergence of the vector $\mathbf{B} = (b_1, b_2, b_3)$.

4.4.2 Show that **(a)** $\nabla \times (\nabla f) = 0$ and **(b)** $\nabla \cdot (\nabla \times \mathbf{A}) = 0$ are consequences of $\mathbf{d}^2 = 0$.

4.4.3 Given that $\mathsf{F} = F_{\alpha\beta} \mathbf{d}x^\alpha \wedge \mathbf{d}x^\beta$, write the two homogeneous Maxwell's equations, $\nabla \cdot \mathbf{B} = 0$ and $\nabla \times \mathbf{E} + \partial \mathbf{B}/\partial t = 0$, in terms of $F_{\alpha\beta}$.

4.4.4 Let $f \in \Lambda^0(\mathbb{R}^3)$ be a function on $\mathbb{R}^3$. Calculate $* \mathbf{d} * \mathbf{d}f$.

4.4.5 Show that current conservation is an automatic consequence of Maxwell's homogeneous equation $\mathbf{d} * \mathsf{F} = (4\pi) * \mathsf{J}$.

4.4.6 Show that $\mathbb{d}P$ is coordinate-independent.

4.4.7 Show that the $\boldsymbol{\omega}_j^k$ in $\mathbb{d}\mathbf{e}_j = \boldsymbol{\omega}_j^k \mathbf{e}_k$ are antisymmetric in their indices.

4.4.8 Show that there is a unique solution for Γ_{ijk} in terms of C_{ijk} if

$$C_{ijk} + C_{jik} = 0 \qquad (1)$$

$$\Gamma_{ijk} + \Gamma_{jik} = 0 \qquad (2)$$

$$\Gamma_{jik} - \Gamma_{kij} = C_{ijk} \qquad (3)$$

4.4.9 Show that $\mathbf{d}\boldsymbol{\Theta} = \boldsymbol{\Omega} \wedge \boldsymbol{\Theta} - \boldsymbol{\Theta} \wedge \boldsymbol{\Omega}$.

4.4.10 Write $\boldsymbol{\varepsilon} \wedge \boldsymbol{\Theta} = 0$ in component form and derive Eq. (4.45c).

4.4.11 Show that the length of the displacement vector $\mathbf{v} = \xi^k \partial_k$ is $g_{ij}\xi^i \xi^j$.

4.4.12 Find the geodesics for the surface of a sphere of radius a having the line element

$$ds^2 = a^2\, d\theta^2 + a^2 \sin^2\theta\, d\varphi^2$$

PROBLEMS

4.1 Show that the mapping $v: \mathscr{V}^* \to \mathbb{R}$ given by $v(\tau) = \tau(v)$ is linear.

4.2 Show that the components of a tensor product are the products of the components of the factors:

$$(\mathbf{U} \otimes \mathbf{T})^{i_1 \ldots i_{r+k}}_{j_1 \ldots j_{s+l}} = U^{i_1 \ldots i_r}_{j_1 \ldots j_s} T^{i_{r+1} \ldots i_{r+k}}_{j_{s+1} \ldots j_{s+l}}$$

4.3 Prove that the linear functional $\mathbb{f}: \mathscr{V} \to \mathbb{R}$ is a linear invariant function.

4.4 Show that

$$\dim \mathscr{S}^r(\mathscr{V}) = \binom{N+r-1}{r} \equiv \frac{(N+r-1)!}{r!(N-1)!}$$

4.5 If $\mathbf{A}$ is skew-symmetric in some pair of variables, show that $\mathbb{S}(\mathbf{A}) = 0$.

4.6 Show whether the following three vectors are linearly dependent or independent:

$$\mathbf{v}_1 = 2\mathbf{e}_1 - \mathbf{e}_2 + 3\mathbf{e}_2 - \mathbf{e}_4$$

$$\mathbf{v}_2 = -\mathbf{e}_1 + 3\mathbf{e}_2 - 2\mathbf{e}_4$$

$$\mathbf{v}_3 = 3\mathbf{e}_1 + 2\mathbf{e}_2 - 4\mathbf{e}_3 + \mathbf{e}_4$$

4.7 Show that $\mathbf{A} \in \Lambda^2(\mathbb{R}^4)$, given by $-7\mathbf{e}_1 \wedge \mathbf{e}_2 + 3\mathbf{e}_1 \wedge \mathbf{e}_3 - 5\mathbf{e}_1 \wedge \mathbf{e}_4 + \mathbf{e}_2 \wedge \mathbf{e}_3 - 4\mathbf{e}_2 \wedge \mathbf{e}_4 + \mathbf{e}_3 \wedge \mathbf{e}_4$, is decomposable. Find $\mathbf{v}_1$ and $\mathbf{v}_2$ such that $\mathbf{A} = \mathbf{v}_1 \wedge \mathbf{v}_2$.

4.8 In one dimension a linear operator acting on a vector simply multiplies that vector by some constant. Show that this constant is independent of the vector chosen. That is, the constant is an intrinsic property of the operator.

4.9 Complete the solution to Exercise 4.1.7.

4.10 Show that $g_{ij} = \mathbf{g}(\mathbf{e}_i, \mathbf{e}_j)$ are the components of the tensor $\mathbf{g} \in \mathscr{S}^2(\mathscr{V}^*)$ in the basis $\{\mathbf{e}_i\}_{i=1}^N$ of $\mathscr{V}$.

4.11 Let $\mathscr{N}(\mathbf{g})$ denote the null space of $\mathbf{g}$ considered as a linear operator, $\mathbb{g}: \mathscr{V} \to \mathscr{V}^*$. Show that $\mathscr{N}(\mathbf{g})$ consists of all vectors $\mathbf{u} \in \mathscr{V}$ such that $\mathbf{g}(\mathbf{u}, \mathbf{v}) = 0$ for all $\mathbf{v} \in \mathscr{V}$. Show also that in the orthonormal basis $\{\mathbf{e}_j\}$ the $\mathbf{e}_i$ with $\mathbf{g}(\mathbf{e}_i, \mathbf{e}_i) = 0$ form a basis of $\mathscr{N}(\mathbf{g})$, and therefore n_0 is the nullity of $\mathbf{g}$.

4.12 Show that $\delta_\pi = \varepsilon_{\pi(1),\pi(2),\ldots,\pi(N)}$.

4.13 Use Eq. (4.15) to show that, for a 3×3 matrix A,

$$\det \mathsf{A} = \frac{1}{3!}[(\operatorname{tr}\mathsf{A})^3 - 3\operatorname{tr}\mathsf{A}\operatorname{tr}(\mathsf{A}^2) + 2\operatorname{tr}\mathsf{A}^3]$$

4.14 Find the index and the signature for the bilinear form $\mathbf{g}$ on $\mathbb{R}^3$, given by

$$\mathbf{g}(\mathbf{v}_1,\mathbf{v}_2) = x_1y_2 + x_2y_1 - y_1z_2 - y_2z_1$$

4.15 Use the linearity of the Hodge star operator and the fact that $\mathbf{v}_i = r_i^j\mathbf{e}_j$ to show that Eq. (4.16a) implies (4.18).

4.16 Show that where there is a sum over an upper index and a lower index, switching the upper index to a lower index, and vice versa, does not change the sum. In other words, $A^iB_i = A_iB^i$.

4.17 Show the following vector identities, using the definition of cross products in terms of ε_{ijk}.

(a) $\mathbf{A} \times \mathbf{A} = 0$

(b) $\nabla\cdot(\nabla \times \mathbf{A}) = 0$

(c) $\nabla\cdot(\mathbf{A} \times \mathbf{B}) = (\nabla \times \mathbf{A})\cdot\mathbf{B} - (\nabla \times \mathbf{B})\cdot\mathbf{A}$

(d) $\nabla \times (\mathbf{A} \times \mathbf{B}) = (\mathbf{B}\cdot\nabla)\mathbf{A} + \mathbf{A}(\nabla\cdot\mathbf{B}) - (\mathbf{A}\cdot\nabla)\mathbf{B} - \mathbf{B}(\nabla\cdot\mathbf{A})$

(e) $\nabla \times (\nabla \times \mathbf{A}) = \nabla(\nabla\cdot\mathbf{A}) - \nabla^2\mathbf{A}$

4.18 Show that any tensor, $\mathbf{T} \in \mathcal{T}_0^2(\mathcal{V})$, can be written as $\mathbf{T} = \mathbf{S} + \mathbf{A}$, where $\mathbf{S} \in \mathcal{S}^2(\mathcal{V})$ is a symmetric tensor and $\mathbf{A} \in \Lambda^2(\mathcal{V})$ is an antisymmetric tensor. Can this be generalized to tensors of arbitrary rank?

4.19 For every $\mathbf{t} \in \mathcal{T}_P(\mathrm{M})$ and every constant function $c \in \mathrm{F}^\infty(P)$, show that $\mathbf{t}(c) = 0$. [Hint: Use parts (i) and (ii) of Definition 4.2.4 and show that $(c\mathbf{t})1 = \mathbf{t}(c) + (c\mathbf{t})1$.]

4.20 Let (x^i) and (y^i) be coordinate systems on a subset U of a manifold M. Let X^i and Y^i be the components of a vector field with respect to the two coordinate systems. Show that

$$Y^i = \sum_j X^j \frac{\partial y}{\partial x^j}$$

4.21 Show that $\mathbf{F} \wedge (*\mathbf{F}) = |\mathbf{B}|^2 - |\mathbf{E}|^2$. Remember that $F_{01} = -F_{10} = -\frac{1}{2}E_x$, and so on.

4.22 With $\mathbf{F} = F_{\alpha\beta}\,\mathbf{d}x^\alpha \wedge \mathbf{d}x^\beta$ and $\mathbf{J} = J_\gamma\,\mathbf{d}x^\gamma$, show that $\mathbf{d}(*\mathbf{F}) = 4\pi(*\mathbf{J})$ takes the following form in components:

$$\frac{\partial F^{\alpha\beta}}{\partial x^\beta} = 4\pi J^\alpha$$

4.23 Interpret Theorem 4.4.3 for $p = 1$ and $p = 2$ on $\mathbb{R}^3$.

4.24 Let A and B be matrices whose elements are one-forms. Show that $\widetilde{(\mathsf{A} \wedge \mathsf{B})} = -\tilde{\mathsf{B}} \wedge \tilde{\mathsf{A}}$.

4.25 Use the symmetries of R_{ijkl} to show that $R_{ijkl} = R_{klij}$.

4.26 Show that in a coordinate basis $\mathbb{d}^2P = 0$ gives

$$\begin{Bmatrix} i \\ jk \end{Bmatrix} + \begin{Bmatrix} i \\ kj \end{Bmatrix} = 0$$

4.27 Show that $[j, ik] + [i, jk] = \partial g_{ij}/\partial x^k$ and $[j, ik] + [j, ki] = 0$ have the unique solution

$$[i, jk] = \frac{1}{2}\left(\frac{\partial g_{ij}}{\partial x^k} + \frac{\partial g_{ik}}{\partial x^j} - \frac{\partial g_{jk}}{\partial x^i}\right)$$

4.28 Find $\mathbf{d}\boldsymbol{\Omega}$ if

$$\boldsymbol{\Omega} = \begin{pmatrix} 0 & \cot\theta\,\boldsymbol{\epsilon}^{\varphi} \\ -\cot\theta\,\boldsymbol{\epsilon}^{\varphi} & 0 \end{pmatrix}$$

4.29 Calculate $\boldsymbol{\Omega}$ from Exercise 4.4.9. Take the exterior derivative of $\boldsymbol{\Omega}$ and show that it is equal to $\boldsymbol{\Omega} \wedge \boldsymbol{\Omega}$.

4.30 Find the curvature of the two-dimensional space whose arc length is given by

$$ds^2 = (dx^1)^2 + (x^1)^2(dx^2)^2$$

4.31 Find the curvature of the three-dimensional space whose line element is given by

$$ds^2 = (dx^1)^2 + (x^1)^2(dx^2)^2 + (dx^3)^2$$

4.32 Find the curvature tensors of the Friedmann and Schwarzschild spaces given in Example 4.4.6.

4.33 Find the geodesics of the metric $ds^2 = (dx^1)^2 + (x^1)^2(dx^2)^2$

4.34 Verify the claim (made in the solution to Exercise 4.4.12) that $A\cos\varphi + B\sin\varphi - \cot\theta = 0$ is a solution of

$$\frac{d^2\theta}{d\varphi^2}\sin\theta - 2\left(\frac{d\theta}{d\varphi}\right)^2\cos\theta - \sin^2\theta\cos\theta = 0$$

4.35 (**a**) Show that in $\mathbb{R}^3$ the composite operator $\mathbf{d} \circ *$ gives the curl of a vector when the vector is written as components of a two-form. (**b**) Similarly, show that $* \circ \mathbf{d}$ is the divergence operator for one-forms. (**c**) Use these results and the procedure of Example 4.4.12 to find expressions for $\nabla \times \mathbf{A}$ and $\nabla \cdot \mathbf{A}$ in curvilinear coordinates.

5

Infinite-Dimensional Vector Spaces (Spaces of Functions)

The basic concepts of finite-dimensional vector spaces introduced in Chapter 2 can readily be generalized to infinite dimensions. The definition of a vector space and the concepts of linear combination, linear independence, basis, subspace, span, and so forth, all carry over to infinite dimensions. However, one thing is crucially different in the new situation, and this difference makes the study of infinite-dimensional vector spaces far richer and decidedly much more nontrivial than the study of finite-dimensional ones. In a finite-dimensional vector space we deal with finite sums, but in infinite-dimensional vector spaces we encounter infinite sums. Thus, we have to concern ourselves with the question of convergence of such sums.

5.1 THE QUESTION OF CONVERGENCE

Any kind of convergence involves the notion of distance. We considered such a notion in the Introduction. In the case of vector spaces a natural candidate for the distance between vectors is the norm, which we encountered in Chapter 2 in the context of inner products. However, the notion of norm does not require the vector space to be an inner product space and is defined as follows.

5.1.1 Definition A vector space $\mathcal{V}$ (not necessarily of finite dimension) is called a *normed linear space* if there exists a mapping $N:\mathcal{V} \to \mathbb{R}$ such that for $|a\rangle, |b\rangle \in \mathcal{V}$

(i) $N(|a\rangle) > 0 \qquad \forall\ |a\rangle \neq 0$

(ii) $N(0) = 0$

(iii) $N(\alpha|\mathrm{a}\rangle) = |\alpha| N(|\mathrm{a}\rangle)$

(iv) $N(|\mathrm{a}\rangle + |\mathrm{b}\rangle) \leqslant N(|\mathrm{a}\rangle) + N(|\mathrm{b}\rangle)$

It is customary to write $N(|\mathrm{a}\rangle) \equiv \|\mathbf{a}\|$, and we will use this notation frequently.[1] An inner product space is automatically normed. We can simply define $\|\mathbf{a}\| \equiv \sqrt{\langle \mathrm{a}|\mathrm{a}\rangle}$, as was done in Chapter 2.

Example 5.1.1

Some examples of normed linear spaces are

(a) $\mathscr{V} = \mathbb{R}$; then, for any $\alpha \in \mathbb{R}$, let $\|\alpha\| = |\alpha|$

(b) $\mathscr{V} = \mathbb{C}$; then, for any $\alpha \in \mathbb{C}$, let

$$\|\alpha\| = |\alpha| = \sqrt{[\mathrm{Re}(\alpha)]^2 + [\mathrm{Im}(\alpha)]^2}$$

(c) (1) $\mathscr{V} = \mathbb{R}^n$; then, for $\boldsymbol{a} = (\alpha_1, \alpha_2, \ldots, \alpha_n)$, let

$$\|\boldsymbol{a}\| = \sum_{i=1}^{n} |\alpha_i|$$

(2) $\mathscr{V} = \mathbb{R}^n$; then, for $\boldsymbol{a} = (\alpha_1, \alpha_2, \ldots, \alpha_n)$, let

$$\|\boldsymbol{a}\| = \sqrt{\alpha_1^2 + \alpha_2^2 + \cdots + \alpha_n^2}$$

(3) $\mathscr{V} = \mathbb{R}^n$; then, for $\boldsymbol{a} = (\alpha_1, \alpha_2, \ldots, \alpha_n)$, let

$$\|\boldsymbol{a}\| = \left(\sum_{i=1}^{n} |\alpha_i|^p\right)^{1/p}$$

It can be shown that all of these norms satisfy the axioms of Definition 5.1.1. In (3) the so-called Minkowski inequality,

$$\left(\sum_{i=1}^{n} |\alpha_i + \beta_i|^p\right)^{1/p} \leqslant \left(\sum_{i=1}^{n} |\alpha_i|^p\right)^{1/p} + \left(\sum_{i=1}^{n} |\beta_i|^p\right)^{1/p}$$

is used (for a proof, see Simmons 1983, p. 218). Also, note that (1) and (2) are special cases of (3) with $p = 1$ and $p = 2$, respectively. ●

The norm function defines a natural distance between vectors. Thus, for $|\mathrm{a}\rangle$, $|\mathrm{b}\rangle \in \mathscr{V}$ we define the distance between them, $d(a, b)$, by

$$d(a, b) \equiv (|\mathrm{a}\rangle - |\mathrm{b}\rangle) \equiv \|a - b\| \tag{5.1}$$

The norm provides a means for testing whether two vectors are close or not: If $\|a - b\|$ is small, we say that $|\mathrm{a}\rangle$ and $|\mathrm{b}\rangle$ are close. Example 5.1.1 showed that several norms may be definable on a given vector space. Thus, the concept of closeness depends on the type of norm used in the vector space, as the following example shows.

[1] The Dirac notation is extremely convenient for inner product spaces but clumsy in other circumstances. For instance, the norm of a vector $|\mathrm{a}\rangle \in \mathscr{V}$ is more conveniently written as $\|\mathbf{a}\|$ than as $\||\mathrm{a}\rangle\|$. Similarly, the distance between $|\mathrm{a}\rangle$ and $|\mathrm{b}\rangle$ is better denoted as $d(a, b)$ than as $d(|\mathrm{a}\rangle, |\mathrm{b}\rangle)$. Whenever it is not confusing and more convenient, the Dirac notation will be dropped.

Thus, $|\psi\rangle$ must be zero, and

$$|\mathrm{f}\rangle = \textstyle\sum_{i=1}^{\infty} |\mathrm{e}_i\rangle\langle \mathrm{e}_i|\mathrm{f}\rangle$$

(ii) $\Rightarrow$ (iii): Since $|\mathrm{f}\rangle = (\sum_{i=1}^{\infty} |\mathrm{e}_i\rangle\langle \mathrm{e}_i|)|\mathrm{f}\rangle$ is true for all $|\mathrm{f}\rangle$, $\sum_{i=1}^{\infty} |\mathrm{e}_i\rangle\langle \mathrm{e}_i| = \mathbb{1}$ must hold.

(iii) $\Rightarrow$ (iv):

$$\langle \mathrm{f}|\mathrm{g}\rangle = \langle \mathrm{f}|\mathbb{1}|\mathrm{g}\rangle = \langle \mathrm{f}|(\textstyle\sum_{i=1}^{\infty} |\mathrm{e}_i\rangle\langle \mathrm{e}_i|)|\mathrm{g}\rangle = \sum_{i=1}^{\infty} \langle \mathrm{f}|\mathrm{e}_i\rangle\langle \mathrm{e}_i|\mathrm{g}\rangle$$

because of the linearity of the inner product.

(iv) $\Rightarrow$ (v): Let $|\mathrm{g}\rangle = |\mathrm{f}\rangle$ in (iv) and recall that $\langle \mathrm{e}_i|\mathrm{f}\rangle = \langle \mathrm{f}|\mathrm{e}_i\rangle^*$.

(v) $\Rightarrow$ (i): Let $|\mathrm{f}\rangle$ be orthogonal to all $|\mathrm{e}_i\rangle$. Then $\langle \mathrm{f}|\mathrm{e}_i\rangle = 0$ for all i. Thus, the sum is zero, and $\|\mathrm{f}\|^2 = 0$ implies that $|\mathrm{f}\rangle = \mathbf{0}$, because only the zero vector has a zero norm. ■

The equality

$$\|\mathbf{f}\|^2 = \langle \mathrm{f}|\mathrm{f}\rangle = \sum_{i=1}^{\infty} |\langle \mathrm{f}|\mathrm{e}_i\rangle|^2 = \sum_{i=1}^{\infty} |f_i|^2 \qquad \text{where } f_i \equiv \langle \mathrm{e}_i|\mathrm{f}\rangle \tag{5.4}$$

is called the *Parseval equality*. The numbers f_i are called *generalized Fourier coefficients*. The relation

$$\sum_{i=1}^{\infty} |\mathrm{e}_i\rangle\langle \mathrm{e}_i| = \mathbb{1} \tag{5.5}$$

is usually called the *completeness relation*, but any other requirement of Proposition 5.1.7 is just as good.

Proposition 5.1.7 makes the next definition natural.

5.1.8 Definition A complete orthonormal sequence $\{|\mathrm{e}_i\rangle\}_{i=1}^{\infty}$ in $\mathcal{H}$ is called a basis in $\mathcal{H}$.

A Hilbert space that has a countable basis, such as the one in Definition 5.1.8, is said to be *separable*. It can be shown (see Simmons 1983, p. 253) that in an inseparable Hilbert space all vectors are orthogonal to one another. Only separable Hilbert spaces will be considered in the book. However, to provide some insight into the peculiarities of inseparable Hilbert spaces, the next section discusses a related subject, the Dirac delta function, which is an example of a general class of mathematical objects called generalized functions, or distributions.

Exercises

5.1.1 Show that

$$|\,\|\mathbf{a}\| - \|\mathbf{b}\|\,| \leqslant \|\mathbf{a} + \mathbf{b}\| \leqslant \|\mathbf{a}\| + \|\mathbf{b}\|$$

and

$$|\,\|\mathbf{a}\| - \|\mathbf{b}\|\,| \leqslant \|\mathbf{a} - \mathbf{b}\| \leqslant \|\mathbf{a}\| + \|\mathbf{b}\|$$

5.1.2 Show that $\{f_k(x)\}$, the sequence of functions defined in Example 5.1.4, is a Cauchy sequence.

5.2 DISTRIBUTIONS (GENERALIZED FUNCTIONS)

Generalized functions are not (really) functions. The natural way to think about them is in terms of linear functionals, which is what we will do later. However, we need motivations, and we want to see how they arise in concrete situations. A natural habitat of these "functions" is the realm of the continuous index. We will therefore investigate this concept, but we have to keep in mind that our investigation will be purely intuitive. (In fact, some of the arguments in the following subsection are wrong!) However, we will obtain enough intuitive feeling for the concept to make it worthwhile to spend a little time on it.

5.2.1 The Continuous Index and the Dirac Delta Function

Let us develop the continuous index step by step. First, we consider the components of a vector $|\mathrm{f}\rangle$ in a basis $\mathrm{B} = \{|\mathrm{i}\rangle\}_{i=1}^{\infty}$ of $\mathscr{H}$ and let $f_i = \langle \mathrm{i}|\mathrm{f}\rangle$ be its ith component.[2] Given the basis B, the set of f_i's determines $|\mathrm{f}\rangle$ uniquely. These f_i's are in the range of the mapping $f: \mathbb{N} \to \mathbb{C}$, from the set of natural numbers $\mathbb{N}$ to the set of complex numbers $\mathbb{C}$, given by $f(i) = f_i$, or

$$f(i) = \langle \mathrm{i}|\mathrm{f}\rangle \tag{5.6}$$

This gives another alternative for representing vectors: An *abstract vector* is a function whose value at the integer i is the ith coordinate of the abstract vector in some specified basis. Such a function can even be graphed, as shown in Fig. 5.1, if, for the time being, the scalars are taken to be real rather than complex.

In Fig. 5.2 the unit length on the i-axis has been decreased so that the points lie closer together. The second step in the abstraction is to let the unit length along the horizontal axis go to zero, or become infinitesimal. In this case the upper limit can be represented by a finite point on the horizontal axis. This corresponds to a horizontal axis that represents a *continuous* variable, say x, and a vertical axis that is a function, $f(x)$, of the continuous variable. The interval between the adjacent integer points on the x-axis is (in the limit) equal to zero. Thus, the range of the variable x is from $x = 0$

[2] The notation $|\mathrm{i}\rangle$ is used here for a member of an orthonormal basis instead of $|\mathrm{e}_i\rangle$ because later we will generalize to continuous indices, and the use of $|\mathrm{i}\rangle$ facilitates this generalization.

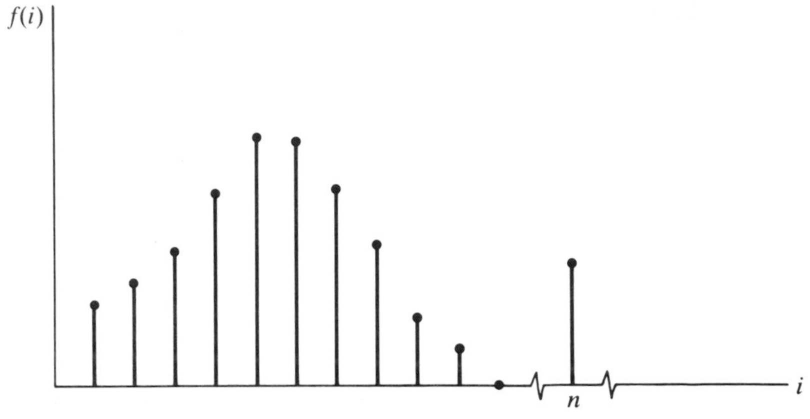

Figure 5.1 The function f plotted for integer values of its argument.

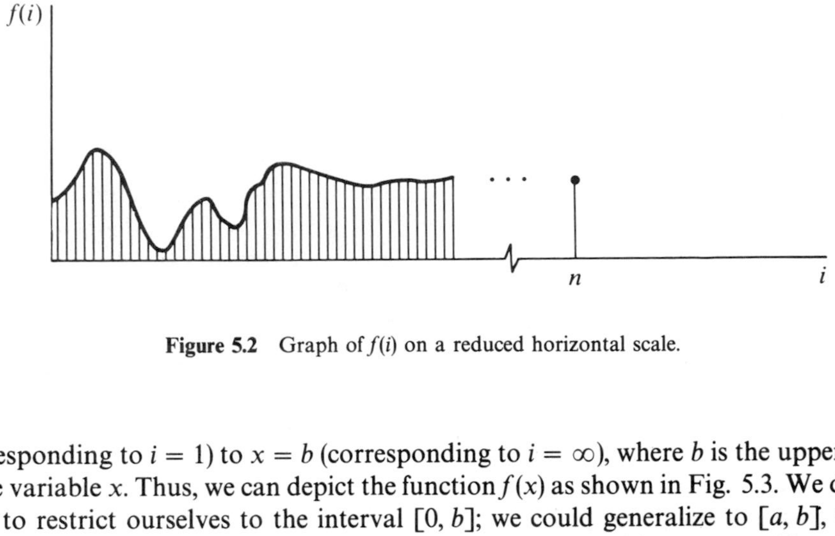

Figure 5.2 Graph of $f(i)$ on a reduced horizontal scale.

(corresponding to $i = 1$) to $x = b$ (corresponding to $i = \infty$), where b is the upper limit of the variable x. Thus, we can depict the function $f(x)$ as shown in Fig. 5.3. We do not have to restrict ourselves to the interval $[0, b]$; we could generalize to $[a, b]$, where a corresponds to $i = 1$ and b corresponds to $i = \infty$, or a could be $-\infty$ and b could be $+\infty$.

How do we express $|\mathrm{f}\rangle$ in terms of the $|\mathrm{x}\rangle$'s? We first divide the interval $[a, b]$ into N equal parts, each of length $\Delta x = (b - a)/N$. We can then write

$$|\mathrm{f}\rangle = \sum_{i=1}^{\infty} f_i|\mathrm{i}\rangle = \sum_{j=1}^{N} \sum_{k=1}^{\infty} f_j(k)|\mathrm{j}, \mathrm{k}\rangle$$

where we have changed the single sum over i into a double sum. The first sum (over j) tells which interval we are in, and the second (infinite) sum adds all the contributions in the jth interval. Let x_j denote the midpoint of the jth interval, and assume that $f(x)$

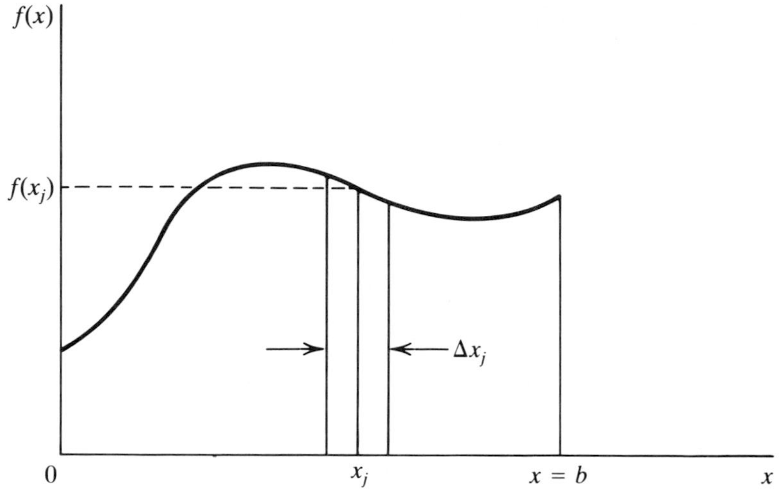

Figure 5.3 The function f plotted against the continuous variable x obtained when the interval between adjacent integral points goes to zero.

does not vary appreciably in that interval. Then $f_j(k) \approx f(x_j)$, and the above sum gives

$$|\mathrm{f}\rangle \approx \sum_{j=1}^{N} f(x_j) \sum_{k=1}^{\infty} |\mathrm{j}, \mathrm{k}\rangle$$

If we were dealing with ordinary numbers, the sum over k would be the number of points in the jth interval, which could be approximated by $w(x_j)\Delta x_j$, where $w(x)$ is a *density function* that tells how many points there are per unit interval.[3] It is tempting to do the same with vectors and write

$$\sum_{k=1}^{\infty} |\mathrm{j}, \mathrm{k}\rangle = w(x_j)\Delta x_j |\mathrm{x}_j\rangle \tag{5.7}$$

where $|\mathrm{x}_j\rangle$ is the "average" of all the vectors in the jth interval. However, we immediately run into trouble, because $|\mathrm{x}_j\rangle$ is orthogonal to all the other vectors in the interval (see the discussion at the end of Section 5.1). This means that it has no components along any of the other vectors and therefore cannot be their "average." This is where intuition clashes with precision. However, we disregard this conflict and continue because the final results are correct and we will obtain some relations that are useful in many circumstances.

[3] This density function, $w(x)$, also called the weight factor, is closely related to the normalization of $|\mathrm{x}\rangle$. The common choice, $w(x) = 1$, is only one of many possible choices for the normalization of $|\mathrm{x}\rangle$ [see Eq. (5.11)].

Using Eq. (5.7), we can write

$$|\mathrm{f}\rangle \approx \sum_{j=1}^{N} f(x_j)w(x_j)\Delta x_j|\mathrm{x}_j\rangle$$

When $N \to \infty$ and $\Delta x_j \to 0$, we obtain

$$|\mathrm{f}\rangle \equiv \int_a^b f(x)w(x)|\mathrm{x}\rangle\, dx \tag{5.8}$$

This is the expansion of $|\mathrm{f}\rangle$ in terms of vectors with a continuous index. Equation (5.6) now becomes

$$f(x) = \langle \mathrm{x}|\mathrm{f}\rangle$$

and

$$|\mathrm{f}\rangle = \int_a^b \langle \mathrm{x}|\mathrm{f}\rangle w(x)|\mathrm{x}\rangle\, dx = \int_a^b w(x)|\mathrm{x}\rangle\langle \mathrm{x}|\mathrm{f}\rangle\, dx$$

$$\equiv \left[\int_a^b |\mathrm{x}\rangle w(x)\langle \mathrm{x}|\, dx\right]|\mathrm{f}\rangle$$

Since this is true for arbitrary $|\mathrm{f}\rangle$, we obtain the completeness relation for a continuous index:

$$\int_a^b |\mathrm{x}\rangle w(x)\langle \mathrm{x}|\, dx = \mathbb{1} \tag{5.9a}$$

This is the generalization of (5.5) to the continuous case. In most (but not all) applications, $w(x)$ is set equal to unity, and (5.9a) becomes

$$\int_a^b |\mathrm{x}\rangle\langle \mathrm{x}|\, dx = \mathbb{1} \tag{5.9b}$$

Although Eq. (5.8) appears innocent, there is a monster sleeping inside it. We should suspect this because of the way (5.8) was derived from (5.7). To get a glimpse of this monster, let us take the inner product of (5.8) with $\langle \mathrm{x}'|$:

$$\langle \mathrm{x}'|\mathrm{f}\rangle = f(x') = \int_a^b f(x)\langle \mathrm{x}'|\mathrm{x}\rangle w(x)\, dx \qquad \forall\ |\mathrm{f}\rangle$$

where x' is assumed to lie in the interval $[a, b]$ [otherwise, $f(x') = 0$, by definition]. Let us denote $\langle \mathrm{x}'|\mathrm{x}\rangle$ by $D(x', x)$, which is a function of x' and x, and write the preceding equation as

$$f(x') = \int_a^b f(x)D(x', x)w(x)\, dx \tag{5.10}$$

This equation, which holds for arbitrary $f(x)$, tells us immediately that $D(x', x)$ is no ordinary function. When the product of two ordinary functions is integrated, the result will not be related to either of the original functions or to its value at a point in the range of integration.

From Eq. (5.10) it can be shown that

(1) $D(x', x) = 0$ if $x \neq x'$

(2) $D(x, x) = \infty$

(3) $D(x', x) = D(x - x') = D(x' - x)$

(4) $\int_a^b w(x)D(x - x')\,dx = 1$

The proof is left as a problem for the reader.

It is customary to define the *Dirac delta function* as

$$\delta(x - x') = w(x)D(x', x) \equiv w(x)\langle \mathrm{x}'|\mathrm{x}\rangle \tag{5.11}$$

We then have the following proposition.

5.2.1 Proposition The Dirac delta function defined in Eqs. (5.10) and (5.11) has the following properties:

(i) $\delta(x - x') = 0$ if $x \neq x'$

(ii) $\delta(0) = \infty$

(iii) $\delta(x) = \delta(-x)$ (that is, the Dirac delta function is even)

(iv) $\int_a^b \delta(x - x')\,dx = 0$ if $x' \notin [a, b]$

(v) $\int_{-\infty}^{\infty} f(x)\delta(x - x')dx = f(x')$; in particular, $\int_{-\infty}^{\infty} \delta(x - x')\,dx = 1$ ■

In Property (v) the interval is taken to be infinite because $\delta(x - x') = 0$ everywhere except at a point. So including all the rest of the real line does not change the integral.

Written in the form

$$\langle \mathrm{x}'|\mathrm{x}\rangle = \frac{1}{w(x)}\delta(x - x')$$

Equation (5.11) is the generalization of the orthonormality relation of vectors to the case of the continuous index.

We can apply all of the above development to complex vector spaces. In that case $f(x)$ will be complex and

$$f^*(x) = \langle \mathrm{x}|\mathrm{f}\rangle^* = \langle \mathrm{f}|\mathrm{x}\rangle$$

Example 5.2.1

An example of a representation of the Dirac delta function is a Gaussian curve whose width approaches zero at the same time and rate that its height approaches infinity, so the area under it remains constant. In fact, we may write

$$\delta(x - x') = \lim_{\varepsilon \to 0} \frac{1}{\sqrt{\varepsilon\pi}} e^{-(x-x')^2/\varepsilon}$$

For any nonzero value of ε, we can easily verify that (see Exercise 5.3.2)

$$\int_{-\infty}^{\infty} \frac{1}{\sqrt{\varepsilon\pi}} e^{-(x-x')^2/\varepsilon}\, dx = 1$$

This relation is independent of ε and therefore still holds in the limit $\varepsilon \to 0$. On the other hand, the graphs of the above Gaussian curves for different ε's reveal a "peaking up" as $\varepsilon \to 0$ (Fig. 5.4). ●

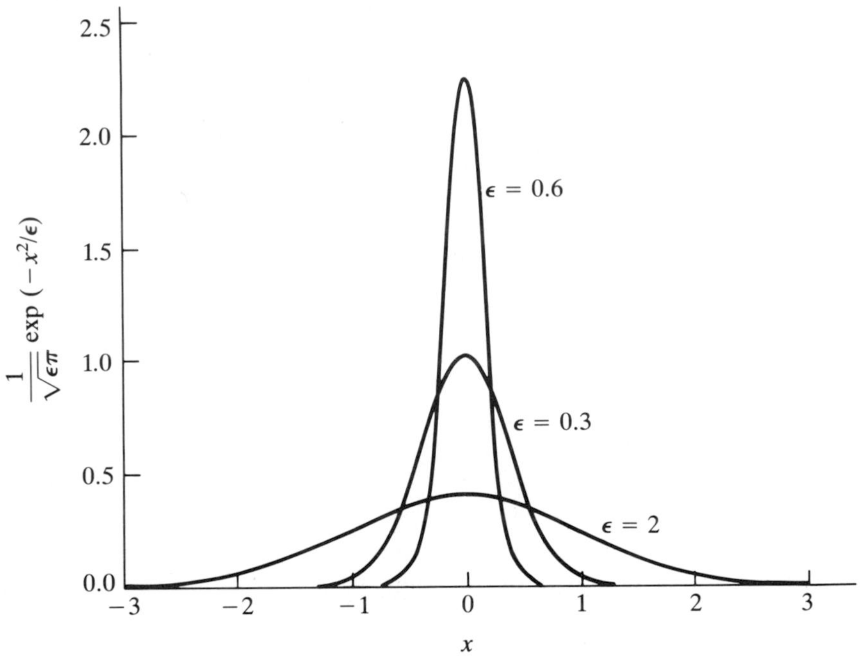

Figure 5.4 The function $e^{-x^2/\varepsilon}/\sqrt{\varepsilon\pi}$ plotted for different values of ε. The smaller ε is, the sharper the peak of the function.

Example 5.2.2

Consider the function $D_T(x - x')$ defined as

$$D_T(x - x') \equiv \frac{1}{2\pi} \int_{-T}^{+T} e^{it(x-x')}\, dt$$

Note that the integration is over t, not x (or x')! The integral is straightforward and yields

$$D_T(x - x') = \frac{1}{2\pi}\left[\frac{e^{it(x-x')}}{i(x-x')}\right]\Bigg|_{-T}^{+T} = \frac{1}{\pi}\left[\frac{\sin T(x-x')}{x-x'}\right]$$

The graph of $D_T(x - x')$ as a function of x for $x' = 0$ and various values of T is shown in Fig. 5.5. Note that the width of the curve decreases as T increases. The area under the curve can be calculated:

$$\int_{-\infty}^{\infty} D_T(x - x')\,dx = \frac{1}{\pi}\int_{-\infty}^{\infty} \frac{\sin T(x - x')}{x - x'}\,dx = \frac{1}{\pi}\int_{-\infty}^{\infty} \frac{\sin y}{y}\,dy = 1$$

This calculation used the result

$$\int_{-\infty}^{\infty} \frac{\sin x}{x}\,dx = \pi$$

which we will obtain in Chapter 7 using the powerful method of the calculus of residues from complex analysis (see Example 7.2.13). Thus, the area is independent of the width, T.

Figure 5.5 shows that $D_T(x - x')$ becomes more and more like the Dirac delta function as T increases. In fact, we have

$$\delta(x - x') = \lim_{T \to \infty} \frac{1}{\pi}\left[\frac{\sin T(x - x')}{x - x'}\right] \tag{1}$$

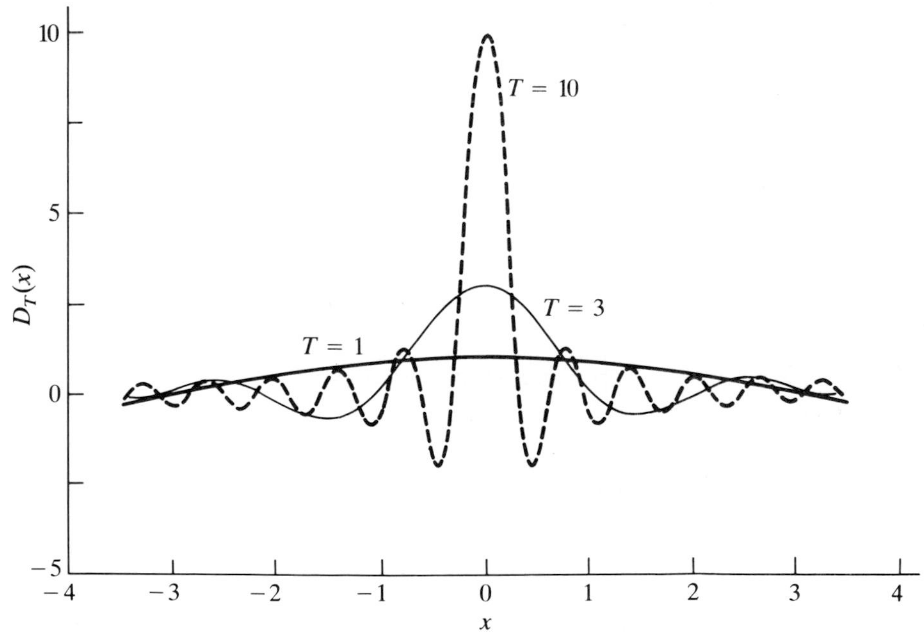

Figure 5.5 The function $\sin Tx/x$ is plotted for different values of T. The larger T is, the sharper the peak of the function.

To see this, we note that for any finite T we can write

$$D_T(x - x') = \frac{T}{\pi}\left[\frac{\sin T(x - x')}{T(x - x')}\right]$$

This shows that *if x is very close to x'*, $T(x - x') \to 0$ and $\sin T(x - x')/T(x - x') \to 1$. Thus, when $x \to x'$, we have

$$D_T(x - x') \approx \frac{T}{\pi}$$

which is large when T is large. This is as expected of a delta function: $\delta(0) = \infty$. On the other hand, the width of $D_T(x - x')$ around x' is given, roughly, by the points at which $D_T(x - x')$ drops to zero: $T(x - x') = \pm\pi$, or $x - x' = \pm\pi/T$. Therefore, the width is given, roughly, by the distance between $x_+ = x' + \pi/T$ and $x_- = x' - \pi/T$, or $\Delta x = x_+ - x_- = 2\pi/T$, which goes to zero as T grows. Again, this is as expected from a delta function.

We see, then, that (1) has all the properties of the Dirac delta function. This suggests writing

$$\delta(x - x') = \frac{1}{2\pi}\int_{-\infty}^{\infty} e^{it(x-x')}\,dt$$ •

Example 5.2.3

A third representation of the Dirac delta function involves the step function, $\theta(x - x')$, which is defined as

$$\theta(x - x') \equiv \begin{cases} 0 & \text{if} \quad x < x' \\ 1 & \text{if} \quad x > x' \end{cases}$$

and is discontinuous at $x = x'$. We can approximate this step function by a continuous function $T_\varepsilon(x - x')$ defined by

$$T_\varepsilon(x - x') \equiv \begin{cases} 0 & \text{if} \quad x \leqslant x' - \varepsilon \\ \dfrac{1}{2\varepsilon}(x - x' + \varepsilon) & \text{if} \quad x' - \varepsilon \leqslant x \leqslant x' + \varepsilon \\ 1 & \text{if} \quad x \geqslant x' + \varepsilon \end{cases}$$

where ε is a small positive number. It is clear that

$$\theta(x - x') = \lim_{\varepsilon \to 0} T_\varepsilon(x - x')$$

Now let us consider the slope of $T_\varepsilon(x - x')$ as a function of x:

$$\frac{dT_\varepsilon(x - x')}{dx} = \begin{cases} 0 & \text{if} \quad x < x' - \varepsilon \\ \dfrac{1}{2\varepsilon} & \text{if} \quad x' - \varepsilon < x < x' + \varepsilon \\ 0 & \text{if} \quad x > x' + \varepsilon \end{cases}$$

Note that the slope is not defined at $x = x' - \varepsilon$ and $x = x' + \varepsilon$. We observe that dT_ε/dx is zero everywhere except when x lies in the interval $[x' - \varepsilon, x' + \varepsilon]$, when it is equal to $1/2\varepsilon$, which goes to infinity as $\varepsilon \to 0$. Here again we see signs of the delta function. In fact, we note that

$$\int_{-\infty}^{\infty} \left(\frac{dT_\varepsilon}{dx}\right) dx = \int_{x'-\varepsilon}^{x'+\varepsilon} \left(\frac{dT_\varepsilon}{dx}\right) dx = \int_{x'-\varepsilon}^{x+\varepsilon} \frac{1}{2\varepsilon}\, dx = 1$$

and the area under the curve of the function dT_ε/dx is 1. It is not surprising, then, to find that

$$\lim_{\varepsilon \to 0} \frac{dT_\varepsilon(x - x')}{dx} = \delta(x - x')$$

Interchanging the differentiation and the limiting process gives

$$\frac{d}{dx}\left[\lim_{\varepsilon \to 0} T_\varepsilon(x - x')\right] = \frac{d}{dx}\theta(x - x') = \delta(x - x') \qquad \bullet$$

5.2.2 Several Continuous Indices

Now that we have some understanding of one continuous index, we can generalize the results to several continuous indices. In the earlier discussion we looked at $f(x)$ as the xth component of some abstract vector $|\mathrm{f}\rangle$, where $f(x)$ is represented by a graph similar to the one in Fig. 5.3. For functions of two variables, x and y, we can think of $f(x, y)$ as the component of an abstract vector $|\mathrm{f}\rangle$ along a basis vector $|\mathrm{x}, \mathrm{y}\rangle$. This basis is a direct generalization of $|\mathrm{x}\rangle$ to more than one continuous index. Of course, the labels x and y are completely independent of each other. Then $f(x, y)$ is defined as

$$f(x, y) \equiv \langle \mathrm{x}, \mathrm{y} | \mathrm{f} \rangle$$

If the range of x is from a to b and that of y is from c to d, then we can generalize the equations of the single variable to

$$|\mathrm{f}\rangle = \int_a^b dx \int_c^d dy\, f(x, y) w(x, y) |\mathrm{x}, \mathrm{y}\rangle$$

$$\int_a^b dx \int_c^d dy\, |\mathrm{x}, \mathrm{y}\rangle w(x, y) \langle \mathrm{x}, \mathrm{y}| = \mathbb{1}$$

$$f(x', y') = \int_a^b dx \int_c^d dy\, f(x, y) w(x, y) \langle \mathrm{x}', \mathrm{y}' | \mathrm{x}, \mathrm{y}\rangle$$

$$\langle \mathrm{x}', \mathrm{y}' | \mathrm{x}, \mathrm{y}\rangle w(x, y) = \delta(x - x')\delta(y - y')$$

A *distribution* is thus a mathematical entity that appears inside an integral in conjunction with a well-behaved test function such that the result of integration is a well-defined number. Depending on the type of test function used, different kinds of distributions can be defined. If we want to include the Dirac delta function plus its derivatives of all orders, then the test functions must have derivatives of all orders; that is, they must be C^∞ functions. In addition, in order for the theory of distributions to work, all the test functions must vanish outside a finite "volume" of $\mathbb{R}^n$ or $\mathbb{C}^n$. We use $C_F^\infty(\mathbb{R}^n)$ or $C_F^\infty(\mathbb{C}^n)$ to denote all such functions defined in $\mathbb{R}^n$ or $\mathbb{C}^n$ (F stands for finite). It is hard to imagine a function that is infinitely differentiable and, at the same time, vanishes outside a finite region. The following example, however, shows that such functions do exist.

Example 5.2.5

Let us define the function $f: \mathbb{R} \to \mathbb{R}$ by

$$f(x) \equiv \begin{cases} \exp(-1/x) & \text{if } x \geqslant 0 \\ 0 & \text{if } x < 0 \end{cases}$$

It is clear that $d^n f/dx^n$ exists for all values of n and all values of x except, possibly, $x = 0$. For $x = 0$ we use the definition of the derivative:

$$\left.\frac{df}{dx}\right|_{x=0} = \lim_{\Delta x \to 0} \frac{f(\Delta x) - f(0)}{\Delta x} = \lim_{\Delta x \to 0} \frac{f(\Delta x)}{\Delta x}$$

$$= \lim_{\Delta x \to 0} \begin{cases} \dfrac{\exp(-1/\Delta x)}{\Delta x} & \text{if } \Delta x > 0 \\ \dfrac{0}{\Delta x} & \text{if } \Delta x < 0 \end{cases} = \lim_{\Delta x \to 0} \begin{cases} \left(\dfrac{1}{\Delta x}\right) e^{-1/\Delta x} & \text{if } \Delta x > 0 \\ 0 & \text{if } \Delta x < 0 \end{cases}$$

Employing L'Hôpital's rule, we get

$$\lim_{\Delta x \to 0} \left(\frac{1}{\Delta x}\right) e^{-1/\Delta x} = 0$$

Thus,

$$\left.\frac{df}{dx}\right|_{x=0} = 0$$

In fact, similar arguments can be used to show that

$$\left.\frac{d^n f}{dx^n}\right|_{x=0} = 0 \qquad \forall\, n$$

Therefore, f is a C^∞ function.

Now let us consider the function $g: \mathbb{R} \to \mathbb{R}$ defined by

$$g(x) = \begin{cases} 0 & \text{if } x \leqslant 0 \text{ or } x \geqslant 1 \\ \exp\left[\dfrac{1}{x(x-1)}\right] & \text{if } 0 < x < 1 \end{cases}$$

This function is C_F^∞ and vanishes outside the interval (0, 1).

We can generalize the above ideas and define

$$h(x; a, b) = \begin{cases} 0 & \text{if } x \leqslant a \text{ or } x \geqslant b \\ \exp\left[\dfrac{1}{(x-a)(x-b)}\right] & \text{if } a < x < b \end{cases}$$

which is C_F^∞ and vanishes outside the interval (a, b). •

The definitive property of distributions concerns the way they combine with test functions to give a number. The test functions used clearly form a vector space over $\mathbb{R}$ or $\mathbb{C}$. In this vector-space language, distributions are linear functionals. The linearity is a simple consequence of the properties of the integral. We, therefore, have the following definition.

5.2.2 Definition A *distribution*, or *generalized function*, is a linear functional on the space of $C_F^\infty(\mathbb{R}^n)$ or $C_F^\infty(\mathbb{C}^n)$. If $f \in C_F^\infty$ and φ is a distribution, then

$$\varphi[f] = \int_{-\infty}^{\infty} \varphi(\mathbf{t}) f(\mathbf{t})\, d^n t$$

It is usually further assumed that φ satisfies certain continuity requirements, but we will not deal with such subtleties.

In Definition 5.2.2 $\mathbf{t} = (t_1, t_2, \ldots, t_n) \in \mathbb{R}^n$ or $\mathbb{C}^n$ and $d^n t = dt_1\, dt_2 \cdots dt_n$. Note that $\varphi[f]$ is a number, as expected of a functional. Another commonly used notation is $\langle \varphi, f \rangle$. This is more appealing not only because φ is a linear functional, in the sense that $\varphi[af + bg] = a\varphi[f] + b\varphi[g]$, but also because the set of all such linear functionals forms a vector space; that is, the linear combination of the φ's is also defined. Thus, $\langle \varphi, f \rangle$ suggests a mutual "democracy" in a natural way.

We now have a shorthand way of writing integrals. For instance, if δ represents the Dirac delta function $\delta(x)$, then

$$\langle \delta, f \rangle = f(0) \tag{5.15a}$$

Similarly,

$$\langle \delta', f \rangle = -f'(0) \tag{5.15b}$$

and, for linear combinations,

$$\langle a\delta + b\delta', f \rangle = af(0) - bf'(0) \tag{5.15c}$$

The disadvantage of such a notation is that it does not differentiate between distributions with different arguments. Thus, Eqs. (5.15) do not indicate whether we are dealing with $\delta(x)$ or $\delta(x - x')$ or $\delta(x^2 - a^2)$ or some other distribution. We will use (5.15) only when the argument is clear from the context.

We also assume that we can generalize the linearity $\langle \varphi, \sum_{i=1}^{n} a_i f_i \rangle = \sum_{i=1}^{n} a_i \langle \varphi, f_i \rangle$ to continuous sums, or integrals. We have

$$\left\langle \varphi, \int_a^b f\,dy \right\rangle = \int_a^b \langle \varphi, f \rangle\,dy$$

or, writing it explicitly,

$$\int_{-\infty}^{\infty} \varphi(x) \int_a^b f(x, y)\,dy\,dx = \int_a^b \int_{-\infty}^{\infty} \varphi(x) f(x, y)\,dx\,dy \tag{5.16a}$$

Similarly,

$$\frac{\partial}{\partial y} \langle \varphi, f \rangle = \left\langle \varphi, \frac{\partial f}{\partial y} \right\rangle \tag{5.16b}$$

Example 5.2.6

An ordinary (continuous) function f can be thought of as a special case of a distribution. The linear functional $f: C_F^\infty(\mathbb{R}) \to \mathbb{R}$ is simply defined by

$$\langle f, g \rangle \equiv f[g] = \int_{-\infty}^{\infty} f(x) g(x)\,dx$$ ●

Example 5.2.7

The interesting applications of distributions (generalized functions) occur when the notion of density is generalized to include not only smooth densities, but also pointlike, linear, and surface densities.

As Exercise 5.2.1 shows, a point charge q located at $\mathbf{r}_0$ can be thought of as having a charge density $\rho(\mathbf{r}) = q\delta(\mathbf{r} - \mathbf{r}_0)$. In the language of linear functionals we interpret ρ as a distribution, $\rho: C_F^\infty \to \mathbb{R}$, which for an arbitrary $C_F^\infty(\mathbb{R}^3)$ function f gives

$$\rho[f] = \langle \rho, f \rangle = q f(\mathbf{r}_0)$$

The "monstrous" character of ρ is hidden in this equation. However, we can detect this character by recalling that the LHS is

$$\iiint \rho(\mathbf{r}) f(\mathbf{r})\,d^3r = \lim_{\substack{N \to \infty \\ \Delta V_i \to 0}} \sum_{i=1}^{N} \rho(\mathbf{r}_i) f(\mathbf{r}_i) \Delta V_i$$

On the RHS of this equation the only volume element that contributes is the one that contains the point $\mathbf{r}_0$; all the rest contribute zero. As $\Delta V_i \to 0$, the only way that the RHS can give a finite number is if $\rho(\mathbf{r}_0) f(\mathbf{r}_0)$ is infinite. Since f is a well-behaved function, $f(\mathbf{r}_0)$ is finite. Thus, $\rho(\mathbf{r}_0)$ is infinite, which implies that $\rho(\mathbf{r})$ is a delta function.

Similarly, let us assume that we have a smooth linear charge density $\lambda(\mathbf{r})$ on a curve γ in $\mathbb{R}^3$. We define a distribution $\rho: C_F^\infty(\mathbb{R}^3) \to \mathbb{R}$ by

$$\langle \rho, f \rangle = \int_\gamma \lambda(\mathbf{r}) f(\mathbf{r})\,dl$$

where dl is an element of the arc of γ and the integral is over the curve. Again the peculiar character of ρ can be detected by writing

$$\langle \rho, f \rangle = \iiint \rho(\mathbf{r}) f(\mathbf{r}) d^3 r = \lim_{\substack{N \to \infty \\ \Delta V_i \to 0}} \sum_{i=1}^{N} \rho(\mathbf{r}_i) f(\mathbf{r}_i) \Delta V_i$$

where ΔV_i is now a cylinder of infinitesimal thickness that is shaped exactly like γ. Again all contributions from the RHS vanish except that of ΔV_i, which contains the charged curve γ. Since $\Delta V_i \to 0$ and $f(\mathbf{r})$ is a good function, $\rho(\mathbf{r})$ must be infinite *for points on* γ and zero everywhere else.

It can also be shown that if $\sigma(\mathbf{r})$ is a smooth surface charge density over a surface S in $\mathbb{R}^3$, then the distribution $\rho: C_F^\infty(\mathbb{R}^3) \to \mathbb{R}$ given by

$$\langle \rho, f \rangle = \iint_S \sigma(\mathbf{r}) f(\mathbf{r}) da$$

is zero everywhere except at points on S, where it is infinite. ●

Example 5.2.7 shows that if a distribution is defined in terms of integrals of lower dimension, then the distribution blows up at points of the integrand. In general, if $\rho: C_F^\infty(\mathbb{R}^n) \to \mathbb{R}$ is defined by

$$\langle \rho, f \rangle = \int_S \sigma(\mathbf{r}) f(\mathbf{r}) d^m r \tag{5.17}$$

where S is a hypersurface, $\sigma(\mathbf{r})$ is a smooth function on S, and $m < n$, then $\rho(\mathbf{r})$ must be infinite for $\mathbf{r} \in S$.

Distributions as limits of sequences of functions. We have seen that the delta function can be thought of as the limit of an ordinary function. This idea can be generalized.

5.2.3 Definition Let $\{\varphi_n(x)\}$ be a sequence of functions such that

$$\lim_{n \to \infty} \int_{-\infty}^{\infty} \varphi_n(x) f(x)\, dx$$

exists for every $f \in C_F^\infty(\mathbb{R})$. Then the sequence is said to converge to the distribution φ, defined by

$$\langle \varphi, f \rangle = \lim_{n \to \infty} \int_{-\infty}^{\infty} \varphi_n(x) f(x)\, dx \qquad \forall f \tag{5.18}$$

This convergence is denoted as $\varphi_n \to \varphi$.

Thus, it can be verified that

$$\frac{n}{\sqrt{\pi}} e^{-n^2x^2} \to \delta(x)$$

and that

$$\frac{1 - \cos nx}{n\pi x^2} \to \delta(x)$$

and so on. The proofs are left as problems.

The derivative of a distribution. The derivative of a distribution is defined as follows.

5.2.4 Definition The derivative of a distribution, φ, is another distribution, φ', defined by

$$\langle \varphi', f \rangle \equiv -\langle \varphi, f' \rangle \qquad \forall f \in \mathrm{C}_{\mathrm{F}}^{\infty} \tag{5.19}$$

Example 5.2.8

We can combine definitions 5.2.3 and 5.2.4 to show that if the functions θ_n are defined as

$$\theta_n(x) \equiv \begin{cases} 0 & \text{if } x < -\dfrac{1}{n} \\ \dfrac{n}{2}x & \text{if } -\dfrac{1}{n} < x < \dfrac{1}{n} \\ 1 & \text{if } x > \dfrac{1}{n} \end{cases}$$

then $\theta_n'(x) \to \delta(x)$.

From (5.19) and the fact that $\theta_n(x)$ can be thought of as a distribution (see Example 5.2.6), we have

$$\langle \theta_n', f \rangle = -\langle \theta_n, f' \rangle$$

which, in terms of integrals, is

$$\begin{aligned}
\int_{-\infty}^{\infty} \theta_n'(x) f(x)\,dx &= -\int_{-\infty}^{\infty} \theta_n(x) \frac{df}{dx}\,dx \\
&= -\left[\int_{-\infty}^{-1/n} \theta_n(x)\frac{df}{dx}\,dx + \int_{-1/n}^{1/n} \theta_n(x)\frac{df}{dx}\,dx + \int_{1/n}^{\infty} \theta_n(x)\frac{df}{dx}\,dx\right] \\
&= -\left(0 + \int_{-1/n}^{1/n} \frac{n}{2}x\,df + \int_{1/n}^{\infty} df\right) \\
&= -\frac{n}{2}\left(xf\Big|_{-1/n}^{1/n} - \int_{-1/n}^{1/n} f\,dx\right) - f(x)\Big|_{1/n}^{\infty}
\end{aligned}$$

For large n we have $1/n \approx 0$ and $f(1/n) \approx f(0)$. Thus,

$$\int_{-\infty}^{\infty} \theta_n'(x) f(x)\, dx \approx -\frac{n}{2}\left[\frac{1}{n} f\left(\frac{1}{n}\right) + \frac{1}{n} f\left(-\frac{1}{n}\right) - f(0)\left(\frac{2}{n}\right)\right] - f(\infty) + f\left(\frac{1}{n}\right) \approx f(0)$$

The approximation becomes equality in the limit $n \to \infty$. Thus,

$$\lim_{n\to\infty} \int_{-\infty}^{\infty} \theta_n'(x) f(x)\, dx = f(0) = \langle \delta, f \rangle$$

and

$$\theta_n' \to \delta$$

Note that $f(\infty) = 0$ because of the assumption that all functions must vanish outside a finite volume. ●

The integral of a distribution. We saw above that the derivative of a distribution is given in terms of the derivative of the test function. It is not surprising, then, that the primitive (or antiderivative, or indefinite integral) of a distribution is also given in terms of the primitive of the test function. In fact, we can turn (5.19) around and write

$$\langle \Phi, f \rangle \equiv -\langle \varphi, F \rangle \tag{5.20a}$$

where Φ and F are primitives of φ and f, respectively. The primitive of f is

$$F(x) = \int_{-\infty}^{x} f(t)\, dt$$

Therefore, even if $f(x) \in C_F^\infty$, F may not vanish at infinity because $\int_{-\infty}^{\infty} f(t)\, dt$ may not be zero. In that case $F \notin C_F^\infty$, and Eq. (5.20a) is not defined for F. However, we can remedy this by the following procedure. Let $g \in C_F^\infty$ be such that $\int_{-\infty}^{\infty} g(x)\, dx = 1$, and let $\int_{-\infty}^{\infty} f(t)\, dt = a$. Then the primitive of $f_0 \equiv f - ag$ belongs to C_F^∞, and (5.20a) is defined. We, therefore, have the following definition.

5.2.5 Definition Let φ be a distribution and $f \in C_F^\infty$. Let $a = \int_{-\infty}^{\infty} f(t)\, dt$ and $g \in C_F^\infty$ such that $\int_{-\infty}^{\infty} g(t)\, dt = 1$. Then the primitive of φ, denoted Φ, is defined as

$$\langle \Phi, f \rangle = \langle \Phi, f_0 \rangle + \langle C, f \rangle \tag{5.20b}$$

where $f_0 \equiv f - ag$, C is a constant (distribution), and $\langle \Phi, f_0 \rangle$ is as defined in (5.20a). The constant C is simply equal to $\langle \Phi, g \rangle$.

Example 5.2.9

Let us find the primitive of $\delta(x)$ using Definition 5.2.5. For $\varphi = \delta$ we have

$$\begin{aligned}\langle \Phi, f\rangle &= \langle \Phi, f_0\rangle + \text{constant} = -\langle \delta, F_0\rangle + \text{constant}\\ &= -F_0(0) + \text{constant} = -\int_{-\infty}^{0} f_0(t)\,dt + \text{constant}\\ &= -\int_{-\infty}^{0} [f(t) - ag(t)]\,dt + \text{constant}\\ &\equiv -\int_{-\infty}^{0} f(t)\,dt + b\int_{-\infty}^{+\infty} f(t)\,dt\end{aligned} \tag{1}$$

where b is a new constant defined by the last step. Rewriting the LHS of (1) in terms of integrals and changing the dummy integration variable from t to x, we get

$$\int_{-\infty}^{\infty} \Phi(x) f(x)\,dx = \int_{-\infty}^{0} (b-1) f(x)\,dx + \int_{0}^{\infty} b f(x)\,dx \qquad \forall f \in C_F^\infty(\mathbb{R})$$

This equation is true for all f if and only if

$$\Phi(x) = \begin{cases} b-1 & \text{if } x < 0 \\ b & \text{if } x > 0 \end{cases}$$

With $b = 1$ we recover the theta function, previously shown to be the primitive of the delta function. •

Changing the variables of integration. We have seen how to deal with $\delta(f(x))$. In the language of distributions expressions such as $\varphi(f(x))$ may not be defined, especially since test functions have been restricted to being C_F^∞. In fact, if we do the most natural thing and define $y = f(x)$ so that the variable of integration changes, we may run into difficulty because x may not be expressible in terms of y. Even if it is, it may not be C_F^∞.

However, if $y = f(x)$ is C_F^∞ and $x = f^{-1}(y)$ is also C_F^∞, then

$$\int_{-\infty}^{\infty} \varphi(f(x)) g(x)\,dx = \int_{-\infty}^{\infty} \varphi(y) g(f^{-1}(y)) \left|\frac{df^{-1}(y)}{dy}\right| dy \tag{5.21}$$

or

$$\langle \psi, g\rangle = \langle \varphi, h\rangle \tag{5.22}$$

where

$$\psi(x) = \varphi(f(x)) \qquad \text{and} \qquad h(y) = g(f^{-1}(y)) \left|\frac{df^{-1}(y)}{dy}\right|$$

In $\mathbb{R}^n$ the Jacobian must be used. Thus, in (5.22),

$$\psi(\mathbf{r}) = \varphi(f(\mathbf{r})) \qquad \text{and} \qquad h(\mathbf{y}) = g(f^{-1}(\mathbf{y})) \left|\frac{\partial(x_1, \ldots, x_n)}{\partial(y_1, \ldots, y_n)}\right|$$

This section discussed continuous indices. The best way to study "functions" resulting from continuous indices, such as the Dirac delta function, is through the notion of distributions, or generalized functions. The remainder of this chapter deals with separable Hilbert spaces, in which a countable orthonormal basis exists. As will be seen, these Hilbert spaces are much more interesting and applicable to many more physical situations than the inseparable ones are.

Exercises

5.2.1 It is sometimes helpful to be able to associate a density with a point particle. For instance, we may wish to represent the mass density and the charge density of a massive point charge q having mass m. Show that

$$\rho_m(x) = m\delta(x - x_0) \qquad \text{and} \qquad \rho_c(x) = q\delta(x - x_0)$$

represent the mass and charge density of a point particle located at $x = x_0$. (This is a one-dimensional situation, which can easily be generalized to more dimensions.)

5.2.2 Write a density function for two point charges, q_1 and q_2, located at $\mathbf{r} = \mathbf{r}_1$ and $\mathbf{r} = \mathbf{r}_2$, respectively.

5.2.3 Show that

$$\delta(\mathbf{r} - \mathbf{r}') = \frac{1}{(2\pi)^3} \iiint e^{i\mathbf{k}\cdot(\mathbf{r}-\mathbf{r}')}\, d^3k$$

5.2.4 Show that

$$\delta(f(x)) = \frac{1}{\left|\dfrac{df}{dx}\right|}\delta(x - x_0)$$

where x_0 is such that $f(x_0) = 0$.

5.2.5 What happens if the function $f(x)$ in Exercise 5.2.4 has more than one zero? Investigate this possibility.

5.2.6 Show that

$$\int_{-\infty}^{\infty} \delta'(x) f(x)\, dx = -f'(0) \qquad \text{where } \delta'(x) \equiv \frac{d}{dx}\delta(x)$$

5.3 THE SPACE OF SQUARE-INTEGRABLE FUNCTIONS

It was mentioned earlier that the collection of all continuous functions defined in $[a, b]$ forms a linear vector space, and Example 5.1.4 showed that this space is not complete. However, if we lift the continuity restriction, then we will have a complete space. Let us study the space of functions in detail.

We can think of a function $f(x)$ as $|\mathrm{f}\rangle$, a member of an abstract (infinite-dimensional) vector space whose "xth" component in the "basis" $\mathrm{B} = \{|\mathrm{x}\rangle | x \in [a, b]\}$ is $f(x)$. In other words, we can write $f(x) = \langle \mathrm{x}|\mathrm{f}\rangle$.

The inner product of two functions, $f(x)$ and $g(x)$, can be "derived" using (5.9) as follows:

$$\langle g|f\rangle = \langle g|\mathbb{1}|f\rangle = \langle g|\left[\int_a^b |x\rangle w(x)\langle x|\,dx\right]|f\rangle$$

$$= \int_a^b \langle g|x\rangle\langle x|f\rangle\, w(x)\,dx$$

$$= \int_a^b g^*(x) f(x) w(x)\,dx$$

If $g(x) = f(x)$, we obtain

$$\langle f|f\rangle = \int_a^b |f(x)|^2 w(x)\,dx \tag{5.23}$$

Functions for which (5.23) is defined are said to be *square-integrable.*

The space of square-integrable functions over the interval $[a, b]$ is denoted by $\mathscr{L}^2_w(a, b)$. In this notation $\mathscr{L}$ stands for Lebesgue, who generalized the notion of the ordinary Riemann integral to cases for which the functions could be highly discontinuous; 2 stands for the power of $f(x)$ in the integral; a and b denote the limits of integration; and w refers to the weight function (a strictly positive real-valued function).[4] When $w(x) = 1$, we use the notation $\mathscr{L}^2(a, b)$.

The space $\mathscr{L}^2_w(a, b)$ is of no significance unless it is complete. The following theorem, due to Riesz and Fischer, proves this completeness (for a proof, see Reed and Simon 1980, Chapter III).

5.3.1 Theorem The space $\mathscr{L}^2_w(a, b)$ is complete. In other words, if there is a sequence of square-integrable functions, $f_1(x), f_2(x), \ldots, f_n(x), \ldots$, such that

$$\lim_{i,j\to\infty} \|f_i - f_j\|^2 \equiv \lim_{i,j\to\infty} \int_a^b |f_i(x) - f_j(x)|^2\, w(x)\,dx = 0$$

then there exists a function $f(x) \in \mathscr{L}^2_w(a, b)$ such that

$$\lim_{i\to\infty} \int_a^b |f_i(x) - f(x)|^2\, w(x)\,dx = 0$$

■

[4] The integral in (5.23) should really be thought of as a Lebesgue integral; however, this book will not distinguish it from the ordinary Riemann integral.

Another complete infinite-dimensional inner product space is the Hilbert space, $\mathcal{H}$. The following theorem shows that the number of complete infinite-dimensional inner product spaces is severely restricted. (For a proof, see Reed and Simon 1980, p. 47.)

5.3.2 Theorem The spaces $\mathcal{H}$ and $\mathcal{L}_w^2(a, b)$ are isomorphic. That is, there exists a one-to-one correspondence between the Hilbert space and the space of square-integrable functions. ■

We have not as yet established the finiteness of $\int_a^b g^*(x) f(x) w(x)\, dx$ in $\mathcal{L}_w^2(a, b)$. We know that

$$\langle \mathrm{f} | \mathrm{f} \rangle = \int_a^b |f(x)|^2 w(x)\, dx$$

and

$$\langle \mathrm{g} | \mathrm{g} \rangle = \int_a^b |g(x)|^2 w(x)\, dx$$

are finite, but we do not know whether

$$\langle \mathrm{g} | \mathrm{f} \rangle = \int_a^b g^*(x) f(x) w(x)\, dx$$

is finite. However, if we use Schwarz inequality,

$$|\langle \mathrm{g} | \mathrm{f} \rangle|^2 \leqslant \langle \mathrm{g} | \mathrm{g} \rangle \langle \mathrm{f} | \mathrm{f} \rangle$$

(which holds for any vector space, finite or infinite), we get

$$\left| \int_a^b g^*(x) f(x) w(x)\, dx \right|^2 \leqslant \left[\int_a^b |g(x)|^2 w(x)\, dx \right] \left[\int_a^b |f(x)|^2 w(x)\, dx \right]$$

of which the LHS is clearly finite.

5.3.1 Generalities Concerning Orthogonal Polynomials

The isomorphism established in Theorem 5.3.2 makes the Hilbert space somewhat more tangible, but it is still fairly abstract. How are we to imagine general square-integrable functions? We need some specific functions to work with, and Weierstrass and Stone come to our rescue with the following theorem (for a proof, see Simmons 1983, pp. 154–161).

5.3.3 Theorem (Stone-Weierstrass Approximation Theorem) The sequence of vectors $|\mathrm{f}_k\rangle$, where $k = 0, 1, 2, \ldots$, having "coordinates"

$$f_k(x) \equiv \langle \mathrm{x} | \mathrm{f}_k \rangle \equiv x^k \qquad \text{where } k = 0, 1, 2, \ldots$$

forms a basis in $\mathscr{L}_w^2(a, b)$. ∎

Thus, any vector $|\mathrm{f}\rangle$ can be written as

$$|\mathrm{f}\rangle = \sum_{k=0}^{\infty} \alpha_k |\mathrm{f}_k\rangle$$

or in a function form that arises from the inner product of this relation with $\langle \mathrm{x}|$,

$$f(x) = \sum_{k=0}^{\infty} \alpha_k f_k(x) = \sum_{k=0}^{\infty} \alpha_k x^k$$

Note that the $|\mathrm{f}_k\rangle$'s are not orthonormal but are linearly independent. Thus, the Gram-Schmidt process can be used to find orthonormal functions—in this case polynomials—that span $\mathscr{L}_w^2(a, b)$. The following example illustrates the construction of the first few of these orthogonal (but not normalized) polynomials.

Example 5.3.1

Let us consider the interval $[-1, +1]$ and try to find linear combinations of x^k that are orthogonal (not necessarily normal). We set $w(x) = 1$ and denote these orthogonal functions by $P_k(x)$. For $k = 0$ we let $P_0(x) = f_0(x) = 1$. For $k = 1$ we want $P_1(x)$ to be orthogonal to $P_0(x)$ in the interval $[-1, 1]$. To find $P_1(x)$ we assume a general form, $P_1(x) = ax + b$, and determine a and b in such a way that $P_1(x)$ is orthogonal to $P_0(x) = 1$.

$$0 = \int_{-1}^{+1} P_1(x) P_0(x)\, dx = \int_{-1}^{1} (ax + b)\, dx = \tfrac{1}{2}ax^2\big|_{-1}^{+1} + 2b = 2b$$

So one of the coefficients, b, is zero. To find the other one, we need some standardization procedure. We "standardize" $P_j(x)$ by requiring that

$$P_j(1) = 1 \qquad \text{for all } j$$

For $j = 1$ this yields

$$P_1(1) = a(1) + 0 = a = 1$$

and we obtain

$$P_1(x) = x$$

To calculate $P_2(x)$, we write $P_2(x) = ax^2 + bx + c$ and determine a, b, and c such that $P_2(x)$ is orthogonal to both $P_0(x)$ and $P_1(x)$ and satisfies the above standardization.

$$0 = \int_{-1}^{1} P_2(x)P_0(x)\,dx = \tfrac{2}{3}a + 2c$$

$$0 = \int_{-1}^{1} P_2(x)P_1(x)\,dx = \tfrac{2}{3}b$$

$$P_2(1) = a + b + c = 1$$

These three equations have the unique solution

$$a = \tfrac{3}{2} \qquad b = 0 \qquad c = -\tfrac{1}{2}$$

Thus,

$$P_2(x) = \tfrac{1}{2}(3x^2 - 1)$$

It can be shown that

$$P_3(x) = \tfrac{1}{2}(5x^3 - 3x)$$

and

$$P_4(x) = \tfrac{1}{8}(35x^4 - 30x^2 + 3)$$

and so on. These polynomials, called *Legendre polynomials,* are solutions of an important differential equation that will be discussed later in this book. In general, the nth polynomial is given by what is called the *Rodriguez formula:*

$$P_n(x) = \frac{1}{2^n n!}\frac{d^n}{dx^n}[(x^2 - 1)^n]$$

•

Example 5.3.1 discusses only one of the many types of so-called *classical orthogonal polynomials,* polynomials that are orthogonal to one another in the specified interval $[a, b]$ using a weight function $w(x)$.

Such polynomials can be produced by starting with the functions $1, x, x^2, \ldots, x^k, \ldots$ and employing the Gram-Schmidt process. However, there is an elegant, but less general, approach that studies all of these polynomials in a unified manner.[5] We will employ this approach.

Consider the functions

$$F_n(x) = \frac{1}{w}\frac{d^n}{dx^n}(ws^n) \qquad \text{for } n = 0, 1, 2, \ldots \tag{5.24}$$

where $F_1(x)$, $w = w(x)$, and $s = s(x)$ satisfy the following conditions:

(1) $F_1(x)$ is a first-degree polynomial in x

(2) $s(x)$ is a polynomial in x of degree less than or equal to 2 and with real roots

(3) $w(x)$ is real, positive, and integrable in the interval $[a, b]$ and satisfies the boundary conditions $w(a)s(a) = w(b)s(b) = 0$

[5] This approach is due to F. G. Tricomi (1955). See also Denery and Krzywicki (1967).

As we will see, these three requirements very severely limit the number of possibilities for $F_n(x)$. In fact, we will see that $F_n(x)$ is a polynomial in x of degree n and that $F_k(x)$, for $k = 0, 1, 2, \ldots$, form a set of orthogonal polynomials with weight $w(x)$ on the interval $[a, b]$. In order to smooth the way for proving these two claims, let us next prove two lemmas.

In the following discussion $p^{(\leftarrow k)}(x)$ denotes a generic polynomial of degree less than or equal to k. For example, $3x^5 - 4x^2 + 5$, $2x + 1$, $-2.4x^4 + 3x^3 - 8x^2 + 2x - 1$, and 2 are all denoted by $p^{(\leftarrow 5)}(x)$ or $p^{(\leftarrow 6)}(x)$ or $p^{(\leftarrow 97)}(x)$ because they all have degrees less than or equal to 5, 6, and 97.

5.3.4 Lemma The following identity holds:

$$\frac{d^m}{dx^m}(ws^n p^{(\leftarrow k)}) = ws^{n-m} p^{(\leftarrow k+m)}$$

Proof. Let $n = 1$ in (5.24), yielding

$$F_1(x) = \frac{1}{w}\frac{d}{dx}(ws) = \frac{1}{w}\frac{dw}{dx}s + \frac{1}{w}w\frac{ds}{dx}$$

or

$$s\frac{dw}{dx} = w\left(F_1(x) - \frac{ds}{dx}\right)$$

Now consider the derivative of $ws^n p^{(\leftarrow k)}$:

$$\begin{aligned}\frac{d}{dx}(ws^n p^{(\leftarrow k)}) &= \frac{dw}{dx}s^n p^{(\leftarrow k)} + nw\frac{ds}{dx}s^{n-1}p^{(\leftarrow k)} + ws^n\frac{dp^{(\leftarrow k)}}{dx} \\ &= ws^{n-1}\left\{\left[F_1(x) + (n-1)\frac{ds}{dx}\right]p^{(\leftarrow k)} + s\frac{dp^{(\leftarrow k)}}{dx}\right\}\end{aligned}$$

However, since $F_1(x)$ and $s(x)$ are, by definition, polynomials of first and (at most) second degree, respectively, the expression inside the curly brackets is of degree less than or equal to $k + 1$. Thus, we can write

$$\frac{d}{dx}(ws^n p^{(\leftarrow k)}) = ws^{n-1} p^{(\leftarrow k+1)}$$

which indicates that n has decreased by 1 while k has increased by 1. It is obvious that

$$\frac{d^2}{dx^2}(ws^n p^{(\leftarrow k)}) = \frac{d}{dx}(ws^{n-1} p^{(\leftarrow k+1)}) = ws^{n-2} p^{(\leftarrow k+2)}$$

and, in general,

$$\frac{d^m}{dx^m}(ws^n p^{(\leftarrow k)}) = ws^{n-m} p^{(\leftarrow k+m)}$$

This is the desired identity. ■

5.3.5 Lemma All of the derivatives $(d^m/dx^m)(ws^n)$, where $m < n$, vanish at $x = a$ and $x = b$.

Proof. Set $k = 0$ in the identity of Lemma 5.3.4 and let the polynomial of degree 0 be $p^{(\leftarrow 0)} = 1$. Then we have

$$\frac{d^m}{dx^m}(ws^n) = ws^{n-m}p^{(\leftarrow m)}$$

When evaluated at $x = a$, this reduces to

$$\left.\frac{d^m}{dx^m}(ws^n)\right|_{x=a} = w(a)s^{n-m}(a)p^{(\leftarrow m)}(a) = 0$$

by Condition (3). Similarly, we have

$$\left.\frac{d^m}{dx^m}(ws^n)\right|_{x=b} = 0$$

and the lemma is proved. ∎

We are now in a position to prove the following important theorem.

5.3.6 Theorem $F_n(x)$ is a polynomial of degree n in x and is orthogonal on the interval $[a, b]$, with weight $w(x)$, to any polynomial $p^{(l)}(x)$ of degree l less than n, so

$$\int_a^b p^{(l)}(x)\,F_n(x)w(x)\,dx = 0 \qquad \text{for } l < n \tag{5.25}$$

Proof. Let us prove the orthogonality first:

$$\int_a^b p^{(l)}(x)\,F_n(x)w(x)\,dx = \int_a^b p^{(l)}(x)\frac{1}{w}\left[\frac{d^n}{dx^n}(ws^n)\right]w\,dx = \int_a^b p^{(l)}(x)\frac{d}{dx}\left[\frac{d^{n-1}}{dx^{n-1}}(ws^n)\right]dx$$

$$= p^{(l)}(x)\frac{d^{n-1}}{dx^{n-1}}(ws^n)\Bigg|_a^b - \int_a^b \frac{dp^{(l)}}{dx}\frac{d^{n-1}}{dx^{n-1}}(ws^n)\,dx$$

$$= 0 - \int_a^b \frac{dp^{(l)}}{dx}\frac{d}{dx}\left[\frac{d^{n-2}}{dx^{n-2}}(ws^n)\right]dx$$

$$= -\frac{dp^{(l)}}{dx}\frac{d^{n-2}}{dx^{n-2}}(ws^n)\Bigg|_a^b + \int_a^b \frac{d^2p^{(l)}}{dx^2}\frac{d^{n-2}}{dx^{n-2}}(ws^n)\,dx$$

$$= 0 + \int_a^b \frac{d^2p^{(l)}}{dx^2}\frac{d^{n-2}}{dx^{n-2}}(ws^n)\,dx$$

Here we have integrated by parts twice and used Lemma 5.3.5 for $m = n - 1$ and $m = n - 2$. We continue integrating by parts (all together l times) to finally obtain

$$\int_a^b p^{(l)}(x) F_n(x) w(x)\, dx = \int_a^b \frac{d^l p^{(l)}}{dx^l} \frac{d^{n-l}}{dx^{n-l}} (ws^n) dx$$

$$= C \int_a^b \frac{d}{dx}\left[\frac{d^{n-l-1}}{dx^{n-l-1}}(ws^n)\right] dx$$

$$= C \frac{d^{n-l-1}}{dx^{n-l-1}}(ws^n)\Big|_a^b = 0$$

where we have used the fact that the lth derivative of a polynomial of degree l is a constant. Note that since $l < n$, $n - l - 1 \geqslant 0$, so the last line of the above proof is well-defined.

To prove the first part of the theorem, we use Lemma 5.3.4 with $k = 0$ and $m = n$ to get

$$\frac{d^n}{dx^n}(ws^n) = ws^0 p^{(\leftarrow n)} = wp^{(\leftarrow n)}$$

or

$$F_n(x) = \frac{1}{w} \frac{d^n}{dx^n}(ws^n) = p^{(\leftarrow n)}$$

This states that $F_n(x)$ is a polynomial of degree less than or equal to n. We want to show that it is in fact of degree precisely equal to n. So we write

$$F_n(x) = p^{(\leftarrow n-1)}(x) + a_n x^n$$

which can always be done (if we include the possibility that a_n could be zero). To show that a_n *cannot be zero*, we first multiply both sides of the above by $F_n(x)$ and integrate over $[a, b]$ with weight function $w(x)$:

$$\int_a^b [F_n(x)]^2 w(x)\, dx = \int_a^b F_n(x) p^{(\leftarrow n-1)}(x) w(x)\, dx + a_n \int_a^b x^n F_n(x) w(x)\, dx$$

The first integral on the RHS vanishes by Eq. (5.25). The LHS is a positive quantity because both $w(x)$ and $[F_n(x)]^2$ are positive. Thus, we have

$$a_n \int_a^b x^n F_n(x) w(x)\, dx \neq 0$$

which indicates, in particular, that $a_n \neq 0$. Thus, $F_n(x)$ is of degree n. ■

It is customary to introduce a normalization constant in the definition of $F_n(x)$ and write

$$F_n(x) = \frac{1}{K_n w} \frac{d^n}{dx^n}(ws^n) \tag{5.26}$$

This equation is called the *generalized Rodriguez formula.* For historical reasons different polynomial functions are normalized differently, which is why K_n is introduced here.

From Theorem 5.3.6 it is clear that the sequence of polynomials $F_0(x)$, $F_1(x)$, $F_2(x), \ldots, F_n(x), \ldots$ forms an orthogonal set of polynomials on $[a, b]$ with weight $w(x)$.

5.3.2 Classification of Orthogonal Polynomials

Let us now investigate the consequences of various choices of $s(x)$ within the limitations imposed by Condition (2). First, we note that if we change x to $\alpha x + \beta$, where α and β are constants, neither the degree of $s(x)$ nor the orthogonality property of $F_n(x)$ will change because the three conditions will still be satisfied. The only change that will occur will be in the limits of integration a and b; these will be fixed to make $s(x)$ and $w(x)$ as simple as possible.

What this means is that α and β are arbitrary ($\alpha \neq 0$, of course) and will be chosen to simplify expressions. In particular, $F_1(x)$, being a polynomial of degree 1, can be chosen to be

$$F_1(x) = -\frac{x}{K_1} \tag{5.27a}$$

corresponding to $\alpha = -1/K_1$ and $\beta = 0$. On the other hand, for $n = 1$, Eq. (5.26) gives

$$F_1(x) = \frac{1}{K_1 w} \frac{d}{dx}(ws) = \frac{s}{K_1 w} \frac{dw}{dx} + \frac{1}{K_1} \frac{ds}{dx} \tag{5.27b}$$

Equating (5.27a) and (5.27b) yields a differential equation for $w(x)$, which is

$$\frac{1}{w} \frac{dw}{dx} = -\frac{x + ds/dx}{s} \tag{5.28}$$

and has the boundary condition

$$w(a)s(a) = w(b)s(b) = 0 \tag{5.29}$$

Imposing Condition (2) and the above boundary condition on (5.28) gives all the orthogonal polynomials of physics. Condition (2) states that $s(x)$ is a polynomial of degree 0, 1, or 2. Let us look at these three possibilities.

If the degree of s is 0, then

$$s(x) = \alpha \neq 0$$

and

$$\frac{ds}{dx} = 0$$

result in the differential equation

$$\frac{1}{w}\frac{dw}{dx} = -\frac{x}{\alpha}$$

The most general solution of this equation is

$$w(x) = Ae^{-x^2/2\alpha}$$

where A is the constant of integration. This weight function must satisfy (5.29). Thus,

$$s(a)w(a) = \alpha w(a) = \alpha Ae^{-a^2/2\alpha}$$

which vanishes only if $\alpha > 0$ and $a = \pm\infty$. Similarly, $s(b)w(b) = 0$ implies that $b = \pm\infty$. But since $b > a$, we are led to the choice

$$a = -\infty \qquad \text{and} \qquad b = +\infty$$

We can choose $\alpha = 1$ and $A = 1$ and make the change of argument from $x/\sqrt{2}$ to x to obtain the following proposition.

5.3.7 Proposition If $s(x)$ is a constant polynomial, then it is possible to choose $s(x) = 1$. In that case

$$w(x) = \exp(-x^2)$$

and

$$a = -\infty \qquad \text{and} \qquad b = +\infty \qquad \blacksquare$$

If the degree of s is 1, we can write very generally

$$s(x) = \alpha(x - \beta)$$

where α and β are arbitrary. Then Eq. (5.28) becomes

$$\frac{1}{w}\frac{dw}{dx} = -\frac{x+\alpha}{\alpha(x-\beta)} = -\frac{x-\beta+\beta+\alpha}{\alpha(x-\beta)} = -\frac{1}{\alpha} - \left(\frac{\alpha+\beta}{\alpha}\right)\frac{1}{x-\beta}$$

or

$$\frac{dw}{w} = -\frac{1}{\alpha}dx - \left(\frac{\alpha+\beta}{\alpha}\right)\frac{dx}{x-\beta}$$

Integrating both sides, we get

$$\ln w = -\frac{x}{\alpha} - \left(\frac{\alpha+\beta}{\alpha}\right)\ln(x-\beta) + \ln A$$

$$= -\frac{x}{\alpha} + \ln[(x-\beta)^{-[(\alpha+\beta)/\alpha]}] + \ln A$$

which can be written as

$$w(x) = A(x-\beta)^{-(1+\beta/\alpha)}e^{-x/\alpha}$$

To impose the boundary condition of (5.29), we form the product

$$s(x)w(x) = A\alpha(x-\beta)^{-\beta/\alpha}e^{-x/\alpha}$$

It is clear that this can be made to vanish at $x = \beta$ *only if* $\beta/\alpha < 0$. On the other hand, the exponential function vanishes at $x = +\infty$ (or $-\infty$) if $\alpha > 0$ (or $\alpha < 0$). Thus, we choose

$$\frac{\beta}{\alpha} < 0 \qquad a = \beta \qquad b = +\infty \qquad \alpha > 0$$

We also make a change of variables,

$$\frac{x-\beta}{\alpha} = y \qquad \text{or} \qquad x = \alpha y + \beta$$

to get

$$\begin{aligned} w(y) &= A(\alpha y)^{-(1+\beta/\alpha)}e^{-y-\beta/\alpha} = Ae^{-\beta/\alpha}\alpha^{-(1+\beta/\alpha)}y^{-(1+\beta/\alpha)}e^{-y} \\ &= By^{\nu}e^{-y} \end{aligned}$$

where B is just another constant, $\nu = -1 - \beta/\alpha > -1$, and the interval is from $a = 0$ to $b = +\infty$. By ignoring the constant B and substituting the more common x for y, we can summarize the above discussion as a proposition.

5.3.8 Proposition If $s(x)$ is a polynomial of degree 1, then it is possible to choose $s(x) = x$, and in that case

$$w(x) = x^{\nu}e^{-x} \qquad \text{where } \nu > -1$$

and

$$a = 0 \qquad \text{and} \qquad b = +\infty$$ ■

If the degree of s is 2, the result is the following proposition. (See Problem 5.18.)

5.3.9 Proposition If $s(x)$ is of degree 2, then it is possible to choose $s(x) = 1 - x^2$, and in that case

$$w = (1-x)^{\nu}(1+x)^{\mu} \qquad \text{where } \nu, \mu > -1$$

and

$$a = -1 \qquad \text{and} \qquad b = +1$$ ■

We can summarize the foregoing discussion by saying that, apart from a trivial linear transformation of the argument, the orthogonal polynomials introduced in Section 5.3.1 can be reduced, up to multiplicative constants, to the three types of polynomials given in Table 5.1.

Specific choices of μ and ν in the Jacobi polynomials give rise to polynomials that have traditionally been known by various names. Legendre polynomials are obtained when $\mu = \nu = 0$; Gegenbauer polynomials correspond to $\mu = \nu = \lambda - \frac{1}{2}$; Tchebichef polynomials of the first kind are those with $\mu = \nu = -\frac{1}{2}$; and Tchebichef

TABLE 5.1 CHARACTERISTICS OF HERMITE, LAGUERRE, AND JACOBI POLYNOMIALS

Interval	$w(x)$	$s(x)$	Polynomial
$(-\infty, +\infty)$	e^{-x^2}	1	Hermite, $H_n(x)$
$[0, \infty)$	$x^\nu e^{-x}$ for $\nu > -1$	x	Laguerre, $L_n^\nu(x)$
$[-1, +1]$	$(1-x)^\nu(1+x)^\mu$ for $\mu, \nu > -1$	$1-x^2$	Jacobi, $P_n^{(\nu,\mu)}(x)$

TABLE 5.2 CHARACTERISTICS OF COMMON JACOBI POLYNOMIALS

μ	ν	$w(x)$	Polynomial
0	0	1	Legendre, $P_n(x)$
$\lambda - \frac{1}{2}$	$\lambda - \frac{1}{2}$	$(1-x^2)^{\lambda-1/2}$ for $\lambda > -\frac{1}{2}$	Gegenbauer, $C_n^\lambda(x)$
$-\frac{1}{2}$	$-\frac{1}{2}$	$(1-x^2)^{-1/2}$	Tchebichef of the first kind, $T_n(x)$
$\frac{1}{2}$	$\frac{1}{2}$	$(1-x^2)^{1/2}$	Tchebichef of the second kind, $U_n(x)$

polynomials of the second kind have $\mu = \nu = +\frac{1}{2}$. Table 5.2 shows these four special polynomials.

Note that the definition of each of the preceding polynomials involves a "standardization," which boils down to a particular choice of K_n in the generalized Rodriguez formula. This choice does not correspond to normalization of the polynomials but is simply a historical convention.

Before studying these orthogonal polynomials separately, let us investigate two more important properties of all of them—the recurrence relations they satisfy and the differential equation of which they are solutions.

Recurrence relations. All orthogonal polynomials (not just the ones discussed in this section) satisfy a recurrence relation; thus, the following considerations apply to all orthogonal polynomials. Let $C_n(x)$ be a set of orthogonal polynomials [of which $F_n(x)$ are members] with the obvious property that

$$\int_a^b C_n(x) p^{(\leftarrow n-1)} w(x)\, dx = 0 \tag{5.30}$$

The following notations will be used in this discussion:

$$k_n = \text{coefficient of } x^n \text{ in } C_n(x) \tag{5.31a}$$

$$k'_n = \text{coefficient of } x^{n-1} \text{ in } C_n(x) \tag{5.31b}$$

$$h_n = \int_a^b [C_n(x)]^2 w(x)\, dx \tag{5.31c}$$

The polynomial

$$C_{n+1}(x) - \left(\frac{k_{n+1}}{k_n}\right)x\, C_n(x)$$

clearly has degree less than or equal to n and, therefore, can be expanded as a linear combination of $C_k(x)$:

$$C_{n+1}(x) - \left(\frac{k_{n+1}}{k_n}\right)xC_n(x) = \sum_{j=0}^{n} a_j C_j(x) \tag{5.32}$$

We multiply both sides of this equation by $C_m(x)w(x)$ and integrate from a to b to obtain

$$\int_a^b C_{n+1}(x)C_m(x)w(x)\,dx - \frac{k_{n+1}}{k_n}\int_a^b xC_m(x)C_n(x)w(x)\,dx = \sum_{j=0}^{n} a_j \int_a^b C_j(x)C_m(x)w(x)\,dx$$

The first integral on the LHS vanishes as long as $m \leqslant n$; the second integral gives zero if $m \leqslant n-2$ [if $m \leqslant n-2$, then $xC_m(x)$ is a polynomial of degree $n-1$, and by Eq. (5.30) the integral must vanish]. Thus, we have

$$\sum_{j=0}^{n} a_j \int_a^b C_j(x)C_m(x)w(x)\,dx = 0 \qquad \text{for } m \leqslant n-2$$

The integral in the sum is zero unless $j = m$, because of the orthogonality of the $C_j(x)$. Therefore, the sum reduces to

$$a_m \int_a^b [C_m(x)]^2\, w(x)\,dx = 0 \qquad \text{for } m \leqslant n-2$$

Since the integral is nonzero, we conclude that

$$a_m = 0 \qquad \text{for } m = 0, 1, 2, \ldots, n-2$$

and Eq. (5.32) reduces to

$$C_{n+1}(x) - \left(\frac{k_{n+1}}{k_n}\right)xC_n(x) = a_{n-1}C_{n-1}(x) + a_nC_n(x)$$

This can be rewritten as

$$C_{n+1}(x) = \left[\left(\frac{k_{n+1}}{k_n}\right)x + a_n\right]C_n(x) + a_{n-1}C_{n-1}(x) \tag{5.33}$$

Exercise 5.3.1 shows that if we define

$$\alpha_n \equiv \frac{k_{n+1}}{k_n} \tag{5.34a}$$

$$\beta_n \equiv \frac{k_{n+1}}{k_n}\left(\frac{k'_{n+1}}{k_{n+1}} - \frac{k'_n}{k_n}\right) \tag{5.34b}$$

and
$$\gamma_n \equiv -\frac{h_n}{h_{n-1}}\left(\frac{\alpha_n}{\alpha_{n-1}}\right) \tag{5.34c}$$

where k_n, k'_n and h_n are as given in Eqs. (5.31), then the recurrence relation can be expressed as

$$C_{n+1}(x) = (\alpha_n x + \beta_n)C_n(x) + \gamma_n C_{n-1}(x) \tag{5.35}$$

It should be emphasized that this recurrence relation is satisfied by all orthogonal polynomials, because none of the special properties associated with $F_n(x)$ were used in its derivation. The only property that was used is the orthogonality relation, Eq. (5.30). The recurrence relation of (5.35) can be used to find other recurrence relations and evaluate integrals involving powers of x. The following examples illustrate these points.

Example 5.3.2

As an application of the recurrence relation of (5.35), let us evaluate the integral

$$I_1 \equiv \int_a^b xC_m(x)C_n(x)\, w(x)\, dx$$

We rewrite the recurrence relation as

$$xC_n(x) = \frac{1}{\alpha_n}C_{n+1}(x) - \frac{\beta_n}{\alpha_n}C_n(x) - \frac{\gamma_n}{\alpha_n}C_{n-1}(x)$$

Substituting in the above integral gives

$$I_1 = \frac{1}{\alpha_n}\int_a^b C_m(x)C_{n+1}(x)w(x)\,dx - \frac{\beta_n}{\alpha_n}\int_a^b C_m(x)C_n(x)w(x)\,dx - \frac{\gamma_n}{\alpha_n}\int_a^b C_m(x)C_{n-1}(x)w(x)\,dx$$

We now use the orthogonality relations among $C_k(x)$ to obtain

$$I_1 = \frac{1}{\alpha_n}\delta_{m,n+1}\int_a^b [C_m(x)]^2 w(x)\,dx - \frac{\beta_n}{\alpha_n}\delta_{m,n}\int_a^b [C_m(x)]^2 w(x)dx - \frac{\gamma_n}{\alpha_n}\delta_{m,n-1}\int_a^b [C_m(x)]^2 w(x)\,dx$$

$$= \left(\frac{1}{\alpha_{m-1}}\delta_{m,n+1} - \frac{\beta_m}{\alpha_m}\delta_{m,n} - \frac{\gamma_{m+1}}{\alpha_{m+1}}\delta_{m,n-1}\right)h_m$$

or
$$I_1 = \begin{cases} h_m/\alpha_{m-1} & \text{if } m = n+1 \\ -\beta_m h_m/\alpha_m & \text{if } m = n \\ -\gamma_{m+1}h_m/\alpha_{m+1} & \text{if } m = n-1 \\ 0 & \text{otherwise} \end{cases}$$

•

Example 5.3.3

Let us use the recurrence relation of (5.35) to find another recurrence relation involving the second power of x. We write the recurrence relation as in Example 5.3.2:

$$xC_n(x) = \frac{1}{\alpha_n}C_{n+1}(x) - \frac{\beta_n}{\alpha_n}C_n(x) - \frac{\gamma_n}{\alpha_n}C_{n-1}(x) \tag{1}$$

We multiply both sides by x and use (1) on the RHS side:

$$\begin{aligned}
x^2C_n(x) &= \frac{1}{\alpha_n}xC_{n+1}(x) - \frac{\beta_n}{\alpha_n}xC_n(x) - \frac{\gamma_n}{\alpha_n}xC_{n-1}(x) \\
&= \frac{1}{\alpha_n}\left[\frac{1}{\alpha_{n+1}}C_{n+2}(x) - \frac{\beta_{n+1}}{\alpha_{n+1}}C_{n+1}(x) - \frac{\gamma_{n+1}}{\alpha_{n+1}}C_n(x)\right] \\
&\quad - \frac{\beta_n}{\alpha_n}\left[\frac{1}{\alpha_n}C_{n+1}(x) - \frac{\beta_n}{\alpha_n}C_n(x) - \frac{\gamma_n}{\alpha_n}C_{n-1}(x)\right] \\
&\quad - \frac{\gamma_n}{\alpha_n}\left[\frac{1}{\alpha_{n-1}}C_n(x) - \frac{\beta_{n-1}}{\alpha_{n-1}}C_{n-1}(x) - \frac{\gamma_{n-1}}{\alpha_{n-1}}C_{n-2}(x)\right]
\end{aligned}$$

Rearranging gives

$$\begin{aligned}
x^2C_n(x) &= \frac{1}{\alpha_n\alpha_{n+1}}C_{n+2}(x) - \left(\frac{\beta_{n+1}}{\alpha_n\alpha_{n+1}} + \frac{\beta_n}{\alpha_n^2}\right)C_{n+1}(x) \\
&\quad - \left(\frac{\gamma_{n+1}}{\alpha_n\alpha_{n+1}} - \frac{\beta_n^2}{\alpha_n^2} + \frac{\gamma_n}{\alpha_n\alpha_{n-1}}\right)C_n(x) \\
&\quad + \left(\frac{\beta_n\gamma_n}{\alpha_n^2} + \frac{\gamma_n\beta_{n-1}}{\alpha_n\alpha_{n-1}}\right)C_{n-1}(x) + \frac{\gamma_n\gamma_{n-1}}{\alpha_n\alpha_{n-1}} + C_{n-2}(x) \\
&\equiv A_nC_{n+2}(x) + B_nC_{n+1}(x) + D_nC_n(x) + E_nC_{n-1}(x) + G_nC_{n-2}(x)
\end{aligned}$$

where the coefficients are appropriately defined. •

Example 5.3.4

We can use the result of Example 5.3.3 to evaluate the integral

$$I_2 \equiv \int_a^b x^2C_m(x)C_n(x)w(x)\,dx$$

Substituting for $x^2C_n(x)$ from Example 5.3.3 gives

$$\begin{aligned}
I_2 &= \int_a^b C_m(x)\,[A_nC_{n+2}(x) + \cdots + G_nC_{n-2}(x)]\,w(x)\,dx \\
&= A_n\int_a^b C_m(x)C_{n+2}(x)w(x)dx + \cdots + G_n\int_a^b C_mC_{n-2}w(x)\,dx
\end{aligned}$$

Using the orthogonality of the C_k's, we get

$$I_2 = A_n \delta_{m,n+2} h_m + \cdots + G_n \delta_{m,n-2} h_m$$
$$= h_m(A_{m-2}\delta_{m,n+2} + B_{m-1}\delta_{m,n+1} + D_m\delta_{m,n} + E_{m+1}\delta_{m,n-1} + G_{m+2}\delta_{m,n-2})$$

or
$$I_2 = \begin{cases} h_m/\alpha_{m-2}\alpha_{m-1} & \text{if } m = n+2 \\ -h_m[\beta_m/\alpha_m\alpha_{m-1} + \beta_{m-1}/\alpha_{m-1}^2] & \text{if } m = n+1 \\ -h_m[\gamma_{m+1}/\alpha_m\alpha_{m+1} - \beta_m^2/\alpha_m^2 + \gamma_m/\alpha_m\alpha_{m-1}] & \text{if } m = n \\ h_m[\beta_{m+1}\gamma_{m+1}/\alpha_{m+1}^2 + \gamma_{m+1}\beta_m/\alpha_m\alpha_{m+1}] & \text{if } m = n-1 \\ h_m\gamma_{m+1}\gamma_{m+2}/\alpha_{m+1}\alpha_{m+2} & \text{if } m = n-2 \\ 0 & \text{otherwise} \end{cases}$$ ●

Differential equation. Let us now seek a differential equation of which $F_n(x)$ are solutions. The differential equation must involve dF_n/dx, which is a polynomial of degree less than or equal to $n - 1$. Using Lemma 5.3.4 with $m = 1$, $n = 1$, and $k = n - 1$ and the polynomial $p^{(\leftarrow k)} = dF_n/dx$, we have

$$\frac{d}{dx}\left(ws\frac{dF_n}{dx}\right) = wp^{(\leftarrow n)}$$

Expanding the generic polynomial on the RHS in terms of the orthogonal polynomials $F_n(x)$, we obtain

$$\frac{d}{dx}\left(ws\frac{dF_n}{dx}\right) = w\sum_{i=1}^{n} \lambda_i F_i(x) \tag{5.36}$$

where λ_i are constants to be determined. Now we multiply both sides by $F_m(x)$ and integrate:

$$\int_a^b F_m \frac{d}{dx}\left(ws\frac{dF_n}{dx}\right)dx = \sum_{i=1}^{n} \lambda_i \int_a^b F_m F_i w\,dx$$

The integrals on the RHS are all zero except when $i = m$, and then the result is simply h_m. For the LHS when $m < n$, we have

$$\int_a^b F_m \frac{d}{dx}\left(ws\frac{dF_n}{dx}\right)dx = F_m ws\frac{dF_n}{dx}\bigg|_a^b - \int_a^b ws\frac{dF_n}{dx}\frac{dF_m}{dx}dx$$

$$= 0 - F_n ws\frac{dF_m}{dx}\bigg|_a^b + \int_a^b F_n \frac{d}{dx}\left(ws\frac{dF_m}{dx}\right)dx$$

$$= \int_a^b wF_n\left[\frac{1}{w}\frac{d}{dx}\left(ws\frac{dF_m}{dx}\right)\right]dx = 0$$

The result is zero because the expression in brackets is a polynomial of degree less than or equal to $m < n$. In the above derivation we used the fact that $w(a)s(a) = 0 = w(b)s(b)$ when integrating by parts twice. Putting everything together, we write

$$0 = \lambda_m h_m \qquad \text{for } m < n$$

or, since $h_m \neq 0$,

$$\lambda_m = 0 \qquad \text{for } m < n$$

Thus, Eq. (5.36) becomes the differential equation of the following proposition.

5.3.10 Proposition The orthogonal polynomials F_n satisfy the differential equation

$$\frac{d}{dx}\left(ws\frac{dF_n}{dx}\right) = w\lambda_n F_n \tag{5.37}$$

where

$$\lambda_n = n\left[K_1\frac{dF_1}{dx} + \frac{1}{2}(n-1)\frac{d^2s}{dx^2}\right]$$

Proof. We only need to evaluate λ_n. To do so, we multiply both sides of (5.37) by $F_n(x)$ and integrate:

$$\int_a^b F_n\frac{d}{dx}\left(ws\frac{dF_n}{dx}\right)dx = \lambda_n\int_a^b w[F_n(x)]^2\,dx = \lambda_n h_n \tag{5.38a}$$

Let us evaluate the LHS:

$$\int_a^b F_n\frac{d}{dx}\left(ws\frac{dF_n}{dx}\right)dx = \int_a^b F_n\left(\frac{d(ws)}{dx}\frac{dF_n}{dx} + ws\frac{d^2F_n}{dx^2}\right)dx$$

$$= \int_a^b wF_n\left\{\left[\frac{1}{w}\frac{d}{dx}(ws)\right]\frac{dF_n}{dx} + s\frac{d^2F_n}{dx^2}\right\}dx$$

But by the generalized Rodriguez formula,

$$\frac{1}{w}\frac{d}{dx}(ws) = K_1F_1$$

We, therefore, have

$$\int_a^b F_n\frac{d}{dx}\left(ws\frac{dF_n}{dx}\right)dx = \int_a^b wF_n\left(K_1F_1\frac{dF_n}{dx} + s\frac{d^2F_n}{dx^2}\right)dx \tag{5.38b}$$

Since $F_1(x)$ is a polynomial of degree 1 and s is a polynomial of degree (at most) 2, we can write them as

$$F_1(x) = \alpha_1 x + \alpha_0$$

and $$s(x) = \alpha_2 x^2 + \beta_1 x + \beta_0$$

On the other hand,

$$\frac{dF_n}{dx} = \frac{d}{dx}(k_n x^n + k'_n x^{n-1} + \cdots) = nk_n x^{n-1} + (n-1)k'_n x^{n-2} + \cdots$$

$$\frac{d^2F_n}{dx^2} = n(n-1)k_n x^{n-2} + (n-1)(n-2)k'_n x^{n-3} + \cdots$$

Because of the orthogonality property of the $F_n(x)$, only the nth power of x in the nth-degree polynomial in the parentheses on the RHS of (5.38b) contributes. Therefore, we can replace that expression in parentheses by the nth degree of x:

$$\left(K_1 F_1 \frac{dF_n}{dx} + s\frac{d^2F_n}{dx^2}\right) \quad \text{becomes} \quad K_1\alpha_1 n k_n x^n + \alpha_2 n(n-1)k_n x^n$$

So (5.38b) becomes

$$\int_a^b F_n \frac{d}{dx}\left(ws\frac{dF_n}{dx}\right)dx = [K_1\alpha_1 n + \alpha_2 n(n-1)]k_n \int_a^b wF_n x^n\, dx$$
$$= [K_1\alpha_1 n + \alpha_2 n(n-1)]h_n$$

where we used Eq. (5.31c) and the facts that $k_n x^n = F_n(x) - p^{(\leftarrow n-1)}$ and that F_n is orthogonal to $p^{(\leftarrow n-1)}$ [Eq. (5.30)]. Substituting this result in (5.38a) gives

$$\lambda_n = K_1\alpha_1 n + \alpha_2 n(n-1)$$

where $$\alpha_1 = \frac{dF_1}{dx} \quad \text{and} \quad \alpha_2 = \frac{1}{2}\frac{d^2s}{dx^2}$$

Thus, we have

$$\lambda_n = n\left[K_1\frac{dF_1}{dx} + \frac{1}{2}(n-1)\frac{d^2s}{dx^2}\right] \tag{5.39}$$

which is the desired result. ■

Before studying specific orthogonal polynomials, let us take a moment to note the generality and elegance of the foregoing discussion. With only a few assumptions [the three conditions in Sec. 5.3.1 and the defining equation, Eq. (5.24)] we have severely restricted the choice of the weight function and, with it, the choice of the interval $[a, b]$. We have nevertheless exhausted the list of the so-called classical orthogonal polynomials (the entries in Table 5.1). What remains is to construct the specific polynomials frequently used in physics.

We have seen that the four parameters K_n, k_n, k'_n, and h_n determine all the properties of the polynomials. Once K_n is fixed by some standardization, we can determine all the other parameters: k_n and k'_n will be given by the generalized Rodriguez formula, and h_n can be calculated as follows:

$$h_n = \int_a^b [F_n(x)]^2 w(x)\,dx = \int_a^b F_n[k_n x^n + k_n' x^{n-1} + \cdots]w\,dx$$

$$= k_n \int_a^b F_n x^n w\,dx = k_n \int_a^b \frac{1}{K_n w}\left[\frac{d^n}{dx^n}(ws^n)\right] x^n w\,dx$$

$$= \frac{k_n}{K_n}\int_a^b x^n \frac{d^n}{dx^n}(ws^n)\,dx = \frac{k_n}{K_n}\int_a^b x^n \frac{d}{dx}\left[\frac{d^{n-1}}{dx^{n-1}}(ws^n)\right]dx$$

Integrating by parts and recalling from Lemma 5.3.5 that

$$\left.\frac{d^m}{dx^m}(ws^n)\right|_{x=a} = 0 = \left.\frac{d^m}{dx^m}(ws^n)\right|_{x=b} \qquad \text{for } m < n$$

we get

$$h_n = \frac{k_n}{K_n}\left[x^n \frac{d^{n-1}}{dx^{n-1}}(ws^n)\Big|_a^b - \int_a^b \frac{d}{dx}(x^n)\frac{d^{n-1}}{dx^{n-1}}(ws^n)\,dx\right]$$

$$= -\frac{k_n}{K_n}\int_a^b \frac{d}{dx}(x^n)\frac{d}{dx}\left[\frac{d^{n-2}}{dx^{n-2}}(ws^n)\right]dx$$

$$= -\frac{k_n}{K_n}\left[\frac{d}{dx}(x^n)\frac{d^{n-2}}{dx^{n-2}}(ws^n)\Big|_a^b - \int_a^b \frac{d^2}{dx^2}(x^n)\frac{d^{n-2}}{dx^{n-2}}(ws^n)\,dx\right]$$

$$= (-1)^2\frac{k_n}{K_n}\int_a^b \frac{d^2}{dx^2}(x^n)\frac{d^{n-2}}{dx^{n-2}}(ws^n)\,dx$$

Clearly, after m integrations by parts we obtain

$$h_n = (-1)^m \frac{k_n}{K_n}\int_a^b \frac{d^m}{dx^m}(x^n)\frac{d^{n-m}}{dx^{n-m}}(ws^n)\,dx$$

When $m = n$, this yields

$$h_n = (-1)^n \frac{k_n}{K_n}\int_a^b \frac{d^n}{dx^n}(x^n)\frac{d^0}{dx^0}(ws^n)\,dx$$

But $\quad \dfrac{d^0}{dx^0}(ws^n) = ws^n \quad$ and $\quad \dfrac{d^n}{dx^n}(x^n) = n!$

So the final result is

$$h_n = \frac{(-1)^n k_n n!}{K_n} \int_a^b ws^n dx \tag{5.40}$$

We will use this result when studying the classical orthogonal polynomials.

Example 5.3.5

Recurrence relations involving derivatives of orthogonal functions are sometimes useful. Let us derive such recurrence relations. Again we start with Eq. (5.35) for $F_n(x)$:

$$F_{n+1}(x) = (\alpha_n x + \beta_n) F_n(x) + \gamma_n F_{n-1}(x) \tag{1}$$

We take the derivative of both sides:

$$\frac{dF_{n+1}}{dx} = \alpha_n F_n + (\alpha_n x + \beta_n)\frac{dF_n}{dx} + \gamma_n \frac{dF_{n-1}}{dx} \tag{2}$$

The second derivative is

$$F''_{n+1} = 2\alpha_n F'_n + (\alpha_n x + \beta_n)F''_n + \gamma_n F''_{n-1} \tag{3}$$

Using the differential equation of (5.37) in the form

$$F''_n + \frac{1}{ws}\frac{d(ws)}{dx} F'_n = \frac{1}{s}\lambda_n F_n$$

or, defining $g(x) \equiv (1/ws)[d(ws)/dx]$, in the form

$$F''_n = -g(x) F'_n + \frac{\lambda_n}{s} F_n$$

we substitute in (3) to obtain

$$-g(x)F'_{n+1} + \frac{\lambda_{n+1}}{s} F_{n+1}$$

$$= 2\alpha_n F'_n + (\alpha_n x + \beta_n)\left[-gF'_n + \frac{\lambda_n}{s}F_n\right] + \gamma_n\left[-gF'_{n-1} + \frac{\lambda_{n-1}}{s}F_{n-1}\right]$$

We substitute (2) in this equation to get

$$-g[\alpha_n F_n + (\alpha_n x + \beta_n)F'_n + \gamma_n F'_{n-1}] + \frac{\lambda_{n+1}}{s} F_{n+1}$$

$$= 2\alpha_n F'_n + (\alpha_n x + \beta_n)\left[-gF'_n + \frac{\lambda_n}{s}F_n\right] + \gamma_n\left[-gF'_{n-1} + \frac{\lambda_{n-1}}{s}F_{n-1}\right]$$

or, after simplifying,

$$2\alpha_n F'_n + \left[\alpha_n g(x) + \frac{\lambda_n}{s}(\alpha_n x + \beta_n)\right]F_n - \frac{\lambda_{n+1}}{s}F_{n+1} + \frac{\gamma_n \lambda_{n-1}}{s}F_{n-1} = 0 \tag{4}$$

This is one form of recurrence relation.

We can get another form of recurrence relation involving derivatives by substituting (1) in (4):

$$2\alpha_n F_n' + \left[\alpha_n g + \frac{\lambda_n}{s}(\alpha_n x + \beta_n)\right] F_n - \frac{\lambda_{n+1}}{s}[(\alpha_n x + \beta_n) F_n + \gamma_n F_{n-1}] + \frac{\gamma_n \lambda_{n-1}}{s} F_{n-1} = 0$$

or

$$2\alpha_n F_n' + \left[\alpha_n g + \frac{(\alpha_n x + \beta_n)}{s}(\lambda_n - \lambda_{n+1})\right] F_n + \frac{\gamma_n}{s}(\lambda_{n-1} - \lambda_{n+1}) F_{n-1} = 0 \qquad (5)$$

We can get a third relation by multiplying Eq. (5) by ws, differentiating both sides, and using (5.37) and the definition of $g(x)$:

$$2\alpha_n w \lambda_n F_n + \frac{d}{dx}\left\{\left[\alpha_n \frac{d}{dx}(ws) + w(\alpha_n x + \beta_n)(\lambda_n - \lambda_{n+1})\right] F_n\right\} + \gamma_n(\lambda_{n-1} - \lambda_{n+1})\frac{d}{dx}(wF_{n-1}) = 0 \qquad (6)$$

Yet another useful relation is obtained by multiplying (4) by ws, differentiating the resulting equation and using (5.37):

$$2\alpha_n w \lambda_n F_n + \frac{d}{dx}\left\{\left[\alpha_n \frac{d}{dx}(ws) + w\lambda_n(\alpha_n x + \beta_n)\right] F_n\right\} - \lambda_{n+1}\frac{d}{dx}(wF_{n+1}) + \gamma_n \lambda_{n-1}\frac{d}{dx}(wF_{n-1}) = 0 \qquad (7)$$

Now we solve (6) for $\gamma_n d/dx(wF_{n-1})$ and substitute the result in (7). After simplification the result is

$$2\alpha_n \lambda_n w F_n + \frac{d}{dx}\left\{\left[\alpha_n \frac{d}{dx}(ws) + (\lambda_n - \lambda_{n-1})(\alpha_n x + \beta_n) w\right] F_n\right\} + (\lambda_{n-1} - \lambda_{n+1})\frac{d}{dx}(wF_{n+1}) = 0 \qquad (8)$$

A final recurrence relation of interest is the following:

$$A_n(x) F_n(x) - \lambda_{n+1}(\alpha_n x + \beta_n)\frac{dw}{dx} F_{n+1} + \gamma_n \lambda_{n-1}\frac{dw}{dx} F_{n-1} + B_n(x)\frac{dF_{n+1}}{dx} + \gamma_n D_n(x) F_{n-1}' = 0 \qquad (9)$$

where

$$A_n(x) \equiv (\alpha_n x + \beta_n)\left[2\alpha_n w \lambda_n + \alpha_n \frac{d^2}{dx^2}(ws) + \frac{dw}{dx}\lambda_n(\alpha_n x + \beta_n)\right] - \alpha_n^2 \frac{d}{dx}(ws)$$

$$B_n(x) \equiv \alpha_n \frac{d}{dx}(ws) - (\alpha_n x + \beta_n) w(x)(\lambda_{n+1} - \lambda_n)$$

$$D_n(x) \equiv (\alpha_n x + \beta_n) w(\lambda_{n-1} - \lambda_n) - \alpha_n \frac{d}{dx}(ws)$$

Equation (9) can be obtained by carrying out the differentiation of the expression in curly brackets in (7), multiplying the resulting equation by $(\alpha_n x + \beta_n)$, and substituting for $(\alpha_n x + \beta_n) F'_n$ from (2). The details are left as a problem for the reader.

All these relations seem to be very complicated. However, the complexity is the price we pay for generality. When we work with specific orthogonal polynomials, the equations simplify considerably. For instance, for Hermite and Legendre polynomials Eq. (5) yields, respectively,

$$H'_n = 2nH_{n-1}$$

and

$$(1 - x^2) P'_n + nxP_n - nP_{n-1} = 0$$

Also, applying (8) to Legendre polynomials gives

$$P'_{n+1} - xP'_n - (n + 1) P_n = 0 \tag{10}$$

Similarly, Eq. (9) yields

$$P'_{n+1} - P'_{n-1} - (2n + 1) P_n = 0 \tag{11}$$

as can easily be verified.

It is possible to find many more recurrence relations by manipulating Eqs. (1), (2), and (4), but we will not go into this. ●

5.3.3 Examples of Orthogonal Polynomials

Let us take a look at some specific examples of orthogonal polynomials that are important in applications.

Hermite polynomials. The Hermite polynomials are denoted by $H_n(x)$ and for historical reasons are standardized such that $K_n = (-1)^n$. Thus, the generalized Rodriguez formula becomes [recall that $s(x) = 1$ and $w(x) = e^{-x^2}$]

$$H_n(x) = (-1)^n e^{x^2} \frac{d^n}{dx^n}(e^{-x^2}) \tag{5.41}$$

It is clear that each time e^{-x^2} is differentiated a factor of $-2x$ is introduced. The highest power of x is therefore given by

$$(-1)^n e^{x^2}(-2x)^n e^{-x^2} = 2^n x^n$$

and we have

$$k_n = 2^n$$

To find k'_n we find it helpful to check first for the evenness or oddness of the polynomial. We substitute $-x$ for x in (5.41) to get

$$H_n(-x) = (-1)^n e^{(-x)^2} \frac{d^n}{d(-x)^n}[e^{-(-x)^2}]$$

$$= (-1)^n e^{x^2}(-1)^n \frac{d^n}{dx^n}(e^{-x^2})$$

If n is even, we have

$$H_n(-x) = (-1)^{(\text{even})} H_n(x) = H_n(x)$$

which means that $H_n(x)$ can have only even powers of x. On the other hand, if n is odd, we get

$$H_n(-x) = -H_n(x)$$

which shows that $H_n(x)$ has only odd powers of x. From this we conclude that the next highest power of x in $H_n(x)$ is not $n-1$ but $n-2$. Thus, the coefficient of x^{n-1} is zero for $H_n(x)$, and we have

$$k'_n = 0$$

We now calculate h_n as given by (5.40). With $s = 1$, $w = e^{-x^2}$, $a = -\infty$, $b = +\infty$ (see Table 5.1), $K_n = (-1)^n$, and $k_n = 2^n$, we obtain

$$h_n = \frac{(-1)^n 2^n n!}{(-1)^n} \int_{-\infty}^{+\infty} e^{-x^2} dx = \sqrt{\pi}\, 2^n n!$$

where we have used

$$\int_{-\infty}^{\infty} e^{-x^2} dx = \sqrt{\pi}$$

a result shown in Exercise 5.3.2.

To find the recurrence relation we need α_n, β_n, and γ_n as defined in Eqs. (5.34):

$$\alpha_n = \frac{k_{n+1}}{k_n} = \frac{2^{n+1}}{2^n} = 2$$

$$\beta_n = \alpha_n\left(\frac{k'_{n+1}}{k_{n+1}} - \frac{k'_n}{k_n}\right) = 0$$

$$\gamma_n = -\frac{h_n}{h_{n-1}}\left(\frac{\alpha_n}{\alpha_{n-1}}\right) = -\frac{\sqrt{\pi}\, 2^n n!}{\sqrt{\pi}\, 2^{n-1}(n-1)!}\left(\frac{2}{2}\right) = -2n$$

Therefore, the recurrence relation of Eq. (5.35) becomes

$$H_{n+1}(x) = 2xH_n(x) - 2nH_{n-1}(x) \tag{5.42}$$

Finally, to write the differential equation satisfied by $H_n(x)$, we first have to find λ_n as given by Eq. (5.39). We note that

$$K_1 = -1 \qquad \frac{d^2 s}{dx^2} = 0 \qquad F_1(x) = (-1)e^{x^2}\frac{d}{dx}(e^{-x^2}) = 2x$$

giving

$$F'_1 = 2$$

which results in

$$\lambda_n = n[(-1)2] = -2n$$

Using all this information in (5.37), we obtain the differential equation

$$\frac{d}{dx}\left(e^{-x^2}\frac{dH_n}{dx}\right) = e^{-x^2}(-2n)H_n$$

or

$$-2xe^{-x^2}H_n' + e^{-x^2}H_n'' + 2ne^{-x^2}H_n = 0$$

Multiplying by e^{x^2} yields

$$\frac{d^2H_n}{dx^2} - 2x\frac{dH_n}{dx} + 2n\,H_n = 0 \tag{5.43}$$

Laguerre polynomials. Laguerre polynomials are denoted by $L_n^\nu(x)$ with $s(x) = x$, $w(x) = x^\nu e^{-x}$ (where $\nu > -1$), $a = 0$, and $b = +\infty$. In this case the standardization is

$$K_n = n!$$

from which we obtain the generalized Rodriguez formula:

$$\begin{aligned} L_n^\nu(x) &= \frac{1}{n!x^\nu e^{-x}}\frac{d^n}{dx^n}(x^\nu e^{-x}x^n) \\ &= \frac{1}{n!}x^{-\nu}e^x\frac{d^n}{dx^n}(x^{n+\nu}e^{-x}) \end{aligned} \tag{5.44}$$

To find k_n we note that differentiating e^{-x} does not introduce any new powers of x but only a factor of -1. Thus, the highest power of x is obtained by leaving $x^{n+\nu}$ alone and differentiating e^{-x} n times. This gives

$$\frac{1}{n!}x^{-\nu}e^x x^{n+\nu}(-1)^n e^{-x} = \frac{(-1)^n}{n!}x^n$$

which yields

$$k_n = \frac{(-1)^n}{n!}$$

We may try to check the evenness or oddness of $L_n^\nu(x)$; however, we immediately see that changing x to $-x$ completely distorts the RHS of (5.44). In fact, $k_n' \neq 0$ in this case, and we can calculate it by noticing that the next highest power of x is obtained by adding the first derivative of $x^{n+\nu}$ n times and multiplying the result by $(-1)^{n-1}$, which comes from differentiating e^{-x}. We obtain

$$\frac{1}{n!}x^{-\nu}e^x[(-1)^{n-1}n(n+\nu)x^{n+\nu-1}e^{-x}] = \frac{(-1)^{n-1}(n+\nu)}{(n-1)!}x^{n-1}$$

resulting in

$$k_n' = \frac{(-1)^{n-1}(n+\nu)}{(n-1)!}$$

Finally, for h_n we get

$$h_n = \frac{(-1)^n[(-1)^n/n!]\,n!}{n!}\int_0^\infty x^\nu e^{-x} x^n\,dx$$

$$= \frac{1}{n!}\int_0^\infty x^{n+\nu} e^{-x}\,dx$$

If ν is not an integer (and it need not be), the integral on the RHS cannot be evaluated by elementary methods. In fact, this integral occurs so frequently in mathematical applications that it is given a special name, the *gamma function*, and written as

$$\Gamma(z+1) \equiv \int_0^\infty x^z e^{-x}\,dx \tag{5.45}$$

By integration by parts we can show that (see Chapter 14 for a discussion of the gamma function)

$$\Gamma(z+1) = z\Gamma(z) = z(z-1)\Gamma(z-1) = \cdots$$

If z is an integer, n, this terminates in $\Gamma(1)$, which is equal to 1, and we get

$$\Gamma(n+1) = n! \tag{5.46}$$

This result shows why the gamma function is sometimes called the *factorial function.*

Now we can write h_n as

$$h_n = \frac{\Gamma(n+\nu+1)}{n!} = \frac{\Gamma(n+\nu+1)}{\Gamma(n+1)} \equiv \frac{(n+\nu)!}{n!}$$

where, by definition, $(n+\nu)! \equiv \Gamma(n+\nu+1)$ for all values of ν.

To find the recurrence relation we first use (5.34) to calculate the relevant parameters:

$$\alpha_n = \frac{k_{n+1}}{k_n} = \frac{(-1)^{n+1}/(n+1)!}{(-1)^n/n!} = -\frac{1}{n+1}$$

$$\beta_n = \alpha_n\left(\frac{k'_{n+1}}{k_{n+1}} - \frac{k'_n}{k_n}\right) = -\frac{1}{n+1}\left[\frac{(-1)^n(n+1+\nu)/n!}{(-1)^{n+1}/(n+1)!} - \frac{(-1)^{n-1}(n+\nu)/(n-1)!}{(-1)^n/n!}\right]$$

$$= \frac{2n+\nu+1}{n+1}$$

$$\gamma_n = -\frac{(n+\nu)!/n!}{(n+\nu-1)!/(n-1)!}\left[\frac{-1/(n+1)}{-1/n}\right] = -\frac{n+\nu}{n+1}$$

or
$$(n+1)P_{n+1}(x) = (2n+1)xP_n(x) - nP_{n-1}(x) \tag{5.49}$$

Now we use $K_1 = -2$, $P_1(x) = -\frac{1}{2}(d/dx)(1-x^2) = x$, $dP_1/dx = 1$, and $d^2s/dx^2 = (d^2/dx^2)(1-x^2) = -2$ to obtain

$$\lambda_n = n[-2 + \tfrac{1}{2}(n-1)(-2)] = -n(n+1)$$

which yields the following differential equation

$$\frac{d}{dx}\left[(1-x^2)\frac{dP_n}{dx}\right] = -n(n+1)P_n \tag{5.50a}$$

This can also be expressed as

$$(1-x^2)P_n'' - 2xP_n' + n(n+1)P_n = 0 \tag{5.50b}$$

Summary of the remaining classical polynomials. Since this discussion of the classical polynomials is already lengthy, we will not go into detail for the remaining ones. Table 5.3 summarizes the results for Jacobi, Gegenbauer, and Tchebichef polynomials.

5.3.4 Expansion in Terms of Orthogonal Polynomials

Having considered the different classical orthogonal polynomials, we can write an arbitrary function, $f \in \mathscr{L}_w^2(a, b)$, as a linear combination, or expansion, of these polynomials. If we denote an orthogonal (not necessarily classical) polynomial by $|\mathrm{C}_k\rangle$ and the given function by $|\mathrm{f}\rangle$, we may write

$$|\mathrm{f}\rangle = \sum_{k=0}^{\infty} a_k|\mathrm{C}_k\rangle \tag{5.51a}$$

where, in general, $a_k \neq \langle \mathrm{C}_k|\mathrm{f}\rangle$ because the $|\mathrm{C}_k\rangle$'s are not normalized to unity. We can find a_k by multiplying the LHS by $\langle \mathrm{C}_i|$ and using the orthogonality of the $|\mathrm{C}_k\rangle$'s:

$$\langle \mathrm{C}_i|\mathrm{f}\rangle = \sum_{k=0}^{\infty} a_k\langle \mathrm{C}_i|\mathrm{C}_k\rangle = a_i\langle \mathrm{C}_i|\mathrm{C}_i\rangle$$

or
$$a_i = \frac{\langle \mathrm{C}_i|\mathrm{f}\rangle}{\langle \mathrm{C}_i|\mathrm{C}_i\rangle} \tag{5.52a}$$

This is written in function form as

$$a_i = \frac{\int_a^b C_i^*(x)f(x)w(x)\,dx}{\int_a^b |C_i(x)|^2 w(x)\,dx} \tag{5.52b}$$

We can also write (5.51a) as a functional relation if we multiply both sides by $\langle \mathrm{x}|$ and use the facts that $\langle \mathrm{x}|\mathrm{f}\rangle = f(x)$ and $\langle \mathrm{x}|\mathrm{C}_k\rangle = C_k(x)$. We obtain

$$f(x) = \sum_{k=0}^{\infty} a_k C_k(x) \tag{5.51b}$$

TABLE 5.3 OTHER ORTHOGONAL POLYNOMIALS

	Jacobi polynomials, $P_n^{(\nu,\mu)}(x)$	Gegenbauer polynomials, $C_n^{\lambda}(x)$
Standardization	$K_n = (-2)^n n!$	$K_n = (-2)^n n! \dfrac{\Gamma(n+\lambda+\frac{1}{2})\Gamma(2\lambda)}{\Gamma(n+2\lambda)\Gamma(\lambda+\frac{1}{2})}$
Constants	$k_n = 2^{-n} \dfrac{\Gamma(2n+\nu+\mu+1)}{\Gamma(n+1)\Gamma(n+\mu+\nu+1)}$	$k_n = \dfrac{2^n}{n!}\left[\dfrac{\Gamma(n+\lambda)}{\Gamma(\lambda)}\right]$
	$k'_n = \dfrac{n(\nu-\mu)}{2n+\nu+\mu} k_n$	$k'_n = 0$
	$h_n = \dfrac{2^{\nu+\mu+1}\Gamma(n+\nu+1)\Gamma(n+\mu+1)}{(2n+\nu+\mu+1)(n!)\Gamma(n+\nu+\mu+1)}$	$h_n = \dfrac{\sqrt{\pi}\,\Gamma(n+2\lambda)\Gamma(\lambda+\frac{1}{2})}{n!(n+\lambda)\Gamma(\lambda)\Gamma(2\lambda)}$
Rodriguez formula	$P_n^{(\nu,\mu)}(x) = \dfrac{(-1)^n}{2^n n!}(1-x)^{-\nu}(1+x)^{-\mu}\dfrac{d^n}{dx^n}[(1-x)^{\nu+n}(1+x)^{\mu+n}]$	$C_n^{\lambda}(x) = \dfrac{(-1)^n\Gamma(n+2\lambda)\Gamma(\lambda+\frac{1}{2})}{2^n n!\,\Gamma(n+\lambda+\frac{1}{2})\Gamma(2\lambda)}(1-x^2)^{-\lambda+1/2}$ $\dfrac{d^n}{dx^n}[(1-x^2)^{n+\lambda-1/2}](1-x^2)$
Differential equation	$(1-x^2)\dfrac{d^2}{dx^2}P_n^{(\nu,\mu)}(x) + [\mu-\nu-(\nu+\mu+2)x]\dfrac{d}{dx}P_n^{(\nu,\mu)}(x)$ $+ n(n+\mu+\nu+1)P_n^{(\nu,\mu)}(x) = 0$	$\dfrac{d^2}{dx^2}C_n^{\lambda} - (2\lambda+1)x\dfrac{dC_n^{\lambda}}{dx} + n(n+2\lambda)C_n^{\lambda}$ $= 0$
Recurrence relation	$2(n+1)(n+\nu+\mu+1)(2n+\nu+\mu)P_{n+1}^{(\nu,\mu)}(x)$ $= (2n+\nu+\mu+1)[(2n+\nu+\mu)(2n+\nu+\mu+2)x+\nu^2-\mu^2]$ $P_n^{(\nu,\mu)}(x) - 2(n+\nu)(n+\mu)(2n+\nu+\mu+2)P_{n-1}^{(\nu,\mu)}(x)$	$(n+1)C_{n+1}^{\lambda} = 2(n+\lambda)xC_n^{\lambda} - (n+2\lambda-1)C_{n-1}^{\lambda}$

(Continued)

TABLE 5.3 (continued)

	Tchebichef polynomials of the first kind, $T_n(x)$	Tchebichef polynomials of the second kind, $U_n(x)$
Standardization	$K_n = (-1)^n \frac{(2n)!}{2^n n!}$	$K_n = \frac{(-1)^n (2n+1)!}{2^n (n+1)!}$
Constants	$k_n = 2^{n-1}$	$k_n = 2^n$
	$k'_n = 0$	$k'_n = 0$
	$h_n = \frac{\pi}{2}$	$h_n = \frac{\pi}{2}$
Rodriguez formula	$T_n(x) = \frac{(-1)^n 2^n n!}{(2n)!} (1-x^2)^{1/2} \frac{d^n}{dx^n} [(1-x^2)^{n-1/2}]$	$U_n(x) = \frac{(-1)^n 2^n (n+1)!}{(2n+1)!} (1-x^2)^{-1/2} \frac{d^2}{dx^2} [(1-x^2)^{n+1/2}]$
Differential equation	$(1-x^2)\frac{d^2 T_n}{dx^2} - x\frac{dT_n}{dx} + n^2 T_n = 0$	$(1-x^2)\frac{d^2 U_n}{dx^2} - 3x\frac{dU_n}{dx} + n(n+2)U_n = 0$
Recurrence relation	$T_{n+1} = 2xT_n - T_{n-1}$	$U_{n+1} = 2xU_n - U_{n-1}$

Let us now consider a few specific examples of the expansion of functions in terms of orthogonal polynomials using (5.51) and (5.52).

Example 5.3.6

Let us expand the function

$$f(x) = \begin{cases} +1 & \text{if } 0 < x < 1 \\ -1 & \text{if } -1 < x < 0 \end{cases}$$

in terms of Legendre polynomials. From (5.51b) we have

$$f(x) = \sum_{k=0}^{\infty} a_k P_k(x)$$

where, by (5.52b),

$$a_k = \frac{\int_{-1}^{1} P_k(x) f(x)\, dx}{\int_{-1}^{1} [P_k(x)]^2 dx} = \frac{\int_{-1}^{1} P_k(x) f(x)\, dx}{h_k} = \frac{2k+1}{2} \int_{-1}^{+1} P_k(x) f(x)\, dx$$

Now we substitute for $f(x)$ in this equation and write

$$a_k = \frac{2k+1}{2}\left[\int_{-1}^{0} P_k(x)(-1)dx + \int_{0}^{1} P_k(x)(+1)dx\right] = \frac{2k+1}{2}\left[\int_{0}^{1} P_k(x)dx - \int_{-1}^{0} P_k(x)dx\right]$$

To proceed we first rewrite the second integral:

$$\int_{-1}^{0} P_k(x)dx = \int_{+1}^{0} P_k(-y)\,d(-y) = -\int_{1}^{0} P_k(-y)dy = \int_{0}^{1} P_k(-x)dx = (-1)^k \int_{0}^{1} P_k(x)dx$$

Here we changed the variable from x to $-y$ in the second step, rewrote the integral using x instead of y in the fourth step, and made use of the evenness-or-oddness property of $P_k(x)$ in the last step. Therefore,

$$a_k = \frac{2k+1}{2}[1 - (-1)^k] \int_{0}^{1} P_k(x)\, dx$$

It is clear from this that $a_k = 0$ if k is even. Thus, only odd polynomials contribute to the expansion. Now, using the result of Exercise 5.3.5 for the integral of Legendre polynomials in the interval [0, 1], we obtain

$$a_k = \frac{(2k+1)(k-1)!}{2^k \left(\frac{k+1}{2}\right)! \left(\frac{k-1}{2}\right)!} (-1)^{(k-1)/2}$$

The first few terms in the expansion can thus be written

$$f(x) = \tfrac{3}{2} P_1(x) - \tfrac{7}{8} P_3(x) + \tfrac{11}{16} P_5(x) - \cdots$$ •

Note that the expansion of Example 5.3.6 gives $f(x)$ in the interval $[-1, +1]$. What happens to $f(x)$ outside this interval is not indicated by the equation. In a sense the expansion could represent many functions that differ from each other outside the interval $[-1, +1]$ but are identical in it.

Note also that $f(x)$ is an odd function; that is, $f(-x) = -f(x)$, as is evident from its definition. Thus, only odd polynomials appear in the expansion of $f(x)$ so as to preserve this property.

The fact that the argument of $P_k(x)$ is restricted to the interval $[-1, +1]$ may seem to indicate that x might be a sine or a cosine function of some angle. In fact, the place where Legendre polynomials appear most naturally is in the solution of partial differential equations that involve the Laplacian when written in spherical coordinates. After the partial differential equation is transformed into three ordinary differential equations using the method of the separation of variables (to be discussed in Chapter 8), the differential equation corresponding to the polar angle θ gives rise to solutions, of which Legendre polynomials are special cases. This differential equation simplifies considerably if the substitution $x = \cos\theta$ is made; in that case the solutions will be Legendre polynomials in x, or in $\cos\theta$. That is why in physical applications the argument of Legendre polynomials lies in the interval $[-1, +1]$.

Example 5.3.7

As another example of the expansion of functions in terms of orthogonal polynomials, let us expand $|x|$ in the interval $[-1, +1]$ in terms of Legendre polynomials. We write

$$|x| = \sum_{k=0}^{\infty} a_k P_k(x)$$

where

$$\begin{aligned} a_k &= \frac{2k+1}{2}\int_{-1}^{1} |x|\, P_k(x)\,dx = \frac{2k+1}{2}\left[\int_{-1}^{0}(-x)P_k(x)\,dx + \int_{0}^{1} xP_k(x)\,dx\right] \\ &= \frac{2k+1}{2}\left[\int_{+1}^{0}(+x)P_k(-x)\,d(-x) + \int_{0}^{1} xP_k(x)\,dx\right] \qquad (1) \\ &= \frac{2k+1}{2}[(-1)^k + 1]\int_{0}^{1} xP_k(x)\,dx \end{aligned}$$

We see that $a_k = 0$ for odd k. This is what we expect because $|x|$ is an even function of x, and, therefore, only the even polynomials must appear in the expansion. To evaluate the integral in (1), we make use of the recurrence relation of Eq. (5.49) and write

$$\int_0^1 xP_k(x)\,dx = \int_0^1\left[\frac{k+1}{2k+1}P_{k+1}(x) + \frac{k}{2k+1}P_{k-1}(x)\right]dx \qquad \text{for } k \geqslant 2$$

Note that since k is even, both $k + 1$ and $k - 1$ are odd, and we can use the result of Exercise 5.3.5 (replacing k with $k + 1$ and $k - 1$, respectively) to write

$$\int_0^1 xP_k(x)\,dx$$

$$= \frac{k+1}{2k+1}(-1)^{k/2}\frac{k!}{2^{k+1}\left(\frac{k+2}{2}\right)!\left(\frac{k}{2}\right)!} + \frac{k}{2k+1}(-1)^{(k-2)/2}\frac{(k-2)!}{2^{k-1}\left(\frac{k}{2}\right)!\left(\frac{k-2}{2}\right)!}$$

$$= (-1)^{k/2}\frac{(k-2)!}{(2k+1)\left(\frac{k}{2}\right)!\left(\frac{k-2}{2}\right)!\,2^{k-1}}\left[\frac{(k+1)(k-1)k}{4\left(\frac{k}{2}+1\right)\frac{k}{2}} - k\right]$$

$$= (-1)^{k/2-1}\frac{(k-2)!}{2^k\left(\frac{k}{2}+1\right)!\left(\frac{k}{2}-1\right)!} \qquad \text{for } k \geqslant 2$$

Thus, (1) yields

$$a_k = (-1)^{k/2-1}\frac{(2k+1)(k-2)!}{2^k\left(\frac{k}{2}+1\right)!\left(\frac{k}{2}-1\right)!} \qquad \text{for } k \geqslant 2$$

For $k = 0$ we use the definition of a_k directly and write

$$a_0 = \frac{1}{2}\int_{-1}^{1} |x|\,P_0(x)\,dx = \frac{1}{2}\int_{-1}^{1} |x|\,dx = \frac{1}{2}$$

If we write $k = 2s$ for $s = 1, 2, 3, \ldots$, we have

$$|x| = \frac{1}{2} + \sum_{s=1}^{\infty} (-1)^{s-1}\frac{(4s+1)(2s-2)!}{4^s(s+1)!(s-1)!}P_{2s}(x)$$

The first few terms in the expansion are

$$|x| = \frac{1}{2} + \frac{5}{8}P_2(x) - \frac{3}{16}P_4(x) + \frac{13}{128}P_6(x) - \cdots$$

●

Example 5.3.8

We can expand the Dirac delta function, $\delta(x)$, in terms of Legendre polynomials. We write

$$\delta(x) = \sum_{n=0}^{\infty} a_n P_n(x)$$

where

$$a_n = \frac{2n+1}{2}\int_{-1}^{1} \delta(x)P_n(x)\,dx = \frac{2n+1}{2}P_n(0)$$

For odd n this will give zero because $P_n(x)$ will be an odd polynomial. Thus, we must evaluate $P_n(0)$ for even n. We use the recurrence relation for $P_n(x)$ evaluated at $x = 0$. Substituting $x = 0$ in (5.49), we get

$$(n+1)P_{n+1}(0) = -nP_{n-1}(0)$$

Substituting $n - 1$ for n (to get a relation involving P_n), we obtain

$$nP_n(0) = -(n-1)P_{n-2}(0)$$

or

$$P_n(0) = -\frac{n-1}{n}P_{n-2}(0) = -\left(\frac{n-1}{n}\right)\left(-\frac{n-3}{n-2}\right)P_{n-4}$$

$$= \cdots = \left(-\frac{n-1}{n}\right)\left(-\frac{n-3}{n-2}\right)\cdots\left(-\frac{n-2m+1}{n-2m+2}\right)P_{n-2m}$$

$$= (-1)^m \frac{(n-1)(n-3)\cdots(n-2m+1)}{n(n-2)(n-4)\cdots(n-2m+2)}P_{n-2m}(0)$$

Thus, if $n = 2m$, we can write

$$P_{2m}(0) = (-1)^m \frac{(2m-1)(2m-3)\cdots(3)(1)}{2m(2m-2)\cdots(4)(2)}P_0(0)$$

Now we "fill the gaps" in the numerator by multiplying it and the denominator by $2m(2m-2)(2m-4)\cdots(4)(2)$, giving

$$P_{2m}(0) = (-1)^m \frac{2m(2m-1)(2m-2)\cdots(3)(2)(1)}{[(2m)2(m-1)2(m-2)\cdots 2(2)2(1)]^2}$$

$$= (-1)^m \frac{(2m)!}{2^{2m}(m!)^2}$$

because $P_0(x) = 1$. Thus, we can write

$$\delta(x) = \sum_{m=0}^{\infty} \frac{4m+1}{2}(-1)^m \frac{(2m)!}{2^{2m}(m!)^2}P_{2m}(x)$$

We can also derive this expansion as follows:

$$\delta(x-x') = w(x)\langle \mathrm{x}|\mathrm{x}'\rangle = \langle \mathrm{x}|\mathbb{1}|\mathrm{x}'\rangle = \langle \mathrm{x}|\left(\sum_{k=0}^{\infty} |\mathrm{f}_k\rangle\langle \mathrm{f}_k|\right)|\mathrm{x}'\rangle$$

$$= \sum_{k=0}^{\infty} f_k^*(x')f_k(x)$$

where $|\mathrm{f}_n\rangle$ form any complete orthonormal set of vectors. We make $P_k(x)$ orthonormal by dividing by

$$h_k^{1/2} \equiv \left\{\int_{-1}^{1} [P_k(x)]^2\,dx\right\}^{1/2} = \sqrt{\frac{2}{2k+1}}$$

We then obtain

$$\delta(x - x') = \sum_{k=0}^{\infty} \frac{P_k(x')}{\sqrt{2/(2k+1)}}\left(\frac{P_k(x)}{\sqrt{2/(2k+1)}}\right) = \sum_{k=0}^{\infty} \frac{2k+1}{2} P_k(x')P_k(x)$$

For $x' = 0$ we get

$$\delta(x) = \sum_{k=0}^{\infty} \frac{2k+1}{2} P_k(0)P_k(x)$$

which agrees with the earlier result. •

5.3.5 Generating Functions

It is possible to generate all orthogonal polynomials of a certain kind from a single function of two variables by differentiating that function repeatedly. Such a function is called a *generating function* and denoted $g(x, t)$. This generating function is required to be expandable in the form

$$g(x, t) = \sum_{n=0}^{\infty} a_n t^n F_n(x) \tag{5.53}$$

so that the nth derivative of $g(x, t)$ with respect to t evaluated at $t = 0$,

$$\left.\frac{\partial^n}{\partial t^n}[g(x, t)]\right|_{t=0} = n!\,a_n F_n(x)$$

gives $F_n(x)$ to within a multiplicative constant. The constant a_n is introduced for convenience in developing a closed form for $g(x, t)$.

We can immediately write a partial differential equation for $g(x, t)$. Since $F_n(x)$ satisfies (5.37), we can write

$$\frac{\partial}{\partial x}\left(ws\frac{\partial g}{\partial x}\right) = \sum_{n=0}^{\infty} a_n t^n \frac{d}{dx}\left(ws\frac{dF_n}{dx}\right) = \sum_{n=0}^{\infty} a_n t^n w\lambda_n F_n$$

Now we use the definition of λ_n in (5.39) to obtain

$$\begin{aligned}\frac{\partial}{\partial x}\left(ws\frac{\partial g}{\partial x}\right) &= w\sum_{n=0}^{\infty} a_n t^n \left[nK_1\frac{dF_1}{dx} + \frac{1}{2}n(n-1)\frac{d^2s}{dx^2}\right]F_n \\ &= wK_1\frac{dF_1}{dx}\sum_{n=0}^{\infty} a_n n t^n F_n + \frac{1}{2}w\frac{d^2s}{dx^2}\sum_{n=0}^{\infty} a_n n(n-1)t^n F_n \\ &= wK_1\frac{dF_1}{dx}t\sum_{n=0}^{\infty} n a_n t^{n-1} F_n + \frac{1}{2}w\frac{d^2s}{dx^2}t^2\sum_{n=0}^{\infty} a_n n(n-1)t^{n-2} F_n\end{aligned}$$

The first sum in the last line of this equation is simply $\partial g/\partial t$, and the second sum is $\partial^2 g/\partial t^2$. Thus, the equation becomes

$$\frac{\partial}{\partial x}\left(ws\frac{\partial g}{\partial x}\right) = wK_1F_1't\frac{\partial g}{\partial t} + \frac{1}{2}ws''t^2\frac{\partial^2 g}{\partial t^2} \tag{5.54}$$

Recall that K_1, F_1', and s'' are all constants.

Equation (5.54) is the desired partial differential equation; however, it is of hardly any use if it cannot be solved in closed form, giving an analytic expression for $g(x, t)$. The generality of (5.54) prevents it from being susceptible to solution in closed form. If we want $g(x, t)$ in closed form, we have to specify the kind of orthogonal polynomial. Thus, the following two examples outline the procedure for obtaining a closed form for the generating functions for Hermite and Legendre polynomials. The remaining generating functions can be derived similarly. We will not do so, but Table 5.4 lists the more important ones. (For a thorough discussion of generating functions, see McBride 1971.)

TABLE 5.4 GENERATING FUNCTIONS FOR SELECTED ORTHOGONAL POLYNOMIALS

Polynomial	Generating function
Hermite, $H_n(x)$	$\exp(-t^2 + 2xt)$
Laguerre, $L_n^\nu(x)$	$\exp[-xt/(1-t)]/(1-t)^{\nu+1}$
Legendre, $P_n(x)$	$(t^2 - 2xt + 1)^{-1/2}$
Tchebichef of the first kind, $T_n(x)$	$(1-t^2)(t^2 - 2xt + 1)^{-1}$
Tchebichef of the second kind, $U_n(x)$	$(t^2 - 2xt + 1)^{-1}$

Example 5.3.9

Let us derive the generating function for Hermite polynomials. We start with the definition

$$g(x, t) \equiv \sum_{n=0}^{\infty} a_n t^n H_n(x) = a_0 H_0(x) + a_1 t H_1(x) + a_2 t^2 H_2(x) + \cdots \tag{1}$$

We differentiate with respect to x (assuming that t is a constant) and use the recurrence relation for Hermite polynomials obtained in Example 5.3.5, to get

$$\frac{dg}{dx} = \sum_{n=0}^{\infty} a_n t^n \frac{dH_n}{dx} = \sum_{n=0}^{\infty} a_n t^n 2n H_{n-1} = 2t \sum_{n=1}^{\infty} n a_n t^{n-1} H_{n-1} \tag{2}$$

Now, if we *define*

$$n a_n \equiv a_{n-1}$$

then (2) yields

$$\begin{aligned}\frac{dg}{dx} &= 2t \sum_{n=1}^{\infty} a_{n-1} t^{n-1} H_{n-1} = 2t(a_0 H_0 + a_1 t H_1 + a_2 t^2 H_2 + \cdots) \\ &= 2tg\end{aligned}$$

We can solve this simple differential equation (recalling that t is just a constant):

$$\frac{dg}{g} = 2t\,dx \quad \Rightarrow \quad \ln g = 2tx + \ln C$$

or

$$g(x, t) = Ce^{2tx}$$

To find C we first note that

$$a_n = \frac{1}{n}a_{n-1} = \frac{1}{n}\left(\frac{1}{n-1}a_{n-2}\right) = \cdots = \frac{1}{n!}a_0 = \frac{1}{n!}$$

for $a_0 = 1$. Next we evaluate $g(x, t)$ in (1) at $x = 0$, which yields

$$g(0, t) = \sum_{n=0}^{\infty} \frac{1}{n!}t^n H_n(0)$$

Using arguments similar to those employed in Example 5.3.8, we can easily show that

$$H_n(0) = \begin{cases} 0 & \text{if } n \text{ is odd} \\ (-1)^m \dfrac{(2m)!}{m!} & \text{if } n = 2m \end{cases}$$

We then obtain

$$g(0, t) = \sum_{m=0}^{\infty} \frac{1}{(2m)!} t^{2m}(-1)^m \frac{(2m)!}{m!} = \sum_{m=0}^{\infty} \frac{(-t^2)^m}{m!} = e^{-t^2}$$

But we also have

$$g(0, t) = C$$

Therefore,

$$g(x, t) = e^{-t^2 + 2tx} = \sum_{n=0}^{\infty} \frac{t^n}{n!} H_n(x)$$

We note that

$$\left.\frac{\partial^n g}{\partial t^n}\right|_{t=0} = H_n(x)$$

so, for instance,

$$\left.\frac{\partial g}{\partial t}\right|_{t=0} = (-2t + 2x)e^{-t^2 + 2tx}|_{t=0} = 2x = H_1(x)$$

$$\left.\frac{\partial^2 g}{\partial t^2}\right|_{t=0} = [-2e^{-t^2 + 2tx} + (-2t + 2x)^2 e^{-t^2 + 2tx}]_{t=0} = -2 + 4x^2 = H_2(x)$$

and so on. •

Example 5.3.10

Deriving the generating function for Legendre polynomials is somewhat complicated. We start with the basic definition of Eq. (5.53):

$$g(x, t) = \sum_{n=0}^{\infty} a_n t^n P_n(x) \tag{1}$$

We differentiate with respect to x:

$$\frac{dg}{dx} = \sum_{n=0}^{\infty} a_n t^n \frac{dP_n}{dx} = a_0 \frac{dP_0}{dx} + a_1 t \frac{dP_1}{dx} + \cdots = a_1 t + \sum_{n=2}^{\infty} a_n t^n P'_n \tag{2}$$

Next we use a recurrence relation to replace P'_n with something without derivatives. We first use Eq. (11) of Example 5.3.5 with $n + 1$ replaced by n and substitute in (2) to obtain

$$\begin{aligned}\frac{dg}{dx} &= a_1 t + \sum_{n=2}^{\infty} a_n t^n [P'_{n-2} + (2n-1)P_{n-1}] \\ &= a_1 t + \frac{d}{dx}\sum_{n=2}^{\infty} a_n t^n P_{n-2} + \sum_{n=2}^{\infty} a_n t^n (2n-1) P_{n-1} \\ &= a_1 t + t^2 \frac{d}{dx}\sum_{n=2}^{\infty} a_n t^{n-2} P_{n-2} + 2\sum_{n=2}^{\infty} n a_n t^n P_{n-1} - t\sum_{n=2}^{\infty} a_n t^{n-1} P_{n-1}\end{aligned} \tag{3}$$

The first and last sums suggest defining a_n as follows:

$$a_n = a_{n-1} = a_{n-2} = \cdots = 1$$

In that case (3) gives

$$\frac{dg}{dx} = t + t^2 \frac{dg}{dx} + 2\sum_{n=2}^{\infty} n t^n P_{n-1} - t(g-1)$$

or

$$(1-t^2)\frac{dg}{dx} + tg = 2\sum_{n=2}^{\infty} n t^n P_{n-1} + 2t \tag{4}$$

We can also substitute the recurrence relation of Eq. (10) in Example 5.3.5 in (2) to get

$$\begin{aligned}\frac{dg}{dx} &= t + \sum_{n=2}^{\infty} t^n (xP'_{n-1} + nP_{n-1}) = t + tx\frac{d}{dx}\left(\sum_{n=2}^{\infty} t^{n-1} P_{n-1}\right) + \sum_{n=2}^{\infty} n t^n P_{n-1} \\ &= t + tx\frac{d}{dx}(g-1) + \sum_{n=2}^{\infty} n t^n P_{n-1} = t + tx\frac{dg}{dx} + \sum_{n=2}^{\infty} n t^n P_{n-1}\end{aligned}$$

or

$$(1-xt)\frac{dg}{dx} = \sum_{n=2}^{\infty} n t^n P_{n-1} + t \tag{5}$$

We see that the RHS of (4) is twice that of (5). Thus, we write

$$(1-t^2)\frac{dg}{dx} + tg = 2(1-xt)\frac{dg}{dx}$$

which gives

$$(t^2 - 2xt + 1)\frac{dg}{dx} = tg$$

or

$$\frac{dg}{g} = \frac{t\,dx}{t^2 - 2xt + 1}$$

Integrating both sides, we obtain

$$\ln g = -\tfrac{1}{2}\ln(t^2 - 2xt + 1) + \ln C$$

or

$$g(x, t) = \frac{C}{\sqrt{t^2 - 2xt + 1}}$$

To find the constant we note that

$$g(1, t) = \frac{C}{1 - t} \qquad \text{for } |t| < 1$$

On the other hand, Eq. (1) gives, using $P_n(1) = 1$,

$$g(1, t) = \sum_{n=0}^{\infty} t^n P_n(1) = \sum_{n=0}^{\infty} t^n = \frac{1}{1 - t} \qquad \text{for } |t| < 1$$

Comparing the last two equations yields $C = 1$, and we have

$$g(x, t) = (t^2 - 2xt + 1)^{-1/2} = \sum_{n=0}^{\infty} t^n P_n(x)$$

We note that, in this case,

$$P_n(x) = \frac{1}{n!}\frac{\partial^n g}{\partial t^n}\bigg|_{t=0}$$

For instance,

$$\frac{\partial g}{\partial t}\bigg|_{t=0} = -\frac{1}{2}(2t - 2x)(t^2 - 2xt + 1)^{-3/2}\bigg|_{t=0} = x = P_1(x)$$

$$\frac{\partial^2 g}{\partial t^2}\bigg|_{t=0} = -1 + 3x^2 = 2P_2(x)$$

and so on. •

Exercises

5.3.1 Calculate a_n and a_{n-1} in Eq. (5.33) in terms of k, k', and h as defined in Eqs. (5.31).

5.3.2 Evaluate the integral $\int_{-\infty}^{\infty} e^{-x^2} dx$.

5.3.3 Show that, for Legendre polynomials,

$$k_n = \frac{2^n \Gamma(n + \frac{1}{2})}{n!\,\Gamma(\frac{1}{2})}$$

5.3.4 Evaluate the integral $\int_{-1}^{+1} (1 - x^2)^n dx$.

5.3.5 Show that

$$\int_0^1 P_k(x)\,dx = \begin{cases} (-1)^{(k-1)/2} \dfrac{(k-1)!}{2^k \left(\dfrac{k+1}{2}\right)! \left(\dfrac{k-1}{2}\right)!} & \text{if } k \text{ is odd} \\ 0 & \text{if } k \text{ is even and greater than or equal to 2} \\ 1 & \text{if } k = 0 \end{cases}$$

5.4 FOURIER SERIES AND TRANSFORMS

The single topic that turns up in all areas of physics is Fourier analysis. It shows up, for example, in classical mechanics and the analysis of normal modes, in electromagnetic theory and the analysis of waves of various frequencies, in noise considerations and thermal physics, in quantum theory and the transformation between momentum and coordinate representations, and in relativistic quantum field theory and creation and annihilation operation formalism.

Because of this importance, the rest of this chapter is devoted to the study of Fourier series and transforms. The best way to begin this study is to invoke a generalization of the Stone-Weierstrass theorem (Theorem 5.3.3), which established the completeness of monomials, x^k, which, in turn, permitted the construction of orthogonal polynomials. The generalization of Theorem 5.3.3 permits us to find another set of orthogonal functions in terms of which we can expand an arbitrary function. This generalization involves polynomials in more than one variable. (For a proof of this theorem, see Simmons 1983, pp. 160 and 161.)

5.4.1 Theorem (Generalized Stone-Weierstrass Theorem) If $f(x_1, x_2, \ldots, x_n)$ is continuous in the domain $a_i \leqslant x_i \leqslant b_i$, then it can be expanded in terms of the monomials $x_1^{k_1} x_2^{k_2} \cdots x_n^{k_n}$, where $k_i \geqslant 0$. ■

We can interpret the generalized Stone-Weierstrass theorem in terms of vectors and their representations as functions, as we did in Section 5.3. In other words, $x_1^{k_1} x_2^{k_2} \cdots x_n^{k_n}$ can be thought of as the coordinates of some abstract vector $|\mathrm{f}_{k_1 k_2 \cdots k_n}\rangle$ along $|\mathbf{r}\rangle = |\mathrm{x}_1, \mathrm{x}_2, \ldots, \mathrm{x}_n\rangle$:

$$\begin{aligned}\langle \mathbf{r} | \mathrm{f}_{k_1 k_2 \cdots k_n}\rangle &\equiv \langle \mathrm{x}_1, \mathrm{x}_2, \ldots, \mathrm{x}_n | \mathrm{f}_{k_1 k_2 \cdots k_n}\rangle \\ &\equiv f_{k_1 k_2 \cdots k_n}(x_1, x_2, \ldots, x_n) = x_1^{k_1} x_2^{k_2} \cdots x_n^{k_n} \end{aligned} \tag{5.55}$$

The theorem then states that the set $\{|\mathrm{f}_{k_1 k_2 \cdots k_n}\rangle\}_{k_i=0}^{\infty}$ represented by Eq. (5.55) forms a basis for the space of continuous functions of the n variables $x_1, x_2, \ldots, x_n$.

5.4.1 Fourier (Trigonometric) Series

Now let us consider functions that are periodic and investigate their expansion in terms of elementary periodic functions. We use the generalized Stone-Weierstrass theorem with two variables, x and y. A function, $f(x, y)$, can be written as a series in powers of x multiplied by powers of y: $f(x, y) = \sum_{k,m} a_{km} x^k y^m$. In this equation x and y could be considered the coordinates in the xy-plane, which, in turn, could be written in terms of polar coordinates r and θ. In that case we obtain[6]

$$f(r, \theta) = \sum_{k,m} a_{km} r^{k+m} \cos^k\theta \sin^m\theta$$

[6]It is common (but sloppy!) to use the same letter for the function even though variables in the argument are replaced by some new variables related to the old ones. Thus, $f(r, \theta)$ really means $f(r\cos\theta, r\sin\theta)$. Thus, if $f(x, y) = 2x^2 + 3y^2$, then $f(r, \theta) \neq 2r^2 + 3\theta^2$, but rather $f(r, \theta) = 2r^2\cos^2\theta + 3r^2\sin^2\theta$.

In particular, if we let $r = 1$, we obtain a function of only θ, $f(\theta) = \sum_{k,m} a_{km} \cos^k \theta \sin^m \theta$, which on substitution of

$$\cos\theta = \tfrac{1}{2}(e^{i\theta} + e^{-i\theta}) \qquad \text{and} \qquad \sin\theta = \frac{1}{2i}(e^{i\theta} - e^{-i\theta})$$

becomes
$$f(\theta) = \sum_{k,m} a_{km}\left(\frac{1}{2}\right)^k (e^{i\theta} + e^{-i\theta})^k \left(\frac{1}{2i}\right)^m (e^{i\theta} - e^{-i\theta})^m$$

Expanding the terms on the RHS and collecting all similar powers together, we obtain

$$f(\theta) = \sum_{n=-\infty}^{+\infty} b_n e^{in\theta} \tag{5.56}$$

where b_n is a constant that depends on a_{km}.

The RHS of (5.56) is periodic with period 2π; thus, it is especially suitable for periodic functions $f(\theta)$ that satisfy the periodicity condition $f(\theta - \pi) = f(\theta + \pi)$.

We can also write (5.56) as

$$\begin{aligned} f(\theta) &= b_0 + \sum_{n=1}^{\infty} (b_n e^{in\theta} + b_{-n} e^{-in\theta}) \\ &= b_0 + \sum_{n=1}^{\infty} [b_n(\cos n\theta + i \sin n\theta) + b_{-n}(\cos n\theta - i \sin n\theta)] \\ &= b_0 + \sum_{n=1}^{\infty} [(b_n + b_{-n}) \cos n\theta + i(b_n - b_{-n}) \sin n\theta] \\ &= b_0 + \sum_{n=1}^{\infty} (A_n \cos n\theta + B_n \sin n\theta) \end{aligned} \tag{5.57}$$

where
$$A_n = b_n + b_{-n} \qquad \text{and} \qquad B_n = i(b_n - b_{-n})$$

are constants to be determined for each $f(\theta)$. If $f(\theta)$ is real, then b_0, A_n, and B_n are also real. Equation (5.56) or (5.57) is called the *Fourier-series expansion* of $f(\theta)$.

Let us now concentrate on the elementary periodic functions, $e^{in\theta}$. We define the vectors $\ldots, |-\mathrm{n}\rangle, \ldots, |-1\rangle, |0\rangle, |1\rangle, \ldots, |\mathrm{n}\rangle, \ldots$ such that their θth components are given by

$$\langle\theta|\mathrm{n}\rangle \equiv \frac{1}{\sqrt{2\pi}} e^{in\theta} \qquad \text{where } \theta \in [-\pi, \pi] \tag{5.58}$$

These functions (vectors), which are assumed to belong to $\mathscr{L}^2(-\pi, \pi)$, are orthonormal, as can easily be verified. It can also be shown that they are complete. In fact, for functions that are continuous on $[-\pi, \pi]$, this completeness is a result of Theorem 5.4.1. It can be shown, however, that the sequence $\{|\mathrm{n}\rangle\}$ defined in (5.58) is also a complete orthonormal sequence for *piecewise* continuous functions on $[-\pi, \pi]$. Therefore, any (periodic) function of θ can be expressed as a linear combination of these orthonormal vectors. This means that any (periodic) $|\mathrm{f}\rangle \in \mathscr{L}^2(-\pi, \pi)$ can be written in vector notation as a linear combination of $|\mathrm{n}\rangle$'s:

$$|\mathrm{f}\rangle = \sum_{n=-\infty}^{\infty} f_n |\mathrm{n}\rangle \tag{5.59a}$$

where

$$f_n = \langle \mathrm{n}|\mathrm{f}\rangle \tag{5.59b}$$

We can write (5.59a) as a functional relation if we take the θth component of both sides and write

$$\langle\theta|\mathrm{f}\rangle = \sum_{n=-\infty}^{\infty} f_n \langle\theta|\mathrm{n}\rangle$$

or

$$f(\theta) = \sum_{n=-\infty}^{\infty} f_n \frac{1}{\sqrt{2\pi}} e^{in\theta} = \frac{1}{\sqrt{2\pi}} \sum_{n=-\infty}^{\infty} f_n e^{in\theta} \tag{5.60a}$$

with f_n given by

$$f_n = \langle \mathrm{n}|\mathbb{1}|\mathrm{f}\rangle = \langle \mathrm{n}|\left(\int_{-\pi}^{\pi} |\theta\rangle\langle\theta|\, d\theta\right)|\mathrm{f}\rangle = \int_{-\pi}^{\pi} d\theta \langle \mathrm{n}|\theta\rangle\langle\theta|\mathrm{f}\rangle$$

$$= \int_{-\pi}^{\pi} d\theta \frac{1}{\sqrt{2\pi}} e^{-in\theta} f(\theta) = \frac{1}{\sqrt{2\pi}} \int_{-\pi}^{\pi} d\theta\, e^{-in\theta} f(\theta) \tag{5.60b}$$

It is important to note that even though $f(\theta)$ may be defined only for $-\pi \leqslant \theta \leqslant \pi$, Eqs. (5.60) extend the domain of the definition of $f(\theta)$ to the interval $(2k-1)\pi \leqslant \theta \leqslant (2k+1)\pi$, for all $k \in \mathbb{Z}$. Thus, if a function is to be represented by (5.60) without any specification of the interval of definition, it must be periodic in θ.

Functions are not necessarily defined on $[-\pi, \pi]$; in fact, they are usually defined on some arbitrary interval $[a, b = a + L]$. As a specific example, let us consider a function, $F(x)$, that is defined in $[a, b = a + L]$ and has period L. We define a new variable,

$$\theta \equiv \frac{2\pi}{L}\left(x - a - \frac{L}{2}\right)$$

such that

$$x = \frac{L}{2\pi}\theta + a + \frac{L}{2}$$

If we substitute this expression in $F(x)$, we obtain a function of θ,

$$F\left(\frac{L}{2\pi}\theta + a + \frac{L}{2}\right) \equiv f(\theta)$$

which is periodic in $[-\pi, \pi]$:

$$f(\theta - \pi) = F\left[\frac{L}{2\pi}(\theta - \pi) + a + \frac{L}{2}\right]$$

$$= F\left(\frac{L}{2\pi}\theta + a\right) = F\left(x - \frac{L}{2}\right)$$

and
$$f(\theta+\pi)=F\left[\frac{L}{2\pi}(\theta+\pi)+a+\frac{L}{2}\right]=F\left(x+\frac{L}{2}\right)$$

The periodicity of $F(x)$ implies that

$$f(\theta-\pi)=f(\theta+\pi)$$

We can thus expand $f(\theta)$ in terms of the elementary functions $(1/\sqrt{2\pi})e^{in\theta}$:

$$f(\theta)=F\left(\frac{L}{2\pi}\theta+a+\frac{L}{2}\right)=\frac{1}{\sqrt{2\pi}}\sum_{n=-\infty}^{\infty}f_n e^{in\theta}$$

Substituting for θ in terms of x, we obtain

$$F(x)=\frac{1}{\sqrt{2\pi}}\sum_{n=-\infty}^{\infty}f_n e^{in[(2\pi/L)(x-a-L/2)]}=\sum_{n=-\infty}^{\infty}F'_n e^{i(2n\pi/L)x}$$

where we have introduced

$$F'_n\equiv\frac{1}{\sqrt{2\pi}}f_n e^{-i(2n\pi/L)(a+L/2)}$$

Using (5.60b), we can write

$$F'_n=\frac{1}{\sqrt{2\pi}}e^{-i(2n\pi/L)(a+L/2)}\frac{1}{\sqrt{2\pi}}\int d\theta\, e^{-in\theta}f(\theta)$$

In terms of x this becomes

$$F'_n=\frac{1}{2\pi}e^{-i(2n\pi/L)(a+L/2)}\int_a^{a+L}\frac{2\pi}{L}dx\,e^{-i(2n\pi/L)(x-a-L/2)}F(x)$$

$$=\frac{1}{L}\int_a^b dx\,e^{-i(2n\pi/L)x}F(x)$$

In the basis $\{|\mathrm{x}\rangle\}$ the orthonormal vectors $|\mathrm{n}\rangle$, which now belong to $\mathscr{L}^2(a,b)$, have components

$$\langle \mathrm{x}|\mathrm{n}\rangle=\frac{1}{\sqrt{L}}e^{i(2n\pi/L)x}$$

It is convenient to redefine F'_n so that it becomes the nth component of $|\mathrm{F}\rangle$ in the orthonormal basis $\{|\mathrm{n}\rangle\}$. We, therefore, write

$$|\mathrm{F}\rangle=\sum_{n=-\infty}^{\infty}F_n|\mathrm{n}\rangle$$

or in the basis $\{|\mathrm{x}\rangle\}$

$$\langle \mathrm{x}|\mathrm{F}\rangle=\sum_{n=-\infty}^{\infty}F_n\langle \mathrm{x}|\mathrm{n}\rangle$$

or

$$F(x) = \frac{1}{\sqrt{L}} \sum_{n=-\infty}^{\infty} F_n e^{i(2n\pi/L)x} \tag{5.61a}$$

with F_n defined as

$$F_n = \langle \mathrm{n} | \mathrm{F} \rangle = \int_a^b dx \langle \mathrm{n} | \mathrm{x} \rangle \langle \mathrm{x} | \mathrm{F} \rangle = \int_a^b dx \frac{1}{\sqrt{L}} e^{-i(2n\pi/L)x} F(x)$$

or

$$F_n = \frac{1}{\sqrt{L}} \int_a^b dx\, e^{-i(2n\pi/L)x} F(x) \tag{5.61b}$$

Equations (5.61) are generalizations of (5.60) to cases where the interval of periodicity is $[a, a + L]$ instead of $[-\pi, \pi]$. The following examples illustrate these points.

Example 5.4.1

In the study of electrical circuitry periodic voltage signals of different shapes are encountered. An example is a square potential of height $\mathcal{E}_0$, "duration" T, and "rest duration" T (Fig. 5.6). The potential as a function of time, $V(t)$, can be expanded as a Fourier series. The interval is $[0, 2T]$ because that is one whole cycle of the potential variation. We, therefore, use (5.61a) and write

$$V(t) = \frac{1}{\sqrt{2T}} \sum_{n=-\infty}^{\infty} V_n e^{i(2n\pi/2T)t} \tag{1}$$

where

$$V_n = \frac{1}{\sqrt{2T}} \int_0^{2T} dt\, e^{-i(2n\pi/2T)t} V(t) \tag{2}$$

The problem is to find V_n. This is easily done by substituting

$$V(t) = \begin{cases} \mathcal{E}_0 & \text{if } 0 \leqslant t \leqslant T \\ 0 & \text{if } T < t \leqslant 2T \end{cases}$$

in (2) to obtain

$$V_n = \frac{\mathcal{E}_0}{\sqrt{2T}} \int_0^T dt\, e^{-i(n\pi/T)t} = \frac{\mathcal{E}_0}{\sqrt{2T}} \left(-\frac{T}{in\pi} \right) [(-1)^n - 1] \qquad \text{where } n \neq 0$$

Thus, we have

$$V_n = \begin{cases} 0 & \text{if } n \text{ is even and } n \neq 0 \\ \dfrac{\sqrt{2T}\mathcal{E}_0}{in\pi} & \text{if } n \text{ is odd} \end{cases}$$

For $n = 0$ we get

$$V_0 = \frac{1}{\sqrt{2T}} \int_0^{2T} dt\, V(t) = \mathcal{E}_0 \sqrt{\frac{T}{2}}$$

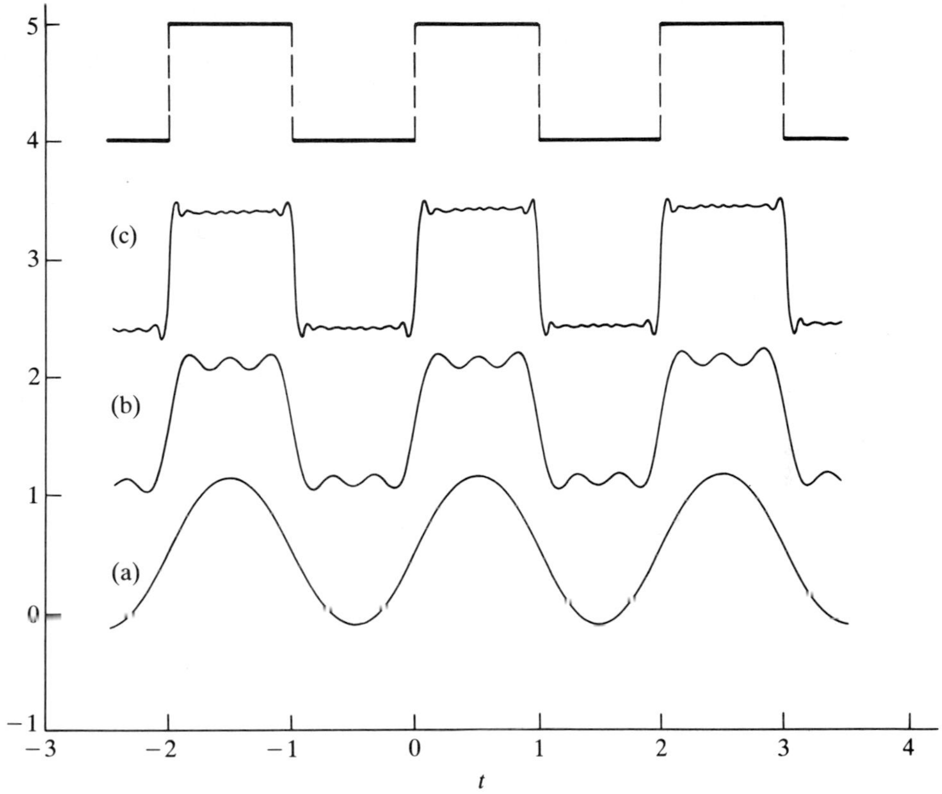

Figure 5.6 The graph of the square wave and its various approximations using Fourier-series expansions with (a) one term, (b) three terms, and (c) eleven terms.

Therefore, we can write

$$V(t) = \frac{1}{\sqrt{2T}}\left[\mathcal{E}_0\sqrt{\frac{T}{2}} + \frac{\sqrt{2T}\mathcal{E}_0}{i\pi}\left(\sum_{\substack{n=-\infty \\ n \text{ odd}}}^{-1} \frac{1}{n} e^{i(n\pi/T)t} + \sum_{\substack{n=1 \\ n \text{ odd}}}^{\infty} \frac{1}{n} e^{i(n\pi/T)t}\right)\right]$$

$$= \mathcal{E}_0\left\{\frac{1}{2} + \frac{1}{i\pi}\left[\sum_{\substack{n=1 \\ n \text{ odd}}}^{\infty}\left(-\frac{1}{n}\right)e^{-i(n\pi/T)t} + \sum_{\substack{n=1 \\ n \text{ odd}}}^{\infty}\frac{1}{n}e^{i(n\pi/T)t}\right]\right\}$$

where we changed n to $-n$ in the first sum. It is now clear that

$$V(t) = \mathcal{E}_0\left\{\frac{1}{2} + \frac{2}{\pi}\sum_{\substack{n=1 \\ n \text{ odd}}}^{\infty}\frac{1}{n}\sin\left(\frac{n\pi t}{T}\right)\right\}$$

Figure 5.6 presents the graphical representation of the above sum when only the first few terms are present. ●

Example 5.4.2

Another frequently used potential is the sawtooth potential (see Fig. 5.7). The equation for $V(t)$ is

$$V(t) = \mathcal{E}_0 \frac{t}{T} \qquad \text{for } 0 \leqslant t \leqslant T$$

and its Fourier representation is

$$V(t) = \frac{1}{\sqrt{T}} \sum_{n=-\infty}^{\infty} V_n e^{i(2n\pi/T)t}$$

where

$$V_n = \frac{1}{\sqrt{T}} \int_0^T dt\, e^{-i(2n\pi/T)t}\, V(t) \tag{1}$$

Note the difference between the above equations, where the period is T, and the corresponding ones in Example 5.4.1, where $2T$ was the period.

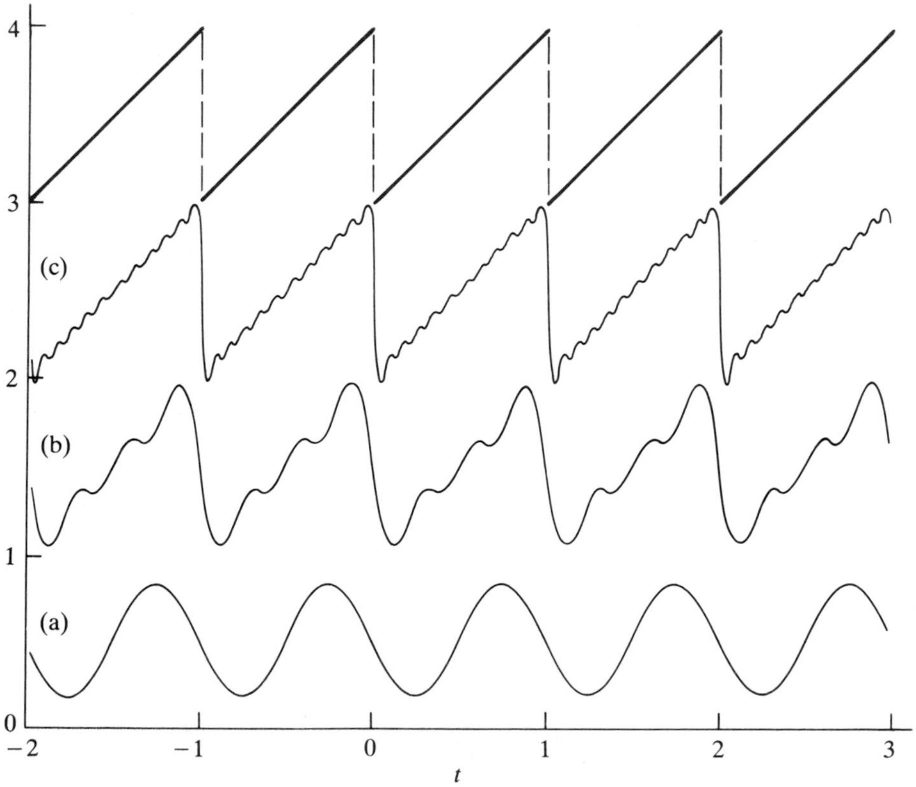

Figure 5.7 The graph of the sawtooth wave and its various approximations using Fourier-series expansions with (a) one term, (b) three terms, and (c) eleven terms.

Substituting for $V(t)$ in the integral in (1) yields

$$V_n = \frac{1}{\sqrt{T}} \int_0^T dt\, e^{-i(2n\pi/T)t} \mathcal{E}_0 \frac{t}{T} = \mathcal{E}_0 T^{-3/2} \int_0^T t e^{-i(2n\pi t/T)}\, dt$$

$$= \mathcal{E}_0 T^{-3/2} \left(\frac{Tt}{-i2n\pi} e^{-i(2n\pi t/T)} \Big|_0^T + \frac{T}{i2n\pi} \int_0^T dt\, e^{-i(2n\pi t/T)} \right)$$

$$= \mathcal{E}_0 T^{-3/2} \left(\frac{T^2}{-i2n\pi} \right) = -\frac{\mathcal{E}_0 \sqrt{T}}{i2n\pi} \qquad \text{for } n \neq 0$$

For $n = 0$ we get

$$V_0 = \frac{1}{\sqrt{T}} \int_0^T dt\, \mathcal{E}_0 \frac{t}{T} = \tfrac{1}{2}\sqrt{T}\mathcal{E}_0$$

Thus,

$$V(t) = \frac{1}{\sqrt{T}} \left[\frac{1}{2}\sqrt{T}\mathcal{E}_0 + \sum_{n=-\infty}^{-1} \left(-\frac{\mathcal{E}_0\sqrt{T}}{i2n\pi} e^{i(2n\pi t/T)} \right) + \sum_{n=1}^{\infty} \left(-\frac{\mathcal{E}_0\sqrt{T}}{i2n\pi} e^{i(2n\pi t/T)} \right) \right]$$

$$= \mathcal{E}_0 \left[\frac{1}{2} - \frac{1}{\pi} \sum_{n=1}^{\infty} \frac{1}{n} \sin\left(\frac{2n\pi t}{T} \right) \right] \tag{2}$$

We write out the first few terms of this expansion:

$$V(t) = \mathcal{E}_0 \left\{ \frac{1}{2} - \frac{1}{\pi} \left[\sin\left(\frac{2\pi t}{T}\right) + \frac{1}{2}\sin\left(\frac{4\pi t}{T}\right) + \frac{1}{3}\sin\left(\frac{6\pi t}{T}\right) + \cdots \right] \right\}$$

Figure 5.7 shows the graph of (2) for the first few terms. •

Example 5.4.3

Let us now expand the following function in a Fourier series:

$$f(\theta) = \theta \qquad \text{for } -\pi \leqslant \theta \leqslant \pi$$

The nth coefficient of the function is

$$f_n = \frac{1}{\sqrt{2\pi}} \int_{-\pi}^{\pi} \theta e^{-in\theta}\, d\theta = \frac{1}{\sqrt{2\pi}} \left(\frac{\theta}{-in} e^{-in\theta} \Big|_{-\pi}^{\pi} + \frac{1}{n^2} e^{-in\theta} \Big|_{-\pi}^{+\pi} \right)$$

$$= \frac{1}{\sqrt{2\pi}} \left[\frac{2\pi}{-in} (-1)^n \right] = \frac{\sqrt{2\pi}}{-in} (-1)^n$$

Also,
$$f_0 = \frac{1}{\sqrt{2\pi}} \int_{-\pi}^{\pi} \theta\, d\theta = 0$$

Thus,
$$\theta = \frac{1}{\sqrt{2\pi}} \left[\sum_{n=-\infty}^{-1} \frac{\sqrt{2\pi}}{-in} (-1)^n e^{in\theta} + \sum_{n=1}^{\infty} \frac{\sqrt{2\pi}}{-in} (-1)^n e^{in\theta} \right]$$
$$= - \sum_{n=1}^{\infty} \frac{(-1)^n}{in} (e^{in\theta} - e^{-in\theta}) = -2 \sum_{n=1}^{\infty} \frac{(-1)^n}{n} \sin n\theta$$

In particular, if we set $\theta = \pi/2$, we obtain
$$\frac{\pi}{4} = - \sum_{n=1}^{\infty} \frac{(-1)^n}{n} \sin \frac{n\pi}{2} = - \sum_{\substack{n=1 \\ n \text{ odd}}}^{\infty} \frac{(-1)^n}{n} (-1)^{(n-1)/2}$$

or, if we substitute $2k + 1$ for n,
$$\frac{\pi}{4} = \sum_{k=0}^{\infty} \frac{1}{2k+1} (-1)^k = 1 - \frac{1}{3} + \frac{1}{5} - \frac{1}{7} + \cdots$$

which is an expansion for π. •

The foregoing examples indicate an important fact about the Fourier series. At the points of discontinuity (for example, $t = T$ in Examples 5.4.1 and 5.4.2 and $\theta = \pm \pi$ in Example 5.4.3), the value of the function is not defined, but the Fourier-series expansion *assigns* it a value—the average of the two values on the right and left of the discontinuity. In Example 5.4.2, for instance, when we substitute $t = T$ in the series, all the sine terms vanish and we obtain
$$V(T) = \tfrac{1}{2} \mathcal{E}_0$$

On the other hand,
$$\lim_{\varepsilon \to 0} [V(T - \varepsilon) + V(T + \varepsilon)] = \lim_{\varepsilon \to 0} \left[\mathcal{E}_0 \frac{T - \varepsilon}{T} + \mathcal{E}_0 \frac{T + \varepsilon}{T} \right] = \mathcal{E}_0$$

so
$$\tfrac{1}{2}[V(T - 0) + V(T + 0)] \equiv \lim_{\varepsilon \to 0} \tfrac{1}{2}[V(T - \varepsilon) + V(T + \varepsilon)] = \tfrac{1}{2}\mathcal{E}_0 = V(T)$$

This is a general property of Fourier series. In fact, the main theorem of Fourier series, which follows, incorporates this property. (For a proof of this theorem, see Courant and Hilbert 1962.)

5.4.2 Theorem The Fourier series of a function $f(\theta)$ that is *piecewise continuous* in the interval $[-\pi, \pi]$ converges to
$$\tfrac{1}{2}[f(\theta + 0) + f(\theta - 0)] \qquad \text{for } -\pi < \theta < \pi$$
$$\tfrac{1}{2}[f(\pi) + f(-\pi)] \qquad \text{for } \theta = \pm \pi$$
■

A convenient rule is that for even (odd) functions only cosine (sine) terms appear in the Fourier expansion. Of course, the even and odd functions are defined on a symmetric interval around the origin. We can show this very generally. We take the interval to be $[-L/2, L/2]$ and rewrite (5.61a) as

$$F(x) = \frac{1}{\sqrt{L}} \sum_{n=-\infty}^{\infty} F_n e^{i(2n\pi/L)x}$$

$$F_n = \frac{1}{\sqrt{L}} \int_{-L/2}^{L/2} F(x) e^{-i(2n\pi/L)x}\, dx$$

Now we change the variable of integration to $y = -x$, obtaining

$$F_n = \frac{1}{\sqrt{L}} \int_{+L/2}^{-L/2} F(-y) e^{i(2n\pi/L)y} (-dy) = \frac{1}{\sqrt{L}} \int_{-L/2}^{L/2} F(-y) e^{i(2n\pi/L)y}\, dy$$

from which we get

$$F_{-n} = \frac{1}{\sqrt{L}} \int_{-L/2}^{L/2} F(-y) e^{-i(2n\pi/L)y}\, dy$$

It is now clear that

$$F_{-n} = \begin{cases} F_n & \text{if } F(-y) = F(y) \text{ or } F \text{ is even} \\ -F_n & \text{if } F(-y) = -F(y) \text{ or } F \text{ is odd} \end{cases} \tag{5.62}$$

Substituting this in the expression for $F(x)$, we obtain

$$F(x) = \frac{1}{\sqrt{L}} \left[F_0 + \sum_{n=1}^{\infty} \left(F_n e^{i(2n\pi/L)x} + F_{-n} e^{-i(2n\pi/L)x} \right) \right]$$

Thus, for even $F(x)$ we get

$$F(x) = \frac{1}{\sqrt{L}} \left[F_0 + 2 \sum_{n=1}^{\infty} F_n \cos\left(\frac{2n\pi}{L} x \right) \right] \tag{5.63a}$$

and for odd $F(x)$ we get

$$F(x) = \frac{1}{\sqrt{L}} \sum_{n=1}^{\infty} 2iF_n \sin\left(\frac{2n\pi}{L} x \right) \tag{5.63b}$$

Note that the constant term is absent in (5.63b) because for $n = 0$ and odd functions (5.62) yields

$$F_0 = -F_0 = 0$$

Equation (5.63b) is disturbing because of the factor i. However, as shown in Exercise 5.4.3, F_n is a purely imaginary number, and iF_n is thus real.

Finally, let us note that when a Fourier series is written in terms of sines and cosines, the coefficients A_n and B_n could be evaluated as follows. We first identify b_n in (5.56) with $f_n/\sqrt{2\pi}$ in (5.60a). Then, from the definitions of A_n and B_n, we have

$$A_n = b_n + b_{-n} = \frac{1}{\sqrt{2\pi}}(f_n + f_{-n})$$

$$= \frac{1}{\sqrt{2\pi}}\left[\frac{1}{\sqrt{2\pi}}\int_{-\pi}^{\pi} d\theta\, e^{-in\theta} f(\theta) + \frac{1}{\sqrt{2\pi}}\int_{-\pi}^{\pi} d\theta\, e^{in\theta} f(\theta)\right]$$

$$= \frac{1}{2\pi}\int_{-\pi}^{\pi} d\theta\,(e^{in\theta} + e^{-in\theta}) f(\theta) = \frac{1}{\pi}\int_{-\pi}^{+\pi} f(\theta)\cos n\theta\, d\theta \tag{5.64a}$$

$$B_n = \frac{1}{\pi}\int_{-\pi}^{\pi} f(\theta)\sin n\theta\, d\theta \tag{5.64b}$$

and

$$b_0 = \frac{1}{\sqrt{2\pi}} f_0 = \frac{1}{2\pi}\int_{-\pi}^{\pi} f(\theta)\, d\theta \equiv \frac{1}{2} A_0$$

So, for a function $f(\theta)$, defined in $[-\pi, +\pi]$, the Fourier series is

$$f(\theta) = \frac{1}{2}A_0 + \sum_{n=1}^{\infty} (A_n \cos n\theta + B_n \sin n\theta) \tag{5.65}$$

where A_n and B_n are as given in (5.64).

For a general function $F(x)$, defined in $[a, a + L]$, the analogue of (5.64) is

$$A_n = \frac{2}{L}\int_a^b F(x)\cos\left(\frac{2n\pi}{L}x\right)dx \tag{5.66a}$$

$$B_n = \frac{2}{L}\int_a^b F(x)\sin\left(\frac{2n\pi}{L}x\right)dx \tag{5.66b}$$

and the expansion is

$$F(x) = \frac{1}{2}A_0 + \sum_{n=1}^{\infty}\left(A_n \cos\frac{2n\pi x}{L} + B_n \sin\frac{2n\pi x}{L}\right) \tag{5.67}$$

Example 5.4.4

Let us expand

$$F(x) = |x| \qquad \text{for } -a \leqslant x \leqslant a$$

Since $|x|$ is even, we expect only cosine terms to be present. So, using (5.67), we write

$$F(x) = \frac{1}{2}A_0 + \sum_{n=1}^{\infty} A_n \cos\frac{2n\pi x}{L}$$

From (5.66a) and the fact that $L = 2a$, we have

$$A_n = \frac{2}{2a}\int_{-a}^{a} |x| \cos\frac{n\pi x}{a}\,dx = \frac{1}{a}\left[\int_{-a}^{b} (-x)\cos\frac{n\pi x}{a}\,dx + \int_{0}^{a} x\cos\frac{n\pi x}{a}\,dx\right] = \frac{2}{a}\int_{0}^{a} x\cos\frac{n\pi x}{a}\,dx$$

Thus,

$$A_0 = \frac{2}{a}\int_0^a x\,dx = a$$

And for $n \neq 0$

$$A_n = \frac{2}{a}\left(\frac{a}{n\pi}x\sin\frac{n\pi x}{a}\Bigg|_0^a - \frac{a}{n\pi}\int_0^a \sin\frac{n\pi x}{a}\,dx\right)$$

$$= \frac{2a}{n^2\pi^2}[(-1)^n - 1]$$

Noting that only odd terms contribute, we have

$$|x| = a\left\{\frac{1}{2} - \frac{4}{\pi^2}\sum_{k=0}^{\infty}\frac{\cos[(2k+1)(\pi x/a)]}{(2k+1)^2}\right\}$$

If we set $x = 0$ [which is a point in the domain of definition of $F(x)$], we obtain

$$0 = \frac{1}{2} - \frac{4}{\pi^2}\sum_{k=0}^{\infty}\frac{1}{(2k+1)^2}$$

or

$$\frac{\pi^2}{8} = \sum_{k=0}^{\infty}\frac{1}{(2k+1)^2} = 1 + \frac{1}{9} + \frac{1}{25} + \cdots \tag{1}$$

This gives another way of evaluating π.

For $x = a$ we can use (1) to obtain

$$F(a) = a\left\{\frac{1}{2} - \frac{4}{\pi^2}\sum_{k=0}^{\infty}\frac{\cos[(2k+1)\pi]}{(2k+1)^2}\right\}$$

$$= a\left\{\frac{1}{2} + \frac{4}{\pi^2}\sum_{k=0}^{\infty}\frac{1}{(2k+1)^2}\right\} = a$$

We see that since $F(x) = |x|$ is continuous at $x = a$, the series yields its true value, as expected. ●

Example 5.4.5 Full-Wave Rectifier

An alternating current is rectified to a direct current by starting with a periodic signal of the form $f(t) = \sin|\omega t|$ and then, by proper arrangements of resistors and capacitors, smoothing out the "bumps" so that the output signal is very nearly a direct voltage.

Let us Fourier-analyze the above signal. Since $f(t)$ is even for $-\pi < \omega t < \pi$, we expect only cosine terms to be present. If, for the time being, we use θ instead of ωt, we can write

$$\sin|\theta| = \frac{1}{2}A_0 + \sum_{n=1}^{\infty} A_n \cos n\theta$$

where

$$A_n = \frac{1}{\pi}\int_{-\pi}^{\pi} \sin|\theta| \cos n\,\theta\, d\theta = \frac{1}{\pi}\left[\int_{-\pi}^{0}(-\sin\theta)\cos n\theta\, d\theta + \int_{0}^{\pi}\sin\theta\cos n\theta\, d\theta\right]$$

$$= \frac{2}{\pi}\int_{0}^{\pi}\sin\theta\cos n\theta\, d\theta$$

$$= \frac{2}{\pi}\int_{0}^{\pi}\frac{1}{2}[\sin(n+1)\theta - \sin(n-1)\theta]\, d\theta$$

$$= \frac{1}{\pi}\left\{-\frac{1}{n+1}[\cos(n+1)\pi - 1] + \frac{1}{n-1}[\cos(n-1)\pi - 1]\right\}$$

$$= -\frac{2}{n^2-1}\left[\frac{(-1)^n + 1}{\pi}\right]$$

Thus,

$$A_n = \begin{cases} -\dfrac{4}{\pi}\left(\dfrac{1}{n^2-1}\right) & \text{for } n \text{ even and } n \neq 0 \\ 0 & \text{for } n \text{ odd} \end{cases}$$

and

$$A_0 = \frac{1}{\pi}\int_{-\pi}^{\pi}\sin|\theta|\, d\theta = \frac{4}{\pi}$$

The expansion, then, yields

$$\sin|\omega t| = \frac{2}{\pi} - \frac{4}{\pi}\sum_{k=1}^{\infty}\frac{\cos 2k\omega t}{4k^2-1}$$ ●

It is instructive to generalize the Fourier series to more than one dimension. This generalization is especially useful in crystallography and solid-state physics, which deal with three-dimensional periodic structures in the form of crystals.

To generalize to N dimensions, we first consider a special case in which an N-dimensional periodic function is a product of one-dimensional periodic functions of the form given by (5.61a). That is, we write

$$F(x_1) = \frac{1}{\sqrt{L_1}} \sum_{n_1=-\infty}^{\infty} F_{n_1} e^{i(2n_1\pi/L_1)x_1}$$

$$F(x_2) = \frac{1}{\sqrt{L_2}} \sum_{n_2=-\infty}^{\infty} F_{n_2} e^{i(2n_2\pi/L_2)x_2}$$

$$\vdots$$

$$F(x_k) = \frac{1}{\sqrt{L_k}} \sum_{n_k=-\infty}^{\infty} F_{n_k} e^{i(2n_k\pi/L_k)x_k}$$

$$\vdots$$

$$F(x_N) = \frac{1}{\sqrt{L_N}} \sum_{n_N=-\infty}^{\infty} F_{n_N} e^{i(2n_N\pi/L_N)x_N}$$

and

$$\begin{aligned} F(\mathbf{r}) &= F(x_1)F(x_2)\cdots F(x_N) \\ &= \frac{1}{\sqrt{L_1L_2\cdots L_N}} \sum_{n_1\ldots n_N=-\infty}^{\infty} F_{n_1}\cdots F_{n_N} e^{i[(2\pi n_1/L_1)x_1+\cdots+(2\pi n_N/L_N)x_N]} \\ &= \frac{1}{V_N^{1/2}} \sum_{\mathbf{n}} F_{\mathbf{n}} e^{i\mathbf{g}_n\cdot\mathbf{r}} \end{aligned} \tag{5.68}$$

where we have used the following new notations:

$$\begin{aligned} F(\mathbf{r}) &\equiv F(x_1)F(x_2)\cdots F(x_N) \\ V_N &\equiv L_1L_2\cdots L_N \\ \mathbf{n} &\equiv (n_1, n_2, \ldots, n_N) \\ F_{\mathbf{n}} &\equiv F_{n_1}F_{n_2}\cdots F_{n_N} \\ \mathbf{g}_n &\equiv 2\pi\left(\frac{n_1}{L_1}, \frac{n_2}{L_2}, \ldots, \frac{n_N}{L_N}\right) \\ \mathbf{r} &\equiv (x_1, x_2, \ldots, x_N) \end{aligned}$$

Although Eq. (5.68) is defined for a particular periodic function in N variables, we can take it as the definition of the Fourier series for *any* periodic function $F(x_1, x_2, \ldots, x_n)$ of N variables. However, to apply (5.68) requires some clarification.

In one dimension the shape of the smallest region of periodicity is unique. It is simply a line segment of length L, for example. In two and more dimensions, however, such regions may have a variety of shapes. For instance, in two dimensions they can be rectangles, hexagons, and so forth. Thus, we let V_N in (5.68) stand for a primitive cell of the N-dimensional lattice. This cell is very important in solid-state physics and (in three dimensions) is called the *Wigner-Seitz cell.*

It is customary to absorb the factor $1/V_N^{1/2}$ into $F_{\mathbf{n}}$ and write

$$F(\mathbf{r}) = \sum_{\mathbf{n}} F_{\mathbf{n}} e^{i\mathbf{g}_n \cdot \mathbf{r}} \tag{5.69a}$$

whose inverse Fourier series is

$$F_{\mathbf{n}} = \frac{1}{V_N} \int_{V_N} F(\mathbf{r}) e^{-i\mathbf{g}_n \cdot \mathbf{r}} \, d^N r \tag{5.69b}$$

where the integral is over a single Wigner-Seitz cell.

Recall that $F(\mathbf{r})$ is a periodic function of $\mathbf{r}$. This means that if $\mathbf{r}$ is changed by ℓ, where ℓ is a vector describing the boundaries of a cell, then we should get the same function:

$$F(\mathbf{r} + \ell) = F(\mathbf{r})$$

When substituted in (5.69a), this yields

$$F(\mathbf{r} + \ell) = \sum_{\mathbf{n}} F_{\mathbf{n}} e^{i\mathbf{g}_n \cdot (\mathbf{r} + \ell)} = \sum_{\mathbf{n}} e^{i\mathbf{g}_n \cdot \ell} F_{\mathbf{n}} e^{i\mathbf{g}_n \cdot \mathbf{r}}$$

which is equal to $F(\mathbf{r})$ if

$$e^{i\mathbf{g}_n \cdot \ell} = 1 \tag{5.70}$$

In three dimensions

$$\ell = m_1 \mathbf{a}_1 + m_2 \mathbf{a}_2 + m_3 \mathbf{a}_3$$

where m_1, m_2, and m_3 are integers and $\mathbf{a}_1$, $\mathbf{a}_2$, and $\mathbf{a}_3$ are *crystal axes* (which are not, generally, orthogonal). On the other hand,

$$\mathbf{g}_n = n_1 \mathbf{b}_1 + n_2 \mathbf{b}_2 + n_3 \mathbf{b}_3$$

where n_1, n_2, and n_3 are integers and $\mathbf{b}_1$, $\mathbf{b}_2$, and $\mathbf{b}_3$ are the *reciprocal lattice vectors* defined by

$$\mathbf{b}_1 = \frac{2\pi(\mathbf{a}_2 \times \mathbf{a}_3)}{\mathbf{a}_1 \cdot (\mathbf{a}_2 \times \mathbf{a}_3)} \qquad \mathbf{b}_2 = \frac{2\pi(\mathbf{a}_3 \times \mathbf{a}_1)}{\mathbf{a}_1 \cdot (\mathbf{a}_2 \times \mathbf{a}_3)} \qquad \mathbf{b}_3 = \frac{2\pi(\mathbf{a}_1 \times \mathbf{a}_2)}{\mathbf{a}_1 \cdot (\mathbf{a}_2 \times \mathbf{a}_3)}$$

It is easily verified that

$$\mathbf{a}_i \cdot \mathbf{b}_j = 2\pi \delta_{ij}$$

Thus,

$$\mathbf{g}_n \cdot \ell = \left(\sum_{i=1}^{3} n_i \cdot \mathbf{b}_i \right) \cdot \left(\sum_{j=1}^{3} m_j \mathbf{a}_j \right) = \sum_{i,j} n_i m_j \mathbf{b}_i \cdot \mathbf{a}_j$$

$$= 2\pi \sum_{i=1}^{3} m_i n_i = 2\pi \text{ (integer)}$$

and (5.70) is satisfied.

5.4.2 The Fourier Transform

Equations (5.61) are valid over the entire real line as long as the function $F(x)$ is periodic. However, most functions encountered in physical applications are defined in some interval $[a, b]$ without repetition beyond that interval. It would be useful if we could also expand such functions in some form of Fourier "series."

One way to do this is to start with the periodic series and then let the period go to infinity while extending the domain of the definition of the function. As a specific case, suppose we are interested in representing a function $f(x)$, which is defined only for the interval $[a, b]$ and is assigned the value zero everywhere else (see Fig. 5.8). To begin with we might try the series representation of Eqs. (5.61), but this will produce a periodic function such that in each period the function looks as is shown in Fig. 5.8. This situation is depicted in Fig. 5.9. Next we may try a function $g_1(x)$ defined in the interval $[a - L_1/2, b + L_1/2]$, where L_1 is an arbitrary positive number:

$$g_1(x) = \begin{cases} 0 & \text{if } a - L_1/2 < x < a \\ f(x) & \text{if } a < x < b \\ 0 & \text{if } b < x < b + L_1/2 \end{cases}$$

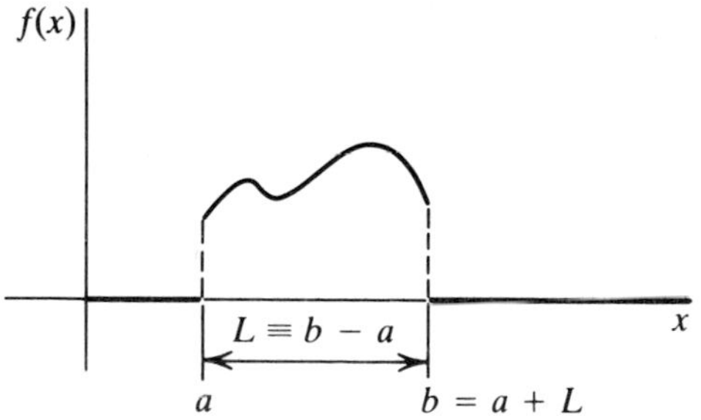

Figure 5.8 The function f defined in the interval $[a, b]$.

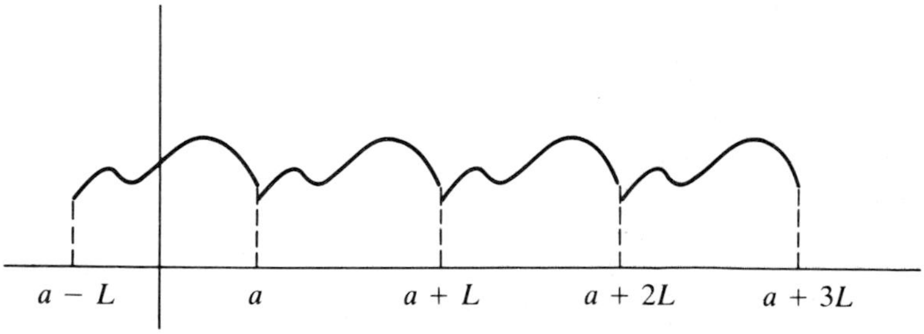

Figure 5.9 The Fourier-series representation of $f(x)$ produces a periodic function with period $L = b - a$, such that in each interval $[a + nL, a + (n + 1)L]$ the function looks like $f(x)$.

Now, by using Eqs. (5.61), we obtain a representation of $g_1(x)$ that is again periodic and is shown in Fig. 5.10. For this function Eqs. (5.61) take the form

$$g_1(x) = \frac{1}{\sqrt{L + L_1}} \sum_{n=-\infty}^{\infty} g_n^{(1)} e^{i[2n\pi/(L + L_1)]x}$$

where
$$g_n^{(1)} = \frac{1}{\sqrt{L + L_1}} \int_{a-L_1/2}^{b+L_1/2} e^{-i[2n\pi/(L + L_1)]x} g_1(x)\, dx$$

Let us try a function $g_2(x)$ defined in the interval $[a - L_2/2, b + L_2/2]$, where $L_2 > L_1$, such that

$$g_2(x) = \begin{cases} 0 & \text{if } a - L_2/2 < x < a \\ f(x) & \text{if } a < x < b \\ 0 & \text{if } b < x < b + L_2/2 \end{cases}$$

The periodic representation of $g_2(x)$ is depicted in Fig. 5.11. We see that increasing the

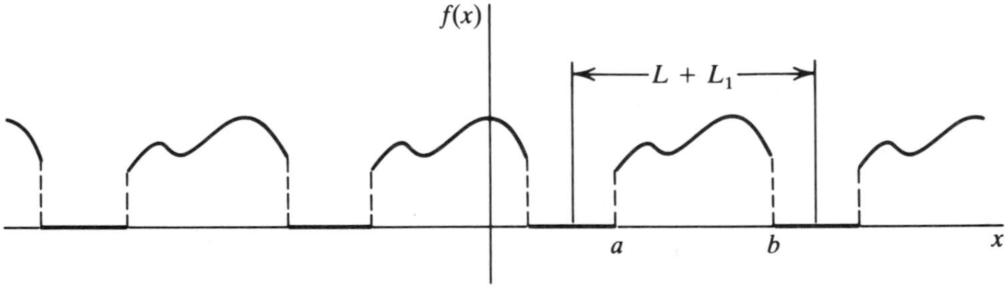

Figure 5.10 The Fourier-series representation of the periodic function $g_1(x)$.

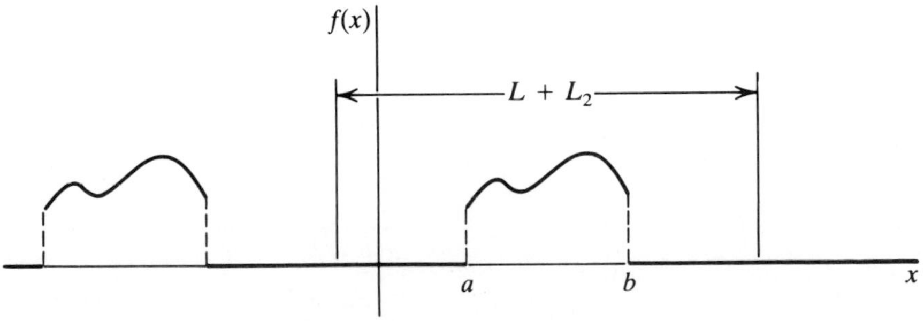

Figure 5.11 The Fourier-series representation of the periodic function $g_2(x)$. Two of the four repetitions of Fig. 5.10 have disappeared.

period from $L + L_1$ to $L + L_2$ makes the different parts of the function become more and more isolated. The function $g_2(x)$ can be represented as

$$g_2(x) = \frac{1}{\sqrt{L + L_2}} \sum_{n=-\infty}^{\infty} g_n^{(2)} e^{i[2n\pi/(L + L_2)]x}$$

where
$$g_n^{(2)} = \frac{1}{\sqrt{L + L_2}} \int_{a - L_2/2}^{b + L_2/2} dx\, e^{-i[2n\pi/(L + L_2)]} g_2(x)$$

In general, we can define a function $g_M(x)$ in the interval $[a - L_M/2, b + L_M/2]$, where $L_M \gg L_1$, such that

$$g_M(x) = \begin{cases} 0 & \text{if } a - L_M/2 < x < a \\ f(x) & \text{if } a < x < b \\ 0 & \text{if } b < x < b + L_M/2 \end{cases}$$

The periodic representation of $g_M(x)$ is of the form

$$g_M(x) = \frac{1}{\sqrt{L + L_M}} \sum_{n=-\infty}^{\infty} g_n^{(M)} e^{i[2n\pi/(L + L_M)]x} \tag{5.71a}$$

where
$$g_n^{(M)} = \frac{1}{\sqrt{L + L_M}} \int_{a - L_M/2}^{b + L_M/2} dx\, e^{-i[2n\pi/(L + L_M)]x} g_M(x) \tag{5.71b}$$

By now it should be clear that if $L_M \to \infty$, we completely isolate the function and stop the repetition. Let us investigate the behavior of Eqs. (5.71) when L_M grows without bound. First, we notice that the quantity $k_n \equiv 2n\pi/(L + L_M)$ appearing in the exponent becomes almost continuous. In other words, as n changes by one unit, k_n changes only slightly. This suggests that the terms in the sum in (5.71a) can be lumped together, giving

$$g_M(x) \approx \sum_{j=-\infty}^{+\infty} \frac{g^{(M)}(k_j)}{\sqrt{L + L_M}} e^{ik_j x} \Delta n_j$$

where we have written $g^{(M)}(k_j)$ instead of $g_{n_j}^{(M)}$. The quantity Δn_j is the width of the jth interval and could be a large number; however, in that interval neither $g^{(M)}(k_j)/\sqrt{L + L_M}$ nor $e^{ik_j x}$ varies appreciably. From the relation

$$k_j \equiv \frac{2n_j\pi}{L + L_M}$$

we obtain

$$\Delta n_j = \frac{L + L_M}{2\pi} \Delta k_j$$

and $g_M(x)$ becomes

$$g_M(x) \approx \sum_{j=-\infty}^{\infty} \left(\frac{g^{(M)}(k_j)}{\sqrt{L+L_M}}\right) e^{ik_j x}\left(\frac{L+L_M}{2\pi}\right)\Delta k_j$$

$$= \frac{1}{\sqrt{2\pi}} \sum_{j=-\infty}^{\infty} \sqrt{\frac{L+L_M}{2\pi}}\, g^{(M)}(k_j) e^{ik_j x}\Delta k_j$$

$$= \frac{1}{\sqrt{2\pi}} \sum_{j=-\infty}^{\infty} \tilde{g}(k_j) e^{ik_j x}\Delta k_j$$

where we introduced $\tilde{g}(k_j)$ defined by

$$\tilde{g}(k_j) \equiv \sqrt{\frac{L+L_M}{2\pi}}\, g^{(M)}(k_j)$$

It is now clear that the preceding sum approaches an integral, and, ignoring the label M on the LHS, we write

$$g(x) = \frac{1}{\sqrt{2\pi}} \int_{-\infty}^{\infty} \tilde{g}(k) e^{ikx}\, dk \tag{5.72a}$$

Using the definition of $\tilde{g}(k_j)$, Eq. (5.71b) becomes

$$\sqrt{\frac{2\pi}{L+L_M}}\,\tilde{g}(k_j) = \frac{1}{\sqrt{L+L_M}} \int_{-\infty}^{\infty} dx\, e^{-ik_j x} g_M(x)$$

We disregard the subscripts j and M and write

$$\tilde{g}(k) = \frac{1}{\sqrt{2\pi}} \int_{-\infty}^{\infty} dx\, e^{-ikx} g(x) \tag{5.72b}$$

Equations (5.72a) and (5.72b) are called the *Fourier integral transforms* of $\tilde{g}(k)$ and $g(x)$, respectively.

The above analysis closely resembles the arguments made in the discussion of the continuous index (see Section 5.2). However, here we ran into no difficulty in going from the (discrete) Fourier series to the (continuous) Fourier transform. The reason for this smooth transition is that we scaled not only the vectors $|\mathbf{n}\rangle$, but also the inner product on those vectors. In each step above we changed the integration interval $[a, b]$ and, with it, the space $\mathcal{L}^2(a, b)$. Therefore, the two rescalings compensated for one another and permitted the smooth transition. In the abstract formalism of Section 5.2, there was no interval $[a, b]$ to adjust, and the inner product was not rescaled to compensate for the scaling of $|\mathrm{n}\rangle \to |\mathrm{x}\rangle$. The analysis here gives us an indication that the final result of the earlier abstract discussion is right even though the derivation has serious flaws.

Example 5.4.6

Let us evaluate the Fourier transform of the function defined as

$$g(x) = \begin{cases} b & \text{if } |x| \leqslant a \text{ and } a > 0 \\ 0 & \text{if } |x| > a \end{cases}$$

From (5.72b) we have

$$\tilde{g}(x) = \frac{1}{\sqrt{2\pi}} \int_{-\infty}^{\infty} dx\, e^{-ikx} g(x) = \frac{b}{\sqrt{2\pi}} \int_{-a}^{+a} e^{-ikx}\, dx$$

$$= \frac{2ab}{\sqrt{2\pi}} \left(\frac{\sin ka}{ka} \right)$$

which is the function encountered in Example 5.2.2.

Let us discuss this result in detail. First, note that if $a \to \infty$, the function $g(x)$ becomes a constant function over the entire real line, and we get

$$\tilde{g}(k) = \frac{2b}{\sqrt{2\pi}} \lim_{a \to \infty} \frac{\sin ka}{k} = \frac{2b}{\sqrt{2\pi}} \pi \delta(k)$$

by Example 5.2.2. This is what we expect for a constant function (see Exercise 5.4.5). Next, let $b \to \infty$ and $a \to 0$ in such a way that $2ab$ = area under $g(x) = 1$. Then $g(x)$ will approach the delta function, and $\tilde{g}(k)$ becomes

$$\tilde{g}(k) = \lim_{\substack{b \to \infty \\ a \to 0}} \frac{2ab}{\sqrt{2\pi}} \frac{\sin ka}{ka} = \frac{1}{\sqrt{2\pi}}$$

which again agrees with the result of Exercise 5.4.5.

Finally, we note that the width of $g(x)$ is

$$\Delta x = 2a$$

and the width of $\tilde{g}(k)$ is roughly the distance, on the k-axis, between the two zeros of $\tilde{g}(k)$ on either side of $k = 0$:

$$\Delta k = k_+ - k_- = \frac{2\pi}{a}$$

We thus note that increasing the width of $g(x)$ results in a decrease of the width of $\tilde{g}(k)$. In other words, when the function is wide, its Fourier transform is sharp. In the limit of infinite width (a constant function), we get infinite sharpness (the δ function). The last two statements are very general. In fact, it can be shown that generally

$$(\Delta x)(\Delta k) \geqslant 1$$

for any function $g(x)$. When both sides of this inequality are multiplied by $\hbar \equiv h/2\pi$, the result is the celebrated *Heisenberg uncertainty relation*, where momentum p is equal to $\hbar k$:

$$(\Delta x)(\Delta p) \geqslant \hbar \equiv \frac{h}{2\pi}$$

where h is the Planck constant and $p = \hbar k$ is the momentum.

Having obtained the transform of $g(x)$, we can write

$$g(x) = \frac{1}{\sqrt{2\pi}} \int_{-\infty}^{\infty} \frac{2b}{\sqrt{2\pi}} \frac{\sin ka}{k} e^{ikx} dk$$

$$= \frac{b}{\pi} \int_{-\infty}^{\infty} \frac{\sin ka}{k} e^{ikx} dk$$

●

Example 5.4.7

Let us evaluate the Fourier transform of a Gaussian:

$$g(x) = ae^{-bx^2} \qquad \text{for } a, b > 0$$

We substitute this in (5.72b) to obtain

$$\tilde{g}(k) = \frac{a}{\sqrt{2\pi}} \int_{-\infty}^{\infty} e^{-b[x^2 + (ik/b)x]} dx$$

Completing the square in the exponent,

$$x^2 + \frac{ik}{b}x = \left(x + \frac{ik}{2b}\right)^2 + \frac{k^2}{4b^2}$$

we obtain

$$\tilde{g}(k) = \frac{a}{\sqrt{2\pi}} e^{-k^2/4b} \int_{-\infty}^{\infty} e^{-b(x + ik/2b)^2} dx$$

To evaluate this integral rigorously, we would have to use techniques developed in complex analysis, which are not introduced until Chapter 7. However, we can ignore the fact that the second term in the exponent is an imaginary number, substitute $y = x + ik/2b$, and write

$$\int_{-\infty}^{\infty} e^{-b(x + ik/2b)^2} dx = \int_{-\infty}^{\infty} e^{-by^2} dy = \sqrt{\frac{\pi}{b}}$$

Thus, we have

$$\tilde{g}(k) = \frac{a}{\sqrt{2b}} e^{-k^2/4b}$$

which is also a Gaussian.

We note again that the width of $g(x)$, which is proportional to $1/\sqrt{b}$, is in inverse relation to the width of $\tilde{g}(k)$, which is proportional to $\sqrt{b}$. Thus, once again we have $(\Delta x)(\Delta k) \approx 1$.

●

Equations (5.72a) and (5.72b) are reciprocals of one another. However, it is not obvious that they are consistent. In other words, if we substitute (5.72a) in the RHS of (5.72b), do we get an identity? Let's try this (note the use of a different dummy variable of integration!):

$$\tilde{g}(k) = \frac{1}{\sqrt{2\pi}} \int_{-\infty}^{\infty} dx\, e^{-ikx} \left[\frac{1}{\sqrt{2\pi}} \int_{-\infty}^{\infty} \tilde{g}(k')\, e^{ik'x}\, dk' \right]$$

$$= \frac{1}{2\pi} \int_{-\infty}^{\infty} dx \int_{-\infty}^{\infty} \tilde{g}(k') e^{ix(k'-k)}\, dk'$$

We now change the order of the two integrations:

$$\tilde{g}(k) = \int_{-\infty}^{\infty} dk' \tilde{g}(k') \left[\frac{1}{2\pi} \int_{-\infty}^{\infty} dx\, e^{i(k'-k)x} \right]$$

But the expression in the square brackets is the delta function (see Example 5.2.2). Thus, we have

$$\tilde{g}(k) = \int_{-\infty}^{\infty} dk' \tilde{g}(k') \delta(k' - k)$$

which is an identity.

As in the case of Fourier series, Eqs. (5.72) are valid even if $g(x)$ and $\tilde{g}(k)$ are piecewise continuous. In that case the Fourier transforms are written as

$$\tfrac{1}{2}[g(x+0) + g(x-0)] = \frac{1}{\sqrt{2\pi}} \int_{-\infty}^{\infty} \tilde{g}(k) e^{ikx}\, dk \tag{5.73a}$$

and

$$\tfrac{1}{2}[\tilde{g}(k+0) + \tilde{g}(k-0)] = \frac{1}{\sqrt{2\pi}} \int_{-\infty}^{\infty} g(x) e^{-ikx}\, dx \tag{5.73b}$$

where each zero on the LHS is an ε that has gone to zero in the limit.

It is useful to generalize Eqs. (5.72) or (5.73) to more than one dimension. The generalization is straightforward, yielding this result:

$$g(x_1, \ldots, x_n) = \frac{1}{(2\pi)^{n/2}} \int_{-\infty}^{\infty} dk_1 \cdots \int_{-\infty}^{\infty} dk_n\, e^{i(k_1 x_1 + \cdots + k_n x_n)} \tilde{g}(k_1, \ldots, k_n)$$

$$\tilde{g}(k_1, \ldots, k_n) = \frac{1}{(2\pi)^{n/2}} \int_{-\infty}^{\infty} dx_1 \cdots \int_{-\infty}^{\infty} dx_n\, e^{-i(k_1 x_1 + \cdots + k_n x_n)} g(x_1, \ldots, x_n)$$

If we introduce the vectors $\mathbf{r} = (x_1, x_2, \ldots, x_n)$ and $\mathbf{k} = (k_1, k_2, \ldots, k_n)$, we obtain

$$g(\mathbf{r}) = \frac{1}{(2\pi)^{n/2}} \int d^n k e^{i\mathbf{k}\cdot\mathbf{r}} \tilde{g}(\mathbf{k}) \tag{5.74a}$$

and

$$\tilde{g}(\mathbf{k}) = \frac{1}{(2\pi)^{n/2}} \int d^n r e^{-i\mathbf{k}\cdot\mathbf{r}} g(\mathbf{r}) \tag{5.74b}$$

Let us now use the abstract notation we used earlier to discuss the preceding results. A Fourier integral transform is obtained from a Fourier series when the discrete index n becomes a continuous index k. This means that we should be able to obtain $|\mathrm{k}\rangle$ from $|\mathrm{n}\rangle$ by some sort of limiting process. It must be emphasized that there are many pitfalls in going from a "discrete" vector to a "continuous" one. We have seen such pitfalls before. However, we again disregard them and look at the completeness relation for $|\mathrm{n}\rangle$'s:

$$\mathbb{1} = \sum_{n=-\infty}^{\infty} |\mathrm{n}\rangle\langle\mathrm{n}|$$

If we lump terms together (n is approaching a continuous index; thus, the difference between $|\mathrm{n}\rangle$ and $|\mathrm{n}+1\rangle$ is infinitesimal), then the above relation can be "approximated" by

$$\mathbb{1} \approx \sum_{j=-\infty}^{\infty} |\mathrm{n}_j\rangle\langle\mathrm{n}_j| \Delta n_j$$

where, as before, Δn_j is the number of integers in the jth interval. Now we define

$$k_j \equiv \frac{2\pi n_j}{L}$$

which yields

$$\Delta n_j = \frac{L}{2\pi} \Delta k_j$$

Substituting this relation in the expression for $\mathbb{1}$, we obtain

$$\begin{aligned}
\mathbb{1} &\approx \sum_{j=-\infty}^{+\infty} |\mathrm{n}_j\rangle\langle\mathrm{n}_j| \frac{L}{2\pi} \Delta k_j \\
&= \sum_j \left(\sqrt{\frac{L}{2\pi}} |\mathrm{n}_j\rangle\right)\left(\sqrt{\frac{L}{2\pi}} \langle\mathrm{n}_j|\right) \Delta k_j \\
&\equiv \sum_{j=-\infty}^{\infty} |\mathrm{k}_j\rangle\langle\mathrm{k}_j| \Delta k_j
\end{aligned}$$

where the vector $|\mathrm{k}_j\rangle$ (not to be confused with the number k_j defined above) is defined by

$$|\mathrm{k}_j\rangle \equiv \sqrt{\frac{L}{2\pi}} |\mathrm{n}_j\rangle$$

The x representation of $|\mathrm{k}_j\rangle$ can be obtained by taking the inner product of $|\mathrm{k}_j\rangle$ with $\langle \mathrm{x}|$:

$$\langle \mathrm{x}|\mathrm{k}_j\rangle = \sqrt{\frac{L}{2\pi}}\langle \mathrm{x}|\mathrm{n}_j\rangle = \sqrt{\frac{L}{2\pi}}\frac{1}{\sqrt{L}}e^{i(2n_j\pi/L)x}$$

$$= \frac{1}{\sqrt{2\pi}}e^{ik_jx}$$

Neglecting the index j, we obtain

$$\langle \mathrm{x}|\mathrm{k}\rangle = \frac{1}{\sqrt{2\pi}}e^{ikx} \tag{5.75}$$

In the limit of infinitesimal Δk_j, the expression for $\mathbb{1}$ turns into an integral:

$$\mathbb{1} = \int_{-\infty}^{\infty} |\mathrm{k}\rangle\langle \mathrm{k}|\,dk \tag{5.76}$$

This is Eq. (5.9a) with $w(k) = 1$. Equation (5.11) yields

$$\langle \mathrm{k}|\mathrm{k}'\rangle = \delta(k - k') \tag{5.77}$$

We can go further and write an integral representation of the delta function:

$$\delta(k - k') = \langle \mathrm{k}|\mathbb{1}|\mathrm{k}'\rangle = \langle \mathrm{k}|\left(\int_{-\infty}^{\infty} |\mathrm{x}\rangle\langle \mathrm{x}|\,dx\right)|\mathrm{k}'\rangle$$

$$= \int_{-\infty}^{\infty} dx\,\langle \mathrm{k}|\mathrm{x}\rangle\langle \mathrm{x}|\mathrm{k}'\rangle$$

$$= \int_{-\infty}^{\infty} dx\frac{1}{\sqrt{2\pi}}e^{-ikx}\frac{1}{\sqrt{2\pi}}e^{ik'x} = \frac{1}{\sqrt{2\pi}}\int_{-\infty}^{\infty} dx\,e^{i(k'-k)x}$$

Obviously, we can also write

$$\delta(x - x') = \frac{1}{2\pi}\int_{-\infty}^{\infty} dk\,e^{i(x-x')k}$$

In fact, we can represent the delta function in terms of any set of orthonormal functions (see Example 5.3.8). Consider a general orthonormal basis $\{|\mathrm{n}\rangle\}$ with the x representation $f_n(x) = \langle \mathrm{x}|\mathrm{n}\rangle$. Using the completeness relation

$$\mathbb{1} = \sum_n |\mathrm{n}\rangle\langle \mathrm{n}|$$

we can write

$$\delta(x-x') = w(x)\langle \mathrm{x}|\mathrm{x}'\rangle = \langle \mathrm{x}|\mathbb{1}|\mathrm{x}'\rangle w(x)$$

$$= \langle \mathrm{x}|\left(\sum_n |\mathrm{n}\rangle\langle \mathrm{n}|\right)|\mathrm{x}'\rangle w(x) = \sum_n \langle \mathrm{x}|\mathrm{n}\rangle\langle \mathrm{n}|\mathrm{x}'\rangle w(x) \tag{5.78}$$

$$= \sum_n f_n(x) f_n^*(x') w(x)$$

This gives the delta function as a sum of products of orthonormal functions and is true for *any* complete set of orthonormal functions. In particular, if we use the trigonometric functions, for which $w(x) = 1$, we obtain

$$\delta(x-x') = \sum_n \frac{1}{\sqrt{L}} e^{i(2n\pi/L)x} \frac{1}{\sqrt{L}} e^{-i(2n\pi/L)x'} = \frac{1}{L}\sum_n e^{i(2n\pi/L)(x-x')}$$

If more than one dimension is involved, we use

$$\delta(\mathbf{k}-\mathbf{k}') = \frac{1}{(2\pi)^n}\int d^n r\, e^{i(\mathbf{k}-\mathbf{k}')\cdot\mathbf{r}} \tag{5.79a}$$

and

$$\delta(\mathbf{r}-\mathbf{r}') = \frac{1}{(2\pi)^n}\int d^n k\, e^{i\mathbf{k}\cdot(\mathbf{r}-\mathbf{r}')} \tag{5.79b}$$

and Eq. (5.75) becomes

$$\langle \mathbf{r}|\mathbf{k}\rangle = \frac{1}{(2\pi)^{n/2}} e^{i\mathbf{k}\cdot\mathbf{r}} \tag{5.80a}$$

with the complex conjugate relation

$$\langle \mathbf{k}|\mathbf{r}\rangle = \frac{1}{(2\pi)^{n/2}} e^{-i\mathbf{k}\cdot\mathbf{r}} \tag{5.80b}$$

Equations (5.79) and (5.80) exhibit a striking resemblance between $|\mathbf{r}\rangle$ and $|\mathbf{k}\rangle$. In fact, any given abstract vector $|\mathrm{f}\rangle$ can be expressed either in terms of its x representation, $\langle \mathbf{r}|\mathrm{f}\rangle = f(\mathbf{r})$, or in terms of its k representation, $\langle \mathbf{k}|\mathrm{f}\rangle = f(\mathbf{k})$. These two representations are completely equivalent, and there is a one-to-one correspondence between the two, given by Eqs. (5.74). The representation that is used in practice is dictated by the physical application. In quantum mechanics, for instance, most of the time the x representation, corresponding to the position, is used, because then the operator equations turn into differential equations that are linear and easier to solve than the corresponding equations with the k representation, which is related to the momentum.

Example 5.4.8 The Fourier Transform of the Coulomb Potential

Let us evaluate the Fourier transform of $V(r)$, the Coulomb potential of two point charges, Q and Q':

$$V(r) = \frac{QQ'}{r}$$

The Fourier transform is important in scattering experiments with atoms and solids. As we will see in the following, the Fourier transform of $V(r)$ is not defined. However, if we work with the Yukawa potential,

$$V_\alpha(r) = \frac{QQ' e^{-\alpha r}}{r} \qquad \text{for } \alpha > 0$$

the Fourier transform is well-defined, and we can take the limit $\alpha \to 0$ to recover the Coulomb potential. Thus, we seek the Fourier transform of $V_\alpha(r)$.

We are dealing with three dimensions and, therefore, may write

$$\tilde{V}_\alpha(\mathbf{k}) = \frac{1}{(2\pi)^{3/2}} \iiint d^3 r\, e^{-i\mathbf{k}\cdot\mathbf{r}} \frac{QQ' e^{-\alpha r}}{r}$$

It is clear from the presence of r that spherical coordinates are appropriate. We are free to pick any direction as the z-axis. A simplifying choice in this case is the direction of k. So we let

$$\mathbf{k} = |\mathbf{k}|\hat{\mathbf{e}}_z \equiv k\hat{\mathbf{e}}_z$$

and

$$\mathbf{k}\cdot\mathbf{r} = kr\cos\theta$$

where θ is the polar angle in spherical coordinates. Now we have

$$\tilde{V}_\alpha(\mathbf{k}) = \frac{QQ'}{(2\pi)^{3/2}} \int_0^\infty r^2\, dr \int_0^\pi \sin\theta\, d\theta \int_0^{2\pi} d\varphi\, e^{-ikr\cos\theta} \frac{e^{-\alpha r}}{r}$$

The φ integration is easy and gives 2π. The θ integration is done next, yielding

$$\int_0^\pi \sin\theta\, e^{-ikr\cos\theta}\, d\theta = \int_{-1}^{1} du\, e^{ikru} = \frac{1}{ikr}(e^{ikr} - e^{-ikr})$$

We thus have

$$\tilde{V}_\alpha(\mathbf{k}) = \frac{QQ'(2\pi)}{(2\pi)^{3/2}} \int_0^\infty dr\, r^2 \frac{e^{-\alpha r}}{r}\left(\frac{1}{ikr}\right)(e^{ikr} - e^{-ikr})$$

$$= \frac{QQ'}{\sqrt{2\pi}}\left(\frac{1}{ik}\right)\int_0^\infty dr(e^{(-\alpha+ik)r} - e^{-(\alpha+ik)r})$$

$$= \frac{QQ'}{\sqrt{2\pi}}\left(\frac{1}{ik}\right)\left(\frac{e^{(-\alpha+ik)r}}{-\alpha+ik}\bigg|_0^\infty + \frac{e^{-(\alpha+ik)r}}{\alpha+ik}\bigg|_0^\infty\right)$$

Note that the factor α comes to our rescue here because it tames the behavior of the exponential at $r \to \infty$. Finally, we have

$$\tilde{V}_\alpha(\mathbf{k}) = \frac{2QQ'}{\sqrt{2\pi}}\left(\frac{1}{k^2 + \alpha^2}\right)$$

The factor α is a measure of the range of the potential. It is clear from $V_\alpha(r)$ that the larger α is, the smaller the range. In fact, it was in response to the short range of nuclear forces that Yukawa introduced α. For electromagnetism, where the range is infinite, α becomes zero and $V_\alpha(r)$ reduces to $V(r)$.

Thus, the Fourier transform of the Coulomb potential is

$$\tilde{V}_{\text{Coul}}(\mathbf{k}) = \frac{2QQ'}{\sqrt{2\pi}}\left(\frac{1}{k^2}\right)$$

Note that this is the Fourier transform of the Coulomb potential energy between two point charges, Q and Q'. If other charge distributions are used, the Fourier transform is different. ●

Example 5.4.9 Form Factors

Example 5.4.8 dealt with the electrostatic potential energy of two point charges. Let us now consider the case where one of the charges is distributed over a finite volume. Then the potential energy is

$$V(\mathbf{r}) = Q\iiint \frac{Q'\rho(\mathbf{r}')}{|\mathbf{r}-\mathbf{r}'|}\,d^3r'$$

where $Q'\rho(\mathbf{r}')$ is the charge density at the point $\mathbf{r}'$. Note that we have normalized $\rho(\mathbf{r}')$ so that $\iiint \rho(\mathbf{r}')d^3r' = 1$. Figure 5.12 shows the geometry of the situation.

Making a change of variables,

$$\mathbf{R} \equiv \mathbf{r}' - \mathbf{r} \quad\Rightarrow\quad \mathbf{r}' = \mathbf{R} + \mathbf{r} \quad\Rightarrow\quad d^3r' = d^3R$$

we get

$$\begin{aligned}\tilde{V}(\mathbf{k}) &= \frac{1}{(2\pi)^{3/2}}\iiint d^3r\, e^{-i\mathbf{k}\cdot\mathbf{r}}V(\mathbf{r}) \\ &= \frac{1}{(2\pi)^{3/2}}\iiint d^3r\, e^{-i\mathbf{k}\cdot\mathbf{r}}Q\int d^3R\,\frac{Q'\rho(\mathbf{r}+\mathbf{R})}{R} \qquad (1)\end{aligned}$$

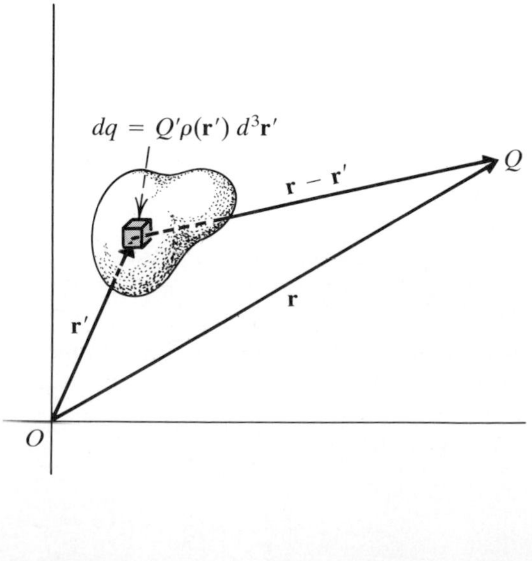

Figure 5.12 The electrostatic potential energy of a point charge Q at $\mathbf{r}$ and a charge distribution can be calculated using the geometry shown here.

Since the differentials d^3r and d^3R contain the information about the number of integrals to be performed, we will use only one integral sign. To evaluate (1) we substitute for $\rho(\mathbf{r}+\mathbf{R})$ in terms of its Fourier transform,

$$\rho(\mathbf{r}+\mathbf{R}) = \frac{1}{(2\pi)^{3/2}} \int d^3k'\,\tilde{\rho}(\mathbf{k}')e^{i\mathbf{k}'\cdot(\mathbf{r}+\mathbf{R})} \tag{2}$$

Note the single integral sign, the combination $\mathbf{r}+\mathbf{R}$ appearing on both sides (as it should), and the use of $\mathbf{k}'$ as the dummy variable of integration. Combining (1) and (2), we obtain

$$\begin{aligned}\tilde{V}(\mathbf{k}) &= \frac{QQ'}{(2\pi)^3}\int d^3r\,d^3R\,d^3k'\,\frac{e^{i\mathbf{k}'\cdot\mathbf{R}}}{R}\,\tilde{\rho}(\mathbf{k}')e^{i\mathbf{r}\cdot(\mathbf{k}'-\mathbf{k})} \\ &= QQ'\int d^3R\,d^3k'\,\frac{e^{i\mathbf{k}'\cdot\mathbf{R}}}{R}\,\tilde{\rho}(\mathbf{k}')\underbrace{\left(\frac{1}{(2\pi)^3}\int d^3r\,e^{i\mathbf{r}\cdot(\mathbf{k}'-\mathbf{k})}\right)}_{\delta(\mathbf{k}'-\mathbf{k})} \\ &= QQ'\int d^3R\,\frac{1}{R}\int d^3k'\,e^{i\mathbf{k}'\cdot\mathbf{R}}\,\tilde{\rho}(\mathbf{k}')\,\delta(\mathbf{k}'-\mathbf{k}) \\ &= QQ'\tilde{\rho}(\mathbf{k})\int d^3R\,\frac{e^{i\mathbf{k}\cdot\mathbf{R}}}{R} \end{aligned} \tag{3}$$

What is nice about this result is that the contribution of the charge distribution, $\tilde{\rho}(\mathbf{k})$, has been completely factored out. The integral, aside from a constant and a change in the sign of $\mathbf{k}$, is simply the Fourier transform of the Coulomb potential energy between two point charges, Q and Q', obtained in Example 5.4.8. We can, therefore, write (3) as

$$\tilde{V}(\mathbf{k}) = (2\pi)^{3/2}\tilde{\rho}(\mathbf{k})\tilde{V}_{\text{Coul}}(-\mathbf{k}) = \frac{4\pi QQ'\tilde{\rho}(\mathbf{k})}{|\mathbf{k}|^2}$$

This equation is important in analyzing the structure of atomic particles. The Fourier transform $\tilde{V}(\mathbf{k})$ is directly measurable in scattering experiments. In a typical experiment a (charged) target is probed with a charged point particle (electron). If the analysis of the scattering data shows a deviation from $1/k^2$ in the behavior of $\tilde{V}(\mathbf{k})$, then it can be concluded that the target particle has a charge distribution. More specifically, a plot of $|\mathbf{k}|^2\tilde{V}(\mathbf{k})$ versus k gives the variation of $\tilde{\rho}(\mathbf{k})$ with k. If the resulting graph is a constant, then $\tilde{\rho}(\mathbf{k})$ is a constant, and the target is a point particle [$\tilde{\rho}(\mathbf{k})$ is a constant for point particles, where $\rho(\mathbf{r}') \propto \delta(\mathbf{r}-\mathbf{r}')$]. If there is any deviation from this constancy, $\tilde{\rho}(\mathbf{k})$ must have a dependence on k, and, correspondingly, the target particle must have a charge distribution.

The above discussion, when generalized to four-dimensional relativistic space-time, was the basis for an argument in favor of the existence of quarks in 1968, when the results of the high-energy scattering of electrons off proton targets at the Stanford Linear Accelerator Center indicated a deviation from constancy for $\tilde{\rho}(\mathbf{k})$ for protons. The function $\tilde{\rho}(\mathbf{k})$ (and its relativistic generalization) is called the *form factor*. ●

Example 5.4.10

Let us investigate the Fourier transform of a periodic function and see if we recover a Fourier series. We can define a periodic function with a periodicity interval L as follows.

Let $g_c(x)$ stand for the single function that is defined on the fundamental cell, $[a, a + L]$; that is, $g_c(x) = 0$ for $x \notin [a, a + L]$. We can write the periodic function $g(x)$, which is a repetition of $g_c(x)$ for all intervals of periodicity:

$$g(x) = \begin{cases} g_c(x) & \text{if } a \leqslant x \leqslant a + L \\ g_c(x - L) & \text{if } a + L \leqslant x \leqslant a + 2L \\ \vdots & \\ g_c(x - jL) & \text{if } a + jL \leqslant x \leqslant a + (j + 1)L \\ \vdots & \end{cases}$$

Since $g_c(x) = 0$ outside the interval, we can write

$$g(x) = \sum_{j=-\infty}^{\infty} g_c(x - jL)$$

Now it is easy to verify that this function is indeed periodic:

$$g(x + L) = \sum_{j=-\infty}^{\infty} g_c(x - jL + L) = \sum_{j=-\infty}^{\infty} g_c[x - (j - 1)L]$$

$$= \sum_{m=-\infty}^{\infty} g_c(x - mL)$$

Here we substituted m for $j - 1$. This equation is identical to the preceding one except for a change of dummy index.

Now that we have written the function in a compact form, we can find its Fourier transform:

$$\tilde{g}(k) = \frac{1}{\sqrt{2\pi}} \int_{-\infty}^{\infty} dx\, e^{-ikx} g(x) = \frac{1}{\sqrt{2\pi}} \int_{-\infty}^{\infty} dx\, e^{-ikx} \sum_{j=-\infty}^{\infty} g_c(x - jL)$$

$$= \frac{1}{\sqrt{2\pi}} \sum_{j=-\infty}^{\infty} \int_{-\infty}^{\infty} dx\, e^{-ikx} g_c(x - jL) \tag{1}$$

But $g_c(x - jL)$ is nonzero only if $a + jL \leqslant x \leqslant a + (j + 1)L$. Thus, (1) reduces to

$$\tilde{g}(k) = \frac{1}{\sqrt{2\pi}} \sum_{j=-\infty}^{\infty} \int_{a+jL}^{a+(j+1)L} dx\, e^{-ikx} g_c(x - jL)$$

Changing the variable of integration to $y = x - jL$, we obtain

$$\tilde{g}(k) = \frac{1}{\sqrt{2\pi}} \sum_j \int_a^{a+L} dy\, e^{-ik(y+jL)} g_c(y)$$

$$= \frac{1}{\sqrt{2\pi}} \sum_j e^{-ikjL} \int_a^{a+L} dy\, e^{-iky} g_c(y) = \frac{1}{\sqrt{2\pi}} \tilde{g}_c(k) \sum_j e^{-ikjL} \tag{2}$$

where $\tilde{g}_c(k)$ is, by definition,

$$\tilde{g}_c(k) \equiv \int_a^{a+L} dy\, e^{-iky}\, g_c(y)$$

Let us rewrite the result of Exercise 5.4.2 for the functions e^{ijkL}:

$$\delta(kL - 2m\pi) = \frac{1}{L}\sum_j e^{-ijkL} \tag{3}$$

As mentioned in that exercise, this equation is not entirely correct. For instance, if $kL = 2l\pi$ for $l \neq m$, then the LHS is zero but the RHS is infinite! The reason is that m is arbitrary and we have to include all m's in the LHS. Thus, we correct Eq. (3) by writing

$$\sum_{m=-\infty}^{\infty} \delta(kL - 2m\pi) = \frac{1}{L}\sum_{j=-\infty}^{\infty} e^{-ijkL}$$

If we substitute this in (2), we obtain

$$\tilde{g}(k) = \frac{1}{\sqrt{2\pi}}\tilde{g}_c(k)L \sum_{m=-\infty}^{\infty} \delta(kL - 2m\pi)$$

We can transform back to "x space" and write

$$\begin{aligned} g(x) &= \frac{1}{\sqrt{2\pi}}\int_{-\infty}^{\infty} dk\, e^{ikx}\,\tilde{g}(k) = \frac{L}{2\pi}\sum_m \int_{-\infty}^{\infty} \tilde{g}_c(k)e^{ikx}\delta(kL - 2m\pi)\,dk \\ &= \frac{L}{2\pi}\sum_m \int_{-\infty}^{\infty} \tilde{g}_c(k)e^{ikx}\frac{1}{L}\delta\left(k - \frac{2m\pi}{L}\right)dk \\ &= \frac{1}{2\pi}\sum_{m=-\infty}^{\infty} \tilde{g}_c\left(\frac{2m\pi}{L}\right)e^{i(2m\pi/L)x} \end{aligned}$$

which is a Fourier series if we identify $(1/2\pi)\tilde{g}(2m\pi/L)$ with the Fourier coefficient of $g(x)$. •

Fourier transforms and derivatives. Probably the most common use of Fourier transforms is in solving differential equations. This is because the derivative operator in **r** space turns into ordinary multiplication in **k** space. For example, if we differentiate $g(\mathbf{r})$ in (5.74a) with respect to x_j, we obtain

$$\begin{aligned} \frac{\partial}{\partial x_j} g(\mathbf{r}) &= \frac{1}{(2\pi)^{n/2}}\int d^n k\,\tilde{g}(\mathbf{k})\frac{\partial}{\partial x_j} e^{i(k_1x_1 + \cdots + k_jx_j + \cdots + k_nx_n)} \\ &= \frac{1}{(2\pi)^{n/2}}\int d^n k\,\tilde{g}(\mathbf{k})(ik_j)e^{i\mathbf{k}\cdot\mathbf{r}} \end{aligned}$$

That is, every time we differentiate with respect to any component of **r**, the corresponding component of **k** "comes down." The n-dimensional gradient is

$$\nabla g(\mathbf{r}) = \frac{1}{(2\pi)^{n/2}} \int d^n k\, \tilde{g}(\mathbf{k})(i\mathbf{k}) e^{i\mathbf{k}\cdot\mathbf{r}}$$

and the n-dimensional Laplacian is

$$\nabla^2 g = \frac{1}{(2\pi)^{n/2}} \int d^n k\, \tilde{g}(\mathbf{k})(-k^2) e^{i\mathbf{k}\cdot\mathbf{r}}$$

where $k^2 = k_1^2 + k_2^2 + \cdots + k_n^2$.

We will use Fourier transforms extensively in solving differential equations in Chapters 11 and 12. Here we can illustrate the above points with a simple example. Consider this ordinary second-order differential equation:

$$C_1 \frac{d^2 y}{dx^2} + C_2 \frac{dy}{dx} + C_3 y = f(x)$$

where C_1, C_2, and C_3 are constants. We can "solve" this equation by simply substituting the following in it:

$$y(x) = \frac{1}{\sqrt{2\pi}} \int dk\, \tilde{y}(k) e^{ikx} \qquad \frac{dy}{dx} = \frac{1}{\sqrt{2\pi}} \int dk\, \tilde{y}(k)(ik) e^{ikx}$$

$$\frac{d^2 y}{dx^2} = \frac{1}{\sqrt{2\pi}} \int dk\, \tilde{y}(k)(-k^2) e^{ikx} \qquad f(x) = \frac{1}{\sqrt{2\pi}} \int dk\, \tilde{f}(k) e^{ikx}$$

This gives

$$\frac{1}{\sqrt{2\pi}} \int dk\, \tilde{y}(k)(-C_1 k^2 + iC_2 k + C_3) e^{ikx} = \frac{1}{\sqrt{2\pi}} \int dk\, \tilde{f}(k) e^{ikx}$$

Equating the coefficient of e^{ikx} on both sides, we obtain

$$\tilde{y}(k) = \frac{\tilde{f}(k)}{-C_1 k^2 + iC_2 k + C_3}$$

If we know $\tilde{f}(k)$ [which can be obtained from $f(x)$], we can calculate $y(x)$ by Fourier-transforming $\tilde{y}(k)$. The resulting integrals are not easy to evaluate. In some cases the methods of Chapter 7 may be helpful; in others numerical integration may be the last resort. However, the real power of the Fourier transform lies in the formal analysis of differential equations.

Fourier transform of a distribution. Exercise 5.4.5 concerns the Fourier transform of the Dirac delta function. We have noticed that such a transform is defined in exactly the same way as ordinary functions. It is convenient to define the Fourier transform of a distribution directly in terms of the action of the distribution as a linear functional.

Let us ignore the distinction between the variables x and k and simply define the Fourier transform of a function $f: \mathbb{R} \to \mathbb{R}$ as

$$\tilde{f}(t_1) = \frac{1}{\sqrt{2\pi}} \int_{-\infty}^{\infty} f(t_2) e^{-it_1t_2}\, dt_2 \tag{5.81}$$

Now we consider two functions, f and g, and note that

$$\langle f, \tilde{g} \rangle \equiv \int_{-\infty}^{\infty} f(x)\tilde{g}(x)\, dx = \int_{-\infty}^{\infty} f(x)\left[\frac{1}{\sqrt{2\pi}} \int_{-\infty}^{\infty} g(y) e^{-ixy}\, dy\right] dx$$

$$= \int_{-\infty}^{\infty} g(y)\left[\frac{1}{\sqrt{2\pi}} \int_{-\infty}^{\infty} f(x) e^{-ixy}\, dx\right] dy$$

$$= \int_{-\infty}^{\infty} g(y)\tilde{f}(y)\, dy = \langle \tilde{f}, g \rangle$$

The following definition is motivated by this equation.

5.4.3 Definition Let φ be a distribution and let f be a C_F^∞ function whose Fourier transform $\tilde{f}$ exists and is also a C_F^∞ function. Then we define the Fourier transform $\tilde{\varphi}$ of φ to be the distribution given by $\langle \tilde{\varphi}, f \rangle = \langle \varphi, \tilde{f} \rangle$.

Example 5.4.11

The Fourier transform of $\delta(x)$ is given by

$$\langle \tilde{\delta}, f \rangle = \langle \delta, \tilde{f} \rangle = \tilde{f}(0) = \frac{1}{\sqrt{2\pi}} \int_{-\infty}^{\infty} f(t_2)\, dt_2$$

$$= \int_{-\infty}^{\infty} \left(\frac{1}{\sqrt{2\pi}}\right) f(t_2)\, dt_2 = \langle 1/\sqrt{2\pi}, f \rangle$$

Thus, $\tilde{\delta} = 1/\sqrt{2\pi}$, as expected. •

Example 5.4.12

The Fourier transform of $\delta(x - x')$ is given by

$$\langle \tilde{\delta}(x - x'), f \rangle = \langle \delta(x - x'), \tilde{f} \rangle = \tilde{f}(x') = \frac{1}{\sqrt{2\pi}} \int_{-\infty}^{\infty} f(t_2) e^{-ix't_2}\, dt_2$$

$$= \int_{-\infty}^{\infty} \left(\frac{1}{\sqrt{2\pi}} e^{-ix't_2}\right) f(t_2)\, dt_2$$

Thus, if $\varphi(x) = \delta(x - x')$, then

$$\tilde{\varphi}(t) = \frac{1}{\sqrt{2\pi}} e^{-ix't}$$ •

Exercises

5.4.1 Show that $e^{in\theta}/\sqrt{2\pi}$, for $n = 0, \pm 1, \pm 2, \ldots$, are orthonormal in the interval $[-\pi, +\pi]$.

5.4.2 Show that, for θ and θ' in $[-\pi, \pi]$,

$$\frac{1}{2\pi} \sum_{n=-\infty}^{\infty} e^{in(\theta-\theta')} = \delta(\theta - \theta')$$

What happens when $\theta' = 2m\pi$ for some integer m?

5.4.3 Show that A_n and B_n in Eq. (5.57) are real for real $f(\theta)$.

5.4.4 Find the Fourier expansion of $\cos kx$ defined in the interval $[-a, +a]$.

5.4.5 What is the Fourier transform of **(a)** the constant function $f(x) = C$ and **(b)** the Dirac delta function?

5.4.6 Evaluate the Fourier transform of

$$g(x) = \begin{cases} b - \dfrac{b}{a}|x| & \text{for } |x| \leqslant a \\ 0 & \text{for } |x| > a \end{cases}$$

5.4.7 Show that **(a)** if $g(x)$ is real, then $\tilde{g}^*(k) = \tilde{g}(-k)$, and **(b)** if $g(x)$ is even (odd), then $\tilde{g}(k)$ is also even (odd).

5.4.8 Use a Fourier transform in three dimensions to find a solution of the Poisson equation,

$$\nabla^2 \phi(\mathbf{r}) = -4\pi\rho(\mathbf{r})$$

5.4.9 Show that $\tilde{\tilde{f}}(t) = f(-t)$.

5.4.10 For $\varphi(x) = \delta'(x - x')$, find $\tilde{\varphi}(y)$.

PROBLEMS

5.1 Show that the *parallelogram law*,

$$\|a + b\|^2 + \|a - b\|^2 = 2\|a\|^2 + 2\|b\|^2$$

holds for an inner product space.

5.2 Jordan and von Neumann have shown that if the parallelogram law (see Problem 5.1) holds in a Banach (complete normed) space, then the space is an inner product space. Consider $\mathscr{L}^1(\mathbb{R})$ consisting of all functions $f(x)$ such that $\int_{-\infty}^{\infty} |f(x)|dx \equiv \|f\|$ is finite. This is clearly a normed vector space. Let f and g be such that there is no x at which both $g(x)$ and $f(x)$ are nonzero. Show that

(a) $\|f \pm g\| = \|f\| + \|g\|$

(b) $\|f+g\|^2 + \|f-g\|^2 = 2(\|f\| + \|g\|)^2$

(c) $\mathscr{L}^1(\mathbb{R})$, with the norm defined as above, is not an inner product space

5.3 Prove the completeness of $\mathbb{C}$, using the completeness of $\mathbb{R}$.

5.4 Prove the statements of Proposition 5.2.1 using the defining equations of the Dirac delta function [Eqs. (5.10) and (5.11)]. [Hint: For $x' \in [a, b]$, break the interval into three parts, $[a, x' - \varepsilon]$, $[x' - \varepsilon, x' + \varepsilon]$, and $[x' + \varepsilon, b]$; assume that $f(x)$ is a "good" function; and note that the two sides of (5.10) are equal for an arbitrary function $f(x)$.]

5.5 For the function f defined in Example 5.2.5, show that the derivatives of all orders exist and vanish at $x = 0$.

5.6 Let $-c < a < b < c$. Construct a C^∞ function that is equal to 1 for $a < x < b$ and equal to 0 for $|x| \geqslant c$. [Hint: Look at functions of the form $H(x) = (1/A)\int_b^x h(t; b, c)\,dt$, where $h(t; b, c)$ is as defined in Example 5.2.5 and $A = H(c)$.]

5.7 Define the distribution $\rho: C^\infty(\mathbb{R}^3) \to \mathbb{R}$ by

$$\langle \rho, f \rangle \equiv \iint_S \sigma(\mathbf{r}) f(\mathbf{r})\, da$$

where $\sigma(\mathbf{r})$ is a smooth function on a smooth surface S in $\mathbb{R}^3$. Show that $\rho(\mathbf{r})$ is zero if $\mathbf{r}$ is not on S and infinite if $\mathbf{r}$ is on S.

5.8 Find the expression for the Dirac delta function in cylindrical coordinates, $\delta(\rho, \varphi, z)$; spherical coordinates, $\delta(r, \theta, \varphi)$; and general curvilinear coordinates, $\delta(q_1, q_2, q_3)$. (Hint: The Dirac delta function in $\mathbb{R}^3$ satisfies $\iiint \delta(\mathbf{r})\, d^3r = 1$.)

5.9 Evaluate the following integrals.

(a) $\displaystyle\int_{-\infty}^{\infty} \delta(x^2 - 5x + 6)(3x^2 - 7x + 2)\,dx$ (b) $\displaystyle\int_{-\infty}^{\infty} \delta(x^2 - \pi^2)\cos x\,dx$

(c) $\displaystyle\int_{0.5}^{\infty} \delta(\sin \pi x)(\tfrac{2}{3})^x\,dx$

5.10 Let $\alpha \in C^\infty(\mathbb{R}^n)$ be a smooth function on $\mathbb{R}^n$, and let φ be a distribution. Show that $\alpha\varphi$ is also a distribution. What is the natural definition for $\alpha\varphi$?

5.11 Let $\varphi(x) = |x|$. Use Eq. (5.19) to find the derivative of φ.

5.12 Show that all of the following sequences of functions approach $\delta(x)$ in the sense of Definition 5.2.3.

(a) $\dfrac{n}{\sqrt{\pi}} e^{-n^2x^2}$ (b) $\dfrac{1 - \cos nx}{\pi n x^2}$ (c) $\dfrac{n}{\pi}\left(\dfrac{1}{1 + n^2x^2}\right)$ (d) $\dfrac{\sin nx}{\pi x}$

[Hint: Approximate $\varphi_n(x)$ for large n and $x \approx 0$, and then evaluate the integral.]

5.13 Show that $\frac{1}{2}[(1 + \tanh(nx)] \to \theta(x)$ *as* $n \to \infty$.

5.14 Let φ be a distribution on $\mathbb{R}$, and let α be a C^∞ function. Defining $\alpha\varphi$ as in Problem 5.10, show that $(\alpha\varphi)' = \alpha'\varphi + \alpha\varphi'$.

5.15 Show that $x\delta'(x) = -\delta(x)$.

5.16 For $f(x) = a_0 + a_1 x + a_2 x^2 + \cdots + a_n x^n$, show that

$$\tilde{f}(t) = \sqrt{2\pi}\,[a_0\delta(t) + ia_1\delta'(t) + \cdots + i^n a_n \delta^{(n)}(t)]$$

where

$$\delta^{(n)}(t) \equiv \frac{d^n}{dt^n}\delta(t)$$

5.17 Following the procedure of Example 5.3.1 and using the fact that

$$\int_{-\infty}^{\infty} e^{-x^2}[H_n(x)]^2\,dx = \sqrt{\pi}\,2^n n!$$

find the first three Hermite polynomials.

5.18 Prove Proposition 5.3.9.

5.19 Use the generalized Rodriguez formula for Hermite polynomials and integration by parts to expand x^{2k} and x^{2k+1} in terms of Hermite polynomials.

5.20 Use the recurrence relation for Hermite polynomials to show that

$$\int_{-\infty}^{\infty} xe^{-x^2} H_n(x)H_m(x)\,dx = \sqrt{\pi}\,2^{n-1} n!\,[\delta_{m,n-1} + 2(n+1)\delta_{m,n+1}]$$

5.21 Apply the general formalism of Example 5.3.5 to find the following recurrence relation for Hermite polynomials:

$$H_n + H'_{n-1} - 2xH_{n-1} = 0$$

5.22 **(a)** Specialize the result of Example 5.3.4 to Hermite polynomials to show that

$$\int_{-\infty}^{\infty} x^2 e^{-x^2}[H_n(x)]^2\,dx = \sqrt{\pi}\,2^n(n + \tfrac{1}{2})n!$$

(b) Evaluate $\int_{-\infty}^{\infty} xe^{-x^2}[H_n(x)]^2\,dx$.

5.23 Use the recurrence relations for Hermite polynomials to show that

$$H_n(0) = \begin{cases} 0 & \text{if } n \text{ is odd} \\ (-1)^m \dfrac{(2m)!}{m!} & \text{if } n = 2m \end{cases}$$

5.24 Use the generating function for Hermite polynomials to obtain

$$\sum_{m,n=0}^{\infty} e^{-x^2} H_m(x)H_n(x)\frac{s^m t^n}{m!n!} = e^{-x^2 + 2x(s+t) - (t^2+s^2)}$$

Then integrate both sides to get

$$\sum_{n=0}^{\infty} \frac{(st)^n}{(n!)^2}\int_{-\infty}^{\infty} e^{-x^2}[H_n(x)]^2\,dx = \sqrt{\pi}\,e^{2st} = \sqrt{\pi}\sum_{n=0}^{\infty}\frac{2^n(st)^n}{n!}$$

Deduce from this the normalization of $H_n(x)$, or

$$\int_{-\infty}^{\infty} e^{-x^2}[H_n(x)]^2\,dx = 2^n\sqrt{\pi}n!$$

5.25 Using the recurrence relation of Eq. (5.42) repeatedly, show that

$$\int_{-\infty}^{\infty} x^k e^{-x^2} H_m(x) H_{m+n}(x)\,dx = \begin{cases} 0 & \text{if } n > k \\ \sqrt{\pi}\,2^m (m+k)! & \text{if } n = k \end{cases}$$

5.26 Given that $P_0(x) = 1$ and $P_1(x) = x$, find $P_2(x)$, $P_3(x)$, and $P_4(x)$ using the recurrence relation for Legendre polynomials, Eq. (5.49).

5.27 Derive Eq. (9) of Example 5.3.5 from Eq. (7).

5.28 Apply the general formalism of Example 5.3.5 to find the following recurrence relation for Legendre polynomials:

$$nP_n - xP'_n + P'_{n-1} = 0$$

5.29 Use the procedure of Example 5.3.5 to obtain the following recurrence relation for Legendre polynomials:

$$(1 - x^2)P'_n - nP_{n-1} + nxP_n = 0$$

5.30 Show that

$$\int_{-1}^{1} x^n P_n(x)\,dx = \frac{2^{n+1}(n!)^2}{(2n+1)!}$$

5.31 Use the generating function for Legendre polynomials to show that $P_n(1) = 1$, $P_n(-1) = (-1)^n$, and $P_n(0) = 0$, for odd n. Also show that $P'_n(1) = \frac{1}{2}n(n+1)$.

5.32 Both electrostatic and gravitational potential energies depend on the expression $1/|\mathbf{r} - \mathbf{r}'|$, where $\mathbf{r}'$ is the position of the source (charge or mass) and $\mathbf{r}$ is the observation point. Let $\mathbf{r}$ lie along the z-axis, and use spherical coordinates and the definition of generating functions to show that

$$\frac{1}{|\mathbf{r} - \mathbf{r}'|} = \frac{1}{r_>} \sum_{n=0}^{\infty} \left(\frac{r_<}{r_>}\right)^n P_n(\cos\theta)$$

where $r_<$ ($r_>$) is the smaller (larger) of $\{r, r'\}$ and θ is the polar angle.

5.33 The electrostatic or gravitational potential energy $\Phi(\mathbf{r})$ is given by

$$\Phi(\mathbf{r}) = k \iiint \frac{\rho(\mathbf{r}')}{|\mathbf{r} - \mathbf{r}'|}\,d^3r'$$

where k is a constant and $\rho(\mathbf{r}')$ is the (charge or mass) density function. Use the result of Problem 5.32 to show that if $\rho(\mathbf{r}')$ is spherically symmetric, then $\Phi(\mathbf{r})$ reduces to the potential energy of a point charge at the origin for $r > r'$.

5.34 Use the generating function for Legendre polynomials and their orthogonality to find the relation

$$\int_{-1}^{1} \frac{dx}{1-2xt+t^2} = \sum_{n=0}^{\infty} t^{2n} \int_{-1}^{1} [P_n(x)]^2\, dx$$

Integrate the LHS, expand the result in powers of t, and compare powers of t on both sides to obtain

$$\int_{-1}^{1} [P_n(x)]^2\, dx = \frac{2}{2n+1}$$

5.35 Evaluate the following integrals using the generating function for Legendre polynomials.

(a) $\displaystyle\int_0^{\pi} \frac{a\cos\theta + b}{\sqrt{a^2 + 2ab\cos\theta + b^2}} \sin\theta\, d\theta$ **(b)** $\displaystyle\int_0^{\pi} \frac{a\cos^2\theta + b\sin^2\theta}{\sqrt{a^2 + 2ab\cos\theta + b^2}} \sin\theta\, d\theta$

5.36 Differentiate $g(x, t) = \sum_{n=0}^{\infty} t^n P_n(x)$ with respect to x and manipulate the resulting expression to obtain

$$(1 - 2tx + t^2) \sum_{n=0}^{\infty} t^n P_n'(x) = t \sum_{n=0}^{\infty} t^n P_n(x)$$

Equate equal powers of t on both sides to derive this recurrence relation:

$$P_{n+1}'(x) + P_{n-1}'(x) - 2xP_n'(x) - P_n(x) = 0$$

5.37 Show that $g(x, t) = g(-x, -t)$ for both Hermite and Legendre polynomials. Also, expand $g(x, t)$ and $g(-x, -t)$ and compare the coefficients of t^n to obtain the so-called *parity relations* for these polynomials:

$$H_n(-x) = (-1)^n H_n(x) \qquad \text{and} \qquad P_n(-x) = (-1)^n P_n(x)$$

5.38 Derive the orthogonality of Legendre polynomials directly from the differential equation they satisfy.

5.39 Apply the general results of Example 5.3.5 to the case of Laguerre polynomials to obtain the following recurrence relations:

$$nL_n^{\nu} - (n + \nu)L_{n-1}^{\nu} - x\frac{dL_n^{\nu}}{dx} = 0$$

$$(n+1)L_{n+1}^{\nu} - (2n + \nu + 1 - x)L_n^{\nu} + (n + \nu)L_{n-1}^{\nu} = 0$$

5.40 From the generating function for Laguerre polynomials, deduce that

$$L_n^{\nu}(0) = \frac{(n+\nu)!}{(n!)(\nu!)}$$

5.41 Show that $L_n(x) \equiv L_n^0(x)$ are orthonormal polynomials.

5.42 Differentiate

$$g(x, t) = \frac{e^{-xt/(1-t)}}{1-t} = \sum_{n=0}^{\infty} t^n L_n(x)$$

with respect to x and compare powers of t to obtain

$$L_n'(0) = -n \qquad \text{and} \qquad L_n''(0) = \tfrac{1}{2}n(n-1)$$

[Hint: Differentiate $1/(1-t) = \sum_{n=0}^{\infty} t^n$ to get an expansion for $(1-t)^{-2}$.]

5.43 Expand e^{-kx} in a series of Laguerre polynomials, $L_n^\nu(x)$. Find the coefficients by using **(a)** The orthogonality of $L_n^\nu(x)$ and **(b)** the generating function.

5.44 Derive the recurrence relations given in Table 5.3 for Jacobi, Gegenbauer, and Tchebichef polynomials.

5.45 Show that $T_n(-x) = (-1)^n T_n(x)$ and $U_n(-x) = (-1)^n U_n(x)$. [Hint: $g(x, t) = g(-x, -t)$.]

5.46 Show that $T_n(1) = 1$, $T_n(-1) = (-1)^n$, $T_{2m}(0) = (-1)^m$, and $T_{2m+1}(0) = 0$ and that $U_n(1) = n + 1$, $U_n(-1) = (-1)^n(n+1)$, $U_{2m}(0) = (-1)^m$, and $U_{2m+1}(0) = 0$.

5.47 Let

$$f(\theta) = \begin{cases} \tfrac{1}{2}(\pi - \theta) & \text{for } 0 < \theta \leqslant \pi \\ -\tfrac{1}{2}(\pi + \theta) & \text{for } -\pi \leqslant \theta < 0 \end{cases}$$

be a periodic function. Show that its Fourier-series expansion is

$$f(\theta) = \sum_{n=1}^{\infty} \frac{1}{n} \sin n\theta$$

5.48 Let $f(x) = x$ be a periodic function defined over the interval $(0, 2a)$. Show that

$$x = a - \frac{2a}{\pi} \sum_{n=1}^{\infty} \frac{1}{n} \sin\left(\frac{n\pi x}{a}\right)$$

5.49 Show that the "approximation" to $\sin(\pi x/a)$ in the interval $[-a, a]$ given by the function

$$f(x) = \begin{cases} 4x(a - x) & \text{for } 0 \leqslant x \leqslant a \\ 4x(a + x) & \text{for } -a \leqslant x \leqslant 0 \end{cases}$$

has the Fourier-series expansion

$$f(x) = \frac{32a^2}{\pi^3} \sum_{n=0}^{\infty} \frac{1}{(2n+1)^3} \sin\left[\frac{(2n+1)\pi y}{a}\right]$$

Draw $f(x)$ and $\sin(\pi x/a)$ between $-a$ and a on the same graph to see why $f(x)$ is called the approximation to $\sin(\pi x/a)$.

5.50 Find the Fourier-series expansion of $f(\theta) = \theta^2$ for $|\theta| < \pi$. Then show that

$$\frac{\pi^2}{6} = \sum_{n=1}^{\infty} \frac{1}{n^2} \qquad \text{and} \qquad \frac{\pi^2}{12} = -\sum_{n=1}^{\infty} \frac{(-1)^n}{n^2}$$

5.51 Find the Fourier-series expansion for

$$f(t) = \begin{cases} \sin \omega t & \text{for } 0 \leqslant t \leqslant \dfrac{\pi}{\omega} \\[2ex] 0 & \text{for } -\dfrac{\pi}{\omega} \leqslant t \leqslant 0 \end{cases}$$

5.52 Let $f(\theta)$ be a periodic function given by $f(\theta) = \sum_{n=-\infty}^{\infty} a_n e^{in\theta}$. Show that

$$\tilde{f}(t) = \sqrt{2\pi} \sum_{n=-\infty}^{\infty} a_n \delta(t-n)$$

5.53 Let

$$f(t) = \begin{cases} \sin \omega_0 t & \text{for } |t| < T \\ 0 & \text{for } |t| > T \end{cases}$$

Show that

$$\tilde{f}(\omega) = \frac{i}{\sqrt{2\pi}} \left\{ \frac{\sin[(\omega - \omega_0)T]}{\omega - \omega_0} - \frac{\sin[(\omega + \omega_0)T]}{\omega + \omega_0} \right\}$$

Verify the uncertainty relation $(\Delta\omega)(\Delta t) \approx 4\pi$.

5.54 If $f(x) = g(x + a)$, show that $\tilde{f}(k) = e^{-iak}\tilde{g}(k)$.

5.55 For $a > 0$ find the Fourier transform of $f(x) = e^{-a|x|}$. Is $\tilde{f}(k)$ symmetric? Is it real? Verify the uncertainty relations.

5.56 A damped harmonic oscillator is given by

$$f(t) = \begin{cases} Ae^{-\alpha t} e^{i\omega_0 t} & \text{for } t > 0 \\ 0 & \text{for } t < 0 \end{cases}$$

Find $\tilde{f}(\omega)$ and show that the frequency distribution, $|\tilde{f}(\omega)|^2$, is given by

$$|\tilde{f}(\omega)|^2 = \frac{A^2}{2\pi} \left[\frac{1}{(\omega - \omega_0)^2 + \alpha^2} \right]$$

5.57 Show that (this is called the convolution theorem)

$$\int_{-\infty}^{\infty} f(x) g(y - x)\, dx = \int_{-\infty}^{\infty} \tilde{f}(k)\tilde{g}(k) e^{-iky}\, dk$$

What will this reduce to when $y = 0$?

5.58 Prove Parseval's relation for Fourier transforms:

$$\int_{-\infty}^{\infty} f(x) g^*(x)\, dx = \int_{-\infty}^{\infty} \tilde{f}(k)\tilde{g}^*(k)\, dk$$

In particular, the norm of a function is invariant under Fourier transformation.

5.59 The Fourier transform of a distribution φ is given by

$$\tilde{\varphi}(t) = \sum_{n=0}^{\infty} \frac{1}{n!} \delta'(t - n)$$

What is $\varphi(x)$? [Hint: Use $\varphi(x) = \tilde{\tilde{\varphi}}(-x)$.]

Part II

COMPLEX ANALYSIS

6

Complex Analysis I: Complex Algebra and Calculus

Complex numbers were developed because there was a need to expand the notion of numbers to include solutions of algebraic equations whose prototype is $x^2 + 1 = 0$. Such developments are not atypical in the history of mathematics. The invention of irrational numbers occurred because of a need for a number that could solve an equation of the form $x^2 - 2 = 0$. Similarly, fractions and rational numbers were offspring of the operations of multiplication and division and the quest for (for example) a number that gives 4 when multiplied by 3, or, equivalently, a number that solves the equation $3x - 4 = 0$.

There is, however, a crucial difference between complex numbers and all the numbers mentioned above: All rational, irrational, and, in general, real numbers correspond to measurable physical quantities. However, there is no single measurable physical quantity that can be described by a complex number. No wonder Leibnitz called a complex number a "flight of God's spirit."

A natural question then is this: What need is there for complex numbers if no physical quantity can be measured in terms of them? The answer is that although no *single* physical quantity can be expressed in terms of complex numbers, a *pair* of physical quantities can be neatly described by a single complex number. For example, a wave that consists of an amplitude and a phase can be concisely described by a complex number, which is a vector in the xy-plane whose length is the amplitude and whose angle with the x-axis gives the phase of the wave. There is no limit to the number of possibilities for the application of complex numbers in physics and engineering. In fact, we have already used complex numbers on many occasions in earlier chapters. In this chapter and the next, we start a systematic and (reasonably) comprehensive study of complex numbers and complex analysis.

6.1 COMPLEX NUMBERS AND THEIR ALGEBRA

We demand a number system broad enough to include solutions to equations such as

$$x^2 + 1 = 0 \qquad \text{or} \qquad x^2 = -1$$

Clearly, the solution(s), x, cannot be real because a real number raised to the second power gives a positive real number, and we want x^2 to be negative. So we broaden the concept of numbers by considering *complex numbers*, which are of the form

$$z = x + iy \tag{6.1}$$

where

$$i \equiv \sqrt{-1} \quad \Rightarrow \quad i^2 = -1 \tag{6.2}$$

is the formal solution of the equation $x^2 + 1 = 0$. It turns out (amazingly!) that we don't need to introduce any other numbers to solve *any* algebraic equation. In fact, the fundamental theorem of algebra states that all roots of any algebraic equation of the form

$$a_n x^n + a_{n-1} x^{n-1} + \cdots + a_1 x + a_0 = 0$$

with arbitrary (real or complex) coefficients $a_n, a_{n-1}, \ldots, a_1, a_0$ are in the complex number system. In this sense, then, the complex number system is the most complete system.

A complex number can be conveniently represented as a point in the xy-plane, called the *complex plane* in this context, as shown in Fig. 6.1. In Eq. (6.1) x is called the *real part* of z, written $\text{Re}(z)$, and y is called the *imaginary part* of z, written $\text{Im}(z)$. Similarly, the horizontal axis in Fig. 6.1 is named the *real axis*, and the vertical axis is named the *imaginary axis*. The set of all complex numbers, or the set of points in the complex plane, is denoted by $\mathbb{C}$.

We can define various operations on $\mathbb{C}$ that are extensions of similar operations on the real number system, $\mathbb{R}$. We must keep in mind that $i^2 = -1$ and that the final form of an equation must be written like Eq. (6.1)—with real and imaginary parts. For

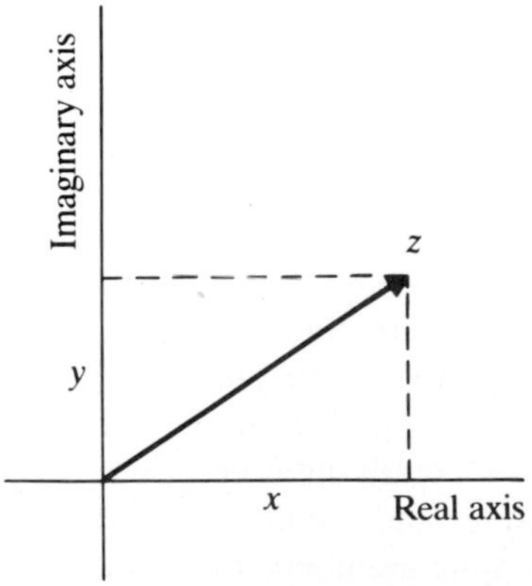

Figure 6.1 The Cartesian representation of the complex number z in the complex plane $\mathbb{C}$.

instance, the sum of two complex numbers, $z_1 = x_1 + iy_1$ and $z_2 = x_2 + iy_2$, is

$$z_1 + z_2 = (x_1 + x_2) + i(y_1 + y_2)$$

This sum can be represented in the complex plane as the *vector* sum of z_1 and z_2, as shown in Fig. 6.2. The product of z_1 and z_2 can be obtained:

$$\begin{aligned} z_1 z_2 &= (x_1 + iy_1)(x_2 + iy_2) = x_1 x_2 + x_1(iy_2) + iy_1 x_2 + (iy_1)(iy_2) \\ &= x_1 x_2 + i(x_1 y_2 + y_1 x_2) + i^2 y_1 y_2 \\ &= x_1 x_2 - y_1 y_2 + i(x_1 y_2 + y_1 x_2) \end{aligned}$$

Thus,

$$\begin{aligned} \text{Re}(z_1 z_2) &= x_1 x_2 - y_1 y_2 \\ \text{Im}\,(z_1 z_2) &= x_1 y_2 + x_2 y_1 \end{aligned} \tag{6.3}$$

To obtain (6.3) we have implicitly used the fact that two complex numbers are equal if and only if their real parts are equal and their imaginary parts are equal.

The factor i in z allows new operations for complex numbers that do not exist for real numbers. One such operation is *complex conjugation*. The *complex conjugate* z^* (or $\bar{z}$) of z is defined as

$$z^* \equiv (x + iy)^* \equiv x - iy \tag{6.4}$$

which is obtained by replacing i with $-i$. We note immediately that

$$zz^* = z^*z = (x + iy)(x - iy) = x^2 + y^2$$

which is a positive real number. The positive square root of zz^* is called the *absolute value*, or *norm*, of z and denoted by $|z|$. Thus, we have

$$|z| = \sqrt{zz^*} = \sqrt{z^*z} = \sqrt{x^2 + y^2} \tag{6.5}$$

This absolute value is, of course, the length of the vector representing z in the complex

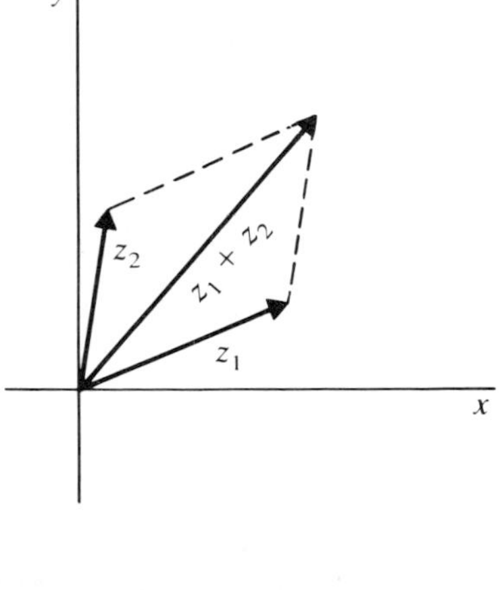

Figure 6.2 The addition of two complex numbers resembles the vector addition of two-dimensional vectors.

plane. Some useful properties of absolute values are as follows:

$$|z_1 z_2| = |z_1||z_2| \tag{6.6a}$$

$$\left|\frac{z_1}{z_2}\right| = \frac{|z_1|}{|z_2|} \tag{6.6b}$$

$$||z_1| - |z_2|| \leqslant |z_1 + z_2| \leqslant |z_1| + |z_2| \tag{6.6c}$$

This last inequality is the triangle inequality encountered in Section 5.1 for norms.

Example 6.1.1

Let us carry out some sample calculations with complex numbers.

(a)
$$\frac{1}{1-i} - \frac{1}{1+i} = i$$

because
$$\frac{1}{1-i} - \frac{1}{1+i} = \frac{1+i-(1-i)}{(1-i)(1+i)} = \frac{2i}{|1-i|^2} = \frac{2i}{2} = i$$

where we used $z = 1 - i$ and $|z|^2 = zz^* = x^2 + y^2$.

(b)
$$(i+1)^{-4} = -\tfrac{1}{4}$$

because
$$(i+1)^{-4} = \frac{1}{(i+1)^4} = \frac{1}{(i+1)^2(i+1)^2} = \frac{1}{(2i)(2i)} = \frac{1}{4i^2} = -\frac{1}{4}$$

(c)
$$\frac{2+i}{3-i} = \tfrac{1}{2}(1+i)$$

because
$$\frac{2+i}{3-i} = \frac{2+i}{3-i}\left(\frac{3+i}{3+i}\right) = \frac{6+5i+i^2}{9-i^2} = \frac{5+5i}{10} = \tfrac{1}{2}(1+i)$$

(d)
$$\left|\frac{2i-1}{i-2}\right| = 1$$

because
$$\left|\frac{2i-1}{i-2}\right| = \sqrt{\frac{2i-1}{i-2}\left(\frac{-2i-1}{-i-2}\right)} = \sqrt{\frac{(2)^2+(-1)^2}{(1)^2+(-2)^2}} = \sqrt{\frac{5}{5}} = 1$$

(e)
$$\left|\frac{z_1}{z_2}\right| = \frac{|z_1|}{|z_2|}$$

because
$$\left|\frac{z_1}{z_2}\right| = \sqrt{\frac{z_1}{z_2}\left(\frac{z_1}{z_2}\right)^*} = \sqrt{\frac{z_1}{z_2}\left(\frac{z_1^*}{z_2^*}\right)} = \sqrt{\frac{z_1 z_1^*}{z_2 z_2^*}} = \sqrt{\frac{|z_1|^2}{|z_2|^2}} = \frac{|z_1|}{|z_2|}$$

(f)
$$|z_1 z_2| = |z_1||z_2|$$

because
$$|z_1 z_2| = \sqrt{(z_1 z_2)(z_1 z_2)^*} = \sqrt{z_1 z_2 z_1^* z_2^*} = \sqrt{(z_1 z_1^*)(z_2 z_2^*)}$$
$$= \sqrt{|z_1|^2 |z_2|^2} = |z_1||z_2|$$

(g) The equation $|z - a| = b$, where a and b are fixed numbers (b real and positive), describes a circle of radius b with center at $a \equiv a_x + ia_y$. This is easily seen because

$$\begin{aligned}|z - a|^2 &= |(x + iy) - (a_x + ia_y)|^2 = |(x - a_x) + i(y - a_y)|^2 \\ &= (x - a_x)^2 + (y - a_y)^2 = b^2\end{aligned}$$

We note that $|z - a|$ is the distance between the two complex numbers z and a. Therefore, $|z - a| = b$, where a is a constant and z is a variable, is the collection of all points z that are at a distance b from a. ●

We can also define the division of two complex numbers using complex conjugation. To find the real and imaginary parts of the quotient, we proceed as follows:

$$\begin{aligned}\frac{z_1}{z_2} = \frac{z_1 z_2^*}{z_2 z_2^*} &= \frac{(x_1 + iy_1)(x_2 - iy_2)}{x_2^2 + y_2^2} = \frac{x_1x_2 + y_1y_2 + i(x_2y_1 - x_1y_2)}{x_2^2 + y_2^2} \\ &= \frac{x_1x_2 + y_1y_2}{x_2^2 + y_2^2} + i\left(\frac{y_1x_2 - x_1y_2}{x_2^2 + y_2^2}\right)\end{aligned}$$

Thus,

$$\mathrm{Re}\left(\frac{z_1}{z_2}\right) = \frac{x_1x_2 + y_1y_2}{x_2^2 + y_2^2} \tag{6.7a}$$

$$\mathrm{Im}\left(\frac{z_1}{z_2}\right) = \frac{y_1x_2 - x_1y_2}{x_2^2 + y_2^2} \tag{6.7b}$$

In particular,

$$\frac{1}{z} = \frac{z^*}{|z|^2} = \frac{x - iy}{x^2 + y^2}$$

Some properties associated with complex conjugation (which were used above) are gathered in the following proposition.

6.1.1 Proposition Complex conjugation satisfies the following identities.

$$\begin{aligned}(z_1 + z_2)^* &= z_1^* + z_2^* \\ (z_1 z_2)^* &= z_1^* z_2^* \\ \left(\frac{z_1}{z_2}\right)^* &= \frac{z_1^*}{z_2^*} \\ (z^*)^* &= z \\ \mathrm{Re}\,(z) &= \tfrac{1}{2}(z + z^*) \\ \mathrm{Im}\,(z) &= \frac{1}{2i}(z - z^*) \\ (z^n)^* &= (z^*)^n\end{aligned}$$

Proof. The proof follows immediately from the definition of complex conjugation and, for the last identity, application of the binomial theorem. ■

In general, we can define the complex conjugate of a function $f(z)$ as

$$(f(z))^* \equiv f(z^*) \tag{6.8}$$

This is equivalent to replacing every i with $-i$ in the expression for $f(z)$. (For details, see the discussion of the Schwarz reflection principle in Chapter 7.)

6.1.1 Polar Coordinates and Complex Numbers

The introduction of polar coordinates in the complex plane makes available a powerful tool with which to facilitate complex manipulations. Figure 6.3 shows a complex number and its polar coordinates. In terms of these polar coordinates, z can be written as

$$z = x + iy = r\cos\theta + ir\sin\theta = r(\cos\theta + i\sin\theta)$$

Assuming that series of complex numbers can be manipulated as those of real numbers can, it can be shown that (see Exercise 6.1.2)

$$e^{i\theta} = \cos\theta + i\sin\theta \tag{6.9}$$

Thus, a complex number can be expressed succinctly as

$$z = re^{i\theta} \tag{6.10a}$$

where

$$r = \sqrt{x^2 + y^2} = \sqrt{zz^*} = |z| \tag{6.10b}$$

and

$$\tan\theta = \frac{y}{x} \tag{6.10c}$$

Note that θ is not uniquely determined; any multiple of 2π can be added to it without affecting z. We can use (6.10) together with

$$x = r\cos\theta \qquad \text{and} \qquad y = r\sin\theta$$

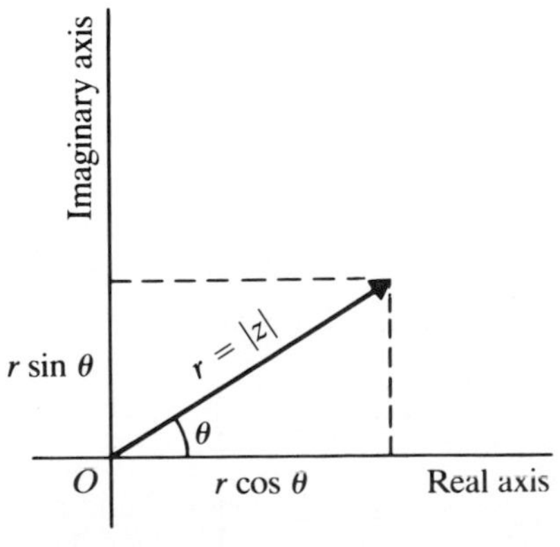

Figure 6.3 The polar representation of the complex number z in the complex plane $\mathbb{C}$.

to convert from Cartesian coordinates to polar coordinates, and vice versa. The coordinate θ is called the *argument* of z, written

$$\theta = \arg(z)$$

The complex conjugate of z in polar coordinates is

$$z^* = x - iy = r\cos\theta - ir\sin\theta = r[\cos(-\theta) + i\sin(-\theta)] = re^{-i\theta}$$

This equation confirms the earlier statement that complex conjugation is equivalent to replacing i with $-i$.

As the following examples show, polar coordinates can be extremely useful.

Example 6.1.2

We can use the polar representation of complex numbers to find some trigonometric identities. In all of the following, $r = 1$.

(a) $$1 = e^{i\theta}e^{-i\theta} = (\cos\theta + i\sin\theta)(\cos\theta - i\sin\theta) = \cos^2\theta + \sin^2\theta$$

(b) $$e^{i(\theta_1+\theta_2)} = \cos(\theta_1 + \theta_2) + i\sin(\theta_1 + \theta_2)$$

On the other hand,

$$e^{i(\theta_1+\theta_2)} = e^{i\theta_1}e^{i\theta_2} = (\cos\theta_1 + i\sin\theta_1)(\cos\theta_2 + i\sin\theta_2)$$

$$= (\cos\theta_1\cos\theta_2 - \sin\theta_1\sin\theta_2) + i(\sin\theta_1\cos\theta_2 + \cos\theta_1\sin\theta_2)$$

Equating the real and imaginary parts of these two equations, we obtain

$$\cos(\theta_1 + \theta_2) = \cos\theta_1\cos\theta_2 - \sin\theta_1\sin\theta_2$$

$$\sin(\theta_1 + \theta_2) = \sin\theta_1\cos\theta_2 + \cos\theta_1\sin\theta_2$$

(c) $$e^{3i\theta} = \cos 3\theta + i\sin 3\theta$$

$$e^{3i\theta} = (e^{i\theta})^3 = (\cos\theta + i\sin\theta)^3 = \cos^3\theta + 3(\cos\theta)^2 i\sin\theta + 3(i\sin\theta)^2\cos\theta$$

$$+ (i\sin\theta)^3 = \cos^3\theta - 3\sin^2\theta\cos\theta + i(3\cos^2\theta\sin\theta - \sin^3\theta)$$

Equating the real and imaginary parts of these two equations gives

$$\cos 3\theta = \cos^3\theta - 3\sin^2\theta\cos\theta$$

$$\sin 3\theta = 3\cos^2\theta\sin\theta - \sin^3\theta$$

•

Example 6.1.3

We can generalize part (c) of Example 6.1.2. From

$$e^{in\theta} = \cos n\theta + i\sin n\theta \qquad \text{and} \qquad e^{in\theta} = (e^{i\theta})^n = (\cos\theta + i\sin\theta)^n$$

we obtain the so-called de Moivré theorem:

$$\cos n\theta + i\sin n\theta = (\cos\theta + i\sin\theta)^n$$

Next we expand the RHS using the binomial expansion:

$$(\cos\theta + i\sin\theta)^n$$
$$= \sum_{k=0}^{n}\binom{n}{k}(i\sin\theta)^k(\cos\theta)^{n-k}$$
$$= \sum_{k\text{ even}}\binom{n}{k}(i\sin\theta)^k(\cos\theta)^{n-k} + \sum_{k\text{ odd}}\binom{n}{k}(i\sin\theta)^k(\cos\theta)^{n-k}$$
$$= \sum_{l=0}^{[n]}\binom{n}{2l}i^{2l}(\sin\theta)^{2l}(\cos\theta)^{n-2l} + \sum_{l=0}^{[n]}\binom{n}{2l+1}i^{2l+1}(\sin\theta)^{2l+1}(\cos\theta)^{n-2l-1}$$
$$= \sum_{l=0}^{[n]}(-1)^l\binom{n}{2l}\sin^{2l}\theta\cos^{n-2l}\theta + i\sum_{l=0}^{[n]}(-1)^l\binom{n}{2l+1}\sin^{2l+1}\theta\cos^{n-2l-1}\theta$$

where
$$\binom{n}{k} \equiv \frac{n!}{k!(n-k)!}$$

and $[n]$ means either n or $n-1$, whichever applies. Equating the real and imaginary parts, we obtain

$$\cos n\theta = \sum_{l=0}^{[n]}(-1)^l\binom{n}{2l}\sin^{2l}\theta\cos^{n-2l}\theta$$
$$\sin n\theta = \sum_{l=0}^{[n]}(-1)^l\binom{n}{2l+1}\sin^{2l+1}\theta\cos^{n-2l-1}\theta \qquad \bullet$$

We can use the polar representation of z and z^* in (6.9) to obtain these useful results:

$$\cos\theta = \tfrac{1}{2}(e^{i\theta} + e^{-i\theta}) \tag{6.11a}$$
$$\sin\theta = \frac{1}{2i}(e^{i\theta} - e^{-i\theta}) \tag{6.11b}$$

6.1.2 The nth Roots of Unity

The exponential nature of polar coordinates makes them especially useful in multiplication, division, and exponentiation. For instance,

$$\frac{z_1}{z_2} = \frac{r_1 e^{i\theta_1}}{r_2 e^{i\theta_2}} = \frac{r_1}{r_2}e^{i(\theta_1-\theta_2)} \tag{6.12a}$$
$$z_1 z_2 = (r_1 e^{i\theta_1})(r_2 e^{i\theta_2}) = r_1 r_2 e^{i(\theta_1+\theta_2)} \tag{6.12b}$$
$$\sqrt{z_1} = \sqrt{r_1 e^{i\theta_1}} = (r_1 e^{i\theta_1})^{1/2} = r_1^{1/2}(e^{i\theta_1})^{1/2} = \sqrt{r_1}e^{i(\theta_1/2)} \tag{6.12c}$$

and so forth. All of these relations have interesting geometric interpretations. For example, (6.12b) says that when you multiply a complex number, z_1, by another complex number, z_2, you dilate its magnitude by a factor r_2 and increase its angle by θ_2. That is, multiplication involves both a dilation and a rotation. In particular, if we multiply a complex number by $e^{i\omega t}$, where t is a real variable (time), we get a vector of constant length in the xy-plane that is rotating with angular velocity ω.

An interesting application of these ideas is the extraction of the nth roots of unity. We want to find all z's satisfying

$$z^n = 1 \tag{6.13}$$

The most general way we can write unity is

$$1 = e^{2i\pi k} \qquad \text{for } k = 0, \pm 1, \pm 2, \ldots$$

Thus, we have

$$z^n = e^{2i\pi k} \qquad \text{for } k = 0, \pm 1, \pm 2, \ldots$$

Taking the nth root of both sides, we obtain

$$z = (e^{2i\pi k})^{1/n} = e^{2i\pi k/n} \qquad \text{for } k = 0, \pm 1, \pm 2, \ldots$$

and the *distinct* roots $\{z_k\}$ are

$$z_k = e^{2i\pi k/n} \qquad \text{for } k = 0, 1, 2, \ldots, n-1 \tag{6.14}$$

We see that there are exactly n nth roots of unity. All of these roots are equally spaced on the unit circle in the complex plane, as shown in Fig. 6.4.

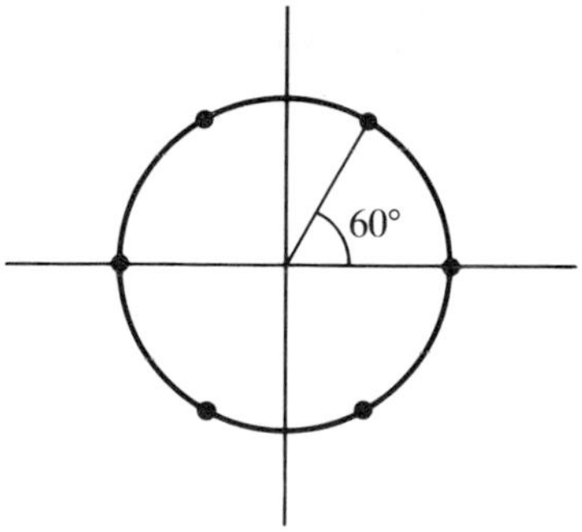

Figure 6.4 The six sixth roots of unity depicted as six equally spaced points on the unit circle.

Example 6.1.4

Various roots of unity are given in the following.
For $n = 2$,

$$z = e^{2\pi ik/2} = e^{\pi ik} \qquad \text{where } k = 0, 1 \quad \Rightarrow \quad z_1 = e^0, z_2 = e^{\pi i} \quad \Rightarrow \quad z_1 = 1, z_2 = -1$$

For $n = 3$,

$$z = e^{2\pi ik/3} \qquad \text{for } k = 0, 1, 2 \quad \Rightarrow \quad z_1 = e^0, z_2 = e^{2\pi i/3}, z_3 = e^{4\pi i/3}$$

$$e^{2\pi i/3} = \cos\frac{2\pi}{3} + i\sin\frac{2\pi}{3} = -\frac{1}{2} + i\frac{\sqrt{3}}{2}$$

$$e^{4\pi i/3} = \cos\frac{4\pi}{3} + i\sin\frac{4\pi}{3} = -\frac{1}{2} - i\frac{\sqrt{3}}{2}$$

We verify that

$$\left(-\frac{1}{2}+i\frac{\sqrt{3}}{2}\right)^3 = -\frac{1}{8}+3\left(-\frac{1}{2}\right)^2 i\frac{\sqrt{3}}{2}+3\left(i\frac{\sqrt{3}}{2}\right)^2\left(-\frac{1}{2}\right)+\left(i\frac{\sqrt{3}}{2}\right)^3$$

$$= -\frac{1}{8}+i\frac{3\sqrt{3}}{8}+\frac{9}{8}-i\frac{3\sqrt{3}}{8}=1$$

and similarly for $\left(-\frac{1}{2}-i\frac{\sqrt{3}}{2}\right)$.

For $n = 4$,

$$z = e^{2\pi ik/4} = e^{\pi ik/2} \qquad \text{where } k = 0, 1, 2, 3$$

$$z_1 = 1 \qquad z_2 = e^{\pi i/2} \qquad z_3 = e^{\pi i} \qquad z_4 = e^{3\pi i/2}$$

or

$$z_1 = 1 \qquad z_2 = i \qquad z_3 = -1 \qquad z_4 = -i$$

For $n = 6$,

$$z = e^{2\pi ik/6} = e^{\pi ik/3} \qquad \text{where } k = 0, 1, \ldots, 5$$

$$z_1 = 1 \qquad z_2 = e^{\pi i/3} \qquad z_3 = e^{2\pi i/3} \qquad z_4 = e^{\pi i} \qquad z_5 = e^{4\pi i/3} \qquad z_6 = e^{5\pi i/3}$$

or

$$z_1 = 1 \qquad z_2 = \tfrac{1}{2}(1+i\sqrt{3}) \qquad z_3 = \tfrac{1}{2}(-1+i\sqrt{3})$$

$$z_4 = -1 \qquad z_5 = -\tfrac{1}{2}(1+i\sqrt{3}) \qquad z_6 = \tfrac{1}{2}(1-i\sqrt{3})$$

•

Example 6.1.5

Let us find the square root of $z = a + ib$ in Cartesian coordinates. The technique we will use also applies to finding the nth root. We first write z in polar form:

$$z = re^{i\theta} \qquad \text{where } r = \sqrt{a^2+b^2},\ \tan\theta = \frac{b}{a}$$

We raise both sides to the $\frac{1}{2}$ power:

$$z^{1/2} = r^{1/2}e^{i\theta/2} = (a^2+b^2)^{1/4}e^{i\theta/2} = (a^2+b^2)^{1/4}\left(\cos\frac{\theta}{2}+i\sin\frac{\theta}{2}\right)$$

We have

$$\cos\frac{\theta}{2} = \left[\frac{1}{2}(1+\cos\theta)\right]^{1/2} = \frac{1}{\sqrt{2}}\left(1+\frac{1}{\sqrt{1+\tan^2\theta}}\right)^{1/2}$$

$$= \frac{1}{\sqrt{2}}\left(1+\frac{a}{\sqrt{a^2+b^2}}\right)^{1/2}$$

Similarly,

$$\sin\frac{\theta}{2} = \frac{1}{\sqrt{2}}\left(1-\frac{a}{\sqrt{a^2+b^2}}\right)^{1/2}$$

Thus,

$$z^{1/2} = (\sqrt{a^2+b^2})^{1/2}\frac{1}{\sqrt{2}}\left[\left(1+\frac{a}{\sqrt{a^2+b^2}}\right)^{1/2} + i\left(1-\frac{a}{\sqrt{a^2+b^2}}\right)^{1/2}\right]$$

$$= \frac{1}{\sqrt{2}}[(\sqrt{a^2+b^2}+a)^{1/2} + i(\sqrt{a^2+b^2}-a)^{1/2}] \qquad (1)$$

We see how complicated a simple square root calculation can become. Of course, when dealing with complex numbers (rather than symbols) in practice, we can directly evaluate r and θ and find the square root in polar coordinates, which is much simpler. However, sometimes we need the analytic—not just the numerical—expression for a square root. Such a need arises, for instance, when considering the propagation of electromagnetic waves in conductors, where the square of the complex index of refraction is given by Maxwell's equations and an *expression* for the index of refraction is desired.

Equation (1) gives only one of the roots. Clearly, there is another square root not included in (1). To find this second root we must replace θ with $\theta + 2k\pi$, for $k = 0, \pm 1, \pm 2, \ldots$, and $\theta/2$ with $\theta/2 + k\pi$, for $k = 0, \pm 1, \pm 2, \ldots$. The distinct roots are obtained when $k = 0, 1$. We thus have

$$\sqrt{z} = r^{1/2}e^{i(\theta/2+k\pi)} = e^{ik\pi}\sqrt{r}e^{i\theta/2} \qquad \text{for } k = 0, 1$$

$$= \pm\sqrt{r}e^{i\theta/2}$$

The other root is simply the negative of that given by (1).

We can generalize this to the nth root. A general Cartesian expression like (1) is prohibitively difficult, however. On the other hand, the polar version is given by

$$z^{1/n} = r^{1/n}e^{i[(\theta+2k\pi)/n]} = e^{2ik\pi/n}r^{1/n}e^{i\theta/n} \qquad \text{for } k = 0, 1, \ldots, n-1$$

•

6.1.3 Alternative Representations of Complex Numbers

The representations of complex numbers that we have been using, $z = x + iy$ and $z = re^{i\theta}$, are the most common in applications. However, other representations are in use. For instance, z can be represented by a pair of real numbers:

$$z \equiv (x, y)$$

The set of all such pairs, together with the operations

$$z_1 + z_2 = (x_1 + x_2, y_1 + y_2)$$

and

$$z_1 z_2 = (x_1, y_1)(x_2, y_2) = (x_1x_2 - y_1y_2, x_1y_2 + x_2y_1)$$

is identical with the set of complex numbers, $z = x + iy$. In particular, the unit complex number, 1_c, is simply

$$1_c \equiv (1, 0)$$

because

$$1_c z = (1, 0)(x, y) = (x, y) = z$$

On the other hand,

$$i = (0, 1)$$

because $$i^2 = (0, 1)(0, 1) = (-1, 0) = -(1, 0) = -1_c$$

There is no difference between the complex identity element, 1_c, and the real identity element, 1.

We see, therefore, that there is no difference between this pair-of-real-numbers approach and the approach adopted previously in this section. Each has its advantages. The approach used earlier is more natural for human computations, but with computers the pair-of-real-numbers approach is preferable because they can easily make room for two-component arrays in their memories.

Other alternatives exist. A third, less common, approach is to make the identifications

$$1 \equiv \begin{pmatrix} 1 & 0 \\ 0 & 1 \end{pmatrix} \quad \text{and} \quad i \equiv \begin{pmatrix} 0 & -1 \\ 1 & 0 \end{pmatrix}$$

and write complex numbers $z = x + iy$ as

$$z = \begin{pmatrix} x & -y \\ y & x \end{pmatrix}$$

In this case we treat them as 2×2 matrices.

Exercises

6.1.1 The complex numbers $z_1 = x_1 + iy_1$ and $z_2 = x_2 + iy_2$ may be regarded as two-dimensional vectors:

$$\mathbf{z}_1 = \hat{\mathbf{e}}_x x_1 + \hat{\mathbf{e}}_y y_1 \quad \text{and} \quad \mathbf{z}_2 = \hat{\mathbf{e}}_x x_2 + \hat{\mathbf{e}}_y y_2$$

Show that

$$z_1^* z_2 = \mathbf{z}_1 \cdot \mathbf{z}_2 + i\hat{\mathbf{e}}_z \cdot (\mathbf{z}_1 \times \mathbf{z}_2)$$

6.1.2 Show that $e^{i\theta} = \cos\theta + i\sin\theta$.

6.1.3 Show that

$$\sum_{k=0}^{n} \cos k\theta = \frac{\sin[(n+1)\theta/2]}{\sin(\theta/2)} \cos\frac{n\theta}{2} \quad \text{and} \quad \sum_{k=0}^{n} \sin k\theta = \frac{\sin[(n+1)\theta/2]}{\sin(\theta/2)} \sin\frac{n\theta}{2}$$

6.1.4 Change each representation of a complex number from Cartesian to polar or vice versa.
(a) $1 + i$ **(b)** $2 + 3i$ **(c)** i **(d)** $2e^{-i\pi/6}$ **(e)** $5e^{i(35^\circ)}$

6.2 FUNCTIONS OF A COMPLEX VARIABLE

A complex function is simply a mapping $f: \mathbb{C} \to \mathbb{C}$ such that f takes in a complex number and gives out (in general) a different complex number. We, therefore, write

$$f(z) = w \tag{6.15}$$

where both z and w are complex numbers. Equation (6.15) can be thought of geometrically as a correspondence between two *complex planes*, the z-plane and the w-plane. The w-plane has a real axis and an imaginary axis, which we can call u and v, respectively. Both u and v are *real functions* of the coordinates of z, which are x and y. Therefore, we may write

$$f(z) = u(x, y) + iv(x, y) \tag{6.16}$$

This equation gives a *unique* point (u, v) in the w-plane for each point (x, y) in the z-plane. Thus, regions of the z-plane are mapped into regions of the w-plane. For instance, a curve in the z-plane gets mapped into a curve in the w-plane. The following example illustrates this point.

Example 6.2.1

Let us investigate the behavior of elementary complex functions. In particular, we will look at the way a line in the z-plane is mapped into various *curves* in the w-plane. Consider a line in the z-plane given by the equation $y = mx$.

(a) Let us start with the simple function $w = f(z) = z^2$. We have

$$w = (x + iy)^2 = x^2 - y^2 + 2ixy$$

and

$$u(x, y) = x^2 - y^2$$

$$v(x, y) = 2xy$$

For $y = mx$ we obtain

$$u = x^2 - m^2x^2 = (1 - m^2)x^2$$

$$v = 2mx^2$$

To find the curve in the w-plane, we eliminate x in these equations for u and v to find v as a function of u:

$$v = 2m\frac{u}{1 - m^2} = \frac{2m}{1 - m^2}u$$

This is a line passing through the origin of the w-plane. Note that if α is the angle the line in the xy-plane makes with the x-axis ($m = \tan\alpha$), then the angle made by the line in the uv-plane with the u-axis is 2α, as shown in Fig. 6.5(a). (Why?)

(b) Now let us consider $w = f(z) = e^z = e^{x+iy}$, which (for now) we define to be $e^x(\cos y + i\sin y)$ with

$$u(x, y) = e^x \cos y$$

$$v(x, y) = e^x \sin y$$

Now, if we substitute $y = mx$, we obtain

$$u = e^x \cos mx$$

$$v = e^x \sin mx$$

However, unlike part (a), here we cannot eliminate x to find v as an explicit function of u. Nevertheless, these are parametric equations of a curve, which we can plot on a uv-plane, as shown in Fig. 6.5(b). ●

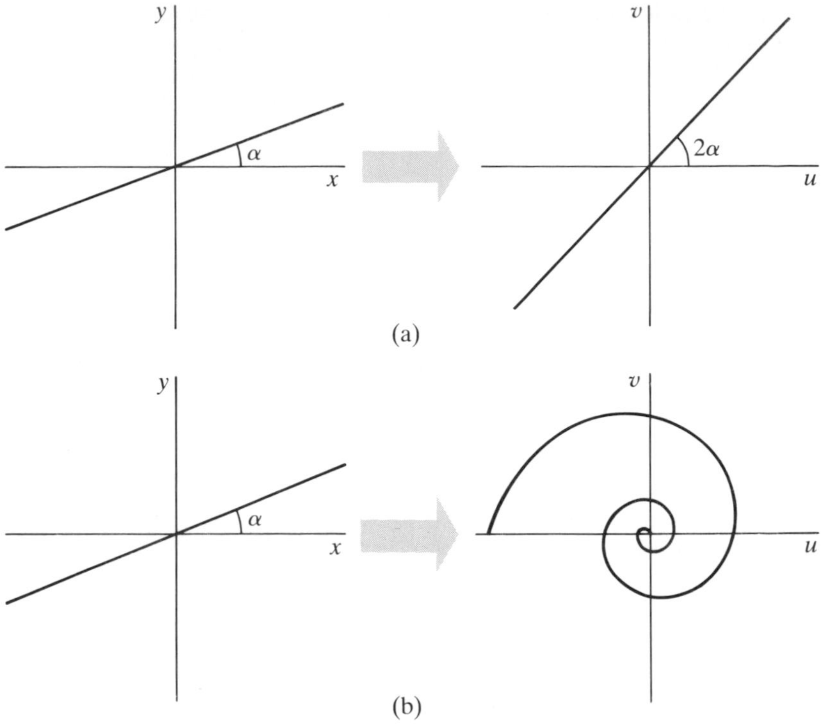

Figure 6.5. (a) The line $y = mx$, making an angle α with the real axis of the z-plane, is mapped onto a line making an angle 2α with the real axis of the w-plane under the mapping $f: \mathbb{C} \to \mathbb{C}$ given by $f(z) = z^2$. (b) The same line is mapped onto a spiral under the mapping $f: \mathbb{C} \to \mathbb{C}$ given by $f(z) = e^z$.

6.2.1 Limits and Continuity

Limits of complex functions are defined in terms of absolute values (norms), as in Chapter 5. We write

$$\lim_{z \to a} f(z) = w_0$$

which means that, given any real number $\varepsilon > 0$, we can find a corresponding real number $\delta > 0$ such that $|f(z) - w_0| < \varepsilon$ whenever $|z - a| < \delta$.

Similarly, we say that a function $f(z)$ is continuous at $z = a$ if

$$\lim_{z \to a} f(z) = f(a)$$

or if there exist $\varepsilon > 0$ and $\delta > 0$ such that $|f(z) - f(a)| < \varepsilon$ whenever $|z - a| < \delta$.

6.2.2 Derivative of a Complex Function

The derivative of a complex function is defined as usual.

6.2.1 Definition Let $f: \mathbb{C} \to \mathbb{C}$ be a complex function. The *derivative of f at z_0*, denoted $(df/dz)_{z_0}$, is

$$\lim_{\Delta z \to 0} \frac{f(z_0 + \Delta z) - f(z_0)}{\Delta z}$$

provided the limit exists and is independent of Δz.

In this definition "independent of Δz" means independent of Δx and Δy (the components of Δz) and, therefore, independent of the direction of approach to z_0. The restrictions of this definition apply to the real case as well. For instance, the derivative of $f(x) = |x|$ at $x = 0$ does not exist because it approaches $+1$ from the right and -1 from the left.

It can easily be shown that all the formal rules of differentiation (for example, for derivatives of products, sums, and ratios of functions) that apply in the real case also apply in the complex case.

Example 6.2.2

Let us examine the derivative of $f(z) = x + 2iy$ at $z = 0$. From Definition 6.2.1 we have

$$\left(\frac{df}{dz}\right)_{z=0} = \lim_{\Delta z \to 0} \frac{f(\Delta z) - f(0)}{\Delta z} = \lim_{\substack{\Delta x \to 0 \\ \Delta y \to 0}} \frac{\Delta x + 2i\,\Delta y}{\Delta x + i\,\Delta y}$$

To find the resulting limit we must choose a path in the z-plane and approach the origin, $z = 0$, along that path. If we choose the x-axis as our path, $\Delta y = 0$, and we get

$$\left(\frac{df}{dz}\right)_{z=0} = \lim_{\Delta x \to 0} \frac{\Delta x + 0}{\Delta x + 0} = 1$$

On the other hand, if we choose the y-axis as our path, we have $\Delta x = 0$, and the limit yields

$$\left(\frac{df}{dz}\right)_{z=0} = \lim_{\Delta y \to 0} \frac{0 + 2i\,\Delta y}{0 + i\,\Delta y} = 2$$

In general, for a line that goes through the origin, $y = mx$, the limit yields

$$\left(\frac{df}{dz}\right)_{z=0} = \lim_{\Delta x \to 0} \frac{\Delta x + 2i(m\,\Delta x)}{\Delta x + im\,\Delta x} = \frac{1 + 2im}{1 + im}$$

$$= \frac{1 + 2m^2 + im}{1 + m^2}$$

This indicates that we get infinitely many values for $(df/dz)_{z=0}$ when we change m arbitrarily; thus, df/dz does not exist at $z = 0$. ●

6.2.3 The Cauchy-Riemann (C-R) Conditions

A question arises naturally at this point: When *does* the limit in Definition 6.2.1 exist? To answer this question we will find the necessary and sufficient conditions for the existence of that limit. It is clear from the definition that differentiability puts a severe restriction on $f(z)$ because it requires the limit to be the same for *all paths* going through z_0. Another important point to keep in mind is that differentiability is a *local* property. To test whether or not a function $f(z)$ is differentiable at z_0, we move away from z_0 by a small amount, Δz, and check the existence of the limit $(df/dz)_{z_0}$.

What are the conditions under which a complex function is differentiable? For $f(z) = u(x, y) + iv(x, y)$, Definition 6.2.1 yields

$$\left(\frac{df}{dz}\right)_{z_0} = \lim_{\substack{\Delta x \to 0 \\ \Delta y \to 0}} \frac{u(x_0 + \Delta x, y_0 + \Delta y) + iv(x_0 + \Delta x, y_0 + \Delta y) - u(x_0, y_0) - iv(x_0, y_0)}{\Delta x + i\Delta y}$$

$$= \lim_{\substack{\Delta x \to 0 \\ \Delta y \to 0}} \frac{u(x_0 + \Delta x, y_0 + \Delta y) - u(x_0, y_0) + i[v(x_0 + \Delta x, y_0 + \Delta y) - v(x_0, y_0)]}{\Delta x + i\Delta y}$$

If this limit is to exist for all paths, it must exist for the two particular paths on which $\Delta y = 0$ (the x-axis) and $\Delta x = 0$ (the y-axis). For the first path we get

$$\left(\frac{df}{dz}\right)_{z_0} = \lim_{\Delta x \to 0} \frac{u(x_0 + \Delta x, y_0) - u(x_0, y_0) + i[v(x_0 + \Delta x, y_0) - v(x_0, y_0)]}{\Delta x}$$

$$= \lim_{\Delta x \to 0} \frac{u(x_0 + \Delta x, y_0) - u(x_0, y_0)}{\Delta x} + i \lim_{\Delta x \to 0} \frac{v(x_0 + \Delta x, y_0) - v(x_0, y_0)}{\Delta x}$$

$$= \left.\frac{\partial u}{\partial x}\right|_{x_0, y_0} + i \left.\frac{\partial v}{\partial x}\right|_{x_0, y_0}$$

where we have used the definition of partial derivatives. For the second path ($\Delta x = 0$) we obtain

$$\left(\frac{df}{dz}\right)_{z_0} = \lim_{\Delta y \to 0} \frac{u(x_0, y_0 + \Delta y) - u(x_0, y_0) + i[v(x_0, y_0 + \Delta y) - v(x_0, y_0)]}{i\Delta y}$$

$$= -i \lim_{\Delta y \to 0} \frac{u(x_0, y_0 + \Delta y) - u(x_0, y_0)}{\Delta y} + \lim_{\Delta y \to 0} \frac{v(x_0, y_0 + \Delta y) - v(x_0, y_0)}{\Delta y}$$

$$= -i \left.\frac{\partial u}{\partial y}\right|_{x_0, y_0} + \left.\frac{\partial v}{\partial y}\right|_{x_0, y_0}$$

If $f(z)$ is to be differentiable at z_0, the equations for the two paths must be equal; that is,

$$\left.\frac{\partial u}{\partial x}\right|_{z_0} + i \left.\frac{\partial v}{\partial x}\right|_{z_0} = -i \left.\frac{\partial u}{\partial y}\right|_{z_0} + \left.\frac{\partial v}{\partial y}\right|_{z_0}$$

Equating the real and imaginary parts of both sides of this equation and ignoring the subscript $z_0(x_0, y_0$ or z_0 is arbitrary), we obtain

$$\frac{\partial u}{\partial x} = \frac{\partial v}{\partial y} \tag{6.17a}$$

and

$$\frac{\partial u}{\partial y} = -\frac{\partial v}{\partial x} \tag{6.17b}$$

These two conditions, which are *necessary* for the differentiability of $f(z)$, are called the *Cauchy-Riemann* (*C-R*) conditions.

An alternative way of writing the C-R conditions is obtained by making the substitutions

$$x = \tfrac{1}{2}(z + z^*) \qquad \text{and} \qquad y = \frac{1}{2i}(z - z^*)$$

in $u(x, y)$ and $v(x, y)$, using the chain rule to write Eqs. (6.17) in terms of z and z^*, substituting the results in

$$\frac{\partial f}{\partial z^*} = \frac{\partial u}{\partial z^*} + i\frac{\partial v}{\partial z^*}$$

and showing that Eqs. (6.17) are equivalent to the single equation (see Exercise 6.2.2)

$$\frac{\partial f}{\partial z^*} = 0$$

This equation says that if f is to be *differentiable*, it must be *independent of* z^*.

If the derivative of f exists, the arguments leading to Eqs. (6.17) imply that it can be expressed as

$$\frac{df}{dz} = \frac{\partial u}{\partial x} + i\frac{\partial v}{\partial x} \tag{6.18a}$$

or

$$\frac{df}{dz} = \frac{\partial v}{\partial y} - i\frac{\partial u}{\partial y} \tag{6.18b}$$

The C-R conditions assure us that these two equations are equivalent.

The following example illustrates the differentiability of complex functions.

Example 6.2.3

Let us determine whether or not certain functions are differentiable.

(a)
$$f(z) = x + 2iy$$

We have already established the nondifferentiability of this function at the origin by considering different paths of approach to the origin. We can show that this function cannot be differentiable at any point in the complex plane. Note that

$$u = x \qquad \text{and} \qquad v = 2y$$

Thus,

$$\frac{\partial u}{\partial x} = 1 \neq \frac{\partial v}{\partial y} = 2$$

and Eq. (6.17a) is not satisfied. Equation (6.17b) is satisfied, but that is not enough.

We can also write $f(z)$ in terms of z and z^*:

$$f(z) = \frac{1}{2}(z + z^*) + 2i\left[\frac{1}{2i}(z - z^*)\right] = \frac{3}{2}z - \frac{1}{2}z^*$$

We see that $f(z)$ has an explicit dependence on z^* and

$$\frac{\partial f}{\partial z^*} = -\frac{1}{2} \neq 0$$

This again shows that $f(z)$ is not differentiable.

(b) $$f(z) = x^2 - y^2 + 2ixy$$

Here $u = x^2 - y^2$ and $v = 2xy$, so we have

$$\frac{\partial u}{\partial x} = 2x = \frac{\partial v}{\partial y}$$

$$\frac{\partial u}{\partial y} = -2y = -\frac{\partial v}{\partial x}$$

Thus, $f(z)$ could be differentiable! Recall that Eqs. (6.17) are *necessary* conditions; we have not shown (but will shortly) that they are also *sufficient*.

Let us check whether $\partial f/\partial z^* = 0$:

$$\begin{aligned} f(z) &= \left[\frac{1}{2}(z + z^*)\right]^2 - \left[\frac{1}{2i}(z - z^*)\right]^2 + 2i\left[\frac{1}{2}(z + z^*)\right]\left[\frac{1}{2i}(z - z^*)\right] \\ &= \tfrac{1}{4}[z^2 + (z^*)^2 + 2zz^*] + \tfrac{1}{4}[z^2 + (z^*)^2 - 2zz^*] + \tfrac{1}{2}[z^2 - (z^*)^2] \\ &= z^2 \end{aligned}$$

There is no z^*-dependence; therefore, $f(z)$ could be differentiable.

(c) Let $u(x, y) = e^x \cos y$ and $v(x, y) = e^x \sin y$. Then

$$\frac{\partial u}{\partial x} = e^x \cos y = \frac{\partial v}{\partial y}$$

$$\frac{\partial u}{\partial y} = -e^x \sin y = -\frac{\partial v}{\partial x}$$

and the C-R conditions are satisfied. Also,

$$f(z) = e^x \cos y + ie^x \sin y = e^x(\cos y + i \sin y) = e^x e^{iy} = e^{x+iy} = e^z$$

and there is no z^*-dependence. •

The requirement for differentiability is very restrictive (the derivative must exist along infinitely many paths). However, the C-R conditions seem deceptively mild (they are derived for only two paths). The two paths are, in fact, true representatives of all paths; that is, the C-R conditions are not only necessary but sufficient, as the following theorem shows.

6.2.2 Theorem The function $f(z) = u(x, y) + iv(x, y)$ is differentiable in a region of the complex plane iff the Cauchy-Riemann conditions,

$$\frac{\partial u}{\partial x} = \frac{\partial v}{\partial y} \qquad \text{and} \qquad \frac{\partial u}{\partial y} = -\frac{\partial v}{\partial x}$$

(or, equivalently, $\partial f/\partial z^* = 0$), are satisfied and all first partial derivatives of u and v are continuous in that region. In that case

$$\frac{df}{dz} = \frac{\partial u}{\partial x} + i\frac{\partial v}{\partial x} = \frac{\partial v}{\partial y} - i\frac{\partial u}{\partial y}$$

Proof. We have already shown that if $f(z)$ is differentiable, then the C-R conditions are satisfied. We now show that if the C-R conditions hold, then the function is differentiable. First, we note that if the derivative exists at all, it must equal (6.18a) or (6.18b). Thus, we have to show that

$$\lim_{\Delta z \to 0} \frac{f(z + \Delta z) - f(z)}{\Delta z} = \frac{\partial u}{\partial x} + i\frac{\partial v}{\partial x}$$

or, equivalently, that

$$\left|\frac{f(z + \Delta z) - f(z)}{\Delta z} - \left(\frac{\partial u}{\partial x} + i\frac{\partial v}{\partial x}\right)\right| < \varepsilon \qquad \text{whenever } |\Delta z| < \delta$$

By definition,

$$f(z + \Delta z) - f(z) = u(x + \Delta x, y + \Delta y) + iv(x + \Delta x, y + \Delta y) - u(x, y) - iv(x, y)$$

Since u and v have continuous first partial derivatives, we can write

$$u(x + \Delta x, y + \Delta y) = u(x, y) + \frac{\partial u}{\partial x}\Delta x + \frac{\partial u}{\partial y}\Delta y + \varepsilon_1 \Delta x + \delta_1 \Delta y$$

$$v(x + \Delta x, y + \Delta y) = v(x, y) + \frac{\partial v}{\partial x}\Delta x + \frac{\partial v}{\partial y}\Delta y + \varepsilon_2 \Delta x + \delta_2 \Delta y$$

where ε_1, ε_2, δ_1, and δ_2 are real numbers that approach zero as Δx and Δy approach zero. Using these expressions, we can write

$$\begin{aligned}
f(z + \Delta z) - f(z) &= u + \frac{\partial u}{\partial x}\Delta x + \frac{\partial u}{\partial y}\Delta y + \varepsilon_1 \Delta x + \delta_1 \Delta y + iv + i\frac{\partial v}{\partial x}\Delta x + i\frac{\partial v}{\partial y}\Delta y \\
&\quad + i\varepsilon_2 \Delta x + i\delta_2 \Delta y - u - iv \\
&= \left(\frac{\partial u}{\partial x} + i\frac{\partial v}{\partial x}\right)\Delta x + i\left(-i\frac{\partial u}{\partial y} + \frac{\partial v}{\partial y}\right)\Delta y + (\varepsilon_1 + i\varepsilon_2)\Delta x \\
&\quad + (\delta_1 + i\delta_2)\Delta y \\
&= \left(\frac{\partial u}{\partial x} + i\frac{\partial v}{\partial x}\right)(\Delta x + i\Delta y) + (\varepsilon_1 + i\varepsilon_2)\Delta x + (\delta_1 + i\delta_2)\Delta y
\end{aligned}$$

where we used the C-R conditions in the last step. Dividing both sides by $\Delta z = \Delta x + i\Delta y$, we get

$$\frac{f(z+\Delta z)-f(z)}{\Delta z} - \left(\frac{\partial u}{\partial x} + i\frac{\partial v}{\partial x}\right) = (\varepsilon_1 + i\varepsilon_2)\frac{\Delta x}{\Delta z} + (\delta_1 + i\delta_2)\frac{\Delta y}{\Delta z}$$

Taking the absolute value of both sides and using the triangle inequality, we obtain

$$\begin{aligned}\left|\frac{f(z+\Delta z)-f(z)}{\Delta z} - \left(\frac{\partial u}{\partial x} + i\frac{\partial v}{\partial x}\right)\right| &= \left|(\varepsilon_1 + i\varepsilon_2)\frac{\Delta x}{\Delta z} + (\delta_1 + i\delta_2)\frac{\Delta y}{\Delta z}\right| \\ &\leqslant \left|(\varepsilon_1 + i\varepsilon_2)\frac{\Delta x}{\Delta z}\right| + \left|(\delta_1 + i\delta_2)\frac{\Delta y}{\Delta z}\right| \\ &= |\varepsilon_1 + i\varepsilon_2|\frac{|\Delta x|}{|\Delta z|} + |\delta_1 + i\delta_2|\frac{|\Delta y|}{|\Delta z|} \\ &\leqslant |\varepsilon_1 + i\varepsilon_2| + |\delta_1 + i\delta_2|\end{aligned}$$

The last step follows from the fact that

$$\frac{|\Delta x|}{|\Delta z|} = \frac{|\Delta x|}{\sqrt{(\Delta x)^2 + (\Delta y)^2}} \leqslant 1$$

$$\frac{|\Delta y|}{|\Delta z|} = \frac{|\Delta y|}{\sqrt{(\Delta x)^2 + (\Delta y)^2}} \leqslant 1$$

The ε and δ terms can be made as small as desired by making Δz small enough. We have thus established that when the C-R conditions hold, the function $f(z)$ is differentiable. ■

6.2.3 Definition A function $f\colon \mathbb{C} \to \mathbb{C}$ is called *analytic* at z_0 if it is differentiable at z_0 and *at all other points* in some neighborhood of z_0. A point at which f is analytic is called a *regular point* of f. A point at which f is not analytic is called a *singular point*, or a *singularity*, of f. A function for which all points in $\mathbb{C}$ are regular points is called an *entire function.*

Example 6.2.4

Let us consider some examples of entire functions.

(a) $$f(z) = z$$

Here $u = x$ and $v = y$; the C-R conditions are easily shown to hold, and

$$\frac{df}{dz} = \frac{\partial u}{\partial x} + i\frac{\partial v}{\partial x} = 1$$

Therefore, the derivative exists at all points of the complex plane.

(b) $$f(z) = z^2$$

Here $u = x^2 - y^2$ and $v = 2xy$; the C-R conditions hold, and

$$\frac{df}{dz} = 2x + i2y = 2z$$

Therefore, $f(z)$ is differentiable at all points.

(c) $$f(z) = z^n \qquad \text{for } n \geqslant 1$$

We use mathematical induction and the fact that the product of two entire functions is an entire function to show that z^n is entire. In fact, we show that

$$\frac{d}{dz}(z^n) = nz^{n-1} \tag{1}$$

We have shown this for $n = 1$ and $n = 2$ in parts (a) and (b). Assume that it is true for k. Then

$$\frac{d}{dz}(z^{k+1}) = \frac{d}{dz}(zz^k) = \frac{dz}{dz}z^k + z\frac{d}{dz}(z^k)$$

↑ by the product rule for differentiation

$$= z^k + z(kz^{k-1}) = (k+1)z^k$$

↑ by part (a) and the induction hypothesis

Thus, (1) is true for all n by mathematical induction. This shows, in particular, that z^n is entire.

(d) $$f(z) = a_0 + a_1 z + a_2 z^2 + \cdots + a_{n-1}z^{n-1} + a_n z^n$$

where a_i are arbitrary constants. That $f(z)$ is entire follows directly from part (c) and the fact that the sum of two entire functions is entire.

Now let us consider some examples of functions that are not entire.

(e) $$f(z) = \frac{1}{z}$$

The derivative can be found to be

$$f'(z) = -\frac{1}{z^2}$$

which does not exist for $z = 0$. Thus, $z = 0$ is a singularity of $f(z)$. However, any other point is a regular point of f.

(f) $$f(z) = |z|^2$$

Using the definition of the derivative, we obtain

$$\frac{\Delta f}{\Delta z} = \frac{|z + \Delta z|^2 - |z|^2}{\Delta z} = \frac{(z + \Delta z)(z^* + \Delta z^*) - zz^*}{\Delta z} = z^* + \Delta z^* + z\frac{\Delta z^*}{\Delta z}$$

For $z = 0$, $\Delta f/\Delta z = \Delta z^*$, which goes to zero as $\Delta z \to 0$. Therefore, $df/dz = 0$ at $z = 0$. However, if $z \neq 0$, the limit of $\Delta f/\Delta z$ will depend on how z is approached. Thus, df/dz does not exist if $z \neq 0$. This shows that $|z|^2$ is differentiable only at $z = 0$ and nowhere else in its neighborhood. It also shows that even if the real and imaginary parts of a complex function have continuous partial derivatives of all orders at a point, the function may not be differentiable there. For the case we are considering,

$$u(x, y) = x^2 + y^2 \qquad \text{and} \qquad v(x, y) = 0$$

have continuous partial derivatives of all orders at every point (x, y), yet $|z|^2$ is not differentiable anywhere (except at $z = 0$).

(g)
$$f(z) = \frac{1}{\sin z}$$

This gives

$$\frac{df}{dz} = -\cos z/(\sin z)^2$$

Thus, f has infinitely many (isolated) singular points:

$$z = \pm n\pi \qquad \text{for } n = 0, 1, 2, \ldots$$

•

Example 6.2.4 shows that any polynomial in z is entire. Exercise 6.2.5 shows that the exponential function e^z is also entire. Therefore, any product and/or sum of polynomials and e^z will also be entire. We can build other entire functions. For instance, e^{iz} and e^{-iz} are entire functions; therefore, the trigonometric functions, defined as follows, are also entire functions:

$$\sin z \equiv \frac{e^{iz} - e^{-iz}}{2i} \qquad \cos z \equiv \frac{e^{iz} + e^{-iz}}{2}$$

It can be shown that $\sin z$ and $\cos z$ have only *real* zeros (see Exercise 6.2.6). The hyperbolic functions can be similarly defined:

$$\sinh z \equiv \frac{e^z - e^{-z}}{2} \qquad \cosh z \equiv \frac{e^z + e^{-z}}{2}$$

Although the sum and the product of entire functions are entire, the ratio, in general, is not. For instance, if $f(z)$ and $g(z)$ are polynomials of degree m and n, respectively, then for $n > 0$ the ratio $f(z)/g(z)$ is not entire, because at the zeros of $g(z)$ (and these zeros always exist) the derivative is not defined.

The functions $u(x, y)$ and $v(x, y)$ of an analytic function have some interesting properties, which the following example investigates.

Example 6.2.5

The family of curves $u(x, y) =$ constant is perpendicular to the family of curves $v(x, y) =$ constant at each point in the region of the complex plane where $f(z) = u + iv$ is analytic.

This can easily be seen by looking at the normal to the curves. The normal to the curve $u(x, y) =$ constant is simply $\nabla u = (\partial u/\partial x, \partial u/\partial y)$ as discussed in Chapter 1. Similarly, the normal to the curve $v(x, y) =$ constant is $\nabla v = (\partial v/\partial x, \partial v/\partial y)$. Taking the dot product of these two normals, we obtain

$$(\nabla u)\cdot(\nabla v) = \frac{\partial u}{\partial x}\frac{\partial v}{\partial x} + \frac{\partial u}{\partial y}\frac{\partial v}{\partial y} = \frac{\partial u}{\partial x}\left(-\frac{\partial u}{\partial y}\right) + \frac{\partial u}{\partial y}\left(\frac{\partial u}{\partial x}\right) = 0$$

by the C-R conditions. •

6.2.4 Analytic Functions and Electrostatics

The real and imaginary parts of an analytic function separately satisfy the two-dimensional *Laplace's equation*:

$$\frac{\partial^2 u}{\partial x^2} + \frac{\partial^2 u}{\partial y^2} = 0 \tag{6.19a}$$

$$\frac{\partial^2 v}{\partial x^2} + \frac{\partial^2 v}{\partial y^2} = 0 \tag{6.19b}$$

This can easily be verified from the C-R conditions. Laplace's equation in three dimensions, $\partial^2\Phi/\partial x^2 + \partial^2\Phi/\partial y^2 + \partial^2\Phi/\partial z^2 = 0$, describes the electrostatic potential, Φ, in a charge-free region of space. In a typical electrostatic problem the potential Φ is given at certain boundaries (usually conducting surfaces) and its value at every point in space is sought. There are numerous techniques for solving such problems (these will be discussed in Chapters 8 and 10). However, some of these problems have a certain degree of symmetry, allowing them to be reduced to two dimensions. In such cases the theory of analytic functions can be extremely helpful.

The symmetry mentioned above is *cylindrical symmetry* where the potential is known to be independent of the z-coordinate (the axis of symmetry). This situation occurs when conductors are cylinders and, if there are charge distributions in certain regions of space, the densities are z-independent. In such cases $\partial\Phi/\partial z = 0$, and the problem reduces to a two-dimensional one.

Functions that satisfy Laplace's equation are called *harmonic functions*. Thus, the electrostatic potential is a three-dimensional harmonic function; however, the potential for a cylindrically symmetric charge distribution and/or boundary condition is a two-dimensional harmonic function, as are the real and imaginary parts of a complex analytic function [see Eqs. (6.19)]. Is there a connection between cylindrically symmetric potentials and analytic functions? Yes.

To find this connection let us consider a long straight filament with a *constant* linear charge density λ. It is shown in elementary textbooks on electricity and magnetism that the electric field $\mathbf{E}$ and the potential Φ (with an appropriate choice of the reference point) are given in cylindrical coordinates by

$$\mathbf{E} = \frac{2\lambda}{\rho}\hat{\mathbf{e}}_\rho$$

and $$\Phi = 2\lambda \ln \rho = 2\lambda \ln [(x^2 + y^2)^{1/2}]$$

Clearly, $\partial^2\Phi/\partial x^2 + \partial^2\Phi/\partial y^2 = 0$ implies that Φ *could* be the real (or imaginary) part of an analytic function. But which analytic function? Example 6.2.5 gives us a clue.

Suppose that u, the real part of the analytic function we are looking for, is the potential function. We have

$$u = \Phi = 2\lambda \ln [(x^2 + y^2)^{1/2}]$$

What should v be? Example 6.2.5 says that the two families of curves u = constant and v = constant are perpendicular to each other at every point. In the case at hand, that u is constant implies that Φ is constant, which, in turn, implies equipotential surfaces. These equipotential surfaces are cylinders, whose intersections with the xy-plane are described by the curves $u(x, y)$ = constant. These curves are given by $x^2 + y^2$ = constant, that is, are circles centered at the origin. Thus, the other family of curves must be rays emanating from the origin (see Fig. 6.6). These are given, in cylindrical coordinates, by

$$v(\rho, \varphi) = g(\varphi)$$

where g is a function to be determined. However, to make $w = u + iv$ analytic (see Exercise 6.2.7), we must choose $v = 2\lambda\varphi$. Then

$$w(z) = 2\lambda \ln \rho + 2i\lambda\varphi = 2\lambda \ln (\rho e^{i\varphi}) = 2\lambda \ln z$$

This is the analytic function we are looking for.

Such a function is called the *complex potential* and is very useful. For instance, it can easily be verified that the magnitude of the electric field is simply given by the magnitude of the derivative of the potential:

$$|\mathbf{E}| = \left|\frac{dw}{dz}\right| \tag{6.20}$$

As the following example shows, this result is general.

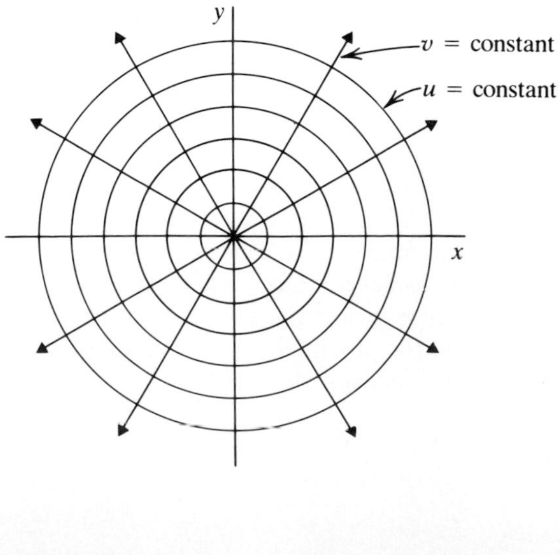

Figure 6.6 The curves of constant u and v for the function $w(z) = \ln z$. Note that the two families of curves are perpendicular to one another at each point in the complex plane.

Example 6.2.6

If $w(z) = u(x, y) + iv(x, y)$ is a complex potential, so $u(x, y) = \Phi$, then

$$\frac{dw}{dz} = \frac{\partial u}{\partial x} + i\frac{\partial v}{\partial x} = \frac{\partial u}{\partial x} - i\frac{\partial u}{\partial y} = -E_x + iE_y$$

and

$$\left|\frac{dw}{dz}\right| = \sqrt{(-E_x)^2 + (E_y^2)} = |\mathbf{E}|$$

because $E_z = -\partial\Phi/\partial z = 0$ by cylindrical symmetry. Thus, the derivative of the complex potential gives the magnitude of the electric field directly. •

Another use of $w(z)$ is in calculating the flux through a surface. Let us consider the simple situation where we want to calculate the flux through a part of a cylinder of radius a and length (in the z-direction) l, which is subtended between the two azimuthal angles φ_1 and φ_2 as shown in Fig. 6.7. It is easy to calculate the flux:

$$\text{flux} = \iint_S \mathbf{E}\cdot\hat{\mathbf{e}}_n\,da = \int_{l_1}^{l_2} dz \int_{\varphi_1}^{\varphi_2} \rho\,d\varphi\,\frac{2\lambda}{\rho} = 2\lambda l(\varphi_2 - \varphi_1)$$

Recalling that $v = 2\lambda\varphi$, we get

$$\frac{\text{flux}}{l} \equiv \text{flux per unit length of } z = v_2 - v_1$$

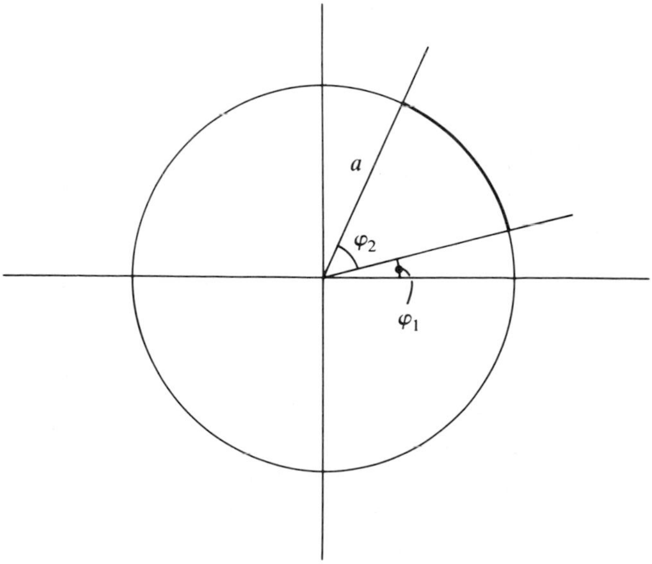

Figure 6.7 A portion of the $u =$ constant curve subtended between two different $v =$ constant curves.

This is also a general result (shown in Exercise 6.2.8), stated in the following proposition.

6.2.4 Proposition Let $w(z) = u(x, y) + iv(x, y)$ be the (analytic) complex potential of a cylindrically symmetric electrostatic charge distribution. Let a curve from (x_1, y_1) to (x_2, y_2) in the xy-plane be the intersection of a cylindrical surface with the xy-plane. Then

$$|\mathbf{E}| = \left|\frac{dw}{dz}\right|$$

$$\begin{array}{l}\text{flux through the segment of the} \\ \text{(cylindrical) surface per unit length} = v(x_2, y_2) - v(x_1, y_1) \\ \text{along the axis of the cylinder}\end{array}$$

The function v is sometimes called the *streamline function.* ∎

When there are two cylindrical conductors such that all electric field lines start from one and end up on the other, we say they form a *capacitor*. These two conductors map two equipotential curves on the xy-plane, $u(x, y) = u_1$ and $u(x, y) = u_2$, where u_1 and u_2 are simply the potentials of the conductors. There are, of course, infinitely many equipotential curves in the xy-plane. Let us start at a point P_1 on $u(x, y) = u_1$ and observe what happens to the values of the streamline function with movement along u_1. For a complete tour of u_1, the change in v will be the total flux, by Proposition 6.2.4. By Gauss's law this total flux is related to the charge per unit length enclosed:

$$\frac{\text{total flux}}{\text{length}} = 4\pi \frac{\text{total charge}}{\text{length}}$$

But total flux per unit length is the change in v. If we denote this total change by $[v]$, we obtain

$$\frac{\text{total charge}}{\text{length}} = \frac{1}{4\pi}[v]$$

We can now easily obtain the capacitance per unit length for the two conductors:

$$c = \frac{C}{l} = \frac{\text{charge per unit length}}{\text{potential difference}} = \frac{\frac{1}{4\pi}[v]}{|u_2 - u_1|}$$

Similarly, the electrostatic energy per unit length is

$$U = \tfrac{1}{2}cV^2 = \frac{1}{8\pi}\left(\frac{[v]}{|u_2 - u_1|}\right)|u_2 - u_1|^2 = \frac{1}{8\pi}[v]|u_2 - u_1|$$

We see that a lot of information is stored in $w(z)$. Both its real part and its imaginary part have physical significance, as exhibited above.

It is useful to know the complex potential function of more than one line of charge. To find such a potential we must first find $w(z)$ for a line charge when it is displaced from the origin. If the line is located at $z_0 = x_0 + iy_0$, then it is easy to show that

$$w(z) = 2\lambda \ln (z - z_0)$$

The superposition principle then immediately gives us the complex potential function of n line charges located at $z_1, z_2, \ldots, z_n$:

$$w(z) = 2 \sum_{k=1}^{n} \lambda_k \ln (z - z_k) \tag{6.21}$$

Example 6.2.7

As an application of Eq. (6.21), let us find the equipotential curves and streamlines for two line charges of equal magnitude and opposite sign located at $y = a$ and $y = -a$ in the xy-plane. Equation (6.21) immediately gives

$$\begin{aligned} w(z) &= 2\lambda \ln(z + ia) - 2\lambda \ln (z - ia) \\ &= 2\lambda \ln \frac{z + ia}{z - ia} \end{aligned}$$

Let us solve for z in terms of w:

$$\frac{z + ia}{z - ia} = e^{w/2\lambda}$$

After some algebra we get

$$\begin{aligned} z = ia \frac{\cosh (w/4\lambda)}{\sinh (w/4\lambda)} &= ia \frac{\cosh\left(\dfrac{u}{4\lambda} + i\dfrac{v}{4\lambda}\right)}{\sinh\left(\dfrac{u}{4\lambda} + i\dfrac{v}{4\lambda}\right)} \\ &= ia \frac{\cosh\dfrac{u}{4\lambda}\cos\dfrac{v}{4\lambda} + i\sinh\dfrac{u}{4\lambda}\sin\dfrac{v}{4\lambda}}{\sinh\dfrac{u}{4\lambda}\cos\dfrac{v}{4\lambda} + i\cosh\dfrac{u}{4\lambda}\sin\dfrac{v}{4\lambda}} \end{aligned}$$

which, after some more algebra, yields

$$z = a \frac{\sin\dfrac{v}{2\lambda} + i\sinh\dfrac{u}{2\lambda}}{\cosh\dfrac{u}{2\lambda} - \cos\dfrac{v}{2\lambda}}$$

Thus,

$$x = \frac{a \sin(v/2\lambda)}{\cosh(u/2\lambda) - \cos(v/2\lambda)} \qquad \text{and} \qquad y = \frac{a \sinh(u/2\lambda)}{\cosh(u/2\lambda) - \cos(v/2\lambda)}$$

Eliminating $v/2\lambda$ from these two equations, which requires still more algebra, we obtain

$$x^2 + y^2 + a^2 - 2ay \coth\frac{u}{2\lambda} = 0$$

or

$$x^2 + \left(y - a\coth\frac{u}{2\lambda}\right)^2 = \frac{a^2}{\sinh^2(u/2\lambda)}$$

Similarly, eliminating $u/2\lambda$ gives

$$\left(x - a\cot\frac{v}{2\lambda}\right)^2 + y^2 = \frac{a^2}{\sin^2(v/2\lambda)}$$

It is now clear that the equipotential surfaces ($u = \text{constant}$) are circles in the xy-plane of radii $a/\sinh(u/2\lambda)$ with centers at $(0, a\coth(u/2\lambda))$. Similarly, the streamlines ($v = \text{constant}$) are also circles of radii $a/\sin(v/2\lambda)$ with centers at $(a\cot(v/2\lambda), 0)$.

Note that equipotential curves with positive potential ($u > 0$) are centered along the positive y-axis, and those with negative potential are centered along the negative y-axis. ●

The formalism of Example 6.2.7 has a very interesting practical application. Let us turn the problem around and ask this question: Given two *arbitrary* circles of radii R_1 and R_2 and potentials u_1 and u_2 separated so that the distance between their centers is D, can we situate two line charges of opposite signs in the xy-plane in such a way that the circles are two of the equipotential curves? If this is possible, then we can find the capacitance between any two cylindrical conductors because we can coincide the conductors (which are equipotential surfaces) with the given circles, find the charge per unit length, λ, and their locations on the y-axis, $\pm a$, in terms of R_1, R_2, and D. The capacitance (per unit length) is then simply

$$c = \frac{\lambda}{|u_2 - u_1|}$$

which will be given in terms of the geometry of the two cylindrical conductors. The following example illustrates this situation. (For further examples, see Smythe 1968.)

Example 6.2.8

Let us find the capacitance per unit length of two cylindrical conductors of radii R_1 and R_2 and with a distance between their centers of D.

We are looking for two line charges with linear charge densities $+\lambda$ and $-\lambda$ such that the two cylinders are two of their equipotential surfaces. From Example 6.2.7 we have

$$R_1 = \frac{a}{\sinh(u_1/2\lambda)} \qquad \text{and} \qquad R_2 = \frac{a}{\sinh(u_2/2\lambda)} \tag{1a}$$

$$y_1 = a\coth(u_1/2\lambda) \qquad \text{and} \qquad y_2 = a\coth(u_2/2\lambda) \tag{1b}$$

where y_1 and y_2 are the locations of the centers of the two conductors on the y-axis (which we assume connects the two centers). We could solve the above equations for λ and a. However, let us directly calculate the capacitance per unit length. We need to find λ in terms of $|u_1 - u_2|$. To do so we consider (see Fig. 6.8)

$$D = |y_1 - y_2| = \left| a\frac{\cosh(u_1/2\lambda)}{\sinh(u_1/2\lambda)} - a\frac{\cosh(u_2/2\lambda)}{\sinh(u_2/2\lambda)} \right|$$
$$= |R_1 \cosh(u_1/2\lambda) - R_2 \cosh(u_2/2\lambda)|$$

and square both sides to get

$$D^2 = R_1^2 \cosh^2(u_1/2\lambda) + R_2^2 \cosh^2(u_2/2\lambda) - 2R_1R_2 \cosh(u_1/2\lambda)\cosh(u_2/2\lambda)$$
$$= R_1^2[1 + \sinh^2(u_1/2\lambda)] + R_2^2[1 + \sinh^2(u_2/2\lambda)] - 2R_1R_2 \cosh(u_1/2\lambda)\cosh(u_2/2\lambda)$$

Now we use $\cosh(a - b) = \cosh a \cosh b - \sinh a \sinh b$ in the last term on the RHS to obtain

$$D^2 = R_1^2 + R_2^2 + R_1^2 \sinh^2\frac{u_1}{2\lambda} + R_2^2 \sinh^2\frac{u_2}{2\lambda} - 2R_1R_2\left[\cosh\left(\frac{u_1}{2\lambda} - \frac{u_2}{2\lambda}\right) + \sinh\frac{u_1}{2\lambda}\sinh\frac{u_2}{2\lambda}\right]$$
$$= R_1^2 + R_2^2 - 2R_1R_2 \cosh\left(\frac{u_1}{2\lambda} - \frac{u_2}{2\lambda}\right) + \left(R_1 \sinh\frac{u_1}{2\lambda} - R_2 \sinh\frac{u_2}{2\lambda}\right)^2 \qquad (2)$$

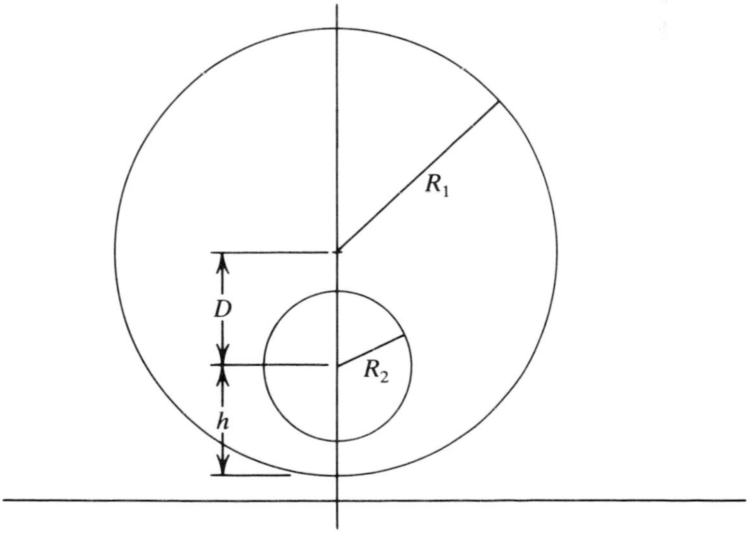

Figure 6.8 When the larger radius R_1 tends to infinity, the geometry approaches that of a cylinder plus a plane.

The last term of (2) is zero by (1a). Thus, we have

$$\cosh\left(\frac{u_1 - u_2}{2\lambda}\right) = \left|\frac{R_1^2 + R_2^2 - D^2}{2R_1 R_2}\right|$$

The absolute value sign on the RHS is necessary to assure positivity of the hyperbolic cosine. The capacitance can now be easily calculated:

$$\left|\frac{u_1 - u_2}{2\lambda}\right| = \left|\cosh^{-1}\left(\left|\frac{R_1^2 + R_2^2 - D^2}{2R_1 R_2}\right|\right)\right|$$

and
$$c = \left|\frac{\lambda}{u_1 - u_2}\right| = \frac{1}{2}\left|\cosh^{-1}\left(\left|\frac{R_1^2 + R_2^2 - D^2}{2R_1 R_2}\right|\right)\right|^{-1}$$

Some special cases are worth mentioning.

(a) For concentric cylinders,

$$D = 0 \quad \Rightarrow \quad c = \frac{1}{2}\left|\ln\left(\frac{R_2}{R_1}\right)\right|^{-1}$$

which can be verified.

(b) For a cylinder and a plane, we let one of the radii, say R_1, go to ∞ while $h \equiv R_1 - D$ remains fixed. Then we immediately get

$$c = \frac{1}{2|\cosh^{-1}(h/R_2)|}$$

(c) For similar cylinders,

$$R_1 = R_2 \quad \Rightarrow \quad c = \frac{1}{2\left|\cosh^{-1}\left(1 - \dfrac{D^2}{2R^2}\right)\right|}$$

which can also be written as

$$c = \frac{1}{4|\cosh^{-1}(D/2R)|}$$

•

6.2.5 Conformal Mappings

The examples in the preceding subsection exhibit the power of applying analytic functions in solving two-dimensional electrostatic problems. However, the real power of analytic functions has a much broader range than those examples indicate. We can get an idea of how useful analytic functions can be by looking at the above methods from a different perspective.

Instead of treating $w(z)$ as a complex potential, let us look at it as a mapping from the z-plane (or xy-plane) to the w-plane (or uv-plane) (see Churchill, Brown, and Verhey 1974). That is, points in the xy-plane are mapped onto points in the uv-plane. In particular, the equipotential curves (circles in the earlier examples) are mapped *onto lines parallel to the v-axis* in the w-plane. This is obvious, since equipotential curves are *defined* by $u =$ constant, which is simply a line parallel to the v-axis. Similarly, the streamlines are mapped onto horizontal lines in the w-plane.

This is an enormous simplification of the geometry. Straight lines, especially when they are parallel to axes, are simpler by far than circles, especially if the circles are not centered around the origin. Thus, if we could somehow transfer the situation to the w-plane, we would have a much simpler problem to solve. Once we solved the problem in the w-plane, we could transfer back to the z-plane and reexpress everything in terms of x and y.

So let us consider two complex "worlds." One is represented by the xy-plane and denoted by z. The other, the "prime world," is represented by z', and its real and imaginary parts by x' and y'. We start in z, where we need to find a physical quantity, such as the electrostatic potential, $\Phi(x, y)$. If the problem is too complicated in the z-world, we transfer it to the z'-world, in which it may be easily soluble; we solve the problem there (in terms of x' and y') and then transfer back to the z-world (x and y). The mapping that relates z and z' *must be cleverly chosen.* Otherwise, there will be no guarantee that the problem will simplify.

Two conditions are necessary for the above strategy to work. First, the differential equation describing the physics must not get more complicated with the transfer to z'. Since in electrostatics Laplace's equation is already of the simplest type, the z'-world must also respect Laplace's equation. Second, and more important, the mapping must preserve the angles between curves. This is necessary, because we want the equipotential curves and the field lines to be perpendicular in both worlds. A mapping that preserves the angle between two curves at a given point is called a *conformal mapping.* We already have such mappings at our disposal, as the following proposition shows.

6.2.5 Proposition Let γ_1 and γ_2 be curves in the complex z-plane that intersect at a point z_0 at an angle α. Let $f: \mathbb{C} \to \mathbb{C}$ be a mapping given by $z' = x' + iy' = f(z)$, which is analytic at z_0. Let γ_1' and γ_2' be the images of γ_1 and γ_2 under this mapping, which intersect at an angle α'.

(i) Then, $\alpha' = \alpha$, that is, the mapping f is *conformal,* if $(dz'/dz)_{z_0} \neq 0$.

(ii) If Φ is harmonic in (x, y), it is also harmonic in (x', y').

Proof. We prove the first part (the second part is a problem involving partial differentiation and the chain rule and is left for the reader). We let

$$z - z_0 = re^{i\theta} \qquad z' - z_0' = r'e^{i\theta'} \qquad \left(\frac{dz'}{dz}\right)_{z_0} = R_0 e^{i\psi_0}$$

Now we consider the ratio

$$\frac{z' - z_0'}{z - z_0} = \frac{r'}{r} e^{i(\theta' - \theta)}$$

and take the limit as $z \to z_0$ on γ_1. The LHS approaches $R_0 e^{i\psi_0}$. The angle on the RHS approaches $\theta'_1 - \theta_1$, as shown in Fig. 6.9. We thus have

$$\psi_0 = \theta'_1 - \theta_1$$

This angle is well-defined because $(dz'/dz)_{z_0}$ is not zero. Therefore, $R_0 \neq 0$ implies that the vector has a nonzero length and makes a well-defined angle with the horizontal axis. Similarly, as $z \to z_0$ on γ_2, we get

$$\psi_0 = \theta'_2 - \theta_2$$

So $\quad \theta'_2 - \theta_2 = \theta'_1 - \theta_1 \quad$ or $\quad \alpha = \theta_2 - \theta_1 = \theta'_2 - \theta'_1 = \alpha'$

Thus, the two angles are equal. ∎

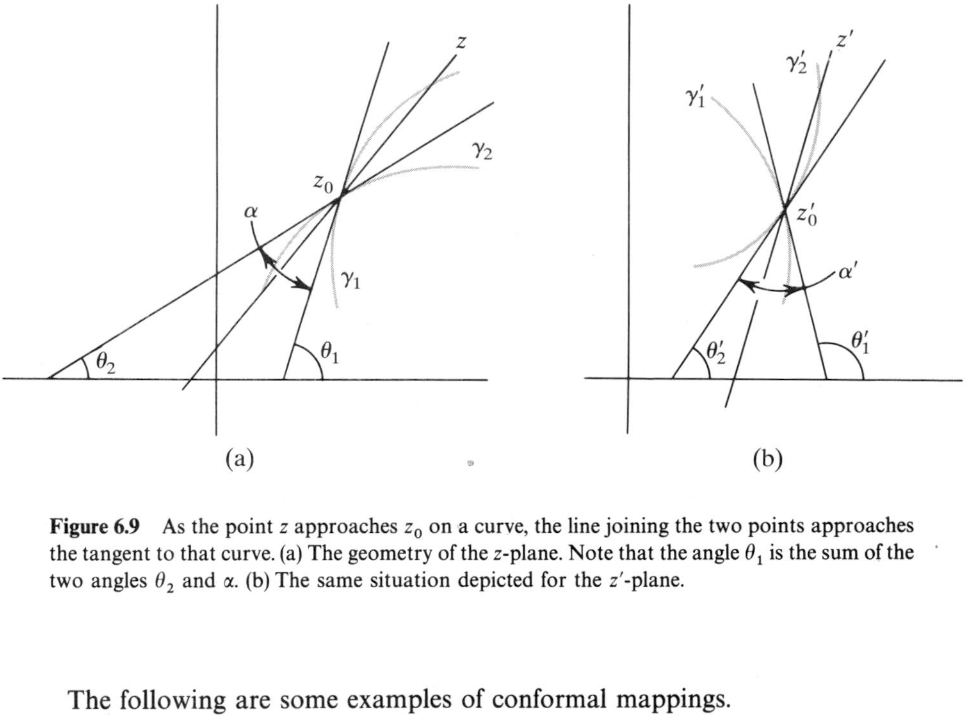

Figure 6.9 As the point z approaches z_0 on a curve, the line joining the two points approaches the tangent to that curve. (a) The geometry of the z-plane. Note that the angle θ_1 is the sum of the two angles θ_2 and α. (b) The same situation depicted for the z'-plane.

The following are some examples of conformal mappings.

(1) $$z' = z + a$$

where a is an arbitrary complex constant. This is simply a *translation* of the z-plane.

(2) $$z' = bz$$

where b is a complex constant. This is a *dilation*, in which distances are dilated by a factor $|b|$. A diagram in the z-plane is mapped onto a *similar* (congruent) diagram in the z'-plane that will be reduced ($|b| < 1$) or enlarged ($|b| > 1$) by a factor of $|b|$.

(3)
$$z' = \frac{1}{z}$$

This is called an *inversion*. Exercise 6.2.9 shows that under such a mapping circles are mapped onto circles or straight lines. A circle in the z-plane with its center at a and having a radius $r \neq |a|$ is mapped onto a circle in the z'-plane with its center at

$$a' = \frac{a^*}{|a|^2 - r^2}$$

and having a radius

$$r' = \frac{r}{||a|^2 - r^2|}$$

If $r = |a|$, then the circle is mapped onto a straight line.

(4) Combining the preceding three examples yields the general mapping

$$z' = \frac{az + b}{cz + d} \tag{6.22}$$

which is conformal if $cz + d \neq 0$ and $dz'/dz \neq 0$. The latter condition yields

$$\frac{dz'}{dz} = \frac{ad - bc}{(cz + d)^2} \neq 0 \quad \Rightarrow \quad ad - bc \neq 0$$

Mappings of the form given in Eq. (6.22) are called *homographic transformations.* A useful property of (6.22) is that it can map an infinite region of the z-plane onto a finite region of the z'-plane. This is clear, since points with very large values of z are mapped onto the neighborhood of the point $z' = a/c$. Of course, this argument goes both ways; (6.22) also maps infinitesimal regions of the z-plane close to $z = -d/c$ onto large regions of the z'-plane. The usefulness of (6.22) is illustrated in the following example.

Example 6.2.9

Consider two cylindrical conductors of equal radius r, held at potentials u_1 and u_2, respectively. Assume that their centers are located on the x-axis at distances a_1 and a_2 from the origin (see Fig. 6.10). Let us find the electrostatic potential produced by such a configuration in the entire xy-plane.

We know from elementary electrostatics that the problem becomes extremely simple if the two cylinders are concentric (and, of course, of different radii). Thus, we try to map the two circles onto two concentric circles in the z'-plane such that the infinite region outside the two circles in the z-plane gets mapped onto the finite annular region between the two concentric circles in the z'-plane. We then (easily) find the potential in the z'-plane and transfer back to the z-plane.

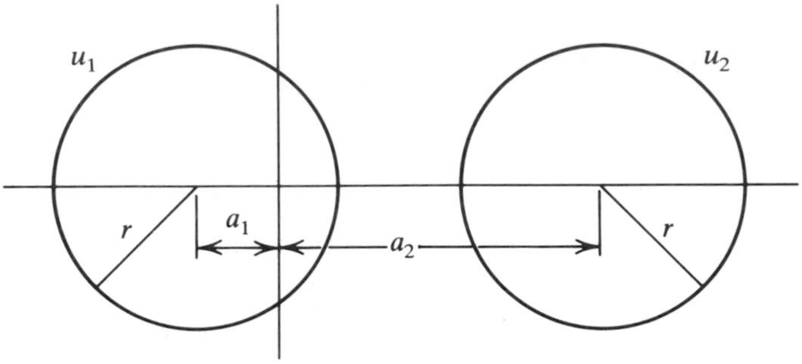

Figure 6.10 The geometry of the problem as viewed from the z-plane.

The most general mapping that may be able to do the job is that given by Eq. (6.22). However, it turns out that we do not have to be this general. In fact, the special case $z' = 1/(z - c)$ in which c is a *real* constant will be sufficient. So let the above equation describe the mapping between z and z'. Then

$$z = \frac{1}{z'} + c$$

and the circles $|z - a_i| = r$, for $i = 1, 2$, will be mapped onto the circles $|z' - a_i'| = r_i'$, for $i = 1, 2$, where (by Exercise 6.2.9)

$$a_i' = \frac{a_i - c}{(a_i - c)^2 - r^2} \qquad \text{and} \qquad r_i' = \frac{r}{|(a_i - c)^2 - r^2|} \qquad \text{for } i = 1, 2 \tag{1}$$

Can we arrange the parameters so that the circles in the z'-plane are concentric, that is, so that $a_1' = a_2'$? The answer is yes. We let $a_i - c = b_i$, set a_1' and a_2' equal, and solve for b_2 in terms of b_1. The result is $b_2 = b_1$ or $b_2 = -r^2/b_1$. The first solution is trivial, but the second gives

$$a_2 - c = -\frac{r^2}{a_1 - c}$$

If we make the simplifying, but generally applicable, assumption that the first cylinder has its center at the origin, then $a_1 = 0$ and

$$a_2 = c + \frac{r^2}{c} \equiv D$$

where D is the distance between the cylinders. We see that c is determined by D. For $a_1 = 0$ we obtain

$$a' \equiv a_1' = a_2' = -\frac{c}{c^2 - r^2}$$

and the geometry of the problem is as shown in Fig. 6.11.

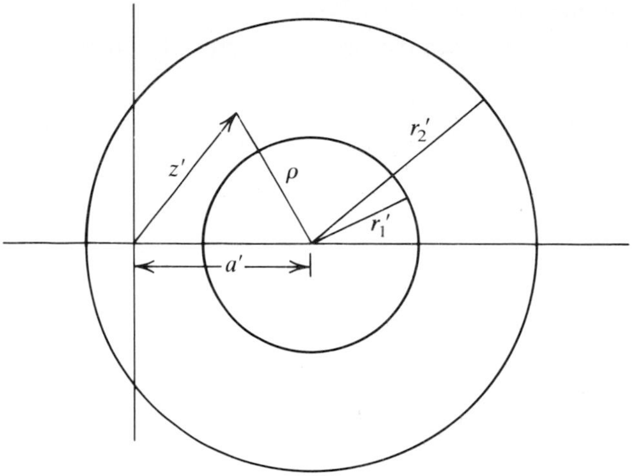

Figure 6.11 The geometry of the problem as viewed from the z'-plane.

For such a geometry the potential at a point in the annular region is given by

$$\Phi' = A \ln \rho + B = A \ln |z' - a'| + B$$

where A and B are real constants determined by the conditions $\Phi'(\rho = r'_1) = u_1$ and $\Phi'(\rho = r'_2) = u_2$, which yield

$$A = \frac{u_1 - u_2}{\ln(r'_1/r'_2)} \qquad \text{and} \qquad B = \frac{u_2 \ln r'_1 - u_1 \ln r'_2}{\ln(r'_1/r'_2)}$$

The potential Φ' is the real part of the complex function

$$F(z') = A \ln(z' - a') + B$$

which is analytic except at $z' = a'$, a point lying outside the region of interest. We can now go back to the z-plane by substituting $z' = 1/(z - c)$ to obtain

$$G(z) = A \ln\left(\frac{1}{z - c} - a'\right) + B$$

whose real part is the potential in the z-plane

$$\begin{aligned}\Phi(x, y) = \operatorname{Re}[G(z)] &= A \ln \left|\frac{1 - a'z + a'c}{z - c}\right| + B \\ &= A \ln \left|\frac{(1 + a'c - a'x) - ia'y}{(x - c) + iy}\right| + B \\ &= \frac{A}{2} \ln \left[\frac{(1 + a'c - a'x)^2 + a'^2 y^2}{(x - c)^2 + y^2}\right] + B\end{aligned}$$

This is the potential we want. •

Exercises

6.2.1 Show that the function $w = 1/z$ maps the straight line $y = \frac{1}{2}$ in the z-plane onto a circle in the w-plane.

6.2.2 Show that Eqs. (6.17) are equivalent to $\partial f/\partial z^* = 0$. What is $\partial f/\partial z$?

6.2.3 Show that, when z is represented by polar coordinates, the derivative of a function $f(z)$ can be written as

$$\frac{df}{dz} = e^{-i\theta}\left(\frac{\partial U}{\partial r} + i\frac{\partial V}{\partial r}\right)$$

where U and V are the real and imaginary parts of $f(z)$ written in polar coordinates. What are the C-R conditions?

6.2.4 Show that $d/dz(\ln z) = 1/z$.

6.2.5 Show that there is a unique function $f(z) = u(x, y) + iv(x, y)$ with the following three properties:
$f(z)$ is single-valued and analytic for all z,
$df(z)/dz = f(z)$, and
$f(z_1 + z_2) = f(z_1)f(z_2)$.
What is $f(z)$?

6.2.6 Show that $\sin z$ and $\cos z$ have only real zeros.

6.2.7 Given that $u = 2\lambda \ln[(x^2 + y^2)^{1/2}]$, show that $v = 2\lambda\theta + \text{constant}$, where u and v are the real and imaginary parts of an analytic function $w(z)$.

6.2.8 Let $w = u + iv$ be the complex potential in which u is the actual electrostatic potential. Show that

$$\text{flux per unit length of the } z\text{-axis} = v(x_2, y_2) - v(x_1, y_1)$$

where (x_2, y_2) and (x_1, y_1) are the end points of the curve representing the intersection of the segment of the cylindrical surface with the xy-plane.

6.2.9 Show that the inversion $z' = 1/z$ maps circles in the z-plane onto circles or lines in the z'-plane.

6.3 INTEGRATION OF COMPLEX FUNCTIONS

The derivative of a complex function is an important concept and, as Section 6.2 demonstrated, provides a powerful tool in physical applications. The concept of integration is even more important. In fact, we will see later in this section that derivatives can be written in terms of integrals. We will study definite integrals of complex functions in detail in this section. The indefinite integral, or primitive, of a complex function is a special case of the definite integral in which the upper limit is a variable.

6.3.1 Definite Integrals of Complex Functions

The definite integral of a complex function is defined in exact analogy to that of a real function. Thus, we have

$$\int_{\alpha_1}^{\alpha_2} f(z)\,dz = \lim_{\substack{N\to\infty \\ \Delta z_i \to 0}} \sum_{i=1}^{N} f(z_i)\Delta z_i$$

where Δz_i is a small segment, situated at z_i, of the curve that connects the complex number α_1 to the complex number α_2 in the z-plane. Since there are infinitely many ways of connecting α_1 and α_2, there is no guarantee that the integral has a unique value. It is possible to obtain different values for the integral for different paths.

We encountered a similar situation in Chapter 1. When we discussed the line integral of a vector field, we saw that some vector fields have integrals that are path-dependent. We noted there that the requirement of path-independence imposes restrictive conditions on a vector field. Let us investigate these ideas in the context of the integral of a complex function.

Let us rewrite the definite integral of a complex function $f(z) = u + iv$ in terms of the real and imaginary parts:

$$\int_{\alpha_1}^{\alpha_2} f(z)\,dz = \int_{\alpha_1}^{\alpha_2} (u + iv)(dx + i\,dy) = \int_{\alpha_1}^{\alpha_2} (u\,dx - v\,dy) + i\int_{\alpha_1}^{\alpha_2} (v\,dx + u\,dy)$$

The two integrals on the RHS look very much like two-dimensional line integrals of vectors. In fact, if we define the two-dimensional vectors

$$\mathbf{A}_1 \equiv (u,\ -v) \qquad \text{and} \qquad \mathbf{A}_2 \equiv (v, u)$$

then those integrals can be expressed as

$$\int_{\alpha_1}^{\alpha_2} f(z)\,dz = \int_{\alpha_1}^{\alpha_2} \mathbf{A}_1 \cdot d\mathbf{l} + i\int_{\alpha_1}^{\alpha_2} \mathbf{A}_2 \cdot d\mathbf{l} \tag{6.23}$$

where $d\mathbf{l} = (dx, dy)$ is an element of displacement along the path from α_1 to α_2. We can now apply the results of Chapter 1 to the two vectors and conclude that (6.23) is path-independent if and only if they have vanishing curls. Since both $\mathbf{A}_1$ and $\mathbf{A}_2$ have only x and y components, which depend only on x and y, we obtain

$$\nabla \times \mathbf{A}_1 = \hat{\mathbf{e}}_z\left(-\frac{\partial v}{\partial x} - \frac{\partial u}{\partial y}\right)$$

$$\nabla \times \mathbf{A}_2 = \hat{\mathbf{e}}_z\left(\frac{\partial u}{\partial x} - \frac{\partial v}{\partial y}\right)$$

both of which will vanish if and only if u and v satisfy the C-R conditions. But this is exactly what is needed for $f(z)$ to be analytic.

Path-independence of an integral is the same as the vanishing of the integral along a closed path. We saw in Chapter 1 that

$$\nabla \times \mathbf{A} = 0 \quad \Rightarrow \quad \oint \mathbf{A} \cdot d\mathbf{l} = 0$$

only if the region of definition of $\mathbf{A}$ is contractable to zero, that is, if $\nabla \times \mathbf{A}$ is indeed zero at *every* point of the region.

The preceding discussion can be encapsulated in an important theorem, called the Cauchy-Goursat theorem. First, however, it is worthwhile to become familiar with some vocabulary used frequently in complex analysis.

(1) A *curve* (see Chapter 4) is a map $\gamma: [a, b] \to \mathbb{C}$, from the real interval $[a, b]$ into the complex plane, given by $z(t) = x(t) + iy(t)$, where $a \leqslant t \leqslant b$.
(2) A *simple arc*, or a Jordan arc, is a curve that does not cross itself [that is, $z(t_1) \neq z(t_2)$ for $t_1 \neq t_2$].
(3) A *simple closed curve*, or a Jordan curve, is a simple arc that crosses at the ends [that is, such that $z(a) = z(b)$].
(4) A *smooth arc* is a curve for which $dz/dt = dx/dt + i\,dy/dt$ exists and is nonzero for $t \in [a, b]$.
(5) A *contour* is an arc consisting of a finite number of smooth arcs joined end to end. When only the initial and final values of $z(t)$ are the same, the contour is said to be a *simple closed contour*.

6.3.1 Theorem (Cauchy-Goursat Theorem) Let $f:\mathbb{C} \to \mathbb{C}$ be analytic on a closed contour C and at all points of the region consisting of the complex plane bounded by C. Then

$$\oint_C f(z)\,dz = 0$$ ■

Example 6.3.1

Let us consider a few examples of definite integrals along different paths.

(a) Let us evaluate the integral

$$I_1 = \int_{\gamma_1} z\,dz$$

where γ_1 is the straight line drawn from the origin to the point (1,2). Along such a line, $y = 2x$ and $z = x + iy = x + 2ix$. In terms of t we have

$$z(t) = t + 2it \qquad \text{where } 0 \leqslant t \leqslant 1$$

Thus,

$$I_1 = \int_{\gamma_1} z\,dz = \int_0^1 (t + 2it)(dt + 2i\,dt)$$

$$= \int_0^1 (-3t\,dt + 4it\,dt)$$

$$= (-3 + 4i)(\tfrac{1}{2}) = -\tfrac{3}{2} + 2i$$

For a different path γ_2, along which $y = 2x^2$, we get

$$z(t) = t + 2it^2 \qquad \text{where } 0 \leqslant t \leqslant 1$$

and
$$I_2 = \int_{\gamma_2} z\,dz = \int_0^1 (t + 2it^2)(dt + 4it\,dt)$$
$$= \int_0^1 (t\,dt - 8t^3\,dt + 6it^2\,dt) = -\tfrac{3}{2} + 2i$$

Therefore, $I_1 = I_2$.

(b) Now let us consider
$$I_3 = \int_{\gamma_1} z^2\,dz$$
with γ_1 as in part (a). Substituting for z in terms of t, we obtain
$$I_3 = \int_0^1 (t + 2it)^2(dt + 2i\,dt) = (1 + 2i)^3 \int_0^1 t^2\,dt = -\frac{11}{3} - \frac{2}{3}i$$

Next we compare I_3 with
$$I'_3 = \int_{\gamma_3} z^2\,dz$$
where γ_3 is as shown in Fig. 6.12. This path can be described by
$$z(t) = \begin{cases} t & \text{for } 0 \leqslant t \leqslant 1 \\ 1 + i(t-1) & \text{for } 1 \leqslant t \leqslant 3 \end{cases}$$

Therefore,
$$I'_3 = \int_0^1 t^2\,dt + \int_1^3 [1 + i(t-1)]^2(i\,dt)$$
$$= \frac{1}{3} + \left(-4 - \frac{2}{3}i\right) = -\frac{11}{3} - \frac{2}{3}i$$

which is identical to I_3.

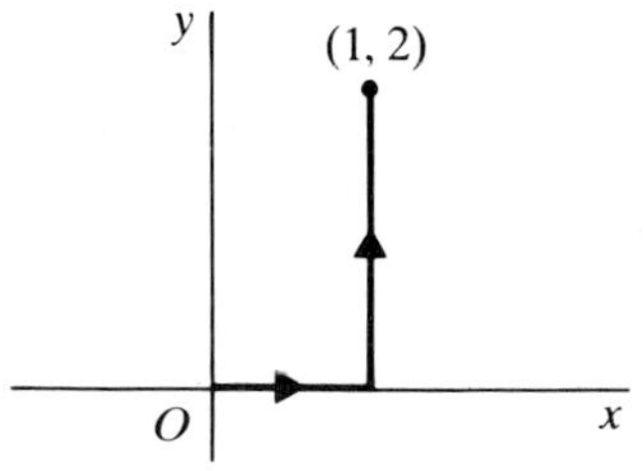

Figure 6.12 The path associated with the integral I'_3.

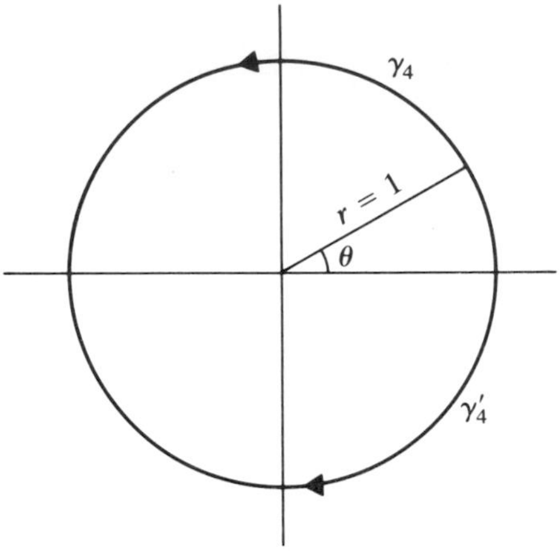

Figure 6.13 The presence of the singular point $z = 0$ within the circle prevents the two integrals along γ_4 and γ'_4 from being equal.

(c) We can show that equality for different paths is not always guaranteed by considering

$$I_4 = \int_{\gamma_4} \frac{1}{z}\, dz$$

where γ_4 is the upper semicircle of unit radius, as shown in Fig. 6.13. A parametric equation for γ_4 can be given in terms of θ:

$$z(\theta) = \cos\theta + i\sin\theta = e^{i\theta} \quad \Rightarrow \quad dz = ie^{i\theta}\, d\theta$$

Thus, we obtain

$$I_4 = \int_0^{\pi} \frac{1}{e^{i\theta}}\, ie^{i\theta}\, d\theta = i\pi$$

On the other hand,

$$I'_4 = \int_{\gamma'_4} \frac{1}{z}\, dz = -\int_{\pi}^{2\pi} \frac{1}{e^{i\theta}}\, ie^{i\theta}\, d\theta = -i\pi$$

Here the two integrals are not equal. From γ_4 and γ'_4 we can construct a counterclockwise contour C, along which the integral of $f(z) = 1/z$ becomes

$$I \equiv \oint_C \frac{1}{z}\, dz = I_4 - I'_4 = 2i\pi$$

That I is not zero is a reflection of the fact that $1/z$ is *not* analytic at all points in the region bounded by the closed contour C. (The Cauchy-Goursat theorem generalizes this fact.) ●

Theorem 6.3.1 applies to more complicated regions than it may appear to suggest. Even if a region contains points at which $f(z)$ is not analytic, those points can be avoided by redefining the region and the contour. Such a procedure requires a convention regarding the direction of "motion" along the contour. This convention is important enough to be stated separately.

6.3.2 Convention When integrating along a closed contour, we agree to move along the contour in such a way that the region enclosed by the contour lies to our left. An integration that follows this convention is called integration *in the positive sense.* Any integration performed otherwise has a minus sign.

For a simple closed contour movement in the counterclockwise direction yields integration in the positive sense. However, as the contour becomes more complicated, this conclusion breaks down. Figure 6.14 shows a complicated path enclosing a region (shaded) in which the integrand is analytic. Note that it is possible to traverse a portion of the region twice in opposite directions without affecting the integral, which may be a sum of integrals for different pieces of the contour. Also note that the "eyes" and "mouth" are traversed clockwise! This is necessary to keep the enclosed region on the left as we move along the contour. A region such as that shown in Fig. 6.14, in which holes are "punched out," is called a *multiply connected region.* In contrast, a *simply connected region* is one in which every simple closed contour encloses only points of the region.

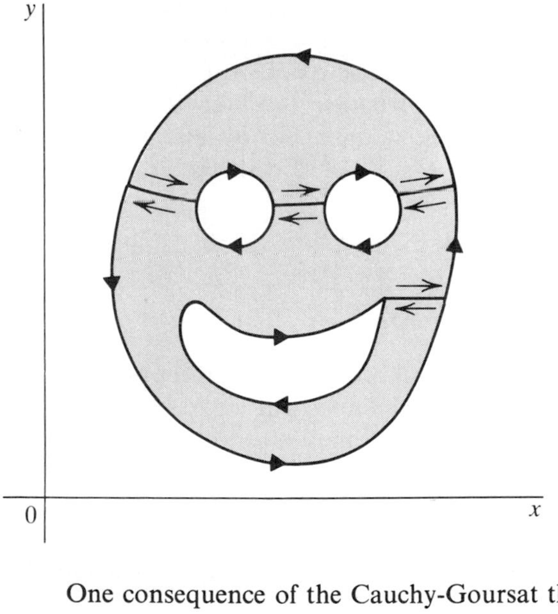

Figure 6.14 A complicated contour can always be broken up into simpler contours. Note that the "eyes" and "mouth" are traversed in the (negative) clockwise direction.

One consequence of the Cauchy-Goursat theorem is an important formula.

6.3.3 Theorem (Cauchy Integral Formula, or CIF) Let f be analytic on and within a simple closed contour C integrated in the positive sense. Let z_0 be any point interior to C. Then

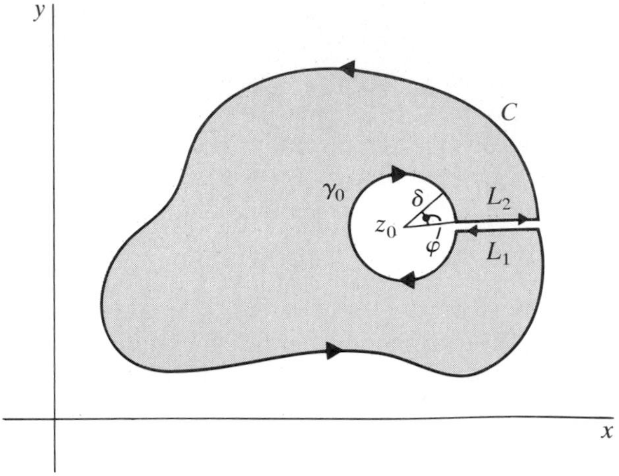

Figure 6.15 The integrand is analytic within and on the boundary of the shaded region. It is always possible to construct contours that exclude all singular points.

$$f(z_0) = \frac{1}{2\pi i} \oint_C \frac{f(z)\,dz}{z - z_0} \tag{6.24}$$

Proof. Consider the shaded region in Fig. 6.15, which is bounded by C, the contour of (6.24); by γ_0, a small circle of infinitesimal radius δ centered at z_0; and by L_1 and L_2, two straight lines infinitesimally close to one another (we can, in fact, assume that L_1 and L_2 are right on top of one another; however, they are separated in the figure for clarity). Let us use C' to denote the union of all these curves.

Since $f(z)/(z - z_0)$ is analytic everywhere on the contour C' and inside the shaded region, we can write

$$0 = \frac{1}{2\pi i} \oint_{C'} \frac{f(z)\,dz}{z - z_0} = \frac{1}{2\pi i}\left[\oint_C \frac{f(z)}{z - z_0}\,dz + \oint_{\gamma_0} \frac{f(z)}{z - z_0}\,dz + \int_{L_1} \frac{f(z)}{z - z_0}\,dz + \int_{L_2} \frac{f(z)}{z - z_0}\,dz \right]$$

The contributions from L_1 and L_2 cancel because they are in opposite directions. Let us evaluate the contribution from the infinitesimal circle, γ_0. First, we note that because $f(z)$ is continuous (differentiability implies continuity) we can write

$$\left| \frac{f(z) - f(z_0)}{z - z_0} \right| = \frac{|f(z) - f(z_0)|}{|z - z_0|} = \frac{|f(z) - f(z_0)|}{\delta} < \frac{\varepsilon}{\delta}$$

for $z \in \gamma_0$, where ε is a small positive number. The Darboux inequality (see Exercise 6.3.2) gives

$$\left| \oint_{\gamma_0} \frac{f(z) - f(z_0)}{z - z_0}\,dz \right| < \frac{\varepsilon}{\delta}\, 2\pi\delta = 2\pi\varepsilon$$

which is translated into

$$\oint_{\gamma_0} \frac{f(z) - f(z_0)}{z - z_0} dz = 0$$

or

$$\oint_{\gamma_0} \frac{f(z)}{z - z_0} dz = \oint_{\gamma_0} \frac{f(z_0)}{z - z_0} dz = f(z_0) \oint_{\gamma_0} \frac{dz}{z - z_0}$$

We can easily calculate the integral on the RHS by noting that $z - z_0 = \delta e^{i\varphi}$ and that γ_0 has a clockwise direction. We obtain

$$\oint_{\gamma_0} \frac{dz}{z - z_0} = -\int_0^{2\pi} \frac{d(\delta e^{i\varphi})}{\delta e^{i\varphi}} = -\int_0^{2\pi} \frac{i\delta e^{i\varphi}\, d\varphi}{\delta e^{i\varphi}} = -2\pi i$$

Putting everything together, we get

$$0 = \frac{1}{2\pi i}\left[\oint_C \frac{f(z)}{z - z_0} dz - 2\pi i f(z_0)\right]$$

and the theorem is established. ■

Example 6.3.2

We can use the CIF to evaluate the integrals

$$I_1 = \oint_{C_1} \frac{z^2\, dz}{(z^2 + 3)^2 (z - i)} \qquad I_2 = \oint_{C_2} \frac{(z^2 - 1)\, dz}{(z - \frac{1}{2})(z^2 - 4)^3} \qquad I_3 = \oint_{C_3} \frac{e^{z/2}\, dz}{(z - i\pi)(z^2 - 20)^4}$$

where C_i is a circle centered at the origin and having a radius r_i such that $r_1 = \frac{3}{2}$, $r_2 = 1$, and $r_3 = 4$.

For I_1 we note that $f_1(z) \equiv z^2/(z^2 + 3)^2$ is analytic within and on C_1. Thus,

$$I_1 = \oint_{C_1} \frac{f_1(z)}{z - i} dz = 2\pi i f_1(i) = 2\pi i \frac{i^2}{(-1 + 3)^2} = -i\frac{\pi}{2}$$

Similarly, $f_2(z) \equiv (z^2 - 1)/(z^2 - 4)^3$ is analytic on and within C_2. Thus, Theorem 6.3.3 gives

$$I_2 = \oint_{C_2} \frac{f_2(z)}{z - \frac{1}{2}} dz = 2\pi i f_2(\tfrac{1}{2}) = \frac{32\pi}{1125} i$$

For the last integral $f_3(z) \equiv e^{z/2}/(z^2 - 20)^4$, and

$$I_3 = \oint_{C_3} \frac{f_3(z)}{z - i\pi} dz = 2\pi i f_3(i\pi) = -\frac{2\pi}{(\pi^2 + 20)^4}$$

●

The Cauchy integral formula gives the value of an analytic function at every point inside a simple closed contour when it is given the value of the function only at points on the contour. It seems as though analytic functions are imprisoned within a contour: They are not free to change inside a region once their value is fixed on the contour enclosing that region.

There is an analogous situation in electrostatics—the determination of the potential at the boundaries, typically conductors, automatically determines it at any other point in the region of space bounded by the conductors. This is the content of the uniqueness theorem used in electrostatic boundary-value problems. However, the electrostatic potential Φ is bound by another condition, Laplace's equation, $\nabla^2\Phi = 0$; and the *combination* of Laplace's equation and the boundary conditions furnishes the uniqueness of Φ.

It seems, on the other hand, as though the mere specification of an analytic function on a contour, without any other condition, is sufficient to determine the function's value at all points enclosed within that contour. This is not the case, however. An analytic function, by its very definition, satisfies another restrictive condition: Its real and imaginary parts separately satisfy Laplace's equation in two dimensions! As discussed earlier, these parts are harmonic functions. Thus, it should come as no surprise that the value of an analytic function at a boundary (contour) determines the function at all points inside the boundary.

6.3.2 Derivatives of Analytic Functions

Theorem 6.3.3 is a very powerful tool for working with analytic functions. One of the applications of this theorem is in evaluating the derivatives of all orders of such functions.

It is convenient to change the dummy integration variable of (6.24) to ξ and write the CIF as

$$f(z) = \frac{1}{2\pi i} \oint_C \frac{f(\xi)\, d\xi}{\xi - z} \tag{6.25}$$

where C is a simple closed contour in the ξ-plane and z is a point within C.

As preparation for defining the derivative of an analytic function, we need the following proposition.

6.3.4 Proposition Let C be a contour (not necessarily closed), and $g(z)$ a continuous function on C. Then the function $f(z)$ defined as

$$f(z) \equiv \frac{1}{2\pi i} \int_C \frac{g(\xi)}{\xi - z}\, d\xi$$

is analytic at any point $z \notin C$.

Proof. We evaluate the derivative of $f(z)$ using the definition of a derivative:

$$\frac{df}{dz} = \lim_{\Delta z \to 0} \frac{f(z + \Delta z) - f(z)}{\Delta z} = \frac{1}{2\pi i} \lim_{\Delta z \to 0} \frac{1}{\Delta z}\left[\int_C \frac{g(\xi)\,d\xi}{\xi - z - \Delta z} - \int_C \frac{g(\xi)\,d\xi}{\xi - z}\right]$$

$$= \frac{1}{2\pi i} \lim_{\Delta z \to 0} \frac{1}{\Delta z} \int_C g(\xi)\left(\frac{1}{\xi - z - \Delta z} - \frac{1}{\xi - z}\right) d\xi$$

$$= \frac{1}{2\pi i} \lim_{\Delta z \to 0} \frac{1}{\Delta z} \int_C g(\xi) \frac{\Delta z}{(\xi - z - \Delta z)(\xi - z)}\, d\xi$$

$$= \frac{1}{2\pi i} \lim_{\Delta z \to 0} \int_C \frac{g(\xi)\,d\xi}{(\xi - z - \Delta z)(\xi - z)}$$

$$= \frac{1}{2\pi i} \int_C \frac{g(\xi)\,d\xi}{(\xi - z)^2}$$

This is defined for all values of z not on C. Thus, $f(z)$ is analytic for $z \notin C$. ■

We can generalize to the nth derivative, obtaining

$$f^{(n)}(z) \equiv \frac{d^n f}{dz^n} = \frac{n!}{2\pi i} \int_C \frac{g(\xi)\,d\xi}{(\xi - z)^{n+1}}$$

We can apply this result to an analytic function expressed by Eq. (6.25), as the following theorem states.

6.3.5 Theorem The derivatives of all orders of an analytic function $f(z)$ exist in the domain of analyticity of the function and are themselves analytic in that domain. The nth derivative of $f(z)$ is given by

$$f^{(n)}(z) \equiv \frac{d^n f}{dz^n} = \frac{n!}{2\pi i} \oint_C \frac{f(\xi)\,d\xi}{(\xi - z)^{n+1}} \tag{6.26}$$

■

Example 6.3.3

Let us apply Eq. (6.26) directly to some simple functions to obtain their derivatives. In all cases we will assume that the contour is a circle of radius r centered at z.

(a) Given $f(z) = K \equiv \text{constant}$, for $n = 1$ we have

$$\frac{df}{dz} = \frac{1}{2\pi i} \oint_C \frac{K}{(\xi - z)^2} d\xi$$

Since ξ is always on the circle C, which is centered at z,

$$\xi - z = re^{i\theta} \qquad \text{and} \qquad d\xi = rie^{i\theta}\, d\theta$$

So we have

$$\frac{df}{dz} = \frac{1}{2\pi i} \int_0^{2\pi} \frac{K}{(re^{i\theta})^2} ire^{i\theta}\, d\theta = \frac{K}{2\pi r} \int_0^{2\pi} e^{-i\theta}\, d\theta = 0$$

That is, the derivative of a constant is zero.

(b) Given $f(z) = z$, for $n = 1$ we have

$$\frac{df}{dz} = \frac{1}{2\pi i} \oint_C \frac{\xi}{(\xi - z)^2} d\xi = \frac{1}{2\pi i} \int_0^{2\pi} \frac{z + re^{i\theta}}{(re^{i\theta})^2} ire^{i\theta}\, d\theta$$

$$= \frac{1}{2\pi}\left(\frac{z}{r} \int_0^{2\pi} e^{-i\theta}\, d\theta + \int_0^{2\pi} d\theta\right) = \frac{1}{2\pi}(0 + 2\pi) = 1$$

(c) Given $f(z) = z^2$, for the first derivative Eq. (6.26) yields

$$\frac{df}{dz} = \frac{1}{2\pi i} \oint_C \frac{\xi^2}{(\xi - z)^2} d\xi = \frac{1}{2\pi i} \int_0^{2\pi} \frac{(z + re^{i\theta})^2}{(re^{i\theta})^2} ire^{i\theta}\, d\theta$$

$$= \frac{1}{2\pi} \int_0^{2\pi} [z^2 + (re^{i\theta})^2 + 2zre^{i\theta}](re^{i\theta})^{-1}\, d\theta$$

$$= \frac{1}{2\pi}\left(\frac{z^2}{r} \int_0^{2\pi} e^{-i\theta}\, d\theta + r \int_0^{2\pi} e^{i\theta}\, d\theta + 2z \int_0^{2\pi} d\theta\right)$$

$$= \frac{1}{2\pi}[0 + 0 + (2z)(2\pi)] = 2z$$

It can be shown that, in general, $(d/dz)z^m = mz^{m-1}$. The proof is left as a problem. ●

Some useful consequences of the CIF. We will make good use of Theorem 6.3.5 later. However, there are a couple of interesting properties of analytic functions that are worth discussing at this point. These are presented in the following propositions.

6.3.6 Proposition The absolute value of an analytic function $f(z)$ cannot have a local maximum within the region of analyticity of the function.

Proof. Let R be the region of analyticity of $f(z)$. Let $z_0 \in R$, and let γ_0 be a small circle of radius δ centered at z_0. Then the CIF gives

$$f(z_0) = \frac{1}{2\pi i} \oint_{\gamma_0} \frac{f(z)}{z_0 - z}\, dz$$

from which we obtain

$$|f(z_0)| = \frac{1}{2\pi} \left| \oint_{\gamma_0} \frac{f(z)}{z_0 - z}\, dz \right|$$

The Darboux inequality (see Exercise 6.3.2) yields

$$|f(z_0)| \leqslant \frac{1}{2\pi} \left(\frac{M}{|z_0 - z|} \right) 2\pi\delta = \frac{1}{2\pi} \left(\frac{M}{\delta} \right) 2\pi\delta = M$$

where M is the maximum value of $|f(z)|$ for $z \in \gamma_0$. This inequality says that there exists at least one point z on the circle γ_0 (the point at which the maximum of $|f(z)|$ is attained) such that $|f(z_0)| \leqslant |f(z)|$. However, γ_0 is chosen to be arbitrarily small. Thus, for *any* point z_0 there will be a point z *arbitrarily close to* z_0 such that $|f(z_0)| \leqslant |f(z)|$. Therefore, there can be no local maximum of $|f(z)|$ anywhere within R. ■

6.3.7 Proposition A bounded entire function is necessarily a constant.

Proof. We show that the derivative of such a function is zero. We consider

$$\frac{df}{dz} = \frac{1}{2\pi i} \oint_C \frac{f(\xi)}{(\xi - z)^2}\, d\xi$$

Since f is an entire function, the closed contour C can be chosen to be a very large circle of radius R with center at z. Taking the absolute value of both sides yields

$$\left|\frac{df}{dz}\right| = \frac{1}{2\pi} \left| \oint_C \frac{f(\xi)}{(\xi - z)^2}\, d\xi \right| \leqslant \frac{1}{2\pi} \max \left| \frac{f(\xi)}{(\xi - z)^2} \right| 2\pi R \leqslant \frac{1}{2\pi} \left(\frac{M}{R^2} \right) 2\pi R = \frac{M}{R}$$

↑ by the Darboux inequality

↑ because f is bounded and $|\xi - z| = R$

It is now clear that as $R \to \infty$, the derivative goes to zero. And $df/dz = 0$ implies that f is constant. ■

Proposition 6.3.7 is a very powerful statement about analytic functions. There are many interesting and nontrivial real functions that are bounded and have derivatives of all orders on the entire real line. For instance, e^{-x^2} and $1/(x^2+1)$ are such functions. In complex analysis, on the other hand, no such freedom exists for analytic functions. Any nontrivial analytic function is either not bounded (it goes to infinity somewhere on the complex plane) or not entire (it is not analytic at at least one point of the complex plane).

Example 6.3.4

A consequence of Proposition 6.3.7 is the fundamental theorem of algebra, which states that any polynomial of degree $n \geqslant 1$ has n roots (some of which may be repeated). In other words, the polynomial

$$p(x) = a_0 + a_1 x + \cdots + a_{n-1}x^{n-1} + a_n x^n \qquad \text{for } n \geqslant 1$$

can be factored completely as

$$p(x) = c(x - z_1)(x - z_2)\cdots(x - z_n)$$

where c is a constant and the z_i are, in general, complex numbers.

To see how Proposition 6.3.7 implies the fundamental theorem of algebra, we let[1]

$$f(z) = \frac{1}{p(z)}$$

and assume that $p(z)$ is never zero for any (finite) $z \in \mathbb{C}$. Then $f(z)$ is bounded and analytic for all $z \in \mathbb{C}$, and Proposition 6.3.7 says that $f(z)$ is a constant. This is a contradiction. Thus, there must be at least one z, say $z = z_1$, for which $p(z)$ is zero. So we can factor out $(z - z_1)$ from $p(z)$ and write

$$p(z) = (z - z_1)q(z)$$

where $q(z)$ is of degree $n - 1$. Applying the above argument to $q(z)$, we have

$$p(z) = (z - z_1)(z - z_2)r(z)$$

where $r(z)$ is of degree $n - 2$. Continuing in this way, we can factor $p(z)$ completely into linear factors. The last polynomial will be a constant (a zero-degree polynomial), which we denote as c. ●

Primitive of an analytic function. The primitive (indefinite integral) of an analytic function can be defined using definite integrals. Let $f: \mathbb{C} \to \mathbb{C}$ be analytic in a region R of the complex plane. Let z_0 and z be two points in R. Then the integral $\int_{z_0}^{z} f(\xi)\,d\xi$ exists and is path-independent. Thus, we can define a function $F(z)$ by

$$F(z) \equiv \int_{z_0}^{z} f(\xi)\,d\xi$$

[1] An algebraic proof of the fundamental theorem of algebra also exists, but it is extremely difficult and is usually presented in advanced books on algebra.

We can show that $F(z)$ is the primitive of $f(z)$. We consider

$$\frac{F(z+\Delta z)-F(z)}{\Delta z}=\frac{1}{\Delta z}\left[\int_{z_0}^{z+\Delta z} f(\xi)\,d\xi-\int_{z_0}^{z} f(\xi)\,d\xi\right]$$

$$=\frac{1}{\Delta z}\left[\int_{z_0}^{z+\Delta z} f(\xi)\,d\xi+\int_{z}^{z_0} f(\xi)\,d\xi\right]$$

$$=\frac{1}{\Delta z}\int_{z}^{z+\Delta z} f(\xi)\,d\xi$$

and form the difference

$$\frac{F(z+\Delta z)-F(z)}{\Delta z}-f(z)=\frac{1}{\Delta z}\int_{z}^{z+\Delta z} f(\xi)\,d\xi-f(z)$$

$$=\frac{1}{\Delta z}\left[\int_{z}^{z+\Delta z} f(\xi)\,d\xi-f(z)\Delta z\right]$$

$$=\frac{1}{\Delta z}\left[\int_{z}^{z+\Delta z} f(\xi)\,d\xi-f(z)\int_{z}^{z+\Delta z} d\xi\right]$$

$$=\frac{1}{\Delta z}\int_{z}^{z+\Delta z}[f(\xi)-f(z)]\,d\xi$$

We now take the absolute value of both sides, use the Darboux inequality, and let $\Delta z \to 0$ to obtain

$$\lim_{\Delta z\to 0}\left|\frac{F(z+\Delta z)-F(z)}{\Delta z}-f(z)\right|=\lim_{\Delta z\to 0}\left|\frac{1}{\Delta z}\int_{z}^{z+\Delta z}[f(\xi)-f(z)]\,d\xi\right|$$

$$\leqslant \lim_{\Delta z\to 0}\frac{1}{|\Delta z|}\max|f(\xi)-f(z)||\Delta z|=0$$

The last equality on the RHS follows from the continuity of f. We have just established the following proposition.

6.3.8 Proposition Let $f\colon \mathbb{C}\to\mathbb{C}$ be analytic in a region R of $\mathbb{C}$. Then at every point $z\in R$ there exists an analytic function $F\colon \mathbb{C}\to\mathbb{C}$ such that

$$\frac{dF}{dz}=f(z)$$

■

Note that in the proof of Proposition 6.3.8 we used only the continuity of f and the fact that the integral was well-defined. These two conditions are sufficient to establish the analyticity of f. The following theorem, due to Morera, states this fact and is the converse of the Cauchy-Goursat theorem, which establishes that analytic functions have vanishing integrals around any closed contour.

6.3.9 Theorem (Morera's Theorem) Let a function $f: \mathbb{C} \to \mathbb{C}$ be continuous in a simply connected region R. If, for each simple closed contour C in R,

$$\oint_C f(\xi)\, d\xi = 0$$

then f is analytic throughout R. ■

Exercises

6.3.1 Evaluate the integral

$$I = \int_\gamma \frac{dz}{z - 1 - i}$$

where γ is (**a**) the line joining $z_1 = 2i$ and $z_2 = 3$, and (**b**) the path from z_1 to $z = 0$ and then from $z = 0$ to z_2.

6.3.2 Let $f(z)$ be a bounded function (not necessarily analytic) along a contour C; that is, there exists $M \in \mathbb{R}^+$ such that

$$|f(z)| \leqslant M \qquad \forall\, z \in \mathbb{C}$$

Show that

$$\left| \int_C f(z)\, dz \right| \leqslant ML$$

where L is the length of the contour C. This is called the *Darboux inequality*.

6.3.3 Let f be analytic within and on the circle γ_0 given by $|z - z_0| = r_0$ and integrated in the positive sense. Show that *Cauchy's inequality* holds:

$$|f^{(n)}(z_0)| \leqslant \frac{n!\,M}{r_0^n}$$

where M is the maximum value of $|f(z)|$ on γ_0.

PROBLEMS

6.1 Express each of the following complex numbers in Cartesian form.

(**a**) $\dfrac{2}{1 - 3i}$ (**b**) $\dfrac{1 - i}{1 + i}$ (**c**) $(1 + i\sqrt{3})^3$

(**d**) $\dfrac{5}{(1 - i)(2 - i)(3 - i)}$ (**e**) $(1 - i)^4$ (**f**) $\dfrac{1 + 2i}{3 - 4i} + \dfrac{2 - i}{5i}$

6.2 Find the modulus (magnitude, length, absolute value) and argument of each of the following complex numbers.

(a) $3i$ **(b)** -2 **(c)** $2-5i$ **(d)** $1+i$ **(e)** $1-i$ **(f)** $5+2i$

6.3 Evaluate the following roots and plot them.

(a) $\sqrt[5]{1+i}$ **(b)** $\sqrt[4]{-1}$ **(c)** $\sqrt[8]{1}$ **(d)** $\sqrt[5]{-32}$

(e) $\sqrt{4+5i}$ **(f)** $\sqrt[3]{-1}$ **(g)** $\sqrt[4]{-16i}$ **(h)** $\sqrt[6]{-1}$

6.4 Prove that $\sqrt{2}|z| \geqslant |\mathrm{Re}(z)| + |\mathrm{Im}(z)|$.

6.5 Prove that z is real iff $z = z^*$.

6.6 Sketch the set of points determined by each of the following conditions.

(a) $|z-2+i| = 2$ **(b)** $|z+2i| \leqslant 4$ **(c)** $|z+i| = |z-i|$

(d) $\mathrm{Im}(\bar{z}+i) = 2$ **(e)** $2z + 3z^* = 1$ **(f)** $z^2 + (z^*)^2 = 2$

6.7 Show that the equation of a circle of radius r centered at z_0 can be written as

$$|z|^2 - 2\mathrm{Re}(zz_0^*) = r^2 - |z_0|^2$$

6.8 Show that

(a) $(-1+i)^7 = -8(1+i)$ **(b)** $(1+i\sqrt{3})^{-10} = 2^{-11}(-1+i\sqrt{3})$

6.9 Find the four roots of $z^4+4=0$ and use them to factor z^4+4 into quadratic factors with real coefficients.

6.10 Given that $z_1 z_2 \neq 0$, show that $\mathrm{Re}(z_1 z_2^*) = |z_1||z_2|$ iff $\arg(z_1) - \arg(z_2) = 2n\pi$, for $n = 0, \pm 1, \pm 2, \ldots$.

6.11 Given that $z_1 z_2 \neq 0$, show that $|z_1 + z_2| = |z_1| + |z_2|$ iff $\arg(z_1) - \arg(z_2) = 2n\pi$, for $n = 0, \pm 1, \pm 2, \ldots$. What does this mean geometrically?

6.12 Assume that $z \neq 1$ and $z^n = 1$. Show that $1 + z + z^2 + \cdots + z^{n-1} = 0$.

6.13 Represent z as (x, y) and find the equations for x and y that result from $z^2 + z + 1 = 0$. Solve those equations to find the roots of the given equation in z. Compare those roots with the ones obtained by directly solving the equation in z.

6.14 Given that

$$z = \begin{pmatrix} x & -y \\ y & x \end{pmatrix}$$

find z^* and show that $zz^* = (\det z) = 1_2$.

6.15 What is the value of i^i?

6.16 Prove the following identities for differentiation by using the definition of derivative (or otherwise) and assuming that the functions are differentiable in a certain region of $\mathbb{C}$.

(a) $\dfrac{d}{dz}(f+g) = \dfrac{df}{dz} + \dfrac{dg}{dz}$ **(b)** $\dfrac{d}{dz}(fg) = \dfrac{df}{dz}g + f\dfrac{dg}{dz}$

(c) $\dfrac{d}{dz}\left(\dfrac{f}{g}\right) = \dfrac{f'(z)g(z) - g'(z)f(z)}{[g(z)]^2}$ where $g(z) \neq 0$

6.17 Show that (**a**) the sum and the product of two entire functions are entire and (**b**) the ratio of two entire functions is analytic everywhere except at the zero(s) of the denominator.

6.18 Establish the following identities.

(**a**) $\text{Re}(\sin z) = \sin x \cosh y$, $\text{Im}(\sin z) = \cos x \sinh y$

(**b**) $\text{Re}(\cos z) = \cos x \cosh y$, $\text{Im}(\cos z) = -\sin x \sinh y$

(**c**) $\text{Re}(\sinh z) = \sinh x \cos y$, $\text{Im}(\sinh z) = \cosh x \sin y$

(**d**) $\text{Re}(\cosh z) = \cosh x \cos y$, $\text{Im}(\cosh z) = \sinh x \sin y$

(**e**) $|\sin z|^2 = \sin^2 x + \sinh^2 y$, $|\cos z|^2 = \cos^2 x + \sinh^2 y$

$|\sinh z|^2 = \sinh^2 x + \sin^2 y$, $|\cosh z|^2 = \sinh^2 x + \cos^2 y$

6.19 Show that there is no $z \in \mathbb{C}$ at which $e^z = 0$.

6.20 Find all the zeros of $\sinh z$ and $\cosh z$.

6.21 Verify the following trigonometric identities.

(**a**) $\cos^2 z + \sin^2 z = 1$

(**b**) $\cos(z_1 + z_2) = \cos z_1 \cos z_2 - \sin z_1 \sin z_2$

(**c**) $\sin(z_1 + z_2) = \sin z_1 \cos z_2 + \cos z_1 \sin z_2$

(**d**) $\sin\left(\dfrac{\pi}{2} - z\right) = \cos z$, $\cos\left(\dfrac{\pi}{2} - z\right) = \sin z$

(**e**) $\cos 2z = \cos^2 z - \sin^2 z$, $\sin^2 z = 2 \sin z \cos z$

(**f**) $\tan(z_1 + z_2) = \dfrac{\sin(z_1 + z_2)}{\cos(z_1 + z_2)} = \dfrac{\tan z_1 + \tan z_2}{1 - \tan z_1 \tan z_2}$

(**g**) $\cos^2 z = \frac{1}{2}(1 + \cos 2z)$, $\sin^2 z = \frac{1}{2}(1 - \cos 2z)$

(**h**) $\tan 2z = \dfrac{2 \tan z}{1 - \tan^2 z}$

6.22 Verify the following hyperbolic identities.

(**a**) $\cosh^2 z - \sinh^2 z = 1$

(**b**) $\cosh(z_1 + z_2) = \cosh z_1 \cosh z_2 + \sinh z_1 \sinh z_2$

(**c**) $\sinh(z_1 + z_2) = \sinh z_1 \cosh z_2 + \cosh z_1 \sinh z_2$

(**d**) $\cosh 2z = \cosh^2 z + \sinh^2 z$, $\sinh 2z = 2 \sinh z \cosh z$

(**e**) $\tanh(z_1 + z_2) = \dfrac{\sinh(z_1 + z_2)}{\cosh(z_1 + z_2)} = \dfrac{\tanh z_1 + \tanh z_2}{1 + \tanh z_1 \tanh z_2}$

(**f**) $\cosh^2 z = \frac{1}{2}(\cosh 2z + 1)$, $\sinh^2 z = \frac{1}{2}(\cosh 2z - 1)$

(**g**) $\tanh 2z = \dfrac{2 \tanh z}{1 + \tanh^2 z}$

6.23 Show that

(a) $\tanh\left(\frac{z}{2}\right) = \frac{\sinh x + i\sin y}{\cosh x + \cos y}$ (b) $\coth\left(\frac{z}{2}\right) = \frac{\sinh x - i\sin y}{\cosh x - \cos y}$

6.24 Where is the function $f(z) = 3z^2 - 2ze^{z^2} + e^{-z}$ differentiable?

6.25 Find all values of z such that

(a) $e^z = -3$ (b) $e^z = 1 + \sqrt{3}i$ (c) $e^{2z-1} = 1$

6.26 Show that $|e^{-3z}| < 1$ iff $\mathrm{Re}(z) > 0$.

6.27 Prove that $\exp(z^*)$ is not analytic anywhere.

6.28 Show that $e^{iz} = \cos z + i\sin z$ for any z.

6.29 Show that both the real and the imaginary parts of an analytic function are harmonic.

6.30 Show that each of the following functions, $u(x, y)$ is harmonic, and find the function's harmonic conjugate, $v(x,y)$, such that $u(x, y) + iv(x, y)$ is analytic.

(a) $x^3 - 3xy^2$ (b) $e^x \cos y$ (c) $\frac{x}{x^2 + y^2}$ where $x^2 + y^2 \neq 0$

(d) $e^{-2y}\cos 2x$ (e) $e^{y^2 - x^2}\cos 2xy$

(f) $e^x(x\cos y - y\sin y) + 2\sinh y \sin x + x^3 - 3xy^2 + y$

6.31 Find the following sums, where α and β are real.

(a) $\cos\alpha + \cos(\alpha + \beta) + \cos(\alpha + 2\beta) + \cdots + \cos(\alpha + n\beta)$

(b) $\sin\alpha + \sin(\alpha + \beta) + \sin(\alpha + 2\beta) + \cdots + \sin(\alpha + n\beta)$

6.32 Prove the following identities.

(a) $\cos^{-1} z = -i\ln(z + \sqrt{z^2 - 1})$ (b) $\sin^{-1} z = -i\ln[i(z + \sqrt{z^2 - 1})]$

(c) $\tan^{-1} z = \frac{1}{2i}\ln\left(\frac{1 + iz}{1 - iz}\right)$ (d) $\cosh^{-1} z = \ln(z + \sqrt{z^2 - 1})$

(e) $\sinh^{-1} z = \ln(z + \sqrt{z^2 + 1})$ (f) $\tanh^{-1} z = \frac{1}{2}\ln\left(\frac{1 + z}{1 - z}\right)$

6.33 Find the curve defined by each of the following equations.

(a) $z = 1 - it$ for $0 \leqslant t \leqslant 2$ (b) $z = t + it^2$ for $-\infty < t < +\infty$

(c) $z = a(\cos t + i\sin t)$ for $\frac{\pi}{2} \leqslant t \leqslant \frac{3\pi}{2}$, $a > 0$ (d) $z = t + \frac{i}{t}$ for $-\infty < t < 0$

6.34 Show that the complex potential for an infinitely long straight wire located at (x_0, y_0) is $w(z) = 2\lambda\ln(z - z_0)$.

6.35 (a) Provide all the missing steps of Example 6.2.7.
(b) Complete the manipulations of Example 6.2.8.

6.36 Let $f: \mathbb{C} \to \mathbb{C}$ be given by $f(z) = z' = x' + iy'$. Show that if f is analytic and $\partial^2\phi/\partial x^2 + \partial^2\phi/\partial y^2 = 0$, then $\partial^2\phi/\partial x'^2 + \partial^2\phi/\partial y'^2 = 0$. (That is, analytic functions map harmonic functions in the z-plane into harmonic functions in the z'-plane.)

6.37 Evaluate (**a**) $\int_0^{\pi/2} e^{it}\,dt$ and (**b**) $\int_0^\infty e^{-zt}\,dt$, where $\mathrm{Re}(z) > 0$.

6.38 Let $f(t) = u(t) + iv(t)$ be a (piecewise) continuous complex-valued function of a real variable t defined in the interval $a \leqslant t \leqslant b$. Show that if $F(t) = U(t) + iV(t)$ is a function such that $dF/dt = f(t)$, then

$$\int_a^b f(t)\,dt = F(b) - F(a)$$

6.39 Find the value of the integral $\int_C [(z + 2)/z]\,dz$, where C is
(**a**) the semicircle $z = 2e^{i\theta}$, for $0 \leqslant \theta \leqslant \pi$
(**b**) the semicircle $z = 2e^{i\theta}$, for $\pi \leqslant \theta \leqslant 2\pi$
(**c**) the circle $z = 2e^{i\theta}$, for $-\pi \leqslant \theta \leqslant \pi$

6.40 Evaluate the integral $\int_C z^m (z^*)^n\,dz$, where m and n are integers and C is the circle $|z| = 1$, taken counterclockwise.

6.41 Let C be the boundary of the square with vertices at the points $z = 0, z = 1, z = 1 + i$, and $z = i$. Show that $\oint_C (5z + 2)\,dz = 0$.

6.42 Assume a counterclockwise direction for C, the contour in Problem 6.41, and evaluate $\oint_C \pi \exp(\pi z^*)\,dz$.

6.43 Let C_1 be a simple closed contour. Deform C_1 into a new contour C_2 in such a way that C_1 does not encounter any singularity of an analytic function $f: \mathbb{C} \to \mathbb{C}$ in the process. Show that $\oint_{C_1} f(z)\,dz = \oint_{C_2} f(z)\,dz$. [That is, the contour can always be deformed into simpler shapes (such as a circle) and the integral evaluated.]

6.44 Use the result of Problem 6.43 to show that

$$\oint_C \frac{dz}{z - 1 - i} = 2\pi i \qquad \text{and} \qquad \oint_C (z - 1 - i)^{m-1}\,dz = 0 \qquad \text{for } n = \pm 1, \pm 2, \ldots$$

when C is the boundary of the square $0 \leqslant x \leqslant 2$, $0 \leqslant y \leqslant 2$, taken counterclockwise.

6.45 Use Eq. (6.26) to show that

$$\frac{dz^m}{dz} = mz^{m-1}$$

6.46 Evaluate

$$\oint_C \frac{dz}{(z^2 - 1)}$$

where C is the circle $|z| = 3$ integrated in the positive sense.

6.47 Let C be the circle $|z| = 3$ integrated in the positive sense. Show that if

$$g(z) = \oint_C \frac{2\xi^2 - \xi - 1}{\xi - z}\,d\xi \qquad \text{where } |z| \neq 3$$

then $g(2) = 10\pi i$. What is the value of $g(z)$ when $|z| > 3$?

6.48 For $g: \mathbb{C} \to \mathbb{C}$ given by

$$g(z) = \oint_C \frac{\xi^3 + 3\xi}{(\xi - z)^3}\,d\xi$$

where C is a simple closed contour integrated in the positive sense, show that $g(1) = 6\pi i$ when C encircles $z = 1$, and $g(1) = 0$ when $z = 1$ is outside C.

6.49 Show that when f is analytic within and on a simple closed contour C and z_0 is not on C, then

$$\oint_C \frac{f'(z)\,dz}{z - z_0} = \oint_C \frac{f(z)\,dz}{(z - z_0)^2}$$

6.50 Let C be the boundary of a square whose sides lie along the lines $x = \pm 3$ and $y = \pm 3$. For the positive sense of integration, evaluate each of the following integrals.

(a) $\displaystyle\oint_C \frac{e^{-z}}{z - i\frac{\pi}{2}}\,dz$ **(b)** $\displaystyle\oint_C \frac{e^3}{z(z^2 + 10)}\,dz$ **(c)** $\displaystyle\oint_C \frac{\cos z}{\left(z - \frac{\pi}{4}\right)(z^2 - 10)}\,dz$

(d) $\displaystyle\oint_C \frac{\tan z}{(z - \alpha)^2}\,dz$ for $-3 < \alpha < 3$ **(e)** $\displaystyle\oint_C \frac{z^2\,dz}{(z^2 - 10)(z - 2)}$

(f) $\displaystyle\oint_C \frac{\sinh z}{z^4}\,dz$ **(g)** $\displaystyle\oint_C \frac{\cosh z}{z^4}\,dz$ **(h)** $\displaystyle\oint_C \frac{\cos z}{z^3}\,dz$

(i) $\displaystyle\oint_C \frac{\cos z}{\left(z - i\frac{\pi}{2}\right)^2}\,dz$ **(j)** $\displaystyle\oint_C \frac{e^z}{(z - i\pi)^2}\,dz$ **(k)** $\displaystyle\oint_C \frac{\cos z}{z + i\pi}\,dz$

(l) $\displaystyle\oint_C \frac{e^z}{z^2 - 5z + 4}\,dz$ **(m)** $\displaystyle\oint_C \frac{\sinh z}{\left(z - i\frac{\pi}{2}\right)^2}\,dz$ **(n)** $\displaystyle\oint_C \frac{\cosh z}{\left(z - i\frac{\pi}{2}\right)^2}\,dz$

6.51 Let C be the circle $|z - i| = 3$ integrated in the positive sense. Find the value of each of the following integrals.

(a) $\displaystyle\oint_C \frac{e^z}{z^2 + \pi^2}\,dz$ **(b)** $\displaystyle\oint_C \frac{\sinh z}{(z^2 + \pi^2)^2}\,dz$ **(c)** $\displaystyle\oint_C \frac{dz}{z^2 + 9}$

(d) $\displaystyle\oint_C \frac{dz}{(z^2 + 9)^2}$ **(e)** $\displaystyle\oint_C \frac{\cosh z}{(z^2 + \pi^2)^3}\,dz$ **(f)** $\displaystyle\oint_C \frac{z^2 + 4 - 3z}{z^2 - 4z + 3}\,dz$

6.52 Show that Legendre polynomials for $|x| < 1$ can be represented as

$$P_n(x) = \frac{(-1)^n}{2^n(2\pi i)} \oint_C \frac{(1 - z^2)^n}{(z - x)^{n+1}}\,dz$$

where C is the unit circle around the origin.

7

Complex Analysis II: Calculus of Residues

One of the most powerful tools made available by complex analysis is the theory of residues, which makes possible the evaluation of certain definite integrals that are impossible to calculate otherwise. The derivation, application, and analysis of this tool constitute the main focus of this chapter. However, it is necessary to develop some preliminary concepts first.

7.1 SERIES OF COMPLEX FUNCTIONS

We discussed the convergence of a sequence in a metric space in Chapter 5. We also saw that the complex number system is a complete space under the norm $\|z\| = |z|$. This section reexamines the convergence of sequences and series of complex numbers and derives some series representations of complex functions.

7.1.1 Convergence of Series

A sequence $\{z_k\}_{k=1}^{\infty}$ of complex numbers is said to converge to a limit z if

$$\lim_{k\to\infty} |z - z_k| = 0$$

In other words, for each positive number ε there exists an integer N such that $|z - z_k| < \varepsilon$ whenever $k \geqslant N$. It is easy to show (see Churchill, Brown, and Verhey 1974) the following proposition.

7.1.1 Proposition Let $\{z_k = x_k + iy_k\}_{k=1}^{\infty}$ and $z = x + iy$. Then $\lim_{k\to\infty} z_k = z$ iff $\lim_{k\to\infty} x_k = x$ *and* $\lim_{k\to\infty} y_k = y$. ■

Series can be converted into sequences by partial summation. For instance, to study the infinite series $\sum_{k=1}^{\infty} z_k$, we form the partial sums

$$Z_n \equiv \sum_{k=1}^{n} z_k$$

and investigate the sequence $\{Z_n\}_{n=1}^{\infty}$. We thus say that the infinite series $\sum_{k=1}^{\infty} z_k$ converges to Z if

$$\lim_{n \to \infty} Z_n = Z$$

These are all definitions we encountered in Chapter 5. We reconsider them here for completeness and to establish some notations.

The following proposition is a direct consequence of Proposition 7.1.1.

7.1.2 Proposition Let $\{z_k = x_k + iy_k\}_{k=1}^{\infty}$ and $Z = X + iY$. Then $\sum_{k=1}^{\infty} z_k = Z$ iff $\sum_{k=1}^{\infty} x_k = X$ and $\sum_{k=1}^{\infty} y_k = Y$. ■

It is often helpful to define the *remainder after n terms* as $R_n \equiv Z - Z_n$. For establishing the convergence of a series, it is sufficient to show that $\lim_{n\to\infty} R_n = 0$.

Example 7.1.1

A series that is used often in analysis is the geometric series

$$Z = \sum_{k=0}^{\infty} z^k$$

Let us show that this series converges to $1/(1 - z)$ for $|z| < 1$.

For a partial sum of n terms, we have

$$Z_n = \sum_{k=0}^{n} z^k = 1 + z + z^2 + \cdots + z^n \tag{1}$$

We multiply (1) by z:

$$zZ_n = z + z^2 + \cdots + z^n + z^{n+1} \tag{2}$$

We subtract (2) from (1) to obtain

$$Z_n - zZ_n = 1 - z^{n+1} \Rightarrow Z_n = \frac{1 - z^{n+1}}{1 - z}$$

We now show that Z_n converges to $1/(1 - z)$ by showing that the remainder after n terms goes to zero as $n \to \infty$. We have

$$R_n = Z - Z_n = \frac{1}{1 - z} - \frac{1 - z^{n+1}}{1 - z} = \frac{z^{n+1}}{1 - z}$$

and

$$\lim_{n\to\infty} |R_n| = \lim_{n\to\infty} \frac{|z|^{n+1}}{|1 - z|} = \frac{1}{|1 - z|} \lim_{n\to\infty} |z|^{n+1} = 0$$

for $|z| < 1$. Thus, $R_n \to 0$ as $n \to \infty$, and we get

$$\sum_{k=0}^{\infty} z^k = \frac{1}{1 - z}$$

●

7.1.2 Taylor and Laurent Series

The expansion of functions in terms of polynomials or monomials is important in calculus and was emphasized in the analysis of Chapter 5. We now apply this concept to analytic functions.

7.1.3 Theorem (Taylor Series) Let f be analytic throughout the interior of a circle C_0 having radius r_0 and centered at z_0. Then at each point z inside C_0

$$f(z) = f(z_0) + f'(z_0)(z - z_0) + \frac{f''(z_0)}{2!}(z - z_0)^2 + \cdots + \frac{f^{(n)}(z_0)}{n!}(z - z_0)^n + \cdots \tag{7.1}$$

That is, the power series converges to $f(z)$ when $|z - z_0| < r_0$.

Proof. From the CIF and the fact that z is inside C_0, we have

$$f(z) = \frac{1}{2\pi i} \oint_{C_0} \frac{f(\xi)}{\xi - z} d\xi$$

On the other hand,

$$\frac{1}{\xi - z} = \frac{1}{\xi - z_0 + z_0 - z} = \frac{1}{(\xi - z_0)\left(1 - \dfrac{z - z_0}{\xi - z_0}\right)}$$

$$= \frac{1}{\xi - z_0} \sum_{k=0}^{\infty} \left(\frac{z - z_0}{\xi - z_0}\right)^k$$

The last equality follows from Example 7.1.1 and the fact that $|(z - z_0)/(\xi - z_0)| < 1$ because z is *inside* the circle C_0 and ξ is *on* it. Substituting in the CIF, we obtain

$$f(z) = \frac{1}{2\pi i} \oint_{C_0} f(\xi) \left[\frac{1}{\xi - z_0} \sum_{k=0}^{\infty} \left(\frac{z - z_0}{\xi - z_0}\right)^k\right] d\xi$$

$$= \sum_{k=0}^{\infty} \frac{1}{2\pi i} \oint_{C_0} f(\xi) \frac{(z - z_0)^k}{(\xi - z_0)^{k+1}} d\xi$$

$$= \sum_{k=0}^{\infty} (z - z_0)^k \left[\frac{1}{2\pi i} \oint_{C_0} \frac{f(\xi)}{(\xi - z_0)^{k+1}} d\xi\right]$$

$$= \sum_{k=0}^{\infty} (z - z_0)^k \frac{f^{(k)}(z_0)}{k!}$$

which is the desired result. The last equality follows from Eq. (6.26). ∎

For $z_0 = 0$ we obtain the Maclaurin series:

$$f(z) = \sum_{k=0}^{\infty} \frac{f^{(k)}(0)}{k!} z^k$$

For both series it is understood that $f^{(0)}(z_0) \equiv f(z_0)$.

The Taylor expansion demands analyticity of the function at all points interior to the circle C_0. On many occasions there may be a point inside C_0 at which the function is not analytic. The Laurent series accommodates such cases.

7.1.4 Theorem (Laurent Series) Let C_1 and C_2 be circles of radii r_1 and r_2, where $r_1 > r_2$, centered at z_0 in the z-plane. Let $f: \mathbb{C} \to \mathbb{C}$ be analytic on C_1 and C_2 and throughout R, the annular region between the two circles. Then, at each point $z \in R, f(z)$ is given by

$$f(z) = \sum_{n=-\infty}^{+\infty} a_n (z - z_0)^n$$

where
$$a_n = \frac{1}{2\pi i} \oint_C \frac{f(\xi)}{(\xi - z_0)^{n+1}} d\xi$$

and C is any contour within R that encircles z_0.

Proof. Let γ be a small closed contour in R and enclosing z, as shown in Fig. 7.1. For the composite contour C' the Cauchy-Goursat theorem gives

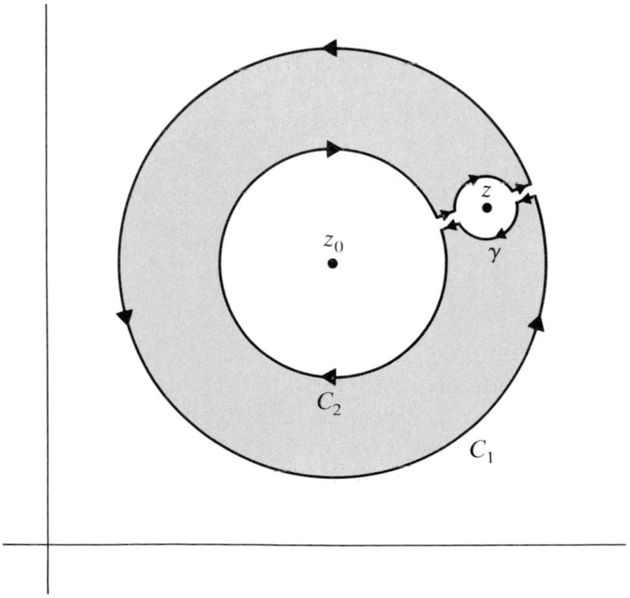

Figure 7.1 The annular region used in developing the Laurent expansion. Gaps are drawn between the straight portions of the contour C' so the direction of motion can be shown.

$\oint_{C'} f(\xi)/(\xi - z)\, d\xi = 0$. However, since C' is composed of three pieces, two of which are traversed in the clockwise direction, we get

$$0 = \oint_{C'} \frac{f(\xi)}{\xi - z} d\xi = -\oint_{\gamma} \frac{f(\xi)}{\xi - z} d\xi - \oint_{C_2} \frac{f(\xi)}{\xi - z} d\xi + \oint_{C_1} \frac{f(\xi)}{\xi - z} d\xi$$

where the γ and C_2 integrations are assumed to be counterclockwise. The γ integral is simply $2\pi i f(z)$ by the CIF. Thus, we obtain

$$2\pi i f(z) = \oint_{C_1} \frac{f(\xi)}{\xi - z} d\xi - \oint_{C_2} \frac{f(\xi)}{\xi - z} d\xi \tag{7.2}$$

Now we use the same trick we used in deriving the Taylor expansion. Since z is located in the annular region, $r_2 < |z - z_0| < r_1$. We have to keep this in mind when expanding the fractions. In particular, for $\xi \in C_1$ we get

$$\frac{1}{\xi - z} = \frac{1}{\xi - z_0 - (z - z_0)} = \frac{1}{(\xi - z_0)\left(1 - \dfrac{z - z_0}{\xi - z_0}\right)}$$

$$= \frac{1}{\xi - z_0} \sum_{n=0}^{\infty} \left(\frac{z - z_0}{\xi - z_0}\right)^n = \sum_{n=0}^{\infty} \frac{(z - z_0)^n}{(\xi - z_0)^{n+1}}$$

and for $\xi \in C_2$ we obtain

$$\frac{1}{\xi - z} = \frac{1}{\xi - z_0 - (z - z_0)} = \frac{1}{-(z - z_0)\left(1 - \dfrac{\xi - z_0}{z - z_0}\right)}$$

$$= -\frac{1}{z - z_0} \sum_{n=0}^{\infty} \left(\frac{\xi - z_0}{z - z_0}\right)^n = -\sum_{n=0}^{\infty} \frac{(\xi - z_0)^n}{(z - z_0)^{n+1}}$$

Substituting the above two expansions in Eq. (7.2) yields

$$2\pi i f(z) = \sum_{n=0}^{\infty} (z - z_0)^n \oint_{C_1} \frac{f(\xi)\, d\xi}{(\xi - z_0)^{n+1}} + \sum_{n=0}^{\infty} \frac{1}{(z - z_0)^{n+1}} \oint_{C_2} f(\xi)(\xi - z_0)^n\, d\xi \tag{7.3}$$

Now we consider an arbitrary contour C in R that encircles z_0. Figure 7.2 shows a region bounded by a contour composed of C_1 and C. In this region $f(\xi)/(\xi - z_0)^{n+1}$ is analytic (because ξ can never equal z_0). Thus, the integral over the composite contour must vanish by the Cauchy-Goursat theorem. By a procedure that should now be familiar, we get

$$\oint_{C_1} \frac{f(\xi)}{(\xi - z_0)^{n+1}} d\xi = \oint_{C} \frac{f(\xi)}{(\xi - z_0)^{n+1}} d\xi$$

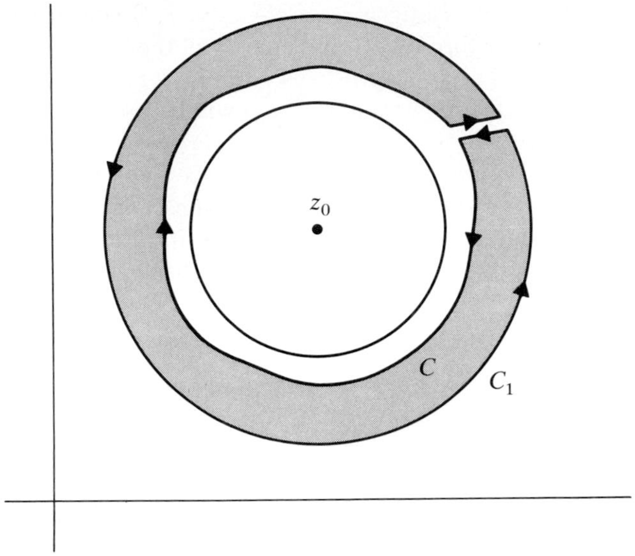

Figure 7.2 The function is analytic in the shaded region. Thus, the integrals around C_1 and C are equal.

Similarly,
$$\oint_{C_2} f(\xi)(\xi - z_0)^n d\xi = \oint_C f(\xi)(\xi - z_0)^n d\xi$$

We let $n + 1 \equiv -m$ in the second sum in Eq. (7.3) to transform that sum into

$$\sum_{m=-1}^{-\infty} \frac{1}{(z - z_0)^{-m}} \oint_C f(\xi)(\xi - z_0)^{-m-1} d\xi = \sum_{m=-\infty}^{-1} (z - z_0)^m \oint_C \frac{f(\xi)}{(\xi - z_0)^{m+1}} d\xi$$

Changing the dummy index back to n gives

$$\sum_{n=-\infty}^{-1} (z - z_0)^n \oint_C \frac{f(\xi)}{(\xi - z_0)^{n+1}} d\xi$$

Substituting this in (7.3) yields

$$2\pi i f(z) = \sum_{n=0}^{\infty} (z - z_0)^n \oint_C \frac{f(\xi)}{(\xi - z_0)^{n+1}} d\xi + \sum_{n=-\infty}^{-1} (z - z_0)^n \oint_C \frac{f(\xi)}{(\xi - z_0)^{n+1}} d\xi$$

We can now combine the sums and divide both sides by $2\pi i$, obtaining

$$f(z) = \sum_{n=-\infty}^{+\infty} (z - z_0)^n \left[\frac{1}{2\pi i} \oint_C \frac{f(\xi)}{(\xi - z_0)^{n+1}} d\xi \right]$$

which is the desired expansion. ■

The Laurent expansion is convergent as long as $r_2 < |z - z_0| < r_1$. In particular, if $r_2 = 0$, that is, if the function is analytic throughout the interior of the larger circle, then a_n will be zero for $n = -1, -2, \ldots, -\infty$, because $f(\xi)/(\xi - z_0)^{n+1}$ will be analytic for negative n. Thus, only positive powers of $(z - z_0)$ will be present in the series, and we will recover Taylor series, as we should.

It is clear that we can expand C_1 and shrink C_2 until we encounter a point at which f is no longer analytic. This is obvious from the construction of the proof, in which only the analyticity in the annular region is important—not its size. Thus, we can include all the possible analytic points by expanding C_1 and shrinking C_2.

Example 7.1.2

Let us expand some functions in terms of series. For entire functions there is no point in the entire complex plane at which they are not analytic. Thus, only positive powers of $(z - z_0)$ will be present, and we will have a Taylor expansion that is valid for all values of z.

(a) Let us expand e^z around $z_0 = 0$. The nth derivative of e^z is e^z. Thus, $f^{(n)}(0) = 1$, and Taylor expansion gives

$$e^z = \sum_{n=0}^{\infty} \frac{f^{(n)}(0)}{n!} z^n = \sum_{n=0}^{\infty} \frac{z^n}{n!}$$

(b) The Maclaurin series for $\sin z$ is obtained by noting that

$$\left.\frac{d^{(n)}}{dz^n}(\sin z)\right|_{z=0} = \begin{cases} 0 & \text{if } n \text{ is even} \\ (-1)^{(n-1)/2} & \text{if } n \text{ is odd} \end{cases}$$

and substituting this in the Taylor expansion:

$$\sin z = \sum_{n\ \text{odd}} (-1)^{(n-1)/2} \frac{z^n}{n!} = \sum_{k=0}^{\infty} (-1)^k \frac{z^{2k+1}}{(2k+1)!}.$$

Similarly, we can obtain

$$\cos z = \sum_{k=0}^{\infty} (-1)^k \frac{z^{2k}}{(2k)!} \qquad \sinh z = \sum_{k=0}^{\infty} \frac{z^{2k+1}}{(2k+1)!} \qquad \cosh z = \sum_{k=0}^{\infty} \frac{z^{2k}}{(2k)!}$$

(c) The function $1/(1 + z)$ is not entire, so the region of its convergence is limited. Let us find the Maclaurin expansion of this function. Starting from the origin ($z_0 = 0$), the function is analytic within all circles of radii $r < 1$. At $r = 1$ we encounter a singularity, the point $z = -1$. Thus, the series converges for all points z for which $|z| < 1$. For such points we have

$$f^{(n)}(0) \equiv \left.\frac{d^n}{dz^n}[(1 + z)^{-1}]\right|_{z=0} = (-1)^n n!$$

Thus,

$$\frac{1}{1+z} = \sum_{n=0}^{\infty} \frac{f^{(n)}(0)}{n!} z^n = \sum_{n=0}^{\infty} (-1)^n z^n$$

Changing z to $-z$ yields

$$\frac{1}{1-z} = \sum_{n=0}^{\infty} (-1)^n(-z)^n = \sum_{n=0}^{\infty} z^n$$

which is the result we obtained in Example 7.1.1. •

Example 7.1.3

Let us expand the following function in a Laurent series:

$$f(z) = \frac{2+3z}{z^2+z^3}$$

We can rewrite this function as

$$f(z) = \frac{1}{z^2}\left(\frac{2+3z}{1+z}\right) = \frac{1}{z^2}\left(3 - \frac{1}{1+z}\right)$$

$$= \frac{1}{z^2}\left(3 - \sum_{n=0}^{\infty}(-1)^n z^n\right) = \frac{1}{z^2}(3 - 1 + z - z^2 + z^3 - \cdots)$$

$\uparrow$ if $|z| < 1$

$$= \frac{2}{z^2} + \frac{1}{z} - 1 + z - z^2 + \cdots$$

This series converges for $0 < |z| < 1$. We note that negative powers of z are also present. In fact, using the notation of Theorem 7.1.4, we have

$$a_{-2} = 2, \quad a_{-1} = 1, \quad a_n = 0 \qquad \text{for } n \leqslant -3$$

and

$$a_n = (-1)^{n+1} \qquad \text{for } n \geqslant 0$$

•

7.1.3 Properties of Series

If the series $\sum_{k=0}^{\infty} z_k$ converges, both the real part, $\sum x_k$, and the imaginary part, $\sum y_k$, of the series also converge. From calculus we know that a necessary condition for the convergence of the real series $\sum_k x_k$ and $\sum_k y_k$ is that $\lim_{k\to\infty} x_k = 0 = \lim_{k\to\infty} y_k$. Thus, for a convergent series of complex numbers,

$$\lim_{k\to\infty} z_k = 0$$

The terms of such a series are, therefore, bounded. Thus, there exists a positive number

$$M \ni |z_k| < M \qquad \forall\, k$$

A complex series is said to converge absolutely, if the following *real* series converges:

$$\sum_{k=0}^{\infty} |z_k| \equiv \sum_{k=0}^{\infty} \sqrt{x_k^2 + y_k^2}$$

Clearly, absolute convergence implies convergence. We now consider an important proposition.

7.1.5 Proposition If the power series $\sum_{k=0}^{\infty} a_k(z-z_0)^k$ converges for $z = z_1$, where $z_1 \neq z_0$, it converges absolutely for every value of z such that $|z - z_0| < |z_1 - z_0|$.

Proof. Since the series converges for $z = z_1$, it is bounded. Thus, there exists a positive number

$$M \ni |a_k(z_1 - z_0)^k| < M$$

and

$$\sum_{k=0}^{\infty} |a_k(z-z_0)^k| = \sum_{k=0}^{\infty} \left| a_k(z_1 - z_0)^k \frac{(z-z_0)^k}{(z_1-z_0)^k} \right|$$

$$= \sum_{k=0}^{\infty} |a_k(z_1 - z_0)^k| \left| \frac{(z-z_0)^k}{(z_1-z_0)^k} \right| \leqslant \sum_{k=0}^{\infty} MC^k$$

$$= M \sum_{k=0}^{\infty} C^k = \frac{M}{1-C}$$

where $C \equiv |(z-z_0)/(z_1-z_0)|$ is a positive real number less than 1. Since the RHS is a finite (positive) number, the series of absolute values converges, and the proof is complete. ■

The largest circle about z_0 such that the power series of Proposition 7.1.5 converges is called the *circle of convergence* of the power series. The proposition implies that the series *cannot converge at any point z_2 outside this circle*, because in that case it would converge at every point inside the larger circle centered at z_0 and having a radius $|z_2 - z_0|$.

Similarly, if the power series $\sum_{k=1}^{\infty} b_k/(z-z_0)^k$ converges for z_1, then it converges for all z such that $|z-z_0|^{-1} < |z_1 - z_0|^{-1}$ or $|z - z_0| > |z_1 - z_0|$, that is, for all points *exterior* to the circle centered at z_0 and passing through z_1. Thus, the region of convergence is the exterior of some circle about z_0.

In determining the convergence of a power series, we consider the behavior of the remainder after n terms. If this remainder approaches zero, we say the series converges. Let us consider the power series

$$S(z) \equiv \sum_{n=0}^{\infty} a_n(z-z_0)^n \tag{7.4}$$

which we assume to be convergent at all points interior to a circle for which $|z - z_0| = r$. This implies that the remainder,

$$R_N(z) \equiv S(z) - \sum_{n=0}^{N-1} a_n(z-z_0)^n = \sum_{n=N}^{\infty} a_n(z-z_0)^n$$

$$= \lim_{m \to \infty} \sum_{n=N}^{m} a_n(z-z_0)^n$$

approaches zero as $N \to \infty$. Therefore, for any $\varepsilon > 0$ there exists an integer N_ε such that

$$|R_N(z)| < \varepsilon \qquad \text{whenever } N > N_\varepsilon$$

In general, the integer N_ε may be dependent on z; that is, for different values of z we may be forced to pick different N_ε's. When N_ε is independent of z, we say that the convergence is *uniform*. The power series of Eq. (7.4) is uniformly convergent, as the following theorem shows.

7.1.6 Theorem Let the power series

$$S(z) = \sum_{n=0}^{\infty} a_n(z - z_0)^n$$

be convergent at all points within its circle of convergence, $|z - z_0| = r$. Then it is *uniformly convergent* at all points interior to that circle.

Proof. Let z' be any fixed point inside the circle of convergence (it can be as far away from z_0 as possible). Define the positive real number Q_N as follows:

$$Q_N \equiv \lim_{m \to \infty} \sum_{n=N}^{m} |a_n(z' - z_0)^n| \tag{7.5}$$

Convergence implies that $Q_N < \varepsilon$ when $N > N_\varepsilon$ for some positive integer N_ε. On the other hand, for any z whose distance from z_0 is smaller than $|z' - z_0|$, we have

$$\left|\sum_{n=N}^{m} a_n(z - z_0)^n\right| \leqslant \sum_{n=N}^{m} |a_n||z - z_0|^n \leqslant \sum_{n=N}^{m} |a_n||z' - z_0|^n$$

$$= \sum_{n=N}^{m} |a_n(z' - z_0)^n| \leqslant Q_N < \varepsilon \qquad \text{whenever } N > N_\varepsilon$$

↑ because all terms of Eq. (7.5) are positive and not all terms are included in the LHS

Note that N_ε is independent of z. It may depend on z', but once we fix z', N_ε can be chosen once and for all. The above inequality is valid for all m; in particular, it is valid as $m \to \infty$, in which case its LHS becomes $R_N(z)$. We have thus established that

$$|R_N(z)| \leqslant Q < \varepsilon \qquad \text{whenever } N > N_\varepsilon$$

where N_ε is *independent of* z. This establishes the uniform convergence of the power series. ■

It is also possible to show that the power series $S(z)$ is continuous for points within its circle of convergence (see Exercise 7.1.2).

By substituting the reciprocal of $(z - z_0)$ in the power series, we can show that if $\sum_{n=1}^{\infty} b_n/(z - z_0)^n$ is convergent in the annulus $r_2 \leqslant |z - z_0| \leqslant r_1$, it is uniformly convergent for all z in that annulus, and the series represents a continuous function of z there.

We have established the continuity of the power series. We can go one step further and establish its analyticity, but we need the following proposition.

7.1.7 Proposition Let C denote any contour interior to the circle of convergence of the power series $S(z) = \sum_{n=0}^{\infty} a_n(z - z_0)^n$. Let $g: \mathbb{C} \to \mathbb{C}$ be continuous on C. Then the series formed by multiplying each term of the power series by $g(z)$ can be integrated term by term:

$$\int_C g(z)S(z)\,dz = \sum_{n=0}^{\infty} a_n \int_C g(z)(z - z_0)^n\,dz \tag{7.6}$$

Proof. We write (7.6) as

$$\int_C g(z)\,S(z)\,dz = \sum_{n=0}^{N-1} a_n \int_C g(z)(z - z_0)^n\,dz + \int_C g(z)R_N(z)\,dz$$

and note that the integral on the LHS exists, because both $S(z)$ and $g(z)$ are continuous. Similarly, the sum on the RHS exists, because it is a finite sum of integrals of continuous functions. Therefore, the last integral on the RHS also exists. The routine argument of showing that

$$\lim_{N\to\infty} \int_C g(z)R_N(z)\,dz = 0$$

is left as a problem. This will complete the proof. ∎

An important consequence of this proposition is obtained when 1 is chosen for $g(z)$ and any simple closed contour for the contour C. Then, for all n, we obtain

$$\oint_C g(z)(z - z_0)^n\,dz = \int_C (z - z_0)^n\,dz = 0 \qquad \forall\ n$$

because $(z - z_0)^n$ is an entire function and the Cauchy-Goursat theorem applies. Thus, Eq. (7.6) gives $\oint_C S(z)\,dz = 0$ for every simple closed contour C within the circle of convergence. The continuity of $S(z)$ and Morera's theorem now imply that $S(z)$ is analytic within its circle of convergence. We have just proved the following theorem.

7.1.8 Theorem A convergent power series represents a function that is analytic at every point within the circle of convergence of the series. ∎

We have noted that the derivative of an analytic function can be expressed in terms of a contour integral (see Theorem 6.3.5). Proposition 7.1.7, on the other hand, enables us to integrate a power series term by term. Combining these two results, we can prove the following theorem.

7.1.9 Theorem A convergent power series can be differentiated term by term; that is,

$$\frac{dS(z)}{dz} = \sum_{n=1}^{\infty} na_n(z - z_0)^{n-1}$$

at each point z interior to the circle of convergence of the power series.

Proof. Since $S(z)$ is analytic, its derivative is given by

$$\frac{dS}{dz} = \frac{1}{2\pi i} \oint_C \frac{S(\xi)}{(\xi - z)^2}\, d\xi$$

where C is a simple closed contour surrounding z and interior to the circle of convergence. This suggests changing z to ξ in Eq. (7.6) and choosing

$$g(\xi) = \frac{1}{2\pi i}\left[\frac{1}{(\xi - z)^2}\right]$$

On the LHS of the resulting equation, we have the derivative of $S(z)$, and the RHS yields

$$\sum_{n=0}^{\infty} a_n \frac{1}{2\pi i} \oint_C \frac{(\xi - z_0)^n}{(\xi - z)^2}\, dz = \sum_{n=0}^{\infty} a_n \frac{d}{dz}(z - z_0)^n = \sum_{n=1}^{\infty} na_n(z - z_0)^{n-1}$$

↑ by Theorem 6.3.5 (applies to the first equality)

Putting the two sides together establishes the theorem. ■

Theorem 7.1.9 and Proposition 7.1.7 are useful in determining the series representations of certain functions from those of simpler functions. The following example illustrates this point.

Example 7.1.4

Given that

$$\frac{1}{1+z} = \sum_{n=0}^{\infty} (-1)^n z^n \tag{1}$$

we can obtain the series expansion for $1/(1 + z)^2$ by differentiating both sides:

$$-\frac{d}{dz}\left(\frac{1}{1+z}\right) = \frac{1}{(1+z)^2} = -\sum_{n=0}^{\infty} (-1)^n \frac{d}{dz} z^n = -\sum_{n=1}^{\infty} (-1)^n n z^{n-1}$$

Differentiating once more, we obtain

$$-\frac{2}{(1+z)^3} = -\sum_{n=1}^{\infty}(-1)^n n\frac{d}{dz}z^{n-1} = -\sum_{n=2}^{\infty}(-1)^n n(n-1)z^{n-2}$$

or

$$\frac{1}{(1+z)^3} = +\frac{1}{2}\sum_{n=2}^{\infty}(-1)^n n(n-1)z^{n-2} = \frac{1}{2}\sum_{n=0}^{\infty}(-1)^n(n+1)(n+2)z^n$$

We can also take the indefinite integral of both sides of (1). This operation is legitimate as long as we confine ourselves to the points lying in the region of analyticity of the series. In this case that region is the interior of the circle $|z| = 1$. The LHS then becomes $\ln(1+z)$, and we can write

$$\ln(1+z) = \sum_{n=0}^{\infty}(-1)^n\frac{z^{n+1}}{n+1} = \sum_{n=1}^{\infty}(-1)^{n+1}\frac{z^n}{n} \qquad \bullet$$

Uniqueness of representations. Taylor and Laurent series allow us to express an analytic function as a power series. For a Taylor series of $f(z)$ the expansion is routine because the coefficient of its nth term is simply $f^{(n)}(z_0)/n!$, where z_0 is the center of the circle of convergence. When a Laurent series is applicable, however, the nth coefficient is a_n, as given in Theorem 7.1.4, which is not, in general, easy to evaluate. Usually a_n can be found by inspection and certain manipulations of other known series. But if we use such an intuitive approach to determine the coefficients, can we be sure that we have obtained *the* Laurent series? The following theorem answers this question.

7.1.10 Theorem If the series $\sum_{n=-\infty}^{\infty} a_n(z-z_0)^n$ converges to $f(z)$ at all points in some annular region about z_0, then it is *the* Laurent series expansion for $f(z)$ in that region.

Proof. We use Theorem 7.1.7, extended to include negative powers of $(z - z_0)$, with

$$g(z) = \frac{1}{2\pi i(z-z_0)^{k+1}} \qquad \text{for } k = 0, \pm 1, \pm 2, \ldots$$

to obtain

$$\frac{1}{2\pi i}\oint_C \frac{f(z)}{(z-z_0)^{k+1}}dz = \frac{1}{2\pi i}\sum_{n=-\infty}^{+\infty} a_n \oint_C \frac{(z-z_0)^n}{(z-z_0)^{k+1}}dz$$

$$= \frac{1}{2\pi i}\sum_{n=-\infty}^{\infty} a_n \oint_C \frac{dz}{(z-z_0)^{k-n+1}}$$

where C is a circle in the annulus. It is easy to verify that

$$\frac{1}{2\pi i}\oint_C \frac{dz}{(z-z_0)^{k-n+1}} = \delta_{kn}$$

This, in turn, says that

$$\frac{1}{2\pi i}\oint_C \frac{f(z)}{(z-z_0)^{k+1}}dz = a_k$$

Thus, the coefficient in the given power series is precisely the coefficient in the Laurent series, and the two must be identical. ■

We will look at some examples that illustrate the abstract ideas developed in the preceding collection of theorems and propositions. However, we can consider a much broader range of examples if we know the arithmetic of power series. The following theorem giving arithmetical manipulations with power series is not difficult to prove (see Churchill, Brown, and Verhey 1974).

7.1.11 Theorem Let the two power series

$$f(z) = \sum_{n=0}^{\infty} a_n(z-z_0)^n \qquad \text{and} \qquad g(z) = \sum_{n=0}^{\infty} b_n(z-z_0)^n$$

be convergent within some circle $|z - z_0| = r$. Then the sum

$$\sum_{n=0}^{\infty} (a_n + b_n)(z-z_0)^n$$

converges to $f(z) + g(z)$, and the product

$$\sum_{m=0}^{\infty}\sum_{n=0}^{\infty} a_n b_m (z-z_0)^{m+n} \equiv \sum_{k=0}^{\infty} c_k(z-z_0)^k$$

converges to $f(z)g(z)$ for z interior to the circle. Furthermore if $g(z) \neq 0$ for some neightborhood of z_0, then the series obtained by long division of numerator by denominator in the following

$$\frac{\sum_{n=0}^{\infty} a_n(z-z_0)^n}{\sum_{m=0}^{\infty} b_m(z-z_0)^m} \equiv \sum_{k=0}^{\infty} d_k(z-z_0)^k$$

converges to $f(z)/g(z)$ in that neighborhood. ■

This theorem, in essence, says that converging infinite series can be manipulated as though they were finite sums (polynomials). Such manipulations are extremely useful when dealing with Taylor and Laurent expansions in which the straightforward calculation of coefficients may be tedious.

The following examples illustrate the power of infinite-series arithmetic.

Example 7.1.5

Let us find the Laurent series for

$$f(z) = \frac{1}{4z - z^2}$$

This function is the ratio of two entire functions. Therefore, by Theorem 7.1.11, it is analytic everywhere except at the zeros of $4z - z^2$, or at $z = 0$ and $z = 4$. For the annular region (here r_2 of Theorem 7.1.4 is zero), where $0 < |z| < 4$, we expand $f(z)$ in the Laurent series around $z = 0$. Instead of actually calculating a_n, we first note that

$$f(z) = \frac{1}{4z(1 - z/4)} = \frac{1}{4z}\left(\frac{1}{1 - z/4}\right)$$

The second factor can be expanded in a geometric series because $|z/4| < 1$:

$$\frac{1}{1 - z/4} = \sum_{n=0}^{\infty} \left(\frac{z}{4}\right)^n = \sum_{n=0}^{\infty} 2^{-2n} z^n \tag{1}$$

Treating $4z$ as a special "series" in which

$$a_n = \begin{cases} 0 & \text{if } n \neq 1 \\ 4 & \text{if } n = 1 \end{cases}$$

dividing (1) by $4z$, and noting that $z = 0$ is the only zero of $4z$, and is excluded from the annular region, we obtain the expansion

$$f(z) = \sum_{n=0}^{\infty} 2^{-2n} \frac{z^n}{4z} = \sum_{n=0}^{\infty} 2^{-2(n+1)} z^{n-1} \qquad 0 < |z| < 4$$

Although we derived this series using manipulations on other series, the uniqueness of series representations assures us that this is *the* Laurent series for the indicated region.

How can we represent $f(z)$ in the region for which $|z| > 4$? This region is exterior to the circle $|z| = 4$, so we expect negative powers of z. To find the Laurent expansion we write

$$f(z) = \frac{1}{-z^2(1 - 4/z)} = -\frac{1}{z^2}\left(\frac{1}{1 - 4/z}\right)$$

and note that $|4/z| < 1$ exterior to the larger circle. Thus, the second factor can be written as a geometric series:

$$\frac{1}{1 - 4/z} = \sum_{n=0}^{\infty} \left(\frac{4}{z}\right)^n = \sum_{n=0}^{\infty} 4^n z^{-n}$$

Dividing by $-z^2$, which is analytic in the region exterior to the larger circle, yields

$$f(z) = \frac{\sum_{n=0}^{\infty} 4^n z^{-n}}{-z^2} = -\sum_{n=0}^{\infty} 4^n z^{-n-2}$$ •

Example 7.1.6

The function

$$f(z) = \frac{z}{(z - 1)(z - 2)}$$

has a Taylor expansion around the origin for $|z| < 1$. To find this expansion, we write

$$f(z) = -\frac{1}{z-1} + \frac{2}{z-2} = \frac{1}{1-z} - \frac{1}{1-z/2} \tag{1}$$

Expanding both fractions in geometric series (both $|z|$ and $|z/2|$ are less than 1), we obtain

$$f(z) = \sum_{n=0}^{\infty} z^n - \sum_{n=0}^{\infty} \left(\frac{z}{2}\right)^n$$

Adding the two series yields, by Theorem 7.1.11,

$$f(z) = \sum_{n=0}^{\infty} (1 - 2^{-n})z^n \qquad \text{for } |z| < 1$$

This is the unique Taylor expansion of $f(z)$ within the circle $|z| = 1$.

For $1 < |z| < 2$ we have a Laurent series. This can be seen by noting that

$$f(z) = \frac{1/z}{1/z - 1} - \frac{1}{1 - z/2} = -\frac{1}{z}\left(\frac{1}{1 - 1/z}\right) - \frac{1}{1 - z/2} \tag{2}$$

Since both fractions on the RHS of (2) converge in the annular region ($|1/z| < 1$, $|z/2| < 1$), we get

$$\begin{aligned} f(z) &= -\frac{1}{z}\sum_{n=0}^{\infty}\left(\frac{1}{z}\right)^n - \sum_{n=0}^{\infty}\left(\frac{z}{2}\right)^n \\ &= -\sum_{n=0}^{\infty} z^{-n-1} - \sum_{n=0}^{\infty} 2^{-n}z^n = -\sum_{n=-1}^{-\infty} z^n - \sum_{n=0}^{\infty} 2^{-n}z^n \\ &\equiv \sum_{n=-\infty}^{\infty} a_n z^n \end{aligned}$$

where $a_n = -1$ for $n < 0$ and $a_n = -2^{-n}$ for $n \geqslant 0$. This is the unique Laurent expansion of $f(z)$ in the given region.

Finally, for $|z| > 2$ we have only negative powers of z. We obtain the expansion in this region by rewriting $f(z)$ as follows:

$$f(z) = -\frac{1/z}{1 - 1/z} + \frac{2/z}{1 - 2/z}$$

Expanding the fractions yields

$$f(z) = -\sum_{n=0}^{\infty} z^{-n-1} + \sum_{n=0}^{\infty} 2^{n+1}z^{-n-1} = \sum_{n=0}^{\infty} (2^{n+1} - 1)z^{-n-1}$$

This is again the unique expansion of $f(z)$ in the region $|z| > 2$. •

Example 7.1.7

Define $f(z)$ as

$$f(z) \equiv \begin{cases} (1 - \cos z)/z^2 & \text{for } z \neq 0 \\ \frac{1}{2} & \text{for } z = 0 \end{cases}$$

We can show that $f(z)$ is an entire function.

Since $1 - \cos z$ and z^2 are entire functions, their ratio is analytic everywhere except at the zeros of its denominator, z^2. The only such zero is $z = 0$. Thus, Theorem 7.1.11 implies that $f(z)$ is analytic everywhere except possibly at $z = 0$. To see the behavior of $f(z)$ at $z = 0$, we look at its Maclaurin series:

$$1 - \cos z = 1 - \sum_{n=0}^{\infty} (-1)^n \frac{z^{2n}}{(2n)!} = \sum_{n=1}^{\infty} (-1)^{n+1} \frac{z^{2n}}{(2n)!}$$

which implies that

$$\frac{1 - \cos z}{z^2} = \sum_{n=1}^{\infty} (-1)^{n+1} \frac{z^{2n-2}}{(2n)!} = \frac{1}{2} - \frac{z^2}{4!} + \frac{z^4}{6!} - \cdots$$

From the LHS we may suspect that the only point at which the series may not converge is $z = 0$. But the expansion on the RHS shows that the value of the series is $\frac{1}{2}$, which, by definition, is $f(0)$. Thus, the series converges for all z, and Theorem 7.1.8 says that $f(z)$ is entire. ●

Example 7.1.8

A Laurent series can give information about the integral of a function around a closed contour in whose interior the function may not be analytic. In fact, Theorem 7.1.4 implies that

$$a_{-1} = \frac{1}{2\pi i} \oint_C f(\xi)\, d\xi \tag{1}$$

Thus, to find the integral of a (nonanalytic) function around a closed contour, we write the Laurent series for the function and read off the coefficient of the $1/(z - z_0)$ term.

As an illustration of this idea, let us evaluate the integral

$$I \equiv \oint_C \frac{dz}{z^2(z-2)} \tag{2}$$

where C is a circle of radius 1 centered at the origin. The function is analytic in the annular region $0 < |z| < 2$. We can, therefore, expand it as a Laurent series in that region. We obtain

$$\frac{1}{z^2(z-2)} = -\frac{1}{2z^2}\left(\frac{1}{1 - z/2}\right) = -\frac{1}{2z^2} \sum_{n=0}^{\infty} \left(\frac{z}{2}\right)^n$$

$$= -\sum_{n=0}^{\infty} \frac{z^{n-2}}{2^{n+1}} = -\frac{1}{2}\left(\frac{1}{z^2}\right) - \frac{1}{4}\left(\frac{1}{z}\right) - \frac{1}{8} - \cdots$$

Thus, $a_{-1} = -\frac{1}{4}$, and (1) gives

$$\oint_C f(z)\, dz = 2\pi i a_{-1} = -i\frac{\pi}{2}$$

Any other way of evaluating the integral is nontrivial. In fact, we will see in the next section that to find certain types of integrals it is necessary to cast them in the form of Eq. (2) and use either Eq. (1) or a related equation. ●

Zeros of analytic functions. Let $f: \mathbb{C} \to \mathbb{C}$ be analytic at z_0. Then, by definition, there exists a neighborhood of z_0 in which f is analytic. In particular, we can find a circle, $|z - z_0| = r$, in whose interior f has the Taylor expansion

$$f(z) = \sum_{n=0}^{\infty} a_n(z - z_0)^n$$

We say that f has a *zero of order* k at z_0 if $f^{(n)}(z_0) = 0$ for $n = 0, 1, \ldots, k - 1$ but $f^{(k)}(z_0) \neq 0$. In that case

$$f(z) = (z - z_0)^k \sum_{n=0}^{\infty} a_{k+n}(z - z_0)^n \qquad \text{where } a_k \neq 0,\ |z - z_0| < r$$

We define $g(z)$ as

$$g(z) \equiv \sum_{n=0}^{\infty} a_{k+n}(z - z_0)^n \qquad \text{where } |z - z_0| < r$$

and note that $g(z_0) = a_k \neq 0$. Convergence of the series on the RHS implies that $g(z)$ is continuous at z_0. Consequently, for each $\varepsilon > 0$, there exists δ such that $|g(z) - a_k| < \varepsilon$ whenever $|z - z_0| < \delta$. If we choose $\varepsilon = |a_k|/2$, then, for some $\delta_0 > 0$,

$$|g(z) - a_k| < \frac{|a_k|}{2} \qquad \text{whenever } |z - z_0| < \delta_0$$

Thus, as long as z is inside the circle $|z - z_0| = \delta_0$, $g(z)$ cannot vanish, because if it did the first inequality would imply that $|a_k| < |a_k|/2$, which is a contradiction. We, therefore, have the following theorem.

7.1.12 Theorem Let $f: \mathbb{C} \to \mathbb{C}$ be analytic at z_0 and $f(z_0) = 0$. Then there exists a neighborhood of z_0 throughout which f has no other zeros unless $f \equiv 0$. Thus, the zeros of an analytic function are *isolated*. ∎

When $k = 1$, we say that z_0 is a *simple zero* of f. To find the order of the zero of a function at a point, we differentiate the function, evaluate the derivative at that point, and continue the process until we obtain a nonzero value for the derivative. Thus, the zeros of $\cos z$, which are $z = (2k + 1)(\pi/2)$, are all simple, because

$$\frac{d}{dz}\cos z\bigg|_{z=(2k+1)(\pi/2)} = -\sin\left[(2k + 1)\frac{\pi}{2}\right] \neq 0$$

To find the order of the zero of

$$f(z) = e^z - 1 - z - \frac{z^2}{2}$$

at $z = 0$, we differentiate $f(z)$ and evaluate $f'(0)$:

$$f'(0) = (e^z - 1 - z)_{z=0} = 0$$

Differentiating again gives

$$f''(0) = (e^z - 1)_{z=0} = 0$$

Differentiating once more yields

$$f'''(0) = (e^z)_{z=0} = 1 \neq 0$$

Thus, the zero is of order 3.

Exercises

7.1.1 For $0 < r < 1$, show that

$$\sum_{k=0}^{\infty} r^k \cos k\theta = \frac{1 - r\cos\theta}{1 + r^2 - 2r\cos\theta} \quad \text{and} \quad \sum_{k=1}^{\infty} r^k \sin k\theta = \frac{r\sin\theta}{1 + r^2 - 2r\cos\theta}$$

7.1.2 Show that if the power series $S(z) = \sum_{n=0}^{\infty} a_n(z - z_0)^n$ converges, then it is a continuous function at points within its circle of convergence.

7.1.3 Show that the following function is entire:

$$f(z) \equiv \begin{cases} \cos z/(z^2 - \pi^2/4) & \text{for } z \neq \pm\pi/2 \\ -1/\pi & \text{for } z = \pm\pi/2 \end{cases}$$

7.1.4 Let $f: \mathbb{C} \to \mathbb{C}$ be analytic at z_0 and $f(z_0) = f'(z_0) = \cdots = f^{(k)}(z_0) = 0$. Show that the following function is analytic at z_0:

$$g(z) \equiv \begin{cases} f(z)/(z - z_0)^{k+1} & \text{if } z \neq z_0 \\ f^{(k+1)}(z_0)/(k+1)! & \text{if } z = z_0 \end{cases}$$

7.1.5 Evaluate the integral $\oint_C dz/z^2 \sin z$, where C is the circle $|z| = 1$ integrated counterclockwise, by writing the Laurent-series expansion and reading off the appropriate coefficient.

7.1.6 Obtain the Laurent-series expansion of $f(z) = \sinh z/z^3$.

7.1.7 Write a series representation of $f(z) = z/(z - 1)$ for all regions of analyticity of $f(z)$.

7.1.8 Show that $\oint_C dz/(e^z - 1) = 2\pi i$, where C is the circle $|z| = 1$.

7.2 CALCULUS OF RESIDUES

This section develops a powerful method for evaluating integrals. In the preceding section we saw examples in which integrals were related to expansion coefficients of Laurent series. Here we will develop a systematic way of evaluating both real and complex integrals.

7.2.1 Residues

Recall that a singular point z_0 of $f: \mathbb{C} \to \mathbb{C}$ is a point at which f fails to be analytic. If, in addition, there is some neighborhood of z_0 in which f is analytic at every point (except of course at z_0 itself), then z_0 is an *isolated singularity* of f. Almost all the singularities we have encountered so far have been isolated singularities. However, we will see later, in discussing multivalued functions, that singularities that are not isolated do exist.

Let z_0 be an isolated singularity of f. Then there exists an $r > 0$ such that within the "annular" region $0 < |z - z_0| < r$ the function f has the Laurent expansion

$$f(z) = \sum_{n=-\infty}^{\infty} a_n(z - z_0)^n \equiv \sum_{n=0}^{\infty} a_n(z - z_0)^n + \frac{b_1}{z - z_0} + \frac{b_2}{(z - z_0)^2} + \cdots$$

where

$$a_n = \frac{1}{2\pi i}\oint_C \frac{f(\xi)}{(\xi - z_0)^{n+1}}\,d\xi \qquad \text{and} \qquad b_n = \frac{1}{2\pi i}\oint_C f(\xi)(\xi - z_0)^{n-1}\,d\xi$$

In particular,

$$b_1 = \frac{1}{2\pi i}\oint_C f(\xi)\,d\xi \tag{7.7}$$

where C is any simple closed contour around z_0, traversed in the positive sense, on and interior to which f is analytic except at the point z_0 itself. The complex number b_1, which is simply related to the integral of $f(z)$ along the contour, is called the *residue* of f at the isolated singular point z_0. It is important to note that the residue is independent of the contour C as long as z_0 is the *only* isolated singular point within C.

We have seen some examples of the application of Eq. (7.7) in evaluating integrals. Let us consider two more.

Example 7.2.1

Let us evaluate the integral

$$\oint_C \frac{\sin z}{(z - \pi/2)^3}\,dz$$

where C is any simple closed contour having $z = \pi/2$ as an interior point.

To evaluate the integral we expand around $z = \pi/2$ and use Eq. (7.7). We note that

$$\sin z = \cos\left(z - \frac{\pi}{2}\right) = \sum_{n=0}^{\infty} (-1)^n \frac{(z - \pi/2)^{2n}}{(2n)!}$$

so

$$\frac{\sin z}{(z - \pi/2)^3} = \sum_{n=0}^{\infty} (-1)^n \frac{(z - \pi/2)^{2n-3}}{(2n)!}$$

$$= (z - \pi/2)^{-3} - \frac{1}{2}\left(\frac{1}{z - \pi/2}\right) + \frac{z - \pi/2}{4!} - \cdots$$

This shows that $b_1 = -\frac{1}{2}$ and, therefore,

$$\oint_C \frac{\sin z}{(z - \pi/2)^3}\,dz = 2\pi i b_1 = -i\pi$$

•

Example 7.2.2

The integral

$$\oint_C \frac{\cos z}{z^2}\,dz$$

where C is the circle $|z| = 1$, is zero because

$$\frac{\cos z}{z^2} = \frac{1}{z^2}\sum_{n=0}^{\infty}(-1)^n\frac{z^{2n}}{(2n)!} = \frac{1}{z^2} - 1 + \frac{z^2}{4!} - \cdots$$

yields $b_1 = 0$ (no $1/z$ term in the Laurent expansion). Therefore, by Eq. (7.7) the integral must vanish.

Similarly, where C is the circle $|z| = 2$, we have

$$\oint_C \frac{e^z}{(z-1)^3}\,dz = i\pi e$$

because

$$e^z = ee^{z-1} = e\sum_{n=0}^{\infty}\frac{(z-1)^n}{n!}$$

and

$$\frac{e^z}{(z-1)^3} = e\sum_{n=0}^{\infty}\frac{(z-1)^{n-3}}{n!} = e\left[\frac{1}{(z-1)^3} + \frac{1}{(z-1)^2} + \frac{1}{2}\left(\frac{1}{z-1}\right) + \frac{1}{3!} + \frac{z}{4!} + \cdots\right]$$

Thus, $b_1 = e/2$. •

It is customary to use $\text{Res}[f(z_0)]$ to denote the residue of f at the isolated singular point z_0. Equation (7.7) can then be written as

$$\oint_C f(z)\,dz = 2\pi i\,\text{Res}[f(z_0)]$$

What if there are several isolated singular points within the simple closed contour C? The following theorem provides the answer.

7.2.1 Theorem (The Residue Theorem) Let C be a positively integrated simple closed contour within and on which a function f is analytic except at a finite number of isolated singular points $z_1, z_2, \ldots, z_m$ interior to C. Then

$$\oint_C f(z)\,dz = 2\pi i \sum_{k=1}^{m} \text{Res}[f(z_k)] \tag{7.8}$$

Proof. Let C_k be the positively traversed circle around z_k. Then Fig. 7.3 and the Cauchy-Goursat theorem yield

$$0 = \oint_{C'} f(z)dz = -\oint_{\text{circles}} f(z)\,dz + \int_{\substack{\text{parallel}\\ \text{lines}}} f(z)\,dz + \oint_C f(z)\,dz$$

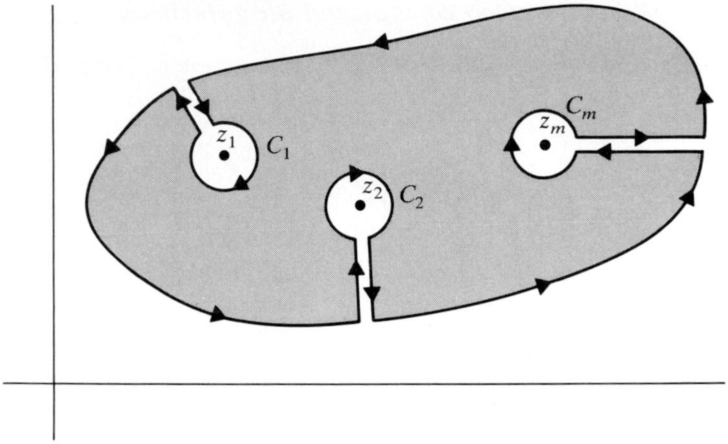

Figure 7.3 Singularities are avoided by encircling them.

where the minus sign on the first integral is due to the negative sense of integration around the circles. The contributions of the parallel lines cancel out, and we obtain

$$\oint_C f(z)\,dz = \sum_{k=1}^{m} \oint_{C_k} f(z)\,dz = \sum_{k=1}^{m} 2\pi i \operatorname{Res}[f(z_k)]$$

where in the last step the definition of residue at z_k has been used. ■

Example 7.2.3

Let us evaluate the integral

$$\oint_C \frac{2z-3}{z(z-1)}\,dz$$

where C is the circle $|z| = 2$. There are two isolated singularities in C, $z_1 = 0$ and $z_2 = 1$. To find $\operatorname{Res}[f(z_1)]$, we expand around the origin:

$$\frac{2z-3}{z(z-1)} = \frac{3}{z} + \frac{-1}{z-1} = \frac{3}{z} + \frac{1}{1-z} = \frac{3}{z} + 1 + z + z^2 + \cdots \qquad \text{for } |z| < 1$$

This gives $\operatorname{Res} f(z_1) = 3$. Similarly, expanding around $z = 1$ gives

$$\frac{2z-3}{z(z-1)} = \frac{3}{z-1+1} - \frac{1}{z-1} = -\frac{1}{z-1} + 3\sum_{n=0}^{\infty}(-1)^n(z-1)^n$$

which yields $\operatorname{Res}[f(z_2)] = -1$. Thus,

$$\oint_C \frac{2z-3}{z(z-1)}\,dz = 2\pi i\{\operatorname{Res}[f(z_1)] + \operatorname{Res}[f(z_2)]\}$$

$$= 2\pi i(3-1) = 4\pi i$$ ●

7.2.2 Classification of Isolated Singularities

Let $f: \mathbb{C} \to \mathbb{C}$ have an isolated singularity at $z = z_0$. Then there exist a real number $r > 0$ and an annular region $0 < |z - z_0| < r$ such that f can be represented by the Laurent series

$$f(z) = \sum_{n=0}^{\infty} a_n(z - z_0)^n + \sum_{n=1}^{\infty} \frac{b_n}{(z - z_0)^n} \tag{7.9}$$

The second sum in Eq. (7.9), involving negative powers of $(z - z_0)$, is called the *principal part* of f at z_0. We can use the principal part to distinguish between three types of isolated singularities. The behavior of the function near the isolated singularity is fundamentally different in each case.

(1) If $b_n = 0$ for all $n \geqslant 1$, z_0 is called a *removable singular point* of f. In this case the Laurent series contains only nonnegative powers of $(z - z_0)$, and setting $f(z_0) = a_0$ makes the function analytic at z_0. For example, the function

$$f(z) = \frac{e^z - 1 - z}{z^2} = \frac{1}{2} + \frac{z}{3!} + \frac{z^2}{4!} + \cdots$$

becomes entire if we set $f(0) = \frac{1}{2}$.

(2) If $b_n = 0$ for all $n > m$ and $b_m \neq 0$, z_0 is called a *pole of order m*. In this case the expansion takes the form

$$f(z) = \sum_{n=0}^{\infty} a_n(z - z_0)^n + \frac{b_1}{z - z_0} + \frac{b_2}{(z - z_0)^2} + \cdots + \frac{b_m}{(z - z_0)^m}$$
$$\text{for } 0 < |z - z_0| < r$$

In particular, if $m = 1$, z_0 is called a *simple pole.*

(3) If the principal part of f at z_0 has an infinite number of nonzero terms, the point z_0 is called an *essential singularity*. A prototype of functions that have essential singularities is

$$\exp\left(\frac{1}{z}\right) = \sum_{n=0}^{\infty} \frac{1}{n!}\left(\frac{1}{z^n}\right)$$

which has an essential singularity at $z = 0$ and a residue of 1 there. To see how strange such functions are, we let a be *any* real number, and consider

$$z = \frac{1}{\ln a + 2n\pi i} \qquad n = 0, \pm 1, \pm 2, \ldots$$

For such a z we have

$$e^{1/z} = e^{\ln a + 2n\pi i} = ae^{2n\pi i} = a \qquad \forall\, n$$

In particular, as $n \to \infty$, z gets arbitrarily close to the origin. Thus, in an arbitrarily small neighborhood of the origin, there are infinitely many points at which the function $\exp(1/z)$ takes on an *arbitrary* value a! This result holds for all functions with essential singularities.

Example 7.2.4

Let us consider some examples of poles of various orders.

(a) The function $(z^2 - 3z + 5)/(z - 1)$ has a Laurent series around $z = 1$ containing only three terms:

$$\frac{z^2 - 3z + 5}{z - 1} = -1 + (z - 1) + \frac{3}{z - 1}$$

Thus, it has a simple pole at $z = 1$, with a residue there of 3.

(b) The function $\sin z/z^6$ has a Laurent series

$$\frac{\sin z}{z^6} = \frac{1}{z^6} \sum_{n=0}^{\infty} (-1)^n \frac{z^{2n+1}}{(2n+1)!} = \sum_{n=0}^{\infty} (-1)^n \frac{z^{2n-5}}{(2n+1)!}$$

$$= \frac{1}{z^5} - \frac{1}{6z^3} + \frac{1}{(5!)z} - \frac{z}{7!} + \cdots$$

The principal part has three terms. The pole, at $z = 0$, is of order 5, and the function has a residue of $\frac{1}{120}$ at $z = 0$.

(c) The function $(z^2 - 5z + 6)/(z - 2)$ has a removable singularity at $z = 2$, because

$$\frac{z^2 - 5z + 6}{z - 2} = \frac{(z - 2)(z - 3)}{(z - 2)} = z - 3 = -1 + (z - 2) \qquad \text{for } z \neq 2$$

and $b_n = 0$ for all n. •

7.2.3 Poles and Residues

The type of isolated singularity that is most important in applications is the second type—poles. For a function that has a pole of order m at z_0, the calculation of residues is routine. Such calculation, in turn, enables us to evaluate most integrals effortlessly. How do we calculate the residue of a function f having a pole of order m at z_0?

It is clear that if f has a pole of order m, then $g: \mathbb{C} \to \mathbb{C}$ defined by

$$g(z) \equiv (z - z_0)^m f(z)$$

is analytic at z_0. Thus, for any simple closed contour C that contains z_0 but no other singular point of f, we have

$$\text{Res}\,[f(z_0)] = \frac{1}{2\pi i} \oint_C f(z)\,dz = \frac{1}{2\pi i} \oint_C \frac{g(z)}{(z - z_0)^m} dz$$

$$= \frac{1}{(m-1)!} g^{(m-1)}(z_0)$$

In terms of f this yields

$$\text{Res}\,[f(z_0)] = \lim_{z \to z_0} \frac{1}{(m-1)!} \frac{d^{m-1}}{dz^{m-1}} [(z - z_0)^m f(z)] \tag{7.10a}$$

For the special, but important, case of a simple pole, we obtain

$$\operatorname{Res}[f(z_0)] = \lim_{z \to z_0} [(z - z_0) f(z)] \tag{7.10b}$$

7.2.4 Evaluation of Definite Integrals Using the Residue Theorem

The most important application of residues is in the evaluation of real definite integrals. It is possible to complexify certain real definite integrals and relate them to contour integrations in the complex plane. We will discuss this method shortly; however, it is helpful to prove a lemma, due to Jordan, first.

7.2.2 Lemma (Jordan's Lemma) Let C_R be a semicircle of radius R in the upper half of the complex plane (UHP) and centered at the origin. Let f be a function that tends uniformly to zero faster than $1/|z|$ for $\arg(z) \in [0, \pi]$ when $|z| \to \infty$. Let α be a non-negative real number. Then

$$\lim_{R \to \infty} I_R \equiv \lim_{R \to \infty} \int_{C_R} e^{i\alpha z} f(z)\, dz = 0$$

Proof. For $z \in C_R$ we can write $z = Re^{i\theta}$, $dz = iRe^{i\theta}\, d\theta$, and

$$i\alpha z = i\alpha(R\cos\theta + iR\sin\theta) = i\alpha R\cos\theta - \alpha R\sin\theta$$

Thus,

$$|I_R| = \left| \int_{C_R} e^{i\alpha z} f(z)\, dz \right| \leqslant \int_{C_R} |e^{i\alpha z}|\, |f(z)|\, |dz|$$

$$= \int_0^\pi |e^{i\alpha R\cos\theta - \alpha R\sin\theta}|\, |f(Re^{i\theta})|\, R\, d\theta$$

$$= \int_0^\pi e^{-\alpha R\sin\theta}\, R\,|f(Re^{i\theta})|\, d\theta$$

By assumption, $R\,|f(Re^{i\theta})| < \varepsilon(R)$ independent of θ, where $\varepsilon(R)$ is an arbitrary positive number that tends to zero as $R \to \infty$. Therefore,

$$|I_R| < \varepsilon(R) \int_0^\pi e^{-\alpha R\sin\theta}\, d\theta \underset{\substack{\uparrow \\ \text{by Problem 7.20}}}{=} 2\,\varepsilon(R) \int_0^{\pi/2} e^{-\alpha R\sin\theta}\, d\theta$$

Furthermore, for $0 \leqslant \theta \leqslant \pi/2$,

$$\sin\theta \geqslant \frac{2\theta}{\pi} \quad \Rightarrow \quad e^{-\alpha R \sin\theta} \leqslant e^{-(2\alpha R/\pi)\theta}$$

Thus,
$$|I_R| < 2\varepsilon(R) \int_0^{\pi/2} e^{-(2\alpha R/\pi)\theta}\, d\theta = \frac{\pi\varepsilon(R)}{\alpha R}(1 - e^{-\alpha R})$$

which implies that $\lim_{R\to\infty} I_R = 0$. ■

Note that Jordan's lemma applies for $\alpha = 0$ as well, because $(1 - e^{-\alpha R}) \to \alpha R$ as $\alpha \to 0$. If $\alpha < 0$, the lemma is still valid if the semicircle C_R is taken in the lower half of the complex plane (LHP) and $f(z)$ goes to zero uniformly for $\pi \leqslant \arg(z) \leqslant 2\pi$.

We are now in a position to apply the residue theorem to the evaluation of definite integrals. The three types of integrals most commonly encountered are discussed separately below. In all cases we assume that Jordan's lemma holds.

Integrals of rational functions. The first type of integral we can evaluate using the residue theorem is of the form

$$I_1 = \int_0^{\infty} \frac{p(x)}{q(x)}\, dx$$

where $p(x)$ and $q(x)$ are real polynomials, $q(x) \neq 0$ for any real x, and $p(x)/q(x)$ is an *even function* of x. We can then write

$$I_1 = \frac{1}{2}\int_{-\infty}^{\infty} \frac{p(x)}{q(x)}\, dx \equiv \frac{1}{2}\lim_{R\to\infty}\int_{-R}^{R} \frac{p(x)}{q(x)}\, dx = \frac{1}{2}\lim_{R\to\infty}\int_{C_x} \frac{p(z)}{q(z)}\, dz$$

where C_x is the (open) contour lying on the real axis from $-R$ to $+R$. Assuming that Jordan's lemma holds, we can close that contour by adding to it the semicircle of radius R. This will not affect the value of the integral because, in the limit $R \to \infty$, the contribution of the integral of the semicircle tends to zero. If $q(z)$ has zeros in the UHP, we may close the contour there. We then get

$$I_1 = \frac{1}{2}\lim_{R\to\infty}\oint_C \frac{p(z)}{q(z)}\, dz = \pi i \sum_{k=1}^{m} \operatorname{Res}\left[\frac{p(z_k)}{q(z_k)}\right]$$

where C is the closed contour composed of the interval $[-R, R]$ and the semicircle C_R and $\{z_k\}_{k=1}^{m}$ are the zeros of $q(z)$ in the UHP. We may instead close the contour in the LHP, in which case

$$I_1 = -\pi i \sum_{j=1}^{l} \operatorname{Res}\left[\frac{p(z_j)}{q(z_j)}\right]$$

where $\{z_j\}_{j=1}^{l}$ are the zeros of $q(z)$ in the LHP. The minus sign indicates that in the LHP we integrate clockwise. Let us look at some examples of the evaluation of this type of integral.

Example 7.2.5

Let us evaluate the integral

$$\int_0^\infty \frac{x^2}{(x^2+1)(x^2+9)}\,dx$$

which, since Jordan's lemma holds (see below), corresponds to the contour integral

$$\frac{1}{2}\oint_C \frac{z^2}{(z^2+1)(z^2+9)}\,dz$$

where C is as shown in Fig. 7.4(a). Note that the contour is integrated in the positive sense. This is always true for the UHP. The singularities of the function in the UHP are the simple poles i and $3i$ corresponding to the simple zeros of the denominator. The residues at these poles are

$$\text{Res}\,[f(i)] = \lim_{z\to i}(z-i)\frac{z^2}{(z-i)(z+i)(z^2+9)} = \frac{i^2}{(2i)(-1+9)} = -\frac{1}{16i}$$

$$\text{Res}\,[f(3i)] = \lim_{z\to 3i}(z-3i)\frac{z^2}{(z^2+1)(z-3i)(z+3i)} = \frac{(3i)^2}{(-9+1)(6i)} = \frac{3}{16i}$$

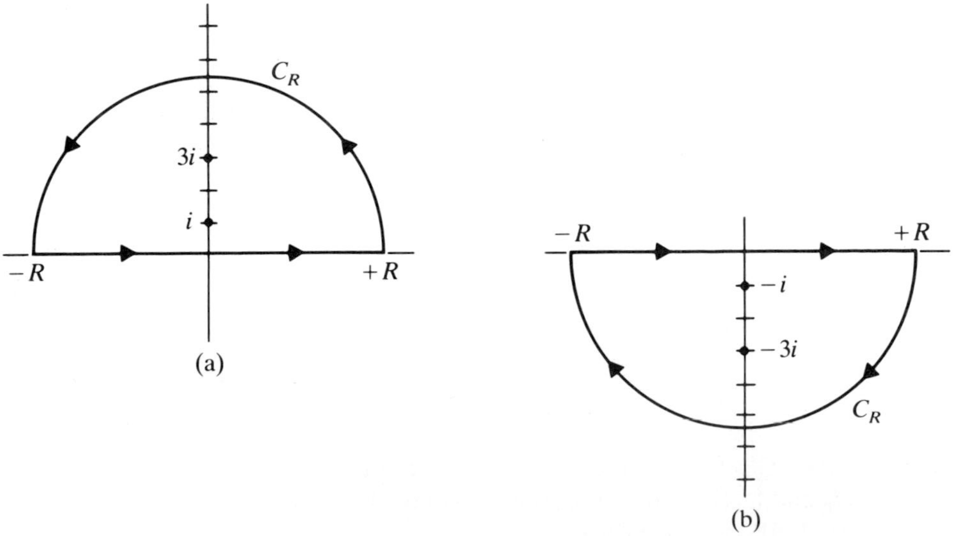

Figure 7.4 (a) The large semicircle is chosen to be in the UHP. (b) The large semicircle is in the LHP. Note that the contour is traversed clockwise.

Thus, we obtain

$$\int_0^\infty \frac{x^2}{(x^2+1)(x^2+9)}\,dx = \frac{1}{2}\oint_C \frac{z^2\,dz}{(z^2+1)(z^2+9)} = i\pi\left(-\frac{1}{16i}+\frac{3}{16i}\right) = \frac{\pi}{8}$$

It is instructive to obtain the same results using the LHP. In this case the contour is as shown in Fig. 7.4(b) and is taken clockwise, so we have to introduce a minus sign for its integration. The singular points are at $z = -i$ and $z = -3i$. These are simple poles at which the residues of the function are

$$\text{Res}\,[f(-i)] = \lim_{z\to -i}(z+i)\frac{z^2}{(z^2+1)(z^2+9)} = \frac{(-i)^2}{(-2i)(8)} = \frac{1}{16i}$$

$$\text{Res}\,[f(-3i)] = \lim_{z\to -3i}(z+3i)\frac{z^2}{(z^2+1)(z^2+9)} = \frac{(-3i)^2}{(-8)(-6i)} = -\frac{3}{16i}$$

Therefore,

$$\int_0^\infty \frac{x^2}{(x^2+1)(x^2+9)}\,dx = \frac{1}{2}\oint_C \frac{z^2\,dz}{(z^2+1)(z^2+9)} = -i\pi\left(\frac{1}{16i}-\frac{3}{16i}\right) = \frac{\pi}{8}$$

To show that Jordan's lemma applies to this integral, we have only to show that $\lim_{R\to\infty} R|f(Re^{i\theta})| = 0$. In this case $\alpha = 0$ because there is no exponential function in the integrand. Thus,

$$R|f(Re^{i\theta})| = R\left|\frac{1}{(R^2e^{2i\theta}+1)(R^2e^{2i\theta}+9)}\right| = \frac{R}{|R^2e^{2i\theta}+1|\,|R^2e^{2i\theta}+9|}$$

However,

$$|R^2e^{2i\theta}+1| = \sqrt{(R^2e^{2i\theta}+1)(R^2e^{\ \ 2i\theta}+1)} = \sqrt{R^4+2R^2\cos 2\theta+1}$$

$$\geqslant \sqrt{R^4-2R^2+1} = R^2-1$$

Similarly, $$|R^2e^{2i\theta}+9| \geqslant R^2-9$$

Therefore, $$R|f(Re^{i\theta})| \leqslant \frac{R^3}{(R^2-1)(R^2-4)} = \varepsilon(R)$$

and $R|f(Re^{i\theta})| \to 0$ as $R\to\infty$. •

Example 7.2.6

Let us now consider a more complicated integral:

$$\int_0^\infty \frac{x^2}{(x^2+1)(x^2+4)^2}\,dx$$

which, in the UHP, turns into

$$\frac{1}{2}\oint_C \frac{z^2}{(z^2+1)(z^2+4)^2}\,dz$$

The poles in the UHP are at $z = i$ and $z = 2i$. The former is a simple pole, and the latter is a pole of order 2. Thus,

$$\text{Res}[f(i)] = \lim_{z\to i}(z-i)\frac{z^2}{(z^2+1)(z^2+4)^2} = \frac{-1}{2i(3)^2} = -\frac{1}{18i}$$

$$\text{Res}[f(2i)] = \frac{1}{(2-1)!}\lim_{z\to 2i}\frac{d}{dz}\left[(z-2i)^2\frac{z^2}{(z^2+1)(z+2i)^2(z-2i)^2}\right]$$

$$= \lim_{z\to 2i}\frac{d}{dz}\left[\frac{z^2}{(z^2+1)(z+2i)^2}\right] = \frac{5}{72i}$$

and
$$\int_0^\infty \frac{x^2\,dx}{(x^2+1)(x^2+4)^2} = i\pi\left(-\frac{1}{18i}+\frac{5}{72i}\right) = \frac{\pi}{72}$$

Closing the contour in the LHP would have yielded the same result. •

Integrals of products of rational functions and trigonometric functions. The second type of integral we can evaluate using the residue theorem is of the form

$$\int_{-\infty}^{\infty}\frac{p(x)}{q(x)}\cos ax\,dx \qquad \text{or} \qquad \int_{-\infty}^{\infty}\frac{p(x)}{q(x)}\sin ax\,dx$$

where a is a real number, $p(x)$ and $q(x)$ are real polynomials in x, and $q(x)$ has no real zeros. These integrals are the real and imaginary parts of

$$I_2 \equiv \int_{-\infty}^{\infty}\frac{p(x)}{q(x)}e^{iax}\,dx$$

The presence of e^{iax} dictates the choice of half-plane: If $a \geqslant 0$, we choose the UHP; otherwise we choose the LHP. We must, of course, have enough powers of x in the denominator to render $R|p(Re^{i\theta})/q(Re^{i\theta})|$ uniformly convergent to zero.

The following examples illustrate the procedure.

Example 7.2.7

Let us evaluate

$$\int_{-\infty}^{\infty}\frac{\cos ax}{(x^2+1)^2}\,dx \qquad \text{where } a \neq 0$$

This integral is the real part of the integral

$$I_2 = \int_{-\infty}^{\infty}\frac{e^{iax}}{(x^2+1)^2}\,dx$$

When $a > 0$, we close in the UHP as advised by Jordan's lemma. Then we proceed as for integrals of rational functions. Thus, we have

$$I_2 = \oint_C \frac{e^{iaz}}{(z^2+1)^2}\,dz = 2\pi i \operatorname{Res}[f(i)] \qquad \text{for } a > 0$$

because there is only one singularity in the UHP, at $z = i$, which is a pole of order 2. We next calculate the residue:

$$\begin{aligned}\operatorname{Res}[f(i)] &= \lim_{z\to i}\frac{d}{dz}\left[(z-i)^2\frac{e^{iaz}}{(z-i)^2(z+i)^2}\right]\\ &= \lim_{z\to i}\frac{d}{dz}[(z+i)^{-2}e^{iaz}] = \lim_{z\to i}[-2(z+i)^{-3}e^{iaz} + ia(z+i)^{-2}e^{iaz}]\\ &= -\frac{2e^{-a}}{(2i)^3} + \frac{iae^{-a}}{(2i)^2}\\ &= \frac{e^{-a}}{4i}(1+a)\end{aligned}$$

Substituting this in the expression for I_2, we obtain

$$I_2 = \frac{\pi}{2}e^{-a}(1+a) \qquad \text{for } a > 0$$

When $a < 0$, we have to close the contour in the LHP, where the pole of order 2 is at $z = -i$ and the contour is taken clockwise. Thus, we get

$$I_2 = \oint_C \frac{e^{iaz}}{(z^2+1)^2}\,dz = -2\pi i \operatorname{Res}[f(-i)]$$

For the residue we obtain

$$\begin{aligned}\operatorname{Res}[f(-i)] &= \lim_{z\to -i}\frac{d}{dz}\left[(z+i)^2\frac{e^{iaz}}{(z-i)^2(z+i)^2}\right] = \lim_{z\to -i}\frac{d}{dz}[(z-i)^{-2}e^{iaz}]\\ &= -\frac{e^{a}}{4i}(1-a)\end{aligned}$$

and the expression for I_2 becomes

$$I_2 = \frac{\pi}{2}e^{a}(1-a) \qquad \text{for } a < 0$$

We can therefore write

$$\int_{-\infty}^{\infty} \frac{\cos ax}{(x^2+1)^2}\,dx = \operatorname{Re}(I_2) = I_2 = \frac{\pi}{2}(1+|a|)e^{-|a|} \qquad \bullet$$

Example 7.2.8

As another example, let us evaluate

$$\int_{-\infty}^{\infty} \frac{x \sin ax}{x^4 + 4}\, dx \qquad \text{where } a \neq 0$$

This is the imaginary part of the integral

$$I_2 = \int_{-\infty}^{\infty} \frac{xe^{iax}}{x^4 + 4}\, dx$$

which, in terms of z and for the closed contour in the UHP, becomes

$$I_2 = \oint_C \frac{ze^{iaz}}{z^4 + 4}\, dz = 2\pi i \sum_{j=1}^{m} \text{Res}\,[f(z_j)] \tag{1}$$

The singularities are determined by the zeros of the denominator:

$$z^4 + 4 = 0 \quad \Rightarrow \quad z^4 = -4 = 4e^{i(\pi + 2n\pi)} \qquad \text{for } n = 0, \pm 1, \pm 2, \cdots$$

or

$$z = \sqrt{2}e^{i(2n+1)\pi/4} = \sqrt{2}\left[\cos\left(\frac{2n+1}{4}\pi\right) + i \sin\left(\frac{2n+1}{4}\pi\right)\right]$$

$$= 1 + i,\ 1 - i,\ -1 + i,\ -1 - i$$

Of these four simple roots only two, $1 + i$ and $-1 + i$, are in the UHP. We now calculate the residues:

$$\text{Res}\,[f(1 + i)] = \lim_{z \to 1+i} (z - 1 - i)\frac{ze^{iaz}}{(z - 1 - i)(z - 1 + i)(z + 1 - i)(z + 1 + i)}$$

$$= \frac{(1 + i)e^{ia(1+i)}}{(2i)(2)(2 + 2i)} = \frac{e^{ia}e^{-a}}{8i}$$

$$\text{Res}\,[f(-1 + i)] = \lim_{z \to -1+i} (z + 1 - i)\frac{ze^{iaz}}{(z + 1 - i)(z + 1 + i)(z - 1 - i)(z - 1 + i)}$$

$$= \frac{(-1 + i)e^{ia(-1+i)}}{(2i)(-2)(-2 + 2i)} = -\frac{e^{-ia}e^{-a}}{8i}$$

Substituting in (1), we obtain

$$I_2 = 2\pi i \frac{e^{-a}}{8i}(e^{ia} - e^{-ia}) = i\frac{\pi}{2}e^{-a} \sin a$$

Thus,

$$\int_{-\infty}^{\infty} \frac{x \sin ax}{x^4 + 4}\, dx = \text{Im}(I_2) = \frac{\pi}{2}e^{-a} \sin a \qquad \text{for } a > 0 \tag{2}$$

For $a < 0$ we note that $-a > 0$, and Eq. (2) yields

$$\int_{-\infty}^{\infty} \frac{x \sin ax}{x^4 + 4}\, dx = -\int_{-\infty}^{\infty} \frac{x \sin [(-a)x]}{x^4 + 4}\, dx = -\frac{\pi}{2} e^{-(-a)} \sin(-a)$$

$$= \frac{\pi}{2} e^{a} \sin a \qquad (3)$$

We can summarize (2) and (3) as

$$\int_{-\infty}^{\infty} \frac{x \sin ax}{x^4 + 4}\, dx = \frac{\pi}{2} e^{-|a|} \sin a$$

•

Integrals involving only trigonometric functions. The third type of integral we can evaluate using the residue theorem involves only trigonometric functions and is typically of the form

$$\int_{0}^{2\pi} F(\sin \theta, \cos \theta)\, d\theta$$

where F is some arbitrary (typically rational) function of $\sin \theta$ and $\cos \theta$. Since θ varies from 0 to 2π, we can consider it an argument of a point z on the unit circle centered at the origin. Then $z = e^{i\theta}$ and $e^{-i\theta} = 1/z$, and we can substitute

$$\cos \theta = \frac{z + 1/z}{2} \qquad \sin \theta = \frac{z - 1/z}{2i} \qquad d\theta = \frac{dz}{iz}$$

in the original integral, obtaining

$$\oint_C F\left(\frac{z^2 - 1}{2iz}, \frac{z^2 + 1}{2z}\right) \frac{dz}{iz}$$

This integral can be evaluated using the method of residues.

The following examples illustrate the use of this procedure.

Example 7.2.9

Let us evaluate the integral

$$\int_{0}^{2\pi} \frac{d\theta}{1 + a \cos \theta} \qquad \text{where } |a| < 1$$

Substituting for $\cos \theta$ and $d\theta$ in terms of z, we obtain

$$\oint_C \frac{dz/iz}{1 + a[(z^2 + 1)/2z]} = \frac{2}{i} \oint_C \frac{dz}{2z + az^2 + a}$$

where C is the unit circle centered at the origin. The singularities of the integrand are the zeros of its denominator. We use the quadratic formula to find the simple zeros of the denominator:

$$z_1 = \frac{-1+\sqrt{1-a^2}}{a} \qquad \text{and} \qquad z_2 = \frac{-1-\sqrt{1-a^2}}{a}$$

For $|a| < 1$ it is clear that z_2 will lie outside C; therefore, it does not contribute to the integral. But z_1 lies inside C, and we obtain

$$\oint_C \frac{dz}{2z + az^2 + a} = 2\pi i \operatorname{Res}[f(z_1)]$$

The residue of the simple pole at z_1 can be calculated:

$$\operatorname{Res}[f(z_1)] = \lim_{z\to z_1} (z - z_1)\frac{1}{a(z-z_1)(z-z_2)} = \frac{1}{a}\left(\frac{1}{z_1 - z_2}\right)$$
$$= \frac{1}{a}\left(\frac{a}{2\sqrt{1-a^2}}\right) = \frac{1}{2\sqrt{1-a^2}}$$

Finally, from this we get

$$\int_0^{2\pi} \frac{d\theta}{1 + a\cos\theta} = \frac{2}{i}\oint \frac{dz}{2z + az^2 + a} = \frac{2}{i}2\pi i\left(\frac{1}{2\sqrt{1-a^2}}\right) = \frac{2\pi}{\sqrt{1-a^2}} \qquad \bullet$$

Example 7.2.10

As another example, let us consider the integral

$$I = \int_0^{\pi} \frac{d\theta}{(a + \cos\theta)^2} \qquad \text{where } a > 1$$

Since $\cos\theta$ is an even function of θ, we may write

$$I = \frac{1}{2}\int_{-\pi}^{+\pi} \frac{d\theta}{(a + \cos\theta)^2}$$

This integration is over a complete cycle around the origin, and we can make the usual substitution:

$$I = \frac{1}{2}\oint_C \frac{dz/iz}{[a + (z^2+1)/2z]^2} = \frac{1}{2i}\oint_C \frac{4z\,dz}{(z^2 + 2az + 1)^2}$$

The denominator has the roots

$$z_1 = a - \sqrt{a^2-1} \qquad \text{and} \qquad z_2 = a + \sqrt{a^2-1}$$

which are both of order 2. The second root is outside the unit circle because $a > 1$. Also, it is easily verified that, for all $a > 1$, z_1 is inside the unit circle. Since z_1 is a pole of order 2, we have

$$\operatorname{Res}[f(z_1)] = \lim_{z \to z_1} \frac{d}{dz}\left[(z - z_1)^2 \frac{z}{(z - z_1)^2 (z - z_2)^2}\right]$$

$$= \lim_{z \to z_1} \frac{d}{dz}[z(z - z_2)^{-2}] = \frac{1}{(z_1 - z_2)^2} - \frac{2z_1}{(z_1 - z_2)^3} = \frac{a}{4(a^2 - 1)^{3/2}}$$

We thus obtain

$$I = \frac{2}{i}\{2\pi i \operatorname{Res}[f(z_1)]\} = \frac{\pi a}{(a^2 - 1)^{3/2}}$$ •

Other manageable integrals. The foregoing three types of definite integrals do not, obviously, exhaust all possibilities. There are other integrals that do not fit into any of the three categories but are still manageable. As the next two examples demonstrate, an ingenious choice of contour allows evaluation of other types of integrals.

Example 7.2.11

Let us evaluate the Gaussian integral

$$I = \int_{-\infty}^{+\infty} e^{iax - bx^2}\, dx \qquad \text{where } a, b \in \mathbb{R},\ b > 0$$

Completing squares in the exponent, we have

$$I = \int_{-\infty}^{\infty} e^{-b[x - (ia/2b)]^2 - (a^2/4b)}\, dx = e^{-a^2/4b} \int_{-\infty}^{\infty} e^{-b[x - (ia/2b)]^2}\, dx$$

$$= e^{-a^2/4b} \lim_{R \to \infty} \int_{-R}^{R} e^{-b[x - (ia/2b)]^2}\, dx$$

If we change the variable of integration to $z = x - ia/2b$, we obtain

$$I = e^{-a^2/4b} \lim_{R \to \infty} \int_{-R - ia/2b}^{R - ia/2b} e^{-bz^2}\, dz$$

Let us now define I_R:

$$I_R \equiv \int_{-R - ia/2b}^{R - ia/2b} e^{-bz^2}\, dz$$

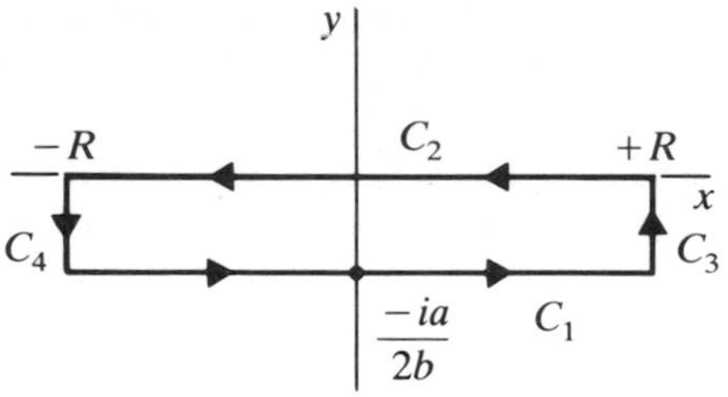

Figure 7.5

This is an integral along a straight line C_1 that is parallel to the x-axis (see Fig. 7.5). We close the contour as shown and note that e^{-bz^2} is analytic throughout the interior of the closed contour. Thus, the contour integral must vanish by the Cauchy-Goursat theorem. So we obtain

$$I_R + \int_{C_3} e^{-bz^2}\,dz + \int_R^{-R} e^{-bx^2}\,dx + \int_{C_4} e^{-bz^2}\,dz = 0$$

Along C_3, $z = R + iy$ and

$$\int_{C_3} e^{-bz^2}\,dz = \int_{-ia/2b}^{0} e^{-b(R+iy)^2}\,i\,dy = ie^{-bR^2}\int_{-ia/2b}^{0} e^{by^2-2ibRy}\,dy$$

which clearly tends to zero as $R \to \infty$. We get a similar result for the integral along C_4. Therefore, we have

$$I_R = \int_{-R}^{R} e^{-bx^2}\,dx$$

and $\lim_{R\to\infty} I_R = \int_{-\infty}^{\infty} e^{-bx^2}\,dx = \sqrt{\pi/b}$ (see Exercise 5.3.2). Finally, we get

$$\int_{-\infty}^{\infty} e^{iax-bx^2} = \sqrt{\frac{\pi}{b}}\,e^{-a^2/4b}$$

•

Example 7.2.12

Let us evaluate

$$I = \int_0^{\infty} \frac{dx}{x^3+1}$$

We cannot extend the lower limit of integration to $-\infty$, as we have done before, because the integrand is not even. To get a hint as to how to close the contour, we study the singularities of the integrand. These are simply the roots of the denominator:

$$z^3 = -1 \quad \Rightarrow \quad z = e^{i(2n+1)\pi/3}$$

or

$$z_1 = e^{i\pi/3} \qquad z_2 = e^{i\pi} = -1 \qquad z_3 = e^{i5\pi/3}$$

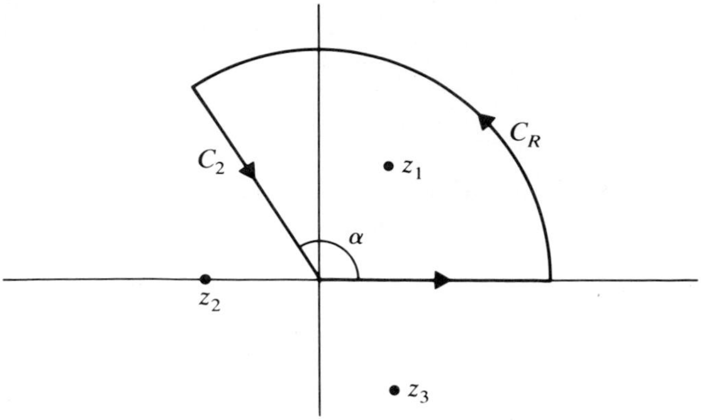

Figure 7.6

These are shown in Fig. 7.6 as well as a contour that has only z_1 as an interior point. We thus have

$$I + \int_{C_R} \frac{dz}{z^3 + 1} + \int_{C_2} \frac{dz}{z^3 + 1} = 2\pi i \operatorname{Res}[f(z_1)] \tag{1}$$

The C_R integral vanishes, as usual. Along C_2, $z = re^{i\alpha}$ with constant α, so

$$\int_{C_2} \frac{dz}{z^3 + 1} = \int_{\infty}^{0} \frac{e^{i\alpha}\, dr}{(re^{i\alpha})^3 + 1} = -e^{i\alpha} \int_{0}^{\infty} \frac{dr}{r^3 e^{3i\alpha} + 1}$$

In particular, if we choose $3\alpha = 2\pi$, we obtain

$$\int_{C_2} \frac{dz}{z^3 + 1} = -e^{i2\pi/3} \int_{0}^{\infty} \frac{dr}{r^3 + 1} = -e^{i2\pi/3} I$$

Substituting this in (1) gives

$$(1 - e^{i2\pi/3})I = 2\pi i \operatorname{Res}[f(z_1)]$$

On the other hand,

$$\operatorname{Res}[f(z_1)] = \lim_{z \to e^{i\pi/3}} (z - e^{i\pi/3}) \frac{1}{(z - e^{i\pi/3})(z + 1)(z - e^{i5\pi/3})}$$

$$= \frac{1}{\left(\frac{3}{2} + i\frac{\sqrt{3}}{2}\right)(i\sqrt{3})}$$

These last two equations yield

$$\left(\frac{3}{2} - i\frac{\sqrt{3}}{2}\right)I = \frac{2\pi i}{\left(\frac{3}{2} + i\frac{\sqrt{3}}{2}\right)(i\sqrt{3})}$$

or
$$I = \frac{2\pi}{\sqrt{3}} \frac{1}{\left(\frac{3}{2}\right)^2 + \left(\frac{\sqrt{3}}{2}\right)^2} = \frac{2\pi}{3\sqrt{3}}$$
•

The principal value of an integral. So far we have discussed only integrals of functions that have no singularities *on* the contour. Let us now investigate the consequences of the presence of singular points on the contour. Consider the integral

$$\int_{-\infty}^{\infty} \frac{f(x)}{x - x_0}\, dx \tag{7.11}$$

where x_0 is a real number and f is analytic at x_0. To avoid x_0, which blows up the integral, we bypass it by indenting the contour as shown in Fig. 7.7. The contour C_0 is simply a semicircle of radius ε. For the contour $\hat{C}$ shown in Fig. 7.7, we have

$$\int_{\hat{C}} \frac{f(z)}{z - x_0}\, dz = \int_{-\infty}^{x_0 - \varepsilon} \frac{f(x)}{x - x_0}\, dx + \int_{x_0 + \varepsilon}^{\infty} \frac{f(x)}{x - x_0}\, dx + \int_{C_0} \frac{f(z)}{z - x_0}\, dz$$

In the limit $\varepsilon \to 0$ the first two terms on the RHS define the *principal value* of the integral in (7.11):

$$\mathrm{P} \int_{-\infty}^{\infty} \frac{f(x)}{x - x_0}\, dx \equiv \lim_{\varepsilon \to 0} \left[\int_{-\infty}^{x_0 - \varepsilon} \frac{f(x)\, dx}{x - x_0} + \int_{x_0 + \varepsilon}^{\infty} \frac{f(x)}{x - x_0}\, dx \right]$$

The integral over the semicircle is calculated by noting that $z - x_0 = \varepsilon e^{i\theta}$ and $dz = i\varepsilon e^{i\theta}\, d\theta$. Thus,

$$\int_{C_0} \frac{f(z)}{z - x_0}\, dz = \int_{\pi}^{0} \frac{f(x_0 + \varepsilon e^{i\theta})}{\varepsilon e^{i\theta}}\, i\varepsilon e^{i\theta}\, d\theta = -i\pi f(x_0)$$

and
$$\int_{\hat{C}} \frac{f(z)}{z - x_0}\, dz = \mathrm{P} \int_{-\infty}^{\infty} \frac{f(x)}{x - x_0}\, dx - i\pi f(x_0) \tag{7.12a}$$

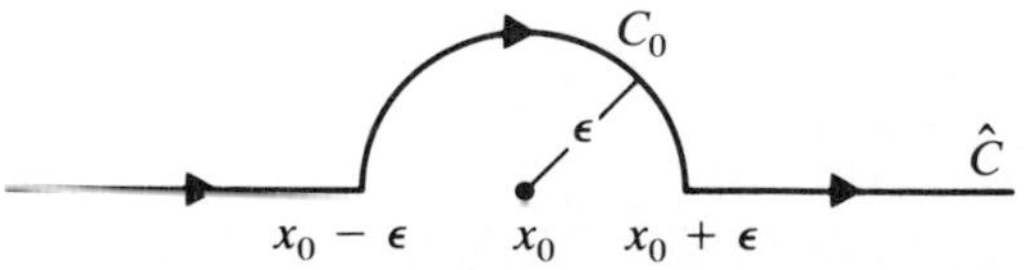

Figure 7.7 The contour $\hat{C}$ that avoids x_0.

On the other hand, if C_0 is taken below the singularity, we obtain

$$\int_{C} \frac{f(z)}{z - x_0}\, dz = \mathrm{P} \int_{-\infty}^{\infty} \frac{f(x)}{x - x_0}\, dx + i\pi f(x_0) \tag{7.12b}$$

We see that the contour integral depends on how the singular point x_0 is avoided. However, the principal value, if it exists, is unique. To calculate this principal value we close the contour by adding a large semicircle to it as before, assuming that the contribution from this semicircle goes to zero by Jordan's lemma. The LHS of (7.12a) or (7.12b) becomes a closed contour integral whose value is given by the residue theorem. We therefore have

$$\mathrm{P} \int_{-\infty}^{\infty} \frac{f(x)}{x - x_0}\, dx = \pm i\pi f(x_0) + 2i\pi \sum_{j=1}^{m} \mathrm{Res}\left[\frac{f(z_j)}{z_j - x_0}\right] \tag{7.13}$$

where the plus sign corresponds to placing the infinitesimal semicircle in the UHP, as shown in Fig. 7.7, and the minus sign corresponds to the other choice. Note that, contrary to the appearance of Eq. (7.13), the principal value (if it exists) has a unique value (see Exercise 7.2.8).

Example 7.2.13

Let us use the principal-value method to evaluate the integral

$$I \equiv \int_0^{\infty} \frac{\sin x}{x}\, dx = \frac{1}{2} \int_{-\infty}^{\infty} \frac{\sin x}{x}\, dx$$

It appears that $x = 0$ is a singular point of the integrand; in reality, however, it is only a removable singularity, as can be verified by Taylor expansion of $\sin x/x$. To make use of the principal-value method, we write

$$I = \frac{1}{2}\,\mathrm{Im}(I') \equiv \frac{1}{2}\,\mathrm{Im}\left(\int_{-\infty}^{\infty} \frac{e^{ix}}{x}\, dx\right) \equiv \frac{1}{2}\,\mathrm{Im}\left(\mathrm{P}\int_{-\infty}^{\infty} \frac{e^{ix}}{x}\, dx\right)$$

We now use Eq. (7.13) with the small circle in the UHP, noting that there are no singularities for e^{ix}/x there. This yields

$$\mathrm{P}\int_{-\infty}^{\infty} \frac{e^{ix}}{x}\, dx = i\pi e^{i(0)} = i\pi$$

Therefore,

$$\int_0^{\infty} \frac{\sin x}{x}\, dx = \frac{1}{2}\,\mathrm{Im}(i\pi) = \frac{\pi}{2}$$

We have already encountered this important result in Example 5.2.2. ●

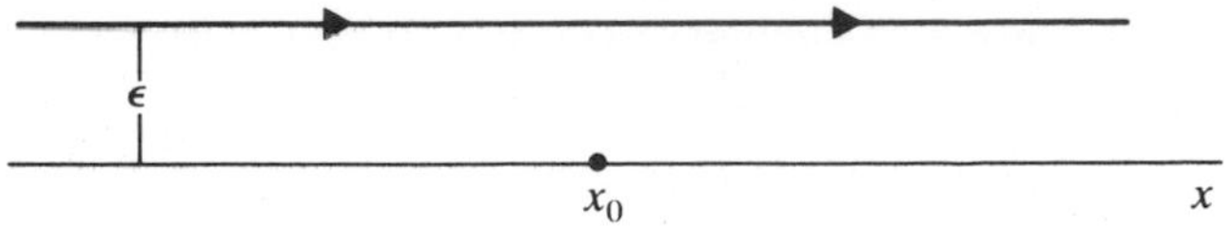

Figure 7.8 The equivalent contour obtained by "stretching" the contour of Figure 7.7.

The principal value of an integral can be written more compactly if we deform the contour in Eq. (7.12a) by stretching it into the shape shown in Fig. 7.8. For small enough ε such a deformation will not change the number of singularities within the infinite closed contour. Thus, the LHS of (7.12a) will not change. We can, therefore, write

$$\int_{\hat{C}} \frac{f(z)}{z - x_0}\,dz = \int_{-\infty + i\varepsilon}^{+\infty + i\varepsilon} \frac{f(z)}{z - x_0}\,dz$$

If we change the variable of integration to $\xi = z - i\varepsilon$, the integral becomes

$$\int_{-\infty}^{\infty} \frac{f(\xi + i\varepsilon)}{\xi + i\varepsilon - x_0}\,d\xi = \int_{-\infty}^{\infty} \frac{f(\xi)\,d\xi}{\xi - x_0 + i\varepsilon} = \int_{-\infty}^{\infty} \frac{f(z)}{z - x_0 + i\varepsilon}\,dz \tag{7.14}$$

where in the last step we changed the dummy integration variable back to z. Note that since f is assumed to be continuous at all points on the contour, $f(\xi + i\varepsilon) \to f(\xi)$ for small ε. The last integral of (7.14) shows that there is no singularity on the new x-axis; we have pushed the singularity to $x - i\varepsilon$. In other words, we have given the singularity on the x-axis a small negative imaginary part. We can thus rewrite (7.12a) as

$$\mathrm{P}\int_{-\infty}^{\infty} \frac{f(x)}{x - x_0}\,dx = i\pi f(x_0) + \int_{-\infty}^{\infty} \frac{f(x)}{x - x_0 + i\varepsilon}\,dx \tag{7.15}$$

where x is used instead of z in the last integral because we are indeed integrating along the new x-axis. A similar argument yields

$$\mathrm{P}\int_{-\infty}^{\infty} \frac{f(x)}{x - x_0}\,dx = -i\pi f(x_0) + \int_{-\infty}^{\infty} \frac{f(x)}{x - x_0 - i\varepsilon}\,dx$$

which can be combined with (7.15) to yield

$$\mathrm{P}\int_{-\infty}^{\infty} \frac{f(x)}{x - x_0}\,dx = \pm i\pi f(x_0) + \int_{-\infty}^{\infty} \frac{f(x)\,dx}{x - x_0 \pm i\varepsilon} \tag{7.16}$$

Example 7.2.14

If there are two singular points on the real axis at x_1 and x_2, we can derive a general formula for the principal value of the function. Assume that the integral is of the form

$$\int_{-\infty}^{\infty} \frac{f(x)}{(x-x_1)(x-x_2)}\,dx \equiv \int_{C_x} \frac{f(z)}{(z-x_1)(z-x_2)}\,dz$$

where C_x is simply the x-axis. Let us avoid x_1 and x_2 by making little semicircles, as before, letting both semicircles be in the UHP (see Fig. 7.9). Without writing the integrands, we can represent the contour integral by

$$\int_{-\infty}^{x_1-\varepsilon} + \int_{C_1} + \int_{x_1+\varepsilon}^{x_2-\varepsilon} + \int_{C_2} + \int_{x_2+\varepsilon}^{\infty} + \int_{C_R} = 2\pi i \sum \text{Res}$$

The principal value of the integral is naturally defined to be

$$\text{P}\int_{-\infty}^{\infty} \cdots = \int_{-\infty}^{x_1-\varepsilon} \cdots + \int_{x_1+\varepsilon}^{x_2-\varepsilon} \cdots + \int_{x_2+\varepsilon}^{\infty} \cdots$$

The contribution from the small semicircles can be calculated:

$$\int_{C_1} \frac{f(z)\,dz}{(z-x_1)(z-x_2)} = \int_{\pi}^{0} \frac{f(x_1+\varepsilon e^{i\theta})}{\varepsilon e^{i\theta}(x_1+\varepsilon e^{i\theta}-x_2)}\, i\varepsilon e^{i\theta}\,d\theta = i\,\frac{f(x_1)}{x_1-x_2}\int_{\pi}^{0} d\theta$$

↑ because $z - x_1 + \varepsilon e^{i\theta}$ on C_1

↑ because ε is small and f is assumed continuous

$$= -i\pi \frac{f(x_1)}{x_1-x_2}$$

Similarly,

$$\int_{C_2} \frac{f(z)\,dz}{(z-x_1)(z-x_2)} = -i\pi \frac{f(x_2)}{x_2-x_1}$$

Putting everything together, we get

$$\text{P}\int_{-\infty}^{\infty} \frac{f(x)\;dx}{(x-x_1)(x-x_2)} - i\pi\frac{f(x_2)-f(x_1)}{x_2-x_1} = 2\pi i \sum \text{Res}$$

If we include the case where both C_1 and C_2 are in the LHP, we get

$$\text{P}\int_{-\infty}^{\infty} \frac{f(x)\,dx}{(x-x_1)(x-x_2)} = \pm i\pi\frac{f(x_2)-f(x_1)}{x_2-x_1} + 2\pi i \sum \text{Res} \tag{1}$$

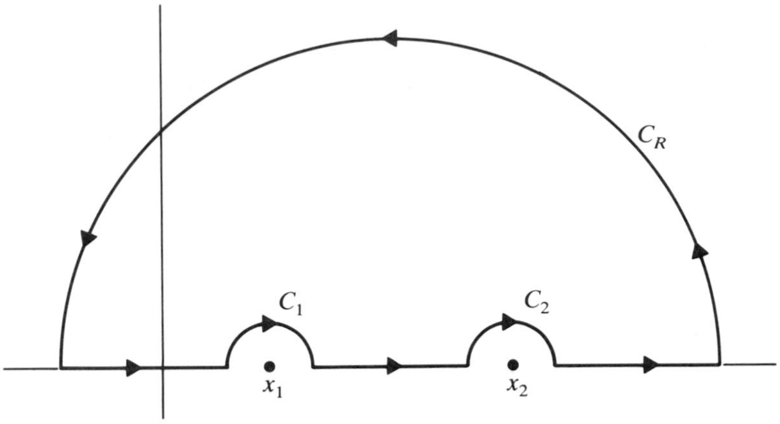

Figure 7.9 One of the four choices for the contour for evaluating the principal value of the integral.

where the plus sign is for the case where C_1 and C_2 are in the UHP and the minus sign for the case where both are in the LHP. In particular, if there exist no other singularities, there will be no contribution from residues, and (1) reduces to

$$\mathrm{P}\int_{-\infty}^{\infty} \frac{f(x)}{(x-x_1)(x-x_2)}\,dx = \pm i\pi \frac{f(x_2)-f(x_1)}{x_2-x_1} \tag{2}$$

We can also obtain the result for the case where the two singularities coincide by taking the limit $x_1 \to x_2$. Then the RHS of (2) becomes a derivative, and we obtain

$$\mathrm{P}\int_{-\infty}^{\infty} \frac{f(x)}{(x-x_0)^2}\,dx = \pm i\pi \left.\frac{df}{dx}\right|_{x=x_0}$$ •

Example 7.2.15

Let us use residues to evaluate the function

$$f(k) = \frac{1}{2\pi i}\int_{-\infty}^{\infty} \frac{e^{ikx}}{x-i\varepsilon}\,dx$$

We have to close the contour by adding a large semicircle. Whether we do this in the UHP or the LHP is dictated by the sign of k: If $k > 0$, we close in the UHP. Thus,

$$f(k) = \frac{1}{2\pi i}\int_C \frac{e^{ikz}}{z-i\varepsilon}\,dz = 2\pi i\,(\text{Res at } i\varepsilon)$$

$$= 2\pi i \lim_{z\to i\varepsilon}\left[(z-i\varepsilon)\frac{1}{2\pi i}\left(\frac{e^{ikz}}{z-i\varepsilon}\right)\right] = e^{-k\varepsilon} \xrightarrow[\varepsilon\to 0]{} 1 \qquad \text{for } k > 0$$

On the other hand, if $k < 0$, we must close in the LHP, in which the function $e^{ikz}/(z - i\varepsilon)$ is analytic. Thus, by the Cauchy-Goursat theorem, the integral vanishes. Therefore, we have

$$f(k) = \begin{cases} 1 & \text{if } k > 0 \\ 0 & \text{if } k < 0 \end{cases}$$

This is precisely the definition of the theta function. Thus, we have obtained an integral representation of that function:

$$\theta(x) = \frac{1}{2\pi i} \int_{-\infty}^{\infty} \frac{e^{ixt}}{t - i\varepsilon} dt$$

●

Exercises

7.2.1 Evaluate the integral

$$\oint_C \frac{e^z \, dz}{(z-1)(z-2)}$$

where C is the circle $|z| = 3$.

7.2.2 Let $h(z)$ be analytic and have a simple zero at $z = z_0$, and let $g(z)$ be analytic there. Let $f(z) = g(z)/h(z)$, and show that

$$\text{Res}\,[f(z_0)] = \frac{g(z_0)}{h'(z_0)}$$

7.2.3 Find the residue of $f(z) = 1/\cos z$ at all its poles.

7.2.4 Evaluate the integral $\int_0^\infty dx/(x^2+1)(x^2+4)$ by closing the contour both in the UHP and in the LHP.

7.2.5 Evaluate the integral $\int_0^\infty dx/(x^2+1)^2(x^2+2)$.

7.2.6 Evaluate the integral $I = \int_0^\infty dx \cos ax/(x^2 + b^2)$.

7.2.7 Evaluate the integral

$$I = \int_0^\pi \frac{\cos 2\theta \, d\theta}{1 - 2a\cos\theta + a^2} \qquad \text{where } |a| < 1$$

7.2.8 Show that Eq. (7.13) gives the same result for both choices of the small semicircle.

7.2.9 Find the principal part of the integral

$$I = \int_{-\infty}^{\infty} \frac{1 - \cos x}{x^2} dx$$

7.2.10 An expression encountered in the study of Green's functions (propagators) is

$$\int_{-\infty}^{\infty} \frac{e^{itx}}{x^2 - k^2} dx$$

where k and t are real constants. What is the principal value of this integral?

7.2.11 Use the result of Example 7.2.15 to show that $d\theta(k)/dk = \delta(k)$.

7.2.12 Evaluate the integral

$$I = \int_{-\infty}^{\infty} \frac{e^{\alpha x}}{1 + e^{x}}\, dx \qquad \text{for } 0 < \alpha < 1$$

7.3 MULTIVALUED FUNCTIONS

The arbitrariness, up to a multiple of 2π, of the angle $\theta \equiv \arg(z)$ in $z = re^{i\theta}$ leads to functions that can take different values at the same point. Consider, for example, the function $f(z) = z^{1/2}$. Writing z in polar coordinates, we obtain

$$f(z) \equiv f(r, \theta) = (re^{i\theta})^{1/2} = \sqrt{r}\, e^{i\theta/2}$$

This shows that for the same z, $z = (r, \theta) = (r, \theta + 2\pi)$, we get two different values, $f(r, \theta)$ and $f(r, \theta + 2\pi) = -f(r, \theta)$.

This may be disturbing at first. After all, the definition of functions (mapping) ensures that for any point in the domain a *unique* image is obtained. Here two different images are obtained for the same z. Riemann found a cure for this complex "double vision"—Riemann sheets. We will discuss these briefly below, but first let us take a closer look at a prototype of multivalued functions.

7.3.1 The Logarithmic Function

Let us consider the natural log function, $\ln z$. For $z = re^{i\theta}$ this is defined as

$$\ln z = \ln r + i\theta \equiv \ln|z| + i\arg(z)$$

where $\arg(z)$ is defined only to within a multiple of 2π; that is, $\arg(z) = \theta + 2n\pi$, for $n = 0, \pm 1, \pm 2, \ldots$.

We can see the peculiar nature of the logarithmic function by considering a closed path around the point $z = 0$, as shown in Fig. 7.10(a). Starting at z_0, we move counterclockwise, noticing the constant increase in the angle θ_0, until we reach the initial point in the z-plane. However, the angle is then $\theta_0 + 2\pi$. Thus, the process of moving around the origin has changed the value of the log function by 2π. Thus,

$$(\ln z_0)_{\text{final}} - (\ln z_0)_{\text{initial}} = 2\pi i$$

because the final z_0 has a different angle than the initial z_0 does. Note that in this process z_0 does not change because

$$(z_0)_{\text{final}} = re^{i(\theta + 2\pi)} = re^{i\theta} = (z_0)_{\text{initial}}$$

7.3.1 Definition A *branch point* of a function $f: \mathbb{C} \to \mathbb{C}$ is a point $z_0 \equiv (r_0, \theta_0) \in \mathbb{C}$ with the property that $f(r_0, \theta_0) \neq f(r_0, \theta_0 + 2\pi)$ for any closed curve C encircling z_0.

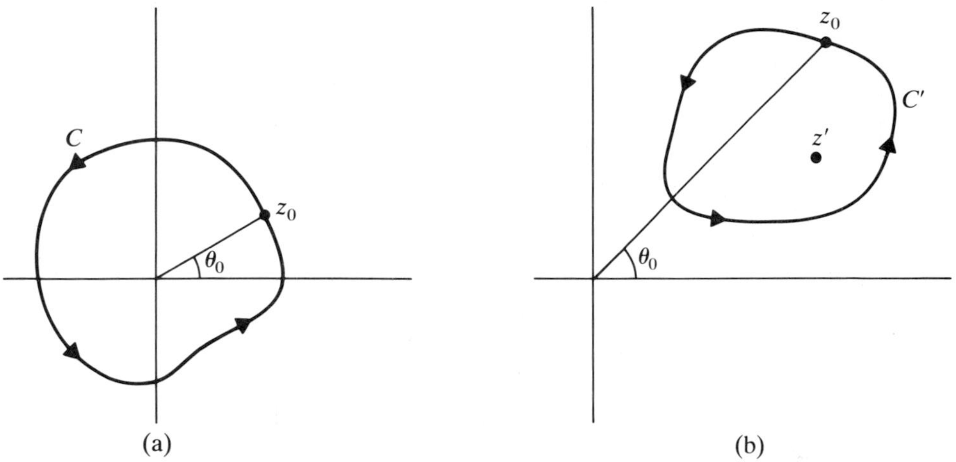

Figure 7.10 (a) The angle θ_0 changes by 2π as z_0 makes a complete circuit around C. (b) The angle θ_0 returns to its original value when z_0 completes the circuit.

Thus, $z = 0$ is a branch point of the logarithmic function. Studying the behavior of $\ln(1/w) = -\ln w$ around $w = 0$ will reveal that the point "at infinity" is also a branch point of $\ln z$. Figure 7.10(b) shows that any other point of the complex plane, such as z', cannot be a branch point because θ_0 does not increase by 2π when C' is traversed completely.

7.3.2 Riemann Surfaces

The idea of a Riemann surface begins with the removal of all points that lie on the line (or any other curve) joining two branch points. For $\ln z$ this means the removal of all points lying on a curve that starts at $z = 0$ and extends all the way to the point at infinity. Such a curve is called a *branch cut*, or simply a *cut*.

Let us concentrate on $\ln z$ and take the cut to be along the negative half of the real axis. Let us also define the functions

$$f_n(z) = f_n(r, \theta) = \ln r + i(\theta + 2n\pi) \qquad \text{for } -\pi < \theta < \pi;\ r > 0;\ n = 0,\ \pm 1, \ldots$$

so $f_n(z)$ takes on the same values for $-\pi < \theta < \pi$ that $\ln z$ takes in the range $(2n - 1)\pi < \theta < (2n + 1)\pi$. We have replaced the multivalued logarithmic function by a series of *different* functions that are analytic in the cut z-plane.

This process of cutting the z-plane and then defining a sequence of functions eliminates the contradiction caused by the existence of branch points, since we are no longer allowed to completely encircle a branch point. A complete encirculation involves crossing the cut, which, in turn, violates the domain of definition of $f_n(z)$.

We have made good progress. We have replaced the (nonanalytic) multivalued function $\ln z$ with a series of analytic (in their domain of the definition) functions $f_n(z)$.

However, there is a problem left: $f_n(z)$ has a discontinuity at the cut. In fact, just above the cut

$$f_n(r, \pi - \varepsilon) = \ln r + i(\pi - \varepsilon + 2n\pi) \qquad \text{where } \varepsilon > 0$$

and just below it

$$f_n(r, -\pi + \varepsilon) = \ln r + i(-\pi + \varepsilon + 2n\pi) \qquad \text{where } \varepsilon > 0$$

and

$$\lim_{\varepsilon \to 0} [f_n(r, \pi - \varepsilon) - f_n(r, -\pi + \varepsilon)] = 2\pi i$$

To cure this we make the observation that the value of $f_n(z)$ just above the cut is the same as the value of $f_{n+1}(z)$ just below the cut. This suggests the following geometrical construction, due to Riemann: Superpose an infinite series of cut complex planes one on top of the other, each plane corresponding to a different value of n. The adjacent planes are connected along the cut such that the upper lip of the cut in the $(n-1)$th plane is connected to the lower lip of the cut in the nth plane. All planes contain the two branch points. That is, the branch points appear as "hinges" at which all the planes are joined. With this geometrical construction, if we cross the cut, we end up in a different plane adjacent to the first one (see Fig. 7.11).

The geometric surface just constructed is called a *Riemann surface*; each plane is called a *Riemann sheet* and denoted R_j, for $j = 0, \pm 1, \pm 2, \ldots$. A single-valued function defined on a Riemann sheet is called a *branch* of the original multivalued function.

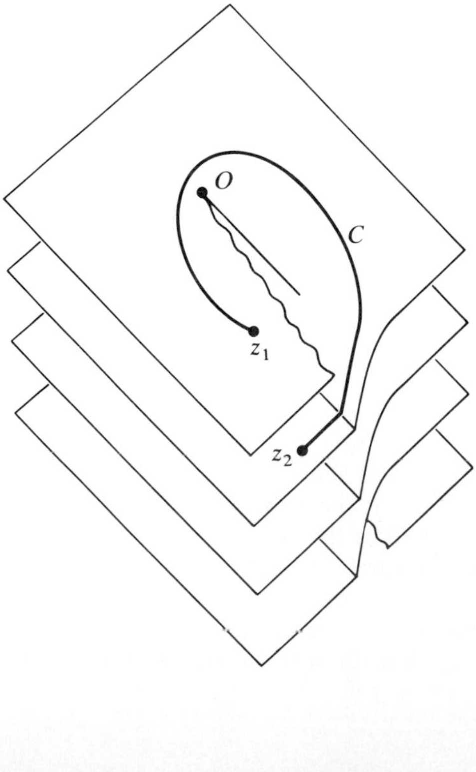

Figure 7.11 A few sheets of the Riemann surface for the logarithmic function. The path C encircling the origin O ends up on the lower sheet. (Adapted from Denery and Krzywicki, *Math for Physicists*, pp. 70 and 71. New York: Harper & Row, 1967.)

We have achieved the following: From a multivalued function we constructed a sequence of single-valued functions, each defined in a single complex plane; from this sequence of functions we have constructed a single complex function defined on a single Riemann surface. Thus, the logarithmic function is analytic throughout the Riemann surface except at the branch points, which are simply the function's singular points.

It is now easy to see the geometrical significance of branch points. A complete cycle around a branch point ends up on another Riemann sheet, where the function takes on a different value. On the other hand, a complete cycle around an ordinary point either never leaves the original sheet or, if it does, comes back to the starting point.

Let us now briefly consider two of the more common multivalued functions and their Riemann surfaces.

The function $f(z) = z^{1/n}$. The only branch points for the function $f(z) = z^{1/n}$ are $z = 0$ and the point at infinity. Defining

$$f_k(z) \equiv r^{1/n} e^{i(\theta + 2k\pi/n)} \qquad \text{for } k = 0, 1, \ldots, n-1;\ 0 < \theta < 2\pi$$

and following the same procedure as for the logarithmic function, we see that there must be n Riemann sheets, labeled $R_0, R_1, \ldots, R_{n-1}$, to the Riemann surface. The lower edge of R_{n-1} is pasted to the upper edge of R_0 along the cut, which is taken to be along the positive real axis. The Riemann surface for $n = 2$ is shown in Fig. 7.12.

It is clear that for any noninteger value of α the function $f(z) = z^\alpha$ has a branch point at $z = 0$ and another at the point at infinity. For irrational α the number of Riemann sheets is infinite.

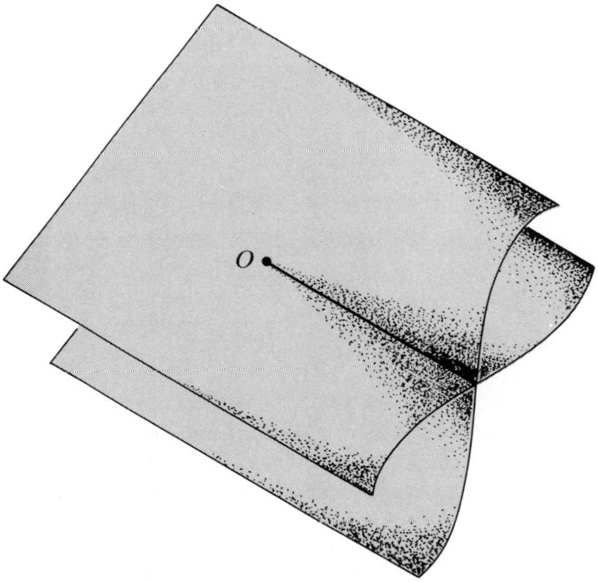

Figure 7.12 The Riemann surface for $f(z) = z^{1/2}$. (Adapted from Denery and Krzywicki, *Math for Physicists*, pp. 70 and 71. New York: Harper & Row, 1967.)

The function $f(z) = (z^2 - 1)^{1/2}$. The branch points for the function $f(z) = (z^2 - 1)^{1/2}$ are at $z_1 = +1$ and $z_2 = -1$ (see Fig. 7.13). Writing $z - 1 = r_1 e^{i\theta_1}$ and $z + 1 = r_2 e^{i\theta_2}$, we have

$$f(z) = (r_1 e^{i\theta_1})^{1/2}(r_2 e^{i\theta_2})^{1/2} = \sqrt{r_1 r_2}\, e^{i[(\theta_1+\theta_2)/2]}$$

The cut is along the real axis from $z = -1$ to $z = +1$. There are two Riemann sheets to the Riemann surface. Clearly, only cycles of 2π involving *one* branch point will cross the cut and, therefore, end up on a different sheet. Any closed curve that has *both* z_1 and z_2 as interior points will remain entirely on the original sheet.

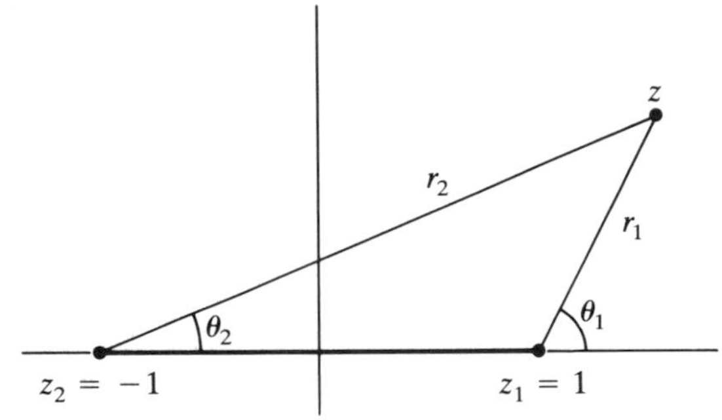

Figure 7.13 The cut is from z_1 to z_2. Paths that encircle only one of the points cross the cut and end up on the other sheet.

Example 7.3.1

Let us evaluate the integral

$$I = \int_0^\infty \frac{x^\alpha}{x^2 + 1}\, dx \qquad \text{for } -1 < \alpha < 1$$

We consider the integral

$$I' = \oint_C \frac{z^\alpha}{z^2 + 1}\, dz$$

where C is as shown in Fig. 7.14 and the cut is taken along the positive real axis. To evaluate the contribution from C_R and C_r, we let ρ stand for either r or R. Then we have

$$I_\rho \equiv \int_{C_\rho} \frac{(\rho e^{i\theta})^\alpha}{(\rho e^{i\theta})^2 + 1}\, i\rho e^{i\theta}\, d\theta = i \int_0^{2\pi} \frac{\rho^{\alpha+1} e^{i(\alpha+1)\theta}}{\rho^2 e^{2i\theta} + 1}\, d\theta$$

It is clear that, since $|\alpha| < 1$, $I_\rho \to 0$ as $\rho \to 0$ or $\rho \to \infty$.

Let us now evaluate the contributions from L_1 and L_2. Note that these do not cancel one another because the value of the function changes above and below the cut. To evaluate the integrals we have to choose a branch of the function. Let us choose that branch on which

$$z^\alpha \equiv |z|^\alpha e^{i\alpha\theta} \qquad \text{for } 0 < \theta < 2\pi$$

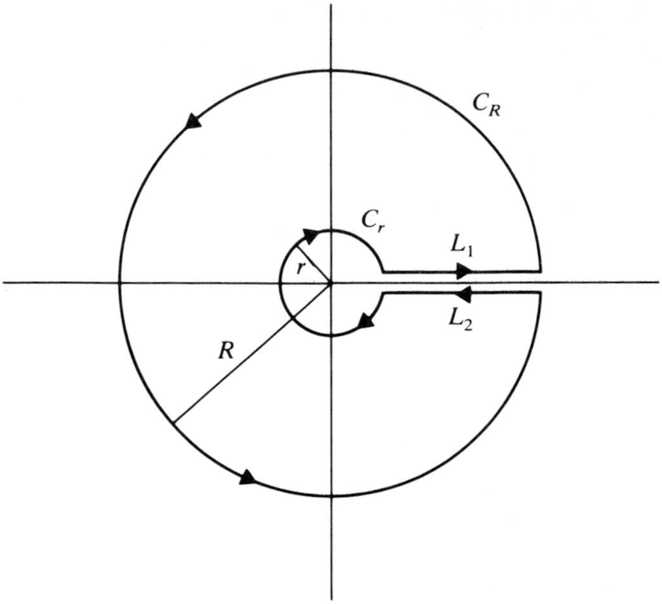

Figure 7.14

Along L_1,

$$\theta \approx 0 \quad \Rightarrow \quad z^\alpha = x^\alpha$$

and along L_2,

$$\theta \approx 2\pi \quad \Rightarrow \quad z^\alpha = (xe^{2\pi i})^\alpha$$

Thus,
$$\oint_C \frac{z^\alpha}{z^2+1}\,dz = \int_0^\infty \frac{x^\alpha}{x^2+1}\,dx + \int_\infty^0 \frac{x^\alpha e^{i2\pi\alpha}}{(xe^{2\pi i})^2+1}\,dx$$

$$= (1 - e^{i2\pi\alpha}) \int_0^\infty \frac{x^\alpha}{x^2+1}\,dx \tag{1}$$

The LHS of this equation can be obtained using the residue theorem. There are two simple poles, at $z = +i$ and $z = -i$. The residues at these points are

$$\text{Res}\,[f(i)] = \lim_{z\to i}(z-i)\frac{z^\alpha}{(z-i)(z+i)} = \frac{i^\alpha}{2i} = \frac{(e^{i\pi/2})^\alpha}{2i}$$

$$\text{Res}\,[f(-i)] = \lim_{z\to -i}(z+i)\frac{z^\alpha}{(z+i)(z-i)} = -\frac{(e^{i3\pi/2})^\alpha}{2i}$$

Thus,
$$\oint_C \frac{z^\alpha}{z^2+1}\,dz = 2\pi i\left(\frac{e^{i\alpha\pi/2}}{2i} - \frac{e^{i3\pi\alpha/2}}{2i}\right) = \pi(e^{i\alpha\pi/2} - e^{i3\alpha\pi/2})$$

Combining this with Eq. (1), we obtain

$$\int_0^\infty \frac{x^\alpha}{x^2+1}\,dx = \frac{\pi(e^{i\alpha\pi/2} - e^{i3\alpha\pi/2})}{e^{i\alpha\pi}(e^{-i\alpha\pi} - e^{i\alpha\pi})}$$

$$= \frac{-2\pi i \sin(\alpha\pi/2)}{-2i\sin(\alpha\pi)} = \frac{\pi}{2}\sec\frac{\alpha\pi}{2}$$

If we had taken a different branch of the function, both the LHS and the RHS of Eq. (1) would have been different, but the final result would still have been the same. ●

Example 7.3.2

Here is another integral involving a branch cut:

$$I = \int_0^\infty \frac{x^{-a}}{x+1}\,dx \qquad \text{for } 0 < a < 1$$

To evaluate this integral we use the zeroth branch of the function and the contour of Fig. 7.14. Thus, writing $z = \rho e^{i\theta}$, we have

$$I' = \oint_C \frac{z^{-\alpha}}{z+1}\,dz$$

$$= \int_0^\infty \frac{\rho^{-a}}{\rho+1}\,d\rho + \int_{C_R} \frac{z^{-a}}{z+1}\,dz + \int_\infty^0 \frac{(\rho e^{2i\pi})^{-a}}{\rho e^{2i\pi}+1}\,d\rho + \int_{C_r} \frac{z^{-a}}{z+1}\,dz$$

$$= 2\pi i \operatorname{Res}[f(-1)] \tag{1}$$

The contributions from both circles vanish by the same argument used in Example 7.3.1. On the other hand,

$$\operatorname{Res}[f(-1)] = \lim_{z\to -1}\left[(z+1)\frac{z^{-a}}{z+1}\right] = (-1)^{-a}$$

For the branch we are using, $-1 = e^{i\pi}$. Thus,

$$\operatorname{Res}[f(-1)] = e^{-i\pi a}$$

The integrals of (1) yield

$$I' = \int_0^\infty \frac{\rho^{-a}}{\rho+1}\,d\rho - e^{-2i\pi a}\int_0^\infty \frac{\rho^{-a}}{\rho+1}\,d\rho = (1-e^{-2i\pi a})I$$

Substituting the above in (1) gives

$$(1-e^{-2i\pi a})I = 2\pi i e^{-i\pi a}$$

or

$$I = \int_0^\infty \frac{x^{-a}}{x+1}\,dx = \frac{\pi}{\sin a\pi} \qquad \text{for } 0 < a < 1$$ ●

Example 7.3.3

Let us evaluate

$$I = \int_0^\infty \frac{\ln x}{x^2 + 1}\, dx$$

We choose the zeroth branch of the logarithmic function, in which $-\pi < \theta < \pi$, and use the contour of Fig. 7.7, closing it with a large circle in the UHP.

Thus, we set $z = \rho e^{i\pi}$ (note that $\rho > 0$), and consider

$$I' = \oint_C \frac{\ln z}{z^2 + 1}\, dz = \int_\infty^\varepsilon \frac{\ln(\rho e^{i\pi})}{(\rho e^{i\pi})^2 + 1} e^{i\pi}\, d\rho + \int_{C_r} \frac{\ln z}{z^2 + 1}\, dz + \int_\varepsilon^\infty \frac{\ln(\rho)}{\rho^2 + 1}\, d\rho + \int_{C_R} \frac{\ln z}{z^2 + 1}\, dz$$

$$= 2\pi i \operatorname{Res}[f(i)] \tag{1}$$

where $z = i$ is the only singularity (and happens to be a simple pole) in the UHP. Now we note that

$$\int_\infty^\varepsilon \frac{\ln(\rho e^{i\pi})}{(\rho e^{i\pi})^2 + 1} e^{i\pi}\, d\rho = \int_\varepsilon^\infty \frac{\ln \rho + i\pi}{\rho^2 + 1}\, d\rho = \int_\varepsilon^\infty \frac{\ln \rho}{\rho^2 + 1}\, d\rho + i\pi \int_\varepsilon^\infty \frac{d\rho}{\rho^2 + 1}$$

The contributions from the circles tend to zero. On the other hand,

$$\operatorname{Res}[f(i)] = \lim_{z \to i} \left[(z - i) \frac{\ln z}{(z - i)(z + i)} \right] = \frac{\ln i}{2i} = \frac{\ln e^{i\pi/2}}{2i} = \frac{\pi}{4}$$

Substituting in (1), we obtain

$$2 \int_\varepsilon^\infty \frac{\ln \rho}{\rho^2 + 1}\, d\rho + i\pi \int_\varepsilon^\infty \frac{d\rho}{\rho^2 + 1} = i\frac{\pi^2}{2}$$

It can also easily be shown [see Problem 7.12(p)] that

$$\int_0^\infty \frac{d\rho}{\rho^2 + 1} = \frac{\pi}{2}$$

Thus, in the limit $\varepsilon \to 0$, we get

$$I = \int_0^\infty \frac{\ln \rho}{\rho^2 + 1}\, d\rho = 0$$

●

Exercises

7.3.1 Show that the point at infinity is not a branch point for $f(z) = (z^2 - 1)^{1/2}$.

7.3.2 Evaluate the integral

$$I = \int_0^\infty \frac{\ln x}{(x^2 + 1)^2}\, dx$$

7.4 ANALYTIC CONTINUATION

Analytic functions have certain unique properties, some of which we have already noted. For instance, the Cauchy integral formula gives the value of an analytic function inside a simple closed contour given only its value on the contour. We have also seen that we can deform the contours of integration as long as we do not encounter any singularities of the function.

Combining these two properties and assuming that $f: \mathbb{C} \to \mathbb{C}$ is analytic within a region $R \subset \mathbb{C}$, we can ask the following question: Is it possible to define f *only in a subset of* R and obtain its values for the rest of R from the values in that subset? We will see in this section that the answer is yes. However, let us first consider the following theorem.

7.4.1 Theorem Let $f_1, f_2: \mathbb{C} \to \mathbb{C}$ be analytic in a region R. If $f_1 = f_2$ in a neighborhood of a point $z \in R$, or if $f_1 = f_2$ for a segment of a curve in R, then $f_1 = f_2 \,\forall z \in R$.

Proof. By Theorem 7.1.12 the zeros of an analytic function are isolated, and if the zeros are not isolated, the function must be identically zero. Applying this to $f_1 - f_2$ immediately yields the result. ■

A consequence of this theorem is the following corollary.

7.4.2 Corollary The behavior of a function that is analytic in a region $R \subset \mathbb{C}$ is completely determined by its behavior in a (small) neighborhood of an arbitrary point in that region. ■

Theorem 7.4.1 can be considered a uniqueness theorem by which the behavior of an analytic function in a small neighborhood of a point $z \in R$ is "telepathically" communicated to distant points of R. We can understand this "telepathy" more clearly if we look at the Taylor-series expansion of the analytic function.

Since f is analytic at all points of R, we can expand $f(z)$ in a Taylor series about an arbitrary point $z_0 \in R$, obtaining

$$f(z) = \sum_{n=0}^{\infty} a_n (z - z_0)^n \tag{7.17a}$$

where $|z - z_0| = r_0$ is the circle of convergence C_0 of f. Now we let $z'_0 \in R$ be another (distant) point and connect z_0 and z'_0 by a continuous path C lying entirely within R. We pick $z_1 \in C$ in such a way that z_1 is inside C_0. The power series of (7.17a) can be differentiated term by term infinitely many times, and the resulting series are all convergent for all values of z within C_0. In particular, all values $f(z_1), f'(z_1), \ldots,$ $f^{(n)}(z_1), \ldots$ are well-defined. Thus, f can be expanded about z_1:

$$f(z) = \sum_{n=0}^{\infty} \frac{f^{(n)}(z_1)}{n!}(z - z_1)^n \qquad \text{for } |z - z_1| < r_1 \tag{7.17b}$$

where r_1 is the radius of convergence of f about z_1. Since f is analytic on C, r_1 is nonzero, and we can always choose z_1 in such a way that its circle of convergence, C_1, has points outside C_0. Thus, knowing the expansion of $f(z)$ *only* for points within C_0 enables us to calculate $f(z)$ for points outside C_0 using (7.17b). Clearly, this process can be repeated until we reach z_0' (see Fig. 7.15).

This process of determining the behavior of an analytic function outside the region in which it was originally defined (in the above case, inside C_0) is called *analytic continuation.* Although there are infinitely many ways of analytically continuing beyond regions of definition, the values of all functions obtained as a result of diverse continuations are the same at any given point. This follows from Corollary 7.4.2.

Let $f_1, f_2 : \mathbb{C} \to \mathbb{C}$ be defined for regions R_1 and R_2, respectively. Suppose that f_1 and f_2 have different functional forms in their respective regions of analyticity. If there is an overlap between R_1 and R_2 and if $f_1 = f_2$ within that overlap, then the (unique) analytic continuation of f_1 into R_2 must be f_2, and vice versa. In fact, we may regard f_1 and f_2 as a single function $f: \mathbb{C} \to \mathbb{C}$ such that

$$f(z) = \begin{cases} f_1(z) & \text{when } z \in R_1 \\ f_2(z) & \text{when } z \in R_2 \end{cases}$$

Clearly, f is analytic for the combined region $R = R_1 \cup R_2$. We then say that f_1 and f_2 are *analytic continuations* of one another.

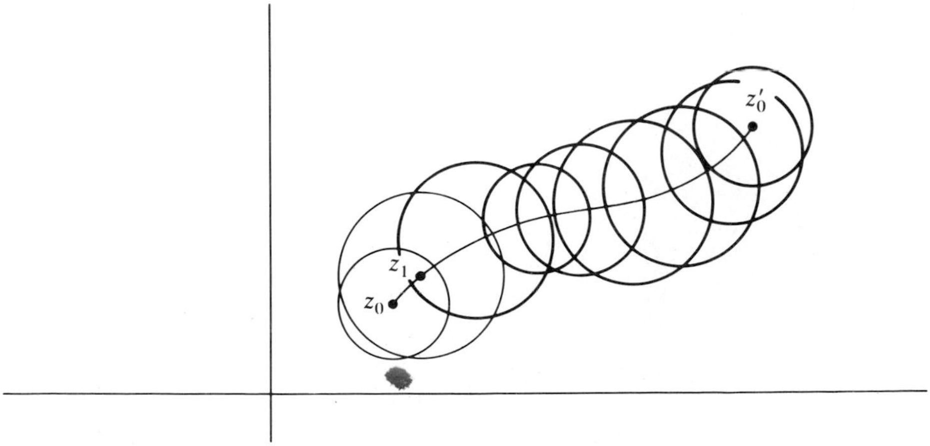

Figure 7.15 The steps involved in the analytic continuation from z_0 to z_0'.

Example 7.4.1

Let us consider the function

$$f_1(z) = \sum_{n=0}^{\infty} z^n$$

which is analytic for $|z| < 1$. We have seen that it converges to $1/(1 - z)$ for $|z| < 1$. Thus, we have

$$f_1(z) = \frac{1}{1 - z} \qquad \text{when } |z| < 1$$

and f_1 is not defined for $|z| \geqslant 1$.

Now let us consider a second function,

$$f_2(z) = \sum_{n=0}^{\infty} (\tfrac{3}{5})^{n+1}(z + \tfrac{2}{3})^n$$

which converges for $|z + \frac{2}{3}| < \frac{5}{3}$. To see what it converges to, we note that

$$f_2(z) = \tfrac{3}{5} \sum_{n=0}^{\infty} [\tfrac{3}{5}(z + \tfrac{2}{3})]^n$$

Thus,
$$f_2(z) = \frac{\frac{3}{5}}{1 - \frac{3}{5}(z + \frac{2}{3})} = \frac{1}{1 - z} \qquad \text{when } |z + \tfrac{2}{3}| < \tfrac{5}{3}$$

We observe that, although $f_1(z)$ and $f_2(z)$ have different functional forms in the two overlapping regions (see Fig. 7.16), they represent the same function, $f(z) = 1/(1 - z)$. We can, therefore, write

$$f(z) = \begin{cases} f_1(z) & \text{when } |z| < 1 \\ f_2(z) & \text{when } |z + \frac{2}{3}| < \frac{5}{3} \end{cases}$$

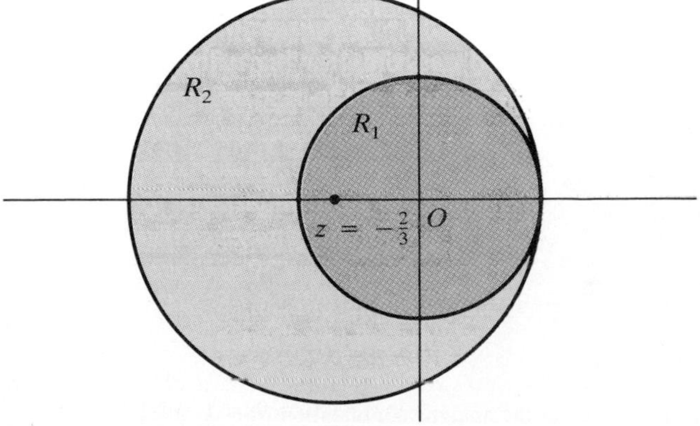

Figure 7.16 The function $f_2(z)$ is the analytic continuation of $f_1(z)$ into a larger region R_2.

and $f_1(z)$ and $f_2(z)$ are analytic continuations of one another. In fact, $f(z) = 1/(1 - z)$ is the analytic continuation of both f_1 and f_2 for all of $\mathbb{C}$ except $z = 1$. Figure 7.16 shows R_i, the region of definition of f_i, for $i = 1, 2$. ●

Example 7.4.2

The function

$$f_1(z) = \int_0^\infty e^{-zt}\, dt$$

exists only if $\text{Re}(z) > 0$, in which case

$$f_1(z) = \frac{1}{z} \qquad \text{for } \text{Re}(z) > 0$$

Its region of definition, R_1, is shown in Fig. 7.17 and is simply the right half-plane.

Now we define f_2 by a geometric series:

$$f_2(z) = i \sum_{n=0}^{\infty} \left(\frac{z+i}{i}\right)^n \qquad \text{where } |z + i| < 1$$

This series converges, within its circle of convergence R_2, to

$$i\frac{1}{1-(z+i)/i} = \frac{1}{z}$$

Thus, we have

$$f(z) \equiv \frac{1}{z} = \begin{cases} f_1(z) & \text{when } z \in R_1 \\ f_2(z) & \text{when } z \in R_2 \end{cases}$$

The two functions are analytic continuations of one another, and $f(z) = 1/z$ is the analytic continuation of both f_1 and f_2 for all $z \in \mathbb{C}$ except $z = 0$. ●

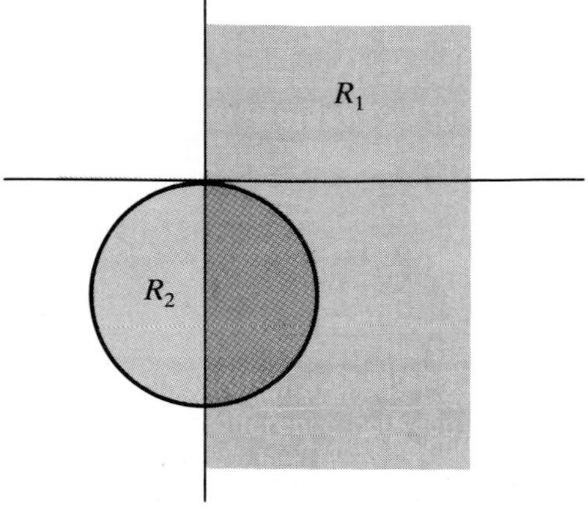

Figure 7.17 The functions $f_1(z)$ and $f_2(z)$ are analytic continuations of one another: f_1 analytically continues f_2 into the right half-plane, and f_2 analytically continues f_1 into the semicircle in the left half-plane.

7.4.1 The Schwarz Reflection Principle

A result that is useful in some physical applications is referred to as dispersion relations. To derive such relations we need to know the behavior of analytic functions on either side of the real axis. This is found using the Schwarz reflection principle, for which we need the following proposition.

7.4.3 Proposition Let f_i be analytic throughout R_i, where $i = 1, 2$. Let B be the boundary between R_1 and R_2 (see Fig. 7.18) and assume that f_1 and f_2 are continuous on B and that

$$f_1(z) = f_2(z) \qquad \text{for } z \in B$$

Then f_1 and f_2 are analytic continuations of one another and together they define a (unique) function

$$f(z) = \begin{cases} f_1(z) & \text{when } z \in R_1 \cup B \\ f_2(z) & \text{when } z \in R_2 \cup B \end{cases}$$

which is analytic throughout the entire region $R_1 \cup R_2 \cup B$.

Proof. Consider an arbitrary closed curve C in $R_1 \cup R_2 \cup B$. If C is entirely in R_1 or R_2, the integral $\oint_C f(z)\,dz$ vanishes. If C is partly in R_1 and partly in R_2, then

$$\oint_C f(z)\,dz = \oint_{C_1} f(z)\,dz + \oint_{C_2} f(z)\,dz = \oint_{C_1} f_1(z)\,dz + \oint_{C_2} f_2(z)\,dz$$

where $C_1 \subset R_1$ and $C_2 \subset R_2$. Note that the contributions from the line segments ab and cd cancel; that is why we can write the integration along C as the sum of two integrations along C_1 and C_2, which are entirely in R_1 and R_2, respectively. The RHS

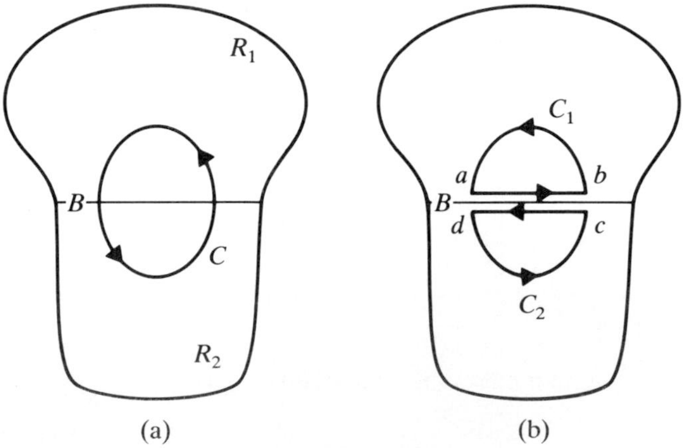

Figure 7.18 (a) Regions R_1 and R_2 separated by the boundary B, and the contour C. (b) The contour C split up into C_1 and C_2.

of the above equation vanishes by the analyticity of f_1 and f_2 in their respective regions. Thus,

$$\oint_C f(z)\,dz = 0 \qquad \forall\, C \subset R_1 \cup R_2 \cup B$$

By Morera's theorem f is analytic in all of $R_1 \cup R_2 \cup B$. Theorem 7.4.1 then implies that f_1 and f_2 are analytic continuations of one another. ■

We are now ready for the Schwarz reflection principle.

7.4.4 Theorem (Schwarz Reflection Principle) Let f be a function that is analytic in a region R, which has a segment of the real axis as part of its boundary B. If $f(z) \in \mathbb{R}$ whenever $\text{Im}(z) = 0$, then the analytic continuation of f into R^* (the mirror image of R with respect to the real axis) exists and is given by

$$g(z) \equiv f^*(z^*) \qquad \text{where } z \in R^*$$

Proof. First, we show that g is analytic in R^*. Let

$$f(z) \equiv f(x, y) = u(x, y) + iv(x, y)$$

Then
$$f(z^*) = f(x, -y) = u(x, -y) + iv(x, -y)$$

and
$$g(z) \equiv U(x, y) + iV(x, y) = u(x, -y) - iv(x, -y)$$

Thus,
$$U(x, y) = u(x, -y) \quad \text{and} \quad V(x, y) = -v(x, -y)$$

and
$$\frac{\partial U}{\partial x} = \frac{\partial u}{\partial x} = \frac{\partial v}{\partial y} = -\frac{\partial v}{\partial(-y)} = \frac{\partial V}{\partial y}$$

$$\frac{\partial U}{\partial y} = -\frac{\partial u}{\partial y} = \frac{\partial v}{\partial x} = -\frac{\partial V}{\partial y}$$

These last two equations are the Cauchy-Riemann conditions for $g(z)$. Thus, g is analytic.

Next, we compare the values of f and g on the real axis:

$$f(x, 0) = u(x, 0) + iv(x, 0)$$

$$g(x, 0) = U(x, 0) + iV(x, 0) = u(x, 0) - iv(x, 0)$$

However, by assumption $f(x, 0)$ is real, implying that $v(x, 0) = 0$ and $f(x, 0) = g(x, 0)$. Proposition 7.4.3 then implies that $f(z)$ and $f^*(z^*) = g(z)$ are analytic continuations of one another. ■

Thus, there exists an analytic function h such that

$$h(z) = \begin{cases} f(z) & \text{when } z \in R \\ f^*(z^*) & \text{when } z \in R^* \end{cases}$$

We note that $h(z^*) = g(z^*) = f^*(z) = h^*(z)$.

7.4.2 Dispersion Relations

Let $f(z)$ be analytic throughout the complex plane except at a cut along the real axis extending from x_0 to infinity. For a point z not on the x-axis, the Cauchy integral formula gives

$$f(z) = \frac{1}{2\pi i} \int_C \frac{f(\xi)}{\xi - z} \, d\xi$$

where C is the contour shown in Fig. 7.19. We assume that f drops to zero fast enough that the contribution from the large circle tends to zero. Then

$$\begin{aligned} f(z) &= \frac{1}{2\pi i} \left[\int_{x_0+i\varepsilon}^{\infty+i\varepsilon} \frac{f(\xi)}{\xi - z} d\xi - \int_{x_0-i\varepsilon}^{\infty-i\varepsilon} \frac{f(\xi)}{\xi - z} d\xi \right] \\ &= \frac{1}{2\pi i} \left[\int_{x_0}^{\infty} \frac{f(x+i\varepsilon)}{x - z + i\varepsilon} dx - \int_{x_0}^{\infty} \frac{f(x-i\varepsilon)}{x - z - i\varepsilon} dx \right] \end{aligned}$$

Since z is not on the real axis, we can ignore the $i\varepsilon$ terms in the denominators,

$$f(z) = \frac{1}{2\pi i} \int_{x_0}^{\infty} \frac{[f(x+i\varepsilon) - f(x-i\varepsilon)]}{x - z} dx$$

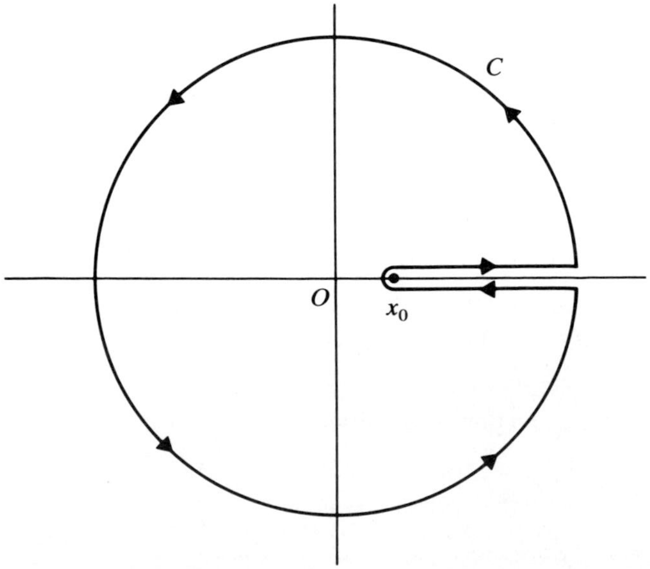

Figure 7.19 The contour used for dispersion relations.

The Schwarz reflection principle in the form $f^*(z) = f(z^*)$ can now be used to yield

$$f(x + i\varepsilon) - f(x - i\varepsilon) = f(x + i\varepsilon) - f^*(x + i\varepsilon) = 2i\,\mathrm{Im}\,[f(x + i\varepsilon)]$$

The final result is

$$f(z) = \frac{1}{\pi}\int_{x_0}^{\infty} \frac{\mathrm{Im}\,[f(x + i\varepsilon)]}{x - z}\,dx$$

This is one form of *dispersion relation.* It expresses the value of a function at any point of the cut complex plane in terms of an integral over the imaginary part of the function on the upper edge of the cut.

Example 7.4.3

Dispersion relations in which all the residues are absent in the UHP can also be obtained from Eq. (7.13). We rewrite (7.13) in the form

$$f(x_0) = \frac{1}{i\pi}\,\mathrm{P}\int_{\infty}^{\infty} \frac{f(x)}{x - x_0}\,dx$$

where we chose the contour with small circle in the UHP. Writing this in terms of real and imaginary parts of f, we obtain

$$\mathrm{Re}\,[f(x_0)] + i\,\mathrm{Im}\,[f(x_0)] = \frac{1}{i\pi}\,\mathrm{P}\int_{-\infty}^{\infty} \frac{\mathrm{Re}\,[f(x)] + i\,\mathrm{Im}\,[f(x)]}{x - x_0}\,dx$$

$$= \frac{1}{\pi}\,\mathrm{P}\int_{-\infty}^{\infty} \frac{\mathrm{Im}\,[f(x)]}{x - x_0}\,dx - i\frac{1}{\pi}\int_{-\infty}^{\infty} \frac{\mathrm{Re}\,[f(x)]}{x - x_0}\,dx$$

From this we get the following dispersion relations

$$\mathrm{Re}\,[f(x_0)] = \frac{1}{\pi}\,\mathrm{P}\int_{-\infty}^{\infty} \frac{\mathrm{Im}\,[f(x)]}{x - x_0}\,dx$$

$$\mathrm{Im}\,[f(x_0)] = -\frac{1}{\pi}\,\mathrm{P}\int_{-\infty}^{\infty} \frac{\mathrm{Re}\,[f(x)]}{x - x_0}\,dx$$

In some applications the imaginary part of f is an odd function of its argument; that is, $\mathrm{Im}\,[f(-x)] = -\mathrm{Im}\,[f(x)]$. Using this property, we have

$$\text{Re}[f(x_0)] = \frac{1}{\pi}\text{P}\int_{-\infty}^{0}\frac{\text{Im}[f(x)]}{x-x_0}dx + \frac{1}{\pi}\text{P}\int_{0}^{\infty}\frac{\text{Im}[f(x)]}{x-x_0}dx$$

$$= \frac{1}{\pi}\text{P}\int_{0}^{\infty}\frac{\text{Im}[f(-x)]}{-x-x_0}dx + \frac{1}{\pi}\text{P}\int_{0}^{\infty}\frac{\text{Im}[f(x)]}{x-x_0}dx$$

$$= \frac{1}{\pi}\text{P}\int_{0}^{\infty}\frac{2x\,\text{Im}[f(x)]}{x^2-x_0^2}dx$$

Thus, sometimes the dispersion relation is written as

$$\text{Re}[f(x_0)] = \frac{2}{\pi}\text{P}\int_{0}^{\infty}\frac{x\,\text{Im}[f(x)]}{x^2-x_0^2}dx$$

•

Example 7.4.4

For dispersion relations to hold a prerequisite is that

$$\lim_{R\to\infty} R|f(\text{Re}^{i\theta})| = 0$$

where R is the radius of the large semicircle in the UHP. If f does not satisfy this prerequisite, it is still possible to obtain a dispersion relation called a *dispersion relation with one subtraction*. This can be done by introducing an extra factor of x in the denominator of the integrand.

We can simply use the result obtained in Example 7.2.14:

$$\frac{f(x_2)-f(x_1)}{x_2-x_1} = \frac{1}{i\pi}\text{P}\int_{-\infty}^{\infty}\frac{f(x)\,dx}{(x-x_1)(x-x_2)}$$

Equating the real and imaginary parts on both sides, we obtain

$$\frac{\text{Re}[f(x_2)]}{x_2-x_1} = \frac{\text{Re}[f(x_1)]}{x_2-x_1} + \frac{1}{\pi}\text{P}\int_{-\infty}^{\infty}\frac{\text{Im}[f(x)]}{(x-x_1)(x-x_2)}dx$$

$$\frac{\text{Im}[f(x_2)]}{x_2-x_1} = \frac{\text{Im}[f(x_1)]}{x_2-x_1} - \frac{1}{\pi}\text{P}\int_{-\infty}^{\infty}\frac{\text{Re}[f(x)]}{(x-x_1)(x-x_2)}dx$$

In particular, if we let $x_1 = 0$ and $x_2 = x_0$, the first of these two equations yields

$$\frac{\text{Re}[f(x_0)]}{x_0} = \frac{\text{Re}[f(0)]}{x_0} + \frac{1}{\pi}\text{P}\int_{-\infty}^{\infty}\frac{\text{Im}[f(x)]}{x(x-x_0)}dx$$

and if $\mathrm{Im}[f(-x)] = -\mathrm{Im}[f(x)]$, we obtain

$$\frac{\mathrm{Re}[f(x_0)]}{x_0} = \frac{\mathrm{Re}[f(0)]}{x_0} + \frac{1}{\pi}\mathrm{P}\left\{\int_{-\infty}^{0} \frac{\mathrm{Im}[f(x)]}{x(x-x_0)}dx + \int_{0}^{\infty} \frac{\mathrm{Im}[f(x)]}{x(x-x_0)}dx\right\}$$

The first integral of this equation can be reexpressed as

$$\int_{\infty}^{0} \frac{\mathrm{Im}[f(-x)]}{-x(-x-x_0)}(-dx) = -\int_{0}^{\infty} \frac{\mathrm{Im}[f(x)]}{x(x+x_0)}dx$$

Substituting this back in the above equation yields

$$\mathrm{Re}[f(x_0)] = \mathrm{Re}[f(0)] + \frac{x_0}{\pi}\mathrm{P}\left\{\int_{0}^{\infty} \mathrm{Im}[f(x)]\left[\frac{-1}{x(x+x_0)} + \frac{1}{x(x-x_0)}\right]dx\right\}$$

or

$$\mathrm{Re}[f(x_0)] = \mathrm{Re}[f(0)] + \frac{2x_0^2}{\pi}\mathrm{P}\int_{0}^{\infty} \frac{dx\,\mathrm{Im}[f(x)]}{x(x^2-x_0^2)} \tag{1}$$

In optics it has been shown (see Bjorken and Drell 1965) that the imaginary part of the forward-scattering amplitude for the frequency, ω, is related, by the so-called optical theorem, to the total cross section for the absorption of light of that frequency:

$$\mathrm{Im}[f(\omega)] = \frac{\omega}{4\pi}\sigma_{\mathrm{tot}}(\omega)$$

Substituting this in (1) yields

$$\mathrm{Re}[f(\omega_0)] = \mathrm{Re}[f(0)] + \frac{\omega_0^2}{2\pi^2}\mathrm{P}\int_{0}^{\infty} \frac{d\omega\,\sigma_{\mathrm{tot}}(\omega)}{\omega^2-\omega_0^2} \tag{2}$$

Thus, the real part of the (coherent) forward scattering of light, that is, the real part of the *index of refraction*, can be computed from Eq. (2) by either measuring or calculating $\sigma_{\mathrm{tot}}(\omega)$, the simpler quantity describing the absorption of light in the medium. Equation (2) is the original Kramers-Kronig relation. ●

Exercises

7.4.1 Use analytic continuation, the analyticity of the exponential, hyperbolic, and trigonometric functions, and the analogous identities for real z to prove the following identities.
(a) $e^z = \cosh z + \sinh z$ **(b)** $\cosh^2 z - \sinh^2 z = 1$ **(c)** $\sin 2z = 2\sin z\cos z$

7.4.2 Show that the function $1/z^2$ represents the analytic continuation into the domain $\mathbb{C} - \{0\}$ of the function defined by

$$\sum_{n=0}^{\infty} (n+1)(z+1)^n \qquad \text{where } |z+1| < 1$$

7.4.3 Find the analytic continuation into $\mathbb{C} - \{i, -i\}$ of

$$f(z) = \int_0^\infty e^{-zt} \sin t \, dt \qquad \text{where } \operatorname{Re}(z) > 0$$

7.5 METHOD OF STEEPEST DESCENT

Let us consider the integral

$$I(\alpha) \equiv \int_C e^{\alpha f(z)} g(z) \, dz \tag{7.18}$$

where $|\alpha|$ is large compared to 1 and f and g are analytic in some region of $\mathbb{C}$ containing the contour C. Since this integral occurs frequently in physical applications, it would be helpful if we could find a general approximation for it that is applicable to all f and g. The fact that $|\alpha|$ is large will be of great help.

By redefining $f(z)$, if necessary, we can assume that α is real and positive [write $\alpha = |\alpha| e^{i\arg(\alpha)}$ and absorb $e^{i\arg(\alpha)}$ into the function $f(z)$].

The exponent of the integrand can be written as

$$\alpha f(z) = \alpha u(x, y) + i\alpha v(x, y)$$

Since α is large and positive, we expect the exponential to be the largest at the maximum of $u(x, y)$. Thus, if we deform the contour so that it passes through a point z_0 at which $u(x, y)$ is maximum, the contribution to the integral may come mostly from the neighborhood of z_0. This opens up the possibility of expanding the exponent about z_0 and keeping the lowest terms in the expansion, which is what we are after. There is one catch, however. Because of the largeness of α, the imaginary part of αf in the exponent will oscillate violently as $v(x, y)$ changes even by a small amount. This oscillation can make the contribution of the real part of $f(z_0)$ negligibly small and render the whole procedure useless. Thus, we want to tame the variation of $e^{i\alpha v(x, y)}$ by making $v(x, y)$ vary as slowly as possible. A necessary (but not sufficient) condition is for the derivative of v to vanish at z_0. This and the fact that the real part is to have a maximum at z_0 lead to

$$\frac{\partial u}{\partial x} + i\frac{\partial v}{\partial x} \equiv \left.\frac{df}{dz}\right|_{z_0} = 0 \tag{7.19}$$

However, we do not stop here but demand that the imaginary part of f be *constant* along C: $\operatorname{Im}[f(z)] = \operatorname{Im}[f(z_0)]$ or $v(x, y) = v(x_0, y_0)$.

Equation (7.19) and the Cauchy-Riemann conditions imply that $\partial u/\partial x = 0 = \partial u/\partial y$ at z_0. Thus, it might appear that z_0 is a true maximum (or minimum) of the surface described by the function $z = u(x, y)$ (here z is the axis, *not* the complex number). There are two ways to show that this is not true. The first is by Proposition 6.3.6, which states that $f(z)$ has no local maximum if f is analytic. The second way is by noting that for the surface to have a maximum (minimum) *both* second derivatives, $\partial^2 u/\partial x^2$ and $\partial^2 u/\partial y^2$, must be negative (positive). But that is impossible because $u(x, y)$ is harmonic, or $\partial^2 u/\partial x^2 + \partial^2 u/\partial y^2 = 0$. A point at which the derivatives vanish but which is neither a maximum nor a minimum is called a *saddle point*. That is why the procedure described below is sometimes called the *saddle point method.*

We are interested in values of z close to z_0. So let us expand $f(z)$ in a Taylor series about z_0, use Eq. (7.19), and keep terms only up to the second term, obtaining

$$f(z) = f(z_0) + \tfrac{1}{2}(z - z_0)^2 f''(z_0) \tag{7.20}$$

Let us assume that $f''(z_0) \neq 0$ (if it does, we go to the next higher term). We define

$$z - z_0 \equiv r_1 e^{i\theta_1} \tag{7.21a}$$

and

$$\tfrac{1}{2}f''(z_0) \equiv r_2 e^{i\theta_2} \tag{7.21b}$$

and substitute in the above expansion to obtain

$$f(z) - f(z_0) = r_1^2 r_2 e^{i(2\theta_1 + \theta_2)}$$

or

$$\mathrm{Re}[f(z) - f(z_0)] = r_1^2 r_2 \cos(2\theta_1 + \theta_2) \tag{7.22a}$$

$$\mathrm{Im}[f(z) - f(z_0)] = r_1^2 r_2 \sin(2\theta_1 + \theta_2) \tag{7.22b}$$

The constancy of $\mathrm{Im}[f(z)]$ implies that $\sin(2\theta_1 + \theta_2) = 0$, or $2\theta_1 + \theta_2 = n\pi$. Thus, for

$$\theta_1 = -\frac{\theta_2}{2} + n\frac{\pi}{2} \qquad \text{where } n = 0, 1, 2, 3$$

the imaginary part of f is constant. Substituting in (7.21a), we obtain

$$z - z_0 = r_1 e^{i(-\theta_2/2 + n\pi/2)}$$

which in Cartesian components becomes

$$x - x_0 = r_1 \cos\left(\frac{n\pi}{2} - \frac{\theta_2}{2}\right)$$

$$y - y_0 = r_1 \sin\left(\frac{n\pi}{2} - \frac{\theta_2}{2}\right)$$

Eliminating r_1 from these two equations yields

$$y - y_0 = \left[\tan\left(\frac{n\pi}{2} - \frac{\theta_2}{2}\right)\right](x - x_0)$$

This is the equation of a line (because θ_2 is constant) passing through $z_0 = (x_0, y_0)$ and making an angle of $(n\pi - \theta_2)/2$ with the real axis. For $n = 0, 2$ we get one line, and for $n = 1, 3$ we get another that is perpendicular to the first (see Fig. 7.20). It is to be emphasized that along these lines the imaginary part of $f(z)$ remains constant.

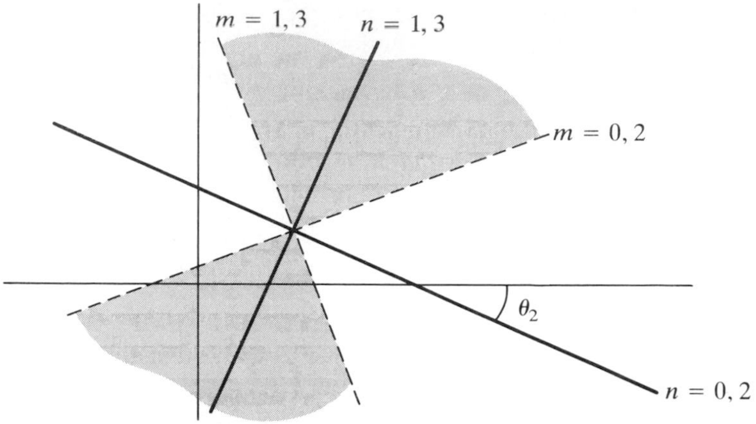

Figure 7.20 Different regions of the z-plane for the method of steepest descent. The shaded regions are those in which $\mathrm{Re}[f(z)] < \mathrm{Re}[f(z_0)]$. Note that only one curve of constant imaginary part passes through such regions.

We can similarly obtain the curves (lines) along which the real part of $f(z)$ is constant. We have

$$\cos(2\theta_1 + \theta_2) = 0 \quad \Rightarrow \quad 2\theta_1 + \theta_2 = \frac{\pi}{2} + n\pi$$

or

$$\theta_1 = -\frac{\theta_2}{2} + \frac{2m+1}{4}\pi \qquad \text{where } m = 0, 1, 2, 3$$

The corresponding lines are given by

$$y - y_0 = \left[\tan\left(\frac{2m+1}{4}\pi - \frac{\theta_2}{2}\right)\right](x - x_0)$$

which are at 45° from the other set of lines (see Fig. 7.20).

We are interested in regions where $\mathrm{Re}(f)$ goes through a relative maximum at z_0. According to Eq. (7.22a), this occurs at points where $\cos(2\theta_1 + \theta_2) < 0$. This, in turn, yields

$$\frac{4k+1}{4}\pi - \frac{\theta_2}{2} \leqslant \theta_1 \leqslant \frac{2k+1}{2}\pi - \frac{\theta_2}{2} \qquad \text{for } k = 0, 1$$

or

$$\frac{2k+1}{2}\pi - \frac{\theta_2}{2} \leqslant \theta_1 \leqslant \frac{4k+3}{4}\pi - \frac{\theta_2}{2} \qquad \text{for } k = 0, 1$$

These regions are shown in Fig. 7.20. It is clear that only one curve of constant $\mathrm{Im}[f(z)]$ passes through these regions. This is generally true because there is only one zero of the sine between two adjacent zeros of the cosine.

We want to choose that path within these regions along which the quantity $|\mathrm{Re}[f(z)] - \mathrm{Re}[f(z_0)]|$ is maximum. Equation (7.22a) determines such a path, on which $\cos(2\theta_1 + \theta_2)$ must be -1. But this corresponds precisely to the path on which $\sin(2\theta_1 + \theta_2) = 0$, that is, the path on which the imaginary part of $f(z)$ is constant. There is only one such path in the region of interest, and the procedure is uniquely determined. Because the descent from the maximum value at z_0 is maximum along such a path, this procedure is called the *method of steepest descent.*

Now that we have determined the contour C_0, let us approximate the integral. First, we note that $\operatorname{Im}[f(z) - f(z_0)] = 0$ along C_0. Thus, $f(z) - f(z_0)$ is real and *negative* along C_0, and we can introduce a real parameter t given, according to Eq. (7.20), by

$$t^2 = -\tfrac{1}{2}(z - z_0)^2 f''(z_0) \tag{7.23}$$

Substituting (7.20) in (7.18) and using (7.23) yields

$$I(\alpha) \approx \int_{C_0} e^{\alpha[f(z_0) - t^2]} g(z)\, dz$$

$$= e^{\alpha f(z_0)} \int_{C_0} e^{-\alpha t^2} g(z)\, dz$$

We can solve for z in terms of t from Eq. (7.23) and substitute in this equation to obtain

$$I(\alpha) \approx e^{\alpha f(z_0)} \int_{-\infty}^{\infty} e^{-\alpha t^2} g(z(t)) \frac{dz}{dt}\, dt \tag{7.24}$$

The extension of the integral limits to infinity does not alter the result significantly because α is assumed large and positive. To evaluate (7.24) we replace $g(z(t))\, dz/dt$ by its Maclaurin series in t,

$$g(z(t))\frac{dz}{dt} = \sum_{k=0}^{\infty} a_k t^k \tag{7.25}$$

Substituting this in (7.24) gives

$$I(\alpha) \approx e^{\alpha f(z_0)} \sum_{k=0}^{\infty} a_k \int_{-\infty}^{\infty} e^{-\alpha t^2} t^k\, dt \tag{7.26}$$

Denoting the integral in the sum by I_k and noting that $I_{2n+1} = 0$ for all n, we have

$$I_{2n+2} = -\frac{\partial}{\partial \alpha} I_{2n} \qquad \text{and} \qquad I_0 = \sqrt{\frac{\pi}{\alpha}}$$

This gives

$$I_{2n} = \sqrt{\pi}\,\frac{(1)(3)(5)\cdots(2n-1)}{2^n \alpha^{(2n+1)/2}} \tag{7.27a}$$

It can also be shown that

$$I_{2n} = \alpha^{-n-1/2}\,\Gamma(n + \tfrac{1}{2}) \qquad \forall\, n \geqslant 0 \tag{7.27b}$$

where $\Gamma(k)$ is the gamma function mentioned in Chapter 5 and discussed in Chapter 14.

We substitute in (7.26) to obtain the final form, the *asymptotic expansion* of $I(\alpha)$:

$$I(\alpha) \approx e^{\alpha f(z_0)} \sum_{n=0}^{\infty} a_{2n} I_{2n} = e^{\alpha f(z_0)} \sum_{n=0}^{\infty} a_{2n}\alpha^{-n-1/2}\,\Gamma(n + \tfrac{1}{2}) \tag{7.28a}$$

In almost all applications only the first term of the above series is retained, giving

$$I(\alpha) \approx e^{\alpha f(z_0)} \sqrt{\frac{\pi}{\alpha}}\, a_0 \tag{7.28b}$$

Example 7.5.1

Let us approximate the integral

$$I(\alpha) = \int_0^{\infty} e^{-z} z^{\alpha}\, dz$$

where α is real. For simplicity, let us assume that $\alpha > 0$. First, we must rewrite the integral in the form of Eq. (7.18). We can do this by noting that $z^{\alpha} = e^{\alpha \ln z}$. Thus, we have

$$I(\alpha) = \int_0^{\infty} e^{\alpha \ln z - z}\, dz = \int_0^{\infty} e^{\alpha(\ln z - z/\alpha)}\, dz$$

and we identify

$$f(z) = \ln z - \frac{z}{\alpha} \qquad \text{and} \qquad g(z) = 1$$

The saddle point is found from

$$\frac{df}{dz} = 0 \quad \Rightarrow \quad \frac{1}{z} - \frac{1}{\alpha} = 0 \quad \Rightarrow \quad z_0 = \alpha$$

We must now choose the path of steepest descent, for which we need $f''(z_0) = -1/\alpha^2$. Thus, Eq. (7.21b) gives

$$\frac{1}{2}\left(-\frac{1}{\alpha^2}\right) = r_2 e^{i\theta_2} = \frac{1}{2\alpha^2}(-1) = \frac{1}{2\alpha^2} e^{i\pi} \quad \Rightarrow \quad \theta_2 = \pi$$

The lines of constant phase are given by

$$\sin(2\theta_1 + \theta_2) = 0 \quad \Rightarrow \quad \sin(2\theta_1 + \pi) = 0 \quad \Rightarrow \quad 2\theta_1 + \pi = n\pi \quad \text{for } n = 0, 1, 2, 3$$

To obtain the steepest descent, we must have

$$\cos(2\theta_1 + \pi) = -1 \quad \Rightarrow \quad \theta_1 = 0, \pi$$

Recalling the definition of θ_1 [Eq. (7.21a)], we can write

$$z - \alpha = r_1 e^{i\theta_1} = \pm r_1 \quad \Rightarrow \quad z = \alpha \pm r_i$$

Thus, z must be on the real line for the approximation to be good. This means that we do not have to deform the contour.

We want to use (7.28b). To find a_0 we use Eq. (7.25), which for small t gives $a_0 = dz/dt$. On the other hand, (7.23) and $f''(\alpha) = -1/\alpha^2$ yield

$$t^2 = \frac{(z-\alpha)^2}{2\alpha^2} \quad \Rightarrow \quad t = \frac{z-\alpha}{\sqrt{2}\alpha} \quad \Rightarrow \quad \frac{dz}{dt} = \sqrt{2}\alpha = a_0$$

Substituting in (7.28b), we obtain

$$I(\alpha) \approx e^{\alpha f(\alpha)} \sqrt{\frac{\pi}{\alpha}} \sqrt{2}\alpha = \sqrt{2\pi}\alpha^{1/2} e^{\alpha(\ln \alpha - 1)}$$

$$= \sqrt{2\pi}\, e^{-\alpha} \alpha^{\alpha + 1/2}$$

Since (as you may have noticed) $I(\alpha) = \Gamma(\alpha + 1)$, this equation is an approximation for the gamma function, called the *Stirling approximation*. We note that if $\alpha = n$, an integer, then

$$\Gamma(n+1) = n! \approx \sqrt{2\pi}\, e^{-n} n^{n+1/2}$$

which is sometimes written as

$$\ln(n!) \approx \ln(\sqrt{2\pi}) - n + (n + \tfrac{1}{2}) \ln n$$

For large enough n this can be further simplified to

$$\ln(n!) \approx n \ln n - n$$

•

Exercise

7.5.1 The Hankel function of the first kind is defined as

$$H_\nu^{(1)}(\alpha) = \frac{i}{\pi} \int_{-\infty}^{0} e^{(\alpha/2)(z - 1/z)} \frac{dz}{z^{\nu+1}}$$

Find the asymptotic expansion of this function. Choose the branch of the function in which $-\pi < \theta \leqslant \pi$.

PROBLEMS

7.1 Show that if $\sum_{n=1}^{\infty} z_n = s$, then $\sum_{n=1}^{\infty} z_n^* = s^*$.

7.2 Show that if the sequence $\{z_n\}_{n=1}^{\infty}$ converges, there exists $M > 0$ such that $|z_n| < M$ for all n.

7.3 Show that if $\lim_{n\to\infty} z_n = z$ and $|z_n| \leqslant M$ for all n, then $|z| \leqslant M$.

7.4 Show that

$$\frac{1}{z^2} = \frac{1}{4} \sum_{n=0}^{\infty} (-1)^n (n+1) \left(\frac{z-2}{2}\right)^n \qquad \text{when } |z-2| < 2$$

7.5 Expand $\sinh z$ in a Taylor series about the point $z = i\pi$.

7.6 What is the largest circle within which the Maclaurin series for $\tanh z$ converges to $\tanh z$?

7.7 Find the (unique) expansion of each of the following functions for the entire region of analyticity.

(a) $\dfrac{1}{(z-2)(z-3)}$ (b) $z\cos(z^2)$ (c) $\dfrac{1}{z^2(1-z)}$ (d) $\dfrac{\sinh z - z}{z^4}$

(e) $\dfrac{1}{(1-z)^3}$ (f) $\dfrac{1}{z^2-1}$ (g) $\dfrac{z^2-4}{z^2-9}$ (h) $\dfrac{1}{(z^2-1)^2}$

7.8 Complete the proof of Proposition 7.1.7.

7.9 Show that the following functions, $f: \mathbb{C} \to \mathbb{C}$, are entire.

(a) $f(z) \equiv \begin{cases} \dfrac{e^{2z}-1}{z^2} - \dfrac{2}{z} & \text{for } z \neq 0 \\ 2 & \text{for } z = 0 \end{cases}$ (b) $f(z) \equiv \begin{cases} \dfrac{\sin z}{z} & \text{for } z \neq 0 \\ 1 & \text{for } z = 0 \end{cases}$

7.10 Obtain the first few nonzero terms of the Laurent-series expansion of each of the following functions. Also find the integral of the function along a small simple closed contour encircling the origin.

(a) $\dfrac{1}{\sin z}$ (b) $\dfrac{1}{1-\cos z}$ (c) $\dfrac{z}{1-\cosh z}$ (d) $\dfrac{z^2}{z-\sin z}$ (e) $\dfrac{z^4}{6z+z^3-6\sinh z}$

7.11 Evaluate each of the following integrals, for which C is the circle $|z| = 3$.

(a) $\oint_C \dfrac{4z-3}{z(z-2)}dz$ (b) $\oint_C \dfrac{e^z}{z(z-i\pi)}dz$ (c) $\oint_C \dfrac{\cos z}{z(z-\pi)}dz$

(d) $\oint_C \dfrac{z^2+1}{z(z-1)}dz$ (e) $\oint_C \dfrac{\cosh z}{z^2+\pi^2}dz$ (f) $\oint_C \dfrac{1-\cos z}{z^2}dz$

(g) $\oint_C \dfrac{\sinh z}{z^4}dz$ (h) $\oint_C z\cos\left(\dfrac{1}{z}\right)dz$ (i) $\oint_C \dfrac{dz}{z^3(z+5)}$

(j) $\oint_C \tan z\, dz$ (k) $\oint_C \dfrac{dz}{\sinh 2z}$ (l) $\oint_C \dfrac{e^z}{z^2}dz$

(m) $\oint_C \dfrac{dz}{z^2\sin z}$

7.12 Evaluate the following integrals.

(a) $\displaystyle\int_0^\infty \frac{2x^2+1}{x^4+5x^2+6}dx$ (b) $\displaystyle\int_0^\infty \frac{dx}{6x^4+5x^2+1}$ (c) $\displaystyle\int_0^\infty \frac{dx}{x^4+1}$

(d) $\displaystyle\int_0^\infty \frac{2x^2-1}{x^6+1}dx$ (e) $\displaystyle\int_0^\infty \frac{dx}{(x^2+4)^2}$ (f) $\displaystyle\int_0^\infty \frac{x^2\,dx}{(x^2+4)^2(x^2+25)}$

(g) $\displaystyle\int_0^\infty \frac{\cos ax}{x^2+1}\,dx$ where $a \neq 0$

(h) $\displaystyle\int_0^\infty \frac{\cos ax\,dx}{(b^2+x^2)^2}$ where $a \neq 0 \neq b$

(i) $\displaystyle\int_0^\infty \frac{\cos x\,dx}{(x^2+a^2)(x^2+b^2)}$

(j) $\displaystyle\int_0^\infty \frac{x^3 \sin ax}{x^6+1}\,dx$

(k) $\displaystyle\int_{-\infty}^\infty \frac{x\,dx}{(x^2+4x+13)^2}$

(l) $\displaystyle\int_0^\infty \frac{x^2\,dx}{(x^2+a^2)^2}$ where $a > 0$

(m) $\displaystyle\int_0^\infty \frac{x^2+1}{x^2+4}\,dx$

(n) $\displaystyle\int_{-\infty}^\infty \frac{x\cos x}{x^2-2x+10}\,dx$

(o) $\displaystyle\int_{-\infty}^\infty \frac{x\sin x}{x^2-2x+10}\,dx$

(p) $\displaystyle\int_0^\infty \frac{dx}{x^2+1}$

(q) $\displaystyle\int_0^\infty \frac{dx}{(x^2+1)^2}$

7.13 Find the Cauchy principal values of the following integrals.

(a) $\displaystyle\int_{-\infty}^\infty \frac{\sin x\,dx}{(x^2+4)(x-1)}$

(b) $\displaystyle\int_{-\infty}^\infty \frac{\cos ax}{1+x^3}\,dx$ where $a \geqslant 0$

(c) $\displaystyle\int_{-\infty}^\infty \frac{x\cos x\,dx}{x^2-5x+6}$

7.14 Evaluate the following integrals.

(a) $\displaystyle\int_0^\infty \frac{x^2-b^2}{x^2+b^2}\left(\frac{\sin ax}{x}\right)dx$

(b) $\displaystyle\int_0^\infty \frac{\sin ax}{x(x^2+b^2)}\,dx$

(c) $\displaystyle\int_0^\infty \frac{\sin ax}{x(x^2+b^2)^2}\,dx$

(d) $\displaystyle\int_0^\infty \frac{\sin^2 x}{x^2}\,dx$

(e) $\displaystyle\int_0^\infty \frac{\cos 2ax - \cos 2bx}{x^2}\,dx$

(f) $\displaystyle\int_0^\infty \frac{\sin^3 x}{x^3}\,dx$

7.15 Evaluate each of the following integrals using the substitution $z = e^{i\theta}$.

(a) $\displaystyle\int_0^{2\pi} \frac{d\theta}{5+4\sin\theta}$

(b) $\displaystyle\int_0^{2\pi} \frac{d\theta}{1+\sin^2\theta}$

(c) $\displaystyle\int_0^{2\pi} \frac{\cos^2 3\theta}{5-4\cos 2\theta}\,d\theta$

(d) $\displaystyle\int_0^{2\pi} \frac{d\theta}{a+\cos\theta}$ where $a > 1$

(e) $\displaystyle\int_0^{2\pi} \frac{d\theta}{(a+b\cos^2\theta)^2}$ where $a, b > 0$

(f) $\displaystyle\int_0^{\pi} \frac{d\phi}{1-2a\cos\phi+a^2}$ where $a \neq \pm 1$

(g) $\displaystyle\int_0^{\pi} \frac{\cos^2 3\phi\,d\phi}{1-2a\cos\phi+a^2}$ where $a \neq \pm 1$

(h) $\displaystyle\int_0^{\pi} e^{\cos\phi}\cos(n\phi - \sin\phi)\,d\phi$ where n is an integer

(i) $\displaystyle\int_0^{\pi} \tan(x+ia)\,dx$ where $a \in \mathbb{R}$

7.16 Derive the integration formula

$$\int_0^\infty e^{-x^2} \cos(2bx)\, dx = \frac{\sqrt{\pi}}{2} e^{-b^2} \qquad \text{where } b \neq 0$$

by integrating the function e^{-z^2} around the rectangular path shown in Fig. 7.21.

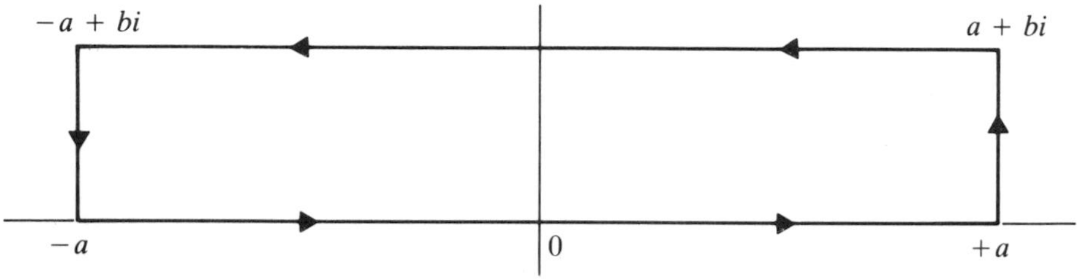

Figure 7.21

7.17 The *beta function* is defined as

$$B(a, b) \equiv \int_0^1 t^{a-1}(1-t)^{b-1}\, dt \qquad \text{where } a > 0, b > 0$$

Make the substitution $t = 1/(x + 1)$ and use the result of Example 7.3.2 to show that

$$B(a, 1 - a) = \frac{\pi}{\sin a\pi}$$

7.18 Find the following integrals for which $a \neq 0$ and $a \in \mathbb{R}$.

(a) $\displaystyle\int_0^\infty \frac{\ln x}{x^2 + a^2} dx$ **(b)** $\displaystyle\int_0^\infty \frac{\ln x}{(x^2 + a^2)^2 \sqrt{x}} dx$ **(c)** $\displaystyle\int_0^\infty \frac{(\ln x)^2\, dx}{x^2 + a^2}$

7.19 Use the contour in Fig. 7.22 to evaluate the following integrals.

(a) $\displaystyle\int_0^\infty \frac{\sin ax}{\sinh x} dx$ **(b)** $\displaystyle\int_0^\infty \frac{x \cos ax}{\sinh x} dx$

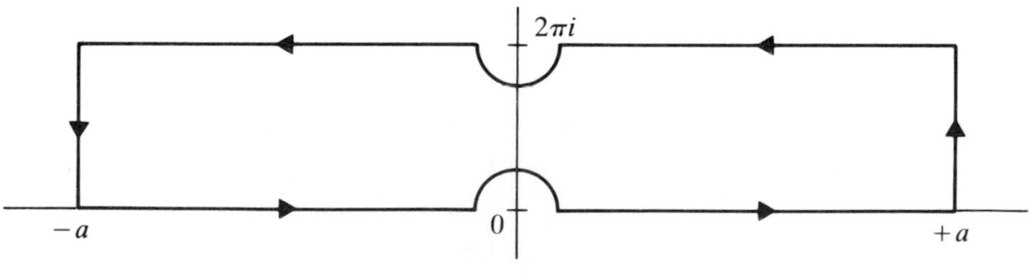

Figure 7.22

7.20 Show that

$$\int_0^{\pi} f(\sin\theta)\,d\theta = 2\int_0^{\pi/2} f(\sin\theta)\,d\theta$$

for an arbitrary function f defined on the interval $[-1, +1]$.

7.21 Expand $f(z) = \sum_{n=0}^{\infty} z^n$ in a Taylor series about $z = a$ (where $|a| < 1$). For what values of a does this expansion permit the function $f(z)$ to be continued analytically?

7.22 The two power series

$$f_1(z) = \sum_{n=1}^{\infty} \frac{z^n}{n} \qquad \text{and} \qquad f_2(z) = i\pi + \sum_{n=1}^{\infty} (-1)^n \frac{(z-2)^n}{n}$$

have no common domain of convergence. Show that they are nevertheless analytic continuations of one another.

7.23 Prove that the functions defined by the two series

$$1 + az + a^2z^2 + \cdots \qquad \text{and} \qquad \frac{1}{1-z} - \frac{(1-a)^z}{(1-z)^2} + \frac{(1-a)^2 z^2}{(1-z)^3} - \cdots$$

are analytic continuations of one another.

7.24 Show that the function

$$f_1(z) = \frac{1}{z^2+1} \qquad \text{where } z \neq \pm i$$

is the analytic continuation into $\mathbb{C} - \{i, -i\}$ of the function

$$f_2(z) = \sum_{n=0}^{\infty} (-1)^n z^{2n} \qquad \text{where } |z| < 1$$

7.25 Find the analytic continuation into $\mathbb{C} - \{0\}$ of the function

$$f(z) = \int_0^{\infty} te^{-zt}\,dt \qquad \text{where } \mathrm{Re}(z) > 0$$

7.26 Find the principal value of the integral

$$\int_{-\infty}^{\infty} \frac{x\sin x}{x^2 - x_0^2}\,dx$$

and evaluate

$$I = \int_{-\infty}^{\infty} \frac{x\sin x\,dx}{(x - x_0 \pm i\varepsilon)(x + x_0 \pm i\varepsilon)}$$

for the four possible choices of signs.

7.27 The Hankel function of the second kind is defined as

$$H_\nu^{(2)}(\alpha) \equiv \frac{1}{\pi i} \int_{-\infty}^{0} e^{(\alpha/2)(z-1/z)} \frac{dz}{z^{\nu+1}}$$

where the limits are understood to be in the LHP. Find the asymptotic expansion of this function.

7.28 Find the asymptotic dependence of the modified Bessel function of the first kind, defined as

$$I_\nu(\alpha) \equiv \frac{1}{2\pi i} \oint_C e^{(\alpha/2)(z+1/z)} \frac{dz}{z^{\nu+1}}$$

where C starts at $-\infty$, circles the origin, and finally goes back to $-\infty$. Thus, the negative real axis is excluded from the domain of analyticity.

7.29 Find the asymptotic dependence of

$$K_\nu(\alpha) \equiv \frac{1}{2} \int_{0}^{\infty} e^{-(\alpha/2)(z+1/z)} \frac{dz}{z^{1-\nu}}$$

called the modified Bessel function of the second kind. The branch cut is taken to be along the positive real axis.

PART III

DIFFERENTIAL EQUATIONS

8

Differential Equations I: Separation of Variables

Physics, the most exact of all sciences, is concerned with predictions. Predictions are based on two factors: the initial information (data), and the law governing the physical process at hand. Thus, knowing what the situation is *here* and *now* (initial data, initial conditions, boundary conditions) enables physicists to predict what it will be *there* and *then*. This ability to predict is based on the intuitive belief that physical quantities, dependent on continuous parameters such as position and time, must be continuous functions of those parameters. Thus, knowledge of the values of those functions at one (initial) point and of how the functions change from one point to a neighboring point (given by the laws of physics) allows the values of the functions at the neighboring point to be predicted. Once the values of the functions are determined at the new point, their values can be predicted for its neighboring points, and the process can continue until a distant point is reached.

In mechanics knowledge of the force acting on a particle of mass m, located at $\mathbf{r}_0$ and moving with momentum $\mathbf{p}_0$ at time t_0, allows its momentum and position at a time $t_0 + \Delta t$ to be predicted. Because $d\mathbf{p}/dt = \mathbf{F}$, by Newton's second law of motion, we have

$$\Delta\mathbf{p} \approx \mathbf{F}(\mathbf{r}_0, t_0)\Delta t$$

and

$$\mathbf{p}(t_0 + \Delta t) = \mathbf{p}_0 + \Delta\mathbf{p} \approx \mathbf{p}_0 + \mathbf{F}(\mathbf{r}_0, t_0)\Delta t$$

Similarly,

$$\mathbf{r}(t_0 + \Delta t) \approx \mathbf{r}_0 + \frac{\mathbf{p}_0}{m}\Delta t$$

The smaller Δt is, the better the prediction will be. In the limit $\Delta t \to 0$ (that is, Δt becomes a differential), an exact result is obtained.

Newton's second law of motion,

$$\frac{d}{dt}\left(m\frac{d\mathbf{r}}{dt}\right) = \mathbf{F}(\mathbf{r}, d\mathbf{r}/dt, t)$$

is an example of an *ordinary differential equation* (ODE). A dependent variable, $\mathbf{r}$, is determined from an equation involving a single independent variable, t, the dependent variable, $\mathbf{r}$, and its various derivatives.

In (point) particle mechanics there is only one independent variable, leading to ODEs. In other areas of physics, however, in which extended objects such as fields are studied, variations with respect to position are also important. Partial derivatives with respect to coordinate variables show up in the differential equations, which are therefore called *partial differential equations* (PDEs). For instance, in electrostatics, where static scalar fields (potential) and vector fields (electrostatic field) are studied, the law is described by Poisson's equation, $\nabla^2\Phi(\mathbf{r}) = -4\pi\rho(\mathbf{r})$, where Φ is the electrostatic potential and $\rho(\mathbf{r})$ is the volume charge density. In the theory of heat transfer, in which the temperature T is a function of position and time, an important relation is the heat equation, which can be written in the simple form $\partial T/\partial t = (k/\rho c)\nabla^2 T$, where k is the heat conductivity, ρ the mass density, and c the heat capacity of the medium. Similarly, the Schrödinger equation, $i\hbar\partial\Psi/\partial t = -(\hbar^2/2m)\nabla^2\Psi + V(\mathbf{r})\Psi$, is important in quantum mechanics, and the wave equation, $\nabla^2\Psi - (1/c^2)\partial^2\Psi/\partial t^2 = 0$, is encountered in many areas of physics.

In fact, except for laws of particle mechanics and electrical circuits, in which the only independent variable is time, almost all laws of physics are described by PDEs. There are various methods of solving PDEs, but by far the most commonly used (and the most powerful) is the separation of variables, by which a PDE is turned into several ODEs. This method, obviously, places special significance on ODEs, and so do this chapter and the next. This chapter will focus on obtaining the ODEs from the most common PDEs of mathematical physics.

8.1 COMMON PARTIAL DIFFERENTIAL EQUATIONS AND THE SEPARATION OF TIME

Let us start with the simplest PDE arising in electrostatic problems, Poisson's equation,

$$\nabla^2\Phi(\mathbf{r}) = -4\pi\rho(\mathbf{r}) \tag{8.1}$$

Note that Poisson's equation is not the most general equation encountered in electrostatics. Equation (8.1) requires some simplifying assumptions, which are explored in the following example.

Example 8.1.1

The two fundamental equations of electrostatics are

$$\nabla \times \mathbf{E} = 0 \qquad \text{and} \qquad \nabla\cdot\mathbf{D} = 4\pi\rho_{\mathrm{f}} \tag{1}$$

The first equation simply says that the electric field $\mathbf{E}$ is conservative. The second equation is the differential form of Gauss's law, which relates the outward flux of the displacement vector $\mathbf{D}$ through a closed surface to the amount of free charges enclosed by that surface.

From $\nabla \times \mathbf{E} = 0$ we infer the existence of an electrostatic potential Φ such that $\mathbf{E} = -\nabla\Phi$. On the other hand, $\mathbf{D} = \mathbf{E} + 4\pi\mathbf{P}$, where $\mathbf{P}$ is the polarization vector. In general, $\mathbf{P}$ is a function of $\mathbf{E}$. Thus, $\mathbf{P} \equiv \mathbf{P}(\mathbf{E}) = \mathbf{P}(-\nabla\Phi)$, and the second equation in (1) yields

$$\nabla\cdot(\mathbf{E} + 4\pi\mathbf{P}) = -\nabla^2\Phi + 4\pi\nabla\cdot\mathbf{P}(\mathbf{E}) = 4\pi\rho_f$$

or

$$\nabla^2\Phi - 4\pi\nabla\cdot\mathbf{P}(\mathbf{E}) = -4\pi\rho_f$$

The general dependence of $\mathbf{P}$ on $\mathbf{E}$ makes this equation very complicated (nonlinear). Only in simple situations, for example, when the medium is a vacuum ($\mathbf{P} = 0$) or linear ($\mathbf{P} = \chi\mathbf{E}$ with χ a constant), do we get Poisson's equation.

For a linear medium we obtain

$$\mathbf{D} = \mathbf{E} + 4\pi\chi\mathbf{E} = (1 + 4\pi\chi)\mathbf{E} \equiv \varepsilon\mathbf{E}$$

and the second equation in (1) becomes

$$\nabla\cdot\mathbf{D} = \nabla\cdot(\varepsilon\mathbf{E}) = \varepsilon\nabla\cdot(-\nabla\Phi) = 4\pi\rho_f$$

or

$$\nabla^2\Phi = -4\pi\frac{\rho_f}{\varepsilon} \equiv -4\pi\rho$$ ●

In vacuum, where $\rho(\mathbf{r}) = 0$, Eq. (8.1) reduces to *Laplace's equation*,

$$\nabla^2\Phi(\mathbf{r}) = 0 \tag{8.2}$$

Many electrostatic problems involve conductors held at certain potentials and situated in vacuum. In the space between such conducting surfaces, the electrostatic potential obeys Eq. (8.2).

Another PDE encountered in mathematical physics is concerned with the flow of heat. Since any kind of flow takes place in time, we expect t to appear in the *heat equation*, the most simplified version of which is

$$\frac{\partial T}{\partial t} = a^2\nabla^2 T \tag{8.3}$$

where T is the (absolute) temperature and a is a constant characterizing the medium in which heat is flowing.

Example 8.1.2

When the temperature of a small volume dV of a substance with mass density ρ changes by dT, the amount of heat exchanged is

$$d^2Q = (\rho\, dV)c\, dT$$

where c is the heat capacity of the substance. The quantities ρ, c, and T may be functions of position; the temperature is also a function of time. Thus, we have

$$d^2Q = \rho(\mathbf{r})c(\mathbf{r})\, dT(\mathbf{r}, t)\, dV$$

If we integrate over the whole volume of the substance, we obtain

$$dQ = \iiint_V \rho(\mathbf{r})\, c(\mathbf{r})\, dT(\mathbf{r}, t)\, dV$$

where dQ is the heat exchanged from the whole volume in the time interval dt and dT is the change in temperature, at the point $\mathbf{r}$, in that time interval. Thus, the amount of heat exchanged per unit time is

$$\frac{dQ}{dt} = \iiint_V \rho(\mathbf{r}) c(\mathbf{r}) \frac{\partial T}{\partial t}\, dV \tag{1}$$

In the absence of a source of heat, this must equal the heat flux through the surface S enclosing the volume V. This heat flux is given by

$$\iint_S k(\nabla T)\cdot \hat{\mathbf{e}}_n\, da$$

Assuming that k, the heat conductivity, is constant, and using the divergence theorem, we obtain

$$\iint_S k(\nabla T)\cdot \hat{\mathbf{e}}_n\, da = k \iiint_V (\nabla^2 T)\, dV$$

Equating the RHS of this with that of (1), making the simplifying assumption that ρ and c are also constant, and assuming that the expressions hold for arbitrary volume V, we get

$$\rho c \frac{\partial T}{\partial t} = k \nabla^2 T$$

or

$$\frac{\partial T}{\partial t} = \left(\frac{k}{\rho c}\right) \nabla^2 T$$

which is Eq. (8.3) with $a^2 = k/\rho c$. •

Probably one of the more important PDEs encountered in mathematical physics is the *wave equation*,

$$\nabla^2 \Psi - \frac{1}{c^2}\frac{\partial^2 \Psi}{\partial t^2} = 0 \tag{8.4}$$

This equation (or its simplification to lower dimensions) is applied to the vibration of strings and drums, the propagation of sound in gases, solids, and liquids, the propagation of disturbances in plasmas, and the propagation of electromagnetic waves. (See Exercise 8.1.1 for a derivation of the wave equation in electrodynamics.)

Without a doubt one of the most important equations in physics is the Schrödinger equation, which can be written in the (unusual) form

$$\nabla^2 \Psi - \frac{2m}{\hbar^2} V(\mathbf{r}) \Psi = -\frac{2im}{\hbar}\frac{\partial \Psi}{\partial t} \tag{8.5}$$

where m is the mass of a microscopic particle, $\hbar$ is Planck's constant divided by 2π, V is the potential energy of the particle, and $|\Psi(\mathbf{r}, t)|^2$ is the probability density of finding the particle at the point $\mathbf{r}$ at time t.

A relativistic generalization of the Schrödinger equation for a free particle of mass m is the Klein-Gordon equation, which, in terms of the natural units ($\hbar = 1 = c$), reduces to

$$\nabla^2\Phi - \frac{\partial^2\Phi}{\partial t^2} = m^2\Phi \tag{8.6}$$

Equations (8.3–8.6) have partial derivatives with respect to time. As a first step toward solving these PDEs and an introduction to similar techniques developed in later sections, let us separate the time variable. We will denote the functions in all four equations by the generic symbol $\Psi(\mathbf{r}, t)$.

The separation of variables starts with separating the $\mathbf{r}$ and t dependences into factors:

$$\Psi(\mathbf{r}, t) \equiv R(\mathbf{r})\, T(t)$$

This factorization permits us to separate the two operations of space differentiation and time differentiation. Thus, for example, $\nabla^2\Psi$ is replaced by $T\nabla^2 R$ and $\partial\Psi/\partial t$ by $R\, dT/dt$. To be as general as possible, we let $\mathbb{L}$ stand for all operations on Ψ except time differentiation, and write all the relevant equations as

$$\mathbb{L}\Psi = \begin{cases} \partial\Psi/\partial t \\ \partial^2\Psi/\partial t^2 \end{cases}$$

Thus, for Eq. (8.5), which involves a first derivative with respect to time,

$$\mathbb{L} = -\frac{\hbar}{2mi}\nabla^2 + \frac{1}{i\hbar}V$$

and for (8.6), which has a second derivative,

$$\mathbb{L} = \nabla^2 - m^2$$

and so on.

With this notation and the above separation, we have

$$\mathbb{L}(RT) = T(\mathbb{L}R) = \begin{cases} R\, dT/dt \\ R\, d^2T/dt^2 \end{cases}$$

Dividing both sides by RT, we obtain

$$\frac{T(\mathbb{L}R)}{RT} = \begin{cases} \dfrac{1}{T}\dfrac{dT}{dt} \\[2ex] \dfrac{1}{T}\dfrac{d^2T}{dt^2} \end{cases}$$

or

$$\frac{1}{R}\mathbb{L}(R) = \begin{cases} \dfrac{1}{T}\dfrac{dT}{dt} \\ \dfrac{1}{T}\dfrac{d^2T}{dt^2} \end{cases} \tag{8.7}$$

It is important to realize that the two R's on the LHS do not cancel because $\mathbb{L}$ is an operator.

Now comes the crucial step in the process of separation of variables. The LHS of Eq. (8.7) is a function of *position only*, and the RHS is a function of *time only*. Since $\mathbf{r}$ and t are independent variables, the only way that (8.7) can hold is for *both sides to be constant*. Denoting this constant generically as α, we obtain

$$\frac{1}{R}\mathbb{L}(R) = \alpha \quad \Rightarrow \quad \mathbb{L}R = \alpha R$$

and

$$\frac{1}{T}\frac{dT}{dt} = \alpha \quad \Rightarrow \quad \frac{dT}{dt} = \alpha T$$

or

$$\frac{1}{T}\frac{d^2T}{dt^2} = \alpha \quad \Rightarrow \quad \frac{d^2T}{dt^2} = \alpha T$$

We have reduced the original time-dependent PDE to an ODE,

$$\frac{dT}{dt} = \alpha T \qquad \text{or} \qquad \frac{d^2T}{dt^2} = \alpha T \tag{8.8a}$$

and a PDE involving only the position variables,

$$(\mathbb{L} - \alpha)R = 0$$

It is clear that the most general form of $\mathbb{L} - \alpha$, arising from Eqs. (8.3–8.6), is

$$\mathbb{L} - \alpha \equiv \nabla^2 + f(\mathbf{r})$$

Therefore, Eqs. (8.3–8.6) are completely equivalent to Eq. (8.8a), and

$$\nabla^2 R + f(\mathbf{r})R = 0 \tag{8.8b}$$

This already includes Laplace's equation as a special case [the case in which $f(\mathbf{r}) \equiv 0$]. To include Poisson's equation, we replace the zero on the RHS by $g(\mathbf{r}) \equiv -4\pi\rho(\mathbf{r})$, obtaining

$$\nabla^2 R + f(\mathbf{r})R = g(\mathbf{r})$$

The foregoing discussion is summarized in this statement: Most (simplified) PDEs encountered in mathematical physics can be reduced to the (very simple) ODE given in Eq. (8.8a) and the PDE given in (8.8b). With the exception of Poisson's equation, in all the foregoing equations the term on the RHS is zero. We will restrict ourselves to this so-called homogeneous case and rewrite (8.8b) as

$$\nabla^2\Psi(\mathbf{r}) + f(\mathbf{r})\Psi(\mathbf{r}) = 0 \tag{8.8c}$$

The inhomogeneous case in which $g(\mathbf{r}) \neq 0$ will be treated in Chapters 11 and 12. The rest of this chapter is devoted to the study of Eq. (8.8c) in various coordinate systems.

Note that there is no *a priori* reason why the basic assumption underlying the separation of variables is legitimate. After all, it is not obvious that we can write $\sin(xt)$ as a product, $f(x)g(t)$. However, in all cases of physical interest the separation of variables works, and the subsequent discussion will assume that it does.

Exercise

8.1.1 Derive the wave equation for **E** and **B**, starting with Maxwell's equations in vacuum:

$$\nabla \cdot \mathbf{E} = 0 \qquad \nabla \cdot \mathbf{B} = 0 \qquad \nabla \times \mathbf{E} = -\frac{1}{c}\frac{\partial \mathbf{B}}{\partial t} \qquad \nabla \times \mathbf{B} = \frac{1}{c}\frac{\partial \mathbf{E}}{\partial t}$$

8.2 SEPARATION IN CARTESIAN COORDINATES

In Cartesian coordinates Eq. (8.8c) becomes

$$\frac{\partial^2 \Psi}{\partial x^2} + \frac{\partial^2 \Psi}{\partial y^2} + \frac{\partial^2 \Psi}{\partial z^2} + f(x, y, z)\Psi = 0$$

The separation method worked for time, and there is no reason it should not work for coordinates. Thus, we assume that we can separate the dependence on various coordinates and write

$$\Psi(x, y, z) = X(x)Y(y)Z(z)$$

Then the above PDE yields

$$YZ\frac{d^2 X}{dx^2} + XZ\frac{d^2 Y}{dy^2} + XY\frac{d^2 Z}{dz^2} + f(x, y, z)XYZ = 0$$

Dividing by XYZ gives

$$\frac{1}{X}\frac{d^2 X}{dx^2} + \frac{1}{Y}\frac{d^2 Y}{dy^2} + \frac{1}{Z}\frac{d^2 Z}{dz^2} + f(x, y, z) = 0$$

This equation is almost separated. The first term is a function of x only, the second of y only, and the third of z only. However, the last term, in general, mixes the coordinates. The only way the separation can become complete is if the last term is also separated, that is, expressed as a sum of three functions, each depending on a *single* coordinate. In such a special case we obtain

$$\frac{1}{X}\frac{d^2 X}{dx^2} + \frac{1}{Y}\frac{d^2 Y}{dy^2} + \frac{1}{Z}\frac{d^2 Z}{dz^2} + f_1(x) + f_2(y) + f_3(z) = 0$$

or

$$\left[\frac{1}{X}\frac{d^2 X}{dx^2} + f_1(x)\right] + \left[\frac{1}{Y}\frac{d^2 Y}{dy^2} + f_2(y)\right] + \left[\frac{1}{Z}\frac{d^2 Z}{dz^2} + f_3(z)\right] = 0$$

The first term on the LHS is dependent on x only, the second on y only, and the third on z only. Since the sum of these three terms is a constant (zero), each term must be a constant. Denoting the constant corresponding to the ith term by α_i, we obtain

$$\frac{1}{X}\frac{d^2X}{dx^2} + f_1(x) = \alpha_1$$

$$\frac{1}{Y}\frac{d^2Y}{dy^2} + f_2(y) = \alpha_2$$

$$\frac{1}{Z}\frac{d^2Z}{dz^2} + f_3(z) = \alpha_3$$

which can be reexpressed as

$$\frac{d^2X}{dx^2} + [f_1(x) - \alpha_1]X = 0 \tag{8.9a}$$

$$\frac{d^2Y}{dy^2} + [f_2(y) - \alpha_2]Y = 0 \tag{8.9b}$$

$$\frac{d^2Z}{dz^2} + [f_3(z) - \alpha_3]Z = 0 \tag{8.9c}$$

$$\alpha_1 + \alpha_2 + \alpha_3 = 0 \tag{8.9d}$$

Equations (8.9) constitute the most general set of ODEs resulting from the separation of the PDE of Eq. (8.8c). Special cases commonly encountered in applications are presented in the following example.

Example 8.2.1

Let us consider a few cases for which Eqs. (8.9) are applicable.

(a) In electrostatics starting with Laplace's equation, in which $f(\mathbf{r}) = 0$, leads to these ODEs:

$$\frac{d^2X}{dx^2} - \alpha_1 X = 0 \qquad \frac{d^2Y}{dy^2} - \alpha_2 Y = 0 \qquad \frac{d^2Z}{dz^2} + (\alpha_1 + \alpha_2)Z = 0 \tag{1}$$

The solutions to (1) are trigonometric or hyperbolic (exponential) functions, determined from the boundary conditions (conducting surfaces).

(b) In quantum mechanics the time-independent Schrödinger equation for a free particle in three dimensions is

$$\nabla^2\Psi + \frac{2mE}{\hbar^2}\Psi = 0 \tag{2}$$

where m, the mass of the particle, E, the energy of the particle, and $\hbar = h/2\pi$ are constant. In this case $f(\mathbf{r}) = 2mE/\hbar^2$ can be grouped with one of the α's, say α_1, and Eqs. (8.9) yield

$$\frac{d^2X}{dx^2} - \left(\alpha_1 - \frac{2mE}{\hbar^2}\right)X = 0 \qquad \frac{d^2Y}{dy^2} - \alpha_2 Y = 0 \qquad \frac{d^2Z}{dz^2} - \alpha_3 Z = 0$$

$$\alpha_1 + \alpha_2 + \alpha_3 = \frac{2mE}{\hbar^2}$$

After time is separated from the coordinate variation, the heat and wave equations also yield an equation similar to (2), so the above ODEs apply to these cases as well.

(c) In quantum mechanics the time-independent Schrödinger equation for a three-dimensional isotropic harmonic oscillator is

$$\nabla^2\Psi - \left(\frac{m^2\omega^2}{\hbar^2}r^2 - \frac{2mE}{\hbar^2}\right)\Psi = 0$$

Thus, $$f(\mathbf{r}) = -\frac{m^2\omega^2}{\hbar^2}r^2 + \frac{2mE}{\hbar^2} = -\frac{m^2\omega^2}{\hbar^2}(x^2 + y^2 + z^2) + \frac{2mE}{\hbar^2}$$

Grouping the constant $2mE/\hbar^2$ with α_1 yields

$$\frac{d^2X}{dx^2} - \frac{m^2\omega^2}{\hbar^2}x^2X = \left(\alpha_1 - \frac{2mE}{\hbar^2}\right)X$$

$$\frac{d^2Y}{dy^2} - \frac{m^2\omega^2}{\hbar^2}y^2Y = \alpha_2 Y$$

$$\frac{d^2Z}{dz^2} - \frac{m^2\omega^2}{\hbar^2}z^2Z = \alpha_3 Z$$

$$\alpha_1 + \alpha_2 + \alpha_3 = \frac{2mE}{\hbar^2}$$

8.3 SEPARATION IN CYLINDRICAL COORDINATES

The general equation, Eq. (8.8c), takes the following form in cylindrical coordinates:

$$\frac{1}{\rho}\frac{\partial}{\partial\rho}\left(\rho\frac{\partial\Psi}{\partial\rho}\right) + \frac{1}{\rho^2}\frac{\partial^2\Psi}{\partial\varphi^2} + \frac{\partial^2\Psi}{\partial z^2} + f(\rho, \varphi, z)\Psi = 0$$

To separate the coordinates, we write

$$\Psi(\rho, \varphi, z) = R(\rho)S(\varphi)Z(z)$$

substitute in the general equation, and divide both sides by RSZ to obtain

$$\frac{1}{R}\left(\frac{1}{\rho}\frac{d}{d\rho}\right)\left(\rho\frac{dR}{d\rho}\right) + \frac{1}{S}\left(\frac{1}{\rho^2}\right)\frac{d^2S}{d\varphi^2} + \frac{1}{Z}\frac{d^2Z}{dz^2} + f(\rho, \varphi, z) = 0$$

If this equation is to be separable, we must have

$$f(\rho, \varphi, z) = f_1(\rho) + \frac{1}{\rho^2}f_2(\varphi) + f_3(z)$$

In that case the equation becomes

$$\left[\frac{1}{R}\left(\frac{1}{\rho}\frac{d}{d\rho}\right)\left(\rho\frac{dR}{d\rho}\right) + f_1(\rho)\right] + \frac{1}{\rho^2}\left[\frac{1}{S}\frac{d^2S}{d\varphi^2} + f_2(\varphi)\right] + \left[\frac{1}{Z}\frac{d^2Z}{dz^2} + f_3(z)\right] = 0$$

The sum of the first two terms is independent of z, so the third term must also be. We thus get

$$\frac{1}{Z}\frac{d^2Z}{dz^2} + f_3(z) = \alpha_1$$

and
$$\left[\frac{1}{R}\left(\frac{1}{\rho}\frac{d}{d\rho}\right)\left(\rho\frac{dR}{d\rho}\right) + f_1(\rho)\right] + \frac{1}{\rho^2}\left[\frac{1}{S}\frac{d^2S}{d\varphi^2} + f_2(\varphi)\right] + \alpha_1 = 0$$

Multiplying this equation by ρ^2 yields

$$\left[\frac{\rho}{R}\frac{d}{d\rho}\left(\rho\frac{dR}{d\rho}\right) + \rho^2 f_1(\rho) + \alpha_1\rho^2\right] + \left[\frac{1}{S}\frac{d^2S}{d\varphi^2} + f_2(\varphi)\right] = 0$$

Since the first term is a function of ρ only and the second a function of φ only, both terms must be constants whose sum vanishes. Thus,

$$\frac{1}{S}\frac{d^2S}{d\varphi^2} + f_2(\varphi) = \alpha_2$$

and
$$\frac{1}{R}\left(\rho\frac{d}{d\rho}\right)\left(\rho\frac{dR}{d\rho}\right) + \rho^2 f_1(\rho) + \alpha_1\rho^2 + \alpha_2 = 0$$

Putting together all of the above, we conclude that when the general equation, Eq. (8.8c), is separable in cylindrical coordinates, it will separate into the following three ODEs:

$$\frac{d^2Z}{dz^2} + [f_3(z) - \alpha_1]Z = 0 \tag{8.10a}$$

$$\frac{d^2S}{d\varphi^2} + [f_2(\varphi) - \alpha_2]S = 0 \tag{8.10b}$$

$$\rho\frac{d}{d\rho}\left(\rho\frac{dR}{d\rho}\right) + \{\rho^2[f_1(\rho) + \alpha_1] + \alpha_2\}R = 0 \tag{8.10c}$$

For the special case where f_1 is identical to zero, we get

$$\rho\frac{d}{d\rho}\left(\rho\frac{dR}{d\rho}\right) + (\alpha_1\rho^2 + \alpha_2)R = 0$$

which is called the *Bessel differential equation*. This equation shows up in electrostatic and heat-transfer problems with cylindrical geometry and in problems involving two-dimensional wave propagation, as in drumheads. (We will solve the Bessel equation in Chapters 9 and 10.)

Example 8.3.1

Consider the electrostatic problem in which the top and bottom surfaces of a circular cylindrical can of length l are grounded (the potential, Φ, is zero there), and the lateral surface is held at a potential V_0.

In such a case Laplace's equation applies. Thus, $f(\mathbf{r}) \equiv 0$, which implies that $f_1 = f_2 = f_3 = 0$. Equation (8.10a) gives

$$\frac{d^2Z}{dz^2} - \alpha_1 Z = 0 \tag{1}$$

which, for $\alpha_1 \neq 0$, has the general solution

$$Z = Ae^{\sqrt{\alpha_1}z} + Be^{-\sqrt{\alpha_1}z}$$

as can easily be verified.

Imposing the boundary condition $\Phi(\rho, \varphi, 0) = 0$ gives

$$R(\rho)S(\varphi)Z(0) = 0$$

Since this is true for arbitrary ρ and φ, we must have $Z(0) = 0$ or $A + B = 0$. Thus, we get

$$Z = A(e^{\sqrt{\alpha_1}z} - e^{-\sqrt{\alpha_1}z}) \tag{2}$$

At $z = l$ the potential must vanish. This implies that $Z(l) = 0$, or

$$Z(l) = A(e^{\sqrt{\alpha_1}l} - e^{-\sqrt{\alpha_1}l}) = 0 \tag{3}$$

For $A = 0$, Eq. (2) gives $Z \equiv 0$, which implies $\Phi = 0$. This is a trivial solution in which we have no interest. We demand that the other factor in Eq. (3) be zero. This yields

$$e^{\sqrt{\alpha_1}l} = e^{-\sqrt{\alpha_1}l}$$

or

$$e^{2\sqrt{\alpha_1}l} = 1$$

If we insist on real values for $\sqrt{\alpha_1}$, the only solution to this equation will be $\alpha_1 = 0$, which, by Eq. (2), is another trivial solution. Thus, we allow imaginary numbers for $\sqrt{\alpha_1}$ and find that $2\sqrt{\alpha_1}l = 2in\pi$, where n is an integer, is a solution. We write

$$\sqrt{\alpha_1} = i\frac{n\pi}{l} \qquad \text{for } n = \pm 1, \pm 2, \ldots$$

or

$$\alpha_1 = -\frac{n^2\pi^2}{l^2} \qquad \text{for } n = \pm 1, \pm 2, \ldots \tag{4}$$

For $\alpha_1 = 0$, Eq. (1) becomes $d^2Z/dz^2 = 0$, which has the general solution $Z = az + b$, where a and b are arbitrary constants.

This result is a typical example of how boundary conditions impose restrictive relations, such as Eq. (4), on the parameters of an equation. Here the simplicity of the differential equation, Eq. (1), leads smoothly to the conditions in Eq. (4). With other, more complicated, equations these conditions are not as easy to obtain.

Another restriction, this time on α_2, is also easily obtainable and comes from Eq. (8.10b). With $f_2 \equiv 0$, this becomes

$$\frac{d^2S}{d\varphi^2} - \alpha_2 S = 0$$

which is analogous to Eq. (1). Thus, a general solution for this equation is

$$S = \begin{cases} Ce^{\sqrt{\alpha_2}\varphi} + De^{-\sqrt{\alpha_2}\varphi} & \text{for } \alpha_2 \neq 0 \\ C'\varphi + D' & \text{for } \alpha_2 = 0 \end{cases} \tag{5}$$

No matter what type of boundary conditions are imposed on the potential Φ, it is clear that Φ must give the same value at φ and at $\varphi + 2\pi$. This means that $\alpha_2 \neq 0$ (we disregard the trivial solution $S = D'$), and $R(\rho)S(\varphi)Z(z) = R(\rho)S(\varphi + 2\pi)Z(z)$, or $S(\varphi) = S(\varphi + 2\pi)$. Substituting the latter in (5) yields

$$Ce^{\sqrt{\alpha_2}\varphi} + De^{-\sqrt{\alpha_2}\varphi} = Ce^{\sqrt{\alpha_2}(\varphi + 2\pi)} + De^{-\sqrt{\alpha_2}(\varphi + 2\pi)}$$

or

$$Ce^{\sqrt{\alpha_2}\varphi}(1 - e^{\sqrt{\alpha_2}2\pi}) + De^{-\sqrt{\alpha_2}\varphi}(1 - e^{-\sqrt{\alpha_2}2\pi}) = 0$$

Multiplying both sides by $e^{\sqrt{\alpha_2}\varphi}$ yields

$$Ce^{2\sqrt{\alpha_2}\varphi}(1 - e^{\sqrt{\alpha_2}2\pi}) + D(1 - e^{-\sqrt{\alpha_2}2\pi}) = 0$$

This must hold for all φ. The only way for that to be true is to have

$$1 - e^{\sqrt{\alpha_2}2\pi} = 0 \qquad \text{and} \qquad 1 - e^{-\sqrt{\alpha_2}2\pi} = 0$$

The most general solution to both of these is

$$\sqrt{\alpha_2} = im \qquad \text{for } m = 0, \pm 1, \pm 2, \ldots$$

or

$$\alpha_2 = -m^2 \qquad \text{for } m = 0, \pm 1, \pm 2, \ldots$$

●

8.4 SEPARATION IN SPHERICAL COORDINATES

By far the most commonly used coordinate system in mathematical physics is the spherical coordinate system. This is because forces, potential energies, and most geometries encountered in nature have a spherical symmetry. The rest of this chapter treats the separation of spherical variables in detail.

8.4.1 Separation of the Angles from the Radial Part of the Laplacian

With Cartesian and cylindrical variables the boundary conditions are important in the nature of the solutions of the ODE obtained from the PDE (the constants α_1 and α_2 are arbitrary; the boundary conditions determine whether they are integers, real numbers, or complex numbers). In almost all applications, however, the angular part of the spherical variables can be separated and studied very generally. This is because the angular part of the Laplacian in the spherical coordinate system is closely related to the operation of rotation and the angular momentum, which are independent of any particular situation.

The separation of the angular part in spherical coordinates can be done in a fashion exactly analogous to the separation of Cartesian and cylindrical coordinates by writing Ψ as a product of three functions, each depending on only one of the variables r, θ, and φ. However, we will use a different approach that is common in quantum mechanical treatments of the Schrödinger equation. This approach, which is based on the operator algebra of Chapters 2 and 3 and is extremely powerful and elegant, gives solutions for the angular part in closed form.

Define the vector operator $\vec{\mathbb{p}}$ as

$$\vec{\mathbb{p}} = -i\nabla \equiv -\sqrt{-1}\nabla$$

so that the jth Cartesian component of $\vec{\mathbb{p}}$ is $\mathbb{p}_j \equiv -i\,\partial/\partial x_j$, for $j = 1, 2, 3$. In quantum mechanics $\vec{\mathbb{p}}$ is the momentum operator. It is easy to verify that

$$[x_j, \mathbb{p}_k] = i\delta_{jk}$$

and

$$[x_j, x_k] = 0 = [\mathbb{p}_j, \mathbb{p}_k]$$

We can also define the angular momentum operator as

$$\vec{\mathbb{L}} = \mathbf{r} \times \vec{\mathbb{p}}$$

This is expressed in components as

$$\mathbb{L}_i = (\mathbf{r} \times \vec{\mathbb{p}})_i = \varepsilon_{ijk} x_j \mathbb{p}_k \qquad \text{for } i = 1, 2, 3$$

where Einstein's summation convention (summing over repeated indices) is utilized. Using the commutation relations above, we obtain (see Exercise 8.4.1)

$$[\mathbb{L}_j, \mathbb{L}_k] = i\varepsilon_{jkl}\mathbb{L}_l \tag{8.11}$$

We will see shortly that $\vec{\mathbb{L}}$ can be written solely in terms of the angles θ and φ. However, there is a single factor of $\vec{\mathbb{p}}$ in the definition of $\vec{\mathbb{L}}$, so if we square $\vec{\mathbb{L}}$, we will get two factors of $\vec{\mathbb{p}}$, which may turn into a Laplacian. In this manner we may be able to write ∇^2 in terms of $\mathbb{L}^2$, which depends only on angles, and perhaps a second term that involves only r. Let us try this:

$$\mathbb{L}^2 \equiv \vec{\mathbb{L}} \cdot \vec{\mathbb{L}} = \sum_{i=1}^{3} \mathbb{L}_i \mathbb{L}_i \equiv \mathbb{L}_i \mathbb{L}_i = \varepsilon_{ijk} x_j \mathbb{p}_k \varepsilon_{imn} x_m \mathbb{p}_n$$

↑ by Einstein's summation convention

$$= \varepsilon_{ijk}\varepsilon_{imn} x_j \mathbb{p}_k x_m \mathbb{p}_n$$

$$= (\delta_{jm}\delta_{kn} - \delta_{jn}\delta_{km}) x_j \mathbb{p}_k x_m \mathbb{p}_n$$

$$= x_j \mathbb{p}_k x_j \mathbb{p}_k - x_j \mathbb{p}_k x_k \mathbb{p}_j$$

We need to write this expression in such a way that factors with the same index are next to each other, to give a dot product. We must also try, when possible, to keep the p factors to the right so that they can operate on functions without intervention from the x factors. We obtain

$$\mathbb{L}^2 = x_j(x_j\mathbb{p}_k - i\delta_{kj})\mathbb{p}_k - (\mathbb{p}_k x_j + i\delta_{kj})x_k\mathbb{p}_j$$

$$= x_j x_j \mathbb{p}_k \mathbb{p}_k - i x_j \mathbb{p}_j - \mathbb{p}_k x_k x_j \mathbb{p}_j - i x_j \mathbb{p}_j$$

$$= x_j x_j \mathbb{p}_k \mathbb{p}_k - 2i x_j \mathbb{p}_j - (x_k \mathbb{p}_k - i\delta_{kk}) x_j \mathbb{p}_j$$

Again using Einstein's summation convention and recalling that $\delta_{kk} = 3$, we can write

$$\mathbb{L}^2 = r^2|\vec{\mathbb{p}}|^2 - 2i\mathbf{r}\cdot\vec{\mathbb{p}} - (\mathbf{r}\cdot\vec{\mathbb{p}})(\mathbf{r}\cdot\vec{\mathbb{p}}) + 3i\mathbf{r}\cdot\vec{\mathbb{p}}$$

which, if we make the substitution $\vec{\mathbb{p}} = -i\boldsymbol{\nabla}$, yields

$$\nabla^2 = -r^{-2}\mathbb{L}^2 + r^{-2}(\mathbf{r}\cdot\boldsymbol{\nabla})(\mathbf{r}\cdot\boldsymbol{\nabla}) + r^{-2}\mathbf{r}\cdot\boldsymbol{\nabla}$$

Letting ∇^2 act on the function $\Psi(r, \theta, \varphi)$, we get

$$\nabla^2\Psi = -\frac{1}{r^2}\mathbb{L}^2\Psi + \frac{1}{r^2}(\mathbf{r}\cdot\boldsymbol{\nabla})(\mathbf{r}\cdot\boldsymbol{\nabla})\Psi + \frac{1}{r^2}\mathbf{r}\cdot\boldsymbol{\nabla}\Psi \tag{8.12}$$

But we note that

$$\mathbf{r}\cdot\boldsymbol{\nabla} = r\hat{\mathbf{e}}_r\cdot\boldsymbol{\nabla} = r\nabla_r \equiv r\frac{\partial}{\partial r}$$

We, thus, get the final form of $\nabla^2\Psi$:

$$\nabla^2\Psi = -\frac{1}{r^2}\mathbb{L}^2\Psi + \frac{1}{r^2}\left(r\frac{\partial}{\partial r}\right)\left(r\frac{\partial\Psi}{\partial r}\right) + \frac{1}{r}\frac{\partial\Psi}{\partial r} \tag{8.13}$$

It is important to note that Eq. (8.12) is a general relation that holds in all coordinate systems. In fact, all the manipulations leading to it were done in Cartesian coordinates. Since it is written in vector notation, however, there is no indication that it was derived using Cartesian coordinates.

Equation (8.13) is the spherical version of (8.12) and is the version we will pursue. We will first make a simplifying, but in most cases nonrestrictive, assumption—that in Eq. (8.8c), the master equation, $f(\mathbf{r})$ is a function of r only. This is certainly true for almost all applications in physics.

In spherical coordinates, then, Eq. (8.8c) becomes

$$-\frac{1}{r^2}\mathbb{L}^2\Psi + \frac{1}{r}\frac{\partial}{\partial r}\left(r\frac{\partial\Psi}{\partial r}\right) + \frac{1}{r}\frac{\partial\Psi}{\partial r} + f(r)\Psi = 0$$

Assuming, for the time being, that $\mathbb{L}^2$ depends only on θ and φ and separating Ψ into a product of two functions, $\Psi(r, \theta, \varphi) = R(r)Y(\theta, \varphi)$, we can rewrite this equation as

$$-\frac{1}{r^2}\mathbb{L}^2(RY) + \frac{1}{r}\frac{\partial}{\partial r}\left[r\frac{\partial}{\partial r}(RY)\right] + \frac{1}{r}\frac{\partial}{\partial r}(RY) + f(r)RY = 0$$

Division by RY yields

$$-\frac{1}{r^2 Y}\mathbb{L}^2(Y) + \frac{1}{rR}\frac{d}{dr}\left(r\frac{dR}{dr}\right) + \frac{1}{rR}\frac{dR}{dr} + f(r) = 0$$

Multiplying by r^2, we finally obtain

$$-\frac{1}{Y}\mathbb{L}^2(Y) + \frac{r}{R}\frac{d}{dr}\left(r\frac{dR}{dr}\right) + \frac{r}{R}\frac{dR}{dr} + r^2 f(r) = 0$$

Using the same argument as in previous sections, we have

$$-\frac{1}{Y}\mathbb{L}^2(Y) = -\alpha$$

and

$$\frac{r}{R}\frac{d}{dr}\left(r\frac{dR}{dr}\right) + \frac{r}{R}\frac{dR}{dr} + r^2 f(r) = \alpha$$

These two equations can be reexpressed as

$$\mathbb{L}^2\, Y(\theta, \varphi) = \alpha Y(\theta, \varphi) \tag{8.14a}$$

and

$$\frac{d^2R}{dr^2} + \frac{2}{r}\frac{dR}{dr} + \left[f(r) - \frac{\alpha}{r^2}\right]R = 0 \tag{8.14b}$$

We will concentrate on the angular part, Eq. (8.14a), leaving the radial part to the general discussion of ODEs in the next chapter. The rest of this subsection will focus on showing that $\mathbb{L}^2$ depends only on θ and φ by calculating $\mathbb{L}_1 \equiv \mathbb{L}_x$, $\mathbb{L}_2 \equiv \mathbb{L}_y$, and $\mathbb{L}_3 = \mathbb{L}_z$ and showing that they are independent of r.

Since $\mathbb{L}_i$ is an operator, we can study its action on an arbitrary function $f(\mathbf{r})$. Thus,

$$\mathbb{L}_i f = -i\varepsilon_{ijk} x_j \nabla_k f \equiv -i\varepsilon_{ijk} x_j \frac{\partial f}{\partial x_k}$$

We can express the Cartesian x_j in terms of r, θ, and φ and use the chain rule to express $\partial f/\partial x_k$ in terms of spherical coordinates. This will give us $\mathbb{L}_i f$ expressed in terms of r, θ, and φ. We want to show that r is absent in the final expression.

We start with

$$x = r\sin\theta\cos\varphi \qquad y = r\sin\theta\sin\varphi \qquad z = r\cos\theta$$

and their inverses,

$$r = (x^2 + y^2 + z^2)^{1/2} \qquad \cos\theta = \frac{z}{r} \qquad \tan\varphi = \frac{y}{x}$$

We can express the Cartesian derivatives, $\partial f/\partial x_k$, in terms of spherical coordinates, for example:

$$\frac{\partial f}{\partial x} = \frac{\partial f}{\partial r}\frac{\partial r}{\partial x} + \frac{\partial f}{\partial \theta}\frac{\partial \theta}{\partial x} + \frac{\partial f}{\partial \varphi}\frac{\partial \varphi}{\partial x}$$

The derivative of one coordinate system with respect to the other can be easily calculated:

$$\frac{\partial r}{\partial x} = \frac{1}{2}(2x)(x^2 + y^2 + z^2)^{-1/2} = \frac{x}{r}$$

Now, differentiating both sides of the equation $\cos\theta = z/r$, we obtain

$$-\sin\theta\frac{\partial \theta}{\partial x} = -\frac{z\,\partial r/\partial x}{r^2} = -\frac{zx}{r^3}$$

from which we get

$$\frac{\partial \theta}{\partial x} = \frac{zx}{r^3 \sin \theta}$$

Similarly,

$$\frac{\partial}{\partial x}(\tan \varphi) = \frac{\partial}{\partial x}\left(\frac{y}{x}\right) \quad \Rightarrow \quad \sec^2 \varphi \frac{\partial \varphi}{\partial x} = -\frac{y}{x^2}$$

or

$$\frac{\partial \varphi}{\partial x} = -\frac{y}{x^2}\cos^2 \varphi$$

Using these expressions for $\partial r/\partial x$, $\partial \theta/\partial x$, and $\partial \varphi/\partial x$ in the equation for $\partial f/\partial x$, we get

$$\frac{\partial f}{\partial x} = \frac{x}{r}\frac{\partial f}{\partial r} + \frac{xz}{r^3 \sin \theta}\frac{\partial f}{\partial \theta} - \frac{y}{x^2}\cos^2 \varphi \frac{\partial f}{\partial \varphi}$$

In exactly the same way, we obtain

$$\frac{\partial f}{\partial y} = \frac{y}{r}\frac{\partial f}{\partial r} + \frac{yz}{r^3 \sin \theta}\frac{\partial f}{\partial \theta} + \frac{1}{x}\cos^2 \varphi \frac{\partial f}{\partial \varphi}$$

$$\frac{\partial f}{\partial z} = \frac{z}{r}\frac{\partial f}{\partial r} + \left(\frac{z^2}{r^3} - \frac{1}{r}\right)\frac{1}{\sin \theta}\frac{\partial f}{\partial \theta}$$

We can now calculate $\mathbb{L}_x$ by letting it act on an arbitrary function:

$$\begin{aligned}
\mathbb{L}_x f &= (y\mathbb{p}_z - z\mathbb{p}_y)f = -i\left(y\frac{\partial f}{\partial z} - z\frac{\partial f}{\partial y}\right) \\
&= -i\left[\frac{yz}{r}\frac{\partial f}{\partial r} + \left(\frac{yz^2}{r^3} - \frac{y}{r}\right)\left(\frac{1}{\sin \theta}\right)\frac{\partial f}{\partial \theta} - \frac{zy}{r}\frac{\partial f}{\partial r} - \frac{yz^2}{r^3 \sin \theta}\frac{\partial f}{\partial \theta} - \frac{z}{x}\cos^2 \varphi \frac{\partial f}{\partial \varphi}\right] \\
&= +i\left(\sin \varphi \frac{\partial}{\partial \theta} + \cot \theta \cos \varphi \frac{\partial}{\partial \varphi}\right)f
\end{aligned}$$

This gives

$$\mathbb{L}_x = i\left(\sin \varphi \frac{\partial}{\partial \theta} + \cot \theta \cos \varphi \frac{\partial}{\partial \varphi}\right) \tag{8.15a}$$

Analogous arguments yield

$$\mathbb{L}_y = i\left(-\cos \varphi \frac{\partial}{\partial \theta} + \cot \theta \sin \varphi \frac{\partial}{\partial \varphi}\right) \tag{8.15b}$$

and

$$\mathbb{L}_z = -i\frac{\partial}{\partial \varphi} \tag{8.15c}$$

We have proved the following proposition.

8.4.1 Proposition When expressed in spherical coordinates, $\mathbb{L}_x$, $\mathbb{L}_y$, and $\mathbb{L}_z$ are operators that depend only on θ and φ. ■

Exercise 8.4.2 demonstrates the procedure to calculate $\mathbb{L}^2$ from Eqs. (8.15). The result is

$$\mathbb{L}^2 = -\frac{1}{\sin\theta}\frac{\partial}{\partial\theta}\left(\sin\theta\frac{\partial}{\partial\theta}\right) - \frac{1}{\sin^2\theta}\frac{\partial^2}{\partial\varphi^2} \tag{8.16}$$

Substitution in Eq. (8.13) yields the Laplacian in spherical coordinates, as derived in Chapter 1.

8.4.2 Construction of the Eigenvalues of $\mathbb{L}^2$

Now that we have $\mathbb{L}^2$ in terms of θ and φ, we could substitute in Eq. (8.14a), separate the θ and φ dependence, and solve the corresponding ODEs. However, there is a much more elegant way of solving (8.14a) algebraically, because it is simply an eigenvalue equation for $\mathbb{L}^2$. In this subsection we will find the eigenvalues and eigenvectors of $\mathbb{L}^2$.

Let us consider $\mathbb{L}^2$ as an abstract operator and write (8.14a) as

$$\mathbb{L}^2|\mathrm{Y}\rangle = \alpha|\mathrm{Y}\rangle$$

where $|\mathrm{Y}\rangle$ is an abstract vector whose θ,φth component can be calculated later. Since $\mathbb{L}^2$ is a differential operator, it does not have a (finite-dimensional) matrix representation. Thus, the determinantal procedure for calculating eigenvalues and eigenfunctions will not work here. Instead we will generate all the eigenvectors and their corresponding eigenvalues by considering a maximum set of commuting operators.

The equation above specifies an eigenvalue, α, and an eigenvector, $|\mathrm{Y}\rangle$. However, there may be more than one $|\mathrm{Y}\rangle$ corresponding to the same α. To distinguish among these *degenerate eigenvectors*, we choose a second operator from among $\{\mathbb{L}_i\}_{i=1}^3$ that commutes with $\mathbb{L}^2$, say $\mathbb{L}_3$. We then diagonalize $\mathbb{L}_3$ in the subspaces corresponding to various α's (see the discussion of spectral decomposition in Chapter 3 for details) and label each $|\mathrm{Y}\rangle$ with the eigenvalue corresponding to $\mathbb{L}_3$. This is possible because both $\mathbb{L}^2$ and $\mathbb{L}_3$ are hermitian operators in the space of square-integrable functions. (The proof is left as a problem.) We continue until there are no more commuting operators. The eigenvectors obtained in this manner will be nondegenerate as far as the action of the operators $\{\mathbb{L}_i\}_{i=1}^3$ is concerned.

We thus look for the maximum set of commuting operators from among $\{\mathbb{L}^2, \mathbb{L}_1, \mathbb{L}_2, \mathbb{L}_3\}$. Exercise 4.1.14 shows that $\mathbb{L}^2$ commutes with all of the $\mathbb{L}_i$'s. However, none of the $\mathbb{L}_i$'s commutes with the rest of them. Thus, the maximum set of commuting operators consists of $\mathbb{L}^2$ and only one of the $\mathbb{L}_i$'s. It is customary to choose $\mathbb{L}_3 \equiv \mathbb{L}_z$ as the partner for $\mathbb{L}^2$, giving $\{\mathbb{L}^2, \mathbb{L}_z\}$ as the maximum set of commuting operators. The eigenvalues of these two operators will label the vector $|\mathrm{Y}\rangle$. Since $[\mathbb{L}^2, \mathbb{L}_z] = 0$, we can choose vectors that are simultaneous eigenvectors of both operators. Let us label these vectors with the eigenvalues of the two operators, obtaining

$$\mathbb{L}^2|\mathrm{Y}_{\alpha,\beta}\rangle = \alpha|\mathrm{Y}_{\alpha,\beta}\rangle \tag{8.17a}$$

$$\mathbb{L}_z|\mathrm{Y}_{\alpha,\beta}\rangle = \beta|\mathrm{Y}_{\alpha,\beta}\rangle \tag{8.17b}$$

The hermiticity of $\mathbb{L}^2$ and $\mathbb{L}_z$ implies the reality of α and β. Next we need to determine the possible values for α and β.

We define two new operators as follows:

$$\mathbb{L}_+ \equiv \mathbb{L}_x + i\mathbb{L}_y$$

$$\mathbb{L}_- \equiv \mathbb{L}_x - i\mathbb{L}_y$$

From these definitions it is easily verified that

$$[\mathbb{L}^2, \mathbb{L}_\pm] = 0 \tag{8.18a}$$

$$[\mathbb{L}_z, \mathbb{L}_\pm] = \pm \mathbb{L}_\pm \tag{8.18b}$$

$$[\mathbb{L}_+, \mathbb{L}_-] = 2\mathbb{L}_z \tag{8.18c}$$

Equation (8.18a) implies that $\mathbb{L}_\pm$ are invariant operators when acting in the subspace corresponding to the eigenvalue α; that is,

$$\mathbb{L}^2(\mathbb{L}_\pm|\mathrm{Y}_{\alpha,\beta}\rangle) = \alpha(\mathbb{L}_\pm|\mathrm{Y}_{\alpha,\beta}\rangle)$$

On the other hand, Eq. (8.18b) yields

$$\begin{aligned}\mathbb{L}_z(\mathbb{L}_+|\mathrm{Y}_{\alpha,\beta}\rangle) &= (\mathbb{L}_z\mathbb{L}_+)|\mathrm{Y}_{\alpha,\beta}\rangle = (\mathbb{L}_+\mathbb{L}_z + \mathbb{L}_+)|\mathrm{Y}_{\alpha,\beta}\rangle \\ &= \mathbb{L}_+\mathbb{L}_z|\mathrm{Y}_{\alpha,\beta}\rangle + \mathbb{L}_+|\mathrm{Y}_{\alpha,\beta}\rangle = \beta\mathbb{L}_+|\mathrm{Y}_{\alpha,\beta}\rangle + \mathbb{L}_+|\mathrm{Y}_{\alpha,\beta}\rangle \\ &= (\beta + 1)(\mathbb{L}_+|\mathrm{Y}_{\alpha,\beta}\rangle)\end{aligned}$$

This indicates that $\mathbb{L}_+|\mathrm{Y}_{\alpha,\beta}\rangle$ has one more unit of the $\mathbb{L}_z$ eigenvalue than $|\mathrm{Y}_{\alpha,\beta}\rangle$ does. In other words, $\mathbb{L}_+$ *raises* the eigenvalue of $\mathbb{L}_z$ by one unit. That is why $\mathbb{L}_+$ is called a *raising operator*. Similarly, $\mathbb{L}_-$ is called a *lowering operator* because

$$\mathbb{L}_z(\mathbb{L}_-|\mathrm{Y}_{\alpha,\beta}\rangle) = (\beta - 1)(\mathbb{L}_-|\mathrm{Y}_{\alpha,\beta}\rangle)$$

We can summarize the above discussion as

$$\mathbb{L}_\pm|\mathrm{Y}_{\alpha,\beta}\rangle = C_\pm|\mathrm{Y}_{\alpha,\beta\pm 1}\rangle$$

where $C_\pm$ are constants to be determined (later) by a suitable normalization.

There are restrictions on and relations between α and β. Before exploring these, let us note that because $\mathbb{L}^2$ is a sum of squares of hermitian matrices, it must be a positive operator (see Chapter 3); that is,

$$\langle \mathrm{a}|\mathbb{L}^2|\mathrm{a}\rangle \geqslant 0 \qquad \forall\, |\mathrm{a}\rangle$$

In particular,

$$0 \leqslant \langle \mathrm{Y}_{\alpha,\beta}|\mathbb{L}^2|\mathrm{Y}_{\alpha,\beta}\rangle = \alpha\langle \mathrm{Y}_{\alpha,\beta}|\mathrm{Y}_{\alpha,\beta}\rangle \equiv \alpha\|\mathrm{Y}_{\alpha,\beta}\|^2$$

Therefore, $\alpha \geqslant 0$

On the other hand, as Exercise 8.4.3 shows,

$$\mathbb{L}^2 = \mathbb{L}_+\mathbb{L}_- + \mathbb{L}_z^2 - \mathbb{L}_z \tag{8.19a}$$

and
$$\mathbb{L}^2 = \mathbb{L}_-\mathbb{L}_+ + \mathbb{L}_z^2 + \mathbb{L}_z \tag{8.19b}$$

Sandwiching both sides of Eq. (8.19a) between $|Y_{\alpha,\beta}\rangle$ and $\langle Y_{\alpha,\beta}|$ yields

$$\langle Y_{\alpha,\beta}|\mathbb{L}^2|Y_{\alpha,\beta}\rangle = \langle Y_{\alpha,\beta}|\mathbb{L}_+\mathbb{L}_-|Y_{\alpha,\beta}\rangle + \langle Y_{\alpha,\beta}|\mathbb{L}_z^2|Y_{\alpha,\beta}\rangle - \langle Y_{\alpha,\beta}|\mathbb{L}_z|Y_{\alpha,\beta}\rangle$$

or
$$\alpha\,\|Y_{\alpha,\beta}\|^2 = \langle Y_{\alpha,\beta}|\mathbb{L}_+\mathbb{L}_-|Y_{\alpha,\beta}\rangle + \beta^2\,\|Y_{\alpha,\beta}\|^2 - \beta\,\|Y_{\alpha,\beta}\|^2$$

Since $\mathbb{L}_+ = (\mathbb{L}_-)^\dagger$,

$$\|\mathbb{L}_-|Y_{\alpha,\beta}\rangle\|^2 = (\alpha - \beta^2 + \beta)\|Y_{\alpha,\beta}\|^2$$

Because of the positivity of norms, this yields

$$\alpha \geqslant \beta^2 - \beta$$

Similarly, (8.19b) gives

$$\alpha \geqslant \beta^2 + \beta$$

Adding these two inequalities gives

$$2\alpha \geqslant 2\beta^2 \quad \Rightarrow \quad |\beta| \leqslant \sqrt{\alpha} \quad \Rightarrow \quad -\sqrt{\alpha} \leqslant \beta \leqslant \sqrt{\alpha}$$

This shows that the values of β are bounded. That is, there exist a maximum β, denoted as β_+, and a minimum β, denoted β_-, beyond which there are no more values of β. This can happen only if

$$\mathbb{L}_+|Y_{\alpha,\beta_+}\rangle = 0$$

$$\mathbb{L}_-|Y_{\alpha,\beta_-}\rangle = 0$$

because if $\mathbb{L}_\pm|Y_{\alpha,\beta_\pm}\rangle$ are not zero, they must have values of β corresponding to $\beta_\pm \pm 1$, which is not allowed by assumption.

Now we sandwich (8.19b) between $\langle Y_{\alpha,\beta_+}|$ and $|Y_{\alpha,\beta_+}\rangle$ and use $\mathbb{L}_+|Y_{\alpha,\beta_+}\rangle = 0$ to obtain

$$\alpha\,\|Y_{\alpha,\beta_+}\|^2 = 0 + \beta_+^2\,\|Y_{\alpha,\beta_+}\|^2 + \beta_+\,\|Y_{\alpha,\beta_+}\|^2$$

or
$$(\alpha - \beta_+^2 - \beta_+)\|Y_{\alpha,\beta_+}\|^2 = 0$$

By definition $|Y_{\alpha,\beta_+}\rangle \neq 0$ (otherwise $\beta_+ - 1$ would be the maximum). Thus, we obtain

$$\alpha = \beta_+^2 + \beta_+$$

An analogous procedure involving (8.19a) yields

$$\alpha = \beta_-^2 - \beta_-$$

We solve these two equations for β_+ and β_-:

$$\beta_+ = \tfrac{1}{2}(-1 \pm \sqrt{1 + 4\alpha})$$

$$\beta_- = \tfrac{1}{2}(1 \pm \sqrt{1 + 4\alpha})$$

The facts that $\beta_+ \geqslant \beta_-$ and $\sqrt{1 + 4\alpha} \geqslant 1$ force us to choose

$$\beta_+ = \tfrac{1}{2}(-1 + \sqrt{1 + 4\alpha})$$

and
$$\beta_- = \tfrac{1}{2}(1 - \sqrt{1 + 4\alpha}) = -\beta_+$$

Starting with $|Y_{\alpha,\beta_+}\rangle$, we can apply $\mathbb{L}_-$ to it repeatedly. In each step we decrease the value of β by one unit. There must be a limit to the number of vectors obtained in this way, because $|\beta| \leqslant \sqrt{\alpha}$. Therefore, there must exist a nonnegative integer k such that

$$(\mathbb{L}_-)^{k+1}|Y_{\alpha,\beta_+}\rangle = 0$$

or
$$\mathbb{L}_-(\mathbb{L}_-^k|Y_{\alpha,\beta_+}\rangle) = 0$$

Thus, $\mathbb{L}_-^k|Y_{\alpha,\beta_+}\rangle$ must be proportional to $|Y_{\alpha,\beta_-}\rangle$. In particular, since $\mathbb{L}_-^k|Y_{\alpha,\beta_+}\rangle$ has a β value equal to $\beta_+ - k$, we have

$$\beta_- = \beta_+ - k$$

Now, using $\beta_- = -\beta_+$ (derived above) yields the most important result:

$$\beta_+ = \frac{k}{2} \equiv j \qquad \text{for } k \in \mathbb{N}$$

And, since $\alpha = \beta_+^2 + \beta_+$,

$$\alpha = j(j+1)$$

This result is important enough to be stated as a theorem.

8.4.2 Theorem The eigenvectors of $\mathbb{L}^2$ can be written as $|Y_{jm}\rangle$, and

$$\mathbb{L}^2|Y_{jm}\rangle = j(j+1)|Y_{jm}\rangle$$

$$\mathbb{L}_z|Y_{jm}\rangle = m|Y_{jm}\rangle$$

where j is a positive integer or half-integer, and m can take a value in the set $\{-j, -j+1, \ldots, j-1, j\}$ of $2j+1$ numbers. ∎

Let us briefly consider the normalization of the eigenvectors. We already know that the $|Y_{jm}\rangle$, being eigenvectors of the hermitian operators $\mathbb{L}^2$ and $\mathbb{L}_z$, are orthogonal. We also demand that they be of unit norm; that is,

$$\langle Y_{j'm'}|Y_{jm}\rangle = \delta_{jj'}\delta_{mm'}$$

This determines $C_\pm$, introduced earlier. Let us consider C_+ first, which is defined by

$$\mathbb{L}_+|Y_{jm}\rangle = C_+|Y_{j,m+1}\rangle$$

The hermitian conjugate of this equation is

$$\langle Y_{jm}|\mathbb{L}_- = C_+^*\langle Y_{j,m+1}|$$

We contract these two equations to get

$$\langle Y_{jm}|\mathbb{L}_-\mathbb{L}_+|Y_{jm}\rangle = |C_+|^2\langle Y_{j,m+1}|Y_{j,m+1}\rangle$$

We use Eq. (8.19b), Theorem 8.4.2, and the above normalization to obtain

$$j(j+1) - m(m+1) = |C_+|^2$$

or

$$|C_+| = \sqrt{j(j+1) - m(m+1)}$$

Adopting the convention that the argument (phase) of the complex number C_+ is zero makes it a positive real number. Thus,

$$C_+ = \sqrt{j(j+1) - m(m+1)}$$

Similarly,

$$C_- = \sqrt{j(j+1) - m(m-1)}$$

Thus, we get

$$\mathbb{L}_+|\mathrm{Y}_{jm}\rangle = \sqrt{j(j+1) - m(m+1)}\,|\mathrm{Y}_{j,m+1}\rangle \tag{8.20a}$$

$$\mathbb{L}_-|\mathrm{Y}_{jm}\rangle = \sqrt{j(j+1) - m(m-1)}\,|\mathrm{Y}_{j,m-1}\rangle \tag{8.20b}$$

The discussion in this subsection is the standard treatment of angular momentum in quantum mechanics. In the context of quantum mechanics, Theorem 8.4.2 states the far-reaching physical result that particles can have integer or half-integer spin. Such a conclusion is tied to the rotation (group) in three dimensions, which, in turn, is an example of a Lie group, or a continuous group of transformation.[1]

8.4.3 Construction of the Eigenvectors of $\mathbb{L}^2$ (Spherical Harmonics)

The treatment in the preceding subsection took place in an abstract vector space. Let us go back to the function space and represent the operators and vectors in terms of θ and φ.

First, let us consider $\mathbb{L}_z$ in the form of a differential operator, as given in Eq. (8.15c). The eigenvalue equation for $\mathbb{L}_z$ becomes

$$-i\frac{\partial}{\partial\varphi}Y_{jm}(\theta,\varphi) = mY_{jm}(\theta,\varphi)$$

We write $Y_{jm}(\theta,\varphi) = P_{jm}(\theta)Q_{jm}(\varphi)$ and substitute in the above equation to obtain

$$\frac{dQ_{jm}}{d\varphi} = imQ_{jm}$$

whose solution is

$$Q_{jm}(\varphi) = C_{jm}e^{im\varphi}$$

[1] Discussion of Lie groups would be too great a digression from PDEs and ODEs. However, it is worth noting that it was the study of differential equations that led Sophus Lie to the investigation of their symmetries and the development of the beautiful branch of mathematics and theoretical physics that bears his name. Thus, the existence of a connection between group theory (rotation, angular momentum) and the differential equation we are trying to solve should not come as a surprise.

where C_{jm} is a constant. Absorbing C_{jm} into P_{jm}, we can write

$$Y_{jm}(\theta, \varphi) = P_{jm}(\theta)e^{im\varphi}$$

In classical physics the functions must give the same value at φ and at $\varphi + 2\pi$. This condition restricts the values of m to integers. In quantum mechanics, on the other hand, it is the absolute values of functions that are relevant, and therefore m can also be half-integers. From now on, we will assume that m is an integer and denote the eigenvectors of $\mathbb{L}^2$ as $Y_{lm}(\theta, \varphi)$, in which l is a *nonnegative integer*. We want to find an analytic expression for $Y_{lm}(\theta, \varphi)$.

For such a task we need analytic (differential) expressions for $\mathbb{L}_\pm$. These can easily be obtained from the expressions for $\mathbb{L}_x$ and $\mathbb{L}_y$ given in Eqs. (8.15). (The straightforward manipulations are left as a problem.) We thus have

$$\mathbb{L}_\pm = e^{\pm i\varphi}\left(\pm\frac{\partial}{\partial\theta} + i\cot\theta\frac{\partial}{\partial\varphi}\right) \tag{8.21}$$

When $\mathbb{L}_+$ acts on $Y_{ll}(\theta, \varphi) \equiv P_{ll}(\theta)e^{il\varphi}$, the result is zero. This leads to the differential equation

$$\left(\frac{\partial}{\partial\theta} + i\cot\theta\frac{\partial}{\partial\varphi}\right)(P_{ll}(\theta)e^{il\varphi}) = 0$$

which, in turn, gives

$$\left(\frac{d}{d\theta} - l\cot\theta\right)P_{ll}(\theta) = 0$$

The solution to this (derived in Exercise 8.4.4) is

$$P_{ll}(\theta) = C_l(\sin\theta)^l$$

The constant is subscripted because each P_{ll} may lead to a different constant of integration. We can now write

$$Y_{ll}(\theta, \varphi) = P_{ll}(\theta)e^{il\varphi} = C_l(\sin\theta)^l e^{il\varphi} \tag{8.22}$$

With $Y_{ll}(\theta, \varphi)$ at our disposal, we can obtain any $Y_{lm}(\theta, \varphi)$ by repeated application of $\mathbb{L}_-$. The following example shows the straightforward, but instructive, manipulation for the case of abstract vectors.

Example 8.4.1

Let us find an expression for $|Y_{lm}\rangle$ by applying $\mathbb{L}_-$ to $|Y_{ll}\rangle$ repeatedly. The action for $\mathbb{L}_-$ is completely described by Eq. (8.20b).

For the first power of $\mathbb{L}_-$, we obtain

$$\mathbb{L}_-|Y_{ll}\rangle = \sqrt{l(l+1) - l(l-1)}|Y_{l,l-1}\rangle = \sqrt{2l}|Y_{l,l-1}\rangle$$

We apply $\mathbb{L}_-$ to this equation, obtaining

$$(\mathbb{L}_-)^2|Y_{ll}\rangle = \sqrt{2l}\mathbb{L}_-|Y_{l,l-1}\rangle = \sqrt{2l}\sqrt{2(2l-1)}|Y_{l,l-2}\rangle$$
$$= \sqrt{2(2l)(2l-1)}|Y_{l,l-2}\rangle$$

Applying $\mathbb{L}_-$ once more yields

$$(\mathbb{L}_-)^3|\mathrm{Y}_{ll}\rangle = \sqrt{2(2l)(2l-1)}\mathbb{L}_-|\mathrm{Y}_{l,l-2}\rangle = \sqrt{2(2l)(2l-1)}\sqrt{6(l-1)}|\mathrm{Y}_{l,l-3}\rangle$$
$$= \sqrt{3!(2l)(2l-1)(2l-2)}|\mathrm{Y}_{l,l-3}\rangle$$

We detect a pattern, giving the following for a general power, k:

$$(\mathbb{L}_-)^k|\mathrm{Y}_{ll}\rangle = \sqrt{k!(2l)(2l-1)\cdots(2l-k+1)}|\mathrm{Y}_{l,l-k}\rangle$$

or

$$(\mathbb{L}_-)^k|\mathrm{Y}_{ll}\rangle = \sqrt{\frac{k!(2l)!}{(2l-k)!}}|\mathrm{Y}_{l,l-k}\rangle$$

If we set $l - k \equiv m$, we get

$$(\mathbb{L}_-)^{l-m}|\mathrm{Y}_{ll}\rangle = \sqrt{\frac{(l-m)!(2l)!}{(l+m)!}}|\mathrm{Y}_{l,m}\rangle$$

which finally yields

$$|\mathrm{Y}_{l,m}\rangle = \sqrt{\frac{(l+m)!}{(l-m)!(2l)!}}(\mathbb{L}_-)^{l-m}|\mathrm{Y}_{ll}\rangle$$ •

In principle, the result of Example 8.4.1 gives all the (abstract) eigenvectors. In practice, however, it is helpful to have a closed form (in terms of derivatives) for just the θ part of $Y_{lm}(\theta, \varphi)$. So, let us apply $\mathbb{L}_-$, as given in Eq. (8.21), to Eq. (8.22) to obtain $\mathrm{Y}_{l,l-1}(\theta, \varphi)$:

$$\mathbb{L}_- Y_{ll} = e^{-i\varphi}\left(-\frac{\partial}{\partial\theta} + i\cot\theta\frac{\partial}{\partial\varphi}\right)[P_{ll}(\theta)e^{il\varphi}]$$
$$= e^{-i\varphi}\left[-\frac{\partial}{\partial\theta} + i\cot\theta(il)\right][P_{ll}(\theta)e^{il\varphi}]$$
$$= (-1)e^{i(l-1)\varphi}\left(\frac{d}{d\theta} + l\cot\theta\right)P_{ll}(\theta)$$

It can be shown (see Exercise 8.4.5) that

$$\left(\frac{d}{d\theta} + n\cot\theta\right)f(\theta) = \frac{1}{(\sin\theta)^n}\frac{d}{d\theta}[(\sin\theta)^n f(\theta)] \tag{8.23}$$

Using this result yields

$$\mathbb{L}_- Y_{ll} = (-1)e^{i(l-1)\varphi}\frac{1}{(\sin\theta)^l}\frac{d}{d\theta}[\sin^l\theta(C_l\sin^l\theta)]$$
$$= (-1)C_l\frac{e^{i(l-1)\varphi}}{(\sin\theta)^l}\frac{d}{d\theta}[(\sin^2\theta)^l]$$
$$\equiv e^{i(l-1)\varphi}P_{l,l-1}(\theta)$$

which shows that

$$P_{l,l-1}(\theta) = (-1)C_l \frac{1}{(\sin\theta)^l}\frac{d}{d\theta}[(\sin^2\theta)^l]$$

We apply $\mathbb{L}_-$ once more to obtain

$$\begin{aligned}(\mathbb{L}_-)^2 Y_{ll} &= \mathbb{L}_-[e^{i(l-1)\varphi}P_{l,l-1}(\theta)] \\ &= e^{-i\varphi}\left(-\frac{\partial}{\partial\theta} + i\cot\theta\frac{\partial}{\partial\varphi}\right)[e^{i(l-1)\varphi}P_{l,l-1}(\theta)] \\ &= e^{-i\varphi}\left[-\frac{\partial}{\partial\theta} - (l-1)\cot\theta\right][e^{i(l-1)\varphi}P_{l,l-1}(\theta)] \\ &= (-1)e^{i(l-2)\varphi}\left[\frac{d}{d\theta} + (l-1)\cot\theta\right]P_{l,l-1}(\theta)\end{aligned}$$

Now we use Eq. (8.23) with $n = l - 1$:

$$(\mathbb{L}_-)^2 Y_{ll} = (-1)e^{i(l-2)\varphi}\frac{1}{(\sin\theta)^{l-1}}\frac{d}{d\theta}[(\sin\theta)^{l-1}P_{l,l-1}]$$

Substituting the above expression for $P_{l,l-1}$, we get

$$\begin{aligned}(\mathbb{L}_-)^2 Y_{ll} &= (-1)^2 C_l e^{i(l-2)\varphi}\frac{1}{(\sin\theta)^{l-1}}\frac{d}{d\theta}\left\{(\sin\theta)^{l-1}\frac{1}{(\sin\theta)^l}\frac{d}{d\theta}[(\sin^2\theta)^l]\right\} \\ &= (-1)^2 C_l \frac{e^{i(l-2)\varphi}}{(\sin\theta)^{l-1}}\frac{d}{d\theta}\left\{\frac{1}{\sin\theta}\frac{d}{d\theta}[(\sin^2\theta)^l]\right\}\end{aligned}$$

Making the substitution $u = \cos\theta$, we obtain

$$(\mathbb{L}_-)^2 Y_{ll} = C_l \frac{e^{i(l-2)\varphi}}{(\sqrt{1-u^2})^{l-2}}\frac{d^2}{du^2}[(1-u^2)^l]$$

It is now clear that

$$(\mathbb{L}_-)^k Y_{ll} = C_l \frac{e^{i(l-k)\varphi}}{(1-u^2)^{(l-k)/2}}\left(\frac{d}{du}\right)^k[(1-u^2)^l]$$

If we let $k = l - m$, we get

$$(\mathbb{L}_-)^{l-m} Y_{ll} = C_l \frac{e^{im\varphi}}{(1-u^2)^{m/2}}\left(\frac{d}{du}\right)^{l-m}[(1-u^2)^l]$$

Making use of Example 8.4.1, we obtain

$$Y_{lm}(\theta,\varphi) = \sqrt{\frac{(l+m)!}{(l-m)!(2l)!}}\,C_l \frac{e^{im\varphi}}{(1-u^2)^{m/2}}\left(\frac{d}{du}\right)^{l-m}[(1-u^2)^l]$$

To specify $Y_{lm}(\theta, \varphi)$ completely, we need to evaluate C_l. Since C_l has no m as index, we set $m = 0$ in the above expression, obtaining

$$Y_{l0}(u, \varphi) = \frac{1}{\sqrt{(2l)!}} C_l \frac{d^l}{du^l}[(1 - u^2)^l]$$

The RHS looks very much like the Legendre polynomials of Chapter 5. In fact,

$$Y_{l0}(u, \varphi) = \frac{C_l}{\sqrt{(2l)!}}(-1)^l 2^l (l!) P_l(u) \equiv A_l P_l(u)$$

Recall that $|Y_{lm}\rangle$ are orthonormal. For the case of $Y_{lm}(\theta, \varphi)$ this translates into

$$\int_0^{2\pi} d\varphi \int_0^{\pi} \sin\theta \, d\theta \, Y^*_{l'm'}(\theta, \varphi) Y_{lm}(\theta, \varphi) = \delta_{ll'}\delta_{mm'}$$

which, in terms of $u = \cos\theta$, becomes

$$\int_0^{2\pi} d\varphi \int_{-1}^{1} Y^*_{l'm'}(u, \varphi) Y_{lm}(u, \varphi) \, du = \delta_{ll'}\delta_{mm'}$$

In particular, for $l = l'$ and $m = m' = 0$, we have

$$\int_0^{2\pi} d\varphi \int_{-1}^{1} |Y_{l0}(u, \varphi)|^2 \, du = 1$$

Substituting for Y_{l0} in terms of Legendre polynomials, we obtain

$$\int_0^{2\pi} d\varphi \int_{-1}^{1} |A_l|^2 [P_l(u)]^2 \, du = 1 = 2\pi |A_l|^2 \int_{-1}^{1} [P_l(u)]^2 \, du = 2\pi |A_l|^2 \frac{2}{2l + 1}$$

↑ by the standardization of $P_l(u)$

$$= \frac{4\pi}{2l + 1} |A_l|^2$$

Setting, by convention, the phase of the complex number A_l equal to zero, we get

$$A_l = \frac{C_l}{\sqrt{(2l)!}}(-1)^l 2^l (l!) = \sqrt{\frac{2l + 1}{4\pi}}$$

Using this relation, we finally obtain

$$Y_{lm}(u, \varphi) = (-1)^l \sqrt{\frac{2l + 1}{4\pi}} \left(\frac{e^{im\varphi}}{2^l l!}\right) \sqrt{\frac{(l + m)!}{(l - m)!}} (1 - u^2)^{-m/2} \left(\frac{d}{du}\right)^{l - m} [(1 - u^2)^l] \tag{8.24}$$

where $u = \cos\theta$. These functions, the eigenfunctions of $\mathbb{L}^2$ and $\mathbb{L}_z$, are called *spherical harmonics* and occur frequently in physical applications for which the Laplacian is expressed in terms of spherical coordinates.

We can immediately read off the θ part of the functions of (8.24):

$$P_{lm}(u) = (-1)^l \sqrt{\frac{2l+1}{4\pi}}\left(\frac{1}{2^l l!}\right)\sqrt{\frac{(l+m)!}{(l-m)!}}(1-u^2)^{-m/2}\left(\frac{d}{du}\right)^{l-m}[(1-u^2)^l]$$

However, this is not the version used in the literature. For historical reasons the *associated Legendre functions*, $P_l^m(u)$, are used. These are defined as follows:

$$P_l^m(u) \equiv (-1)^m \sqrt{\frac{(l+m)!}{(l-m)!}}\sqrt{\frac{4\pi}{2l+1}}P_{lm}(u)$$

$$= (-1)^{l+m}\frac{(l+m)!}{(l-m)!}\left[\frac{(1-u^2)^{-m/2}}{2^l l!}\right]\left(\frac{d}{du}\right)^{l-m}[(1-u^2)^l] \tag{8.25a}$$

Thus,

$$Y_{lm}(\theta,\varphi) = (-1)^m\left\{\frac{2l+1}{4\pi}\left[\frac{(l-m)!}{(l+m)!}\right]\right\}^{1/2} P_l^m(\cos\theta)e^{im\varphi} \tag{8.25b}$$

We generated the spherical harmonics starting with $Y_{ll}(\theta, \varphi)$ and applying the lowering operator, $\mathbb{L}_-$. We could have started with $Y_{l,-l}(\theta,\varphi)$ instead and applied the raising operator, $\mathbb{L}_+$. This procedure is identical to the other and is outlined below.

We first note that

$$|\mathrm{Y}_{l,-m}\rangle = \sqrt{\frac{(l+m)!}{(l-m)!(2l)!}}(\mathbb{L}_+)^{l-m}|\mathrm{Y}_{l,-l}\rangle \tag{8.26}$$

(This can be obtained following the steps of Example 8.4.1.) Next, we use $\mathbb{L}_-|\mathrm{Y}_{l,-l}\rangle = 0$ in differential form to obtain

$$\left(\frac{d}{d\theta} - l\cot\theta\right)P_{l,-l}(\theta) = 0$$

which has the same form as the differential equation for P_{ll}. Thus, the solution is

$$P_{l,-l} = C_l'(\sin\theta)^l$$

and

$$Y_{l,-l}(\theta,\varphi) \equiv P_{l,-l}(\theta)e^{-il\varphi} = C_l'(\sin\theta)^l e^{-il\varphi}$$

Applying $\mathbb{L}_+$ repeatedly yields

$$(\mathbb{L}_+)^k Y_{l,-l}(u,\varphi) = C_l'\frac{(-1)^k e^{-i(l-k)\varphi}}{(1-u^2)^{(l-k)/2}}\left(\frac{d}{du}\right)^k[(1-u^2)^l]$$

where $u = \cos\theta$. Substituting $k = l - m$ and using Eq. (8.26) gives

$$Y_{l,-m}(u,\varphi) = \sqrt{\frac{(l+m)!}{(l-m)!(2l)!}}C_l' e^{-im\varphi}\frac{(-1)^{l-m}}{(1-u^2)^{m/2}}\left(\frac{d}{du}\right)^{l-m}[(1-u^2)^l]$$

The constant C_l' can be determined as before. In fact, for $m = 0$ we get exactly the same result as before, so we expect C_l' to be identical to C_l. Thus,

$$Y_{l,\,-m}(u, \varphi) = (-1)^{l+m}\sqrt{\frac{2l+1}{4\pi}}\left(\frac{e^{-im\varphi}}{2^l l!}\right)\sqrt{\frac{(l+m)!}{(l-m)!}}(1-u^2)^{-m/2}\left(\frac{d}{du}\right)^{l-m}[(1-u^2)^l] \tag{8.27}$$

Comparison with (8.24) yields

$$Y_{l,\,-m}(\theta, \varphi) = (-1)^m Y_{lm}^*(\theta, \varphi) \tag{8.28a}$$

and, using the definition $Y_{l,\,-m}(\theta, \varphi) \equiv P_{l,\,-m}(\theta)e^{-im\varphi}$ and the first part of Eq. (8.25a), we obtain

$$P_l^{-m}(\theta) = (-1)^m \frac{(l-m)!}{(l+m)!} P_l^m(\theta) \tag{8.28b}$$

The first few of the spherical harmonics with positive m are given below. Those with negative m can be obtained using Eq. (8.28a).

For $l = 0$, $\quad Y_{0,0} = \dfrac{1}{\sqrt{4\pi}}$

For $l = 1$, $\quad Y_{1,1} = -\sqrt{\dfrac{3}{8\pi}}\, e^{i\varphi}\sin\theta \qquad Y_{1,0} = \sqrt{\dfrac{3}{4\pi}}\cos\theta$

For $l = 2$, $\quad Y_{2,2} = \sqrt{\dfrac{15}{32\pi}}\, e^{2i\varphi}\sin^2\theta \qquad Y_{2,1} = -\sqrt{\dfrac{15}{8\pi}}\, e^{i\varphi}\sin\theta\cos\theta$

$$Y_{2,0} = \sqrt{\frac{5}{16\pi}}(3\cos^2\theta - 1)$$

For $l = 3$, $\quad Y_{3,3} = -\sqrt{\dfrac{35}{64\pi}}\, e^{3i\varphi}\sin^3\theta \qquad Y_{3,2} = \sqrt{\dfrac{105}{32\pi}}\, e^{2i\varphi}\sin^2\theta\cos\theta$

$$Y_{3,1} = -\sqrt{\frac{21}{64\pi}}\, e^{i\varphi}\sin\theta(5\cos^2\theta - 1)$$

$$Y_{3,0} = \sqrt{\frac{7}{16\pi}}(5\cos^3\theta - 3\cos\theta)$$

Having found the spherical harmonics, we can derive the differential equation they obey. From Eqs. (8.14a), (8.16), and (8.25b) and the fact that $\alpha = l(l+1)$ for some integer l, we obtain

$$-\frac{1}{\sin\theta}\frac{\partial}{\partial\theta}\left(\sin\theta\frac{\partial}{\partial\theta}\right)(P_l^m(\cos\theta)e^{im\varphi}) - \frac{1}{\sin^2\theta}\frac{\partial^2}{\partial\varphi^2}[P_l^m(\cos\theta)e^{im\varphi}] = l(l+1)P_l^m e^{im\varphi}$$

which gives

$$\frac{1}{\sin\theta}\frac{d}{d\theta}\left(\sin\theta\frac{dP_l^m}{d\theta}\right) - \frac{m^2}{\sin^2\theta}P_l^m + l(l+1)P_l^m = 0$$

As before, we let $u = \cos\theta$ to obtain

$$\frac{d}{du}\left[(1-u^2)\frac{dP_l^m}{du}\right] + \left[l(l+1) - \frac{m^2}{1-u^2}\right]P_l^m = 0$$

This is called the *associated Legendre differential equation.* Its solutions, the associated Legendre functions, are given in closed form in Eq. (8.25a).

For $m = 0$ we have

$$\frac{d}{du}\left[(1-u^2)\frac{dP_l^0}{du}\right] + l(l+1)P_l^0 = 0$$

which is the Legendre differential equation whose solutions, again given by (8.25a) with $m = 0$, are the Legendre polynomials encountered in Chapter 5. When $m = 0$, the spherical harmonics become φ-independent. This corresponds to a physical situation in which there is an explicit azimuthal symmetry. In such cases (when it is obvious that the physical property in question does not depend on φ) only a Legendre polynomial, depending only on $\cos\theta$, will multiply the radial function.

8.4.4 Expansion of Angular Functions

The orthonormality of $Y_{lm}(\theta, \varphi)$ suggests the expansion of functions of θ and φ in terms of spherical harmonics. The fact that these functions are complete will be discussed in a general way in Chapter 10. Assuming completeness for now, we write

$$f(\theta, \varphi) = \sum_{l=0}^{\infty}\sum_{m=-l}^{+l} a_{lm} Y_{lm}(\theta, \varphi) \tag{8.29a}$$

To find a_{lm} we multiply both sides by $Y^*_{l'm'}(\theta, \varphi)$ and integrate over the solid angle. The result, obtained by using the orthonormality relation, is

$$a_{lm} = \int_0^{2\pi} d\varphi \int_0^{\pi} \sin\theta\, f(\theta, \varphi)\, Y^*_{lm}(\theta, \varphi)\, d\theta \tag{8.29b}$$

The addition theorem for spherical harmonics. An important consequence of the above expansion is called the addition theorem for spherical harmonics. Consider two unit vectors $\hat{\mathbf{e}}_r$ and $\hat{\mathbf{e}}_{r'}$ making spherical angles (θ, φ) and (θ', φ'), respectively (see Fig. 8.1). Let γ be the angle between the two vectors. Then it can easily be shown that

$$\cos\gamma = \cos\theta\cos\theta' + \sin\theta\sin\theta'\cos(\varphi - \varphi')$$

Now consider the Legendre polynomial $P_l(\cos\gamma)$, which is clearly a function of θ and

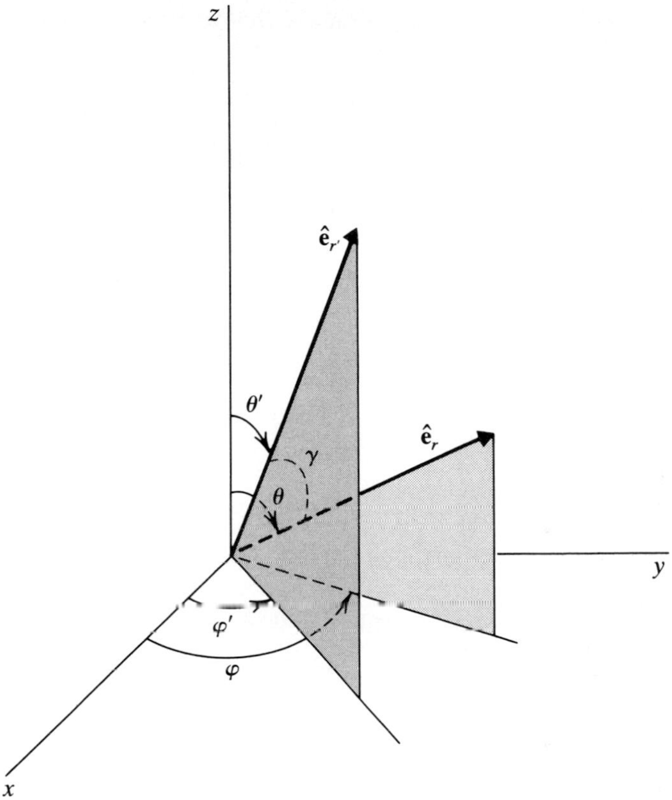

Figure 8.1 The unit vectors $\hat{\mathbf{e}}_r$ and $\hat{\mathbf{e}}_{r'}$ with the spherical angles of each and the angle γ between them.

φ. According to (8.29), it can be expanded as a linear combination of $Y_{lm}(\theta, \varphi)$. Since $P_l(\cos\gamma)$ is also a function of θ' and φ', we expect the coefficients a_{lm} to be functions of θ' and φ'. Thus, quite generally, we have

$$P_l(\cos\gamma) = \sum_{k=0}^{\infty} \sum_{m=-k}^{+k} a_{km}(\theta', \varphi') Y_{km}(\theta, \varphi)$$

The addition theorem states that

$$a_{km} = \delta_{kl} \frac{4\pi}{2l+1} Y^*_{lm}(\theta', \varphi')$$

so

$$P_l(\cos\gamma) = \frac{4\pi}{2l+1} \sum_{m=-l}^{l} Y^*_{lm}(\theta', \varphi') Y_{lm}(\theta, \varphi) \tag{8.30}$$

To prove this theorem, we make the observation that $\cos\gamma = \cos\theta$ for $\theta' = 0$. On the other hand,

$$\mathbb{L}^2 P_l(\cos\theta) = \mathbb{L}^2 Y_{l0}(\theta, \varphi) = l(l+1)P_l(\cos\theta)$$

and $P_l(\cos\theta)$ is an eigenvector of $\mathbb{L}^2$ with eigenvalue $l(l+1)$. We also note that $P_l(\cos\gamma)$ is obtained from $P_l(\cos\theta)$ by rotating the axes. But the operator $\mathbb{L}^2 = \vec{\mathbb{L}}\cdot\vec{\mathbb{L}}$ is a scalar product of two vector operators. Invariance of the scalar product under rotation implies invariance of $\mathbb{L}^2$. In abstract operator notation this is expressed by $\mathbb{R}\mathbb{L}^2\mathbb{R}^{-1} = \mathbb{L}^2$, where $\mathbb{R}$ is the rotation operator. Thus, using abstract vectors, we can write

$$\mathbb{L}^2|l, 0\rangle = l(l+1)|l, 0\rangle$$

and

$$\mathbb{R}\mathbb{L}^2|l, 0\rangle = l(l+1)\mathbb{R}|l, 0\rangle$$

or

$$\mathbb{R}\mathbb{L}^2\mathbb{R}^{-1}(\mathbb{R}|l, 0\rangle) = l(l+1)(\mathbb{R}|l, 0\rangle) \quad \Rightarrow \quad \mathbb{L}^2(\mathbb{R}|l, 0\rangle) = l(l+1)(\mathbb{R}|l, 0\rangle)$$

Thus, the rotated vector $\mathbb{R}|l, 0\rangle$ has the same l value as the original vector. Translated into the language of the function space, this means that the rotated Legendre polynomial $P_l(\cos\gamma)$ has the same l value as before. Thus, in the expansion of $P_l(\cos\gamma)$ we should have only the $k = l$ term. This reduces the sum to

$$P_l(\cos\gamma) = \sum_{m=-l}^{l} A_m(\theta', \varphi')\, Y_{lm}(\theta, \varphi) \tag{8.31}$$

where $a_{lm} \equiv A_m$ (because l is fixed). We want to show that

$$A_m(\theta', \varphi') = \frac{4\pi}{2l+1} Y^*_{lm}(\theta', \varphi')$$

We will do this in several steps.

First, we note that

$$Y_{lm}(0, \varphi) = \delta_{m0} Y_{l0}(0, \varphi) = \delta_{m0}\sqrt{\frac{2l+1}{4\pi}}\, P_l(1) = \delta_{m0}\sqrt{\frac{2l+1}{4\pi}}$$

This can be seen from (8.25a) and the fact that, unless $m = 0$, the factor $(1 - u^2)$ will be left in all terms resulting from differentiation of $(1 - u^2)^l$.

Second, we let $\theta = 0$ on both sides of (8.29a) to obtain

$$f(\theta, \varphi)|_{\theta=0} = \sum_{l=0}^{\infty} a_{l0}\sqrt{\frac{2l+1}{4\pi}}$$

In particular, if there is only one l term in the sum, we have

$$f_l(\theta, \varphi)|_{\theta=0} = \sqrt{\frac{2l+1}{4\pi}}\, a_{l0}$$

By (8.29b),

$$a_{l0} = \iint d\Omega\, f_l(\theta, \varphi) \sqrt{\frac{2l+1}{4\pi}}\, P_l(\cos\theta)$$

where the commonly used notation $d\Omega = \sin\theta\, d\theta\, d\varphi$ is used. The angles (θ, φ) refer to a particular choice of axes. Let us use (γ, β) for the angles in an arbitrary coordinate system. Then

$$a_{l0} = \sqrt{\frac{2l+1}{4\pi}} \iint d\Omega\, f_l(\gamma, \beta) P_l(\cos\gamma)$$

Third, we multiply both sides of (8.31) by $Y^*_{lm}(\theta, \varphi)$ and integrate to get

$$A_m(\theta', \varphi') = \iint d\Omega\; Y^*_{lm}(\theta, \varphi) P_l(\cos\gamma)$$

The last two equations are the same if we set up the following identity:

$$f_l(\gamma, \beta) \equiv \sqrt{\frac{4\pi}{2l+1}}\; Y^*_{lm}[\theta(\gamma, \beta, \theta', \varphi'), \varphi(\gamma, \beta, \theta', \varphi')]$$

We, therefore, have

$$A_m(\theta', \varphi') \equiv a_{l0} = \sqrt{\frac{4\pi}{2l+1}} f_l(\gamma, \beta)\bigg|_{\gamma=0}$$

$$= \frac{4\pi}{2l+1}\; Y^*_{lm}[\theta(\gamma, \beta, \theta', \varphi'), \varphi(\gamma, \beta, \theta', \varphi')]_{\gamma=0}$$

But $\gamma = 0$ implies that $\theta = \theta'$ and $\varphi = \varphi'$, as is clear from Fig. 8.1. Thus,

$$A_m(\theta', \varphi') = \frac{4\pi}{2l+1}\; Y^*_{lm}(\theta', \varphi')$$

which is the result we want.

The addition theorem is particularly useful in the expansion of the frequently occurring expression $1/|\mathbf{r} - \mathbf{r}'|$. For definiteness we assume $|\mathbf{r}'| \equiv r' < |\mathbf{r}| \equiv r$. Then, introducing $t = r'/r$, we have

$$\frac{1}{|\mathbf{r} - \mathbf{r}'|} = \frac{1}{[r^2 + (r')^2 - 2rr'\cos\gamma]^{1/2}} = \frac{1}{r}\left[\frac{1}{(1 + t^2 - 2t\cos\gamma)^{1/2}}\right]$$

Recalling the generating function for Legendre polynomials from Chapter 5 and using the addition theorem, we get

$$\frac{1}{|\mathbf{r} - \mathbf{r}'|} = \frac{1}{r}\sum_{l=0}^{\infty} t^l P_l(\cos\gamma) = \sum_{l=0}^{\infty} \frac{(r')^l}{r^{l+1}} \left(\frac{4\pi}{2l+1}\right) \sum_{m=-l}^{l} Y^*_{lm}(\theta', \varphi')\, Y_{lm}(\theta, \varphi)$$

$$= 4\pi \sum_{l=0}^{\infty} \sum_{m=-l}^{l} \frac{1}{2l+1}\left[\frac{(r')^l}{r^{l+1}}\right] Y^*_{lm}(\theta', \varphi')\, Y_{lm}(\theta, \varphi)$$

It is clear that if $r < r'$, we can expand in terms of the ratio r/r'. It is therefore customary to use $r_<$ to denote the smaller and $r_>$ the larger of the two radii r and r'. Then the above equation is written as

$$\frac{1}{|\mathbf{r} - \mathbf{r}'|} = 4\pi \sum_{l=0}^{\infty} \sum_{m=-l}^{l} \frac{1}{2l+1} \left(\frac{r_<^l}{r_>^{l+1}} \right) Y_{lm}^*(\theta', \varphi') Y_{lm}(\theta, \varphi) \tag{8.32}$$

This equation is very useful in the study of Coulomb potentials.

The general solution in spherical coordinates. It is useful to have a general formula for the solutions of Eq. (8.8c) in a spherical coordinate system. We have already reduced the equation to two equivalent equations, (8.14a) and (8.14b). Rewriting (8.14b) as

$$\frac{1}{r^2}\frac{d}{dr}\left(r^2 \frac{dR}{dr} \right) + \left[f(r) - \frac{l(l+1)}{r^2} \right] R = 0$$

and making the substitution $R = u/r$ gives

$$\frac{d^2 u}{dr^2} + \left[f(r) - \frac{l(l+1)}{r^2} \right] u = 0$$

In general, $f(r)$ contains a constant that has a label, an index such as n that can take infinitely many (integer) values. Thus, u has two labels: n and l. Recalling that $\Psi(\mathbf{r}) = R(r) Y(\theta, \varphi)$, we note that for any particular n, l, and m, we have the solution

$$\Psi_{nlm}(\mathbf{r}) = R_{nl}(r) Y_{lm}(\theta, \varphi) = \frac{u_{nl}}{r} Y_{lm}(\theta, \varphi)$$

and a general solution is of the form

$$\Psi(\mathbf{r}) = \sum_{n,l,m} C_{nlm} \frac{u_{nl}(r)}{r} Y_{lm}(\theta, \varphi)$$

Exercises

8.4.1 For the angular momentum operator $\mathbb{L}_i = \varepsilon_{ijk} x_j \mathbb{p}_k$, show that the following commutation relations hold:

$$[\mathbb{L}_j, \mathbb{L}_k] = i\varepsilon_{jkl} \mathbb{L}_l$$

8.4.2 Obtain an expression for $\mathbb{L}^2$ in terms of θ and φ.

8.4.3 Show that

$$\mathbb{L}^2 = \mathbb{L}_+ \mathbb{L}_- + \mathbb{L}_z^2 - \mathbb{L}_z \quad \text{and} \quad \mathbb{L}^2 = \mathbb{L}_- \mathbb{L}_+ + \mathbb{L}_z^2 + \mathbb{L}_z$$

8.4.4 Find the solution to $dP/d\theta - (l \cot \theta)P = 0$.

8.4.5 Prove the following differential identity:

$$\left(\frac{d}{d\theta} + n \cot \theta \right) f(\theta) = \frac{1}{(\sin \theta)^n} \frac{d}{d\theta} [(\sin \theta)^n f(\theta)]$$

PROBLEMS

8.1 By applying the *operator* $[x_j, \mathbb{p}_k]$ to an arbitrary function $f(\mathbf{r})$, show that $[x_j, \mathbb{p}_k] = i\delta_{jk}$.

8.2 Use the definitive relation $\mathbb{L}_i = \varepsilon_{ijk} x_j \mathbb{p}_k$ to show that $x_j \mathbb{p}_k - x_k \mathbb{p}_j = \varepsilon_{ijk} \mathbb{L}_i$. (In both of these expressions a sum over the repeated indices is understood.)

8.3 Evaluate $\partial f/\partial y$ and $\partial f/\partial z$ in spherical coordinates.

8.4 Calculate $\mathbb{L}_y$ and $\mathbb{L}_z$ in terms of spherical coordinates.

8.5 Substitute the expression for $\mathbb{L}^2$ in terms of θ and φ in Eq. (8.13) and show that this yields the Laplacian in spherical coordinates, as given in Chapter 1.

8.6 Show that $\mathbb{L}^2$, $\mathbb{L}_1$, $\mathbb{L}_2$, and $\mathbb{L}_3$ are hermitian operators in the space of square-integrable functions.

8.7 Verify the following commutation relations:

$$[\mathbb{L}^2, \mathbb{L}_\pm] = 0 \qquad [\mathbb{L}_z, \mathbb{L}_\pm] = \pm \mathbb{L}_\pm \qquad [\mathbb{L}_+, \mathbb{L}_-] = 2\mathbb{L}_z$$

8.8 Show that $\mathbb{L}_\pm |\mathrm{Y}_{\alpha\beta}\rangle$ has the same eigenvalue for $\mathbb{L}^2$ as $|\mathrm{Y}_{\alpha\beta}\rangle$ does. Also show that $\mathbb{L}_- |\mathrm{Y}_{\alpha\beta}\rangle$ has $\beta - 1$ as its eigenvalue for $\mathbb{L}_z$.

8.9 Let $\mathscr{V}$ be an inner product space. Let $\mathbb{H} \in \mathscr{L}(\mathscr{V})$ be hermitian. Show that, for any $|\mathrm{a}\rangle \in \mathscr{V}$, $\langle \mathrm{a}|\mathbb{H}^2|\mathrm{a}\rangle \geqslant 0$.

8.10 Show that $|\mathrm{Y}_{\alpha\beta_\pm}\rangle$ cannot be zero.

8.11 Show that if $|\mathrm{Y}_{jm}\rangle$ are normalized to unity, then, with proper choice of phase,

$$\mathbb{L}_- |\mathrm{Y}_{jm}\rangle = \sqrt{j(j+1) - m(m-1)}\, |\mathrm{Y}_{j,m-1}\rangle$$

8.12 Starting with $\mathbb{L}_x$ and $\mathbb{L}_y$, derive the following expression for $\mathbb{L}_\pm$:

$$\mathbb{L}_\pm = e^{\pm i\varphi}\left(\pm \frac{\partial}{\partial\theta} + i\cot\theta \frac{\partial}{\partial\varphi}\right)$$

8.13 Perform the calculation leading to an expression for $(\mathbb{L}_-)^2 Y_{ll}$ when the substitution $u = \cos\theta$ is made.

8.14 Derive Eq. (8.26).

8.15 Show that

$$(\mathbb{L}_+)^k Y_{l,-l}(u, \varphi) = C_l' \frac{(-1)^k e^{i(l-k)\varphi}}{(1-u^2)^{(l-k)/2}} \left(\frac{d}{du}\right)^k [(1-u^2)^l]$$

8.16 Derive Eqs. (8.28).

8.17 Show that

$$\sum_{m=-l}^{+l} |Y_{lm}(\theta, \varphi)|^2 = \frac{2l+1}{4\pi}$$

Verify this for $l = 1$ and $l = 2$.

8.18 Show that the addition theorem for spherical harmonics can be written as

$$P_l(\cos\gamma) = P_l(\cos\theta)P_l(\cos\theta') + 2\sum_{m=1}^{l} \frac{(l-m)!}{(l+m)!} P_l^m(\cos\theta)P_l^m(\cos\theta')\cos[m(\varphi - \varphi')]$$

9

Differential Equations II: Ordinary Differential Equations

The discussion of the preceding chapter clearly singled out ordinary differential equations (ODEs), especially those of second order, as objects requiring special attention. This is because most of the common partial differential equations (PDEs) of mathematical physics can be separated into ODEs (of second order). Therefore, the rest of this part of the book is devoted to the study of these equations.

The most general ODE can be expressed as

$$F\left(x, y, \frac{dy}{dx}, \frac{d^2y}{dx^2}, \ldots, \frac{d^ny}{dx^n}\right) = 0 \tag{9.1}$$

in which $F: \mathbb{R}^{n+2} \to \mathbb{R}$ is a real-valued function on $\mathbb{R}^{n+2}$. When F depends explicitly and nontrivially on d^ny/dx^n, Eq. (9.1) is an *nth-order ODE*. An ODE is said to be *linear* if the part of the function F that includes y and all its derivatives is linear in y. Thus, the most general nth-order linear ODE is

$$p_0(x)y + p_1(x)\frac{dy}{dx} + p_2(x)\frac{d^2y}{dx^2} + \cdots + p_n(x)\frac{d^ny}{dx^n} = q(x) \qquad \text{for } p_n \neq 0 \tag{9.2}$$

where $\{p_i(x)\}_{i=0}^n$ and $q(x)$ are functions of the independent variable x. Equation (9.2) is said to be *homogeneous* if $q(x) \equiv 0$; otherwise, it is said to be *inhomogeneous* and $q(x)$ is called the *inhomogeneous term*. It is customary, and convenient, to define a linear (differential) operator $\mathbb{L}$ by

$$\mathbb{L} \equiv p_0(x) + p_1(x)\frac{d}{dx} + p_2(x)\frac{d^2}{dx^2} + \cdots + p_n(x)\frac{d^n}{dx^n} \tag{9.3}$$

and write Eq. (9.2) as

$$\mathbb{L}[y] = q(x) \tag{9.4}$$

A *solution* of Eq. (9.1) or (9.4) is a function $f: \mathbb{R} \to \mathbb{R}$ such that

$$F(x, f(x), f'(x), \ldots, f^{(n)}(x)) = 0$$

or

$$\mathbb{L}[f] = q(x)$$

for all x in the domain of definition of f. The solution of a differential equation may not exist if we put too many restrictions on it. For instance, if we demand that $f: \mathbb{R} \to \mathbb{R}$ be differentiable too many times, we may not be able to find a solution, as the following example shows.

Example 9.0.1

The most general solution of

$$\frac{dy}{dx} = |x|$$

with the property $f(0) = 0$, is

$$y = f(x) = \begin{cases} \frac{1}{2}x^2 & \text{if } x \geqslant 0 \\ -\frac{1}{2}x^2 & \text{if } x \leqslant 0 \end{cases}$$

This function is continuous and has the first derivative $f'(x) = |x|$, which is also continuous at $x = 0$. However, if we demand that its second derivative also be continuous at $x = 0$, we cannot find a solution because

$$f''(x) = \begin{cases} +1 & \text{if } x > 0 \\ -1 & \text{if } x < 0 \end{cases}$$

That is, $f''(x)$ is not continuous at $x = 0$. If we want $f'''(x)$ to exist at $x = 0$, then we have to expand the notion of a function to include distributions (discussed in Chapter 5). •

Overrestricting a solution for a differential equation results in its absence, but underrestricting it allows multiple solutions. To strike a balance between these two extremes, we agree to ask a solution to be as many times differentiable as plausible and to satisfy certain *initial conditions*. For an nth-order DE such initial conditions are commonly equivalent (but not restricted) to a specification of the function and of its first $n - 1$ derivatives. This sort of specification is based on the following theorem.

9.0.1 Theorem (The Implicit Function Theorem) Let $F: \mathbb{R}^{n+1} \to \mathbb{R}$, given by $F(x_1, \ldots, x_{n+1}) \in \mathbb{R}$, have continuous partial derivatives up to the kth order at a neighborhood of a point $P_0 \equiv (r_1, r_2, \ldots, r_{n+1})$ in $\mathbb{R}^{n+1}$. Let $(\partial F/\partial x_{n+1})|_{r_{n+1}} \neq 0$. Then there exists a unique function $G: \mathbb{R}^n \to \mathbb{R}$ which is continuously differentiable k times at the neighborhood of P_0 such that $x_{n+1} = G(x_1, x_2, \ldots, x_n)$ for all points $P \equiv (x_1, \ldots, x_{n+1})$ in the neighborhood of P_0 and $F(x_1, x_2, \ldots, x_n, G(x_1, \ldots, x_n)) = 0$. ■

Theorem 9.0.1 simply says that, under certain (mild) conditions, we can "solve" for one of the independent variables in $F(x_1, x_2, \ldots, x_{n+1}) = 0$ in terms of the others. (For a proof of this theorem, see any book on advanced calculus, such as Courant 1936.)

Application of this theorem to Eq. (9.1) leads to

$$\frac{d^n y}{dx^n} = G\left(x, y, \frac{dy}{dx}, \ldots, \frac{d^{n-1}y}{dx^{n-1}}\right)$$

provided that F satisfies the conditions of the theorem. If we know the value of the solution $y = f(x)$, where $f: \mathbb{R} \to \mathbb{R}$, and its derivatives up to order $n-1$, we can evaluate its nth derivative using this equation. In addition, we can calculate the derivatives of all orders (assuming they exist) by differentiating this equation. This allows us to expand the solution into a Taylor series. Thus, at least for solutions that have derivatives of all orders, knowledge of the value of a solution and its first $n-1$ derivatives at a point x_0 determines that solution at a neighboring point x.

We will not study the type of general ODE of Eq. (9.1) or even the simpler linear version of Eq. (9.2). We will only briefly study general ODEs of the first order, and then concentrate on linear ODEs of the second order.

9.1 FIRST-ORDER DIFFERENTIAL EQUATIONS

A general first-order differential equation (FODE) is of the form

$$F(x, y, y') = 0 \tag{9.5a}$$

We can find y' (the derivative of y) in terms of a function of x and y if the function $F(x_1, x_2, x_3)$ is differentiable with respect to its third argument and $\partial F/\partial x_3 \neq 0$. In that case we have

$$\frac{dy}{dx} = y' = G(x, y) \tag{9.5b}$$

which is said to be a *normal FODE*.

Example 9.1.1

There are three special cases of Eq. (9.5b) that lead immediately to a solution.

(a) If $G(x, y)$ is independent of y, then

$$\frac{dy}{dx} = g(x) \quad \Rightarrow \quad dy = g(x)\,dx$$

and the most general solution can be written as

$$y = f(x) = C + \int_a^x g(t)\,dt \qquad \text{where } C = f(a) \tag{1}$$

by the *fundamental theorem of calculus*. This theorem simply states that if $g:[a, b] \to \mathbb{R}$ is continuous on $[a, b]$, then for any $C \in \mathbb{R}$, there exists a unique solution $f:[a, b] \to \mathbb{R}$ of $dy/dx = g(x)$ such that $f(a) = C$. Such a solution is given by (1).

(b) If $G(x, y)$ is independent of x, then

$$\frac{dy}{dx} = h(y) \quad \Rightarrow \quad \frac{dy}{h(y)} = dx \qquad \text{where } h(y) \neq 0$$

and the implicit function

$$F(x, y) \equiv \int_C^y \frac{dt}{h(t)} - x + a = 0$$

embodies a solution. That is, $F(x, y) = 0$ can be solved by the implicit function theorem [assuming its conditions are satisfied, that is, if $\partial F/\partial y = 1/h(y)$ is continuous]. Note that $y|_{x=a} \equiv f(a) = C$.

(c) The third special case is really a generalization of the first two. If $G(x, y) = g(x)h(y)$, then

$$\frac{dy}{dx} = g(x)h(y) \quad \Rightarrow \quad \frac{dy}{h(y)} = g(x)\,dx$$

and

$$\int_C^y \frac{dt}{h(t)} = \int_a^x g(s)\,ds$$

is an implicit solution. •

As Example 9.1.1 shows, the solutions to a FODE are usually obtained in an implicit form, as a function $u:\mathbb{R}^2 \to \mathbb{R}$ such that the solution y can be found by solving $u(x, y) = 0$ for y. Included in $u(x, y)$ is an arbitrary constant related to the initial conditions. The equation $u(x, y) = 0$ defines a curve in the xy-plane, which depends on the (hidden) constant in $u(x, y)$. Since different constants give rise to different curves, it is convenient to separate the constant and write $u(x, y) = C$. This leads to the concept of an integral of a differential equation.

9.1.1 Definition An *integral* of a normal FODE [Eq. (9.5b)] is a function $u:\mathbb{R}^2 \to \mathbb{R}$ such that $u(x, f(x))$ is a constant for all possible values of x whenever $y = f(x)$ is a solution of the DE.

The integrals of differential equations are encountered frequently in physics. If x is replaced by t (time), then the differential equation describes the motion of a physical system, and a solution, $y = f(t)$, can be written implicitly as $u(t, y) = C$, where u is an integral of the differential equation. The equation $u(t, y) = C$ describes a curve in the ty-plane on which the value of the function $u(t, y)$ remains unchanged for all t. Thus, $u(t, y)$ is called a *constant of the motion*.

Example 9.1.2

A situation frequently encountered in mechanics is that where a point particle is in motion under the influence of a force depending on position only. Denoting the position by x and the velocity by v, we have, by Newton's second law,

$$m\frac{dv}{dt} = F(x) \tag{1a}$$

Using the chain rule, $dv/dt = dv/dx\, dx/dt = dv/dx\, v$, we obtain

$$mv\frac{dv}{dx} = F(x) \quad \Rightarrow \quad mv\,dv = F(x)\,dx \tag{1b}$$

which is easily integrated to

$$\tfrac{1}{2}mv^2 = \int F(x)\,dx + C \equiv -V(x) + C \tag{2}$$

where the *potential energy*,

$$V(x) = -\int F(x)\,dx$$

has been introduced as the indefinite integral of the (negative of the) force. We can write (2) as

$$\tfrac{1}{2}mv^2 + V(x) = C \tag{3}$$

Thus, the integral of (1) is

$$u(x, v) = \tfrac{1}{2}mv^2 + V(x)$$

which is the expression for the energy (or the Hamiltonian) of the one-dimensional motion of a particle experiencing the potential $V(x)$. If v is a solution of (1), then $u(x, v) = \text{constant}$. Since a solution of Eq. (1) describes a *possible motion* of the particle, Eq. (3) implies that the energy of a particle does not change in the course of its motion. The latter statement is the conservation of (mechanical) energy.

A special and important case is the motion of a harmonic oscillator, for which $V(x) = \frac{1}{2}kx^2$ and for which Eq. (3) becomes

$$\tfrac{1}{2}mv^2 + \tfrac{1}{2}kx^2 = C$$

or

$$\frac{x^2}{2C/k} + \frac{v^2}{2C/m} = 1$$

The latter, in general, is an ellipse in the xv-plane. It must be emphasized that the shape of the curve in the xv-plane has nothing to do with the shape of the curve describing the path of motion of the particle. ●

9.1.1 Integrating Factors

Let $M, N: \mathbb{R}^2 \to \mathbb{R}$ be continuous functions in a domain D of $\mathbb{R}^2$. This is denoted as $M, N \in \mathrm{C}^{(0)}(\mathrm{D})$. [Recall that $\mathrm{C}^{(n)}(\mathrm{D})$ stands for the collection of functions whose first

n derivatives are continuous on D.] Most of the time the superscript (0) is omitted. The differential $M\,dx + N\,dy$ is *exact* if the line integral

$$\int_{P_1}^{P_2} [M(x, y)\,dx + N(x, y)\,dy]$$

from $P_1 \in \text{D}$ to $P_2 \in \text{D}$, is independent of the path joining the two (arbitrary) points. We have seen examples of exact differentials in Chapters 1 and 4. The above condition is equivalent to saying that the closed line integral of the integrand vanishes. Discussion in Chapters 1 and 4 then implies that if the domain D is contractable to a point (or, in the language of Chapter 6, is simply connected), then a necessary and sufficient condition for exactness is that the curl of the vector $\mathbf{A} = (M, N, 0)$ vanishes. The vector $\mathbf{A}$ is then conservative, and we can define a potential function v such that $\mathbf{A} = \nabla v = (\partial v/\partial x, \partial v/\partial y, 0)$, or

$$dv = (\nabla v)\cdot d\mathbf{l} = \frac{\partial v}{\partial x}dx + \frac{\partial v}{\partial y}dy = M\,dx + N\,dy$$

This shows that $M = \partial v/\partial x$ and $N = \partial v/\partial y$. It is now possible to state the following proposition.

9.1.2 Proposition If $M(x, y)\,dx + N(x, y)\,dy$ is an exact differential dv in a domain $\text{D} \subset \mathbb{R}^2$, then $v(x, y)$ is an integral of the DE

$$M(x, y) + N(x, y)\frac{dy}{dx} = 0 \tag{9.6}$$

whose solutions are of the form $v(x, y) = C$.

Proof. If $y = f(x)$ is a solution of $v(x, y) = C$, then

$$0 = \frac{dC}{dx} = \frac{dv}{dx} \equiv \frac{\partial v}{\partial x} + \frac{\partial v}{\partial y}\frac{dy}{dx} = M(x, y) + N(x, y)\frac{dy}{dx}$$

Thus, $y = f(x)$, obtained by solving $v(x, y) = C$, is a solution of (9.6), and $v(x, y) = C$ is a (implicit) solution of (9.6). ■

Not very many FODEs are exact. However, there are many that can be turned into exact FODEs by multiplication by a suitable function. Such a function, if it exists, is called an *integrating factor*. Thus, if the differential $M(x, y)\,dx + N(x, y)\,dy$ is not exact, but

$$q(x, y)M(x, y)\,dx + q(x, y)N(x, y)\,dy = dv$$

then $q(x, y)$ is an integrating factor for the differential equation

$$M(x, y) + N(x, y)\frac{dy}{dx} = 0$$

whose solution is then $v(x, y) = C$. This is clear because if $y = f(x)$ solves $v(x, y) = C$, then

$$0 = \frac{dv}{dx} = q\left(M + N\frac{dy}{dx}\right) \quad \Leftrightarrow \quad M + N\frac{dy}{dx} = 0$$

assuming that $q \neq 0$.

Integrating factors are not unique, as the following example illustrates and Exercise 9.1.1 shows in general.

Example 9.1.3

The differential $x\,dy - y\,dx$ is not exact. Let us see if we can find a function $q(x, y)$ such that

$$dv = qx\,dy - qy\,dx \tag{1}$$

for some $v: \mathbb{R}^2 \to \mathbb{R}$. We assume that the domain D is contractable to a point (or, equivalently, simply connected). Then a necessary and sufficient condition for (1) to hold is

$$\frac{\partial}{\partial x}(qx) = \frac{\partial}{\partial y}(-qy)$$

or

$$x\frac{\partial q}{\partial x} + q = -y\frac{\partial q}{\partial y} - q \tag{2}$$

(a) We assume that q is a function of x only and try to find a solution to Eq. (2). With this assumption (2) reduces to

$$x\frac{dq}{dx} + q = -q \quad \Rightarrow \quad x\frac{dq}{dx} = 2q$$

which has a solution

$$q = C/x^2 \qquad \text{where } x \neq 0$$

In this case we get

$$dv = C\left(\frac{1}{x}dy - \frac{y}{x^2}dx\right) = C\,d\left(\frac{y}{x}\right) = d\left(\frac{Cy}{x}\right) \qquad \text{where } x \neq 0$$

Thus, as long as $x \neq 0$, any function C/x^2, with arbitrary C, is an integrating factor for

$$x\,dy - y\,dx = 0$$

This integrating factor leads to the solution

$$v = \frac{Cy}{x} = \text{constant} \quad \Rightarrow \quad y = mx \qquad \text{where } x \neq 0 \tag{3}$$

The condition $x \neq 0$ removes the entire y-axis from the domain of definition of (3). If the domain is to be simply connected, it has to be completely contained in either the right half-plane (RHP) or the left half-plane (LHP). Thus, the largest such domain is either the RHP or the LHP, but *not* both.

(b) Now let us assume that q is a function of y only. This leads to the integrating factor

$$q = \frac{C}{y^2} \qquad \text{where } y \neq 0$$

for $x\,dy - y\,dx = 0$. In this case $v = Cx/y$ is the integral of the DE, again leading to the solution given in Eq. (3). Here the largest domain is either the upper half-plane (UHP) or the lower half-plane (LHP), but not both.

It is important to realize that, although there are many integrating factors, the solution to the DE $x\,dy - y\,dx = 0$ is *unique* (subject to certain initial conditions). This is clear from the solutions derived here and in part (a). The uniqueness of the solution, subject to initial conditions, is a general property of all differential equations and will be discussed later.

(c) It can easily be verified that

$$q = \frac{C}{x^2 + y^2} \qquad \text{where } (x, y) \neq (0, 0)$$

is also an integrating factor leading to the integral

$$v = \arctan\left(\frac{y}{x}\right)$$

which gives rise to the solution

$$\arctan\left(\frac{y}{x}\right) = \text{constant} \quad \Rightarrow \quad \frac{y}{x} = m$$

as before. ●

In fact, the following theorem can be proved (see Rubenstein 1969, pp. 11 and 12).

9.1.3 Theorem If the DE

$$M(x, y)dx + N(x, y)\,dy = 0$$

has a unique solution in the interval $[a, b]$, then it has infinitely many integrating factors. The general integrating factor is given by

$$v(x, y) = q(x, y)\,F(u)$$

where $u(x, y) = C$ is the general solution of the DE, $F(u)$ is an arbitrary differentiable function, and

$$q(x, y) \equiv \frac{\partial u/\partial x}{M} = \frac{\partial u/\partial y}{N}$$

is a particular integrating factor. ■

9.1.2 First-Order Linear Differential Equations

The most general first-order linear differential equation (FOLDE) is obtained from Eq. (9.2) by setting $n = 1$. This gives

$$p_1(x)\,y' + p_0(x)y = q(x) \tag{9.7a}$$

or, equivalently,

$$(p_0 y - q)\,dx + p_1\,dy = 0 \tag{9.7b}$$

By Theorem 9.1.3, if Eq. (9.7b) has a (unique) solution, then it must have at least one integrating factor. Let $\mu(x, y)$ be the integrating factor. Then there exists $v(x, y)$ such that

$$dv = \mu(p_0 y - q)dx + \mu p_1\,dy = 0$$

A necessary and sufficient condition for this to hold is

$$\frac{\partial}{\partial y}[\mu(p_0 y - q)] = \frac{\partial}{\partial x}(\mu p_1)$$

To simplify the problem, let us assume that μ is a function of x only (we are looking for any integrating factor, not the most general one). Then the above condition leads to

$$\mu p_0 = \frac{d}{dx}(\mu p_1)$$

which has a solution

$$\mu(x) = \mu(x_0)\frac{p_1(x_0)}{p_1(x)}\exp\left[\int_{x_0}^{x}\frac{p_0(t)}{p_1(t)}\,dt\right]$$

where x_0 is any arbitrary point at which μ and p_1 are defined. We now rewrite dv as

$$\begin{aligned} dv &= \left[\frac{d}{dx}(\mu p_1)y\right]dx + \mu p_1\,dy - \mu q\,dx \\ &= d(\mu p_1 y) - \mu q\,dx = d\left[\mu p_1 y - \int_{x_1}^{x}\mu(t)q(t)\,dt\right] = 0 \end{aligned}$$

which yields

$$\mu p_1 y - \int_{x_1}^{x}\mu(t)q(t)\,dt = C$$

where x_1 is an arbitrary point at which $\mu(x_1)\,p_1(x_1)\,y(x_1) = C$. We thus obtain the general solution

$$y = f(x) = \frac{1}{\mu(x)p_1(x)}\left[C + \int_{x_1}^{x}\mu(t)q(t)\,dt\right] \tag{9.8}$$

and along with it the following theorem.

9.1.4 Theorem Any FOLDE of the form

$$p_1(x)y' + p_0(x)y = q(x)$$

in which p_0, p_1, and q are continuous functions in the interval $[a, b]$, has a general solution

$$y = f(x) = \frac{1}{\mu(x)\,p_1(x)}\left[C + \int_{x_1}^{x} \mu(t)q(t)\,dt\right]$$

where C is an arbitrary constant [determined by $y(x_1)$],

$$\mu(x) = \mu(x_0)\frac{p_1(x_0)}{p_1(x)}\exp\left[\int_{x_0}^{x} \frac{p_0(t)}{p_1(t)}\,dt\right]$$

and x_0 and x_1 are arbitrary points in the interval $[a, b]$. ■

Example 9.1.4

In an electric circuit with a resistance R and a capacitance C, Kirchhoff's law gives rise to the equation

$$R\frac{dQ}{dt} + \frac{Q}{C} = V(t)$$

where $V(t)$ is the time-dependent voltage and Q is the charge on the capacitor. This is a simple FOLDE with $p_1(t) = R$, $p_0(t) = 1/C$, and $q(t) \equiv V(t)$. The integrating factor is

$$\mu(t) = \mu(t_0)\exp\left[\int_{t_0}^{t} \frac{p_0(s)}{p_1(s)}\,ds\right] = \mu(t_0)\exp\left[\frac{1}{RC}(t - t_0)\right] \equiv Ae^{t/RC}$$

which yields

$$Q(t) = \frac{1}{\mu(t)p_1(t)}\left[B + \int_{t_1}^{t} Ae^{s/RC}\,V(s)\,ds\right]$$

$$= \frac{B}{AR}e^{-t/RC} + \frac{e^{-t/RC}}{R}\int_{t_1}^{t} e^{s/RC}\,V(s)\,ds$$

Usually t_1 is taken to be zero. Then $Q(0) \equiv Q_0$ is the initial charge, given by (for $t_1 = 0$; the integral vanishes at $t = 0$)

$$Q_0 = \frac{B}{AR}$$

Thus, the charge at time t is given by

$$Q(t) = Q_0 e^{-t/RC} + \frac{1}{R}e^{-t/RC}\int_{0}^{t} e^{s/RC}\,V(s)\,ds$$ •

9.1.3 Some Nonlinear First-Order Differential Equations and Their Solutions

Although FOLDEs can always be solved, yielding solutions as given in Theorem 9.1.4, no general rule can be applied to solve a general FODE. There are various techniques that can be applied to special FODEs, however. This subsection enumerates some of these methods for some well-known FODEs.

Separable first-order differential equations. A separable FODE is of the form

$$\frac{dy}{dx} = \frac{g(x)}{h(y)}$$

With minor differences, this was discussed in Example 9.1.1. The (implicit) solution is

$$\int^{y} h(t)\,dt = \int^{x} g(t)\,dt + C$$

where the lower limit of integration and C are arbitrary.

Homogeneous first-order differential equations. The general form of a homogeneous FODE is

$$\frac{dy}{dx} = w\left(\frac{y}{x}\right)$$

where $w: \mathbb{R} \to \mathbb{R}$ is a continuous function on some interval of $\mathbb{R}$. A solution can be obtained by substituting

$$u = \frac{y}{x} \quad \Rightarrow \quad y = xu \quad \Rightarrow \quad y' = u + xu'$$

yielding

$$u + xu' = w(u) \quad \Rightarrow \quad u' = \frac{w(u) - u}{x}$$

which is separable with $g(t) = 1/t$ and $h(t) = 1/[w(t) - t]$. Thus, a general solution is given by

$$\int^{u} \frac{dt}{w(t) - t} = \int^{x} \frac{dt}{t} + C$$

or

$$x = A \exp\left[\int^{y/x} \frac{dt}{w(t) - t}\right] \qquad \text{where } A = e^{-C}$$

Exact first-order differential equations. An exact FODE, discussed earlier, is of the form

$$M(x, y)\,dx + N(x, y)\,dy = 0$$

where there exists a function v such that $dv = M\,dx + N\,dy = 0$. The solution is given simply by $v(x, y) = C$.

Bernoulli's first-order differential equation. Bernoulli's FODE is of the form

$$y' + p(x)y + q(x)y^n = 0 \qquad \text{where } n \neq 1$$

Since the equation contains a power of the independent variable, we make the substitution

$$y = u^r \qquad \text{where } r \neq 0$$

and choose r, if possible, to simplify it. Substituting this in the original form yields

$$u' + \frac{p(x)}{r}u + \frac{q(x)}{r}u^{nr-r+1} = 0$$

If we could make the exponent of u in the last term equal to one, we would have a very simple FOLDE to solve. But this would require that $nr - r = 0$, which has no solution for r. So we do the next best thing and demand

$$nr - r + 1 = 0 \quad \Rightarrow \quad r = \frac{1}{1-n}$$

This reduces the equation in u to

$$u' + (1-n)\,p(x)u + (1-n)q(x) = 0$$

which is a FOLDE.

Lagrange's first-order differential equation. Lagrange's FODE can be expressed as

$$y - xp(y') - q(y') = 0$$

To solve this equation we let $y' = t$ and consider x as a function of t. Then

$$\frac{dx}{dt} = \frac{dx}{dy}\frac{dy}{dt} = \frac{1}{y'}\frac{dy}{dt} = \frac{1}{t}\frac{dy}{dt}$$

On the other hand, the original FODE yields

$$y = xp(t) + q(t)$$

Thus,

$$\frac{dx}{dt} = \frac{1}{t}\left[\frac{dx}{dt}p(t) + x\frac{dp}{dt} + \frac{dq}{dt}\right]$$

or

$$[t - p(t)]\frac{dx}{dt} - p'(t)x = q'(t)$$

We must consider two cases:

(1) When $p(t) = t$, the resulting DE is called Clairot's equation. Then $p'(t) = 1$, and

$$x = -q'(t)$$
$$y = xp(t) + q(t) = -tq'(t) + q(t)$$

This is a parametric solution.

(2) When $t - p(t) \neq 0$, the resulting equation in $x(t)$ is a FOLDE that can be solved to obtain x as a function of t. Again we obtain a parametric solution:

$$x = x(t)$$
$$y = xp(t) + q(t) = x(t)p(t) + q(t)$$

9.1.4 Existence and Uniqueness of Solutions of First-Order Differential Equations

For completeness this subsection summarizes some of the ideas involved in the proof of the existence and uniqueness of the solutions to FODEs. (For proofs, see the excellent book by Birkhoff and Rota 1978.)

9.1.5 Definition A function $F(x, y)$ satisfies a *Lipschitz condition* in a domain $D \subset \mathbb{R}^2$ when, for some finite constant L (Lipschitz constant), it satisfies the inequality

$$|F(x, y_1) - F(x, y_2)| \leqslant L|y_1 - y_2|$$

for all points (x, y_1) and (x, y_2) in D having the same x-coordinate.

9.1.6 Proposition Let $F \in C^{(1)}(D)$; that is, let F be continuously differentiable in any bounded closed convex domain D. Then it satisfies a Lipschitz condition there, with $L = \sup_D |\partial F/\partial y|$. ■

The language used in this proposition can be translated as follows: First, continuous differentiability means simply that $\partial F/\partial x$ and $\partial F/\partial y$ are continuous functions of x and y. Next, a bounded domain is a region in the xy-plane that can be enclosed in a circle of finite radius. Third, for these purposes, a closed domain is a region in the xy-plane that includes the points on its boundary. Finally, $\sup_D |\partial F/\partial y|$ means the maximum possible value that $|\partial F/\partial y|$ attains as x and y vary over the domain D. Here sup stands for *supremum*, which, disregarding certain mathematical subtleties, is the same as "maximum."

Let us apply these results to the solutions of the general FODE $y' = F(x, y)$.

9.1.7 Theorem (Uniqueness Theorem) Let $f(x)$ and $g(x)$, where $x \in [a, b]$, be any two solutions of the FODE $y' = F(x, y)$ in a domain D, where F satisfies a Lipschitz condition. Then

$$|f(x) - g(x)| \leqslant e^{L|x-a|}|f(a) - g(a)|$$

In particular, the FODE has at most one solution curve passing through the point $(a, c) \in \mathrm{D}$. ■

The final conclusion of Theorem 9.1.7 is an easy consequence of the requirement $f(a) = g(a) = c$. The theorem says that if there is *a* solution, $y = f(x)$, to $y' = F(x, y)$, satisfying $f(a) = c$, then it is *the* solution. However, the theorem says nothing about the existence of such a solution. The following theorem bridges this gap.

9.1.8 Theorem (Local Existence Theorem) Suppose that the function $F(x, y)$ is defined and continuous in the closed domain $|y - c| \leqslant K$, $|x - a| \leqslant N$ and satisfies a Lipschitz condition there. Let $M = \sup|F(x, y)|$ in this domain. Then the differential equation $y' = F(x, y)$ has a unique solution, $y = f(x)$, satisfying $f(a) = c$ and defined on the interval $|x - a| \leqslant \min(N, K/M)$. ■

The uniqueness requirement in Theorem 9.1.8 restricts the existence to a local region. If we lift this uniqueness requirement, we obtain another theorem.

9.1.9 Theorem (Peano's Existence Theorem) If the function $F(x, y)$ is continuous for $|y - c| \leqslant K$ and $|x - a| \leqslant N$, and if $|F(x, y)| \leqslant M$ there, then the differential equation $y' = F(x, y)$ has *at least* one solution, $y = f(x)$, defined for $|x - a| \leqslant \min(N, K/\mathrm{M})$ and satisfying the initial condition $f(a) = c$. ■

Exercises

9.1.1 Show that if the DE $M(x, y)\,dx + N(x, y)\,dy = 0$ has one integrating factor q, then it has infinitely many integrating factors.

9.1.2 Find the charge on a capacitor in an R-C circuit when a voltage, $V(t) = V_0 \cos \omega t$, is applied to it for a period $T > 0$ and then removed.

9.1.3 Find a solution to the *linear fractional differential equation*

$$\frac{dy}{dx} = \frac{a_1 x + a_2 y}{b_1 x + b_2 y} \qquad \text{where } a_1 b_2 \neq a_2 b_1$$

9.2 GENERAL PROPERTIES OF SECOND-ORDER LINEAR DIFFERENTIAL EQUATIONS

The majority of differential equations encountered in physics, both partial and ordinary, lead to a SOLDE when the so-called nonlinear terms are approximated out. (This was shown in Chapter 8 in great detail). Thus, the remainder of this chapter is devoted to a discussion of SOLDEs.

The most general SOLDE is

$$p_2(x)\frac{d^2y}{dx^2} + p_1(x)\frac{dy}{dx} + p_0(x)y = p_3(x) \tag{9.9a}$$

Dividing by $p_2(x)$ reduces this to the *normal form,*

$$y'' + p(x)y' + q(x)y = r(x) \qquad \text{where } p = \frac{p_1}{p_2},\ q = \frac{p_0}{p_2},\ r = \frac{p_3}{p_2} \tag{9.9b}$$

Equation (9.9b) is equivalent to (9.9a) if $p_2(x) \neq 0$. The points at which $p_2(x)$ vanishes are called the *singular points* of the differential equation.

There is a crucial difference between the singular points of linear differential equations and those of nonlinear differential equations. For a nonlinear differential equation such as $(x^2 - y)y' = x^2 + y^2$, the curve $y = x^2$ is the collection of singular points. This makes it impossible to construct solutions $y = f(x)$ that are defined on an interval $\mathrm{I} = [a, b]$ of the x-axis because for any $x \in \mathrm{I}$, there is a $y = x^2$ for which the differential equation is undefined. On the other hand, linear differential equations do not have this problem because the coefficients of the derivatives are functions of x only. Therefore, all the singular curves are *vertical*. Thus, we have the following definition.

9.2.1 Definition The normal form of a SOLDE [Eq. (9.9b)] is *regular* on an interval $[a, b]$ of the x-axis if $p(x)$, $q(x)$, and $r(x)$ are continuous on $[a, b]$. A solution of a normal SOLDE is a *twice-differentiable* function $y = f(x)$, which satisfies the SOLDE at every point of $[a, b]$.

It is clear that any function that satisfies Eqs. (9.9) must necessarily be twice differentiable, and that is all that is demanded of the solutions. Any higher-order differentiability requirement may be too restrictive, as was pointed out in the preceding section. Most solutions to a normal SOLDE, however, are automatically differentiable many more than two times.

9.2.1 Linearity, Superposition, and Uniqueness

If we introduce the differential operator,

$$\mathbb{L} = p_2\frac{d^2}{dx^2} + p_1\frac{d}{dx} + p_0 \tag{9.10}$$

then Eq. (9.9a) can be written as

$$\mathbb{L}[y] = p_3 \tag{9.11}$$

It is clear that $\mathbb{L}$ is a *linear* operator because d/dx is linear, as are all powers of it. Thus, for constants α and β,

$$\mathbb{L}[\alpha y_1 + \beta y_2] = \alpha\mathbb{L}[y_1] + \beta\mathbb{L}[y_2]$$

In particular, if y_1 and y_2 are two solutions of (9.11), then

$$\mathbb{L}[y_1 - y_2] = 0$$

That is, the difference between any two solutions of a SOLDE is a solution of the *homogeneous* equation obtained by setting $p_3 = 0$.

An immediate consequence of the linearity of $\mathbb{L}$ is the following lemma.

9.2.2 Lemma If $\mathbb{L}[u] = r(x)$, $\mathbb{L}[v] = s(x)$, α and β are constants, and $w = \alpha u + \beta v$, then $\mathbb{L}[w] = \alpha r(x) + \beta s(x)$. ■

The proof of this lemma is trivial, but the result describes the fundamental property of linear operators. When $r = s = 0$, that is, when we are dealing with homogeneous equations, the lemma says that if u and v are two arbitrary solutions of the homogeneous SOLDE (HSOLDE), then any *linear combination* of u and v is also a solution. This is called the *superposition principle.*

We saw in Chapter 8 that based on physical intuition, we expect to be able to predict the behavior of a physical system if we know the differential equation obeyed by that system and equally importantly, the initial data. A prediction is not a prediction unless it is *unique.* A situation in which given initial data, called *initial conditions,* lead to multiple predictions (multiple solutions to differential equations) is unacceptable physically. Thus, on the basis of physical intuition alone, we expect a unique solution to any differential equation subject to given initial conditions.[1]

This expectation for linear equations is borne out in the language of mathematics in the form of an existence theorem and a uniqueness theorem. We consider the latter next, leaving the former until Section 9.3.1.

9.2.3 Theorem (Uniqueness Theorem) If p and q are continuous on $[a, b]$, then at most one solution $y = f(x)$ of Eq. (9.9b) can satisfy the initial conditions $f(a) = c_1$ and $f'(a) = c_2$, where c_1 and c_2 are arbitrary constants.

Proof. Let f_1 and f_2 be two solutions satisfying the given initial conditions. Then their difference, $f \equiv f_1 - f_2$, satisfies the homogeneous equation with $r(x) = 0$. The initial condition that $f(x)$ satisfies is clearly $f(a) = 0 = f'(a)$. Now consider the nonnegative function $u(x) \equiv [f(x)]^2 + [f'(x)]^2$, which, by its definition, satisfies the initial condition $u(a) = 0$. Differentiating $u(x)$, we get

$$\begin{aligned} u'(x) &= 2f'f + 2f'f'' = 2f'(f + f'') = 2f'(f - pf' - qf) \\ &= -2p(f')^2 + 2(1 - q)ff' \end{aligned}$$

Since $(f \pm f')^2 \geqslant 0$, it follows that $2|ff'| \leqslant f^2 + f'^2$. Thus,

$$\begin{aligned} 2(1 - q)ff' &\leqslant 2|(1 - q)ff'| = 2|(1 - q)||ff'| \\ &\leqslant |(1 - q)|(f^2 + f'^2) \leqslant (1 + |q|)(f^2 + f'^2) \end{aligned}$$

[1] Physical intuition also tells us that if the initial conditions are changed by an infinitesimal amount, then the solutions will be changed infinitesimally. Thus, the solutions of linear differential equations are said to be continuous functions of the initial conditions. Nonlinear differential equations can have completely different solutions for two initial conditions that are infinitesimally close. Since initial conditions cannot be specified with mathematical precision in practice, nonlinear differential equations lead to unpredictable solutions, or chaos. This subject has received much attention in recent years but cannot be covered here.

and, therefore,

$$u'(x) \leqslant |u'(x)| = |-2pf'^2 + 2(1-q)ff'| \leqslant 2|p|f'^2 + (1+|q|)(f^2+f'^2)$$
$$= [1 + |q(x)|]f^2 + [1 + |q(x)| + 2|p(x)|]f'^2$$

Now let $K \equiv 1 + \max[|q(x)| + 2|p(x)|]$, where the maximum is taken over $[a, b]$. Then we obtain

$$u'(x) \leqslant K(f^2 + f'^2) = Ku(x) \quad \forall\, x \in [a, b]$$

Using the result of Exercise 9.2.1 yields

$$u(x) \leqslant u(a)e^{K(x-a)} \quad \forall\, x \in [a, b]$$

The vanishing of $u(a)$ implies the vanishing of $u(x)$ because u cannot be negative. Thus, $u(x) = f^2(x) + f'^2(x) = 0$ implies

$$f(x) = 0 = f'(x) \quad \Rightarrow \quad f_1(x) - f_2(x) = 0 \qquad \forall\, x \in [a, b]. \qquad \blacksquare$$

Theorem 9.2.3 can be applied to any HSOLDE to find its most general solution. In particular, lef $f_1(x)$ and $f_2(x)$ be any two solutions of

$$y'' + p(x)y' + q(x) = 0 \tag{9.12}$$

Assume that the two vectors $\mathbf{v}_1 = (f_1(x_0), f_1'(x_0))$ and $\mathbf{v}_2 = (f_2(x_0), f_2'(x_0))$ in $\mathbb{R}^2$ are linearly independent for some $x_0 \in [a, b]$. Let $g(x)$ be another solution to Eq. (9.12). The vector $(g(x_0), g'(x_0))$ can be written as a linear combination of $\mathbf{v}_1$ and $\mathbf{v}_2$, giving the two equations

$$g(x_0) = c_1 f_1(x_0) + c_2 f_2(x_0)$$
$$g'(x_0) = c_1 f_1'(x_0) + c_2 f_2'(x_0)$$

Consider the function

$$u(x) \equiv g(x) - c_1 f_1(x) - c_2 f_2(x)$$

which satisfies Eq. (9.12) and the initial conditions $u(x_0) = u'(x_0) = 0$. Theorem 9.2.3 says that $u(x) \equiv 0$ or $g(x) = c_1 f_1(x) + c_2 f_2(x)$, and we have proved the following theorem.

9.2.4 Theorem Let f_1 and f_2 be two solutions of the HSOLDE $y'' + py' + qy = 0$, where $p, q \in C^{(0)}[a, b]$. If $(f_1(x_0), f_1'(x_0))$ and $(f_2(x_0), f_2'(x_0))$ are linearly independent vectors of $\mathbb{R}^2$ for some $x_0 \in [a, b]$, then every solution $g(x)$ of this HSOLDE is equal to some linear combination $g(x) = c_1 f_1(x) + c_2 f_2(x)$ of f_1 and f_2 with constant coefficients c_1 and c_2. $\blacksquare$

9.2.2 The Wronskian

The two solutions $f_1(x)$ and $f_2(x)$ in Theorem 9.2.4 have the property that any other solution $g(x)$ can be expressed as a linear combination of them. Borrowing the terminology of vector spaces, we call f_1 and f_2 a *basis of solutions* of the HSOLDE. To form a basis of solutions, f_1 and f_2 must be linearly independent. The linear independence of functions was discussed in Chapter 5, but it is important to note that the linear dependence or independence of a number of functions, $f_1, f_2, \ldots, f_n\colon [a, b] \to \mathbb{R}$, is a concept that must hold for all $x \in [a, b]$. Thus, if $\alpha_1, \alpha_2, \ldots, \alpha_n \in \mathbb{R}$ can be found such that

$$\alpha_1 f_1(x_0) + \alpha_2 f_2(x_0) + \cdots + \alpha_n f_n(x_0) = 0$$

for some $x_0 \in [a, b]$, it does not mean that $f_1, \ldots, f_n$ are linearly dependent. Linear dependence requires that the equality hold for all $x \in [a, b]$. In fact, we must write

$$\alpha_1 f_1 + \alpha_2 f_2 + \cdots + \alpha_n f_n = \mathbb{0}$$

where $\mathbb{0}$ is the *zero function* whose value is zero for *all* $x \in [a, b]$.

The nature of the linear relation between f_1 and f_2 can be determined by their Wronskian.

9.2.5 Definition The *Wronskian* of any two differentiable functions $f_1(x)$ and $f_2(x)$ is

$$W(f_1, f_2; x) \equiv f_1(x) f_2'(x) - f_2(x) f_1'(x) = \det\begin{pmatrix} f_1(x) & f_1'(x) \\ f_2(x) & f_2'(x) \end{pmatrix}$$

Applying Definition 9.2.5 to the solutions of Eq. (9.12) gives a proposition.

9.2.6 Proposition The Wronskian of any two solutions of Eq. (9.12) satisfies

$$W(f_1, f_2; x) = W(f_1, f_2; a) e^{-\int_a^x p(t)\,dt} \tag{9.13}$$

Proof. Differentiating Eq. (9.13) and substituting from (9.12) yields a FOLDE for $W(f_1, f_2; x)$, which can be easily solved. The details are left as a problem. ■

An important consequence of Proposition 9.2.6 is that the Wronskian of any two solutions of Eq. (9.12) is always positive *or* always negative *or* always zero. Thus, if the Wronskian vanishes at one point in $[a, b]$, it vanishes at all points in $[a, b]$.

The real importance of the Wronskian is revealed by the following theorem.

9.2.7 Theorem Two solutions f_1 and f_2 of a HSOLDE [Eq. (9.12)] are linearly dependent if and only if their Wronskian vanishes.

Proof. We can show trivially that if f_1 and f_2 are linearly dependent, then their Wronskian vanishes. Conversely, let us assume that $W(f_1, f_2; x)$ vanishes for $x = x_1$. Then the two vectors $(f_1(x_1), f_1'(x_1))$ and $(f_2(x_1), f_2'(x_1))$ are linearly dependent in $\mathbb{R}^2$. We can thus write $f_1(x_1) = k f_2(x_1)$ and $f_1'(x_1) = k f_2'(x_1)$. The function $f(x) \equiv f_1(x) - k f_2(x)$ is a solution of (9.12) and satisfies the initial conditions $f(x_1) = f'(x_1) = 0$. By the proof of Theorem 9.2.3, f must be zero. Thus $f_1(x) = k f_2(x)$ for all $x \in [a, b]$, and f_1 and f_2 are linearly dependent. ■

As the proof of this theorem shows, linear dependence always leads to the vanishing of the Wronskian whether or not the two functions are solutions of the HSOLDE. However, the vanishing of the Wronskian does not lead to the establishment of linear dependence in general. Only if the two functions are *also* solutions of Eq. (9.12) does the vanishing of the Wronskian imply linear dependence.

Example 9.2.1

Let $f_1(x) = x$ and $f_2(x) = |x|$ for $x \in [-1,1]$. These two functions are linearly independent in the given interval, because $\alpha_1 x + \alpha_2|x| = 0$ *for all* x if and only if $\alpha_1 = \alpha_2 = 0$. The Wronskian, on the other hand, vanishes for all $x \in [-1, +1]$:

$$W(f_1, f_2; x) = x\frac{d}{dx}(|x|) - |x|\frac{d}{dx}(x)$$

$$= x\frac{d}{dx}(|x|) - |x| = \begin{cases} x - x = 0 & \text{if } x > 0 \\ -x - (-x) = 0 & \text{if } x < 0 \end{cases}$$

Thus, it is possible for two general functions to have a vanishing Wronskian without being linearly dependent. ●

Example 9.2.2

The Wronskian can be generalized to n functions. The Wronskian of the functions $f_1, f_2, \ldots, f_n$ is

$$W(f_1,f_2,\ldots,f_n;x) = \det\begin{pmatrix} f_1(x) & f_1'(x) & \cdots & f_1^{(n-1)}(x) \\ f_2(x) & f_2'(x) & \cdots & f_2^{(n-1)}(x) \\ \vdots & \vdots & & \vdots \\ f_n(x) & f_n'(x) & \cdots & f_n^{(n-1)}(x) \end{pmatrix}$$

If the functions are linearly dependent, then

$$W(f_1, f_2, \ldots, f_n; x) \equiv 0$$

For instance, it is clear that e^x, e^{-x}, and $\sinh x$ are linearly dependent. Thus, we expect

$$W(e^x,e^{-x},\sinh x; x) = \det\begin{pmatrix} e^x & e^x & e^x \\ e^{-x} & -e^{-x} & e^{-x} \\ \sinh x & \cosh x & \sinh x \end{pmatrix}$$

to vanish, as it is easily seen (the first and last column are the same). ●

A second solution to the HSOLDE. If we know one solution to Eq. (9.12), we can use the Wronskian to obtain a second linearly independent solution. Let $W(x) \equiv W(f_1,f_2;x)$ be the Wronskian of the two solutions f_1 and f_2. Then, by definition and Proposition 9.2.6, we have

$$f_1(x)f_2'(x) - f_2(x)f_1'(x) = W(x) = W(a)e^{-\int_a^x p(t)dt}$$

Given $f_1(x)$, this is a FOLDE in $f_2(x)$, which can be solved by the method of Section 9.1. In fact, $1/f_1^2(x)$ is an integrating factor, so dividing both sides by $f_1^2(x)$ gives

$$\frac{d}{dx}\left[\frac{f_2(x)}{f_1(x)}\right] = \frac{W(x)}{f_1^2(x)}$$

or

$$\frac{f_2(x)}{f_1(x)} = c + \int_a^x \frac{W(s)}{f_1^2(s)}ds = c + \int_a^x \frac{1}{f_1^2(s)}[W(a)e^{-\int_a^s p(t)dt}]ds$$

where c is an arbitrary constant of integration. Thus,

$$f_2(x) = f_1(x)\left\{c + W(a)\int_a^x \frac{1}{f_1^2(s)}[e^{-\int_a^s p(t)dt}]ds\right\} \tag{9.14}$$

Note that $W(a)$ is another arbitrary (nonzero) constant; we do not have to know $W(x)$ (this would require knowledge of f_2, which we are trying to calculate!) to obtain $W(a)$. It is customary to set $c = 0$ because that gives a term that is proportional to $f_1(x)$.

Example 9.2.3

(a) A solution to the SOLDE $y'' - k^2y = 0$ is e^{kx}. To find a second solution, we let $c = 0$ and $W(a) = 1$ in Eq. (9.14). Since $p(x) = 0$, we have

$$f_2(x) = e^{kx}\left(0 + \int_a^x \frac{ds}{e^{2ks}}\right) = e^{kx}\left(-\frac{1}{2k}e^{-2ks}\Big|_a^x\right)$$

$$= -\frac{1}{2k}e^{-kx} + \frac{e^{-2ka}}{2k}e^{kx}$$

which leads directly to the choice of e^{-kx} as a second solution.

(b) When $k^2 \to -k^2$, the differential equation $y'' + k^2y = 0$ has $\sin kx$ as a solution. With $c = 0$, $a = \pi/2k$, and $W(\pi/2k) = 1$, we get

$$f_2(x) = \sin kx\left(0 + \int_{\pi/2k}^x \frac{ds}{\sin^2 ks}\right) = -\sin kx \cot ks\,\Big|_{\pi/2k}^x$$

$$= -\cos kx$$

(c) For the solutions in part (a),

$$W(x) = \det\begin{pmatrix} e^{kx} & ke^{kx} \\ e^{-kx} & -ke^{-kx}\end{pmatrix} = -2k$$

and for those in part (b),

$$W(x) = \det\begin{pmatrix} \sin kx & k\cos kx \\ \cos kx & -k\sin kx\end{pmatrix} = -k$$

Both Wronskians are constant. This is a special case of the general result that the Wronskian of any two linearly independent solutions of $y'' + q(x)y = 0$ is constant.

●

Most special functions used in mathematical physics are solutions of SOLDEs. The behavior of these functions at certain points is determined by the physics of the particular problem. In most situations physical expectation leads to a preference for one particular solution over the other. For example, although there are two linearly independent solutions to the Legendre equation,

$$\frac{d}{dx}\left[(1-x^2)\frac{dy}{dx}\right] + n(n+1)y = 0$$

the solution that is most frequently encountered is a Legendre polynomial, $P_n(x)$, which was discussed in Chapter 5. The other solution can be obtained by solving the Legendre equation or by using Eq. (9.14). The following example discusses this point.

Example 9.2.4

The Legendre equation,

$$\frac{d}{dx}\left[(1-x^2)\frac{dy}{dx}\right] + n(n+1)y = 0$$

can be reexpressed as

$$\frac{d^2y}{dx^2} - \frac{2x}{1-x^2}\frac{dy}{dx} + \frac{n(n+1)}{1-x^2}y = 0$$

This is an HSOLDE with

$$p(x) = -\frac{2x}{1-x^2} \qquad \text{and} \qquad q(x) = \frac{n(n+1)}{1-x^2}$$

One solution of this HSOLDE is the well-known Legendre polynomial, $P_n(x)$. Using this as our input and employing Eq. (9.14), we can generate another set of solutions.

Let $Q_n(x)$ stand for the linearly independent "partner" of $P_n(x)$. Then, setting $c = 0$ in Eq. (9.14) yields

$$\begin{aligned} Q_n(x) &= W(a)P_n(x)\int_a^x \frac{1}{P_n^2(s)}\left(e^{-\int_a^s[-2t/(1-t^2)]dt}\right)ds \\ &= W(a)P_n(x)\int_a^x \frac{1}{P_n^2(s)}\left(\frac{1-a^2}{1-s^2}\right)ds \\ &= A_nP_n(x)\int_a^x \frac{ds}{(1-s^2)P_n^2(s)} \end{aligned}$$

where A_n is an arbitrary constant determined by standardization and a is an arbitrary point in the interval $[-1, +1]$. For instance, for $n = 0$, $P_0 = 1$, and we obtain

$$Q_0(x) = A_0 \int_a^x \frac{ds}{1-s^2} = A_0\left[\frac{1}{2}\ln\left(\frac{1+x}{1-x}\right) - \frac{1}{2}\ln\left(\frac{1+a}{1-a}\right)\right]$$

The standard form of $Q_0(x)$ is obtained by setting $A_0 = 1$ and $a = 0$:

$$Q_0(x) = \frac{1}{2}\ln\left(\frac{1+x}{1-x}\right) \qquad \text{for } |x| < 1$$

Similarly, since $P_1(x) = x$,

$$Q_1(x) = A_1 x \int_a^x \frac{ds}{s^2(1-s^2)} = A_1 x\left[-\frac{1}{x} + \frac{1}{a} + \frac{1}{2}\ln\left(\frac{1+x}{1-x}\right) - \frac{1}{2}\ln\left(\frac{1+a}{1-a}\right)\right]$$

$$= Ax + Bx\ln\left(\frac{1+x}{1-x}\right) + C \qquad \text{for } |x| < 1$$

Here standardization gives $A = 0$, $B = \frac{1}{2}$, and $C = -1$. Thus,

$$Q_1(x) = \frac{1}{2}x \ln\left(\frac{1+x}{1-x}\right) - 1$$

●

The general solution to an ISOLDE. Inhomogeneous SOLDEs (ISOLDEs) can be most elegantly discussed in terms of Green's functions, the subject of Chapter 11. At this point, however, we can determine the most general solution of an ISOLDE.

Let $g(x)$ be a particular solution of

$$\mathbb{L}[y] \equiv y'' + py' + qy = r(x) \tag{9.15}$$

and let $h(x)$ be any other solution of this equation. Then $h(x) - g(x)$ satisfies Eq. (9.12) and can be written as a linear combination of a basis of solutions $f_1(x)$ and $f_2(x)$, leading to the following equation:

$$h(x) = c_1 f_1(x) + c_2 f_2(x) + g(x) \tag{9.16}$$

Thus, if we have a *particular* solution of the ISOLDE of (9.15) and two basis solutions of the HSOLDE, then the most *general* solution of (9.15) can be expressed as a linear combination of the two basis solutions and the particular solution.

We know how to find a second solution to the HSOLDE once we know one solution. We now show that knowing one such solution can also allow us to find a particular solution to the ISOLDE. The method we use is called the *method of variation of constants.* This method can also be used to find a second solution to the HSOLDE (see Exercise 9.2.3).

Let $g(x)$ be the desired solution to Eq. (9.15). If we write it as $g(x) \equiv f_1(x)v(x)$, then $v(x)$ satisfies the SOLDE

$$v'' + \left(p + \frac{2f_1'}{f_1}\right)v' = \frac{r}{f_1}$$

Let $u = v'$. Then u satisfies the FOLDE

$$u' + \left(p + \frac{2f_1'}{f_1}\right)u = \frac{r}{f_1}$$

which has the integrating factor

$$\mu(x) = cf_1^2(x)\exp\left[\int_a^x p(t)\,dt\right] = \frac{f_1^2(x)}{W(x)}$$

and a solution of the form

$$u = \frac{W(x)}{f_1^2(x)}\left[c + \int_a^x \frac{f_1(t)r(t)}{W(t)}\,dt\right]$$

But knowledge of the second solution, f_2, given by (9.14), allows us to write

$$\frac{W(x)}{f_1^2(x)} = \frac{d}{dx}\left(\frac{f_2}{f_1}\right)$$

Thus, setting $c = 0$ (we are interested in a particular solution) and noting that $u = dv/dx$ gives

$$\frac{dv}{dx} = \frac{d}{dx}\left(\frac{f_2}{f_1}\right)\int_a^x \frac{f_1(t)r(t)}{W(t)}\,dt$$

$$= \frac{d}{dx}\left[\frac{f_2}{f_1}\int_a^x \frac{f_1(t)r(t)}{W(t)}\,dt\right] - \frac{f_2(x)}{f_1(x)}\left[\frac{f_1(x)r(x)}{W(x)}\right]$$

This leads to the particular solution:

$$g(x) = f_2(x)\int_a^x \frac{f_1(t)r(t)}{W(t)}\,dt - f_1(x)\int_a^x \frac{f_2(t)r(t)}{W(t)}\,dt \tag{9.17}$$

The following proposition summarizes the above discussion.

9.2.8 Proposition The knowledge of only *one* solution to an HSOLDE is enough to construct the most general solution to both the HSOLDE and its corresponding ISOLDE. ■

The separation and comparison theorems. The Wronskian can be used to derive some properties of the graphs of solutions of HSOLDEs. One such property is described by a theorem due to Sturm and concerns the relative position of the zeros of two linearly independent solutions of an HSOLDE.

9.2.9 Theorem (The Separation Theorem) The zeros of two linearly independent solutions $f_1(x)$ and $f_2(x)$ of the HSOLDE of Eq. (9.12) occur alternately.

Proof. We have to show that a zero of f_1 exists between any two zeros of f_2. The linear independence of f_1 and f_2 implies that $W(f_1, f_2) \neq 0$ for any $x \in [a, b]$. Let $x_i \in [a, b]$ be a zero of f_2; that is, let $f_2(x_i) = 0$. Then

$$\begin{aligned} 0 \neq W(f_1, f_2; x_i) &= f_1(x_i)f'_2(x_i) - f_2(x_i)f'_1(x_i) \\ &= f_1(x_i)f'_2(x_i) \end{aligned}$$

Thus, $f_1(x_i) \neq 0$ and $f'_2(x_i) \neq 0$. Suppose that x_1 and x_2, where $x_2 > x_1$, are two successive zeros of f_2. Since f_2 is continuous in $[a, b]$ and $f'_2(x_1) \neq 0$, f_2 has to be either increasing $[f'_2(x_1) > 0]$ or decreasing $[f'_2(x_1) < 0]$ at x_1. For f_2 to be zero at the next point, x_2, $f'_2(x_2)$ must have the *opposite* sign from $f'_2(x_1)$. We proved earlier that the sign of $W(f_1, f_2)$ does not change in $[a, b]$. The above equation then says that $f_1(x_1)$ and $f_1(x_2)$ also have opposite signs. The continuity of f_1 then implies that

$$\exists\, x_1 < x_0 < x_2 \ni f_1(x_0) = 0$$

A similar argument shows that there exists one zero of f_2 between any two zeros of f_1. ■

Example 9.2.5

Two linearly independent solutions of $y'' + y = 0$ are $\sin x$ and $\cos x$. The separation theorem suggests that the zeros of $\sin x$ and $\cos x$ must alternate, a fact known from elementary trigonometry. In fact, the zeros of $\cos x$ occur at odd multiples of $\pi/2$, and those of $\sin x$ occur at even multiples of $\pi/2$. ●

Another result due to Sturm is known as the comparison theorem (for a proof, see Birkhoff and Rota 1978, p. 38).

9.2.10 Theorem (The Comparison Theorem) Let $f(x)$ and $g(x)$ be nontrivial solutions of $u'' + p(x)u = 0$ and $v'' + q(x)v = 0$, respectively, where $p(x) \geqslant q(x)$ for all $x \in [a, b]$. Then $f(x)$ vanishes at least once between any two zeros of $g(x)$, unless $p(x) \equiv q(x)$ and f is a constant multiple of g. ■

The form of the differential equations used in the comparison theorem is not restrictive because any HSOLDE can be cast in this form (as Exercise 9.2.4 shows). A useful special case of this theorem is given as a corollary (the easy but instructive proof is left as a problem).

9.2.11 Corollary If $q(x) \leqslant 0$, then no nontrivial solution of the differential equation $v'' + q(x)v = 0$ can have more than one zero. ■

Example 9.2.6

It should be clear from the preceding discussion that the oscillations of the solutions of $v'' + q(x)v = 0$ are mostly determined by the sign and magnitude of $q(x)$. For $q(x) \leqslant 0$ there is no oscillation; that is, there is no solution that changes sign more than once. Now suppose that $q(x) \geqslant k^2 > 0$ for some real k. Then, by Theorem 9.2.10, any solution of $v'' + q(x)v = 0$ must have at least one zero between any two successive zeros of the solution $\sin kx$ of $u'' + k^2 u = 0$. This means that any solutions of $v'' + q(x)v = 0$ has a zero in any interval of length π/k if $q(x) \geqslant k^2 > 0$.

Let us apply this to the Bessel equation,

$$y'' + \frac{1}{x}y' + \left(1 - \frac{n^2}{x^2}\right)y = 0$$

We can eliminate the y' term by substituting $v/\sqrt{x}$ for y. This transforms the Bessel equation into

$$v'' + \left(1 - \frac{4n^2 - 1}{4x^2}\right)v = 0$$

We compare this, for $n = 0$, with $u'' + u = 0$, which has a solution $u = \cos x$, and conclude that each interval of length π of the positive x-axis contains at least one zero of any solution of order zero ($n = 0$) of the Bessel equation. Thus, in particular, the zeroth Bessel function, usually denoted as $J_0(x)$, has a zero in each interval of length π of the x-axis.

On the other hand, for $4n^2 - 1 > 0$, or $n > \frac{1}{2}$, we have $1 > [1 - (4n^2 - 1)/4x^2]$. This implies that $\sin x$ has *at least* one zero between any two successive zeros of the Bessel functions of order greater than $\frac{1}{2}$. It follows that such a Bessel function can have *at most* one zero between any two successive zeros of $\sin x$ (or in each interval of length π on the positive x-axis).

Applying Corollary 9.2.11 to $v'' - v = 0$, whose most general solution is $c_1 e^x + c_2 e^{-x} \equiv f(x)$, we note that

$$f(x) = 0 \quad \Rightarrow \quad x = \tfrac{1}{2} \ln\left| -\frac{c_2}{c_1}\right|$$

This (real) x (if it exists) is unique. •

9.2.3 Adjoint Differential Operators

Adjoint operators were discussed in detail in the context of finite-dimensional vector spaces in Chapters 2 and 3. In particular, the importance of self-adjoint, or hermitian, operators was clearly spelled out by the spectral decomposition theorem. A consequence of that theorem is the completeness of the eigenvectors of a hermitian operator—the fact that an arbitrary vector can be expressed as a linear combination of the (orthonormal) eigenvectors of a hermitian operator.

Self-adjoint differential operators are equally important because their "eigenfunctions" also form complete orthogonal sets (see Chapter 10). This subsection will generalize the concept of the adjoint to the case of a differential operator (of second degree). Let us begin with a definition.

9.2.12 Definition The HSOLDE

$$\mathbb{L}[y] \equiv p_2(x)y'' + p_1(x)y' + p_0(x)y = 0 \tag{9.18a}$$

is said to be *exact* if and only if

$$p_2 y'' + p_1 y' + p_0 y = \frac{d}{dx}[A(x)y' + B(x)y] \tag{9.18b}$$

for all $y \in \mathrm{C}^{(2)}[a, b]$ and for some $A,\ B \in \mathrm{C}^{(1)}[a, b]$. An *integrating factor* for Eq. (9.18a) is a function $\mu(x)$ such that $\mu\mathbb{L}[y] = 0$ is exact.

If an integrating factor exists, then Eq. (9.18a) reduces to

$$\frac{d}{dx}[A(x)y' + B(x)y] = 0 \quad \Rightarrow \quad A(x)y' + B(x)y = C$$

which is a FOLDE with a constant inhomogeneous term. Even the ISOLDE corresponding to (9.18a) can be solved, because

$$\mu\mathbb{L}[y] = \mu(x)r(x) \quad \Rightarrow \quad \frac{d}{dx}[A(x)y' + B(x)y] = \mu(x)r(x)$$

$$\Rightarrow \quad A(x)y' + B(x)y = \int^x \mu(t)r(t)\,dt$$

which is a general FOLDE. Thus, the existence of an integrating factor completely solves a SOLDE. It is, therefore, important to know whether or not a SOLDE is exact. The following proposition gives a criterion for the exactness of a SOLDE.

9.2.13 Proposition The SOLDE of Eq. (9.18a) is exact if and only if $p_2'' - p_1' + p_0 = 0$.

Proof. The SOLDE is exact if and only if (9.18b) holds, which is true if and only if

$$p_2 = A \qquad p_1 = A' + B \qquad \text{and} \qquad p_0 = B'$$

These equations are equivalent to

$$p_2'' = A'' \qquad p_1' = A'' + B' \qquad \text{and} \qquad p_0 = B'$$

which in turn are equivalent to the single equation $p_2'' - p_1' + p_0 = 0$. ■

A general SOLDE is clearly not exact. Can we make it exact by multiplying it by an integrating factor as we did with a FOLDE? The following proposition contains the answer.

9.2.14 Proposition A function $\mu \in C^{(2)}(a, b)$ is an integrating factor for the SOLDE of Eq. (9.18a) if and only if it is a solution of the HSOLDE

$$\mathbb{M}[\mu] \equiv (p_2\mu)'' - (p_1\mu)' + p_0\mu = 0 \tag{9.19a}$$

Proof. Equation (9.19a) is an immediate consequence of Proposition 9.2.13. ■

We can expand Eq. (9.19a) to obtain the equivalent equation

$$p_2\mu'' + (2p_2' - p_1)\mu' + (p_2'' - p_1' + p_0)\mu = 0 \tag{9.19b}$$

The operator $\mathbb{M}$, given by

$$\mathbb{M} \equiv p_2 \frac{d^2}{dx^2} + (2p_2' - p_1)\frac{d}{dx} + (p_2'' - p_1' + p_0) \tag{9.20}$$

is called the *adjoint* of the operator $\mathbb{L}$, denoted $\mathbb{M} \equiv \mathbb{L}^\dagger$. The reason for the use of the word "adjoint" will be made clear below.

Proposition 9.2.14 affirms the existence of an integrating factor. However, it can be obtained only by solving Eq. (9.19a) or (9.19b), which is at least as bad as solving the original differential equation! In contrast the integrating factor for a FOLDE can be obtained by a mere integration (see Theorem 9.1.4).

Although integrating factors for SOLDEs are not as powerful as their counterparts for FOLDEs, they can facilitate the study of SOLDEs. Let us first note that the adjoint of the adjoint of a differential operator $\mathbb{L}$ is the original operator: $(\mathbb{L}^\dagger)^\dagger = \mathbb{L}$ (see Exercise 9.2.5). This suggests that if v is an integrating factor of $\mathbb{L}[u] = 0$, then u is an integrating factor of $\mathbb{M}[v] \equiv \mathbb{L}^\dagger[v] = 0$. In particular, multiplying (9.18a) (in which y is replaced by u) by v, and (9.19a) (in which μ is replaced by v) by u, and subtracting the result, we obtain

$$v\mathbb{L}[u] - u\mathbb{M}[v] = (vp_2)u'' - u(p_2v)'' + (vp_1)u' + u(p_1v)'$$

which can be simplified to

$$v\mathbb{L}[u] - u\mathbb{M}[v] = \frac{d}{dx}[p_2vu' - (p_2v)'u + p_1uv] \tag{9.21a}$$

Integrating this from a to b yields

$$\int_a^b \{v\mathbb{L}[u] - u\mathbb{M}[v]\}\,dx = [p_2vu' - (p_2v)'u + p_1uv]\big|_a^b \tag{9.21b}$$

Equations (9.21) are called the *Lagrange identity*. Equation (9.21b) embodies the reason for calling $\mathbb{M}$ the adjoint of $\mathbb{L}$. If we consider u and v as abstract vectors $|u\rangle$ and $|v\rangle$, $\mathbb{L}$ and $\mathbb{M}$ as operators in a Hilbert space, and define the inner product as

$$\langle u|v\rangle = \int_a^b u^*(x)v(x)\,dx$$

then (9.21b) can be written as

$$\begin{aligned}\langle v|\mathbb{L}|u\rangle - \langle u|\mathbb{M}|v\rangle &= \langle u|\mathbb{L}^\dagger|v\rangle^* - \langle u|\mathbb{M}|v\rangle \\ &= [p_2 v u' - (p_2 v)' u + p_1 u v]\big|_a^b\end{aligned}$$

If the RHS is zero, then

$$\langle u|\mathbb{L}^\dagger|v\rangle^* = \langle u|\mathbb{M}|v\rangle \qquad \forall\ |u\rangle, |v\rangle$$

and, since all these operators and functions are real,

$$\mathbb{L}^\dagger = \mathbb{M}$$

Like a self-adjoint finite-dimensional vector space, a self-adjoint differential operator merits special consideration. For $\mathbb{L}^\dagger = \mathbb{M}$ to be equal to $\mathbb{L}$, we must have [see Eqs. (9.18a) and (9.19b)]

$$2p_2' - p_1 = p_1$$

$$p_2'' - p_1' + p_0 = p_0$$

The first equation gives

$$p_2' = p_1$$

which also solves the second equation. If this condition holds, then we can write (9.18a) as

$$\mathbb{L}[y] = p_2 y'' + p_2' y' + p_0 y = (p_2 y')' + p_0 y = 0$$

or

$$\mathbb{L}[y] = \frac{d}{dx}\left[p_2(x)\frac{dy}{dx}\right] + p_0(x)y = 0$$

Of course, not all SOLDEs are self-adjoint. Can we make them so? Let us multiply both sides of (9.18a) by a function $h(x)$, to be determined later. We get

$$h(x)p_2(x)y'' + h(x)p_1(x)y' + h(x)p_0(x)y = 0$$

We choose $h(x)$ so that

$$\frac{d}{dx}(hp_2) = hp_1$$

or

$$p_2\frac{dh}{dx} + h(p_2' - p_1) = 0$$

This is a separable FODE with the solution

$$h(x) = \frac{1}{p_2} \exp\left[\int^x \frac{p_1(t)}{p_2(t)}\, dt \right]$$

We have just proved the following theorem.

9.2.15 Theorem The SOLDE for Eq. (9.18) is self-adjoint if and only if it has the form

$$\frac{d}{dx}\left[p_2(x) \frac{dy}{dx} \right] + p_0(x) y = 0$$

If it is not self-adjoint, it can be made so by multiplying through by

$$h(x) = \frac{1}{p_2} \exp\left[\int^x \frac{p_1(t)}{p_2(t)}\, dt \right]$$ ■

Example 9.2.7

(a) The Legendre equation in normal form,

$$y'' - \frac{2x}{1 - x^2} y' + \frac{\lambda}{1 - x^2} y = 0$$

is not self-adjoint because

$$p_2' = (1)' = 0 \neq -\frac{2x}{1 - x^2}$$

However, we get a self-adjoint version if we multiply through by $1 - x^2$:

$$(1 - x^2) y'' - 2xy' + \lambda y = 0$$

or

$$\frac{d}{dx}\left[(1 - x^2) \frac{dy}{dx} \right] + \lambda y = 0$$

(b) Similarly, the normal form of the Bessel equation,

$$y'' + \frac{1}{x} y' + \left(1 - \frac{n^2}{x^2}\right) y = 0$$

is not self-adjoint, but multiplying through by x yields

$$\frac{d}{dx}\left(x \frac{dy}{dx} \right) + \left(x - \frac{n^2}{x} \right) y = 0$$

which is clearly self-adjoint. ●

9.2.4 The Riccati Equation

It is possible to transform a SOLDE into a first-order nonlinear differential equation by making the substitution $v = y'/y$. The equation

$$y'' + p(x)y' + q(x)y = 0$$

turns into

$$v' + v^2 + p(x)v + q(x) = 0 \tag{9.22a}$$

which is a first-order nonlinear equation called the *Riccati equation.* It can be further simplified to (see Exercise 9.2.4)

$$u' + u^2 + S(x) = 0 \qquad \text{where } S(x) = -\tfrac{1}{2}p' - \tfrac{1}{4}p^2 + q \tag{9.22b}$$

Thus, the Riccati substitution reduces a SOLDE to a first-order quadratic differential equation. Once either form of the Riccati equation is solved, the original function can be recovered using

$$y(x) = C\exp\left[\int^x v(t)\,dt\right]$$

Exercises

9.2.1 Let $u(x)$ be a differentiable function satisfying the differential inequality

$$u'(x) \leqslant Ku(x) \qquad \text{for } x \in [a, b]$$

where K is a constant. Show that

$$u(x) \leqslant u(a)e^{K(x-a)}$$

9.2.2 Show that if (f_1, f_1') and (f_2, f_2') for Eq. (9.12) are linearly dependent at one point, then f_1 and f_2 are linearly dependent at all $x \in [a, b]$.

9.2.3 Use the method of variation of constants to obtain $f_2(x)$ from $f_1(x)$ for an HSOLDE.

9.2.4 Show that $y'' + p(x)y' + q(x)y = 0$ can be cast in the form

$$u'' + S(x)u = 0$$

by an appropriate functional transformation. What is $S(x)$ in terms of $p(x)$ and $q(x)$?

9.2.5 Show that the adjoint of $\mathbb{M}$ given in (9.19b) is the original $\mathbb{L}$.

9.3 POWER-SERIES SOLUTIONS OF SECOND-ORDER LINEAR DIFFERENTIAL EQUATIONS

The theory of functions is one of the richest branches of mathematics, focusing on the endless variety of objects known as functions. The simplest kind of function is a polynomial, which is obtained by performing the simple algebraic operations of addition and multiplication on the independent variable x. The next in complexity are the trigonometric functions, which are obtained by geometric methods. If we demand a simplistic, intuitive approach to functions, the list ends there. It was only with the advent of derivatives, integrals, and differential equations that a vastly rich variety of functions exploded into existence in the eighteenth and nineteenth centuries.

For instance, e^x, nonexistent before the invention of calculus, can be thought of as that function which equals its derivative, or the function that solves

$$\frac{dy}{dx} = y$$

This leads to an implicit solution

$$\int_1^y \frac{dt}{t} = x + C$$

The LHS can be used as a *definition* of $\ln y$ and, in fact, is in some calculus texts. (Problem 9.20 outlines arguments leading to the identification of $\int_1^x dt/t$ with $\ln x$.)

Although this definition of a function seems a bit artificial, for most applications it is the *only* way to define a function. For instance, the error function, used in statistics, is defined as

$$\text{er}[f(x)] \equiv \frac{1}{\sqrt{\pi}} \int_0^x e^{-t^2}\, dt$$

Such a function cannot be expressed in terms of elementary functions. Similarly, functions such as

$$\int_x^\infty \frac{\sin t}{t}\, dt \qquad \int_0^{\pi/2} \sqrt{1 - x^2 \sin^2 t}\, dt \qquad \int_0^{\pi/2} \frac{dt}{\sqrt{1 - x^2 \sin^2 t}} \qquad \text{and so on}$$

are encountered frequently in applications. None of these functions can be expressed in terms of other well-known functions.

A better way of studying such functions is to study the differential equation they satisfy. In fact, the majority of functions encountered in mathematical physics obey the HSOLDE

$$p_2(x)y'' + p_1(x)y' + p_0(x)y = 0$$

in which $p_i(x)$ are elementary functions, mostly polynomials (of at most degree 2). Of course, to specify functions completely, appropriate boundary conditions are necessary. For instance, the error function mentioned above satisfies the HSOLDE

$$y'' + 2xy' = 0$$

with the boundary conditions $y(0) = 0$ and $y'(0) = 1/\sqrt{\pi}$.

The natural tendency to resist the idea of a function as a solution of a SOLDE is mostly due to the abstract nature of differential equations. After all, it is easier to imagine constructing functions by simple multiplications or with simple geometric figures that have been around for centuries. The following beautiful example (see Birkhoff and Rota 1978, pp. 85–87) should overcome this resistance and convince the skeptic that differential equations contain all the information about a function.

Example 9.3.1

We can show that the solutions to $y'' + y = 0$ have all the properties we expect of $\sin x$ and $\cos x$. Let us denote the two linearly independent solutions of this equation by $C(x)$ and $S(x)$. To specify these functions completely, we set

$$C(0) = 1 \qquad C'(0) = 0$$

$$S(0) = 0 \qquad S'(0) = 1$$

We claim that this information is enough to identify $C(x)$ and $S(x)$ as $\cos x$ and $\sin x$, respectively.

First, let us show that the solutions exist and are well-behaved functions. With $C(0)$ and $C'(0)$ given, the equation $y'' + y = 0$ can generate all derivatives of $C(x)$ at zero:

$$C''(0) = -C(0) = -1$$

$$C'''(0) = -C'(0) = 0$$

$$C^{(4)}(0) = -C''(0) = +1$$

and, in general,

$$C^{(n)}(0) = \begin{cases} 0 & \text{for odd } n \\ (-1)^k & \text{for } n = 2k, \text{ where } k = 0, 1, \ldots \end{cases}$$

Thus, the (real) Taylor expansion of $C(x)$ is

$$C(x) = \sum_{k=0}^{\infty} (-1)^k \frac{x^{2k}}{(2k)!} \tag{1}$$

Similarly,

$$S(x) = \sum_{k=0}^{\infty} (-1)^k \frac{x^{2k+1}}{(2k+1)!} \tag{2}$$

A simple ratio test on the series representation of $C(x)$ yields

$$\lim_{k\to\infty} \frac{a_{k+1}}{a_k} \equiv \lim_{k\to\infty} \frac{(-1)^{k+1} \dfrac{x^{2(k+1)}}{(2k+2)!}}{(-1)^k \dfrac{x^{2k}}{(2k)!}} = \lim_{k\to\infty} \frac{-x^2}{(2k+2)(2k+1)} = 0$$

which shows that the series for $C(x)$ converges for all values of x. Similarly, the series for $S(x)$ is also convergent. Thus, we are dealing with well-defined finite-valued functions.

Let us now enumerate and prove some properties of $C(x)$ and $S(x)$.

(i) $$C'(x) = -S(x)$$

We prove this relation by differentiating

$$C''(x) + C(x) = 0$$

and writing the result as $[C'(x)]'' + C'(x) = 0$ to make evident the fact that $C'(x)$ is also a solution. Since $C'(0) = 0$ and $[C'(0)]' = C''(0) = -1$, the *unique* solution must be $C'(x) = -S(x)$. Similarly, $S'(x) = C(x)$.

(ii) $$C^2(x) + S^2(x) = 1$$

Since the $p(x)$ term is absent from the SOLDE, Proposition 9.2.6 implies that the Wronskian of $C(x)$ and $S(x)$ is constant. On the other hand,

$$W(C, S; x) = C(x)S'(x) - C'(x)S(x) = C^2(x) + S^2(x)$$
$$= C^2(0) + S^2(0) = 1$$

↑ because W is constant

(iii) $$S(a + x) = S(a)C(x) + C(a)S(x)$$

The use of the chain rule easily shows that $S(a + x)$ is a solution of the equation $y'' + y = 0$. Thus, it can be written as a linear combination of $C(x)$ and $S(x)$ [which are linearly independent by (ii)]:

$$S(a + x) = AS(x) + BC(x) \tag{3}$$

This is a functional identity, which, for $x = 0$, becomes

$$S(a) = BC(0) = B \tag{4}$$

If we differentiate both sides of (3), we get

$$C(a + x) = AS'(x) + BC'(0) = AC(x) - BS(x) \tag{5}$$

which, for $x = 0$, gives $C(a) = A$. Substituting Eqs. (4) and (5) in (1) yields the desired identity. A similar argument leads to $C(a + x) = C(a)C(x) - S(a)S(x)$.

(iv) Periodicity of $C(x)$ and $S(x)$

Let x_0 be the smallest positive real number such that $S(x_0) = C(x_0)$. Then property (ii) implies that $C(x_0) = S(x_0) = 1/\sqrt{2}$. On the other hand,

$$S(x_0 + x) = S(x_0)C(x) + C(x_0)S(x) = C(x_0)C(x) + S(x_0)S(x)$$
$$= C(x_0)C(x) - S(x_0)S(-x) = C(x_0 - x)$$

↑ because, by (2), $S(x)$ is an odd function of x ↑ by (iii)

This is true for all x; in particular, for $x = x_0$ it yields

$$S(2x_0) = C(0) = 1 \quad \Rightarrow \quad C(2x_0) = 0$$

Using property (iii) once more, we get

$$S(2x_0 + x) = S(2x_0)C(x) + C(2x_0)S(x) = C(x)$$

$$C(2x_0 + x) = C(2x_0)C(x) - S(2x_0)S(x) = -S(x)$$

Substituting $x = 2x_0$ yields

$$S(4x_0) = C(2x_0) = 0$$

$$C(4x_0) = -S(2x_0) = -1$$

Continuing in this manner, we can easily obtain

$$S(8x_0 + x) = S(x)$$

$$C(8x_0 + x) = C(x)$$

which prove the periodicity of $S(x)$ and $C(x)$ and show that their period is $8x_0$.

It is even possible to determine x_0. This determination is left as a problem, but the result is

$$x_0 = \int_0^{1/\sqrt{2}} \frac{dt}{\sqrt{1-t^2}}$$

A numerical calculation will show that this is $\pi/4$. •

Convinced that infinite-series solutions of differential equations have as much information as the functions they represent, we are ready to find such solutions.

9.3.1 The Method of Undetermined Coefficients (The Frobenius Method)

A proper treatment of SOLDEs requires the medium of complex analysis and will be undertaken later in this chapter. At this point, however, we want a *formal* infinite-series solution to the SOLDE

$$y'' + p(x)y' + q(x)y = 0$$

where $p(x)$ and $q(x)$ are real and analytic. This means that $p(x)$ and $q(x)$ can be represented by *convergent* power series in some interval (a, b). [The interesting case where $p(x)$ and $q(x)$ may have singularities will be treated in the context of complex solutions.]

The general procedure is to assume expansions of the form

$$p(x) = \sum_{k=0}^{\infty} a_k x^k \qquad q(x) = \sum_{k=0}^{\infty} b_k x^k \tag{9.23}$$

for the coefficient functions, and a trial series

$$y = \sum_{k=0}^{\infty} c_k x^k \tag{9.24}$$

for the solution. Then these expansions are substituted in the SOLDE and the coefficients of powers of x on the LHS are equated to zero. This yields

$$y' = \sum_{k=1}^{\infty} kc_k x^{k-1} = \sum_{k=0}^{\infty} (k+1)c_{k+1}x^k$$

and

$$y'' = \sum_{k=1}^{\infty} (k+1)kc_{k+1}x^{k-1} = \sum_{k=0}^{\infty} (k+1)(k+2)c_{k+2}x^k$$

Thus,

$$\begin{aligned} p(x)y' &= \sum_{k=0}^{\infty}\sum_{m=0}^{\infty} (k+1)c_{k+1}x^k a_m x^m \\ &= \sum_{k,m} (k+1)a_m c_{k+1} x^{k+m} \end{aligned}$$

Let $k + m \equiv n$. Then one of the sums, say m, cannot exceed n. Thus,

$$p(x)y' = \sum_{n=0}^{\infty}\sum_{m=0}^{n} [(n-m+1)a_m c_{n-m+1}]x^n$$

Similarly,

$$q(x)y = \sum_{n=0}^{\infty}\sum_{m=0}^{n} (b_m c_{n-m})x^n$$

Substituting these sums and the series for y'' in the SOLDE, we obtain

$$\sum_{n=0}^{\infty}\left\{(n+1)(n+2)c_{n+2} + \sum_{m=0}^{n} [(n-m+1)a_m c_{n-m+1} + b_m c_{n-m}]\right\}x^n = 0$$

For this to be true for all x, all coefficients of powers of x must vanish. This gives

$$(n+1)(n+2)c_{n+2} = -\sum_{m=0}^{n} [(n-m+1)a_m c_{n-m+1} + b_m c_{n-m}] \qquad \text{for } n \geqslant 0$$

or

$$c_{n+1} = -\frac{\sum_{m=0}^{n-1} [(n-m)a_m c_{n-m} + b_m c_{n-m-1}]}{n(n+1)} \qquad \text{for } n \geqslant 1 \tag{9.25}$$

If we know c_0 and c_1, for instance, from boundary conditions, we can determine c_n, where $n \geqslant 2$, uniquely from (9.25). This, in turn, gives a unique power-series expansion for y, and we have the following theorem.

9.3.1 Theorem (The Existence Theorem) For any SOLDE of the form $y'' + p(x)y' + q(x)y = 0$ with analytic coefficient functions given by (9.23), there exists a unique power series, given by (9.24), which formally satisfies the SOLDE for each choice of c_0 and c_1. ∎

This theorem merely states the existence of a formal power series and says nothing about its convergence. The following example will demonstrate that convergence does not necessarily follow.

Example 9.3.2

Let us find the formal power-series solution for $x^2y' - y + x = 0$.

Let $y = \sum_{n=0}^{\infty} c_n x^n$. Then $y' = \sum_{n=0}^{\infty} c_{n+1}(n+1)x^n$, and substitution gives

$$\sum_{n=0}^{\infty} c_{n+1}(n+1)x^{n+2} - \sum_{n=0}^{\infty} c_n x^n + x = 0$$

or

$$\sum_{n=0}^{\infty} c_{n+1}(n+1)x^{n+2} - c_0 - c_1 x - \sum_{n=2}^{\infty} c_n x^n + x = 0$$

We see that

$$c_0 = 0 \qquad c_1 = 1$$

and

$$(n+1)c_{n+1} = c_{n+2} \qquad \text{for } n \geqslant 0$$

Thus, we have the recursion relation

$$nc_n = c_{n+1} \qquad \text{for } n \geqslant 1$$

Its unique solution is $c_n = (n-1)!$, which yields the solution

$$y(x) = x + x^2 + (2!)x^3 + (3!)x^4 + \cdots + (n-1)!x^n + \cdots$$

This series is not convergent for any nonzero x. •

As we will see later, for *normal* SOLDEs the power series of (9.24) converges to an analytic function. The SOLDE solved in Example 9.3.2 is not normal.

Example 9.3.3

As an application of Theorem 9.3.1, let us consider the Legendre equation in its normal form,

$$y'' - \frac{2x}{1-x^2}y' + \frac{\lambda}{1-x^2}y = 0$$

where λ is a constant. For $|x| < 1$ both p and q are analytic, and

$$p(x) = -2x \sum_{m=0}^{\infty} (x^2)^m = \sum_{m=0}^{\infty} (-2)x^{2m+1}$$

$$q(x) = \lambda \sum_{m=0}^{\infty} (x^2)^m = \sum_{m=0}^{\infty} \lambda x^{2m}$$

Thus,

$$a_m = \begin{cases} 0 & \text{if } m \text{ is even} \\ -2 & \text{if } m \text{ is odd} \end{cases} \qquad \text{and} \qquad b_m = \begin{cases} \lambda & \text{if } m \text{ is even} \\ 0 & \text{if } m \text{ is odd} \end{cases}$$

We want to substitute for a_m and b_m in Eq. (9.25) to find c_{n+1}. It is convenient to consider two cases: when n is even and when n is odd. First, we let $n = 2r + 1$, for some integer r. It is then easy to verify that the use of (9.25) yields

$$(2r+1)(2r+2)c_{2r+2} = \sum_{m=0}^{r} (4r - 4m - \lambda)c_{2(r-m)} \tag{1}$$

With $r \to r + 1$, this becomes

$$(2r + 3)(2r + 4)c_{2r+4} = \sum_{m=0}^{r+1} (4r + 4 - 4m - \lambda)c_{2(r+1-m)}$$

$$= (4r + 4 - \lambda)c_{2r+2} + \sum_{m=1}^{r+1} (4r + 4 - 4m + \lambda)c_{2(r+1-m)}$$

$$= (4r + 4 - \lambda)c_{2r+2} + \sum_{m=0}^{r} (4r - 4m - \lambda)c_{2(r-m)}$$

↑ by a change of the dummy index

$$= (4r + 4 - \lambda)c_{2r+2} + (2r + 1)(2r + 2)c_{2r+2}$$

↑ by (1)

$$= [4r + 4 - \lambda + (2r + 1)(2r + 2)]c_{2r+2}$$

Now we let $2r + 2 \equiv k$ to obtain

$$(k + 1)(k + 2)c_{k+2} = [k(k + 1) - \lambda]c_k$$

or

$$c_{k+2} = \frac{k(k + 1) - \lambda}{(k + 1)(k + 2)} c_k \qquad \text{for even } k$$

It is not difficult to show that this equation also holds for odd k. Thus, we can write

$$c_{n+2} = \frac{n(n + 1) - \lambda}{(n + 1)(n + 2)} c_n \tag{2}$$

For arbitrary c_0 and c_1, we obtain two independent solutions, one of which has only even powers of x and the other only odd powers. The ratio test,

$$\lim_{n\to\infty} \frac{c_{n+2}x^{n+2}}{c_n x^n} = \lim_{n\to\infty} \frac{n(n + 1) - \lambda}{(n + 1)(n + 2)} x^2 = x^2$$

shows that the series is divergent for $x = \pm 1$ unless $\lambda = l(l + 1)$ for some positive integer l. In that case the infinite series becomes a *polynomial*, the Legendre polynomial encountered in Chapter 5!

Equation (2) could have been obtained by substituting Eqs. (9.23) and (9.24) directly into the Legendre equation in its normal form. The roundabout way to (2) taken here shows the generality of Eq. (9.25). With specific differential equations it is generally better to substitute (9.23) and (9.24) directly. •

Example 9.3.4

We studied Hermite polynomials in Chapter 5 in the context of classical orthogonal polynomials. Let us see how they arise in physics.

The one-dimensional Schrödinger equation for a particle in a potential, $V(x)$, is obtained from the classical energy relation $E = p^2/2m + V(x)$ by replacing E with $i\hbar\, \partial/\partial t$

and p with $-i\hbar\,\partial/\partial x$ and letting the resulting operator act on the wave function $\Psi(x, t)$. Thus,

$$i\hbar\frac{\partial\Psi}{\partial t} = -\frac{\hbar^2}{2m}\frac{\partial^2\Psi}{\partial x^2} + V(x)\Psi$$

Writing $\Psi(x, t) = u(x)T(t)$ separates the dependence on time and gives

$$T(t) = \exp\left(-i\frac{Et}{\hbar}\right)$$

where E is the energy of the particle. It also yields

$$\frac{d^2u}{dx^2} - \frac{2m}{\hbar^2}V(x)u + \frac{2m}{\hbar^2}Eu = 0$$

For a harmonic oscillator, $V(x) = \frac{1}{2}kx^2 \equiv \frac{1}{2}m\omega^2x^2$, and

$$u'' - \frac{m^2\omega^2}{\hbar^2}x^2u + \frac{2m}{\hbar^2}Eu = 0$$

Substituting $u(x) = H(x)\exp(-m\omega x^2/2\hbar)$ and then making the change of variables $x = (1/\sqrt{m\omega/\hbar})y$ yields

$$H'' - 2yH' + \lambda H = 0 \qquad \text{where } \lambda = \frac{2E}{\hbar\omega} - 1 \tag{1}$$

This is the Hermite differential equation in normal form. Thus, we assume the expansion

$$H(y) = \sum_{n=0}^{\infty} c_n y^n$$

which yields

$$H'(y) = \sum_{n=1}^{\infty} nc_n y^{n-1} = \sum_{n=0}^{\infty} (n+1)c_{n+1}y^n$$

$$H''(y) = \sum_{n=1}^{\infty} n(n+1)c_{n+1}y^{n-1} = \sum_{n=0}^{\infty} (n+1)(n+2)c_{n+2}y^n$$

Substituting in (1) gives

$$\sum_{n=0}^{\infty} [(n+1)(n+2)c_{n+2} + \lambda c_n]y^n - 2\sum_{n=0}^{\infty} (n+1)c_{n+1}y^{n+1} = 0$$

or
$$2c_2 + \lambda c_0 + \sum_{n=0}^{\infty} [(n+2)(n+3)c_{n+3} + \lambda c_{n+1} - 2(n+1)c_{n+1}]y^{n+1} = 0$$

Setting the coefficients of powers of y equal to zero, we obtain

$$c_2 = -\frac{\lambda}{2}c_0$$

$$c_{n+3} = \frac{2(n+1) - \lambda}{(n+2)(n+3)}c_{n+1} \qquad \text{for } n \geqslant 0$$

or, replacing n with $n - 1$,

$$c_{n+2} = \frac{2n - \lambda}{(n+1)(n+2)} c_n \tag{2}$$

The ratio test gives

$$\lim_{n\to\infty} \frac{c_{n+2}x^{n+2}}{c_n x^n} = \lim_{n\to\infty} \frac{2n - \lambda}{(n+1)(n+2)} x^2 = 0 \qquad \forall\ x$$

Thus, the infinite series whose coefficients obey the recursive relation in (2) converges for all x. However, on physical grounds [the demand that $\lim_{x\to\infty} u(x) = 0$; see Exercise 9.3.2], the series must be truncated. This happens only if $\lambda = 2l$ for some integer l, and in that case we obtain a polynomial, the Hermite polynomial of order l. A consequence of such a truncation is the quantization of harmonic oscillator energy. This can easily be seen:

$$2l = \frac{2E}{\hbar\omega} - 1 \quad \Rightarrow \quad E = (l + \tfrac{1}{2})\hbar\omega$$

Two solutions are generated from Eq. (2), one including only even powers and the other only odd powers. These are clearly linearly independent. Thus, knowledge of c_0 and c_1 determines the general solution of the HSOLDE in (1). ●

The preceding two examples show how certain special functions used in mathematical physics are obtained in an analytic way, by solving a differential equation. We saw in Chapter 8 how to obtain Legendre polynomials by algebraic methods (see the discussion of spherical harmonics). It is instructive to solve the harmonic oscillator problem using algebraic methods, as the following example demonstrates.

Example 9.3.5

The Hamiltonian of a harmonic oscillator is

$$\mathbb{H} = \frac{\mathbb{p}^2}{2m} + \frac{1}{2} m\omega^2 x^2$$

where $\mathbb{p} = -i\hbar d/dx$ is the momentum operator. Let us find the eigenvectors and eigenvalues of $\mathbb{H}$.

We define the operator

$$\mathbb{a} \equiv \sqrt{\frac{m\omega}{2\hbar}}\, x + i \frac{\mathbb{p}}{\sqrt{2m\hbar\omega}}$$

and its adjoint (note that x and $\mathbb{p}$ are hermitian)

$$\mathbb{a}^\dagger \equiv \sqrt{\frac{m\omega}{2\hbar}}\, x - i \frac{\mathbb{p}}{\sqrt{2m\hbar\omega}}$$

Using the commutation relation

$$[x, \mathbb{p}] = i\hbar\mathbb{1}$$

we can show that

$$[\mathbb{a}, \mathbb{a}^\dagger] = \mathbb{1}$$

and that

$$\mathbb{H} = \hbar\omega \mathbb{a}^\dagger \mathbb{a} + \tfrac{1}{2}\hbar\omega\mathbb{1} \tag{1}$$

Furthermore,[2]

$$[\mathbb{H}, \mathbb{a}] = \hbar\omega[\mathbb{a}^\dagger\mathbb{a} + \tfrac{1}{2}, \mathbb{a}] = \hbar\omega[\mathbb{a}^\dagger, \mathbb{a}]\mathbb{a} = -\hbar\omega\mathbb{a} \tag{2a}$$

Similarly,

$$[\mathbb{H}, \mathbb{a}^\dagger] = \hbar\omega\mathbb{a}^\dagger \tag{2b}$$

Let $|u_E\rangle$ be the eigenvector corresponding to the eigenvalue E:

$$\mathbb{H}|u_E\rangle = E|u_E\rangle$$

Equations (2a) and (2b) give

$$\mathbb{H}\mathbb{a}|u_E\rangle = (\mathbb{a}\mathbb{H} - \hbar\omega\mathbb{a})|u_E\rangle = (E - \hbar\omega)\mathbb{a}|u_E\rangle$$

and, similarly,

$$\mathbb{H}\mathbb{a}^\dagger|u_E\rangle = (E + \hbar\omega)\mathbb{a}^\dagger|u_E\rangle$$

Thus, $\mathbb{a}|u_E\rangle$ is an eigenvector of $\mathbb{H}$, with eigenvalue $E - \hbar\omega$, and $\mathbb{a}^\dagger|u_E\rangle$ is another, with eigenvalue $E + \hbar\omega$. That is why $\mathbb{a}^\dagger$ and $\mathbb{a}$ are called the *raising* and *lowering* (or creation and annihilation) *operators*, respectively. We can write

$$\mathbb{a}|u_E\rangle = c_E|u_{E - \hbar\omega}\rangle$$

By applying $\mathbb{a}$ repeatedly, we obtain states of lower and lower energies. But there is a limit to this—because $\mathbb{H}$ is a positive operator, it cannot have a negative eigenvalue. Thus, there must exist a *ground state*, $|u_0\rangle$, such that

$$\mathbb{a}|u_0\rangle = 0$$

The energy of this ground state (or the eigenvalue corresponding to $|u_0\rangle$) can be obtained:

$$\mathbb{H}|u_0\rangle = (\hbar\omega\mathbb{a}^\dagger\mathbb{a} + \tfrac{1}{2}\hbar\omega)|u_0\rangle = \tfrac{1}{2}\hbar\omega|u_0\rangle$$

Repeated application of the raising operator yields both higher-level states and eigenvalues. We thus define $|u_n\rangle$ by

$$(\mathbb{a}^\dagger)^n|u_0\rangle = c_n|u_n\rangle \tag{3}$$

where c_n is a normalizing constant. The energy associated with $|u_n\rangle$ is

$$E_n = (n + \tfrac{1}{2})\hbar\omega$$

which is what we obtained in Example 9.3.4.

To find c_n, we demand orthonormality for the $|u_n\rangle$. Taking the inner product of (3) with itself, we can show (see Problem 9.24) that

$$|c_n|^2 = n|c_{n-1}|^2 \quad \Rightarrow \quad |c_n|^2 = n!|c_0|^2$$

[2] From here on, the unit operator $\mathbb{1}$ will not be shown explicitly.

which for $|c_0| = 1$ and real c_n yields

$$c_n = \sqrt{n!}$$

It follows, then, that

$$|\mathrm{u}_n\rangle = \frac{1}{\sqrt{n!}}(\mathbb{a}^\dagger)^n|\mathrm{u}_0\rangle \tag{4}$$

On the other hand, in terms of functions and derivative operators, $\mathbb{a}|\mathrm{u}_0\rangle = 0$ gives

$$\left(\sqrt{\frac{m\omega}{2\hbar}}\,x + \sqrt{\frac{\hbar}{2m\omega}}\frac{d}{dx}\right)u_0(x) = 0$$

which has the solution

$$u_0(x) = c\exp\left(-\frac{m\omega x^2}{2\hbar}\right)$$

Normalizing $u_0(x)$, gives

$$1 = \langle \mathrm{u}_0|\mathrm{u}_0\rangle = c^2\int_{-\infty}^{\infty}[u_0(x)]^2\,dx = c^2\left(\frac{\hbar\pi}{m\omega}\right)^{1/2}$$

Thus,

$$u_0(x) = \left(\frac{m\omega}{\hbar\pi}\right)^{1/4} e^{-(m\omega/2\hbar)x^2}$$

We can now write (4) in terms of differential operators:

$$u_n(x) = \frac{1}{\sqrt{n!}}\left(\frac{m\omega}{\hbar\pi}\right)^{1/4}\left(\sqrt{\frac{m\omega}{2\hbar}}\,x - \sqrt{\frac{\hbar}{2m\omega}}\frac{d}{dx}\right)^n e^{-(m\omega/2\hbar)x^2}$$

Defining a new variable, $y = \sqrt{m\omega/\hbar}\,x$, transforms this equation into

$$u_n\left(\sqrt{\frac{\hbar}{m\omega}}\,y\right) = \left(\frac{m\omega}{\hbar\pi}\right)^{1/4}\frac{1}{2^{n/2}\sqrt{n!}}\left(y - \frac{d}{dy}\right)^n e^{-y^2/2}$$

From this and the definition of Hermite polynomials given in Example 9.3.4, a general formula for $H_n(x)$ can be obtained. In particular, if we note that (see Problem 9.24)

$$e^{y^2/2}\left(y - \frac{d}{dy}\right)e^{-y^2/2} = -e^{y^2}\frac{d}{dy}e^{-y^2}$$

and, in general,

$$e^{y^2/2}\left(y - \frac{d}{dy}\right)^n e^{-y^2/2} = (-1)^n e^{y^2}\frac{d^n}{dy^n}e^{-y^2}$$

we recover the generalized Rodriguez formula of Chapter 5. •

Other differential equations, such as the Bessel and Laguerre equations, are important in physics. We will study them later, using complex analytical methods.

The following important theorem ends this section (for a proof, see Birkhoff and Rota 1978, p. 95).

9.3.2 Theorem For any choice of c_0 and c_1 the radius of convergence of any power-series solution of the form of (9.24) for the normal HSOLDE $y'' + p(x)y' + q(x)y = 0$ and defined by the recursion relation of (9.25) is at least as large as the smaller of the two radii of convergence of the two series in (9.23). ■

In particular, if $p(x)$ and $q(x)$ are analytic in an interval around $x = 0$, then the solution of the normal HSOLDE is also analytic in a neighborhood of $x = 0$.

Exercises

9.3.1 The Laplace equation in electrostatics is $\nabla^2\Phi = 0$. When it is separated in spherical coordinates, its radial part leads to

$$\frac{d}{dx}\left[x^2\frac{dy}{dx}\right] - n(n+1)y = 0 \qquad \text{for } n \geqslant 0$$

Use the method of undetermined coefficients to find the two independent solutions of this ODE. (You have to include negative powers of x as well as positive powers!)

9.3.2 Show that the function defined by

$$f(x) = \sum_{n=0}^{\infty} c_n x^n, \; c_{n+2} = \frac{2n - \lambda}{(n+1)(n+2)} c_n$$

goes to infinity at least as fast as e^{x^2} does. In other words, show that $\lim_{x\to\infty} f(x)e^{-x^2} \neq 0$. (Hint: Consider only even powers of x.)

9.4 LINEAR DIFFERENTIAL EQUATIONS WITH CONSTANT COEFFICIENTS

The most general nth-order linear differential equation (NOLDE) with constant coefficients can be written as

$$\mathbb{L}[y] \equiv y^{(n)} + a_{n-1}y^{(n-1)} + \cdots + a_1 y' + a_0 y = r(x) \tag{9.26a}$$

The corresponding homogeneous NOLDE (HNOLDE) is obtained by setting $r(x) = 0$. Let us consider such a homogeneous case first.

9.4.1 The Homogeneous Case

The solution to the HNOLDE

$$\mathbb{L}[y] = y^{(n)} + a_{n-1}y^{(n-1)} + \cdots + a_1 y' + a_0 y = 0 \tag{9.26b}$$

can be found by making the *exponential substitution* $y = e^{\lambda x}$, which results in the equation

$$\mathbb{L}[e^{\lambda x}] = (\lambda^n + a_{n-1}\lambda^{n-1} + \cdots + a_1\lambda + a_0)e^{\lambda x} = 0$$

This is satisfied only if λ is a root of the *characteristic polynomial*

$$p(\lambda) \equiv \lambda^n + a_{n-1}\lambda^{n-1} + \cdots + a_1\lambda + a_0$$

which by the fundamental theorem of algebra (see Chapter 7) can be written as

$$p(\lambda) = (\lambda - \lambda_1)^{k_1}(\lambda - \lambda_2)^{k_2} \cdots (\lambda - \lambda_m)^{k_m} \tag{9.27}$$

The λ_j are the distinct roots of $p(\lambda)$ and have multiplicity k_j.

It is convenient to introduce $\mathbb{D} \equiv d/dx$ and define the operator

$$\mathbb{L} \equiv p(\mathbb{D}) = \mathbb{D}^n + a_{n-1}\mathbb{D}^{n-1} + \cdots + a_1\mathbb{D} + a_0$$

Since $\mathbb{D} - \mu$ and $\mathbb{D} - \lambda$ commute for arbitrary constants μ and λ, we can unambiguously factor out the above, obtaining

$$\mathbb{L} = p(\mathbb{D}) = (\mathbb{D} - \lambda_1)^{k_1}(\mathbb{D} - \lambda_2)^{k_2} \cdots (\mathbb{D} - \lambda_m)^{k_m} \tag{9.28}$$

In preparation for finding the most general solution for Eq. (9.26b), we note that, for any integer $r > 0$,

$$(\mathbb{D} - \lambda)(x^r e^{\lambda x}) = r x^{r-1} e^{\lambda x}$$

and, in general,

$$(\mathbb{D} - \lambda)^k(x^r e^{\lambda x}) = r(r-1)\cdots(r-k+1)x^{r-k}e^{\lambda x}$$

In particular,

$$(\mathbb{D} - \lambda)^k(x^r e^{\lambda x}) = 0 \qquad \text{if } k > r \tag{9.29}$$

Thus, the set of functions $\{x^r e^{\lambda_j x}\}_{r=0}^{k_j-1}$ are all solutions of (9.26b).

The roots λ_j are generally complex. If the coefficients $\{a_i\}_{i=0}^{n-1}$ are *real*, then whenever $x^r e^{\lambda x}$ is a solution, so is $x^r e^{\lambda^* x}$. Thus, writing $\lambda_j = \alpha_j + i\beta_j$ and using the linearity of $\mathbb{L}$, we conclude that

$$x^r e^{\alpha_j x}\cos\beta_j x \qquad \text{and} \qquad x^r e^{\alpha_j x}\sin\beta_j x \qquad \text{where } r = 0, 1, \ldots, k_j - 1$$

are all solutions of (9.26b).

It is easily proved that the functions $x^r e^{\lambda_j x}$ are linearly independent (see Exercise 9.4.1). On the other hand, Eq. (9.27) indicates that the set $\{x^r e^{\lambda_j x}\}$, where $r = 0, 1, \ldots, k_j - 1$ and $j = 1, 2, \ldots, m$, contains exactly n elements. We have thus shown that there are at least n linearly independent solutions for the HNOLDE of (9.26b). In fact, it can be shown (see Birkhoff and Rota 1978, p. 65) that there are exactly n linearly independent solutions.

9.4.1 Theorem Let $\{\lambda_j\}_{j=1}^m$ be the roots of the characteristic polynomial of the real HNOLDE of Eq. (9.26b), and let the respective roots have multiplicities $\{k_j\}_{j=1}^m$. Then the functions $x^r e^{\lambda_j x}$, where $r = 0, 1, \ldots, k_j - 1$, are a basis of solutions of (9.26b). ■

Example 9.4.1

The three-dimensional motion of a particle under the influence of a central force can be reduced to a one-dimensional problem as follows. First, the xy-plane is taken to be the plane formed by the initial velocity of the particle and the line joining it to the center of force, which is taken to be the origin. Since there is no force acting out of that plane, the problem has been reduced to two dimensions. Next, the equations of motion are written in polar coordinates (see Chapter 1):

$$m\frac{d^2r}{dt^2} - mr\left(\frac{d\theta}{dt}\right)^2 = F(r)$$

$$mr\frac{d^2\theta}{dt^2} + 2m\left(\frac{dr}{dt}\right)\left(\frac{d\theta}{dt}\right) = 0$$

The second equation reduces to (the dot means the derivative with respect to time)

$$\frac{d}{dt}(mr^2\dot{\theta}) = 0$$

This is conservation of angular momentum, $L = mr^2\dot{\theta}$. Substituting back in the first equation yields

$$m\ddot{r} - \frac{L^2}{mr^3} = F(r) \qquad (1)$$

Let $u \equiv 1/r$. Then

$$r = \frac{1}{u} \quad \Rightarrow \quad \dot{r} = -\frac{1}{u^2}\frac{du}{dt} = -\frac{1}{u^2}\frac{du}{d\theta}\dot{\theta} = -r^2\dot{\theta}\frac{du}{d\theta} = -\frac{L}{m}\frac{du}{d\theta}$$

Similarly, $\ddot{r} = -(L^2u^2/m^2)d^2u/d\theta^2$. Substituting in (1) gives

$$\frac{d^2u}{d\theta^2} + u = -\frac{m}{L^2u^2}F\left(\frac{1}{u}\right)$$

For the Kepler problem this equation is easy to solve because

$$F(r) = \frac{GMm}{r^2} \quad \Rightarrow \quad F\left(\frac{1}{u}\right) = GMmu^2$$

and we have

$$\frac{d^2u}{d\theta^2} + u = -\frac{GMm^2}{L^2} \qquad (2)$$

Let $v = u + GMm^2/L^2$. Then Eq. (2) becomes

$$\frac{d^2v}{d\theta^2} + v = 0$$

The characteristic polynomial is $\lambda^2 + 1$, whose roots are $\lambda = \pm i$. These simple roots give rise to the linearly independent solutions $v = \sin\theta$ and $v = \cos\theta$. The general solution can therefore be expressed as

$$v = c_1 \sin\theta + c_2 \cos\theta$$

or

$$v = A\cos(\theta - \theta_0) \equiv u + \frac{GMm^2}{L^2}$$

or, finally,

$$\frac{1}{r} = A\cos(\theta - \theta_0) - \frac{GMm^2}{L^2}$$

This equation of a conic section embraces Kepler's three laws. •

Example 9.4.2

An equation that is used in both mechanics and circuit theory is

$$\frac{d^2y}{dt^2} + a\frac{dy}{dt} + by = 0 \qquad \text{for } a, b > 0 \tag{1}$$

Its characteristic polynomial is

$$p(\lambda) = \lambda^2 + a\lambda + b$$

which has the roots

$$\lambda_1 = \tfrac{1}{2}(-a + \sqrt{a^2 - 4b}) \qquad \text{and} \qquad \lambda_2 = \tfrac{1}{2}(-a - \sqrt{a^2 - 4b})$$

We can distinguish three cases, depending on the relative sizes of a and b.

(a) $a^2 > 4b$ (overdamped)

Let $\gamma \equiv \frac{1}{2}\sqrt{a^2 - 4b}$. Then the most general solution is

$$y(t) = e^{-at/2}(c_1 e^{\gamma t} + c_2 e^{-\gamma t})$$

Since $a > 2\gamma$, this solution starts at $y = c_1 + c_2$ at $t = 0$ and continuously decreases; so, as $t \to \infty$, $y(t) \to 0$.

(b) $a^2 = 4b$ (critically damped)

In this case we have a multiple root. Thus, the general solution is

$$y(t) = c_1 t e^{-at/2} + c_2 e^{-at/2}$$

This solution starts at $y(0) = c_2$ at $t = 0$, reaches a maximum (or minimum) at $t = 2/a - c_2/c_1$, and subsequently decays exponentially to zero.

(c) $a^2 < 4b$ (damped oscillatory)

Let $\omega \equiv \frac{1}{2}\sqrt{4b - a^2}$. Then $\lambda_i = -a/2 + i\omega$ and $\lambda_2 = \lambda_1^*$. The roots are complex, and the most general solution is thus of the form

$$\begin{aligned} y(t) &= e^{-at/2}(c_1 \cos \omega t + c_2 \sin \omega t) \\ &\equiv Ae^{-at/2}\cos(\omega t + \alpha) \end{aligned}$$

Thus, the solution is a harmonic motion with the *decaying* amplitude $A\exp(-at/2)$. Note that if $a = 0$, the amplitude does not decay. That is why a is called the *damping factor* (or the damping constant).

These equations describe either a mechanical system oscillating, with no external force, in a viscous (dissipative) fluid, or an electrical circuit consisting of a resistance R, an inductance L, and a capacitance C. For mechanical oscillators, $a = r/m$ and $b = k/m$, where r is the dissipative constant related to the dissipative force and the velocity by $f = rv$, and k is simply the spring constant (a measure of the stiffness of the spring). For RLC circuits, $a = R/L$ and $b = 1/LC$. Thus, the damping factor depends on the relative magnitudes of R and L. On the other hand, the frequency $\omega = \sqrt{b - (a/2)^2} = \sqrt{1/LC - R^2/4L^2}$ depends on all three elements. In particular, for $R \geqslant 4L/C$ the circuit does not oscillate. ●

9.4.2 The Inhomogeneous Case and the Transfer Function

An interesting application of NOLDEs takes place when there is a driving force acting on a physical system. This driving force is simply the inhomogeneous term of the NOLDE. The best way to solve such an INOLDE in its most general form is by using Fourier transforms and Green's functions, as we will do in Chapter 11. For the particular, but important, case in which the inhomogeneous term is a product of polynomials and exponentials, the solution can be found in closed form. This subsection shows how this is done.

We assume that the inhomogeneous term in Eq. (9.26a) is of the form

$$r(x) = \sum_k p_k(x)e^{\lambda_k x}$$

where $p_k(x)$ are polynomials and λ_k are (complex) constants. The most general solution of (9.26a) is a linear combination of a basis of solutions (as given in Theorem 9.4.1) and a *particular* solution of the NOLDE. We need to find the latter. Because $\mathbb{L}$ is a linear operator, it is clear that if y_1 is a particular solution of $\mathbb{L}[y] = r_1(x)$ and y_2 of $\mathbb{L}[y] = r_2(x)$, then $y_1 + y_2$ is a solution of $\mathbb{L}[y] = r_1(x) + r_2(x)$. Thus, no generality is lost if we restrict $r(x)$ to

$$r(x) = p(x)e^{\lambda x}$$

where $p(x)$ is a polynomial.

It is easy to verify that, for any $f \in \mathrm{C}^{(1)}[a, b]$, we have

$$(\mathbb{D} - \lambda)[e^{\lambda x}f(x)] = e^{\lambda x}f'(x)$$

In particular, if $p(x)$ is a polynomial of degree k, then

$$(\mathbb{D} - \lambda)(u) = e^{\lambda x}p(x)$$

has a solution of the form $u = e^{\lambda x}q(x)$, where $q(x)$ is a polynomial of degree $k + 1$ that is the primitive of $p(x)$. It is equally easy to show that if $\lambda_1 \neq \lambda$, then

$$(\mathbb{D} - \lambda_1)[e^{\lambda x}f(x)] = e^{\lambda x}[(\lambda - \lambda_1)f(x) + f'(x)] \tag{9.30}$$

and, therefore,

$$(\mathbb{D} - \lambda_1)(u) = e^{\lambda x}p(x) \tag{9.31}$$

has a solution of the form $u = e^{\lambda x}q(x)$, where $q(x)$ is a polynomial of degree k.

Applying Eqs. (9.30) and (9.31) repeatedly leads to a theorem.

9.4.2 Theorem The NOLDE $\mathbb{L}[y] = e^{\lambda x} S(x)$, where $S(x)$ is a polynomial, has the particular solution $e^{\lambda x} q(x)$, where $q(x)$ is also a polynomial. The degree of $q(x)$ equals that of $S(x)$ unless $\lambda = \lambda_j$, a root of the characteristic polynomial of $\mathbb{L}$ [Eq. (9.27)]. If $\lambda = \lambda_j$ has multiplicity k_j, then the degree of $q(x)$ exceeds that of $S(x)$ by k_j. ■

Once we know the form of the particular solution of the NOLDE, we can find the coefficients in the polynomial of the solution by substituting in the NOLDE and matching the powers on both sides.

Example 9.4.3

Let us find the most general solutions for two differential equations subject to the boundary conditions $y(0) = 0$ and $y'(0) = 1$.

(a)
$$y'' + y = xe^x \tag{1}$$

The characteristic polynomial is $\lambda^2 + 1$, whose roots are $\lambda_1 = i$ and $\lambda_2 = -i$. Thus, a basis of solutions is $\{\cos x, \sin x\}$. To find the particular solution we note that λ (the coefficient of x in the exponential part of the inhomogeneous term) is 1, which is not either of the roots λ_1 and λ_2. Thus, the particular solution is of the form $q(x)e^x$, where $q(x) = Ax + B$ is of degree 1 [same degree as that of $S(x) = x$]. We now substitute $u = (Ax + B)e^x$ in (1) and find A and B:

$$u' = Ae^x + (Ax + B)e^x$$

$$u'' = 2Ae^x + (Ax + B)e^x$$

Substituting in (1) gives

$$Axe^x + (2A + B)e^x + (Ax + B)e^x = xe^x$$

Matching the coefficients, we have

$$2A = 1 \qquad \text{and} \qquad 2A + 2B = 0 \quad \Rightarrow \quad A = \tfrac{1}{2} = -B$$

Thus, the most general solution is

$$y = c_1 \cos x + c_2 \sin x + \tfrac{1}{2}(x - 1)e^x$$

Imposing the given boundary conditions yields

$$0 = y(0) = c_1 - \tfrac{1}{2} \quad \Rightarrow \quad c_1 = \tfrac{1}{2}$$

$$1 = y'(0) = c_2$$

Thus,
$$y = \tfrac{1}{2}\cos x + \sin x + \tfrac{1}{2}(x - 1)e^x$$

is the unique solution.

(b)
$$y'' - y = xe^x \tag{2}$$

Here $p(\lambda) = \lambda^2 - 1$, and the roots are $\lambda_1 = 1$ and $\lambda_2 = -1$. A basis of solutions is $\{e^x, e^{-x}\}$. To find the particular solution, we note that $S(x) = x$ and $\lambda = 1 = \lambda_1$. Theorem 9.4.2 then implies that $q(x)$ is of degree 2 (because λ_1 is a simple root). We, therefore, try

$q(x) = Ax^2 + Bx + C$, which leads to the particular solution $u = (Ax^2 + Bx + C)e^x$. Taking the derivatives and substituting in (2) yields two equations,

$$4A = 1 \qquad \text{and} \qquad A + B = 0$$

whose solution is $A = -B = \frac{1}{4}$. Note that C is not determined, because Ce^x is a solution of the homogeneous DE corresponding to (2), and when $\mathbb{L}$ is applied to u, it eliminates the term Ce^x. Another way of looking at the situation is to note that the most general solution to (2) is of the form

$$y = C_1 e^x + C_2 e^{-x} + (\tfrac{1}{4}x^2 - \tfrac{1}{4}x + C)e^x$$

The term Ce^x could be absorbed in $C_1 e^x$. We, therefore, set $C = 0$, apply the boundary conditions, and find the unique solution

$$y = \tfrac{5}{4}\sinh x + \tfrac{1}{4}(x^2 - x)e^x$$ •

The inhomogeneous differential equation (IDE) $\mathbb{L}[y] = r(x)$ can be thought of as a machine (or a black box) that produces a function $y(x)$ when a function $r(x)$ is fed into it. In this context, it is more natural to look at the "inverse" of $\mathbb{L}$, which is an operator $\mathbb{M}$ such that $\mathbb{M}[r] = u$. However, in general, $\mathbb{M}$ does not exist because it is possible to have different functions $u(x)$ for a given $r(x)$. We have encountered such a situation, where certain terms of the particular solution could not be determined uniquely because they were solutions of the HDE. However, with certain restrictions (to be discussed later) we can make $u(x)$ unique to a given $r(x)$.

The interpretation of the IDE as a machine is common in the study of electrical or acoustic filters. A signal, the function $r(x)$, is sent into the filter, and a second function, $u(x)$, is received as an output. In such a context, by far the most important input signal is a sinusoidal function of the general form

$$r(t) = A\cos(\omega t + \alpha)$$

or, in complex notation, with $A \equiv |b|$ and $\arg(b) \equiv \alpha$,

$$r(t) = \operatorname{Re}(be^{i\omega t})$$

where ω is a constant (the angular frequency), t represents time (the independent variable), and A, b, and α are constants. Assuming that $i\omega$ is not a root of $p(\lambda)$, the characteristic polynomial of $\mathbb{L}$, Theorem 9.4.2 suggests a particular solution, $U = C(\omega)e^{i\omega t}$, where $C(\omega)$ is a (ω-dependent) constant that can be determined by substituting in $\mathbb{L}[U] = be^{i\omega t}$:

$$\mathbb{L}[C(\omega)e^{i\omega t}] = be^{i\omega t} \quad \Rightarrow \quad C(\omega) = \frac{b}{p(i\omega)}$$

Writing $C(\omega) = \rho(\omega)e^{i\gamma(\omega)}$, $b = Ae^{i\alpha}$, and $p(i\omega) = R(\omega)e^{i\theta(\omega)}$, we obtain

$$\rho(\omega) = \frac{A}{R(\omega)} \qquad \text{and} \qquad \gamma(\omega) = \alpha - \theta(\omega)$$

The real solution, $u(t) = \text{Re}[U(t)]$, will then be

$$u(t) = \text{Re}[C(\omega)e^{i\omega t}] = \rho(\omega)\cos[\omega t + \gamma(\omega)]$$
$$= \frac{A}{R(\omega)}\cos[\omega t + \alpha - \theta(\omega)] \tag{9.32}$$

The function $C(\omega)$ is called the *transfer function* associated with the linear operator $\mathbb{L}$. Equation (9.32) shows that the output, $u(t)$, has the same frequency as the input. It also indicates that the amplitude of $u(t)$ is frequency-dependent, making it possible to obtain large output amplitudes by varying the frequency until $R(\omega)$ is minimum. This is the phenomenon of resonance in AC circuits.

Example 9.4.4

Let us consider Eq. (1) of Example 9.4.2 and, for definiteness, take the case of damped oscillation. In this case, $4b > a^2$ and

$$\omega_1 \equiv \sqrt{b - \left(\frac{a}{2}\right)^2} \equiv \omega_0\sqrt{1 - \frac{a^2}{4\omega_0^2}}$$

where $\omega_0 \equiv \sqrt{b}$ is called the *natural frequency* of the system.

The characteristic polynomial is $p(\lambda) = \lambda^2 + a\lambda + b$. Thus,

$$p(i\omega) = -\omega^2 + i\omega a + b = (\omega_0^2 - \omega^2) + i\omega a$$

and

$$R(\omega) = \sqrt{(\omega_0^2 - \omega^2)^2 + \omega^2 a^2}$$

$$\theta(\omega) = \tan^{-1}\left(\frac{\omega a}{\omega_0^2 - \omega^2}\right)$$

The amplitude of the output signal, sometimes called the *gain function*, is

$$\rho(\omega) = \frac{A}{R(\omega)} = \frac{A}{\sqrt{(\omega_0^2 - \omega^2)^2 + \omega^2 a^2}}$$

The minimum of the denominator occurs at $\omega = \omega_0$, that is, when the driving frequency equals the natural frequency. In such a situation we have

$$\rho(\omega) = \frac{A}{\omega_0 a}$$

which shows that the output signal will have a large amplitude when a is small. In the limit $a \to 0$ the amplitude becomes infinite, which is bothersome. However, we note that if $a = 0$, then, at resonance, $i\omega_0$ will be a root of the characteristic polynomial $p(\lambda) = \lambda^2 + \omega_0^2$. This contradicts the assumption that $i\omega$ is not a root of $p(\lambda)$.

We have considered only the particular solution, $u(t)$, because the most general solution

$$y(t) = Be^{-at/2}\cos(\omega_1 t + \beta) + u(t)$$

in which B and β are constants, eventually reduces to $u(t)$. The first term on the RHS, the *transient term*, decays to zero. The rate of this decay is determined by the *time constant*

$2/a$, a measure of the length of the interval during which the amplitude of the transient term drops to $1/e$ of its initial value. ●

Example 9.4.4 shows (at least for the special case of an ISOLDE) that when there is no dissipative force (when $a = 0$), the gain function $\rho(\omega)$ will be infinite for a particular frequency. This can be seen for the general case by noting that if $i\omega$ is a root of the characteristic polynomial, say a simple root for convenience, then Theorem 9.4.2 suggests a solution of the form $u(t) = (a_1 t + a_2)e^{i\omega t}$. This clearly shows that the amplitude, $a_1 t + a_2$, "blows up" as time progresses.

The importance of the sinusoidal signal becomes apparent when we realize that any periodic signal $r(t)$ can be expanded in a Fourier series,

$$R(t) = \sum_{n=-\infty}^{\infty} b_n e^{in\omega t}$$

where ω is the fundamental frequency and $r(t) \equiv \mathrm{Re}[R(t)]$. The linearity of $\mathbb{L}$ suggests the solution $u(t) = \mathrm{Re}[U(t)]$, where

$$U(t) = \sum_{n=-\infty}^{\infty} C_n(\omega) e^{in\omega t}$$

Substituting in $\mathbb{L}[U] = R(t)$ gives

$$\sum_{n=-\infty}^{\infty} C_n(\omega) p(in\omega) e^{in\omega t} = \sum_{n=-\infty}^{\infty} b_n e^{in\omega t}$$

Since the $e^{in\omega t}$ are orthonormal, we get

$$C_n(\omega) = \frac{b_n}{p(in\omega)}$$

and

$$u(t) = \mathrm{Re}\left[\sum_{n=-\infty}^{\infty} \frac{b_n e^{in\omega t}}{p(i\omega n)}\right] \tag{9.33}$$

Thus, $u(t)$ is also periodic and has the same fundamental frequency as $r(t)$.

9.4.3 Linear Systems with Constant Coefficients

Systems of FODEs in more than one variable, such as

$$\begin{aligned}
\frac{dx_1}{dt} &= F_1(x_1, x_2, \ldots, x_n; t) \\
\frac{dx_2}{dt} &= F_2(x_1, x_2, \ldots, x_n; t) \\
&\vdots \\
\frac{dx_n}{dt} &= F_n(x_1, x_2, \ldots, x_n; t)
\end{aligned} \tag{9.34}$$

are important because *any* normal differential equation of nth order,

$$\frac{d^n y}{dt^n} = F(y, y', \ldots, y^{(n-1)}; t)$$

can be reduced to a system of n first-order differential equations as given by (9.34) by defining $x_1 = y$, $x_2 = y'$, . . . , $x_n = y^{(n-1)}$. This gives the system

$$\begin{aligned} \frac{dx_1}{dt} &= x_2 \\ \frac{dx_2}{dt} &= x_3 \\ &\vdots \\ \frac{dx_{n-1}}{dt} &= x_n \\ \frac{dx_n}{dt} &= F(x_1, x_2, \ldots, x_n; t) \end{aligned} \tag{9.35}$$

We will not undertake a general study of (9.34) or (9.35), because that would carry us too far into the formal theory of differential equations. We will, however, discuss the simplest version of (9.34), in which all the F_i are linear functions of x_j with constant coefficients.

Let us find a solution for the system

$$\begin{aligned} \frac{dx_1}{dt} &= a_{11}x_1 + a_{12}x_2 + \cdots + a_{1n}x_n + b_1(t) \\ \frac{dx_2}{dt} &= a_{21}x_1 + a_{22}x_2 + \cdots + a_{2n}x_n + b_2(t) \\ &\vdots \\ \frac{dx_n}{dt} &= a_{n1}x_1 + a_{n2}x_2 + \cdots + a_{nn}x_n + b_n(t) \end{aligned} \tag{9.36}$$

We can reexpress this system of equations in matrix form. With

$$\mathbf{X}(t) = \begin{pmatrix} x_1(t) \\ x_2(t) \\ \vdots \\ x_n(t) \end{pmatrix} \qquad \mathsf{A} = \begin{pmatrix} a_{11} & a_{12} & \cdots & a_{1n} \\ a_{21} & a_{22} & \cdots & a_{2n} \\ \vdots & \vdots & & \vdots \\ a_{n1} & a_{n2} & \cdots & a_{nn} \end{pmatrix} \qquad \text{and} \qquad \mathbf{B}(t) = \begin{pmatrix} b_1(t) \\ b_2(t) \\ \vdots \\ b_n(t) \end{pmatrix}$$

we can write (9.36) as

$$\frac{d\mathbf{X}}{dt} = \mathsf{A}\mathbf{X} + \mathbf{B} \tag{9.37}$$

Let us first consider the homogeneous case:

$$\frac{d\mathbf{X}}{dt} = \mathsf{A}\mathbf{X} \tag{9.38}$$

To find a solution, let $x_i(t) = c_i e^{\lambda t}$, or in matrix notation

$$\mathbf{X}(t) = \mathbf{C}e^{\lambda t} \qquad \text{where } \mathbf{C} \equiv \begin{pmatrix} c_1 \\ c_2 \\ \vdots \\ c_n \end{pmatrix}$$

Substituting in (9.38) gives

$$\lambda \mathbf{C} e^{\lambda t} = \mathsf{A}\mathbf{C}e^{\lambda t}$$

or

$$(\mathsf{A} - \lambda \mathsf{1})\mathbf{C} = 0 \tag{9.39}$$

We have thus reduced the problem to the standard one of finding eigenvalues and eigenvectors for matrices, a subject discussed at great length in Chapter 3. It follows that if λ_i is any eigenvalue of the matrix A and $\mathbf{C}_i$ is its corresponding eigenvector, then

$$\mathbf{X}_i(t) \equiv \mathbf{C}_i e^{\lambda_i t} \equiv \begin{pmatrix} c_{i1} \\ c_{i2} \\ \vdots \\ c_{in} \end{pmatrix} e^{\lambda_i t} \tag{9.40}$$

is a solution of (9.38).

These solutions are linearly independent if and only if

$$\det[\mathbf{X}_1(t), \ldots, \mathbf{X}_n(t)] \equiv \det \begin{pmatrix} c_{11}e^{\lambda_1 t} & c_{21}e^{\lambda_2 t} & \cdots & c_{n1}e^{\lambda_n t} \\ c_{12}e^{\lambda_1 t} & c_{22}e^{\lambda_2 t} & \cdots & c_{n_2}e^{\lambda_n t} \\ \vdots & \vdots & & \vdots \\ c_{1n}e^{\lambda_1 t} & c_{2n}e^{\lambda_2 t} & \cdots & c_{nn}e^{\lambda_n t} \end{pmatrix}$$

does not vanish. As long as $\lambda_1, \lambda_2, \ldots, \lambda_n$ are all distinct, this will be true. Thus, if the λ_i are all distinct, the most general solution for (9.38) is

$$\mathbf{X}(t) = \sum_{i=1}^{n} \mathbf{C}_i \mathbf{e}^{\lambda_i t} = \sum_{i=1}^{n} \begin{pmatrix} c_{i1} \\ c_{i2} \\ \vdots \\ c_{in} \end{pmatrix} e^{\lambda_i t}$$

The following theorem covers the case of nonsimple eigenvalues (for a proof, see Rubenstein 1969, pp. 82–86).

9.4.3 Theorem Let the $n \times n$ matrix A have λ_0 as an eigenvalue of multiplicity r. Then there exist r linearly independent solutions of $d\mathbf{X}/dt = \mathrm{A}\mathbf{X}$ corresponding to λ_0. These solutions are of the form

$$e^{\lambda_0 x}(\mathbf{V}_0 + x\mathbf{V}_1 + \cdots + x^k\mathbf{V}_k)$$

where $k < r$ and the $\mathbf{V}_i$ are constant *vectors*. ■

Example 9.4.5

Consider the linear system

$$\begin{aligned} \dot{x}_1 &= x_2 \\ \dot{x}_2 &= x_3 \\ \dot{x}_3 &= -2x_1 - 5x_2 - 4x_3 \end{aligned} \tag{1}$$

where the dots on the x_i indicate differentiation with respect to time (t). In matrix form (1) becomes

$$\dot{\mathbf{X}} = \mathrm{A}\mathbf{X} \qquad \text{where } \mathrm{A} = \begin{pmatrix} 0 & 1 & 0 \\ 0 & 0 & 1 \\ -2 & -5 & -4 \end{pmatrix} \tag{2}$$

The eigenvalues of A are easily found to be $\lambda_1 = -1$ with multiplicity 2 and $\lambda_2 = -2$ with multiplicity 1.

The eigenvector for λ_2 is

$$a\begin{pmatrix} 1 \\ -2 \\ 4 \end{pmatrix} \tag{3}$$

where a is an arbitrary constant.

For λ_1 we use Theorem 9.4.3 and write a solution of the form

$$\mathbf{X} = e^{-t}(\mathbf{V}_0 + \mathbf{V}_1 t)$$

where $\mathbf{V}_0$ and $\mathbf{V}_1$ are constant vectors. Differentiating this and substituting in (2), we get

$$\dot{\mathbf{X}} = e^{-t}(\mathbf{V}_1 - \mathbf{V}_0 - \mathbf{V}_1 t) = \mathrm{A}\mathbf{X} = e^{-t}\mathrm{A}(\mathbf{V}_0 + \mathbf{V}_1 t)$$

which leads to a system of equations:

$$\begin{aligned} \mathrm{A}\mathbf{V}_0 &= \mathbf{V}_1 - \mathbf{V}_0 \\ \mathrm{A}\mathbf{V}_1 &= -\mathbf{V}_1 \end{aligned}$$

The second equation of this system is an eigenvalue problem corresponding to $\lambda_1 = -1$. Solving it gives

$$\mathbf{V}_1 = b\begin{pmatrix} 1 \\ -1 \\ 1 \end{pmatrix}$$

where b is an arbitrary constant. Substituting this in the first equation of the system and writing $\mathbf{V}_0$ as the column vector (a_1, a_2, a_3) yields

$$\begin{pmatrix} 1 & 1 & 0 \\ 0 & 1 & 1 \\ -2 & -5 & -3 \end{pmatrix} \begin{pmatrix} a_1 \\ a_2 \\ a_3 \end{pmatrix} = \begin{pmatrix} b \\ -b \\ b \end{pmatrix}$$

This equation has two linearly independent solutions, which can be chosen to be

$$\mathbf{V}_0^{(1)} = b \begin{pmatrix} 0 \\ 1 \\ -2 \end{pmatrix} \quad \text{and} \quad \mathbf{V}_0^{(2)} = b \begin{pmatrix} 1 \\ 0 \\ -1 \end{pmatrix}$$

These give rise to the two solutions

$$\mathbf{X}^{(1)}(t) \equiv e^{-t}(\mathbf{V}_0^{(1)} + \mathbf{V}_1 t)$$

$$\mathbf{X}^{(2)}(t) = e^{-t}(\mathbf{V}_0^{(2)} + \mathbf{V}_1 t)$$

The most general solution to (1) is then

$$\mathbf{X}(t) = C_1 \begin{pmatrix} 1 \\ -2 \\ 4 \end{pmatrix} e^{-2t} + C_2 \left[\begin{pmatrix} 0 \\ 1 \\ -2 \end{pmatrix} + \begin{pmatrix} 1 \\ -1 \\ 1 \end{pmatrix} t \right] e^{-t} + C_3 \left[\begin{pmatrix} 1 \\ 0 \\ -1 \end{pmatrix} + \begin{pmatrix} 1 \\ -1 \\ 1 \end{pmatrix} t \right] e^{-t}$$

●

To solve Eq. (9.37), the inhomogeneous equation, we need only one of its particular solutions. To obtain such a solution we employ the method of variation of coefficients. This involves taking n linearly independent solutions, $\{\mathbf{X}_i\}_{i=1}^n$, of the *homogeneous* equation and writing

$$\mathbf{X}(t) = C_1(t)\mathbf{X}_1(t) + \cdots + C_n(t)\mathbf{X}_n(t) \tag{9.41}$$

where the $C_i(t)$ are to be determined. Differentiating this equation, we obtain

$$\dot{\mathbf{X}} = C_1\dot{\mathbf{X}}_1 + \cdots + C_n\dot{\mathbf{X}}_n + \dot{C}_1\mathbf{X}_1 + \cdots + \dot{C}_n\mathbf{X}_n$$

Applying the matrix A to (9.41) yields

$$\mathsf{A}\mathbf{X} = C_1\mathsf{A}\mathbf{X}_1 + \cdots + C_n\mathsf{A}\mathbf{X}_n = C_1\dot{\mathbf{X}}_1 + \cdots + C_n\dot{\mathbf{X}}_n$$

because by assumption the $\mathbf{X}_i$ are solutions of the homogeneous equation. Substituting the above in (9.37) gives

$$\dot{C}_1\mathbf{X}_1 + \dot{C}_2\mathbf{X}_2 + \cdots + \dot{C}_n\mathbf{X}_n = \mathbf{B} \tag{9.42}$$

This is a system of n equations (remember that the $\mathbf{X}_i$ are n-vectors) in n unknowns (the $\dot{C}_i$), which always has a solution because the $\mathbf{X}_i$ are linearly independent, implying that the matrix of coefficients, whose columns are the independent vectors $\mathbf{X}_i$, is invertible. Once $\{\dot{C}_i\}_{i=1}^n$ is determined, the $\dot{C}_i$ can be integrated to obtain C_i, which in turn determine the particular solution $\mathbf{X}(t)$ in (9.41).

A result from Chapter 2 has a bearing on linear systems with constant coefficients. Example 2.3.10 showed that an operator equation,

$$\frac{d\mathbb{U}}{dt} = \mathbb{H}\mathbb{U}(t)$$

in which $\mathbb{H}$ is independent of t, has the solution

$$\mathbb{U}(t) = e^{\mathbb{H}t}\mathbb{U}(0)$$

Applying this idea to the vector equation of (9.38) yields a proposition.

9.4.4 Proposition The most general solution of the linear system of FODEs $d\mathbf{X}/dt = \mathsf{A}\mathbf{X}$, in which A is a constant matrix, is

$$\mathbf{X}(t) = e^{\mathsf{A}t}\mathbf{X}(0) \qquad \blacksquare$$

In general, $e^{\mathsf{A}t}$ is not easy to calculate. In fact, in Chapter 3 we saw that calculation of any function of an operator (matrix) involves diagonalization, calculation of projection operators, and so forth.

Exercises

9.4.1 Show that the functions $x^r e^{\lambda_j x}$, where $r = 0, 1, 2, \ldots, k$, are linearly independent.

9.4.2 Solve $y'' - 2y' + y = xe^x$ subject to the boundary conditions $y(0) = 0$, $y'(0) = 1$.

9.4.3 Rewrite the SOLDE $\ddot{y} - 2\dot{y} + y = 0$ as a linear system of FODEs and solve the system.

9.4.4 Find a particular solution for $\ddot{y} - 2\dot{y} + y = te^t$ by turning it into a linear system.

9.5 COMPLEX DIFFERENTIAL EQUATIONS

We have familiarized ourselves with some useful techniques for finding solutions to differential equations. One powerful method that leads to formal solutions is the power series method. We also stated a theorem which guarantees the convergence of the solution of the power series within a circle whose size is at least as large as the smallest of the circles of convergence of the coefficient functions.

Thus, the convergence of the solution is related to the convergence of the coefficient functions. What about the *nature* of the convergence, or the analyticity of the solution? Is it related to the analyticity of the coefficient functions? If so, how? Are the singular points of the coefficients also singular points of the solution? If so, is the nature of the singularities the same? This section attempts to answer these questions.

Any question of analyticity is best handled in the complex plane. An important reason for this is the property of analytic continuation discussed in Chapter 7. The differential equation $du/dx = u^2$ has a solution $u = -1/x$ for all x except $x = 0$. Thus, we have to "puncture" the real line by removing $x = 0$ from it. Then we have two solutions, because the domain of definition of $u = -1/x$ is not connected on the real line (technically, the definition of a function includes its domain as well as the rule for going from the domain to the range). In addition, if we confine ourselves to the real line, there is no way that we can connect the $x > 0$ region to the $x < 0$ region. On the other hand, in the complex plane the same equation, $dw/dz = w^2$, has the complex solution $w = -1/z$, which is analytic everywhere except at $z = 0$. Puncturing the complex plane does not destroy the connectivity of the region of definition of w. Thus, the solution in the $x > 0$ region can be analytically continued to the solution in the $x < 0$ region by going around the origin.

The purpose of this section is to investigate the analytic properties of the solutions of some well-known SOLDEs in mathematical physics. We begin with a result from differential equation theory (for a proof, see Birkhoff and Rota 1978, p. 223).

9.5.1 Proposition (Continuation Principle) The function obtained by analytic continuation of any solution of an analytic differential equation along any path in the complex plane is a solution of the analytic continuation of the differential equation along the same path. ■

An analytic differential equation is one with analytic coefficient functions. This proposition makes it possible to find a solution in one region of the complex plane and then continue it analytically. The following example shows how the singularities of the coefficient functions affect the behavior of the solution.

Example 9.5.1

Let us consider the FODE

$$\frac{dw}{dz} - \frac{\gamma}{z} w = 0 \qquad \text{for } \gamma \in \mathbb{R}$$

The coefficient function $p(z) \equiv -\gamma/z$ has a simple pole at $z = 0$. The solution to the FODE is easily found to be

$$w = z^{\gamma}$$

Thus, depending on whether γ is a nonnegative integer, a negative integer, $-m$, or a noninteger, at $z = 0$ the solution has a regular point, a pole of order m, or a branch point, respectively. •

This example shows that the singularities of the solution can be worse or better than those of the coefficient functions, depending on the parameters in the differential equation.

9.5.1 General Analytic Properties of Complex Differential Equations

To prepare for discussing the analytic properties of the solutions of SOLDEs, let us consider some general properties of differential equations from a complex analytical point of view.

Complex first-order linear differential equations. In the homogeneous FOLDE

$$\frac{dw}{dz} + p(z)w = 0 \tag{9.43}$$

$p(z)$ is assumed to have only isolated singular points. It follows that $p(z)$ can be expanded about a point z_0, which may be a singularity of $p(z)$, as a Laurent series in some annular region $r_1 < |z - z_0| < r_2$:

$$p(z) = \sum_{n=-\infty}^{\infty} a_n(z - z_0)^n \qquad \text{where } r_1 < |z - z_0| < r_2$$

The solution to (9.43), as given in Theorem 9.1.4 with $q = 0$, is

$$\begin{aligned} w(z) &= \exp\left[-\int p(z)dz\right] \\ &= C\exp\left[-\int a_{-1}\frac{dz}{z - z_0} - \sum_{n=0}^{\infty} a_n \int (z - z_0)^n dz - \sum_{n=2}^{\infty} a_{-n} \int (z - z_0)^{-n} dz\right] \\ &= C\exp\left[-a_{-1}\ln(z - z_0) - \sum_{n=0}^{\infty} \frac{a_n}{n+1}(z - z_0)^{n+1} + \sum_{n=1}^{\infty} \frac{a_{-n-1}}{n}(z - z_0)^{-n}\right] \end{aligned}$$

We can write this solution as

$$w(z) = C(z - z_0)^{\alpha} g(z) \tag{9.44}$$

where $\alpha \equiv -a_{-1}$ and $g(z)$ is an analytic *single-valued* function in the annular region $r_1 < |z - z_0| < r_2$. This follows from the fact that the exponential of an analytic function is analytic.

Depending on the nature of the singularity of $p(z)$ at z_0, the solutions given by (9.44) have different classifications. For instance, if $p(z)$ has a removable singularity (if $a_{-n} = 0\ \forall n \geqslant 1$), the solution is $Cg(z)$, which is analytic. In this case we say that the FOLDE [Eq. (9.43)] has a *removable singularity* at z_0. If $p(z)$ has a simple pole at z_0 (if $a_{-1} \neq 0$ and $a_{-n} = 0\ \forall n \geqslant 2$), then, in general, the solution has a branch point at z_0. In this case we say that the FOLDE has a *regular singular point*. Finally, if $p(z)$ has a pole of order $m > 1$, then the solution will have an essential singularity (see Problem 9.34). In this case the FOLDE is said to have an *irregular singular point*.

To arrive at the solution given by (9.44), we had to solve the FOLDE. Since higher-order differential equations are not as easily solved, it is desirable to obtain such a solution through other considerations. The following example sets the stage for this endeavor.

Example 9.5.2

A FOLDE has a unique solution, to within a multiplicative constant, given by Theorem 9.1.4. Thus, with a given solution $w(z)$, any other solution must be of the form $Cw(z)$. Let z_0 be a singularity of $p(z)$, and let $z - z_0 = re^{i\theta}$. Start at a point z and circle z_0 so that $\theta \to \theta + 2\pi$. Even though $p(z)$ may have a simple pole at z_0, the solution may have a *branch point* there. This is clear from the general solution, where α may be a noninteger. Thus, $\tilde{w}(z) \equiv w(z_0 + re^{i(\theta+2\pi)})$ may be different from $w(z)$. However, Proposition 9.5.1 implies that $\tilde{w}(z)$ is also a solution to the FOLDE. Thus, there exists a C such that $\tilde{w}(z) = Cw(z)$. Define $\alpha \in \mathbb{C}$ by $C \equiv e^{2\pi i\alpha}$. Then the function

$$g(z) = (z - z_0)^{-\alpha} w(z)$$

is *single-valued* around z_0. In fact,

$$\begin{aligned} g(z_0 + re^{i(\theta+2\pi)}) &= (re^{i(\theta+2\pi)})^{-\alpha} w(z_0 + re^{i(\theta+2\pi)}) \\ &= (z - z_0)^{-\alpha} e^{-2\pi i\alpha} e^{2\pi i\alpha} w(z) = (z - z_0)^{-\alpha} w(z) = g(z) \end{aligned}$$

This argument shows that a solution, $w(z)$, of the FOLDE of Eq. (9.43) can be written as $w(z) = (z - z_0)^{\alpha} g(z)$, where $g(z)$ is single-valued. •

The circuit matrix. The method used in Example 9.5.2 can be generalized to obtain a similar result for the NOLDE

$$\mathbb{L}[w] = \frac{d^n w}{dz^n} + p_{n-1}(z)\frac{d^{n-1}w}{dz^{n-1}} + \cdots + p_1(z)\frac{dw}{dz} + p_0(z)w = 0 \tag{9.45}$$

where all the $p_i(z)$ are analytic in $r_1 < |z - z_0| < r_2$.

Let $\{w_j(z)\}_{j=1}^n$ be a basis of solutions of (9.45), and let $z - z_0 = re^{i\theta}$. Start at z and analytically continue the functions $w_j(z)$ one complete turn to $\theta + 2\pi$. Let $\tilde{w}_j(z) \equiv \tilde{w}_j(z_0 + re^{i\theta}) \equiv w_j(z_0 + re^{i(\theta+2\pi)})$. Then it is not hard to show that $\{\tilde{w}_j(z)\}_{j=1}^n$ is also a basis of solutions. On the other hand, $\tilde{w}_j(z)$ can be expressed as a linear combination of $w_j(z)$. Thus,

$$\tilde{w}_j(z) = w_j(z_0 + re^{i(\theta+2\pi)}) = \sum_{k=1}^{n} a_{jk} w_k(z)$$

The matrix $\mathsf{A} \equiv (a_{jk})$, called the *circuit matrix* of the NOLDE, is invertible because it transforms one basis into another. Therefore, it has nonzero eigenvalues. We let $\lambda \neq 0$ be one such eigenvalue, and choose a column vector,

$$\mathbf{C} = \begin{pmatrix} c_1 \\ c_2 \\ \vdots \\ c_n \end{pmatrix}$$

such that $\tilde{\mathsf{A}}\mathbf{C} = \lambda\mathbf{C}$, that is, $\mathbf{C}$ is an eigenvector of $\tilde{\mathsf{A}}$ with eigenvalue λ (note that A and $\tilde{\mathsf{A}}$ have the same set of eigenvalues). At least one such eigenvector always exists, because the characteristic polynomial of $\tilde{\mathsf{A}}$ has at least one root, for which it is always possible to find a $\mathbf{C}$ such that $\tilde{\mathsf{A}}\mathbf{C} = \lambda\mathbf{C}$. Now we let

$$w(z) = \sum_{j=1}^{n} c_j w_j(z)$$

Clearly, this $w(z)$ is a solution of (9.45), and

$$\begin{aligned}\tilde{w}(z) \equiv w(z_0 + re^{i(\theta+2\pi)}) &= \sum_{j=1}^{n} c_j w_j(z_0 + re^{i(\theta+2\pi)}) \\ &= \sum_{j=1}^{n} c_j \sum_{k=1}^{n} a_{jk} w_k(z) \\ &= \sum_{j,k} (\tilde{A})_{kj} c_j w_k(z) = \sum_{k=1}^{n} \lambda c_k w_k(z) \\ &= \lambda w(z)\end{aligned}$$

If we define α by $\lambda = e^{2\pi i\alpha}$, then

$$w(z_0 + re^{i(\theta+2\pi)}) = e^{2\pi i\alpha} w(z)$$

Now we write $f(z) \equiv (z - z_0)^{-\alpha} w(z)$. Following the argument used in Example 9.5.2, we get $f(z_0 + re^{i(\theta+2\pi)}) = f(z)$; that is, $f(z)$ is single-valued around z_0. We thus have the following theorem.

9.5.2 Theorem Any NOLDE of the form of (9.45) with analytic coefficient functions in $r_1 < |z - z_0| < r_2$ admits a solution of the form

$$w(z) = (z - z_0)^{\alpha} f(z)$$

where $f(z)$ is single-valued around z_0 in $r_1 < |z - z_0| < r_2$. ■

An isolated singular point, z_0, near which an analytic function $w(z)$ can be written as $w(z) = (z - z_0)^{\alpha} f(z)$, where $f(z)$ is single-valued and analytic in the punctured neighborhood of z_0, is called a *simple branch point* of $w(z)$. The arguments leading to Theorem 9.5.2 imply that a solution with a simple branch point exists if and only if the vector **C** whose components appear in $w(z) = \sum_{j=1}^{n} c_j w_j(z)$ is an eigenvector of $\tilde{\mathsf{A}}$, the transpose of the circuit matrix. Thus, there are as many solutions with simple branch points as there are linearly independent eigenvectors.

9.5.2 Complex Second-Order Linear Differential Equations

Let us now consider the SOLDE

$$w'' + p(z)w' + q(z)w = 0$$

Given two linearly independent solutions, $w_1(z)$ and $w_2(z)$, we form the matrix A and try to diagonalize it. There are three possible outcomes:

(1) The matrix A is diagonalizable, and we can find two eigenvectors, $F(z)$ and $G(z)$, corresponding, respectively, to two distinct eigenvalues, λ_1 and λ_2. This means that

$$F(z_0 + re^{i(\theta+2\pi)}) = \lambda_1 F(z) \qquad \text{and} \qquad G(z_0 + re^{i(\theta+2\pi)}) = \lambda_2 G(z)$$

Defining $\lambda_1 = e^{2\pi i\alpha}$ and $\lambda_2 = e^{2\pi i\beta}$, we get, as Theorem 9.5.2 suggests,

$$F(z) = (z - z_0)^\alpha f(z) \qquad \text{and} \qquad G(z) = (z - z_0)^\beta g(z)$$

The set $\{F(z), G(z)\}$ is called a *canonical basis* for the SOLDE.

(2) The matrix A is diagonalizable, and the two eigenvalues are the same. In this case both $F(z)$ and $G(z)$ have the same constant α:

$$F(z) = (z - z_0)^\alpha f(z) \qquad \text{and} \qquad G(z) = (z - z_0)^\alpha g(z)$$

(3) We cannot find *two* eigenvectors. This corresponds to the case where A is not diagonalizable (it is not a normal matrix, in the language of Chapter 3). However, we can always find one eigenvector, so A has only one eigenvalue, λ. We let $w_1(z)$ be the solution of the form $(z - z_0)^\alpha f(z)$, where $f(z)$ is single-valued and $\lambda = e^{2\pi i\alpha}$. The existence of such a solution is guaranteed by Theorem 9.5.2. Let $w_2(z)$ be any other linearly independent solution (Theorem 9.2.4 ensures the existence of such a second solution). Then

$$w_2(z_0 + re^{i(\theta + 2\pi)}) = aw_1(z) + bw_2(z)$$

The circuit matrix will be

$$\mathsf{A} = \begin{pmatrix} \lambda & 0 \\ a & b \end{pmatrix}$$

which has eigenvalues λ and b. Since A is assumed to have only one eigenvalue (otherwise the first outcome would occur), it must be true that $b = \lambda$. This reduces A to

$$\mathsf{A} = \begin{pmatrix} \lambda & 0 \\ a & \lambda \end{pmatrix} \qquad \text{where } a \neq 0$$

The condition $a \neq 0$ is necessary to distinguish this case from the second outcome. Now we analytically continue $h(z) \equiv w_2(z)/w_1(z)$ one whole turn around z_0, obtaining

$$h(z_0 + re^{i(\theta + 2\pi)}) = \frac{w_2(z_0 + re^{i(\theta + 2\pi)})}{w_1(z_0 + re^{i(\theta + 2\pi)})} = \frac{aw_1(z) + \lambda w_2 z}{\lambda w_1(z)}$$

$$= \frac{a}{\lambda} + \frac{w_2(z)}{w_1(z)} = \frac{a}{\lambda} + h(z)$$

It then follows that the function

$$g_1(z) \equiv h(z) - \frac{a}{2\pi i\lambda} \ln z$$

is single-valued in $r_1 < |z - z_0| < r_2$. Thus, $w_2(z) = h(z)w_1(z)$ can be written as

$$w_2(z) = w_1(z)g_1(z) + \frac{a}{2\pi i\lambda}(\ln z)w_1(z)$$

If we redefine $g_1(z)$ and $w_2(z)$ as $(2\pi i\lambda/a)g_1(z)$ and $(2\pi i\lambda/a)w_2(z)$, respectively, we have a theorem.

9.5.3 Theorem If $p(z)$ and $q(z)$ are analytic in $r_1 < |z - z_0| < r_2$, then the SOLDE $w'' + p(z)w' + q(z)w = 0$ admits a basis of solutions in the neighborhood of the singular point z_0, and the solutions have the form

$$w_1(z) = (z - z_0)^\alpha f(z)$$

$$w_2(z) = (z - z_0)^\beta g(z)$$

or, in *exceptional cases* (there are not two eigenvectors), the form

$$w_1(z) = (z - z_0)^\alpha f(z)$$

$$w_2(z) = w_1(z)[g_1(z) + \ln z]$$

The functions $f(z)$, $g(z)$, and $g_1(z)$ are analytic and single-valued in $r_1 < |z - z_0| < r_2$. ■

This theorem allows us to separate the branch point z_0 from the rest of the solutions. However, even though $f(z)$, $g(z)$, and $g_1(z)$ are analytic in the annular region $r_1 < |z - z_0| < r_2$, they may very well have poles of arbitrary orders *at* z_0. Can we also separate the poles? In general, we cannot; however, when $p(z)$ and $q(z)$ have poles of certain order, we can. To do this we need the following definition.

9.5.4 Definition A SOLDE of the form $w'' + p(z)w' + q(z)w = 0$ that is analytic in $0 < |z - z_0| < r$ has a *regular singular point* at z_0 if $p(z)$ has at worst a simple pole there and $q(z)$ has at worst a pole of order 2 there.

In the neighborhood of a regular singular point z_0, the coefficient functions $p(z)$ and $q(z)$ have the power-series expansions

$$p(z) = \frac{a_{-1}}{z - z_0} + \sum_{k=0}^{\infty} a_k(z - z_0)^k$$

$$q(z) = \frac{b_{-2}}{(z - z_0)^2} + \frac{b_{-1}}{z - z_0} + \sum_{k=0}^{\infty} b_k(z - z_0)^k$$

Multiplying both sides of the first equation by $(z - z_0)$ and both sides of the second by $(z - z_0)^2$, we obtain

$$(z - z_0)p(z) \equiv P(z) \equiv \sum_{k=0}^{\infty} A_k(z - z_0)^k$$

and

$$(z - z_0)^2 q(z) \equiv Q(z) \equiv \sum_{k=0}^{\infty} B_k(z - z_0)^k$$

It is convenient to multiply the SOLDE of the definition by $(z - z_0)^2$ and write it as

$$\mathbb{L}[w] \equiv (z - z_0)^2 w'' + (z - z_0)P(z)w' + Q(z)w = 0 \tag{9.46}$$

We now claim that a solution of the form

$$w(z) = (z - z_0)^\nu \left[1 + \sum_{k=1}^{\infty} C_k(z - z_0)^k\right] \equiv (z - z_0)^\nu \sum_{k=0}^{\infty} C_k(z - z_0)^k \quad \text{where } C_0 = 1 \tag{9.47}$$

formally solves the SOLDE. Let us substitute this in (9.46) to obtain

$$\sum_{n=0}^{\infty} \left\{(n + \nu)(n + \nu - 1)C_n + \sum_{k=0}^{n} [(k + \nu)A_{n-k} + B_{n-k}]C_k\right\}(z - z_0)^{n+\nu} = 0$$

which results in the recursion relation

$$(n + \nu)(n + \nu - 1)C_n = -\sum_{k=0}^{n} [(k + \nu)A_{n-k} + B_{n-k}]C_k \qquad \text{for } n = 0, 1, 2, \ldots \tag{9.48a}$$

For $n = 0$ this leads to what is known as the *indicial equation* for the exponent ν:

$$I(\nu) \equiv \nu(\nu - 1) + A_0\nu + B_0 = 0 \tag{9.49}$$

The roots of this equation are called the *characteristic exponents* of z_0, and $I(\nu)$ is called its *indicial polynomial.* In terms of this polynomial (9.48a) can be expressed as

$$I(\nu + n)C_n = -\sum_{k=0}^{n-1} [(k + \nu)A_{n-k} + B_{n-k}]C_k \qquad \text{for } n = 1, 2, \cdots \tag{9.48b}$$

Equation (9.49) determines what values of ν are possible, and Eq. (9.48b) gives $C_1, C_2, C_3, \ldots$, which in turn determine $w(z)$. However, if the indicial polynomial vanishes at $n + \nu$ for some positive integer n, that is, if $n + \nu$, as well as ν, is a root of (9.49), then $I(\nu + n) = 0 = I(\nu)$, and (9.48b) will not be defined.

If ν_1 and ν_2 are characteristic exponents of the indicial equation and $\text{Re}(\nu_1) > \text{Re}(\nu_2)$, then a solution for ν_1 always exists. A solution for ν_2 also exists if and only if $\nu_1 - \nu_2 \neq n$ for some (positive) integer n. In particular, if z_0 is an *ordinary* point [a point at which both $p(z)$ and $q(z)$ are analytic], then only one solution is determined by (9.48b). (Why?)

The foregoing discussion is summarized in the following theorem.

9.5.5 Theorem If the differential equation $w'' + p(z)w' + q(z)w = 0$ has a regular singular point at $z = z_0$, then at least one power series of the form of (9.47) formally satisfies the equation. If ν_1 and ν_2 are the characteristic exponents of z_0, then there are two linearly independent formal solutions unless $\nu_1 - \nu_2$ is an integer. ■

Example 9.5.3

Let us consider some differential equations commonly occurring in physics.

(a) The Bessel equation is

$$w'' + \frac{1}{z}w' + \left(1 - \frac{\alpha^2}{z^2}\right)w = 0$$

In this case $z_0 = 0$, $A_0 = 1$, and $B_0 = -\alpha^2$. Thus, the indicial equation is

$$\nu(\nu - 1) + \nu - \alpha^2 = 0$$

and its solutions are $\nu_1 = -\alpha$ and $\nu_2 = +\alpha$. Therefore, there are two linearly independent solutions to the Bessel equation unless $\nu_2 - \nu_1 = 2\alpha$ is an integer, or α is either an integer or a half-integer.

(b) For the Coulomb potential, $f(r) = \beta/r$, the most general radial equation [Eq. (8.14b)] reduces to

$$w'' + \frac{2}{z}w' + \left(\frac{\beta}{z} - \frac{\alpha}{z^2}\right)w = 0$$

The point $z = 0$ is a regular singular point at which $A_0 = 2$ and $B_0 = -\alpha$. The indicial polynomial is

$$I(\nu) = \nu^2 + \nu - \alpha$$

with characteristic exponents

$$\nu_1 = -\tfrac{1}{2} - \tfrac{1}{2}\sqrt{1 + 4\alpha} \quad \text{and} \quad \nu_2 = -\tfrac{1}{2} + \tfrac{1}{2}\sqrt{1 + 4\alpha}$$

There are two independent solutions unless $\nu_2 - \nu_1 = \sqrt{1 + 4\alpha}$ is an integer. In practice, $\alpha = l(l + 1)$, where l is some integer; so $\nu_2 - \nu_1 = 2l + 1$. Thus, only one solution is obtained.

(c) The hypergeometric differential equation is

$$w'' + \frac{\gamma - (\alpha + \beta + 1)z}{z(1 - z)}w' - \frac{\alpha\beta}{z(1 - z)}w = 0$$

A substantial number of functions in mathematical physics are solutions of this remarkable equation, with appropriate values for α, β, and γ. The regular singular points are $z = 0$ and $z = 1$. At $z = 0$, $A_0 = \gamma$, $B_0 = 0$, and the indicial polynomial is $I(\nu) = \nu(\nu + \gamma - 1)$, whose roots are $\nu_1 = 0$ and $\nu_2 = 1 - \gamma$. Unless γ is an integer, we have two formal solutions, corresponding to ν_1 and ν_2. ●

It is shown in differential equation theory (see Birkhoff and Rota 1978, pp. 240–242) that, as long as $\nu_1 - \nu_2$ is not an integer, the series solution of Theorem 9.5.5 is convergent for a neighborhood of z_0. What happens to the exceptional case when $\nu_1 - \nu_2$ *is* an integer?

It is convenient to translate the coordinate axes so that the point z_0 coincides with the origin. This will save us some writing, because instead of powers of $(z - z_0)$, we will have powers of z.

Assume that $v_1 - v_2$ is an integer, and let v_1 be the root having the larger real part. Then $v_2 = v_1 - n$ for some $n \geqslant 0$, and a solution of the form

$$w_1 = z^{v_1} f(z) \equiv z^{v_1}\left(1 + \sum_{k=1}^{\infty} C_k z^k\right)$$

exists in the region $0 < |z| < r$ for some $r > 0$. Now we define $w(z) \equiv w_1(z)h(z) = z^{v_1} f(z)h(z)$, and substitute in the SOLDE to obtain a FOLDE in h',

$$h'' + \left(p + \frac{2w_1'}{w_1}\right)h' = 0$$

or, by substituting $w_1'/w_1 = v_1/z + f'/f$, the equivalent FOLDE

$$h'' + \left(\frac{2v_1}{z} + 2\frac{f'}{f} + p\right)h' = 0 \tag{9.50}$$

This DE in h' has a regular singular point at $z = 0$ because $f(0) = 1 \neq 0$ and $p(z)$ has at worst a simple pole there. It can be shown that the coefficient of the z^{-1} term in the Laurent-series expansion of the term multiplying h' in (9.50) is $n + 1$ and that

$$h(z) = \begin{cases} \ln z + g_1(z) & \text{if } n = 0 \\ C \ln z + z^{-n} g_2(z) & \text{if } n \neq 0 \end{cases} \tag{9.51}$$

where $g_1(z)$ and $g_2(z)$ are analytic at $z = 0$ (see Exercise 9.5.1). The following theorem collects these results.

9.5.6 Theorem Suppose that the characteristic exponents of a SOLDE with a regular singular point at $z = 0$ are v_1 and v_2. If $v_1 - v_2$ is not an integer, there exists a basis of solutions of the form of (9.47) with $v = v_1$ or $v = v_2$. On the other hand, if $v_2 = v_1 - n$, where n is a nonnegative integer, then there exists a canonical basis of solutions of the form

$$w_1 = z^{v_1}\left(1 + \sum_{k=1}^{\infty} a_k z^k\right)$$

$$w_2 = z^{v_2}\left(1 + \sum_{k=1}^{\infty} b_k z^k\right) + C w_1 \ln z$$

where the power series are convergent in a neighborhood of $z = 0$.

Proof. Except for the form of w_2 given in the theorem, the proof was completed above. The derivation of this form of w_2 is left as a simple problem (Problem 9.39). ∎

9.5.3 Fuchsian Differential Equations

In many cases of physical interest, the behavior of the solution of a SOLDE at infinity is important. For instance, as Example 9.3.4 and Exercise 9.3.2 show, a function that describes the probability density of a particle in quantum mechanics must tend to zero as the distance from the center of the binding force increases.

We have seen that the behavior of a solution is determined by the behavior of the DE, or, more specifically, by the behavior of its coefficient functions. To determine the behavior at infinity, we substitute $z = 1/t$ in the SOLDE

$$\frac{d^2w}{dz^2} + p(z)\frac{dw}{dz} + q(z)w = 0 \tag{9.52}$$

and obtain

$$\frac{d^2v}{dt^2} + \left[\frac{2}{t} - \frac{1}{t^2}r(t)\right]\frac{dv}{dt} + \frac{1}{t^4}S(t)v(t) = 0 \tag{9.53}$$

where $$v(t) \equiv w\left(\frac{1}{t}\right) \qquad r(t) \equiv p\left(\frac{1}{t}\right) \qquad S(t) \equiv q\left(\frac{1}{t}\right)$$

Clearly, as $z \to \infty$, $t \to 0$. Thus, we are interested in the behavior of (9.53) for $t = 0$. We assume that both $r(t)$ and $S(t)$ are analytic at $t = 0$. Equation (9.53) clearly shows, however, that the solution $v(t)$ may still have singularities at $t = 0$ (or $z = \infty$) because of the extra terms appearing in the coefficient functions.

We assume that infinity is a regular singular point of (9.52), by which we mean that $t = 0$ is a regular singular point of (9.53). Therefore, in the Taylor expansions of $r(t)$ and $S(t)$ (because they are analytic), the first (constant) term of $r(t)$ and the first two terms of $S(t)$ must be zero. Thus, we write

$$r(t) = a_1 t + a_2 t^2 + \cdots + a_n t^n + \cdots = \sum_{k=1}^{\infty} a_k t^k$$

and $$S(t) = b_2 t^2 + b_3 t^3 + \cdots + b_n t^n + \cdots = \sum_{k=2}^{\infty} b_k t^k$$

For $p(z)$ and $q(z)$ we must have expressions of the form

$$\begin{aligned} p(z) &= \frac{a_1}{z} + \frac{a_2}{z^2} + \cdots + \frac{a_n}{z^n} + \cdots = \sum_{k=1}^{\infty} \frac{a_k}{z^k} \\ q(z) &= \frac{b_2}{z^2} + \cdots + \frac{b_n}{z^n} + \cdots = \sum_{k=2}^{\infty} \frac{b_k}{z^k} \end{aligned} \tag{9.54}$$

for $|z| \to \infty$.

When $z = \infty$ is a regular singular point of Eq. (9.52), or, equivalently, $t = 0$ is a regular singular point of (9.53), it follows from Theorem 9.5.6 that there exist solutions of the form

$$v_1(t) = t^{\alpha}\left(1 + \sum_{k=1}^{\infty} C_k t^k\right) \qquad \text{for } \alpha = \nu_1, \nu_2$$

or, in terms of z,

$$w_1(z) = z^{-\alpha}\left(1 + \sum_{k=1}^{\infty} \frac{C_k}{z^k}\right) \qquad \text{for } \alpha = \nu_1, \nu_2 \tag{9.55}$$

if $\nu_1 - \nu_2$ is not an integer. Here ν_1 and ν_2 are the characteristic exponents at $t = 0$ of (9.53), whose indicial polynomial is easily found to be

$$\nu(\nu - 1) + (2 - a_1) + b_2 = 0$$

If $\nu_1 - \nu_2$ is an integer, there still exist solutions of the form of (9.55), but the second one may contain a logarithmic term.

A homogeneous differential equation with single-valued analytic coefficient functions is called a *Fuchsian differential equation* (FDE) if it has *only* regular singular points in the extended complex plane, $\mathbb{C} \cup \{\infty\}$ (the complex plane including the point at infinity).

Since one particular kind of FDE describes a large class of nonelementary functions encountered in mathematical physics, it is instructive to classify various kinds of FDEs. A fact that is used in such a classification is that complex functions whose only singularities in the extended complex plane are poles are rational functions (ratios of polynomials; see Exercise 9.5.2). We thus expect FDEs to have only rational functions as coefficients.

Let us start with second-order Fuchsian differential equations (SOFDEs). First, we consider the case where the equation has at most two regular singular points at z_1 and z_2. We introduce the variable $\xi = (z - z_1)/(z - z_2)$. The regular singular points at z_1 and z_2 are mapped onto the points $\xi_1 = 0$ and $\xi_2 = \infty$, respectively, in the extended ξ-plane. Equation (9.52) becomes

$$\frac{d^2u}{d\xi^2} + \Phi(\xi)\frac{du}{d\xi} + \Theta(\xi)u = 0 \tag{9.56}$$

where u, Φ, and Θ are functions of ξ obtained when z is expressed in terms of ξ in $w(z)$, $p(z)$, and $q(z)$, respectively. Since $\xi = 0$ is at most a simple pole of $\Phi(\xi)$, we must have

$$\Phi(\xi) = \frac{a_1}{\xi} + \sum_{k=0}^{\infty} \alpha_k \xi^k$$

But $\xi = \infty$ is also a regular singular point. According to (9.54), it follows that $\alpha_k = 0$ for $k = 0, 1, 2, \ldots$. Thus, $\Phi(\xi) = a_1/\xi$. Similarly, we have $\Theta(\xi) = b_2/\xi^2$. Thus, a SOFDE with two regular singular points is equivalent to the DE

$$w'' + \frac{a_1}{z}w' + \frac{b_2}{z^2}w = 0$$

Multiplying both sides by z^2, we obtain

$$z^2w'' + a_1zw' + b_2w = 0$$

which is the second-order *Euler differential equation.* A general nth-order Euler differential equation is equivalent to a NOLDE with constant coefficients (see Problem 9.32). Thus, a SOFDE with two regular singular points is equivalent to a SOLDE with constant coefficients and produces nothing new.

The simplest SOFDE whose solutions may include nonelementary functions is, therefore, one having three regular singular points, at z_1, z_2, and z_3, for example. By the transformation

$$\xi = \frac{(z - z_1)(z_3 - z_2)}{(z - z_2)(z_3 - z_1)}$$

we can map z_1, z_2, and z_3 onto $\xi_1 = 0$, $\xi_2 = \infty$, and $\xi_3 = 1$. Thus, we assume that the three regular singular points are at $z = 0$, $z = 1$, and $z = \infty$. It can be shown (see Exercise 9.5.3) that the most general $p(z)$ and $q(z)$ are

$$p(z) = \frac{A_1}{z} + \frac{B_1}{z - 1} \qquad \text{and} \qquad q(z) = \frac{A_2}{z^2} + \frac{B_2}{(z - 1)^2} - \frac{A_3}{z(z - 1)}$$

We thus have the following theorem.

9.5.7 Theorem The most general SOFDE with three regular singular points can be transformed into the form

$$w'' + \left(\frac{A_1}{z} + \frac{B_1}{z - 1}\right)w' + \left[\frac{A_2}{z^2} + \frac{B_2}{(z - 1)^2} - \frac{A_3}{z(z - 1)}\right]w = 0 \tag{9.57}$$

where A_1, A_2, A_3, B_1, and B_2 are constants. This equation is called the *Riemann differential equation.* ■

We can write the Riemann DE in terms of pairs of characteristic exponents, (λ_1, λ_2), (μ_1, μ_2), and (ν_1, ν_2), belonging to the singular points 0, 1, and ∞, respectively. The indicial equations are easily found to be

$$\lambda(\lambda - 1) + A_1\lambda + A_2 = 0$$

$$\mu(\mu - 1) + B_1\mu + B_2 = 0$$

and

$$\nu^2 + (1 - A_1 - B_1)\nu + A_2 + B_2 - A_3 = 0$$

By writing the indicial equations as $(\lambda - \lambda_1)(\lambda - \lambda_2) = 0$, and so forth, and comparing coefficients, we can find the following relations:

$$A_1 = 1 - \lambda_1 - \lambda_2 \qquad A_2 = \lambda_1\lambda_2$$

$$B_1 = 1 - \mu_1 - \mu_2 \qquad B_2 = \mu_1\mu_2$$

$$A_1 + B_1 = \nu_1 + \nu_2 + 1 \qquad A_2 + B_2 - A_3 = \nu_1\nu_2$$

These equations lead easily to the *Riemann identity*

$$\lambda_1 + \lambda_2 + \mu_1 + \mu_2 + \nu_1 + \nu_2 = 1 \tag{9.58}$$

Substituting these results in (9.57) gives another theorem.

9.5.8 Theorem A SOFDE with three regular singular points in the extended complex plane is equivalent to the Riemann DE,

$$w'' + \left(\frac{1-\lambda_1-\lambda_2}{z} + \frac{1-\mu_1-\mu_2}{z-1}\right)w' + \left(\frac{\lambda_1\lambda_2}{z^2} + \frac{\mu_1\mu_2}{(z-1)^2} + \frac{\nu_1\nu_2-\lambda_1\lambda_2-\mu_1\mu_2}{z(z-1)}\right)w = 0 \tag{9.59}$$

which is *uniquely* determined by the pairs of characteristic exponents at each singular points. The characteristic exponents satisfy the Riemann identity, Eq. (9.58). ■

The uniqueness of the Riemann DE allows us to derive identities for solutions and reduce the independent parameters of (9.59) from five to three. We first note that if $w(z)$ is a solution of (9.59) corresponding to (λ_1, λ_2), (μ_1, μ_2), and (ν_1, ν_2), then the function $v(z) = z^\lambda w(z)$, which also has branch points at 0, 1, and ∞, is a solution corresponding to $(\lambda_1 + \lambda, \lambda_2 + \lambda)$, (μ_1, μ_2), and $(\nu_1 - \lambda, \nu_2 - \lambda)$. This can be seen from Theorem 9.5.6 and Eq. (9.55). More generally, the function

$$v(z) = z^\lambda (z-1)^\mu w(z)$$

has branch points at $z = 0, 1, \infty$ [because $w(z)$ does]; therefore, it is a solution of the Riemann DE. Its pairs of characteristic exponents are

$$(\lambda_1 + \lambda, \lambda_2 + \lambda) \qquad (\mu_1 + \mu, \mu_2 + \mu) \qquad \text{and} \qquad (\nu_1 - \lambda - \mu, \nu_2 - \lambda - \mu)$$

In particular, if we let $\lambda = -\lambda_1$ and $\mu = -\mu_1$, then the pairs reduce to

$$(0, \lambda_2 - \lambda_1) \qquad (0, \mu_2 - \mu_1) \qquad \text{and} \qquad (\nu_1 + \lambda_1 + \mu_1, \nu_2 + \lambda_1 + \mu_1)$$

If we define $\alpha \equiv \nu_1 + \lambda_1 + \mu_1$, $\beta \equiv \nu_2 + \lambda_1 + \mu_1$, and $\gamma \equiv 1 - \lambda_2 + \lambda_1$ and use (9.58), we can write these pairs as

$$(0, 1-\gamma) \qquad (0, \gamma - \alpha - \beta) \qquad \text{and} \qquad (\alpha, \beta)$$

which yield the Riemann DE [upon substitution of $\lambda_1 = \mu_1 = 0$, $\lambda_2 = 1 - \gamma$, $\mu_2 = \gamma - \alpha - \beta$, $\nu_1 = \alpha$, and $\nu_2 = \beta$ in (9.59)]

$$w'' + \left(\frac{\gamma}{z} + \frac{1-\gamma+\alpha+\beta}{z-1}\right)w' + \frac{\alpha\beta}{z(z-1)}w = 0$$

This important equation is commonly written

$$z(1-z)w'' + [\gamma - (1+\alpha+\beta)z]w' - \alpha\beta w = 0 \tag{9.60}$$

and is called the *hypergeometric differential equation* (HDE). We will study this equation next.

9.5.4 The Hypergeometric Function

The two characteristic exponents of Eq. (9.60) at $z = 0$ are 0 and $1 - \gamma$. It follows from Theorem 9.5.6 that there exists an analytic solution at $z = 0$. Let us denote this solution by $F(\alpha, \beta; \gamma; z)$ and write

$$F(\alpha, \beta; \gamma; z) = \sum_{k=0}^{\infty} a_k z^k \qquad \text{where } a_0 = 1$$

Substituting in (9.60), we obtain the recurrence relation

$$a_{k+1} = \frac{(\alpha + k)(\beta + k)}{(k + 1)(\gamma + k)} a_k \qquad \text{for } k \geqslant 0$$

These coefficients can be determined successively if $\gamma \neq 0, -1, -2, \ldots$. We get

$$\begin{aligned} F(\alpha, \beta; \gamma; z) &= 1 + \sum_{k=1}^{\infty} \frac{\alpha(\alpha + 1)\cdots(\alpha + k - 1)\beta(\beta + 1)\cdots(\beta + k - 1)}{k!\,\gamma(\gamma + 1)\cdots(\gamma + k + 1)} z^k \\ &= \frac{\Gamma(\gamma)}{\Gamma(\alpha)\Gamma(\beta)} \sum_{k=0}^{\infty} \frac{\Gamma(\alpha + k)\Gamma(\beta + k)}{\Gamma(k + 1)\Gamma(\gamma + k)} z^k \end{aligned} \tag{9.61}$$

The series in (9.61) is called the *hypergeometric series*, because $F(1, \beta; \beta; z)$ is simply the geometric series. Some of the elementary properties of the hypergeometric function are discussed in the following examples.

Example 9.5.4

Many of the properties of the hypergeometric function can be obtained directly from the HDE [Eq. (9.60)]. For instance, differentiating (9.60) and letting $v = w'$, we obtain

$$z(1 - z)v'' + [\gamma + 1 - (\alpha + \beta + 3)z]v' - (\alpha + 1)(\beta + 1)v = 0$$

which shows that

$$F'(\alpha, \beta; \gamma; z) = C\,F(\alpha + 1, \beta + 1; \gamma + 1; z)$$

The constant C can be determined by differentiating (9.61), setting $z = 0$ in the result, and noting that $F(\alpha + 1, \beta + 1; \gamma + 1; 0) = 1$. Then we obtain

$$F'(\alpha, \beta; \gamma; z) = \frac{\alpha\beta}{\gamma} F(\alpha + 1, \beta + 1; \gamma + 1; z)$$

On the other hand, making the substitution $w = z^{1-\gamma}u\,(\gamma \neq 1)$ in (9.60), we obtain

$$z(1 - z)u'' + [\gamma_1 - (\alpha_1 + \beta_1 + 1)z]u' - \alpha_1 \beta_1 u = 0$$

where $\alpha_1 = \alpha - \gamma + 1$, $\beta_1 = \beta - \gamma + 1$, and $\gamma_1 = 2 - \gamma$. Thus,

$$u = F(\alpha - \gamma + 1, \beta - \gamma + 1; 2 - \gamma; z)$$

and u is, therefore, analytic at $z = 0$. This leads to an interesting result. Provided that γ is not an integer, the two functions

$$w_1(z) \equiv F(\alpha, \beta; \gamma; z)$$

and
$$w_2(z) = z^{1-\gamma} F(\alpha - \gamma + 1, \beta - \gamma + 1; 2 - \gamma; z)$$

form a canonical basis of solutions to the HDE at $z = 0$. This follows from Theorem 9.5.6 and the fact that $(0, 1 - \gamma)$ are the pair of characteristic exponents at $z = 0$.

A third relation can be obtained by making the substitution $w = (1 - z)^{\gamma - \alpha - \beta} u$. This leads to a hypergeometric equation for u with $\alpha_1 = \gamma - \alpha, \beta_1 = \gamma - \beta$, and $\gamma_1 = \gamma$. We, therefore, have the identity

$$F(\alpha, \beta; \gamma; z) = (1 - z)^{\gamma - \alpha - \beta} F(\gamma - \alpha, \gamma - \beta; \gamma; z)$$

To obtain the canonical basis at $z = 1$, we make the substitution $t = 1 - z$, and note that the result is again the HDE with $\alpha_1 = \alpha$, $\beta_1 = \beta$, and $\gamma_1 = \alpha + \beta - \gamma + 1$. It follows from $w_1(z)$ and $w_2(z)$ that

$$w_3(z) = F(\alpha, \beta; \alpha + \beta - \gamma + 1; 1 - z)$$

and
$$w_4(z) = (1 - z)^{\gamma - \alpha - \beta} F(\gamma - \beta, \gamma - \alpha; \gamma - \alpha - \beta + 1; 1 - z)$$

These functions form a canonical basis of solutions to the HDE at $z = 1$.

A symmetry of the hypergeometric function that is easily obtained from the HDE is

$$F(\alpha, \beta; \gamma; z) = F(\beta, \alpha; \gamma; z)$$

The six functions $F(\alpha \pm 1, \beta; \gamma; z), F(\alpha, \beta \pm 1; \gamma; z)$, and $F(\alpha, \beta; \gamma \pm 1; z)$ are called hypergeometric functions *contiguous* to $F(\alpha, \beta; \gamma; z)$. We can use the series representation in (9.61) to derive the following relations among these contiguous hypergeometric functions:

$$[\gamma - 2\alpha - (\beta - \alpha)z]\, F(\alpha, \beta; \gamma; z) + \alpha(1 - z)\, F(\alpha + 1, \beta; \gamma; z) - (\gamma - \alpha)\, F(\alpha - 1, \beta; \gamma; z) = 0$$

$$(\gamma - \alpha - 1)\, F(\alpha, \beta; \gamma; z) + \alpha F(\alpha + 1, \beta; \gamma; z) - (\gamma - 1)\, F(\alpha, \beta; \gamma - 1; z) = 0$$

●

Example 9.5.5

Example 9.5.4 showed how to obtain the basis of solutions at $z = 1$ from the regular solution to the HDE $z = 0$, which is $F(\alpha, \beta; \gamma; z)$. We can show that the basis of solutions at $z = \infty$ can also be obtained from the hypergeometric function.

Equation (9.55) suggests a function of the form

$$v(z) = z^r F\left(\alpha_1, \beta_1; \gamma_1; \frac{1}{z}\right) \equiv z^r w\left(\frac{1}{z}\right) \tag{1}$$

where r, α_1, β_1, and γ_1 are to be determined. It is convenient to let $t \equiv 1/z$, $dw/dt \equiv \dot{w}$, and $d^2w/dt^2 \equiv \ddot{w}$. Then, since w is a solution of the HDE, we have

$$\ddot{w} = -\frac{[\gamma - (\alpha + \beta + 1)t]\dot{w}}{t(1 - t)} + \frac{\alpha\beta}{t(1 - t)} w$$

In terms of $z = 1/t$, this is

$$\ddot{w} = \frac{[\gamma z^2 - (\alpha + \beta + 1)z]\dot{w}}{1 - z} - \frac{\alpha\beta z^2}{1 - z} w$$

Now we differentiate (1) with respect to z and use

$$\frac{df}{dz} = \left(\frac{df}{dt}\right)\frac{dt}{dz} = -\left(\frac{1}{z^2}\right)\frac{df}{dt}$$

to obtain

$$v' \equiv \frac{dv}{dz} = rz^{r-1}w - z^{r-2}\dot{w}$$

$$v'' \equiv \frac{d^2v}{dz^2} = r(r-1)z^{r-2}w - 2(r-1)z^{r-3}\dot{w} + z^{r-4}\ddot{w}$$

Multiplying the equation for v'' by $z(1-z)$ gives

$$z(1-z)v'' = r(r-1)z^{r-1}(1-z)w - 2(r-1)z^{r-2}(1-z)\dot{w} + z^{r-3}(1-z)\ddot{w} \tag{2}$$

Evaluating $\dot{w}$ from the equation for v' gives

$$\dot{w} = z^{2-r}(rz^{r-1}w - v') = z^{2-r}\left(\frac{rv}{z} - v'\right)$$

and substituting for w, $\dot{w}$, and $\ddot{w}$ in terms of v in (2) yields

$$z(1-z)v'' + [1-\alpha-\beta-2r-(2-\gamma-2r)z]v' - \left[r^2 - r + r\gamma - \frac{1}{z}(r+\alpha)(r+\beta)\right]v = 0 \tag{3}$$

This reduces to the HDE if $r = -\alpha$ or $r = -\beta$. For $r = -\alpha$ the parameters become $\alpha_1 = \alpha$, $\beta_1 = 1 + \alpha - \gamma$, and $\gamma_1 = \alpha - \beta + 1$, which yield

$$v_1(z) = z^{-\alpha}F\left(\alpha, 1+\alpha-\gamma; \alpha-\beta+1; \frac{1}{z}\right)$$

For $r = -\beta$ the parameters are $\alpha_1 = \beta$, $\beta_1 = 1 + \beta - \gamma$, and $\gamma_1 = \beta - \alpha + 1$. Thus,

$$v_2(z) = z^{-\beta}F\left(\beta, 1+\beta-\gamma; \beta-\alpha+1; \frac{1}{z}\right)$$

These two equations form a basis of solutions for the HDE that are valid about $z = \infty$.

●

As the preceding examples suggest, it is possible to obtain many relations among the hypergeometric functions with different parameters and independent variables. In fact, the nineteenth-century mathematician Kummer showed that there are 24 different (but linearly dependent, of course) solutions to the HDE. These are collectively known as *Kummer's solutions*, and six of them were derived in the two foregoing examples. Another important relation (shown in Exercise 9.5.5) is that if $F(\alpha, \beta; \gamma; z)$ is a solution of the HDE, then so is

$$z^{\alpha-\gamma}(1-z)^{\gamma-\alpha-\beta}F\left(\gamma-\alpha, 1-\alpha; 1-\alpha+\beta; \frac{1}{z}\right) \tag{9.62}$$

Many of the functions that occur in mathematical physics are related to the hypergeometric function. Even some of the common elementary functions can be expressed in terms of the hypergeometric function with appropriate parameters. For example, when $\beta = \gamma$, we obtain

$$F(\alpha, \beta; \beta; z) = \sum_{k=0}^{\infty} \frac{\Gamma(\alpha + k)}{\Gamma(\alpha)\Gamma(k+1)} z^k = (1-z)^{-\alpha}$$

Similarly,

$$F(\tfrac{1}{2}, \tfrac{1}{2}; \tfrac{3}{2}; z^2) = \frac{1}{z} \sin^{-1} z$$

$$F(1, 1; 2; -z) = \frac{1}{z} \ln(1+z)$$

However, the real power of the hypergeometric function is that it encompasses almost all of the nonelementary functions encountered in physics. Let us look briefly at a few of these.

Jacobi functions. Jacobi functions are solutions of the DE

$$(1-x^2)\frac{d^2u}{dx^2} + [\beta - \alpha - (\alpha + \beta + 2)x]\frac{du}{dx} + \lambda(\lambda + \alpha + \beta + 1)u = 0 \qquad (9.63)$$

Defining $x \equiv 1 - 2z$ changes (9.63) into the HDE with parameters $\alpha_1 = \lambda$, $\beta_1 = \lambda + \alpha + \beta + 1$, and $\gamma_1 = 1 + \alpha$. The solutions of (9.63), called the *Jacobi functions of the first kind*, are, with appropriate normalization,

$$P_\lambda^{(\alpha,\beta)}(z) = \frac{\Gamma(\lambda + \alpha + 1)}{\Gamma(\lambda + 1)\Gamma(\alpha + 1)} F\left(-\lambda, \lambda + \alpha + \beta + 1; 1 + \alpha; \frac{1-z}{2}\right) \qquad (9.64)$$

When $\lambda = n$, where n is a nonnegative integer, $P_n^{(\alpha,\beta)}(z)$ is a polynomial of degree n with the following expansion:

$$P_n^{(\alpha,\beta)}(z) = \frac{\Gamma(n + \alpha + 1)}{\Gamma(n+1)\Gamma(n + \alpha + \beta + 1)} \sum_{k=0}^{n} \frac{\Gamma(n + \alpha + \beta + k + 1)}{\Gamma(\alpha + k + 1)} \left(\frac{z-1}{2}\right)^k$$

These are the Jacobi polynomials discussed in Chapter 5. In fact, the DE satisfied by $P_n^{(\alpha,\beta)}(x)$ in Chapter 5 is identical to (9.63). Note that the transformation $x = 1 - 2z$ translates the points $z = 0$ and $z = 1$ to the points $x = 1$ and $x = -1$, respectively. Thus the regular singular points of the Jacobi functions of the first kind are at ± 1 and ∞.

A second, linearly independent, set of solutions of (9.63) is obtained by using (9.62). These are called the *Jacobi functions of the second kind*:

$$Q_\lambda^{(\alpha,\beta)}(z) =$$

$$\frac{2^{\lambda+\alpha+\beta}\Gamma(\lambda + \alpha + 1)\Gamma(\lambda + \beta + 1)}{\Gamma(2\lambda + \alpha + \beta + 2)(z-1)^{\lambda+\alpha+1}(z+1)^\beta} F\left(\lambda + \alpha + 1, \lambda + 1; 2\lambda + \alpha + \beta + 2; \frac{2}{1-z}\right) \qquad (9.65)$$

Gegenbauer (ultraspherical) functions. Gegenbauer functions, or *ultraspherical functions*, are special cases of Jacobi functions for which $\alpha = \beta = \mu - \frac{1}{2}$. They are defined by

$$C_\lambda^\mu(z) \equiv \frac{\Gamma(\lambda + 2\mu)}{\Gamma(\lambda + 1)\Gamma(2\mu)} F\left(-\lambda, \lambda + 2\mu; \mu + \tfrac{1}{2}; \frac{1-z}{2} \right) \tag{9.66}$$

Note the change in the normalization constant. Linearly independent Gegenbauer functions "of the second kind" can be obtained from (9.65) by the substitution $\alpha = \beta = \mu - \frac{1}{2}$.

Legendre functions. Another special case of the Jacobi functions is obtained when $\alpha = \beta = 0$. These are called *Legendre functions of the first kind*:

$$P_\lambda(z) \equiv P_\lambda^{(0,0)}(z) = F\left(-\lambda, \lambda + 1; 1; \frac{1-z}{2} \right) \tag{9.67}$$

Legendre functions of the second kind are obtained by letting $\alpha = \beta = 0$ in (9.65), yielding

$$Q_\lambda(z) \equiv Q_\lambda^{(0,0)}(z) = \frac{2^\lambda [\Gamma(\lambda + 1)]^2}{\Gamma(2\lambda + 2)(z-1)^{\lambda+1}} F\left(\lambda + 1, \lambda + 1; 2\lambda + 2; \frac{2}{1-z} \right)$$

Other functions derived from the Jacobi functions are obtained similarly (see Chapter 5).

9.5.5 Confluent Hypergeometric Functions; Bessel Functions

The transformation $x = 1 - 2z$ translates the regular singular points of the HDE by a finite amount. Consequently, the new functions still have two regular singular points, $z = \pm 1$, in the complex plane, $\mathbb{C}$. In some physical cases of importance, only the origin, corresponding to $r = 0$ in spherical coordinates (typically the location of the source of a central force), is the singular point. If we want to obtain a differential equation consistent with such a case, we have to "push" the singular point $z = 1$ to infinity. This can be achieved by making the substitution $t = rz$ in the HDE and taking the limit $r \to \infty$. The substitution yields

$$\frac{d^2 w}{dt^2} + \left(\frac{\gamma}{t} + \frac{1 - \gamma + \alpha + \beta}{t - r} \right) \frac{dw}{dt} + \frac{\alpha\beta}{t(t-r)} w = 0 \tag{9.68}$$

If we blindly take the limit $r \to \infty$ with α, β, and γ remaining finite, Eq. (9.68) reduces to the elementary FODE

$$\frac{d^2 w}{dt^2} + \frac{\gamma}{t} \frac{dw}{dt} = 0$$

To obtain a nonelementary DE, we need to manipulate the parameters, to let some of them tend to infinity. We want γ to remain finite, because it occurs in the first term of the coefficient of dw/dt. We, therefore, let β or α tend to infinity. The result will be the same either way because α and β appear symmetrically in the equation. It is customary to let $\beta = r \to \infty$. In that case Eq. (9.68) becomes (note that t is *finite*)

$$\frac{d^2w}{dt^2} + \left(\frac{\gamma}{t} - 1\right)\frac{dw}{dt} - \frac{\alpha}{t}w = 0$$

Multiplying by t and changing the independent variable back to z yields

$$zw''(z) + (\gamma - z)w'(z) - \alpha w(z) = 0 \tag{9.69}$$

This is called the *confluent hypergeometric differential equation* (CHDE).

It is important to note that the point at infinity is no longer a regular singular point of Eq. (9.69). However, we can still seek solutions of the CHDE in the form of power series. Since $z = 0$ is still a regular singular point of the CHDE, we can obtain expansions about that point. The characteristic exponents are 0 and $1 - \gamma$, as before. Thus, there is an analytic solution to the CHDE at the origin, which is called the *confluent hypergeometric function* and denoted $\Phi(\alpha, \gamma; z)$. Since $z = 0$ is the only (finite) singularity of the CHDE, $\Phi(\alpha, \gamma; z)$ is an entire function.

We can obtain the series expansion of $\Phi(\alpha, \gamma; z)$ directly from (9.61) and the fact that

$$\Phi(\alpha, \gamma; z) = \lim_{\beta \to \infty} F\left(\alpha, \beta; \gamma; \frac{z}{\beta}\right)$$

The result is

$$\Phi(\alpha, \gamma; z) = \frac{\Gamma(\gamma)}{\Gamma(\alpha)} \sum_{k=0}^{\infty} \frac{\Gamma(\alpha + k)}{\Gamma(\gamma + k)\Gamma(k + 1)} z^k \tag{9.70}$$

This is called the *confluent hypergeometric series.*

A second solution of the CHDE can be obtained, as for the HDE. If $1 - \gamma$ is not an integer, then (see Example 9.5.4)

$$z^{1-\gamma} \lim_{\beta \to \infty} F\left(\beta - \gamma + 1, \alpha - \gamma + 1; 2 - \gamma; \frac{z}{\beta}\right)$$

$$= z^{1-\gamma} \lim_{\beta \to \infty} F\left(\alpha - \gamma + 1, \beta - \gamma + 1; 2 - \gamma; \frac{z}{\beta}\right)$$

↑ because F is symmetric in its first two parameters

$$= z^{1-\gamma}\Phi(\alpha - \gamma + 1, 2 - \gamma; z)$$

Thus, any solution of the CHDE can be written as a linear combination of $\Phi(\alpha, \gamma; z)$ and $z^{1-\gamma}\Phi(\alpha - \gamma + 1, 2 - \gamma; z)$.

Example 9.5.6

The time-independent Schrödinger equation, in units in which $\hbar = 1$, is

$$-\frac{1}{2m}\nabla^2\Psi + V(r)\Psi = E\Psi$$

For the case of hydrogenlike atoms, $V(r) = -Ze^2/r$, where Z is the atomic number, and the equation reduces to

$$\nabla^2\Psi + \left(2mE + \frac{2mZe^2}{r}\right)\Psi = 0$$

The radial part of this equation is given by Eq. (8.14b) with $f(r) = 2mE + 2mZe^2/r$. Defining $u \equiv rR(r)$, we may write

$$\frac{d^2u}{dr^2} + \left(\lambda + \frac{a}{r} - \frac{b}{r^2}\right)u = 0 \tag{1}$$

where $\lambda \equiv 2mE$, $a = 2mZe^2$, and $b = l(l+1)$. This equation can be further simplified by defining $r \equiv kz$ (k is an arbitrary constant to be determined later), yielding

$$\frac{d^2u}{dz^2} + \left(\lambda k^2 + \frac{ak}{z} - \frac{b}{z^2}\right)u = 0$$

Choosing $\lambda k^2 = -\frac{1}{4}$ yields

$$\frac{d^2u}{dz^2} + \left(-\frac{1}{4} + \frac{a/2\sqrt{-\lambda}}{z} - \frac{b}{z^2}\right)u = 0$$

Equations of this form can be transformed into the CHDE by making the substitution

$$u(z) = z^{\mu}e^{-\nu z}f(z)$$

It then follows that (with $a' \equiv a/2\sqrt{-\lambda}$)

$$\frac{d^2f}{dz^2} + \left(\frac{2\mu}{z} - 2\nu\right)\frac{df}{dz} + \left[-\frac{1}{4} + \frac{\mu(\mu-1)}{z^2} - \frac{2\mu\nu}{z} + \frac{a'}{z} - \frac{b}{z^2} + \nu^2\right]f = 0 \tag{2}$$

Choosing $\nu^2 = \frac{1}{4}$ and $\mu(\mu - 1) = b$ reduces (2) to

$$f'' + \left(\frac{2\mu}{z} - 2\nu\right)f' - \frac{2\mu\nu - a'}{z}f = 0$$

which is of the form of (9.69).

On physical grounds, we expect $u(z) \to 0$ as $z \to \infty$. Therefore, $\nu = \frac{1}{2}$. Similarly,

$$b = l(l+1) \quad \Rightarrow \quad \mu = -l \quad \text{or} \quad \mu = l + 1$$

Again on physical grounds, we demand that $u(0)$ be finite (the wave function must not "blow up" at $r = 0$). This implies that $\mu = l + 1$. We thus obtain

$$f'' + \left(\frac{2\mu}{z} - 1\right)f' - \frac{\mu - a'}{z}f = 0 \qquad \text{where } \mu = l + 1$$

Multiplying by z gives

$$zf'' + (2\mu - z)f' - (\mu - a')f = 0$$

Comparing this with (9.69) shows that f is proportional to $\Phi(\mu - a', 2\mu; z)$. Thus, the solution of (1) can be written as

$$u = Cz^{l+1}e^{(-1/2)z}\Phi(l + 1 - a', 2l + 2; z)$$

An argument similar to that used in solving Exercise 9.3.2 shows that the product $e^{-1/2z}\Phi(l + 1 - a', 2l + 2; z)$ will be infinite unless the power series representing Φ terminates (becomes a polynomial). This takes place only if

$$l + 1 - a' = -N \tag{3}$$

for some integer $N \geqslant 0$. In that case we define the *Laguerre* polynomials

$$L_n^j(z) \equiv \frac{\Gamma(n + j + 1)}{\Gamma(n + 1)\Gamma(j + 1)}\Phi(-N, j + 1; z) \qquad \text{where } j = 2l + 1$$

Condition (3) is the quantization rule for the energy levels of a hydrogenlike atom. Recalling the definition of a', we obtain $a' = a/2\sqrt{-\lambda} = N + l + 1 \equiv n$, where $n \geqslant 1$, which yields

$$\lambda = -\frac{a^2}{4n^2} = -\frac{4m^2Z^2e^4}{4n^2} \qquad \text{for } n \geqslant 1$$

or

$$E = -Z^2\left(\frac{mc^2}{2}\right)\alpha\frac{1}{n^2} \qquad \text{for } n \geqslant 1$$

where $\alpha = e^2/\hbar c = 1/137$ is the *fine structure constant*. By substituting for all constants, we can write this equation in a form that gives the energy in electron volts:

$$E = -\frac{13.6Z^2}{n^2}\,(\text{eV}) \qquad \text{for } n \geqslant 1 \tag{4}$$

The solutions to (1), on the other hand, are

$$R_{n,l}(r) \equiv \frac{u_{n,l}(r)}{r} = Cr^l e^{-Zr/na_0}\,\Phi\left(-n + l + 1, 2l + 2; \frac{2Zr}{na_0}\right)$$

and the most general solution is

$$\Psi(\mathbf{r}) = \sum_{n=0}^{\infty}\sum_{l=0}^{n-1}\sum_{m=-l}^{+l} C_{nlm}Y_{lm}(\theta, \varphi)r^l e^{-Zr/na_0}\Phi\left(-n + l + 1, 2l + 2; \frac{2Zr}{na_0}\right)$$

where $a_0 = \hbar^2/me^2 = 0.529 \times 10^{-8}$ cm is the Bohr radius. •

The Bessel equation, introduced in Chapter 8, is usually written as

$$w'' + \frac{1}{z}w' + \left(1 - \frac{\nu^2}{z^2}\right)w = 0 \tag{9.71}$$

As in Example 9.5.6, the substitution $u = z^{\mu} e^{-\eta z} f(z)$ transforms (9.71) into

$$f'' + \left(\frac{2\mu + 1}{z} - 2\eta\right) f' + \left[\frac{\mu^2 - \nu^2}{z^2} - \frac{\eta(2\mu + 1)}{z} + \eta^2 + 1\right] f = 0$$

which, if we set $\mu = \nu$ and $\eta = i$, reduces to

$$f'' + \left(\frac{2\nu + 1}{z} - 2i\right) f' - \frac{(2\nu + 1)i}{z} f = 0$$

Making the further substitution $2iz = t$, we obtain

$$t\frac{d^2 f}{dt^2} + [(2\nu + 1) - t]\frac{df}{dt} - (\nu + \tfrac{1}{2}) f = 0$$

which is of the form of (9.69) with $\alpha = \nu + \frac{1}{2}$ and $\gamma = 2\nu + 1$.

Thus, solutions of the Bessel equation [Eq. (9.71)] can be written as constant multiples of $z^{\nu} e^{-iz} \Phi(\nu + \frac{1}{2}, 2\nu + 1; 2iz)$. With proper normalization, we define the *Bessel function of the first kind and of order* ν as

$$J_{\nu}(z) \equiv \frac{1}{\Gamma(\nu + 1)} \left(\frac{z}{2}\right)^{\nu} e^{-iz} \Phi(\nu + \tfrac{1}{2}, 2\nu + 1; 2iz) \tag{9.72}$$

Using Eq. (9.70) and the expansion for e^{-iz}, we can show that

$$J_{\nu}(z) = \left(\frac{z}{2}\right)^{\nu} \sum_{k=0}^{\infty} \frac{(-1)^k}{k!\,\Gamma(\nu + k + 1)} \left(\frac{z}{2}\right)^{2k} \qquad \text{where } \nu \geqslant 0 \tag{9.73}$$

Choosing ν to be nonnegative ensures that $J_{\nu}(z)$ is well-defined at $z = 0$. Thus, as mentioned earlier in a broader sense, $J_{\nu}(z)$ is an entire function if $\nu \geqslant 0$. The second linearly independent solution can be obtained as usual and is proportional to

$$\begin{aligned} &\left(\frac{z}{2}\right)^{\nu} e^{-iz} z^{1-(2\nu+1)} \Phi(\nu + \tfrac{1}{2} - (2\nu + 1) + 1, 2 - (2\nu + 1); 2iz) \\ &\qquad = (C)\left(\frac{z}{2}\right)^{-\nu} e^{-iz} \Phi(-\nu + \tfrac{1}{2}, -2\nu + 1; 2iz) \\ &\qquad = (C) J_{-\nu}(z) \end{aligned}$$

provided that $1 - \gamma \equiv 1 - (2\nu + 1) = -2\nu$ is not an integer. When ν is an integer, $J_{-n}(z) = (-1)^n J_n(z)$ (see Exercise 9.5.6). Thus, when ν is a noninteger, the most general solution is of the form $AJ_{\nu}(z) + BJ_{-\nu}(z)$.

How do we find a second linearly independent solution when ν is an integer, n? We first define

$$Y_{\nu}(z) \equiv \frac{[J_{\nu}(z) \cos(\nu\pi) - J_{-\nu}(z)]}{\sin(\nu\pi)} \tag{9.74}$$

called the *Bessel function of the second kind*, or the *Neumann function.* For noninteger ν this is simply a linear combination of the two linearly independent solutions J_{ν} and

$J_{-\nu}$. For integer ν the function is indeterminate. Therefore, we use l' Hôpital's rule and define

$$Y_n(z) \equiv \lim_{\nu \to n} Y_\nu(z) = \frac{1}{\pi} \lim_{\nu \to n} \left[\frac{\partial J_\nu}{\partial \nu} - (-1)^n \frac{\partial J_{-\nu}}{\partial \nu} \right]$$

Differentiating (9.73) with respect to ν, we obtain

$$\frac{\partial J_\nu}{\partial \nu} = J_\nu \ln\left(\frac{z}{2}\right) - \left(\frac{z}{2}\right)^\nu \sum_{k=0}^{\infty} (-1)^k \frac{\Psi(\nu + k + 1)}{k!\Gamma(\nu + k + 1)} \left(\frac{z}{2}\right)^{2k}$$

where $\Psi(z) \equiv (d/dz) \ln \Gamma(z)$. Similarly,

$$\frac{\partial J_{-\nu}}{\partial \nu} = -J_{-\nu} \ln\left(\frac{z}{2}\right) + \left(\frac{z}{2}\right)^{-\nu} \sum_{k=0}^{\infty} \frac{\Psi(-\nu + k + 1)}{k!\Gamma(-\nu + k + 1)} \left(\frac{z}{2}\right)^{2k}$$

Substituting these expressions for $\partial J_\nu / \partial \nu$ and $\partial J_{-\nu}/\partial \nu$ in the definition of $Y_n(z)$, we obtain

$$\begin{aligned} Y_n(z) = \frac{2}{\pi} J_n(z) \ln\left(\frac{z}{2}\right) &- \frac{1}{\pi}\left(\frac{z}{2}\right)^n \sum_{k=0}^{\infty} (-1)^k \frac{\Psi(n + k + 1)}{k!\Gamma(n + k + 1)} \left(\frac{z}{2}\right)^{2k} \\ &- \frac{1}{\pi}\left(\frac{z}{2}\right)^{-n} (-1)^n \sum_{k=0}^{\infty} (-1)^k \frac{\Psi(k - n + 1)}{k!\Gamma(k - n + 1)} \left(\frac{z}{2}\right)^{2k} \end{aligned} \tag{9.75}$$

Since $Y_\nu(z)$ is linearly independent of $J_\nu(z)$ for any ν, integer or noninteger, it is convenient to consider $\{J_\nu(z), Y_\nu(z)\}$ as a basis of solutions for the Bessel equation.

Another basis of solutions is defined as

$$H_\nu^{(1)}(z) \equiv J_\nu(z) + iY_\nu(z) \qquad \text{and} \qquad H_\nu^{(2)}(z) \equiv J_\nu(z) - iY_\nu(z) \tag{9.76}$$

which are called *Bessel functions of the third kind*, or *Hankel functions*.

Replacing z by iz in the Bessel equation yields

$$\frac{d^2 w}{dz^2} + \frac{1}{z}\frac{dw}{dz} - \left(1 + \frac{\nu^2}{z^2}\right) w = 0$$

whose basis of solutions consists of multiples of $J_\nu(iz)$ and $J_{-\nu}(iz)$. Thus, the *modified Bessel functions of the first kind* are defined as

$$I_\nu(z) \equiv e^{-i(\pi/2)\nu} J_\nu(iz) = \left(\frac{z}{2}\right)^\nu \sum_{k=0}^{\infty} \frac{1}{k!\Gamma(k + \nu + 1)} \left(\frac{z}{2}\right)^{2k}$$

Similarly, the *modified Bessel functions of the third kind* are defined as

$$K_\nu(z) = \frac{\pi}{2 \sin \nu\pi} [I_{-\nu}(z) - I_\nu(z)]$$

When ν is an integer, n, $I_n = I_{-n}$, and K_n is indeterminate. Thus, we define $K_n(z)$ as $\lim_{\nu \to n} K_\nu(z)$. This gives

$$K_n(z) = \frac{(-1)^n}{2} \lim_{\nu \to n} \left[\frac{\partial I_{-\nu}(z)}{\partial \nu} - \frac{\partial I_\nu(z)}{\partial \nu} \right]$$

which has the power-series representation

$$K_n(z) = (-1)^{n+1} I_n(z) \ln\left(\frac{z}{2}\right) + \frac{1}{2}(-1)^n \left(\frac{z}{2}\right)^n \sum_{k=0}^{\infty} \frac{\Psi(k+n+1)}{k!(n+k)!}\left(\frac{z}{2}\right)^{2k}$$
$$+ \frac{1}{2}(-1)^n \left(\frac{z}{2}\right)^{-n} \sum_{k=0}^{\infty} \frac{\Psi(k-n+1)}{k!\Gamma(k-n+1)}\left(\frac{z}{2}\right)^{2k}$$

We can obtain a recurrence relation for solutions of the Bessel equation as follows. If $Z_\nu(z)$ is a solution of order ν, then (see Problem 9.56)

$$Z_{\nu+1} = C_1 z^\nu \frac{d}{dz}[z^{-\nu} Z_\nu(z)] \qquad \text{and} \qquad Z_{\nu-1} = C_2 z^{-\nu} \frac{d}{dz}[z^\nu Z_\nu(z)]$$

If the constants are chosen in such a way that $Z_{\nu+1}$ and $Z_{\nu-1}$ satisfy (9.73), then $C_1 = -1$ and $C_2 = 1$. Carrying out the differentiation of the equations for $Z_{\nu+1}$ and $Z_{\nu-1}$, we obtain

$$Z_{\nu+1} = \left(\frac{\nu}{z}\right) Z_\nu - \frac{dZ_\nu}{dz} \tag{9.77a}$$

$$Z_{\nu-1} = \left(\frac{\nu}{z}\right) Z_\nu + \frac{dZ_\nu}{dz} \tag{9.77b}$$

Adding these two equations yields the recursion relation

$$Z_{\nu-1}(z) + Z_{\nu+1}(z) = \frac{2\nu}{z} Z_\nu(z) \tag{9.78}$$

where $Z_\nu(z)$ can be any of the three kinds of Bessel functions.

The next chapter will show, in the context of Sturm-Liouville systems, that Bessel functions are orthogonal functions, and, therefore, that other appropriate functions can be expanded in terms of them.

Exercises

9.5.1 Show that the FOLDE given in Eq. (9.50) has the solution given in (9.51).

9.5.2 Show that a function that has *only* poles in the extended complex plane must be a rational function.[3]

9.5.3 Show that Eq. (9.57) represents the most general SOFDE.

9.5.4 Show that the elliptic function of the first kind, defined as

$$K(z) \equiv \int_0^{\pi/2} \frac{d\theta}{\sqrt{1 - z\sin^2\theta}}$$

can be expressed as $(\pi/2)F(\frac{1}{2}, \frac{1}{2}; 1; z)$.

[3] This exercise does not really belong here and should be in Chapter 7. However, because of its relevance to this material and also because it provides an excuse to review complex analysis, it appears here.

9.5.5 Show that if $F(\alpha, \beta; \gamma; z)$ is a solution of the HDE, then so is

$$z^{\alpha-\gamma}(1-z)^{\gamma-\alpha-\beta}F\left(\gamma-\alpha, 1-\alpha; 1+\alpha-\beta; \frac{1}{z}\right)$$

9.5.6 Show that $J_{-n}(z) = (-1)^n J_n(z)$.

9.5.7 In a potential-free region the radial equation of (8.14b) reduces to a DE in which $f(r) = \lambda$, a constant. Write the solutions of that DE in terms of Bessel functions.

9.5.8 Theorem 9.5.6 states that if the difference between characteristic exponents of a SOLDE is an integer, then there *may* be a $\ln z$ term in the second solution. However, this term will be multiplied by a constant. Only if this constant is nonzero will there be a logarithmic divergence at $z = 0$. In solutions of the Bessel equation with $\nu = n + \frac{1}{2}$, where n is a nonnegative integer, the log term is absent. This can be seen by showing that $Z_{n+1/2}(z)$ and $Z_{-n-1/2}(z)$ are linearly independent solutions. Evaluate the Wronskian of the solutions to the Bessel equation and show that it vanishes only if ν is an integer. [Hint: Consider the value of the Wronskian for small z, and use the formula $\Gamma(\nu)\Gamma(1-\nu) = \pi/\sin \nu\pi$.]

9.5.9 Use the recursion relation of (9.78) to prove that

$$\left(\frac{1}{z}\frac{d}{dz}\right)^m [z^{-\nu}Z_\nu(z)] = z^{\nu-m}Z_{\nu-m}(z)$$

9.5.10 Write $J_{1/2}(z)$ and $J_{-1/2}(z)$ in terms of elementary functions.

9.5.11 From the results of Exercises 9.5.9 and 9.5.10, derive the relation

$$J_{-n-1/2}(z) = \left(\frac{2}{\pi}\right)^{1/2} z^{n+1/2}\left(\frac{1}{z}\frac{d}{dz}\right)^n\left(\frac{\cos z}{z}\right)$$

9.5.12 Use the relations

$$\frac{d}{dz}[z^\nu J_\nu(z)] = z^\nu J_{\nu-1}(z) \qquad \text{and} \qquad \frac{d}{dz}[z^{-\nu}J_\nu(z)] = -z^{-\nu}J_{\nu+1}(z) \tag{1}$$

to obtain

(a) $$\int z^{\nu+1}J_\nu(z)\,dz = z^{\nu+1}J_{\nu+1}(z) \qquad \text{and} \qquad \int z^{-\nu+1}J_\nu(z)\,dz = -z^{-\nu+1}J_{\nu-1}(z)$$

(b) $$\int z^{\mu+1}J_\nu(z)\,dz = z^{\mu+1}J_{\nu+1}(z) + (\mu-\nu)z^\mu J_\nu(z) - (\mu^2-\nu^2)\int z^{\mu-1}J_\nu(z)\,dz$$

(c) Evaluate $\int z^3 J_0(z)\,dz$.

PROBLEMS

9.1 Use direct differentiation to show that the function given in Eq. (9.8) satisfies the FOLDE of Eq. (9.7).

9.2 Let $u(x, y) = C$ be a solution of the DE $M\,dx + N\,dy = 0$. Show that **(a)** $(\partial u/\partial x)/M = (\partial u/\partial y)/N$, and **(b)** $\mu(x, y) \equiv (\partial u/\partial x)/M$ is an integrating factor for the DE. **(c)** Given an arbitrary differentiable function $F: \mathbb{R} \to \mathbb{R}$, show that $v(x, y) = \mu(x, y)F(u)$ is also an integrating factor for the DE.

9.3 Analyze the capacitor's charge in an RC circuit in which a constant potential V_0 is applied for a time $T > 0$ and then disconnected. Consider the cases where $t < T$ and $t > T$.

9.4 Find all functions $f(x)$ whose definite integral from 0 to x equals the square of their reciprocal.

9.5 **(a)** Let $p_1 u' + p_0 u = 0$ be a homogeneous FOLDE in u. Solve it. (Note that it is separable and can easily be integrated.)
(b) Consider $p_1 y' + p_0 y = q$. Let $y = uv$, and obtain an equation for v. Solve this equation, and obtain a general solution for $p_1 y' + p_0 y = q$. This is the method of variation of parameters, which can also be used for the second-order equations.

9.6 A falling body in air has a motion approximately described by the DE

$$m\frac{dv}{dt} = mg - bv$$

where $v = dx/dt$ is the velocity of the body. Find this velocity as a function of time (t) assuming that the object starts from rest.

9.7 Prove Proposition 9.2.6.

9.8 Let f and g be two differentiable functions that are linearly dependent. Show that their Wronskian vanishes. (Note that f and g need not be solutions of the homogeneous SOLDE.)

9.9 Show that the solutions to the SOLDE $y'' + q(x)y = 0$ have a constant Wronskian.

9.10 Find a general integral relation for $G_n(x)$, the linearly independent "partner" of the Hermite polynomial $H_n(x)$. Specialize this to $n = 0, 1$. Is it possible to find $G_0(x)$ and $G_1(x)$ in terms of elementary functions?

9.11 For each pair of solutions and their DE, calculate the Wronskian, and give a solution satisfying the initial conditions $y(0) = 2$ and $y'(0) = 1$.
(a) $\cos x$ and $\sin x$; $y'' + y = 0$ **(b)** e^{-x} and e^{-3x}; $y'' + 4y' + 3y = 0$
(c) x and e^x; $y'' + \dfrac{x}{1-x}y' - \dfrac{1}{1-x}y = 0$

9.12 Let f_1, f_2, and f_3 be any three solutions of the $y'' + py' + qy = 0$. Show that

$$\det\begin{pmatrix} f_1 & f_1' & f_1'' \\ f_2 & f_2' & f_2'' \\ f_3 & f_3' & f_3'' \end{pmatrix} = 0$$

9.13 For the HSOLDE $y'' + py' + qy = 0$, show that

$$p = -\frac{f_1 f_2'' - f_2 f_1''}{w(f_1, f_2)} \qquad \text{and} \qquad q = \frac{f_1' f_2'' - f_2' f_1''}{w(f_1, f_2)}$$

(Thus, knowing two solutions of an HSOLDE allows us to construct the DE.)

9.14 Let f_1, f_2, and f_3 be three solutions of the third-order linear differential equation

$$y''' + p_2(x)y'' + p_1(x)y' + p_0(x)y = 0$$

Derive a FODE satisfied by the Wronskian

$$w(x) \equiv \det \begin{pmatrix} f_1 & f_1' & f_1'' \\ f_2 & f_2' & f_2'' \\ f_3 & f_3' & f_3'' \end{pmatrix}$$

9.15 Prove Corollary 9.2.11. (Hint: Consider the solution $v = 1$ of the DE $v'' = 0$.)

9.16 Show that if $u(x)$ and $v(x)$ are solutions of the self-adjoint DE $(pu')' + qu = 0$, then *Abel's identity*, $p(uv' - vu') = \text{constant}$, holds.

9.17 Reduce each DE to self-adjoint form.

(a) $x^2y'' + xy' + y = 0$ **(b)** $y'' + y' \tan x = 0$

9.18 Reduce the self-adjoint DE $(py')' + qy = 0$ to $u'' + S(x)u = 0$ by an appropriate change of the dependent variable. What is $S(x)$? Apply this reduction to the Legendre equation,

$$[(1 - x^2)y']' + n(n + 1)y = 0$$

and show that

$$S(x) = \frac{[1 + n(n + 1) - n(n + 1)x^2]}{(1 - x^2)^2}$$

Now use this result to show that every solution of the Legendre equation has at least $(2n + 1)/\pi$ zeros on $(-1, +1)$.

9.19 Obtain the Riccati equation from the HSOLDE. Use an appropriate functional transformation to reduce the result to the simple form given in Eq. (9.22b):

$$u' + u^2 + S(x) = 0$$

9.20 Define the function $L(x) \equiv \int_1^x dt/t$.

(a) Show that $L(1) = 0$.

(b) Show that $\displaystyle\int_\alpha^x \frac{dt}{t} = \int_1^{x/\alpha} \frac{dt}{t}$, and thus $L\left(\frac{x}{\alpha}\right) = L(x) - L(\alpha)$.

(c) Show that $L(\alpha x) = L(\alpha) + L(x)$.

(d) Let x_0 be the point at which $L(x_0) = 1$. By comparing certain appropriate integrands, show that $2 < x_0 < 3$.

(These properties strongly suggest that $L(x)$ is really the natural log function.)

9.21 For the function $S(x)$ defined in Example 9.3.1, let $S^{-1}(x)$ be the inverse. Show that

$$\frac{d}{dx}[S^{-1}(x)] = \frac{1}{\sqrt{1 - x^2}}$$

and conclude that

$$S^{-1}(x) = \int_0^x \frac{dt}{\sqrt{1 - t^2}}$$

9.22 **(a)** Derive Eq. (1) of Example 9.3.3.

(b) Derive Eq. (2) of Example 9.3.3 by direct substitution.

(c) Let $\lambda = l(l+1)$. Calculate the Legendre polynomials $P_l(x)$ for $l = 0, 1, 2, 3$, subject to the condition that $P_l(1) = 1$.

9.23 Use Eq. (2) of Example 9.3.4 to generate the first three Hermite polynomials. Use the normalization

$$\int_{-\infty}^{\infty} [H_n(x)]^2 e^{-x^2} dx = \sqrt{\pi} 2^n n!$$

to determine the arbitrary constant.

9.24 Refer to Example 9.3.5 for this problem.

(a) Derive the commutation relation $[\mathbb{a}, \mathbb{a}^\dagger] = \mathbb{1}$.

(b) Show that the Hamiltonian can be written as given in Eq. (1).

(c) Derive the commutation relation $[\mathbb{a}, (\mathbb{a}^\dagger)^n] = n(\mathbb{a}^\dagger)^{n-1}$.

(d) Take the inner product of $(\mathbb{a}^\dagger)^n|\mathrm{u}_0\rangle = C|\mathrm{u}_n\rangle$ with itself to show that $|C_n|^2 = n|C_{n-1}|^2$. From this, conclude that $|C_n|^2 = n!|C_0|^2$.

(e) For any function $f(y)$, show that

$$\left(y - \frac{d}{dy}\right)(e^{y^2/2} f) = -e^{y^2/2}\frac{df}{dy} \tag{1}$$

Apply $\left(y - \frac{d}{dy}\right)$ repeatedly to both sides of (1) to obtain

$$\left(y - \frac{d}{dy}\right)^n (e^{y^2/2} f) = (-1)^n e^{y^2/2} \frac{d^n f}{dy^n}$$

(f) Choose an appropriate $f(y)$ in part (e) and show that

$$e^{y^2/2}\left(y - \frac{d}{dy}\right)^n e^{-y^2/2} = (-1)^n e^{y^2} \frac{d^n}{dy^n}(e^{-y^2})$$

9.25 Solve $y'' + xy = 0$ (Airy's differential equation) by the power-series method. Show that the radius of convergence for both independent solutions is infinite. Use the comparison theorem to show that for $x > 0$ these solutions have infinitely many zeros, but for $x < 0$ they can have at most one zero.

9.26 Define $\sinh x$ and $\cosh x$ as the solutions of $y'' = y$ satisfying the boundary conditions $y(0) = 0$, $y'(0) = 1$ and $y(0) = 1$, $y'(0) = 0$, respectively. Show that

(a) $\cosh^2 x - \sinh^2 x = 1$ **(b)** $\cosh(-x) = \cosh x$

(c) $\sinh(-x) = -\sinh x$ **(d)** $\sinh(a + x) = \sinh a \cosh x + \cosh a \sinh x$

9.27 Show that $(Ax^2 + B)y'' + Cxy' + Dy = 0$ has as a solution a polynomial of degree n iff $An^2 + (C - A)n + D = 0$.

9.28 Show that the function defined by

$$f_n(x) \equiv \frac{d^n}{dx^n}[(1 - x^2)^n]$$

satisfies the Legendre equation, $(1 - x^2)y'' - 2xy' + n(n+1)y = 0$.

9.29 Find a basis of real solutions for each DE.

(a) $y'' + 5y' + 6 = 0$ **(b)** $y''' + 6y'' + 12y' + 8y = 0$

(c) $\dfrac{d^4y}{dx^4} = y$ **(d)** $\dfrac{d^4y}{dx^4} = -y$

9.30 Solve the following initial value problems.

(a) $\dfrac{d^4y}{dx^4} = y,\ y(0) = y'(0) = y'''(0),\ y''(0) = 1$

(b) $\dfrac{d^4y}{dx^4} + \dfrac{d^2y}{dx^2} = 0,\ y(0) = y''(0) = y'''(0) = 0,\ y'(0) = 1$

(c) $\dfrac{d^4y}{dx^4} = 0,\ y(0) = y'(0) = y''(0) = 0,\ y'''(0) = 2$

9.31 Find the general solution of each equation.

(a) $y'' = xe^x$ **(b)** $y'' - 4y' + 4y = x^2$

(c) $y'' + y = \sin x \sin 2x$ **(d)** $y'' - y = (1 + e^{-x})^2$

(e) $y'' - y = e^x \sin 2x$ **(f)** $y^{(6)} - y^{(4)} = x^2$

(g) $y'' - 4y' + 4 = e^x + xe^{2x}$ **(h)** $y'' + y = e^{2x}$

9.32 Consider the Euler equation,

$$x^n y^{(n)} + a_{n-1}x^{n-1}y^{(n-1)} + \cdots + a_1 xy' + a_0 y = r(x)$$

Substitute $x = e^t$ and show that such a substitution reduces this to a DE with constant coefficients. In particular, solve

$$x^2y'' - 4xy' + 6y = x$$

9.33 Solve the following systems of equations.

(a) $\dot{x} = x + y + e^{\sqrt{2}t}$
$\dot{y} = x - y$

(b) $\dot{x} + ax - by = e^t$
$\dot{y} - ay + bx = e^t$
where $a^2 - b^2 = 1$

(c) $\dot{x} = -x + y + z$
$\dot{y} = x - y + z$
$\dot{z} = x + y - z$

(d) $\dot{x} = y$
$\dot{y} = z$
$\dot{z} = 2x - 5y + 4z$

In each case show how the system can be reduced to a single DE of higher order. (Hint: Differentiate the equations as many times as necessary, then eliminate all functions and their derivatives except one.)

9.34 Show that if $p(z) = 1/z^2$, then the solution of $w' + p(z)w = 0$ has an essential singularity at $z = 0$.

9.35 Let $\{w_j(z)\}_{j=1}^n$ be a basis of solutions for the NOLDE of Eq. (9.45). Show that $\{w_j(z_0 + re^{i(\theta+2\pi)})\}_{j=1}^n$ is also a basis of solutions.

9.36 Derive the recursion relation of (9.48a) and reexpress it in terms of the indicial polynomial, as in (9.48b).

9.37 Show that only one solution of $w'' + p(z)w' + q(z)w = 0$ can be determined for an ordinary point. What is the characteristic exponent associated with this solution?

9.38 Find the indicial polynomial, characteristic exponents, and recursion relation at both of the regular singular points of the Legendre equation,

$$w'' - \frac{2z}{1 - z^2} w' + \frac{\alpha}{1 - z^2} w = 0$$

What is A_k for the point $z = +1$?

9.39 Derive the function $w_2(z)$ of Theorem 9.5.6.

9.40 Show that the substitution $z = 1/t$ transforms (9.52) into (9.53).

9.41 Obtain the indicial polynomial of Eq. (9.53).

9.42 Using equations (9.49) and (9.53), derive the indicial equation for the Riemann DE [Eq. (9.57)].

9.43 Show that the transformation

$$v(z) = z^{\lambda}(z - 1)^{\mu} w(z)$$

changes the pairs of characteristic exponents for the Riemann DE to

$$(\lambda_1 + \lambda, \lambda_2 + \lambda) \qquad (\mu_1 + \mu, \mu_2 + \mu) \qquad \text{and} \qquad (\nu_1 - \mu - \lambda, \nu_2 - \mu - \lambda)$$

9.44 Derive all the results of Example 9.5.4.

9.45 By differentiating Eq. (9.61), show that

$$\frac{d^n}{dz^n} F(\alpha, \beta; \gamma; z) = \frac{\Gamma(\alpha + n)\Gamma(\beta + n)\Gamma(\gamma)}{\Gamma(\alpha)\Gamma(\beta)\Gamma(\gamma + n)} F(\alpha + n, \beta + n; \gamma + n; z)$$

9.46 Use direct substitution in the series of (9.61) to show that

$$F(-\alpha, \beta; \beta; -z) = (1 + z)^{\alpha}$$

$$F(\tfrac{1}{2}, \tfrac{1}{2}; \tfrac{3}{2}; z^2) = \frac{1}{z} \sin^{-1} z$$

and

$$F(1, 1; 2; -z) = \frac{1}{z} \ln(1 + z)$$

9.47 Provide all the missing steps in Example 9.5.5.

9.48 Provide all the missing steps in the solution of Exercise 9.5.5.

9.49 Derive Eq. (9.64) from Eq. (9.63).

9.50 Derive Eq. (9.65).

9.51 Show that $z = \infty$ is not a regular singular point of the CHDE.

9.52 Derive Eq. (9.70) from Eq. (9.61).

9.53 **(a)** Show that the Weber-Hermite equation,

$$\frac{d^2 u}{dz^2} + (\nu + \tfrac{1}{2} - \tfrac{1}{4} z^2) u = 0$$

can be transformed into the CHDE. [Hint: Make the substitution $u(z) = \exp(-\tfrac{1}{4} z^2) v(z)$.]

(b) The linear combination defined as

$$\Psi(\alpha, \gamma; z) \equiv \frac{\Gamma(1-\gamma)}{\Gamma(\alpha-\gamma+1)}\Phi(\alpha, \gamma; z) + \frac{\Gamma(\gamma-1)}{\Gamma(\alpha)} z^{1-\gamma}\,\Phi(\alpha-\gamma+1, 2-\gamma; z)$$

is also a solution of the CHDE. Show that the Hermite polynomials can be written as

$$H_n\left(\frac{z}{\sqrt{2}}\right) = 2^n \Psi\left(-\frac{n}{2}, \frac{1}{2}; \frac{z^2}{2}\right)$$

9.54 Verify that the error function

$$\mathrm{Er}\,[f(z)] \equiv \int_0^z e^{-t^2}\,dt$$

satisfies the relation

$$\mathrm{Er}\,[f(z)] = z\Phi(\tfrac{1}{2}, \tfrac{3}{2}; -z^2)$$

9.55 Derive (9.73) from (9.70), (9.72), and the expansion of the exponential. Check your answer by obtaining the same result directly from (9.71) using the method of undetermined coefficients discussed in Section 9.3.

9.56 If Z_ν is a solution of the Bessel equation of order ν, show that $z^\nu(d/dz)[z^{-\nu}Z_\nu(z)]$ is a solution of order $\nu+1$ and $z^{-\nu}(d/dz)[z^\nu Z_\nu(z)]$ is a solution of order $\nu-1$.

9.57 Show that, for integer ν, a second solution to the Bessel equation exists and can be written as

$$Y_n(z) = J_n(z)[f_n(z) + K_n \ln z]$$

where $f_n(z)$ is analytic about $z=0$. [Hint: Use Theorem 9.5.6 and the fact that $J_n(z)$ is entire.]

9.58 **(a)** Show that the Wronskian, $W(J_\nu, Z; z)$, of J_ν and any other solution Z of the Bessel equation, satisfies the equation

$$\frac{d}{dz}[zW(J_\nu, Z; z)] = 0$$

(b) For some constant A, show that

$$\frac{d}{dz}\left[\frac{Z(z)}{J_\nu(z)}\right] = \frac{W(z)}{J_\nu^2(z)} = \frac{A}{zJ_\nu^2(z)}$$

(c) Show that the general second solution of the Bessel equation can be written as

$$Z_\nu(z) = J_\nu(z)\left[B + A\int \frac{dz}{zJ_\nu^2(z)}\right]$$

9.59 Spherical Bessel functions are defined by

$$f_l(z) \equiv \sqrt{\frac{\pi}{2}}\left(\frac{Z_{l+1/2}(z)}{\sqrt{z}}\right)$$

Let $f_l(z)$ denote a spherical Bessel function "of some kind." By direct differentiation and substitution in the Bessel equation, show that

(a) $\dfrac{d}{dz}[z^{l+1}f_l(z)] = z^{l+1}f_{l-1}(z)$ (b) $\dfrac{d}{dz}[z^{-l}f_l(z)] = -z^{-l}f_{l+1}(z)$

(c) Combine the results of parts (a) and (b) to derive the recursion relations

$$f_{l-1}(z) + f_{l+1}(z) = \frac{2l+1}{z}f_l(z) \quad \text{and} \quad lf_{l-1}(z) - (l+1)f_{l+1}(z) = (2l+1)\frac{df_l}{dz}$$

9.60 Show that $W(J_\nu, Y_\nu; z) = 2/\pi z$.

9.61 Show that

(a) $Y_{n+1/2}(z) = (-1)^{n+1}J_{-n-1/2}(z)$ and $Y_{-n-1/2}(z) = (-1)^n J_{n+1/2}(z)$

(b) $Y_{-\nu}(z) = (\sin \nu\pi)J_\nu(z) + (\cos \nu\pi)Y_\nu(z) = \dfrac{J_\nu(z) - \cos \nu\pi\, J_{-\nu}(z)}{\sin \nu\pi}$

(c) $Y_{-n}(z) = (-1)^n Y_n(z)$ in the limit $\nu \to n$

9.62 Show that $W(H_\nu^{(1)}, H_\nu^{(2)}; z) = -4i/(\pi z)$.

9.63 Use the recurrence relation for the Bessel function, $J_\nu(z)$, to show that $J_1(z) = -J_0'(z)$.

9.64 The generating function for Bessel functions is $\exp[\frac{1}{2}z(t - t^{-1})]$. To see this, rewrite the function as $\exp(zt/2)\exp(-z/2t)$, expand both factors, and write the product as powers of t^n. Now show that the coefficient of t^n is simply $J_n(z)$. Finally, use $J_{-n}(z) = (-1)^n J_n(z)$ to derive the formula

$$\exp[\tfrac{1}{2}z(t - t^{-1})] = \sum_{n=-\infty}^{\infty} t^n J_n(z)$$

9.65 Show that $J_{-1/2}(z) = (2/\pi z)^{1/2}\cos z$.

9.66 Use Eq. (9.77a) and a procedure similar to that used for Exercise 9.5.9 to derive the formula

$$\left(\frac{1}{z}\frac{d}{dz}\right)^m [z^{-\nu}Z_\nu(z)] = (-1)^m z^{-\nu-m}Z_{\nu+m}(z)$$

Then show that

$$J_{n+1/2}(z) = \left(\frac{2}{\pi}\right)^{1/2} z^{n+1/2}\left(-\frac{1}{z}\frac{d}{dz}\right)^n\left(\frac{\sin z}{z}\right)$$

9.67 Make the substitutions $z = \beta t^\gamma$ and $w = t^\alpha u$ to transform the Bessel equation,

$$z^2\frac{d^2w}{dz^2} + z\frac{dw}{dz} + (z^2 - \nu^2)w = 0$$

into

$$t^2\frac{d^2u}{dt^2} + (2\alpha + 1)t\frac{du}{dt} + (\beta^2\gamma^2t^{2\gamma} + \alpha^2 - \nu^2\gamma^2)u = 0$$

Now show that Airy's differential equation, $d^2u/dt^2 - tu = 0$, has solutions of the form $J_{1/3}(\frac{2}{3}it^{3/2})$ and $J_{-1/3}(\frac{2}{3}it^{3/2})$.

9.68 Show that the general solution of

$$\frac{d^2w}{dt^2} + \frac{e^{2t} - \nu^2}{t^4}w = 0$$

is

$$w = t[AJ_\nu(e^{1/t}) + BY_\nu(e^{1/t})]$$

9.69 Transform $dw/dz + w^2 + z^m = 0$ by making the substitution $w = (d/dz)\ln v$. Now make the further substitutions

$$v = u\sqrt{z} \qquad \text{and} \qquad t = \frac{2}{m+2} z^{1+(1/2)m}$$

to show that the new DE can be transformed into a Bessel equation of order $1/(m+2)$.

9.70 Starting with the relation

$$\exp\left[\tfrac{1}{2}x(t - t^{-1})\right]\exp\left[\tfrac{1}{2}y(t - t^{-1})\right] = \exp\left[\tfrac{1}{2}(x + y)(t - t^{-1})\right]$$

and the fact that the exponential function is the generating function for $J_n(z)$, prove the "addition theorem" for Bessel functions:

$$J_n(x + y) = \sum_{k=-\infty}^{+\infty} J_k(x)\, J_{n-k}(y)$$

10

Sturm-Liouville Systems

Chapters 8 and 9 discussed the properties of solutions of the linear differential operator (DO), $\mathbb{L} = p_2(x)\, d^2/dx^2 + p_1(x)d/dx + p_0(x)$, from an analytic standpoint, focusing on the power-series expansions of solutions, the convergence of this series, the singular points of the solutions, and so on. This chapter emphasizes the algebraic aspect of SOLDEs. Of course, the analytic aspect cannot be abandoned for a differential operator, because by its very nature it is smoothly varying, unlike matrices, for instance, which are discrete.

Specifically, this chapter will focus on the eigenvectors and eigenvalues of the abovementioned linear DO. This is particularly important because, as was noted in Chapter 8, the separation of PDEs always results in expressions of the form

$$\frac{1}{u}\,\mathbb{L}[u] + \lambda = 0$$

where u is a function of a single variable, λ is, *a priori*, an arbitrary constant, and $\mathbb{L}$ is a second-order DO.

10.1 THE STURM-LIOUVILLE EQUATION

The equation given above can be written as

$$\mathbb{L}[u] + \lambda u = 0 \tag{10.1a}$$

or

$$p_2(x)\frac{d^2u}{dx^2} + p_1(x)\frac{du}{dx} + p_0(x)u + \lambda u = 0 \tag{10.1b}$$

This is an eigenvalue equation for the operator $\mathbb{L}$. If we use Theorem 9.2.15 and multiply (10.1b) by

$$w(x) \equiv \frac{1}{p_2(x)} \exp\left[\int^x \frac{p_1(t)}{p_2(t)} dt \right]$$

then it becomes self-adjoint, for real λ, and can be written as

$$\frac{d}{dx}\left[p(x)\frac{du}{dx} \right] + [\lambda w(x) - q(x)]u = 0 \qquad (10.2a)$$

where $p(x) \equiv w(x)p_2(x)$ and $q(x) \equiv -p_0(x)w(x)$. Equation (10.2a) is the standard form of the *Sturm-Liouville (S-L) equation*, which we will study in this chapter. Denoting d/dx by $\mathbb{D}$ and $\mathbb{L}$ by $\mathbb{D}[p(x)\mathbb{D}] - q(x)$, we can write (10.2a) in the abbreviated form

$$\mathbb{D}[p\mathbb{D}u] - qu + \lambda wu \equiv \mathbb{L}[u] + \lambda w(x)u = 0 \qquad (10.2b)$$

The S-L problem is to find all the eigenvalues and (nontrivial) eigenfunctions of the S-L equation. We assume that λ is real and that p, q, and w are continuous in $[a, b]$, and we let $u(x)$ be, in general, a complex-valued function of the real variable x [however, in all the following discussion $u(x)$ will turn out to be real]. A *regular S-L equation* is one in which $p(x)$ and $w(x)$ are positive for all $x \in [a, b]$. The S-L problem is not defined unless the *end point conditions*, or boundary conditions, are also specified. A general boundary condition (BC) is usually written as two separate and linearly independent combinations of values of u and u' at the two end points. Such a condition will be stated later in this chapter. For now, we consider the following definition.

10.1.1 Definition A regular S-L equation [Eq. (10.2a) or (10.2b)] on the finite closed interval $[a, b]$, along with the separated boundary conditions

$$\alpha u(a) + \alpha' u'(a) = 0 \quad \text{and} \quad \beta u(b) + \beta' u'(b) = 0$$

where α, α', β, and β' are given constants, is called a *regular S-L system*. The two trivial cases for which $\alpha = \alpha' = 0$ and $\beta = \beta' = 0$ are excluded.

Another type of boundary condition that is commonly used is the *periodic BC*. This is especially well-suited to cases in which the coefficient functions of the S-L equation are periodic in $[a, b]$ with period $b - a$. In such cases the BCs

$$u(a) = u(b) \quad \text{and} \quad u'(a) = u'(b)$$

are imposed on the eigenfunctions of the S-L equation.

Example 10.1.1

(a) The S-L system consisting of the S-L equation $d^2u/dt^2 + \omega^2 u = 0$ in the interval $[0, T]$ with the separated BCs $u(0) = 0$ and $u(T) = 0$ has the eigenfunctions

$$u_n(t) = \sin \frac{n\pi}{T} t \qquad \text{for } n = 1, 2, \ldots$$

and the eigenvalues

$$\lambda_n \equiv \omega_n^2 = \left(\frac{n\pi}{T}\right)^2 \qquad \text{for } n = 1, 2, \ldots$$

Note that $n = 0$ is excluded because it leads to the trivial solution $u_0(t) \equiv 0$.

(b) Let the S-L equation be the same as in part (a) but in the interval $[-T, +T]$ and with the periodic BCs $u(-T) = u(T)$ and $u'(-T) = u'(T)$. We note that periodic BCs are applicable here because the coefficient functions $p(t) = 1 = w(t)$ and $q(t) = 0$ are (trivially) periodic in $[-T, +T]$. The eigenfunctions are 1, $\sin(n\pi t/T)$, and $\cos(n\pi t/T)$, where n is a positive integer. Note that there is a degeneracy here in the sense that there are two linearly independent eigenfunctions having the same eigenvalue, $(n\pi/T)^2$.

(c) The Bessel equation for a given fixed ν^2 is

$$u'' + \frac{1}{x}u' + \left(k^2 - \frac{\nu^2}{x^2}\right)u = 0 \qquad \text{where } a \leqslant x \leqslant b$$

and can be turned into an S-L system if we multiply it by

$$w(x) = \frac{1}{p_2(x)}\exp\left[\int^x \frac{p_1(t)}{p_2(t)}\,dt\right] = \exp\left[\int^x \frac{dt}{t}\right] = x$$

Then we can write

$$\frac{d}{dx}\left(x\frac{du}{dx}\right) + \left(k^2 x - \frac{\nu^2}{x}\right)u = 0$$

which is of the form of Eq. (10.2a) with $p = w = x$, $\lambda = k^2$, and $q(x) = \nu^2/x$. If $a > 0$, we can obtain a regular S-L system by applying appropriate separated BCs. •

The treatment of finite-dimensional vector spaces in Chapter 3 showed that the eigenvectors of a self-adjoint (hermitian) operator span the vector space. On the other hand, the discussion of infinite-dimensional Hilbert spaces in Chapter 5 made no mention of operators and eigenvalues. We can now make a connection between these two chapters.

In Chapter 3 we saw that two eigenvectors of a hermitian operator corresponding to different eigenvalues are automatically orthogonal. Here we have $\mathbb{L}$, a hermitian (self-adjoint) DO. Are its eigenfunctions orthogonal? Let us investigate this point.

In Chapter 9 we derived the Lagrange identity for a self-adjoint operator $\mathbb{L}$:

$$u\mathbb{L}[v] - v\mathbb{L}[u] = \frac{d}{dx}\{p(x)[u(x)v'(x) - v(x)u'(x)]\} \tag{10.3}$$

If we specialize this identity to the S-L equation of (10.2a) with $u = u_1$ corresponding to eigenvalue λ_1 and $v = u_2$ corresponding to eigenvalue λ_2, we obtain for the LHS

$$\begin{aligned} u_1\mathbb{L}[u_2] - u_2\mathbb{L}[u_1] &= u_1(-\lambda_2 w u_2) + u_2(\lambda_1 w u_1) \\ &= (\lambda_1 - \lambda_2) w u_1 u_2 \end{aligned}$$

Integrating both sides of (10.3) then yields

$$(\lambda_1 - \lambda_2)\int_a^b wu_1u_2dx = \{p(x)[u_1(x)u_2'(x) - u_2(x)u_1'(x)]\}_a^b$$

It can be easily shown that the RHS of this equation vanishes if u_1 and u_2 satisfy either a separated or a periodic BC. In the latter case the additional requirement that $p(a) = p(b)$ is imposed. We thus have

$$(\lambda_1 - \lambda_2)\int_a^b w(x)u_1(x)u_2(x)dx = 0$$

In particular, if $\lambda_1 \neq \lambda_2$,

$$\int_a^b w(x)u_1(x)u_2(x)dx = 0$$

and the two eigenfunctions $u_1(x)$ and $u_2(x)$, corresponding to two *different* eigenvalues, are orthogonal if we define the inner product as

$$\langle \mathrm{u}|\mathrm{v}\rangle \equiv \int_a^b w(x)u(x)v(x)dx \qquad (10.4)$$

This is exactly how the inner product was defined in Chapter 5. Here, however, u is not complex-conjugated because the operator $\mathbb{L}$ is *real.* The complex conjugation is necessary only if $\mathbb{L}$ is complex, that is, if $p(x)$ or $q(x)$ is a complex function of real x. Since all S-L systems occurring in physics are real, we can use the definition given in Eq. (10.4). The foregoing results can be summarized in a theorem.

10.1.2 Theorem Eigenfunctions of a regular S-L system or an S-L equation with periodic BCs are orthogonal and have the weight function $w(x)$. ■

Example 10.1.2

(a) The regular S-L system $u'' + \lambda u = 0$, $x \in [0,\pi]$, has eigenfunctions $u_n(x) = \sin nx$, where $n = 1, 2, \ldots$. For this system $w(x) = 1$, and Theorem 10.1.2 states that

$$\int_0^\pi \sin nx \sin mx \, dx = 0 \qquad \text{if } n \neq m$$

which is known from the theory of Fourier series (Chapter 5).

(b) The regular S-L system $u'' + \lambda u = 0$, $x \in [-\pi, \pi]$, has eigenfunctions 1, $\cos nx$, $\sin nx$, where $n = 1, 2, \ldots$. Here again Theorem 10.1.2 simply says that

$$\int_{-\pi}^{\pi} \sin nx = 0 = \int_{-\pi}^{\pi} \cos nx\, dx \qquad \forall\ n$$

$$\int_{-\pi}^{\pi} \sin nx \sin mx\, dx = 0 = \int_{-\pi}^{\pi} \cos nx \cos mx\, dx \qquad \text{for } n \neq m$$

$$\int_{-\pi}^{\pi} \sin nx \cos mx\, dx = 0 \qquad \forall \text{ integers } m,\ n$$

●

A regular S-L system is too restrictive for applications where either a or b or both may be infinite or where either a or b may be a singular point of the S-L equation. A *singular S-L system* is one for which one or more of the following conditions hold:

(i) The interval $[a,b]$ stretches to infinity in either or both directions.
(ii) Either p or w vanishes at one or both end points a and b.
(iii) The function $q(x)$ is not continuous in $[a,b]$.
(iv) Any one of the functions $p(x)$, $q(x)$, and $w(x)$ is singular at a or b.

If the interval $[a,b]$ is restricted in such a way that the eigenfunctions become square-integrable with weight function $w(x)$ (see Chapter 5), then it is clear that the inner product of (10.4) is always finite. Thus, the eigenfunctions of a singular S-L system are orthogonal if the RHS of (10.3) vanishes.

Example 10.1.3

(a) Bessel functions, $J_\nu(x)$ are entire functions. Thus, they are square-integrable in the interval $[0,b]$ for any finite positive b. For *fixed* ν the DE

$$r^2 \frac{d^2u}{dr^2} + r\frac{du}{dr} + (k^2r^2 - \nu^2)u = 0 \tag{1}$$

transforms into the Bessel equation,

$$x^2 \frac{d^2u}{dx^2} + x\frac{du}{dx} + (x^2 - \nu^2)u = 0$$

if we make the substitution $kr = x$. Thus, the solution to (1) that is regular at $r = 0$ and corresponds to the eigenvalue k^2 can be written as

$$u_k(r) \equiv J_\nu(kr)$$

Thus, for two different eigenvalues, k^2 and k'^2, the eigenfunctions are orthogonal if the boundary term of (10.3) vanishes, that is, if

$$\{r[J_\nu(kr)J_\nu'(k'r) - J_\nu(k'r)J_\nu'(kr)]\}_0^b$$

vanishes, which will occur if and only if

$$J_\nu(kb)J'_\nu(k'b) - J_\nu(k'b)J'_\nu(kb) = 0$$

A common choice is to take

$$J_\nu(kb) = 0 = J_\nu(k'b)$$

that is, to take both kb and $k'b$ as (different) roots of the Bessel function of order ν. [That more than one (in fact, infinitely many) such root can be found is discussed later in this chapter.] We thus have

$$\int_0^b r J_\nu(k_i r) J_\nu(k_j r) dr = 0$$

if k_i and k_j are different roots of $J_\nu(kb) = 0$.

(b) The Legendre equation,

$$\frac{d}{dx}\left[(1 - x^2)\frac{du}{dx}\right] + \lambda u = 0 \qquad \text{where } -1 < x < 1$$

is already self-adjoint. Thus, $w(x) = 1$, and $p(x) = 1 - x^2$. The eigenfunctions of this singular S-L system [singular because $p(1) = p(-1) = 0$] are regular at the end points $x = \pm 1$ and are the Legendre polynomials $P_n(x)$ corresponding to $\lambda = n(n + 1)$. The boundary term of (10.3) clearly vanishes at $a = -1$ and $b = +1$. Since $P_n(x)$ are square-integrable on $[-1, +1]$, we obtain

$$\int_{-1}^{+1} P_n(x) P_m(x) dx = 0 \qquad \text{if } m \neq n$$

•

Example 10.1.4

The Hermite DE is

$$u'' - 2xu' + \lambda u = 0 \tag{1}$$

It is transformed into an S-L system if we multiply it by

$$w(x) = \exp\left[\int^x (-2t)dt\right] = e^{-x^2}$$

The resulting S-L equation is

$$\frac{d}{dx}\left[e^{-x^2}\frac{du}{dx}\right] + \lambda e^{-x^2} u = 0 \tag{2}$$

The boundary term corresponding to the two eigenfunctions $u_1(x)$ and $u_2(x)$ having the respective eigenvalues λ_1 and $\lambda_2 \neq \lambda_1$ is

$$\{e^{-x^2}[u_1(x)u'_2(x) - u_2(x)u'_1(x)]\}_a^b$$

This vanishes for arbitrary u_1 and u_2 (because they are polynomials, as shown below) if $a = -\infty$ and $b = +\infty$.

The function u is an eigenfunction of (2) corresponding to eigenvalue λ if and only if it is a solution of (1). Solutions of this DE corresponding to $\lambda = 2n$ are the Hermite polynomials $H_n(x)$ discussed in Chapter 5. We can therefore write

$$\int_{-\infty}^{\infty} e^{-x^2} H_n(x) H_m(x)\,dx = 0 \qquad \text{if } m \neq n$$

This is the orthogonality relation for Hermite polynomials derived in Chapter 5. •

10.2 PROPERTIES OF STURM-LIOUVILLE SYSTEMS

The study of S-L systems is one of the most important areas of mathematical physics because an S-L equation produces orthogonal functions, which are highly in demand in, for example, the theory of Hilbert spaces. It is beyond the scope of this book to provide complete coverage of S-L systems. We will consider examples from different areas of physics and a few important theorems, stated without proof. (See Birkhoff and Rota 1978 or Hellwig 1967 for proofs and a thorough discussion.) Let us start with the following theorem.

10.2.1 Theorem Any regular S-L system has an infinite sequence of real eigenvalues, which can be ordered as $\lambda_0 < \lambda_1 < \lambda_2 < \cdots < \lambda_n \ldots$, where $\lim_{n\to\infty} \lambda_n = \infty$. The eigenfunction $u_n(x)$ corresponding to eigenvalue λ_n has exactly n zeros in the interval $a < x < b$. ■

Example 10.2.1

The regular S-L system $u'' + \lambda u = 0$ on $[0, \pi]$ with the BCs $u(0) = 0 = u(\pi)$ has the eigenfunctions $u_n(x) = \sin nx$ and the eigenvalues $\lambda_n = n^2$, for $n = 1, 2, \ldots$. We can rewrite these eigenvalues and eigenfunctions as

$$\lambda_n = (n+1)^2 \qquad \text{for } n = 0, 1, 2, \ldots$$

$$u_n(x) = \sin(n+1)x \qquad \text{for } n = 0, 1, 2, \ldots$$

This kind of labeling allows a zeroth eigenvalue and eigenfunction, as implied by Theorem 10.2.1.

It is clear that $\lim_{n\to\infty} \lambda_n = \infty$. Also, the zeros of $u_n(x)$ are given by

$$\sin(n+1)x = 0 \quad \Rightarrow \quad (n+1)x = m\pi \qquad \text{for } m = 1, 2, \ldots, n$$

or

$$x_m = \frac{m}{n+1}\pi \qquad \text{for } m = 1, 2, \ldots, n$$

This shows that in the *open* interval $0 < x < \pi$, that is, excluding 0 and π, we have exactly n zeros. •

10.2.1 Asymptotic Behavior for Large Eigenvalues

The S-L problem is central to the solution of many DEs in mathematical physics. In some cases the S-L equation has direct bearing on the physics. For example, the eigenvalue λ may correspond to the orbital angular momentum of an electron in an atom (see the treatment of spherical harmonics in Chapter 8) or to the energy levels of a particle experiencing a potential (see Examples 9.3.4 and 9.5.6). In many cases, then, it is worthwhile to gain some knowledge of the behavior of an S-L system in the limit of large λ (high angular momentum or high energy).

It is convenient to transform the S-L equation into a simpler and more tractable form. This is done by making the *Liouville substitution* (see Exercise 10.2.1), consisting of

$$u(x) = v(t)[p(x)w(x)]^{-1/4} \qquad \text{and} \qquad t = \int_a^x \sqrt{\frac{w(s)}{p(s)}}\,ds \tag{10.5}$$

The substitution transforms the S-L equation into

$$\frac{d^2v}{dt^2} + [\lambda - \hat{q}(t)]v = 0 \tag{10.6a}$$

where

$$\hat{q}(t) = \frac{q(x(t))}{w(x(t))} + [p(x(t))w(x(t))]^{-1/4}\frac{d^2}{dt^2}[(pw)^{1/4}] \tag{10.6b}$$

Example 10.2.2

For the Bessel equation, $(d/dx)(x\,du/dx) + (k^2x - \nu^2/x)u = 0$, we have

$$p(x) = x = w(x) \qquad \text{and} \qquad q(x) = \frac{\nu^2}{x}$$

Thus, Eq. (10.5) gives

$$v(t) = u(x(t))[x(t)\,x(t)]^{1/4} = \sqrt{x}u(x) \tag{1}$$

and

$$t = \int^x \sqrt{\frac{w(s)}{p(s)}}\,ds = \int^x \sqrt{\frac{s}{s}}\,ds = x$$

From (10.6) we obtain

$$\hat{q}(t) = \frac{\nu^2/t}{t} + [t^2]^{-1/4}\frac{d^2}{dt^2}[t^{1/2}] = \frac{\nu^2 - \frac{1}{4}}{t^2}$$

and

$$\frac{d^2v}{dt^2} + \left[k^2 - \frac{\nu^2 - \frac{1}{4}}{t^2}\right]v = 0$$

We can obtain an interesting result when $\nu = \frac{1}{2}$. In that case we have

$$\frac{d^2v}{dt^2} + k^2v = 0$$

which has solutions of the form $\cos kt$ and $\sin kt$. Equation (1) then gives

$$J_{1/2}(kt) = A\frac{\sin kt}{\sqrt{t}} \qquad \text{or} \qquad J_{1/2}(kt) = B\frac{\cos kt}{\sqrt{t}}$$

However, since $J_\nu(x)$ is analytic at $x = 0$, we must have

$$J_{1/2}(kx) = A\frac{\sin kx}{\sqrt{x}}$$

which is the result obtained in Exercises 9.5.9–9.5.11.

The preceding discussion made use of the fact, explained in Example 10.1.3, that the solutions to the Bessel equation corresponding to eigenvalue k are written as $Z_\nu(kx)$. ●

Since any S-L equation can be transformed into (10.6), we can consider only S-L systems of the following form:

$$u'' + [\lambda - q(x)]u \equiv u'' + Q(x)u = 0 \qquad \text{where } Q = \lambda - q \tag{10.7a}$$

with separated BCs

$$\alpha u(a) + \alpha' u'(a) = 0 \qquad \text{and} \qquad \beta u(b) + \beta' u'(b) = 0 \tag{10.7b}$$

Let us assume that $Q(x) > 0$ for all $x \in [a, b]$, that is, $\lambda > q(x)$. This is reasonable since we are interested in very large λ.

The study of the system of (10.7) is simplified if we make the *Prüfer substitution*, consisting of

$$u = RQ^{-1/4}\sin\phi \qquad \text{and} \qquad u' = RQ^{1/4}\cos\phi \tag{10.8}$$

where $R(x,\lambda)$ and $\phi(x,\lambda)$ are λ-dependent functions of x. This substitution transforms the S-L equation of (10.7a) into a pair of equations (see Exercise 10.2.2):

$$\frac{d\phi}{dx} = \sqrt{\lambda - q(x)} - \frac{q'}{4[\lambda - q(x)]}\sin 2\phi \tag{10.9a}$$

$$\frac{dR}{dx} = \frac{Rq'}{4(\lambda - q)}\cos 2\phi \tag{10.9b}$$

The function $R(x,\lambda)$ is assumed to be positive because any negativity of u can be transferred to the phase $\phi(x,\lambda)$. Also, R cannot be zero at any point $a \leqslant x \leqslant b$, because both u and u' would vanish at that point, and, by the argument used in the proof of Theorem 9.2.3, $u(x) \equiv 0$.

Equations (10.9) are very useful in discussing the asymptotic behavior of solutions of S-L systems both when $\lambda \to \infty$ and when $x \to \infty$. First, we need to consider a symbol that is used often in analysis.

It is often useful to have a notation for the behavior of a function $f(x,\lambda)$ for large λ and all values of x. If the function remains bounded for all values of x as $\lambda \to \infty$, we can write

$$f(x,\lambda) = O(1)$$

Intuitively, this means that as λ gets larger and larger, the magnitude of the function $f(x, \lambda)$ remains of order 1. In other words, for no value of x is $\lim_{\lambda \to \infty} f(x,\lambda)$ infinite. If the function $g(x, \lambda) \equiv \lambda^n f(x, \lambda)$ is of order 1, that is, if

$$g(x, \lambda) \equiv \lambda^n f(x, \lambda) = O(1)$$

then, we can write

$$f(x, \lambda) = \frac{O(1)}{\lambda^n}$$

This means that as λ tends to infinity, $f(x,\lambda)$ goes to zero as fast as $1/\lambda^n$ does. Sometimes this is written as $f(x, \lambda) = O(\lambda^{-n})$.

Some properties of $O(1)$ are as follows;

(i) $O(1) + O(1) = O(1)$, and $O(1)O(1) = O(1)$.

(ii) For finite a and b, $\int_a^b O(1)dx = O(1)$.

(iii) If r and s are real numbers with $r \leqslant s$, then

$$\frac{O(1)}{\lambda^r} + \frac{O(1)}{\lambda^s} = \frac{O(1)}{\lambda^r}$$

(iv) If $g(x)$ is any bounded function of x, then a Taylor-series expansion yields

$$[\lambda + g(x)]^r = \lambda^r\left[1 + \frac{g(x)}{\lambda}\right]^r = \lambda^r\left\{1 + r\frac{g(x)}{\lambda} + \frac{r(r-1)}{2}\left[\frac{g(x)}{\lambda}\right]^2 + \frac{O(1)}{\lambda^3}\right\}$$

$$= \lambda^r + rg(x)\lambda^{r-1} + \frac{r(r-1)}{2}[g(x)]^2\lambda^{r-2} + O(1)\lambda^{r-3}$$

$$= \lambda^r + rg(x)\lambda^{r-1} + O(1)\lambda^{r-2}$$

$$= \lambda^r + O(1)\lambda^{r-1}$$

$$= O(1)\lambda^r$$

[Sometimes $O(1)\lambda^s$ is written as $O(\lambda^s)$.]

Returning to Eqs. (10.9) and expanding the RHS's using Property (iv), we obtain

$$\frac{d\phi}{dx} = \sqrt{\lambda} + \frac{O(1)}{\sqrt{\lambda}} + \frac{O(1)}{\lambda} = \sqrt{\lambda} + \frac{O(1)}{\sqrt{\lambda}}$$

$$\frac{dR}{dx} = \frac{O(1)}{\lambda}$$

Taylor-series expansion of $\phi(x, \lambda)$ and $R(x,\lambda)$ about $x = a$ then yields

$$\phi(x,\lambda) = \phi(a,\lambda) + \left(\frac{d\phi}{dx}\right)(x - a) + \cdots$$

$$R(x,\lambda) = R(a,\lambda) + \left(\frac{dR}{dx}\right)(x - a) + \cdots$$

Thus, in the limit $\lambda \to \infty$, we obtain

$$\phi(x, \lambda) = \phi(a, \lambda) + \sqrt{\lambda}(x - a) + \frac{O(1)}{\sqrt{\lambda}} \tag{10.10a}$$

$$R(x, \lambda) = R(a, \lambda) + \frac{O(1)}{\lambda} \tag{10.10b}$$

These results are useful in determining the behavior of λ_n for large n. To do so, we use (10.7b) and (10.8) to write

$$-\frac{\alpha}{\alpha'} = \frac{u'(a)}{u(a)} = \frac{R(a,\lambda)Q^{1/4}(a,\lambda)\cos[\phi(a,\lambda)]}{R(a,\lambda)Q^{-1/4}(a,\lambda)\sin[\phi(a,\lambda)]}$$

where we have assumed that $\alpha' \neq 0$. If $\alpha' = 0$, we can take the ratio α'/α, which is defined because at least one of the α's must be different from zero. Let $A = -\alpha/\alpha'$ and write

$$\cot[\phi(a,\lambda)] = \frac{A}{\sqrt{Q}} = \frac{A}{\sqrt{\lambda - q(a)}}$$

Similarly,

$$\cot[\phi(b,\lambda)] = \frac{B}{\sqrt{\lambda - q(b)}}$$

where $B = -\beta/\beta'$. Let us concentrate on the nth eigenvalue and write the first of the preceding two equations as

$$\phi(a, \lambda_n) = \cot^{-1}\frac{A}{\sqrt{\lambda_n - q(a)}}$$

For large λ_n the argument of $\cot^{-1}$ is small. Therefore, we can expand the RHS in a Taylor series about zero. Keeping only the lowest orders, we get

$$\cot^{-1}\frac{A}{\sqrt{\lambda_n - q(a)}} = \cot^{-1}(0) - \frac{A}{\sqrt{\lambda_n - q(a)}} + \cdots = \cot^{-1}(0) + \frac{O(1)}{\sqrt{\lambda_n}}$$

or

$$\phi(a, \lambda_n) = \frac{\pi}{2} + \frac{O(1)}{\sqrt{\lambda_n}} \tag{10.11a}$$

Similarly,

$$\phi(b, \lambda_n) = \frac{\pi}{2} + n\pi + \frac{O(1)}{\sqrt{\lambda_n}} \tag{10.11b}$$

The term $n\pi$ appears in (10.11b) because, by Theorem 10.2.1, the nth eigenfunction has n zeros between a and b. Since $u = RQ^{-1/4}\sin\phi$, this means that $\sin\phi$ must go through n zeros as x goes from a to b. Thus, at $x = b$ the phase ϕ must be $n\pi$ larger than it is at $x = a$.

Substituting $x = b$ in (10.10a), with $\lambda \to \lambda_n$, and using (10.11), we obtain

$$\frac{\pi}{2} + n\pi + \frac{O(1)}{\sqrt{\lambda_n}} = \frac{\pi}{2} + \frac{O(1)}{\sqrt{\lambda_n}} + \sqrt{\lambda_n}(b - a) + \frac{O(1)}{\sqrt{\lambda_n}}$$

or
$$\sqrt{\lambda_n}(b-a) = n\pi + \frac{O(1)}{\sqrt{\lambda_n}} \tag{10.12}$$

One consequence of this result is that, for large λ_n,

$$\lim_{n\to\infty} n\lambda_n^{-1/2} = \frac{b-a}{\pi}$$

Thus, $\sqrt{\lambda_n} = C_n n$, where $\lim_{n\to\infty} C_n = \pi/(b-a)$, and Eq. (10.12) can be rewritten as

$$\sqrt{\lambda_n} = \frac{n\pi}{b-a} + \frac{O(1)}{C_n n} = \frac{n\pi}{b-a} + \frac{O(1)}{n} \tag{10.13}$$

This equation describes the asymptotic behavior of eigenvalues. The following theorem, stated without proof, describes the asymptotic behavior of eigenfunctions.

10.2.2 Theorem Let $\{u_n(x)\}_{n=0}^{\infty}$ be the normalized eigenfunctions of the regular S-L system given by Eqs. (10.7) with $\alpha'\beta' \neq 0$. Then, for $n \to \infty$,

$$u_n(x) = \sqrt{\frac{2}{b-a}}\cos\frac{n\pi(x-a)}{b-a} + \frac{O(1)}{n} \tag{10.14}$$

■

Example 10.2.3

Let us derive an asymptotic formula for the Legendre polynomials, $P_n(x)$. We first make the Liouville substitution, as given in (10.5), to transform the Legendre DE,

$$\frac{d}{dx}\left[(1-x^2)\frac{dP_n}{dx}\right] + n(n+1)P_n = 0$$

into
$$\frac{d^2v}{dt^2} + [\lambda_n - \hat{q}(t)]v = 0 \qquad \text{where } \lambda_n \equiv n(n+1) \tag{1}$$

Here $p(x) = 1 - x^2$ and $w(x) = 1$, so

$$t = \int^x \frac{ds}{\sqrt{1-s^2}} = \cos^{-1}x \quad \Rightarrow \quad x(t) = \cos t$$

and
$$P_n(x(t)) = v(t)[1 - x^2(t)]^{-1/4} = v(t)(\sin t)^{-1/2} \tag{2}$$

In (1)

$$\hat{q}(t) = (1-x^2)^{-1/4}\frac{d^2}{dt^2}[(1-x^2)^{1/4}]$$
$$= (\sin t)^{-1/2}\frac{d^2}{dt^2}[(\sin t)^{1/2}] = -\frac{1}{4}\left(1 + \frac{1}{\sin^2 t}\right)$$

For large n we can neglect $\hat{q}(t)$, make the approximation $\lambda_n = n^2 + n \approx (n + \frac{1}{2})^2$, and write

$$\frac{d^2 v}{dt^2} + (n + \tfrac{1}{2})^2 v = 0$$

The general solution is

$$v(t) = A\cos[(n + \tfrac{1}{2})t + \alpha]$$

where A and α are arbitrary constants yet to be determined. Substituting this solution in (2) yields

$$P_n(\cos t) = \frac{A}{\sqrt{\sin t}}\cos[(n + \tfrac{1}{2})t + \alpha]$$

To determine α we note that $P_n(0) = 0$ if n is odd. Thus, if we let $t = \pi/2$, the cosine part becomes

$$\cos\left[n\frac{\pi}{2} + \frac{\pi}{4} + \alpha\right]$$

which vanishes for odd n if and only if $\alpha = -\pi/4$. Thus, the general asymptotic formula for Legendre polynomials is

$$P_n(\cos t) = \frac{A}{(\sin t)^{1/2}}\cos\left[\left(n + \frac{1}{2}\right)t - \frac{\pi}{4}\right] \qquad \bullet$$

10.2.2 Asymptotic Behavior for Large x

Liouville and Prüfer substitutions are very useful in investigating the behavior of the solutions of S-L systems for large x. Rather than pursue this for the general case, however, we will consider only the important case of the Bessel equation.

Let $Z_\nu(kx)$ be a solution of the Bessel equation,

$$\frac{d}{dx}\left(x\frac{du}{dx}\right) + \left(k^2 x - \frac{\nu^2}{x}\right)u = 0$$

Then, according to Example 10.2.2, the Liouville substitution $v(x) = \sqrt{x}Z_\nu(kx)$ transforms the Bessel equation into

$$\frac{d^2 v}{dx^2} + \left(k^2 - \frac{\nu^2 - \frac{1}{4}}{x^2}\right)v = 0 \tag{10.15}$$

We want to look into the behavior of $v(x)$ for large values of x.

The best way to do this is to look at the equations obtained from the Prüfer substitution, which for the Bessel equation are

$$\frac{d\phi}{dx} = \sqrt{k^2 - \frac{a}{x^2}} + \frac{a\sin 2\phi}{2(k^2x^3 - ax)}$$

$$\frac{dR}{dx} = -\frac{Ra\cos 2\phi}{2(k^2x^3 - ax)}$$

where $a \equiv \nu^2 - \frac{1}{4}$. For large x these equations reduce to

$$\phi' = k\left(1 - \frac{a}{2k^2x^2}\right) + \frac{O(1)}{x^3}$$

$$\frac{R'}{R} = \frac{O(1)}{x^3}$$

Integrating the first of these two equations between x and $b > x$ yields

$$\phi(b) - \phi(x) = kb - kx - \frac{a}{2kx} + \frac{a}{2kb} + \frac{O(1)}{x^2}$$

Keeping x fixed and letting $b \to \infty$, we observe that $\phi(b) - kb$ tends to a finite value, which can be represented as ϕ_∞. Then, for $b \to \infty$, we obtain

$$\phi(x) = \phi_\infty + kx + \frac{a}{2kx} + \frac{O(1)}{x^2} \tag{10.16a}$$

The same integration on the equation for R'/R yields

$$R(x) = R_\infty + \frac{O(1)}{x^2} \tag{10.16b}$$

where $R_\infty = \lim_{b\to\infty} R(b)$.

Substituting Eqs. (10.16) and

$$Q^{-1/4} = \left(k^2 - \frac{a}{x^2}\right)^{-1/4} = k^{-1/2} + \frac{O(1)}{x^2}$$

in Eq. (10.8), we obtain

$$v(x) = \left[R_\infty + \frac{O(1)}{x^2}\right]\left[k^{-1/2} + \frac{O(1)}{x^2}\right]\sin\left[\phi_\infty + kx + \frac{a}{2kx} + \frac{O(1)}{x^2}\right]$$

The identity

$$\sin\left[\alpha + \frac{O(1)}{x^2}\right] = \sin\alpha + \frac{O(1)}{x^2}$$

can be easily obtained by expanding the LHS. Using it and $kx_\infty \equiv \pi/2 - \phi_\infty$, we get

$$Z_\nu(kx) \equiv \frac{v(x)}{\sqrt{x}} = \frac{R_\infty}{\sqrt{kx}}\cos\left(kx - kx_\infty + \frac{\nu^2 - \frac{1}{4}}{2kx}\right) + \frac{O(1)}{x^{5/2}} \tag{10.17}$$

The constants R_∞ and ϕ_∞ uniquely determine $Z_\nu(kx)$.

For the Bessel functions, $J_\nu(x)$, it can be shown (see Section 11.3.2) that

$$kx_\infty = \left(\nu + \frac{1}{2}\right)\frac{\pi}{2} \qquad \text{and} \qquad R_\infty = \sqrt{\frac{2}{\pi}}$$

For the Neumann functions, $Y_\nu(x)$, it can similarly be shown that

$$kx_\infty = \left(\nu + \frac{3}{2}\right)\frac{\pi}{2} \qquad \text{and} \qquad R_\infty = \sqrt{\frac{2}{\pi}}$$

We can, therefore, write

$$J_\nu(x) = \sqrt{\frac{2}{\pi x}}\cos\left[x - \left(\nu + \frac{1}{2}\right)\frac{\pi}{2} + \frac{\nu^2 - \frac{1}{4}}{2x}\right] + \frac{O(1)}{x^{5/2}}$$

$$Y_\nu(x) = \sqrt{\frac{2}{\pi x}}\sin\left[x - \left(\nu + \frac{1}{2}\right)\frac{\pi}{2} + \frac{\nu^2 - \frac{1}{4}}{2x}\right] + \frac{O(1)}{x^{5/2}}$$

These two relations easily yield the asymptotic expressions for the Hankel functions:

$$H_\nu^{(1)}(x) \equiv J_\nu(x) + iY_\nu(x) = \sqrt{\frac{2}{\pi x}}e^{i[x-(\nu+1/2)(\pi/2)+(\nu^2-1/4)/2x]} + \frac{O(1)}{x^{5/2}}$$

$$H_\nu^{(2)}(x) \equiv J_\nu(x) - iY_\nu(x) = \sqrt{\frac{2}{\pi x}}e^{-i[x-(\nu+1/2)(\pi/2)+(\nu^2-1/4)/2x]} + \frac{O(1)}{x^{5/2}}$$

If the last term in the exponent, which vanishes as $x \to \infty$, is ignored, the asymptotic expression for $H_\nu^{(1)}(x)$ matches what was obtained in Chapter 7 using the method of steepest descent.

The same procedure is used for other DEs. First, the DE is transformed into the form of Eq. (10.6) by the Liouville substitution; then the Prüfer substitution of Eq. (10.8) is made to obtain two DEs of the form of (10.9). Solving Eqs. (10.9) when $x \to \infty$ determines the behavior of ϕ and R and, subsequently, of u.

Exercises

10.2.1 The Liouville substitution starts with a change of both the independent and dependent variables

$$u = y(x)v \qquad \text{and} \qquad t = \int^x h(s)\,ds$$

Determine y and h such that the general S-L equation,

$$\frac{d}{dx}\left[p(x)\frac{du}{dx}\right] + [\lambda w(x) - q(x)]\,u = 0$$

is transformed into (10.6).

10.2.2 Derive Eqs. (10.9) from (10.7a) using the defining equations of (10.8).

10.3 EXPANSIONS IN TERMS OF EIGENFUNCTIONS

The fact that, at least for spherical coordinates, the solution to a PDE can always be expanded in terms of products of certain indexed functions was mentioned at the end of Chapter 8. Let us elaborate on this fact.

Central to such an expansion is the question of the completeness of the eigenfunctions of the S-L system. We have already noted that all the PDEs we have discussed lead to S-L systems. Solving these systems results in infinitely many discrete eigenvalues and their corresponding orthogonal eigenfunctions. But what can be done with these eigenfunctions, and how are they related to the desired solution of a PDE? The natural tendency is to express each factor of the separated solution as an (infinite) linear combination of the eigenfunctions of the corresponding S-L system. However, for this to work, the set of eigenfunctions of the S-L system must be complete.

Any question concerning the completeness of an infinite-dimensional (Hilbert) space is nontrivial. A theorem that is the most important result of this chapter shows that the eigenfunctions of an S-L system are indeed complete. Before considering this theorem, we need to be familiar with some notation. First, recall that $\mathscr{L}^2_w(a,b)$ denotes the space of square-integrable functions with weight function $w(x)$ in the interval $[a,b]$. Next, the separated BCs of Definition 10.1.1 and the periodic BC are generalized to

$$R_1 u = 0 \qquad \text{and} \qquad R_2 u = 0$$

where
$$R_j u \equiv \alpha_{j1} u(a) + \alpha_{j2} u'(a) + \alpha_{j3} u(b) + \alpha_{j4} u'(b) \qquad \text{for } j = 1, 2 \tag{10.18}$$

and $\alpha_{11}, \alpha_{12}, \ldots, \alpha_{24}$ are numbers such that the rank of the following matrix is 2:

$$\mathsf{a} \equiv \begin{pmatrix} \alpha_{11} & \alpha_{12} & \alpha_{13} & \alpha_{14} \\ \alpha_{21} & \alpha_{22} & \alpha_{23} & \alpha_{24} \end{pmatrix}$$

The separated BCs correspond to the case for which $\alpha_{11} = \alpha$, $\alpha_{12} = \alpha'$, $\alpha_{23} = \beta$, and $\alpha_{24} = \beta'$, with all other α_{ij} zero. Similarly, the periodic BC is a special case for which $\alpha_{11} = -\alpha_{13} = \alpha_{22} = -\alpha_{24} = 1$, with all other α_{ij} zero. It is easy to verify that the rank of the matrix a is 2 for these two special cases. Let

$$\mathscr{U} \equiv \{u(x) \mid u \in C^{(2)}(a,b);\ R_j u = 0, \quad \text{for } j = 1, 2\} \tag{10.19}$$

be a subspace of $\mathscr{L}^2_w(a,b)$. Finally, to assure the vanishing of the RHS of the Lagrange identity the following equality must hold:

$$p(b) \det\begin{pmatrix} \alpha_{11} & \alpha_{12} \\ \alpha_{21} & \alpha_{22} \end{pmatrix} = p(a) \det\begin{pmatrix} \alpha_{13} & \alpha_{14} \\ \alpha_{23} & \alpha_{24} \end{pmatrix} \tag{10.20}$$

We are now ready to consider the theorem (for a proof, see Chapter 7 of Hellwig 1967).

10.3.1 Theorem (The Completeness of the Eigenfunctions of an S-L System) The eigenfunctions $\{u_n(x)\}_{n=1}^{\infty}$ of an S-L system consisting of the S-L equation $(pu')' + (\lambda w - q)u = 0$ and the BCs of (10.18) form a complete basis for the subspace $\mathscr{U}$ of $\mathscr{L}^2_w(a,b)$ described in (10.19). The eigenvalues are real and countably infinite and each one has a multiplicity of at most 2. They can be ordered according to size $\lambda_1 \leqslant \lambda_2 \leqslant \ldots$, and their only limit point is $+\infty$. ■

Note that (10.20) contains both separated and periodic BCs as special cases (Exercise 10.3.1). Thus, all the eigenfunctions discussed so far are covered by Theorem 10.3.1. Second, the orthogonality of eigenfunctions corresponding to different eigenvalues and the fact that there are infinitely many distinct eigenvalues assure the existence of infinitely many eigenfunctions. Third, the eigenfunctions form a basis of $\mathcal{U}$ and not the whole $\mathcal{L}^2_w(a, b)$. Only those functions $u \in \mathcal{L}^2_w(a, b)$ that satisfy the BC in (10.18) are expandable in terms of $u_n(x)$. Finally, the last statement of Theorem 10.3.1 is a repetition of part of Theorem 10.2.1 but is included because the conditions under which Theorem 10.3.1 holds are more general than those applying to Theorem 10.2.1.

Chapter 5 discussed orthonormal functions in detail and showed how other functions could be expanded in terms of them. The discussion covered various orthogonal polynomials, the recurrence relations they satisfy, the generalized Rodriguez formula for evaluating them, and even the DE of which they are solutions. However, the procedure used in Chapter 5 is *ad hoc* from a logical standpoint. After all, the orthogonal polynomials were logical consequences of attempts on the part of nineteenth-century mathematician-physicists to solve PDEs. In the struggle to solve the PDEs of physics using the separation of variables, those mathematicians came across various ODEs of the second order, all of which were recognized later as S-L systems. From a logical standpoint, therefore, this chapter should precede Chapter 5. But the order of the chapters was based on clarity and ease of presentation and the fact that the machinery of differential equations is a prerequisite for Chapter 10.

Theorem 10.3.1 is the important link between the algebraic concepts developed in Chapter 5 and the analytic machinery of differential equation theory. This theorem puts at our disposal concrete mathematical functions that are calculable to any desired accuracy on a computer and can serve as basis functions for all the expansions described in Chapter 5. The remainder of this chapter is devoted to solving some PDEs of mathematical physics using the separation of variables and Theorem 10.3.1.

10.3.1 Separation in Cartesian Coordinates

Problems expressible in Cartesian coordinates typically include those for which the boundaries are rectangular boxes or planes.

Example 10.3.1 Conducting Box

Consider a rectangular conducting box with sides a, b, and c (Fig. 10.1). All faces are held at zero potential except the top face, whose potential is given by $f(x, y)$. Let us find the potential at all points inside the box.

The relevant PDE for this situation is Laplace's equation, $\nabla^2\Phi = 0$. Writing $\Phi(x, y, z)$ as a product of three functions, we get

$$\Phi(x, y, z) = X(x)\, Y(y)\, Z(z)$$

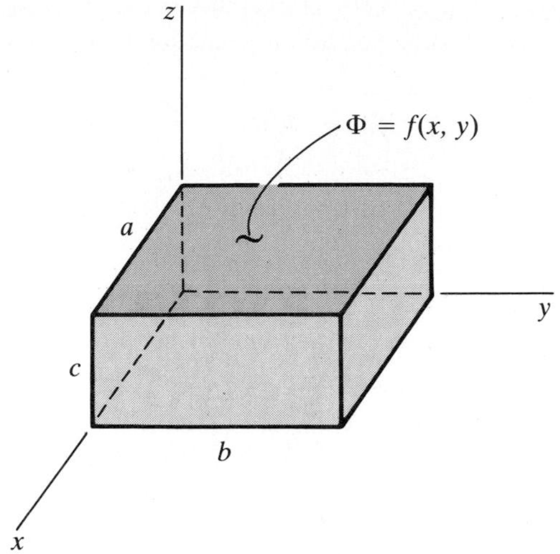

Figure 10.1 A rectangular conducting box of which one face is held at the potential $f(x, y)$ and the other faces are grounded.

which yields three ODEs (see Chapter 8):

$$\frac{d^2X}{dx^2} + \lambda^{(1)}X = 0 \tag{1}$$

$$\frac{d^2Y}{dy^2} + \lambda^{(2)}Y = 0 \tag{2}$$

$$\frac{d^2Z}{dz^2} + \lambda^{(3)}Z = 0 \tag{3}$$

where $$\lambda^{(1)} + \lambda^{(2)} + \lambda^{(3)} = 0$$

The vanishing of Φ at $x = 0$ and $x = a$ means that

$$\Phi(0, y, z) = X(0)Y(y)Z(z) = 0 \quad \Rightarrow \quad X(0) = 0$$

$$\Phi(a, y, z) = X(a)Y(y)Z(z) = 0 \quad \Rightarrow \quad X(a) = 0$$

We thus obtain an S-L system,

$$\frac{d^2X}{dx^2} + \lambda^{(1)}X = 0 \qquad X(0) = 0 = X(a)$$

whose BC is neither separated nor periodic, but satisfies (10.18) with $\alpha_{11} = \alpha_{23} = 1$ and all other α_{ij} zero. This S-L system has the eigenvalues and eigenfunctions

$$\lambda_n^{(1)} = \left(\frac{n\pi}{a}\right)^2 \qquad \text{and} \qquad X_n(x) \equiv \sin\left(\frac{n\pi}{a}x\right) \qquad \text{for } n = 1, 2, \ldots$$

Similarly, Eq. (2) leads to

$$\lambda_m^{(2)} = \left(\frac{m\pi}{b}\right)^2 \qquad \text{and} \qquad Y_m(x) \equiv \sin\left(\frac{m\pi}{b}y\right) \qquad \text{for } m = 1, 2, \ldots$$

On the other hand, Eq. (3) does not lead to an S-L system because the BC for the top of the box does not fit (10.18). However, we can find a solution for that equation. The substitution

$$\gamma_{mn}^2 \equiv \left(\frac{n\pi}{a}\right)^2 + \left(\frac{m\pi}{b}\right)^2$$

changes Eq. (3) to

$$\frac{d^2Z}{dz^2} - \gamma_{mn}^2 Z = 0$$

The solution of this, consistent with $Z(0) = 0$, is

$$Z(z) = C\,\sinh(\gamma_{mn} z)$$

We note that, for example, $X(x)$ is a function satisfying $R_1 X = 0 = R_2 X$. Thus, by Theorem 10.3.1, it can be written as a linear combination of $X_n(x)$:

$$X(x) = \sum_{n=1}^{\infty} A_n \sin\left(\frac{n\pi}{a}x\right)$$

Similarly,
$$Y(y) = \sum_{m=1}^{\infty} B_m \sin\left(\frac{m\pi}{b}y\right)$$

Consequently, the most general solution can be expressed as

$$\Phi(x, y, z) = X(x)\,Y(y)Z(z) = \sum_{n=1}^{\infty}\sum_{m=1}^{\infty} A_{mn} \sin\left(\frac{n\pi}{a}x\right)\sin\left(\frac{m\pi}{b}y\right)\sinh(\gamma_{mn} z)$$

where $A_{mn} \equiv A_n B_m C$.

To specify Φ completely, we must determine the arbitrary constants A_{mn}. This is done by imposing the remaining BC, $\Phi(x, y, c) = f(x, y)$, yielding the identity

$$f(x, y) = \sum_{n=1}^{\infty}\sum_{m=1}^{\infty} A_{mn} \sin\left(\frac{n\pi}{a}x\right)\sin\left(\frac{m\pi}{b}y\right)\sinh(\gamma_{mn} c)$$
$$\equiv \sum_{n=1}^{\infty}\sum_{m=1}^{\infty} B_{mn} \sin\left(\frac{n\pi}{a}x\right)\sin\left(\frac{m\pi}{b}y\right)$$

where $B_{mn} \equiv A_{mn}\sinh(\gamma_{mn} c)$. This is a two-dimensional Fourier series (see Chapter 5) whose coefficients are given by

$$B_{mn} = \frac{4}{ab}\int_0^a dx \int_0^b dy\, f(x, y) \sin\left(\frac{n\pi}{a}x\right)\sin\left(\frac{m\pi}{b}y\right)$$ •

Example 10.3.2 Steady-State Diffusion

When the transfer (diffusion) of heat takes place with the temperature independent of time, the process is known as steady-state heat transfer. The diffusion equation, $\partial T/\partial t = a^2\nabla^2 T$, becomes the Laplacian equation, $\nabla^2 T = 0$, and the technique of Example 10.3.1 can be used. It is easy to see that the diffusion equation allows us to perform any linear transformation on T, such as $T \to aT + b$, and still satisfy that equation. This implies that T can be measured on any of the three common scales (Kelvin, Celsius, and Fahrenheit).

Let us consider a rectangular heat-conducting plate with sides of lengths a and b. Three of the sides are held at $T = 0$, and the fourth side has a temperature variation $T = f(x)$ (see Fig. 10.2) The flat faces are insulated, so they cannot lose heat to the surroundings. Assuming a steady-state heat transfer, let us calculate the variation of T over the plate.

The problem is two-dimensional. The separation of variables leads to

$$T(x, y) = X(x)Y(y)$$

$$\frac{d^2X}{dx^2} + \lambda^{(1)}X = 0 \tag{1}$$

$$\frac{d^2Y}{dy^2} + \lambda^{(2)}Y = 0 \tag{2}$$

$$\lambda^{(1)} + \lambda^{(2)} = 0$$

Equation (1) and the periodic BCs $T(0, y) = T(a, y) = 0$ form an S-L system whose eigenfunctions and eigenvalues are

$$X_n(x) = \sin\left(\frac{n\pi}{a}x\right) \qquad \lambda_n^{(1)} = \left(\frac{n\pi}{a}\right)^2 \qquad \text{for } n = 1, 2, \ldots$$

Thus, according to Theorem 10.3.1, a general $X(x)$ can be written as

$$X(x) = \sum_{n=1}^{\infty} A_n \sin\left(\frac{n\pi}{a}x\right)$$

Equation (2), on the other hand, does not form an S-L system. However, we can solve the equation

$$\frac{d^2Y}{dy^2} - \left(\frac{n\pi}{a}\right)^2 Y = 0$$

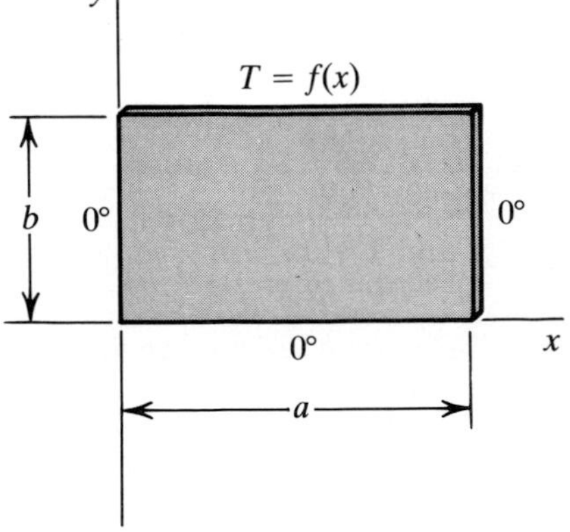

Figure 10.2 A heat-conducting rectangular plate.

to obtain the general solution

$$Y = Ae^{(n\pi/a)y} + Be^{-(n\pi/a)y}$$

Since $T(x, 0) = 0$, we must have $Y(0) = 0$. This implies that $A + B = 0$, which, in turn, reduces the solution to $Y = C \sinh(n\pi y/a)$.

Thus, the most general solution, consistent with the three BCs $T(0, y) = T(a, y) = T(x, 0) = 0$, is

$$T(x, y) = \sum_{n=1}^{\infty} B_n \sin\left(\frac{n\pi}{a}x\right) \sinh\left(\frac{n\pi}{a}y\right)$$

The fourth BC gives a Fourier series,

$$f(x) = \sum_{n=1}^{\infty} \left[B_n \sinh\left(\frac{n\pi}{a}b\right)\right] \sin\left(\frac{n\pi}{a}x\right) \equiv \sum_{n=1}^{\infty} C_n \sin\left(\frac{n\pi}{a}x\right)$$

whose coefficients can be determined from

$$C_n \equiv B_n \sinh\left(\frac{n\pi}{a}b\right) = \frac{2}{a}\int_0^a f(x) \sin\left(\frac{n\pi}{a}x\right) dx$$

In particular, if the fourth side is held at the constant temperature T_0, then

$$C_n = \frac{2T_0}{a}\left(\frac{a}{n\pi}\right)[1 - (-1)^n] = \begin{cases} \dfrac{4T_0}{n\pi} & \text{if } n \text{ is odd} \\ 0 & \text{if } n \text{ is even} \end{cases}$$

and we obtain

$$T(x, y) = \frac{4T_0}{\pi} \sum_{k=0}^{\infty} \frac{1}{2k+1} \frac{\sinh[(2k+1)\pi y/a]}{\sinh[(2k+1)\pi b/a]} \sin[(2k+1)\pi x/a] \tag{3}$$

On the other hand, if the temperature variation of the fourth side is of the form $f(x) = T_0 \sin(\pi x/a)$, then

$$C_n = \frac{2T_0}{a}\int_0^a \sin\left(\frac{\pi x}{a}\right) \sin\left(\frac{n\pi x}{a}\right) dx = \frac{2T_0}{a}\left(\frac{a}{2}\right)\delta_{n,1} = T_0\delta_{n,1}$$

and

$$B_n = \frac{C_n}{\sinh\left(\dfrac{n\pi b}{a}\right)} = \frac{T_0}{\sinh\left(\dfrac{\pi b}{a}\right)}\delta_{n,1}$$

and we have

$$T(x, y) = T_0 \frac{\sinh(\pi y/a)}{\sinh(\pi b/a)} \sin\left(\frac{\pi x}{a}\right) \tag{4}$$

Only one term of the series survives, because the variation on the fourth side happens to be one of the harmonics of the expansion.

Note that the temperature variations given by (3) and (4) are independent of the material of the plate because we are dealing with a steady state. The conductivity of the material is a factor only in the process of heat transfer, while *attaining* the steady state.

Once equilibrium has been reached, the distribution of temperature will be the same for all materials. •

The preceding two examples concerned themselves with static (time-independent) situations. Most cases of physical interest, however, are dynamic. Thus, it is important to include time-dependence in the equations.

Example 10.3.3 Thin Rod with Ends Held at $T = 0$

Let us consider a one-dimensional conducting rod with one end at the origin, $x = 0$, and the other at $x = b$. The two ends are held at $T = 0$. Initially, at $t = 0$, we assume a temperature distribution on the rod given by the function $f(x)$. We want to calculate the temperature at time t at any point x on the rod.

We have to solve this dynamic equation:

$$\frac{\partial T}{\partial t} = a^2 \nabla^2 T \equiv a^2 \frac{\partial^2 T}{\partial x^2}$$

A separation of variables, $T(t, x) = g(t) X(x)$, leads to two ODEs:

$$\frac{dg}{dt} + a^2 \lambda g = 0 \tag{1}$$

$$\frac{d^2 X}{dx^2} + \lambda X = 0 \tag{2}$$

The BCs $T(t, 0) = 0 = T(t, b)$ imply that $X(0) = 0 = X(b)$. Thus, Eq. (2) constitutes an S-L system whose solutions are

$$\lambda_n = \left(\frac{n\pi}{b}\right)^2 \qquad X_n(x) = \sin\left(\frac{n\pi}{b} x\right) \qquad \text{for } n = 1, 2, \ldots$$

By Theorem 10.3.1, therefore, we have

$$X(x) = \sum_{n=1}^{\infty} A_n \sin\left(\frac{n\pi}{b} x\right)$$

The solution to Eq. (1) is also simply obtained:

$$g(t) = Ce^{-a^2 (n\pi/b)^2 t}$$

This leads to a general solution of the form

$$T(t, x) = \sum_{n=1}^{\infty} B_n e^{-(n\pi a/b)^2 t} \sin\left(\frac{n\pi}{b} x\right)$$

The initial condition, $f(x) = T(0, x)$, yields

$$f(x) = \sum_{n=1}^{\infty} B_n \sin\left(\frac{n\pi}{b} x\right)$$

from which we can calculate the coefficients:

$$B_n = \frac{2}{b} \int_0^b f(x) \sin\left(\frac{n\pi}{b} x\right) dx$$ •

Example 10.3.4

For the plate of Example 10.3.2, let all sides be held at $T = 0$. Assume that at time $t = 0$ the temperature has a distribution function $f(x, y)$. Let us find the variation of temperature for all points (x, y) at all times $t > 0$.

The diffusion equation for this problem is

$$\frac{\partial T}{\partial t} = k^2\left(\frac{\partial^2 T}{\partial x^2} + \frac{\partial^2 T}{\partial y^2}\right)$$

where k is used instead of a because a is the length of one of the sides of the plate. A separation of variables, $T(x, y, t) = X(x)Y(y)g(t)$, leads to three DEs:

$$\frac{d^2 X}{dx^2} + \lambda^{(1)} X = 0 \tag{1}$$

$$\frac{d^2 Y}{dy^2} + \lambda^{(2)} Y = 0 \tag{2}$$

$$\frac{dg}{dt} + k^2(\lambda^{(1)} + \lambda^{(2)})g = 0 \tag{3}$$

The BCs $T(0, y, t) = T(a, y, t) = T(x, 0, t) = T(x, b, t) = 0$, together with Eqs. (1) and (2), give rise to two S-L systems. The solutions to both of these are easily found:

$$\lambda_n^{(1)} = \left(\frac{n\pi}{a}\right)^2 \qquad X_n(x) = \sin\left(\frac{n\pi}{a}x\right)$$

$$\lambda_m^{(2)} = \left(\frac{m\pi}{b}\right)^2 \qquad Y_m(y) = \sin\left(\frac{m\pi}{b}y\right)$$

These give rise to the general solutions (by Theorem 10.3.1)

$$X(x) = \sum_{n=1}^{\infty} A_n \sin\left(\frac{n\pi}{a}x\right)$$

$$Y(y) = \sum_{m=1}^{\infty} B_m \sin\left(\frac{m\pi}{b}y\right)$$

With $\gamma_{mn} \equiv k^2\pi^2(n^2/a^2 + m^2/b^2)$, the solution to Eq. (3) can be expressed as

$$g(t) = Ce^{-\gamma_{mn}t}$$

Putting everything together, we obtain

$$T(x, y, t) = \sum_{n=1}^{\infty}\sum_{m=1}^{\infty} A_{mn} e^{-\gamma_{mn}t} \sin\left(\frac{n\pi}{a}x\right)\sin\left(\frac{m\pi}{b}y\right)$$

where $A_{mn} \equiv CA_nB_m$ is an arbitrary constant. To determine it, we impose the initial condition $T(x, y, 0) = f(x, y)$. This yields

$$f(x, y) = \sum_{m,n} A_{mn} \sin\left(\frac{n\pi}{a}x\right)\sin\left(\frac{m\pi}{b}y\right)$$

which determines the coefficients A_{mn}:

$$A_{mn} = \frac{4}{ab}\int_0^a dx \int_0^b dy\, f(x, y) \sin\left(\frac{n\pi}{a}x\right) \sin\left(\frac{m\pi}{b}y\right)$$ •

Example 10.3.5 Particle in a Box

The behavior of an atomic particle of mass μ confined in a rectangular box with sides a, b, and c (an infinite three-dimensional potential well) is governed by the Schrödinger equation for a free particle,

$$i\hbar \frac{\partial \psi}{\partial t} = -\frac{\hbar^2}{2\mu}\left(\frac{\partial^2 \psi}{\partial x^2} + \frac{\partial^2 \psi}{\partial y^2} + \frac{\partial^2 \psi}{\partial z^2}\right)$$

and the BC that $\psi(x, y, z, t)$ vanishes at all sides of the box for all time.

A separation of variables, $\psi(x, y, z, t) = X(x)Y(y)Z(z)T(t)$, yields the following ODEs:

$$\frac{d^2X}{dx^2} + \lambda^{(1)}X = 0$$

$$\frac{d^2Y}{dy^2} + \lambda^{(2)}Y = 0$$

$$\frac{d^2Z}{dz^2} + \lambda^{(3)}Z = 0$$

$$\frac{dT}{dt} + i\omega T = 0$$

where

$$\omega \equiv \frac{\hbar}{2\mu}(\lambda^{(1)} + \lambda^{(2)} + \lambda^{(3)})$$

The spatial equations, together with the BCs

$$\psi(0, y, z, t) = \psi(a, y, z, t) = 0 \quad \Rightarrow \quad X(0) = 0 = X(a)$$

$$\psi(x, 0, z, t) = \psi(x, b, z, t) = 0 \quad \Rightarrow \quad Y(0) = 0 = Y(b)$$

$$\psi(x, y, 0, t) = \psi(x, y, c, t) = 0 \quad \Rightarrow \quad Z(0) = 0 = Z(c)$$

lead to three S-L systems, whose solutions are easily found:

$$X_n(x) = \sin\left(\frac{n\pi}{a}x\right) \qquad \lambda^{(1)} = \left(\frac{n\pi}{a}\right)^2 \qquad \text{for } n = 1, 2, \ldots$$

$$Y_m(x) = \sin\left(\frac{m\pi}{b}y\right) \qquad \lambda^{(2)} = \left(\frac{m\pi}{b}\right)^2 \qquad \text{for } m = 1, 2, \ldots$$

$$Z_l(x) = \sin\left(\frac{l\pi}{c}z\right) \qquad \lambda^{(3)} = \left(\frac{l\pi}{c}\right)^2 \qquad \text{for } l = 1, 2, \ldots$$

The time equation, on the other hand, has a solution of the form

$$T = Ce^{-i\omega_{lmn}t}$$

where
$$\omega_{lmn} \equiv \frac{\hbar}{2\mu}\left[\left(\frac{n\pi}{a}\right)^2 + \left(\frac{m\pi}{b}\right)^2 + \left(\frac{l\pi}{c}\right)^2\right]$$

The solution of the Schrödinger equation that is consistent with the BCs is, therefore,

$$\psi(x, y, z, t) = \sum_{l;m,n=1}^{\infty} A_{lmn} e^{-i\omega_{lmn}t} \sin\left(\frac{n\pi}{a}x\right) \sin\left(\frac{m\pi}{b}y\right) \sin\left(\frac{l\pi}{c}z\right)$$

The constants A_{lmn} are determined by the initial shape of the wave function, $\psi(x, y, z, 0)$.

It is worth mentioning that the energy of the particle is

$$E = \hbar\omega_{lmn} = \frac{\hbar^2}{2\mu}k^2$$

where
$$k^2 \equiv \pi^2\left(\frac{n^2}{a^2} + \frac{m^2}{b^2} + \frac{l^2}{c^2}\right)$$

is the wave number. Clearly, the particle's energy is quantized.

Each set of three positive integers n, m, and l represents a state of the particle. For a cube, $a = b = c \equiv L$, the energy of the particle is

$$E = \frac{\hbar^2\pi^2}{2\mu L^2}(n^2 + m^2 + l^2) = \frac{\hbar^2\pi^2}{2\mu V^{2/3}}(n^2 + m^2 + l^2) \tag{1}$$

where $V \equiv L^3$ is the volume of the box. The ground state is (1, 1, 1) and has energy $E = 3\hbar^2\pi^2/2\mu V^{2/3}$ and is nondegenerate (only one state corresponds to this energy). However, the higher-level states are degenerate. For instance, the three states (1, 1, 2), (1, 2, 1), and (2, 1, 1) all correspond to the same energy, $E = 6\hbar^2\pi^2/2\mu V^{2/3}$. States corresponding to larger values of n, m, and l are even more degenerate.

Equation (1) can be written as

$$n^2 + m^2 + l^2 = R^2$$

where $R^2 \equiv 2\mu E V^{2/3}/\hbar^2\pi^2$. This looks like the equation of a sphere in the *nml*-space. If R is large, the number of states contained within the sphere of radius R (the number of states with energy less than or equal to E) is simply the volume of the first octant of the sphere. If N is the number of such states, we have

$$\begin{aligned} N &= \frac{1}{8}\left(\frac{4\pi}{3}\right)R^3 = \frac{\pi}{6}\left(\frac{2\mu E V^{2/3}}{\pi^2\hbar^2}\right)^{3/2} \\ &= \frac{\pi}{6}\left(\frac{2\mu E}{\pi^2\hbar^2}\right)^{3/2} V \end{aligned}$$

Thus the density of states (the number of states per unit volume) is then

$$n \equiv \frac{N}{V} = \frac{\pi}{6}\left(\frac{2\mu}{\pi^2\hbar^2}\right)^{3/2} E^{3/2} \tag{2}$$

This is an important formula in solid-state physics, because the energy E is (with minor modifications required by the spin of the particle) the Fermi energy. If the Fermi energy is denoted E_F, Eq. (2) gives

$$E_F = \alpha n^{2/3}$$

where α is some constant. •

In the preceding examples the time variation is given by a first derivative. Thus, as far as time is concerned, we have a FODE. It follows that the specification of the physical quantity of interest (temperature T or Schrödinger wave function ψ) is sufficient to determine the solution uniquely.

A second kind of time-dependent PDE occurring in physics is the wave equation which involves time derivatives of the second order and is a SODE in time. Thus, there are two arbitrary parameters in the general solution. To determine these two arbitrary parameters, we expect two initial conditions.

Example 10.3.6

The simplest kind of wave equation is that in one-dimension, for example, for a wave propagating on a rope. Such a wave equation can be written as

$$\frac{\partial^2 \psi}{\partial x^2} = \frac{1}{c^2}\frac{\partial^2 \psi}{\partial t^2}$$

where c is the speed of wave propagation. For a rope this speed is related to the tension τ and linear mass density ρ by $c = \sqrt{\tau/\rho}$.

Let us assume that the rope has length a and is fastened at both ends (located at $x = 0$ and $x = a$). This means that the "displacement" ψ is zero at $x = 0$ and at $x = a$.

A separation of variables, $\psi(x, t) = X(x)T(t)$, leads to two ODEs:

$$\frac{d^2 X}{dx^2} + \lambda X = 0 \tag{1}$$

$$\frac{d^2 T}{dt^2} + \lambda c^2 T = 0 \tag{2}$$

Equation (1) and the spatial BC define an S-L system whose solutions (eigenvalues and eigenfunctions) are

$$\lambda_n = \left(\frac{n\pi}{a}\right)^2 \qquad X_n = \sin\left(\frac{n\pi}{a}x\right) \qquad \text{for } n = 1, 2, \ldots$$

Equation (2) has a general solution of the form

$$T = A_n \cos \omega_n t + B_n \sin \omega_n t$$

where $\omega_n \equiv cn\pi/a$ and A_n and B_n are arbitrary constants. The general solution is thus

$$\psi(x, t) = \sum_{n=1}^{\infty} (A_n \cos \omega_n t + B_n \sin \omega_n t) \sin\left(\frac{n\pi}{a}x\right)$$

Further specifying the initial shape of the rope as $\psi(x, 0) = f(x)$ gives a Fourier series,

$$f(x) = \sum_{n=1}^{\infty} A_n \sin\left(\frac{n\pi}{a} x\right)$$

which determines A_n. What about B_n? Physically, the shape of the wave is not enough to define the problem uniquely. It is possible that the rope, while having the required initial shape, may be in motion of some sort. Thus, we must also know the velocity shape, which means specifying the function $\partial\psi/\partial t$ at $t = 0$. If it is given that

$$\left.\frac{\partial\psi}{\partial t}\right|_{t=0} = g(x)$$

then

$$g(x) = \sum_{n=1}^{\infty} \omega_n B_n \sin\left(\frac{n\pi}{a} x\right)$$

and B_n is also determined.

A unique determination of ψ requires specification not only of the initial wave shape but also of the "velocity shape."

The various frequencies ω_n, where $n = 1, 2, \ldots$, are referred to as *modes of oscillation*. Thus, in general, a solution is a linear superposition of infinitely many modes. In practice, it is possible to "excite" one mode or, with appropriate initial conditions, a finite number of modes with some degree of significance. •

In practice, for traveling waves rather than standing waves, specification of the wave shape and velocity shape is not as important as the mode of propagation. Thus, for instance, in the theory of wave guides, after the time variation is separated, a particular time variation, such as $e^{+i\omega t}$, and a particular direction for the propagation of the wave, usually the z-axis, are chosen.

Thus, if u denotes a component of the electric or the magnetic field, we can write $u(x, y, z, t) = \psi(x, y)e^{i(\omega t \pm kz)}$, where k is a constant, the wave number. The wave equation then reduces to

$$\frac{\partial^2\psi}{\partial x^2} + \frac{\partial^2\psi}{\partial y^2} + \left(\frac{\omega^2}{c^2} - k^2\right)\psi = 0$$

Introducing $\gamma^2 \equiv \omega^2/c^2 - k^2$ and $\mathbf{\nabla}_t \equiv (\partial/\partial x, \partial/\partial y)$ and writing the above equation in terms of the full vectors, we obtain

$$(\nabla_t^2 + \gamma^2)\begin{Bmatrix}\mathbf{E}\\ \mathbf{B}\end{Bmatrix} = 0 \tag{10.21a}$$

where

$$\begin{Bmatrix}\mathbf{E}\\ \mathbf{B}\end{Bmatrix} = \begin{Bmatrix}\mathbf{E}(x, y)\\ \mathbf{B}(x, y)\end{Bmatrix} e^{i(\omega t \pm kz)} \tag{10.21b}$$

These are the basic equations used in the study of electromagnetic wave guides and resonant cavities.

Example 10.3.7 Rectangular Wave Guides

Maxwell's equations in conjunction with Eq. (10.21b) give the transverse components (components perpendicular to the propagation direction) $\mathbf{E}_t$ and $\mathbf{B}_t$ in terms of the longitudinal components E_z and B_z (see Lorrain, Corson, and Lorrain 1988, Chapter 33):

$$\gamma^2 \mathbf{E}_t = \nabla_t\left(\frac{\partial E_z}{\partial z}\right) - i\frac{\omega}{c}\hat{\mathbf{e}}_z \times (\nabla_t B_z) \tag{1}$$

$$\gamma^2 \mathbf{B}_t = \nabla_t\left(\frac{\partial B_z}{\partial z}\right) + i\frac{\omega}{c}\hat{\mathbf{e}}_z \times (\nabla_t E_z) \tag{2}$$

where $\gamma^2 = \omega^2/c^2 - k^2$ and ∇_t is the two-dimensional gradient operator in the transverse plane.

Three types of guided waves are usually studied.

(1) Transverse magnetic (TM) waves have $B_z = 0$ everywhere. The BC on $\mathbf{E}$ demands that E_z vanish at the conducting walls of the guide.

(2) Transverse electric (TE) waves have $E_z = 0$ everywhere. The BC on B_z requires that $\partial B_z/\partial n$, where $\partial/\partial n$ is a directional derivative normal to the wall, vanish at the walls.

(3) Transverse electromagnetic (TEM) waves have $B_z = 0 = E_z$. For a nontrivial solution Eqs. (1) and (2) demand that $\gamma^2 = 0$. This form resembles a free wave with no boundaries.

We will discuss the TM mode briefly (see any book on electromagnetic theory for further details). The basic equations in this mode are

$$(\nabla_t^2 + \gamma^2)E_z = 0$$

$$B_z = 0$$

$$\gamma^2 \mathbf{E}_t = \nabla_t\left(\frac{\partial E_z}{\partial z}\right)$$

$$\gamma^2 \mathbf{B}_t = i\frac{\omega}{c}\hat{\mathbf{e}}_z \times (\nabla_t E_z)$$

For a wave guide with a rectangular cross section with sides a and b in the x and the y direction, respectively, we have

$$\frac{\partial^2 E_z}{\partial x^2} + \frac{\partial^2 E_z}{\partial y^2} + \gamma^2 E_z = 0$$

A separation of variables, $E_z(x, y) \equiv X(x)Y(y)$, leads to two S-L systems,

$$\frac{d^2 X}{dx^2} + \lambda^{(1)} X = 0 \qquad X(0) = X(a) = 0$$

$$\frac{d^2 Y}{dy^2} + \lambda^{(2)} Y = 0 \qquad Y(0) = Y(b) = 0$$

where $\gamma^2 = \lambda^{(1)} + \lambda^{(2)}$. These equations have the solutions

$$X_n(x) = \sin\left(\frac{n\pi}{a}x\right) \qquad \lambda_n^{(1)} \equiv \left(\frac{n\pi}{a}\right)^2 \qquad \text{for } n = 1, 2, \ldots$$

$$Y_m(y) = \sin\left(\frac{m\pi}{b}y\right) \qquad \lambda_m^{(2)} \equiv \left(\frac{m\pi}{b}\right)^2 \qquad \text{for } m = 1, 2, \ldots$$

The wave number is given by

$$k_{mn} = \sqrt{\frac{\omega^2}{c^2} - \left(\frac{n\pi}{a}\right)^2 - \left(\frac{m\pi}{b}\right)^2}$$

which has to be real if the wave is to propagate (an imaginary k leads to exponential decay or growth along the z-axis). Thus, there is a *cut-off frequency*,

$$\omega_{mn} \equiv c\sqrt{\left(\frac{n\pi}{a}\right)^2 + \left(\frac{m\pi}{b}\right)^2} \qquad \text{for } m, n \geqslant 1$$

below which the wave cannot propagate through the wave guide. Thus, for a TM wave, the lowest frequency that can propagate along a rectangular wave guide is

$$\omega_{11} = \frac{\pi c\sqrt{a^2 + b^2}}{ab}$$

The most general solution for E_z is, therefore,

$$E_z = \sum_{m,n=1}^{\infty} A_{mn} \sin\left(\frac{n\pi}{a}x\right)\sin\left(\frac{m\pi}{b}y\right)e^{i(\omega t \pm k_{mn}z)}$$

The constants A_{mn} are arbitrary and can be determined from the initial shape of the wave, but that is not commonly done. Once E_z is found, the other components can be calculated using Eqs. (1) and (2). •

10.3.2 Separation in Cylindrical Coordinates

When the geometry of the boundaries is cylindrical, the appropriate coordinate system is the cylindrical one. This always leads to Bessel functions "of some kind."

Example 10.3.8 Conducting Cylindrical Can

Consider a cylindrical conducting can of radius a and height h (Fig. 10.3). The potential $V(\rho, \varphi)$ varies at the top face, and the lateral surface and the bottom face are grounded. Let us find the electrostatic potential at all points inside the can.

A separation of variables, $\Phi(\rho, \varphi, z) = R(\rho)S(\varphi)Z(z)$, transforms Laplace's equation, $\nabla^2\Phi = 0$, into three ODEs:

$$\frac{d^2R}{d\rho^2} + \frac{1}{\rho}\frac{dR}{d\rho} + \left(k^2 - \frac{m^2}{\rho^2}\right)R = 0$$

$$\frac{d^2S}{d\varphi^2} + m^2 S = 0$$

$$\frac{d^2Z}{dz^2} - k^2 Z = 0$$

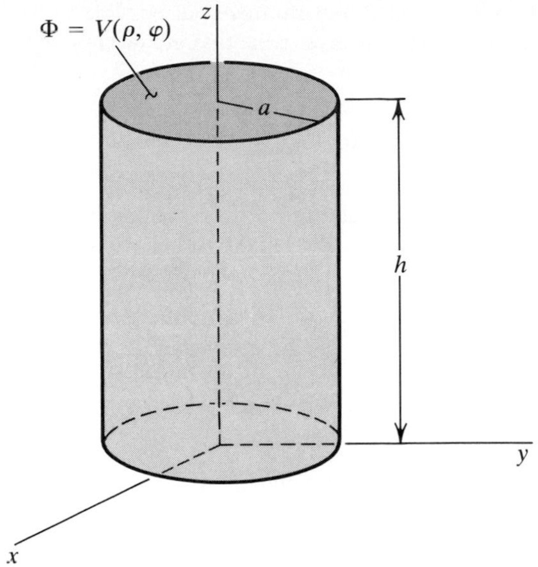

Figure 10.3 A conducting cylindrical can whose top has a potential given by $V(\rho, \varphi)$ with the rest of the surface grounded.

The first of these is the Bessel equation, whose general solution can be written as

$$R(\rho) = AJ_m(k\rho) + BY_m(k\rho)$$

The second DE has the general solution

$$S(\varphi) = C \cos m\varphi + D \sin m\varphi$$

The periodicity of Φ (and, therefore, of S) in φ implies that m is an integer. Finally the third DE has a general solution of the form

$$Z = Ee^{kz} + Fe^{-kz}$$

We note that none of the three ODEs lead to an S-L system because the BCs associated with them do not satisfy (10.18). However, we can still solve the problem by imposing the given BCs.

The fact that the potential must be finite everywhere inside the can (including at $\rho = 0$) forces B to vanish because the Neumann function $Y_m(k\rho)$ is not defined at $\rho = 0$. On the other hand, we want Φ to vanish at $\rho = a$. This gives

$$J_m(ka) = 0$$

which demands that ka be a root of the Bessel function of order m. Denoting by x_{mn} the nth zero of the Bessel function of order m, we have

$$ka = x_{mn} \quad \Rightarrow \quad k \equiv \frac{x_{mn}}{a} \qquad \text{for } n = 1, 2, \ldots$$

Similarly, the vanishing of Φ at $z = 0$ implies that

$$E = -F \qquad \text{and} \qquad Z = E \sinh\left(\frac{x_{mn} z}{a}\right)$$

We can now multiply R, S, and Z and sum over all possible values of m and n, keeping in mind that negative values of m give terms that are linearly dependent on the corresponding positive values. The result is

$$\Phi(\rho, \varphi, z) = \sum_{m=0}^{\infty} \sum_{n=1}^{\infty} J_m\left(\frac{x_{mn}}{a}\rho\right) \sinh\left(\frac{x_{mn}}{a} z\right)(A_{mn} \cos m\varphi + B_{mn} \sin m\varphi) \tag{1}$$

where A_{mn} and B_{mn} are constants to be determined by the remaining BC.

To find these constants we use the orthogonality of the trigonometric and Bessel functions. If we let $z = h$, then (1) reduces to

$$V(\rho, \varphi) = \sum_{m=0}^{\infty} \sum_{n=1}^{\infty} J_m\left(\frac{x_{mn}\rho}{a}\right) \sinh\left(\frac{x_{mn}}{a} h\right)(A_{mn} \cos m\varphi + B_{mn} \sin m\varphi)$$

from which we obtain

$$A_{mn} = \frac{2}{\pi a^2 J_{m+1}^2(x_{mn}) \sinh\left(x_{mn}\dfrac{h}{a}\right)} \int_0^{2\pi} d\varphi \int_0^a d\rho\, \rho V(\rho, \varphi) J_m\left(\frac{x_{mn}}{a}\rho\right) \cos m\varphi$$

and

$$B_{mn} = \frac{2}{\pi a^2 J_{m+1}^2(x_{mn}) \sinh\left(x_{mn}\dfrac{h}{a}\right)} \int_0^{2\pi} d\varphi \int_0^a d\rho\, \rho V(\rho, \varphi) J_m\left(\frac{x_{mn}}{a}\rho\right) \sin m\varphi$$

using a result derived in Exercise 10.3.7:

$$\int_0^a \rho J_m^2\left(\frac{x_{mn}}{a}\rho\right) d\rho = \frac{a^2}{2} J_{m+1}^2(x_{mn})$$

For the special but important case of azimuthal symmetry, for which $V(\rho, \varphi)$ is independent of φ, we obtain

$$A_{mn} = \frac{4\delta_{m,0}}{a^2 J_1^2(x_{0n}) \sinh\left(x_{0n}\dfrac{h}{a}\right)} \int_0^a d\rho\, \rho V(\rho) J_0\left(\frac{x_{0n}}{a}\rho\right)$$

$$B_{mn} = 0$$

●

Example 10.3.8 illustrates a typical problem whose solution is given in terms of the so-called Fourier-Bessel series,

$$\Phi(\rho, \varphi, z) = \sum_{m=0}^{\infty} \sum_{n=1}^{\infty} J_m(k_{mn}\rho) \sinh(k_{mn}z)(A_{mn} \cos m\varphi + B_{mn} \sin m\varphi) \tag{10.22}$$

where $k_{mn} \equiv x_{mn}/a$ and x_{mn} is the nth zero of the Bessel function of order m. Some of these zeros are given in Table 10.1.

TABLE 10.1 SOME VALUES OF x_{mn} AND y_{mn}, THE ZEROS OF THE BESSEL FUNCTIONS $J_m(x)$ AND THE NEUMANN FUNCTIONS $y_m(x)$, RESPECTIVELY

n	x_{0n}	x_{1n}	x_{2n}	x_{3n}	x_{4n}	x_{5n}	x_{6n}	x_{7n}	x_{8n}
1	2.405	3.832	5.136	6.380	7.588	8.771	9.936	11.086	12.225
2	5.520	7.016	8.417	9.761	11.065	12.339	13.589	14.821	16.038
3	8.654	10.173	11.620	13.015	14.373	15.700	17.004	18.288	19.555
4	11.791	13.324	14.796	16.223	17.616	18.980	20.321	21.642	22.945
5	14.931	16.471	17.960	19.409	20.827	22.218	23.586	24.935	26.267
	y_{0n}	y_{1n}	y_{2n}	y_{3n}	y_{4n}	y_{5n}	y_{6n}	y_{7n}	y_{8n}
1	0.894	2.197	3.384	4.527	5.645	6.747	7.838	8.920	9.995
2	3.958	5.430	6.794	8.098	9.362	10.597	11.811	13.008	14.190
3	7.086	8.596	10.023	11.396	12.730	14.034	15.314	16.574	17.818
4	10.222	11.749	13.210	14.623	16.000	17.347	18.671	19.974	21.261
5	13.361	14.897	16.379	17.818	19.224	20.603	21.958	23.294	24.613

The reason we obtained discrete values for k is the demand that Φ vanish at $\rho = a$. If we let $a \to \infty$, then k will be a continuous variable, and instead of a sum over k, we will obtain an integral. This is completely analogous to the transition from a Fourier series to a Fourier transform, but we will not pursue it further.

Example 10.3.9

Consider a circular heat-conducting plate of radius a whose temperature at time $t = 0$ has a distribution function $f(\rho, \varphi)$. Let us find the variation of T for all points (ρ, φ) on the plate for time $t > 0$ when the edge is kept at $T = 0$.

This is a two-dimensional problem involving the heat equation,

$$\frac{\partial T}{\partial t} = k^2 \nabla^2 T = k^2 \left[\frac{1}{\rho} \frac{\partial}{\partial \rho} \left(\rho \frac{\partial T}{\partial \rho} \right) + \frac{1}{\rho^2} \frac{\partial^2 T}{\partial \varphi^2} \right]$$

A separation of variables, $T(\rho, \varphi, t) = R(\rho)S(\varphi)g(t)$, leads to the following ODEs:

$$\frac{dg}{dt} = k^2 \lambda^{(1)} g$$

$$\frac{d^2 S}{d\varphi^2} + \lambda^{(2)} S = 0$$

$$\frac{d^2 R}{d\rho^2} + \frac{1}{\rho} \frac{dR}{d\rho} - \left(\frac{\lambda^{(2)}}{\rho^2} + \lambda^{(1)} \right) R = 0$$

To obtain exponential decay rather than growth for the temperature, we demand that $\lambda^{(1)} \equiv -b^2 < 0$. To ensure periodicity, we must have $\lambda^{(2)} \equiv m^2$, where m is an integer.

This leads to the following solutions:

$$g(t) = Ae^{-k^2b^2t}$$

$$S(\varphi) = B\cos m\varphi + C\sin m\varphi$$

$$R(\rho) = DJ_m(b\rho)$$

The other linear independent solution of the Bessel equation is absent because $\rho = 0$ is in the region of interest.

If the temperature is to be zero at $\rho = a$, we must have

$$R(a) = 0 \quad \Rightarrow \quad J_m(ba) = 0 \quad \Rightarrow \quad b = \frac{x_{mn}}{a}$$

It follows that the general solution can be written as

$$T(\rho, \varphi, t) = \sum_{m=0}^{\infty} \sum_{n=1}^{\infty} e^{-k^2(x_{mn}/a)^2t} J_m\left(\frac{x_{mn}}{a}\rho\right)(A_{mn}\cos m\varphi + B_{mn}\sin m\varphi)$$

The coefficients A_{mn} and B_{mn} can be determined as in Example 10.3.8. •

Example 10.3.10 Particle in a Cylindrical Can

Let us consider the cylindrical analogue of the situation of Example 10.3.5. For a cylindrical can of length L and radius a in which an atomic particle of mass μ is confined, the relevant Schrödinger equation in cylindrical coordinates is

$$i\frac{\partial\psi}{dt} = -\frac{\hbar}{2\mu}\left[\frac{1}{\rho}\frac{\partial}{\partial\rho}\left(\rho\frac{\partial\psi}{\partial\rho}\right) + \frac{1}{\rho^2}\frac{\partial^2\psi}{\partial\varphi^2} + \frac{\partial^2\psi}{\partial z^2}\right]$$

Let us solve this equation subject to the BC that $\psi(\rho, \varphi, z, t)$ vanishes at the sides of the can.

A separation of variables, $\psi(\rho, \varphi, z, t) = R(\rho)S(\varphi)Z(z)T(t)$, yields

$$\frac{d^2S}{d\varphi^2} + \lambda^{(1)}S = 0$$

$$\frac{d^2Z}{dz^2} + \lambda^{(2)}Z = 0 \tag{1}$$

$$\frac{d^2R}{d\rho^2} + \frac{1}{\rho}\frac{dR}{d\rho} + \left[\frac{2\mu}{\hbar}\omega - \lambda^{(2)} - \frac{\lambda^{(1)}}{\rho^2}\right]R = 0 \tag{2}$$

The periodicity of ψ in φ gives $\lambda^{(1)} = m^2$. Equation (1) along with the periodic BCs on Z constitutes an S-L system with solutions

$$Z(z) = \sin\left(\frac{k\pi}{L}z\right) \qquad \lambda_k^{(2)} = \left(\frac{k\pi}{L}\right)^2 \qquad \text{for } k = 1, 2, \ldots$$

If we let $\frac{2\mu}{\hbar}\omega - \left(\frac{k\pi}{L}\right)^2 \equiv b^2$, then Eq. (2) becomes

$$\frac{d^2R}{d\rho^2} + \frac{1}{\rho}\frac{dR}{d\rho} + \left(b^2 - \frac{m^2}{\rho^2}\right)R = 0$$

and the solution that is well-behaved at $\rho = 0$ is $J_m(b\rho)$. Since $R(a) = 0$, we obtain the quantization condition

$$ba = x_{mn} \quad \Rightarrow \quad b = \frac{x_{mn}}{a} \qquad \text{for } n = 1, 2, \ldots$$

Thus, the energy eigenvalues are

$$E_{kmn} \equiv \hbar\omega_{kmn} = \frac{\hbar^2}{2\mu}\left[\left(\frac{k\pi}{L}\right)^2 + \frac{x_{mn}^2}{a^2}\right]$$

and the general solution can be written as

$$\psi(\rho, \varphi, z, t) = \sum_{\substack{k,n=1\\m=0}}^{\infty} e^{-i\omega_{kmn}t} J_m\left(\frac{x_{mn}}{a}\rho\right)\sin\left(\frac{k\pi}{L}z\right)(A_{kmn}\cos m\varphi + B_{kmn}\sin m\varphi) \quad \bullet$$

Waves on a circular drumhead are historically important because the study of them was one of the first instances in which Bessel functions appeared. The following example considers such waves.

Example 10.3.11 A Vibrating Membrane

The wave equation for a two-dimensional vibrating system, such as a drumhead, is

$$\frac{\partial^2\psi}{\partial x^2} + \frac{\partial^2\psi}{\partial y^2} = \frac{1}{c^2}\frac{\partial^2\psi}{\partial t^2}$$

where c, the velocity of wave propagation, is given by $c = \sqrt{\tau/\sigma}$, in which τ is the tension per unit length and σ is the surface mass density of the membrane.

For a circular membrane over a cylinder, we need to express the above wave equation in polar coordinates, ρ and φ. This yields

$$\frac{1}{\rho}\frac{\partial}{\partial\rho}\left(\rho\frac{\partial\psi}{\partial\rho}\right) + \frac{1}{\rho^2}\frac{\partial^2\psi}{\partial\varphi^2} = \frac{1}{c^2}\frac{\partial^2\psi}{\partial t^2}$$

which, after separation of variables, reduces to

$$S = A\cos m\varphi + B\sin m\varphi \qquad \text{for } m = 0, 1, 2, \ldots$$

$$T = A'\cos\omega t + B'\sin\omega t$$

$$\frac{d^2R}{d\rho^2} + \frac{1}{\rho}\frac{dR}{d\rho} + \left(\frac{\omega^2}{c^2} - \frac{m^2}{\rho^2}\right)R = 0$$

The solution of this last equation, which is defined for $\rho = 0$ and vanishes at $\rho = a$, is

$$R = CJ_m\left(\frac{x_{mn}}{a}\rho\right) \qquad \text{where } \frac{\omega}{c} = \frac{x_{mn}}{a} \text{ and } n = 1, 2, \ldots$$

This shows that only the following frequencies propagate:

$$\omega_{mn} \equiv \frac{c}{a}x_{mn}$$

If we assume an initial shape for the membrane, given by $f(\rho, \varphi)$, and an initial velocity of zero, then $B' = 0$, and the general solution is

$$\psi(\rho, \varphi, t) = \sum_{\substack{m=0 \\ n=1}}^{\infty} J_m\left(\frac{x_{mn}}{a}\rho\right)\cos\left(\frac{cx_{mn}}{a}t\right)(A_{mn}\cos m\varphi + B_{mn}\sin m\varphi)$$

where

$$A_{mn} = \frac{2}{\pi a^2 J_{m+1}^2(x_{mn})}\int_0^{2\pi} d\varphi \int_0^a d\rho\, \rho f(\rho, \varphi) J_m\left(\frac{x_{mn}}{a}\rho\right)\cos m\varphi$$

$$B_{mn} = \frac{2}{\pi a^2 J_{m+1}^2(x_{mn})}\int_0^{2\pi} d\varphi \int_0^a d\rho\, \rho f(\rho, \varphi) J_m\left(\frac{x_{mn}}{a}\rho\right)\sin m\varphi$$

In particular, if the initial displacement of the membrane is independent of φ, then only the term with $m = 0$ contributes, and we get

$$\psi(\rho, \varphi, t) = \sum_{n=1}^{\infty} A_n J_0\left(\frac{x_{0n}}{a}\rho\right)\cos\left(\frac{cx_{0n}}{a}t\right)$$

where

$$A_n = \frac{4}{a^2 J_1^2(x_{0n})}\int_0^a d\rho\, \rho f(\rho) J_0\left(\frac{x_{0n}}{a}\rho\right)$$

Clearly, the wave does not develop any φ-dependence at later times. •

An important application of the wave equation concerns cylindrical wave guides, considered in the following example.

Example 10.3.12 Cylindrical Wave Guides

For a TM wave propagating along the z-axis in a hollow circular conductor, we have (see Example 10.3.7)

$$\frac{1}{\rho}\frac{\partial}{\partial\rho}\left(\rho\frac{\partial E_z}{\partial\rho}\right) + \frac{1}{\rho^2}\frac{\partial^2 E_z}{\partial\varphi^2} + \gamma^2 E_z = 0$$

The separation $E_z = R(\rho)S(\varphi)$ yields $S(\varphi) = A\cos m\varphi + B\sin m\varphi$ and

$$\frac{d^2R}{d\rho^2} + \frac{1}{\rho}\frac{dR}{d\rho} + \left(\gamma^2 - \frac{m^2}{\rho^2}\right)R = 0$$

The solution to this equation, which is regular at $\rho = 0$ and vanishes at $\rho = a$, is

$$R = CJ_m\left(\frac{x_{mn}}{a}\rho\right) \quad \text{and} \quad \gamma = \frac{x_{mn}}{a}$$

Recalling the definition of γ, we obtain

$$\gamma^2 = \frac{x_{mn}^2}{a^2} = \frac{\omega^2}{c^2} - k^2 \quad \Rightarrow \quad k = \sqrt{\frac{\omega^2}{c^2} - \frac{x_{mn}^2}{a^2}}$$

This gives the cut-off frequency:

$$\omega_{mn} = \frac{c}{a} x_{mn}$$

The solution for the azimuthally symmetric case ($m = 0$) is

$$E_z = \sum_{n=1}^{\infty} A_n J_0\left(\frac{x_{0n}}{a}\rho\right) e^{i(\omega t \pm k_n z)} \qquad \text{and} \qquad B_z = 0$$

where

$$k_n \equiv \sqrt{\frac{\omega^2}{c^2} - \frac{x_{0n}^2}{a^2}}$$

•

There are many variations on the theme of Bessel functions. We have encountered Bessel functions of three kinds, as well as modified Bessel functions. Another variation encountered in applications leads to what are known as Kelvin functions, introduced in the following example.

Example 10.3.13 Current Distribution in a Circular Wire

Consider the flow of charges in an infinitely long wire with a circular cross section of radius a. The relevant equation can be obtained as follows.

We start with Maxwell's equations:

$$\nabla \cdot \mathbf{E} = 0 \qquad \nabla \cdot \mathbf{B} = 0$$

$$\nabla \times \mathbf{E} = -\frac{1}{c}\frac{\partial \mathbf{B}}{\partial t} \qquad \nabla \times \mathbf{B} = \frac{4\pi}{c}\mathbf{j} + \frac{1}{c}\frac{\partial \mathbf{E}}{\partial t} \tag{1}$$

Taking the curl of the third equation gives

$$\nabla \times (\nabla \times \mathbf{E}) = -\frac{1}{c}\frac{\partial}{\partial t}(\nabla \times \mathbf{B}) = -\frac{1}{c}\frac{\partial}{\partial t}\left[\frac{4\pi}{c}\mathbf{j} + \frac{1}{c}\frac{\partial \mathbf{E}}{\partial t}\right] \tag{2}$$

For ordinary frequencies, $|4\pi \mathbf{j}| \gg |\partial \mathbf{E}/\partial t|$. Thus, we drop the second term of the RHS. The LHS can be written as

$$\nabla \times (\nabla \times \mathbf{E}) = \nabla(\nabla \cdot \mathbf{E}) - \nabla^2 \mathbf{E} = -\nabla^2 \mathbf{E} \tag{3}$$

Using Ohm's law, $\mathbf{j} = \sigma \mathbf{E}$, in (3) and substituting the result in (2) yields the equation with which we work:

$$\nabla^2 \mathbf{j} - \frac{4\pi\sigma}{c^2}\frac{\partial \mathbf{j}}{\partial t} = 0$$

We make the simplifying assumptions that the wire is along the z-axis and that there is no turbulence, so $\mathbf{j}$ is also in the z direction. We further assume that $\mathbf{j}$ is independent of φ and z in the cylindrical coordinate system and that its time-dependence is given by $e^{-i\omega t}$. Then we get

$$\frac{1}{\rho}\frac{d}{d\rho}\left(\rho \frac{dj}{d\rho}\right) + \tau^2 j = 0 \tag{4}$$

where $$\tau^2 = i\frac{4\pi\sigma\omega}{c^2} \equiv \frac{2i}{\delta^2}$$

and $\delta = c/\sqrt{2\pi\sigma\omega}$ is called the *skin depth.*

The *Kelvin equation* is usually given as

$$\frac{d^2w}{dx^2} + \frac{1}{x}\frac{dw}{dx} - ik^2w = 0 \tag{5}$$

If we substitute $x = \sqrt{it}/k$, it becomes

$$\frac{d^2w}{dt^2} + \frac{1}{t}\frac{dw}{dt} + w = 0$$

which is a Bessel equation of order zero. If the solution is to be regular at $x = 0$, then the only choice is

$$w(t) = J_0(t) = J_0(e^{-i\pi/4}kx)$$

This is the *Kelvin function* for Eq. (5). It is usually written as

$$J_0(e^{-i\pi/4}kx) \equiv \text{ber}(kx) + i[\text{bei}(kx)]$$

where ber and bei stand for "Bessel real" and "Bessel imaginary," respectively. If we substitute $z = e^{-i\pi/4}kx$ in the expansion for $J_0(z)$ and separate the real and the imaginary parts of the expansion, we obtain

$$\text{ber}(x) = 1 - \frac{(x/2)^4}{(2!)^2} + \frac{(x/2)^8}{(4!)^2} - \cdots$$

$$\text{bei}(x) = \frac{(x/2)^2}{(1!)^2} - \frac{(x/2)^6}{(3!)^2} + \frac{(x/2)^{10}}{(5!)^2} - \cdots$$

Equation (4) written in the form

$$\frac{d^2\jmath}{d\rho^2} + \frac{1}{\rho}\frac{d\jmath}{d\rho} + i\frac{2}{\delta^2}\jmath = 0$$

is the complex conjugate of (5) with $k^2 \equiv 2/\delta^2$. Thus, its solution is

$$\jmath(\rho) = AJ_0(e^{i\pi/4}k\rho) \equiv A\left\{\text{ber}\left(\frac{\sqrt{2}}{\delta}\rho\right) - i\left[\text{bei}\left(\frac{\sqrt{2}}{\delta}\rho\right)\right]\right\}$$

We can compare the value of the current density at ρ with its value at the surface $\rho = a$:

$$\left|\frac{\jmath(\rho)}{\jmath(a)}\right| = \left[\frac{\text{ber}^2\left(\frac{\sqrt{2}}{\delta}\rho\right) + \text{bei}^2\left(\frac{\sqrt{2}}{\delta}\rho\right)}{\text{ber}^2\left(\frac{\sqrt{2}}{\delta}a\right) + \text{bei}^2\left(\frac{\sqrt{2}}{\delta}a\right)}\right]^{1/2}$$

For low frequencies δ is large, which implies that $\sqrt{2}\rho/\delta$ is small; thus, $\text{ber}(\sqrt{2}\,\rho/\delta) \approx 1$ and $\text{bei}(\sqrt{2}\,\rho/\delta) \approx 0$, which imply that $|\jmath(\rho)/\jmath(a)| \approx 1$, and the current density is almost *uniform*. For higher frequencies the ratio of the current densities starts at a value less than 1 at $\rho = 0$ and increases to 1 at $\rho = a$. The starting value depends on the frequency. For very large frequencies the starting value is almost zero (see Marion and Heald 1980).

●

10.3.3 Separation in Spherical Coordinates

Recall that the most general PDE encountered in physical applications can be separated, in spherical coordinates, into

$$\mathbb{L}^2 Y(\theta, \varphi) = l(l+1) Y(\theta, \varphi) \tag{10.23a}$$

and

$$\frac{d^2R}{dr^2} + \frac{2}{r}\frac{dR}{dr} + \left[f(r) - \frac{l(l+1)}{r^2}\right]R = 0 \tag{10.23b}$$

We discussed the first of these two equations in great detail in Chapter 8 [as Eq. (8.14a)]. In particular, we constructed $Y_{lm}(\theta, \varphi)$ in such a way that they formed an orthonormal sequence. However, that construction was purely algebraic and did not say anything about the completeness of $Y_{lm}(\theta, \varphi)$.

With Theorem 10.3.1 at our disposal, we can separate (10.23a) into two ODEs by writing $Y_{lm}(\theta, \varphi) = P_{lm}(\theta) S_m(\varphi)$. We obtain

$$\frac{d^2 S_m}{d\varphi^2} + m^2 S_m = 0$$

$$\frac{d}{dx}\left[(1 - x^2)\frac{dP_{lm}}{dx}\right] + \left[l(l+1) - \frac{m^2}{1 - x^2}\right]P_{lm} = 0$$

where $x = \cos\theta$. These are both S-L systems satisfying the conditions of Theorem 10.3.1. Thus, the S_m are orthogonal among themselves and form a complete set for $\mathcal{L}^2[0, 2\pi]$. Similarly, for any fixed m, $P_{lm}(x)$ form a complete orthogonal set for $\mathcal{L}^2[-1, +1]$ (actually for the subset of $\mathcal{L}[-1, +1]$ that satisfies the same BC as the P_{lm} do at $x = \pm 1$). Thus, the (tensor) products $Y_{lm}(x, \varphi) = P_{lm}(x) S_m(\varphi)$ form a complete orthogonal sequence in the (Cartesian product) set $[-1, +1] \times [0, 2\pi]$, which, in terms of spherical angles, is the unit sphere, $0 \leqslant \theta \leqslant \pi$, $0 \leqslant \varphi \leqslant 2\pi$.

Let us consider some specific examples of expansion in the spherical coordinate system, starting with the simplest case—Laplace's equation when $f(r) = 0$.

Example 10.3.14 Electrostatics

For charge-free electrostatic problems Laplace's equation, $\nabla^2\Phi = 0$, gives (10.23b) with $f(r) = 0$:

$$\frac{d^2R}{dr^2} + \frac{2}{r}\frac{dR}{dr} - \frac{l(l+1)}{r^2}R = 0$$

Multiplying by r^2 reduces this to the Euler equation, which leads to the following SOLDE with constant coefficients after we substitute $r = e^t$ and use the chain rule and the fact that $dt/dr = 1/r$:

$$\frac{d^2R}{dt^2} + \frac{dR}{dt} - l(l+1)R = 0$$

This has a characteristic polynomial $p(\lambda) = \lambda^2 + \lambda - l(l+1)$ with roots $\lambda_1 = l$ and $\lambda_2 = -(l+1)$. Thus, a general solution is of the form

$$R(t) = Ae^{\lambda_1 t} + Be^{\lambda_2 t} = A(e^t)^l + B(e^t)^{-l-1}$$

or, in terms of r,

$$R(r) = Ar^l + Br^{-l-1}$$

Thus, the most general solution of Laplace's equation is

$$\Phi(r, \theta, \varphi) = \sum_{l=0}^{\infty} \sum_{m=-l}^{l} (A_{lm}r^l + B_{lm}r^{-l-1})Y_{lm}(\theta, \varphi)$$

For regions containing the origin the finiteness of Φ implies that $B_{lm} = 0$. Denoting the potential in such regions by Φ_{in}, we obtain

$$\Phi_{\text{in}}(r, \theta, \varphi) = \sum_{l=0}^{\infty} \sum_{m=-l}^{l} A_{lm}r^l Y_{lm}(\theta, \varphi)$$

Similarly, for regions including $r = \infty$, we have

$$\Phi_{\text{out}}(r, \theta, \varphi) = \sum_{l=0}^{\infty} \sum_{m=-l}^{l} B_{lm}r^{-l-1} Y_{lm}(\theta, \varphi)$$

To determine A_{lm} and B_{lm}, we need to invoke appropriate BCs. In particular, for a sphere of radius a on which the potential is given by $V(\theta, \varphi)$, we have

$$V(\theta, \varphi) \equiv \Phi_{\text{in}}(a, \theta, \varphi) = \sum_{l=0}^{\infty} \sum_{m=-l}^{l} A_{lm}a^l Y_{lm}(\theta, \varphi)$$

Multiplying by $Y^*_{l'm'}(\theta, \varphi)$ and integrating over the solid angle $d\Omega = \sin\theta \, d\theta \, d\varphi$, we obtain

$$A_{lm} = \frac{1}{a^l} \iint d\Omega \; V(\theta, \varphi) \, Y^*_{lm}(\theta, \varphi)$$

Similarly,

$$B_{lm} = a^{l+1} \iint d\Omega \; V(\theta, \varphi) \, Y^*_{lm}(\theta, \varphi)$$

In particular, if $V(\theta, \varphi)$ is independent of φ, only the components for which $m = 0$ are nonzero, and we have

$$A_{l0} = \frac{2\pi}{a^l} \int_0^{\pi} \sin\theta \; V(\theta) \, Y^*_{l0}(\theta) \, d\theta = \frac{2\pi}{a^l} \sqrt{\frac{2l+1}{4\pi}} \int_0^{\pi} \sin\theta \; V(\theta) P_l(\cos\theta) \, d\theta$$

which yields

$$\Phi_{\text{in}}(r, \theta) = \sum_{l=0}^{\infty} \left(\frac{2l+1}{2}\right) A_l \left(\frac{r}{a}\right)^l P_l(\cos\theta)$$

where
$$A_l = \int_0^\pi \sin\theta\, V(\theta) P_l(\cos\theta)\, d\theta$$

Similarly,
$$\Phi_{\text{out}}(r, \theta) = \sum_{l=0}^{\infty} \left(\frac{2l+1}{2}\right) A_l \left(\frac{a}{r}\right)^{l+1} P_l(\cos\theta)$$

As the alert reader may have noticed, all electrostatic problems have an exact counterpart among steady-state diffusion problems. This is because both situations obey Laplace's equation. •

The next most difficult case is that for which $f(r)$ is a constant. The diffusion equation, the wave equation, and the Schrödinger equation for a free particle give rise to such a case after time is separated from the rest of the variables.

The *Helmholtz equation* is

$$\nabla^2\psi + k^2\psi = 0 \tag{10.24}$$

and its radial part is

$$\frac{d^2R}{dr^2} + \frac{2}{r}\frac{dR}{dr} + \left[k^2 - \frac{l(l+1)}{r^2}\right]R = 0 \tag{10.25}$$

(This equation was discussed in Exercise 9.5.7.) The solutions are *spherical Bessel functions*, generically denoted as $z_l(x)$ and given by

$$z_l(x) \equiv \sqrt{\frac{\pi}{2}}\frac{Z_{l+1/2}(x)}{\sqrt{x}} \tag{10.26}$$

where $Z_\nu(x)$ is a solution of the Bessel equation of order ν.

A general solution of (10.25) can therefore be written as

$$R_l(r) = Aj_l(kr) + By_l(kr)$$

If the origin is included in the region of interest, then we must set $B = 0$. For such a case the solution to the Helmholtz equation is

$$\psi_k(r, \theta, \varphi) = \sum_{l=0}^{\infty}\sum_{m=-l}^{l} A_{lm} j_e(kr) Y_{lm}(\theta, \varphi) \tag{10.27}$$

The subscript k indicates that ψ is a solution of (10.24), which has k^2 as its constant.

Example 10.3.15 Particle in a Three-Dimensional Infinite Potential Well

The time-independent Schrödinger equation for a particle in an infinite potential well of radius a is

$$-\frac{\hbar^2}{2\mu}\nabla^2\psi = E\psi$$

with the BC $\psi(a, \theta, \varphi) = 0$. Here E is the energy of the particle and μ is its mass. We rewrite the Schrödinger equation as

$$\nabla^2\psi + \frac{2\mu E}{\hbar^2}\psi = 0$$

If $k^2 \equiv 2\mu E/\hbar^2$, we can immediately write the radial solution

$$R_l(r) = Aj_l(kr) = Aj_l(\sqrt{2\mu E}\,r/\hbar)$$

The vanishing of ψ at a implies that $j_l(\sqrt{2\mu E}\,a/\hbar) = 0$, or

$$\frac{\sqrt{2\mu E}\,a}{\hbar} = X_{ln} \qquad \text{for } n = 1, 2, \ldots$$

where X_{ln} is the nth zero of $j_l(x)$. Thus, the energy is quantized as

$$E_{ln} = \frac{\hbar^2 X_{ln}^2}{2\mu a^2} \qquad \text{for } l = 0, 1, \ldots; n = 1, 2, \ldots$$

The general solution to the Schrödinger equation is

$$\psi(r, \theta, \varphi) = \sum_{\substack{n=1\\ l=0}}^{\infty} \sum_{m=-l}^{l} A_{nlm} j_l\left(X_{ln}\frac{r}{a}\right) Y_{lm}(\theta, \varphi)$$ •

A particularly useful consequence of Eq. (10.27) is the expansion of a plane wave in terms of spherical Bessel functions. It is easily verified that if k is a vector, with $\mathbf{k}\cdot\mathbf{k}$ equal to k^2 in (10.24), then $e^{i\mathbf{k}\cdot\mathbf{r}}$ is a solution of (10.24). Thus, $e^{i\mathbf{k}\cdot\mathbf{r}}$ can be expanded as in (10.27). Assuming that $\mathbf{k}$ is along the z-axis, we get $\mathbf{k}\cdot\mathbf{r} = kr\cos\theta$, which is independent of φ. Thus, only the terms of (10.27) for which $m = 0$ will survive in such a case. Thus, we may write

$$e^{ikr\cos\theta} = \sum_{l=0}^{\infty} A_l j_l(kr) P_l(\cos\theta)$$

It can be shown (see Exercise 10.3.13) that $A_l = i^l(2l + 1)$. Thus, we have

$$e^{ikr\cos\theta} = \sum_{l=0}^{\infty} (2l + 1) i^l j_l(kr) P_l(\cos\theta) \tag{10.28}$$

For an arbitrary direction of $\mathbf{k}$, $\mathbf{k}\cdot\mathbf{r} = kr\cos\gamma$, where γ is the angle between $\mathbf{k}$ and $\mathbf{r}$. Thus, we may write

$$e^{i\mathbf{k}\cdot\mathbf{r}} = \sum_{l=0}^{\infty} (2l + 1) i^l j_l(kr) P_l(\cos\gamma)$$

Using the addition theorem for spherical harmonics, Eq. (8.30), we finally obtain

$$e^{i\mathbf{k}\cdot\mathbf{r}} = 4\pi \sum_{l=0}^{\infty} \sum_{m=-l}^{l} i^l j_l(kr) Y_{lm}^*(\theta', \varphi') Y_{lm}(\theta, \varphi) \tag{10.29}$$

where θ' and φ' are the angles of $\mathbf{k}$ and θ and φ are those of $\mathbf{r}$.

Such a decomposition of plane waves into components with definite orbital angular momenta is extremely useful when working with scattering theory for waves and particles.

Exercises

10.3.1 Show that separated and periodic BCs are special cases of the equality in (10.20).

10.3.2 Find the potential inside a cube with sides of length a when the top side is held at a constant potential V_0 with all other sides grounded (zero potential).

10.3.3 A long heat-conducting plate of width b is stretched along the positive x-axis with one corner at (0, 0) and the other at (0, b). The side of width b is held at temperature T_0, and the two long sides are held at $T = 0$. The two flat faces are insulated. Find the temperature variation of the plate, assuming equilibrium.

10.3.4 The two ends of a thin heat-conducting bar are held at $T = 0$. Initially, the first half of the bar is held at $T = T_0$, and the second half is held at $T = 0$. The lateral surface of the bar is then thermally insulated. Find the temperature distribution for all time.

10.3.5 The midpoint of the rope of Example 10.3.6 is raised a distance $a/2$, measured perpendicular to the tense rope, and released from rest. What is the subsequent wave function?

10.3.6 Find a general solution for the electromagnetic wave propagation in a *resonant cavity*, a rectangular box of sides $0 \leqslant x \leqslant a$, $0 \leqslant y \leqslant b$, and $0 \leqslant z \leqslant d$ with perfectly conducting walls. Discuss the modes the cavity can accommodate.

10.3.7 Derive the following normalization of the Bessel functions:

$$\int_0^a \rho J_\nu\left(\frac{x_{\nu n}}{a}\rho\right) J_\nu\left(\frac{x_{\nu l}}{a}\rho\right) d\rho = \frac{a^2}{2}[J_{\nu+1}(x_{\nu n})]\delta_{nl}$$

10.3.8 Find the potential of a cylindrical conducting can whose top is held at a constant potential V_0 while the rest is grounded.

10.3.9 Find the temperature of a circular conducting plate of radius a at all points of its surface for all time $t > 0$, assuming that its edge is held at $T = 0$ and initially its surface from the center to $a/2$ is in contact with a heat bath of temperature T_0.

10.3.10 Find the modes and the corresponding fields of a cylindrical resonant cavity of length L and radius a. Discuss the lowest TM mode.

10.3.11 Find the electrostatic potential both inside and outside a conducting sphere of radius a held at the constant potential V_0.

10.3.12 Find the electrostatic potential both inside and outside a sphere of radius a whose upper half is held at the constant potential V_0 and lower half is held at $-V_0$.

10.3.13 Derive the plane wave expansion, Eq. (10.28). (Hint: Use the orthogonality of Legendre polynomials and take the limit $k \to 0$.)

PROBLEMS

10.1 Show that the Liouville substitution given by (10.5) transforms regular S-L systems into regular S-L systems and separated and periodic BCs into separated and periodic BCs.

10.2 Let $u_1(x)$ and $u_2(x)$ be transformed into $v_1(t)$ and $v_2(t)$ by the Liouville substitution. Show that the inner product on $[a, b]$ with weight function $w(x)$ is transformed into the inner product on $[0, c]$ with unit weight, where $c = \int_a^b \sqrt{w/p}\, dx$.

10.3 Find the electrostatic potential inside a cube with sides of length a if all faces are grounded except the top, which is held at a potential given by

(a) $\frac{V_0}{a}x$ where $0 \leqslant x \leqslant a$ **(b)** $\frac{V_0}{a}y$ where $0 \leqslant y \leqslant a$

(c) $\frac{V_0}{a^2}xy$ where $0 \leqslant x, y \leqslant a$ **(d)** $V_0 \sin\left(\frac{\pi}{a}x\right)$ where $0 \leqslant x \leqslant a$

10.4 The lateral faces of a cube are grounded, and its top and bottom faces are held at potentials $f_1(x, y)$ and $f_2(x, y)$, respectively.
(a) Find a general expression for the potential inside the cube (as in Example 10.3.1).
(b) Find the potential if the top is held at V_0 volts and the bottom at $-V_0$ volts.

10.5 Find the potential inside a semi-infinite cylindrical conductor whose cross section is a square with sides of length a. All sides are grounded except the square side, which is held at the constant potential V_0.

10.6 Find the temperature distribution of a rectangular plate (see Fig. 10.4) with sides of lengths a and b if three sides are held at $T = 0$ and the fourth side has a temperature variation given by

(a) $\frac{T_0}{a}x$ where $0 \leqslant x < a$ **(b)** $\frac{T_0}{a^2}x(x-a)$ where $0 \leqslant x \leqslant a$

(c) $\frac{T_0}{a}\left|x - \frac{a}{2}\right|$ where $0 < x < a$ **(d)** $T = 0$ where $0 \leqslant x \leqslant a$

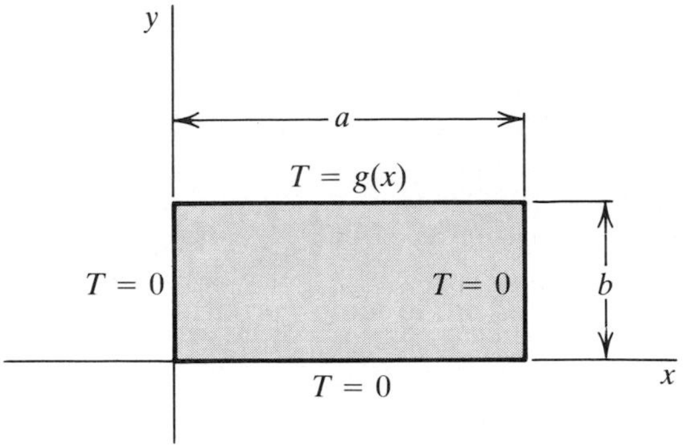

Figure 10.4

10.7 Repeat Exercise 10.3.3 with the temperature of the short side held at each of the following:

(a) $T = \begin{cases} 0^\circ & \text{if } 0 < y < \frac{b}{2} \\ T_0 & \text{if } \frac{b}{2} < y < b \end{cases}$ **(b)** $T = \frac{T_0}{b}y$ where $0 \leqslant y < b$

(c) $T = T_0 \cos\left(\frac{\pi}{b}y\right)$ where $0 \leqslant y \leqslant b$ **(d)** $T = T_0 \sin\left(\frac{\pi}{b}y\right)$ where $0 \leqslant y \leqslant b$

10.8 The two ends of a thin heat-conducting bar of length b are held at $T = 0$. The lateral surface of the bar is thermally insulated. Find the temperature distribution at all times if initially it is given by

(a) $T(0, x) = \begin{cases} T_0 \text{ for the middle third of the bar} \\ 0 \text{ for the other two thirds} \end{cases}$

(b) $T(0, x) = \left|x - \dfrac{b}{2}\right| - \dfrac{b}{2}$ where $0 \leqslant x \leqslant b$

(c) $T(0, x) = \cos\left(\dfrac{\pi}{b}x\right)$ where $0 < x < b$

(d) $T(0, x) = \begin{cases} 0 & \text{if } 0 \leqslant x \leqslant \dfrac{b}{3} \\ T_0 \sin\left(\dfrac{3\pi}{b}x - \pi\right) & \text{if } \dfrac{b}{3} \leqslant x \leqslant \dfrac{2b}{3} \\ 0 & \text{if } \dfrac{2b}{3} \leqslant x \leqslant b \end{cases}$

10.9 Repeat Example 10.3.3 assuming that the end at $x = 0$ is held at T_0 and the end at $x = b$ at $-T_0$. (Hint: The solution corresponding to $\lambda = 0$ is essential and cannot be excluded.)

10.10 Repeat Problem 10.8 for the case where the initial temperature distribution is given by

(a) $T(0, x) = -\dfrac{2T_0}{b}x + T_0$ where $0 \leqslant x \leqslant b$

(b) $T(0, x) = -\dfrac{2T_0}{b^2}x^2 + T_0$ where $0 \leqslant x \leqslant b$

(c) $T(0, x) = \dfrac{T_0}{b}x + T_0$ where $0 \leqslant x < b$

10.11 Determine $T(x, y, t)$ for the rectangular plate of Example 10.3.4 if initially the lower left quarter is held at T_0 and the rest of the plate is held at $T = 0$.

10.12 All sides of the plate of Example 10.3.2 are held at $T = 0$. Find the temperature distribution for all time if the initial temperature distribution is given by

(a) $T(x, y, 0) = \begin{cases} T_0 & \text{if } \frac{1}{4}a \leqslant x \leqslant \frac{3}{4}a \text{ and } \frac{1}{4}b \leqslant y \leqslant \frac{3}{4}b \\ 0 & \text{otherwise} \end{cases}$

(b) $T(x, y, 0) = \dfrac{T_0}{ab}xy$ where $0 \leqslant x < a$ and $0 \leqslant y < b$

(c) $T(x, y, 0) = \dfrac{T_0}{a}x$ where $0 \leqslant x < a$ and $0 < y < b$

10.13 Repeat Example 10.3.4 with the temperatures of the sides equal to T_1, T_2, T_3, and T_4. (Hint: You must include $\lambda^{(i)} = 0$ solutions.)

10.14 A string of length a fastened at both ends has an initial velocity of zero and is given an initial displacement as shown in the following figures. Find $\psi(x, t)$ in each case.

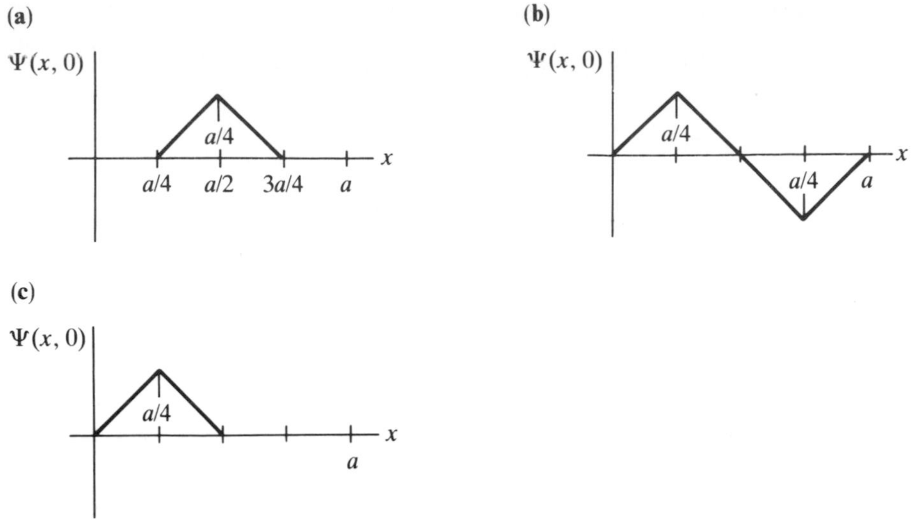

10.15 Repeat Problem 10.14 assuming that the initial displacement is zero and the initial velocity distribution is given by each figure.

10.16 Repeat Problem 10.15 with the initial velocity distribution given by

(a) $g(x) = \begin{cases} V_0 & \text{if } 0 \leqslant x \leqslant \dfrac{a}{2} \\ 0 & \text{if } \dfrac{a}{2} < x \leqslant a \end{cases}$ **(b)** $g(x) = \begin{cases} V_0 \sin \dfrac{2\pi x}{a} & \text{if } 0 \leqslant x \leqslant \dfrac{a}{2} \\ 0 & \text{if } \dfrac{a}{2} < x \leqslant a \end{cases}$

10.17 A string of length a is fixed at the left end, and the right end moves with displacement $A \sin \omega t$. Find $\psi(x, t)$ and a consistent set of initial conditions for the displacement and the velocity.

10.18 Repeat Example 10.3.8 if the can has a semi-infinite length and the base is held at the potential $V(\rho, \varphi)$. The lateral surface is grounded.

10.19 Repeat Problem 10.18 with the potential of the bottom surface given by

(a) $V = \dfrac{V_0}{a} y$ **(b)** $V = \dfrac{V_0}{a} x$ **(c)** $V = \dfrac{V_0}{a^2} xy$

[Hint: Use the integral identity $\int z^{\nu+1} J_\nu(z)\, dz = z^{\nu+1} J_{\nu+1}(z)$.]

10.20 Find the steady-state temperature distribution, $T(\rho, \varphi, z)$, in a semi-infinite solid cylinder of radius a if the temperature distribution of the base is $f(\rho, \varphi)$ and the lateral surface is held at $T = 0°$.

10.21 Find the steady-state temperature distribution of a solid cylinder with a height and radius of 10, assuming that the base and the lateral surface are at $T = 0$ and the top is at $T = 100°$.

10.22 A flat circular plate of radius a is initially at a temperature T_0. From $t = 0$ on, the temperature of the circumference is held at $T = 0$. Find the temperature distribution for all time.

10.23 Repeat Problem 10.22 with the temperature distribution at $t = 0$ given by

(a) $T(\rho, \varphi, 0) = \dfrac{T_0}{a} y$ **(b)** $T(\rho, \varphi, 0) = \dfrac{T_0}{a} x$ **(c)** $T(\rho, \varphi, 0) = \dfrac{T_0}{a^2} xy$

10.24 Two identical long conducting half-cylindrical shells of radius a are glued together in such a way that they are insulated from one another. One half-cylinder is held at potential V_0 and the other is gounded. Find the potential at any point inside the resulting cylinder. (Hint: Separate Laplace's equation in two dimensions.)

10.25 Find the equation for a vibrating rectangular membrane with sides of lengths a and b rigidly fastened on all sides. For $a = b$, show that a given mode frequency may have more than one solution.

10.26 A linear charge distribution of uniform density λ extends along the z-axis from $z = -b$ to $z = b$. Show that the electrostatic potential, $\Phi(r, \theta, \varphi)$, at any point $r > b$ is given by

$$\Phi(r, \theta, \varphi) = 2\lambda \sum_{k=0}^{\infty} \frac{(b/r)^{2k+1}}{2k+1} P_{2k}(\cos\theta)$$

(Hint: Consider a point on the z-axis at a distance $r > b$ from the origin. Solve the simple problem using $\Phi = \int dq/r$, and compare the result with the infinite series to obtain the unknown coefficients.)

10.27 Find the electrostatic potential inside a sphere of radius a if the top hemisphere is gounded and the bottom hemisphere is maintained at a constant potential V_0.

10.28 A sphere of radius a is maintained at a temperature T_0. The sphere is inside a large heat-conducting mass. Find the expressions for the steady-state temperature distribution both inside and outside the sphere.

10.29 The upper half of a heat-conducting sphere of radius a has $T = 100\,°\text{C}$; the lower half is maintained at $T = -100\,°\text{C}$. The whole sphere is inside an infinitely large mass of heat-conducting material. Find the steady-state temperature distribution inside and outside the sphere.

10.30 Find the steady-state temperature distribution inside a sphere of radius a when the surface temperature is given by

(a) $T_0 \cos^2\theta$ **(b)** $T_0 \cos^4\theta$ **(c)** $T_0(\cos\theta - \cos^3\theta)$ **(d)** $T_0|\cos\theta|$

10.31 Find the electrostatic potential both inside and outside a conducting sphere of radius a when the sphere is maintained at a potential given by

(a) $V_0(\cos\theta - 3\sin^2\theta)$ **(b)** $V_0(5\cos^3\theta - 3\sin^2\theta)$

(c) $\begin{cases} V_0 \cos\theta \text{ for the upper hemisphere} \\ 0 \text{ for the lower hemisphere} \end{cases}$

10.32 Find the steady-state temperature distribution *inside* a hemisphere of radius a if the curved surface is held at T_0 and the flat surface at $T = 0$. (Hint: Imagine completing the sphere and maintaining the lower hemisphere at a temperature such that the overall surface temperature distribution is an *odd* function about $\theta = \pi/2$.)

10.33 Find the steady-state temperature distribution in a spherical shell of inner radius R_1 and outer radius R_2 when the inner surface has a constant temperature T_1 and the outer surface has a constant temperature T_2.

Part IV

OPERATORS, GREEN'S FUNCTIONS, AND INTEGRAL EQUATIONS

11

Operators in Hilbert Spaces and Green's Functions

So far, the treatment of differential equations, with the exception of SOLDEs with constant coefficients, has not considered inhomogeneous equations. At this point, however, we can put into use one of the most elegant pieces of machinery in higher mathematics to solve inhomogeneous differential equations. This machinery, which harmoniously blends algebraic and analytical techniques, is called the Green's function method and is the central theme of this part of this book.

11.1 INTRODUCTION

We can solve an abstract vector-operator equation in an N-dimensional vector space $\mathcal{V}$,

$$\mathbb{A}|\mathrm{u}\rangle = |\mathrm{v}\rangle$$

where $\mathbb{A} \in \mathcal{L}(\mathcal{V})$ and $|\mathrm{u}\rangle, |\mathrm{v}\rangle \in \mathcal{V}$, by selecting a basis $\mathrm{B} = \{|\mathrm{a}_i\rangle\}_{i=1}^N$ for $\mathcal{V}$, writing the equation in matrix form, and solving the resulting system of N linear equations. This produces the components of the solution $|\mathrm{u}\rangle$ in B. If components in another basis B′ are desired, they can be obtained using the matrix transformation R connecting B and B′ (see Chapter 3).

There is a standard formal procedure for obtaining the matrix equation. It is convenient to choose an orthonormal basis $\mathrm{B} = \{|\mathrm{e}_i\rangle\}_{i=1}^N$ for $\mathcal{V}$ and refer all components to this basis. Contracting both sides of the equation $\mathbb{A}|\mathrm{u}\rangle = |\mathrm{v}\rangle$ with $\langle \mathrm{e}_i|$ and inserting $\mathbb{1} = \sum_{j=1}^N |\mathrm{e}_j\rangle\langle \mathrm{e}_j|$ between $\mathbb{A}$ and $|\mathrm{u}\rangle$ gives

$$\sum_{j=1}^N \langle \mathrm{e}_i|\mathbb{A}|\mathrm{e}_j\rangle\langle \mathrm{e}_j|\mathrm{u}\rangle = \langle \mathrm{e}_i|\mathrm{v}\rangle \qquad \text{for } i = 1, 2, \ldots, N$$

or
$$\sum_{j=1}^{N} A_{ij}u_j = v_i \qquad \text{for } i = 1, 2, \ldots, N \tag{11.1a}$$

where $A_{ij} \equiv \langle e_i|\mathbb{A}|e_j\rangle$, $u_j \equiv \langle e_j|u\rangle$, and $v_i \equiv \langle e_i|v\rangle$. Equation (11.1a) is simply a system of N linear equations in N unknowns $\{u_j\}_{j=1}^{N}$, which can be solved to obtain the solution(s) of the original equation in B.

An important case is that where $\mathbb{A}$ is a diagonal in B, that is, $A_{ij} = \lambda_i\delta_{ij}$, where $\{\lambda_i\}_{i=1}^{N}$ is the set of eigenvalues of $\mathbb{A}$. For this special situation we have

$$\lambda_i u_i = v_i \qquad \text{for } i = 1, 2, \ldots, N \qquad \text{(no sum over } i) \tag{11.1b}$$

This equation has a *unique* solution (for arbitrary v_i) if and only if $\lambda_i \neq 0$ for all i. In that case

$$u_i = \frac{v_i}{\lambda_i} \qquad \text{for } i = 1, 2, \ldots, N$$

In particular, if $v_i = 0$ for all i, that is, when Eq. (11.1b) is homogeneous, the unique solution is the trivial solution. On the other hand, when some of the λ_i are zero, there may be no solution to Eq. (11.1b), but the homogeneous equation has a *nontrivial* solution (u_i need not be zero). Recalling (from Chapter 3) that an operator $\mathbb{A}$ is invertible if and only if none of its eigenvalues is zero, we have a proposition.

11.1.1 Proposition The operator $\mathbb{A} \in \mathcal{L}(\mathcal{V})$ is invertible if and only if the homogeneous equation $\mathbb{A}|u\rangle = 0$ has no nontrivial solutions. ■

Let us now apply the procedure just described to infinite-dimensional vector spaces, in particular, for the case of a continous index. We can, formally at least, proceed as for the N-dimensional space. Thus, using the notation commonly used for the continuous case, we want to find the solutions of

$$\mathbb{K}|u\rangle = |f\rangle \tag{11.2}$$

Following the procedure used above, we obtain

$$\langle x|\mathbb{K}\left(\int_a^b |y\rangle w(y)\langle y|\,dy\right)|u\rangle = \int_a^b \langle x|\mathbb{K}|y\rangle w(y)\langle y|u\rangle\,dy = \langle x|f\rangle$$

where we have used the results obtained in Section 5.2. Writing this result in continuous argument notation, we have

$$\int_a^b K(x, y)w(y)u(y)\,dy = f(x) \tag{11.3}$$

which is the continuous analogue of (11.1). Here $[a, b]$ is the interval on which functions are defined. We note that the indices have turned into continuous arguments, and the sum has turned into an integral. The operator $\mathbb{K}$ in (11.2) that leads to an equation such as (11.3) is called an *integral operator* (IO), and the "matrix element" $K(x, y)$ is said to be its *kernel*.

If $\mathbb{K}$ has an inverse, $\mathbb{G}$, Eq. (11.2) can be solved for $|u\rangle$. Multiplying both sides by $\mathbb{G}$ yields $|u\rangle = \mathbb{G}|f\rangle$, which in integral form is

$$u(x) = \int_a^b G(x, y)w(y)f(y)\,dy$$

Thus, the problem of solving (11.2) reduces to finding the function $G(x, y)$, the inverse of $K(x, y)$. The nice thing about $G(x, y)$ is that once we have it, we can solve Eq. (11.2) for *arbitrary* $f(x)$. This is particularly handy when $\mathbb{K}$ is a differential operator (DO). But how does a DO arise from an IO such as $\mathbb{K}$?

The discussion of the discrete case mentioned the possibility of the operator $\mathbb{A}$ being diagonal in the given basis B. Let us do the same with (11.3); that is (noting that x and y are indices for K), let us assume that $K(x, y) = 0$ for $x \neq y$. Such operators are called *local operators*. Thus, the contribution to the integral comes only at the *point* where $x = y$. If $K(x, y)$ is finite at this point, and the functions $w(y)$ and $u(y)$ are well-behaved there, the LHS of (11.3) will vanish, and we will get inconsistencies. It follows then that

$$K(x, y) = \begin{cases} 0 & \text{if } x \neq y \\ \infty & \text{if } x = y \end{cases}$$

Thus, $K(x, y)$ has the behavior of a delta function. Letting

$$K(x, y) \equiv L(x)\frac{\delta(x - y)}{w(x)}$$

and substituting in (11.3) yields

$$L(x)u(x) = f(x)$$

In the discrete case λ_i is merely an indexed number; its continuous analogue, $L(x)$, may represent merely a function. However, the fact that x is a continuous variable (index) gives rise to other possibilities for $L(x)$, which do not exist for the discrete case. For instance, $L(x)$ could be a DO, an operator that takes $u(x)$ on a "cruise" to neighboring points and then returns it to x with a derivative in front of it. Such a journey is possible for $u(x)$ because x is a continuous variable, on which the analytic process of taking a limit can be defined. For the discrete case, u_i must "hop" from i to $i + 1$ and then back to i. Such a difference (as opposed to differential) process is *not local*, it involves not only i but also $i + 1$. The "point" i does not have a (infinitesimally close) neighbor. The derivative, although defined by a limiting process involving neighboring points, is a local operator. Thus, we can speak of the derivative of a function *at a point*.

This essential difference between discrete operators and continuous operators makes the latter far richer in possibilities for applications. In particular, if $L(x)$ is considered a differential operator, the equation $L(x)u(x) = f(x)$ leads directly to the fruitful area of differential equation theory. In that context $G(x, y)$, the inverse of $L(x)$, is called the *Green's function* for $\mathbb{L}$. We will discuss Green's functions in great detail later, but first let us briefly consider operators in Hilbert spaces.

11.2 OPERATORS IN HILBERT SPACES

The concept of an operator on a Hilbert space is extremely subtle. Even the elementary characteristics of operators, such as the operation of hermitian conjugation, cannot generally be defined on the whole Hilbert space.

In finite-dimensional vector spaces there is a one-to-one correspondence between operators and matrices. So, in some sense, the study of operators reduces to a study of matrices, which are collections of real or complex numbers. Although we have already noted an analogy between matrices and kernels, a whole new realm of questions arises when A_{ij} goes to $K(x, y)$—questions about the continuity of $K(x, y)$ in both its arguments, about the limit of $K(x, y)$ as x and/or y approach the "end points" of the interval on which $\mathbb{K}$ is defined, about the boundedness and "compactness" of $\mathbb{K}$, and so on. Such subtleties are not unexpected. After all, when we tried to generalize concepts of finite-dimensional vector spaces to infinite dimensions in Chapter 5, we encountered some difficulties. There we were concerned about vectors only; the generalization of operators is more complicated by orders of magnitude.

This section outlines some of the general features of operator theory. Let us begin by discussing an important class of operators—the integral operators.

11.2.1 Integral Operators

An integral operator (IO) is defined in terms of its kernel. Let $K(x, y)$ be a (complex-valued) continuous function defined on the square $a \leqslant x \leqslant b$, $a \leqslant y \leqslant b$. It induces a transformation on the space of the so-called Reimann-integrable functions, denoted by $\mathscr{R}(a, b)$, such that for $f \in \mathscr{R}(a, b)$ we have

$$\mathbb{K}f(x) \equiv g(x) \equiv \int_a^b K(x, y) f(y)\, dy$$

This equation is called an *integral transform*. Note that $w(x) = 1$.

The *adjoint integral operator* is defined in analogy with the finite-dimensional case. Thus, in Dirac notation we define $\langle \mathrm{x}|\mathbb{K}^\dagger|\mathrm{y}\rangle = \langle \mathrm{y}|\mathbb{K}|\mathrm{x}\rangle^*$, which, in terms of kernels, is $K^\dagger(x, y) = [K(y, x)]^*$. The kernel $K^\dagger(x, y)$ gives rise to an integral transform:

$$\mathbb{K}^\dagger f(x) = \int_a^b K^\dagger(x, y) f(y)\, dy = \int_a^b K^*(y, x) f(y)\, dy$$

An integral transform $K: \mathscr{R}(a, b) \to \mathscr{R}(a, b)$, is said to be hermitian if

$$\langle \mathbb{K}\mathrm{f}|\mathrm{g}\rangle = \langle \mathrm{f}|\mathbb{K}|\mathrm{g}\rangle \qquad \forall\, f, g \in \mathscr{R}(a, b)$$

This leads directly to the condition (see Exercise 11.2.1)

$$K(x, y) = K^*(y, x)$$

In particular, for real-valued functions and operators, this condition reduces to

$$K(x, y) = K(y, x)$$

and the kernel is called a *symmetric kernel.*

Example 11.2.1

Let us consider some examples of integral transforms.

(a) The *Fourier transform* is familiar from the discussion in Chapter 5. The kernel is not hermitian but symmetric (as a complex-valued function):

$$K(x, y) = e^{ixy}$$

(b) The *Laplace transform* is used frequently in electrical engineering. Its kernel is

$$K(x, y) = e^{-xy}$$

which, evidently, is symmetric and hermitian (x and y are real).

(c) The *Euler transform* has the kernel

$$K(x, y) = (x - y)^{\nu}$$

If ν is an even integer, the kernel becomes symmetric.

(d) The *Mellin transform* has the kernel

$$K(x, y) = G(x^y)$$

where G is an arbitrary function. Most of the time $K(x, y)$ is taken to be simply x^y.

(e) The *Hankel transform* has the kernel

$$K(x, y) = yJ_n(xy)$$

where J_n is the nth-order Bessel function.

(f) A transform that is useful in connection with the Bessel equation has the kernel

$$K(x, y) = \left(\frac{x}{2}\right)^{\nu} e^{(y - x^2/4y)}$$ ●

Unless otherwise indicated, the kernels considered here will be real and symmetric. All functions are assumed to be real, and the inner product is given by

$$\langle f|g \rangle = \int_a^b f(x)g(x)\,dx$$

which incorporates the assumption that $w(x) = 1$.

As we did in finite dimensions, we can define the eigenvectors and eigenvalues of an integral operator. The function $f \in \mathscr{R}(a, b)$ is called an *eigenfunction* of $\mathbb{K}$ with eigenvalue λ if and only if $f \not\equiv 0$ and

$$\mathbb{K} f(x) = \lambda f(x) \tag{11.4a}$$

or

$$\int_a^b K(x, y) f(y)\, dy = \lambda f(x) \tag{11.4b}$$

It can easily be verified that the eigenvectors corresponding to distinct eigenvalues of a symmetric integral transform K are orthogonal.

11.2.2 Separable Kernels

The eigenvalue equations, Eqs. (11.4), can be easily solved if $K(x, y)$ is *separable*, that is, if there exist functions $\{h_i(x)\}_{i=1}^n$, continuous on the interval $[a, b]$, such that

$$K(x, y) = \sum_{i,j=1}^{n} c_{ij} h_i(x) h_j(y)$$

with $c_{ij} = c_{ji}$. Substituting this in (11.4b), we obtain

$$\sum_{i,j=1}^{n} c_{ij} h_i(x) \int_a^b h_j(y) f(y)\, dy = \lambda f(x)$$

Defining $a_j \equiv \int_a^b h_j(y) f(y)\, dy$ and substituting it in the preceding equation gives

$$\lambda f(x) = \sum_{i,j=1}^{n} c_{ij} h_i(x) a_j \tag{11.5}$$

Substituting this back in the definition of a_j yields

$$\lambda a_j = \int_a^b h_j(y)[\lambda f(y)]\, dy = \int_a^b h_j(y) \left[\sum_{i,k=1}^{n} c_{ik} h_i(y) a_k \right] dy$$

$$= \sum_{i,k=1}^{n} \left[c_{ik} \int_a^b h_j(y) h_i(y)\, dy \right] a_k \equiv \sum_{k=1}^{n} m_{jk} a_k$$

where

$$m_{jk} \equiv \sum_{i=1}^{n} c_{ik} \int_a^b h_j(y) h_i(y)\, dy$$

This is an eigenvalue equation for the $n \times n$ matrix M with elements m_{ij}. Once the eigenvectors and the eigenvalues are found, we can substitute them in (11.5) and obtain $f(x)$. It is important to note that (11.5) has a solution,

$$f(x) = \frac{1}{\lambda} \sum_{i,j=1}^{n} c_{ij} h_i(x) a_j$$

if and only if λ is a nonzero eigenvalue of M.

It is clear that there are only a finite number of eigenvalues. Let $\{\lambda_k\}_{k=1}^{m}$ be the set of nonzero eigenvalues and $\{f_k(x)\}_{k=1}^{m}$ the set of corresponding eigenfunctions. Then we can write

$$f_k(x) = \frac{1}{\lambda_k} \sum_{i,j=1}^{n} c_{ij} h_i(x) a_j^{(k)}$$

where $a_j^{(k)}$ is the jth component of the kth eigenvector. Clearly, if there is more than one linearly independent eigenvector, there will be more than one $f_k(x)$.

On the other hand, if $\mathbb{K}g(x) = 0$ for $g \in \mathscr{R}(a, b)$, then $g(x)$ is an eigenvector of $\mathbb{K}$ with eigenvalue zero. Since all of $\{\lambda_k\}_{k=1}^{m}$ are different from zero, $g(x)$ must be orthogonal to all $f_k(x)$. We thus have the following theorem (see Roach 1970, pp. 86 and 87, for refinements).

11.2.1 Theorem Let $K(x, y)$ be a nonzero, symmetric, separable kernel. Then there exist a set of orthonormal functions, $\{f_k(x)\}_{k=1}^{m}$, in $\mathscr{R}(a, b)$ and a set of nonzero real scalars, $\{\lambda_k\}_{k=1}^{m}$, such that $\mathbb{K}f_k(x) = \lambda_k f_k(x)$ for $k = 1, 2, \ldots, m$. If $g \in \mathscr{R}(a, b)$ is orthogonal to all f_k, then $\mathbb{K}g(x) = 0$. The scalars λ_k are the only nonzero eigenvalues of $\mathbb{K}$, and there is only a finite number of linearly independent eigenvectors corresponding to each eigenvalue. ■

Example 11.2.2

Let us find the nonzero eigenvalues and corresponding eigenfunctions of the kernel $K(x, y) = 1 + \sin(x + y)$ for $-\pi \leqslant x, y \leqslant \pi$.

We are seeking functions f and scalars λ satisfying $\mathbb{K}f(x) = \lambda f(x)$, or

$$\int_{-\pi}^{\pi} K(x, y) f(y)\, dy = \lambda f(x) \tag{1}$$

Expanding $\sin(x + y)$ and substituting the result in (1), we obtain

$$\int_{-\pi}^{\pi} [1 + \sin x \cos y + \cos x \sin y] f(y)\, dy = \lambda f(x)$$

or

$$a_1 + a_2 \sin x + a_3 \cos x = \lambda f(x) \tag{2}$$

where

$$a_1 \equiv \int_{-\pi}^{\pi} f(y)\, dy \qquad a_2 \equiv \int_{-\pi}^{\pi} \cos y f(y)\, dy \qquad a_3 \equiv \int_{-\pi}^{\pi} \sin y f(y)\, dy$$

We can write

$$\lambda a_1 = \int_{-\pi}^{\pi} \lambda f(y)\,dy = \int_{-\pi}^{\pi} (a_1 + a_2 \sin y + a_3 \cos y)\,dy = 2\pi a_1$$

Similarly, $$\lambda a_2 = \pi a_3 \qquad \text{and} \qquad \lambda a_3 = \pi a_2 \tag{3}$$

If $a_1 \neq 0$, we get $\lambda = 2\pi$, which, when substituted in (3), yields $a_2 = a_3 = 0$. We thus have, as a first solution,

$$\lambda_1 = 2\pi \qquad \text{and} \qquad a^{(1)} = a\begin{pmatrix}1\\0\\0\end{pmatrix}$$

where a is an arbitrary constant. Equation (2) now gives

$$a_1 = \lambda_1 f_1(x) \quad \Rightarrow \quad f_1(x) = \text{constant}$$

On the other hand, if $\lambda \neq 2\pi$, $a_1 = 0$. Then Eq. (3) gives

$$\lambda = \pm\,\pi \quad \text{and} \quad a_2 = \pm\, a_3$$

For $\lambda \equiv \lambda_2 \equiv \pi$, Eq. (2) gives

$$f(x) \equiv f_2(x) = (\text{constant})(\sin x + \cos x)$$

For $\lambda = \lambda_3 \equiv -\pi$, it yields

$$f(x) = f_3(x) = (\text{constant})(\sin x - \cos x)$$

The normalized eigenfunctions are

$$f_1(x) = \frac{1}{\sqrt{2\pi}}$$

$$f_2(x) = \frac{1}{\sqrt{2\pi}}(\sin x + \cos x)$$

and $$f_3(x) = \frac{1}{\sqrt{2\pi}}(\sin x - \cos x)$$

Direct substitution in (1) easily verifies that f_1, f_2, and f_3 are eigenfunctions of $\mathbb{K}$ with the eigenvalues shown. •

Theorem 11.2.1 indicates that there is a finite number of eigenfunctions for each nonzero eigenvalue. There may be an infinite number of eigenfunctions for a zero eigenvalue, as the following example demonstrates.

Example 11.2.3

Let $K(x, y) = \sin x \sin y$ for $-\pi \leqslant x, y \leqslant \pi$. Then, for $\lambda = 0$, the eigenvalue equation is

$$\sin x \int_{-\pi}^{\pi} \sin y f(y)\,dy = \lambda f(x) = 0$$

Since there are infinitely many functions orthogonal to $\sin y$ [for example, $f(y) = \sin ny$, where $n = 2, 3, \ldots$], we conclude that there are infinitely many solutions to the above eigenvalue equation. ●

11.2.3 Eigenvalues and Inverses

The set of eigenvalues of an operator acting in a finite-dimensional vector space is finite; it consists of the roots of the characteristic polynomial of the operator. On the other hand, in Chapter 10 we encountered the Sturm-Liouville operator, $\mathbb{L}$, which has an infinite number of eigenvalues (see Theorem 10.3.1). Once an infinite number of eigenvalues are allowed, nothing can prevent the set of eigenvalues from becoming a *continuum*. Thus, in general, an operator may have both discrete and continuous eigenvalues. Continuity brings with it both the agony and the ecstasy of analysis.

The inverse of an operator. The question as to whether or not an operator $\mathbb{A}$ in a finite-dimensional vector space is invertible is succinctly answered by the value of its determinant: $\mathbb{A}$ is invertible if and only if $\det \mathbb{A} \neq 0$.

In infinite-dimensional (Hilbert) spaces there is no determinant. How can we tell whether or not an operator in a Hilbert space is invertible? The exploitation of the connection between invertibility and eigenvalues at the beginning of this chapter led to Proposition 11.1.1, which can be proved for an operator acting on any vector space, finite or infinite.

Consider the equation $\mathbb{A}|u\rangle = 0$ in the Hilbert space $\mathcal{H}$. Note that, in general, neither the domain nor the range of $\mathbb{A}$ is the whole of $\mathcal{H}$. If $\mathbb{A}$ is invertible, then the only solution to the equation $\mathbb{A}|u\rangle = 0$ is $|u\rangle = 0$. On the other hand, assuming that the equation has no nontrivial solution implies that the null space of $\mathbb{A}$ consists of only the zero vector. Thus,

$$\mathbb{A}|u_1\rangle = \mathbb{A}|u_2\rangle \quad \Rightarrow \quad \mathbb{A}(|u_1\rangle - |u_2\rangle) = 0 \quad \Rightarrow \quad |u_1\rangle - |u_2\rangle = \mathbf{0}$$

This shows that $\mathbb{A}$ is injective, that is, is a bijective linear mapping from the domain of $\mathbb{A}$, $D(\mathbb{A})$, onto the range of $\mathbb{A}$, or $\mathbb{A}(\mathcal{H})$. Therefore, $\mathbb{A}$ must have an inverse.

The foregoing discussion can be expressed as follows. If $\mathbb{A}|u\rangle = 0$, then (by the definition of eigenvectors) $\lambda = 0$ is an eigenvalue of $\mathbb{A}$ iff $|u\rangle \neq \mathbf{0}$. Thus, if $\mathbb{A}|u\rangle = 0$ has no nontrivial solution, then zero cannot be an eigenvalue of $\mathbb{A}$. This can also be stated as a theorem.

11.2.2 Theorem An operator $\mathbb{A}$ on a Hilbert space $\mathcal{H}$ has an inverse $\mathbb{A}^{-1}$ iff $\lambda = 0$ is not an eigenvalue of $\mathbb{A}$. ■

Bear in mind that $\mathbb{A}$ is defined only on a subset of $\mathcal{H}$, namely $D(\mathbb{A})$. Similarly, $\mathbb{A}^{-1}$ is defined only on the range of $\mathbb{A}$, which is $\mathbb{A}(\mathcal{H})$.

The eigenvalues of an arbitrary operator $\mathbb{A}$ are studied in terms of the operator $\mathbb{A} - \lambda\mathbb{1}$.

11.2.3 Definition Let $\mathbb{A} \in \mathscr{L}(\mathscr{H})$. A complex number λ is said to be in the *resolvent set* $\rho(\mathbb{A})$ if $\mathbb{A} - \lambda\mathbb{1}$ is a bijection. The operator

$$\mathbb{R}_\lambda(\mathbb{A}) \equiv (\mathbb{A} - \lambda\mathbb{1})^{-1}$$

is called the *resolvent* of $\mathbb{A}$ at λ. If $\lambda \notin \rho(\mathbb{A})$, then λ is said to be in the *spectrum* of $\mathbb{A}$, denoted $\sigma(\mathbb{A})$.

A rigorous study of the spectrum of $\mathbb{A}$ requires knowledge as to whether or not $\mathbb{R}_\lambda(\mathbb{A})$ is bounded. In fact, the precise version of Definition 11.2.3 specifies that $\mathbb{R}_\lambda(\mathbb{A})$ is bounded.

11.2.4 Definition An operator $\mathbb{B} \in \mathscr{L}(\mathscr{H})$ is said to be *bounded* if there exists a nonnegative real number M such that

$$\|\mathbb{B}\mathbf{u}\| \leqslant M\|\mathbf{u}\| \qquad \forall\ |\mathrm{u}\rangle \in \mathscr{H}$$

The smallest such M is called the *norm* of $\mathbb{B}$, denoted $\|\mathbb{B}\|$.

For most operators of interest, such as self-adjoint operators, the spectrum of the operator consists of its eigenvalues and is sometimes called the *point spectrum* (see Reed and Simon 1980, Chapter VI, for exceptions and further discussion).

Example 11.2.4

Let us consider a few illustrations of the above discussion.

(a) Let $\mathbb{A}$ be the self-adjoint operator $-i\,d/dx$, and let $\mathscr{H} = \mathscr{L}^2(-\infty, \infty) \equiv \mathscr{L}^2(\mathbb{R})$. Then, as can easily be shown (see Chapter 3), $\sigma(\mathbb{A}) = \mathbb{R}$; that is, every real number is an eigenvalue of $\mathbb{A}$. The eigenfunction $u_\lambda(x)$ corresponding to the eigenvalue λ is $e^{-i\lambda x}$. Note that $u_\lambda(x) \notin \mathscr{L}^2(\mathbb{R})$; however, $u_\lambda(x)$ can be "approximated" to any desired accuracy by a suitable wave packet (see Richtmeyer 1978, Chapter 8).

(b) Let $\mathscr{H} = \mathscr{L}^2(\mathbb{R})$, and let $\mathbb{A}$ be the operator that multiplies a function $f(x)$ by x: $\mathbb{A}f(x) = xf(x)$. Again $\sigma(\mathbb{A}) = \mathbb{R}$, and every function is an eigenfunction.

(c) Let $\mathscr{H} = \mathscr{L}^2(\mathbb{R})$ and $\mathbb{A} = -d^2/dx^2$. The function $u(x) \in \mathscr{L}^2(\mathbb{R})$ is an eigenfunction corresponding to eigenvalue λ iff

$$-\frac{d^2u}{dx^2} = \lambda u \quad \Rightarrow \quad \frac{d^2u}{dx^2} + \lambda u = 0$$

For this equation to have a solution that is finite at $\pm\infty$, λ must be positive. Thus, $\sigma(\mathbb{A}) = (0, \infty)$. Here, as in part (a), $u_\lambda(x) \notin \mathscr{L}^2(\mathbb{R})$.

(d) Let $\mathbb{A} = (1 - x^2)(d^2/dx^2) - 2x(d/dx)$ be defined on $\mathscr{L}^2[-1, +1]$. Then $\sigma(\mathbb{A}) = \{l(l+1) \mid l \text{ is a nonnegative integer}\}$, and the eigenfunctions are simply the Legendre polynomials. ●

A self-adjoint operator $\mathbb{H}$ has only real eigenvalues; therefore, $\sigma(\mathbb{H}) \subset \mathbb{R}$. The eigenvalues of a unitary operator $\mathbb{U}$ have magnitude 1; therefore, $\sigma(\mathbb{U})$ is the unit circle, or $\{\lambda \in \mathbb{C} \mid |\lambda| = 1\}$. On the other hand, $\rho(\mathbb{H})$ consists of the UHP $(y > 0)$, the LHP

($y < 0$), and possibly part of the real axis as well, and $\rho(\mathbb{U})$ consists of all points interior and exterior to the unit circle and possibly some points on the circle.

Useful properties of the resolvent. There are two important properties of the resolvent of an operator that are extremely useful in analyzing the spectrum of the operator.

Recall that if $\lambda \in \rho(\mathbb{A})$, the operator $\mathbb{A} - \lambda\mathbb{1}$ is invertible. In fact, $\mathbb{R}_\lambda(\mathbb{A}) = (\mathbb{A} - \lambda\mathbb{1})^{-1}$. Similarly, $\mathbb{R}_\mu(\mathbb{A}) = (\mathbb{A} - \mu\mathbb{1})^{-1}$. Let us assume that $\lambda \neq \mu$ and take the difference between the two resolvents. By formally manipulating the difference, we obtain

$$\begin{aligned}\mathbb{R}_\lambda(\mathbb{A}) - \mathbb{R}_\mu(\mathbb{A}) &= \frac{1}{\mathbb{A} - \lambda\mathbb{1}} - \frac{1}{\mathbb{A} - \mu\mathbb{1}} \\ &= \frac{\mathbb{A} - \mu\mathbb{1} - \mathbb{A} + \lambda\mathbb{1}}{(\mathbb{A} - \lambda\mathbb{1})(\mathbb{A} - \mu\mathbb{1})} = \frac{(\lambda - \mu)\mathbb{1}}{(\mathbb{A} - \lambda\mathbb{1})(\mathbb{A} - \mu\mathbb{1})} \\ &= (\lambda - \mu)\frac{1}{\mathbb{A} - \lambda\mathbb{1}}\left(\frac{1}{\mathbb{A} - \mu\mathbb{1}}\right) = (\lambda - \mu)\mathbb{R}_\lambda(\mathbb{A})\mathbb{R}_\mu(\mathbb{A})\end{aligned}$$

This can be rewritten as

$$\frac{\mathbb{R}_\lambda(\mathbb{A}) - \mathbb{R}_\mu(\mathbb{A})}{\lambda - \mu} = \mathbb{R}_\lambda(\mathbb{A})\mathbb{R}_\mu(\mathbb{A}) \tag{11.6}$$

(For a more rigorous derivation of this result, see the solution to Exercise 11.2.4.)

To obtain the second property of the resolvent, we formally differentiate $\mathbb{R}_\lambda(\mathbb{A})$ with respect to λ and evaluate the result at $\lambda = \mu$:

$$\frac{d}{d\lambda}\mathbb{R}_\lambda(\mathbb{A}) = \frac{d}{d\lambda}[(\mathbb{A} - \lambda\mathbb{1})^{-1}] = (-1)(-\mathbb{1})(\mathbb{A} - \lambda\mathbb{1})^{-2} = [\mathbb{R}_\lambda(\mathbb{A})]^2$$

Differentiating both sides of this equation, we get

$$\frac{d^2}{d\lambda^2}\mathbb{R}_\lambda(\mathbb{A}) = 2\frac{d\mathbb{R}_\lambda}{d\lambda}[\mathbb{R}_\lambda(\mathbb{A})] = 2[\mathbb{R}_\lambda(\mathbb{A})]^3$$

In general,

$$\frac{d^n}{d\lambda^n}\mathbb{R}_\lambda(\mathbb{A}) = n![\mathbb{R}_\lambda(\mathbb{A})]^{n+1}$$

Thus,

$$\left.\frac{d^n}{d\lambda^n}\mathbb{R}_\lambda(\mathbb{A})\right|_{\lambda=\mu} = n!\mathbb{R}_\mu^{n+1}(\mathbb{A})$$

If the Taylor-series expansion exists, we may write

$$\mathbb{R}_\lambda(\mathbb{A}) = \sum_{n=0}^{\infty}\frac{(\lambda - \mu)^n}{n!}\left.\frac{d^n}{d\lambda^n}\mathbb{R}_\lambda(\mathbb{A})\right|_{\lambda=\mu} = \sum_{n=0}^{\infty}(\lambda - \mu)^n\mathbb{R}_\mu^{n+1}(\mathbb{A}) \tag{11.7}$$

which is the second property of the resolvent.

Strictly speaking we should consider (11.7) sandwiched between two arbitrary vectors in the domain of $\mathbb{R}_\lambda$, which would give

$$\langle u|\mathbb{R}_\lambda(\mathbb{A})|v\rangle = \sum_{n=0}^{\infty} (\lambda - \mu)^n \langle u|\mathbb{R}_\mu^{n+1}(\mathbb{A})|v\rangle$$

in which both sides become ordinary functions of λ. Furthermore, (11.7) was derived by formally differentiating $\mathbb{R}_\lambda(\mathbb{A})$, which is not strictly correct. For a correct derivation, see Chapter VI of Reed and Simon (1980) or Chapter 8 of Richtmeyer (1978).

11.2.4 The Spectral Decomposition of a Matrix from an Analytical Viewpoint

Because analytic methods are powerful when applied to operator algebras and because they are the only tool available for the study of the infinite-dimensional case, this subsection reconsiders the spectral decomposition of a matrix (discussed in Chapter 3) from an analytical viewpoint.

Let A be an arbitrary (not necessarily hermitian) $N \times N$ matrix. Let $\{\lambda_i\}_{i=1}^r$ be the roots of its characteristic polynomial, $p(\mu) = \det(\mathsf{A} - \mu 1)$. Let $\lambda \in \mathbb{C}$ and $|\lambda| > |\lambda_i|$ for all $i = 1, 2, \ldots, r$. It is then possible to expand $\mathsf{R}_\lambda(\mathsf{A})$ as follows:

$$\mathsf{R}_\lambda(\mathsf{A}) \equiv \frac{1}{\mathsf{A} - \lambda 1} = -\frac{1}{\lambda}\left[\frac{1}{1 - \left(\frac{\mathsf{A}}{\lambda}\right)}\right]$$

$$= -\frac{1}{\lambda}\sum_{n=0}^{\infty}\left(\frac{\mathsf{A}}{\lambda}\right)^n = -\frac{1}{\lambda}\left(1 + \frac{\mathsf{A}}{\lambda} + \frac{\mathsf{A}^2}{\lambda^2} + \cdots\right) \tag{11.8}$$

This expansion makes sense only if

$$\left\|\frac{\mathsf{A}}{\lambda}\right\| = \frac{\|\mathsf{A}\|}{|\lambda|} < 1$$

where $\|\mathsf{A}\|$ is as defined in Definition 11.2.4. The λ we have chosen assures this inequality (see Reed and Simon 1980, Chapter VI).

Equation (11.8) is the Laurent expansion of $\mathsf{R}_\lambda(\mathsf{A})$. We can immediately read off the residue of $\mathsf{R}_\lambda(\mathsf{A})$, which is the coefficient of $1/\lambda$:

$$\operatorname{Res}[\mathsf{R}_\lambda(\mathsf{A})] = -1$$

or

$$-\frac{1}{2\pi i}\oint_\Gamma \mathsf{R}_\lambda(\mathsf{A})\,d\lambda = 1$$

where Γ is a circle with its center at the origin and a radius large enough to encompass all the eigenvalues of A. A similar argument shows that

$$-\frac{1}{2\pi i}\oint_\Gamma \lambda\mathsf{R}_\lambda(\mathsf{A})\,d\lambda = \mathsf{A}$$

and, in general,

$$-\frac{1}{2\pi i}\oint_{\Gamma} \lambda^n \mathsf{R}_\lambda(\mathsf{A})\,d\lambda = \mathsf{A}^n \qquad \text{for } n = 0, 1, \ldots$$

Expanding the function $f(\mathsf{A})$ in a power series and using the above integral for each power, we get

$$-\frac{1}{2\pi i}\oint_{\Gamma} f(\lambda)\mathsf{R}_\lambda(\mathsf{A})\,d\lambda = f(\mathsf{A}) \tag{11.9}$$

Writing this equation in the form

$$-\frac{1}{2\pi i}\oint_{\Gamma} \frac{f(\lambda)}{\lambda\mathsf{1} - \mathsf{A}}\,d\lambda = f(\mathsf{A})$$

makes it recognizable as the generalization of the Cauchy integral formula to operator-valued functions.

To use any of the above integral formulas, we must know the analytic behavior of $\mathsf{R}_\lambda(\mathsf{A})$. By Theorem 3.4.7,

$$[\mathsf{R}_\lambda(\mathsf{A})]_{ij} = [(\mathsf{A} - \lambda\mathsf{1})^{-1}]_{ij} = \frac{[\widetilde{\mathsf{C}(\mathsf{A} - \lambda\mathsf{1})}]_{ij}}{\det(\mathsf{A} - \lambda\mathsf{1})} \equiv \frac{C_{ij}(\lambda)}{p(\lambda)}$$

where $C_{ij}(\lambda)$ is the cofactor of the ijth element of the matrix $\mathsf{A} - \lambda\mathsf{1}$ and $p(\lambda)$ is the characteristic polynomial of A. Clearly, $C_{ij}(\lambda)$ is a polynomial of degree $N - 1$, and $p(\lambda)$ is of degree N. Thus, $[\mathsf{R}_\lambda(\mathsf{A})]_{ij}$ is a rational function of λ. It follows that $\mathsf{R}_\lambda(\mathsf{A})$ has only poles as singularities. The poles are simply the zeros of the denominator, or the eigenvalues of A.

We can distort the contour Γ in such a way that it consists of small circles γ_j that encircle the isolated eigenvalues λ_j (Fig. 11.1). Then, with $f(\mathsf{A}) = \mathsf{1}$, Eq. (11.9) yields

$$\mathsf{1} = -\frac{1}{2\pi i}\sum_{j=1}^{r}\oint_{\gamma_j} \mathsf{R}_\lambda(\mathsf{A})\,d\lambda \equiv \sum_{j=1}^{r} \mathsf{P}_j \tag{11.10a}$$

where

$$\mathsf{P}_j \equiv -\frac{1}{2\pi i}\oint_{\gamma_j} \mathsf{R}_\lambda(\mathsf{A})\,d\lambda \tag{11.10b}$$

It can be shown (see Exercise 11.2.6) that $\{\mathsf{P}_j\}_{j=1}^{r}$ is a set of orthogonal *idempotents*, by which is meant that

$$\mathsf{P}_i\mathsf{P}_j = \begin{cases} 0 & \text{if } i \neq j \\ \mathsf{P}_i & \text{if } i = j \end{cases}$$

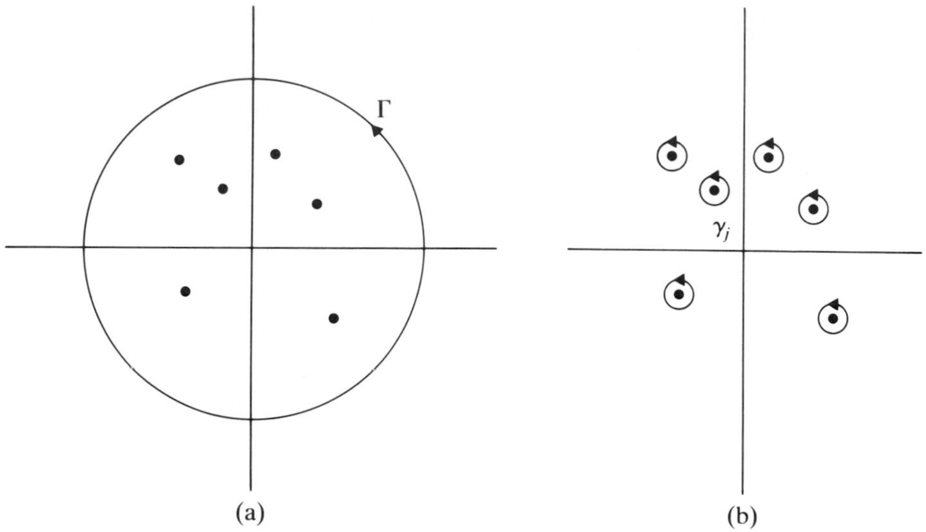

Figure 11.1 (a) The large circle Γ and (b) the distorted contour consisting of the small circles γ_j.

Thus, (11.10a) is a resolution of identity, as specified in the spectral decomposition theorem in Chapter 3.

Now we let $f(\mathsf{A}) = \mathsf{A}$ in (11.9), deform the contour as above, and write

$$\begin{aligned} \mathsf{A} &= -\frac{1}{2\pi i}\sum_{j=1}^{r}\oint_{\gamma_j}\lambda \mathsf{R}_\lambda(\mathsf{A})\,d\lambda \\ &= -\frac{1}{2\pi i}\sum_{j=1}^{r}\left[\lambda_j\oint_{\gamma_j}\mathsf{R}_\lambda(\mathsf{A})\,d\lambda + \oint_{\gamma_j}(\lambda-\lambda_j)\mathsf{R}_\lambda(\mathsf{A})\,d\lambda\right] \\ &\equiv \sum_{j=1}^{r}(\lambda_j\mathsf{P}_j + \mathsf{D}_j) \end{aligned} \tag{11.11}$$

where
$$\mathsf{D}_j \equiv -\frac{1}{2\pi i}\oint_{\gamma_j}(\lambda-\lambda_j)\mathsf{R}_\lambda(\mathsf{A})\,d\lambda$$

It can be shown (see Exercise 11.2.7) that

$$D_j^n = -\frac{1}{2\pi i}\oint_{\gamma_j}(\lambda-\lambda_j)^n\mathsf{R}_\lambda(\mathsf{A})\,d\lambda$$

In particular, since $\mathsf{R}_\lambda(\mathsf{A})$ has only poles as singularities, there exists a positive integer m such that $\mathsf{D}_j^m = 0$.

We have not yet made any assumptions about A. If we assume that A is hermitian, then $\mathsf{R}_\lambda(\mathsf{A})$ will have simple poles (see Exercise 11.2.8). It follows that $(\lambda - \lambda_j)\mathsf{R}_\lambda(\mathsf{A})$ will be analytic at λ_j for all $j = 1, 2, \ldots, r$, and $\mathsf{D}_j = 0$ in Eq. (11.11). We thus have

$$\mathsf{A} = \sum_{j=1}^{r} \lambda_j \mathsf{P}_j$$

which is the spectral decomposition discussed in Chapter 3. This finishes the proof of Theorem 3.6.12 except for showing that the P_j are hermitian (see Exercise 11.2.9).

Example 11.2.5

The most general 2 × 2 hermitian matrix is of the form

$$\mathsf{A} = \begin{pmatrix} a_{11} & a_{12} \\ a_{12}^* & a_{22} \end{pmatrix}$$

where a_{11} and a_{22} are real numbers. Thus,

$$\mathsf{A} - \lambda 1 = \begin{pmatrix} a_{11} - \lambda & a_{12} \\ a_{12}^* & a_{22} - \lambda \end{pmatrix}$$

and
$$\det(\mathsf{A} - \lambda 1) = \lambda^2 - (a_{11} + a_{22})\lambda + a_{11}a_{22} - |a_{12}|^2$$

which has roots

$$\lambda_1 \equiv \tfrac{1}{2}\left[a_{11} + a_{22} - \sqrt{(a_{11} - a_{22})^2 + 4|a_{12}|^2}\right]$$

and
$$\lambda_2 \equiv \tfrac{1}{2}\left[a_{11} + a_{22} + \sqrt{(a_{11} - a_{22})^2 + 4|a_{12}|^2}\right] \tag{1}$$

The inverse of $\mathsf{A} - \lambda 1$ can immediately be written:

$$\begin{aligned} \mathsf{R}_\lambda(\mathsf{A}) = (\mathsf{A} - \lambda 1)^{-1} &= \frac{1}{\det(\mathsf{A} - \lambda 1)} \begin{pmatrix} a_{22} - \lambda & -a_{12} \\ -a_{12}^* & a_{11} - \lambda \end{pmatrix} \\ &= \frac{1}{(\lambda - \lambda_1)(\lambda - \lambda_2)} \begin{pmatrix} a_{22} - \lambda & -a_{12} \\ -a_{12}^* & a_{11} - \lambda \end{pmatrix} \end{aligned} \tag{2}$$

We want to show that (2) has only simple poles. Two cases arise:

(1) If $\lambda_1 \neq \lambda_2$, then it is clear that $\mathsf{R}_\lambda(\mathsf{A})$ has simple poles.

(2) If $\lambda_1 = \lambda_2$, it *appears* that $\mathsf{R}_\lambda(\mathsf{A})$ may have a pole of order 2. However, note that if $\lambda_1 = \lambda_2$, then the square root in Eq. (1) must vanish. This happens iff $a_{11} = a_{22} \equiv a$ *and* $a_{12} = 0$. It then follows that $\lambda_1 = \lambda_2 = \lambda$ and

$$\mathsf{R}_\lambda(\mathsf{A}) = \frac{1}{(\lambda - a)^2} \begin{pmatrix} a - \lambda & 0 \\ 0 & a - \lambda \end{pmatrix}$$

This clearly shows that $\mathsf{R}_\lambda(\mathsf{A})$ has only simple poles in this case.

We have thus demonstrated directly that the resolvent of a 2 × 2 hermitian matrix can have only simple poles. ●

If A is not hermitian, $\mathsf{D}_j \neq 0$; however, D_j is still a *nilpotent*. That is, $\mathsf{D}_j^m = 0$ for some positive integer m. This property and Eq. (11.11) can be used to show that A can be cast into a *Jordan canonical form* via a similarity transformation. That is, there exists an $N \times N$ matrix S such that

$$\mathsf{SAS}^{-1} \equiv \mathsf{J} \equiv \begin{pmatrix} \mathsf{J}_1 & 0 & 0 \ldots 0 \\ 0 & \mathsf{J}_2 & 0 \ldots 0 \\ \vdots & \vdots & \vdots \quad\; \vdots \\ 0 & 0 & 0 \ldots \mathsf{J}_k \end{pmatrix}$$

where J_i is of the form

$$\mathsf{J}_i = \begin{pmatrix} \lambda & 1 & 0 & 0 \ldots 0 & 0 \\ 0 & \lambda & 1 & 0 \ldots 0 & 0 \\ 0 & 0 & \lambda & 1 \ldots 0 & 0 \\ \vdots & \vdots & \vdots & \vdots \quad\; \vdots & \vdots \\ 0 & 0 & 0 & 0 \ldots \lambda & 1 \\ 0 & 0 & 0 & 0 \ldots 0 & \lambda \end{pmatrix}$$

in which λ is one of the eigenvalues of A. Different J_i may contain the same eigenvalues of A. For a discussion of the Jordan canonical form of a matrix, see Birkhoff and MacLane (1977), Denery and Krzywicki (1967), or Halmos (1958).

11.2.5 The Spectral Decomposition of a Self-Adjoint Operator

The discussion of the preceding subsection can be generalized to a self-adjoint operator acting on a Hilbert space. First, we define

$$\mathbb{E}_j \equiv \sum_{i=1}^{j} \mathbb{P}_i = \mathbb{P}_1 + \mathbb{P}_2 + \cdots + \mathbb{P}_j$$

and, with $\mathbb{E}_0 \equiv \mathbb{0}$,

$$\Delta\mathbb{E}_j \equiv \mathbb{E}_j - \mathbb{E}_{j-1} = \mathbb{P}_j$$

We then write

$$\mathbb{1} = \sum_j \Delta\mathbb{E}_j \qquad \text{and} \qquad \mathbb{A} = \sum_j \lambda_j \Delta\mathbb{E}_j$$

The generalization to a continuum now becomes possible via a change of Δ to d and Σ to $\int$. In particular, the equation for $\mathbb{A}$ becomes

$$\mathbb{A} = \int_{-\infty}^{\infty} \lambda \, d\mathbb{E}_\lambda \tag{11.12}$$

where λ is a real parameter (the eigenvalue of the self-adjoint $\mathbb{A}$).

Equation (11.12) should really be written

$$\langle \mathrm{u}|\mathbb{A}|\mathrm{v}\rangle = \int_{-\infty}^{\infty} \lambda \langle \mathrm{u}|d\mathbb{E}_\lambda|\mathrm{v}\rangle \equiv \int_{-\infty}^{\infty} \lambda d(\langle \mathrm{u}|\mathbb{E}_\lambda|\mathrm{v}\rangle)$$

It can be shown (see Richtmeyer 1978, Chapter 9) that

$$\mathbb{E}_\lambda = \frac{1}{2\pi i}\int_C \mathbb{R}_\xi(\mathbb{A})\, d\xi$$

where C is an appropriate contour, and that

$$\mathbb{R}_\lambda(\mathbb{A}) = \int_{-\infty}^{\infty} \frac{1}{\xi - \lambda}\, d\mathbb{E}_\xi$$

and that the $\mathbb{E}_\lambda$, where $-\infty < \lambda < \infty$, are, in some sense, "projection operators." In such a context Eq. (11.12) is the analogue of spectral decomposition for the self-adjoint operator $\mathbb{A}$ on a Hilbert space. A detailed discussion of such matters involves difficult questions of the convergence of the series and integrals and is beyond the scope of this book.

11.2.6 Eigenvalues of a Symmetric Integral Operator

In Section 11.2.2 we saw that a symmetric separable integral operator (IO) has only a finite number of eigenvalues. What about a general symmetric IO? This subsection gives a few important results concerning them. The first of these is a theorem (for a proof, see Roach 1970, pp. 91 and 92).

11.2.5 Theorem The number of distinct nonzero eigenvalues of a symmetric IO is countable. For each such eigenvalue there exist at most a finite number of linearly independent eigenfunctions. ■

Recall that a countable set has either a finite number of elements or an infinite number of elements that can be put in a one-to-one correspondence with the integers. Thus, the eigenvalues of an IO can be labeled with integers k. This is in contrast to a general self-adjoint operator, as discussed in the preceding subsection. The set $\{\lambda_k\}_{k=-\infty}^{\infty}$ is called the *spectrum* of the integral operator $\mathbb{K}$. Of course, the set of λ_k's may have a finite number of nonzero elements, in which case the kernel of the operator will be separable. It may even be the case that none of the λ_k are different from zero. The set may even be empty for a general IO; that is, the (nonsymmetric) IO may have no eigenvalue at all.

What kind of operators have nonzero eigenvalues? Let us first consider the finite-dimensional case. If $\mathbb{A} \in \mathcal{L}(\mathcal{V})$ is a symmetric operator on an N-dimensional vector space $\mathcal{V}$, then its eigenvalues can be found by solving its characteristic polynomial, $p(\lambda) \equiv \det(\mathbb{A} - \lambda \mathbb{1})$, of degree N. As long as $N \geqslant 1$, this polynomial has at least one root. Furthermore, for a symmetric (hermitian) operator, at least one of the roots is different from zero. This is because a symmetric operator $\mathbb{A}$ can be diagonalized; that is, there exists a basis in which $\mathbb{A}$ is represented by a diagonal matrix whose elements are simply the eigenvalues of $\mathbb{A}$. If all the eigenvalues are zero, the matrix reduces to the zero matrix. Transforming the zero matrix to any other basis yields the zero matrix. Thus, $\mathbb{A}$ will be represented by the zero matrix in all bases. This implies that $\mathbb{A}$ is the *zero operator*. We, therefore, conclude that a nonzero symmetric operator in a finite-dimensional vector space has at least one nonzero eigenvalue.

With appropriate modifications the foregoing discussion can be generalized to the infinite-dimensional case. In particular, real symmetric IOs must have nonzero eigenvalues if the IOs are not identically zero (see Stakgold 1979, Chapter 5).

We are interested in a spectral decomposition for a symmetric IO. Specifically, we want to express the kernel as a (infinite) sum of products of functions. To do so, we again look at the finite-dimensional case.

Recall, from Chapter 3 or subsection 11.2.4, that a hermitian (or real symmetric) operator $\mathbb{A}$ can be expressed as $\mathbb{A} = \sum_{i=1}^{r} \lambda_i \mathbb{P}_i$ where λ_i are eigenvalues of $\mathbb{A}$ and $\mathbb{P}_i$ are projection operators onto the subspace $\mathcal{M}_{\lambda_i}$ spanned by the eigenvectors corresponding to λ_i. Given a basis $\{|e_k^{(i)}\rangle\}_{k=1}^{m_i}$ for $\mathcal{M}_{\lambda_i}$, we can write

$$\mathbb{P}_i = \sum_{k=1}^{m_i} |e_k^{(i)}\rangle\langle e_k^{(i)}|$$

and

$$\mathbb{A} = \sum_{i=1}^{r} \lambda_i \sum_{k=1}^{m_i} |e_k^{(i)}\rangle\langle e_k^{(i)}|$$

If we collect all the $|e_k^{(i)}\rangle$ for different i and k, we obtain a basis for the whole vector space, which we denote $B = \{|e_j\rangle\}_{j=1}^{N}$. If we also allow for repetition of λ's, we can write

$$\mathbb{A} = \sum_{j=1}^{N} \lambda_j |e_j\rangle\langle e_j|$$

Using Theorem 11.2.5, we can write the formal generalization of the above to a symmetric IO:

$$\mathbb{K} = \sum_{j} \lambda_j |f_j\rangle\langle f_j|$$

"Sandwiching" this between $\langle x|$ and $|y\rangle$ produces

$$K(x, y) = \sum_{j} \lambda_j f_j(x) f_j(y) \tag{11.13}$$

which is the form we want. Equation (11.13) is, in general, an infinite sum. However, it is possible that all but a finite number of terms in the sum will be zero. Then the kernel will be separable. This can be stated as a proposition.

11.2.6 Proposition Let $\{\lambda_j\}_{j=1}^m$ be the nonzero eigenvalues and $\{f_j\}_{j=1}^m$ the corresponding orthonormal eigenfunctions of a symmetric integral operator $\mathbb{K}$. If $\mathbb{K}$ has no other nonzero eigenvalues, then

$$K(x, y) = \sum_{j=1}^{m} \lambda_j f_j(x) f_j(y)$$

That is, K is separable. ■

This proposition is the converse of Theorem 11.2.1. It is also the analogue of the spectral decomposition theorem for symmetric IOs having a finite number of nonzero eigenvalues (separable IOs). The sum in the proposition is considered an infinite sum in which all but a finite number of the coefficients (λ_j) are zero.

If the number of nonzero eigenvalues of $\mathbb{K}$ is infinite, then the sum of (11.13) converges *uniformly* and *absolutely* only if $K(x, y)$ is positive semidefinite, or

$$\langle \mathrm{f}|\mathbb{K}|\mathrm{f}\rangle \geqslant 0 \qquad \forall f \in \mathscr{R}(a, b)$$

It is only for such a kernel that Eq. (11.13) makes sense.

Exercises

11.2.1 Show that if $\mathbb{K}$ satisfies $\langle \mathbb{K}\mathrm{f}|\mathrm{g}\rangle = \langle \mathrm{f}|\mathbb{K}\mathrm{g}\rangle$, for all $f, g \in \mathscr{R}(a, b)$, then the kernel of $\mathbb{K}$ satisfies $K(x, y) = K^*(y, x)$.

11.2.2 Find the nonzero eigenvalues and corresponding eigenfunctions of the kernel $K(x, y) = xy$ in the interval $a \leqslant x, y \leqslant b$.

11.2.3 Show that the momentum operator $-i\,d/dx$ is not bounded. [Hint: Consider $\mathscr{L}^2(0, \mathrm{a})$ for $a > 0$, and find a counterexample.]

11.2.4 Derive Eq. (11.6).

11.2.5 For a hermitian $\mathbb{A}$, show that

$$\|\mathbb{R}_\lambda(\mathbb{A})|\mathrm{u}\rangle\| \leqslant \frac{1}{|\mathrm{Im}\,(\lambda)|}\|\mathbf{u}\|$$

11.2.6 Using the definition (11.10b), show that $\mathbb{P}_j$ are orthogonal idempotents. [Hint: Write $\mathbb{P}_j^2$ as a double integral and use (11.6).]

11.2.7 Use the procedure of Exercise 11.2.6 to show that

$$\mathbb{D}_j^n = -\frac{1}{2\pi i}\oint_{\gamma_j} (\lambda - \lambda_j)^n \mathbb{R}_\lambda(\mathbb{A})\, d\lambda$$

(Hint: Use mathematical induction.)

11.2.8 Show that for a hermitian matrix A, the resolvent, $\mathsf{R}_\lambda(\mathsf{A})$, has only simple poles. (Hint: Multiply the inequality of Exercise 11.2.5 by $|\lambda - \lambda_j|$, and show that the LHS is finite even when $\lambda \to \lambda_j$.)

11.2.9 Show that the $\mathbb{P}_j$ in Eqs. (11.10) are hermitian when $\mathbb{A}$ is. [Hint: Take the hermitian conjugate of Eq. (11.10b).]

11.3 INTEGRAL TRANSFORMS AND DIFFERENTIAL EQUATIONS

The discussion in Chapter 9 introduced a general method of solving differential equations by power series, also called the Frobenius method, which gives a solution that converges within a circle of convergence. In general, this circle of convergence may be small; however, the function represented by the power series can be analytically continued using methods presented in Chapter 7. This section introduces another method which uses integral transforms (IOs) and incorporates the analytic continuation automatically. Thus, instead of being an infinite power series, the solution $u(z)$ of a DE is represented by an integral transform,

$$u(z) = \int_C K(z, t)v(t)\,dt$$

where the contour C, the kernel $K(z, t)$, and the function $v(t)$ are to be determined.

Let $\mathbb{L}_z$ be a DO in z. We want to determine $u(z)$ such that

$$\mathbb{L}_z[u] = 0$$

or, equivalently, such that

$$\int_C (\mathbb{L}_z[K(z, t)])v(t)\,dt = 0$$

Suppose that we can find $\mathbb{M}_t$, a DO in the variable t such that

$$\mathbb{L}_z[K(z, t)] = \mathbb{M}_t[K(z, t)]$$

Then the DE becomes

$$\int_C (\mathbb{M}_t[K(z, t)])v(t)\,dt = 0$$

If $\mathbb{M}_t^\dagger$ is the adjoint of $\mathbb{M}_t$ and C has a and b as initial and final points (a and b may be equal), then the Lagrange identity [see Eq. (9.21b)] yields

$$0 = \mathbb{L}_z[u] = \int_a^b K(z, t)\mathbb{M}_t^\dagger[v(t)]\,dt + Q(K, v)\Big|_a^b$$

where $Q(K, v)$ is the surface term. Thus, if $v(t)$ and the contour C (or a and b) are chosen so that

$$Q(K, v)\big|_a^b = 0 \qquad \text{and} \qquad \mathbb{M}_t^\dagger[v] = 0 \tag{11.14}$$

then the problem is solved. The trick is to find an $\mathbb{M}_t$ such that (11.14) is easier to solve than the original equation, $\mathbb{L}_z[u] = 0$. This in turn is facilitated by a clever choice of the kernel, $K(z, t)$.

This section discusses how to solve some common differential equations of mathematical physics using the general idea presented above.

11.3.1 Integral Representations of the Hypergeometric Function and the Confluent Hypergeometric Function

Recall that for the hypergeometric differential equation the differential operator (DO) is

$$\mathbb{L}_z = z(1-z)\frac{d^2}{dz^2} + [\gamma - (\alpha+\beta+1)z]\frac{d}{dz} - \alpha\beta$$

For such operators, whose coefficient functions are polynomials, the proper choice for $K(z, t)$ is the Euler kernel, $(z-t)^s$. Applying $\mathbb{L}_z$ to this kernel, we obtain

$$\begin{aligned}\mathbb{L}_z[K(z,t)] &= z(1-z)s(s-1)(z-t)^{s-2} + [\gamma-(\alpha+\beta+1)z]s(z-t)^{s-1} - \alpha\beta(z-t)^s \\ &= \{z^2[-s(s-1) - s(\alpha+\beta+1) - \alpha\beta] + z[s(s-1) + s\gamma + st(\alpha+\beta+1) \\ &\quad + 2\alpha\beta t] - \gamma st - \alpha\beta t^2\}(z-t)^{s-2} \qquad (11.15)\end{aligned}$$

Note that, except for a multiplicative constant, $K(z, t)$ is symmetric in z and t. This suggests that the general form of $\mathbb{M}_t$ may be chosen to be the same as that of $\mathbb{L}_z$ except for the interchange of z and t. If we can manipulate the parameters in such a way that $\mathbb{M}_t$ becomes simple, then we have a chance of solving the problem. For instance, if $\mathbb{M}_t$ has the form of $\mathbb{L}_z$ with the constant term absent, then the hypergeometric DE effectively reduces to a FODE (in dv/dt). We can exploit this possibility.

Equation (11.15) shows that the action of $\mathbb{L}_z$ on $K(z, t)$ generates, aside from a general common factor $(z-t)^{s-2}$, a quadratic term in t whose t^2 term has the coefficient $-\alpha\beta$. The symmetry of t and z implies that, since $\mathbb{L}_z[K(z,t)] = \mathbb{M}_t[K(z,t)]$, exactly the same quadratic form is produced when $\mathbb{M}_t$ acts on $K(z, t)$. With $\mathbb{M}_t$, however, we are free to choose s in such a way that the z^2 term will be absent. Then the constant term will be eliminated, and the DE involving $\mathbb{M}_t$ will be easily solved. Thus, setting the coefficient of z^2 equal to zero, we get

$$s^2 + s(\alpha+\beta) + \alpha\beta = 0 \quad \Rightarrow \quad s = -\alpha \text{ or } s = -\beta$$

If we choose $s = -\alpha$ ($s = -\beta$ leads to a different representation), we obtain

$$\begin{aligned}\mathbb{L}_z[K(z,t)] = \mathbb{M}_t[K(z,t)] &= \{z[\alpha(\alpha+1) - \alpha\gamma + \alpha t(\beta-\alpha-1)] + \alpha\gamma t - \alpha\beta t^2\}(z-t)^{-\alpha-2} \\ &= \{(z-t)[\alpha(\alpha+1) - \alpha\gamma + \alpha t(\beta-\alpha-1)] + t[\alpha(\alpha+1) - \alpha\gamma \\ &\quad + \alpha t(\beta-\alpha-1)] + \alpha\gamma t - \alpha\beta t^2\}(z-t)^{-\alpha-2} \\ &= [\alpha(\alpha+1) - \alpha\gamma + \alpha t(\beta-\alpha-1)](z-t)^{-\alpha-1} + \alpha(\alpha+1)(t-t^2)(z-t)^{-\alpha-2} \\ &= \left\{(t-t^2)\frac{d^2}{dt^2} + [\alpha+1-\gamma+t(\beta-\alpha-1)]\frac{d}{dt}\right\}(z-t)^{-\alpha}\end{aligned}$$

We thus have

$$\mathbb{M}_t = (t-t^2)\frac{d^2}{dt^2} + [\alpha+1-\gamma+t(\beta-\alpha-1)]\frac{d}{dt} \qquad (11.16a)$$

which yields the DE

$$\mathbb{M}_t[u] = (t - t^2)\frac{d^2u}{dt^2} + [\alpha + 1 - \gamma + t(\beta - \alpha - 1)]\frac{du}{dt} = 0$$

However, we are interested in the adjoint equation, which is easily obtained by referring to Eq. (9.19a):

$$\mathbb{M}_t^\dagger[v] = \frac{d^2}{dt^2}[(t - t^2)v(t)] - \frac{d}{dt}\{[\alpha - \gamma + 1 + t(\beta - \alpha - 1)]v(t)\} = 0 \qquad (11.16b)$$

The solution to this equation is (see Exercise 11.3.1)

$$v(t) = Ct^{\alpha-\gamma}(t - 1)^{\gamma-\beta-1}$$

We also need the surface term, $Q(K, v)$, in the Lagrange identity. We can calculate this (see Exercise 11.3.2):

$$Q(K, v) = C\alpha t^{\alpha-\gamma+1}(t - 1)^{\gamma-\beta}(z - t)^{-\alpha-1}$$

Finally, we need a specification of the contour. For different contours we will get different solutions. The contour chosen must, of course, have the property that $Q(K, v)$ vanishes as a result of the integration. There are two possibilities: either the contour is closed [$a = b$ in (11.14)] or $a \neq b$ but $Q(K, v)$ takes on the same value at a and at b.

Let us consider the second of these possibilities. Clearly, $Q(K, v)$ vanishes at $t = 1$ *if* $\text{Re}(\gamma) > \text{Re}(\beta)$. Also, as $t \to \infty$,

$$Q(K, v) \to (-1)^{-\alpha-1}C\alpha t^{\alpha-\gamma+1}t^{\gamma-\beta}t^{-\alpha-1} = (-1)^{-\alpha-1}C\alpha t^{-\beta}$$

which vanishes if $\text{Re}(\beta) > 0$. We thus take $a = 1$ and $b = \infty$ and assume that $\text{Re}(\gamma) > \text{Re}(\beta) > 0$. It then follows that (note the interchange of z and t in the first factor)

$$u(z) = C'\int_1^\infty (t - z)^{-\alpha}t^{\alpha-\gamma}(t - 1)^{\gamma-\beta-1}\,dt \qquad (11.17)$$

The constant C' can be determined to be $\Gamma(\gamma)/\Gamma(\beta)\Gamma(\gamma - \beta)$ (see Exercise 11.3.3). Therefore,

$$u(z) \equiv F(\alpha, \beta; \gamma; z) = \frac{\Gamma(\gamma)}{\Gamma(\beta)\Gamma(\gamma - \beta)}\int_1^\infty (t - z)^{-\alpha}t^{\alpha-\gamma}(t - 1)^{\gamma-\beta-1}\,dt$$

Changing the variable of integration from t to $1/t$ yields what is called the *Euler formula* for the hypergeometric function:

$$F(\alpha, \beta; \gamma; z) = \frac{\Gamma(\gamma)}{\Gamma(\beta)\Gamma(\gamma - \beta)}\int_0^1 (1 - tz)^{-\alpha}t^{\beta-1}(1 - t)^{\gamma-\beta-1}\,dt \qquad (11.18)$$

In deriving (11.18), we had to assume that $|z| < 1$ (see Exercise 11.3.3). We can now analytically continue $F(\alpha, \beta; \gamma; z)$ to all regions of the complex z-plane in which the integral is well-defined. The only term in the integral that may cause problems is $(1 - tz)^{-\alpha}$. For arbitrary α this has two branch points, one at $z = 1/t$ and the other at $z = \infty$. Therefore, we cut the z-plane from $z_1 = 1/t$, a point on the positive real axis, to $z_2 = \infty$. Since $0 \leqslant t \leqslant 1$, z_1 is somewhere in the interval $[1, \infty]$. To ensure that the cut is applicable for all values of t, we take $z_1 = 1$ and cut the plane along the positive real axis. It follows that Eq. (11.18) is well-behaved as long as

$$0 < \arg(1 - z) < 2\pi \tag{11.19}$$

We can also choose a different contour, which, in general, will lead to a different solution. For example, as Exercise 11.3.4 shows, the contour from $t = 0$ to $t = 1$ yields the solution

$$w(z) = C\int_0^1 (z - t)^{-\alpha} t^{\alpha-\gamma}(t - 1)^{\gamma-\beta-1}\, dt$$

which, with the appropriate choice of the constant C, yields

$$\begin{aligned} w(z) &= z^{-\alpha}F(\alpha, \alpha - \gamma + 1; \alpha - \beta + 1; 1/z) \\ &= \frac{\Gamma(\alpha - \beta + 1)}{\Gamma(\gamma - \beta)\Gamma(\alpha - \gamma + 1)}\int_0^1 (z - t)^{-\alpha} t^{\alpha-\gamma}(1 - t)^{\gamma-\beta-1}\, dt \end{aligned}$$

This is the solution obtained in Example 9.5.5.

Having obtained the integral representation of the hypergeometric function, we can readily obtain the integral representation of the confluent hypergeometric function by taking the proper limit.

It was shown in Section 9.5.5 that

$$\Phi(\alpha, \gamma; z) = \lim_{\beta \to \infty} F\left(\alpha, \beta; \gamma; \frac{z}{\beta}\right)$$

Thus, by taking the limit of (11.18), we should be able to obtain $\Phi(\alpha, \gamma; z)$. The presence of $\Gamma(\beta)$ and $\Gamma(\gamma - \beta)$ complicates things, but, on the other hand, the symmetry of the hypergeometric function is helpful. Thus, we have

$$\begin{aligned} \Phi(\alpha, \gamma; z) &= \lim_{\beta \to \infty} F\left(\alpha, \beta; \gamma; \frac{z}{\beta}\right) = \lim_{\beta \to \infty} F\left(\beta, \alpha; \gamma; \frac{z}{\beta}\right) \\ &= \lim_{\beta \to \infty} \frac{\Gamma(\gamma)}{\Gamma(\alpha)\Gamma(\gamma - \alpha)}\int_0^1 \left(1 - \frac{tz}{\beta}\right)^{-\beta} t^{\alpha-1}(1 - t)^{\gamma-\alpha-1}\, dt \end{aligned}$$

Employing the identity

$$\lim_{\beta \to \infty} \left(1 - \frac{tz}{\beta}\right)^{-\beta} - e^{tz}$$

in the above equation for $\Phi(\alpha, \gamma; z)$ yields the final result:

$$\Phi(\alpha, \gamma; z) = \frac{\Gamma(\gamma)}{\Gamma(\alpha)\Gamma(\gamma - \alpha)} \int_0^1 e^{tz} t^{\alpha - 1}(1 - t)^{\gamma - \alpha - 1}\, dt \qquad (11.20)$$

Note that the condition $\mathrm{Re}(\gamma) > \mathrm{Re}(\alpha) > 0$ must still hold here.

Integral transforms are particularly useful in determining the asymptotic behavior of functions. The following example discusses such behavior of the confluent hypergeometric function.

Example 11.3.1

Let us use the integral representation in a determination of the asymptotic behavior of $\Phi(\alpha, \gamma; z)$ for $z \to \infty$.

We write Eq. (11.20) as

$$\Phi(\alpha, \gamma; z) = \frac{\Gamma(\gamma)}{\Gamma(\alpha)\Gamma(\gamma - \alpha)}\left[- \int_{-\infty}^{0} e^{zt} t^{\alpha - 1}(1 - t)^{\gamma - \alpha - 1}\, dt + \int_{-\infty}^{1} e^{zt} t^{\alpha - 1}(1 - t)^{\gamma - \alpha - 1}\, dt \right]$$

We substitute $-t/z$ for t in the first integral and $1 - t/z$ for t in the second, transforming the equation into

$$\Phi(\alpha, \gamma; z) = \frac{\Gamma(\gamma)}{\Gamma(\alpha)\Gamma(\gamma - \alpha)}\left[- \int_{\infty}^{0} e^{-t}\left(-\frac{t}{z}\right)^{\alpha - 1}\left(1 + \frac{t}{z}\right)^{\gamma - \alpha - 1}\left(-\frac{dt}{z}\right) \right.$$

$$\left. + \int_{\infty}^{0} e^{z(1 - t/z)}\left(1 - \frac{t}{z}\right)^{\alpha - 1}\left(\frac{t}{z}\right)^{\gamma - \alpha - 1}\left(-\frac{dt}{z}\right)\right]$$

$$= \frac{\Gamma(\gamma)}{\Gamma(\alpha)\Gamma(\gamma - \alpha)}\left[(-z)^{-\alpha}\int_0^{\infty} e^{-t} t^{\alpha - 1}\left(1 + \frac{t}{z}\right)^{\gamma - \alpha - 1} dt \right.$$

$$\left. + z^{\alpha - \gamma} e^{z} \int_0^{\infty} e^{-t} t^{\gamma - \alpha - 1}\left(1 - \frac{t}{z}\right)^{\alpha - 1} dt\right]$$

Assuming that $z \to \infty$ along the positive real axis, the second term within the brackets will dominate because of the exponential factor. Also, the integral in the second term approaches $\Gamma(\gamma - \alpha)$ as $z \to \infty$ (see Chapter 14). Thus, as $z \to \infty$,

$$\Phi(\alpha, \gamma; z) \to \frac{\Gamma(\gamma)}{\Gamma(\alpha)} z^{\alpha - \gamma} e^{z}$$

•

11.3.2 Integral Representation and Asymptotic Behavior of Bessel Functions

Choosing the kernel, the contour, and the function $v(t)$ that lead to an integral representation of a function is an art, and the nineteenth century produced many masters at it. A particularly popular theme in such endeavors was the Bessel equation and Bessel functions. This section considers the integral representations of Bessel functions.

As was mentioned in Example 11.2.1, a useful kernel for the Bessel equation is

$$K(z, t) = \left(\frac{z}{2}\right)^{\nu} \exp\left(t - \frac{z^2}{4t}\right)$$

When the Bessel DO,

$$\mathbb{L}_z \equiv \frac{d^2}{dz^2} + \frac{1}{z}\frac{d}{dz} + \left(1 - \frac{\nu^2}{z^2}\right)$$

acts on $K(z, t)$, it yields

$$\mathbb{L}_z K(z, t) = \left(-\frac{\nu+1}{t} + 1 + \frac{z^2}{4t^2}\right)\left(\frac{z}{2}\right)^{\nu} e^{t - z^2/4t} = \left(\frac{d}{dt} - \frac{\nu+1}{t}\right) K(z, t)$$

Thus, $\mathbb{M}_t = d/dt - (\nu + 1)/t$, and Eq. (9.19a) gives

$$\mathbb{M}_t^{\dagger}[v(t)] = -\frac{dv}{dt} - \frac{\nu+1}{t} v = 0$$

whose solution, including the arbitrary constant of integration k, is

$$v(t) = kt^{-\nu-1}$$

When we substitute this solution and the kernel in the surface term of the Lagrange identity, Eq. (9.21), we obtain

$$Q(K, v) = p_1 K v = k\left(\frac{z}{2}\right)^{\nu} t^{-\nu-1} e^{t - z^2/4t}$$

A contour in the t-plane that ensures the vanishing of $Q(K, v)$ *for all values* of v starts at $t = -\infty$, comes to the origin, encircles it, and finally goes back to $t = -\infty$ (see Fig. 11.2). Such a contour is possible because of the factor e^t in the expression for $Q(K, v)$. We thus can write

$$J_{\nu}(z) = k\left(\frac{z}{2}\right)^{\nu} \int_C t^{-\nu-1} e^{t - z^2/4t}\, dt \tag{11.21}$$

Note that the integrand has a cut along the negative real axis due to the factor $t^{-\nu-1}$. If ν is an integer, the cut shrinks to a pole at $t = 0$.

The constant k must be determined in such a way that the above expression for $J_{\nu}(z)$ agrees with the series representation obtained in Chapter 9. It can be shown (see

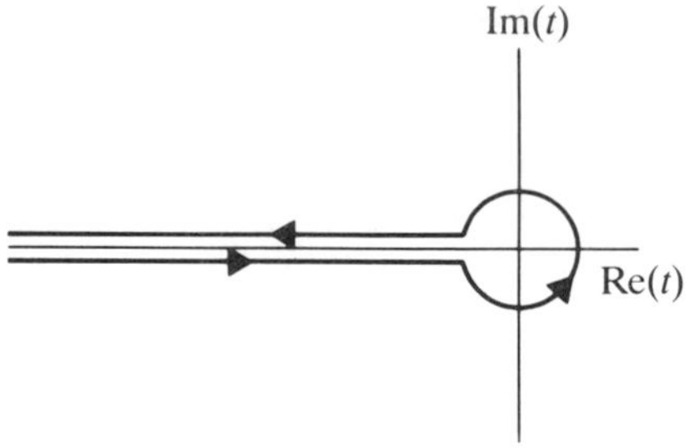

Figure 11.2 The contour C in the t-plane used in evaluating $J_\nu(z)$.

Exercise 11.3.5) that $k = 1/2\pi i$. Thus, we have

$$J_\nu(z) = \frac{1}{2\pi i}\left(\frac{z}{2}\right)^\nu \int_C t^{-\nu-1} e^{t - z^2/4t}\, dt$$

It is more convenient to take the factor $(z/2)^\nu$ into the integral, introduce a new integration variable, $u = 2t/z$, and rewrite the preceding equation as

$$J_\nu(z) = \frac{1}{2\pi i}\int_C u^{-\nu-1} e^{(z/2)(u - 1/u)}\, du \tag{11.22}$$

This result is valid as long as $\text{Re}\,(zu) < 0$ when $u \to -\infty$ on the negative real axis; that is, $\text{Re}\,(z)$ must be positive for (11.22) to work.

An interesting result can be obtained from (11.22) when ν is an integer. In that case the only singularity will be at the origin, so the contour can be taken to be a circle about the origin. This yields

$$J_n(z) = \frac{1}{2\pi i}\int_C u^{-n-1}\, e^{(z/2)(u - 1/u)}\, du$$

which, by Theorem 7.1.4, is the nth coefficient of the Laurent-series expansion of $\exp[(z/2)(u - 1/u)]$ about the origin. We thus have this important result:

$$e^{(z/2)(t - 1/t)} = \sum_{n=-\infty}^{+\infty} J_n(z) t^n \tag{11.23}$$

The function $\exp[(z/2)(t - 1/t)]$ is, therefore, appropriately called the *generating function* for Bessel functions of integer order. Equation (11.23) can be useful in deriving relations for such Bessel functions.

Example 11.3.2

Let us rewrite the LHS of Eq. (11.23) as $\exp(zt/2)\exp(-z/2t)$, expand the exponentials, and collect terms to obtain

$$e^{(z/2)(t-1/t)} = e^{zt/2}e^{-z/2t} = \sum_{m=0}^{\infty} \frac{1}{m!}\left(\frac{zt}{2}\right)^m \sum_{n=0}^{\infty} \frac{1}{n!}\left(-\frac{z}{2t}\right)^n$$

$$= \sum_{m=0}^{\infty} \sum_{n=0}^{\infty} \frac{(-1)^n}{m!n!}\left(\frac{z}{2}\right)^{m+n} t^{m-n} \tag{1}$$

If we let $m - n = k$, change the summation over m to k, and note that k goes from $-\infty$ to $+\infty$, we get

$$e^{(z/2)(t-1/t)} = \sum_{k=-\infty}^{\infty} \sum_{n=0}^{\infty} \frac{(-1)^n}{(n+k)!n!}\left(\frac{z}{2}\right)^{2n+k} t^k$$

$$= \sum_{k=-\infty}^{\infty} \left[\left(\frac{z}{2}\right)^k \sum_{n=0}^{\infty} \frac{(-1)^n}{\Gamma(n+k+1)\Gamma(n+1)}\left(\frac{z}{2}\right)^{2n}\right] t^k \tag{2}$$

Comparing (2) with (11.23) yields the expansion for the Bessel function:

$$J_k(z) = \left(\frac{z}{2}\right)^k \sum_{n=0}^{\infty} \frac{(-1)^n}{\Gamma(n+k+1)\Gamma(n+1)}\left(\frac{z}{2}\right)^{2n}$$

We can also obtain a recurrence relation for $J_n(z)$. Differentiating both sides of (11.23) with respect to t yields

$$\frac{z}{2}\left(1 + \frac{1}{t^2}\right) e^{(z/2)(t-1/t)} = \sum_{n=-\infty}^{\infty} nJ_n(z)t^{n-1} \tag{3}$$

Using (11.23) on the LHS gives

$$\sum_{n=-\infty}^{\infty} \left(\frac{z}{2} + \frac{z}{2t^2}\right) J_n(z)t^n = \sum_{n=-\infty}^{\infty} \frac{z}{2} J_n(z)t^n + \sum_{n=-\infty}^{\infty} \frac{z}{2} J_n(z)t^{n-2}$$

$$= \frac{z}{2} \sum_{n=-\infty}^{\infty} J_{n-1}(z)t^{n-1} + \frac{z}{2} \sum_{n=-\infty}^{\infty} J_{n+1}(z)t^{n-1} \tag{4}$$

$$= \frac{z}{2} \sum_{n=-\infty}^{\infty} [J_{n-1}(z) + J_{n+1}(z)]t^{n-1}$$

where we substituted $n - 1$ for n in the first sum and $n + 1$ for n in the second. Equating powers of t in Eqs. (3) and (4), we get

$$nJ_n(z) = \frac{z}{2}[J_{n-1}(z) + J_{n+1}(z)]$$

which was obtained differently in Chapter 9 [see Eq. (9.78)]. •

We can start with (11.22) and obtain other integral representations of Bessel functions by making different substitutions. For instance, we can let $u = e^w$ and assume that the circle of the contour C has unit radius (this is possible because in inflating ε to 1, we do not encounter any singularity). The contour C' in the w-plane is determined as follows. Write $u = re^{i\theta}$ and $w = x + iy$, so

$$re^{i\theta} = e^x e^{iy} \quad \Rightarrow \quad r = e^x \quad \text{and} \quad e^{i\theta} = e^{iy}$$

Along the first part of C, $\theta = -\pi$ and r goes from ∞ to 1. Thus, along the corresponding part of C', $y = -\pi$ and x goes from ∞ to 0. On the circle of C, $r = 1$ and θ goes from $-\pi$ to $+\pi$. Thus, along the corresponding part of C', $x = 0$ and y goes from $-\pi$ to $+\pi$. Finally, on the last part of C', $y = \pi$ and x goes from 0 to ∞. Therefore, the contour C' in the w-plane is as shown in Fig. 11.3.

Substituting $u = e^w$ in (11.22) yields

$$J_\nu(z) = \frac{1}{2\pi i}\int_{C'} e^{z\sinh w - \nu w}\,dw \qquad \text{where } \mathrm{Re}(z) > 0 \tag{11.24}$$

which can be transformed into (see Exercise 11.3.6)

$$J_\nu(z) = \frac{1}{\pi}\int_0^\pi \cos(\nu\theta - z\sin\theta)\,d\theta - \frac{\sin\nu\pi}{\pi}\int_0^\infty e^{-\nu t - z\sinh t}\,dt \qquad \text{where } \mathrm{Re}(z) > 0 \tag{11.25}$$

A special case is that where n is an integer. Then

$$J_n(z) = \frac{1}{\pi}\int_0^\pi \cos(n\theta - z\sin\theta)\,d\theta$$

In particular,

$$J_0(z) = \frac{1}{\pi}\int_0^\pi \cos(z\sin\theta)\,d\theta$$

We can use the integral representation for $J_\nu(z)$ to find the integral representation for Bessel functions of other kinds. For instance, to obtain the integral repres-

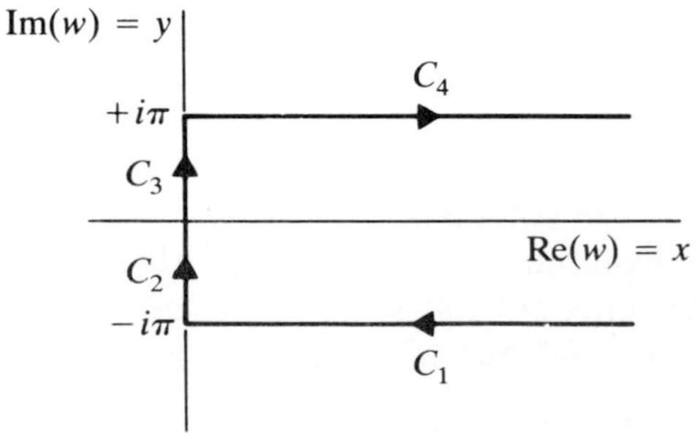

Figure 11.3 The contour C' in the w-plane used in evaluating $J_\nu(z)$.

entation for the Neumann function $Y_\nu(z)$, we use Eq. (9.74):

$$Y_\nu(z) = (\cot \nu\pi) J_\nu(z) - \frac{1}{\sin \nu\pi} J_{-\nu}(z)$$

$$= \frac{\cot \nu\pi}{\pi} \int_0^\pi \cos(\nu\theta - z\sin\theta)\, d\theta - \frac{\cos \nu\pi}{\pi} \int_0^\infty e^{-\nu t - z\sinh t}\, dt$$

$$- \frac{1}{\pi \sin \nu\pi} \int_0^\pi \cos(\nu\theta + z\sin\theta)\, d\theta - \frac{1}{\pi} \int_0^\infty e^{\nu t - z\sinh t}\, dt \qquad \text{where } \mathrm{Re}(z) > 0$$

Substituting $\pi - \theta$ for θ in the third integral on the RHS and substituting the result and Eq. (11.25) in $H_\nu^{(1)}(z) = J_\nu(z) + i Y_\nu(z)$ yields

$$H_\nu^{(1)}(z) = \frac{1}{\pi} \int_0^\pi e^{i(z\sin\theta - \nu\theta)}\, d\theta + \frac{1}{i\pi} \int_0^\infty e^{\nu t - z\sinh t}\, dt + \frac{e^{-i\nu\pi}}{i\pi} \int_0^\infty e^{-\nu t - z\sinh t}\, dt$$

$$\text{where } \mathrm{Re}(z) > 0$$

This can easily be shown to result from integrating along the contour C'' of Fig. 11.4. Thus, we have

$$H_\nu^{(1)}(z) = \frac{1}{i\pi} \int_{C''} e^{z\sinh w - \nu w}\, dw \qquad \text{where } \mathrm{Re}(z) > 0$$

By changing i to $-i$, we can show that

$$H_\nu^{(2)}(z) = -\frac{1}{i\pi} \int_{C'''} e^{z\sinh w - \nu w}\, dw \qquad \text{where } \mathrm{Re}(z) > 0$$

where C''' is the mirror image of C'' about the real axis.

As mentioned before, integral representations are particularly useful for determining the asymptotic behavior of functions. For Bessel functions we can consider two kinds of limits. Assuming that both ν and $z \equiv x$ are real, we have $\nu \to \infty$ or

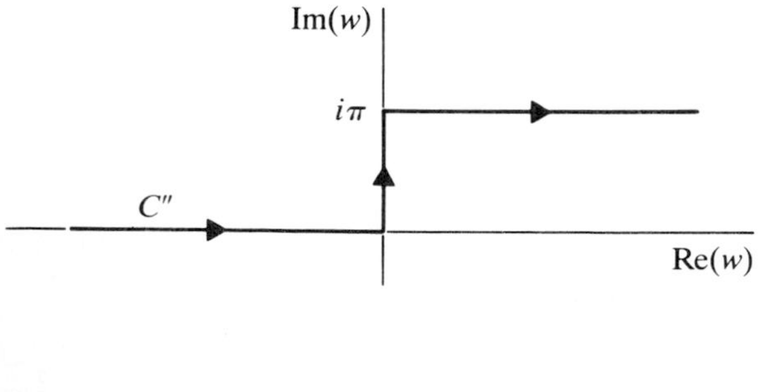

Figure 11.4 The contour C'' in the w-plane used in evaluating $H_\nu^{(1)}(z)$.

$x \to \infty$. First, let us consider the behavior of $J_\nu(x)$ of large order ($\nu \to \infty$). We assume that $\nu > x$ and write

$$\nu \equiv x \cosh w_0 \qquad \text{where } w_0 > 0$$

Then Eq. (11.24) yields

$$J_\nu(x) = \frac{1}{2\pi i} \int_{C'} e^{x(\sinh w - w \cosh w_0)}\, dw$$

In the language of Section 7.5 (with w replacing z), we have

$$g(w) = 1 \qquad \text{and} \qquad f(w) = \sinh w - w \cosh w_0$$

The saddle point is obtained from $df/dw = 0$ or $\cosh w = \cosh w_0$. Thus, $w = \pm w_0 + 2in\pi$, for $n = 0, 1, 2, \ldots$. Since the contour C' lies in the RHP, we choose w_0 as the saddle point.

We now have to find a path going through w_0 and along which $\operatorname{Im}[f(w)] = \operatorname{Im}[f(w_0)] = 0$. Writing $w = u + iv$, we have

$$\begin{aligned} f(w) &= \sinh u \cosh iv + \cosh u \sinh iv - (u + iv)(\cosh w_0) \\ &= \sinh u \cos v - u \cosh w_0 + i(\cosh u \sin v - v \cosh w_0) \end{aligned}$$

Thus, $\operatorname{Im}[f(w)] = 0$ determines the path

$$\cosh u = \frac{v \cosh w_0}{\sin v}$$

shown in Fig. 11.5. It is clear that the contour C' of Eq. (11.24) can be deformed into C_0 (see Fig. 11.3).

We now define the real variable [see Eqs. (7.20) and (7.23)]

$$t^2 \equiv \sinh w_0 - w_0 \cosh w_0 - (\sinh w - w \cosh w_0)$$

Expanding it about w_0 gives, up to the fourth order,

$$t^2 = -\frac{\sinh w_0}{2!}(w - w_0)^2 - \frac{\cosh w_0}{3!}(w - w_0)^3 - \frac{\sinh w_0}{4!}(w - w_0)^4 + \cdots \tag{11.26}$$

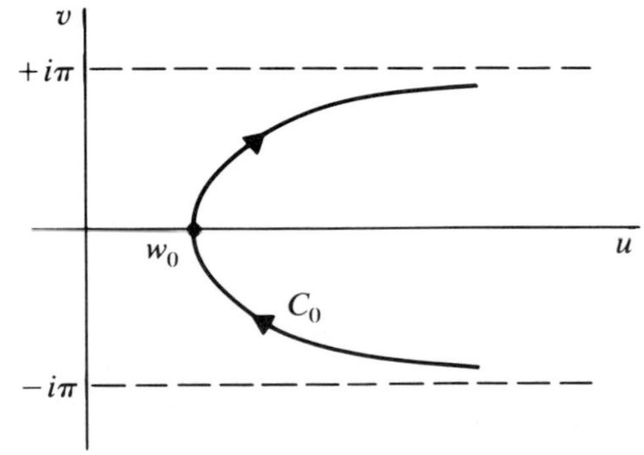

Figure 11.5 The contour C_0 in the w-plane used in evaluating $J_\nu(x)$ for large values of ν.

However, Eq. (7.25) indicates that we need to express w as a power series in t. To find such a power series, let us write

$$w(t) = w_0 + \sum_{m=1}^{\infty} b_m t^m$$

and approximate the series by

$$w(t) - w_0 \approx b_1 t + b_2 t^2 + b_3 t^3$$

Substituting this in (11.26) and setting the coefficients of the same powers of t from both sides equal, we obtain

$$-\frac{\sinh w_0}{2} b_1^2 = 1$$

$$(-\sinh w_0) b_1 b_2 - \frac{\cosh w_0}{6} b_1^3 = 0$$

$$-\frac{\sinh w_0}{2} b_2^2 - (\sinh w_0) b_1 b_3 - \frac{\cosh w_0}{2} b_1^2 b_2 - \frac{\sinh w_0}{24} b_1^4 = 0$$

whose solutions are

$$b_1 = \pm i\left(\frac{2}{\sinh w_0}\right)^{1/2} \qquad b_2 = \frac{\coth w_0}{3\sinh w_0} \qquad b_3 = \frac{b_1^3}{24}\left(\frac{5}{3}\coth^2 w_0 - 1\right)$$

The fact that

$$\frac{\pi}{2} = \arg\left(\lim_{t \to 0} \frac{dw}{dt}\right) = \arg(b_1)$$

requires us to choose the b_1 that has the plus sign.

Differentiating $w(t)$ and comparing with Eq. (7.25) yields

$$a_0 = b_1 = i\left(\frac{2}{\sinh w_0}\right)^{1/2}$$

$$a_1 = 2b_2 = \frac{2\coth w_0}{3\sinh w_0}$$

$$a_2 = 3b_3 = \frac{b_1^3}{8}\left(\frac{5}{3}\coth^2 w_0 - 1\right)$$

If we substitute the above in Eq. (7.28a), we obtain the asymptotic formula valid for $w_0 \to \infty$:

$$J_{x\cosh w_0}(x) \approx \frac{1}{2\pi i} e^{x(\sinh w_0 - w_0 \cosh w_0)}\left(a_0\sqrt{\frac{\pi}{x}} + a_2\frac{\sqrt{\pi}}{2x^{3/2}} + \cdots\right)$$

$$= \frac{e^{x(\sinh w_0 - w_0\cosh w_0)}}{(2\pi x \sinh w_0)^{1/2}}\left[1 + \frac{1}{8x\sinh w_0}\left(1 - \frac{5}{3}\coth^2 w_0\right) + \cdots\right]$$

Let us consider the asymptotic behavior for large x, which was briefly considered in Chapter 10. It is convenient to consider the Hankel functions $H_\nu^{(i)}(x)$, for $i = 1, 2$. The contours C'' and C''' involve both the positive and the negative real axis; therefore, it is convenient, assuming that $x > \nu$, to write $\nu = x\cos\beta$ so that

$$H_\nu^{(1)}(x) = \frac{1}{i\pi}\int_{C''} e^{x(\sinh w - w\cos\beta)}\,dw$$

The saddle points are given by the solutions to $\cosh w = \cos\beta$, which are $w_0 = \pm i\beta$. Choosing $w_0 = +i\beta$, we note that the contour along which $\text{Im}\,[\sinh w - w\cos\beta] =$ constant $= \text{Im}\,[\sinh w_0 - w_0\cos\beta]$ is given by

$$\cosh u = \frac{\sin\beta + (v - \beta)\cos\beta}{\sin v}$$

This contour is shown in Fig. 11.6. The rest of the procedure is exactly the same as that for $J_\nu(x)$ described above. In fact, to obtain the expansion for $H_\nu^{(1)}(x)$, we simply replace w_0 by $i\beta$. The result is

$$H_\nu^{(1)}(x) \approx \left(\frac{2}{i\pi x\sin\beta}\right)^{1/2} e^{i(x\sin\beta - \nu\beta)}\left[1 + \frac{1}{8ix\sin\beta}\left(1 + \frac{5}{3}\cot^2\beta\right) + \cdots\right]$$

When $x \gg \nu$, $\beta \approx \pi/2$, and we have

$$H_\nu^{(1)}(x) \approx \sqrt{\frac{2}{\pi x}}\, e^{i[x - \nu(\pi/2) - (\pi/4)]}\left(1 + \frac{1}{8ix}\right)$$

which, with $1/x \to 0$, is what we obtained in Exercise 7.5.1.

The other saddle point at $-i\beta$ gives the other Hankel function, with the asymptotic limit

$$H_\nu^{(2)}(x) \approx \sqrt{\frac{2}{\pi x}}\, e^{-i[x - \nu(\pi/2) - (\pi/4)]}\left(1 - \frac{1}{8ix}\right)$$

We can use the expressions for $H_\nu^{(1)}(x)$ and $H_\nu^{(2)}(x)$ to write the asymptotic forms of $J_\nu(x)$ and $Y_\nu(x)$ for large x:

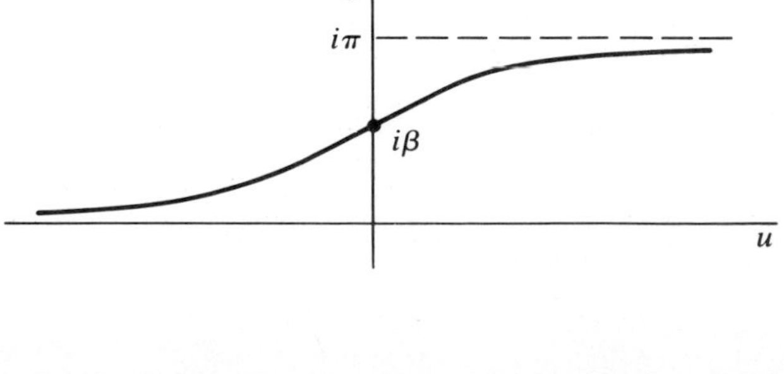

Figure 11.6 The contour in the w-plane used in evaluating $H_\nu^{(1)}(x)$ in the limit of large x.

$$J_\nu(x)=\frac{1}{2}[H_\nu^{(1)}(x)+H_\nu^{(2)}(x)]\approx\sqrt{\frac{2}{\pi x}}\left[\cos\left(x-\nu\frac{\pi}{2}-\frac{\pi}{4}\right)+\frac{1}{8x}\sin\left(x-\nu\frac{\pi}{2}-\frac{\pi}{4}\right)+\cdots\right]$$

$$Y_\nu(x)=\frac{1}{2i}[H_\nu^{(1)}(x)-H_\nu^{(2)}(x)]\approx\sqrt{\frac{2}{\pi x}}\left[\sin\left(x-\nu\frac{\pi}{2}-\frac{\pi}{4}\right)-\frac{1}{8x}\cos\left(x-\nu\frac{\pi}{2}-\frac{\pi}{4}\right)+\cdots\right]$$

In the limit $1/x \to 0$, these tend to the results obtained in Section 10.2.2.

Exercises

11.3.1 Solve Eq. (11.16) and find $v(t)$.

11.3.2 Calculate the surface term in the Lagrange identity for $\mathbb{M}_t$ and $\mathbb{M}_t^\dagger$ of Eqs. (11.16).

11.3.3 Determine the constant C' in Eq. (11.17), the solution to the hypergeometric DE. [Hint: Expand $(t-z)^{-\alpha}$, and use

$$\mathrm{B}(a,b)\equiv\frac{\Gamma(a)\Gamma(b)}{\Gamma(a+b)}=\int_1^\infty t^{-a-b}(t-1)^{b-1}\,dt$$

derived in Chapter 14.]

11.3.4 Taking the contour of integration of $u(z)$ to be from $t=0$ to $t=1$ and making the appropriate choice of the constant C, derive the solution

$$w(z)=z^{-\alpha}F\left(\alpha,\alpha-\gamma+1;\alpha-\beta+1;\frac{1}{z}\right)=\frac{\Gamma(\alpha-\beta+1)}{\Gamma(\gamma-\beta)\Gamma(\alpha-\gamma+1)}\int_0^1(z-t)^{-\alpha}t^{\alpha-\gamma}(1-t)^{\gamma-\beta-1}\,dt$$

11.3.5 **(a)** Write the contour integral of Eq. (11.21) for each of the three pieces of the contour of Fig. 11.2. Note that $\arg(t)=-\pi$ as t comes from $-\infty$ and $\arg(t)=+\pi$ as t goes to $-\infty$. Obtain a real integral from 0 to ∞.

(b) Use the relation $\Gamma(z)\Gamma(1-z)=\pi/\sin\pi z$, obtained in Chapter 14, to show that

$$\Gamma(-z)=-\frac{\pi}{\Gamma(z+1)\sin\pi z}$$

(c) Expand the function $\exp(z^2/4t)$ in the integral of part (a), and show that the contour integral reduces to

$$-2i\sin\nu\pi\sum_{n=0}^\infty\left(\frac{z}{2}\right)^{2n}\frac{\Gamma(-n-\nu)}{\Gamma(n+1)}$$

(d) Use the result of part (c) in part (b), and compare the result with the series expansion of $J_\nu(z)$ in Chapter 9 to finally arrive at $k=1/2\pi i$.

11.3.6 By integrating along C_1, C_2, C_3, and C_4 of Fig. 11.3, derive Eq. (11.25).

11.4 GREEN'S FUNCTIONS IN ONE DIMENSION

This section addresses the primary subject of this chapter, Green's functions in one dimension, that is, Green's functions of ordinary differential equations.

Consider the ODE

$$\mathbb{L}_x[u] = f(x)$$

where $\mathbb{L}_x$ is a general linear differential operator. In the abstract Dirac notation this can be formally written as

$$\mathbb{L}|\mathrm{u}\rangle = |\mathrm{f}\rangle$$

If $\mathbb{L}$ has an inverse $\mathbb{L}^{-1} \equiv \mathbb{G}$, the solution can be formally written as $|\mathrm{u}\rangle = \mathbb{L}^{-1}|\mathrm{f}\rangle \equiv \mathbb{G}|\mathrm{f}\rangle$. Multiplying it by $\langle\mathrm{x}|$ and inserting $\mathbb{1} = \int dy|\mathrm{y}\rangle w(y)\langle\mathrm{y}|$ between $\mathbb{G}$ and $|\mathrm{f}\rangle$ gives

$$u(x) = \int dy\, G(x,y)\, w(y) f(y) \tag{11.27}$$

Thus, once we know $G(x,y)$, we can find the solution, $u(x)$, in an integral form. But how do we find $G(x,y)$?

By the definition of an inverse, we have $\mathbb{L}\mathbb{G} = \mathbb{1}$. Sandwiching both sides of this between $\langle\mathrm{x}|$ and $|\mathrm{y}\rangle$ and using $\mathbb{1} = \int dx'|\mathrm{x}'\rangle w(x')\langle\mathrm{x}'|$ between $\mathbb{L}$ and $\mathbb{G}$ yields

$$\int dx'\, L(x,x')w(x')G(x',y) = \langle\mathrm{x}|\mathrm{y}\rangle = \frac{\delta(x-y)}{w(x)}$$

In particular, if $\mathbb{L}$ is a local differential operator (see Section 11.1) $L(x,x') \equiv [\delta(x-x')/w(x)]\mathbb{L}_x$ and we obtain

$$\mathbb{L}_x G(x,y) = \frac{\delta(x-y)}{w(x)} \tag{11.28a}$$

Sometimes it is assumed that $w(x) \equiv 1$. In that case we have

$$\mathbb{L}_x G(x,y) = \delta(x-y) \tag{11.28b}$$

In either case $G(x,y)$ is called the *Green's function* (GF) for the DO $\mathbb{L}_x$.

It is important to emphasize from the outset that $\mathbb{L}_x$ might not be defined for all $\mathbb{R}$. Thus, we speak of the domain of $\mathbb{L}$ as some interval, such as $[a,b]$, of $\mathbb{R}$. Second, and more important, a complete specification of $\mathbb{L}_x$ requires some initial (or boundary) conditions. Therefore, we expect $G(x,y)$ to depend on such initial conditions as well. Third, we note that when $\mathbb{L}_x$ is applied to (11.27), we get

$$\mathbb{L}_x u(x) = \int dy[L_x G(x,y)]w(y)f(y) = \int dy \frac{\delta(x-y)}{w(x)} w(y) f(y) = f(x)$$

indicating that $u(x)$ is indeed a solution of the original ODE. Fourth, Eqs. (11.28), involving the generalized function (or distribution in the language of Chapter 5) $\delta(x-y)$, are meaningful only in the same context. Thus, we look on $G(x,y)$ not as an ordinary function but as *a distribution*. Finally, Eq. (11.27) is assumed to hold for an arbitrary (well-behaved) function.

11.4.1 Calculation of Some Green's Functions

This subsection presents some examples of calculating $G(x,y)$ for very simple $\mathbb{L}_x$'s. Later we will see how to obtain Green's functions for the general second-order linear differential operator.

Example 11.4.1

The simplest DO is $\mathbb{L}_x \equiv d/dx$. Let us find its GF.

We need to find a distribution such that its derivative is the Dirac delta function:

$$\frac{d}{dx}G(x,y) = \delta(x-y)$$

In Chapter 5 we encountered such a distribution—the step function $\theta(x-y)$. Thus,

$$G(x,y) = \theta(x-y) + \alpha(y)$$

where $\alpha(y)$ is the "constant" of integration. •

Example 11.4.1 did not include a boundary (or initial) condition. We may therefore call the GF we obtained an indefinite GF. Let us see how boundary conditions affect the resulting GF.

Example 11.4.2

Let us solve

$$\frac{du}{dx} = f(x) \qquad \text{where } x \in [0, \infty),\ u(0) = 0$$

A general solution of this DE is given by (11.27) and Example 11.4.1:

$$u(x) = \int_0^\infty \theta(x-y)f(y)\,dy + \int_0^\infty \alpha(y)f(y)dy$$

The factor $\theta(x-y)$ in the first term on the RHS chops off the integral at x. It follows that

$$u(x) = \int_0^x f(y)dy + \int_0^\infty \alpha(y)f(y)dy$$

The BC gives

$$0 = u(0) = 0 + \int_0^\infty \alpha(y)\,f(y)\,dy$$

The only way that this can be satisfied for arbitrary $f(y)$ is if $\alpha(y)$ is zero. Thus,

$$G(x,y) = \theta(x-y)$$

and

$$u(x) = \int_0^\infty \theta(x-y)f(y)dy = \int_0^x f(y)\,dy$$

This is like killing a fly with a sledge hammer! We could have obtained the result by a simple integration. However, the roundabout way outlined here illustrates some important features of GFs that will be discussed later. (Note that the BC introduced here is very special. What happens if it is changed to $u(0) = a$? Exercise 11.4.1 answers this.) ●

Example 11.4.3

A more complicated DO is $\mathbb{L}_x \equiv d^2/dx^2$. Let us find its indefinite GF.

Integrating

$$\frac{d^2}{dx^2} G(x, y) = \delta(x - y)$$

once with respect to x gives

$$\frac{d}{dx} G(x,y) = \theta(x - y) + \alpha(y)$$

A second integration yields

$$G(x,y) = \int dx\, \theta(x - y) + x\alpha(y) + \eta(y)$$

where α and η are arbitrary functions of y and the integral is an indefinite integral to be evaluated below.

Let $\Omega(x,y)$ be the primitive of $\theta(x - y)$; that is

$$\frac{d\Omega}{dx} = \theta(x - y) = \begin{cases} 1 & \text{if } x > y \\ 0 & \text{if } x < y \end{cases} \tag{1}$$

The solution to this equation is

$$\Omega(x,y) = \begin{cases} x + a(y) & \text{if } x > y \\ b(y) & \text{if } x < y \end{cases}$$

Note that we have not defined $\Omega(x,y)$ at $x = y$. It will become clear below that $\Omega(x,y)$ is continuous at $x = y$. We can rewrite $\Omega(x,y)$ as

$$\Omega(x,y) = [x + a(y)]\theta(x - y) + b(y)\theta(y - x) \tag{2}$$

To determine $a(y)$ and $b(y)$, we differentiate (2) and compare with (1):

$$\begin{aligned} \frac{d\Omega}{dx} &= \theta(x - y) + [x + a(y)]\,\delta(x - y) - b(y)\,\delta(x - y) \\ &= \theta(x - y) + [x - b(y) + a(y)]\,\delta(x - y) \end{aligned} \tag{3}$$

where we have used

$$\frac{d}{dx}\theta(x - y) = -\frac{d}{dx}\theta(y - x) = \delta(x - y)$$

For Eq. (3) to agree with (1), we must have

$$[x - b(y) + a(y)]\,\delta(x - y) = 0$$

Recalling from Chapter 5 that $x\delta(x) \equiv 0$, we conclude that

$$a(y) - b(y) = -y$$

Substituting this in the expression for $\Omega(x,y)$ yields

$$\Omega(x,y) = (x-y)\theta(x-y) + b(y)[\theta(x-y) + \theta(y-x)]$$

But $\theta(x) + \theta(-x) \equiv 1$; therefore,

$$\Omega(x,y) = (x-y)\theta(x-y) + b(y)$$

This shows that $\Omega(x,y)$ is continuous at $x = y$, a property that is a consequence of the relation $x\delta(x) = 0$. We can now write

$$G(x,y) = (x-y)\theta(x-y) + x\alpha(y) + \beta(y)$$

where $\beta(y) \equiv \eta(y) + b(y)$. ●

The GF in Example 11.4.3 has two arbitrary functions, $\alpha(y)$ and $\beta(y)$, which are the result of underspecification of $\mathbb{L}_x$ (a full specification of $\mathbb{L}_x$ requires BCs). The following example remedies the situation by specifying some BCs.

Example 11.4.4

Let us calculate the GF of

$$\mathbb{L}_x[u] = \frac{d^2u}{dx^2} = f(x)$$

subject to the BC $u(a) = u(b) = 0$ where $[a,b]$ is the interval on which $\mathbb{L}_x$ is defined.

Example 11.4.3 gives us the (indefinite) GF for $\mathbb{L}_x$. Using that, we can write

$$\begin{aligned} u(x) &= \int_a^b (x-y)\theta(x-y)f(y)\,dy + x\int_a^b \alpha(y)f(y)\,dy + \int_a^b \beta(y)f(y)\,dy \\ &= \int_a^x (x-y)f(y)\,dy + x\int_a^b \alpha(y)f(y)\,dy + \int_a^b \beta(y)f(y)\,dy \end{aligned}$$

Applying the BCs yields

$$0 = u(a) = a\int_a^b \alpha(y)f(y)\,dy + \int_a^b \beta(y)f(y)\,dy \tag{1}$$

and

$$0 = u(b) = \int_a^b (b-y)f(y)\,dy + b\int_a^b \alpha(y)f(y)\,dy + \int_a^b \beta(y)f(y)\,dy \tag{2}$$

From these two relations it is possible to determine $\alpha(y)$ and $\beta(y)$. Substituting for the last integral on the RHS of (2) from (1), we obtain

$$0 = \int_a^b [b - y + b\alpha(y) - a\alpha(y)]f(y)\,dy$$

Since this must hold for arbitrary $f(y)$, we conclude that

$$b - y + (b - a)\alpha(y) \equiv 0 \quad \Rightarrow \quad \alpha(y) = -\frac{b - y}{b - a}$$

Substituting for $\alpha(y)$ in Eq. (1) yields

$$0 = \int_a^b \left[-\frac{a(b - y)}{b - a} + \beta(y) \right] f(y)\,dy$$

from which we obtain

$$\beta(y) = \frac{a(b - y)}{b - a}$$

Insertion of $\alpha(y)$ and $\beta(y)$ in the expression for $G(x, y)$ obtained in Example 11.4.3 gives

$$G(x, y) = (x - y)\theta(x - y) + (x - a)\frac{y - b}{b - a} \qquad \text{where } a \leqslant x;\ y \leqslant b$$

It is striking that, because $a - y \leqslant 0$,

$$G(a, y) = (a - y)\theta(a - y) = 0$$

and, because $\theta(b - y) = 1$ for all $y \leqslant b$,

$$G(b, y) = (b - y)\theta(b - y) + (b - a)\frac{y - b}{b - a} = 0$$

These two equations reveal the important fact that as a function of x, $G(x, y)$ satisfies the same (homogeneous) BC as the solution of the DE. This is a general property that will be discussed later. ●

In all the preceding examples the BCs were very simple. Specifically, the value of the solution and/or its derivative at the boundary points was zero. What if the BCs are not so simple? In particular, how can we handle a case where $u(a)$ [or $u'(a)$] and $u(b)$ [or $u'(b)$] are not zero?

Consider a general differential operator $\mathbb{L}_x$ and the differential equation

$$\mathbb{L}_x[u] = f(x)$$

subject to the BCs

$$u(a) = a_1 \qquad \text{and} \qquad u(b) = b_1$$

We claim that we can reduce this system to the case where $u(a) = u(b) = 0$. Recall from

Chapter 9 that the most general solution to $\mathbb{L}_x[u] = f(x)$ is of the form

$$u = u_h + u_i$$

where u_h, the solution to the homogeneous equation, satisfies $\mathbb{L}_x u_h = 0$ and contains the arbitrary parameters inherent in solutions of differential equations. For instance, if $\mathbb{L}_x$ is a second-order linear differential operator (SOLDO) with two linearly independent solutions, v and w, then

$$u_h(x) = C_1 v(x) + C_2 w(x)$$

On the other hand, u_i is any solution of the inhomogeneous DE.

If we demand that $u_h(a) = a_1$ and $u_h(b) = b_1$, then u_i satisfies the system

$$\mathbb{L}_x[u_i] = f(x)$$
$$u_i(a) = u_i(b) = 0$$

which is of the type discussed in the preceding examples. When $\mathbb{L}_x$ is a SOLDO, we can put all the machinery of Chapter 9 to work to obtain $v(x)$, $w(x)$, and, therefore, $u_h(x)$. The problem then reduces to a DE for which the BCs are homogeneous; that is, the value of the solution and/or its derivative is zero at the boundary points.

Example 11.4.5

Let us assume that $\mathbb{L}_x = d^2/dx^2$. Calculation of u_h is trivial:

$$\mathbb{L}_x[u_h] = 0 \quad \Rightarrow \quad \frac{d^2 u_h}{dx^2} = 0 \quad \Rightarrow \quad u_h(x) = C_1 x + C_2$$

To evaluate C_1 and C_2 we impose the BCs $u_h(a) = a_1$ and $u_h(b) = b_1$:

$$C_1 a + C_2 = a_1$$
$$C_1 b + C_2 = b_1$$

This gives

$$C_1 = \frac{b_1 - a_1}{b - a} \qquad \text{and} \qquad C_2 = \frac{a_1 b - a b_1}{b - a}$$

The inhomogeneous equation defines a problem identical to that of Example 11.4.4. Thus, we can immediately write

$$u_i(x) = \int_a^b G(x,y) f(y)\, dy$$

where
$$G(x,y) = (x - y)\theta(x - y) + (x - a)\frac{y - b}{b - a}$$

Thus, the general solution is

$$u(x) = \frac{b_1 - a_1}{b - a}x + \frac{a_1 b - a b_1}{b - a} + \int_a^x (x - y) f(y)\, dy + \frac{(x - a)}{b - a}\int_a^b (y - b) f(y)\, dy \qquad \bullet$$

Example 11.4.5 shows that an inhomogeneous DE with inhomogeneous BCs can be separated into two DEs, one homogeneous with inhomogeneous BCs and the other inhomogeneous with homogeneous BCs. The foregoing examples indicate that solutions of DEs can be succinctly written in terms of GFs that automatically incorporate the BCs as long as the latter are homogeneous. Can a GF also give the solution to a DE with inhomogeneous BCs? The following example shows that if we regard $\mathbb{L}_x$ as an operator in the generalized sense of Chapter 5, that is, as an operator *acting on generalized functions* (distributions), then the determination of $G(x,y)$ is sufficient to give the solution to a DE with inhomogeneous BCs.

This conclusion is motivated by the fact that $u_{\text{h}}(x)$ does not belong to the domain of the DO defined by the system $\mathbb{L}_x[u] = f(x)$, $u(a) = u(b) = 0$, and thus we do not ordinarily know how the DO, with the homogeneous BC as part of its definition, acts on $u_{\text{h}}(x)$. However, as outlined in Chapter 5, distributions bypass this difficulty. There we saw how to "differentiate" a discontinuous function at its point of discontinuity. If $\varphi(x)$ is such a function and $f(x)$ is a "good" function, then the derivative of φ is given by

$$\langle \varphi', f \rangle \equiv -\langle f', \varphi \rangle \qquad \forall\, f \in C^{\infty}[a,b]$$

where

$$\langle f, \varphi \rangle \equiv \int_a^b f(x)\varphi(x)\,dx$$

Example 11.4.6

Let us solve the system of Example 11.4.5 using the method just described. We note that $\mathbb{L}_x \equiv d^2/dx^2$ is self-adjoint (see Theorem 9.2.15). Thus, assuming that u satisfies the homogeneous BC $u(a) = u(b) = 0$, we have

$$\langle \mathbb{L}_x u, v \rangle = \langle u, \mathbb{L}_x v \rangle \tag{1}$$

We use (1) to *define* how $\mathbb{L}_x$ acts on *any* function v. Integrating the LHS by parts yields

$$\int_a^b \frac{d^2u}{dx^2} v\,dx = \left(\frac{du}{dx}\right)v\bigg|_a^b - \int_a^b \frac{du}{dx}\frac{dv}{dx}\,dx = u'(b)v(b) - u'(a)v(a) - uv'|_a^b + \int_a^b u\frac{d^2v}{dx^2}\,dx$$

Since $u(x)$ satisfies the homogeneous BC, we get

$$\langle \mathbb{L}_x u, v \rangle = \langle u, \mathbb{L}_x v \rangle = \int_a^b uv''\,dx + v(b)u'(b) - v(a)u'(a)$$

$$= \int_a^b u[v'' - v(b)\delta'(x-b) + v(a)\delta'(x-a)]\,dx$$

Therefore, we define the generalized operator as

$$\mathbb{L}_x v \equiv v'' - v(b)\delta'(x-b) + v(a)\delta'(x-a)$$

Note that if v belongs to the domain of $\mathbb{L}_x$, that is, if $v(a) = 0 = v(b)$, then $\mathbb{L}_x$ reduces to d^2/dx^2, as it should.

The problem of Example 11.4.5 has been reduced to finding a function $v(x)$ such that [note that u has been replaced by v here and that $v'' = f(x)$]

$$\mathbb{L}_x v = f(x) - b_1 \delta'(x - b) + a_1 \delta'(x - a)$$

Writing $v = v_{\text{i}} + v_{\text{h}}$, in which

$$v_{\text{i}} = \int_a^b G(x,y)\, f(y)\, dy$$

which, similar to $G(x,y)$, satisfies the homogeneous BCs $v_{\text{i}}(a) = v_{\text{i}}(b) = 0$, reduces the problem to finding v_{h} such that

$$\mathbb{L}_x v_{\text{h}} = -b_1 \delta'(x - b) + a_1 \delta'(x - a) \tag{2}$$

The derivatives of the delta function on the RHS give us a clue. Could we manipulate the GF to obtain v_{h}? After all, $G(x,y)$ satisfies

$$\mathbb{L}_x G(x,y) = \delta(x - y) \tag{3}$$

and if we differentiate both sides of this we obtain the derivative of the delta function. We can differentiate with respect to x or y. Either works in this case because $\mathbb{L}_x$ and d/dx commute, so

$$\frac{d}{dx}[\mathbb{L}_x G(x,y)] = \mathbb{L}_x \left[\frac{d}{dx} G(x,y)\right] = \delta'(x - y)$$

and a proper linear combination of dG/dx evaluated at $y = b$ and $y = a$ would give us a solution for v_{h}. However, in general, $\mathbb{L}_x$ and d/dx do not commute because of the coefficient functions of $\mathbb{L}_x$. Thus, we differentiate both sides of (3) with respect to y and note that

$$\mathbb{L}_x \left\{ b_1 \left[\frac{\partial G}{\partial y}\right]_{y=b} \right\} = -b_1 \delta'(x - b)$$

$$\mathbb{L}_x \left\{ -a_1 \left[\frac{\partial G}{\partial y}\right]_{y=b} \right\} = a_1 \delta'(x - a)$$

The linearity of $\mathbb{L}_x$ now implies that the following satisfies (2):

$$v_{\text{h}}(x) \equiv b_1 \left[\frac{\partial G}{\partial y}\right]_{y=b} - a_1 \left[\frac{\partial G}{\partial y}\right]_{y=a}$$

We see, therefore, that a determination of $G(x,y)$ that satisfies the homogeneous BC is sufficient to obtain the solution to the DE even if the BCs are inhomogeneous. We do not have to separate the problem into two parts as we did in Example 11.4.5. Later we will see that this is a general property of GFs and is not peculiar to that for $\mathbb{L}_x = d^2/dx^2$.

Let us now calculate $v_{\text{h}}(x)$ using the GF obtained in Example 11.4.4:

$$G(x,y) = (x - y)\,\theta(x - y) + (x - a)\frac{y - b}{b - a}$$

We have

$$\left(\frac{\partial G}{\partial y}\right)_{y=b} = -\theta(x-b) - (x-b)\delta(x-b) + \frac{x-a}{b-a} = \frac{x-a}{b-a}$$

$$\left(\frac{\partial G}{\partial y}\right)_{y=a} = -\theta(x-a) - (x-a)\delta(x-a) + \frac{x-a}{b-a} = \frac{x-a}{b-a} - 1 = \frac{x-b}{b-a}$$

Thus,
$$v_h = b_1\frac{x-a}{b-a} - a_1\frac{x-b}{b-a} = \frac{b_1 - a_1}{b-a}x + \frac{ba_1 - ab_1}{b-a}$$

which is the same as the homogeneous solution derived in Example 11.4.5. •

11.4.2 Formal Considerations Concerning Green's Functions

The discussion and examples of the preceding subsection hint at the power of Green's functions. The elegance of such a function becomes apparent from the realization that it contains all the information about the solutions of a DE for any type of BCs (see Example 11.4.6). Therefore, it is time to give the discussion of these functions some formal flavor.

Green's functions are inverses of differential operators. Therefore, it is important to have a clear understanding of the latter. An nth-order linear differential operator (NOLDO) is defined via the following theorem (for a proof, see Birkhoff and Rota 1978, Chapter 6).

11.4.1 Theorem Let

$$\mathbb{L}_x = p_n(x)\frac{d^n}{dx^n} + p_{n-1}(x)\frac{d^{n-1}}{dx^{n-1}} + \cdots + p_1(x)\frac{d}{dx} + p_0(x) \tag{11.29a}$$

where $p_n(x) \not\equiv 0$ in $\mathrm{I} \equiv [a,b]$. Let $x_0 \in \mathrm{I}$, and let $\gamma_1, \gamma_2, \ldots, \gamma_n$ be given numbers and $f(x)$ be a given piecewise continuous function on I. Then the initial value problem (IVP)

$$\mathbb{L}_x[u] = f \qquad \text{for } x \in \mathrm{I}$$
$$u(x_0) = \gamma_1,\ u'(x_0) = \gamma_2, \ldots, u^{(n-1)}(x_0) = \gamma_n \tag{11.29b}$$

has one and only one solution. ■

This is simply the existence and uniqueness theorem for a NOLDE. Equation (11.29b) is referred to as the *IVP with data* $\{f(x); \gamma_1, \ldots, \gamma_n\}$. This theorem is used to define $\mathbb{L}_x$. Part of that definition are *the BCs* that the solutions to $\mathbb{L}_x$ must satisfy.

A particularly important BC is the homogeneous one in which $\gamma_1 = \gamma_2 = \cdots = \gamma_n = 0$. In such a case it can be shown (see Exercise 11.4.3) that the only nontrivial solution of the homogeneous DE $\mathbb{L}_x[u] = 0$ is $u \equiv 0$. Theorem 11.2.2 then tells us that $\mathbb{L}_x$ is invertible; that is, there is a unique operator $\mathbb{G}$ such that $\mathbb{L}\mathbb{G} = \mathbb{1}$, or, in the language of "components,"

$$\mathbb{L}_x G(x,y) = \frac{\delta(x-y)}{w(x)}$$

We thus have the following theorem.

11.4.2 Theorem The DO $\mathbb{L}_x$ of Eq. (11.29a) associated with the IVP with data $\{f(x); 0, 0, \ldots, 0\}$ is invertible; that is, $\exists$ a function $G(x,y)$ such that

$$\mathbb{L}_x G(x,y) = \frac{\delta(x-y)}{w(x)} \qquad \blacksquare$$

The importance of homogeneous BCs can now be appreciated. Theorem 11.4.2 is the reason why we had to impose homogeneous BCs to obtain the GF in all the examples of Section 11.4.1.

The BCs in (11.29b) clearly are not the only ones that can be used. The most general linear BCs encountered in differential operator theory are

$$\mathbb{R}_1[u] \equiv \alpha_{11}u(a) + \cdots + \alpha_{1n}u^{(n-1)}(a) + \beta_{11}u(b) + \cdots + \beta_{1n}u^{(n-1)}(b) = \gamma_1$$

$$\vdots$$

$$\mathbb{R}_n[u] \equiv \alpha_{n1}u(a) + \cdots + \alpha_{nn}u^{(n-1)}(a) + \beta_{n1}u(b) + \cdots + \beta_{nn}u^{(n-1)}(b) = \gamma_n \tag{11.30}$$

The n row vectors $(\alpha_{11}, \ldots, \alpha_{1n}, \beta_{11}, \ldots, \beta_{1n}), \ldots, (\alpha_{n1}, \ldots, \alpha_{nn}, \beta_{n1}, \ldots, \beta_{nn})$ are assumed to be independent (in particular, no row is identical to zero). We refer to $\mathbb{R}_1$, $\mathbb{R}_2, \ldots, \mathbb{R}_n$ as *boundary functionals* because, for each smooth enough function u, they give a number γ_i. The DO of (11.29a) and the BCs of (11.30) together form a *boundary value problem* (BVP). The DE $\mathbb{L}_x[u] = f$ subject to the BCs of (11.30) is a *BVP with data* $\{f(x); \gamma_1, \gamma_2, \ldots, \gamma_n\}$.

We note that $\mathbb{R}_i$, for $i = 1, 2, \ldots, n$, are linear functionals; that is, $\mathbb{R}_i[u_1 + u_2] = \mathbb{R}_i[u_1] + \mathbb{R}_i[u_2]$ and $\mathbb{R}_i[\alpha u] = \alpha\mathbb{R}_i[u]$. Since $\mathbb{L}_x$ is also linear, we conclude that the *superposition principle* applies to the system consisting of $\mathbb{L}_x[u] = f$ and the BCs of (11.30), which is sometimes denoted as $(\mathbb{L}; \mathbb{R}_1, \ldots, \mathbb{R}_n)$. If u satisfies the BVP with data $\{f; \gamma_1, \ldots, \gamma_n\}$ and v satisfies the BVP with data $\{g; \mu_1, \ldots, \mu_n\}$, then $\alpha u + \beta v$ satisfies the BVP with data $\{\alpha f + \beta g; \alpha\gamma_1 + \beta\mu_1, \ldots, \alpha\gamma_n + \beta\mu_n\}$. If u and v both satisfy the BVP with data $\{f; \gamma_1, \gamma_2, \ldots, \gamma_n\}$, then $u - v$ satisfies the BVP with data $\{0; 0, 0, \ldots, 0\}$, which is called the *completely homogeneous problem.*

Unlike the IVP, the BVP with data $\{0; 0, \ldots, 0\}$ may have a nontrivial solution. If the completely homogeneous problem has no nontrivial solution, then the BVP with data $\{f; \gamma_1, \gamma_2, \ldots, \gamma_n\}$ has at most one solution (a solution exists for any data). On the other hand, if the completely homogeneous problem has nontrivial solutions, then the BVP with data $\{f; \gamma_1, \gamma_2, \ldots, \gamma_n\}$ either has no solutions or has more than one solution (see Stakgold 1979, pp. 203–204).

It is usually assumed that a differential operator acts on a Hilbert space, such as $\mathscr{L}^2_w(a, b)$. On the other hand, not all functions in $\mathscr{L}^2_w(a, b)$ satisfy the BCs necessary for defining $\mathbb{L}_x$. Thus, the functions for which the operator is defined (those that satisfy the BCs) form a subset of $\mathscr{L}^2_w(a,b)$ called the *domain* of $\mathbb{L}$ and denoted $D(\mathbb{L})$. From a formal standpoint it is extremely important to distinguish among functions that have different domains. For instance, the derivative operator has completely different properties when defined on $\mathscr{L}^2_w(-\pi,\pi)$ as opposed to $\mathscr{L}^2_w(-\infty, +\infty)$.

Example 11.4.7

The momentum operator $\mathbb{p} = -i\,d/dx$ when acting on $\mathscr{L}^2(a,b)$ has the property that

$$\langle \mathrm{u}|\mathbb{p}|\mathrm{v}\rangle \equiv \int_a^b u^*(x)\left[-i\frac{d}{dx}v(x)\right]dx = -iu^*(x)v(x)\Big|_a^b + i\int_a^b v(x)\frac{du^*}{dx}\,dx$$

$$= -iu^*(b)v(b) + iu^*(a)v(a) + \langle \mathrm{v}|\mathrm{p}|\mathrm{u}\rangle^* \tag{1}$$

In particular, $\mathbb{p}$ will be hermitian if and only if the boundary term vanishes, or $u^*(b)v(b) = u^*(a)v(a)$.

Thus, the domain of the hermitian operator $\mathbb{p}$ consists of all functions u in $\mathscr{L}^2(a,b)$ such that $u(a) = u(b)$. For instance, $\sin x$ belongs to the domain of $\mathbb{p}$ when it acts on $\mathscr{L}^2(-\pi,\pi)$. On the other hand, all functions belonging to $\mathscr{L}^2(-\infty, +\infty)$ automatically vanish at $a = -\infty$ and $b = +\infty$ by the definition of $\mathscr{L}^2(-\infty, +\infty)$. However, $\sin x$ does not belong to $\mathscr{L}^2(-\infty, +\infty)$ or, therefore, to the domain of the hermitian $\mathbb{p}$ as it acts on $\mathscr{L}^2(-\infty, +\infty)$. ●

The adjoint of a differential operator plays a central role in studying Green's functions. It is, therefore, important to give it a precise meaning. For generality the following definition assumes that u, v, and $\mathbb{L}_x$ are not necessarily real.

11.4.3 Definition Let $\mathbb{L}_x$ be the DO of Eq. (11.29a). Suppose there exists a DO $\mathbb{L}_x^\dagger$ with the property that, for arbitrary $u \in D(\mathbb{L}_x)$ and $v \in D(\mathbb{L}_x^\dagger)$,

$$w\{v^*(\mathbb{L}_x[u]) - u(\mathbb{L}_x^\dagger[v])^*\} = \frac{d}{dx}[Q(u,v^*)]$$

where $Q(u,v)$, called the *conjunct* of the functions u and v, depends on u, v, and their derivatives of up to order $n-1$. The DO $\mathbb{L}_x^\dagger$ is then called the *formal adjoint* of $\mathbb{L}_x$. If $\mathbb{L}_x^\dagger = \mathbb{L}_x$, then $\mathbb{L}_x$ is said to be *formally self-adjoint*. If furthermore, $D(\mathbb{L}_x) = D(\mathbb{L}_x^\dagger)$, then $\mathbb{L}_x$ is said to be *self-adjoint* or *hermitian*.

The relation given in Definition 11.4.3 involving the conjunct is a generalization of Lagrange identity and can also be written in integral form:

$$\int_a^b dx\, w\{v^*(\mathbb{L}_x[u])\} - \int_a^b dx\, w\{u(\mathbb{L}_x^\dagger[v])^*\} = Q(u,v^*)|_a^b \tag{11.31}$$

This form is sometimes called the *generalized Green's identity*.

Second-order linear differential operators. Since second-order linear differential operators (SOLDOs) are general enough for most physical applications, we will concentrate on them.

Because homogeneous BCs are important in constructing Green's functions, let us first consider such BCs of the form

$$\begin{aligned} \mathbb{R}_1[u] &\equiv \alpha_{11}u(a) + \alpha_{12}u'(a) + \beta_{11}u(b) + \beta_{12}u'(b) = 0 \\ \mathbb{R}_2[u] &\equiv \alpha_{21}u(a) + \alpha_{22}u'(a) + \beta_{21}u(b) + \beta_{22}u'(b) = 0 \end{aligned} \tag{11.32}$$

where it is assumed, as usual, that $(\alpha_{11}, \alpha_{12}, \beta_{11}, \beta_{12})$ and $(\alpha_{21}, \alpha_{22}, \beta_{21}, \beta_{22})$ are linearly independent.

If, as is customary, we define the inner product as

$$\langle \mathrm{u}|\mathrm{v}\rangle = \int_a^b dx\, w(x)\, u^*(x)\, v(x)$$

then Eq. (11.31) can be formally written as

$$\langle \mathrm{v}|\mathbb{L}|\mathrm{u}\rangle = \langle \mathrm{u}|\mathbb{L}^\dagger|\mathrm{v}\rangle^* + Q(u,v^*)|_a^b$$

This would coincide with the usual definition of the adjoint if the surface term vanishes, that is, if

$$Q(u,v^*)|_{x=b} = Q(u,v^*)|_{x=a} \tag{11.33}$$

Let us investigate the conditions under which this is true.

Assuming the BCs of (11.32), we want to find the restrictions imposed on $v(x)$ such that (11.33) is true (see Exercise 11.4.5). It can be shown that v must satisfy homogeneous BCs similar to those of (11.32):

$$\begin{aligned} \mathbb{B}_1[v^*] &\equiv \gamma_{11}v^*(a) + \gamma_{12}v'^*(a) + \delta_{11}v^*(b) + \delta_{12}v'^*(b) = 0 \\ \mathbb{B}_2[v^*] &\equiv \gamma_{21}v^*(a) + \gamma_{22}v'^*(a) + \delta_{21}v^*(b) + \delta_{22}v'^*(b) = 0 \end{aligned} \tag{11.34}$$

These homogeneous BCs are said to be *adjoint* to those of (11.32). Obviously, these BCs are not the same as those of (11.32). Thus, the domain of a differential operator need not be the same as that of its adjoint.

Example 11.4.8

Let $\mathbb{L}_x = d^2/dx^2$ and

$$\mathbb{R}_1[u] \equiv \alpha u(a) - u'(a) = 0 \qquad \text{and} \qquad \mathbb{R}_2[u] \equiv \beta u(b) - u'(b) = 0 \tag{1}$$

We want to find $Q(u, v^*)$ and the adjoint BCs for v. By repeated integration by parts [or by using Eq. (9.21a)], we find

$$Q(u,v^*) = u'v^* - uv'^*$$

For the surface term to vanish, we must have

$$u'(a)v^*(a) - u(a)v'^*(a) = u'(b)v^*(b) - u(b)v'^*(b)$$

Substituting from (1) in this equation, we get

$$u(a)\,[\alpha v^*(a) - v'^*(a)] = u(b)\,[\beta v^*(b) - v'^*(b)]$$

which holds for arbitrary u if and only if

$$\mathbb{B}_1[v^*] \equiv \alpha v^*(a) - v'^*(a) = 0 \qquad \text{and} \qquad \mathbb{B}_2[v^*] \equiv \beta v^*(b) - v'^*(b) = 0 \qquad (2)$$

This is a special case in which the adjoint BCs are the same as the original BCs (except for the complex conjugation of v).

To see that the original BCs and their adjoint need not be the same, we consider

$$\mathbb{R}_1[u] = u'(a) - \alpha u(b) = 0 \qquad \text{and} \qquad \mathbb{R}_2[u] = \beta u(a) - u'(b) = 0 \qquad (3)$$

From which we obtain

$$u(b)\,[\alpha v^*(a) + v'^*(b)] = u(a)[\,\beta v^*(b) + v'^*(a)]$$

Thus,

$$\mathbb{B}_1[v^*] \equiv \alpha v^*(a) + v'^*(b) = 0 \qquad \text{and} \qquad \mathbb{B}_2[v^*] \equiv \beta v^*(b) + v'^*(a) = 0 \qquad (4)$$

which is not the same as (3). Boundary conditions such as those in (1) and (2), in which each equation contains the function and its derivative evaluated at one point, are called *unmixed BCs*. On the other hand, (3) and (4) are *mixed BCs*. •

It is sometimes convenient to write (11.32) and (11.34) as

$$\mathbb{R}[u] \equiv \begin{pmatrix} \mathbb{R}_1[u] \\ \mathbb{R}_2[u] \end{pmatrix} = (\boldsymbol{\alpha} \quad \boldsymbol{\beta})\begin{pmatrix} \mathbf{U}_a \\ \mathbf{U}_b \end{pmatrix} = 0 \qquad \text{where } \mathbf{U}_a \equiv \begin{pmatrix} u(a) \\ u'(a) \end{pmatrix},\ \mathbf{U}_b \equiv \begin{pmatrix} u(b) \\ u'(b) \end{pmatrix} \qquad (11.35)$$

and

$$\mathbb{B}[v^*] \equiv \begin{pmatrix} \mathbb{B}_1[v^*] \\ \mathbb{B}_2[v^*] \end{pmatrix} \equiv (\boldsymbol{\gamma} \quad \boldsymbol{\delta})\begin{pmatrix} \mathbf{V}_a^* \\ \mathbf{V}_b^* \end{pmatrix} = 0 \qquad \text{where } \mathbf{V}_a^* \equiv \begin{pmatrix} v^*(a) \\ v'^*(a) \end{pmatrix},\ \mathbf{V}_b^* \equiv \begin{pmatrix} v^*(b) \\ v'^*(b) \end{pmatrix} \qquad (11.36)$$

respectively, where $\boldsymbol{\alpha}$, $\boldsymbol{\beta}$, $\boldsymbol{\gamma}$, and $\boldsymbol{\delta}$ are 2×2 matrices whose entries are obvious, and multiplication of row and column vectors is defined as block multiplication.

Self-adjoint second-order linear differential operators. In Chapter 9 we showed that a SOLDO satisfies the generalized Green's identity with $w(x) = 1$. In fact, since u and v in Eq. (9.21b) are real, that equation is identical to (11.31) if we set $w = 1$ and

$$Q(u,v) = p_2 v u' - (p_2 v)' u + p_1 u v \qquad (11.37)$$

Also, we have seen that any SOLDO can be made (formally) self-adjoint. Thus, let us consider the formally self-adjoint SOLDO

$$\mathbb{L}_x = \mathbb{L}_x^\dagger = \frac{d}{dx}\left(p \frac{d}{dx}\right) + q$$

where both $p(x)$ and $q(x)$ are real functions and the inner product is defined with weight $w = 1$. If we are interested in formally self-adjoint operators with respect to a general weight $w > 0$, we can construct them as follows. We first note that if $\mathbb{L}_x$ is formally self-adjoint with respect to a weight of unity, then $(1/w)\,\mathbb{L}_x$ is self-adjoint with respect to weight w. Next, we note that $\mathbb{L}_x$ is formally self-adjoint for all functions q, in particular, wq. Now we define

$$\mathbb{L}_x^{(w)} \equiv \frac{d}{dx}\left(p\frac{d}{dx}\right) + qw$$

Thus, $\mathbb{L}_x^{(w)}$ is formally self-adjoint with respect to a weight of unity, and, therefore,

$$\mathbb{L}_x \equiv \frac{1}{w}\mathbb{L}_x^{(w)} = \frac{1}{w}\frac{d}{dx}\left(p\frac{d}{dx}\right) + q \tag{11.38a}$$

is formally self-adjoint with respect to weight $w(x) > 0$.

For SOLDOs that are formally self-adjoint with respect to weight w, the conjunct given in (11.37), with u and v real, reduces to

$$Q(u,v) = p(x)w(x)(vu' - uv') \tag{11.38b}$$

Thus, the surface term in the generalized Green's identity vanishes if and only if

$$p(b)w(b)[v(b)u'(b) - u(b)v'(b)] = p(a)w(a)[v(a)u'(a) - u(a)v'(a)] \tag{11.39}$$

The DO becomes self-adjoint (or hermitian) if u and v satisfy both Eq. (11.39) and *the same BCs*. It can easily be shown that the following four types of BCs on $u(x)$ assure the validity of (11.39) and, therefore, define a hermitian operator $\mathbb{L}_x$ given by (11.38):

(1) The *Dirichlet BCs*: $u(a) = u(b) = 0$

(2) The *Neumann BCs*: $u'(a) = u'(b) = 0$

(3) *General unmixed BCs*: $\alpha u(a) - u'(a) = \beta u(b) - u'(b) = 0$

(4) *Periodic BCs*: $u(a) = u(b)$ and $u'(a) = u'(b)$

11.4.3 Green's Functions for Second-Order Linear Differential Operators

We are now in a position to find the Green's function for a SOLDO. Recall that if $\mathbb{L}$ is a differential operator, its inverse $\mathbb{G}$, called the *Green's operator*, is defined by

$$\mathbb{L}\mathbb{G} = \mathbb{1}$$

As outlined at the beginning of this section, this equation leads to the DE

$$\mathbb{L}_x G(x,y) = \frac{\delta(x-y)}{w(x)} \tag{11.40}$$

where $G(x,y)$ is the Green's function for $\mathbb{L}_x$. It is clear that $\mathbb{G}$ is the formal inverse of $\mathbb{L}$. If $\mathbb{G}$ is to exist, $\mathbb{L}$ must be invertible. Let us look into the invertibility of $\mathbb{L}$.

First, note that a complete specification of $\mathbb{L}_x$ requires not only knowledge of $p_0(x)$, $p_1(x)$, and $p_2(x)$, its coefficient functions, but also knowledge of the BCs imposed on the solutions. The most general BCs for a SOLDO are of the type given in (11.30) with $n = 2$. Thus, to specify $\mathbb{L}_x$ uniquely, we consider the system $(\mathbb{L}; \mathbb{R}_1, \mathbb{R}_2)$ with data $(f; \gamma_1, \gamma_2)$. This system defines a unique BVP:

$$\mathbb{L}_x[u] \equiv p_2(x)\frac{d^2u}{dx^2} + p_1(x)\frac{du}{dx} + p_0(x)u = f(x) \tag{11.41a}$$

$$\mathbb{R}_i[u] = \gamma_i, \qquad \text{for } i = 1, 2 \tag{11.41b}$$

A necessary condition for $\mathbb{L}_x$ to be invertible is that the *homogeneous* DE $\mathbb{L}_x[u] = 0$ must have only the solution $u \equiv 0$. For $u \equiv 0$ to be the *only* solution, it must be *a* solution. This means that it must meet all the conditions in Eqs. (11.41). In particular, since $\mathbb{R}_i$ are linear functionals of u, we must have

$$\mathbb{R}_i[0] = 0$$

This can be stated as a lemma.

11.4.4 Lemma A necessary condition for the DO defined in (11.41a) to be invertible is that $\gamma_i = 0$; that is, the BCs of (11.41b) must be *homogeneous.* ∎

Thus, to study Green's functions we must restrict ourselves to problems with homogeneous BCs. This, at first, may seem restrictive, since not all problems have homogeneous BCs. Can we solve the others by the Green's function method? The answer is yes, as will be shown later in this section.

The above discussion clearly indicates that the Green's function of $\mathbb{L}_x$, being its "inverse," is defined only if we consider the system $(\mathbb{L}; \mathbb{R}_1, \mathbb{R}_2)$ with data $(f; 0, 0)$. If the Green's function exists, it must satisfy (11.40), in which $\mathbb{L}_x$ acts on $G(x,y)$. But part of the definition of $\mathbb{L}_x$ are the BCs imposed on the solutions. Thus, if the LHS of (11.40) is to make any sense, $G(x,y)$ must also satisfy those same BCs. We, therefore, have the following definition.

11.4.5 Definition The GF of a DO $\mathbb{L}_x$ is a function $G(x,y)$ that satisfies both the DE

$$\mathbb{L}_x G(x,y) = \frac{\delta(x-y)}{w(x)}$$

and, as a function of x, the homogeneous BCs

$$\mathbb{R}_i[G] = 0 \qquad \text{for } i = 1, 2$$

where the $\mathbb{R}_i$ are defined as in Eq. (11.30).

It is convenient to study the Green's function for the adjoint of $\mathbb{L}_x$ simultaneously. Denoting this by $g(x,y)$, we have

$$\mathbb{L}_x^\dagger g(x,y) = \frac{\delta(x-y)}{w(x)} \tag{11.42a}$$

$$\mathbb{B}_i[g] = 0 \qquad \text{for } i = 1, 2 \tag{11.42b}$$

where $\mathbb{B}_i$ are the boundary functionals adjoint to $\mathbb{R}_i$ and given in (11.34). The function $g(x,y)$ is known as the *adjoint Green's function* associated with the DE of (11.41a).

We can now use (11.41) and (11.42) to find the solutions to

$$\mathbb{L}_x[u] = f(x) \qquad \mathbb{R}_i[u] = 0 \quad \text{for } i = 1, 2 \tag{11.43a}$$

and

$$\mathbb{L}_x^\dagger[v] = h(x) \qquad \mathbb{B}_i[v^*] = 0 \quad \text{for } i = 1, 2 \tag{11.43b}$$

With $v(x) = g(x,y)$ in (11.31), whose RHS is assumed to be zero, we get

$$\int_a^b wg^*(x,y)\,\mathbb{L}_x[u]\,dx = \int_a^b wu(x)(\mathbb{L}_x^\dagger[g])^*\,dx$$

Using (11.42a) on the RHS and (11.43a) on the LHS, we obtain

$$u(y) = \int_a^b g^*(x,y)\,w(x)f(x)\,dx$$

Similarly, with $u(x) = G(x,y)$, Eq. (11.31) gives

$$v^*(y) = \int_a^b G(x,y)\,w(x)h^*(x)\,dx$$

or, since $w(x)$ is a (positive) real function,

$$v(y) = \int_a^b w(x)\,G^*(x,y)\,h(x)\,dx$$

These equations for $u(y)$ and $v(y)$ are not what we expect [see, for instance, Eq. (11.27)]. However, if we take into account certain properties of Green's functions that we will discuss next, these equations become plausible.

Properties of Green's functions. Let us rewrite Eq. (11.31), with the RHS equal to zero, as

$$\int_a^b dt\, w(t)\,\{v^*(t)(\mathbb{L}_t[u])\} = \int_a^b dt\, w(t)\,\{u(t)(\mathbb{L}_t^\dagger[v])^*\} \tag{11.44}$$

This is sometimes called *Green's identity*. Substituting $u(t) \equiv G(t,y)$ and $v(t) = g(t,x)$ in (11.44) gives

$$\int_a^b dt\, w(t) g^*(t,x) \frac{\delta(t-y)}{w(t)} = \int_a^b dt\, w(t) G(t,y) \frac{\delta(t-x)}{w(t)}$$

or

$$g^*(y,x) = G(x,y)$$

A consequence of this identity is that $G(x,y)$ must satisfy the adjoint BCs with respect to its *second* argument.

If, for the time being, we assume that the Green's function associated with a system $(\mathbb{L}; \mathbb{R}_1, \mathbb{R}_2)$ is unique, then, since for a hermitian differential operator $\mathbb{L}_x$ and $\mathbb{L}_x^\dagger$ are identical and u and v both satisfy the same BCs, we must have

$$G(x, y) = g(x, y)$$

or, using $g^*(y,x) = G(x,y)$, we get

$$G(x,y) = G^*(y,x)$$

In particular, if the coefficient functions of $\mathbb{L}_x$ are all real, $G(x, y)$ will be real, and we have

$$G(x,y) = G(y,x)$$

This means that G is a symmetric function of its two arguments.

The last property has to do with the continuity of $G(x,y)$ and its derivative at $x = y$. For a SOLDO, we have

$$\mathbb{L}_x G(x,y) = p_2(x) \frac{\partial^2 G}{\partial x^2} + p_1(x) \frac{\partial G}{\partial x} + p_0(x) G = \frac{\delta(x-y)}{w(x)}$$

where p_0, p_1, and p_2 are assumed to be real and continuous in the interval $\mathrm{I} = [a,b]$, and $w(x)$ and $p_2(x)$ are assumed to be greater than 0 for all $x \in \mathrm{I}$. We multiply both sides of the DE by

$$h(x) \equiv \frac{\mu(x)}{p_2(x)} \qquad \text{where } \mu(x) \equiv \exp\left[\int_a^x \frac{p_1(t)}{p_2(t)} dt\right]$$

noting that $d\mu/dx = (p_1/p_2)\mu$. This transforms the DE into

$$\frac{\partial}{\partial x}\left[\mu(x) \frac{\partial G(x,y)}{\partial x}\right] + \frac{p_0(x)\mu(x)}{p_2(x)} G(x,y) = \frac{\mu(y)}{p_2(y) w(y)} \delta(x-y)$$

Integrating this equation gives

$$\mu(x) \frac{\partial G(x,y)}{\partial x} + \int_a^x \frac{p_0(t)\mu(t)}{p_2(t)} G(t,y)\, dt = \frac{\mu(y)}{p_2(y) w(y)} \theta(x-y) + \text{constant} \qquad (11.45)$$

because the primitive of $\delta(x-y)$ is $\theta(x-y)$ (see Chapter 5).

Consider the case where $p_0 = 0$, for which the Green's function can be denoted by $G_0(x,y)$. Then Eq. (11.45) becomes

$$\mu(x)\frac{\partial G_0}{\partial x} = \frac{\mu(y)}{p_2(y)w(y)}\theta(x-y) + C_1$$

which, since μ, p_2, and w are continuous on I and $\theta(x-y)$ has a discontinuity only at $x = y$, indicates that $\partial G_0/\partial x$ is continuous everywhere on I except at $x = y$. However, $G_0(x,y)$ itself is continuous at $x = y$, as shown in Exercise 11.4.6.

Next, we write

$$G(x,y) = G_0(x,y) + H(x,y)$$

and apply $\mathbb{L}_x$ to both sides. This gives

$$\frac{\delta(x-y)}{w(x)} = \left(p_2\frac{d^2}{dx^2} + p_1\frac{d}{dx}\right)G_0 + p_0G_0 + \mathbb{L}_xH(x,y)$$

$$= \frac{\delta(x-y)}{w(x)} + p_0G_0 + \mathbb{L}_xH(x,y)$$

or

$$p_2\frac{d^2H}{dx^2} + p_1\frac{dH}{dx} + p_0H = -p_0G_0$$

The continuity of G_0, p_0, p_1, and p_2 on I implies the continuity of H, because a discontinuity in H implies a delta function discontinuity in dH/dx, which is impossible because there are no delta functions in the equation for H. Since both G_0 and H are continuous, G must also be continuous on I.

We can now calculate the jump in $\partial G/\partial x$ at $x = y$. We denote the jump as $\Delta G'(y)$ and define it as follows:

$$\Delta G'(y) \equiv \lim_{\varepsilon\to 0}\left[\left.\frac{\partial G(x,y)}{\partial x}\right|_{x=y+\varepsilon} - \left.\frac{\partial G(x,y)}{\partial x}\right|_{x=y-\varepsilon}\right]$$

Dividing (11.45) by $\mu(x)$ and taking the above limit for all terms, we obtain

$$\Delta G'(y) + \lim_{\varepsilon\to 0}\left[\frac{1}{\mu(y+\varepsilon)}\int_a^{y+\varepsilon}\frac{p_0(t)\mu(t)}{p_2(t)}G(t,y)\,dt - \frac{1}{\mu(y-\varepsilon)}\int_a^{y-\varepsilon}\frac{p_0(t)\mu(t)}{p_2(t)}G(t,y)dt\right]$$

$$= \frac{\mu(y)}{p_2(y)w(y)}\lim_{\varepsilon\to 0}\left[\frac{\theta(+\varepsilon)}{\mu(y+\varepsilon)} - \frac{\theta(-\varepsilon)}{\mu(y-\varepsilon)}\right] + 0$$

The second term on the LHS is zero because all functions are continuous at y. The limit on the RHS is simply $1/\mu(y)$. We, therefore, obtain

$$\Delta G'(y) = \frac{1}{p_2(y)w(y)} \tag{11.46}$$

Construction and uniqueness of Green's functions. We are now in a position to calculate the Green's function for a general SOLDO and show that it is unique.

11.4.6 Theorem Consider the system $(\mathbb{L}; \mathbb{R}_1, \mathbb{R}_2)$ with data $(f(x); 0, 0)$, where $\mathbb{L}_x$ is a SOLDO. If the homogeneous DE $\mathbb{L}_x[u] = 0$ has no nontrivial solution, then the GF associated with the given system exists and is unique. The solution of the system is

$$u(x) = \int_a^b dy\, w(y)\, G(x,y) f(y)$$

and is also unique.

Proof. The GF satisfies the DE

$$\mathbb{L}_x G(x,y) = 0$$

for all $x \in [a,b]$ except $x = y$. We thus divide $[a,b]$ into two intervals, $I_1 = [a,y)$ and $I_2 = (y,b]$, and note that a general solution to the above homogeneous DE can be written as a linear combination of a basis of solutions, for example, u_1 and u_2. Thus, we can write the solution of the DE as

$$\begin{aligned} G_l(x,y) &\equiv c_1 u_1(x) + c_2 u_2(x) \qquad \text{for } x \in I_1 \\ G_r(x,y) &\equiv d_1 u_1(x) + d_2 u_2(x) \qquad \text{for } x \in I_2 \\ G(x,y) &= \begin{cases} G_l(x,y) & \text{if } x \in I_1 \\ G_r(x,y) & \text{if } x \in I_2 \end{cases} \end{aligned} \tag{11.47}$$

where c_1, c_2, d_1, and d_2 are, in general, functions of y. To determine $G(x,y)$ we must determine four unknowns. We also have four relations: the continuity of G, the jump of $\partial G/\partial x$ at $x = y$, and the two BCs $\mathbb{R}_1[G] = \mathbb{R}_2[G] = 0$.

The continuity of G gives

$$c_1(y)u_1(y) + c_2(y)u_2(y) = d_1(y)u_1(y) + d_2(y)u_2(y)$$

The jump of $\partial G/\partial x$ at $x = y$ yields

$$c_1(y)u_1'(y) + c_2(y)u_2'(y) - d_1(y)u_1'(y) - d_2(y)u_2'(y) = -\frac{1}{p_2(y)w(y)}$$

Introducing $b_1 \equiv c_1 - d_1$ and $b_2 \equiv c_2 - d_2$ changes the two preceding equations to

$$b_1 u_1 + b_2 u_2 = 0$$

and
$$b_1 u_1' + b_2 u_2' = -\frac{1}{p_2 w}$$

These equations have a unique solution iff

$$\det\begin{pmatrix} u_1 & u_2 \\ u_1' & u_2' \end{pmatrix} \neq 0$$

But the determinant is simply the Wronskian of the two *independent* solutions u_1 and u_2 and, therefore, cannot be zero. Thus, $b_1(y)$ and $b_2(y)$ are determined in terms of u_1, u_1', u_2, u_2', p_2, and w.

We now define

$$h(x,y) \equiv \begin{cases} b_1(y)u_1(x) + b_2(y)u_2(x) & \text{if } x \in \mathrm{I}_1 \\ 0 & \text{if } x \in \mathrm{I}_2 \end{cases}$$

so that

$$G(x,y) = h(x,y) + d_1(y)u_1(x) + d_2(y)u_2(x)$$

We have reduced the number of unknowns to two, d_1 and d_2.

Imposing the BCs gives two more relations:

$$\mathbb{R}_1[G] = \mathbb{R}_1[h] + d_1\mathbb{R}_1[u_1] + d_2\mathbb{R}_1[u_2] = 0$$

$$\mathbb{R}_2[G] = \mathbb{R}_2[h] + d_1\mathbb{R}_2[u_1] + d_2\mathbb{R}_2[u_2] = 0$$

Can we solve these equations and determine d_1 and d_2 uniquely? We can, if

$$\det\begin{pmatrix} \mathbb{R}_1[u_1] & \mathbb{R}_1[u_2] \\ \mathbb{R}_2[u_1] & \mathbb{R}_2[u_2] \end{pmatrix} \neq 0$$

It can be shown that this determinant is nonzero (see Exercise 11.4.7).

Having found the unique $\{b_i,d_i\}_{i=1}^2$, we can calculate c_i uniquely, substitute all of them in (11.47), and obtain the *unique* $G(x,y)$. That $u(x)$ is also unique can be shown similarly. ■

Example 11.4.9

Let us calculate the GF for $\mathbb{L}_x = d^2/dx^2$ with BCs $u(a) = u(b) = 0$. We note that $\mathbb{L}_x[u] = 0$ with the given BCs has no nontrivial solution (verify this). Thus, the GF exists. The DE for $G(x,y)$ is

$$\frac{d^2G}{dx^2} = 0 \qquad \text{for } x \neq y$$

whose solutions are

$$G(x,y) = \begin{cases} c_1x + c_2 & \text{for } a \leqslant x < y \\ d_1x + d_2 & \text{for } y < x \leqslant b \end{cases} \tag{1}$$

Continuity at $x = y$ gives

$$c_1y + c_2 = d_1y + d_2$$

or, defining $b_i \equiv c_i - d_i$,

$$b_1y + b_2 = 0 \tag{2}$$

The discontinuity of dG/dx at $x = y$ gives

$$d_1 - c_1 = \frac{1}{p_2w} = 1 \tag{3}$$

assuming that $w = 1$. Equations (2) and (3) give

$$b_1 = -1 \qquad \text{and} \qquad b_2 = y$$

On the other hand, $G(x,y)$ must satisfy the given BCs. Thus, $G(a,y) = 0 = G(b,y)$. Since $a \leqslant y$ and $b \geqslant y$, we obtain

$$c_1 a + c_2 = 0$$

$$d_1 b + d_2 = 0$$

or, substituting $c_i = b_i + d_i$,

$$ad_1 + d_2 = a - y$$

$$bd_1 + d_2 = 0$$

The solution to these equations is

$$d_1 = \frac{y-a}{b-a} \qquad \text{and} \qquad d_2 = -\frac{b(y-a)}{b-a}$$

With b_1, b_2, d_1, and d_2 as given above, we find

$$c_1 = b_1 + d_1 = -\frac{b-y}{b-a}$$

$$c_2 = b_2 + d_2 = a\frac{b-y}{b-a}$$

Writing (1) as

$$G(x,y) = (c_1 x + c_2)\theta(y-x) + (d_1 x + d_2)\theta(x-y)$$

and using the identity $\theta(y-x) = 1 - \theta(x-y)$, we get

$$G(x,y) = c_1 x + c_2 - (c_1 x + c_2)\theta(x-y) + (d_1 x + d_2)\theta(x-y)$$

$$= c_1 x + c_2 - (b_1 x + b_2)\theta(x-y)$$

Using the values found for the b's and c's, we obtain

$$G(x,y) = (a-x)\left(\frac{b-y}{b-a}\right) + (x-y)\theta(x-y)$$

which is the same as the GF obtained in Example 11.4.4. ●

Example 11.4.10

Let us find the GF for $\mathbb{L}_x = d^2/dx^2 + 1$ with the BCs $u(0) = u(\pi/2) = 0$. The general solution of $\mathbb{L}_x[u] = 0$ is

$$u(x) = A \sin x + B \cos x$$

If the BCs are imposed, we get $u \equiv 0$. Thus, $G(x,y)$ exists. The general form of $G(x,y)$ is

$$G(x,y) = \begin{cases} c_1 \sin x + c_2 \cos x & \text{for } 0 \leqslant x < y \\ d_1 \sin x + d_2 \cos x & \text{for } y < x \leqslant \pi/2 \end{cases} \tag{1}$$

The continuity of G at $x = y$ gives

$$b_1 \sin y + b_2 \cos y = 0 \tag{2}$$

with $b_i \equiv c_i - d_i$. The discontinuity of the derivative of G at $x = y$ gives

$$b_1 \cos y - b_2 \sin y = -1 \tag{3}$$

where we have set $w(x) = 1$. Solving (2) and (3) yields

$$b_1 = -\cos y \qquad \text{and} \qquad b_2 = \sin y$$

The BCs give

$$G(0,y) = 0 \quad \Rightarrow \quad c_2 = 0 \quad \Rightarrow \quad d_2 = -b_2 = -\sin y$$

$$G(\pi/2,y) = 0 \quad \Rightarrow \quad d_1 = 0 \quad \Rightarrow \quad c_1 = b_1 = -\cos y$$

Substituting in (1) gives

$$G(x,y) = \begin{cases} -\cos y \ \sin x & x < y \\ -\sin y \ \cos x & x > y \end{cases}$$

or, using the theta function,

$$\begin{aligned} G(x,y) &= (-\cos y \sin x)\theta(y - x) - (\sin y \cos x)\theta(x - y) \\ &= -(\cos y \sin x)[1 - \theta(x - y)] - (\sin y \cos x)\theta(x - y) \\ &= -\cos y \sin x + \theta(x - y)\sin(x - y) \end{aligned} \tag{4}$$

It is instructive to verify directly that Eq. (4) satisfies $\mathbb{L}_x[G] = \delta(x - y)$:

$$\begin{aligned} \mathbb{L}_x[G] &= -\cos y \, \mathbb{L}_x[\sin x] + \mathbb{L}_x[\theta(x - y)\sin(x - y)] \\ &= 0 + \frac{d^2}{dx^2}[\theta(x - y)\sin(x - y)] + \theta(x - y)\sin(x - y) \\ &= \frac{d}{dx}[\delta(x - y)\sin(x - y) + \theta(x - y)\cos(x - y)] + \theta(x - y)\sin(x - y) \end{aligned}$$

The first term vanishes because the sine vanishes at the only point where the delta function is nonzero (see Chapter 5). Thus, we have

$$\mathbb{L}_x[G] = [\delta(x - y)\cos(x - y) - \theta(x - y)\sin(x - y)] + \theta(x - y)\sin(x - y) = \delta(x - y)$$

because the delta function demands that $x = y$, at which point $\cos(x - y) = 1$. ●

The existence and uniqueness of the Green's function, $G(x,y)$, in conjunction with its properties and its adjoint, imply the existence and uniqueness of the adjoint Green's function, $g(x,y)$. Using this fact, we can show that the condition for the nonexistence of a nontrivial solution for $\mathbb{L}_x[u] = 0$ is also a necessary condition for the existence of $G(x,y)$. That is, if $G(x,y)$ exists, then $\mathbb{L}_x[u] = 0$ implies that $u = 0$.

Suppose $G(x,y)$ exists; then $g(x,y)$ also exists. In Green's identity, let $v = g(x,y)$. This gives an identity:

$$\int_a^b w(x)\, g^*(x,y)\, (\mathbb{L}_x[u])dx = \int_a^b w(x)u(x)\, (\mathbb{L}_x^\dagger[g])^* dx = \int_a^b w(x)u(x)\frac{\delta(x - y)}{w(x)}dx = u(y)$$

In particular, if $\mathbb{L}_x[u] = 0$, then $u(y) = 0$ for all y, which implies that $u \equiv 0$. We have proved the following proposition.

11.4.7 Proposition The DE $\mathbb{L}_x[u] = 0$ implies that $u \equiv 0$ iff the GF corresponding to $\mathbb{L}_x$ and the homogeneous BCs exist. ■

It is sometimes stated that the Green's function of a SOLDO with *constant* coefficients depends on the difference $x - y$. This statement is motivated by the observation that if $u(x)$ is a solution of

$$\mathbb{L}_x[u] \equiv a_2 \frac{d^2u}{dx^2} + a_1 \frac{du}{dx} + a_0 u = f(x)$$

then $u(x - y)$ is the solution of

$$a_2 \frac{d^2u}{dx^2} + a_1 \frac{du}{dx} + a_0 u = f(x - y)$$

if a_0, a_1, and a_2 are constant. Thus, if $G(x)$ is a solution of $\mathbb{L}_x[G] = \delta(x)$ [again assuming that $w(x) = 1$], then it seems that the solution of $\mathbb{L}_x[G] = \delta(x - y)$ is simply $G(x - y)$. *This is clearly wrong*, as Examples 11.4.9 and 11.4.10 show. The reason is, of course, the BCs. The fact that $G(x - y)$ satisfies the right DE does not guarantee that it also satisfies the right BCs. Exercise 11.4.8, however, shows that the conjecture is true for a homogeneous IVP.

Inhomogeneous boundary conditions. So far we have concentrated on problems with homogeneous BCs, $\mathbb{R}_i[u] = 0$, for $i = 1, 2$. What if the BCs are inhomogeneous? It turns out that the Green's function method, even though it was derived for homogeneous BCs, solves this kind of problem as well! In fact, we have seen an illustration of this in Example 11.4.6. The secret of this success is the generalized Green's identity.

Suppose we are interested in solving the DE

$$\mathbb{L}_x[u] = f(x)$$

with

$$\mathbb{R}_i[u] = \gamma_i \qquad \text{for } i = 1, 2$$

and we have the GF for $\mathbb{L}_x$ (with homogeneous BCs, of course). We can substitute $v = g(x,y) = G^*(y,x)$ in the generalized Green's identity and use the DE to obtain

$$\int_a^b w(x) G(y,x) f(x)\, dx - \int_a^b w(x) u(x) (\mathbb{L}_x^\dagger[g(x,y)])^* \, dx = Q(u, G(y,x))\big|_{x=a}^{x=b}$$

or, using $\mathbb{L}_x^\dagger[g(x,y)] = \delta(x - y)/w(y)$,

$$u(y) = \int_a^b w(x) G(y,x) f(x)\, dx - Q(u, G(y,x))\big|_{x=a}^{x=b}$$

Let us evaluate the surface term. We can solve the equations for $\mathbb{R}_i$ for two of the four quantities $u(a)$, $u'(a)$, $u(b)$, and $u'(b)$ in terms of the other two. Let us assume that $\mathbf{U}_b$ is found in terms of $\mathbf{U}_a$ [see Eq. (1) of Exercise 11.4.5]. Thus, solving for $u(b)$ and $u'(b)$, we obtain, very generally

$$\mathbf{U}_b = \mathsf{\Gamma}\mathbf{U}_a + \mathsf{M}\boldsymbol{\gamma}$$

where $\mathsf{\Gamma}$ and M are matrices obtained when solving $\mathbf{U}_b \equiv \begin{pmatrix} u(b) \\ u'(b) \end{pmatrix}$ in terms of $\mathbf{U}_a$. The first term on the RHS is the same as is obtained for the homogeneous case because exactly the same row operations are performed on the LHS of the equations for the BCs even when $\gamma_1 = \gamma_2 = 0$. The second term is present because of the corresponding operations on the γ_i's. A result obtained in Exercise 11.4.5 is

$$Q(u, v^*)|_a^b = \tilde{\mathbf{U}}_b\mathsf{A}(b)\mathbf{V}_b^* - \tilde{\mathbf{U}}_a\mathsf{A}(a)\mathbf{V}_a^*$$

Substituting for $\mathbf{U}_b$ in this equation yields [using part (a) of Exercise 11.4.5 in the last step]

$$\begin{aligned} Q(u,v^*)|_a^b &= (\tilde{\mathbf{U}}_a\tilde{\mathsf{\Gamma}} + \tilde{\boldsymbol{\gamma}}\tilde{\mathsf{M}})\mathsf{A}(b)\mathbf{V}_b^* - \tilde{\mathbf{U}}_a\mathsf{A}(a)\mathbf{V}_a^* \\ &= \tilde{\mathbf{U}}_a\{\tilde{\mathsf{\Gamma}}\mathsf{A}(b)\mathbf{V}_b^* - \mathsf{A}(a)\mathbf{V}_a^*\} + \tilde{\boldsymbol{\gamma}}\tilde{\mathsf{M}}\mathsf{A}(b)\mathbf{V}_b^* \\ &= \tilde{\boldsymbol{\gamma}}\tilde{\mathsf{M}}\mathsf{A}(b)\mathbf{V}_b^* \end{aligned} \tag{11.48}$$

Recall that

$$\mathbf{V}_b^* \equiv \begin{pmatrix} v^*(b) \\ v'^*(b) \end{pmatrix} = \begin{pmatrix} g^*(b,y) \\ \dfrac{\partial}{\partial x}g^*(x,y)|_{x=b} \end{pmatrix} = \begin{pmatrix} G(y,b) \\ \dfrac{\partial}{\partial x}G(y,x)|_{x=b} \end{pmatrix}$$

That is, $Q(u,v^*)|_a^b$ is given entirely in terms of G and its derivative and the constants γ_1 and γ_2. The fact that G and $\partial G/\partial x$ appear to be evaluated at $x = b$ is due to the simplifying (but harmless!) assumption used in Exercise 11.4.5, where the choice is made to write $u(b)$ and $u'(b)$ in terms of $u(a)$ and $u'(a)$. Of course, this may not be possible; then we have to find another pair of the four quantities in terms of the other two, and any two of the four quantities $G(y,a)$, $(\partial G(y,x)/\partial x)|_{x=a}$, $G(y,b)$, and $(\partial G(y,x)/\partial x)|_{x=b}$ may appear as components of $\mathbf{V}_b^*$. In any case Eq. (11.48) is still valid; however, $\mathbf{V}_b^*$ involves G and $\partial G/\partial x$ evaluated at either a or b. We can now write

$$u(y) = \int_a^b w(x)G(y,x)f(x)dx - \tilde{\boldsymbol{\gamma}}\,\tilde{\mathsf{M}}\mathsf{A}\mathbf{V}^* \tag{11.49}$$

where the subscript b has been removed from $\mathbf{V}_b^*$. This equation shows that u can be determined completely once we know $G(x,y)$, even though the BCs are *inhomogeneous*.

Equation (11.49) emphasizes the fact that $u(y)$ depends on γ_1, γ_2, G, and $\partial G/\partial x$. In practice, there is no need to calculate $\tilde{\mathsf{M}}$ and A to insert in this equation. We can use the expression for $Q(u,v^*)$ obtained from the Lagrange identity of Chapter 9 and evaluate it at b and a. This, in general, involves evaluating u and G and their derivatives at a and b. We know how to handle the evaluation of G because we can actually construct it (if it exists). But what about the evaluation of u? Two of the four quantities corresponding to u can be evaluated in terms of the other two and inserted in the expression for $Q(u,v^*)|_a^b$. Equation (11.48) then guarantees that the coefficients of the other two terms will be zero. Thus, we can simply drop all the terms in $Q(u,v^*)|_a^b$ containing a factor of the other two terms.

Specifically, we use the conjunct for a formally self-adjoint SOLDO [see Eq. (11.38b)] to obtain

$$u(y) = \int_a^b w(x)G(y,x)\,f(x)dx - \left\{p(x)w(x)\left[G(y,x)\frac{du}{dx} - u(x)\frac{\partial G}{\partial x}(y,x)\right]\right\}_{x=a}^{x=b}$$

Interchanging x and y gives

$$u(x) = \int_a^b w(y)G(x,y)\,f(y)dy + \left\{p(y)w(y)\left[u(y)\frac{\partial G(x,y)}{\partial y} - G(x,y)\frac{du}{dy}\right]\right\}_{y=a}^{y=b} \tag{11.50}$$

This equation is only valid for a self-adjoint SOLDO. That is, using it requires casting the SOLDO into a self-adjoint form (a process that is always possible, in light of Theorem 9.2.15).

By setting $f(x) = 0$, we can also obtain the solution to the homogeneous DE $\mathsf{L}_x[u] = 0$ that satisfies the inhomogeneous BCs.

Example 11.4.11

Let us find the solution of $d^2u/dx^2 = f(x)$ subject to the BCs $u(a) = \gamma_1$ and $u(b) = \gamma_2$. The GF for this problem has been calculated in Examples 11.4.5 and 11.4.9. Let us begin by calculating the surface term in (11.50). We have $p(y) = 1$, and we set $w(y) = 1$, then the surface term becomes

$$\begin{aligned}\text{surface term} &= u(b)\frac{\partial G}{\partial y}\bigg|_{y=b} - G(x,b)u'(b) - u(a)\frac{\partial G}{\partial y}\bigg|_{y=a} + G(x,a)u'(a)\\ &= \gamma_2\frac{\partial G}{\partial y}\bigg|_{y=b} - \gamma_1\frac{\partial G}{\partial y}\bigg|_{y=a} + G(x,a)u'(a) - G(x,b)u'(b)\end{aligned}$$

That the unwanted terms are zero can be seen by observing that $G(x,a) = g^*(a,x) = (g(a,x))^*$, but $g(a,x) = 0$ because the adjoint GF, $g(x,y)$, satisfies the BCs adjoint to the homogeneous BCs (obtained when $\gamma_i = 0$). In this particular and simple case, the BCs happen to be self-adjoint (Dirichlet BCs). Thus, $u(a) = u(b) = 0$ implies that $g(a,x) = g(b,x) = 0$ for all $x \in [a,b]$. (In a more complicated case the coefficient of $u'(a)$ would be more complicated, but still zero.) Thus, we finally have

$$\text{surface term} = \gamma_2\frac{\partial G}{\partial y}\bigg|_{y=b} - \gamma_1\frac{\partial G}{\partial y}\bigg|_{y=a}$$

Now, using the expression for $G(x,y)$ obtained in Examples 11.4.5 and 11.4.9, we get

$$\frac{\partial G}{\partial y} = -\frac{a-x}{b-a} - \theta(x-y) - \underbrace{(x-y)\delta(x-y)}_{\equiv 0}$$

$$= \frac{x-a}{b-a} - \theta(x-y)$$

Thus,

$$\left.\frac{\partial G}{\partial y}\right|_{y=b} = \frac{x-a}{b-a}$$

$$\left.\frac{\partial G}{\partial y}\right|_{y=a} = \frac{x-a}{b-a} - 1 = \frac{x-b}{b-a}$$

Substituting in (11.50), we get

$$u(x) = \int_a^b G(x,y)\, f(y) + \frac{\gamma_2 - \gamma_1}{b-a}x + \frac{b\gamma_1 - a\gamma_2}{b-a}$$

(Compare this with the result obtained in Example 11.4.5.) •

Green's functions have a very simple and enlightening physical interpretation. An inhomogeneous DE such as $\mathbb{L}_x[u] = f(x)$ can be interpreted as a black box ($\mathbb{L}_x$) that determines a physical quantity (u) when there is a source (f) producing the physical quantity. For instance, electrostatic potential is a physical quantity whose source is charge, a magnetic field has an electric current as its source, displacements and velocities have forces as their sources, and so forth.

Applying this interpretation and assuming that $w(x) = 1$, we have $G(x,y)$ as the physical quantity, evaluated at x when its source $\delta(x-y)$ is located at y. To be more precise, let us say that the strength of the source is S_1 and it is located at y_1; then the source becomes $S_1\delta(x-y_1)$. The physical quantity, the Green's function, is then $S_1G(x,y_1)$, because of the linearity of $\mathbb{L}_x$. If $G(x,y)$ is a solution of $\mathbb{L}_x u = \delta(x-y)$, then $S_1G(x,y_1)$ is a solution of $\mathbb{L}_x u = S_1\delta(x-y_1)$. If there are many sources located at $\{y_i\}_{i=1}^N$ with corresponding strengths $\{S_i\}_{i=1}^N$, then the source f as a function of x becomes

$$f(x) = \sum_{i=1}^N S_i\delta(x-y_i)$$

and the corresponding physical quantity $u(x)$ becomes

$$u(x) = \sum_{i=1}^N S_iG(x,y_i)$$

Since the source S_i is located at y_i, it is more natural to define a function $S(x)$ and write $S_i = S(y_i)$. When the number of point sources goes to infinity and y_i becomes a smooth continuous variable, the sums become integrals, and we have

$$f(x) = \int_a^b S(y)\delta(x - y)dy$$

$$u(x) = \int_a^b S(y)G(x,y)dy$$

The first integral shows that $S(x) \equiv f(x)$. Thus, the second integral becomes

$$u(x) = \int_a^b f(y)G(x,y)dy$$

which is precisely what we obtained formally.

Exercises

11.4.1 Using the GF method, solve the DE $\mathbb{L}_x u(x) \equiv du/dx = f(x)$ subject to the BC $u(0) = a$.

11.4.2 Solve the problem of Example 11.4.4 subject to the BCs $u(a) = u'(a) = 0$. Show that the corresponding GF also satisfies these BCs.

11.4.3 Show that the IVP with data $\{0; 0, 0, \ldots, 0\}$ has only $u \equiv 0$ as a solution.

11.4.4 The DO

$$\mathbb{L}_x^{(n)} \equiv \sum_{k=0}^{n} p_k(x)\frac{d^k}{dx^k}$$

is said to be *exact* if there exists a DO

$$\mathbb{M}_x^{(n-1)} \equiv \sum_{k=0}^{n-1} a_k(x)\frac{d^k}{dx^k}$$

such that

$$\mathbb{L}_x^{(n)}[u] = \frac{d}{dx}(\mathbb{M}_x^{(n-1)}[u]) \qquad \forall\, u \in \mathrm{C}^{(n)}(a,b)$$

(a) Show that $\mathbb{L}_x^{(n)}$ is exact iff

$$\sum_{m=0}^{n} (-1)^m \frac{d^m p_m}{dx^m} = 0$$

(b) Show that there exists an integrating factor for $\mathbb{L}_x^{(n)}$, that is, a function $\mu(x)$ such that $\mu(x)\mathbb{L}_x^{(n)}$ is exact iff $\mu(x)$ satisfies the DE

$$\mathbb{N}_x^{(n)}[\mu] \equiv \sum_{m=0}^{n} (-1)^m \frac{d^m}{dx^m}(\mu p_m) = 0$$

The DO $\mathbb{N}_x^{(n)}$ is the formal adjoint of $\mathbb{L}_x^{(n)}$.

Note that this exercise generalizes the ideas discussed in Section 9.2.3.

11.4.5 Investigate the restrictions that must be imposed on v so that the surface term vanishes when Eq. (11.32) holds. Since the two row vectors $(\alpha_{11}, \ldots, \beta_{12})$ and $(\alpha_{21}, \ldots, \beta_{22})$ are linearly independent, two of the quantities $u(a)$, $u(b)$, $u'(a)$, and $u'(b)$ can be written in terms of the other two. To simplify the discussion, assume that $u(b)$ and $u'(b)$ are linear combinations of $u(a)$ and u'(a).

(a) Write the most general form of $Q(u, v^*)$. Defining the column vectors

$$\mathbf{U}_{x_0} \equiv \begin{pmatrix} u(x_0) \\ u'(x_0) \end{pmatrix} \quad \text{and} \quad \mathbf{V}^*_{x_0} \equiv \begin{pmatrix} v^*(x_0) \\ v'^*(x_0) \end{pmatrix} \tag{1}$$

show that the surface term gives

$$\tilde{\mathbf{U}}_b \mathsf{A}(b) \mathbf{V}^*_b = \tilde{\mathbf{U}}_a \mathsf{A}(a) \mathbf{V}^*_a$$

where $\mathsf{A}(x)$ is a 2×2 matrix function of x.

(b) Define the 2×2 matrices

$$\boldsymbol{\alpha} \equiv \begin{pmatrix} \alpha_{11} & \alpha_{12} \\ \alpha_{21} & \alpha_{22} \end{pmatrix} \quad \text{and} \quad \boldsymbol{\beta} = \begin{pmatrix} \beta_{11} & \beta_{12} \\ \beta_{21} & \beta_{22} \end{pmatrix}$$

and show that (11.32) can be expressed as

$$\boldsymbol{\alpha} \mathbf{U}_a = -\boldsymbol{\beta} \mathbf{U}_b \tag{2}$$

(c) Show that v^* must also satisfy a condition similar to (2) [or (11.32)].

11.4.6 Show that $G_0(x, y)$, the GF for the operator

$$p_2(x)\frac{d^2}{dx^2} + p_1(x)\frac{d}{dx}$$

is continuous at $x = y$.

11.4.7 Assuming that $\mathbb{L}_x[u] = 0$ has no nontrivial solution, show that the matrix

$$\mathsf{R} \equiv \begin{pmatrix} \mathbb{R}_1[u_1] & \mathbb{R}_1[u_2] \\ \mathbb{R}_2[u_1] & \mathbb{R}_2[u_2] \end{pmatrix} \tag{1}$$

where u_1 and u_2 are independent solutions of $\mathbb{L}_x[u] = 0$ and $\mathbb{R}_i$ are the boundary functionals, has a nonzero determinant. (Hint: Consider the system of homogeneous linear equations $\alpha\mathbb{R}_1[u_1] + \beta\mathbb{R}_1[u_2] = 0$ and $\alpha\mathbb{R}_2[u_1] + \beta\mathbb{R}_2[u_2] = 0$.)

11.4.8 Let $\mathbb{L}_x$ be a SOLDO with constant coefficients and the BCs $u(a) = u'(a) = 0$. Find the GF and show that it is a function of $x - y$.

11.5 EIGENFUNCTION EXPANSION OF GREEN'S FUNCTIONS

We have seen that Green's functions, taken in a broad sense, are inverses of differential operators. Inverses of operators in a Hilbert space are best studied in terms of resolvents. This is because if an operator $\mathbb{A}$ has an inverse, zero is not one of its eigenvalues, and

$$\mathbb{R}_0(\mathbb{A}) \equiv \mathbb{R}_\lambda(\mathbb{A})|_{\lambda=0} \equiv (\mathbb{A} - \lambda\mathbb{1})^{-1}|_{\lambda=0} = \mathbb{A}^{-1}$$

Thus, it is instructive to discuss the resolvent of a differential operator. We will consider only the case where the eigenvalues are discrete, for example, when $\mathbb{L}_x$ is a Sturm-Liouville operator.

Formally, we have

$$(\mathbb{L} - \lambda \mathbb{1})\mathbb{R}_\lambda(\mathbb{L}) \equiv \mathbb{1}$$

This leads to the DE

$$(\mathbb{L}_x - \lambda)\,\mathrm{R}_\lambda(x,y) = \frac{\delta(x-y)}{w(x)}$$

where $\mathrm{R}_\lambda(x,y) \equiv \langle \mathrm{x}|\mathbb{R}_\lambda(\mathbb{L})|\mathrm{y}\rangle$. The DE simply says that $\mathrm{R}_\lambda(x,y)$ is the Green's function for the operator $\mathbb{L}_x - \lambda$. So we can rewrite the equation as

$$(\mathbb{L}_x - \lambda)G_\lambda(x,y) = \frac{\delta(x-y)}{w(x)}$$

where $\mathbb{L}_x - \lambda$ is a DO having some homogeneous BCs. The GF $G_\lambda(x,y)$ exists if and only if $(\mathbb{L}_x - \lambda)[u] = 0$ has no nontrivial solution, which is true only if $\mathbb{L}_x[u] = \lambda u$ has no nontrivial solution, which in turn is true only if λ is not an eigenvalue of $\mathbb{L}_x$. We choose the BCs in such a way that $\mathbb{L}_x$ becomes hermitian.

If $\mathbb{L}_x$ is a hermitian SOLDO, then the system $\mathbb{L}_x[u] = \lambda u$, $\{\mathbb{R}_i[u] = 0\}_{i=1}^2$ can be considered an S-L system. Let $\{\lambda_n\}_{n=0}^\infty$ be the eigenvalues of $\mathbb{L}_x$ and $u_n^{(k)}(x)$ the corresponding eigenfunctions. The index k distinguishes the linearly independent vectors corresponding to the same eigenvalue λ_n. These eigenfunctions form a complete set for the subspace of the Hilbert space that consists of those functions that satisfy the same BCs as the $u_n^{(k)}(x)$ do. In particular, $G_\lambda(x,y)$ can be expanded in terms of $u_n^{(k)}(x)$. The expansion coefficients are, of course, functions of y. Thus, we can write

$$G_\lambda(x,y) = \sum_k \sum_{n=0}^\infty a_n^{(k)}(y)u_n^{(k)}(x)$$

where

$$a_n^{(k)}(y) \equiv \int_a^b w(x)u_n^{*(k)}(x)G_\lambda(x,y)dx$$

Using Green's identity, Eq. (11.44), and the fact that λ_n is real, we have

$$\lambda_n a_n^{(k)}(y) = \int_a^b w(x)[\lambda_n u_n^{(k)}(x)]^* G_\lambda(x,y)dx$$

$$= \int_a^b w(x)G_\lambda(x,y)\{\mathbb{L}_x[u_n^{(k)}(x)]\}^* dx$$

$$= \int_a^b w(x)[u_n^{(k)}(x)]^* \mathbb{L}_x[G_\lambda(x,y)]dx$$

↑ by Green's identity

$$= \int_a^b w(x)\,[u_n^{(k)}(x)]^* \left[\frac{\delta(x-y)}{w(x)} + \lambda G_\lambda(x,y)\right] dx$$

$$= [u_n^{(k)}(y)]^* + \lambda \int_a^b w(x) u_n^{*(k)}(x) G_\lambda(x,y)\, dx$$

$$= u_n^{*(k)}(y) + \lambda a_n^{(k)}(y)$$

Thus,

$$a_n^{(k)}(y) = \frac{u_n^{*(k)}(y)}{\lambda_n - \lambda}$$

and the expansion for the Green's function is

$$G_\lambda(x,y) = \sum_k \sum_{n=0}^{\infty} \frac{u_n^{*(k)}(y) u_n^{(k)}(x)}{\lambda_n - \lambda} \tag{11.51}$$

This expansion is valid as long as $\lambda \neq \lambda_n$ for any $n = 0, 1, 2, \ldots$. But this is precisely the condition that assures the existence of an inverse for $\mathbb{L} - \lambda \mathbb{1}$.

An interesting result is obtained from Eq. (11.51) if λ is considered a complex variable. Then $G_\lambda(x,y)$ has simple poles (infinitely many of them) at $\{\lambda_n\}_{n=0}^{\infty}$. The residue at the pole λ_n is $-\sum_k u_n^{*(k)}(y) u_n^{(k)}(x)$. If C_m is a contour in whose interior the poles $\{\lambda_n\}_{n=0}^{m}$ are located, then, by the residue theorem, we have

$$\frac{1}{2\pi i} \oint_{C_m} G_\lambda(x,y)\, d\lambda = -\sum_k \sum_{n=0}^{m} u_n^{*(k)}(y) u_n^{(k)}(x)$$

In particular, if we let $m \to \infty$, we obtain

$$\frac{1}{2\pi i} \oint_{C_\infty} G_\lambda(x,y)\, d\lambda = -\sum_k \sum_{n=0}^{\infty} u_n^{*(k)}(y) u_n^{(k)}(x)$$

$$\underset{\uparrow}{=} -\frac{\delta(x-y)}{w(x)} \tag{11.52}$$

by the completeness
of $u_n^{(k)}(x)$ and Chapter 5

where C_∞ is any contour that includes all the eigenvalues $\{\lambda_n\}_{n=0}^{\infty}$. Equation (11.52) is the infinite-dimensional analogue of Eq. (11.9) with $f(\mathbb{A}) = \mathbb{1}$.

Example 11.5.1

Consider the DO $\mathbb{L}_x = d^2/dx^2$ with BCs $u(0) = u(a) = 0$. This is an S-L operator with the following eigenvalues and normalized eigenfunctions:

$$\lambda_n = \left(\frac{n\pi}{a}\right)^2 \quad \text{and} \quad u_n(x) = \sqrt{\frac{2}{a}} \sin\left(\frac{n\pi}{a}x\right) \quad \text{for } n = 1, 2, \ldots$$

Equation (11.51) becomes

$$G_\lambda(x,y) = -\frac{2}{a}\sum_{n=0}^{\infty} \frac{\sin(n\pi x/a)\sin(n\pi y/a)}{\lambda - (n\pi/a)^2}$$

which leads to

$$\frac{1}{2\pi i}\oint_{C_\infty}\left[-\frac{2}{a}\sum_{n=0}^{\infty}\frac{\sin(n\pi x/a)\sin(n\pi y/a)}{\lambda-(n\pi/a)^2}\right]d\lambda$$

$$= -\frac{1}{2\pi i}\left(\frac{2}{a}\right)\sum_{n=0}^{\infty}\sin\left(\frac{n\pi}{a}x\right)\sin\left(\frac{n\pi}{a}y\right)\oint_C \frac{d\lambda}{\lambda-(n\pi/a)^2}$$

$$= -\frac{1}{2\pi i}\left(\frac{2}{a}\right)\sum_{n=0}^{\infty}\sin\left(\frac{n\pi}{a}x\right)\sin\left(\frac{n\pi}{a}y\right)\left\{2\pi i \operatorname{Res}\left[\frac{1}{\lambda-(n\pi/a)^2}\right]_{\lambda=(n\pi/a)^2}\right\}$$

$$= -\frac{2}{a}\sum_{n=0}^{\infty}\sin\left(\frac{n\pi}{a}x\right)\sin\left(\frac{n\pi}{a}y\right)$$

The RHS is recognizable as $-\delta(x-y)$. •

If zero is not an eigenvalue of $\mathbb{L}_x$, Eq. (11.51) yields

$$G(x,y) \equiv G_0(x,y) = \sum_k \sum_{n=0}^{\infty} \frac{u_n^{*(k)}(y)u_n^{(k)}(x)}{\lambda_n} \qquad (11.53)$$

which is an expression for the Green's function of $\mathbb{L}_x$ in terms of its eigenvalues and eigenfunctions.

PROBLEMS

11.1 Show that $\mathbb{P}_i\mathbb{P}_j = 0$ if $i \neq j$, where $\mathbb{P}_i$ is as defined in (11.10b).

11.2 Show that

$$\mathbb{R}_\lambda(\mathbb{A}) = -\frac{1}{2\pi i}\oint_\Gamma \frac{\mathbb{R}_\xi(\mathbb{A})}{\xi-\lambda}d\xi$$

11.3 Find the nonzero eigenvalues and the corresponding eigenfunctions for each kernel.

(a) $K(x,y) = 1 + \cos(x-y)$, where $-\pi \leqslant x, y \leqslant \pi$

(b) $K(x,y) = \sin x \cos y$, where $-\pi \leqslant x, y \leqslant \pi$

(c) $K(x,y) = \sin x \cos y$, where $0 \leqslant x, y \leqslant \pi/2$

(d) $K(x,y) = x + y$, where $a \leqslant x, y \leqslant b$

(e) $K(x,y) = 1 + xy$, where $a \leqslant x, y \leqslant b$

(f) $K(x,y) = e^{x+y}$, where $a \leqslant x, y \leqslant b$

11. ... ogonal functions on $[a, b]$. Show that ... ues.

... transform always has zero as an eigen-

11.5 ... $|f_j\rangle\langle f_j|$ with eigenvalue 0 when it is an ... m.

11.6

11.7

$$\ldots 2i \sum_{n=0}^{\infty} J_{2n+1}(z) \sin[(2n + 1)\theta]$$

$$\ldots {}^{\theta}\, d\theta$$

11.8 ... $H_\nu^{(2)}(z)$ given in Section 11.3.2.

11.9 ... to show that the Fourier transform

$$G(t) = \frac{1}{2\pi i} \int_{-i\infty+\alpha} F(\omega) t^{-\omega}\, d\omega \quad \text{where } F(\omega) = \int_0^\infty G(t) t^{\omega - 1}\, dt$$

11.10 The Laplace transform of a function $f(t)$ is defined as

$$\mathbb{L}[f](s) \equiv \int_0^\infty e^{-st} f(t)\, dt$$

Show that the Laplace transform of

(a) $f(t) = 1$ is $1/s$, where $s > 0$

(b) $f(t) = \exp(\omega t)$, for $t > 0$, is $1/(s - \omega)$, where $s > \omega$

(c) $f(t) = \cosh \omega t$ is $s/(s^2 - \omega^2)$

(d) $f(t) = \sinh \omega t$ is $\omega/(s^2 - \omega^2)$

(e) $f(t) = \cos \omega t$ is $s/(s^2 + \omega^2)$

(f) $f(t) = \sin \omega t$ is $\omega/(s^2 + \omega^2)$

(g) $f(t) = t^n$ is $\Gamma(n + 1)/s^{n+1}$, where $s > 0$, $n > -1$

11.11 Evaluate the integral

$$f(t) = \int_0^\infty \frac{\sin \omega t}{\omega}\, d\omega$$

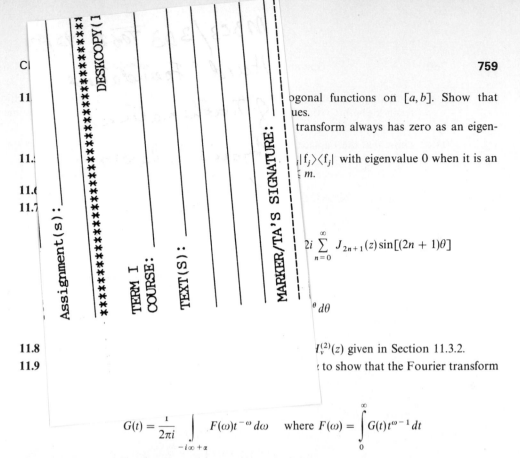

by finding the Laplace transform and changing the order of integration. Express the result for both $t > 0$ and $t < 0$ in terms of the theta function. (You will need some results from problem 11.10.)

11.12 Show that the Laplace transform of the derivative of a function is given by

$$\mathbb{L}[F'](s) = s\mathbb{L}[F](s) - F(0)$$

Similarly, show that for the second derivative the transform is

$$\mathbb{L}[F''](s) = s^2\mathbb{L}[F](s) - sF(0) - F'(0)$$

Use these results to solve the differential equation

$$u''(t) + \omega^2 u(t) = 0$$

subject to the boundary conditions $u(0) = a$, $u'(0) = 0$.

11.13 Determine the formal adjoint of each of the following (**a**) as a differential operator and (**b**) as an operator, that is, including the BCs. Which DOs are formally self-adjoint? Which operators are formally self-adjoint (the surface terms vanish)? Which operators are hermitian?

(**a**) $\mathbb{L}_x = d^2/dx^2$ in $[0, 1]$ with BCs $u(0) = u'(0) = 0$

(**b**) $\mathbb{L}_x = d^2/dx^2 + 1$ in $[0,1]$ with BCs $u(0) = u(1) = 0$

(**c**) $\mathbb{L}_x = d/dx + 1$ in $[0, \infty)$ with BC $u(0) = 0$

(**d**) $\mathbb{L}_x = d^3/dx^3 - \sin x\, d/dx + 3$ in $[0, \pi]$ with BCs $u(0) = 0$, $u'(0) = 0$, $u''(0) - 4u(\pi) = 0$

11.14 Show that the Dirichlet, Neumann, general unmixed, and periodic BCs make the following formally self-adjoint SOLDO self-adjoint (hermitian):

$$\mathbb{L}_x = \frac{1}{\omega}\frac{d}{dx}\left(p\frac{d}{dx}\right) + q$$

11.15 Using a procedure similar to that described in the text for SOLDOs, show that for the FOLDO $\mathbb{L}_x = p_1 d/dx + p_0$

(**a**) the indefinite GF is

$$G(x,y) \equiv \frac{\mu(y)}{p_1(y)w(y)}\left[\frac{\theta(x-y)}{\mu(x)}\right] + C(y)$$

where

$$\mu(x) = \exp\left[\int^x \frac{p_0(t)}{p_1(t)}\,dt\right]$$

(**b**) and the GF itself is discontinuous at $x = y$ with

$$\lim_{\varepsilon\to 0}[G(y+\varepsilon, y) - G(y-\varepsilon, y)] = \frac{1}{p_1(y)w(y)}$$

(**c**) For the homogeneous BC

$$\mathbb{R}[u] \equiv \alpha_1 u(a) + \alpha_2 u'(a) + \beta_1 u(b) + \beta_2 u'(b) = 0$$

construct $G(x,y)$ and show that

$$G(x,y) = \frac{1}{p_1(y)w(y)v(y)}v(x)\theta(x-y) + C(y)v(x)$$

where $v(x)$ is any solution to the homogeneous DE $\mathbb{L}_x[v] = 0$ and

$$C(y) = \frac{\beta_1 v(b) + \beta_2 v'(b)}{\mathbb{R}[v]p_1(y)w(y)v(y)}$$

and $\mathbb{R}[v] \neq 0$.

(d) Show directly that $\mathbb{L}_x[G] = \delta(x-y)/w(x)$.

11.16 Let $\mathbb{L}_x$ be a NOLDO with constant coefficients. Show that if $u(x)$ satisfies $\mathbb{L}_x[u] = f(x)$, then $u(x-y)$ satisfies $\mathbb{L}_x[u] = f(x-y)$. (Note that no BCs are specified.)

11.17 Find the GF for $\mathbb{L}_x = d^2/dx^2 + 1$ with BCs $u(0) = u'(0) = 0$. Show that it can be written as a function of $x - y$ only.

11.18 For all the GFs in examples and exercises in Section 9.4, answer the following: Is the GF symmetric in its arguments? If not, try to rewrite it in a form that is. Is this always possible? Why or why not?

11.19 Find the GF for $\mathbb{L}_x = d^2/dx^2 + k^2$ with BCs $u(0) = u(a) = 0$.

11.20 Find the GF for $\mathbb{L}_x = d^2/dx^2 - k^2$ with BCs $u(\infty) = u(-\infty) = 0$.

11.21 Find the GF for $\mathbb{L}_x = (d/dx)(x\,d/dx)$ given the condition that $G(x, y)$ is finite at $x = 0$ and vanishes at $x = 1$.

11.22 Evaluate the GF and the solutions for each of the following DEs in the interval $[0, 1]$.

(a) $u'' - k^2u = f$; $u(0) - u'(0) = a$, $u(1) = b$

(b) $u'' = f$; $u(0) = u'(0) = 0$

(c) $u'' + 6u' + 9u = 0$; $u(0) = 0$, $u'(0) = 1$

(d) $u'' + \omega^2 u = f(x)$, for $x > 0$; $u(0) = a$, $u'(0) = b$

(e) $u^{(4)} = f$; $u(0) = 0$, $u'(0) = 2u'(1)$, $u(1) = a$, $u''(0) = 0$

11.23 Use eigenfunction expansion of the GF to solve the BVP $u'' = x$, $u(0) = 0$, $u(1) - 2u'(1) = 0$.

12

Green's Functions in More Than One Dimension

The extensive study of Green's functions in one dimension in the last chapter should have made manifest the power and elegance of their use in solving inhomogeneous differential equations. If the differential equation has a (unique) solution, the GF exists and contains all the information necessary to build up a solution. The solution results from operating on the inhomogeneous term with an integral operator whose kernel is the appropriate Green's function [see, for example, Eq. (11.27)].

One crucial property of the Green's function is that it must satisfy some BCs. In fact, *its very existence depends* on the type of BCs imposed. We encountered two types of problems in solving ODEs. The first, called initial value problems (IVPs), involves fixing (for an nth-order DE) the value of the solution and its first $n-1$ derivatives at a fixed point. Then the ODE, if it is sufficiently well-behaved, will determine the values of the solution, in a *unique* way, in the neighborhood of the fixed point. Because of this uniqueness, Green's functions always exist for IVPs.

The second type of problems, called boundary value problems (BVPs), consists of (when the DE is second order) determining a relation between the solution and its derivative evaluated at the boundaries of the interval $[a, b]$. These boundaries are relations that we denoted by $\mathbb{R}_i[u] = \gamma_i$, where $i = 1, 2$. In this case the existence and uniqueness of the Green's function are not guaranteed.

There is a fundamental (topological) difference between a boundary in one dimension and a boundary in two and more dimensions. In one dimension a boundary consists of *only two points*; in m dimensions, $m \geqslant 2$, a boundary has *infinitely many points*. The boundary of a region in $\mathbb{R}^2$ is a closed curve, in $\mathbb{R}^3$ it is a closed *surface*, and in $\mathbb{R}^m$ it is a *hypersurface*. This fundamental difference makes the study of Green's functions in m dimensions more complicated, but also richer and more relevant, and intricately intertwined with the study of BVPs for PDEs. A thorough study of this subject is beyond the scope of this book. However, the basics of PDE theory are presented in the first section of this chapter in preparation for a discussion of m-dimensional Green's functions.

12.1 PROPERTIES OF PARTIAL DIFFERENTIAL EQUATIONS

This section presents certain facts and properties of PDEs, in particular, how BCs affect their solutions. We will note the important difference between ODEs and PDEs: The existence of a solution to a PDE satisfying a given BC depends on the *type* of the PDE.

12.1.1 The Cauchy Problem

When there is more than one variable, the innocent procedure of denoting a partial derivative of arbitrary order becomes a monstrous problem. To minimize such a problem, we can use some special notation.

If $i \equiv (i_1, i_2, \ldots, i_m)$ is an m-tuple whose components are *nonnegative integers*, we define

$$[i]_m \equiv \sum_{k=1}^{m} i_k$$

$$\frac{\partial^{[i]_m}}{\partial x^{[i]_m}} \equiv \frac{\partial^{[i]_m}}{\partial x_1^{i_1} \partial x_2^{i_2} \cdots \partial x_m^{i_m}} \equiv \frac{\partial^{i_1 + i_2 + \cdots + i_m}}{\partial x_1^{i_1} \partial x_2^{i_2} \cdots \partial x_m^{i_m}} \tag{12.1}$$

Thus, for $m = 1$, $[i]_1 = i_1$, and

$$\frac{\partial^{[i]_1}}{\partial x^{[i]_1}} = \frac{\partial^{i_1}}{\partial x_1^{i_1}}$$

is simply the i_1th derivative with respect to the only variable, x_1. For $m = 2$ we have

$$[i]_2 = i_1 + i_2$$

and

$$\frac{\partial^{[i]_2}}{\partial x^{[i]_2}} = \frac{\partial^{i_1 + i_2}}{\partial x_1^{i_1} \partial x_2^{i_2}}$$

That is, the partial derivative is of order $i_1 + i_2$, with i_1 factors with respect to x_1 and i_2 factors with respect to x_2. For example, if $[i]_2 = 3$, then

$$\frac{\partial^{[i]_2}}{\partial x^{[i]_2}} \equiv \frac{\partial^3}{\partial x^3} \equiv \frac{\partial^3}{\partial x_1^{i_1} \partial x_2^{i_2}}$$

could be any of the following partial derivatives:

$$\frac{\partial^3}{\partial x_1^3} \qquad \frac{\partial^3}{\partial x_1^2 \partial x_2} \qquad \frac{\partial^3}{\partial x_1 \partial x_2^2} \qquad \frac{\partial^3}{\partial x_2^3}$$

For higher values of m and $[i]_m$, the number of possible partial derivatives increases rapidly.

Finally, a multicomponent indexed parameter, such as $a_{i_1 i_2 \ldots i_m}$, is simply denoted by a_i; and $\sum_i^{[i]_m}$ means a summation over all possible values of $i_1, i_2, \ldots, i_m$ such that the sum $i_1 + i_2 + \cdots + i_m$ is equal to the value given to $[i]_m$. Thus, for $[i]_2 = 3$ we have

$$\sum_i^{[i]_2} a_i \frac{\partial^{[i]_2} f}{\partial x^{[i]_2}} \equiv \sum_{\substack{i_1, i_2 \\ i_1 + i_2 = 3}} a_{i_1 i_2} \frac{\partial^3 f}{\partial x_1^{i_1} \partial x_2^{i_2}}$$

$$= a_{30} \frac{\partial^3 f}{\partial x_1^3} + a_{21} \frac{\partial^3 f}{\partial x_1^2 \partial x_2} + a_{12} \frac{\partial^3 f}{\partial x_1 \partial x_2^2} + a_{03} \frac{\partial^3 f}{\partial x_2^3}$$

The most general PDE in m variables can be written as

$$F\left(x_j, u, \left\{\frac{\partial^{[i]_m} u}{\partial x^{[i]_m}}\right\}_{[i]_m = 1}^{M}\right) = 0$$

in which F is an appropriate multivariable function, $\{x_j\}_{j=1}^m$ is the set of independent variables, $u: \mathbb{R}^m \to \mathbb{R}$ is a real-valued, M times differentiable function of m independent variables, and M is a positive integer, called the *order* of the PDE.

A *quasilinear*, or *semilinear*, *PDE* of order M is of the form

$$\sum_i^M a_i \frac{\partial^M u}{\partial x^{[i]_m}} + F\left(x_j, u, \left\{\frac{\partial^{[i]_m} u}{\partial x^{[i]_m}}\right\}_{[i]_m = 1}^{M-1}\right) = 0$$

where $a_i \equiv a_{i_1 i_2 \ldots i_m}$ are functions of $\{x_j\}_{j=1}^m$. For example,

$$\sum_i^2 a_i \frac{\partial^2 u}{\partial x^{[i]_m}} + F\left(x_j, u, \left\{\frac{\partial u}{\partial x^{[i]_m}}\right\}\right) = 0$$

is a second-order partial differential equation (SOPDE) in m variables that can be rewritten as

$$\sum_{j,k=1}^m b_{jk}(x) \frac{\partial^2 u}{\partial x_j \partial x_k} + F\left(x_j, u, \frac{\partial u}{\partial x_1}, \frac{\partial u}{\partial x_2}, \ldots, \frac{\partial u}{\partial x_m}\right) = 0$$

In this equation $b_{jk}(x) \equiv b_{jk}(x_1, x_2, \ldots, x_m)$ are real-valued functions of m variables. In particular, if $m = 2$, with $x_1 \equiv x$ and $x_2 \equiv y$, we have the following as the most general quasilinear SOPDE in two variables:

$$a(x, y) \frac{\partial^2 u}{\partial x^2} + b(x, y) \frac{\partial^2 u}{\partial y^2} + c(x, y) \frac{\partial^2 u}{\partial x \partial y} + F\left(x, y, u, \frac{\partial u}{\partial x}, \frac{\partial u}{\partial y}\right) = 0$$

A *linear PDE* of order M in m variables is of the form

$$\mathbb{L}_{\mathbf{x}}[u] = f(x_1, x_2, \ldots, x_m) \tag{12.2a}$$

where

$$\mathbb{L}_{\mathbf{x}} \equiv \sum_{[i]_m \equiv k = 1}^{M} \sum_i^{[i]_m = k} a_i^{(k)}(x_1, \ldots, x_m) \frac{\partial^k}{\partial x^{[i]_m}} \tag{12.2b}$$

The *principal part* of $\mathbb{L}_{\mathbf{x}}$ is

$$\mathbb{L}_{\mathrm{p}} \equiv \sum_{i}^{[i]_m = M} a_i^{(M)}(x_1, \ldots, x_m) \frac{\partial^M}{\partial x^{[i]_m}} \tag{12.2c}$$

The coefficients $a_i^{(k)}$ and the inhomogeneous (or source) term f are assumed to be continuous functions of their arguments.

We will concentrate on linear PDEs and consider Eqs. (12.2) as an IVP with appropriate initial data. The most direct generalization of the IVP of differential equation theory is to specify the values of u and all its normal derivatives of order less than or equal to $M - 1$ on a *hypersurface* Γ of dimension $m - 1$. This type of initial data is called *Cauchy data*, and the resulting IVP is known as the *Cauchy problem* for $\mathbb{L}_{\mathbf{x}}$. Note that the tangential derivatives do not come into play here, because once we know the values of u on Γ, we can evaluate u on two neighboring points on Γ, take the limit as the points get closer and closer, and evaluate the tangential derivatives.

12.1.2 Characteristic Hypersurfaces

In contrast to the IVP in one dimension, the Cauchy problem for arbitrary Cauchy data may not have a solution, or if it does, the solution may not be unique. The existence and uniqueness of the solution depend crucially on the hypersurface Γ and on the type of PDE. We assume that Γ is smooth. A technical definition of a smooth hypersurface (manifold) is found in Chapter 4. Here "smooth" means simply that it is possible to introduce coordinate systems at every point $P \in \Gamma$.

Consider a point $P \in \Gamma$. Introduce $m - 1$ coordinates $\xi_2, \xi_3, \ldots, \xi_m$ to label points *on* Γ. Choose, by translation if necessary, coordinates in such a way that P is the origin, with coordinates $(0, 0, \ldots, 0)$. The ξ's are called *tangential coordinates*. Now let $v = \xi_1$ stand for the remaining coordinate normal to Γ. Usually ξ_i is taken to be the ith coordinate of the projection of the point on Γ onto the hyperplane tangent to Γ at P.

As long as we do not move too far away from P, the Cauchy data on Γ can be written as

$$u(0, \xi_2, \ldots, \xi_m), \frac{\partial u}{\partial v}(0, \xi_2, \ldots, \xi_m), \ldots, \frac{\partial^{M-1} u}{\partial v^{M-1}}(0, \xi_2, \ldots, \xi_m)$$

Using the chain rule,

$$\frac{\partial u}{\partial x_i} = \sum_{j=1}^{m} \frac{\partial u}{\partial \xi_j} \frac{\partial \xi_j}{\partial x_i} \qquad \text{where } \xi_1 \equiv v$$

we can also determine the first $M - 1$ derivatives of u with respect to x_i. The fundamental question is whether we can determine u uniquely using the above Cauchy data and the DE. To motivate the answer, let's look at the analogous problem in one dimension.

Consider the Mth-order linear DE

$$\mathbb{L}_x[u] = a_M(x)\frac{d^M u}{dx^M} + \cdots + a_1(x)\frac{du}{dx} + a_0(x)u = f(x) \tag{12.3}$$

with the following given initial data at x_0: $\{u(x_0), u'(x_0), \ldots, u^{(M-1)}(x_0)\}$. If the coefficients $\{a_k(x)\}_{k=0}^{M}$ and $f(x)$ are continuous and if $a_M(x_0) \neq 0$, Theorem 11.4.1 implies that there exists a unique solution to the IVP in a neighborhood of x_0.

For $a_M(x_0) \neq 0$ we can determine $u^{(M)}(x_0)$ from the DE:

$$u^{(M)}(x_0) = \frac{1}{a_M(x_0)}[f(x_0) - a_{M-1}(x_0)u^{(M-1)}(x_0) - \cdots - a_1(x_0)u'(x_0) - a_0(x_0)u(x_0)]$$

The initial data and a knowledge of $f(x_0)$ give $u^{(M)}(x_0)$ uniquely. Having found $u^{(M)}(x_0)$, we can calculate, with arbitrary accuracy (by choosing Δx small enough), the following set of *new* initial data at $x_1 = x_0 + \Delta x$:

$$u(x_1) = u(x_0) + u'(x_0)\Delta x, \ldots, u^{(M-1)}(x_1) = u^{(M-1)}(x_0) + u^{(M)}(x_0)\,\Delta x$$

Using these new initial data and Theorem 11.4.1, we are assured of a unique solution at x_1. Since $a_M(x)$ is assumed to be continuous for x_1 close enough to x_0, $a_M(x_1) \neq 0$, and it is possible to find newer initial data at $x_2 = x_1 + \Delta x$. The process can continue until we reach a singularity of the DE, a point where $a_M(x)$ vanishes. We can thus construct the unique solution of the IVP on an interval (x_0, b) as long as $a_M(x)$ does not vanish anywhere in $[x_0, b]$. This procedure is analogous to the one used in the analytic continuation of a complex function in Chapter 7.

For $a_M(x_0) = 0$, however, we cannot calculate $u^{(M)}(x_0)$ unambiguously. In such a case the LHS of (12.3) is *completely* determined from the initial data. If the LHS happens to be equal to $f(x_0)$, then there exist infinitely many solutions for $u^{(M)}(x_0)$; if the LHS is not equal to $f(x_0)$, there are no solutions. The difficulty can be stated in another way, which is useful for generalization to the m-dimensional case: $\mathbb{L}_x[u]$ itself is determined from the initial data.

Let us now return to the question of constructing u and investigate conditions under which the Cauchy problem may have a solution. We follow the same steps as for the IVP. To construct the solution numerically for points near P but away from Γ (since the function is completely determined on Γ, not only its Mth derivative but derivatives of all orders are known on Γ), we must be able to calculate $\partial^M u/\partial v^M$ at P. This is not possible if the coefficient of $\partial^M u/\partial v^M$ in $\mathbb{L}_{\mathbf{x}}[u]$ is zero when $x_1, \ldots, x_m$ is written in terms of $v, \xi_2, \ldots, \xi_m$. This motivates the following definition.

12.1.1 Definition If $\mathbb{L}_{\mathbf{x}}[u]$ can be evaluated at a point $P \in \Gamma$ from the Cauchy data alone, the hypersurface Γ is called *characteristic for* $\mathbb{L}_{\mathbf{x}}$ *at* P. If Γ is characteristic for all $P \in \Gamma$, then Γ is said to be a *characteristic hypersurface* for $\mathbb{L}_{\mathbf{x}}$.

Thus, if Γ is characteristic at P, the coefficient of $\partial^M u/\partial v^M$ vanishes in the expression for $\mathbb{L}_{\mathbf{x}}[u]$. In fact, we have the following theorem.

12.1.2 Theorem Let Γ be a smooth $(m-1)$-dimensional hypersurface. Let $\mathbb{L}_\mathbf{x}[u]$ be an Mth-order linear PDE in m variables. Then Γ is characteristic at $P \in \Gamma$ iff the coefficient of $\partial^M u/\partial v^M$ vanishes when $\mathbb{L}_\mathbf{x}$ is expressed in terms of the normal-tangential coordinate system $(v, \xi_2, \ldots, \xi_m)$. ■

The following corollary is a useful consequence of Theorem 12.1.2.

12.1.3 Corollary The hypersurface Γ is *not* characteristic at P iff all Mth-order partial derivatives of u with respect to $\{x_i\}_{i=1}^m$ are unambiguously determined at P from the Cauchy data on Γ and the DE. ■

In the one-dimensional case the difficulty arose when $a_M(x_0) = 0$. In the language being used here, we could call x_0 a "characteristic point." This makes sense because in this special case $m = 1$, and the hypersurfaces can only be of dimension $m - 1 = 0$. Thus, we can say that in the neighborhood of a characteristic point, the IVP has no well-defined solution.

For the general case where $m > 1$, we can similarly say that the Cauchy problem has no well-defined solution in the neighborhood of P if P happens to lie on a characteristic hypersurface of $\mathbb{L}$. Thus, it is of utmost importance to determine the characteristic hypersurfaces of PDEs.

Example 12.1.1

Let us consider the first-order PDE in two variables

$$\mathbb{L}[u] \equiv a(x, y)\frac{\partial u}{\partial x} + b(x, y)\frac{\partial u}{\partial y} + F(x, y, u) = 0 \tag{1}$$

This would be linear if $F(x, y, u) \equiv c(x, y)u + d(x, y)$. In its present form it is quasilinear. For this discussion the form of $F(x, y, u)$ is irrelevant.

We wish to find the characteristic hypersurfaces (in this case, curves) of $\mathbb{L}$. The Cauchy data consist of a simple determination of u on Γ. We use Corollary 12.1.3 to find the characteristic curves. This involves deriving relations that ensure a determination (or lack of it) of $\partial u/\partial x$ and $\partial u/\partial y$ at $P \equiv (x, y)$. Using an obvious notation, the PDE of (1) gives

$$-F(P, u(P)) = a(P)\frac{\partial u}{\partial x}(P) + b(P)\frac{\partial u}{\partial y}(P)$$

On the other hand, if $Q \equiv (x + dx, y + dy)$ lies on the curve Γ, then

$$u(Q) - u(P) = dx\frac{\partial u}{\partial x}(P) + dy\frac{\partial u}{\partial y}(P)$$

The Cauchy data determine the LHS of both of the preceding equations. Treating these equations as a system of two linear equations in two unknowns, $\partial u/\partial x(P)$ and $\partial u/\partial y(P)$, we conclude that the system has a unique solution if and only if the matrix of coefficients is invertible. Thus, Γ *is* a characteristic curve if and only if

$$\det\begin{pmatrix} dx & dy \\ a(P) & b(P) \end{pmatrix} = b(P)\,dx - a(P)\,dy = 0$$

or, assuming that $a(x, y) \neq 0$,

$$\frac{dy}{dx} = \frac{b(x, y)}{a(x, y)}$$

Solving this FODE yields y as a function of x, thus determining the characteristic curve. Note that a general solution of this FODE involves an arbitrary constant, resulting in a *family* of characteristic curves. •

12.1.3 Two-Dimensional Second-Order Partial Differential Equations and Their Boundary Conditions

Example 12.1.1 determined the characteristic curves of a quasilinear first-order PDE. Because second-order partial differential equations (SOPDEs) are important, we will now consider the same problem for a SOPDE in two variables. The most general quasilinear SOPDE is

$$\mathbb{L}[u] \equiv a(x, y)\frac{\partial^2 u}{\partial x^2} + 2b(x, y)\frac{\partial^2 u}{\partial x\,\partial y} + c(x, y)\frac{\partial^2 u}{\partial y^2} + F\left(x, y, u, \frac{\partial u}{\partial x}, \frac{\partial u}{\partial y}\right) = 0 \tag{12.4}$$

As in Example 12.1.1, we use Corollary 12.1.3 to determine the characteristic curves of $\mathbb{L}$. We seek conditions under which all second-order partial derivatives of u can be determined from the DE and the Cauchy data, which are values of u and all its first derivatives on Γ. Consider a point $Q \equiv (x + dx, y + dy)$ close to $P \equiv (x, y)$. We can write

$$\frac{\partial u}{\partial x}(Q) - \frac{\partial u}{\partial x}(P) = dx\,\frac{\partial^2 u}{\partial x^2}(P) + dy\,\frac{\partial^2 u}{\partial x\,\partial y}(P)$$

$$\frac{\partial u}{\partial y}(Q) - \frac{\partial u}{\partial y}(P) = dx\,\frac{\partial^2 u}{\partial x\,\partial y}(P) + dy\,\frac{\partial^2 u}{\partial y^2}(P)$$

$$-F\left(P, u(P), \frac{\partial u}{\partial x}(P), \frac{\partial u}{\partial y}(P)\right) = a(P)\frac{\partial^2 u}{\partial x^2}(P) + 2b(P)\frac{\partial^2 u}{\partial x\,\partial y}(P) + c(P)\frac{\partial^2 u}{\partial y^2}(P)$$

This system of three linear equations in the three unknowns, $(\partial^2 u/\partial x^2)(P)$, $(\partial^2 u/\partial x\,\partial y)(P)$, and $(\partial^2 u/\partial y^2)(P)$, has a unique solution if and only if the determinant of the coefficients is nonzero. Thus, Γ *is* a characteristic curve if and only if

$$\det\begin{pmatrix} dx & dy & 0 \\ 0 & dx & dy \\ a(P) & 2b(P) & c(P) \end{pmatrix} = 0$$

or

$$a(x, y)(dy)^2 - 2b(x, y)\,dx\,dy + c(x, y)(dx)^2 = 0$$

It then follows, assuming that $a(x, y) \neq 0$, that

$$\frac{dy}{dx} = \frac{b \pm \sqrt{b^2 - ac}}{a} \tag{12.5}$$

There are three cases to consider:

(1) If $b^2 - ac < 0$, Eq. (12.5) has no solution, which implies that no characteristic curves exist at P. Equation (12.4) is then said to be *elliptic*. Thus, the Laplace equation in two dimensions, $\partial^2 u/\partial x^2 + \partial^2 u/\partial y^2 = 0$, is elliptic because $b^2 - ac = -1$. In fact, it is elliptic in the whole plane or, put another way, has no characteristic curve in the entire xy-plane. This may lead us to believe that the Cauchy problem for the Laplace equation in two dimensions has a unique solution. However, even though the absence of a characteristic hypersurface at P is a necessary condition for the existence of a solution to the Cauchy problem, it is *not* sufficient. Exercise 12.1.1 presents a Cauchy problem that is ill-posed, meaning that the solution at any fixed point is not a continuous function of the initial data. Satisfying this condition is required of a well-posed problem on both mathematical and physical grounds.

(2) If $b^2 - ac > 0$, Eq. (12.5) has two solutions; that is, there are two characteristic curves passing through P. Equation (12.4) is then said to be *hyperbolic*. The wave equation, $\partial^2 u/\partial x^2 - \partial^2 u/\partial t^2 = 0$, is such an equation in the entire xt-plane.

(3) If $b^2 - ac = 0$, Eq. (12.4) is said to be *parabolic*. In this case there is only one characteristic curve at P. The one-dimensional diffusion equation, $\partial u/\partial t - a\,\partial^2 u/\partial x^2 = 0$ is parabolic in the whole xt-plane.

The question of what type of BCs to use to obtain a unique solution for a PDE is a very intricate mathematical problem. As Exercise 12.1.1 shows, even though all conditions are ripe for the Cauchy problem to have a unique solution for the two-dimensional Laplace equation, in reality it does not have a well-posed solution. On the other hand, examples in Chapter 10 that deal with electrostatic potentials and temperatures lead us to believe that a specification of the solution u on a *closed* curve gives a unique solution. This has a sound physical basis. After all, specifying the temperature (or electrostatic potential) on a closed surface should be enough to give us information about the temperature (or electrostatic potential) in the region close to the curve. A BC in which the value of the solution is given on a closed curve is called a *Dirichlet boundary condition*, and the associated problem, a *Dirichlet boundary value problem*.

There is another type of BC, which, on physical grounds, is appropriate for the Laplace equation in two dimensions. This condition is based on the fact that if the surface charge on a conductor is specified, then the electrostatic potential in the vicinity of the conductor should be determined uniquely. The surface charge on a conductor is proportional to the value of the electric field on the conductor. The electric field, on the other hand, is the normal derivative of the potential. A BC in which the value of the normal derivative of the solution is specified on a closed curve is called a *Neumann boundary condition*, and the associated problem, a *Neumann boundary value problem*.

Thus, at least in two dimensions and on physical grounds alone, either a Dirichlet BVP or a Neumann BVP is a well-posed problem for the Laplace equation.

For the heat (or diffusion) equation we are given an initial temperature distribution on a bar along the x-axis with end points held at constant temperatures. For a bar with end points at $x = a$ and $x = b$, this is equivalent to the data $u(0, x)$, $u(t, a) = T_1$, and $u(t, b) = T_2$. These are not Cauchy data, so we need not worry about characteristic curves. The boundary curve consists of three parts:

(1) $t = 0$, for $a \leqslant x \leqslant b$

(2) $t > 0$, for $x = a$

(3) $t > 0$, for $x = b$

In the xt-plane these form an open rectangle consisting of $\overline{ab}$ as one side and vertical lines at a and b as the other two. The problem is to determine u on the side that closes the curve, that is, on the side $a \leqslant x \leqslant b$ at $t > 0$.

The wave equation requires specification of both u and $\partial u/\partial t$ at $t = 0$. The displacement of the boundaries of the waving medium, a taut rope, for example, must also be specified. Again the curve is open, as for the diffusion case, but the initial data are Cauchy. Thus, for the wave equation we do have a Cauchy problem with Cauchy data specified on an open curve. Since the curve, the open rectangle, is not a characteristic curve of the wave equation, the Cauchy problem is well-posed.

12.1.4 Second-Order Partial Differential Equations in m Dimensions and Their Boundary Conditions

Because of their importance in mathematical physics, the rest of this chapter is devoted to SOPDEs. This subsection classifies SOPDEs and the BCs associated with them.

The most general quasilinear SOPDE in m variables can be written as

$$\sum_{j,k=1}^{m} A_{jk}(x_1, \ldots, x_m) \frac{\partial^2 u}{\partial x_j \partial x_k} + F\left(x_1, \ldots, x_m, u, \frac{\partial u}{\partial x_1}, \ldots, \frac{\partial u}{\partial x_m}\right) = 0 \tag{12.6}$$

where A_{jk} is assumed to be symmetric in i and j. If it is not, then the symmetry of the mixed second derivative is used to write its coefficient in the symmetric combination $(1/2)\,(A_{jk} + A_{kj})$. We choose a new set $\{y_k\}_{k=1}^{m}$, given by

$$y_k = \sum_{l=1}^{m} R_{kl} x_l \qquad \text{where } k = 1, 2, \ldots, m \tag{12.7}$$

where R is, for now, an arbitrary constant matrix. Thus,

$$\frac{\partial y_k}{\partial x_l} = R_{kl} \qquad \text{and} \qquad \frac{\partial^2 y_k}{\partial x_n \partial x_l} = 0$$

Defining the matrices

$$U^{(x)}_{mn} \equiv \frac{\partial^2 u}{\partial x_m \partial x_n} \quad \text{and} \quad U^{(y)}_{kj} \equiv \frac{\partial^2 u}{\partial y_k \partial y_j}$$

we can write (with summation over repeated indices understood)

$$U^{(x)}_{mn} = U^{(y)}_{kj} \frac{\partial y_k}{\partial x_m} \frac{\partial y_j}{\partial x_n} = U^{(y)}_{kj} R_{km} R_{jn} = (\tilde{\mathsf{R}} \mathsf{U}^{(y)} \mathsf{R})_{mn}$$

The sum in (12.6) is

$$\text{sum} \equiv A_{jk}(\mathbf{x}) U^{(x)}_{jk} = A_{kj}(\mathbf{x}) U^{(x)}_{jk} = (\mathsf{A}(\mathbf{x}) \mathsf{U}^{(x)})_{kk} = \text{tr}[\mathsf{A}(\mathbf{x}) \mathsf{U}^{(x)}]$$

which, in terms of y, becomes

$$\text{sum} = \text{tr}(\mathsf{A}(\mathbf{x}) \tilde{\mathsf{R}} \mathsf{U}^{(y)} \mathsf{R}) = \text{tr}[\mathsf{R} \mathsf{A}(\mathbf{x}) \tilde{\mathsf{R}} \mathsf{U}^{(y)}]$$

At a particular *fixed point*, $\mathbf{x}_0 \equiv (x_{01}, x_{02}, \ldots, x_{0m})$, the matrix $\mathsf{A}(\mathbf{x})$ has (real) numbers (not functions) as elements. Since A is real and symmetric, it can be diagonalized by an orthogonal matrix. Let R, which up to now has been arbitrary, be that orthogonal matrix. Then $\mathsf{R}\mathsf{A}(\mathbf{x}_0)\tilde{\mathsf{R}}$ will be diagonal with elements, say

$$(\mathsf{R}\mathsf{A}(\mathbf{x}_0)\tilde{\mathsf{R}})_{jk} \equiv b_j(\mathbf{x}_0)\delta_{jk} \equiv a_j(\mathbf{y}_0)\delta_{jk} \qquad \text{(no sum)}$$

where $\mathbf{x}_0$ has been replaced by $\mathbf{y}_0$ using (the inverse of) Eq. (12.7). Here b_j is simply the jth eigenvalue of A. At the point $\mathbf{x}_0$, we now have

$$\begin{aligned} \text{sum} &= \text{tr}(\mathsf{R}\mathsf{A}(\mathbf{x}_0)\tilde{\mathsf{R}} U^{(y_0)}) = (\mathsf{R}\mathsf{A}(\mathbf{x}_0)\tilde{\mathsf{R}})_{jk} U^{(y_0)}_{kj} \\ &= \sum_{j,k} a_j(\mathbf{y}_0)\delta_{jk} U^{(y_0)}_{kj} = \sum_{j=1}^{m} a_j(\mathbf{y}_0) U^{(y_0)}_{jj} \\ &= \sum_{j=1}^{m} a_j(\mathbf{y}_0) \frac{\partial^2 u}{\partial y_j^2} \end{aligned}$$

We have thus shown that *at the point* $\mathbf{x}_0$, it is possible to transform (12.6), which has cross derivatives, into

$$\sum_{j=1}^{m} a_j(\mathbf{y}_0) \left(\frac{\partial^2 u}{\partial y_j^2} \right)_{\mathbf{y}=\mathbf{y}_0} + F\left(\mathbf{y}_0, u, \frac{\partial u}{\partial \mathbf{y}}(\mathbf{y}_0) \right) = 0 \tag{12.8}$$

where $\mathbf{y}_0$ and $(\partial u/\partial \mathbf{y})(\mathbf{y}_0)$ are abbreviations for $(y_{01}, y_{02}, \ldots, y_{0m})$ and $((\partial u/\partial y_1)|_{\mathbf{y}=\mathbf{y}_0}, (\partial u/\partial y_2)|_{\mathbf{y}=\mathbf{y}_0}, \ldots, (\partial u/\partial y_m)|_{\mathbf{y}=\mathbf{y}_0})$, respectively.

Equation (12.8) can be used to classify SOPDEs as follows:

(1) Equation (12.6) is said to be of the *elliptic type at* $\mathbf{x}_0$ if all the coefficients $a_j(\mathbf{y}_0)$ in (12.8) are nonzero and have the same sign.

(2) Equation (12.6) is said to be of the *ultrahyperbolic type at* $\mathbf{x}_0$ if all $a_j(\mathbf{y}_0)$ in (12.8) are nonzero but do not have the same sign. If only one of the coefficients has a sign different from the rest, the equation is said to be of the *hyperbolic type*.

(3) Equation (12.6) is said to be of the *parabolic type at* $\mathbf{x}_0$ if at least one of the coefficients $a_j(\mathbf{y}_0)$ in (12.8) is zero.

If a SOPDE is of a given type at every point of its domain, it is said to be of that given type. In particular, if the coefficients a_j are constants, the type of the PDE does not change from point to point.

The type of a SOPDE in two dimensions has been defined in two ways. Exercise 12.1.2 shows that the two are equivalent, at least in the special case of constant coefficients.

As indicated before, the question of the dependence of the BCs on the type of PDE under discussion is a very intricate one. The preceding subsection introduced three types of BCs appropriate for the three types of PDEs, based solely on physical arguments. We can generalize that discussion to m dimensions. We know, from encounters with physical problems, that such a generalization from $m = 2$ to $m = 3$ works. Thus, we can make the following correspondence, denoted by $\leftrightarrow$, between a SOPDE with m variables and the appropriate BCs:

(1) Elliptic SOPDE $\leftrightarrow$ Dirichlet or Neumann BCs on a closed hypersurface
(2) Hyperbolic SOPDE $\leftrightarrow$ Cauchy data on an open hypersurface
(3) Parabolic SOPDE $\leftrightarrow$ Dirichlet or Neumann BCs on an open hypersurface

Exercises

12.1.1 Solve the Cauchy problem for the two-dimensional Laplace equation subject to the Cauchy data $u(0, y) = 0$, $(\partial u/\partial x)(0, y) = \varepsilon \sin ky$, where ε and k are constants. Show that the solution does not vary continuously as the Cauchy data vary.

12.1.2 Show that the two definitions of types of SOPDEs discussed in the text are equivalent for a SOPDE in two variables,

$$a\frac{\partial^2 u}{\partial x^2} + 2b\frac{\partial^2 u}{\partial x\,\partial y} + c\frac{\partial^2 u}{\partial y^2} + F\left(x, y, u, \frac{\partial u}{\partial x}, \frac{\partial u}{\partial y}\right) = 0$$

where a, b, and c are constants.

12.2 GREEN'S FUNCTIONS AND DELTA FUNCTIONS IN HIGHER DIMENSIONS

This section will discuss some of the characteristics of Green's functions in higher dimensions. These characteristics are related to the formal partial differential operator associated with the Green's function and also to the delta functions.

Using the formal idea of several continuous indices, we can turn the operator equation

$$\mathbb{L}\mathbb{G} = \mathbb{1}$$

into the PDE

$$\mathbb{L}_{\mathbf{x}} G(\mathbf{x}, \mathbf{y}) = \frac{\delta(\mathbf{x} - \mathbf{y})}{w(\mathbf{x})} \qquad (12.9a)$$

where $\mathbf{x} \equiv (x_1, \ldots, x_m)$, $\mathbf{y} \equiv (y_1, \ldots, y_m)$, $w(\mathbf{x})$ is a weight function that is usually set equal to one, $\mathbb{L}_\mathbf{x}$ is a SOPDO in the $\mathbf{x}$ variables, and, in *Cartesian coordinates only,*

$$\delta(\mathbf{x} - \mathbf{y}) = \delta(x_1 - y_1)\delta(x_2 - y_2) \cdots \delta(x_m - y_m) \tag{12.9b}$$

In most applications Cartesian coordinates are not the most convenient to use. Therefore, it is helpful to express (12.9) in other coordinate systems. In particular, it is helpful to know how the delta function transforms under a general coordinate transformation.

12.2.1 Coordinate Transformation of the Delta Function

Let

$$x_i \equiv f_i(\xi_1, \xi_2, \ldots, \xi_m)$$

be a coordinate transformation from $\{x_i\}_{i=1}^m$ to $\{\xi_j\}_{j=1}^m$. Let P be a point whose coordinates are $\mathbf{a} \equiv (a_1, \ldots, a_m)$ and $\boldsymbol{\alpha} \equiv (\alpha_1, \ldots, \alpha_m)$ in the x and ξ coordinate systems, respectively. Let J be the Jacobian of the transformation, that is, the absolute value of the determinant of a matrix whose elements are $\partial x_i/\partial \xi_j$. Consider a function $F(\mathbf{x}) \equiv F(x_1, \ldots, x_m)$. By the definition of the delta function, we have

$$\int d^m x\, F(\mathbf{x})\delta(\mathbf{x} - \mathbf{a}) = F(\mathbf{a})$$

Expressing this equation in terms of the ξ coordinate system and recalling that $d^m x = J d^m \xi$, we obtain

$$\int d^m \xi\, J F(f_1(\boldsymbol{\xi}), f_2(\boldsymbol{\xi}), \ldots, f_m(\boldsymbol{\xi}))\delta(f_1(\boldsymbol{\xi}) - a_1) \cdots \delta(f_m(\boldsymbol{\xi}) - a_m)$$

$$= F(a_1, \ldots, a_m) = F(f_1(\boldsymbol{\alpha}), \ldots, f_m(\boldsymbol{\alpha})) \tag{12.10}$$

where we have used $a_i = f_i(\boldsymbol{\alpha})$. Introducing the notation $H(\boldsymbol{\xi}) \equiv F(f_1(\boldsymbol{\xi}), \ldots, f_m(\boldsymbol{\xi}))$, we rewrite (12.10) as

$$\int d^m \xi\, J H(\boldsymbol{\xi})\delta(f_1(\boldsymbol{\xi}) - f_1(\boldsymbol{\alpha})) \cdots \delta(f_m(\boldsymbol{\xi}) - f_m(\boldsymbol{\alpha})) = H(\boldsymbol{\alpha})$$

On the other hand,

$$\int d^m \xi\, H(\boldsymbol{\xi})\delta(\xi_1 - \alpha_1) \cdots \delta(\xi_m - \alpha_m) = H(\boldsymbol{\alpha})$$

We, therefore, obtain

$$J\delta(f_1(\boldsymbol{\xi}) - f_1(\boldsymbol{\alpha})) \cdots \delta(f_m(\boldsymbol{\xi}) - f_m(\boldsymbol{\alpha})) = \delta(\xi_1 - \alpha_1) \cdots \delta(\xi_m - \alpha_m)$$

or, in more compact notation,

$$J\delta(\mathbf{x}(\boldsymbol{\xi}) - \mathbf{a}(\boldsymbol{\alpha})) = \delta(\boldsymbol{\xi} - \boldsymbol{\alpha})$$

or, simply,

$$J\delta(\mathbf{x} - \mathbf{a}) = \delta(\boldsymbol{\xi} - \boldsymbol{\alpha}) \tag{12.11}$$

It is, of course, understood that $J \neq 0$ at P.

Example 12.2.1

For the spherical coordinate system

$$x_1 \equiv x = r \sin\theta \cos\varphi \equiv \xi_1 \sin\xi_2 \cos\xi_3$$

$$x_2 \equiv y = r \sin\theta \sin\varphi \equiv \xi_1 \sin\xi_2 \sin\xi_3$$

$$x_3 \equiv z = r \cos\theta \equiv \xi_1 \cos\xi_2$$

and the Jacobian is the absolute value of

$$\det \begin{pmatrix} \sin\xi_2 \cos\xi_3 & \xi_1 \cos\xi_2 \cos\xi_3 & -\xi_1 \sin\xi_2 \sin\xi_3 \\ \sin\xi_2 \sin\xi_3 & \xi_1 \cos\xi_2 \sin\xi_3 & \xi_1 \sin\xi_2 \cos\xi_3 \\ \cos\xi_2 & -\xi_1 \sin\xi_2 & 0 \end{pmatrix}$$

$$= \cos\xi_2(\xi_1^2 \sin\xi_2 \cos\xi_2) + \xi_1 \sin\xi_2(\xi_1 \sin^2\xi_2) = \xi_1^2 \sin\xi_2 \equiv r^2 \sin\theta$$

Equation (12.11) gives

$$\delta(x - x_0)\delta(y - y_0)\delta(z - z_0) = \frac{\delta(r - r_0)\delta(\theta - \theta_0)\delta(\varphi - \varphi_0)}{r_0^2 \sin\theta_0}$$

where (x_0, y_0, z_0) and $(r_0, \theta_0, \varphi_0)$ are the coordinates of the same point in the two coordinate systems. ●

What happens when $J = 0$ at P? A point at which the Jacobian vanishes is called a *singular point* of the transformation. Thus, all points on the z-axis, including the origin, are singular points in Example 12.2.1. Since J is a determinant, its vanishing at a point indicates that, in the vicinity of that point, the transformation is not invertible, that is, is not one to one. Thus, in the transformation from Cartesian to spherical coordinates, the point $(0, 0, -5)$ transforms into $(5, 180°, \varphi)$, where φ is arbitrary. Similarly, the point $(0, 0, 0)$ in the Cartesian coordinate system goes to $(0, \theta, \varphi)$ in the spherical system, with θ and φ arbitrary. A coordinate whose value is not determined at a singular point is called an *ignorable coordinate* at that point. Thus, at the origin both θ and φ are ignorable.

Among the ξ coordinates, let $\{\xi_i\}_{i=k+1}^m$ be ignorable at $P \equiv (a_1, a_2, \ldots, a_m)$. This means that any function such as F, when expressed in terms of ξ coordinates, will be independent of the ignorable coordinates. Thus, a reexamination of Eq. (12.10) shows that (see Exercise 12.2.1)

$$\delta(\mathbf{x} - \mathbf{a}) = \frac{1}{|J_k|}\delta(\xi_1 - \alpha_1) \cdots \delta(\xi_k - \alpha_k) \tag{12.12a}$$

where

$$J_k \equiv \int d\xi_{k+1}\, d\xi_{k+2} \cdots d\xi_m\, J \tag{12.12b}$$

The foregoing discussion can be stated as a proposition.

12.2.1 Proposition Let $x_i \equiv f_i(\boldsymbol{\xi})$, where $i = 1, \ldots, m$, be a transformation from Cartesian coordinates x_i to curvilinear coordinates ξ_j. Let J be the transformation

Jacobian. Let P be a singular point of the transformation. Let $\xi_{k+1}, \xi_{k+2}, \ldots, \xi_m$ be the ignorable coordinates at P, whose coordinates in x_i are $(a_1, \ldots, a_m) \equiv \mathbf{a}$ and in ξ_j are $(\alpha_1, \alpha_2, \ldots, \alpha_m)$. Then, defining $J_k \equiv \int d\xi_{k+1}\, d\xi_{k+2} \cdots d\xi_m\, J$, we have

$$\delta(\mathbf{x} - \mathbf{a}) = \frac{1}{|J_k|}\delta(\xi_1 - \alpha_1)\delta(\xi_2 - \alpha_2) \cdots \delta(\xi_k - \alpha_k)$$ ■

In particular, if the transformation is one to one, $k = m$ and $J_m \equiv J$, and we recover (12.11).

Example 12.2.2

In two dimensions the transformation between Cartesian and polar coordinates is given by

$$x_1 \equiv x = r\cos\theta \equiv \xi_1 \cos\xi_2$$

$$x_2 \equiv y = r\sin\theta \equiv \xi_1 \sin\xi_2$$

with the Jacobian

$$J = \det\begin{pmatrix} \partial x_1/\partial\xi_1 & \partial x_1/\partial\xi_2 \\ \partial x_2/\partial\xi_1 & \partial x_2/\partial\xi_2 \end{pmatrix} = \det\begin{pmatrix} \cos\xi_2 & -\xi_1\sin\xi_2 \\ \sin\xi_2 & \xi_1\cos\xi_2 \end{pmatrix} = \xi_1 = r$$

which vanishes at the origin ($r = 0$). The angle θ is the only ignorable coordinate at the origin. Thus, $k = 1$, and

$$J_1 = \int_0^{2\pi} J\, d\theta = \int_0^{2\pi} r\, d\theta = 2\pi r$$

This yields

$$\delta(\mathbf{x}) \equiv \delta(x)\delta(y) = \frac{\delta(r)}{2\pi r}$$

In three dimensions the transformation between Cartesian and spherical coordinates is given in Example 12.2.1, along with its Jacobian, which is $J = r^2\sin\theta$. This vanishes at the origin no matter what the values of θ and φ are. We thus have two ignorable coordinates at the origin ($k = 1$), over which we integrate to obtain

$$J_1 = \int_0^{2\pi} d\varphi \int_0^{\pi} d\theta\, r^2 \sin\theta = 4\pi r^2$$

Thus, we have

$$\delta(x)\delta(y)\delta(z) = \frac{\delta(r)}{4\pi r^2}$$ •

12.2.2 Spherical Coordinates in m Dimensions

In discussing Green's functions in m dimensions, a particular curvilinear coordinate system will prove useful. This system is the generalization of spherical coordinates in three dimensions. The *m-dimensional spherical coordinate system* is defined as

$$x_1 = r\sin\theta_1 \sin\theta_2 \cdots \sin\theta_{m-1}$$
$$x_k = r\sin\theta_1 \sin\theta_2 \cdots \sin\theta_{m-k}\cos\theta_{m-k+1} \qquad \text{where } 2 \leqslant k \leqslant m-1 \qquad (12.13)$$
$$x_m = r\cos\theta_1$$

(Note that for $m = 3$ the first two coordinates are out of order compared to their usual definitions.)

It is not hard to show (see Example 12.2.3) that the Jacobian of the transformation given by (12.13) is

$$J = r^{m-1}(\sin\theta_1)^{m-2}(\sin\theta_2)^{m-3}\cdots(\sin\theta_k)^{m-k-1}\cdots\sin\theta_{m-2} \qquad (12.14)$$

and that the volume element in terms of these coordinates is

$$d^m x = J\,dr\,d\theta_1\,d\theta_2\cdots d\theta_{m-1} = r^{m-1}\,dr\,d\Omega_m \qquad (12.15a)$$

where $$d\Omega_m \equiv (\sin\theta_1)^{m-2}(\sin\theta_2)^{m-3}\cdots\sin\theta_{m-2}\,d\theta_1\,d\theta_2\cdots d\theta_{m-1} \qquad (12.15b)$$

is the element of the m-dimensional solid angle.

Example 12.2.3

For $m = 4$ we have

$$x_1 = r\sin\theta_1\sin\theta_2\sin\theta_3$$
$$x_2 = r\sin\theta_1\sin\theta_2\cos\theta_3$$
$$x_3 = r\sin\theta_1\cos\theta_2$$
$$x_4 = r\cos\theta_1$$

The Jacobian is given by

$$J = \det\begin{pmatrix} \sin\theta_1\sin\theta_2\sin\theta_3 & r\cos\theta_1\sin\theta_2\sin\theta_3 & r\sin\theta_1\cos\theta_2\sin\theta_3 & r\sin\theta_1\sin\theta_2\cos\theta_3 \\ \sin\theta_1\sin\theta_2\cos\theta_3 & r\cos\theta_1\sin\theta_2\cos\theta_3 & r\sin\theta_1\cos\theta_2\cos\theta_3 & -r\sin\theta_1\sin\theta_2\sin\theta_3 \\ \sin\theta_1\cos\theta_2 & r\cos\theta_1\cos\theta_2 & -r\sin\theta_1\sin\theta_2 & 0 \\ \cos\theta_1 & -r\sin\theta_1 & 0 & 0 \end{pmatrix}$$
$$= r^3\sin^2\theta_1\sin\theta_2$$

We can list the Jacobians we have found for various values of m:

m	J
2	r
3	$r^2\sin\theta_1$
4	$r^3\sin^2\theta_1\sin\theta_2$

These results clearly generalize to Eq. (12.14) for any $m \geqslant 2$. •

As Exercise 12.2.2 shows, the total solid angle in m dimensions is

$$\Omega_m = \frac{2\pi^{m/2}}{\Gamma(m/2)} \qquad (12.16)$$

An interesting result that is easily obtainable is an expression of the delta function in terms of spherical coordinates at the origin. Since $r = 0$, Eq. (12.14) shows that all the angles are ignorable. Thus, we have

$$J_1 = \int J\,d\theta_1\,d\theta_2 \cdots d\theta_{m-1} = r^{m-1} \int d\Omega_m = r^{m-1}\Omega_m$$

which yields

$$\delta(\mathbf{x}) \equiv \delta(x_1) \cdots \delta(x_m) = \frac{\delta(r)}{\Omega_m r^{m-1}} = \frac{\Gamma(m/2)\delta(r)}{2\pi^{m/2} r^{m-1}} \tag{12.17}$$

12.2.3 Green's Function for the Laplacian

With the machinery developed above, we can easily obtain the Green's function for the Laplacian. We will ignore questions of BCs and simply develop a function that satisfies

$$\nabla^2 G(\mathbf{x}, \mathbf{y}) = \delta(\mathbf{x} - \mathbf{y})$$

With no loss of generality we let $\mathbf{y} = 0$; that is, we translate the axes so that $\mathbf{y}$ becomes the new origin. Then we have

$$\nabla^2 G(\mathbf{x}) = \delta(\mathbf{x})$$

In spherical coordinates this becomes

$$\nabla^2 G(\mathbf{x}) = \frac{\delta(r)}{\Omega_m r^{m-1}} \tag{12.18a}$$

Since the RHS is a function of r only, we expect G to behave in the same way. We now have to express ∇^2 in terms of spherical coordinates. In general, this is difficult; however, for a function of r only we can easily find ∇^2. Note that in Cartesian coordinates $\nabla^2 = \sum_{i=1}^{m} \partial^2/\partial x_i^2$. For a function $F(r)$ that depends on r only, we have

$$\frac{\partial F}{\partial x_i} = \frac{\partial F}{\partial r}\frac{\partial r}{\partial x_i} = \frac{\partial F}{\partial r}\frac{x_i}{r}$$

This can easily be verified by noting that $r = \sqrt{x_1^2 + x_2^2 + \cdots + x_m^2}$. Differentiating once more gives

$$\frac{\partial^2 F}{\partial x_i^2} = \frac{\partial^2 F}{\partial r^2}\frac{x_i^2}{r^2} + \frac{\partial F}{\partial r}\left(\frac{1}{r} - \frac{x_i^2}{r^3}\right)$$

which results in

$$\begin{aligned}\nabla^2 F(r) &= \sum_{i=1}^{m} \frac{\partial^2 F}{\partial x_i^2} = \frac{\partial^2 F}{\partial r^2} + \frac{m-1}{r}\frac{\partial F}{\partial r} \\ &= \frac{1}{r^{m-1}}\frac{\partial}{\partial r}\left(r^{m-1}\frac{\partial F}{\partial r}\right)\end{aligned}$$

For the Green's function, therefore, we get

$$\frac{d}{dr}\left(r^{m-1}\frac{dG}{dr}\right) = \frac{\delta(r)}{\Omega_m} \tag{12.18b}$$

The solution, for $m \geqslant 3$, is (see Exercise 12.2.3)

$$\begin{aligned} G &= -\frac{1}{(m-2)\Omega_m}\left(\frac{1}{r^{m-2}}\right) \\ &= -\frac{\Gamma(m/2)}{2(m-2)\pi^{m/2}}\left(\frac{1}{r^{m-2}}\right) \qquad \text{for } m \geqslant 3 \end{aligned} \tag{12.19a}$$

We can restore the vector $\mathbf{y}$, at which we placed the origin, by noting that $r = |\mathbf{r}| = |\mathbf{x} - \mathbf{y}|$. Thus, we get

$$\begin{aligned} G(\mathbf{x}, \mathbf{y}) &= -\frac{\Gamma(m/2)}{2(m-2)\pi^{m/2}}\left(\frac{1}{|\mathbf{x}-\mathbf{y}|^{m-2}}\right) \\ &= -\frac{\Gamma(m/2)}{2(m-2)\pi^{m/2}}\left[\sum_{i=1}^{m}(x_i - y_i)^2\right]^{-(m-2)/2} \qquad \text{for } m \geqslant 3 \end{aligned} \tag{12.19b}$$

For $m = 2$ we have

$$G(\mathbf{x}, \mathbf{y}) = \frac{1}{2\pi}\ln(|\mathbf{x}-\mathbf{y}|) = \frac{1}{4\pi}\ln[(x_1 - y_1)^2 + (x_2 - y_2)^2] \tag{12.19c}$$

Having found the Green's function for the Laplacian, we can use the usual method to find a solution to the inhomogeneous equation, the *Poisson equation,*

$$\nabla^2 u = -\rho(\mathbf{x})$$

Thus,

$$\begin{aligned} u(\mathbf{x}) &= -\int d^m y\, G(\mathbf{x}, \mathbf{y})\rho(\mathbf{y}) \\ &= \frac{\Gamma(m/2)}{2(m-2)\pi^{m/2}}\int d^m y \frac{\rho(\mathbf{y})}{|\mathbf{x}-\mathbf{y}|^{m-2}} \end{aligned}$$

In particular, for $m = 3$ we obtain

$$u(\mathbf{x}) = \frac{1}{4\pi}\int d^3 y \frac{\rho(\mathbf{y})}{|\mathbf{x}-\mathbf{y}|}$$

which is recognizable as the electrostatic potential due to a charge density $\rho(\mathbf{y})$.

Exercises

12.2.1 Prove Eqs. (12.12). First, note that the RHS of Eq. (12.10) is a function of $\alpha_1, \alpha_2, \ldots, \alpha_k$ only. This means that

$$F(f_1(\xi), f_2(\xi), \ldots, f_m(\xi))|_{\xi=\alpha} \equiv F(\alpha_1, \ldots, \alpha_k) \tag{1}$$

(a) Rewrite Eq. (12.10) by separating the integral into two parts, one involving $\xi_1, \ldots, \xi_k$ and the other $\xi_{k+1}, \ldots, \xi_m$. Compare the RHS with the LHS and show that

$$\int d\xi_{k+1} \cdots d\xi_m \, J\delta(\mathbf{x} - \mathbf{a}) = \delta(\xi_1 - \alpha_1) \cdots \delta(\xi_k - \alpha_k) \tag{2}$$

(b) Show that $\delta(\mathbf{x} - \mathbf{a})$ is independent of $\{\xi_j\}_{j=k+1}^{m}$ by assuming otherwise and showing that the identity in (2) is violated.

(c) Take the delta function out of the integral and arrive at (12.12a). Complete the proof by showing that $J_k \neq 0$ at $(\alpha_1, \ldots, \alpha_k)$.

12.2.2 Find the total solid angle Ω_m in m dimensions.

12.2.3 Find the m-dimensional Green's function for the Laplacian by taking the following steps.

(a) Solve (12.18b) for G assuming that $r \neq 0$ and demanding that $G(r) \to 0$ as $r \to \infty$ (this can be done only for $m \geqslant 3$).

(b) Use the divergence theorem in m dimensions and (12.18a) to show that

$$\int_S \frac{dG}{dr} \, da = 1$$

where S is a spherical hypersurface of radius r. Now use this and the result of part (a) to find the remaining constant of integration.

12.3 FORMAL DEVELOPMENT OF GREEN'S FUNCTIONS IN m DIMENSIONS

The preceding section was devoted to discussion of the Green's function for the Laplacian with no mention of the BCs. This section will develop a formalism that not only works for more general operators, but also incorporates the BCs.

12.3.1 General Properties of Green's Functions in m Dimensions

Basic to any study of Green's functions is Green's identity, whose one-dimensional version we encountered in Chapter 11 [Eq. (11.31)]. Here, we generalize it to m dimensions. Suppose there exist two differential operators, $\mathbb{L}_{\mathbf{x}}$ and $\mathbb{L}_{\mathbf{x}}^{\dagger}$, which, for any two functions u and v, satisfy the following relation

$$\begin{aligned} v^*\mathbb{L}_{\mathbf{x}}[u] - u(\mathbb{L}_{\mathbf{x}}^{\dagger}[v])^* &= \nabla \cdot \mathbf{Q}(u, v^*) \\ &\equiv \sum_{i=1}^{m} \frac{\partial}{\partial x_i} Q_i(u, v^*) \end{aligned} \tag{12.20}$$

The differential operator $\mathbb{L}_{\mathbf{x}}^{\dagger}$ is, as in the one-dimensional case, called the formal adjoint of $\mathbb{L}_{\mathbf{x}}$. Integrating (12.20) over a closed domain D in $\mathbb{R}^m$, whose boundary we denote (as is customary in the mathematics literature) by ∂D, and using the divergence theorem, we obtain

$$\int_{D} d^m x \{v^* \mathbb{L}_{\mathbf{x}}[u] - u(\mathbb{L}_{\mathbf{x}}^{\dagger}[v])^*\} = \int_{\partial D} \mathbf{Q}\cdot\hat{\mathbf{e}}_n \, da \tag{12.21}$$

where $\hat{\mathbf{e}}_n$ is an m-dimensional unit vector normal to ∂D and da is an element of "area" of the m-dimensional hypersurface ∂D. Equation (12.21) is the generalized Green's identity for m dimensions. Note that the weight function is set equal to one for simplicity.

The differential operator $\mathbb{L}_{\mathbf{x}}$ is said to be formally self-adjoint if the RHS of (12.21), the surface term, vanishes. In such a case, we necessarily have

$$\mathbb{L}_{\mathbf{x}} = \mathbb{L}_{\mathbf{x}}^{\dagger} \tag{12.22}$$

as in one dimension. Equation (12.22) is a necessary condition for the surface term to vanish because u and v are, by assumption, arbitrary. $\mathbb{L}_{\mathbf{x}}$ is called hermitian if (12.22) is satisfied and the domains of $\mathbb{L}_{\mathbf{x}}$ and $\mathbb{L}_{\mathbf{x}}^{\dagger}$, determined by the vanishing of the surface term, are identical.

Example 12.3.1

For the operator

$$\mathbb{L}_{\mathbf{x}} \equiv \nabla^2 + \mathbf{b}\cdot\nabla + c$$

in which $\{b_i\}_{i=1}^m$ and c are functions of $\{x_i\}_{i=1}^m$, the identity

$$\nabla\cdot[v\nabla u - u\nabla v + \mathbf{b}uv] = v\nabla^2 u - u\nabla^2 v + v\mathbf{b}\cdot(\nabla u) + u\nabla\cdot(\mathbf{b}v)$$

shows that

$$\mathbb{L}_{\mathbf{x}}^{\dagger}[v] \equiv \nabla^2 v - \nabla\cdot(\mathbf{b}v) + cv$$

and

$$\mathbf{Q}(u, v^*) = \mathbf{Q}(u, v) = v\nabla u - u\nabla v + \mathbf{b}uv$$

The generalized Green's identity is then

$$\int_{D} d^m x(v\mathbb{L}_{\mathbf{x}}[u] - u\mathbb{L}_{\mathbf{x}}^{\dagger}[v]) = \int_{\partial D} (v\nabla u - u\nabla v + uv\mathbf{b})\cdot\hat{\mathbf{e}}_n \, da$$

A necessary condition for $\mathbb{L}_{\mathbf{x}}$ to be self-adjoint is [see Eq. (12.22)]

$$\mathbb{L}_{\mathbf{x}}[u] = \mathbb{L}_{\mathbf{x}}^{\dagger}[u] \qquad \forall u$$

or

$$\nabla^2 u + \mathbf{b}\cdot\nabla u + cu = \nabla^2 u - \nabla\cdot(\mathbf{b}u) + cu$$

This gives the condition

$$\mathbf{b}\cdot\nabla u = -\nabla\cdot(\mathbf{b}u) = -(\nabla\cdot\mathbf{b})u - \mathbf{b}\cdot\nabla u \tag{1}$$

or

$$2\mathbf{b}\cdot\nabla u + u(\nabla\cdot\mathbf{b}) = 0 \qquad \forall u \tag{2}$$

For $u = 1$ Eq. (2) gives $\nabla\cdot\mathbf{b} = 0$, reducing Eq. (2) to

$$\mathbf{b}\cdot\nabla u = 0$$

Letting $u = x_i$ gives $b_i = 0$, from which we conclude that

$$\mathbb{L}_{\mathbf{x}} = \nabla^2 + c(\mathbf{x})$$

is formally self-adjoint. •

We can use Eq. (12.21) to study the PDE

$$\mathbb{L}_{\mathbf{x}}[u] = f(\mathbf{x}) \tag{12.23}$$

and its formal adjoint

$$\mathbb{L}_{\mathbf{x}}^{\dagger}[v] = h(\mathbf{x}) \tag{12.24}$$

As in one dimension, we let $G(\mathbf{x}, \mathbf{y})$ and $g(\mathbf{x}, \mathbf{y})$ denote the Green's functions for $\mathbb{L}_{\mathbf{x}}$ and $\mathbb{L}_{\mathbf{x}}^{\dagger}$, respectively. Let us assume that the BCs are such that the surface term in (12.21) vanishes. Then we get Green's identity

$$\int_D d^m x\, v^* \mathbb{L}_{\mathbf{x}}[u] = \int_D d^m x\, u(\mathbb{L}_{\mathbf{x}}^{\dagger}[v])^* \tag{12.25}$$

If we let $u = G(\mathbf{x}, \mathbf{t})$ and $v = g(\mathbf{x}, \mathbf{y})$, where $\mathbf{t}, \mathbf{y} \in \mathrm{D}$, Eq. (12.25) yields

$$\int_D d^m x\, g^*(\mathbf{x}, \mathbf{y})\delta(\mathbf{x} - \mathbf{t}) = \int_D d^m x\, G(\mathbf{x}, \mathbf{t})\delta(\mathbf{x} - \mathbf{y})$$

or

$$g^*(\mathbf{t}, \mathbf{y}) = G(\mathbf{y}, \mathbf{t}) \tag{12.26a}$$

In particular, when $\mathbb{L}_{\mathbf{x}}$ is formally self-adjoint, we have

$$G^*(\mathbf{t}, \mathbf{y}) = G(\mathbf{y}, \mathbf{t}) \tag{12.26b}$$

If all the coefficient functions of $\mathbb{L}_{\mathbf{x}}$ are real, G will be real and (12.26b) will reduce to

$$G(\mathbf{t}, \mathbf{y}) = G(\mathbf{y}, \mathbf{t}) \tag{12.26c}$$

That is, the Green's function will be symmetric.

If we let $v = g(\mathbf{x}, \mathbf{y})$ and use (12.23) in (12.25), we get

$$u(\mathbf{y}) = \int_D d^m x\, g^*(\mathbf{x}, \mathbf{y}) f(\mathbf{x})$$

which, using (12.26a) and interchanging $\mathbf{x}$ and $\mathbf{y}$, becomes

$$u(\mathbf{x}) = \int_D d^m y\, G(\mathbf{x}, \mathbf{y}) f(\mathbf{y})$$

It can similarly be shown that

$$v(\mathbf{x}) = \int_D d^m y\, g(\mathbf{x}, \mathbf{y}) h(\mathbf{y})$$

12.3.2 The Fundamental (Singular) Solutions

The inhomogeneous term of the differential equation to which $G(\mathbf{x}, \mathbf{y})$ is a solution is the delta function, $\delta(\mathbf{x} - \mathbf{y})$. It would be surprising if $G(\mathbf{x}, \mathbf{y})$ did not "take notice" of this catastrophic source term and adapt itself to behave differently at $\mathbf{x} = \mathbf{y}$ than at any other "ordinary" point. We noted the singular behavior of the Green's function at $x = y$ in one dimension when we proved Theorem 11.4.6. There we introduced $h(x, y)$, which was discontinuous at $x = y$, as a part of the Green's function. Similarly, when we discussed the Green's functions for the Laplacian in two and m dimensions earlier in this chapter, we noted that they behaved singularly at $\mathbf{r} = 0$ or $\mathbf{x} = \mathbf{y}$.

Next to the Laplacian in difficulty is the elliptic PDO discussed in Example 12.3.1:

$$\mathbb{L}_{\mathbf{x}} \equiv \nabla^2 + q(\mathbf{x})$$

Substituting this in (12.21) and using the expression for $\mathbf{Q}$ given in that example, we obtain

$$\int_{\mathrm{D}} d^m x \{v \mathbb{L}_{\mathbf{x}}[u] - u(\mathbb{L}_{\mathbf{x}}[v])\} = \int_{\partial \mathrm{D}} (v \hat{\mathbf{e}}_{\mathrm{n}} \cdot \nabla u - u \hat{\mathbf{e}}_{\mathrm{n}} \cdot \nabla v)\, da$$

Letting $v = G(\mathbf{x}, \mathbf{y})$ and denoting $\hat{\mathbf{e}}_{\mathrm{n}} \cdot \nabla$ by $\partial/\partial n$ gives

$$\int_{\mathrm{D}} d^m x [G(\mathbf{x}, \mathbf{y}) \mathbb{L}_{\mathbf{x}} u - u \mathbb{L}_{\mathbf{x}} G(\mathbf{x}, \mathbf{y})] = \int_{\partial \mathrm{D}} \left[G(\mathbf{x}, \mathbf{y}) \frac{\partial u}{\partial n} - u \frac{\partial G}{\partial n}(\mathbf{x}, \mathbf{y}) \right] da \qquad (12.27)$$

We want to use (12.27) to find out about the behavior of $G(\mathbf{x}, \mathbf{y})$ as $|\mathbf{x} - \mathbf{y}| \to 0$. Therefore, assuming that $\mathbf{y} \in \mathrm{D}$, we divide the domain D into two parts: one is an infinitesimal hypersphere S_ε with radius ε and center at $\mathbf{y}$; the other is the rest of D. Instead of D we use the region $\mathrm{D}' \equiv \mathrm{D} - \mathrm{S}_\varepsilon$. The following facts are easily deduced for D′:

(1) $\mathbb{L}_{\mathbf{x}} G(\mathbf{x}, \mathbf{y}) = 0$ because $\mathbf{x} \neq \mathbf{y}$ in D′

(2) $\displaystyle \int_{\mathrm{D}} = \lim_{\varepsilon \to 0} \int_{\mathrm{D}'}$

(3) $\partial \mathrm{D}' = \partial \mathrm{D} \cup \partial \mathrm{S}_\varepsilon$

Suppose that we are interested in finding a solution to

$$\mathbb{L}_{\mathbf{x}}[u] \equiv [\nabla^2 + q(\mathbf{x})]u(\mathbf{x}) = f(\mathbf{x})$$

subject to certain, as yet unspecified, BCs. Using the three facts listed above, Eq. (12.27) yields

$$\int_{\mathrm{D}} d^m x\,[G\mathbb{L}_{\mathbf{x}} u - u\mathbb{L}_{\mathbf{x}} G] = \lim_{\varepsilon\to 0}\int_{\mathrm{D}'} d^m x\,[G\mathbb{L}_x u - u\mathbb{L}_x G]$$

$$= \lim_{\varepsilon\to 0}\int_{\mathrm{D}'} d^m x\, G(\mathbf{x},\mathbf{y}) f(\mathbf{x}) \equiv \int_{\mathrm{D}} d^m x\, G(\mathbf{x},\mathbf{y}) f(\mathbf{x})$$

$$= \int_{\partial\mathrm{D}} \left(G\frac{\partial u}{\partial n} - u\frac{\partial G}{\partial n}\right) da + \int_{\partial\mathrm{S}_\varepsilon} \left(G\frac{\partial u}{\partial n} - u\frac{\partial G}{\partial n}\right) da$$

Assuming that the BCs are such that the integral over $\partial\mathrm{D}$ vanishes (recall from Chapter 11 that, at least for one dimension, this is a necessary condition for the existence of Green's functions), we finally obtain

$$\int_{\mathrm{D}} d^m x\, G(\mathbf{x},\mathbf{y}) f(\mathbf{x}) = \int_{\partial\mathrm{S}_\varepsilon} \left(G\frac{\partial u}{\partial n} - u\frac{\partial G}{\partial n}\right) da$$

For an m-dimensional sphere, $da = r^{m-1}\,d\Omega_m$, which for S_ε reduces to $\varepsilon^{m-1}\,d\Omega_m$. Substituting in the preceding equation yields

$$\int_{\mathrm{D}} d^m x\, G(\mathbf{x},\mathbf{y}) f(\mathbf{x}) = \int_{\partial\mathrm{S}_\varepsilon} \left(G\frac{\partial u}{\partial n} - u\frac{\partial G}{\partial n}\right)\varepsilon^{m-1}\,d\Omega_m$$

We would like the RHS to be $u(\mathbf{y})$. This will be the case if

$$\lim_{\varepsilon\to 0}\int_{\partial\mathrm{S}_\varepsilon} G(\mathbf{x},\mathbf{y})\frac{\partial u}{\partial n}\varepsilon^{m-1}\,d\Omega_m = 0 \qquad \text{and} \qquad \lim_{\varepsilon\to 0}\int_{\partial\mathrm{S}_\varepsilon} u\frac{\partial G}{\partial n}\varepsilon^{m-1}\,d\Omega_m = u(\mathbf{y})$$

for *arbitrary* u. This will happen only if

$$\lim_{r\to 0} G(\mathbf{y}+\mathbf{r},\mathbf{y})r^{m-1} = 0 \tag{12.28a}$$

and

$$\lim_{r\to 0}\frac{\partial G}{\partial r}(\mathbf{y}+\mathbf{r},\mathbf{y})r^{m-1} = \text{constant} \tag{12.28b}$$

A solution to these two equations is

$$G(\mathbf{x},\mathbf{y}) = \begin{cases} -\dfrac{F(\mathbf{x},\mathbf{y})}{2\pi}\ln(|\mathbf{x}-\mathbf{y}|) + H(\mathbf{x},\mathbf{y}) & \text{if } m = 2 \\[2ex] -\dfrac{1}{(m-2)\Omega_m}\left[\dfrac{F(\mathbf{x},\mathbf{y})}{|\mathbf{x}-\mathbf{y}|^{m-2}}\right] + H(\mathbf{x},\mathbf{y}) & \text{if } m \geqslant 3 \end{cases} \tag{12.29}$$

where $H(\mathbf{x},\mathbf{y})$ and $F(\mathbf{x},\mathbf{y})$ are well-behaved at $\mathbf{x}=\mathbf{y}$. The introduction of these functions is necessary because Eqs. (12.28) determine the behavior of $G(\mathbf{x},\mathbf{y})$ only when $\mathbf{x} \approx \mathbf{y}$. Such behavior does not uniquely determine $G(\mathbf{x}, \mathbf{y})$. For instance, $e^{|\mathbf{x}-\mathbf{y}|} \ln(|\mathbf{x} - \mathbf{y}|)$ and $\ln(|\mathbf{x} - \mathbf{y}|)$ behave in the same way as $|\mathbf{x} - \mathbf{y}| \to 0$.

Equation (12.29) shows that, at least for $\mathbb{L}_\mathbf{x} = \nabla^2 + q(\mathbf{x})$, the Green's function consists of two parts. The first part determines the singular behavior of the Green's function as $\mathbf{x} \to \mathbf{y}$. The nature of this singularity (how badly the GF "blows up" as $\mathbf{x} \to \mathbf{y}$) is extremely important, because it is a prerequisite for being able to write the solution in terms of an integral representation with the Green's function as its kernel. Because of their importance in such representations, the first terms on the RHS of (12.29) are called the *fundamental solutions* of the differential equation, or the *singular part* of the Green's function.

What about the second part of the Green's function? What role does it play in obtaining a solution? So far we have been avoiding consideration of BCs. Here $H(\mathbf{x}, \mathbf{y})$ can help. We choose $H(\mathbf{x}, \mathbf{y})$ in such a way that $G(\mathbf{x}, \mathbf{y})$ satisfies the appropriate BCs. Let us discuss this in greater detail and generality.

If BCs are ignored, the Green's function for a SOPDO $\mathbb{L}_\mathbf{x}$ cannot be determined uniquely. In particular, if $G(\mathbf{x}, \mathbf{y})$ is a Green's function, that is, if $\mathbb{L}_\mathbf{x} G(\mathbf{x}, \mathbf{y}) = \delta(\mathbf{x} - \mathbf{y})$, then so is $G(\mathbf{x}, \mathbf{y}) + H(\mathbf{x}, \mathbf{y})$ as long as $H(\mathbf{x}, \mathbf{y})$ is a solution of the homogeneous equation, $\mathbb{L}_\mathbf{x} H(\mathbf{x}, \mathbf{y}) = 0$. Thus, we can break the Green's function into two parts as follows:

$$G(\mathbf{x}, \mathbf{y}) \equiv G_s(\mathbf{x}, \mathbf{y}) + H(\mathbf{x}, \mathbf{y}) \tag{12.30a}$$

Here $G_s(\mathbf{x}, \mathbf{y})$ is the singular part of the Green's function, satisfying

$$\mathbb{L}_\mathbf{x} G_s(\mathbf{x}, \mathbf{y}) = \delta(\mathbf{x} - \mathbf{y}) \tag{12.30b}$$

and $H(\mathbf{x}, \mathbf{y})$ is the *regular* part of the Green's function, satisfying

$$\mathbb{L}_\mathbf{x} H(\mathbf{x}, \mathbf{y}) = 0 \tag{12.30c}$$

Neither G_s nor H (nor G, therefore) is unique. However, if we impose the appropriate BCs, which depend on the type of $\mathbb{L}_\mathbf{x}$, then $G(\mathbf{x}, \mathbf{y})$ will be unique.

To be more specific, let us assume that we want to find a Green's function for $\mathbb{L}_\mathbf{x}$ that vanishes at the boundary ∂D; that is, we wish to find $G(\mathbf{x}, \mathbf{y})$ such that $G(\mathbf{x}_b, \mathbf{y}) = 0$, where $\mathbf{x}_b$ is an arbitrary point of the boundary. All that is required is to find a $G_s(\mathbf{x}, \mathbf{y})$ satisfying Eq. (12.30b) and an $H(\mathbf{x}, \mathbf{y})$ satisfying Eq. (12.30c) with the BC that $H(\mathbf{x}_b, \mathbf{y}) = -G_s(\mathbf{x}_b, \mathbf{y})$. The latter problem, involving a homogeneous differential equation, can be handled by the methods of Chapter 10. Since any discussion of BCs is tied to the type of the PDE, the various types are discussed in the following section.

12.4 GREEN'S FUNCTIONS FOR THE THREE TYPES OF PARTIAL DIFFERENTIAL EQUATIONS

This section considers the Green's functions for elliptic, parabolic, and hyperbolic equations that satisfy the BCs appropriate for each type of PDE, as outlined at the end of Section 12.1.

12.4.1 Elliptic Equations

The most general linear PDE in m variables that is of the elliptic type was discussed in Section 12.1. We will not discuss this general case because all elliptic PDOs encountered in mathematical physics are of a much simpler nature. In fact, the self-adjoint elliptic PDO of the form

$$\mathbb{L}_{\mathbf{x}} \equiv \nabla^2 + q(\mathbf{x})$$

is sufficiently general for purposes of this discussion.

Recall from Section 12.1 that the BCs associated with an elliptic PDE are of two types, Dirichlet and Neumann. Let us consider these separately.

The Dirichlet boundary value problem. A Dirichlet BVP consists of an elliptic PDE together with a Dirichlet BC, such as the PDE

$$\mathbb{L}_{\mathbf{x}}[u] \equiv \nabla^2 u + q(\mathbf{x})u = f(\mathbf{x}) \qquad \text{for } \mathbf{x} \in \mathrm{D} \tag{12.31a}$$

together with the BC

$$u(\mathbf{x}_b) = g(\mathbf{x}_b) \qquad \text{for } \mathbf{x}_b \in \partial\mathrm{D} \equiv \mathrm{S} \tag{12.31b}$$

where $g(\mathbf{x}_b)$ is a given function defined on the closed hypersurface S.

The Green's function for the Dirichlet BVP must satisfy the *homogeneous* BC, for the same reason as for the one-dimensional Green's function. Thus, the Dirichlet Green's function, denoted by $G_\mathrm{D}(\mathbf{x}, \mathbf{y})$, must satisfy

$$\mathbb{L}_{\mathbf{x}}[G_\mathrm{D}(\mathbf{x}, \mathbf{y})] = \delta(\mathbf{x} - \mathbf{y})$$

$$G_\mathrm{D}(\mathbf{x}_b, \mathbf{y}) = 0, \qquad \mathbf{x}_b \in \mathrm{S} \equiv \partial\mathrm{D}$$

This, along with the discussion in Section 12.3.2 concerning the singular and regular parts of the Green's function, leads to a definition.

12.4.1 Definition The Green's function associated with the self-adjoint elliptic operator $\mathbb{L}_{\mathbf{x}}$ in the domain D is a function $G_\mathrm{D}(\mathbf{x}, \mathbf{y})$ satisfying the following conditions:

(i) $\mathbb{L}_{\mathbf{x}}[G_\mathrm{D}(\mathbf{x}, \mathbf{y})] = \delta(\mathbf{x} - \mathbf{y})$

(ii) $G_\mathrm{D}(\mathbf{x}_b, \mathbf{y}) = 0, \qquad \forall\, \mathbf{x}_b \in \partial\mathrm{D}$

(iii) $G_\mathrm{D}(\mathbf{x}, \mathbf{y})$ can be written as $G_\mathrm{D}^{(s)}(\mathbf{x}, \mathbf{y}) + H(\mathbf{x}, \mathbf{y})$, where $G_\mathrm{D}^{(s)}$ is any fundamental solution of $\mathbb{L}_{\mathbf{x}} G_\mathrm{D}^{(s)}(\mathbf{x}, \mathbf{y}) = \delta(\mathbf{x} - \mathbf{y})$ and $H(\mathbf{x}, \mathbf{y})$ is that regular solution of the Dirichlet problem in D which satisfies the BC

$$H(\mathbf{x}_b, \mathbf{y}) = -G_\mathrm{D}^{(s)}(\mathbf{x}, \mathbf{y})$$

Using Eqs. (12.31) and the properties of $G_\mathrm{D}(\mathbf{x}, \mathbf{y})$ in Eq. (12.27) and switching $\mathbf{x}$ and $\mathbf{y}$, we obtain

$$u(\mathbf{x}) = \int_\mathrm{D} d^m y\, G_\mathrm{D}(\mathbf{x}, \mathbf{y}) f(\mathbf{y}) + \int_{\partial\mathrm{D}} g(\mathbf{y}_b) \frac{\partial G}{\partial n_y}(\mathbf{x}, \mathbf{y}_b)\, da \tag{12.32}$$

where $\partial/\partial n_y$ indicates normal differentiation with respect to the second argument.

Some special cases of (12.32) are worthy of mention. The first is $u(\mathbf{x}_b) = 0$, the solution to the inhomogeneous DE satisfying the homogeneous BC. We obtain this by substituting zero for $g(\mathbf{x}_b)$ in (12.32), which yields

$$u(\mathbf{x}) = \int_{\mathrm{D}} d^m y\, G_{\mathrm{D}}(\mathbf{x}, \mathbf{y}) f(\mathbf{y})$$

The second special case is when the DE is homogeneous, that is, when $f(\mathbf{x}) = 0$, but the BC is inhomogeneous. This yields

$$u(\mathbf{x}) = \int_{\partial \mathrm{D}} g(\mathbf{x}_b) \frac{\partial G}{\partial n}(\mathbf{x}_b, \mathbf{y})\, da$$

which is again very similar to the one-dimensional case. Finally, the solution to the homogeneous DE with the homogeneous BC is simply $u \equiv 0$, referred to as the zero solution. This is consistent with physical intuition. If the function is zero on the boundary and there is no source, $f(\mathbf{x})$, to produce any "disturbance," we expect no nontrivial solution.

Example 12.4.1

Let us find the Green's function for the three-dimensional Laplacian, $\mathbb{L}_{\mathbf{x}} \equiv \nabla^2$, satisfying the Dirichlet BC that $G_{\mathrm{D}}(\boldsymbol{\rho}, \mathbf{y}) = 0$ for $\boldsymbol{\rho}$ on the xy-plane. Here D is the upper half-space ($z > 0$) and ∂D is the xy-plane.

It is more convenient to use $\mathbf{r} \equiv (x, y, z)$ and $\mathbf{r}' \equiv (x', y', z')$ instead of $\mathbf{x}$ and $\mathbf{y}$, respectively. Using (12.19) as $G_{\mathrm{D}}^{(s)}$, the singular part of G_{D}, we can write

$$G_{\mathrm{D}}(\mathbf{r}, \mathbf{r}') = -\frac{1}{4\pi\,|\mathbf{r} - \mathbf{r}'|} + H(\mathbf{r}, \mathbf{r}')$$

$$= -\frac{1}{4\pi} \frac{1}{\sqrt{(x - x')^2 + (y - y')^2 + (z - z')^2}} + H(x, y, z; x', y', z')$$

The requirement that G_{D} vanish on the xy-plane gives

$$H(x, y, 0; x', y', z') = \frac{1}{4\pi} \frac{1}{\sqrt{(x - x')^2 + (y - y')^2 + z'^2}}$$

This fixes the dependence of H on x and y. On the other hand, $\nabla^2 H = 0$ in D implies that the form of H must be the same as that of $G_{\mathrm{D}}^{(s)}$ because, except at $\mathbf{r} = \mathbf{r}'$, the latter does satisfy the Laplace equation. Thus, we have two choices for the z-dependence: $(z - z')^2$ and $(z + z')^2$. The former gives $G \equiv 0$, which is a trivial solution. Thus, we must choose

$$H(x, y, z; x', y', z') = \frac{1}{4\pi} \frac{1}{\sqrt{(x - x')^2 + (y - y')^2 + (z + z')^2}}$$

Note that, with $\mathbf{r}'' \equiv (x', y', -z')$, this equation satisfies

$$\nabla^2 H(\mathbf{r}; \mathbf{r}') = -\delta(\mathbf{r} - \mathbf{r}'')$$

From this, it may appear that H does not satisfy the homogeneous DE, as it should. However, $\mathbf{r}''$ is outside D. Thus, $\mathbf{r} \neq \mathbf{r}''$ as long as $\mathbf{r} \in \mathrm{D}$, and H does satisfy the homogeneous DE in D. The Green's function for the given Dirichlet BC is, therefore,

$$G_{\mathrm{D}}(\mathbf{r}, \mathbf{r}') = -\frac{1}{4\pi}\left(\frac{1}{|\mathbf{r} - \mathbf{r}'|} - \frac{1}{|\mathbf{r} - \mathbf{r}''|}\right)$$

where $\mathbf{r}''$ is the *reflection of* $\mathbf{r}'$ in the xy-plane.

This result has a direct physical interpretation. If determining the solution of the Laplace equation is considered a problem in electrostatics, then $G_{\mathrm{D}}^{(s)}(\mathbf{r}, \mathbf{r}')$ is simply the potential at $\mathbf{r}$ of a unit point charge located at $\mathbf{r}'$, and $G_{\mathrm{D}}(\mathbf{r}, \mathbf{r}')$ is the potential of two point charges of opposite signs, one at $\mathbf{r}'$ and the other at the mirror image of $\mathbf{r}'$. The fact that the two charges are equidistant from the xy-plane assures the vanishing of the potential there. The introduction of image charges to assure the vanishing of $G_{\mathrm{D}}(\mathbf{r}, \mathbf{r}')$ on $\partial\mathrm{D}$ is very common in electrostatics and is known as the *method of images*. Thus, the Dirichlet problem for the Laplacian reduces to finding appropriate point charges *outside* D that guarantee the vanishing of the potential on $\partial\mathrm{D}$. For simple geometries, such as the one discussed in this example, determination of the magnitudes and locations of such image charges is easy, rendering the method extremely useful. For complicated geometries it is necessary to resort to other methods.

Having found the Green's function, we can pose the general Dirichlet BVP:

$$\nabla^2 u = -\rho(\mathbf{r}) \qquad \text{and} \qquad u(x, y, 0) \equiv g(x, y) \qquad \text{for } z > 0$$

The solution is

$$u(x, y, z) = \frac{1}{4\pi}\int_{-\infty}^{\infty} dx' \int_{-\infty}^{\infty} dy \int_{0}^{\infty} dz'\, \rho(x', y', z')\left[\frac{1}{\sqrt{(x - x')^2 + (y - y')^2 + (z - z')^2}} - \frac{1}{\sqrt{(x - x')^2 + (y - y')^2 + (z + z')^2}}\right] + \int_{-\infty}^{\infty} dx' \int_{-\infty}^{\infty} dy'\, g(x', y')\left.\frac{\partial G_{\mathrm{D}}}{\partial z}\right|_{z=0} \tag{1}$$

A typical application consists of introducing a number of charges in the vicinity of an infinite conducting sheet, which is held at a constant potential V_0. If there are N charges, $\{q_i\}_{i=1}^{N}$, at $\mathbf{r}_1, \mathbf{r}_2, \ldots, \mathbf{r}_N$, then

$$\rho(\mathbf{x}, \mathbf{y}, \mathbf{z}) = \sum_{i=1}^{N} q_i\,\delta(\mathbf{r} - \mathbf{r}_i)$$

and

$$g(x, y) = \text{constant} = V_0$$

Therefore,

$$u(x, y, z) = \frac{1}{4\pi}\sum_{i=1}^{N}\left(\frac{q_i}{|\mathbf{r} - \mathbf{r}_i|} - \frac{q_i}{|\mathbf{r} - \mathbf{r}'_i|}\right) + V_0\int_{-\infty}^{\infty} dx' \int_{-\infty}^{\infty} dy' \left.\frac{\partial G_{\mathrm{D}}}{\partial z}\right|_{z=0} \tag{2}$$

where $\mathbf{r}_i \equiv (x_i, y_i, z_i)$ and $\mathbf{r}'_i \equiv (x_i, y_i, -z_i)$. That the double integral in (2) is unity can be seen by direct integration or by noting that the sum vanishes when $z = 0$. On the other hand, $u(x, y, 0) = V_0$. Thus, the solution becomes

$$u(x, y, z) = \frac{1}{4\pi} \sum_{i=1}^{N} \left(\frac{q_i}{|\mathbf{r} - \mathbf{r}_i|} - \frac{q_i}{|\mathbf{r} - \mathbf{r}_i'|} \right) + V_0$$

•

Example 12.4.2

The method of images is also applicable when the boundary is a sphere. Inside a sphere of radius a with center at the origin, we wish to solve this Dirichlet BVP:

$$\nabla^2 u = -\rho(r, \theta, \varphi) \qquad \text{for } r < a$$

$$u(a, \theta, \varphi) \equiv g(\theta, \varphi)$$

The GF satisfies

$$\nabla^2 G_D(r, \theta, \varphi; r', \theta', \varphi') = \delta(\mathbf{r} - \mathbf{r}') \qquad \text{for } r < a \tag{1}$$

$$G_D(a, \theta, \varphi; r', \theta', \varphi') = 0 \tag{2}$$

Thus, G_D can again be interpreted as potentials of point charges, of which one is in the sphere and the others are outside.

We write $G_D = G_D^{(s)} + H$ and choose H in such a way that Eq. (2) is satisfied. Since $G_D^{(s)}$ satisfies (1), H must satisfy the Laplace equation. If we let

$$H(\mathbf{r}, \mathbf{r}'') = \frac{-k}{4\pi|\mathbf{r} - \mathbf{r}''|}$$

where k is a constant to be determined and $\mathbf{r}''$ is *outside* the sphere, then $\nabla^2 H$ will vanish *inside* the sphere. The problem has been reduced to finding k and $\mathbf{r}''$ (the location of the image charge). We can write

$$G_D(\mathbf{r}, \mathbf{r}') = \frac{-1}{4\pi|\mathbf{r} - \mathbf{r}'|} + \frac{k}{4\pi|\mathbf{r} - \mathbf{r}''|}$$

Assuming that $\mathbf{r}''$ is in the same direction as $\mathbf{r}'$, we want to choose $\mathbf{r}''$ so that

$$\left(\frac{1}{|\mathbf{r} - \mathbf{r}'|} \right)_{r=a} = \left(\frac{k}{|\mathbf{r} - \mathbf{r}''|} \right)_{r=a}$$

or

$$k(|\mathbf{r} - \mathbf{r}'|)_{r=a} = (|\mathbf{r} - \mathbf{r}''|)_{r=a}$$

This shows that k must be positive. Squaring both sides and expanding the result yields

$$k^2(a^2 + r'^2 - 2ar' \cos\gamma) = a^2 + r''^2 - 2ar'' \cos\gamma \tag{3}$$

where γ is the angle between $\mathbf{r}$ and $\mathbf{r}'$, and we have assumed that $\mathbf{r}'$ and $\mathbf{r}''$ are in the same direction. For (3) to hold for arbitrary γ, we must have

$$k^2 r' = r'' \tag{4}$$

Then

$$k^2(a^2 + r'^2) = a^2 + r''^2 \tag{5}$$

Equations (4) and (5) yield

$$k^4 r'^2 - k^2(a^2 + r'^2) + a^2 = 0$$

whose positive solutions are

$$k = 1, \frac{a}{r'}$$

The first choice implies that $r'' = r'$, which is impossible because r'' must be outside the sphere. We thus choose $k = a/r'$, which gives $\mathbf{r}'' = (a^2/r'^2)\mathbf{r}'$. With this value for $\mathbf{r}''$, we have

$$G_D(\mathbf{r}, \mathbf{r}') = -\frac{1}{4\pi}\left[\frac{1}{|\mathbf{r} - \mathbf{r}'|} - \frac{ar'}{|r'^2\mathbf{r} - a^2\mathbf{r}'|}\right] \tag{6}$$

Substituting this in Eq. (12.32) and noting that $\partial G/\partial n_y = (\partial G/\partial r')_{r'=a}$ yields

$$\begin{aligned} u(\mathbf{r}) = {} & \frac{1}{4\pi}\int_0^a r'^2\,dr' \int_0^\pi \sin\theta'\,d\theta' \int_0^{2\pi}\left(\frac{1}{|\mathbf{r} - \mathbf{r}'|} - \frac{ar'}{|r'^2\mathbf{r} - a^2\mathbf{r}'|}\right)\rho(\mathbf{r}') \\ & + \frac{a(a^2 - r^2)}{4\pi}\int_0^{2\pi} d\varphi' \int_0^\pi \sin\theta'\,d\theta' \frac{g(\theta', \varphi')}{|\mathbf{r} - \mathbf{a}|^3} \end{aligned} \tag{7}$$

where $\mathbf{a} \equiv (a, \theta', \varphi')$ is a vector from the origin to a point on the sphere. For the Laplace equation $\rho(\mathbf{r}') = 0$, and (7) yields

$$u(\mathbf{r}) = \frac{a(a^2 - r^2)}{4\pi}\int_0^{2\pi} d\varphi' \int_0^\pi \sin\theta'\,d\theta' \frac{g(\theta', \varphi')}{|\mathbf{r} - \mathbf{a}|^3} \tag{8}$$

It can be shown that if $g(\theta', \varphi') = V_0 =$ constant, then $u(\mathbf{r}) = V_0$. This fact is shown in electromagnetic theory. If the potential on a sphere is kept constant, the potential inside the sphere will be constant and equal to the potential at the surface. The proof is left as a problem. ●

The Neumann boundary value problem. The Neumann BVP is not as simple as the Dirichlet BVP. This is because the normal derivative of the solution must be determined for the Neumann BVP. But the normal derivative is related to the Laplacian through the divergence theorem. Thus, the BC and the DE are tied together, and unless we impose some *solvability conditions*, we may have no solution at all. These points are illustrated clearly if we consider the Laplacian operator.

Consider the Neumann BVP

$$\nabla^2 u = f(\mathbf{x}) \qquad \text{for } \mathbf{x} \in \mathrm{D}$$

$$\frac{\partial u}{\partial n} = g(\mathbf{x}) \qquad \text{for } \mathbf{x} \in \partial\mathrm{D}$$

Integrating the first equation over D and using the divergence theorem, we obtain

$$\int_{\mathrm{D}} f(x)\,d^m x = \int_{\mathrm{D}} \nabla\cdot(\nabla u)\,d^m x = \int_{\partial\mathrm{D}} \hat{\mathbf{e}}_n\cdot\nabla u\,da = \int_{\partial\mathrm{D}} \frac{\partial u}{\partial n}\,da$$

This shows that we cannot arbitrarily assign values of $\partial u/\partial n$ on the boundary. In particular, if the BC is homogeneous, as in the case of Green's functions, the RHS is zero:

$$\int_D f(\mathbf{x})\, d^m x = 0$$

This relation is restrictive on the DE and not on the solution, and is a solvability condition, as mentioned above. To satisfy this condition, it is necessary to subtract from the inhomogeneous term its average value over the region D. Thus, if V_D is the volume of the region D, then

$$\nabla^2 u = f(\mathbf{x}) - \bar{f}$$

where

$$\bar{f} \equiv \frac{1}{V_D} \int_D d^m x\, f(\mathbf{x})$$

ensures that the Neumann BVP is solvable. In particular, the inhomogeneous term for the Green's function is not simply $\delta(\mathbf{x}-\mathbf{y})$ but $\delta(\mathbf{x}-\mathbf{y}) - \bar{\delta}$, where

$$\bar{\delta} \equiv \frac{1}{V_D} \int_D d^m x\, \delta(\mathbf{x}-\mathbf{y}) = \frac{1}{V_D} \qquad \text{if } \mathbf{y} \in D$$

Thus, the Green's function for the Neumann BVP, $G_N(\mathbf{x}, \mathbf{y})$, satisfies

$$\nabla^2 G_N(\mathbf{x}, \mathbf{y}) = \delta(\mathbf{x}-\mathbf{y}) - \frac{1}{V_D}$$

$$\frac{\partial G_N}{\partial n}(\mathbf{x}, \mathbf{y}) = 0 \qquad \text{for } \mathbf{x} \in \partial D$$

Applying Green's identity, Eq. (12.27), we get

$$\int_D d^m y\, G_N(\mathbf{x}, \mathbf{y}) f(\mathbf{y}) - \int_D u\left[\delta(\mathbf{x}-\mathbf{y}) - \frac{1}{V_D}\right] d^m y = \int_{\partial D} G_N(\mathbf{x}, \mathbf{y}) \frac{\partial u}{\partial n}\, da$$

which leads to the representation

$$u(\mathbf{x}) = \int_D d^m y\, G(\mathbf{x}, \mathbf{y}) f(\mathbf{y}) - \int_{\partial D} G_N(\mathbf{x}, \mathbf{y}) \frac{\partial u}{\partial n}\, da + \bar{u} \tag{12.33}$$

where

$$\bar{u} = \frac{1}{V_D} \int_D d^m x\, u(\mathbf{x})$$

is the average value of u in D. Equation (12.33) is valid only for the Laplacian operator, although a similar result can be obtained for a general self-adjoint

SOLPDO with constant coefficients. We will not pursue that result, however, since it is of little practical use.

Throughout the discussion so far we have assumed that D is bounded; that is, we have considered points inside D with BCs on the boundary ∂D specified. This is called an *interior BVP*. In many physical situations we are interested in points outside D. We are then dealing with an *exterior BVP*. In dealing with such a problem, we must specify the behavior of the Green's function at infinity. In most cases the physical situation dictates such behavior. For instance, for the case of an exterior Dirichlet BVP, where

$$u(\mathbf{x}) = \int_{\mathrm{D}} G_{\mathrm{D}}(\mathbf{x}, \mathbf{y}) f(\mathbf{y})\, d^m y + \int_{\partial \mathrm{D}} u(\mathbf{y}_b) \frac{\partial G_{\mathrm{D}}}{\partial n_y}(\mathbf{x}, \mathbf{y}_b)\, da$$

and it is desired that $u(\mathbf{x}) \to 0$ as $|\mathbf{x}| \to \infty$, the vanishing of $G_{\mathrm{D}}(\mathbf{x}, \mathbf{y})$ at infinity guarantees that the second integral vanishes, as long as ∂D is a finite hypersurface. To guarantee the disappearance of the first integral, we must demand that $G_{\mathrm{D}}(\mathbf{x}, \mathbf{y})$ tend to zero faster than $f(\mathbf{y})\, d^m y$ tends to infinity. For most cases of physical interest, the calculation of the exterior Green's functions is not conceptually different from that of the interior ones. However, the algebra may be more involved.

Later we will develop general methods for finding the Green's functions for certain partial differential operators that satisfy appropriate BCs. At this point, let us simply mention what are called *mixed BCs* for elliptic PDEs. A general mixed BC is of the form

$$\alpha(\mathbf{x})\, u(\mathbf{x}) + \beta(\mathbf{x}) \frac{\partial u}{\partial n}(\mathbf{x}) = \gamma(\mathbf{x}) \qquad \text{for } \mathbf{x} \in \partial \mathrm{D} \tag{12.34}$$

Exercise 12.4.1 examines the conditions that the GF must satisfy in such a case.

12.4.2 Parabolic Equations

Elliptic partial differential equations arise in static problems, where the solution is independent of time. As was mentioned in Chapters 8 and 10, there are two major time-dependent equations: the diffusion (or heat) equation (which turns into the Schrödinger equation if t is changed to it; thus, the following discussion incorporates the Schrödinger equation as well) and the wave equation. The former is a parabolic PDE and the latter a hyperbolic PDE. This subsection examines the diffusion equation, which is of the form

$$\nabla^2 u = a^2 \frac{\partial u}{\partial t}$$

By changing t to t/a^2, we can write the equation as

$$\mathbb{L}_{\mathbf{x},t}[u] \equiv \left(\frac{\partial}{\partial t} - \nabla^2\right) u(\mathbf{x}, t) = 0$$

We wish to calculate the Green's function associated with $\mathbb{L}_{\mathbf{x},t}$ and the homogeneous BCs. Because of the time variable, we must also specify the solution at $t = 0$. Thus, we consider the BVP

$$\begin{aligned} \mathbb{L}_{\mathbf{x},t}[u] \equiv \left(\frac{\partial}{\partial t} - \nabla^2\right) u = 0 \qquad & \text{for } \mathbf{x} \in \mathrm{D} \\ u(\mathbf{x}_\mathrm{b}, t) = 0 \qquad & \text{for } \mathbf{x}_\mathrm{b} \in \partial\mathrm{D} \\ u(\mathbf{x}, 0) = h(\mathbf{x}) \qquad & \text{for } \mathbf{x} \in \mathrm{D} \end{aligned} \tag{12.35}$$

To find a solution to (12.35), we can use a method that turns out to be useful for evaluating Green's functions in general—the method of eigenfunctions. Let $\{u_n\}_{n=1}^{\infty}$ be the eigenfunctions of ∇^2 with eigenvalues $\{-\lambda_n\}_{n=1}^{\infty}$. Let the BC be $u_n(\mathbf{x}_\mathrm{b}) = 0$ for $\mathbf{x}_\mathrm{b} \in \partial\mathrm{D}$. Then

$$\begin{aligned} \nabla^2 u_n(\mathbf{x}) + \lambda_n u_n(\mathbf{x}) = 0 \qquad & \text{for } n = 1, 2, \ldots ; \mathbf{x} \in \mathrm{D} \\ u_n(\mathbf{x}_\mathrm{b}) = 0 \qquad & \text{for } \mathbf{x}_\mathrm{b} \in \partial\mathrm{D} \end{aligned} \tag{12.36}$$

Equations (12.36) constitute a Sturm-Liouville problem in m dimensions. We saw in Chapter 10 (for $m = 3$) how such a problem can be reduced to m S-L problems in one dimension and how each S-L system produces orthogonal eigenfunctions. Applying the results of Chapter 10, we assume that (12.36) has a solution with $\{u_n\}_{n=1}^{\infty}$ a complete orthonormal set.

Because of the completeness of u_n, we can write

$$u(\mathbf{x}, t) = \sum_{n=1}^{\infty} C_n(t)\, u_n(\mathbf{x}) \tag{12.37}$$

This is possible because at each specific value of t, $u(\mathbf{x}, t)$ is a function of $\mathbf{x}$ and, therefore, can be written as a linear combination of the *same set*, $\{u_n\}_{n=1}^{\infty}$. Since the function $u(\mathbf{x}, t)$ is in terms of the same u_n's, the coefficients must clearly depend on time. The coefficients $C_n(t)$ are given by

$$C_n(t) = \int_\mathrm{D} u(\mathbf{x}, t)\, u_n(\mathbf{x})\, d^m x$$

To calculate $C_n(t)$, we differentiate this equation with respect to time and use (12.35) to obtain

$$\begin{aligned} \dot{C}_n(t) \equiv \frac{dC_n}{dt} &= \int_\mathrm{D} \frac{\partial u}{\partial t}(\mathbf{x}, t)\, u_n(\mathbf{x})\, d^m x \\ &= \int_\mathrm{D} [\nabla^2 u(\mathbf{x}, t)]\, u_n(\mathbf{x})\, d^m x \end{aligned}$$

Using Green's identity for the operator ∇^2 yields

$$\int_D [u_n \nabla^2 u - u \nabla^2 u_n]\, d^m x = \int_{\partial D} \left(u_n \frac{\partial u}{\partial n} - u \frac{\partial u_n}{\partial n} \right) da$$

Since both u and u_n vanish on ∂D, the RHS is zero, and we get

$$\dot{C}_n(t) = \int_D u \nabla^2 u_n\, d^m x = -\lambda_n \int_D u(\mathbf{x}, t) u_n(\mathbf{x})\, d^m x = -\lambda_n C_n(t)$$

This has the solution

$$C_n(t) = C_n(0) e^{-\lambda_n t}$$

where

$$C_n(0) = \int_D u(\mathbf{y}, 0) u_n(\mathbf{y})\, d^m y = \int_D h(\mathbf{y}) u_n(\mathbf{y})\, d^m y$$

Substituting the expression for $C_n(t)$ in (12.37), we get

$$u(\mathbf{x}, t) = \sum_{n=1}^{\infty} \left[\int_D h(\mathbf{y}) u_n(\mathbf{y})\, d^m y \right] e^{-\lambda_n t} u_n(\mathbf{x})$$

$$= \int_D \left[\sum_{n=1}^{\infty} e^{-\lambda_n t} u_n(\mathbf{x}) u_n(\mathbf{y}) \right] h(\mathbf{y})\, d^m y$$

This becomes a representation of a Green's function if we define

$$G(\mathbf{x}, \mathbf{y}; t) \equiv \sum_{n=1}^{\infty} e^{-\lambda_n t} u_n(\mathbf{x}) u_n(\mathbf{y})$$

If we also demand that $u(\mathbf{x}, t) = 0$ for $t < 0$, then we have

$$G(\mathbf{x}, \mathbf{y}, t) \equiv \sum_{n=1}^{\infty} e^{-\lambda_n t} u_n(\mathbf{x}) u_n(\mathbf{y}) \theta(t)$$

or, more generally,

$$G(\mathbf{x}, \mathbf{y}; t - \tau) \equiv \sum_{n=1}^{\infty} e^{-\lambda_n (t-\tau)} u_n(\mathbf{x}) u_n(\mathbf{y}) \theta(t - \tau) \qquad (12.38)$$

Note this important property:

$$\lim_{\tau \to t} G(\mathbf{x}, \mathbf{y}; t - \tau) = \sum_{n=1}^{\infty} u_n(\mathbf{x}) u_n(\mathbf{y}) = \delta(\mathbf{x} - \mathbf{y})$$

This is usually written as

$$G(\mathbf{x}, \mathbf{y}; 0^+) = \delta(\mathbf{x} - \mathbf{y}) \qquad (12.39)$$

Note also that

$$\mathbb{L}_{\mathbf{x},t} G(\mathbf{x}, \mathbf{y}; t - \tau) = \left(\frac{\partial}{\partial t} - \nabla^2\right) \sum_{n=1}^{\infty} e^{-\lambda_n(t-\tau)} u_n(\mathbf{x}) u_n(\mathbf{y}) \theta(t - \tau)$$

$$= \sum_{n=1}^{\infty} (-\lambda_n) e^{-\lambda_n(t-\tau)} u_n(\mathbf{x}) u_n(\mathbf{y}) \theta(t - \tau)$$

$$+ \sum_{n=1}^{\infty} e^{-\lambda_n(t-\tau)} u_n(\mathbf{x}) u_n(\mathbf{y}) \delta(t - \tau)$$

$$- \sum_{n=1}^{\infty} e^{-\lambda_n(t-\tau)} [\nabla^2 u_n(\mathbf{x})] u_n(\mathbf{y}) \theta(t - \tau)$$

The first and the last terms on the RHS cancel each other because of (12.36). The middle term is nonzero only if $t = \tau$; in that case the sum gives $\delta(\mathbf{x} - \mathbf{y})$. Thus,

$$\mathbb{L}_{\mathbf{x},t} G(\mathbf{x}, \mathbf{y}; t - \tau) = \delta(\mathbf{x} - \mathbf{y}) \delta(t - \tau) \tag{12.40}$$

This is precisely what we expect for the Green's function of an operator in the variables $\mathbf{x}$ and t. Another property of $G(\mathbf{x}, \mathbf{y}; t - \tau)$ is that it vanishes on ∂D.

Having found the Green's function and noted its properties, we are in a position to solve the inhomogeneous equation

$$\mathbb{L}_{\mathbf{x},t} [u] = f(\mathbf{x}, \mathrm{t}) \qquad \text{for } \mathbf{x} \in \mathrm{D} \tag{12.41a}$$

subject to the BCs

$$\begin{aligned} u(\mathbf{x}_b, t) &= g(\mathbf{x}_b, t) \qquad &\text{for } \mathbf{x}_b \in \partial\mathrm{D} \\ u(\mathbf{x}, 0) &= h(\mathbf{x}) \qquad &\text{for } \mathbf{x} \in \mathrm{D} \end{aligned} \tag{12.41b}$$

Experience with similar but simpler problems indicates that to make any progress toward a solution, we must come up wth a form of Green's identity involving $\mathbb{L}_{\mathbf{x},t}$ and its adjoint. It is easy to show that

$$v(\mathbf{x}, t) \mathbb{L}_{\mathbf{x},t} u(\mathbf{x}, t) - u(\mathbf{x}, t) \mathbb{L}^{\dagger}_{\mathbf{x},t} v(\mathbf{x}, t) = \frac{\partial}{\partial t}(uv) - \boldsymbol{\nabla} \cdot (v \boldsymbol{\nabla} u - u \boldsymbol{\nabla} v) \tag{12.42a}$$

where $\mathbb{L}^{\dagger}_{\mathbf{x},t} = -\partial/\partial t - \nabla^2$.

Now consider the $(m + 1)$-dimensional "cylinder," one of whose bases is at $t = \varepsilon$, where ε is a small positive number. This base is at the m-dimensional hyperplane $\mathbb{R}^m$. The other base is at $t = \tau - \varepsilon$ and is a duplicate of $\mathrm{D} \subset \mathbb{R}^m$ (see Fig. 12.1). Let a^μ, where $\mu = 0, 1, \ldots, m$, be the components of an $(m + 1)$-dimensional vector $\mathbf{a} \equiv (a^0, a^1, \ldots, a^m)$. Define an inner product by

$$\mathbf{a} \cdot \mathbf{b} \equiv a^\mu b_\mu \equiv a^0 b^0 - a^1 b^1 - a^2 b^2 - \cdots - a^m b^m \equiv a^0 b^0 - \mathbf{a} \cdot \mathbf{b}$$

Define the vector $\mathbf{Q}$ by

$$Q_0 = uv \qquad \mathbf{Q} = v \boldsymbol{\nabla} u - u \boldsymbol{\nabla} v$$

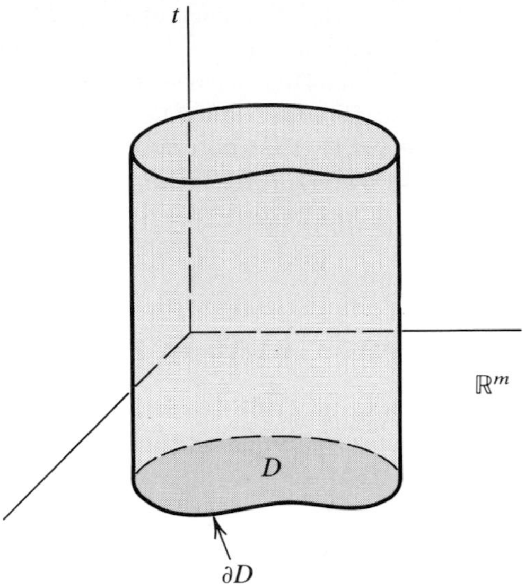

Figure 12.1 The "cylinder" used in evaluating the GF for the diffusion and wave equations. Note that the bases are not simply planes but hyperplanes (that is, spaces such as $\mathbb{R}^m$).

Then (12.42a) can be reexpressed as

$$v\mathbb{L}_{\mathbf{x},t}u - u\mathbb{L}^{\dagger}_{\mathbf{x},t}v = \sum_{\mu=0}^{m}\frac{\partial Q^{\mu}}{\partial x^{\mu}} \tag{12.42b}$$

We recognize the RHS as a divergence in the $(m+1)$-dimensional space. Denoting the volume of the $(m+1)$-dimensional cylinder by $\mathscr{D}$ and its boundary by $\partial\mathscr{D}$ and integrating (12.42b) over $\mathscr{D}$, we obtain

$$\begin{aligned}\int_{\mathscr{D}}(v\mathbb{L}_{\mathbf{x},t}u - u\mathbb{L}^{\dagger}_{\mathbf{x},t}v)\,d^{m+1}x &= \int_{\mathscr{D}}\sum_{\mu}\frac{\partial Q^{\mu}}{\partial x^{\mu}}\,d^{m+1}x\\ &= \int_{\partial\mathscr{D}}\sum_{\mu=0}^{m}Q^{\mu}n_{\mu}\,ds\end{aligned} \tag{12.43}$$

where ds is an element of "area" of $\partial\mathscr{D}$. Note that the divergence theorem was used in the last step. The LHS is an integration over t and $\mathbf{x}$, which can immediately be written as

$$\int_{\mathscr{D}}(v\mathbb{L}_{\mathbf{x},t}u - u\mathbb{L}^{\dagger}_{\mathbf{x},t}v)\,d^{m+1}x = \int_{\varepsilon}^{\tau-\varepsilon}dt\int_{\mathrm{D}}d^{m}x\,(v\mathbb{L}_{\mathbf{x},t}u - u\mathbb{L}^{\dagger}_{\mathbf{x},t}v)$$

The RHS of (12.43), on the other hand, can be split into three parts: a base at $t=\varepsilon$, a base at $t=\tau-\varepsilon$, and the lateral surface. The base at $t=\varepsilon$ is simply the region D, whose outward-pointing normal is in the negative t direction. Thus, $n_0 = -1, n_1 = n_2 = \cdots = n_m = 0$. The base at $t = \tau - \varepsilon$ is also the region D; however, its normal is in the positive t direction. Thus, $n_0 = +1, n_i = 0$, for $i = 1, \ldots, m$. The element of "area" for these two bases is simply $d^m x$. The unit normal to the lateral surface has no time component and is simply the unit normal to the boundary of D. The element of "area" for the lateral surface is $dt\, da$, where da is an element of "area" for ∂D. Putting everything together, we can write (12.43) as

$$\int_{\varepsilon}^{\tau-\varepsilon} dt \int_{D} d^m x\, (v \mathbb{L}_{\mathbf{x},t} u - u \mathbb{L}^{\dagger}_{\mathbf{x},t} v)$$

$$= \int_{D} (-Q^0)\Big|_{t=\varepsilon} d^m x + \int_{D} (Q^0|_{t=\tau-\varepsilon})\, d^m x - \int_{\partial D} da \int_{\varepsilon}^{\tau-\varepsilon} dt\, \mathbf{Q}\cdot\hat{\mathbf{e}}_n$$

The minus sign for the last term is due to the definition of the inner product. Substituting for Q yields

$$\int_{\varepsilon}^{\tau-\varepsilon} dt \int_{D} d^m x\, (v \mathbb{L}_{\mathbf{x},t} u - u \mathbb{L}^{\dagger}_{\mathbf{x},t} v)$$

$$= -\int_{D} u(\mathbf{x},\varepsilon) v(\mathbf{x},\varepsilon)\, d^m x + \int_{D} u(\mathbf{x},\tau-\varepsilon) v(\mathbf{x},\tau-\varepsilon)\, d^m x - \int_{\partial D} da \int_{\varepsilon}^{\tau-\varepsilon} dt \left(v\frac{\partial u}{\partial n} - u\frac{\partial v}{\partial n}\right) \tag{12.44}$$

Letting v be $g(\mathbf{x}, \mathbf{y}; t-\tau)$, the GF associated with the adjoint operator, gives

$$\int_{\varepsilon}^{\tau-\varepsilon} dt \int_{D} d^m x\, [g(\mathbf{x},\mathbf{y}; t-\tau) f(\mathbf{x},t) - u(\mathbf{x},t)\,\delta(\mathbf{x}-\mathbf{y})\,\delta(t-\tau)]$$

$$= -\int_{D} u(\mathbf{x},\varepsilon) g(\mathbf{x},\mathbf{y}; \varepsilon-\tau)\, d^m x + \int_{D} u(\mathbf{x},\tau-\varepsilon) g(\mathbf{x},\mathbf{y}; -\varepsilon)\, d^m x$$

$$- \int_{\partial D} da \int_{\varepsilon}^{\tau-\varepsilon} dt \left[g(\mathbf{x}_b, \mathbf{y}; t-\tau)\frac{\partial u}{\partial n} - u(\mathbf{x}_b, t)\frac{\partial g}{\partial n}\right] \tag{12.45}$$

We now use the following facts:

(1) In the second integral on the LHS, $\delta(t-\tau) = 0$ because t can never be equal to τ in the range of integration.

(2) Using Eq. (12.26a), the symmetry property of the Green's function (note that $\mathbb{L}_{\mathbf{x},t}$ is real), we have

$$g(\mathbf{x}, \mathbf{y}; t - \tau) = G(\mathbf{y}, \mathbf{x}; \tau - t)$$

where we have used the fact that t and τ are the time components of $\mathbf{x}$ and $\mathbf{y}$, respectively.

In particular, by (12.39),

$$g(\mathbf{x}, \mathbf{y}; -\varepsilon) = G(\mathbf{y}, \mathbf{x}; \varepsilon) = \delta(\mathbf{x} - \mathbf{y})$$

(3) The function $g(\mathbf{x}, \mathbf{y}; t - \tau)$ satisfies the same homogeneous BC as $G(\mathbf{x}, \mathbf{y}; t - \tau)$. Thus, $g(\mathbf{x}_b, \mathbf{y}; t - \tau) = 0$ for $\mathbf{x}_b \in \partial D$.

Substituting all the above in (12.45) and then taking the limit $\varepsilon \to 0$, we obtain

$$\int_0^\tau dt \int_D d^m x\, G(\mathbf{y}, \mathbf{x}; \tau - t) f(\mathbf{x}, t)$$

$$= -\int_D u(\mathbf{x}, 0)\, G(\mathbf{y}, \mathbf{x}; \tau)\, d^m x + u(\mathbf{y}, \tau) + \int_{\partial D}\int_0^\tau dt\, da\, u(\mathbf{x}_b, t) \frac{\partial G}{\partial n_y}(\mathbf{y}, \mathbf{x}; \tau - t)$$

Switching $\mathbf{x}$ and $\mathbf{y}$ and t and τ, we finally get

$$\mathbf{u}(\mathbf{x}, t) = \int_0^t d\tau \int_D d^m y\, G(\mathbf{x}, \mathbf{y}; t - \tau) f(\mathbf{y}, \tau) + \int_D u(\mathbf{y}, 0)\, G(\mathbf{x}, \mathbf{y}; t)\, d^m y$$

$$- \int_0^t \int_{\partial D} d\tau\, da\, u(\mathbf{y}_b, \tau) \frac{\partial}{\partial n_y} [G(\mathbf{x}, \mathbf{y}_b; t - \tau)] \qquad (12.46)$$

where $\partial/\partial n_y$ in the last integral means normal differentiation with respect to the second argument of the Green's function.

Equation (12.46) gives the complete solution to Eqs. (12.41), the BVP associated with a parabolic PDE. If $f(\mathbf{y}, \tau) = 0$ and u vanishes on the hypersurface ∂D, then Eq. (12.46) gives

$$u(\mathbf{x}, t) = \int_D u(\mathbf{y}, 0)\, G(\mathbf{x}, \mathbf{y}; t)\, d^m y \qquad (12.47)$$

which is the solution to the BVP of Eqs. (12.35), which led to the general Green's function of (12.38).

Equation (12.47) lends itself beautifully to a physical interpretation. The RHS can be thought of as an IO with kernel $G(\mathbf{x}, \mathbf{y}; t)$. This IO acts on $u(\mathbf{y}, 0)$ and gives $u(\mathbf{x}, t)$; that is, given the shape of the solution at $t = 0$, the IO produces the shape for all subsequent time. That is why $G(\mathbf{x}, \mathbf{y}; t)$ is called the *evolution operator*, or *propagator*.

12.4.3 Hyperbolic Equations

The hyperbolic equation we will discuss is the wave equation

$$\left(\nabla^2 - \frac{1}{c^2}\frac{\partial^2}{\partial t^2}\right)u = 0$$

Redefining the variable t transforms this into

$$\mathbb{L}_{\mathbf{x},t}[u] \equiv \left(\frac{\partial^2}{\partial t^2} - \nabla^2\right)u = 0 \tag{12.48}$$

We wish to calculate the Green's function for $\mathbb{L}_{\mathbf{x},t}$ subject to appropriate BCs. Let us proceed as we did for a parabolic equation and write

$$G(\mathbf{x}, \mathbf{y}; t) = \sum_{n=1}^{\infty} C_n(\mathbf{y}; t)u_n(\mathbf{x}) \tag{12.49a}$$

$$C_n(\mathbf{y}; t) = \int_{\mathbf{D}} G(\mathbf{x}, \mathbf{y}; t)u_n(\mathbf{x})\, d^m x \tag{12.49b}$$

where $u_n(\mathbf{x})$ are orthonormal functions satisfying

$$\nabla^2 u_n(\mathbf{x}) + \lambda_n u_n(\mathbf{x}) = 0 \tag{12.50}$$

and certain, as yet unspecified, BCs. As usual, we expect G to satisfy

$$\mathbb{L}_{\mathbf{x},t}[G] = \left(\frac{\partial^2}{\partial t^2} - \nabla^2\right)G(\mathbf{x}, \mathbf{y}; t - \tau) = \delta(\mathbf{x} - \mathbf{y})\delta(t - \tau) \tag{12.51}$$

Substituting (12.49a) in (12.51) with $\tau = 0$ and using (12.50) gives

$$\sum_{n=1}^{\infty}\left\{\frac{\partial^2}{\partial t^2}[C_n(\mathbf{y}; t)]\right\}u_n(\mathbf{x}) + \sum_{n=1}^{\infty} C_n(\mathbf{y}; t)\lambda_n u_n(\mathbf{x}) = \delta(\mathbf{x} - \mathbf{y})\delta(t)$$

Using $\delta(\mathbf{x} - \mathbf{y}) = \sum_{n=1}^{\infty} u_n(\mathbf{x})u_n(\mathbf{y})$ on the RHS, we obtain

$$\sum_{n=1}^{\infty}\left\{\frac{\partial^2 C_n}{\partial t^2}(\mathbf{y}; t) + \lambda_n C_n(\mathbf{y}; t)\right\}u_n(\mathbf{x}) = \sum_{n=1}^{\infty}[u_n(\mathbf{y})\delta(t)]u_n(\mathbf{x})$$

The orthonormality of $u_n(\mathbf{x})$ gives

$$\frac{\partial^2 C_n}{\partial t^2}(\mathbf{y}; t) + \lambda_n C_n(\mathbf{y}; t) = u_n(\mathbf{y})\delta(t)$$

which shows that $C_n(\mathbf{y}; t)$ is separable. In fact,

$$C_n(\mathbf{y}; t) = u_n(\mathbf{y})T_n(t)$$

where

$$\left(\frac{d^2}{dt^2} + \lambda_n\right)T_n(t) = \delta(t)$$

This equation describes a one-dimensional Green's function and can be solved using the methods of Chapter 11. Assuming that $T_n(t) = 0$ for $t \leqslant 0$, we obtain

$$T_n(t) = \frac{\sin \omega_n t}{\omega_n} \theta(t)$$

where $\omega_n^2 = \lambda_n$. Substituting all the above results in (12.49a), we obtain

$$G(\mathbf{x}, \mathbf{y}; t) = \sum_{n=1}^{\infty} u_n(\mathbf{x}) u_n(\mathbf{y}) \frac{\sin \omega_n t}{\omega_n} \theta(t)$$

or, more generally,

$$G(\mathbf{x}, \mathbf{y}; t - \tau) = \sum_{n=1}^{\infty} u_n(\mathbf{x}) u_n(\mathbf{y}) \frac{\sin \omega_n (t - \tau)}{\omega_n} \theta(t - \tau) \tag{12.52}$$

We note that

$$G(\mathbf{x}, \mathbf{y}; 0^+) = 0 \tag{12.53a}$$

and

$$\frac{\partial}{\partial t} G(\mathbf{x}, \mathbf{y}; t)\bigg|_{t \to 0+} = \delta(\mathbf{x} - \mathbf{y}) \tag{12.53b}$$

as can easily be verified.

With the Green's function for the operator $\mathbb{L}_{\mathbf{x},t}$ of Eq. (12.48) at our disposal, we can attack the BVP given by

$$\begin{aligned}
\left(\frac{\partial^2}{\partial t^2} - \nabla^2\right) u(\mathbf{x}, t) &= f(\mathbf{x}, t) && \text{for } \mathbf{x} \in \mathrm{D} \\
u(\mathbf{x}_b, t) &= h(\mathbf{x}_b, t) && \text{for } \mathbf{x}_b \in \partial\mathrm{D} \\
u(\mathbf{x}, 0) &= \phi(\mathbf{x}) && \text{for } \mathbf{x} \in \mathrm{D} \\
\frac{\partial u}{\partial t}(\mathbf{x}, t)\bigg|_{t=0} &= \psi(\mathbf{x}) && \text{for } \mathbf{x} \in \mathrm{D}
\end{aligned} \tag{12.54}$$

As in the case of the parabolic equation, we first derive an appropriate expression of Green's identity. This can be done by noting that

$$v \mathbb{L}_{\mathbf{x},t}\, u - u \mathbb{L}_{\mathbf{x},t}\, v = \frac{\partial}{\partial t}\left(u \frac{\partial v}{\partial t} - v \frac{\partial u}{\partial t}\right) - \boldsymbol{\nabla} \cdot (u \boldsymbol{\nabla} v - v \boldsymbol{\nabla} u)$$

Thus, $\mathbb{L}_{\mathbf{x},t}$ is formally self-adjoint. Furthermore, we can identify

$$Q_0 = u \frac{\partial v}{\partial t} - v \frac{\partial u}{\partial t} \qquad \text{and} \qquad \mathbf{Q} = u \boldsymbol{\nabla} v - v \boldsymbol{\nabla} u$$

Following the procedure used for the parabolic case step by step, we can easily derive a Green's identity and show that

$$u(\mathbf{x}, t) = \int_0^t d\tau \int_D d^m y\, G(\mathbf{x}, \mathbf{y}; t-\tau) f(\mathbf{y}, \tau)$$
$$+ \int_D \left[\psi(\mathbf{y}) G(\mathbf{x}, \mathbf{y}; t) - \phi(\mathbf{y}) \frac{\partial}{\partial t} G(\mathbf{x}, \mathbf{y}; t)\right] d^m y$$
$$- \int_0^t d\tau \int_{\partial D} da\, h(\mathbf{y}_b, \tau) \frac{\partial}{\partial n_y} [G(\mathbf{x}, \mathbf{y}_b; t-\tau] \qquad (12.55)$$

The details are left as a problem.

For the homogeneous PDE with the homogeneous BC $h = 0 = \psi$, we get

$$u(\mathbf{x}, t) = -\int_D \phi(\mathbf{y}) \frac{\partial}{\partial t} G(\mathbf{x}, \mathbf{y}; t)\, d^m y$$

Note the difference between this equation and Eq. (12.47). Here the propagator is the time derivative of the Green's function. There is another difference between hyperbolic and parabolic equations. When the solution to a parabolic equation vanishes on the boundary and is initially zero, and the PDE is homogeneous $[f(\mathbf{x}, t) = 0]$, the solution must be zero. This is clear from Eq. (12.46). On the other hand, Eq. (12.55) indicates that under the same circumstance, there may be a nonzero solution for a hyperbolic equation if

$$\left.\frac{\partial u}{\partial t}\right|_{t=0} = \psi(\mathbf{x})$$

is nonzero. In such a case we obtain

$$u(\mathbf{x}, t) = \int_D \psi(\mathbf{y})\, G(\mathbf{x}, \mathbf{y}; t)\, d^m y$$

This difference in the two types of equations is due to the fact that hyperbolic equations have second-order time derivatives. Thus, the initial shape of a solution is not enough to uniquely specify it. The initial velocity profile is also essential. We saw examples of this in Chapter 10.

Exercises

12.4.1 Find the BC that the GF must satisfy in order for the solution u to be representable in terms of the GF when the BC on u is mixed, as in Eq. (12.34). Consider a self-adjoint SOLPDO of the elliptic type, and assume either $\alpha(\mathbf{x})$ or $\beta(\mathbf{x})$ is different from zero for $\mathbf{x} \in \partial D$.

12.4.2 Find the Dirichlet GF for a circle of radius a centered at the origin. [Hint: Use

$$H(\mathbf{r}, \mathbf{r}'') = -\frac{1}{2\pi} \ln (|\mathbf{r} - \mathbf{r}''|) - \frac{1}{2\pi} \ln [f(\mathbf{r}'')]$$

for the regular part of the GF and determine $f(\mathbf{r}'')$ and $\mathbf{r}''$.] Use this GF to write the solution of the Dirichlet BVP.

12.5 TECHNIQUES FOR CALCULATING GREEN'S FUNCTIONS

The discussion of Green's functions in this chapter has been very formal. We have yet to see a concrete example of a Green's function, except in Section 12.2. The main purpose of this section is to bridge the gap between formalism and concrete applications. Several powerful techniques are used in obtaining Green's functions, but we will discuss only three: the Fourier transform technique, the eigenfunction expansion technique, and the operator technique.

12.5.1 The Fourier Transform Technique

Recall that any Green's function can be written as a sum of a singular part and a regular part:

$$G(\mathbf{x}, \mathbf{y}) = G_s(\mathbf{x}, \mathbf{y}) + H(\mathbf{x}, \mathbf{y})$$

The singular part satisfies

$$\mathbb{L}_{\mathbf{x}}[G_s] = \delta(\mathbf{x} - \mathbf{y})$$

and the regular part, which obeys $\mathbb{L}_{\mathbf{x}}[H] = 0$, is chosen in such a way that $G(\mathbf{x}, \mathbf{y})$ satisfies the appropriate BC. Since we have already discussed homogeneous equations in detail in Chapter 10, we will not evaluate $H(\mathbf{x}, \mathbf{y})$ in this subsection but will calculate the singular parts of various Green's functions.

Since BCs play no role with respect to $G_s(\mathbf{x}, \mathbf{y})$, the Fourier transform technique (FTT), which involves integration over all space, can be utilized. The FTT has a drawback—it does not work if the coefficient functions are not constants. For physical applications, however, this does not matter, because almost all PDEs of mathematical physics have constant coefficients.

Let us consider the most general SOLPDE with constant coefficients,

$$\mathbb{L}_{\mathbf{x}}[u] = f(\mathbf{x})$$

where
$$\mathbb{L}_{\mathbf{x}} \equiv a_0 + \sum_{j=1}^{m} a_j \frac{\partial}{\partial x_j} + \sum_{j,k=1}^{m} b_{jk} \frac{\partial^2}{\partial x_j \partial x_k} \tag{12.56a}$$

and a_0, a_j, and b_{jk} are constants. The corresponding Green's function has a singular part that satisfies

$$\mathbb{L}_{\mathbf{x}} G_s(\mathbf{x}, \mathbf{y}) = \delta(\mathbf{x} - \mathbf{y}) \tag{12.56b}$$

The FTT starts with assuming a Fourier integral representation in the variable $\mathbf{x}$ for the singular part and for the delta function:

$$G_s(\mathbf{x}, \mathbf{y}) = \frac{1}{(2\pi)^{m/2}} \int d^m k\, \tilde{G}_s(\mathbf{k}, \mathbf{y}) e^{i\mathbf{k}\cdot\mathbf{x}}$$

$$\delta(\mathbf{x} - \mathbf{y}) = \frac{1}{(2\pi)^m} \int d^m k\, e^{i\mathbf{k}\cdot(\mathbf{x}-\mathbf{y})}$$

Substituting from these two formulas in (12.56b), we get

$$\begin{aligned} \mathbb{L}_{\mathbf{x}} G_s(\mathbf{x}, \mathbf{y}) &= \frac{1}{(2\pi)^{m/2}} \int d^m k\, \tilde{G}_s(\mathbf{k}, \mathbf{y}) \mathbb{L}_{\mathbf{x}}[e^{i\mathbf{k}\cdot\mathbf{x}}] \\ &= \frac{1}{(2\pi)^{m/2}} \int d^m k\, \tilde{G}_s(\mathbf{k}, \mathbf{y}) \left[a_0 + \sum_{j=1}^m a_j(ik_j) - \sum_{j,l=1}^m b_{jl} k_j k_l \right] e^{i\mathbf{k}\cdot\mathbf{x}} \\ &= \frac{1}{(2\pi)^m} \int d^m k\, e^{i\mathbf{k}\cdot(\mathbf{x}-\mathbf{y})} \end{aligned}$$

Thus,

$$\tilde{G}_s(\mathbf{k}, \mathbf{y}) = \frac{1}{(2\pi)^{m/2}} \left(\frac{e^{-i\mathbf{k}\cdot\mathbf{y}}}{a_0 + i \sum_{j=1}^m a_j k_j - \sum_{j,l=1}^m b_{jl} k_j k_l} \right) \tag{12.57a}$$

Substituting this in the integral representation for $G_s(\mathbf{x}, \mathbf{y})$ gives

$$G_s(\mathbf{x}, \mathbf{y}) = \frac{1}{(2\pi)^m} \int d^m k \frac{e^{i\mathbf{k}\cdot(\mathbf{x}-\mathbf{y})}}{a_0 + i \sum_{j=1}^m a_j k_j - \sum_{j,l=1}^m b_{jl} k_j k_l} \tag{12.57b}$$

If we can evaluate the integral in (12.57b), we can find $G_s(\mathbf{x}, \mathbf{y})$.

The following examples apply Eqs. (12.57) to specific problems. Note that (12.57b) clearly indicates that $G_s(\mathbf{x}, \mathbf{y})$ is a function of $\mathbf{x} - \mathbf{y}$. This point was mentioned in Chapter 11 where it was noted that the only obstacle to writing a Green's function as a function of $x - y$ is the BCs. The BCs play no part in an evaluation of the singular part of the Green's function, however. Thus, we can write

$$G_s(\mathbf{x}, \mathbf{y}) = G_s(\mathbf{x} - \mathbf{y}) \equiv G_s(\mathbf{r})$$

where $\mathbf{r} \equiv \mathbf{x} - \mathbf{y}$.

Example 12.5.1

Let us calculate the singular part of the Green's function for the Laplacian in m dimensions, where $m > 2$. (We did this in Section 12.2 using a different method.)

With $a_0 = 0$, $a_j = 0$, $b_{jl} = \delta_{jl}$, and $\mathbf{r} = \mathbf{x} - \mathbf{y}$, Eq. (12.57b) reduces to

$$G_s(\mathbf{r}) = \frac{1}{(2\pi)^m} \int d^m k \frac{e^{i\mathbf{k}\cdot\mathbf{r}}}{-k^2} \tag{1}$$

where $k^2 = k_1^2 + k_2^2 + \cdots + k_m^2 \equiv \mathbf{k}\cdot\mathbf{k}$. To integrate (1) we choose spherical coordinates in the m-dimensional k-space. Furthermore, to simplify calculations we let the k_m-axis lie along $\mathbf{r}$ so that $\mathbf{r} \equiv (0, 0, \ldots, |\mathbf{r}|)$ and $\mathbf{k}\cdot\mathbf{r} = k|\mathbf{r}|\cos\theta_1$. Substituting this in (1) and writing $d^m k$ in spherical coordinates yields

$$G_s(\mathbf{r}) = \frac{-1}{(2\pi)^m}\int \frac{e^{ik|\mathbf{r}|\cos\theta_1}}{k^2} k^{m-1}(\sin\theta_1)^{m-2}(\sin\theta_2)^{m-3}\cdots\sin\theta_{m-2}\,dk\,d\theta_1\,d\theta_2\cdots d\theta_{m-1} \tag{2}$$

We note that $d\Omega_m = (\sin\theta_1)^{m-2}d\theta_1\,d\Omega_{m-1}$, which is clear from Eq. (12.15b). Thus, after integration over the angles $\theta_2, \theta_3, \ldots, \theta_{m-1}$, Eq. (2) becomes

$$G_s(\mathbf{r}) = -\frac{1}{(2\pi)^m}\Omega_{m-1}\int_0^\infty k^{m-3}\,dk\int_0^\pi (\sin\theta_1)^{m-2}e^{ik|\mathbf{r}|\cos\theta_1}\,d\theta_1$$

The inner integral can be looked up in an integral table (see, for example, Gradshteyn and Ryzhik 1965, p. 482):

$$\int_0^\pi (\sin\theta)^{m-2}e^{ik|\mathbf{r}|\cos\theta}\,d\theta = \sqrt{\pi}\left(\frac{2}{kr}\right)^{m/2-1}\Gamma\left(\frac{m-1}{2}\right)J_{m/2-1}(kr) \qquad \text{for } m \neq 2$$

Substituting this and Eq. (12.15b) in the preceding equation gives

$$G_s(\mathbf{r}) = -\frac{1}{(2\pi)^{m/2}}\left(\frac{1}{r^{m/2-1}}\right)\int_0^\infty k^{m/2-2}J_{m/2-1}(kr)\,dk \qquad \text{for } m > 2$$

Using the result (see Gradshteyn and Ryzhik, p. 684)

$$\int_0^\infty x^\mu J_\nu(ax)\,dx = 2^\mu a^{-\mu-1}\frac{\Gamma\left(\dfrac{\mu+\nu+1}{2}\right)}{\Gamma\left(\dfrac{\nu-\mu+1}{2}\right)}$$

we obtain

$$G_s(\mathbf{r}) = -\frac{\Gamma\left(\dfrac{m}{2}-1\right)}{4\pi^{m/2}}\left(\frac{1}{r^{m-2}}\right) \qquad \text{for } m > 2$$

which agrees precisely with (12.19a) since $\Gamma(m/2) = (m/2-1)\Gamma(m/2-1)$. •

Example 12.5.2

Let us find the singular part of the Green's function for the (modified) Helmholz operator $\nabla^2 - \mu^2$. In this case Eq. (12.57b) reduces to

$$G_s(\mathbf{r}) = \frac{-1}{(2\pi)^m}\int d^m k\,\frac{e^{i\mathbf{k}\cdot\mathbf{r}}}{\mu^2 + k^2}$$

Following the same procedure as in Example 12.5.1, we find

$$G_s(\mathbf{r}) = -\frac{\Omega_{m-1}}{(2\pi)^m}\int \frac{k^{m-1}\,dk}{\mu^2+k^2}\int_0^{\pi}(\sin\theta_1)^{m-2}e^{ikr\cos\theta_1}\,d\theta_1$$

$$= -\frac{\Omega_{m-1}}{(2\pi)^m}\sqrt{\pi}\left(\frac{2}{r}\right)^{m/2-1}\Gamma\left(\frac{m-1}{2}\right)\int_0^{\infty}\frac{k^{m/2}}{\mu^2+k^2}J_{m/2-1}(kr)\,dk$$

Here we can use the integral formula (see Gradshteyn and Ryzhik 1965, pp. 686 and 952):

$$\int_0^{\infty}\frac{J_\nu(bx)x^{\nu+1}}{(x^2+a^2)^{\eta+1}}dx = \frac{a^{\nu-\eta}b^{\eta}}{2^{\eta}\Gamma(\eta+1)}K_{\nu-\eta}(ab)$$

where

$$K_\nu(z) \equiv \frac{i\pi}{2}e^{i\nu\pi/2}H_\nu^{(1)}(iz)$$

We obtain

$$G_s(\mathbf{r}) = -\frac{\Omega_{m-1}}{(2\pi)^m}\sqrt{\pi}\left(\frac{2}{r}\right)^{m/2-1}\Gamma\left(\frac{m-1}{2}\right)\mu^{m/2-1}\frac{\pi}{2}e^{im\pi/4}H_{m/2-1}^{(1)}(i\mu r)$$

which simplifies to

$$G_s(\mathbf{r}) = -\frac{\pi/2}{(2\pi)^{m/2}}\left(\frac{\mu}{r}\right)^{m/2-1}e^{im\pi/4}H_{m/2-1}^{(1)}(i\mu r) \tag{1}$$

It can be shown (see Exercise 12.5.1) that for $m = 3$ this reduces to

$$G_s(\mathbf{r}) = -\frac{e^{-\mu r}}{4\pi r}$$

which is the Yukawa potential due to a unit charge (see Chapter 5).

We can easily obtain the GF for $\nabla^2 + \mu^2$ by substituting $\pm i\mu$ for μ in (1). The result is

$$G_s(\mathbf{r}) = i^{m+1}\frac{\pi/2}{(2\pi)^{m/2}}\left(\frac{\mu}{r}\right)^{m/2-1}H_{m/2-1}^{(1)}(\pm\mu r) \tag{2}$$

For $m = 3$ this yields

$$G_s(\mathbf{r}) = -\frac{e^{\pm i\mu r}}{4\pi r}$$

The two signs in the exponent correspond to the so-called incoming and outgoing "waves." ●

When dealing with parabolic and hyperbolic equations, we will find it convenient to consider the "different" variable (usually t) as the zeroth coordinate function. In the Fourier transform we then use $\omega \equiv -k_0$ and write

$$G_s(\mathbf{r}, t) \equiv \frac{1}{(2\pi)^{(m+1)/2}}\int_{-\infty}^{\infty}d\omega\int d^m k\,\tilde{G}_s(\mathbf{k},\omega)e^{i(\mathbf{k}\cdot\mathbf{r}-\omega t)} \tag{12.58a}$$

and
$$\delta(\mathbf{r})\delta(t) = \frac{1}{(2\pi)^{m+1}} \int_{-\infty}^{\infty} d\omega \int d^m k e^{i(\mathbf{k}\cdot\mathbf{r} - \omega t)} \tag{12.58b}$$

where $\mathbf{r}$ is the m-dimensional displacement vector.

Example 12.5.3

Let us calculate $G_s(\mathbf{r}, t)$ for the m-dimensional diffusion operator $\partial/\partial t - \nabla^2$, in which ∇^2 is the m-dimensional Laplacian.

Substituting from (12.58) in

$$\left(\frac{\partial}{\partial t} - \nabla^2\right) G_s(\mathbf{r}, t) = \delta(\mathbf{r})\delta(t)$$

we immediately obtain

$$G_s(\mathbf{r}, t) = \frac{1}{(2\pi)^{m+1}} \int_{-\infty}^{\infty} d\omega \int d^m k \frac{1}{-i\omega + k^2} e^{i(\mathbf{k}\cdot\mathbf{r} - \omega t)}$$

$$= \frac{i}{(2\pi)^{m+1}} \int d^m k e^{i\mathbf{k}\cdot\mathbf{r}} \int_{-\infty}^{\infty} d\omega \frac{e^{-i\omega t}}{\omega + ik^2} \tag{1}$$

where, as usual, $k^2 = k_1^2 + k_2^2 + \cdots + k_m^2$. The ω integration can be done using the calculus of residues developed in Chapter 7. The integrand has a simple pole at $\omega = -ik^2$, that is, in the lower half of the complex ω-plane. To integrate we must know the sign of t. If $t > 0$, the exponential factor dictates that the contour be closed in the LHP, where there is a pole and, therefore, a contribution to the residues. On the other hand, if $t < 0$, the contour must be closed in the UHP. The integral is then zero because there are no poles in the UHP. We must, therefore, introduce a step function, $\theta(t)$, in the Green's function.

Evaluating the residue, we easily obtain

$$\int_{-\infty}^{\infty} d\omega \frac{e^{-i\omega t}}{\omega + ik^2} = -2\pi i \operatorname{Res}\left(\frac{e^{-i\omega t}}{\omega + ik^2}\right)\bigg|_{\omega = -ik^2}$$

$$= -2\pi i e^{-i(-ik^2)t} = -2\pi i e^{-k^2 t}$$

(The minus sign arises because of clockwise contour integration in the LHP.) Substituting this in Eq. (1), using spherical coordinates in which the last k-axis is along $\mathbf{r}$, and integrating over all angles except θ_1, we obtain

$$G_s = \theta(t) \frac{\Omega_{m-1}}{(2\pi)^m} \int_0^{\infty} k^{m-1} dk e^{-k^2 t} \int_0^{\pi} (\sin\theta_1)^{m-2} e^{ikr\cos\theta_1} d\theta_1$$

$$= \theta(t) \frac{\Omega_{m-1}}{(2\pi)^m} \sqrt{\pi} \left(\frac{2}{r}\right)^{m/2-1} \Gamma\left(\frac{m-1}{2}\right) \int_0^{\infty} k^{m/2} e^{-k^2 t} J_{m/2-1}(kr)\, dk$$

For the θ_1 integration we used the result quoted in Example 12.5.1.

Using the integral formula (see Gradshteyn and Ryzhik 1965, pp. 716 and 1058)

$$\int_0^\infty x^\mu e^{-\alpha x^2} J_\nu(\beta x)dx = \frac{\beta^\nu \Gamma\left(\frac{\mu+\nu+1}{2}\right)}{2^{\nu+1}\alpha^{(1/2)(\mu+\nu+1)}\Gamma(\nu+1)} \Phi\left(\frac{\mu+\nu+1}{2}, \nu+1; -\frac{\beta^2}{4\alpha}\right)$$

we obtain

$$G_s(\mathbf{r}, t) = \theta(t)\frac{2\pi^{(m-1)/2}}{(2\pi)^m}\sqrt{\pi}\left(\frac{2}{r}\right)^{m/2-1}\frac{r^{m/2-1}}{2^{m/2}t^{m/2}}\Phi\left(\frac{m}{2}, \frac{m}{2}; \frac{-r^2}{4t}\right) \tag{2}$$

Equation (9.70), the power-series expansion for the confluent hypergeometric function Φ, shows that $\Phi(\alpha, \alpha; z) = e^z$. Substituting this result in (2) and simplifying, we finally obtain

$$G_s(\mathbf{r}, t) = \frac{e^{-r^2/4t}}{(4\pi t)^{m/2}}\theta(t)$$

●

Example 12.5.4

Let us find the singular part of the Green's function for the wave equation operator, $\partial^2/\partial t^2 - \nabla^2$. The difference between this example and the preceding one is that here the time derivative is second-order.

Thus, instead of Eq. (1) of Example 12.5.3, we start with

$$G_s(\mathbf{r}, t) = -\frac{1}{(2\pi)^{m+1}}\int d^m k e^{i\mathbf{k}\cdot\mathbf{r}}\int_{-\infty}^{\infty} d\omega \frac{e^{-i\omega t}}{\omega^2 - k^2} \tag{1}$$

The ω integration can be done using the method of residues. Since the singularities of the integrand, $\omega = \pm k$, are on the real axis, it seems reasonable to use the principal value as the true value of the integral. This, in turn, depends on the sign of t. If $t > 0$, we have to close the contour in the LHP to avoid the explosion of the exponential. Example 7,2.14, then, gives

$$\mathrm{P}\int_{-\infty}^{\infty} d\omega \frac{e^{-i\omega t}}{(\omega - k)(\omega + k)} = -i\pi\frac{e^{-ikt} - e^{ikt}}{2k} = -\pi\frac{\sin kt}{k} \qquad \text{for } t > 0$$

For $t < 0$ we must close the contour in the UHP, and this gives

$$\mathrm{P}\int_{-\infty}^{\infty} d\omega \frac{e^{-i\omega t}}{(\omega - k)(\omega + k)} = +i\pi\frac{e^{-ikt} - e^{ikt}}{2k} = \pi\frac{\sin kt}{k} \qquad \text{for } t < 0$$

Combining these two equations, we have

$$\mathrm{P}\int_{-\infty}^{\infty} d\omega \frac{e^{-i\omega t}}{\omega^2 - k^2} = -\pi\frac{\sin kt}{k}[\theta(t) - \theta(-t)]$$

$$\equiv -\pi\frac{\sin kt}{k}\varepsilon(t)$$

where
$$\varepsilon(t) \equiv \theta(t) - \theta(-t) = \begin{cases} 1 & \text{if } t > 0 \\ -1 & \text{if } t < 0 \end{cases}$$

Substituting in (1) yields

$$G_s(\mathbf{r}, t) = \frac{\varepsilon(t)\Omega_{m-1}}{2(2\pi)^m} \int_0^\infty k^{m-1}\, dk \frac{\sin kt}{k} \int_0^\pi (\sin\theta_1)^{m-1} e^{ikr\cos\theta_1}\, d\theta_1$$

where we have integrated over all angles except θ_1. Using the result for the θ_1 integration quoted in Example 12.5.1 and simplifying, we finally obtain

$$G_s(\mathbf{r}, t) = \frac{\varepsilon(t)}{2(2\pi)^{m/2} r^{m/2-1}} \int_0^\infty k^{m/2-1} J_{m/2-1}(kr) \sin kt\, dk \tag{2}$$

As Exercise 12.5.3 shows, the Green's function as given by Eq. (2) satisfies only the homogeneous wave equation, which has no delta function on the RHS. The reason for this is that the principal value of an integral, although reasonable mathematically, may not reflect the physical situation. In fact, the Green's function in (2) contains two pieces corresponding to the two different contours of integration.

It turns out that physically interesting Green's functions are obtained, not from the principal value, but from giving small imaginary parts to the poles. Thus, replacing the ω integral with a contour integral for which the two poles are pushed in the LHP and using the method of residues, we obtain

$$I_{\text{up}} \equiv \int_{-\infty}^{\infty} \frac{e^{-i\omega t}}{\omega^2 - k^2}\, d\omega \equiv \int_{C_1} \frac{e^{-izt}}{z^2 - k^2}\, dz = \begin{cases} 0 & \text{if } t < 0 \\ \dfrac{2\pi}{k}\sin kt & \text{if } t > 0 \end{cases}$$

$$= \frac{2\pi}{k}\theta(t)\sin kt$$

The zero is due to the fact that for $t < 0$ the contour must be closed in the UHP where there are no poles inside C_1. Substituting this in (1) and working through as before, we obtain what is called the *retarded Green's function*:

$$G_s^{(\text{ret})}(\mathbf{r}, t) = \frac{\theta(t)}{(2\pi)^{m/2} r^{m/2-1}} \int_0^\infty k^{m/2-1} J_{m/2-1}(kr) \sin kt\, dk \tag{3}$$

On the other hand, replacing the ω integral with the contour integral C_2 for which the poles are pushed in the UHP yields what is called the *advanced Green's function*:

$$G_s^{(\text{adv})}(\mathbf{r}, t) = -\frac{\theta(-t)}{(2\pi)^{m/2} r^{m/2-1}} \int_0^\infty k^{m/2-1} J_{m/2-1}(kr) \sin kt\, dk \tag{4}$$

Before discussing the physical significance of these two functions for the case where $m = 3$, let us examine the behavior of (2), (3), and (4) for t close to $\pm r$.

Unlike the elliptic and parabolic equations discussed earlier, the integral over k is not a function but a *distribution*, which can best be seen by noting that the following is an even function of α (see Gradshteyn and Ryzhik 1965, p. 712):

$$\int_0^\infty e^{-\alpha x} J_\nu(\beta x) x^\nu dx = \frac{(2\beta)^\nu \Gamma(\nu + \frac{1}{2})}{\sqrt{\pi}(\alpha^2 + \beta^2)^{\nu+1/2}} \qquad \text{for } \operatorname{Re}(\alpha) > |\operatorname{Im}(\beta)|$$

Thus, writing

$$\sin kt = \frac{e^{ikt} - e^{-ikt}}{2i}$$

and adding a small negative number to the exponential, we easily find that (2), (3), and (4) are zero as long as $r^2 \neq t^2$. To see what happens when $t = \pm r$, we take the difference between these two integrals:

$$I_\varepsilon^+ \equiv \int_0^\infty e^{-(-it+\varepsilon)k} J_\nu(kr) k^\nu dk = \frac{(2r)^\nu \Gamma(\nu + \frac{1}{2})}{\sqrt{\pi}} [(-it + \varepsilon)^2 + r^2]^{-(\nu+1/2)}$$

$$I_\varepsilon^- \equiv \int_0^\infty e^{-(it+\varepsilon)k} J_\nu(kr) k^\nu dk = \frac{(2r)^\nu \Gamma(\nu + \frac{1}{2})}{\sqrt{\pi}} [(it + \varepsilon)^2 + r^2]^{-(\nu+1/2)}$$

The result, to the first order in ε, is

$$I_\varepsilon^+ - I_\varepsilon^- = \frac{2i(2\nu + 1)\varepsilon t (2r)^\nu \Gamma(\nu + \frac{1}{2})}{\sqrt{\pi}(r^2 - t^2)^{\nu+3/2}}$$

Substituting this in one of the expressions for G_s, say (2), with $\nu = m/2 - 1$, yields

$$G_s(\mathbf{r}, t; \varepsilon) \approx \frac{i(m-1)\Gamma\left(\frac{m-1}{2}\right) t\varepsilon(t)}{2(\pi)^{[(m+1)/2]}} \left[\frac{\varepsilon}{(r^2 - t^2)^{[(m+1)/2]}}\right] \tag{5}$$

This clearly shows that as long as $r^2 \neq t^2$,

$$\lim_{\varepsilon \to 0} G_s(\mathbf{r}, t; \varepsilon) = 0$$

However, if $r^2 = t^2$, then the denominator of the last factor in (5) vanishes and $G_s(\mathbf{r}, t; \varepsilon)$ blows up. This is behavior like that of a delta function.

To evaluate the GF, we need to find an expression for the integral in (2), (3), or (4). Denoting this integral by $I^{(\nu)}$, we have

$$I^{(\nu)} = \frac{1}{2i} \lim_{\varepsilon \to \infty} (I_\varepsilon^+ - I_\varepsilon^-)$$

$$= \frac{(2r)^\nu \Gamma(\nu + 1/2)}{2i\sqrt{\pi}} \lim_{\varepsilon \to 0} \left\{ \frac{1}{[r^2 + (-it + \varepsilon)^2]^{\nu+1/2}} - \frac{1}{[r^2 + (it + \varepsilon)^2]^{\nu+1/2}} \right\}$$

In our case, $\nu = m/2 - 1$. Thus, the above expression yields

$$I^{(m/2-1)} = \frac{(2r)^{m/2-1}\Gamma\left(\frac{m-1}{2}\right)}{2i\sqrt{\pi}} \lim_{\varepsilon\to 0}\left\{\frac{1}{[r^2+(-it+\varepsilon)^2]^{(m-1)/2}} - \frac{1}{[r^2+(it+\varepsilon)^2]^{(m-1)/2}}\right\} \tag{6}$$

At this point, it is convenient to discuss separately the two cases for which m is odd and m is even. Let us derive the expression for odd m (the even case is left for Problem 12.27). When m is odd, $m - 1$ is even. Thus, we can define the integer $n = (m - 1)/2$ and write (6) as

$$I^{(n)} = \frac{(2r)^{n-1/2}\Gamma(n)}{2i\sqrt{\pi}} \lim_{\varepsilon\to 0}\left\{\frac{1}{[r^2+(-it+\varepsilon)^2]^n} - \frac{1}{[r^2+(it+\varepsilon)^2]^n}\right\} \tag{7}$$

Let $u \equiv r^2 + (-it + \varepsilon)^2$; then we can write

$$\frac{1}{u^n} = \frac{(-1)^{n-1}}{(n-1)!}\frac{d^{n-1}}{du^{n-1}}\left(\frac{1}{u}\right)$$

Noting that the following identity holds for an arbitrary function of u,

$$\frac{\partial f}{\partial r} = \frac{df}{du}\frac{\partial u}{\partial r} = \frac{df}{du}2r \quad \Rightarrow \quad \frac{df}{du} = \frac{1}{2r}\frac{\partial f}{\partial r}$$

we can write $d/du = (1/2r)\,\partial/\partial r$ and

$$\frac{1}{u^n} = \frac{(-1)^{n-1}}{(n-1)!}\left(\frac{1}{2r}\frac{\partial}{\partial r}\right)^{n-1}\left(\frac{1}{u}\right)$$

or

$$\frac{1}{[r^2+(\pm it+\varepsilon)^2]^n} = \frac{1}{(n-1)!}\left(-\frac{1}{2r}\frac{\partial}{\partial r}\right)^{n-1}\left[\frac{1}{r^2+(\pm it+\varepsilon)^2}\right]$$

Therefore, Eq. (7) can be written as

$$I^{(n)} = \frac{(2r)^{n-1/2}\Gamma(n)}{2i\sqrt{\pi}}\frac{1}{(n-1)!}\left(-\frac{1}{2r}\frac{\partial}{\partial r}\right)^{n-1}\left\{\lim_{\varepsilon\to 0}\left[\frac{1}{r^2+(-it+\varepsilon)^2} - \frac{1}{r^2+(it+\varepsilon)^2}\right]\right\} \tag{8}$$

But the limit in (8) is precisely that of the three-dimensional case evaluated in Exercise 12.5.2. It is found that

$$I^{(1)} = \frac{(2r)^{1/2}}{2i\sqrt{\pi}}\lim_{\varepsilon\to 0}\left[\frac{1}{r^2+(-it+\varepsilon)^2} - \frac{1}{r^2+(it+\varepsilon)^2}\right] = -\sqrt{\frac{\pi}{2r}}[\delta(t+r) - \delta(t-r)]$$

which gives

$$\lim_{\varepsilon\to 0}\left[\frac{1}{r^2+(-it+\varepsilon)^2} - \frac{1}{r^2+(it+\varepsilon)^2}\right] = -\frac{i\pi}{r}[\delta(t+r) - \delta(t-r)]$$

Substituting this in (8) and noting that $\Gamma(n) = (n-1)!$, we obtain

$$I^{(n)} = -\frac{\sqrt{\pi}(2r)^{n-1/2}}{2}\left(-\frac{1}{2r}\frac{\partial}{\partial r}\right)^{n-1}\left\{\frac{1}{r}[\delta(t+r) - \delta(t-r)]\right\}$$

Employing this result in (3) and (4) yields

$$G_s^{(\text{ret})}(\mathbf{r}, t) = \frac{1}{4\pi}\left(-\frac{1}{2\pi r}\frac{\partial}{\partial r}\right)^{n-1}\left[\frac{\delta(t-r)}{r}\right] \qquad \text{for } n = \frac{m-1}{2} \tag{9a}$$

$$G_s^{(\text{adv})}(\mathbf{r}, t) = \frac{1}{4\pi}\left(-\frac{1}{2\pi r}\frac{\partial}{\partial r}\right)^{n-1}\left[\frac{\delta(t+r)}{r}\right] \qquad \text{for } n = \frac{m-1}{2} \tag{9b}$$

The theta functions are not needed in (9) because the arguments of the delta functions already meet the restrictions imposed by the theta functions.

The two functions in (9) have an interesting physical interpretation. Green's functions are propagators (of signals of some sort), and $G_s^{(\text{ret})}(\mathbf{r}, t)$ is capable of propagating signals only for positive times. On the other hand, $G_s^{(\text{adv})}(\mathbf{r}, t)$ can propagate only in the negative time direction. Thus, if initially ($t = 0$) a signal is produced (by appropriate BCs), both $G_s^{(\text{ret})}$ and $G_s^{(\text{adv})}$ work to propagate it in their respective time directions. It may seem that $G_s^{(\text{adv})}$ is useless because *every* signal propagates forward in time. This is true, however, only for classical events. In relativistic quantum field theory antiparticles are interpreted mathematically as moving in the negative time direction! Thus, we cannot simply ignore $G_s^{(\text{adv})}$. In fact, the correct propagator to choose in this theory is a linear combination of $G_s^{(\text{adv})}$ and $G_s^{(\text{ret})}$, called the *Feynman propagator* (see Bjorken and Drell 1965). ●

Example 12.5.4 shows a subtle difference between Green's functions for second-order differential operators in one dimension and those for such operators in higher dimensions. We saw in Chapter 11 that the former are continuous functions in the interval on which they are defined. Here we see not only that the Green's functions in higher dimensions are not continuous functions but that they are not even *functions* in the ordinary sense; they contain a delta function as a factor. Thus, in general, Green's functions in higher dimensions must be considered as distributions (generalized functions).

Example 12.5.5

Let us find the GF for the Helmholtz operator $\nabla^2 + \mu^2$ in two dimensions with a circular boundary of radius a.

We can either find G_s directly or use Eq. (2) of Example 12.5.2 with $m = 2$. Either way the result is

$$G_s(\mathbf{r}, \mathbf{r}') = -\frac{i}{4}H_0^{(1)}(\mu|\mathbf{r} - \mathbf{r}'|)$$

which yields

$$G(\mathbf{r}, \mathbf{r}') = -\frac{i}{4}H_0^{(1)}(\mu|\mathbf{r} - \mathbf{r}'|) + H(\mathbf{r}, \mathbf{r}')$$

The BC $G(\mathbf{a}, \mathbf{r}') = 0$, in which $\mathbf{a}$ is a vector from the origin to the circular boundary, requires that H satisfy the BC

$$H(\mathbf{a}, \mathbf{r}') = \frac{i}{4} H_0^{(1)}(\mu|\mathbf{a} - \mathbf{r}'|) \tag{1}$$

Recall that $H(\mathbf{r}, \mathbf{r}')$ satisfies the homogeneous Helmholtz equation $(\nabla^2 + \mu^2)H = 0$. A separation of variables (see Chapters 8 and 10) and the fact that H is regular at $\mathbf{r} = \mathbf{r}'$ yield

$$H(\mathbf{r}, \mathbf{r}') = \sum_{n=0}^{\infty} J_n(\mu r)[a_n(\mathbf{r}')\cos n\theta + b_n(\mathbf{r}')\sin n\theta]$$

Because the coefficients a_n and b_n are constants with respect to $\mathbf{r}$ only, the argument $\mathbf{r}'$ appears for them. Equation (1) now yields

$$\sum_{n=0}^{\infty} J_n(\mu a)[a_n(\mathbf{r}')\cos n\theta + b_n(\mathbf{r}')\sin n\theta] = \frac{i}{4} H_0^{(1)}(\mu\sqrt{a^2 + r'^2 - 2ar'\cos(\theta - \theta')})$$

The LHS is a Fourier-series expansion of the RHS. The coefficients can be evaluated as usual:

$$a_0(\mathbf{r}') = \frac{i}{8\pi J_0(\mu a)} \int_0^{2\pi} H_0^{(1)}(\mu\sqrt{a^2 + r'^2 - 2ar'\cos(\theta - \theta')})\,d\theta$$

$$a_n(\mathbf{r}') = \frac{i}{4\pi J_n(\mu a)} \int_0^{2\pi} H_0^{(1)}(\mu\sqrt{a^2 + r'^2 - 2ar'\cos(\theta - \theta')})\cos n\theta\,d\theta$$

$$b_n(\mathbf{r}') = \frac{i}{4\pi J_n(\mu a)} \int_0^{2\pi} H_0^{(1)}(\mu\sqrt{a^2 + r'^2 - 2ar'\cos(\theta - \theta')})\sin n\theta\,d\theta$$

These equations completely determine $H(\mathbf{r}, \mathbf{r}')$ and, therefore, $G(\mathbf{r}, \mathbf{r}')$. •

The procedure outlined in Example 12.5.5 can be applied to a large number of problems of practical importance. Problems discussed in Chapters 8 and 10 illustrate the technique sufficiently.

12.5.2 The Eigenfunction Expansion Technique

Suppose that the differential operator $\mathbb{L}_{\mathbf{x}}$, defined in a domain D with boundary ∂D, has discrete eigenvalues $\{\lambda_n\}_{n=1}^{\infty}$ with corresponding orthonormal eigenfunctions $\{u_m(\mathbf{x})\}_{m=1}^{\infty}$. Of course, these two sets may not be in one-to-one correspondence. There may be degeneracies; for example, one λ_n may correspond to several $u_m(\mathbf{x})$'s. Assume that the $u_m(\mathbf{x})$'s satisfy the same BCs as the Green's function to be defined below.

Now consider the operator $\mathbb{L}_{\mathbf{x}} - \lambda\mathbb{1}$, where λ is different from all λ_n. Then, as in the one-dimensional case, this operator is invertible, and we can define its Green's function by

$$(\mathbb{L}_{\mathbf{x}} - \lambda)G_\lambda(\mathbf{x}, \mathbf{y}) = \delta(\mathbf{x} - \mathbf{y})$$

where the weight function is set equal to one. The completeness of $\{u_n(\mathbf{x})\}_{n=1}^{\infty}$ implies that

$$\delta(\mathbf{x} - \mathbf{y}) = \langle \mathbf{x}|\mathbf{y}\rangle = \langle \mathbf{x}|\mathbb{1}|\mathbf{y}\rangle = \langle \mathbf{x}|\left(\sum_{n=1}^{\infty} |\mathrm{u}_n\rangle\langle \mathrm{u}_n|\right)|\mathbf{y}\rangle$$

$$= \sum_{n=1}^{\infty} u_n(\mathbf{x})u_n^*(\mathbf{y})$$

and

$$G_\lambda(\mathbf{x}, \mathbf{y}) = \sum_{n=1}^{\infty} a_n(\mathbf{y})u_n(\mathbf{x})$$

where $a_n(\mathbf{y})$ are the "coefficients" of the expansion of $G_\lambda(\mathbf{x}, \mathbf{y})$, which by assumption satisfies the same BCs as the u_n's and, therefore, is expandable in terms of them. Substituting these two expansions in the given differential equation yields

$$(\mathbb{L}_{\mathbf{x}} - \lambda)\left[\sum_{n=1}^{\infty} a_n(\mathbf{y})u_n(\mathbf{x})\right] = \sum_{n=1}^{\infty} u_n(\mathbf{x})u_n^*(\mathbf{y})$$

or using $\mathbb{L}_{\mathbf{x}}u_n(\mathbf{x}) = \lambda_n u_n(\mathbf{x})$,

$$\sum_{n=1}^{\infty} (\lambda_n - \lambda)a_n(\mathbf{y})u_n(\mathbf{x}) = \sum_{n=1}^{\infty} u_n(\mathbf{x})u_n^*(\mathbf{y})$$

The orthonormality of the u_n's gives

$$a_n(\mathbf{y}) = \frac{u_n^*(\mathbf{y})}{\lambda_n - \lambda}$$

Thus,

$$G_\lambda(\mathbf{x}, \mathbf{y}) = \sum_{n=1}^{\infty} \frac{u_n(\mathbf{x})u_n^*(\mathbf{y})}{\lambda_n - \lambda} \tag{12.59}$$

In particular, if zero is not an eigenvalue of $\mathbb{L}_{\mathbf{x}}$, the Green's function for $\mathbb{L}_{\mathbf{x}}$ can be written as

$$G(\mathbf{x}, \mathbf{y}) = G_0(\mathbf{x}, \mathbf{y}) = \sum_{n=1}^{\infty} \frac{u_n(\mathbf{x})u_n^*(\mathbf{y})}{\lambda_n} \tag{12.60}$$

This is an expansion of the Green's function in terms of the eigenfunctions of $\mathbb{L}_{\mathbf{x}}$.

It is instructive to consider a formal interpretation of Eq. (12.60). Recall that the spectral decomposition theorem permits us to write, for an operator $\mathbb{A}$ with eigenvalues λ_i and projection operators $\mathbb{P}_i$,

$$f(\mathbb{A}) = \sum_i f(\lambda_i)\mathbb{P}_i$$

Assuming the form

$$\mathbb{P}_i \equiv \sum_j |\mathrm{u}_j^{(i)}\rangle\langle \mathrm{u}_j^{(i)}|$$

where $|\mathrm{u}_j^{(i)}\rangle$ is the jth eigenfunction corresponding to the eigenvalue λ_i, and summing over all eigenfunctions, we may write

$$f(\mathbb{A}) = \sum_n f(\lambda_n)|\mathrm{u}_n\rangle\langle \mathrm{u}_n|$$

Here n counts all the eigenfunctions corresponding to the different eigenvalues. Now, if we use a particular function, $f(\mathbb{A}) \equiv \mathbb{A}^{-1}$, we get

$$\mathbb{G} \equiv \mathbb{A}^{-1} = \sum_n \lambda_n^{-1}|\mathrm{u}_n\rangle\langle \mathrm{u}_n| = \sum_n \frac{|\mathrm{u}_n\rangle\langle \mathrm{u}_n|}{\lambda_n}$$

or, in "matrix element" form,

$$G(\mathbf{x}, \mathbf{y}) \equiv \langle \mathbf{x}|\mathbb{G}|\mathbf{y}\rangle = \sum_n \frac{\langle \mathbf{x}|\mathrm{u}_n\rangle\langle \mathrm{u}_n|\mathbf{y}\rangle}{\lambda_n} = \sum_n \frac{u_n(\mathbf{x})u_n^*(\mathbf{y})}{\lambda_n}$$

This last expression coincides with the RHS of (12.60).

Although the preceding development is admittedly formal, it shows the connection between algebraic operators (introduced in Chapters 2 and 3) and differential operators.

Equations (12.59) and (12.60) demand that the $u_n(\mathbf{x})$ form a complete discrete orthonormal set. We encountered many examples of such eigenfunctions in discussing Sturm-Liouville systems in Chapter 10. All the S-L systems there were, of course, one-dimensional. Here we are generalizing the S-L system to m dimensions. This is not a limitation, however, because the separation of variables reduces an m-dimensional PDE to m one-dimensional ODEs. If the BCs are appropriate, the m ODEs will be m S-L systems. A review of Chapter 10 will reveal that homogeneous BCs always lead to S-L systems. In fact, Theorem 10.3.1 is evidence for this claim. Since Green's functions must also satisfy homogeneous BCs, expansions such as those of (12.59) and (12.60) become possible.

Example 12.5.6

As a concrete example, let us obtain an eigenfunction expansion of the GF of the two-dimensional Laplacian, $\nabla^2 = \partial^2/\partial x^2 + \partial^2/\partial y^2$, inside the rectangular region $0 \leqslant x \leqslant a$, $0 \leqslant y \leqslant b$ with Dirichlet BCs. Since the GF vanishes at the boundary, the eigenvalue problem becomes

$$\nabla^2 u = \lambda u \qquad u = 0 \text{ on } \partial D$$

A separation of variables, $u(x, y) = X(x)Y(y)$, gives

$$\frac{d^2X}{dx^2} + \alpha X = 0 \qquad X(0) = X(a) = 0 \tag{1}$$

$$\frac{d^2Y}{dy^2} - (\alpha + \lambda)Y = 0 \qquad Y(0) = Y(b) = 0 \tag{2}$$

The solution to (1) is

$$X(x) = A\sin\left(\frac{n\pi}{a}x\right) \qquad \alpha = \left(\frac{n\pi}{a}\right)^2 \qquad \text{for } n = 1, 2, \ldots$$

On the other hand, Eq. (2) will have a solution only if

$$b\sqrt{\alpha + \lambda} = im\pi \quad \Rightarrow \quad \lambda = -\left[\left(\frac{n\pi}{a}\right)^2 + \left(\frac{m\pi}{b}\right)^2\right] \qquad \text{for } m, n = 1, 2, \ldots$$

In that case the solution is

$$Y(y) = B\sin\left(\frac{m\pi}{b}y\right) \qquad \text{for } m = 1, 2, \ldots$$

We thus obtain the orthogonal eigenfunctions

$$u_{mn}(x, y) \equiv X(x)\,Y(y) = A_{mn}\sin\left(\frac{n\pi}{a}x\right)\sin\left(\frac{m\pi}{b}y\right)$$

whose corresponding eigenvalues are

$$\lambda_{mn} = -\left[\left(\frac{n\pi}{a}\right)^2 + \left(\frac{m\pi}{b}\right)^2\right]$$

To normalize the u_{mn}'s, we must choose A_{mn} properly. We have

$$1 = \int_0^a dx \int_0^b dy [u_{mn}(x, y)]^2 = A_{mn}^2 \int_0^a \sin^2\left(\frac{n\pi}{a}x\right)dx \int_0^b \sin^2\left(\frac{mn}{b}y\right)dy$$

$$= (A_{mn})^2\left(\frac{a}{2}\right)\left(\frac{b}{2}\right)$$

and

$$A_{mn} = \frac{2}{\sqrt{ab}}$$

Inserting all the above in Eq. (12.60) and noting that the eigenfunctions and eigenvalues are doubly indexed, we obtain

$$G(\mathbf{r}, \mathbf{r}') \equiv G(x, y; x', y') = \sum_{m,n=1}^{\infty} \frac{u_{mn}(x, y)u_{mn}(x', y')}{\lambda_{mn}}$$

$$= -\frac{4}{ab}\sum_{m,n=1}^{\infty} \frac{\sin\left(\frac{n\pi}{a}x\right)\sin\left(\frac{m\pi}{b}y\right)\sin\left(\frac{n\pi}{a}x'\right)\sin\left(\frac{m\pi}{b}y'\right)}{\left(\frac{n\pi}{a}\right)^2 + \left(\frac{m\pi}{b}\right)^2}$$

Note the change from **x** to **r** and **y** to **r**′. Note also that the λ_{mn} are never zero; thus, $G(\mathbf{r}, \mathbf{r}')$ is well-defined. ●

In Example 12.5.6, $\lambda = 0$ was not an eigenvalue of $\mathbb{L}_{\mathbf{x}}$. This must be true when a Green's function is expanded in terms of eigenfunctions. In physical applications certain conditions (which have nothing to do with the BCs) exclude the zero eigenvalue automatically when they are applied to the Green's function. For instance, the condition that the Green's function remain finite at the origin is severe enough to exclude the zero eigenvalue.

Example 12.5.7

Let us consider the two-dimensional Dirichlet BVP $\nabla^2 u = f$, $u = 0$ on a circle of radius a. If we consider only the BCs and ask whether $\lambda = 0$ is an eigenvalue, the answer will be yes, as the following argument shows.

The most general solution to $\nabla^2 u = 0$ in polar coordinates can be obtained by the method of separation of variables:

$$u(\rho, \varphi) = A + B\ln\rho + \sum_{n=1}^{\infty}(b_n\rho^n + b'_n\rho^{-n})\cos n\varphi + \sum_{n=1}^{\infty}(c_n\rho^n + c'_n\rho^{-n})\sin n\varphi \tag{1}$$

Invoking the BC gives

$$0 = u(a, \varphi) = A + B\ln a + \sum_{n=1}^{\infty}(b_n a^n + b'_n a^{-n})\cos n\varphi + \sum_{n=1}^{\infty}(c_n a^n + c'_n a^{-n})\sin n\varphi$$

which holds for arbitrary φ if and only if

$$A = -B\ln a \qquad b'_n = -b_n a^{2n} \qquad c'_n = -c_n a^{2n}$$

Substituting in (1) gives

$$u(\rho, \varphi) = B\ln\left(\frac{\rho}{a}\right) + \sum_{n=1}^{\infty}\left(\rho^n - \frac{a^{2n}}{\rho^n}\right)(b_n\cos n\varphi + c_n\sin n\varphi) \tag{2}$$

Thus, if we demand nothing beyond the BCs, $\nabla^2 u = 0$ will have a nontrivial solution, given by (2). That is, $\lambda = 0$ is an eigenvalue of the equation $\nabla^2 u = \lambda u$.

Physical reality, however, demands that $u(\rho, \varphi)$ be well-behaved at the origin. This condition sets B, b'_n, and c'_n of Eq. (1) equal to zero. The BCs then make the remaining coefficients in (1) vanish. Thus, the demand that $u(\rho, \varphi)$ be well-behaved at $\rho = 0$ turns the situation completely around and ensures the nonexistence of a zero eigenvalue for ∇^2, which in turn guarantees the existence of a GF. •

12.5.3 The Operator Technique

The basic idea behind the operator technique (OT) is as follows: Suppose that $\mathbb{L}_1$ and $\mathbb{L}_2$ are two commuting operators, and we are seeking $(\mathbb{L}_1 + \mathbb{L}_2)^{-1}$. Since $\mathbb{L}_2$ commutes with $\mathbb{L}_1$, it can be regarded as a constant as far as operations with (and on) $\mathbb{L}_1$ are concerned. In particular,

$$(\mathbb{L}_1 + \mathbb{L}_2)\mathbb{G} = \mathbb{1}$$

can be regarded as an operator equation in $\mathbb{L}_1$ *alone*, with $\mathbb{L}_2$ treated as a constant:

$$\mathbb{L}_1\mathbb{G} + \mathbb{L}_2\mathbb{G} = \mathbb{1}$$

If $\mathbb{L}_1$ (and $\mathbb{L}_2$) are differential operators, then this equation can be written as

$$\mathbb{L}_1 G(\mathbf{x}, \mathbf{y}) + \mathbb{L}_2 G(\mathbf{x}, \mathbf{y}) = \delta(\mathbf{x} - \mathbf{y})$$

In most cases $\mathbb{L}_1$ depends only on a subset of $\{x_i\}_{i=1}^m$. Let $\mathbf{x}_1$ denote this subset, and let $\mathbf{x}_2$ denote the remainder of the coordinates. Then we can write

$$\delta(\mathbf{x} - \mathbf{y}) = \delta(\mathbf{x}_1 - \mathbf{y}_1)\delta(\mathbf{x}_2 - \mathbf{y}_2)$$

If $G_1(\mathbf{x}_1, \mathbf{y}_1, \mathbf{x}_2, \mathbf{y}_2; k)$ denotes the Green's function for $\mathbb{L}_1 + k$, where k is a constant, we obtain

$$G(\mathbf{x}, \mathbf{y}) = G_1(\mathbf{x}_1, \mathbf{y}_1, \mathbf{x}_2, \mathbf{y}_2; \mathbb{L}_2)\delta(\mathbf{x}_2 - \mathbf{y}_2) \tag{12.61}$$

This can be easily verified by noting that

$$\begin{aligned}(\mathbb{L}_1 + \mathbb{L}_2)G(\mathbf{x}, \mathbf{y}) &= [(\mathbb{L}_1 + \mathbb{L}_2)G_1(\mathbf{x}_1, \mathbf{y}_1, \mathbf{x}_2, \mathbf{y}_2; \mathbb{L}_2)]\delta(\mathbf{x}_2 - \mathbf{y}_2)\\ &= \delta(\mathbf{x}_1 - \mathbf{y}_1)\delta(\mathbf{x}_2 - \mathbf{y}_2) = \delta(\mathbf{x} - \mathbf{y})\end{aligned}$$

Once G_1 is found as a function of $\mathbb{L}_2$, it can operate on $\delta(\mathbf{x}_2 - \mathbf{y}_2)$ to give the desired Green's function.

The following example illustrates the technique.

Example 12.5.8

Let us redo Example 12.5.6 using the OT.

Let $\mathbb{L}_1 = \partial^2/\partial x^2$ and $\mathbb{L}_2 = \partial^2/\partial y^2$. Then the GF satisfies

$$\begin{aligned}(\mathbb{L}_1 + \mathbb{L}_2)G(x, y; x', y') &= \delta(x - x')\delta(y - y')\\ G(0, y; x', y') = G(a, y; x', y') &= G(x, 0; x', y') = G(x, b; x', y') = 0\end{aligned} \tag{1}$$

Let us assume that $\mathbb{L}_2$ is a constant (we can do this because $[\mathbb{L}_1, \mathbb{L}_2] = 0$) and concentrate on the one-dimensional problem

$$\begin{aligned}\frac{d^2G}{dx^2} + \mathbb{L}_2 G &= \delta(x - x')\delta(y - y')\\ G(0, y; x', y') &= G(a, y; x', y') = 0\end{aligned} \tag{2}$$

where $\mathbb{L}_2$, y, x', and y' are considered to be simply parameters. Eqs. (2) can be solved by the methods of Chapter 11 (in particular, see Problem 11.19, where $\mathbb{L}_2 \equiv k^2$). The solution is

$$G_1(x, y, x', y'; \mathbb{L}_2) = \begin{cases}(\mathbb{k} \sin \mathbb{k}a)^{-1} \sin \mathbb{k}x \sin \mathbb{k}(a - x') & \text{if } 0 \leqslant x < x'\\ (\mathbb{k} \sin \mathbb{k}a)^{-1} \sin \mathbb{k}(a - x) \sin \mathbb{k}x' & \text{if } x' < x \leqslant a\end{cases} \tag{3}$$

where $\mathbb{k} = \sqrt{\mathbb{L}_2}$.

To find the full Green's function, we must apply the *operator* $G_1(x, y, x', y'; \mathbb{L}_2)$ to $\delta(y - y')$. The best way to do this is to express $\delta(y - y')$ in terms of the eigenfunctions of $\mathbb{L}_2$ and then apply G_1 to thc resulting expression. These eigenfunctions, and their corresponding eigenvalues, are easily found by solving the S-L system

$$\mathbb{L}_2 u \equiv \frac{d^2u}{dy^2} = \lambda u \qquad u(0) = u(b) = 0$$

The normalized solutions are

$$u_n = \sqrt{\frac{2}{b}}\sin\left(\frac{n\pi}{b}y\right) \qquad \lambda_n = -\left(\frac{n\pi}{b}\right)^2 \qquad \text{for } n = 1, 2, \ldots$$

The completeness of the eigenfunctions implies that

$$\delta(y - y') = \frac{2}{b}\sum_{n=1}^{\infty} \sin\left(\frac{n\pi}{b}y\right)\sin\left(\frac{n\pi}{b}y'\right)$$

This, together with the fact that

$$f(\mathbb{L}_2)u_n(y) = f(\lambda_n)u_n(y)$$

yields
$$f(\mathbb{L}_2)\delta(y - y') = \frac{2}{b}\sum_{n=1}^{\infty} f\left(-\frac{n^2\pi^2}{b^2}\right)\sin\left(\frac{n\pi}{b}y\right)\sin\left(\frac{n\pi}{b}y'\right)$$

In particular,

$$f(\mathbb{k})\delta(y - y') = f(\sqrt{\mathbb{L}_2})\delta(y - y') = \frac{2}{b}\sum_{n=1}^{\infty} f\left(i\frac{n\pi}{b}\right)\sin\left(\frac{n\pi}{b}y\right)\sin\left(\frac{n\pi}{b}y'\right)$$

Using this and (3) in (12.61) gives

$$G(x, y; x', y') = G_1(x, y, x', y'; \mathbb{L}_2)\delta(y - y')$$

$$= \begin{cases} \dfrac{2}{b}\displaystyle\sum_{n=1}^{\infty} \dfrac{\sin\left(\frac{in\pi}{b}x\right)\sin\left[\frac{in\pi}{b}(a - x')\right]\sin\left(\frac{n\pi}{b}y\right)\sin\left(\frac{n\pi}{b}y'\right)}{\frac{in\pi}{b}\sin\left(\frac{in\pi}{b}a\right)} & \text{if } 0 \leqslant x < x' \\[2ex] \dfrac{2}{b}\displaystyle\sum_{n=1}^{\infty} \dfrac{\sin\left[\frac{in\pi}{b}(a - x)\right]\sin\left(\frac{in\pi}{b}x'\right)\sin\left(\frac{n\pi}{b}y\right)\sin\left(\frac{n\pi}{b}y'\right)}{\frac{in\pi}{b}\sin\left(\frac{in\pi}{b}a\right)} & \text{if } x' < x \leqslant a \end{cases}$$

which can be simplified to

$$G = \begin{cases} \dfrac{2}{\pi}\displaystyle\sum_{n=1}^{\infty} \dfrac{\sinh\left(\frac{n\pi}{b}x\right)\sinh\left[\frac{n\pi}{b}(a - x')\right]\sin\left(\frac{n\pi}{b}y\right)\sin\left(\frac{n\pi}{b}y'\right)}{n\sinh\left(\frac{n\pi a}{b}\right)} & \text{if } 0 \leqslant x < x' \\[2ex] \dfrac{2}{\pi}\displaystyle\sum_{n=1}^{\infty} \dfrac{\sinh\left[\frac{n\pi}{b}(a - x)\right]\sinh\left(\frac{n\pi}{b}x'\right)\sin\left(\frac{n\pi}{b}y\right)\sin\left(\frac{n\pi}{b}y'\right)}{n\sinh\left(\frac{n\pi a}{b}\right)} & \text{if } x' < x \leqslant a \end{cases}$$

Although this result appears to differ from that of Example 12.5.6, it can be shown that the two are equivalent. The proof is left as a problem. ●

Sometimes it is more convenient to break an operator into more than two parts. For instance, the operator may be written as $\mathbb{L}_1 + \mathbb{L}_2 + \mathbb{L}_3$, where $[\mathbb{L}_i, \mathbb{L}_j] = 0$ for all $i, j = 1, 2, 3$. Then we can consider $\mathbb{L}_2$ and $\mathbb{L}_3$ as numbers, work out the problem for $\mathbb{L}_1$, get a Green's function of the form $G_1(\mathbf{x}_1, \mathbf{y}_1; \mathbf{x}_2, \mathbf{y}_2; \mathbb{L}_2, \mathbb{L}_3)$, and obtain the full Green's function from

$$G(\mathbf{x}, \mathbf{y}) = G_1(\mathbf{x}_1, \mathbf{y}_1; \mathbf{x}_2, \mathbf{y}_2; \mathbb{L}_2, \mathbb{L}_3)\delta(\mathbf{x}_2 - \mathbf{y}_2)$$

where $\mathbf{x}_2$ and $\mathbf{y}_2$ are the variables on which $\mathbb{L}_2$ and $\mathbb{L}_3$ act.

Example 12.5.9

Let us consider the Dirichlet BVP for the operator $\nabla^2 - k^2$ in the region

$$\mathrm{D} \equiv \{(x, y, z) | 0 \leqslant x \leqslant a, 0 \leqslant y \leqslant b, -\infty < z < +\infty\}$$

Thus, there are no boundaries in the z direction. However, we demand that the GF be defined at $z = \pm\infty$.

The GF satisfies

$$(\nabla^2 - k^2)G = \delta(\mathbf{r} - \mathbf{r}')$$

$$G(0, y, z) = G(a, y, z) = G(x, 0, z) = G(x, b, z) = 0 \tag{1}$$

Writing $\mathbb{L}_1 = \partial^2/\partial z^2$, $\mathbb{L}_2 = \partial^2/\partial x^2$, and $\mathbb{L}_3 = \partial^2/\partial y^2$, we have

$$\frac{d^2 G}{dz^2} - (k^2 - \mathbb{L}_2 - \mathbb{L}_3)G = \delta(x - x')\delta(y - y')\delta(z - z')$$

If we define

$$\mu^2 \equiv k^2 - \mathbb{L}_2 - \mathbb{L}_3$$

then we have a one-dimensional problem:

$$\frac{d^2 G_1}{dz^2} - \mu^2 G_1 = \delta(z - z')$$

$$G_1(z = -\infty) = G_1(z = +\infty) = 0$$

Its solution is easily found to be (see Problem 11.20)

$$G_1(z, z'; x, x', y, y'; \mathbb{L}_2, \mathbb{L}_3) = -\frac{e^{-\mu|z-z'|}}{2\mu}$$

and the full GF is given by

$$G(\mathbf{r}, \mathbf{r}') = \left(-\frac{e^{-\mu|z-z'|}}{2\mu}\right)\delta(x - x')\delta(y - y') \tag{2}$$

As before, we express the delta functions as series of cigenfunctions of $\mathbb{L}_2$ and $\mathbb{L}_3$. It can easily be shown that the normalized eigenfunctions of $\mathbb{L}_2$ satisfying $u(0) = u(a) = 0$ are

$$u_n^{(2)}(x) = \sqrt{\frac{2}{a}}\sin\left(\frac{n\pi}{a}x\right) \qquad \text{for } n = 1, 2, \ldots$$

and the corresponding eigenvalues are

$$\lambda_n^{(2)} = -\left(\frac{n\pi}{a}\right)^2 \qquad \text{for } n = 1, 2, \ldots$$

Similarly, the eigenfunctions and eigenvalues of $\mathbb{L}_3$ are

$$u_m^{(3)}(y) = \sqrt{\frac{2}{b}}\sin\left(\frac{m\pi}{b}y\right) \qquad \text{and} \qquad \lambda_m^{(3)} = -\left(\frac{m\pi}{b}\right)^2 \qquad \text{for } m = 1, 2, \ldots$$

Therefore, we can write

$$\delta(x - x')\delta(y - y') = \frac{4}{ab}\sum_{m,n=1}^{\infty} \sin\left(\frac{n\pi}{a}x\right)\sin\left(\frac{n\pi}{a}x'\right)\sin\left(\frac{m\pi}{b}y\right)\sin\left(\frac{m\pi}{b}y'\right)$$

Inserting this in Eq. (2) yields

$$G(\mathbf{r}, \mathbf{r}') = \frac{2}{ab}\sum_{m,n=1}^{\infty} \frac{e^{-\mu_{mn}|z - z'|}}{\mu_{mn}} \sin\left(\frac{n\pi}{a}x\right)\sin\left(\frac{n\pi}{a}x'\right)\sin\left(\frac{m\pi}{b}y\right)\sin\left(\frac{m\pi}{b}y'\right) \qquad (3)$$

where

$$\mu_{mn}^2 \equiv k^2 - \lambda_n^{(2)} - \lambda_m^{(3)} = k^2 + \left(\frac{n\pi}{a}\right)^2 + \left(\frac{m\pi}{b}\right)^2$$

From (3) we can also obtain the GF for the Helmholtz operator $\nabla^2 + k^2$ in the domain D. This is accomplished by letting $k \to ik$. •

In Examples 12.5.8 and 12.5.9 we picked out one operator and treated the rest as constants. There is, of course, no *a priori* reason for such a choice. For instance, we could have let $\mathbb{L}_1 = \partial^2/\partial y^2$ in Example 12.5.8. Then the Green's function would be

$$G = \begin{cases} \dfrac{2}{\pi}\displaystyle\sum_{n=1}^{\infty} \dfrac{\sinh\left(\frac{n\pi}{a}y\right)\sinh\left[\frac{n\pi}{a}(b - y')\right]\sin\left(\frac{n\pi}{a}x\right)\sin\left(\frac{n\pi}{a}x'\right)}{n\sinh\left(\frac{n\pi}{a}b\right)} & \text{if } 0 \leqslant y < y' \\[2ex] \dfrac{2}{\pi}\displaystyle\sum_{n=1}^{\infty} \dfrac{\sinh\left[\frac{n\pi}{a}(b - y)\right]\sinh\left(\frac{n\pi}{a}y'\right)\sin\left(\frac{n\pi}{a}x\right)\sin\left(\frac{n\pi}{a}x'\right)}{n\sinh\left(\frac{n\pi}{a}b\right)} & \text{if } y' < y \leqslant b \end{cases} \qquad (12.62)$$

This is, of course, completely equivalent to the function obtained in Example 12.5.8. Both Green's functions converge for all values of x and y in the rectangular region $0 \leqslant x \leqslant a$, $0 \leqslant y \leqslant b$ except at $(x, y) = (x', y')$; however, the *rate* of convergence may be different for the two. We can see this by looking at (12.62) with $y > y'$, for definiteness. We note that the series is simply a Fourier-series expansion, which we may write as

$$G = \sum_{n=1}^{\infty} a_n(y, y'; x')\sin\left(\frac{n\pi}{a}x\right)$$

where a_n is the expansion coefficient, which happens to be dependent on certain parameters. The rate of convergence of this series is determined by the rate at which the coefficients go to zero for large n. For such values of n we have

$$a_n \approx \frac{\frac{1}{2}\exp\left[\frac{n\pi}{a}(b-y)\right]\frac{1}{2}\exp\left[\frac{n\pi}{a}y'\right]}{\frac{1}{2}n\exp\left(\frac{n\pi}{a}b\right)}\sin\left(\frac{n\pi}{a}x'\right)$$

$$= \frac{e^{(n\pi/a)(y'-y)}}{2n}\sin\left(\frac{n\pi}{a}x'\right)$$

It is clear from this expression that, regardless of the value of x', a_n tends to zero quickly as long as y is away from y'. Thus, if we are interested in $G(x, y; x', y')$ at points (x, y) whose y-coordinates are far from y', then the appropriate expansion is that given by (12.62), that is, an expansion in terms of x eigenfunctions. On the other hand (and this can be shown in exact analogy with the above), if the Green's function is to be calculated for a point (x, y) whose x-coordinate is far away from the singular point (x', y'), then the appropriate expansion is the one given in Example 12.5.8.

This discussion of the OT has concentrated on Cartesian coordinates, for which the full operator, say ∇^2, is a simple sum of derivative operators with constant coefficients. In other coordinate systems the procedure is not as straightforward (as Examples 12.5.8 and 12.5.9 may have suggested). In such cases it may be possible to define a set of self-adjoint (differential) operators, $\{\mathbb{M}_i\}_{i=1}^k$, such that $[\mathbb{M}_i, \mathbb{M}_j] = 0$ for all $i, j = 1, \ldots, k$ and $\mathbb{L} = \mathbb{L}_1\mathbb{M}_1 + \cdots + \mathbb{L}_k\mathbb{M}_k$, where $\mathbb{L}$ is the full operator and $\{\mathbb{L}_i\}_{i=1}^k$ consists of the differential operators acting on variables on which the $\mathbb{M}_i$ have no action. Then we can choose a complete set of eigenfunctions $\{u_j(\mathbf{y})\}_{j=1}^\infty$ such that

$$\mathbb{M}_i u_j(\mathbf{y}) = \lambda_{ij} u_j(\mathbf{y})$$

where $\mathbf{y}$ is the set of variables on which the $\mathbb{M}_i$ act. As a minor change in notation, let us take $\mathbf{r}$ and $\mathbf{r}'$ as the arguments of the full Green's function and expand it in terms of $u_j(\mathbf{y})$:

$$G(\mathbf{r}, \mathbf{r}') = \sum_{n=1}^{\infty} g_n(\mathbf{x}, \mathbf{r}')u_n(\mathbf{y}) \tag{12.63}$$

Here $\mathbf{x}$ is the collection of variables on which the $\mathbb{L}_i$ act. We can also write

$$\delta(\mathbf{r}-\mathbf{r}') = \frac{\delta(\mathbf{x}-\mathbf{x}')\delta(\mathbf{y}-\mathbf{y}')}{J_1(\mathbf{x}')J_2(\mathbf{y}')}$$

$$= \frac{\delta(\mathbf{x}-\mathbf{x}')}{J_1(\mathbf{x}')}\sum_{n=1}^{\infty} u_n(\mathbf{y})u_n^*(\mathbf{y}') \tag{12.64}$$

where $J_1(\mathbf{x}')J_2(\mathbf{y}') \equiv J(\mathbf{x}', \mathbf{y}')$ is the Jacobian of the transformation from Cartesian coordinates $(\mathbf{r}, \mathbf{r}')$ to curvilinear coordinates $(\mathbf{x}, \mathbf{y}; \mathbf{x}', \mathbf{y}')$.

We note that the part of the Jacobian depending on the **y** variables has disappeared. This is due to the fact that $J_2(\mathbf{y})$ is really the weight function for integrals over **y**. Substituting (12.63) and (12.64) in $\mathbb{L}G = \delta(\mathbf{r} - \mathbf{r}')$ gives

$$\begin{aligned}\mathbb{L}\mathbb{G} &= (\mathbb{L}_1\mathbb{M}_1 + \mathbb{L}_2\mathbb{M}_2 + \cdots + \mathbb{L}_k\mathbb{M}_k) \sum_{n=1}^{\infty} g_n(\mathbf{x}, \mathbf{r}')u_n(\mathbf{y}) \\ &= \sum_{n=1}^{\infty} [(\mathbb{L}_1\mathbb{M}_1 + \cdots + \mathbb{L}_k\mathbb{M}_k)u_n(\mathbf{y})]g_n(\mathbf{x}, \mathbf{r}') \\ &= \sum_{n=1}^{\infty} [(\lambda_{1n}\mathbb{L}_1 + \cdots + \lambda_{kn}\mathbb{L}_k)g_n(\mathbf{x}, \mathbf{r}')]u_n(\mathbf{y}) \\ &= \sum_{n=1}^{\infty} \left[\frac{\delta(\mathbf{x} - \mathbf{x}')}{J_1(\mathbf{x}')} u_n^*(\mathbf{y}')\right]u_n(\mathbf{y})\end{aligned}$$

The orthonormality of the $u_n(\mathbf{y})$ gives

$$(\lambda_{1n}\mathbb{L}_1 + \lambda_{2n}\mathbb{L}_2 + \cdots + \lambda_{kn}\mathbb{L}_k)g_n(\mathbf{x}, \mathbf{x}', \mathbf{y}') = \frac{u_n^*(\mathbf{y}')}{J_1(\mathbf{x}')}\delta(\mathbf{x} - \mathbf{x}') \tag{12.65}$$

This equation clearly shows that all the **y** variables have been eliminated. By applying the same procedure to the operators on the LHS, we can further reduce the number of variables until we finally obtain a one-dimensional operator, to which we can apply the procedure of Chapter 11.

Example 12.5.10

Let us consider the Laplacian in spherical coordinates,

$$\nabla^2 u = \frac{1}{r^2}\frac{\partial}{\partial r}\left(r^2\frac{\partial u}{\partial r}\right) + \frac{1}{r^2 \sin\theta}\left[\frac{\partial}{\partial\theta}\left(\sin\theta\frac{\partial u}{\partial\theta}\right) + \frac{\partial^2 u}{\partial\varphi^2}\right]$$

If we introduce

$$\begin{aligned}&\mathbb{M}_1 u = u &&\mathbb{L}_1 u = \frac{1}{r^2}\frac{\partial}{\partial r}\left(r^2\frac{\partial u}{\partial r}\right) \\ &\mathbb{M}_2 u = \frac{1}{\sin\theta}\left[\frac{\partial}{\partial\theta}\left(\sin\theta\frac{\partial u}{\partial\theta}\right) + \frac{\partial^2 u}{\partial\varphi^2}\right] &&\mathbb{L}_2 u = \frac{1}{r^2}u\end{aligned} \tag{1}$$

the Laplacian becomes

$$\nabla^2 = \mathbb{L}_1\mathbb{M}_1 + \mathbb{L}_2\mathbb{M}_2$$

where $[\mathbb{M}_1, \mathbb{M}_2] = 0$ because $\mathbb{M}_1$ is simply the identity operator. The mutual eigenfunctions of $\mathbb{M}_1$ and $\mathbb{M}_2$ are those of $\mathbb{M}_2$. But $\mathbb{M}_2$ is simply (the negative of) the angular momentum operator discussed in Chapter 8, whose eigenfunctions are the spherical harmonics. We thus have

$$\mathbb{M}_2 Y_{lm}(\theta, \varphi) = -l(l+1)Y_{lm}(\theta, \varphi)$$

Let us write the Green's function as in Eq. (12.63):

$$G(\mathbf{r}, \mathbf{r}') = \sum_{l,m} g_{lm}(r; r', \theta', \varphi') Y_{lm}(\theta, \varphi)$$

We also write the delta function as

$$\delta(\mathbf{r} - \mathbf{r}') = \frac{\delta(r - r')\delta(\theta - \theta')\delta(\varphi - \varphi')}{r'^2 \sin\theta'}$$

$$= \frac{\delta(r - r')}{r'^2} \sum_{l,m} Y_{lm}(\theta, \varphi) Y^*_{lm}(\theta', \varphi')$$

where we have used the completeness of the spherical harmonics,

$$\frac{\delta(\theta - \theta')\delta(\varphi - \varphi')}{\sin\theta'} = \sum_{l,m} Y_{lm}(\theta, \varphi) Y^*_{lm}(\theta', \varphi')$$

Substituting all of the above in $\nabla^2 G = \delta(\mathbf{r} - \mathbf{r}')$, we obtain

$$\nabla^2 G = (\mathbb{L}_1\mathbb{M}_1 + \mathbb{L}_2\mathbb{M}_2) \sum_{l,m} g_{lm}(r; r', \theta', \varphi') Y_{lm}(\theta, \varphi)$$

$$= \sum_{l,m} \{[\mathbb{L}_1 - l(l+1)\mathbb{L}_2] g_{lm}(r; r', \theta', \varphi')\} Y_{lm}(\theta, \varphi)$$

$$= \frac{\delta(r - r')}{r'^2} \sum_{l,m} Y^*_{lm}(\theta', \varphi') Y_{lm}(\theta, \varphi)$$

The orthogonality of the $Y_{lm}(\theta, \varphi)$ yields

$$[\mathbb{L}_1 - l(l+1)] g_{lm}(r; r', \theta', \varphi') = \frac{\delta(r - r')}{r'^2} Y^*_{lm}(\theta', \varphi')$$

This shows that the angular part of g_{lm} is simply $Y^*_{lm}(\theta', \varphi')$. Separating this from the dependence on r and r' and substituting for $\mathbb{L}_1$ and $\mathbb{L}_2$, we obtain

$$\frac{1}{r^2}\frac{d}{dr}\left(r^2 \frac{dg_{lm}}{dr}\right) - \frac{l(l+1)}{r^2} g_{lm} = \frac{\delta(r - r')}{r^2} \tag{2}$$

where g_{lm} is a function of r and r' only. The techniques of Chapter 11 can be employed to solve Eq. (2) (see Exercise 12.5.5). •

As you may have noticed, there is a striking resemblance between the OT and the separation of variables. The OT is in fact a fancy way of implementing the separation of variables in which the operator aspect is emphasized.

Exercises

12.5.1 Show that for $m = 3$ the expression for $G_s(\mathbf{r})$ given by Eq. (1) of Example 12.5.2 reduces to

$$G_s(\mathbf{r}) = -\frac{e^{-\mu r}}{4\pi r}$$

12.5.2 Find the singular part of the retarded GF and the advanced GF for the wave equation in three dimensions.

12.5.3 Show that the GF [Eq. (2) of Example 12.5.4] derived from the principal value of the ω integration for the wave equation in three dimensions satisfies only the homogeneous PDE.

12.5.4 Find the eigenfunction expansion of the GF for the Dirichlet BVP for the Laplacian operator in two dimensions for which the region of interest is the interior of a circle of radius a.

12.5.5 Complete the calculations of Example 12.5.10 and find the GF for the Laplacian with Dirichlet BCs on two concentric spheres of radii a and b, where $a < b$. Consider the case where $a \to 0$ and $b \to \infty$, and compare the result with the singular part of the GF for the Laplacian.

PROBLEMS

12.1 Find the characteristic curves for the two-dimensional wave equation and the two-dimensional diffusion equation.

12.2 Find the characteristic curves for $\mathbb{L}[u] = \partial u/\partial x$.

12.3 Show that the x_i in (12.13) describe an m-dimensional sphere of radius r, that is, $\sum_{i=1}^{m} x_i^2 = r^2$.

12.4 Find the volume of an m-dimensional sphere.

12.5 Find the GF for the Dirichlet BVP in two dimensions if D is the UHP and ∂D is the x-axis.

12.6 Use the method of images to find the GF for the interior of a "sphere" of radius a in two and three dimensions.

12.7 Derive Eqs. (12.53).

12.8 Derive Eq. (12.55) using the procedure outlined for parabolic equations.

12.9 Use the FTT to find the singular part of the GF for $\partial/\partial t - \partial^2/\partial x^2$ and $\partial/\partial t - \partial^2/\partial x^2 - \partial^2/\partial y^2 - \partial^2/\partial z^2$. Compare your results with that obtained in Example 12.5.3.

12.10 Show directly that both $G^{(\text{ret})}$ and $G^{(\text{adv})}$ satisfy $\nabla^2 G = \delta(\mathbf{r})\delta(t)$ in three dimensions.

12.11 Derive Eq. (7) of Example 12.4.2 from Eq. (6).

12.12 Using Eq. (8) of Example 12.4.2, show that if $g(\theta', \varphi') = V_0$, the potential at any point inside the sphere is V_0.

12.13 Repeat Example 12.5.1 for $m = 2$.

12.14 Consider a rectangular box with sides a, b, and c located in the first octant with one corner at the origin. Let D denote the inside of this box. (**a**) Show that zero cannot be an eigenvalue of the Laplacian operator with the Dirichlet BCs on ∂D. (**b**) Find the GF for this Dirichlet BVP.

12.15 Show that the results in Examples 12.5.6 and 12.5.8 are equivalent.

12.16 Find the GF for the Helmholtz equation $(\nabla^2 + k^2)u = 0$ on the rectangle $0 \leqslant x \leqslant a$, $0 \leqslant y \leqslant b$.

12.17 Find the singular part of the one-dimensional GF for the operator $a\,d^2/dx^2 + b$, where $a > 0$ and $b < 0$.

12.18 Calculate the GF of the two-dimensional Laplacian operator appropriate for Neumann BCs on the rectangle $0 \leqslant x \leqslant a$, $0 \leqslant y \leqslant b$.

12.19 Find the three-dimensional Dirichlet GF for the static Klein-Gordon operator $\nabla^2 - k^2$ in the half-space $z \geqslant 0$.

12.20 Find the three-dimensional Neumann GF for $\nabla^2 - k^2$ in the half-space $z \leqslant 0$.

12.21 Show that the two-dimensional Dirichlet GF for the two-dimensional static Klein-Gordon operator $\nabla^2 - k^2$ in the infinite strip $0 \leqslant x \leqslant a$, $-\infty < y < +\infty$ is

$$G(x, y; x', y') = -\sum_{n=1}^{\infty} \frac{e^{-\lambda_n|y-y'|}}{\lambda_n} \sin\left(\frac{n\pi}{a}x\right) \sin\left(\frac{n\pi}{a}x'\right)$$

where $\lambda_n^2 = k^2 + (n\pi/a)^2$ and it is assumed that $G(x, y; x', y') \to 0$ as $|y| \to \infty$.

12.22 Use the operator technique to calculate the two-dimensional Dirichlet GF for the two-dimensional operator $\nabla^2 - k^2$ on the rectangle $0 \leqslant x \leqslant a$, $0 \leqslant y \leqslant b$. Also obtain an eigenfunction expansion for this GF.

12.23 Use the operator technique to find the three-dimensional Dirichlet GF for the Laplacian in a circular cylinder of radius a and height h.

12.24 Calculate the singular part of the GF for the three-dimensional free Schrödinger operator

$$i\hbar\frac{\partial}{\partial t} - \frac{\hbar^2}{2\mu}\nabla^2$$

where $\hbar$ and μ are constants.

12.25 Use the operator technique to show that the GF for the Helmholtz operator $\nabla^2 + k^2$ in three dimensions is

$$G(\mathbf{r}, \mathbf{r}') = -ik \sum_{l=0}^{\infty} \sum_{m=-l}^{l} j_l(kr_<) h_l(kr_>) Y_{lm}(\theta, \varphi) Y_{lm}^*(\theta', \varphi')$$

where $r_<$ ($r_>$) is the smaller (larger) of r and r' and j_l and h_l are the spherical Bessel and Hankel functions, respectively. No explicit BCs are assumed except that there is regularity at $r = 0$ and that $G(\mathbf{r}, \mathbf{r}') \to 0$ for $|\mathbf{r}| \to \infty$. Now obtain the identity

$$\frac{e^{ik|\mathbf{r}-\mathbf{r}'|}}{4\pi|\mathbf{r}-\mathbf{r}'|} = ik \sum_{l,m} j_l(kr_<) h_l(kr_>) Y_{lm}(\theta, \varphi) Y_{lm}^*(\theta', \varphi')$$

12.26 Use the result of Problem 12.25 to derive the plane wave expansion [see Eq. (10.29)]

$$e^{i\mathbf{k}\cdot\mathbf{r}} = 4\pi \sum_{l,m} i^l j_l(kr) Y_{lm}(\theta, \varphi) Y_{lm}^*(\theta', \varphi')$$

where θ' and φ' are assumed to be the angular coordinates of $\mathbf{k}$. [Hint: Let $|\mathbf{r}'| \to \infty$ and use

$$|\mathbf{r} - \mathbf{r}'| = (\mathbf{r}'^2 + \mathbf{r}^2 - 2\mathbf{r}\cdot\mathbf{r}')^{1/2} \to r' - \frac{\mathbf{r}'\cdot\mathbf{r}}{r'} \equiv r' - \frac{\mathbf{k}\cdot\mathbf{r}}{k}$$

and the asymptotic formula for $h_l^{(1)}(z)$ for large z, $h_l^{(1)}(z) \to (1/z)e^{i[z+(l+1)(\pi/2)]}$.]

12.27 Calculate the retarded GF for the wave operator in two dimensions and show that it is equal to

$$G_s^{(\mathrm{ret})}(\mathbf{r}, t) = \frac{\theta(t)}{2\pi\sqrt{t^2 - r^2}}$$

Now use this result to obtain the GF for any even number of dimensions:

$$G_s^{(\mathrm{ret})}(\mathbf{r}, t) = \frac{\theta(t)}{2\pi}\left(-\frac{1}{2\pi r}\frac{\partial}{\partial r}\right)^{n-1}\left[\frac{1}{\sqrt{t^2 - r^2}}\right] \qquad \text{for } n = \frac{m}{2}$$

13

Integral Equations

Consider the following problem. Let the point mass m be moving on a perfectly smooth curve in the yz-plane under the influence of gravity, which acts in the negative z direction. Let the equation of the curve be given by $y = F(z)$. The conservation of energy gives

$$\tfrac{1}{2}m\dot{z}^2 + \tfrac{1}{2}m\dot{y}^2 + mgz = E$$

Substituting $\dot{y} = (dF/dz)\dot{z}$ and solving for $\dot{z}$ yields

$$\dot{z} \equiv \frac{dz}{dt} = \frac{\sqrt{\dfrac{2E}{m} - 2gz}}{\sqrt{1 + (dF/dz)^2}} \equiv \frac{\sqrt{E/mg - z}}{u(z)} \tag{13.1}$$

where
$$u(z) \equiv \sqrt{1 + (dF/dz)^2/2g}$$

If the initial speed of the point mass is zero, then E/mg is simply the initial height, z_0. Let us solve (13.1) for time:

$$t = -\int_h^{z_0} \frac{u(z)}{\sqrt{z_0 - z}}\, dz$$

where h is a fixed height the particle reaches at time t. For simplicity, we take h to be zero and rewrite the above equation as

$$t = -\int_0^z \frac{u(\eta)}{\sqrt{z - \eta}}\, d\eta \equiv -f(z) \tag{13.2}$$

The dependence of t on the initial height has been emphasized by taking the latter to be z.

Given $f(z)$, that is, given the dependence of the time of travel as a function of the initial height for various initial heights, can we determine $u(z)$, and, therefore, the curve? Such a question takes (13.2) to be an *integral equation* (IE), an equation in which the integral of an unknown function is given and which is used to find the function itself.

13.1 CLASSIFICATION OF INTEGRAL EQUATIONS

Equation (13.2), called *Abel's integral equation*, is only one example of many kinds of integral equations. Some other examples follow.

Example 13.1.1

The Fourier transform of a function,

$$\tilde{f}(k) = \frac{1}{\sqrt{2\pi}} \int_{-\infty}^{\infty} dx\, e^{-ikx} f(x)$$

can be considered an IE in which $\tilde{f}(k)$ is given and by which $f(x)$ is to be found. The solution, of course, is simply the inverse transform:

$$f(x) = \frac{1}{\sqrt{2\pi}} \int_{-\infty}^{\infty} dk\, e^{ikx} \tilde{f}(k)$$ •

Example 13.1.2

Let us consider an eigenvalue equation with Dirichlet BCs:

$$\begin{aligned} \nabla^2 u &= \lambda u \qquad &&\text{for } \mathbf{x} \in \mathrm{D} \\ u(\mathbf{x}_b) &= 0 \qquad &&\text{for } \mathbf{x}_b \in \partial \mathrm{D} \end{aligned} \tag{1}$$

Regarding λu as the inhomogeneous term of the DE, we can "solve" (1) immediately:

$$u(\mathbf{x}) = \lambda \int_{\mathrm{D}} d^m y\, G(\mathbf{x}, \mathbf{y})\, u(\mathbf{y})$$

where $G(\mathbf{x}, \mathbf{y})$ is the GF appropriate for the BCs. This is an IE in which the unknown function u appears both inside and outside the integral sign. •

Example 13.1.3

The Dirichlet BVP in m dimensions,

$$\begin{aligned} \mathbb{L}_{\mathbf{x}}[u] &= 0 \qquad &&\text{for } \mathbf{x} \in \mathrm{D} \\ u(\mathbf{x}_b) &= f(\mathbf{x}_b) \qquad &&\text{for } \mathbf{x}_b \in \partial \mathrm{D} \end{aligned} \tag{1}$$

can be converted into an IE in $m - 1$ dimensions.

Let $G(\mathbf{x}, \mathbf{y})$ be the GF for $\mathbb{L}_\mathbf{x}$ with Dirichlet BCs. The $(m-1)$-dimensional IE equivalent to (1) is

$$\int_{\partial D} G(\mathbf{x}_b, \mathbf{y})\eta(\mathbf{y})\, da_y = f(\mathbf{x}_b) \qquad \text{for } \mathbf{x}_b \in \partial D$$

where $\eta(\mathbf{y})$ is the unknown function. If we can find $\eta(\mathbf{y})$, the solution to the original BVP will be

$$u(\mathbf{x}) = \int_{\partial D} G(\mathbf{x}, \mathbf{y})\eta(\mathbf{y})\, da_y \tag{2}$$

because for $\mathbf{x} \in D$ *not on the boundary*

$$\mathbb{L}_\mathbf{x}[u(\mathbf{x})] = \int_{\partial D} \mathbb{L}_\mathbf{x} G(\mathbf{x}, \mathbf{y})\eta(\mathbf{y})\, da_y = \int_{\partial D} \delta(\mathbf{x}-\mathbf{y})\eta(\mathbf{y})\, da_y = 0$$

On the other hand, if $\mathbf{x} \in \partial D$, then $u(\mathbf{x})$ as given in (2) will reduce to $u(\mathbf{x}_b) = f(\mathbf{x}_b)$, as it should. ●

Integral equations can be divided into two major groups. Those that have a variable limit of integration are called *Volterra equations.* Abel's integral equation is an example of this group. If the limits of integration are constants, the integral equation is called a *Fredholm equation.* If the unknown function appears only inside the integral, the integral equation is said to be of the *first kind.* Integral equations having the unknown function outside the integral are said to be of the *second kind.* The four kinds of equations can be written as follows.

$$\int_a^x K(x, t)u(t)\, dt = f(x) \qquad \text{Volterra equation of the first kind}$$

$$\int_a^b K(x, t)u(t)\, dt = f(x) \qquad \text{Fredholm equation of the first kind}$$

$$u(x) + \int_a^x K(x, t)u(t)\, dt = f(x) \qquad \text{Volterra equation of the second kind}$$

$$u(x) + \int_a^b K(x, t)u(t)\, dt = f(x) \qquad \text{Fredholm equation of the second kind}$$

In each case $K(x, t)$ is the kernel of the integral equation. If its kernel is changed to $\theta(x-t)K(x, t)$, a Volterra equation becomes a Fredholm equation. Thus, we will discuss only Fredholm equations.

We will try to solve the Fredholm equation

$$u(x) - \lambda \int_a^b K(x, t)u(t)\,dt = f(x) \tag{13.3a}$$

which we can abbreviate as

$$|u\rangle - \lambda\mathbb{K}|u\rangle = |f\rangle \tag{13.3b}$$

This equation is of the second kind, which is the only kind we will discuss. Integral equations of the first kind are much more difficult to handle but, fortunately, arise rarely in physical applications. The constant λ is introduced for convenience here. Later it will be identified as the eigenvalue of $\mathbb{K}$.

It is worth mentioning that just as there are partial differential equations (involving more than one independent variable), there are integral equations in higher dimensions. The generalization of (13.3a) to m dimensions yields

$$u(\mathbf{x}) - \lambda \int_{\mathrm{D}} d^m y\, K(\mathbf{x}, \mathbf{y})u(\mathbf{y}) = f(\mathbf{x}) \tag{13.4}$$

where D is a given region in $\mathbb{R}^m$ and $\mathbf{x}$ and $\mathbf{y}$ are m-dimensional vectors.

13.2 NEUMANN-SERIES SOLUTIONS

There is a systematic way of solving Eq. (13.3b) that involves successive approximation of the solution. Writing the equation as

$$|u\rangle = |f\rangle + \lambda\mathbb{K}|u\rangle \tag{13.5a}$$

we can interpret it as follows. The difference between $|u\rangle$ and $|f\rangle$ is $\lambda\mathbb{K}|u\rangle$. If $\lambda\mathbb{K}$ is absent, the two vectors $|u\rangle$ and $|f\rangle$ will be equal. The effect of $\lambda\mathbb{K}$ is to change $|u\rangle$ in such a way that when the result is added to $|f\rangle$, it gives $|u\rangle$. As our initial approximation, therefore, we take $|u\rangle$ to be equal to $|f\rangle$ and write

$$|u_0\rangle = |f\rangle$$

where the index reminds us of the order (in this case zeroth, because $\lambda\mathbb{K} = 0$) of the approximation. To find a better approximation, we substitute for $|u\rangle$ in the RHS of (13.5a) and obtain

$$|u_1\rangle = |f\rangle + \lambda\mathbb{K}|f\rangle \tag{13.5b}$$

Still a better approximation is achieved if we substitute this expression for $|u_1\rangle$ in (13.5a):

$$|u_2\rangle = |f\rangle + \lambda\mathbb{K}(|f\rangle + \lambda\mathbb{K}|f\rangle) = |f\rangle + \lambda\mathbb{K}|f\rangle + \lambda^2\mathbb{K}^2|f\rangle$$

The procedure is now clear. Once $|u_n\rangle$, the nth approximation, is obtained, we can get $|u_{n+1}\rangle$ by substituting on the RHS of (13.5a).

Before continuing, let us write the above equations in integral form. For (13.5b) we have

$$u_1(x) = f(x) + \lambda \int_a^b K(x, t) f(t)\, dt$$

Substituting this in (13.3a) yields

$$\begin{aligned} u_2(x) &= f(x) + \lambda \int_a^b ds\, K(x, s) u_1(s) \\ &= f(x) + \lambda \int_a^b ds\, K(x, s)\left[f(s) + \lambda \int_a^b K(s, t) f(t)\, dt \right] \qquad (13.5c) \\ &= f(x) + \lambda \int_a^b ds\, K(x, s) f(s) + \lambda^2 \int_a^b dt \left[\int_a^b K(x, s) K(s, t)\, ds \right] f(t) \end{aligned}$$

If we define

$$K^2(x, t) \equiv \int_a^b K(x, s) K(s, t)\, ds$$

Equation (13.5c) becomes

$$u_2(x) = f(x) + \lambda \int_a^b ds\, K(x, s) f(s) + \lambda^2 \int_a^b dt\, K^2(x, t) f(t)$$

Substituting this in the RHS of (13.3a) yields

$$u_3(x) = f(x) + \lambda \int_a^b dt\, K(x, t) f(t) + \lambda^2 \int_a^b dt\, K^2(x, t) f(t) + \lambda^3 \int_a^b dt\, K^3(x, t) f(t)$$

where

$$K^3(x, t) \equiv \int_a^b ds_1 \int_a^b ds_2\, K(x, s_1) K(s_1, s_2) K(s_2, t)$$

Similar expressions can be derived for $u_4(x)$, $u_5(x)$, and so forth. The integrals expressing various "powers" of K can be obtained using Dirac notation and vectors with continuous indices, as discussed in Chapter 5. Thus, for instance,

$$
\begin{aligned}
K^2(x,t) &\equiv \langle \mathrm{x}|\mathbb{K}^2|\mathrm{t}\rangle = \langle \mathrm{x}|\mathbb{K}\mathbb{1}\mathbb{K}|\mathrm{t}\rangle \\
&= \langle \mathrm{x}|\mathbb{K}\left(\int_a^b |\mathrm{s}\rangle\langle \mathrm{s}|\, ds\right)\mathbb{K}|\mathrm{t}\rangle \\
&= \int_a^b \langle \mathrm{x}|\mathbb{K}|\mathrm{s}\rangle\langle \mathrm{s}|\mathbb{K}|\mathrm{t}\rangle\, ds = \int_a^b K(x,s)K(s,t)\, ds
\end{aligned}
$$

We can always use this technique to convert an equation in kets into an equation in functions and integrals. Therefore, we can concentrate on Eq. (13.3b) and its various approximations.

Continuing to the nth-order approximation, we easily obtain

$$
\begin{aligned}
|\mathrm{u}_n\rangle &= |\mathrm{f}\rangle + \lambda\mathbb{K}|\mathrm{f}\rangle + \cdots + \lambda^n\mathbb{K}^n|\mathrm{f}\rangle \\
&= \sum_{j=0}^{n} (\lambda\mathbb{K})^j|\mathrm{f}\rangle
\end{aligned} \tag{13.6a}
$$

which is written in integral form as

$$
u_n(x) = \sum_{j=0}^{n} \lambda^j \int_a^b K^j(x,t)f(t)\, dt \tag{13.6b}
$$

Here $K^j(x,t)$ is defined iteratively by

$$
K^0(x,t) = \delta(x-t) \qquad K^j(x,t) = \int_a^b K(x,s)K^{j-1}(s,t)\, ds \tag{13.6c}
$$

What we hope is that as $n \to \infty$, $|\mathrm{u}_n\rangle \to |\mathrm{u}\rangle$. We, therefore, tentatively write

$$
|\mathrm{u}\rangle = \sum_{j=0}^{\infty} (\lambda\mathbb{K})^j|\mathrm{f}\rangle \tag{13.7a}
$$

and

$$
u(x) = \sum_{j=0}^{\infty} \lambda^j \int_a^b K^j(x,t)f(t)\, dt \tag{13.7b}
$$

The series in Eqs. (13.7) is called the *Neumann series.*

First, we note that Eqs. (13.7) *formally* satisfy Eqs. (13.3). Substituting (13.7a), for example, in (13.3b), we obtain

$$
\begin{aligned}
|\mathrm{u}\rangle - \lambda\mathbb{K}|\mathrm{u}\rangle &= \sum_{j=0}^{\infty} (\lambda\mathbb{K})^j|\mathrm{f}\rangle - \lambda\mathbb{K}\sum_{j=0}^{\infty} (\lambda\mathbb{K})^j|\mathrm{f}\rangle \\
&= |\mathrm{f}\rangle + \sum_{j=1}^{\infty} (\lambda\mathbb{K})^j|\mathrm{f}\rangle - \sum_{j=0}^{\infty} (\lambda\mathbb{K})^{j+1}|\mathrm{f}\rangle \\
&= |\mathrm{f}\rangle
\end{aligned}
$$

because the last two sums cancel. Second, the power series in (13.7a) can be obtained directly from (13.3b) by noting that the latter can be written as

$$(\mathbb{1} - \lambda\mathbb{K})|u\rangle = |f\rangle$$

which has a "solution"

$$|u\rangle = (\mathbb{1} - \lambda\mathbb{K})^{-1}|f\rangle \equiv \left(\frac{1}{\mathbb{1} - \lambda\mathbb{K}}\right)|f\rangle$$

Equation (13.7a) is simply the geometric-series expansion of $1/(\mathbb{1} - \lambda\mathbb{K})$. Finally, Eqs. (13.7) have meaning only if the series converges in a sense to be described next.

The series in (13.7a) converges if its norm is finite. Because of the triangle inequality, we can write

$$\left\| \sum_{j=0}^{\infty} (\lambda\mathbb{K})^j |f\rangle \right\| \leqslant \sum_{j=0}^{\infty} \|(\lambda\mathbb{K})^j |f\rangle\|$$

Thus, if the series (of numbers) on the RHS here converges, the series in (13.7a) also converges. Since the RHS involves a vector, $|f\rangle$, the convergence depends on the nature of $|f\rangle$. This means that we must test for the convergence of the series for every new $|f\rangle$. It would be nice if we could come up with a test that is *independent* of $|f\rangle$. We can do this—by using the concept of the norm of an operator (see Definition 11.2.4). Recall that a bounded operator $\mathbb{A} \in \mathscr{L}(\mathscr{V})$ is one with the property that, for some $M \geqslant 0$,

$$\|\mathbb{A}|u\rangle\| \leqslant M\|\mathbf{u}\| \qquad \forall |u\rangle \in \mathscr{V}$$

The smallest such M is simply the norm of $\mathbb{A}$. Thus, we have

$$\|\mathbb{A}|u\rangle\| \leqslant \|\mathbb{A}\|\,\|\mathbf{u}\|$$

The corresponding inequality for $\mathbb{K}$ can be obtained as follows. Considering $K(x, t)$ as a square-integrable function of x and t, that is, assuming that

$$\int_a^b dx \int_a^b [K(x, t)]^2\, dt \equiv B^2 < \infty$$

we can regard

$$[\mathbb{K}f(x)]^2 \equiv \left[\int_a^b K(x, t) f(t)\, dt\right]^2$$

as the inner product of K (as a function of t alone) and f. We can now apply the Schwarz inequality to this inner product and write

$$[g(x)]^2 \equiv [\mathbb{K}f(x)]^2 \leqslant \|f\|^2 \left\{\int_a^b [K(x, t)]^2\, dt\right\}$$

Integrating both sides with respect to x, we obtain

$$\int_a^b [g(x)]^2\, dx \leqslant \|f\|^2 \int_a^b dx \int_a^b dt\, [K(x,t)]^2\, dt$$

or

$$\|g\|^2 \equiv \|\mathbb{K}|\mathrm{f}\rangle\|^2 \leqslant \|f\|^2 B^2$$

which finally yields

$$\|\mathbb{K}|\mathrm{f}\rangle\| \leqslant \|f\| B \tag{13.8a}$$

This can be easily generalized to (see Exercise 13.2.1)

$$\|\mathbb{K}^n|\mathrm{f}\rangle\| \leqslant \|f\| B^n \tag{13.8b}$$

Going back to the series in Eqs. (13.7), we write

$$\left\| \sum_{j=0}^{\infty} (\lambda\mathbb{K})^j|\mathrm{f}\rangle \right\| \leqslant \sum_{j=0}^{\infty} \|(\lambda\mathbb{K})^j|\mathrm{f}\rangle\| = \sum_{j=0}^{\infty} |\lambda|^j \|\mathbb{K}^j|\mathrm{f}\rangle\|$$

$$\leqslant \sum_{j=0}^{\infty} |\lambda|^j B^j \|f\| = \|f\| \sum_{j=0}^{\infty} (|\lambda|B)^j = \frac{\|f\|}{1-|\lambda|B}$$

if $|\lambda|B < 1$. The expression on the RHS is meaningful if $\|f\| < \infty$, that is, if $f(x)$ is square-integrable, which we assume.

The above information can be summarized in a theorem.

13.2.1 Theorem A Fredholm equation of the second kind, Eq. (13.3), has a unique solution given by a uniformly convergent Neumann series,

$$u(x) = \sum_{j=0}^{\infty} \lambda^j \int_a^b dt\, K^j(x,t) f(t)$$

provided that

(i) f is square-integrable
(ii) $\int_a^b dx \int_a^b dt [K(x,t)]^2 \equiv B^2 < \infty$
(iii) $|\lambda|B < 1$ ■

Example 13.2.1

Consider the IE

$$u(x) - \int_0^1 K(x,t)u(t)\, dt = x$$

where

$$K(x,t) = \begin{cases} x & \text{if } 0 \leqslant x < t \\ t & \text{if } t < x \leqslant 1 \end{cases}$$

Here $\lambda = 1$; therefore, a Neumann-series solution exists if $B < 1$. Let us calculate B^2. It is convenient to write K in terms of the theta function:

$$K(x, t) = x\theta(t - x) + t\theta(x - t)$$

This gives

$$K^2(x, t) = x^2\theta(t - x) + t^2\theta(x - t)$$

because $\theta^2(x - t) = \theta(x - t)$ and $\theta(x - t)\theta(t - x) = 0$. Thus, we have

$$B^2 = \int_0^1 dx \int_0^1 dt\, K^2(x, t) = \int_0^1 dt \int_0^t dx\, x^2 + \int_0^1 dx \int_0^x dt\, t^2$$

$$= \int_0^1 dt\left(\frac{t^3}{3}\right) + \int_0^1 dx\left(\frac{x^3}{3}\right) = \frac{1}{6}$$

giving $B = 1/\sqrt{6}$, which is less than 1. The Neumann series converges and we have

$$u(x) = \sum_{j=0}^{\infty} \int_0^1 K^j(x, t) f(t)\, dt$$

$$= \sum_{j=0}^{\infty} \int_0^1 K^j(x, t)\, t\, dt \equiv \sum_{j=0}^{\infty} u_j(x)$$

The first few terms are evaluated as follows:

$$u_0(x) = \int_0^1 K^0(x, t) t\, dt = \int_0^1 \delta(x - t) t\, dt = x$$

$$u_1(x) = \int_0^1 K(x, t) t\, dt = \int_0^1 [x\theta(t - x) + t\theta(x - t)]\, t\, dt$$

$$= x\int_x^1 t\, dt + \int_0^x t^2\, dt = \frac{x}{2} - \frac{x^3}{6}$$

$$u_2(x) = \int_0^1 K^2(x, t) t\, dt = \int_0^1 [x^2\theta(t - x) + t^2\theta(x - t)]\, t\, dt$$

$$= -\frac{x^4}{4} + \frac{x^2}{2}$$

Adding these terms, we obtain an approximation for $u(x)$ that is valid for $0 \leqslant x \leqslant 1$:

$$u(x) \approx u_0(x) + u_1(x) + u_2(x) = \frac{3x}{2} + \frac{x^2}{2} - \frac{x^3}{6} - \frac{x^4}{4} \qquad \bullet$$

Exercises

13.2.1 Show that

$$\| \mathbb{K}^n f \| \leqslant B^n \| f \|$$

where $$B^2 \equiv \int_a^b dx \int_a^b dt\,[K(x,t)]^2 \qquad \text{and} \qquad \| f \|^2 \equiv \int_a^b [f(x)]^2\,dx$$

(Hint: Use mathematical induction.)

13.2.2 Use the method of successive approximations to solve this Volterra equation:

$$u(x) = 1 + \lambda \int_0^x u(t)\,dt$$

13.3 THE FREDHOLM ALTERNATIVE

This discussion of integral equations has, so far, centered around expressing their solutions in terms of series, assuming that the solutions exist. Let us now take up the question of their existence.

It is convenient, and necessary, to discuss the integral equation and its adjoint concurrently. Thus, we will consider the four equations

$$|\mathrm{u}\rangle - \lambda \mathbb{K} |\mathrm{u}\rangle = |\mathrm{f}\rangle \tag{13.9a}$$

$$|\mathrm{u}\rangle - \lambda \mathbb{K} |\mathrm{u}\rangle = 0 \tag{13.9b}$$

$$|\mathrm{v}\rangle - \lambda^* \mathbb{K}^\dagger |\mathrm{v}\rangle = |\mathrm{h}\rangle \tag{13.10a}$$

$$|\mathrm{v}\rangle - \lambda^* \mathbb{K}^\dagger |\mathrm{v}\rangle = 0 \tag{13.10b}$$

where $\mathbb{K}^\dagger$, the adjoint of $\mathbb{K}$, is defined by

$$\mathbb{K}^\dagger v(x) \equiv \int_a^b K^*(t,x) v(t)\,dt$$

and $f(x)$ and $h(x)$ are arbitrary square-integrable functions.

Under what conditions do Eqs. (13.9) and (13.10) have solutions? This question is not only legitimate but required, because the functions $f(x)$ and $h(x)$ are not completely arbitrary. For instance, in the case of a point mass sliding on the surface $y = g(z)$, discussed at the beginning of this chapter, the function $f(z)$, which represents the travel time, cannot be smaller than $\sqrt{2z/g}$, the free-fall time. Thus, if $|f(z)| < |\sqrt{2z/g}|$, Eq. (13.2) will have no solution. The question of the existence of a solution to an integral equation is best answered by first looking at separable kernels, also called *degenerate kernels.*

13.3.1 Degenerate Kernels

Recall that a degenerate (or separable, in the language of Chapter 11) kernel can be written as

$$K(x, t) = \sum_{j=1}^{n} \phi_j(x)\psi_j^*(t) \tag{13.11}$$

where ϕ_j and ψ_j are assumed to be linearly independent. This assumption places no restriction on $K(x, t)$ because if $\{\phi_j\}$ and $\{\psi_j\}$ are linearly dependent, they can always be expressed as a linear combination of a smaller set of linearly independent functions chosen from each group. Substituting from (13.11) in the integral versions of (13.9a) and (13.10b), we obtain

$$u(x) - \lambda \sum_{j=1}^{n} \phi_j(x) \int_a^b \psi_j^*(t)u(t)\,dt = f(x)$$

$$v(x) - \lambda^* \sum_{j=1}^{n} \psi_j(x) \int_a^b \phi_j^*(t)v(t)\,dt = 0$$

If we define

$$u_j \equiv \int_a^b \psi_j^*(t)u(t)\,dt \qquad \text{and} \qquad v_j \equiv \int_a^b \phi_j^*(t)v(t)\,dt \tag{13.12}$$

the preceding equations become

$$u(x) - \lambda \sum_{j=1}^{n} u_j\phi_j(x) = f(x) \tag{13.13a}$$

$$v(x) - \lambda^* \sum_{j=1}^{n} v_j\psi_j(x) = 0 \tag{13.13b}$$

If we multiply (13.13a) by $\psi_i^*(x)$ and integrate over x, we get

$$u_i - \lambda \sum_{j=1}^{n} a_{ij}u_j = f_i \qquad \text{for } i = 1, 2, \ldots, n \tag{13.14}$$

where

$$a_{ij} \equiv \int_a^b \psi_i^*(t)\phi_j(t)\,dt$$

$$f_i \equiv \int_a^b \psi_i^*(t)f(t)\,dt$$

We can now determine the u_i by solving the system of linear equations given by (13.14). Once the u_i are determined, we can substitute them in (13.13a) and solve for $u(x)$:

$$u(x) = \lambda \sum_{j=1}^{n} u_j \phi_j(x) + f(x) \tag{13.15}$$

Thus, for a degenerate kernel the Fredholm problem reduces to a system of linear equations.

To investigate this system we let $|U\rangle$, $|F\rangle$, and $\mathbb{A}$ stand for finite-dimensional vectors with components u_i, f_i, and the operator corresponding to (the matrix) a_{ij}, respectively. Equation (13.14) can then be written as

$$|U\rangle - \lambda \mathbb{A} |U\rangle = |F\rangle \tag{13.16}$$

Furthermore, multiplying (13.13b) by $\phi_i^*(x)$, integrating over x, and using notation similar to that used in (13.16), we get

$$|V\rangle - \lambda^* \mathbb{A}^\dagger |V\rangle = 0 \tag{13.17}$$

It is convenient to study (13.16) and (13.17) together, to seek conditions under which these equations have solutions. The existence of their solutions depends, in the first place, on the determinant of

$$\mathbb{B} \equiv \mathbb{1} - \lambda \mathbb{A}$$

If det $\mathbb{B} \neq 0$, then, clearly,

$$\det \mathbb{B}^\dagger \equiv \det(\mathbb{1} - \lambda^* \mathbb{A}^\dagger) \neq 0$$

and we have

$$|U\rangle = \mathbb{B}^{-1} |F\rangle$$

and

$$|V\rangle = 0$$

If, on the other hand, det $\mathbb{B} = 0$, then Eq. (13.17) will have nontrivial solutions and Eq. (13.16) will have solutions only if certain conditions hold.

To investigate these conditions, let us take the adjoint of Eq. (13.17) and "multiply" the result by $|U\rangle$:

$$\langle V|U\rangle - \lambda \langle V|\mathbb{A}|U\rangle = 0$$

or

$$\langle V|(|U\rangle - \lambda \mathbb{A}|U\rangle) = 0$$

But, according to Eq. (13.16), the vector in parentheses is simply $|F\rangle$. We thus have

$$\langle V|F\rangle = 0$$

Therefore, for $|U\rangle$ to be a solution of Eq. (13.16), that is, for (13.16) to have a solution, its inhomogeneous term $|F\rangle$ must necessarily be orthogonal to the solution of the homogeneous adjoint equation. Is this condition also sufficient? That is, if $|F\rangle$ is orthogonal to the solution of the homogeneous adjoint equation, are we guaranteed a solution to Eq. (13.16)? The answer is yes (for a proof, see Exercise 13.3.1). Thus, if $\det(\mathbb{1} - \lambda\mathbb{A}) = 0$, we will have a solution if and only if

$$0 = \langle \mathrm{F}|\mathrm{V}\rangle \equiv \lambda^*\langle \mathrm{F}|\mathrm{V}\rangle \equiv \lambda^* \sum_{i=1}^{n} v_i f_i^* = \lambda^* \sum_{i=1}^{n} \left(\left(\int_a^b \phi_i^*(x)v(x)\,dx\right)\left(\int_a^b \psi_i(t)f^*(t)\,dt\right)\right)$$

$$= \lambda^* \int_a^b dt\, f^*(t) \int_a^b \left(\sum_{i=1}^{n} \phi_i(x)\psi_i^*(t)\right)^* v(x)\,dx$$

$$= \int_a^b dt\, f^*(t)\left[\lambda^* \int_a^b K^*(x,t)v(x)\,dx\right]$$

$$= \int_a^b dt\, f^*(t)\lambda^*\mathbb{K}^\dagger v(t)$$

$$= \int_a^b dt\, f^*(t)v(t)$$

The last equality follows from Eq. (13.10b). The above shows that $\langle \mathrm{F}|\mathrm{V}\rangle = 0$ is equivalent to the orthogonality of the functions $f(x)$ and $v(x)$.

The foregoing discussion can be summarized as a theorem.

13.3.1 Theorem (The Fredholm Alternative) Let $\mathbb{K}$ be an integral operator with a degenerate kernel, and let λ be a nonzero complex number. If Eqs. (13.9b) and (13.10b) (the homogeneous equations) have only trivial solutions, then Eqs. (13.9a) and (13.10a) (the inhomogeneous equations) have unique solutions $|\mathrm{u}\rangle$ and $|\mathrm{v}\rangle$, respectively, for all $|\mathrm{f}\rangle, |\mathrm{h}\rangle \in \mathcal{L}^2(a, b)$.

If the homogeneous equations have nontrivial solutions, that is, if the null spaces of $\mathbb{1} - \lambda\mathbb{K}$ and $\mathbb{1} - \lambda^*\mathbb{K}^\dagger$ are nontrivial, then the dimensions of these null spaces are equal, and the inhomogeneous equations have solutions iff $\langle \mathrm{v}|\mathrm{f}\rangle = 0$ and $\langle \mathrm{u}|\mathrm{h}\rangle = 0$. ■

Although a proof of this theorem is outlined above only for Eqs. (13.9a) and (13.10b), the proof for Eqs. (13.9b) and (13.10a) is exactly the same. Note also that, in general, $\mathcal{N}(\mathrm{B}) \neq \mathcal{N}(\mathrm{B}^\dagger)$; therefore, it is essential to pair the equations correctly, pairing the inhomogeneous equation with the homogeneous adjoint equation (see Exercise 13.3.1).

Example 13.3.1

Let us solve

$$u(x) = x + \lambda \int_a^b K(x, t)u(t)\,dt$$

where $K(x, t) \equiv xt$.

Let us first find the Neumann-series solution. We note that

$$B^2 = \int_a^b\int_a^b x^2 t^2\,dx\,dt = \tfrac{1}{9}(b^3 - a^3)^2$$

and the Neumann series converges if

$$|\lambda|(b^3 - a^3) < 3$$

Assuming that this condition holds, we have

$$u(x) = x + \sum_{n=1}^{\infty} \lambda^n \int_a^b K^n(x, t)t\,dt$$

It can easily be shown that

$$K^n(x, t) = \int_a^b K^{n-1}(x, s)K(s, t)\,ds = xt\,B^{n-1}$$

from which we obtain

$$\int_a^b K^n(x, t)t\,dt = xB^{n-1}\tfrac{1}{3}(b^3 - a^3) = xB^n$$

Substituting this in the expression for $u(x)$ yields

$$\begin{aligned} u(x) &= x + x\lambda B \sum_{n=1}^{\infty} \lambda^{n-1}B^{n-1} \\ &= x\left(1 + \lambda B\frac{1}{1 - \lambda B}\right) = \frac{x}{1 - \lambda B} \\ &= \frac{3x}{3 - \lambda(b^3 - a^3)} \end{aligned}$$

Now let us use the Fredholm alternative. In this case Eq. (13.11) reduces to $K(x, t) = xt$. Thus,

$$u(x) = x + \lambda x \int_a^b t\,u(t)\,dt \equiv x + \lambda x A = x(1 + \lambda A) \tag{1}$$

Multiplying both sides by x and integrating, we obtain

$$\int_a^b xu(x)\,dx = (1 + \lambda A)\int_a^b x^2\,dx = (1 + \lambda A)B \tag{2}$$

The LHS is simply A. Solving Eq. (2) for A yields

$$A = \frac{B}{1 - \lambda B}$$

which, when substituted in Eq. (1), gives

$$u(x) = x\left(1 + \frac{\lambda B}{1 - \lambda B}\right) = \frac{x}{1 - \lambda B}$$

This solution is the same as the first one we obtained. However, no series was involved here, and, therefore, no assumption is necessary concerning $|\lambda|B$. •

Example 13.3.2

As a slightly more complicated example, let us solve

$$u(x) - \lambda\int_0^1 (1 + xt)u(t)\,dt = x \tag{1}$$

The kernel, $K(x, t) = 1 + xt$, is degenerate with

$$\phi_1(x) = 1 \qquad \phi_2(x) = x$$
$$\psi_1(t) = 1 \qquad \psi_2(t) = t$$

This gives the matrix

$$\mathsf{A} = \begin{pmatrix} 1 & \frac{1}{2} \\ \frac{1}{2} & \frac{1}{3} \end{pmatrix}$$

(a) Let us assume that $\lambda = 1$. In that case

$$\mathsf{B} = 1 - \lambda\mathsf{A} = \begin{pmatrix} 0 & -\frac{1}{2} \\ -\frac{1}{2} & \frac{2}{3} \end{pmatrix}$$

has a nonzero determinant. Thus, B^{-1} exists, and

$$\mathsf{B}^{-1} = \begin{pmatrix} -\frac{8}{3} & -2 \\ -2 & 0 \end{pmatrix}$$

With

$$f_1 = \int_0^1 \psi_1^*(t)f(t)\,dt = \int_0^1 t\,dt = \tfrac{1}{2} \qquad \text{and} \qquad f_2 = \int_0^1 \psi_2^*(t)f(t)\,dt = \int_0^1 t^2\,dt = \tfrac{1}{3}$$

we obtain

$$\begin{pmatrix} u_1 \\ u_2 \end{pmatrix} = \mathsf{B}^{-1}\begin{pmatrix} f_1 \\ f_2 \end{pmatrix} = \begin{pmatrix} -2 \\ -1 \end{pmatrix}$$

Equation (13.15) then gives

$$u(x) = u_1\phi_1(x) + u_2\phi_2(x) + x = -2$$

(b) Now let us take $\lambda = 8 + 2\sqrt{13}$. Then

$$\mathsf{B} = 1 - \lambda\mathsf{A} = -\begin{pmatrix} 7 + 2\sqrt{13} & 4 + \sqrt{13} \\ 4 + \sqrt{13} & (5 + 2\sqrt{13})/3 \end{pmatrix}$$

and $\det \mathsf{B} = 0$. We thus have a solution only if $f(x) \equiv x$ is orthogonal to the null space of $\mathsf{B}^\dagger$. To determine a basis for this null space, we have to find vectors $|\mathrm{v}\rangle$ such that $\mathsf{B}^\dagger|\mathrm{v}\rangle = 0$. Since B is real and symmetric, $\mathsf{B}^\dagger = \mathsf{B}$, and we must solve

$$\begin{pmatrix} 7 + 2\sqrt{13} & 4 + \sqrt{13} \\ 4 + \sqrt{13} & (5 + 2\sqrt{13})/3 \end{pmatrix}\begin{pmatrix} a_1 \\ a_2 \end{pmatrix} = 0$$

The solution to this equation is a multiple of

$$|\mathrm{v}\rangle \equiv \begin{pmatrix} 3 \\ -2 - \sqrt{13} \end{pmatrix}$$

If the integral equation, Eq. (1), is to have a solution, we must have

$$\langle \mathrm{v}|\mathrm{F}\rangle = 0 \quad \Rightarrow \quad (3 \quad -2 - \sqrt{13})\begin{pmatrix} \frac{1}{2} \\ \frac{1}{3} \end{pmatrix} = 0$$

which is impossible. Therefore, there is no solution. •

It may be slightly disturbing that the functions $\phi_i(x)$ and $\psi_i(t)$ appearing in a degenerate kernel are arbitrary to within a multiplicative constant. After all, we can multiply $\phi_j(x)$ by a nonzero constant, such as γ_j, and divide $\psi_j(t)$ by the same constant, and get the same kernel. Such a change clearly alters the matrices A and B and, therefore, seems likely to change the solution, $u(x)$. That this is not so is demonstrated in Exercise 13.3.2. In fact, it can be shown quite generally that the transformations described above do not change the solution. The proof is left as a simple problem.

13.3.2 Hilbert-Schmidt Kernels

The degenerate kernels discussed in the preceding subsection comprise a special type, making integral equations easily solvable. What about other kernels? Clearly, we do not expect an integral equation to have a solution for *all* kernels in existence. Thus, we want integral equations to have kernels that are more general than a degenerate kernel, yet restricted enough to guarantee a solution.

The essential restricting property imposed on a kernel (or its associated integral operator) is compactness. An operator on a Hilbert space $\mathscr{H}$ is said to be *compact* (or completely continuous) if it transforms any sequence of bounded vectors in $\mathscr{H}$ (a sequence of vectors whose norms are all smaller than a given finite positive number) into a sequence that has a convergent subsequence. For finite-dimensional vector spaces compactness is not essential, because any bounded sequence of vectors is a Cauchy sequence (verify this), and, as shown in Chapter 5, a Cauchy sequence always converges in a finite-dimensional vector space. Compactness is important only in infinite-dimensional spaces.

We will not be dealing with compact operators in general but with a special class of such operators. Nevertheless, the following two theorems are presented to indicate the significant role compact operators play with respect to integral equations and operator theory. (For proofs of these theorems, see Byron and Fuller 1969, Chapter 9, and Reed and Simon 1980, Chapter VI.)

13.3.2 Theorem The Fredholm alternative holds for any compact operator acting on a Hilbert space. ■

13.3.3 Theorem Let $\mathbb{K}$ be a hermitian compact operator on the Hilbert space $\mathcal{H}$. Then there exists an orthonormal basis $\{|u_n\rangle\}$ in $\mathcal{H}$ such that $\mathbb{K}|u_n\rangle = \lambda_n|u_n\rangle$ and $\lambda_n \to 0$ as $n \to \infty$. ■

The next group of integral equations we will consider are those that can be approximated by separable kernels.

13.3.4 Definition An integral operator $\mathbb{K}$, acting on $\mathcal{L}^2(a, b)$, whose kernel $K(x, t)$ is square-integrable, that is, for which

$$\int_a^b dx \int_a^b dt\,|K(x, t)|^2 < \infty$$

is called a *Hilbert-Schmidt* (*H-S*) *operator*, and the kernel is called a *Hilbert-Schmidt* (*H-S*) *kernel.*

An H-S operator is bounded; that is, its norm is finite. In fact, we have the following proposition.

13.3.5 Proposition Let $\mathbb{K}$ be an H-S operator with kernel $K(x, t)$. Then

$$\|\mathbb{K}\| \leqslant \left[\int_a^b dt \int_a^b dx\,|K(x, t)|^2\right]^{1/2}$$

Proof. Let $u \in \mathcal{L}^2(a, b)$. Then, by the Schwarz inequality,

$$|\mathbb{K}u(x)|^2 = \left|\int_a^b dt\, K(x, t)u(t)\right|^2 \leqslant \int_a^b |u(t)|^2\, dt \int_a^b |K(x, t)|^2\, dt$$

Integrating over x yields

$$\int_a^b |\mathbb{K}u(x)|^2\,dx \leqslant \int_a^b |u(t)|^2\,dt \int_a^b dt \int_a^b dx\,|K(x,t)|^2$$

Using the definition of the norm, we obtain

$$\|\mathbb{K}u\|^2 \leqslant \|u\|^2 \int_a^b dt \int_a^b dx\,|K(x,t)|^2$$

For an operator, $\|\mathbb{K}\|$ is the smallest number among real numbers M for which the inequality $\|\mathbb{K}u\| \leqslant \|u\|\,M$ holds. Since the inequality holds for the double integral, $\|\mathbb{K}\|$ cannot be larger than the double integral. ■

The importance of H-S operators is shown by the following theorem.

13.3.6 Theorem An H-S operator can be approximated, in the norm, to arbitrary accuracy by a suitable degenerate kernel.

Proof. Let $\{e_i(x)\}_{i=1}^{\infty}$ be an orthonormal basis for $\mathscr{L}^2(a,b)$. Since $K(x,t)$ is square-integrable with respect to both its arguments, we have

$$K(x,t) = \sum_{i,j=1}^{\infty} k_{ij} e_i(x) e_j^*(t)$$

where

$$k_{ij} = \int_a^b dt \int_a^b dx\, e_i^*(x) e_j(t) K(x,t)$$

Parseval's equality gives

$$\sum_{i,j=1}^{\infty} |k_{ij}|^2 = \int_a^b dx \int_a^b dt\,|K(x,t)|^2 < \infty$$

which shows that the infinite double sum converges. We define the operator $\mathbb{K}_n$, whose kernel is $K_n(x,t) \equiv \sum_{i=1}^{n} k_{ij} e_i(x) e_j^*(t)$. It is clear that $\mathbb{A} \equiv \mathbb{K} - \mathbb{K}_n$ is an H-S operator. In fact, denoting the kernel of $\mathbb{A}$ by $K_{\mathbb{A}}(x,t)$, we obtain

$$\int_a^b dx \int_a^b dt\,|K_{\mathbb{A}}(x,t)|^2 = \int_a^b dx \int_a^b dt\,|K(x,t) - \sum_{i,j=1}^{n} k_{ij} e_i(x) e_j^*(t)|^2$$

$$= \int_a^b dx \int_a^b dt \left| \sum_{i,j=n+1}^{\infty} k_{ij} e_i(x) e_j^*(t) \right|^2 = \sum_{i,j=n+1}^{\infty} |k_{ij}|^2 < \infty$$

Using Proposition 13.3.5, we get

$$\| \mathbb{K} - \mathbb{K}_n \| \equiv \| \mathbb{A} \| \leqslant \int_a^b dx \int_a^b dt \, |K_{\mathrm{A}}(x, t)|^2 = \sum_{i,j=n+1}^{\infty} |k_{ij}|^2$$

The convergence of the double sum in Parseval's equality implies that the sum on the RHS of this equation tends to zero as n approaches infinity. Thus, by making n large enough, we can make $\| \mathbb{K} - \mathbb{K}_n \|$ as small as we please. ■

It is possible to push the foregoing proof a little farther (see Stakgold 1979, Chapter 6) by showing that $\mathbb{K}_n$ is compact and that $\lim_{n\to\infty} \mathbb{K}_n$ is compact. The result is another theorem.

13.3.7 Theorem An H-S operator is compact. ■

Theorem 13.3.6 opens up the possibility of reducing an H-S integral equation (an IE with an H-S kernel) to a system of n linear equations, as outlined in Section 13.3.1. Of course, for this linear system to be a good representation of the original integral equation, n must be as large as possible. This requirement makes this procedure especially suited to computers.

Those H-S operators whose kernels are self-adjoint (symmetric, in the case of real functions) are of particular importance because Theorems 13.3.3 and 13.3.7 imply that

$$K(x, t) = \sum_{i=1}^{\infty} \lambda_i u_i(x) u_i^*(t)$$

where $\{\lambda_i\}_{i=1}^{\infty}$ and $\{u_i(x)\}_{i=1}^{\infty}$ are the (real) eigenvalues and corresponding orthonormal eigenfunctions of the self-adjoint H-S operator $\mathbb{K}$. Moreover, any function in a Hilbert space can be expanded as a linear combination of $\{u_i(x)\}_{i=1}^{\infty}$. Thus, the hermitian H-S integral equation

$$u(x) = \lambda \int_a^b K(x, t) u(t) \, dt + f(x)$$

can be written as

$$\sum_{i=1}^{\infty} a_i u_i(x) = \lambda \int_a^b \left[\sum_{j=1}^{\infty} \lambda_j u_j(x) u_j^*(t) \right] \left[\sum_{i=1}^{\infty} a_i u_i(t) \right] dt + \sum_{i=1}^{\infty} b_i u_i(x)$$

where $$a_i \equiv \int_a^b u_i^*(s) u(s) \, ds \qquad \text{and} \qquad b_i \equiv \int_a^b u_i^*(s) f(s) \, ds$$

are the expansion coefficients of u and f, respectively. Using the orthonormality of the

$u_i(t)$, the preceding equation can be written as

$$\sum_{i=1}^{\infty} a_i u_i(x) = \lambda \sum_{i=1}^{\infty} \lambda_i a_i u_i(x) + \sum_{i=1}^{\infty} b_i u_i(x)$$

The linear independence of the u_i yields

$$(1 - \lambda\lambda_i)a_i = b_i \qquad \text{for } i = 1, 2, \ldots \tag{13.18}$$

If λ is not the reciprocal of any of the eigenvalues λ_i, then

$$a_i = \frac{b_i}{1 - \lambda\lambda_i} \qquad \text{for } i = 1, 2, \ldots$$

and the solution, $u(x) = \sum_{i=1}^{\infty} a_i u_i(x)$, will be given by

$$u(x) = \sum_{i=1}^{\infty} \frac{b_i}{1 - \lambda\lambda_i} u_i(x) = \sum_{i=1}^{\infty} \frac{\int_a^b u_i^*(s) f(s)\, ds}{1 - \lambda\lambda_i} u_i(x) \tag{13.19}$$

Thus, if we know the eigenvalues and eigenfunctions of $\mathbb{K}$, we can solve an integral equation having an H-S kernel.

If $\lambda = \lambda_r$ for some integer r, then b_r must vanish; that is, $f(x)$ *must* be orthogonal to $u_r(x)$ (otherwise the equation has no solution). Assuming that this condition holds, we can write

$$u(x) = a_r u_r(x) + \sum_{\substack{i=1 \\ i \neq r}}^{\infty} \frac{\int_a^b u_i^*(s)\, f(s)\, ds}{1 - \lambda\lambda_i} u_i(x) \tag{13.20}$$

where a_r is an undetermined constant.

Example 13.3.3

We can use the above procedure to solve

$$u(x) = 3 \int_{-1}^{1} K(x,t) u(t)\, dt + x^2$$

where

$$K(x,t) \equiv \sum_{l=0}^{\infty} \frac{u_l(x) u_l(t)}{(\sqrt{2})^l}$$

$$u_l(x) \equiv \sqrt{\frac{2l+1}{2}}\, P_l(x)$$

and $P_l(x)$ is a Legendre polynomial.

We first note that $\{u_l\}$ is an orthonormal set of functions, that $K(x, t)$ is real and symmetric (therefore, hermitian), and that

$$\int_{-1}^{+1} dt \int_{-1}^{+1} dx\, |K(x,t)|^2 = \int_{-1}^{1} dt \int_{-1}^{1} dx \sum_{l,m=0}^{\infty} \frac{1}{(\sqrt{2})^l} \left[\frac{1}{(\sqrt{2})^m}\right] [u_m(x) u_l(x) u_m(t) u_l(t)]$$

$$= \sum_{l=0}^{\infty} \frac{1}{2^l} = 2 < \infty$$

Thus, $K(x,t)$ is an H-S kernel. It is clear that the u_l are eigenfunctions of $K(x,t)$ with eigenvalues $1/(\sqrt{2})^l$. Since $3 \neq 1/(\sqrt{2})^l$ for any integer l, we can use Eq. (13.19) to write

$$u(x) = \sum_{l=0}^{\infty} \frac{\int_{-1}^{1} u_l(s)s^2\,ds}{1 - [3/(\sqrt{2})^l]} u_l(x)$$

But $\int_{-1}^{1} u_l(s)s^2\,ds = 0$ for $l \geqslant 3$. For $l \leqslant 2$ we have

$$\int_{-1}^{1} u_0(s)s^2\,ds = \sqrt{\frac{1}{2}} \int_{-1}^{1} P_0(s)s^2\,ds = \frac{\sqrt{2}}{3}$$

$$\int_{-1}^{1} u_1(s)s^2\,ds = 0$$

and

$$\int_{-1}^{1} u_2(s)s^2\,ds = \sqrt{\frac{5}{2}} \int_{-1}^{1} P_2(s)s^2\,ds = \frac{2}{3}\sqrt{\frac{2}{5}}$$

Thus,

$$u(x) = \frac{\frac{\sqrt{2}}{3}}{-2} u_0(x) + \frac{\frac{2}{3}\sqrt{\frac{2}{5}}}{-\frac{1}{2}} u_2(x) = \tfrac{1}{2} - 2x^2$$

●

Example 13.3.4

The kernel $K(x,t) = e^{-xt}$ can be "expanded" as

$$e^{-xt} \equiv \sum_{m=0}^{\infty} \frac{(-xt)^m}{m!} = \sum_{m=0}^{\infty} \frac{(-x)^m}{\sqrt{m!}} \left(\frac{t^m}{\sqrt{m!}}\right)$$

Thus, using the procedure described above to solve the following integral equation may be tempting:

$$u(x) = \lambda \int_{0}^{\infty} e^{-xt} u(t)dt + f(x)$$

However, caution is necessary, because the procedure works only for H-S operators, and e^{-xt} is *not* an H-S kernel because

$$\int_0^\infty dt \int_0^\infty dx\, |e^{-xt}|^2 = \int_0^\infty dt \int_0^\infty dx\, e^{-2xt}$$

$$= \int_0^\infty dt \left(-\frac{1}{2t}\right)$$

which is not finite! ●

Exercises

13.3.1 Let $\mathcal{V}$ be an N-dimensional inner product space. Let $\mathbb{B} \in \mathcal{L}(\mathcal{V})$ be a linear operator. Consider the two equations

$$\mathbb{B}|U\rangle = |F\rangle \qquad \text{and} \qquad \mathbb{B}^\dagger|V\rangle = 0 \tag{1}$$

Let $\mathcal{N}(\mathbb{B})$ and $\mathcal{N}(\mathbb{B}^\dagger)$ be the null spaces of $\mathbb{B}$ and $\mathbb{B}^\dagger$, respectively, and $\mathcal{M}_\perp$ the collection of vectors orthogonal to $\mathcal{N}(\mathbb{B})$.

(a) Show that $\mathcal{M}_\perp$ is isomorphic to the range of $\mathbb{B}$, which is $\mathbb{B}(\mathcal{V})$. (Hint: Consider the restriction of $\mathbb{B}$ to $\mathcal{M}_\perp$.)

(b) Show that $\mathcal{V} = \mathcal{M}_\perp \oplus \mathcal{N}(\mathbb{B})$ and $\mathcal{V} = \mathbb{B}(\mathcal{V}) \oplus \mathcal{N}(\mathbb{B})$, and thus $\mathcal{M}_\perp = \mathbb{B}(\mathcal{V})$.

(c) Show that $\mathcal{N}(\mathbb{B}) = \mathcal{N}(\mathbb{B}^\dagger)$

(d) Prove that, for det $\mathbb{B} = 0$, there exists a solution to (1) iff

$$\langle V|F\rangle = 0 \qquad \forall\ |V\rangle \in \mathcal{N}(\mathbb{B}^\dagger)$$

13.3.2 Repeat part (a) of Example 13.3.2 using

$$\phi_1(x) = \tfrac{1}{2} \qquad \phi_2(x) = x$$

$$\psi_1(t) = 2 \qquad \psi_2(t) = t$$

so that $K(x,t)$ is still $\phi_1(x)\psi_1(t) + \phi_2(x)\psi_2(t)$.

13.3.3 For what values of λ does the following integral equation have a solution?

$$u(x) = \lambda \int_0^\pi \sin(x + t)\, u(t)\, dt + x$$

What is that solution?

13.4 INTEGRAL EQUATIONS AND GREEN'S FUNCTIONS

Integral equations are best applied in combination with Green's functions. In fact, we can use a Green's function to show that a differential equation can be turned into an integral equation. If the kernel of this integral equation is symmetric and H-S, then the problem lends itself to the approximation methods described at the end of Section 13.3.2.

Let $\mathbb{L}_\mathbf{x}$ be a SOPDO in m variables. We are interested in solving the following SOPDE subject to some BCs:

$$\mathbb{L}_\mathbf{x}[u] + \lambda V(\mathbf{x})u(\mathbf{x}) = f(\mathbf{x}) \tag{13.21}$$

Here λ is an arbitrary constant and $V(\mathbf{x})$ is a well-behaved function of $(x_1, \ldots, x_m)$. Transferring the second term on the LHS to the RHS and then treating the RHS as an inhomogeneous term, we can write the "solution" to (13.21) as

$$u(\mathbf{x}) = H(\mathbf{x}) + \int_D d^m y\, G(\mathbf{x},\mathbf{y})\,[f(\mathbf{y}) - \lambda V(\mathbf{y})u(\mathbf{y})]$$

where D is the domain of $\mathbb{L}_\mathbf{x}$ and $G(\mathbf{x},\mathbf{y})$ is the Green's function for $\mathbb{L}_\mathbf{x}$ with some, as yet unspecified, BCs. The function $H(\mathbf{x})$ is a solution to the homogeneous equation $\mathbb{L}_{\mathbf{x}}[u] = 0$, and it is present to guarantee the appropriate BCs. If the BCs are homogeneous (or the same as those for the GF), then $H(\mathbf{x}) = 0$, and we can write

$$u(\mathbf{x}) = -\lambda \int_{\mathbf{D}} d^m y\, G(\mathbf{x},\mathbf{y})\, V(\mathbf{y})u(\mathbf{y}) + \int_{\mathbf{D}} d^m y\, G(\mathbf{x},\mathbf{y})\, f(\mathbf{y}) \tag{13.22a}$$

On the other hand, if $u(\mathbf{x})$ satisfies an inhomogeneous BC, we can choose $G(\mathbf{x},\mathbf{y})$ to be a convenient Green's function (for instance, the singular part of the full GF) and $H(\mathbf{x})$ to be an arbitrary solution to the homogeneous PDE, and write

$$u(\mathbf{x}) = H(\mathbf{x}) - \lambda \int_{\mathbf{D}} d^m y\, G(\mathbf{x},\mathbf{y})\, V(\mathbf{y})u(\mathbf{y}) + \int_{\mathbf{D}} d^m y\, G(\mathbf{x},\mathbf{y})\, f(\mathbf{y}) \tag{13.22b}$$

Then we adjust $H(\mathbf{x})$ to fit the BCs.

We note that Eqs. (13.22) can both be written as

$$u(\mathbf{x}) = F(\mathbf{x}) - \lambda \int_{\mathbf{D}} d^m y\, G(\mathbf{x},\mathbf{y})\, V(\mathbf{y})u(\mathbf{y}) \tag{13.23}$$

With homogeneous BCs $F(\mathbf{x})$ is given; in the inhomogeneous case it has an adjustable part. Equation (13.23) is an m-dimensional Fredholm equation, whose solution can be obtained in the form of a Neumann series.

13.4.1 Using Green's Functions to Find Neumann-Series Solutions

To find a Neumann-series solution we can follow the procedure described in Section 13.2. However, there is an alternative method that is more direct. We can substitute the expression for u given by the RHS of Eq. (13.23) in the integral of that equation. To do so we write

$$u(\mathbf{y}_1) = F(\mathbf{y}_1) - \lambda \int_{\mathbf{D}} d^m y_2\, G(\mathbf{y}_1,\mathbf{y}_2)\, V(\mathbf{y}_2)u(\mathbf{y}_2)$$

change the variable of integration in (13.23) from $\mathbf{y}$ to $\mathbf{y}_1$ for later convenience, and substitute the preceding equation in (13.23) to get

$$\begin{aligned} u(\mathbf{x}) &= F(\mathbf{x}) - \lambda \int_{\mathbf{D}} d^m y_1\, G(\mathbf{x},\mathbf{y}_1)\, V(\mathbf{y}_1)\left[F(\mathbf{y}_1) - \lambda \int_{\mathbf{D}} d^m y_2\, G(\mathbf{y}_1,\mathbf{y}_2)\, V(\mathbf{y}_2)u(\mathbf{y}_2)\right] \\ &= F(\mathbf{x}) - \lambda \int_{\mathbf{D}} d^m y_1\, G(\mathbf{x},\mathbf{y}_1)\, V(\mathbf{y}_1) F(\mathbf{y}_1) \\ &\quad + (-\lambda)^2 \int_{\mathbf{D}} d^m y_1 \int_{\mathbf{D}} d^m y_2 [G(\mathbf{x},\mathbf{y}_1)\, V(\mathbf{y}_1)][G(\mathbf{y}_1,\mathbf{y}_2)\, V(\mathbf{y}_2)]u(\mathbf{y}_2) \end{aligned}$$

Substituting for $u(\mathbf{y}_2)$, we have

$$u(\mathbf{x}) = F(\mathbf{x}) - \lambda \int_D d^m y_1 \, G(\mathbf{x}, \mathbf{y}_1) \, V(\mathbf{y}_1) \, F(\mathbf{y}_1)$$

$$+ (-\lambda)^2 \int_D d^m y_1 \int_D d^m y_2 [G(\mathbf{x},\mathbf{y}_1) V(\mathbf{y}_1)][G(\mathbf{y}_1,\mathbf{y}_2) V(\mathbf{y}_2)] \, F(\mathbf{y}_2)$$

$$+ (-\lambda)^3 \int_D d^m y_1 \int_D d^m y_2 \int_D d^m y_3 [G(\mathbf{x},\mathbf{y}_1) V(\mathbf{y}_1)][G(\mathbf{y}_1,\mathbf{y}_2) V(\mathbf{y}_2)] \, [G(\mathbf{y}_2,\mathbf{y}_3)$$

$$\times \, V(\mathbf{y}_3)] u(\mathbf{y}_3)$$

Let us use a new notation:

$$K(\mathbf{x},\mathbf{y}) \equiv G(\mathbf{x},\mathbf{y}) \, V(\mathbf{y})$$

$$K^n(\mathbf{x},\mathbf{y}) \equiv \int_D d^m t K^{n-1}(\mathbf{x},\mathbf{t}) K(\mathbf{t},\mathbf{y}) \qquad \text{for } n \geqslant 2$$

Iterating N times, we can write

$$u(\mathbf{x}) = F(\mathbf{x}) + \sum_{n=1}^{N-1} (-\lambda)^n \int_D d^m y_n K^n(\mathbf{x},\mathbf{y}_n) F(\mathbf{y}_n) + (-\lambda)^N \int_D d^m y_N K^N(\mathbf{x},\mathbf{y}_N) u(\mathbf{y}_N)$$

Finally, letting $N \to \infty$, we obtain

$$u(\mathbf{x}) = F(\mathbf{x}) + \sum_{n=1}^{\infty} (-\lambda)^n \int_D d^m y_n K^n(\mathbf{x},\mathbf{y}_n) F(\mathbf{y}_n) \tag{13.24a}$$

Except for the fact that here the integrations are in m variables, Eq. (13.24a) is the same as Eq. (13.7b). In exact analogy, therefore, we abbreviate (13.24a) as

$$|\mathrm{u}\rangle = |\mathrm{F}\rangle + \sum_{n=1}^{\infty} (-\lambda)^n \mathbb{K}^n |\mathrm{F}\rangle \tag{13.24b}$$

Equations (13.24) have meaning only if the Neumann series converges. However, the condition

$$|\lambda| \left[\int_D d^m y \int_D d^m x \, |K(\mathbf{x},\mathbf{y})|^2 \right]^{1/2} < 1 \tag{13.25}$$

guarantees the convergence of the series, as in the one-dimensional case.

Example 13.4.1

The time-independent Schrödinger equation,

$$-\frac{\hbar^2}{2\mu} \nabla^2 \Psi + V(\mathbf{r}) \Psi = E \Psi$$

can be rewritten as

$$(\nabla^2 + k^2)\Psi - \frac{2\mu}{\hbar^2} V(\mathbf{r})\Psi = 0$$

where $k^2 \equiv 2\mu E/\hbar^2$. Comparing this equation with (13.21), we recognize $\mathbb{L}_{\mathbf{x}} \equiv \nabla^2 + k^2$ and $\lambda = -2\mu/\hbar^2$. Assuming that there are no boundaries, we can choose $F(\mathbf{r}) \equiv Ae^{i\mathbf{k}\cdot\mathbf{r}}$ as a homogeneous solution and take the singular part (obtained in Example 12.5.2) of the Green's function:

$$G_s(\mathbf{r}, \mathbf{r}') = -\frac{1}{4\pi}\left(\frac{e^{ik|\mathbf{r}-\mathbf{r}'|}}{|\mathbf{r}-\mathbf{r}'|}\right)$$

Then we can write (13.23) as

$$\Psi(\mathbf{r}) = Ae^{i\mathbf{k}\cdot\mathbf{r}} - \frac{\mu}{2\pi\hbar^2}\int_{\mathbb{R}^3} d^3r' \frac{e^{ik|\mathbf{r}-\mathbf{r}'|}}{|\mathbf{r}-\mathbf{r}'|} V(\mathbf{r}')\Psi(\mathbf{r}') \tag{1}$$

This will have a convergent series as a solution, provided that

$$\frac{\mu}{2\pi\hbar^2}\left(\int_{\mathbb{R}^3} d^3r \int_{\mathbb{R}^3} d^3r' \left|\frac{e^{ik|\mathbf{r}-\mathbf{r}'|}}{|\mathbf{r}-\mathbf{r}'|} V(\mathbf{r}')\right|^2\right)^{1/2} < 1$$

To evaluate the integral, we change $\mathbf{r}$ to $\boldsymbol{\rho}$, defined by $\boldsymbol{\rho} \equiv \mathbf{r} - \mathbf{r}'$. Then the integral becomes

$$\int_{\mathbb{R}^3} |V(\mathbf{r}')|^2 d^3r' \int_{\mathbb{R}^3} d^3\rho \frac{e^{-2\mathrm{Im}(k)\rho}}{\rho^2} = 4\pi \frac{1}{-2\mathrm{Im}(k)} e^{-2\mathrm{Im}(k)\rho}\Big|_0^\infty \int_{\mathbb{R}^3} |V(\mathbf{r}')|^2 d^3r'$$

$$= \frac{2\pi}{\mathrm{Im}(k)} \int_{\mathbb{R}^3} |V(\mathbf{r}')|^2 d^3r'$$

and the series converges if

$$\int_{\mathbb{R}^3} |V(\mathbf{r}')|^2 d^3r' < \frac{2\pi\hbar^4\,\mathrm{Im}(k)}{\mu^2}$$

However, for most cases of interest, for example, in scattering problems, $E > 0$ and $\mathrm{Im}(k) = 0$. Therefore, this inequality will not hold. On the other hand, scattering problems are concerned with the scattering amplitude $f(\theta, \varphi)$, defined by

$$f(\theta, \varphi) \equiv -\frac{\mu}{2\pi^2\hbar^2}\int_{\mathbb{R}^3} d^3r\, e^{-i\mathbf{k}_f\cdot\mathbf{r}} V(\mathbf{r})\Psi(\mathbf{r}) \tag{2}$$

where $\hbar\mathbf{k}_f$ is the final momentum of the scattered particle (wave), whose direction in spherical coordinates is given by θ and φ. This resembles an inner product with weight function $V(\mathbf{r})$. In fact, if we define $\Phi_f(\mathbf{r}) \equiv e^{i\mathbf{k}_f\cdot\mathbf{r}}$, the integral in Eq. (2) can be written simply as $\langle\Phi_f|\Psi\rangle$.

With the scattering amplitude as the goal and the new scalar product as a means to get there, we no longer care about $\Psi(\mathbf{r})$. In fact, Eq. (2) shows that only in regions where $V(\mathbf{r})$ is significant do we need to know $\Psi(\mathbf{r})$. Interpreting V as a weight function and recalling that $\mathbb{1} = \int d^m x |\mathbf{x}\rangle V(\mathbf{x})\langle\mathbf{x}|$, we write Eq. (13.23) as an abstract vector-operator equation,

$$|\mathrm{u}\rangle = |\mathrm{F}\rangle - \lambda\mathbb{G}|\mathrm{u}\rangle$$

for which the condition of (13.25) becomes

$$\frac{\mu}{2\pi\hbar^2}\left[\int_{\mathbb{R}^3} d^3r \int_{\mathbb{R}^3} d^3r'\, V(\mathbf{r})V(\mathbf{r}')|G(\mathbf{r},\mathbf{r}')|^2\right]^{1/2} < 1 \tag{3}$$

We evaluate the integral in (3) for the important case of central potentials, where $V(\mathbf{r})$ is a function of $r = |\mathbf{r}|$ only. Writing out the Green's function gives

$$\iiint r^2 \sin\theta\, dr\, d\theta\, d\varphi \iiint r'^2 \sin\theta'\, dr'\, d\theta'\, d\varphi' \frac{e^{-2\,\mathrm{Im}(k)|\mathbf{r}-\mathbf{r}'|}}{|\mathbf{r}-\mathbf{r}'|^2} V(r)\, V(r')$$

For integration with respect to $\mathbf{r}'$, $\mathbf{r}$ is fixed; let it be along z'. Then

$$|\mathbf{r}-\mathbf{r}'|^2 = r^2 + r'^2 - 2rr'\cos\theta'$$

Thus, the φ, φ', and θ integrations can be done immediately, and they give $8\pi^2$. Introducing $\eta \equiv \cos\theta'$ reduces the integral to

$$8\pi^2 \iiint_{\mathbb{R}^3} r^2 r'^2\, dr\, dr'\, d\eta \frac{e^{-2\mathrm{Im}(k)\sqrt{r^2+r'^2-2rr'\eta}}}{r^2+r'^2-2rr'\eta} V(r)\, V(r')$$

We introduce new variables of integration,

$$u \equiv r + r'$$

$$v \equiv r - r'$$

and

$$w \equiv \sqrt{r^2 + r'^2 - 2rr'\eta}$$

with inverse transformations

$$r = \tfrac{1}{2}(u+v) \qquad r' = \tfrac{1}{2}(u-v) \qquad \eta = \frac{u^2+v^2-2w^2}{u^2-v^2}$$

whose Jacobian is easily found to be

$$J = \frac{2w}{u^2-v^2}$$

This gives

$$dr\, dr'\, d\eta = J\, du\, dv\, dw = \frac{2w}{u^2-v^2} du\, dv\, dw$$

Thus, the integral becomes

$$8\pi^2 \iiint \frac{(u^2-v^2)^2}{16}\left(\frac{2w}{u^2-v^2}\right) du\, dv\, dw \frac{e^{-2\mathrm{Im}(k)w}}{w^2} V(r)\, V(r')$$

From the definitions of u, v, and w, we note that $-w \leqslant v \leqslant w$, $0 \leqslant w \leqslant u$, and $0 \leqslant u \leqslant \infty$. Therefore, the final result for Eq. (3) is

$$\left(\frac{\mu}{2\hbar^2}\right)^2 \int_0^\infty du \int_0^u dw \int_{-w}^{w} dv(u^2 - v^2)\frac{e^{-2\text{Im}(k)w}}{w} V[\tfrac{1}{2}(u+v)]\, V[\tfrac{1}{2}(u-v)] < 1 \tag{4}$$

For example, in the case of a Yukawa potential with strength g^2, we have $V(r) = g^2 e^{-\alpha r}/r$. This gives

$$V[\tfrac{1}{2}(u+v)]\, V[\tfrac{1}{2}(u-v)] = g^4 \frac{e^{-(\alpha/2)(u+v)}}{\frac{1}{2}(u+v)}\left[\frac{e^{-(\alpha/2)(u-v)}}{\frac{1}{2}(u-v)}\right]$$

$$= 4g^4 \frac{e^{-\alpha u}}{u^2 - v^2}$$

and Eq. (4) becomes

$$\left(\frac{\mu g^2}{\hbar^2}\right)^2 \int_0^\infty du\, e^{-\alpha u} \int_0^u dw \frac{e^{-2\text{Im}(k)w}}{w} \int_{-w}^{w} dv < 1$$

These integrals are elementary and give

$$\left(\frac{\mu g^2}{\hbar^2 \alpha}\right)^2 \frac{2}{1 + \dfrac{2\text{Im}(k)}{\alpha}} < 1$$

Assuming that $\text{Im}(k) = 0$ gives

$$g^2 < \frac{\hbar^2 \alpha}{\sqrt{2}\mu}$$

Therefore, for weak enough potentials, the Neumann series converges.

Note that the condition given by (13.25) presupposes square-integrability for functions. However, the free wave $F = Ae^{i\mathbf{k}\cdot\mathbf{r}}$ used in $\Psi(\mathbf{r})$ is square-integrable if and only if $\langle F|F\rangle < \infty$, or

$$|A|^2 \int_{\mathbb{R}^3} e^{-i\mathbf{k}\cdot\mathbf{r}} e^{i\mathbf{k}\cdot\mathbf{r}} V(\mathbf{r}) d^3r = |A|^2 \int_{\mathbb{R}^3} V(\mathbf{r}) d^3r < \infty$$

That is, $\int V(\mathbf{r}) d^3r$ must be finite. For most potentials this is the case; for the important Coulomb potential, however, it is not. For this potential it is necessary to use a wave packet instead of $Ae^{i\mathbf{k}\cdot\mathbf{r}}$. ●

Example 13.4.2

Sometimes it is useful to consider nonlocal potentials. For such potentials the Schrödinger equation is

$$-\frac{\hbar^2}{2\mu}\nabla^2\Psi + \int_{\mathbb{R}^3} V(\mathbf{r},\mathbf{r}')\Psi(\mathbf{r}') d^3r' = E\Psi(\mathbf{r})$$

Then, for scattering problems [see Eq. (1) of Example 13.4.1], we have

$$\Psi(\mathbf{r}) = Ae^{i\mathbf{k}\cdot\mathbf{r}} - \frac{\mu}{2\pi\hbar^2}\int_{\mathbb{R}^3} d^3r' \frac{e^{ik|\mathbf{r}-\mathbf{r}'|}}{|\mathbf{r}-\mathbf{r}'|}\int_{\mathbb{R}^3} d^3r''\, V(\mathbf{r},\mathbf{r}'')\Psi(\mathbf{r}'') \tag{1}$$

For a separable potential, for which $V(\mathbf{r}',\mathbf{r}'') \equiv -g^2 U(\mathbf{r})U(\mathbf{r}'')$, we can solve (1) exactly. We substitute for $V(\mathbf{r}',\mathbf{r}'')$ in (1) to obtain

$$\Psi(\mathbf{r}) = Ae^{i\mathbf{k}\cdot\mathbf{r}} + \frac{\mu g^2}{2\pi\hbar^2}\int_{\mathbb{R}^3} d^3r' \frac{e^{ik|\mathbf{r}-\mathbf{r}'|}}{|\mathbf{r}-\mathbf{r}'|}U(\mathbf{r}')\int_{\mathbb{R}^3} d^3r''\, U(\mathbf{r}'')\Psi(\mathbf{r}'') \tag{2}$$

Defining the following quantities

$$Q(\mathbf{r}) \equiv \frac{\mu g^2}{2\pi\hbar^2}\int_{\mathbb{R}^3} d^3r' \frac{e^{ik|\mathbf{r}-\mathbf{r}'|}}{|\mathbf{r}-\mathbf{r}'|}U(\mathbf{r}')$$

$$C \equiv \int_{\mathbb{R}^3} U(\mathbf{r}'')\Psi(\mathbf{r}'')d^3r'' \tag{3}$$

and substituting them in (2) yields

$$\Psi(\mathbf{r}) = Ae^{i\mathbf{k}\cdot\mathbf{r}} + CQ(\mathbf{r})$$

Multiplying both sides of this equation by $U(\mathbf{r})$ and integrating over $\mathbb{R}^3$, we get

$$\begin{aligned} C &= A\int_{\mathbb{R}^3} U(\mathbf{r})e^{i\mathbf{k}\cdot\mathbf{r}}d^3r + C\int_{\mathbb{R}^3} U(\mathbf{r})Q(\mathbf{r})d^3r \\ &= (2\pi)^{3/2}A\tilde{U}(-\mathbf{k}) + C\int_{\mathbb{R}^3} U(\mathbf{r})Q(\mathbf{r})d^3r \end{aligned}$$

Solving for C, we obtain

$$C = \frac{(2\pi)^{3/2}A\tilde{U}(-\mathbf{k})}{1 - \int_{\mathbb{R}^3} U(\mathbf{r})Q(\mathbf{r})d^3r}$$

from which the solution can be calculated:

$$\Psi(\mathbf{r}) = Ae^{i\mathbf{k}\cdot\mathbf{r}} + \frac{(2\pi)^{3/2}A\tilde{U}(-\mathbf{k})}{1 - \int_{\mathbb{R}^3} U(\mathbf{r}')Q(\mathbf{r}')d^3r'}Q(\mathbf{r}) \tag{4}$$

In principle, $\tilde{U}(-\mathbf{k})$ [the Fourier transform of $U(\mathbf{r})$] and $Q(\mathbf{r})$ can be calculated once the functional form of $U(\mathbf{r})$ is known. Equations (3) and (4) give the solution to the Schrödinger equation in closed form. ●

13.4.2 Feynman Diagrams

We will briefly discuss an intuitive physical interpretation of the Neumann series due to Feynman. Although Feynman developed this *diagrammatic technique* for quantum electrodynamics, it has been useful in other areas of physics, such as statistical physics and many-body theories.

In most cases of interest, the SOPDE of Eq. (13.21) is homogeneous, so $f(\mathbf{x}) = 0$. Then $\mathbb{L}_\mathbf{x}$ and $V(\mathbf{x})$ are called the *free operator* and the *interacting potential*, respectively, and the solution to $\mathbb{L}_\mathbf{x}[u] = 0$ is called the *free solution* and denoted $u_f(\mathbf{x})$.

Let us start with Eq. (13.23) written as

$$u(\mathbf{x}) = u_f(\mathbf{x}) - \lambda \int_{\mathbb{R}^m} d^m y \, G_0(\mathbf{x},\mathbf{y}) V(\mathbf{y}) u(\mathbf{y}) \tag{13.26}$$

where $G_0(\mathbf{x},\mathbf{y})$ stands for the Green's function for the free operator $\mathbb{L}_\mathbf{x}$. The full Green's function, that is, that for $\mathbb{L}_\mathbf{x} + V$, will be denoted $G(\mathbf{x},\mathbf{y})$. Moreover, as is usually the case, the region D is taken to be all of $\mathbb{R}^m$. This implies that no boundary conditions are imposed on $u(\mathbf{x})$, which in turn permits us to use the singular part of the Green's function in the integral. Because of the importance of the full Green's function, we are interested in finding a series for $G(\mathbf{x}, \mathbf{y})$ in terms of $G_0(\mathbf{x}, \mathbf{y})$, since this is supposed to be known. To obtain such a series we note that

$$u(\mathbf{x}) = \int d^m z \, G(\mathbf{x}, \mathbf{z}) u(\mathbf{z})$$

$$u_f(\mathbf{x}) = \int d^m z \, G_0(\mathbf{x},\mathbf{z}) u_f(\mathbf{z})$$

Furthermore, if $\mathbf{z}$ is chosen correctly, for instance, in such a way that its time component tends to $-\infty$ (before any interaction takes place), then $u(\mathbf{t}) \equiv u_f(\mathbf{t})$, and we have

$$u(\mathbf{x}) = \int d^m z \, G(\mathbf{x},\mathbf{z}) u_f(\mathbf{z})$$

Substituting for $u(\mathbf{x})$ and $u_f(\mathbf{x})$ in (13.26) yields

$$\int d^m z \, G(\mathbf{x},\mathbf{z}) u_f(\mathbf{z}) = \int d^m z \, G_0(\mathbf{x}, \mathbf{z}) \, u_f(\mathbf{z}) - \lambda \int d^m y \, G_0(\mathbf{x}, \mathbf{y}) V(\mathbf{y}) \int d^m z \, G(\mathbf{y},\mathbf{z}) u_f(\mathbf{z})$$

Since u_f is arbitrary, we finally obtain

$$G(\mathbf{x},\mathbf{z}) = G_0(\mathbf{x},\mathbf{z}) - \lambda \int d^m y \, G_0(\mathbf{x},\mathbf{y}) V(\mathbf{y}) \, G(\mathbf{y},\mathbf{z}) \tag{13.27}$$

This equation is the analogue of (13.23) and, just like that equation, is susceptible to a Neumann-series expansion. The result is

$$G(\mathbf{x},\mathbf{y}) = G_0(\mathbf{x},\mathbf{y}) + \sum_{n=1}^{\infty} (-\lambda)^n \int_{\mathbb{R}^m} d^m y_n K^n(\mathbf{x},\mathbf{y}_n) G_0(\mathbf{y}_n,\mathbf{y})$$

$$= G_0(\mathbf{x},\mathbf{y}) - \lambda \int_{\mathbb{R}^m} d^m y_1\, G_0(\mathbf{x},\mathbf{y}_1) V(\mathbf{y}_1) G_0(\mathbf{y}_1,\mathbf{y})$$

$$+ (-\lambda)^2 \int_{\mathbb{R}^m} d^m y_1 \int_{\mathbb{R}^m} d^m y_2\, G_0(\mathbf{x},\mathbf{y}_1) V(\mathbf{y}_1) G_0(\mathbf{y}_1,\mathbf{y}_2) V(\mathbf{y}_2) G_0(\mathbf{y}_2,\mathbf{y})$$

$$+ (-\lambda)^3 \int_{\mathbb{R}^m} d^m y_1 \int_{\mathbb{R}^m} d^m y_2 \int_{\mathbb{R}^m} d^m y_3 \big[G_0(\mathbf{x},\mathbf{y}_1) V(\mathbf{y}_1) G_0(\mathbf{y}_1,\mathbf{y}_2) V(\mathbf{y}_2)$$

$$\times G_0(\mathbf{y}_2,\mathbf{y}_3) V(\mathbf{y}_3) G_0(\mathbf{y}_3,\mathbf{y}) \big] + \cdots \tag{13.28}$$

Feynman's idea is to consider $G(\mathbf{x},\mathbf{y})$ as a propagator between points $\mathbf{x}$ and $\mathbf{y}$ and $G_0(\mathbf{x},\mathbf{y})$ as a *free propagator*. The first term on the RHS of (13.28) is simply a free propagation from $\mathbf{x}$ to $\mathbf{y}$. Diagrammatically, it is represented by a line joining the points $\mathbf{x}$ and $\mathbf{y}$ [see Fig. 13.1(a)]. The second term is a free propagation from $\mathbf{x}$ to $\mathbf{y}_1$, interaction at $\mathbf{y}_1$ (also called a *vertex*) with a potential $-\lambda V(\mathbf{y}_1)$, and subsequent

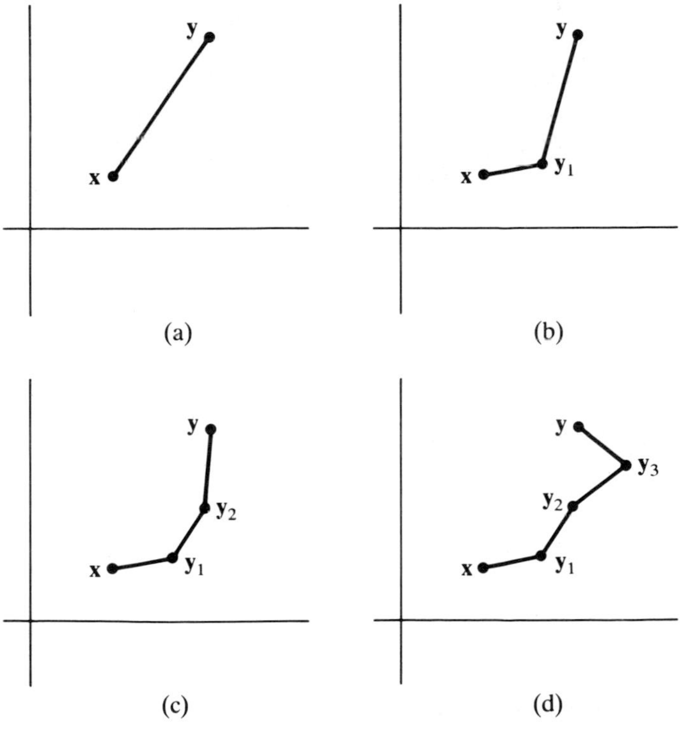

Figure 13.1 Contributions to the full propagator in (a) the zeroth order, (b) the first order, (c) the second order, and (d) the third order. At each vertex there is a factor of $-\lambda V$ and an integration over all values of the variable of that vertex.

propagation to $\mathbf{y}$ [see Fig. 13.1(b)]. According to the third term, the particle or wave [represented by $u_f(\mathbf{x})$] propagates freely from $\mathbf{x}$ to $\mathbf{y}_1$, interacts at $\mathbf{y}_1$ with the potential $-\lambda V(\mathbf{y}_1)$, propagates freely from $\mathbf{y}_1$ to $\mathbf{y}_2$, interacts for a second time with the potential $-\lambda V(\mathbf{y}_2)$, and finally propagates freely from $\mathbf{y}_2$ to $\mathbf{y}$ [see Fig. 13.1(c)]. The interpretation of the rest of the series in (13.28) is now clear: The nth-order term of the series has n vertices between $\mathbf{x}$ and $\mathbf{y}$ with a factor $-\lambda V(\mathbf{y}_k)$ and an integration over $\mathbf{y}_k$ at vertex k, and between any two consecutive vertices, $\mathbf{y}_k$ and $\mathbf{y}_{k+1}$, there is a factor of the free propagator $G_0(\mathbf{y}_k, \mathbf{y}_{k+1})$.

Feynman diagrams are used extensively in relativistic quantum field theory, in which $m = 4$, corresponding to the four-dimensional space-time. In this context λ is determined by the strength of the interaction. For quantum electrodynamics, for instance, λ is the fine structure constant, $e^2/\hbar c = 1/137$.

The basic idea of Feynman diagrams described above only scratches the surface of the very complex subject of interacting quantum fields. A careful analysis of that subject must take into account the operator nature of $u_f(\mathbf{x})$, the fact that $V(\mathbf{x})$ itself is a product of operators, the commutativity or anticommutativity of various operators, the singular nature of most integrals encountered and how to cure such singularities (this is related to the renormalization of the theory), and so on. The scope of this book does not allow any treatment of these topics.

Exercises

13.4.1 Consider the Schrödinger equation in one dimension:

$$-\frac{\hbar^2}{2\mu}\frac{d^2\Psi}{dx^2} + V(x)\Psi(x) = E\Psi(x)$$

(a) Write the equivalent IE for the bound states ($E < 0$) assuming that the free solution (for $V = 0$) remains finite as $x \to \pm\infty$.

(b) Find the eigenvalues (E) and the corresponding eigenfunctions when $V(x)$ is an attractive delta function potential:

$$V(x) \equiv -V_0\delta(x - a) \qquad \text{where } V_0 > 0$$

13.4.2 Regard the Fourier transform,

$$\mathscr{F}[f](\mathbf{x}) \equiv \frac{1}{\sqrt{2\pi}}\int_{-\infty}^{\infty} e^{ixy} f(y)dy$$

as an integral operator.

(a) Show that $\mathscr{F}^2[f](x) = f(-x)$.

(b) Deduce that, therefore, the only eigenvalues of this IO are $\lambda = \pm 1, \pm i$.

(c) Let $f(x)$ be any even function of x. Show that the appropriate choice of α can make $u = f + \alpha\mathscr{F}[f]$ an eigenfunction of $\mathscr{F}$. (This shows that the eigenvalues of $\mathscr{F}$ have infinite multiplicity.)

PROBLEMS

13.1 It is possible to multiply the functions $\phi_j(x)$ by γ_j and $\psi_j(t)$ by $1/\gamma_j$ and get the same degenerate kernel, $K(x,t) = \sum_{j=1}^{n} \phi_j(x)\psi_j^*(t)$. Show that such arbitrariness, although affecting the matrices A and B, does not change the solution of the Fredholm problem:

$$u(x) - \lambda \int_a^b K(x,t)u(t)dt = f(x)$$

13.2 Repeat Exercise 13.3.3 using a Neumann-series expansion. Under what condition is the series convergent?

13.3 Show, by direct substitution, that the solution found in Example 13.3.3 does satisfy the IE.

13.4 Solve $u(x) = \frac{1}{2}\int_{-1}^{1}(x+t)u(t)dt + x$.

13.5 Solve $u(x) = \lambda\int_0^1 xt\,u(t)dt + x$ using the Neumann-series method (for what values of λ is the series convergent?), the degenerate kernel technique, and the H-S technique (Eqs. (13.19) and (13.20)].

13.6 Solve $u(x) = \lambda\int_0^\infty K(x,t)\,u(t)dt + x^\alpha$, where $\alpha \neq -1, -2, -3, \ldots,$ and $K(x,t) = e^{-(x+t)}$. For what values of λ does the IE have a solution?

13.7 Use the method of successive approximations to solve the Volterra equation $u(x) = \lambda\int_0^x u(t)dt$. Then derive a DE equivalent to the Volterra equation (make sure to include the initial condition), and solve it. Do the solutions agree?

13.8 Solve the integral form of the Schrödinger equation for an attractive double delta function potential,

$$V(x) = -V_0[\delta(x-a_1) + \delta(x-a_2)]$$

Find the eigenfunctions and obtain a transcendental equation for the eigenvalues (see Exercise 13.4.1).

13.9 Using the integral form of the Schrödinger equation in three dimensions, show that an attractive delta potential, $V(\mathbf{r}) = -V_0\delta(\mathbf{r}-\mathbf{a})$, does not have a bound state ($E < 0$). Contrast this with the result of Exercise 13.4.1.

13.10 By taking the Fourier transform of both sides of the integral form of the Schrödinger equation, show that, for bound-state problems ($E < 0$), the equation in "momentum space" can be written as

$$\tilde{\psi}(\mathbf{p}) = -\frac{2\mu}{(2\pi)^{3/2}\hbar^2}\left(\frac{1}{\kappa^2+p^2}\right)\int \tilde{V}(\mathbf{p}-\mathbf{q})\tilde{\psi}(\mathbf{q})\,d^3q$$

where $\kappa^2 = -2\mu E/\hbar^2$.

13.11 Show that the IE associated with the damped harmonic oscillator

$$\frac{d^2x}{dt^2} + 2\gamma\frac{dx}{dt} + \omega_0^2 x = 0$$

having the BCs $x(0) = x_0$, $(dx/dt)_{t=0} = 0$ can be written in either of the following forms.

(a) $x(t) = x_0 - \frac{\omega_0^2}{2\gamma}\int_0^t (1 - e^{-2\gamma(t-t')})\, x(t')dt'$

(b) $x(t) = x_0 \cos\omega_0 t + \frac{2\gamma x_0}{\omega_0}\sin\omega_0 t - 2\gamma\int_0^t \cos\omega_0(t - t')x(t')dt'$

(Hint: Take $\omega_0^2 x$ or $2\gamma\dot{x}$, respectively, as the inhomogeneous term.)

13.12 (a) Show that for scattering problems ($E > 0$) the integral form of the Schrödinger equation in one dimension is

$$\psi(x) = e^{ikx} - \frac{i\mu}{\hbar^2 k}\int_{-\infty}^{\infty} e^{ik|x-y|}V(y)\psi(y)dy$$

(b) Divide $(-\infty, +\infty)$ into three regions $\{R_i\}_{i=1}^3$, where $R_1 \equiv (-\infty, -a)$, $R_2 \equiv (-a, +a)$, and $R_3 \equiv (a, \infty)$. Let $\psi_i(x)$ be $\psi(x)$ in region R_i. Assume that the potential $V(x)$ vanishes in R_1 and R_3. Show that

$$\psi_1(x) = e^{ikx} - \frac{i\mu}{\hbar^2 k}e^{-ikx}\int_{-a}^{a} e^{iky}V(y)\psi_2(y)dy$$

$$\psi_2(x) = e^{ikx} - \frac{i\mu}{\hbar^2 k}\int_{-a}^{a} e^{ik|x-y|}V(y)\psi_2(y)dy$$

$$\psi_3(x) = e^{ikx} - \frac{i\mu}{\hbar^2 k}e^{ikx}\int_{-a}^{a} e^{-iky}V(y)\psi_2(y)dy$$

(This shows that determining the wave function in regions where there is no potential requires the wave function in the region where the potential acts.)

13.13 Let

$$V(x) = \begin{cases} V_0 & \text{if } |x| < a \\ 0 & \text{if } |x| > a \end{cases}$$

and find $\psi_2(x)$ of Problem 13.12 by the method of successive approximations. Show that the nth term is less than $(2\mu V_0 a/\hbar^2 k)^{n-1}$, so the Neumann series will converge if

$$\frac{2V_0 a}{\hbar v} < 1$$

where v is the velocity and $\mu v = \hbar k$ is the momentum of the wave. Therefore, for large velocities, the Neumann-series expansion is valid.

13.14 Write the bound-state Schrödinger integral equation for a nonlocal potential, noting that

$$G(\mathbf{r}, \mathbf{r}') = \frac{e^{-\kappa|\mathbf{r}-\mathbf{r}'|}}{|\mathbf{r}-\mathbf{r}'|}$$

and $\kappa^2 = -2\mu E/\hbar^2$, where μ is the mass of the bound particle. The homogeneous solution is zero, as is always the case with bound states. (See Example 13.4.2.)

(a) Assuming that the potential is separable and of the form

$$V(\mathbf{r},\mathbf{r}') = -g^2 U(\mathbf{r}) U(\mathbf{r}')$$

show that a solution to the Schrödinger equation exists iff

$$\frac{\mu g^2}{2\pi\hbar^2}\int d^3r \int d^3r' \frac{e^{-\kappa|\mathbf{r}-\mathbf{r}'|}}{|\mathbf{r}-\mathbf{r}'|} U(\mathbf{r}) U(\mathbf{r}') = 1 \tag{1}$$

(b) Assume that $U(\mathbf{r})$ is a central potential depending on r only, Show that (1) can be written as

$$\frac{\pi\mu g^2}{2\hbar^2}\int_0^\infty du \int_0^u dw \int_{-w}^{w} dv (u^2 - v^2) e^{-\kappa w} U[\tfrac{1}{2}(u+v)]\, U[\tfrac{1}{2}(u-v)] = 1 \tag{2}$$

(Hint: See Example 13.4.1.)

(c) Taking $U(r) = e^{-\alpha r}/r$, show that the condition in (2) becomes

$$\frac{4\pi\mu g^2}{\alpha\hbar^2}\left[\frac{1}{(\alpha+\kappa)^2}\right] = 1 \tag{3}$$

(d) Since $\kappa > 0$, prove that (3) has a unique solution only if $g^2 > \hbar^2\alpha^2/(4\pi\mu)$, in which case the bound-state energy is

$$E = -\frac{\hbar^2}{2\mu}\left[\left(\frac{4\pi\mu g^2}{\alpha\hbar^2}\right)^{1/2} - \alpha\right]^2$$

Part V

SPECIAL TOPICS

14

Gamma and Beta Functions

In the preceding chapters we have used properties of gamma and beta functions on numerous occasions. This short chapter derives some useful relations involving these functions.

14.1 THE GAMMA FUNCTION AND ITS DERIVATIVE

The gamma function is a generalization of factorials (which are defined only for positive integers) to the system of complex numbers. Consider the integral

$$I(\alpha) \equiv \int_0^\infty e^{-\alpha t}\,dt = \frac{1}{\alpha}$$

Note that

$$\frac{d^n I}{d\alpha^n} = \int_0^\infty \left[\frac{\partial^n}{\partial \alpha^n}(e^{-\alpha t})\right] dt = (-1)^n \int_0^\infty t^n e^{-\alpha t}\,dt$$

$$= \frac{d^n}{d\alpha^n}\left(\frac{1}{\alpha}\right) = (-1)^n \frac{n!}{\alpha^{n+1}}$$

For $\alpha = 1$ and n a positive integer, this gives

$$\int_0^\infty t^n e^{-t}\,dt = n!$$

This fact motivates the generalization

$$\Gamma(z) \equiv \int_0^\infty t^{z-1} e^{-t}\, dt \qquad \text{for } \mathrm{Re}(z) > 0 \tag{14.1}$$

where Γ is called the *gamma* (or *factorial*) *function*. It is also called *Euler's integral of the second kind*. It is clear, from its definition, that

$$\Gamma(n + 1) = n! \tag{14.2}$$

if n is a positive integer. The restriction that $\mathrm{Re}(z) > 0$ assures the convergence of the integral. (Such convergence is shown in Exercise 14.1.1.)

An immediate consequence of Eq. (14.1) is obtained by integration by parts:

$$\Gamma(z + 1) = z\Gamma(z) \tag{14.3}$$

This also leads to Eq. (14.2) by iteration.

Another consequence is the analyticity of $\Gamma(z)$. Differentiating Eq. (14.3) with respect to z, we obtain

$$\frac{d\Gamma(z+1)}{dz} = \Gamma(z) + z\frac{d\Gamma(z)}{dz}$$

Thus, $d\Gamma(z)/dz$ exists and is finite if and only if $d\Gamma(z + 1)/dz$ is finite (recall that $z \neq 0$). (Exercise 14.1.2 shows this.) Therefore, $\Gamma(z)$ is analytic. However, it is not analytic for all complex numbers. To see the singularities of $\Gamma(z)$, we note that

$$\Gamma(z + n) = z(z + 1)(z + 2) \cdots (z + n - 1)\Gamma(z)$$

or

$$\Gamma(z) = \frac{\Gamma(z+n)}{z(z+1)(z+2)\cdots(z+n-1)}$$

The numerator is analytic as long as $\mathrm{Re}(z + n) > 0$, or $\mathrm{Re}(z) > -n$. Thus, for $\mathrm{Re}(z) > -n$, the singularities of $\Gamma(z)$ are the poles at $z = 0, -1, -2, \ldots, -n + 1$. Since n is arbitrary, we conclude that $\Gamma(z)$ is analytic at all $z \in \mathbb{C}$ except at $z = 0, -1, -2, \ldots$, where $\Gamma(z)$ has simple poles.

A useful result is obtained by setting $z = \frac{1}{2}$ in Eq. (14.1):

$$\Gamma(\tfrac{1}{2}) = \sqrt{\pi} \tag{14.4}$$

This can be obtained by making the substitution $u = \sqrt{t}$. With the aid of the gamma function we can write double factorials of odd numbers as

$$(2k - 1)!! \equiv (2k - 1)(2k - 3)(2k - 5) \cdots 5 \cdot 3 \cdot 1 = \frac{2^k}{\sqrt{\pi}}\Gamma\left(\frac{2k+1}{2}\right)$$

(See Section 5.3.)

We can derive an expression for the logarithmic derivative of the gamma function that involves an infinite series. First, we need a general result that holds for *meromorphic functions* (functions that have only poles as singularities). Assume that $f(z)$ has simple poles at $\{z_j\}_{j=1}^N$, where N could be infinity. Then, if $z \neq z_j$ for all j, the residue theorem yields

$$\frac{1}{2\pi i}\int_{C_n} \frac{f(\xi)}{\xi - z}\, d\xi = f(z) + \sum_{j=1}^{n} \operatorname{Res}\left(\frac{f}{\xi - z}\right)_{\xi = z_j}$$

where C_n is a circle containing the first n poles, and it is assumed that $0 \leqslant |z_1| \leqslant |z_2| \leqslant \cdots \leqslant |z_n| \leqslant \cdots$. Since the poles of f are assumed to be simple, we have

$$\begin{aligned}\operatorname{Res}\left(\frac{f}{\xi - z}\right)_{\xi = z_j} &= \lim_{\xi \to z_j} (\xi - z_j)\frac{f(\xi)}{\xi - z} \\ &= \frac{1}{z_j - z}\lim_{\xi \to z_j} (\xi - z_j) f(\xi) \\ &= \frac{1}{z_j - z}\operatorname{Res}\,[f(\xi)]_{\xi = z_j} \equiv \frac{r_j}{z_j - z}\end{aligned}$$

where r_j is, by definition, the residue of $f(\xi)$ at $\xi = z_j$. Substituting in the preceding equation gives

$$\frac{1}{2\pi i}\int_{C_n} \frac{f(\xi)}{\xi - z}\, d\xi = f(z) + \sum_{j=1}^{n} \frac{r_j}{z_j - z}$$

Evaluating this equation for $z = 0$ (assumed to be none of the poles) and subtracting, we can write

$$f(z) - f(0) = \frac{z}{2\pi i}\int_{C_n} \frac{f(\xi)\, d\xi}{\xi(\xi - z)} + \sum_{j=1}^{n} r_j\left(\frac{1}{z - z_j} + \frac{1}{z_j}\right)$$

If $\lim_{|\xi| \to \infty} |f(\xi)| < \infty$, the integral vanishes for an infinite circle, and we have

$$f(z) = f(0) + \sum_{j=1}^{N} r_j\left(\frac{1}{z - z_j} + \frac{1}{z_j}\right) \tag{14.5}$$

This is called a *Mittag-Leffler expansion* of the meromorphic function f (with simple poles).

Now we let $g(z)$ be an entire function with simple zeros. Then $(dg/dz)/g(z)$ is meromorphic with simple poles at $\{z_j\}_{j=1}^N$, for example. It can easily be shown that the function is bounded for $|z| \to \infty$, and its residues are all unity. Equation (14.5) thus reduces to

$$\frac{d}{dz}[\ln g(z)] = c + \sum_{j=1}^{N} \left(\frac{1}{z - z_j} + \frac{1}{z_j}\right) \tag{14.6}$$

whose solution (derived in Exercise 14.1.3) is

$$g(z) = g(0)e^{cz}\prod_{j=1}^{N}\left(1 - \frac{z}{z_j}\right)e^{z/z_j} \tag{14.7}$$

where
$$c \equiv \frac{(dg/dz)|_{z=0}}{g(0)}$$

and it is assumed that $z_j \neq 0$ for all j.

Returning to the gamma function, we note that $1/\Gamma(z+1)$ is an entire function with simple zeros at $\{-k\}_{k=1}^{\infty}$. Equation (14.7) gives

$$\frac{1}{\Gamma(z+1)} = e^{\gamma z} \prod_{k=1}^{\infty} \left(1 + \frac{z}{k}\right) e^{-z/k}$$

where γ is a constant to be determined. Using Eq. (14.3), we obtain

$$\frac{1}{\Gamma(z)} = z e^{\gamma z} \prod_{k=1}^{\infty} \left(1 + \frac{z}{k}\right) e^{-z/k} \tag{14.8}$$

To determine γ we evaluate Eq. (14.8) for $z = 1$:

$$e^{-\gamma} = \prod_{k=1}^{\infty} \left(1 + \frac{1}{k}\right) e^{-1/k}$$

or

$$\gamma = \sum_{k=1}^{\infty} \left[\frac{1}{k} - \ln\left(1 + \frac{1}{k}\right)\right] \tag{14.9}$$

The constant γ is called the *Euler-Mascheroni constant*, and its value is $\gamma = 0.57721566\ldots$.

Differentiating the logarithm of both sides of Eq. (14.8), we obtain

$$\frac{d}{dz} \ln [\Gamma(z)] = -\frac{1}{z} - \gamma + \sum_{k=1}^{\infty} \left(\frac{1}{k} - \frac{1}{z+k}\right) \tag{14.10}$$

In Example 7.5.1 we derived another property of the gamma function—its asymptotic form. The result, valid for large z, is

$$\Gamma(z+1) \approx \sqrt{2\pi}\, e^{-z} z^{z+1/2} \tag{14.11}$$

Other properties of the gamma function are derivable from the results presented here. Those derivations are left as problems.

Exercises

14.1.1 Show that the integral in Eq. (14.1) converges.

14.1.2 Show that $d\Gamma(z+1)/dz$ exists and is finite by establishing the following.

(**a**) $|\ln t| \leqslant t + 1/t$ for $t > 0$ (Hint: For $t \geqslant 1$, show that $t - \ln t$ is a monotonically increasing function. For $t < 1$, make the substitution $t = 1/s$.)

(**b**) Use the result from part (a) in the integral for $d\Gamma(z+1)/dz$ to show that $|d\Gamma(z+1)/dz|$ is finite.

14.1.3 Derive Eq. (14.7) from Eq. (14.6).

14.2 THE BETA FUNCTION

The *beta function*, or *Euler's integral of the first kind*, is defined for complex numbers a and b as follows:

$$B(a, b) \equiv \int_0^1 t^{a-1}(1-t)^{b-1}\, dt \qquad \text{where } \operatorname{Re}(a), \operatorname{Re}(b) > 0 \tag{14.12}$$

By changing t to $1/t$, we can write

$$B(a, b) \equiv \int_1^\infty t^{-a-b}(t-1)^{b-1}\, dt \tag{14.13}$$

Since $0 \leqslant t \leqslant 1$ in Eq. (14.12), we can define θ by $t = \sin^2\theta$. This gives

$$B(a, b) = 2\int_0^{\pi/2} \sin^{2a-1}\theta \cos^{2b-1}\theta\, d\theta \tag{14.14}$$

This relation can be used to establish a connection between the gamma and beta functions. We note that

$$\Gamma(a) = \int_0^\infty t^{a-1}e^{-t}\, dt = 2\int_0^\infty x^{2a-1}e^{-x^2}\, dx$$

where in the last step we changed the variable to $x = \sqrt{t}$. Thus, we can write

$$\begin{aligned}\Gamma(a)\Gamma(b) &= 4\int_0^\infty x^{2a-1}e^{-x^2}\, dx \int_0^\infty y^{2b-1}e^{-y^2}\, dy \\ &= 4\int_0^\infty\int_0^\infty dx\, dy\, e^{-(x^2+y^2)}x^{2a-1}y^{2b-1}\end{aligned}$$

Changing to polar coordinates yields

$$\begin{aligned}\Gamma(a)\Gamma(b) &= 4\int_0^\infty r\, dr \int_0^{\pi/2} d\theta\, e^{-r^2}(r\cos\theta)^{2a-1}(r\sin\theta)^{2b-1} \\ &= \int_0^\infty d(r^2)\, e^{-r^2}(r^2)^{a+b-1}\left(2\int_0^{\pi/2} \sin^{2b-1}\theta \cos^{2a-1}\theta\, d\theta\right) \\ &= \Gamma(a+b)B(b, a)\end{aligned}$$

or

$$\mathrm{B}(b, a) \equiv \mathrm{B}(a,b) = \frac{\Gamma(a)\Gamma(b)}{\Gamma(a + b)} \tag{14.15}$$

The symmetry of $\mathrm{B}(a, b)$ in its arguments follows from the RHS of (14.15) or its definition [Eq. (14.12)].

Let us now establish an important relation used in earlier chapters:

$$\Gamma(z)\Gamma(1 - z) = \frac{\pi}{\sin \pi z} \tag{14.16}$$

With $a = z$ and $b = 1 - z$, Eq. (14.15) gives

$$\Gamma(z)\Gamma(1 - z) = \mathrm{B}(z, 1 - z) = 2 \int_0^{\pi/2} \sin^{2z-1}\theta \cos^{-2z+1}\theta \, d\theta$$

$$= 2 \int_0^{\pi/2} \tan^{2z-1}\theta \, d\theta \qquad \text{for } \mathrm{Re}(z), \mathrm{Re}(1 - z) > 0$$

If we let $u = \tan\theta$, then we have

$$\Gamma(z)\Gamma(1 - z) = 2 \int_0^{\infty} \frac{u^{2z-1}}{u^2 + 1} \, du \qquad \text{for } 0 < \mathrm{Re}(z) < 1$$

The integral was evaluated in Example 7.3.1. Using the result obtained there, we immediately get Eq. (14.16), valid for $0 < \mathrm{Re}(z) < 1$. By analytic continuation we then generalize (14.16) to values of z for which both sides are analytic.

Example 14.2.1

As an illustration of the use of Eq. (14.16), let us show that $\Gamma(z)$ can also be written as

$$\frac{1}{\Gamma(z)} = \frac{1}{2\pi i} \int_C \frac{e^t}{t^z} \, dt \tag{1}$$

where C is the contour shown in Fig. 14.1. Because of the factor t^z, the integrand has a cut along the negative real axis. The contour is chosen on the Riemann sheet where $-\pi < \arg(z) < \pi$.

It is clear that if we want to obtain Eq. (14.1) from (1), we have to make a change of variable, $w \equiv -t = e^{-i\pi}t$. This changes the contour C in the t-plane to C' in the w-plane (see Fig. 14.2). For $1 - z$, in terms of w, Eq. (1) gives

$$\frac{1}{\Gamma(1 - z)} = -\frac{e^{-i\pi(z-1)}}{2\pi i} \int_{C'} e^{-w} w^{z-1} \, dw \tag{2}$$

The contour C' is composed of C_1, for which $\arg(w) = 0$; C_2, whose contribution can be

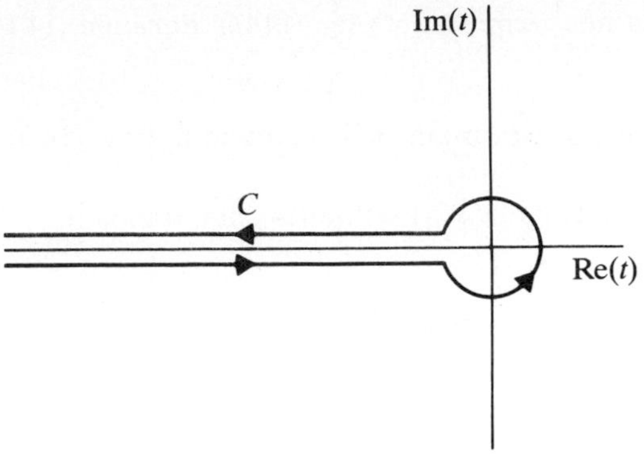

Figure 14.1 The contour C used in evaluating the reciprocal of the gamma function.

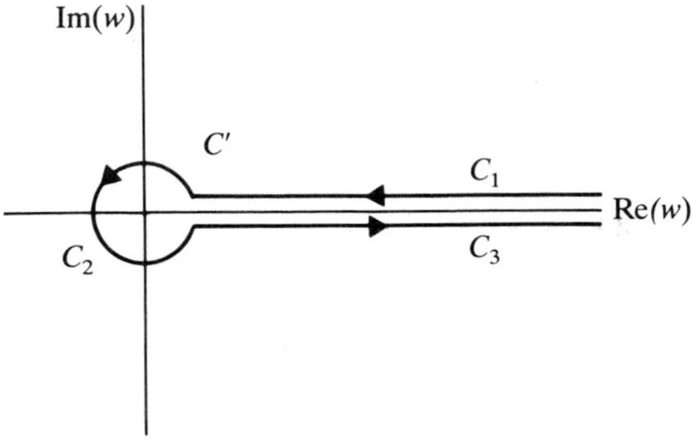

Figure 14.2 The transformed contour C' obtained from C by the transformation $w = \exp(-i\pi)t$.

shown to vanish as the radius of the circle goes to zero; and C_3, for which $\arg(w) = 2\pi$. We can, therefore, write Eq. (2) as

$$
\begin{aligned}
\frac{1}{\Gamma(1-z)} &= -\frac{e^{-i\pi(z-1)}}{2\pi i}\left(\int_{\infty}^{0} e^{-r} r^{z-1}\,dr + \int_{0}^{\infty} e^{-r} e^{2\pi i(z-1)} r^{z-1}\,dr\right) \\
&= -\frac{1}{2\pi i}\left(e^{i\pi(z-1)} - e^{-i\pi(z-1)}\right)\int_{0}^{\infty} e^{-r} r^{z-1}\,dr \\
&= -\frac{1}{\pi}\sin[\pi(z-1)]\int_{0}^{\infty} e^{-r} r^{z-1}\,dr \\
&= \frac{\sin \pi z}{\pi}\int_{0}^{\infty} e^{-r} r^{z-1}\,dr \\
&= \frac{1}{\Gamma(z)\Gamma(1-z)}\int_{0}^{\infty} e^{-r} r^{z-1}\,dr
\end{aligned}
$$

In the last step we used (the reciprocal of) Eq. (14.16). Equation (14.1) now follows immediately. ●

Another useful relation can be obtained by combining Eqs. (14.3) and (14.16):

$$\Gamma(z)\Gamma(1-z) = \Gamma(z)\Gamma(-z+1) = \Gamma(z)(-z)\Gamma(-z) = \frac{\pi}{\sin \pi z}$$

Thus,

$$\Gamma(z)\Gamma(-z) = -\frac{\pi}{z \sin \pi z} \tag{14.17}$$

This result was used in Chapter 9 (see Exercise 9.5.8).

Once we know $\Gamma(x)$ for positive values of real x, we can use (14.17) to find $\Gamma(x)$ for $x < 0$. Thus, for instance,

$$\Gamma(\tfrac{1}{2}) = \sqrt{\pi} \quad \Rightarrow \quad \Gamma(-\tfrac{1}{2}) = -2\sqrt{\pi}$$

Equation (14.17) also shows that the gamma function has simple poles wherever z is a negative integer.

PROBLEMS

14.1 Derive Eq. (14.3) from Eq. (14.1).

14.2 Show that $\Gamma(\frac{1}{2}) = \sqrt{\pi}$.

14.3 Show that

$$\Gamma(z) = \int_0^1 \left[\ln\left(\frac{1}{t}\right)\right]^{z-1} dt \qquad \text{for } \mathrm{Re}(z) > 0$$

14.4 Show that

$$\int_0^\infty e^{-x^\alpha} dx = \Gamma\left(\frac{\alpha+1}{\alpha}\right)$$

14.5 Consider the function $f(z) = (1+z)^\alpha$.

(a) Show that

$$\left.\frac{d^n f}{dz^n}\right|_{z=0} = \frac{\Gamma(\alpha+1)}{\Gamma(\alpha-n+1)}$$

and use it to derive the relation

$$(1+z)^\alpha = \sum_{n=0}^\infty \binom{\alpha}{n} z^n$$

where

$$\binom{\alpha}{n} = \frac{\alpha!}{n!(\alpha-n)!} \equiv \frac{\Gamma(\alpha+1)}{n!\,\Gamma(\alpha-n+1)}$$

(**b**) Show that for general complex numbers a and b we can formally write

$$(a+b)^\alpha = \sum_{n=0}^{\infty} \binom{\alpha}{n} a^n b^{\alpha - n}$$

(**c**) Show that if α is a positive integer m, the series in part (b) truncates at $n = m$.

14.6 Prove that the residue of $\Gamma(z)$ at $z = -k$ is $r_k = (-1)^k/k!$. [Hint: Iterate Eq. (14.3) to find an expression for $\Gamma(z+n)$; then write $\Gamma(z)$ in terms of $\Gamma(z+n)$. Choosing n appropriately, calculate the residue and obtain the result.]

14.7 Derive the following relation for $z = x + iy$:

$$|\Gamma(z)| = \Gamma(x) \prod_{k=0}^{\infty} \left[1 + \frac{y^2}{(x+k)^2}\right]^{-1/2}$$

14.8 Using the definition of $\mathrm{B}(a, b)$, Eq. (14.12), show that $\mathrm{B}(a, b) = \mathrm{B}(b, a)$.

14.9 Integrate Eq. (1) of Example 14.2.1 by parts and derive Eq. (14.3).

14.10 For positive integers n, show that $\Gamma(\frac{1}{2} - n)\Gamma(\frac{1}{2} + n) = (-1)^n \pi$.

14.11 Show that

(**a**) $\mathrm{B}(a, b) = \mathrm{B}(a+1, b) + \mathrm{B}(a, b+1)$

(**b**) $\mathrm{B}(a, b+1) = \left(\dfrac{b}{a+b}\right)\mathrm{B}(a, b)$

(**c**) $\mathrm{B}(a, b)\mathrm{B}(a+b, c) = \mathrm{B}(b, c)\mathrm{B}(a, b+c)$

14.12 Verify that

$$\int_{-1}^{1} (1+t)^a (1-t)^b \, dt - 2^{a+b+1} \mathrm{B}(a+1, b+1)$$

14.13 Show that the volume of the solid formed by the surface $z = x^a y^b$, the xy-, yz-, and xz-planes, and the plane parallel to the z-axis and going through the points $(0, y_0)$ and $(x_0, 0)$ is

$$\frac{x_0^{a+1} y_0^{b+1}}{a+b+2} \mathrm{B}(a+1, b+1)$$

14.14 Derive this relation:

$$\int_0^\infty \frac{\sinh^a x}{\cosh^b x} dx = \frac{1}{2} \mathrm{B}\left(\frac{a+1}{2}, \frac{b-a}{2}\right) \qquad \text{for } -1 < a < b$$

(Hint: Let $t = \tanh^2 x$.)

15

Numerical Methods

Very few problems in physics and applied mathematics can be solved analytically. Moreover, only through idealizations and extra (sometimes nonrealistic) assumptions can a real problem be turned into a problem that is susceptible to analytic solutions. Such idealizations are, of course, important for a basic understanding of a problem. For example, the idealization that the earth and the moon are perfect spheres was extremely helpful in Newton's derivation of the universal law of gravitation.

However, for practical situations, where extra assumptions may render the problem too idealistic, analytic solutions can, at best, be only a "rough draft" of the real solution. In such cases all the relevant parameters must be included, a process that renders the analytic method of solution obsolete. Such cases demand a numerical solution. With computers becoming almost a supermarket item, numerical solutions are ever more feasible.

This chapter offers a brief discussion of the most common numerical methods used in applied mathematics. For a more detailed treatment, see the vast literature on the subject.

15.1 ROOTS OF EQUATIONS

Let us begin by looking at one of the simplest and oldest numerical problems—the solution of an arbitrary equation. The problem can be posed as follows: Given a real-valued function $f(x)$, for what, if any, real-valued r does $f(r) = 0$ hold?

The method that is commonly used to find r (called the *root* of f) is *Newton's method*. The best way to understand this method is by looking at the graph of $f(x)$.

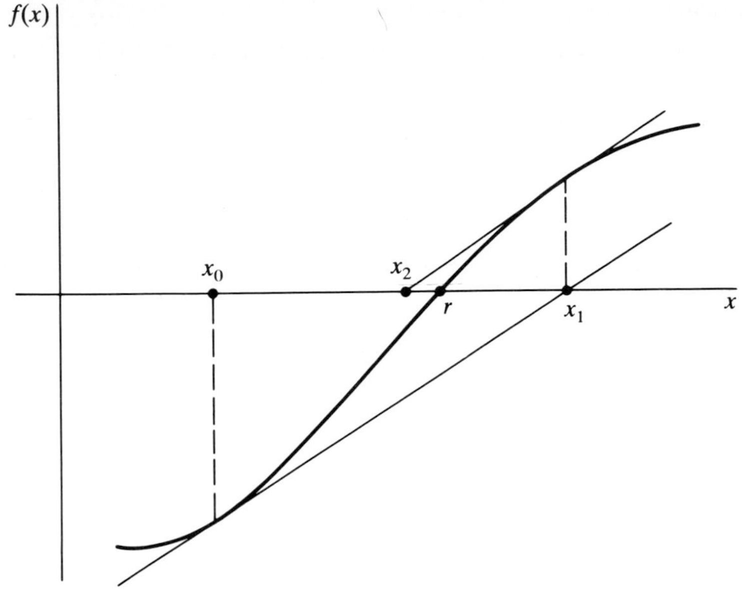

Figure 15.1 Newton's method for finding the root r of $f(x)$.

Figure 15.1 shows a portion of the graph of f close to one of its roots, r. We start with x_0, a zeroth guess at the root. If we are lucky and $f(x_0) = 0$, we are done. Otherwise, we find the x-intercept of the tangent line at $(x_0, f(x_0))$, which is given by

$$y - f(x_0) = f'(x_0)(x - x_0)$$

We call this intercept x_1. Thus, assuming that $f'(x_0) \neq 0$ (otherwise, we choose another x_0), we have

$$0 - f(x_0) = f'(x_0)(x_1 - x_0) \quad \Rightarrow \quad x_1 = x_0 - \frac{f(x_0)}{f'(x_0)}$$

and x_1 is a new, and hopefully better, guess. The equation of the tangent line at $(x_1, f(x_1))$ is

$$y - f(x_1) = f'(x_1)(x - x_1)$$

which leads to the new estimate

$$x_2 = x_1 - \frac{f(x_1)}{f'(x_1)}$$

This process can go on indefinitely until we reach the desired level of accuracy. Thus, for the nth estimate, obtainable from the $(n - 1)$th estimate, we have

$$x_n = x_{n-1} - \frac{f(x_{n-1})}{f'(x_{n-1})} \tag{15.1}$$

Example 15.1.1

The cube root of 2 can be determined as follows. We let $f(x) = x^3 - 2$ and use Newton's method to estimate the root of $f(x)$.

Equation (15.1) becomes

$$x_n = x_{n-1} - \frac{x_{n-1}^3 - 2}{3x_{n-1}^2} = \frac{2}{3}x_{n-1} + \frac{2}{3x_{n-1}^2} \tag{1}$$

Since $1 < 2^{1/3} < 2$, we take $x_0 = 1.5$. Substituting consecutively in (1) gives

$$x_1 = \frac{2}{3}(1.5) + \frac{2}{3(1.5)^2} = 1.296296296$$

$$x_2 = \frac{2}{3}x_1 + \frac{2}{3x_1^2} = 1.260932225$$

$$x_3 = \frac{2}{3}x_2 + \frac{2}{3x_2^2} = 1.259921861$$

$$x_4 = \frac{2}{3}x_3 + \frac{2}{3x_3^2} = 1.25992105$$

$$x_5 = \frac{2}{3}x_4 + \frac{2}{3x_4^2} = 1.25992105$$

Since $x_4 = x_5$, to nine significant figures, we cannot further improve the estimate. Thus, we can say that to nine significant figures, or to eight decimal places, $2^{1/3} = 1.25992105$.

●

It is assumed that with each new iteration of Eq. (15.1), we get closer to the actual root. Two situations in which this assumption does not hold, and, therefore, for which (15.1) is not applicable, are shown in Fig. 15.2.

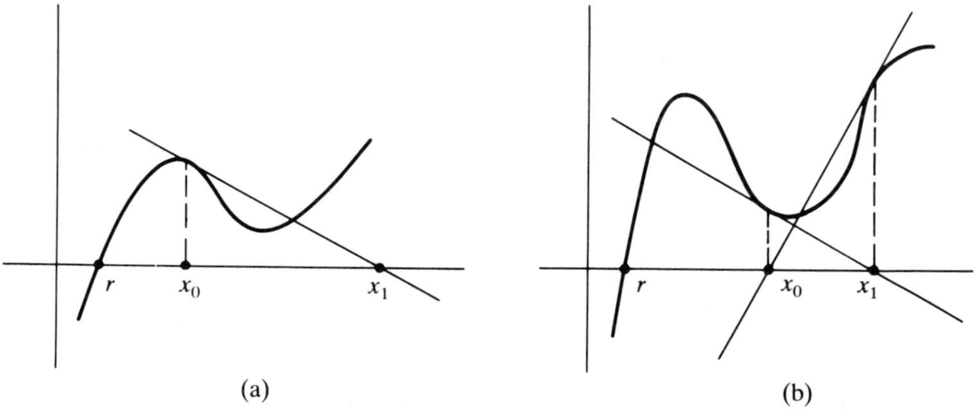

Figure 15.2 (a) Newton's method fails here because x_1 is far removed from r (perhaps it is closer to another zero of the function). (b) Here the estimate oscillates between x_0 and x_1 and will never approach r.

15.2 THE USE OF OPERATORS IN NUMERICAL ANALYSIS

Very broadly speaking, in numerical calculations infinities and zeros are replaced with finite values. For example, instead of taking the actual limit for

$$e = \lim_{n \to \infty} \sum_{k=0}^{n} \frac{1}{k!}$$

we take $n = 10$, or 20, or any desired finite number. Similarly, to find the derivative of a function $f(x)$ at x_0 numerically, we use

$$f'(x_0) \approx \frac{f(x_0 + h) - f(x_0)}{h}$$

where h is a finite (but small) number.

Many analytical problems involve certain limiting operations on functions. For example, differentiation involves the difference between values of a function at two neighboring points as the two points get closer and closer; similarly, integration is a limit of a sum. Since such limiting operations are impossible for numerical calculations, they are *approximated* by other operators, which are discussed in this section.

15.2.1 Finite-Difference Operators

In numerical analysis a function is considered as a table with two columns. The first column lists the (discrete) values of the independent variable x_i, and the second column lists the value of the function f at various values of that variable. Generally, $f_i \equiv f(x_i)$ is the value of f at x_i.

Three (discrete) operators that are in use in numerical analysis are the *forward difference operator* Δ, the *backward difference operator* ∇ (not to be confused with the gradient), and the *central difference operator* δ. These are defined as follows:

$$\Delta f_i \equiv f_{i+1} - f_i \tag{15.2}$$

$$\nabla f_i \equiv f_i - f_{i-1} \tag{15.3}$$

$$\delta f_i \equiv f_{i+1/2} - f_{i-1/2} \tag{15.4}$$

Equation (15.4) has only theoretical significance because a half-step is not used in the tabulation of functions or in computer calculations. Typically, the data are equally spaced, so $x_{i+1} - x_i = \Delta x = h$ is the same for all i. Then $f_{i \pm 1} = f(x_i \pm \Delta x)$, and we define $f_{i \pm 1/2} \equiv f(x_i \pm \Delta x/2)$.

We can define products of the three operators. In particular, Δ^2 is given by

$$\begin{aligned} \Delta^2 f_i &\equiv \Delta(\Delta f_i) = \Delta(f_{i+1} - f_i) \\ &= \Delta f_{i+1} - \Delta f_i = f_{i+2} - f_{i+1} - (f_{i+1} - f_i) \\ &= f_{i+2} - 2f_{i+1} + f_i \end{aligned} \tag{15.5}$$

Similarly,

$$\nabla^2 f_i = f_i - 2f_{i-1} + f_{i-2} \tag{15.6}$$

and
$$\delta^2 f_i = f_{i+1} - 2f_i + f_{i-1} \tag{15.7}$$

We note that

$$\delta^2 f_i = (f_{i+1} - f_i) - (f_i - f_{i-1}) = \Delta f_i - \nabla f_i = (\Delta - \nabla) f_i$$

which yields

$$\delta^2 = \Delta - \nabla$$

This shows that the three operators are related. More relations will be developed later.

From now on we will assume that all the data are equally spaced, with a spacing equal to h. This, in particular, implies that

$$f_{i+r} = f(x_i + rh)$$

It is convenient to introduce the *shifting operator*,

$$\mathbb{E} f(x) \equiv f(x+h) \tag{15.8}$$

and the *averaging operator*,

$$\mu f(x) \equiv \frac{1}{2}\left[f\left(x + \frac{h}{2}\right) + f\left(x - \frac{h}{2}\right)\right] \tag{15.9}$$

Note that, for any positive integer n,

$$\mathbb{E}^n f(x) \equiv f(x + nh)$$

Generalizing this to any real number α, we can write

$$\mathbb{E}^\alpha f(x) \equiv f(x + \alpha h) \tag{15.10}$$

All the other finite-difference operators can be written in terms of E:

$$\Delta = \mathbb{E} - \mathbb{1} \tag{15.11a}$$

$$\nabla = \mathbb{1} - \mathbb{E}^{-1} \tag{15.11b}$$

$$\delta = \mathbb{E}^{1/2} - \mathbb{E}^{-1/2} \tag{15.11c}$$

$$\mu = \tfrac{1}{2}(\mathbb{E}^{1/2} + \mathbb{E}^{-1/2}) \tag{15.11d}$$

Example 15.2.1

Equations (15.11) can be used to derive the relation given earlier for δ, Δ, and ∇:

$$\begin{aligned}\delta^2 &= (\mathbb{E}^{1/2} - \mathbb{E}^{-1/2})(\mathbb{E}^{1/2} - \mathbb{E}^{-1/2}) = \mathbb{E} - \mathbb{1} - \mathbb{1} + \mathbb{E}^{-1} \\ &= \Delta - \nabla\end{aligned}$$

Other relations can also be obtained, for example:

$$\begin{aligned}\Delta\nabla &= (\mathbb{E} - \mathbb{1})(\mathbb{1} - \mathbb{E}^{-1}) = \mathbb{E} - \mathbb{1} - \mathbb{1} + \mathbb{E}^{-1} \\ &= \Delta - \nabla\end{aligned}$$

Thus, we have

$$\delta^2 = \Delta\nabla$$

which can also be obtained directly:

$$\Delta\nabla f_i = \Delta(f_i - f_{i-1}) = \Delta f_i - \Delta f_{i-1} = f_{i+1} - f_i - (f_i - f_{i-1}) = f_{i+1} - 2f_i + f_{i-1}$$

This agrees with Eq. (15.7). ●

Equations (15.11a) and (15.11b) can be rewritten as

$$\mathbb{E} = \mathbb{1} + \Delta \tag{15.12a}$$

$$\mathbb{E} = (\mathbb{1} - \nabla)^{-1} \tag{15.12b}$$

We can obtain a useful formula for the shifting operator when it acts on polynomials of degree n or less. We note that

$$\mathbb{1} - \nabla^{n+1} = (\mathbb{1} - \nabla)(\mathbb{1} + \nabla + \nabla^2 + \cdots + \nabla^n)$$

But ∇^{n+1} annihilates all polynomials of degree n or less (see Exercise 15.2.1). Therefore, for such polynomials, we have

$$\mathbb{1} = (\mathbb{1} - \nabla)(\mathbb{1} + \nabla + \nabla^2 + \cdots + \nabla^n)$$

which shows that

$$\mathbb{E} = (\mathbb{1} - \nabla)^{-1} = \mathbb{1} + \nabla + \nabla^2 + \cdots + \nabla^n$$

Letting $n \to \infty$, we can write

$$\mathbb{E} = (\mathbb{1} - \nabla)^{-1} = \sum_{k=0}^{\infty} \nabla^k \tag{15.13}$$

for polynomials of *any* degree.

Example 15.2.2

Numerical interpolation illustrates the use of the formulas derived above. Suppose that we are given a table of the values of a function. As long as we are interested in exactly the x_i given in the table, there is no problem. However, what if we want the value of the function for an x located between two entries? In such a case we use the following procedure.

Assume that the values of the function f are given for $x_1, x_2, \ldots, x_i, \ldots$, and we are interested in the value for x such that $x_i < x < x_{i+1}$. This corresponds to f_{i+r}, where $0 < r < 1$. We have

$$f_{i+r} = \mathbb{E}^r f_i = (\mathbb{1} + \Delta)^r f_i = \left[\sum_{k=0}^{\infty} \frac{\Gamma(r+1)}{\Gamma(k+1)\Gamma(r-k+1)} \Delta^k\right] f_i$$

$$= \left(\mathbb{1} + r\Delta + \frac{r(r-1)}{2}\Delta^2 + \frac{r(r-1)(r-2)}{3!}\Delta^3 + \cdots\right) f_i \tag{1}$$

This equation is exact as long as all the terms are kept. In practice, however, the infinite sum is truncated, and only a finite number of terms are kept.

If only two terms are kept, we have

$$f_{i+r} \approx (\mathbb{1} + r\Delta)f_i = f_i + r(f_{i+1} - f_i)$$
$$= (1 - r)f_i + rf_{i+1} \qquad (2)$$

In particular, for $r = \frac{1}{2}$, Eq. (2) yields

$$f_{i+1/2} \approx \tfrac{1}{2}(f_i + f_{i+1})$$

which simply says that the value at the midpoint is approximately equal to the average of the values at the end points.

If the third term of the series in Eq. (1) is also retained, then

$$f_{i+r} \approx \left[\mathbb{1} + r\Delta + \frac{r(r-1)}{2}\Delta^2\right]f_i = f_i + r\Delta f_i + \frac{r(r-1)}{2}\Delta^2 f_i$$
$$= f_i + r(f_{i+1} - f_i) + \frac{r(r-1)}{2}(f_{i+2} - 2f_{i+1} + f_i)$$
$$= \frac{(2-r)(1-r)}{2}f_i + r(2-r)f_{i+1} + \frac{r(r-1)}{2}f_{i+2} \qquad (3)$$

For $r = \frac{1}{2}$, that is, at the midpoint between x_i and x_{i+1}, Eq. (3) yields

$$f_{i+1/2} \approx \tfrac{3}{8}f_i + \tfrac{3}{4}f_{i+1} - \tfrac{1}{8}f_{i+2}$$

Equation (3) gives a better approximation than (2) does. However, it involves not only the two points on either side of x but also a relatively distant point, x_{i+2}.

If we were to retain terms up to Δ^k, then f_{i+r} would be given in terms of $f_i, f_{i+1}, \ldots, f_{i+k}$ and the result would be more accurate than (3). Thus, the more information we have about the behavior of a function at distant points, the better we can approximate it at $x \in (x_i, x_{i+1})$.

The foregoing analysis was based on forward interpolation. We may want to use backward interpolation, where f_{i-r} is sought, for $0 < r < 1$. In such a case we use the backward difference operator:

$$f_{i-r} = \mathbb{E}^{-r}f_i = (\mathbb{E}^{-1})^r f_i = (\mathbb{1} - \nabla)^r f_i$$
$$= \left[\sum_{k=0}^{\infty} \frac{(-1)^k \Gamma(r+1)}{\Gamma(k+1)\Gamma(r-k+1)}\nabla^k\right]f_i$$
$$= \left(\mathbb{1} - r\nabla + \frac{r(r-1)}{2}\nabla^2 - \frac{r(r-1)(r-2)}{3!}\nabla^3 + \cdots\right)f_i \qquad \bullet$$

Example 15.2.3

Let us check the conclusion made in Example 15.2.2 using a calculator and a specific function, sin x. With the help of a calculator, we can make the following table:

x	sin x
0	0
0.1	0.0998334166
0.2	0.1986693308
0.3	0.2955202067

Suppose that we want to find sin(0.15) by interpolation. Using Eq. (2) of Example 15.2.2 with $r = \frac{1}{2}$, we obtain

$$\sin(0.15) \approx \tfrac{1}{2}[\sin(0.1) + \sin(0.2)] = 0.1492513737$$

On the other hand, using Eq. (3) with $r = \frac{1}{2}$, yields

$$\sin(0.15) \approx \tfrac{3}{8}\sin(0.1) + \tfrac{3}{4}\sin(0.2) - \tfrac{1}{8}\sin(0.3)$$
$$= 0.1494994959$$

The value of sin(0.15) obtained using a calculator is 0.1494381325. It is clear that Eq. (3) gives a better estimate than Eq. (2). ●

15.2.2 The Differentiation and Integration Operators

The two most important operations of mathematical physics can be written in terms of the finite-difference operators. Let $\mathbb{D}$ and $\mathbb{J}$ stand for differentiation and integration, or, more precisely, define

$$\mathbb{D}f(x) \equiv \frac{df}{dx}(x) \equiv f'(x)$$
$$\mathbb{J}f(x) \equiv \int_x^{x+h} f(x)\,dx \equiv F(x+h) - F(x) \tag{15.14}$$

where $F(x)$ is the primitive of $f(x)$.

Assuming that $\mathbb{D}^{-1}$ exists, we note that $f(x) = \mathbb{D}^{-1}[f'(x)]$. This shows that $\mathbb{D}^{-1}$ is the operation of antidifferentiation, or indefinite integration. That is,

$$\mathbb{D}^{-1}f(x) = F(x) + \text{constant}$$

On the other hand,

$$\Delta F(x) \equiv F(x+h) - F(x) \equiv \mathbb{J}f(x)$$

These two equations show that

$$\Delta\mathbb{D}^{-1} = \mathbb{J} \quad \Rightarrow \quad \Delta = \mathbb{J}\mathbb{D} = \mathbb{D}\mathbb{J} = \mathbb{E} - \mathbb{1} \tag{15.15}$$

The fact that $[\mathbb{J}, \mathbb{D}] = 0$, which can easily be verified, was used here.

To obtain a relation between $\mathbb{D}$ and the forward difference operator, we use Taylor expansion:

$$f(x+h) = f(x) + hf'(x) + \frac{h^2}{2!}f''(x) + \cdots$$
$$= \left[\mathbb{1} + h\mathbb{D} + \frac{h^2}{2!}\mathbb{D}^2 + \cdots\right]f(x) = [e^{h\mathbb{D}}]f(x)$$

On the other hand,

$$f(x+h) \equiv \mathbb{E}f(x)$$

We therefore have an important identity:

$$\mathbb{E} = e^{h\mathbb{D}} \tag{15.16}$$

This yields $h\mathbb{D} = \ln \mathbb{E}$, or

$$\mathbb{D} = \frac{1}{h}\ln(\mathbb{1} + \Delta) = \frac{1}{h}\left[\Delta - \frac{\Delta^2}{2} + \frac{\Delta^3}{3} - \frac{\Delta^4}{4} + \cdots\right] \tag{15.17}$$

This is only one form of expression for $\mathbb{D}$. Others express $\mathbb{D}$ in terms of other difference operators. For instance, $\delta = \mathbb{E}^{1/2} - \mathbb{E}^{-1/2}$ yields

$$\delta = e^{(1/2)h\mathbb{D}} - e^{(-1/2)h\mathbb{D}} = 2\sinh(\tfrac{1}{2}h\mathbb{D})$$

or

$$\tfrac{1}{2}h\mathbb{D} = \sinh^{-1}\left(\frac{\delta}{2}\right)$$

Expanding $\sinh^{-1}(\delta/2)$ in a Taylor series yields

$$\mathbb{D} = \frac{2}{h}\left[\frac{\delta}{2} - \frac{1}{3!}\left(\frac{\delta}{2}\right)^3 + \frac{3^2}{5!}\left(\frac{\delta}{2}\right)^5 - \frac{3^2 5^2}{7!}\left(\frac{\delta}{2}\right)^7 + \cdots\right] \tag{15.18}$$

Similarly,

$$\mathbb{D} = \frac{1}{h}[-\ln \mathbb{E}^{-1}] = -\frac{1}{h}\ln(\mathbb{1} - \nabla)$$

$$= \frac{1}{h}\left[\nabla + \frac{\nabla^2}{2} + \frac{\nabla^3}{3} + \cdots\right] \tag{15.19}$$

If we are interested in higher derivatives, we can raise $\mathbb{D}$ to various powers. For example, in terms of the forward difference operator, we have

$$\mathbb{D}^2 = \frac{1}{h^2}\left[\Delta - \frac{\Delta^2}{2} + \frac{\Delta^3}{3} - \frac{\Delta^4}{4} + \cdots\right]^2$$

$$= \frac{1}{h^2}[\Delta^2 - \Delta^3 + \tfrac{11}{12}\Delta^4 - \tfrac{5}{6}\Delta^5 - \cdots] \tag{15.20}$$

Having found $\mathbb{D}$ as a power series in a difference operator, we can evaluate $\mathbb{D}^{-1}$ by long division. For example, we can obtain

$$\mathbb{D}^{-1} = \frac{h}{\ln(\mathbb{1} + \Delta)} = \frac{h}{\Delta - \dfrac{\Delta^2}{2} + \dfrac{\Delta^3}{3} - \dfrac{\Delta^4}{4} + \cdots}$$

$$= h\Delta^{-1}\frac{\mathbb{1}}{\mathbb{1} - \dfrac{\Delta}{2} + \dfrac{\Delta^2}{3} - \dfrac{\Delta^3}{4} + \cdots}$$

$$= h\Delta^{-1}\left(\mathbb{1} + \frac{\Delta}{2} - \frac{\Delta^2}{12} + \frac{\Delta^3}{24} - \cdots\right)$$

This leads to an important relation:

$$\mathbb{J} = \Delta\mathbb{D}^{-1} = h\left(\mathbb{1} + \frac{\Delta}{2} - \frac{\Delta^2}{12} + \frac{\Delta^3}{24} - \cdots\right) \tag{15.21}$$

The operator $\mathbb{J}$ is a special case of a more general operator defined by

$$\mathbb{J}_\alpha f(x) \equiv \int_x^{x+\alpha h} f(x)\,dx \tag{15.22}$$

This definition leads immediately to the formula

$$\mathbb{J}_\alpha = (\mathbb{E}^\alpha - \mathbb{1})\mathbb{D}^{-1} \tag{15.23}$$

For $\alpha = 1$, Eq. (15.21) is recovered. For numerical integration it is sometimes useful to choose $\alpha \neq 1$. In such cases Eq. (15.23) is used to derive an appropriate formula for $\mathbb{J}_\alpha$ in terms of a difference operator.

Exercises

15.2.1 Show that when the forward difference operator is applied to a polynomial, the degree of the polynomial is reduced by 1. (Hint: Consider x^n first.) Then show that Δ^{n+1} annihilates all polynomials of degree n or less. Assume equal spacing.

15.2.2 Write expressions for $\mathbb{E}^{1/2}$, Δ, ∇, and μ in terms of δ.

15.3 TRUNCATION ERROR

There are two kinds of errors that occur in numerical calculations. The first, called *roundoff errors*, occur because, with a few exceptions, when a real number is represented in decimal (or binary) form, an infinite number of digits is required. For example, $\frac{1}{3} = 0.3333\ldots$, $\sqrt{2} = 1.414\ldots$, and $\pi = 3.14159\ldots$, where the dots indicate that there are infinitely many more numbers that have been left out of the representation. Roundoff errors take place when (inevitably) such an infinite sequence of digits is replaced by some finite number of digits. For a typical calculation done on a computer, the programmer has very little control over such errors. However, double or triple precision does reduce roundoff errors.

The second kind of errors, called *truncation errors*, are more relevant to this discussion because they are controlled by the programmer. A truncation error takes place when an infinite sum is replaced by a finite sum. Such an error can be reduced by simply adding more terms to the finite sum (at the cost of longer computer time) or by judiciously choosing the terms of the infinite sum (so that it converges faster).

It is important to be able to estimate the error invoked when an infinite series is truncated to a finite sum. Such an estimate is best described using the order function discussed in Chapter 10. For example, suppose that this series is truncated at the third term:

$$e^{0.01} = \sum_{n=0}^{\infty} \frac{(0.01)^n}{n!} = 1 + 0.01 + \frac{(0.01)^2}{2!} + \cdots$$

Then we say that the error involved is $O((0.01)^3)$, or $O(10^{-6})$. This, of course, does not mean that the error is exactly 10^{-6}, but that it is of the order of, or approximately, 10^{-6}. We will not go into any detail concerning error analysis; orders of magnitude for errors are sufficient for the purposes at hand.

As was noted in Section 15.2, all operators of interest can be approximated by expressions involving the finite-difference operators. In general, this yields an infinite power series of a finite-difference operator. Since the infinite sum is truncated in practical calculations, it is important to estimate the error involved in such a truncation. We can obtain the order of magnitude of this error if we know the error for each power of the difference operator. The forward difference operator can be written as

$$\begin{aligned}\Delta &= \mathbb{E} - \mathbb{1} = e^{h\mathbb{D}} - \mathbb{1} \\ &= h\mathbb{D} + \frac{(h\mathbb{D})^2}{2} + \cdots = O(h)\end{aligned}$$

because the largest term of the series is of order h. Recalling the properties of the function O, we note that $\Delta^n = O(h^n)$. This shows that truncating a power series in Δ at the nth term produces an error equal to $O(h^{n+1})$, written $\mathrm{e} = O(h^{n+1})$. Also, it can be shown that $\nabla^n = O(h^n)$ and $\delta^n = O(h^n)$.

Example 15.3.1

By writing $\mathbb{D} = (1/h)\Delta$, we are invoking an error whose order of magnitude is $(1/h)O(h^2) = h$, because we have truncated the infinite series of Eq. (15.17) at the Δ term. On the other hand, we can write

$$\mathbb{D} = \frac{1}{h}\left(\Delta - \frac{\Delta^2}{2}\right) \qquad \mathrm{e} = O(h^2) \tag{1}$$

meaning that the error involved is of order h^2 when the expression for $\mathbb{D}$ is truncated at Δ^2.

Similarly, Eq. (15.21) can be written as

$$\mathbb{J} = h\left(\mathbb{1} + \frac{\Delta}{2} - \frac{\Delta^2}{12}\right) \qquad \mathrm{e} = O(h^4) \tag{2}$$

●

One way to minimize the truncation error is to make the step size h as small as possible and keep as many terms as is reasonable. In practice, however, this will increase roundoff error.

Example 15.3.2

Let us calculate cos(0.1), considering $\cos x$ to be $(d/dx)(\sin x)$ and using the table of Example 15.2.3, with $e = O(0.01)$. Applying Eq. (1) of Example 15.3.1, we get

$$\mathbb{D}f_i = \frac{1}{h}\left(\Delta - \frac{\Delta^2}{2}\right)f_i = \frac{1}{h}(\Delta f_i - \tfrac{1}{2}\Delta^2 f_i)$$

$$= \frac{1}{h}[f_{i+1} - f_i - \tfrac{1}{2}(f_{i+2} - 2f_{i+1} + f_i)]$$

$$= \frac{1}{2h}(-f_{i+2} + 4f_{i+1} - 3f_i)$$

For the case at hand, $h = 0.1$ and $f(x) = \sin x$. Thus,

$$\cos(0.1) \approx \frac{1}{0.2}[-0.2955202067 + 4(0.1986693308) - 3(0.0998334166)]$$

$$= 0.9982843335 \qquad e = O(0.01)$$

In comparison, the value obtained directly from a calculator is 0.9950041653. •

Exercise

15.3.1 (**a**) Write expressions for $\mathbb{D}^3$ in terms of Δ with $e = O(h)$ and $e = O(h^2)$. (**b**) For $e = O(h^2)$ approximate the third derivative of a function f at x_i and write it in terms of its values at the neighboring discrete points.

15.4 NUMERICAL INTEGRATION

Suppose that we are interested in numerically calculating the integral

$$I = \int_a^b f(x)\,dx$$

Let $x_0 \equiv a$ and $x_N \equiv b$, and divide the interval $[a, b]$ into N equal parts, each of length $h = (b - a)/N$. The method commonly used in calculating integrals is to find the integral $\int_{x_i}^{x_i+\alpha h} f(x)\,dx$, where α is a suitable number, and then add all such integrals to find I. Thus, we have

$$I = \int_{x_0}^{x_0+\alpha h} f(x)\,dx + \int_{x_0+\alpha h}^{x_0+2\alpha h} f(x)\,dx + \cdots + \int_{x_0+(M-1)\alpha h}^{x_0+M\alpha h} f(x)\,dx$$

where M is a suitably chosen number. In fact, since $x_N = x_0 + Nh$, $M\alpha = N$. We next employ Eq. (15.23) to get

$$I = \mathbb{J}_\alpha f_0 + \mathbb{J}_\alpha f_\alpha + \cdots + \mathbb{J}_\alpha f_{(M-1)\alpha} = \mathbb{J}_\alpha\left(\sum_{k=0}^{M-1} f_{k\alpha}\right) \tag{15.24}$$

where $f_{k\alpha} \equiv f(x_0 + k\alpha h)$. We thus need an expression for $\mathbb{J}_\alpha$ to evaluate the integral. Such an expression can be derived (see Exercise 15.4.1):

$$\mathbb{J}_\alpha = h\frac{(\mathbb{1} + \Delta)^\alpha - \mathbb{1}}{\ln(\mathbb{1} + \Delta)} = \alpha h\left[\mathbb{1} + \frac{\alpha}{2}\Delta + \frac{\alpha(2\alpha - 3)}{12}\Delta^2 + \frac{\alpha(\alpha - 2)^2}{24}\Delta^3 + \frac{\alpha(6\alpha^3 - 45\alpha^2 + 110\alpha - 90)}{720}\Delta^4 + \cdots\right] \tag{15.25}$$

Equations (15.24) and (15.25) give the desired evaluation of the integral.

Let us note a few things about this method. First, once h has been set, the function can be evaluated only at $x_0 + nh$, where n is an integer. This means that f_n is only given for integer n. Thus, in the sum in (15.24) $k\alpha$ must be an integer. Since k is an integer, we conclude that α *must be an integer*. Second, since $N = M\alpha$ for some integer M, we must choose N to be a multiple of α. Third, if we are to be able to evaluate $\mathbb{J}_\alpha f_{(M-1)\alpha}$, the last term in (15.24), $\mathbb{J}_\alpha$ cannot have powers of Δ higher than α because $\Delta^n f_{(M-1)\alpha}$ contains a term of the form

$$f[x_0 + (M - 1)\alpha h + nh] = f[x_N + (n - \alpha)h]$$

which, for $n > \alpha$, gives f at a point beyond the upper limit $b = x_N$. Thus, in the power-series expansion of $\mathbb{J}_\alpha$, we must make sure that no power of Δ beyond α is retained. Fourth, the smaller the error in each individual term in (15.24), the smaller the overall error for I.

There are several specific $\mathbb{J}_\alpha$'s commonly used in numerical integration. We will consider these next.

15.4.1 The Trapezoidal Rule

The trapezoidal rule specifies that $\alpha = 1$. According to the remarks above, we therefore retain terms up to the first power in the expansion of $\mathbb{J}_\alpha$. Thus, $\mathbb{J}_1 \approx h(\mathbb{1} + \frac{1}{2}\Delta)$, or

$$\mathbb{J}_1 = h(\mathbb{1} + \tfrac{1}{2}\Delta) \qquad \mathrm{e} = O(h^3)$$

Substituting this in Eq. (15.24), we obtain

$$I = h(\mathbb{1} + \tfrac{1}{2}\Delta)\left(\sum_{k=0}^{N-1} f_k\right) = h\sum_{k=0}^{N-1}[f_k + \tfrac{1}{2}(f_{k+1} - f_k)]$$

$$= \frac{h}{2}\sum_{k=0}^{N-1}(f_k + f_{k+1}) \qquad \mathrm{e} = O(h^2) \tag{15.26}$$

This is sometimes written as

$$I = \frac{h}{2}(f_0 + 2f_1 + 2f_2 + \cdots + 2f_{N-1} + f_N) \qquad \mathrm{e} = O(h^2)$$

Note that $\mathrm{e} = O(h^2)$ because each term has an error of $O(h^3)$. Thus, the total error is $N[O(h^3)]$. But $N = O(1/h)$, so the total error is

$$O(1/h)O(h^3) = O(h^3/h) = O(h^2)$$

15.4.2 Simpson's One-Third Rule

Simpson's one-third rule sets $\alpha = 2$. Thus, we have to retain up to the Δ^2 term. However, for $\alpha = 2$ the third power of Δ also disappears in Eq. (15.25). We thus get an extra "power" of accuracy for free! Therefore, Eq. (15.25) yields

$$\mathbb{J}_2 = 2h(\mathbb{1} + \Delta + \tfrac{1}{6}\Delta^2) \qquad e = O(h^5)$$

which shows that the order of magnitude of the error is smaller than expected by 1. Because of this, Simpson's one-third rule is very popular for numerical integrations. Substituting the expression for $\mathbb{J}_2$ in (15.24) yields

$$\begin{aligned} I &= \frac{h}{3}\sum_{k=0}^{N/2-1} (6\mathbb{1} + 6\Delta + \Delta^2) f_{2k} \\ &= \frac{h}{3}\sum_{k=0}^{N/2-1} [6f_{2k} + 6(f_{2k+1} - f_{2k}) + f_{2k+2} - 2f_{2k+1} + f_{2k}] \\ &= \frac{h}{3}\sum_{k=0}^{N/2-1} [f_{2k+2} + 4f_{2k+1} + f_{2k}] \qquad e = O(h^4) \end{aligned} \tag{15.27}$$

It is understood, of course, that N is an even integer. Because of the sum, the power of h in the error has been reduced here by 1. Equation (15.27) is sometimes written as

$$I = \frac{h}{3}(f_0 + 4f_1 + 2f_2 + 4f_3 + 2f_4 + 4f_5 + 2f_6 + \cdots + 2f_{N-2} + 4f_{N-1} + f_N) \quad e = O(h^4)$$

The factor $\frac{1}{3}$ gives this method its name.

15.4.3 Simpson's Three-Eighths Rule

According to *Simpson's three-eighths rule*, we set $\alpha = 3$, retain terms up to Δ^3, and write

$$\begin{aligned} \mathbb{J}_3 &= 3h(\mathbb{1} + \tfrac{3}{2}\Delta + \tfrac{3}{4}\Delta^2 + \tfrac{1}{8}\Delta^3) \\ &= \frac{3h}{8}(8\mathbb{1} + 12\Delta + 6\Delta^2 + \Delta^3) \qquad e = O(h^5) \end{aligned}$$

Substituting in Eq. (15.24), we get

$$\begin{aligned} I &= \frac{3h}{8}\sum_{k=0}^{N/3-1} (8\mathbb{1} + 12\Delta + 6\Delta^2 + \Delta^3) f_{3k} \\ &= \frac{3h}{8}\sum_{k=0}^{N/3-1} [8f_{3k} + 12(f_{3k+1} - f_{3k}) + 6(f_{3k+2} - 2f_{3k+1} + f_{3k}) \\ &\quad + (f_{3k+3} - 3f_{3k+2} + 3f_{3k+1} - f_{3k})] \\ &= \frac{3h}{8}\sum_{k=0}^{N/3-1} (f_{3k+3} + 3f_{3k+2} + 3f_{3k+1} + f_{3k}) \qquad e = O(h^4) \end{aligned} \tag{15.28}$$

which is usually written as

$$I = \frac{3h}{8}(f_0 + 3f_1 + 3f_2 + 2f_3 + 3f_4 + 3f_5 + 2f_6 + \cdots + 2f_{N-3}$$

$$+ 3f_{N-2} + 3f_{N-1} + f_N) \qquad e = O(h^4)$$

15.4.4 Bounds on Errors

We can calculate the upper and lower limits of the error involved in approximating an integral by a sum as described in the preceding subsections. Let us do so for the trapezoidal rule.

First, we calculate the error for a single interval of length h. With $x_i + h \equiv t$, the trapezoidal rule gives

$$I = \int_{x_i}^{t} f(x)\,dx = \frac{h}{2}[f(x_i) + f(t)] + e_i(t)$$

where $e_i(t)$ is the "exact" error corresponding to the interval $[x_i, t]$. We rewrite this as

$$e_i(t) = \int_{x_i}^{t} f(x)\,dx - \frac{t - x_i}{2}[f(x_i) + f(t)]$$

Differentiating with respect to t yields

$$e_i'(t) = f(t) - \frac{1}{2}[f(x_i) + f(t)] - \frac{t - x_i}{2} f'(t)$$

Similarly,
$$e_i''(t) = -\frac{t - x_i}{2} f''(t)$$

With $(f''_{\min})_i$ and $(f''_{\max})_i$ as, respectively, the minimum and the maximum values of $f''(t)$ in the interval $[x_i, t]$, we can write

$$\frac{t - x_i}{2}(f''_{\min})_i \leqslant -e_i''(t) \leqslant \frac{t - x_i}{2}(f''_{\max})_i$$

Antidifferentiating once gives

$$\frac{(t - x_i)^2}{4}(f''_{\min})_i \leqslant -e_i'(t) \leqslant \frac{(t - x_i)^2}{4}(f''_{\max})_i$$

Antidifferentiating once more yields

$$\frac{(t - x_i)^3}{12}(f''_{\min})_i \leqslant -e_i(t) \leqslant \frac{(t - x_i)^3}{12}(f''_{\max})_i$$

or, since $t - x_i = h$,

$$\frac{h^3}{12}(f''_{\min})_i \leqslant -e_i \leqslant \frac{h^3}{12}(f''_{\max})_i$$

Having found the bound for a single interval, we can proceed to the whole interval $[a, b]$. Let $f''_{\min}$ and $f''_{\max}$ represent, respectively, the minimum and the maximum of f'' in the entire interval $[a, b]$. Then, clearly,

$$\frac{h^3}{12} f''_{\min} \leqslant -e_i \leqslant \frac{h^3}{12} f''_{\max} \qquad \forall i$$

Adding all the errors for various i's, we finally obtain the inequality

$$\frac{h^3}{12} N f''_{\min} \leqslant -e \leqslant \frac{h^3}{12} N f''_{\max}$$

Since $N = (b - a)/h$, we finally get the following limits for the error for the trapezoidal rule:

$$\frac{h^2(b-a)}{12} f''_{\min} \leqslant -e \leqslant \frac{h^2(b-a)}{12} f''_{\max} \tag{15.29}$$

Some books define e to be the approximation minus the exact result; in that case the minus sign in front of the e in Eq. (15.29) disappears.

The limits of the errors for Simpson's rules are harder to determine. However, it can be shown very generally that (see Hildebrand 1987, Chapter 3)

$$\int_{x_0}^{x_0+nh} f(x)\,dx = h \sum_{k=0}^{n} c_k f(x_k) + e_n(\xi) \tag{15.30a}$$

where ξ is a number between x_0 and $x_0 + nh$,

$$c_k = \int_0^n \frac{s(s-1)\cdots(s-k+1)(s-k-1)\cdots(s-n)}{k(k-1)\cdots(k-k+1)(k-k-1)\cdots(k-n)}\,ds \tag{15.30b}$$

and

$$e_n(\xi) = \begin{cases} \dfrac{h^{n+2} f^{(n+1)}(\xi)}{(n+1)!} \displaystyle\int_0^n s(s-1)\cdots(s-n)\,ds & \text{if } n \text{ is odd} \\[2ex] \dfrac{h^{n+3} f^{(n+2)}(\xi)}{(n+2)!} \displaystyle\int_0^n \left(s - \frac{n}{2}\right) s(s-1)\cdots(s-n)\,ds & \text{if } n \text{ is even} \end{cases} \tag{15.30c}$$

Note that in (15.30b) the term $(s - k)$ is absent from the numerator and the term $(k - k) = 0$ is absent from the denominator

Applying the formula of (15.30a) to Simpson's one-third rule (see Exercise 15.4.2) gives

$$e_2(\xi) = -\frac{h^5 f^{(iv)}(\xi)}{90}$$

yielding the following bounds on the total error for numerical integration by that rule:

$$\frac{h^4(b-a)}{90} f^{(iv)}_{\min} \leqslant -e \leqslant \frac{h^4(b-a)}{90} f^{(iv)}_{\max} \tag{15.31}$$

Example 15.4.1

Let us use Simpson's one-third rule with four intervals to evaluate the integral

$$I = \int_0^1 e^x\, dx$$

With $h = 0.25$, $N = 4$, and $f(x) = e^x$, Eq. (15.27) yields

$$I \approx \frac{0.25}{3}(1 + 4e^{0.25} + 2e^{0.5} + 4e^{0.75} + e)$$

$$= \frac{0.25}{3}(20.62) = 1.71832$$

This is very close to the result obtained with a calculator, $e - 1 = 1.71828$. Note that the approximate value is larger. Thus, if we write

$$I = I_{\text{approx}} + e_2^{\text{tot}}(\xi)$$

we should obtain a negative value for $e_2^{\text{tot}}(\xi)$. This agrees with the expression

$$e_2^{\text{tot}}(\xi) = Ne_2(\xi) = -4\left[\frac{(0.25)^5 e^\xi}{90}\right]$$

because e^ξ is always positive. Since $0 \leqslant \xi \leqslant 1$, we have $1 \leqslant e^\xi \leqslant e$, which yields

$$\frac{(0.25)^5(4)}{90} \leqslant -e_2^{\text{tot}} \leqslant \frac{(0.25)^5(4)}{90} e$$

or

$$4.34 \times 10^{-5} \leqslant -e_2^{\text{tot}} \leqslant 1.18 \times 10^{-4}$$

●

Exercises

15.4.1 Find an expression for $\mathbb{J}_\alpha$ in powers of Δ. Retain all terms up to the fourth power.

15.4.2 Show that Eq. (15.30a) reduces to the trapezoidal rule when $n = 1$ and to Simpson's one-third rule when $n = 2$. Find the error term for each case.

15.5 NUMERICAL SOLUTIONS OF DIFFERENTIAL EQUATIONS

Numerical methods probably find their widest application in the solution of differential equations. There are a variety of methods having various degrees of simplicity of use and accuracy. This section considers a few representative ones as applied to the solution of ODEs.

We saw in Chapter 9 that an ODE of a higher order can always be written as a system of first-order ODEs in several variables. We, therefore, consider a FODE of the form

$$\dot{x} = \frac{dx}{dt} = f(x, t)$$

in which f is a well-behaved function of two real variables.

Two general types of problems are encountered in applications. An initial value problem (IVP) gives $x(t)$ at an initial time t_0 and asks for the value of x at other times. The second type, the boundary value problem (BVP), applies only to differential equations of higher order than first. A second-order BVP specifies the value of $x(t)$ and/or $\dot{x}(t)$ at one or more points and asks for x or $\dot{x}$ at other values of t.

15.5.1 Use of the Backward Difference Operator

Let us consider the IVP

$$\dot{x} = f(x, t) \qquad x(t_0) = x_0 \tag{15.32}$$

A numerical solution of (15.32) means an evaluation of the function $x(t)$ at $\{t_k\}_{k=1}^N$, where $t_k = t_0 + kh$ and h is a convenient spacing. That is, the problem is to find $\{x_k = x(t_0 + kh)\}_{k=1}^N$, given x_0 and the FODE in (15.32).

Let us begin by integrating (15.32) between t_n and $t_n + h$:

$$x(t_n + h) - x(t_n) = \int_{t_n}^{t_n+h} \dot{x}(t)\,dt$$

Changing the variable of integration to $s = (t - t_n)/h$ and using the integral formula $\int a^s ds = a^s/\ln a$ yields

$$x_{n+1} - x_n = h\int_0^1 \dot{x}(t_n + sh)\,ds = h\int_0^1 [\mathbb{E}^s\dot{x}(t_n)]\,ds$$

$$= h\left(\int_0^1 \mathbb{E}^s\,ds\right)\dot{x}(t_n) = h\left(\frac{\mathbb{E} - 1}{\ln \mathbb{E}}\right)\dot{x}(t_n) \tag{15.33a}$$

Since a typical situation involves calculating x_{n+1} from the values of $x(t)$ and $\dot{x}(t)$ at preceding steps, we want an expression in which the RHS contains such preceding terms. This suggests expressing $\mathbb{E}$ in terms of the backward difference operator: $\mathbb{E} = (\mathbb{1} - \nabla)^{-1}$. Then Eq. (15.33a) becomes

$$x_{n+1} - x_n = h\left(\int_0^1 (\mathbb{1} - \nabla)^{-s}\, ds\right)\dot{x}_n$$

$$= h\left[\int_0^1 \sum_{k=0}^{\infty} \frac{\Gamma(-s+1)\, ds}{k!\,\Gamma(-s-k+1)}(-\nabla)^k\right]\dot{x}_n$$

$$\equiv h\left(\sum_{k=1}^{\infty} a_k \nabla^k\right)\dot{x}_n \qquad (15.33b)$$

where

$$a_k \equiv \frac{(-1)^k}{k!}\int_0^1 \frac{\Gamma(-s+1)}{\Gamma(-s-k+1)}\, ds$$

$$= \frac{1}{k!}\int_0^1 s(s+1)(s+2)\cdots(s+k-1)\, ds \qquad (15.33c)$$

Keeping the first few terms of the expansion yields

$$x_{n+1} \approx x_n + h\left(\mathbb{1} + \frac{\nabla}{2} + \frac{5\nabla^2}{12} + \frac{3\nabla^3}{8} + \frac{251\nabla^4}{720} + \frac{95\nabla^5}{288} + \cdots\right)\dot{x}_n \qquad (15.34)$$

Because of the presence of ∇ in Eq. (15.34), to find the value of $x(t)$ at t_{n+1}, we need to know $x(t)$ and $\dot{x}(t)$ only at points $t_0, t_1, \ldots, t_n$. Because of this, (15.34) is called a formula of *open type*. In contrast, in formulas of *closed type*, the RHS contains values at t_{n+1} as well. We can obtain a formula of closed type by changing $\mathbb{E}^s\dot{x}_n$ to its equivalent form, $\mathbb{E}^{s-1}\dot{x}_{n+1}$, giving

$$x_{n+1} = x_n + h\left(\int_0^1 \mathbb{E}^{s-1}\, ds\right)\dot{x}_{n+1} = x_n + h\left(\int_0^1 (\mathbb{1} - \nabla)^{-s+1}\, ds\right)\dot{x}_{n+1}$$

$$= x_n + h\left(\sum_{k=0}^{\infty} b_k \nabla^k\right)\dot{x}_{n+1} \qquad (15.35a)$$

where

$$b_k \equiv \frac{(-1)^k}{k!}\int_0^1 \frac{\Gamma(-s+2)}{\Gamma(-s-k+2)}\, ds = \frac{1}{k!}\int_0^1 (s-1)s(s+1)\cdots(s+k-2)\, ds \qquad (15.35b)$$

Keeping the first few terms, we obtain

$$x_{n+1} \approx x_n + h\left(1 - \frac{\nabla}{2} - \frac{\nabla^2}{12} - \frac{\nabla^3}{24} - \frac{19\nabla^4}{720} - \frac{3\nabla^5}{160} - \cdots\right)\dot{x}_{n+1} \tag{15.36}$$

which involves evaluation at t_{n+1} on the RHS.

If, instead of integrating (15.32) from t_n to $t_n + h$, we integrate it from $t_n - ph$ to $t_n + h$ (the reason for doing so will be made clear shortly), we obtain

$$x_{n+1} = x_{n-p} + h \int_{-p}^{1} \dot{x}(t_n + sh)\, ds$$

After manipulation very similar to that in the foregoing derivations, this yields a formula of open type:

$$x_{n+1} = x_{n-p} + h \sum_{k=0}^{\infty} a_k^{(p)} \nabla^k \dot{x}_n \tag{15.37}$$

where
$$a_k^{(p)} \equiv \frac{(-1)^k}{k!} \int_{-p}^{1} \frac{\Gamma(-s+1)}{\Gamma(-s-k+1)}\, ds$$

Similarly, for a formula of closed type, we obtain

$$x_{n+1} = x_{n-p} + h \sum_{k=0}^{\infty} b_k^{(p)} \nabla^k \dot{x}_{n+1} \tag{15.38}$$

where
$$b_k^{(p)} \equiv \frac{(-1)^k}{k!} \int_{-p}^{1} \frac{\Gamma(-s+2)}{\Gamma(-s-k+2)}\, ds$$

The error committed when the series is truncated at $k = M$ for the two types of formula is

$$e_M^{(p)}(\xi) = h^{M+2} \alpha_{M+1}^{(p)} x^{(M+2)}(\xi) \tag{15.39}$$

where
$$\alpha_k^{(p)} \equiv \begin{cases} a_k^{(p)} & \text{open type} \\ b_k^{(p)} & \text{closed type} \end{cases}$$

and
$$t_{n+1} > \xi > t_{n-M}$$

If we write the first few terms of (15.37) and (15.38) for $p = 1$ and $p = 3$, we obtain

$$x_{n+1} \approx x_{n-1} + h(2 + 0\nabla + \tfrac{1}{3}\nabla^2 + \tfrac{1}{3}\nabla^3 + \tfrac{29}{90}\nabla^4 + \cdots)\dot{x}_n$$

$$x_{n+1} \approx x_{n-3} + h(4 - 4\nabla + \tfrac{8}{3}\nabla^2 + 0\nabla^3 + \tfrac{14}{45}\nabla^4 + \cdots)\dot{x}_n$$

$$x_{n+1} \approx x_{n-1} + h(2 - 2\nabla + \tfrac{1}{3}\nabla^2 + 0\nabla^3 - \tfrac{1}{90}\nabla^4 - \cdots)\dot{x}_{n+1}$$

$$x_{n+1} \approx x_{n-3} + h(4 - 8\nabla + \tfrac{20}{3}\nabla^2 - \tfrac{8}{3}\nabla^3 + \tfrac{14}{45}\nabla^4 - 0\nabla^5 - \cdots)\dot{x}_{n+1}$$

These equations show why we chose $p = 1$ and $p = 3$. The coefficient of ∇^p or ∇^{p+2} is zero. Thus, retaining terms up to the $(p - 1)$th power of ∇ automatically gives us an accuracy of h^{p+2}. Thus is the advantage of using nonzero values of p and the reason we considered such cases. We can use (15.39) to calculate the errors. This gives the following for the open-type formula:

$$x_{n+1} = x_{n-1} + 2h\dot{x} + \frac{h^3}{3}\dddot{x}(\xi) \tag{15.40a}$$

$$x_{n+1} = x_{n-3} + 4h(\dot{x}_n - \nabla\dot{x}_n + \tfrac{2}{3}\nabla^2\dot{x}_n) + \frac{14h^5}{45}\frac{d^5x}{dt^5}(\xi)$$

Similarly, for the closed-type formula,

$$x_{n+1} = x_{n-1} + 2h(\dot{x}_{n+1} - \nabla\dot{x}_{n+1} + \tfrac{1}{6}\nabla^2\dot{x}_{n+1}) - \frac{h^5}{90}\frac{d^5x}{dt^5}(\xi)$$

$$\begin{aligned} x_{n+1} = x_{n-3} &+ 4h(\dot{x}_{n+1} - 2\nabla\dot{x}_{n+1} + \tfrac{5}{3}\nabla^2\dot{x}_{n+1} - \tfrac{2}{3}\nabla^3\dot{x}_{n+1} \\ &+ \tfrac{7}{90}\nabla^4\dot{x}_{n+1}) - \frac{8h^7}{945}\frac{d^7x}{dt^7}(\xi) \end{aligned} \tag{15.40b}$$

Equations (15.40b) demonstrate why formulas of the closed type are used at all, even though they require extra effort to find values at t_{n+1}. The reason for the popularity of such formulas is the smallness of the error involved. Comparing errors of the same order in (15.40a) and (15.40b), we note that the coefficient of the error term is much smaller in the closed formula than in the open one ($\frac{1}{90}$ versus $\frac{14}{45}$).

Starting the solution. All the formulas derived in this section involve powers of ∇ operating on $\dot{x}_n$ or $\dot{x}_{n+1}$. This means that to find x_{n+1} we must know the values of $\dot{x}_k$ for $k \leqslant n+1$. However, $\dot{x} = f(x, t)$ implies that $\dot{x}_k = f(x_k, t_k)$. Thus, knowledge of $\dot{x}_k$ requires knowledge of x_k. Therefore, to find x_{n+1} we must know not only the values of $\dot{x}$ but also the values of $x(t)$ at t_k for $k \leqslant n+1$. In particular, we cannot start with $n = 0$ in the formulas given by Eqs. (15.40) because we would get negative indices for x_n on the RHS due to the powers of ∇. This means that the first few values of x_k must be obtained using a different method.

One common method of starting the solution is to use a Taylor-series expansion:

$$x_k \equiv x(t_0 + kh) = x_0 + h\dot{x}_0 k + \frac{h^2\ddot{x}_0}{2}k^2 + \cdots + \frac{h^m \dfrac{d^m x}{dt^m}(t_0)}{m!}k^m + \cdots \tag{15.41}$$

where

$$\begin{aligned} \dot{x}_0 &= f(x_0, t_0) \\ \ddot{x}_0 &= \frac{d\dot{x}}{dt}(t_0) = \left.\frac{\partial f}{\partial x}\right|_{x_0, t_0}\dot{x}_0 + \left.\frac{\partial f}{\partial t}\right|_{t_0, x_0} \\ \dddot{x}_0 &= \frac{d\ddot{x}}{dt}(t_0) = \left.\frac{\partial^2 f}{\partial x^2}\right|_{x_0, t_0}\dot{x}_0^2 + 2\left.\frac{\partial^2 f}{\partial t\,\partial x}\right|_{x_0, t_0}\dot{x}_0 + \left.\frac{\partial f}{\partial x}\right|_{x_0, t_0}\ddot{x}_0 + \left.\frac{\partial^2 f}{\partial t^2}\right|_{x_0, t_0} \\ &\;\;\vdots \end{aligned}$$

For the general case it is clear that the derivatives required for the RHS of Eq. (15.41) involve very complicated expressions. The following example therefore illustrates the procedure for a specific case.

Example 15.5.1

Let us use the procedure described in this subsection to solve

$$\frac{dx}{dt} + x + e^t x^2 = 0 \qquad x(0) \equiv x_0 = 1 \tag{1}$$

We can obtain a Taylor-series expansion for x by noting that

$$\dot{x} = -x - e^t x^2$$

$$\ddot{x} = -\dot{x} - 2x\dot{x}e^t - x^2 e^t$$

$$\dddot{x} = \frac{d\ddot{x}}{dt} = -\ddot{x} - 2\dot{x}^2 e^t - 2x\ddot{x}e^t - 4x\dot{x}e^t - x^2 e^t$$

$$\vdots$$

Therefore,

$$\begin{aligned} \dot{x}_0 &= -x_0 - x_0^2 \\ \ddot{x}_0 &= -\dot{x}_0 - 2x_0\dot{x}_0 - x_0^2 \\ \dddot{x}_0 &= -\ddot{x}_0 - 2\dot{x}_0^2 - 2x_0\ddot{x}_0 - 4x_0\dot{x}_0 - x_0^2 \\ &\vdots \end{aligned} \tag{2}$$

Substituting $x_0 = 1$ in the first equation of (2) yields $\dot{x}_0 = -2$. Substituting this and $x_0 = 1$ in the second equation gives $\ddot{x}_0 = 5$. Continuing in this way, we can obtain derivatives of all orders. Up to the fifth order we have

$$\dot{x}_0 = -2 \qquad \ddot{x}_0 = 5 \qquad \dddot{x}_0 = -16 \qquad \left(\frac{d^4x}{dt^4}\right)_0 = 65 \qquad \left(\frac{d^5x}{dt^5}\right)_0 = -326$$

Substituting these values in a Taylor-series expansion with $h = 0.1$ yields

$$x_k = 1 - 0.2k + 0.025k^2 - 0.0027k^3 + (2.7 \times 10^{-4})k^4 - (2.7 \times 10^{-5})k^5 + \cdots$$

Thus,

$$x_1 = 0.82254 \qquad x_2 = 0.68186 \qquad \text{and} \qquad x_3 = 0.56741$$

The corresponding values of $\dot{x}$ are

$$\dot{x}_1 = -x_1 - e^{t_1}x_1^2 = -1.57026$$

$$\dot{x}_2 = -x_2 - e^{t_2}x_2^2 = -1.24973$$

and

$$\dot{x}_3 = -x_3 - e^{t_3}x_3^2 = -1.00200$$

●

Once the starting values are obtained, either a formula of open type or one of closed type is used to find the next x value. Only formulas of open type will be discussed here. However, as mentioned earlier, the accuracy of closed-type formulas is better. The price of having x_{n+1} on the RHS is that using closed-type formulas requires *estimating* x_{n+1}, putting it in the RHS, and finding an improved estimate for x_{n+1}. This continues until no further improvement in the estimate is achieved.

Using formulas of open type. The use of open-type formulas involves simple substitution of the known quantities $x_0, x_1, \ldots, x_n$ on the RHS to obtain x_{n+1}. The master equation (for $p = 0$) is Eq. (15.34). The number of powers of ∇ that are retained gives rise to different methods. For instance, when *no* power of ∇ is retained, the method is called *Euler's method*, for which we use

$$x_{n+1} \approx x_n + h\dot{x}_n \qquad e(\xi) = h^2 \ddot{x}(\xi)$$

A more commonly used method is *Adam's method*, for which all powers of ∇ up to and including the third are retained. We then have

$$x_{n+1} \approx x_n + h(1 + \tfrac{1}{2}\nabla + \tfrac{5}{12}\nabla^2 + \tfrac{3}{8}\nabla^3)\dot{x}_n \qquad e(\xi) = \frac{251}{720} h^5 \frac{d^5 x}{dt^5}(\xi)$$

or, in terms of values of $\dot{x}$,

$$x_{n+1} \approx x_n + \frac{h}{24}(55\dot{x}_n - 59\dot{x}_{n-1} + 37\dot{x}_{n-2} - 9\dot{x}_{n-3}) \qquad e(\xi) = \frac{251}{720} h^5 \frac{d^5 x}{dt^5}(\xi) \tag{15.42}$$

Recall that $\dot{x}_k = f(x_k, t_k)$. Thus, if we know the values x_n, x_{n-1}, x_{n-2}, and x_{n-3}, we can obtain x_{n+1}.

Example 15.5.2

Knowing x_0, x_1, x_2, and x_3, we can calculate x_4 for Eq. (1) of Example 15.5.1.

$$x_4 \approx x_3 + \frac{0.1}{24}(55\dot{x}_3 - 59\dot{x}_2 + 37\dot{x}_1 - 9\dot{x}_0)$$

$$= 0.56741 + \frac{0.1}{24}[55(-1.002) - 59(-1.24973) + 37(-1.57026) - 9(-2)]$$

$$= 0.47793$$

With x_4 at our disposal, we can evaluate $\dot{x}_4 = -x_4 - x_4^2 e^{t_4}$, and substitute it in

$$x_5 \approx x_4 + \frac{0.1}{24}(55\dot{x}_4 - 59\dot{x}_3 + 37\dot{x}_2 - 9\dot{x}_1)$$

to find x_5, and so on. ●

A crucial fact about such methods is that every value obtained is in error by some amount, and using such values to obtain new values propagates the error. Thus, error can accumulate rapidly and make approximations worse at each step. Discussion of error propagation is common in the literature (see, for example, pp. 267–268 of Hildebrand 1987).

15.5.2 The Runge-Kutta Method

The FODE of Eq. (15.32) leads to a unique Taylor series,

$$x(t_0 + h) = x_0 + h\dot{x}_0 + \frac{h^2}{2}\ddot{x}_0 + \cdots$$

where $\dot{x}_0$, $\ddot{x}_0$, and all the rest of the derivatives can be evaluated by differentiating $\dot{x} = f(x, t)$. Thus, theoretically, the Taylor series gives the solution (for $t_0 + h$; but $t_0 + 2h$, $t_0 + 3h$, and so on can be obtained similarly). However, in practice, the Taylor series converges slowly and the accuracy involved is not high. Thus one resorts to other methods of solution such as described earlier.

Another method, known as the *Runge-Kutta method*, replaces the Taylor series

$$x_{n+1} = x_n + h\dot{x}_n + \frac{h^2}{2}\ddot{x}_n + \frac{h^3}{3!}\dddot{x}_n + \cdots \tag{15.43}$$

with

$$\begin{aligned} x_{n+1} = x_n &+ h[\alpha_0 f(x_n, t_n) + \alpha_1 f(x_n + b_1 h, t_n + \mu_1 h) \\ &+ \alpha_2 f(x_n + b_2 h, t_n + \mu_2 h) + \cdots + \alpha_p f(x_n + b_p h, t_n + \mu_p h)] \end{aligned} \tag{15.44}$$

where $\{\alpha_k, \mu_k, b_k\}_{k=1}^{p}$ and α_0 are constants chosen so that if the RHS of (15.44) were expanded in powers of the spacing h, the coefficients of a certain number of the leading terms would agree with the corresponding expansion coefficients of the RHS of (15.43).

It is convenient to express the b's as linear combinations of preceding values of f,

$$\begin{aligned} hb_1 &\equiv \lambda_{10} k_0 \\ hb_2 &\equiv \lambda_{20} k_0 + \lambda_{21} k_1 \\ &\vdots \\ hb_m &\equiv \lambda_{m0} k_0 + \lambda_{m1} k_1 + \cdots + \lambda_{m,m-1} k_{m-1} \end{aligned}$$

where the k_r are recursively defined as

$$\begin{aligned} k_0 &= hf(x_n, t_n) \\ k_r &= hf(x_n + \lambda_{r0} k_0 + \lambda_{r1} k_1 + \cdots + \lambda_{r,r-1} k_{r-1}, t_n + \mu_r h) \end{aligned}$$

Then Eq. (15.44) gives

$$x_{n+1} = x_n + \alpha_0 k_0 + \alpha_1 k_1 + \cdots + \alpha_p k_p$$

where α_i, μ_i, and λ_{ij} are to be determined.

In general, the determination of these constants is extremely tedious. Let us consider the very simple case where $p = 1$, and let $\lambda \equiv \lambda_{01}$ and $\mu \equiv \mu_1$. Then we obtain

$$x_{n+1} = x_n + \alpha_0 k_0 + \alpha_1 k_1 \tag{15.45a}$$

where $\qquad k_0 = hf(x_n, t_n) \qquad$ and $\qquad k_1 = hf(x_n + \lambda k_0, t_n + \mu h) \qquad$ (15.45b)

The Taylor expansion for k_1, a function of two variables, is

$$k_1 = h[f + (\mu h f_t + \lambda k_0 f_x) + \tfrac{1}{2}(\mu^2 h^2 f_{tt} + 2\lambda\mu h k_0 f_{tx} + \lambda^2 k_0^2 f_{xx}) + O(h^3)]$$

where $f \equiv f(x_n, t_n)$, $f_t \equiv (\partial f/\partial t)(x_n, t_n)$, $f_{tt} \equiv (\partial^2 f/\partial t^2)(x_n, t_n)$, $f_{tx} \equiv (\partial^2 f/\partial t\,\partial x)\,(x_n, t_n)$, and $f_{xx} \equiv (\partial^2 f/\partial x^2)(x_n, t_n)$. Expressed in powers of h, with $k_0 = hf$, this becomes

$$k_1 = hf + h^2(\mu f_t + \lambda f f_x) + \frac{h^3}{2}(\mu^2 f_{tt} + 2\lambda\mu f f_{xt} + \lambda^2 f^2 f_{xx}) + O(h^4)$$

Substituting in Eq. (15.45a), we get

$$x_{n+1} = x_n + h(\alpha_0 + \alpha_1)f + h^2\alpha_1(\mu f_t + \lambda f f_x) + \frac{h^3}{2}\alpha_1(\mu^2 f_{tt} + 2\lambda\mu f f_{xt} + \lambda^2 f^2 f_{xx}) + O(h^4) \tag{15.46}$$

On the other hand, with

$$\dot{x} = f$$

$$\ddot{x} = \frac{df}{dt} = \frac{\partial f}{\partial x}\frac{dx}{dt} + \frac{\partial f}{\partial t} = \dot{x} f_x + f_t = f f_x + f_t$$

$$\dddot{x} = f_{tt} + 2f f_{xt} + f^2 f_{xx} + f_x(f f_x + f_t)$$

Equation (15.43) gives

$$x_{n+1} = x_n + hf + \frac{h^2}{2}(f f_x + f_t) + \frac{h^3}{6}[f_{tt} + 2f f_{xt} + f^2 f_{xx} + f_x(f f_x + f_t)] + O(h^4) \tag{15.47}$$

If we demand that (15.46) and (15.47) agree up to the h^2 term (we cannot demand agreement for h^3 or higher because of overspecification), then we must have

$$\alpha_0 + \alpha_1 = 1 \qquad \alpha_1 \mu = \tfrac{1}{2} \qquad \alpha_1 \lambda = \tfrac{1}{2}$$

There are only three equations for four unknowns. Therefore, there will be an arbitrary parameter β in terms of which the unknowns can be written:

$$\alpha_0 = 1 - \beta \qquad \alpha_1 = \beta \qquad \mu = \frac{1}{2\beta} \qquad \lambda = \frac{1}{2\beta}$$

Substituting these values in Eqs. (15.45) gives

$$x_{n+1} = x_n + h\left[(1-\beta)f(x_n, t_n) + \beta f\left(x_n + \frac{hf}{2\beta}, t_n + \frac{h}{2\beta}\right)\right] + O(h^3)$$

This formula becomes useful if we let $\beta = \frac{1}{2}$. Then $t_n + h/2\beta = t_n + h = t_{n+1}$, which makes evaluation of the second term in square brackets convenient. For $\beta = \frac{1}{2}$ we have

$$x_{n+1} = x_n + \frac{h}{2}[f(x_n, t_n) + f(x_n + hf, t_{n+1})] + O(h^3) \tag{15.48}$$

What is nice about this equation is that it needs no starting up! We can plug in the known quantities t_n, t_{n+1}, and x_n on the RHS and find x_{n+1} starting with $n = 0$. However, the result is not very accurate, and we cannot make it any more accurate by demanding agreement for higher powers of h, because, as mentioned earlier, such a demand overspecifies the unknowns.

Formulas that give more accurate results can be obtained by retaining terms beyond $p = 1$. Thus, for $p = 2$, if we write

$$x_{n+1} = x_n + \alpha_0 k_0 + \alpha_1 k_1 + \alpha_2 k_2$$

there will be eight unknowns (three α's, three λ_{ij}'s, and two μ's), and the demand for agreement between the Taylor expansion and the expansion of f up to h^3 will yield only six equations. Therefore, there will be two arbitrary parameters whose specification results in various formulas. The details of this kind of algebraic derivation are very messy, so we will merely consider two specific formulas.

One such formula, due to Kutta, is

$$x_{n+1} = x_n + \tfrac{1}{6}(k_0 + 4k_1 + k_2) + O(h^4) \tag{15.49a}$$

where

$$\begin{aligned} k_0 &= hf(x_n, t_n) \\ k_1 &= hf(x_n + \tfrac{1}{2}k_0, t_n + \tfrac{1}{2}h) \\ k_2 &= hf(x_n + 2k_1 - k_0, t_n + h) \end{aligned} \tag{15.49b}$$

A second formula, due to Heun, has the form

$$x_{n+1} = x_n + \tfrac{1}{4}(k_0 + 3k_2) + O(h^4)$$

where

$$\begin{aligned} k_0 &= hf(x_n, t_n) \\ k_1 &= hf(x_n + \tfrac{1}{3}k_0, t_n + \tfrac{1}{3}h) \\ k_2 &= hf(x_n + \tfrac{2}{3}k_1, t_n + \tfrac{2}{3}h) \end{aligned}$$

These two formulas are of about the same order of accuracy, but each possesses some advantages over the other.

Example 15.5.3

Let us solve Eq. (1) of Example 15.5.1 using the Runge-Kutta method with Eqs. (15.49). With $t_0 = 0$, $x_0 = 1$, and $h = 0.1$, Eqs. (15.49) give

$$k_0 = 0.1(-2) = -0.2$$

$$k_1 = 0.1\{-[1 - \tfrac{1}{2}(0.2)] - [1 - \tfrac{1}{2}(0.2)]^2 e^{(1/2)(0.1)}\} = -0.17515$$

$$k_2 = 0.1\{-[1 + 2(0.17515) + 0.2] - [1 + 2(0.17515) + 0.2]^2 e^{0.1}\} = -0.16476$$

$$x_1 = x_0 + \tfrac{1}{6}(k_0 + 4k_1 + k_2) = 1 + \tfrac{1}{6}[-0.2 - 4(0.17515) - 0.16476] = 0.82244$$

Substituting this x_1 and $t_1 = t_0 + h = 0.1$ and $h = 0.1$ in (15.49b) yields

$$k_0 = -0.15700 \qquad k_1 = -0.13870 \qquad k_2 = -0.13040$$

leading to

$$x_2 = 0.82244 + \tfrac{1}{6}[-0.15700 - 4(0.13870) - 0.13040]$$
$$= 0.68207$$

We similarly obtain

$$x_3 = 0.56964 \qquad \text{and} \qquad x_4 = 0.47858$$

Solving the FODE analytically gives the exact result

$$x(t) = \frac{e^{-t}}{1+t}$$

Table 15.1 compares the values obtained in Examples 15.5.1 and 15.5.2, those obtained here, and the exact values to five decimal places. It is clear that the Runge-Kutta method is more accurate than the method discussed earlier. ●

The accuracy of the Runge-Kutta method (illustrated in Example 15.5.3 for a particular case) and the fact that it requires no start-up procedure make it one of the most popular methods for solving differential equations.

The Runge-Kutta method can be made more accurate by using higher values of p. For instance, a formula that is used for $p = 3$ is

$$x_{n+1} = x_n + \tfrac{1}{6}(k_0 + 2k_1 + 2k_2 + k_3) + O(h^5) \tag{15.50a}$$

where

$$k_0 = hf(x_n, t_n) \qquad k_1 = hf(x_n + \tfrac{1}{2}k_0, t_n + \tfrac{1}{2}h)$$
$$k_2 = hf(x_n + \tfrac{1}{2}k_1, t_n + \tfrac{1}{2}h) \qquad k_3 = hf(x_n + k_2, t_n + h) \tag{15.50b}$$

15.5.3 Solving Higher-Order Equations, Initial Value Problems, and Boundary Value Problems

Any nth-order differential equation is equivalent to n first-order differential equations in $n + 1$ variables. Thus, for instance, the most general SODE, $F(\ddot{x}, \dot{x}, x, t) = 0$, can be

TABLE 15.1 SOLUTIONS TO THE DE OF EXAMPLE 15.5.1 OBTAINED IN THREE WAYS

t	Analytical solution	Runge-Kutta solution	Adam's method solution
0	1	1	1
0.1	0.82258	0.82244	0.82254
0.2	0.68228	0.68207	0.68186
0.3	0.56986	0.56964	0.56741
0.4	0.47880	0.47858	0.47793

reduced to two FODEs as follows. We solve for $\ddot{x}$, obtaining

$$\ddot{x} = G(\dot{x}, x, t)$$

we define $\dot{x} \equiv u$ and write

$$\dot{u} = G(u, x, t)$$

and

$$\dot{x} = u$$

These two equations are completely equivalent to the original SODE. Thus, it is appropriate to discuss numerical solutions of systems of FODEs in several variables. The discussion here will be limited to systems consisting of two equations because such a system is equivalent to the SODE that is most common in applications. The generalization to several equations is not difficult.

Consider the system of equations

$$\frac{dx}{dt} = f(x, u, t) \qquad \frac{du}{dt} = g(x, u, t) \tag{15.51a}$$

with thc initial conditions

$$x(t_0) \equiv x_0 \qquad u(t_0) \equiv u_0 \tag{15.51b}$$

Using an obvious generalization of Eqs. (15.50), we can write

$$\begin{aligned} x_{n+1} &= x_n + \tfrac{1}{6}(k_0 + 2k_1 + 2k_2 + k_3) + O(h^5) \\ u_{n+1} &= u_n + \tfrac{1}{6}(m_0 + 2m_1 + 2m_2 + m_3) + O(h^5) \end{aligned} \tag{15.52a}$$

where

$$\begin{aligned} k_0 &= hf(x_n, u_n, t_n) \\ k_1 &= hf(x_n + \tfrac{1}{2}k_0, u_n + \tfrac{1}{2}m_0, t_n + \tfrac{1}{2}h) \\ k_2 &= hf(x_n + \tfrac{1}{2}k_1, u_n + \tfrac{1}{2}m_1, t_n + \tfrac{1}{2}h) \\ k_3 &= hf(x_n + k_2, u_n + m_2, t_n + h) \end{aligned} \tag{15.52b}$$

and

$$\begin{aligned} m_0 &= hg(x_n, u_n, t_n) \\ m_1 &= hg(x_n + \tfrac{1}{2}k_0, u_n + \tfrac{1}{2}m_0, t_n + \tfrac{1}{2}h) \\ m_2 &= hg(x_n + \tfrac{1}{2}k_1, u_n + \tfrac{1}{2}m_1, t_n + \tfrac{1}{2}h) \\ m_3 &= hg(x_n + k_2, u_n + m_2, t_n + h) \end{aligned} \tag{15.52c}$$

These formulas give the values of the dependent variables x and u at $t_0, t_1, \ldots, t_N$.

The formulas of Eqs. (15.52) are more general than is necessary to solve a general SODE, since, as mentioned above, such a SODE is equivalent to the system $\dot{x} = u$,

$\dot{u} = g(x, u, t)$. Thus, $f(x, u, t) \equiv u$, and Eqs. (15.52) specialize to

$$k_0 = hu_n = h\dot{x}_n$$

$$k_1 = h(u_n + \tfrac{1}{2}m_0) = h\dot{x}_n + \frac{h}{2}m_0 \tag{15.53a}$$

$$k_2 = h\dot{x}_n + \tfrac{1}{2}hm_1$$

$$k_3 = h\dot{x}_n + hm_2$$

and

$$x_{n+1} = x_n + h\dot{x}_n + \frac{h}{6}(m_0 + m_1 + m_2) + O(h^5)$$

$$\dot{x}_{n+1} = \dot{x}_n + \tfrac{1}{6}(m_0 + 2m_1 + 2m_2 + m_3) + O(h^5) \tag{15.53b}$$

where

$$m_0 = hg(x_n, \dot{x}_n, t_n)$$

$$m_1 = hg(x_n + \tfrac{1}{2}h\dot{x}_n, \dot{x}_n + \tfrac{1}{2}m_0, t_n + \tfrac{1}{2}h)$$

$$m_2 = hg(x_n + \tfrac{1}{2}h\dot{x}_n + \tfrac{1}{4}hm_0, \dot{x}_n + \tfrac{1}{2}m_1, t_n + \tfrac{1}{2}h) \tag{15.53c}$$

$$m_3 = hg(x_n + h\dot{x}_n + \tfrac{1}{2}hm_1, \dot{x}_n + m_2, t_n + h)$$

Given the IVP $\ddot{x} = g(x, \dot{x}, t)$, $x(0) = x_0$, $\dot{x}(0) = \dot{x}_0$, we can use Eqs. (15.53b) successively to generate $x_1, x_2, \ldots, x_N$ and, therefore, solve the problem.

Example 15.5.4

The IVP $\ddot{x} + x = 0$, $x(0) = 0$, $\dot{x}(0) = 1$ clearly has the analytic solution $x(t) = \sin t$. However, let us use Eqs. (15.53) to illustrate the Runge-Kutta method of solving SODEs and compare the result with the exact solution.

For this problem $g(x, \dot{x}, t) = -x$. Therefore, we can easily calculate the m's

$$m_0 = -hx_n$$

$$m_1 = -h(x_n + \tfrac{1}{2}h\dot{x}_n)$$

$$m_2 = -h(x_n + \tfrac{1}{2}h\dot{x}_n - \tfrac{1}{4}h^2x_n)$$

$$m_3 = -h[x_n + h\dot{x}_n - \tfrac{1}{2}h^2(x_n + \tfrac{1}{2}h\dot{x}_n)]$$

These lead to expressions for x_{n+1} and $\dot{x}_{n+1}$:

$$x_{n+1} = x_n + h\dot{x}_n - \frac{h^2}{6}\left(3x_n + h\dot{x}_n - \frac{h^2}{4}x_n\right) \tag{1}$$

$$\dot{x}_{n+1} = \dot{x}_n - \frac{h}{6}\left[6x_n + 3h\dot{x}_n - h^2\left(x_n + \frac{h}{4}\dot{x}_n\right)\right] \tag{2}$$

Starting with $x_0 = 0$ and $\dot{x}_0 = 1$, we can generate x_1, x_2, and so on by using Eqs. (1) and (2) successively. The results for $x_1, x_2, \ldots, x_{10}$ with $h = 0.1$ are given to five significant figures in Table 15.2. Note that up to x_5 there is complete agreement with the exact result. This indicates the high accuracy of the Runge-Kutta method once more. ●

TABLE 15.2 COMPARISON OF THE RUNGE-KUTTA AND EXACT SOLUTIONS TO THE SODE OF EXAMPLE 15.5.4

t	Runge-Kutta solution	$\sin t$
0.1	0.09983	0.09983
0.2	0.19867	0.19867
0.3	0.29552	0.29552
0.4	0.38942	0.38942
0.5	0.47943	0.47943
0.6	0.56466	0.56464
0.7	0.64425	0.64422
0.8	0.71741	0.71736
0.9	0.78342	0.78333
1.0	0.84161	0.84147

Note that the Runge-Kutta method lends itself readily to use in computer programs. Because Eqs. (15.53) do not require any start-ups, they can be used directly to generate solutions to any IVP involving a SODE.

Another, more direct, method of solving higher-order differential equations is to substitute $\mathbb{D} = -(1/h)\ln(\mathbb{1} - \nabla)$ for the derivative operator in the differential equation, expand in terms of ∇, and keep an appropriate number of terms. Exercise 15.5.1 illustrates this point for a linear SODE.

For FODEs we encounter only IVPs: The initial value $x(t_0)$ is given, and $x(t)$ for $t > t_0$ is sought. For higher-order differential equations we also encounter BVPs: Two independent linear combinations of the values of x and $\dot{x}$ are given. (We have denoted these as $\mathbb{R}_i[u] = \gamma_i$, for $i = 1, 2$ in Chapters 9, 10, and 11.)

The case of a linear differential equation can be handled using the superposition principle. Assume that we want to solve

$$\ddot{x} + p(t)\dot{x} + q(t)x = r(t) \qquad \text{for } a \leqslant t \leqslant b \tag{15.54a}$$

with the BCs

$$x(a) = A \qquad x(b) = B \tag{15.54b}$$

Such a BVP can be reduced to two IVPs as follows. Let $u(t)$ be *any* solution of the (underdetermined) IVP

$$\ddot{u} + p\dot{u} + qu = r \qquad u(a) = A$$

and let $v(t)$ be *any* (nontrivial) solution of the (also underdetermined) IVP

$$\ddot{v} + p\dot{v} + qv = 0 \qquad v(a) = 0$$

Then, for *any* constant C, the function

$$x(t) = u(t) + Cv(t)$$

satisfies (15.54a) and the first BC of (15.54b). To satisfy the second BC, we determine C such that

$$x(b) = u(b) + Cv(b) = B$$

If $u(b) = B$, then $u(t)$ is the solution. Thus, we assume that $u(b) \neq B$. Then

$$C = \frac{B - u(b)}{v(b)}$$

is the constant. Note that $v(b) \neq 0$, because $v(b) = 0$ and $v(a) = 0$ imply that $v \equiv 0$.

In order to use the numerical methods of solving differential equations, we have to assign an initial slope to u and to v. Such assigning of slopes is almost completely arbitrary; the only restriction is that $\dot{v}(a) \neq 0$ (why?). A convenient choice is $\dot{u}(a) = 0$ and $\dot{v}(a) = 1$. This reduces the original BVP to two IVPs, which can be solved by the methods described earlier.

Superposition does not apply to nonlinear cases, but other methods of solution exist. Let us now consider one that is common.

Consider the nonlinear BVP

$$\ddot{x} = g(x, \dot{x}, t)$$

$$x(a) = A \qquad x(b) = B$$

and an associated IVP

$$\ddot{u} = g(u, \dot{u}, t)$$

$$u(a) = A \qquad \dot{u}(a) = \alpha$$

For any value of α we can solve the IVP numerically. Using $u(t, \alpha)$ to denote the solution $u(t)$ corresponding to α, we demand that

$$u(b, \alpha) = B \tag{15.55}$$

Thus, the problem reduces to solving Eq. (15.55) for α. We can accomplish this using Newton's method with $f(\alpha) \equiv u(b, \alpha) - B$. However, since $f'(\alpha)$ is not known, we need to approximate it by the ratio of the difference. Thus, using Eq. (15.1), we get

$$\alpha_{n+1} = \alpha_n - \frac{f(\alpha_n)}{\left[\dfrac{f(\alpha_n) - f(\alpha_{n-1})}{\alpha_n - \alpha_{n-1}}\right]}$$

Substituting for $f(\alpha)$ and simplifying yields

$$\alpha_{n+1} = \alpha_n - \frac{(\alpha_n - \alpha_{n-1})[u(b, \alpha_n) - B]}{u(b, \alpha_n) - u(b, \alpha_{n-1})} \tag{15.56}$$

In practice, the IVP is usually solved for two or more values of α; then Eq. (15.56) is used to obtain improved values of α. This process is iterated until Eq. (15.55) is satisfactorily approximated (assuming that the iteration converges).

A third method of solving linear BVPs numerically involves changing the differential equation into a difference equation (see Exercise 15.5.1), substituting for various values of n, and solving the resulting system of linear equations. The best way to appreciate this method is to consider an example.

Example 15.5.5

Consider this SOLDE and BCs:

$$\ddot{x} + p(t)\dot{x} + q(t)x = r(t)$$

$$x(0) = 0 \qquad x(5) = 0$$

It can be shown (see Exercise 15.5.1) that, up to $O(h^2)$, this is equivalent to the difference equation

$$(2 - \tfrac{3}{2}hp_n + h^2 q_n)x_n - (5 - 2hp_n)x_{n-1} + (4 - \tfrac{1}{2}hp_n)x_{n-2} - x_{n-3} = h^2 r_n \quad \text{where } n \geqslant 3$$

If we assume, for simplicity, that $h = 1$, then the maximum n will be 5. Thus, for $n = 3, 4$, and 5, we obtain

$$\begin{aligned}(2 - \tfrac{3}{2}p_3 + q_3)x_3 - (5 - 2p_3)x_2 + (4 - \tfrac{1}{2}p_3)x_1 - 0 &= r_3 \\ (2 - \tfrac{3}{2}p_4 + q_4)x_4 - (5 - 2p_4)x_3 + (4 - \tfrac{1}{2}p_4)x_2 - x_1 &= r_4 \\ 0 - (5 - 2p_5)x_4 + (4 - \tfrac{1}{2}p_5)x_3 - x_2 &= r_5\end{aligned} \tag{1}$$

These constitute three equations in the four unknowns $x_1, \ldots, x_4$. We need one more equation, which can be obtained by simply determining x_1 using a Taylor-series expansion. ●

This third method can clearly be generalized to any linear differential equation with any linear BCs. Thus, BCs of the form $\mathbb{R}_i[u] = \gamma_i$, for $i = 1, 2, \ldots, N$ (where N is the order of the differential equation) simply provide additional linear equations that can be combined with Eqs. (1) of Example 15.5.5. Of course, for small h we expect to get a large number of simultaneous linear equations. However, there are various packages for solving linear equations on (personal) computers that can handle a large (but with some limit) number of simultaneous linear equations.

This treatment of basic techniques of numerical calculations has been nowhere near complete. Many topics that have not been touched on are of importance in numerical analysis, including error analysis, error propagation, interpolations, partial differential equations, and so on. Such topics are covered by a wide variety of books on numerical analysis. Two books worth mentioning are the classic by Hildebrand (1987) and the work by Press et al. (1987). The latter is full of computer programs for numerically solving many problems in mathematical physics.

Exercise

15.5.1 Approximate a general SOLDE by a difference equation involving terms up to the third order of ∇.

PROBLEMS

15.1 Use a hand-held calculator or computer to find a root to six decimal places for each of the transcendental equations $x = \tan x$ and $x^2 = e^x$.

15.2 For each equation, find a root to five significant figures.

(**a**) $x + \ln x = 5$ (**b**) $x^5 + 3x - 5 = 0$ (**c**) $x^3 = 3x - 2$ (**d**) $\cos x = x$

15.3 Show that δ^{n+1} or ∇^{n+1} annihilates $p(x)$, a polynomial of degree n.

15.4 Show that all of the finite-difference operators commute with one another.

15.5 Verify the following identities.

(**a**) $\nabla\mathbb{E} = \delta\mathbb{E}^{1/2} = \Delta$ (**b**) $\nabla + \Delta = 2\mu\delta$ (**c**) $\mathbb{E}^{-1/2} = \mu - \dfrac{\delta}{2}$ (**d**) $2\mu\delta = \mathbb{E} - \mathbb{E}^{-1}$

15.6 Find an expression for $\mathbb{D}^2$ in terms of ∇ and δ.

15.7 Derive Eqs. (15.18), (15.20), (15.21), and (15.23).

15.8 Use the backward difference operator to write an expression for each of the following.

(**a**) $\mathbb{D}^2 f_i$ with $\text{e} = O(h)$ (**b**) $\mathbb{D}^2 f_i$ with $\text{e} = O(h^3)$
(**c**) $\mathbb{D}^3 f_i$ with $\text{e} = O(h)$ (**d**) $\mathbb{D}^4 f_i$ with $\text{e} = O(h^2)$

15.9 Use a table of Bessel functions (see Table 10.1) to calculate the second derivative of an entry in the table with $\text{e} = O(h^3)$.

15.10 Find the error $e_3(\xi)$ for Simpson's three-eighths rule.

15.11 Evaluate the following integrals numerically, using six steps with the trapezoidal rule, Simpson's one-third rule, and Simpson's three-eighths rule. Compare with the exact result when possible. Also, find the bounds on the error in each case.

(**a**) $\displaystyle\int_0^5 x^3\,dx$ (**b**) $\displaystyle\int_0^2 e^{-x^2}\,dx$ (**c**) $\displaystyle\int_0^{1.2} xe^x \cos x\,dx$

(**d**) $\displaystyle\int_1^4 \frac{dx}{x}$ (**e**) $\displaystyle\int_1^4 \ln x\,dx$ (**f**) $\displaystyle\int_1^2 e^{x^2} \sin x\,dx$

(**g**) $\displaystyle\int_{-1}^1 \frac{dx}{1+x^2}$ (**h**) $\displaystyle\int_0^1 xe^{x^2}\,dx$ (**i**) $\displaystyle\int_0^1 e^x \tan x\,dx$

15.12 Solve any five FODEs in Chapter 9 using Eqs. (15.42) and (15.50). Compare the results with the exact solutions obtained in that chapter.

15.13 Write a computer program that solves the following differential equations by (**a**) Adam's method [Eq. (15.42)] and (**b**) the Runge-Kutta method [Eqs. (15.50)].

$$\dot{x} = t - x^2 \qquad x(0) = 1$$

$$\dot{x} = t + \sin x \qquad x(0) = \pi/2$$

$$\dot{x} = e^{-xt} \qquad x(0) = 1$$

$$\dot{x} = \sin xt \qquad x(0) = 1$$

$$\dot{x} = x^2 t^2 + 1 \qquad x(0) = 1$$

15.14 Solve the following IVPs numerically, with $h = 0.1$. Find the first ten values of x.

(**a**) $\ddot{x} + 0.2\dot{x}^2 + 10x = 20t \qquad x(0) = 0, \dot{x}(0) = 0$

(**b**) $\ddot{x} + 4x = t^2 \qquad x(0) = 1, \dot{x}(0) = 0$

(**c**) $\ddot{x} + \dot{x} + x = 0 \qquad x(0) = 2, \dot{x}(0) = 0$

(**d**) $t\ddot{x} + \dot{x} + xt = 0 \qquad x(0) = 1, \dot{x}(0) = 0$

(**e**) $\ddot{x} + \dot{x} + x^2 = t \qquad x(0) = 1, \dot{x}(0) = 0$

(**f**) $\ddot{x} + xt = 0 \qquad x(0) = 0, \dot{x}(0) = 1$

(**g**) $\ddot{x} + \sin x = t \qquad x(0) = \pi/2, \dot{x}(0) = 0$

REFERENCES

ABRAHAM, R., J. MARSDEN, and T. RATIU, *Manifolds, Tensor Analysis, and Applications.* Reading, MA: Addison-Wesley, 1983.

AHLFORS, L., *Complex Analysis*, 2nd ed. New York: McGraw-Hill, 1966.

BIRKHOFF, G., and S. MACLANE, *Modern Algebra*, 4th ed. New York: Macmillan, 1977.

BIRKHOFF, G., and G.-C. ROTA, *Ordinary Differential Equations*, 3rd ed. New York: John Wiley, 1978.

BISHOP, R., and S. GOLDBERG, *Tensor Analysis on Manifolds.* New York: Dover, 1980.

BJORKEN, J., and S. DRELL, *Relativistic Quantum Fields.* New York: McGraw-Hill, 1965.

BYRON, F., and R. FULLER, *Mathematics of Classical and Quantum Physics.* Reading, MA: Addison-Wesley, 1969.

CHURCHILL, R., J. BROWN, and R. VERHEY, *Complex Variables and Applications*, 3rd ed. New York: McGraw-Hill, 1974.

COURANT, R., *Differential and Integral Calculus*, vol. 2. New York: John Wiley, 1936.

COURANT, R., and D. HILBERT, *Methods of Mathematical Physics.* New York: Interscience, 1962.

DENERY, P., and A. KRZYWICKI, *Mathematics for Physicists.* New York: Harper & Row, 1967.

FLANDERS, H., *Differential Forms.* New York: Academic Press, 1963.

GRADSHTEYN, I., and I. RYZHIK, *Table of Integrals, Series, and Products.* New York: Academic Press, 1965.

HALMOS, P., *Finite-Dimensional Vector Spaces*, 2nd ed. Princeton, NJ: D. Van Nostrand, 1958.

HELLWIG, G., *Differential Operators of Mathematical Physics.* Reading, MA: Addison-Wesley, 1967.

HILDEBRAND, F., *Introduction to Numerical Analysis*, 2nd ed. New York: Dover Publications, 1987.

LORRAIN, P., D. CORSON, and F. LORRAIN, *Electromagnetic Fields and Waves*, 3rd ed. New York: W. H. Freeman, 1988.

MARISON, J., and M. HEALD, *Classical Electromagnetic Radiation*, 2nd ed. New York: Academic Press, 1980.

MCBRIDE, E., *Obtaining Generating Functions.* New York: Springer-Verlag, 1971.

MISNER, C., K. THORNE, and J. WHEELER, *Gravitation.* San Francisco: W. H. Freeman, 1973.

PALEY, H., and M. WEICHSEL, *Abstract Algebra.* New York: Holt Rinehart and Winston, 1966.

PRESS, W., B. FLANNERY, S. TEUKOLSKY, and W. VETTERLING, *Numerical Recipes.* Cambridge, England: Cambridge University Press, 1987.

REED, M., and B. SIMON, *Functional Analysis.* New York: Academic Press, 1980.

RICHTMEYER, R., *Principles of Advanced Mathematical Physics*, vol. 1. New York: Springer-Verlag, 1978.

ROACH, G., *Green's Functions.* London: Van Nostrand Reinhold, 1970.

RUBINSTEIN, Z., *A Course in Ordinary and Partial Differential Equations.* New York and London: Academic Press, 1969.

SIMMONS, G., *Topology and Modern Analysis.* Malabar, FL: Krieger Publishing, 1983.

SMYTHE, W., *Static and Dynamic Electricity*, 3rd ed. New York: McGraw-Hill, 1968.

STAKGOLD, I., *Green's Functions and Boundary Value Problems.* New York: Wiley-Interscience, 1979.

TRICOMI, F. G., *Vorlesungen über Orthogonalreihen.* Berlin: Springer, 1955.

VARADARAJAN, V. *Lie Groups, Lie Algebras, and Their Representations.* New York: Springer-Verlag, 1984.

WEYL, H., *Classical Groups.* Princeton, NJ: Princeton University Press, 1946.

INDEX